Life in the Open Ocean

Life in the Open Ocean

The Biology of Pelagic Species

Joseph J. Torres
Professor Emeritus
University of South Florida, College of Marine Science
St. Petersburg, FL, USA

Thomas G. Bailey
Center Director (retired)
Caribbean Marine Research Center
Jupiter, FL, USA

This edition first published 2022
© 2022 John Wiley & Sons Ltd

All rights reserved. No part of this publication may be reproduced, stored in a retrieval system, or transmitted, in any form or by any means, electronic, mechanical, photocopying, recording or otherwise, except as permitted by law. Advice on how to obtain permission to reuse material from this title is available at http://www.wiley.com/go/permissions.

The right of Joseph J. Torres and Thomas G. Bailey to be identified as the authors of this work has been asserted in accordance with law.

Registered Offices
John Wiley & Sons, Inc., 111 River Street, Hoboken, NJ 07030, USA
John Wiley & Sons Ltd, The Atrium, Southern Gate, Chichester, West Sussex, PO19 8SQ, UK

Editorial Office
9600 Garsington Road, Oxford, OX4 2DQ, UK

For details of our global editorial offices, customer services, and more information about Wiley products visit us at www.wiley.com.

Wiley also publishes its books in a variety of electronic formats and by print-on-demand. Some content that appears in standard print versions of this book may not be available in other formats.

Limit of Liability/Disclaimer of Warranty
The contents of this work are intended to further general scientific research, understanding, and discussion only and are not intended and should not be relied upon as recommending or promoting scientific method, diagnosis, or treatment by physicians for any particular patient. In view of ongoing research, equipment modifications, changes in governmental regulations, and the constant flow of information relating to the use of medicines, equipment, and devices, the reader is urged to review and evaluate the information provided in the package insert or instructions for each medicine, equipment, or device for, among other things, any changes in the instructions or indication of usage and for added warnings and precautions. While the publisher and authors have used their best efforts in preparing this work, they make no representations or warranties with respect to the accuracy or completeness of the contents of this work and specifically disclaim all warranties, including without limitation any implied warranties of merchantability or fitness for a particular purpose. No warranty may be created or extended by sales representatives, written sales materials or promotional statements for this work. The fact that an organization, website, or product is referred to in this work as a citation and/or potential source of further information does not mean that the publisher and authors endorse the information or services the organization, website, or product may provide or recommendations it may make. This work is sold with the understanding that the publisher is not engaged in rendering professional services. The advice and strategies contained herein may not be suitable for your situation. You should consult with a specialist where appropriate. Further, readers should be aware that websites listed in this work may have changed or disappeared between when this work was written and when it is read. Neither the publisher nor authors shall be liable for any loss of profit or any other commercial damages, including but not limited to special, incidental, consequential, or other damages.

Library of Congress Cataloging-in-Publication Data

Names: Torres, Joseph J., 1950– author. | Bailey, Thomas G., 1945– author.
  | Wiley-Blackwell (Firm), publisher.
Title: Life in the open ocean : the biology of pelagic species / Joseph J Torres, Thomas G Bailey.
Description: Hoboken, NJ : Wiley-Blackwell, 2022. | Includes
  bibliographical references and index.
Identifiers: LCCN 2021021994 (print) | LCCN 2021021995 (ebook) | ISBN
  9781405145299 (cloth) | ISBN 9781119840305 (adobe pdf) | ISBN
  9781119840312 (epub)
Subjects: LCSH: Marine biology. | Oceanography.
Classification: LCC QH91 .T67 2022 (print) | LCC QH91 (ebook) | DDC
  577.7–dc23
LC record available at https://lccn.loc.gov/2021021994
LC ebook record available at https://lccn.loc.gov/2021021995

Cover Design: Wiley
Cover Image: © Danté Fenolio

Set in 9.5/12.5pt STIXTwoText by Straive, Pondicherry, India
Printed and bound by CPI Group (UK) Ltd, Croydon, CR0 4YY

From JJT – To my family, near and far – especially mom and dad and my lovely wife Linda for her incredible patience with me and our other family member: the book.

From TGB – To my lovely wife, Jan for her love and support through the years.

From Both of Us

To our mentors, Charlotte Mangum, Jim Childress, George Somero, and Bruce Robison, a constant source of inspiration to this day.

And especially to Linda M. Torres, our gatekeeper, editor, and Angel of Detail, for keeping us on the straight and narrow, and without whom this book would still be "in preparation."

Contents

Preface *xviii*
Acknowledgments *xx*

1 Physics and the Physical Environment *1*
The Vastness of the Open Ocean *2*
The Properties of Water *2*
  Density *4*
  Viscosity *5*
  Reynolds Number *6*
  Drag *7*
Temperature *8*
The Oceans and Ocean Basins *9*
  Ocean Circulation *10*
    Surface Currents: Ocean Gyres and Geostrophic Flow *10*
    Ocean Gyres and Geostrophic Flow *15*
    Upwelling *16*
    Deep-Ocean Circulation *16*
    Water Masses *18*
Oxygen *24*
Pressure *24*
Sound *26*
Light *29*
  Absorption and Scattering *31*
Traditional Depth Zones in the Ocean *33*
Concluding Thoughts *34*
References *35*

2 Physiological Accommodation to Environmental Challenges *36*
Temperature *37*
  Terms *38*
  Temperature Effects on Survival: The Tolerance Polygon *39*
  Temperature Effects on Rate Processes – The Q_{10} Approximation *41*
  Patterns of Thermal Acclimation *43*
  Climatic Adaptation in Ectotherms *44*
  Temperature Compensation via Changes in Enzyme Concentration:
  The Quantitative Strategy for Short-term Change *47*

Compensation via Changes in Enzyme Quality – Isozymes, Allozymes, and Temperature Adaptation   47
What Properties of Enzymes Can Be Changed?   51
Lipids and Temperature   53
A Membrane Primer   54
Pressure   59
  Early Work   60
  Later Work   61
    Whole Animal Work   63
    Molecular Mechanisms of Adaptation to Pressure   64
  Pressure and Membranes   68
Oxygen   69
  Severity of Oxygen Minima, "Dead Zones," and the Intertidal   71
  Adaptations to Oxygen Minima   71
    The Aerobic Strategy   71
Salinity   75
Depth   77
Concluding Thoughts   83
References   84

3 The Cnidaria   89

Introduction   89
Classification   89
  History   89
  Classification Schemes   90
Phylum Cnidaria   91
  Subphylum Medusozoa   91
  Subphylum Anthozoa   91
  Subphylum Myxozoa   91
The Hydromedusae   92
  Morphology Basics   92
  Morphological Detail and Life Histories   95
The Scyphomedusae   99
  Basic Characteristics   99
  Morphological Detail and Life Histories   101
    General   101
    Coronatae   101
    Semaeostomae   102
    Rhizostomae   104
The Cubomedusae   105
Foraging Strategies   105
  General Considerations   106
  The Cnidae   107
  Venoms   108
  Interaction with Prey   109
    Direct Interception   110
    Encounter Zone   110
    The Model   114

Swimming and Hunting Behavior   *115*
  Water Flow and Swimming   *115*
  Attraction Between Predator and Prey   *116*
  Diets, Feeding Rates, and Impacts on Prey Populations   *117*
  Rogue Hydroids: Predatory Polyps in the Midwater   *119*
  Feeding in the Cubomedusae   *120*
 Locomotion   *121*
  The Mesoglea   *123*
  Nerve Nets and Nervous Control of Swimming   *124*
  Senses and Sensory Mechanisms   *125*
 The Siphonophores   *127*
  Terminology and Affinities of Siphonophore "Persons"   *128*
  Whole Animal Organization   *134*
  Life Histories   *137*
  The Siphonophore Conundrum   *137*
  Feeding   *138*
   Fishing Behavior   *138*
   Digestion   *139*
   Diets and Selectivity   *139*
   Ecological Importance   *141*
   Locomotion   *141*
   Buoyancy   *143*
   Vertical Distribution   *144*
   Diurnal Vertical Migration   *146*
   Geographical Distribution   *147*
   Organization and Sensory Mechanisms   *147*
   Epithelial Conduction vs. Neural Conduction   *148*
 The Cnidaria Formerly Known as Chondrophora   *150*
  Classification   *150*
  Characteristics of the Chondrophoran Medusa   *151*
  Evolution Within the Chondrophora   *151*
  Feeding in the Chondrophora   *151*
  Locomotion   *151*
 References   *153*

4 The Ctenophora   *159*
 Introduction   *159*
 Classification   *159*
  History   *159*
  Classification Schemes   *160*
 Ctenophore Basics   *161*
 Morphology   *162*
  Cydippida   *162*
  Lobata   *165*
  Cestida   *168*
  Beroida   *168*
  Platyctenida   *168*
  Ganeshida   *171*

     Thalassocalycida   173
  Foraging Strategies   173
    General Considerations   173
    Interaction with Prey   173
      The Cydippids   173
      The Lobates   174
      The Cestids   175
      The Beroids   175
      The Platyctenids   178
      The Ganeshids   178
      The Thalassocalycids   178
      Specialists   178
    Diet, Feeding Rates, and Impacts on Prey Populations   178
  Ctenophores as Invasive Species   179
  Digestion   181
  Nerves and Sense Organs: Coordination and Conduction   181
  Locomotion   182
  Distribution   185
  Ctenophores and Evolution   186
  References   189

5 **The Nemertea**   192
  Introduction   192
  Classification   193
    History   193
    Of Germ Layers and Body Cavities   194
    Hydrostatic Skeletons   197
    Classification   200
  Morphology   200
    Proboscis Apparatus   202
    The Pelagic Body Form and Locomotion   204
    Color   205
  Nervous System   205
    Sense Organs   207
  Circulatory System   207
  Excretory System   209
  Digestive System   211
  Reproduction   213
    Development   215
  Foraging Strategies   215
  Vertical and Geographic Distributions   215
  References   217

6 **The Annelida**   219
  Introduction   219
  History   221

Classification *222*
Phylum Annelida *224*
  Class Polychaeta *224*
    Subclass Echiura *225*
    Subclass Errantia *226*
  Class Clitellata *226*
    Subclass Oligochaeta *227*
    Subclass Hirudinea *228*
  The Pelagic Polychaetes *228*
  Polychaete Subclass Errantia *229*
    Order Phyllodocida *229*
    Polychaete Subclass Sedentaria *229*
    Order Terebellida *231*
Morphology *231*
  General *231*
  External Anatomy *233*
    The Head Region *234*
    The Trunk or Metastomial Region *236*
  Internal Anatomy *239*
Excretory System *241*
  Pelagic Species *243*
The Nervous System *244*
  Sense Organs *244*
Circulatory System *246*
  Pelagic Species *249*
  Gas Exchange *249*
  Pelagic Species *249*
Digestive System *250*
Reproduction *251*
  Epitoky *252*
  Synchronicity *252*
  The Pelagic Species *254*
    Tomopteris *254*
    Alciopini and Lopadorrynchidae *254*
  Development *255*
Locomotion *255*
Foraging Strategies *256*
  The Hunters *256*
    Diets *258*
  The Suspension Feeders *258*
  Poeobiidae, Chaetopteridae *259*
Distributions *259*
  Geographical *259*
  Vertical *267*
Bioluminescence *267*
References *268*

7 The Crustacea 273

Introduction 273
Arthropod Classification 273
  History 273
  Subphylum Crustacea 276
  Subphylum Hexapoda 277
  Subphylum Myriapoda 277
  Subphylum Chelicerata 277
Panarthropoda Phyla 277
  Phylum Onychophora 277
  Phylum Tardigrada 277
Synopsis of Universal Arthropod Characteristics 278
The Crustacea 280
  Characteristics 281
  Classification 281
Subphylum Crustacea 283
  Class Remipedia 283
  Class Cephalocarida 283
  Class Branchiopoda 283
  Class Copepoda 283
  Class Thecostraca 283
  Class Tantulocarida 283
  Class Mystacocarida 283
  Class Branchiura 283
  Class Pentastomida 284
  Class Ostracoda 284
  Class Malacostraca 284
    Subclass Phyllocarida 284
    Subclass Hoplocarida 284
    Subclass Eumalacostraca 284
Crustacean Systems 292
  Integument and Molting 292
    Integument 292
    Molting 293
  Joints and Appendages 296
    Joints 296
    Appendages 297
  Excretory System 297
    Extra-renal Mechanisms 301
    How the System Works 303
    Nitrogen Excretion 303
  The Nervous System and Sensory Mechanisms 303
    The Central Nervous System 303
    Sensory Modalities 305
    Photoreception 305
    Mechanoreception 308
    Chemoreception 310

Circulatory and Respiratory Systems *311*
    Circulation and Oxygen Transport in the Blood: Hemocyanin *317*
  Digestive System *318*
  Basic Development *319*
  The Micronektonic Crustacea *319*
  The Pelagic Eucarida *321*
    Order Euphausiacea *321*
    Ecological Factors *339*
    Order Decapoda *346*
    Infraorder Anomura; Superfamily Galatheoidea; Family Munididae; Genera *Pleuroncodes, Munida,* and *Cervimunida* *378*
    Order Amphionidacea *379*
    Superorder Peracarida *380*
    Orders Lophogastrida and Mysida *380*
    Order Amphipoda *397*
    Cameo Players *420*
  References *424*

8 The Mollusca *439*

  Introduction *439*
  Classification *440*
    History *440*
    The Pelagic Molluscs *441*
  Phylum Mollusca *442*
    Class Caudofoveata *442*
    Class Solenogastres *442*
    Class Monoplacophora *442*
    Class Polyplacophora *442*
    Class Scaphopoda *443*
    Class Bivalvia *444*
    Class Gastropoda *445*
    Class Cephalopoda *445*
    Body Organization *445*
  The Gastropoda *445*
    Classification *445*
    Class Gastropoda *449*
      Classification Below Subclass Given for Pelagic Species Only *449*
    Gastropod Systems and Structures *451*
      The Digestive Tract *451*
      Circulation *454*
      Respiration *457*
      Excretion *461*
      Shell Formation *464*
      The Nervous System *466*
      Sensory Mechanisms *469*
    The Pelagic Gastropods: Anatomy and Habits *475*
      The Janthinid Snails *475*

The Heteropods   477
The Pteropods   491
Order Pteropoda   492
The Nudibranchs   516
The Cephalopoda   525
Classification   528
Basic Anatomy of the Major Cephalopod Groups   529
General   529
The Nautilida   529
The Sepiida and Spirulida   531
The Myopsida and Oegopsida   532
The Octopodiformes   536
Cephalopod Systems   540
Feeding and Digestion   540
Circulation   544
Gas Exchange   544
Excretion   549
Nervous System and Sensory Mechanisms   552
Locomotion and Buoyancy   569
Life Histories   575
Reproduction and Development   578
Vertical Distribution and Migration   582
Geographic Distribution   588
References   589

9 The Chordata   603

Introduction   603
Deuterostomes and the Phylogenetic Toolkit   604
Classification   607
Subphylum Tunicata   607
Class Ascidiacea (2935)   607
Class Appendicularia (68)   607
Class Thaliacea (78)   607
Basic Anatomy and Life History   609
The Ascidians   609
The Pyrosomes   610
The Salps   613
The Doliolids   618
The Appendicularia   625
The Appendicularian House   629
Tunicate Systems   634
Locomotion and Buoyancy   634
Pyrosomes   634
Salps   635
Doliolids   637
Appendicularia   638
Nervous Systems and Sensory Mechanisms   638

Pyrosomes *639*
　　　Doliolids *639*
　　　Salps *641*
　　　Appendicularia *643*
　　Gas Exchange, Circulation, and Excretion *645*
　　Trophic Role *647*
　　Bioluminescence *653*
　　Predators, Parasites and Other Interactions *654*
　　Geographic and Vertical Distributions *656*
　References *660*

10 The Fishes *669*

Introduction *669*
The Deep-Sea Groups *672*
A Brief History of Fishes *674*
　The Jawless Fishes *674*
　The Jawed Fishes *676*
　　Teleosts *676*
　　Elasmobranchs *677*
　　Holocephali *677*
The Classes of Living Fishes *678*
　Class Myxini *678*
　Class Petromyzonti *679*
　Class Elasmobranchii *682*
　Class Holocephali *691*
　Class Coelacanthi *693*
　Class Dipneusti *695*
　Class Cladistii *695*
　Class Actinopterygii *695*
　　Subclass Chondrostei *696*
　　Subclass Holostei *696*
　　Subclass Teleostei *696*
Fish Systems *758*
　Basic Anatomy *758*
　　External Features and Terms *758*
　　Skull and Skeleton *758*
　Feeding and Digestion *760*
　　Food Acquisition, the Three Dominant Modes: Ramming, Sucking, and Biting *760*
　　Food Sorting: The "Pharyngeal Jaws" *764*
　　Digestion: The Alimentary Canal *765*
　Circulation, Respiration, and Excretion *767*
　　Circulation *767*
　　Gas-Exchange in the Teleosts and Elasmobranchs *771*
　　Unidirectional Flow and Countercurrent Exchange: Maximizing the Concentration Gradient *774*
　　Blood and Oxygen at the Respiratory Surface *775*

CO$_2$ Transport–Far Different from O$_2$ 780
The Bohr and Haldane Effects 781
Secretion of Gases into the Swimbladder of Fishes 782
Nitrogen Excretion 785
Osmotic and Ionic Regulation 785
Introduction 785
Osmosis and Diffusion 787
The gills 790
Esophagus 791
Stomach 793
Intestine 793
Locomotion 793
Musculature 793
Red and White Muscle 795
Drag and Swimming Costs 796
Maximum Swim Speeds 797
Endothermy 799
Warm-Brained Billfishes 799
Swimming in Mesopelagic Fishes 801
Buoyancy 801
The Nervous System 804
Anatomy and Basics 804
The Brain 806
Cranial Nerves 807
Sensory Mechanisms 809
Sensory Modalities 809
Photoreception 811
Mechanoreception 815
The Inner Ear and Sound Reception 818
Electroreception 820
Chemoreception: Olfaction and Gustation 822
Camouflage, Bioluminescence, Photophores 824
Camouflage 824
Bioluminescence and Photophores 824
References 829

11 Communities 845
Introduction 845
The Gulf of Mexico 846
The Northern California Current 847
The Antarctic 852
System Comparisons 858
The Decapods and Mysids 876
The Euphausiids 881
The Myctophids 882
Non-myctophid Fishes 883

The Cephalopods *884*
      Gelatinous Zooplankton and Amphipods *885*
   Concluding Observations *887*
   Physical and Biological Factors that Change *887*
      Mean Annual Temperature *887*
      Seasonal Cycling *887*
      Annual Production *887*
      Current Patterns *887*
   References *888*

12 **Energetics** *893*
   Introduction *893*
   A Model Energy Budget *894*
   Digestibility of Biomolecules *896*
   Energy Value of Biomolecules *897*
   Measuring Metabolic Rate *898*
      Oxygen Consumption Rate-Modifying Factors *898*
      Activity *899*
      Experimental Protocol *900*
      Routine Metabolic Rate *900*
      Animal Size as a Modifier of Metabolism *901*
   Life History Strategies *902*
   Metabolism and Composition of Pelagic Species *914*
      Metabolism of Euphausiids, Decapods, Mysids, and Amphipods *929*
      Proximate Composition of Pelagic Decapods, Mysids, and Euphausiids *930*
         Terminology *931*
         Trends with Depth of Occurrence *931*
         Seasonal Changes *932*
         Trends Across Systems *932*
      Metabolism of Mesopelagic Fishes *933*
      Proximate Composition of Pelagic Fishes *934*
         Trends with Depth of Occurrence *934*
         Trends Across Systems *934*
      Energy and Life History in the Midwater Fauna *935*
         Midwater Fishes *936*
         The Cephalopods *938*
         Pteropods, Nemerteans, Annelids, Salps, and Pyrosomes *938*
         The Cnidaria and Ctenophora *939*
   Conclusions *939*
   References *940*

Appendix A Classification of the Chordata *945*

Glossary *959*

Index *963*

Preface

Most of the planet earth (over 60% of it) is deep ocean. Within the oceanic realm are two basic ecosystems, the ocean bottom, a two-dimensional environment containing creatures that creep, crawl, burrow, or lie in wait for prey, and the immense, three-dimensional pelagic region that lies above it, the largest living space on the planet, containing the swimmers and drifters. The deep ocean bottom has been the focus of a lot of excitement over the last 50 years, with many expeditions to the fabulous communities inhabiting the hydrothermal vents at our planet's oceanic ridges. More recently, the thrilling discovery of deep coral reefs off Australia has captured the public eye, showing that still more oceanic discoveries may yet await us.

More fascinating yet are the communities of marine animals that inhabit the oceans' pelagic realm, and the creatures' adaptations to an environment devoid of barriers to movement in three-dimensional space. Many people are familiar with the term "plankton," the tiny plants and animals that drift with the ocean currents. More are familiar with the large pelagic species such as tuna, sharks, and swordfish, not only from pictures or fishing trips but also from the dinner table. The large, highly capable swimming species like tuna and sharks are termed "nekton." In between the tiny drifters and the strong swimmers are an entire community of animals that are familiar mainly to oceanographers but are the critical link between the small and the large. Animals in the intermediate community are not as capable at swimming as the tunas but are better at it than the small zooplankton. Collectively, the creatures are known as the micronekton and macrozooplankton, and they make up one of the largest animal communities on planet earth.

The nekton, micronekton, and macrozooplankton include a variety of different animal groups. Several different families of fishes are represented, many with unusual adaptations such as light organs like fireflies, huge gapes to allow them to swallow prey larger than themselves, and large tubular eyes. Among the invertebrates are shrimp similar to the ones we enjoy in shrimp cocktails and other crustaceans that can produce clouds of biological light or live inside jellyfishes. Among the jellies are species larger than a meter across and those that can double their population size in a matter of days by reproducing asexually.

The blue water pelagic community is truly fascinating. The problem is, to learn about the wonders we have briefly described above, you currently need to access many sources. *Life in the Open Ocean: The Biology of Pelagic Species* gathers the information available on the wide array of taxa making up the pelagic community and presents it as one cohesive whole. It is a synthesis of the information available on the biology of all the groups you will see if you tow a scientific trawl net between the surface and bathypelagic depths. When you bring the net up and look at your catch, you are looking at a community of coexisting species with each group having its own way of solving the problems posed by nature. The book combines basic information about the different animal groups as well as their different strategies for solving natures' challenges.

Topics covered in the book include basic physical oceanography, properties of water, physical variables that covary with depth such as light, temperature, and pressure and how animals have adapted to cope with each. Animal groups covered in depth (no pun intended) include the Cnidaria, or stinging jellies, the ctenophores or comb jellies, pelagic nemerteans, pelagic annelids, the Crustacea, the Mollusca including the "swimming snails" and cephalopods, the invertebrate chordates, including the salps, pyrosomes, and larvaceans, and lastly, the incredible fishes, focusing on the micronekton but also including the sharks, tunas, mackerels and mahi-mahi.

Within each of the animal phyla the pelagic groups are identified and detailed coverage is provided for classification and history, internal and external anatomy, vertical and geographic distributions, locomotion and buoyancy, foraging strategies, feeding and digestion, bioluminescent systems and their function, reproduction and development, respiration, excretion, nervous systems, heart and circulation and all sensory mechanisms: vision, mechanoreception (touch, balance, and vibration) and chemoreception (smell and taste).

Life in the Open Ocean: The Biology of Pelagic Species is written so that it can be used as a textbook at the advanced undergraduate or graduate level of instruction, and as a reference for those interested in marine biology including professors, interested undergraduates, and perhaps for High School teachers teaching at the AP level. It is our fondest hope that it will make open-ocean biology considerably more accessible, increasing its visibility and its presence in college-level science curricula.

Joseph J. Torres
Thomas G. Bailey

Acknowledgments

Our heartfelt thanks to Dr. Dante Fenolio for the use of his fabulous images in *Life in the Open Ocean: The Biology of Pelagic Species*. They enabled us to show structures such as eyes, photophores, barbels and lures as well as whole animal appearance with freshly captured specimens, a boon for descriptions and a treat for the reader. As we progressed through the animal groups, specific questions arose that our longtime colleagues gave us help with. Advice on all things fish came from Dr. Tracey Sutton; questions on neurophysiology and bioluminescence were covered by Drs. Tamara Frank and Edie Widder; crustacean help came from Drs. Scott Burghart, Robin Ross, and Langdon Quetin; Claudia Mills gave us help with classification in cnidarians and ctenophores; and Eileen Hofmann and John Klinck helped with Antarctic physical oceanography. Special thanks to Bruce Robison and Alice Alldredge for the opportunity to dive WASP and Deep Rover many years ago. JJT would like to thank George Somero for a great sabbatical year and much biochemical advice over the years and Jim Childress for being the deep-sea sage that he is. Many thanks to our friend Ms. Cynthia Brown who obtained many important publications for us as head of the interlibrary loan office at USF St. Petersburg.

Study of open-ocean biology requires going to sea in ships, and running a trawling or diving program at sea requires a team. As one progresses from being a participant in early days to principal investigator, field-team leader, and chief scientist, the teams change, as do the ships, the nets, and if you're really lucky, the submersibles. But, also if you're lucky, you get to keep a few colleagues for several years. Foremost among those for JJT was his longtime research associate Joe Donnelly, colleague over two decades, and his co-principal investigator Dr. Tom Hopkins, peerless zooplankton biologist and expert on the zooplankton fauna in three oceanic systems. For TGB his tireless research associate and good friend Gary Owen as well the crews of the ships and Johnson-Sea-Link submersibles at the Harbor Branch Oceanographic Institution provided invaluable support throughout his research career. All our colleagues mentioned above were with us on multiple cruises as well. JJT would like to thank my many graduate students, most of whom went to sea with me several times, some to the Antarctic, some to the Gulf of Mexico, some to the Caribbean, and some to Cariaco Basin. All are remembered here, and the reader will see their names many times in the text. In several cases, we were one science party in a multidisciplinary science team, examples of which were the AMERIEZ program discussed in the text, and the Southern Ocean GLOBEC program. Lastly, all the science teams were aboard research vessels, and the captains and crews of those vessels, and in the Antarctic our science liaison officers, made everything possible. Thanks to all.

Lastly, the authors would like to thank the National Science Foundation and NOAA's National Undersea Research Program for funding our research over many years. Without their support, the research reported in here would have never happened, and not only ours but that of the multiple other labs whose research is cited in the book. Special recognition to Dr. G. Richard Harbison, exceptional gelatinous zooplankton biologist and never-ending source of good humor who has crossed the rainbow bridge. We miss him.

1

Physics and the Physical Environment

For many beginning the study of oceanic fauna, the ocean itself is a fairly mysterious place. We know that it is vast, deeper in some places than others, and that the deep sea is cold and dark. What is less clear is how physical factors vary over the global ocean and why they are the way they are. The purpose of this first chapter is to briefly describe the physical factors impacting pelagic (open ocean) animal life and how those factors are distributed in the world ocean in the horizontal and vertical planes. Physical factors play an important role in shaping the adapted characteristics of animal life, particularly physiological characteristics, and by virtue of being physical factors they vary predictably in space and time.

One of the main purposes of this book is to give the reader an appreciation of pelagic communities, with as many of the players being treated as possible and with an accent on the community as a whole. Oceanic communities are constrained to water masses, identifiable (sometimes very large!) parcels of water, because the species comprising those communities live out their life histories in a discrete region and those regions have predictable characteristics to which life has become adapted. Adjacent oceanic regions that harbor fundamentally different communities presumably must differ enough physically and be separated enough from a biological perspective, that selection can change species composition. Physical factors play a big part in that selection process.

The physical factors limiting the distributions of open-ocean species are temperature, oxygen, light, and pressure. Salinity, an important variable in estuarine systems, is of far less importance in the open ocean. Salinities in the open sea vary from approximately 33 parts per thousand (ppt or ‰, a 3.3% salt solution) to 38 ppt (a 3.8% solution), which is not a sufficient fluctuation to act as an important selective pressure on pelagic fauna. However, salinity does act indirectly to influence oceanic communities as it is an important operator in ocean circulation and the formation of water masses. And it does vary enough to be useful in identifying water masses when plotted against temperature in a T-S diagram, discussed later in this chapter.

An individual animal's interactions with the open-ocean environment are governed not only by temperature and pressure but also by the properties of water as a fluid. How fast a shark sinks relative to a jellyfish is within the province of basic fluid dynamics, as are the forces acting on the swimming individuals as they make their way quickly or slowly through the fluid medium.

Life in the Open Ocean: The Biology of Pelagic Species, First Edition. Joseph J. Torres and Thomas G. Bailey.
© 2022 John Wiley & Sons Ltd. Published 2022 by John Wiley & Sons Ltd.

Physics and the Physical Environment

Since the distribution and characteristics of oceanic life are dictated partly by the characteristics of the physical environment, it is imperative that one have a fundamental understanding of how that environment varies with location and of some of the principles that govern the variability. This chapter will cover enough of the basics to provide an understanding of the physical environment of the open sea and the biological interaction with it.

The Vastness of the Open Ocean

A few facts about the open ocean are important to help put the vastness of the oceans in perspective. To appreciate the total living space available to pelagic fauna, we need to consider both the ocean's surface area and the volume beneath the surface: the ocean's horizontal and vertical extent. The ocean basins cover 71% of the planet and their average depth is 3800 m (Sverdrup et al. 1942). Since the Earth is a sphere with a radius of 6371 km, its total surface area is calculated at 5.1×10^8 km^2. The surface area of the world's oceans at 3.6×10^8 km^2 is ~71% of the total surface area of our planet. Consider volume. The average depth of the ocean basins is ~3800 m so the volume of water contained in the ocean is ~1.335×10^9 km^3. The volume of the moon is 2.2×10^{10} km^3; the volume of the world's oceans could form a body over half (60% in fact) the size of our moon. Thus the oceans contain an immense volume of water that affords habitat to the creatures that live within it.

We can arrive at a similar volume measurement for terrestrial systems by assuming that biologically useful space on land is defined by the height of a tall tree (0.05 km). Using 29% of the total surface area of the Earth for the terrestrial ecosystems and multiplying that by the height of the tree, we arrive at one useful-volume estimate of 7.4×10^6 km^3, which is only 0.5% of the volume in the oceans (cf. Herring 2002).

It is interesting to look at extremes. The highest point on Earth above sea level is Mt Everest at 8.55 km. The lowest point on Earth is in the Pacific Ocean, the Challenger Deep in the Marianas Trench near the Philippines at 10.93 km below sea level.

The total living space in the pelagic realm can also be compared with that in the terrestrial biome by evaluating surface area. The Hypsographic Curve of the world in Figure 1.1 expresses graphically the amounts of the planet's surface, by elevation, above and below sea level. From the Curve it is easy to see that over 60% of the Earth's surface is covered by water of depths greater than 1000 m. Since 97% of all the water on Earth is contained in the oceans, the open sea at 1000 m deep and below is therefore the Earth's "average environment"!

Freshwater in rivers and lakes accounts for about 1% of the Earth's water; glaciers, polar ice, and groundwater contain about 2% more. The remainder of the Earth's water is in the vastness of the oceans.

The Properties of Water

The physical properties of water do much to shape the Earth's climate, while the density of water literally influences the shape of marine species. Water exhibits many unusual characteristics. It is the only compound that can be found in all three phases (solid, liquid, and gas) at the temperatures encountered at the Earth's surface. Water has a high specific heat, which means that it must absorb a great deal of heat energy for its temperature to rise by 1°C (1.0 cal g^{-1} at 10°C). Because it can absorb a lot of heat, it

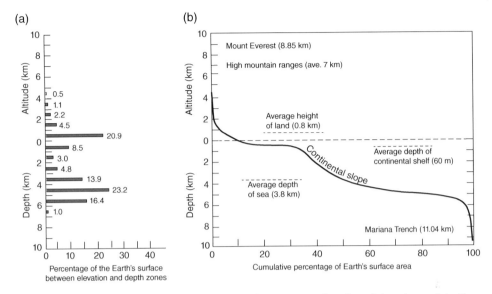

Figure 1.1 The Earth's surface area. (a) Earth's surface area as a function of elevation and depth. (b) The Hypsographic Curve; cumulative percentage of earth's surface area as a function of elevation and depth. *Source:* Gage and Tyler (1991), figure 2.4 (p. 12). Reproduced with the permission of Cambridge University Press.

can also deliver a lot of heat to regions far removed from where the heat was absorbed, and ocean currents such as the Gulf Stream do just that. The climate of Great Britain, for example, would be far colder than it is were it not for the warm maritime influence of the Gulf Stream.

When water changes its physical state, a great deal of energy is gained or lost. When ice is formed, $80\,cal\,g^{-1}$ is liberated as the latent heat of fusion. This property is particularly important at the poles during the freezing and melting of sea ice. Similarly, $540\,cal\,g^{-1}$ must be applied to cause liquid water to evaporate (latent heat of evaporation) and an equal amount of energy is released upon condensation. Water's freezing (0 °C) and boiling (100 °C) points are far higher than those of closely related compounds: H_2S, e.g. boils at about −59 °C.

Water has a high thermal conductivity $(0.587\,W\,m^{-1}\,°K^{-1}$ at 10 °C), so heat diffuses readily through it. When well mixed, like the wind-mixed surface layer of the ocean, large volumes of water can be quite homogeneous in temperature. Water has a high surface tension, which makes it fairly "sticky." Small volumes will form drops, capillary action will cause water to readily invade a small tube, and small water-dwelling animals must contend with a fairly viscous environment.

Water's most unusual characteristic by far is that it is less dense in its solid form than in its liquid phase. It is unique in the physical world in that regard, and it is because of that distinctive characteristic that ice floats. Pure water reaches its maximum density at 4 °C. Seawater does not reach maximum density until −3.5 °C (Vogel 1981), which is below its freezing point at −1.9 °C and ice crystals have already begun to form. When seawater freezes, the salt is excluded as a brine; the ice itself is fundamentally salt-free.

Density

As Vogel (1981) observed, the concept of mass when applied to the inherent shapelessness of fluids is a bit awkward and for practical purposes is replaced by density. The density of pure water is about 830 times that of air, or about 1000 kg m^{-3} at 0 °C and atmospheric pressure. It only varies by about 0.8% in density over the biological range of temperatures (0–40 °C) despite our attention to the fact that its maximum density is above its freezing point. Density of water is even less sensitive to pressure; it increases by only 0.5% with each kilometer of depth (Denny 1993).

Salinity alters the density of water considerably more than does temperature or pressure. The difference in density between freshwater and seawater is substantial. Salinity itself is the amount of dissolved material expressed in g per kg of seawater. The material consists mainly of salts; the principal salt is sodium chloride. The nicely intuitive definition of salinity as g per kg, or parts per thousand, has been replaced in some circles by the introduction of practical salinity. Practical salinity (S_p; UNESCO 1983) is the ratio of the electrical conductivity of a seawater sample and to a standard solution of potassium chloride. Since it is a ratio, practical salinity has no units. It is very close to actual salinity, though. To get back to actual salinity in g per kg from practical salinity, you use the following equation: $S = 1.00510 S_p$ (Bearman 1989).

The fact that water is far denser than air has its good points and bad points from the perspective of a swimming animal. On the plus side, it means that much less structural investment is required to support the weight of an organism in water than on land. A popular analogy compares a tree and a kelp of equal height above the substrate. Clearly, the kelp has far less energy invested in its 0.05 m diameter stipe than the tree has in its 0.5 m trunk. The principle works equally well for a jellyfish elegantly trailing its tentacles in the ocean or piled up in a soggy mass on the beach.

The aquatic medium provides buoyant support according to the difference in density between the body and the medium in which it is immersed. The weight of an object in water is described by the equation

$$\text{Weight} = (\rho - \rho_w) g V,$$

where ρ (rho) is the density of the object, ρ_w is the density of water, g is the acceleration of gravity (9.8 m s^{-2}), and V is the volume of the object in question. The expression $(\rho - \rho_w)$ is the effective density or ρ_e of our submerged body and determines whether it will float, remain suspended, or sink. Its effective weight in water is thus $g \cdot \rho_e V$, and it follows logically that a body will weigh more in air than in water, usually by 5- to 50-fold (Denny 1993). Do not be guilty of synonymizing weight and mass. Mass is a scalar quantity measured in kilograms; weight is a force that is measured in newtons. You will notice that the mass of an object in water does not change, but its weight does.

Seawater at a salinity of 35‰ has a density of 1024 kg m^{-3}, meaning that marine species enjoy more buoyant support from their medium than do their freshwater counterparts. Knowing what we know about relative weights in air and water, neutral buoyancy for marine species will be achieved with a density equal to that of seawater. Let us compare the density of some common biological materials. Mollusk shells at 2700 kg m^{-3} are quite dense, providing protection and support for the soft tissues beneath but also assuring that they are most useful in bottom-dwelling species. Cow bones are also quite dense, 2060 kg m^{-3}, providing the skeletal support needed by a heavy animal in air. Neither structure is appropriate for a species concerned with remaining suspended in mid-water, so the likelihood of cows invading the marine environment remains low. In contrast, muscle is 1050–1080 kg m^{-3}, only about 5% higher than the density of seawater. Fats are slightly less dense than seawater, 915–945 kg m^{-3} so they provide a source of static lift for marine species. It is instructive to note that small changes in an animal's density can confer big advantages to its weight in water but would do little to affect its weight in air. The energetic advantages of neutral buoyancy have done much to influence how pelagic species are put together. In succeeding chapters, we shall explore buoyancies and mechanisms for achieving neutral buoyancy in open-ocean taxa.

Viscosity

The first characteristic of fluids that must be appreciated for an understanding of viscosity is the "no-slip condition" with respect to solids. That is, at the interface between a solid and a fluid flowing over it, the velocity of the fluid is zero. A zero-velocity boundary layer is created, whose thickness depends on the velocity of the fluid flow. At the solid–fluid boundary, fluids stick to solids absolutely. Any object in a flow thus creates a shear, as the fluid particles at the no-slip boundary must be moving at a different velocity than those at a distance from the body in the flow.

The resistance to shear is the dynamic viscosity of the fluid, and it is best described using the classical "flat plate" analogy. Imagine two plates of negligible thickness oriented parallel to one another and with a fluid between them (Figure 1.2). The bottom plate is stationary, and the movable plate is pushed forward with a force that results in a constant velocity. Since the fluid at the boundary of each plate has zero velocity due to the no-slip rule, the fluid between them is deformed or sheared, and a uniform velocity gradient develops between them in response to the constant pushing. The force needed to maintain the constant velocity depends upon the velocity itself, upon the area of the plate because the amount of fluid that needs to be moved is a function of the size of plate, and upon the viscosity of the fluid, or how easily the fluid is deformed. The resulting equation is:

$$F = US\left(\frac{\mu}{l}\right) \text{ or } \mu = \frac{Fl}{US} \quad (1.1)$$

where U is velocity, S is the area of the plate, μ is viscosity, and l is the distance between the plates. Think of the flat plates as the

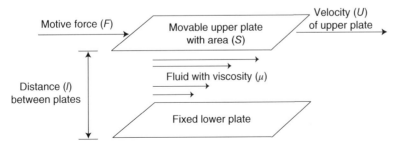

Figure 1.2 Dynamic viscosity (μ). A fluid's "stickiness" or resistance to shear. Expressed mathematically as $\mu = Fl/US$.

bottom and top card in a deck of playing cards. The fluid is all the cards between them, and the stickiness between the cards determines how easy it is to deform the deck. Viscosity is that stickiness. The viscosity (dynamic viscosity) is an important property of fluids because it determines how easy it is to move them and to move through them.

A second type of viscosity is quite important in understanding flow around and through objects: the kinematic viscosity or υ. It is the ratio of dynamic viscosity (μ) to density (ρ):

$$\upsilon = \frac{\mu}{\rho} \quad (1.2)$$

Kinematic viscosity is considerably less easy to grasp on an intuitive level, but it relates two important properties of a fluid that will be significant to us in examining the locomotion of open-ocean fauna. Viscosity and density have much to do with patterns of flow around an organism. On the one hand, viscosity measures how adjacent particles retard a fellow fluid particle's movement when it encounters a body in a flow. On the other, density is a measure of how likely it is that a fluid particle will keep moving. The ratio of the two forces, inertial and viscous, is the subject of our next topic, the Reynolds number.

Reynolds Number

Osborne Reynolds observed that a dye stream introduced into a liquid flowing through a pipe would yield a nice linear (laminar) flow or a turbulent disturbed one depending upon three characteristics of the liquid and one of the pipe. The velocity of the flow, the density of the liquid, the viscosity of the liquid, and the diameter of the pipe determined whether the flow was laminar or turbulent. Manipulating any one of the four variables was equally effective in changing the characteristics of the flow. The relationship between those variables is described in the equation for Reynolds number:

$$\mathrm{Re} = Ul\left(\frac{\rho}{\mu}\right) \quad (1.3)$$

where U is the velocity of the flow, l is the diameter of the pipe, and ρ and μ are by now familiar as density and viscosity, respectively. You will also note that the ratio of density and viscosity gives us the inverse of the kinematic viscosity, which can then be put in the denominator:

$$\mathrm{Re} = \frac{Ul}{\upsilon} \quad (1.4)$$

A Reynolds number of 2000 marks the transition between laminar and turbulent flow. The diameter of the pipe (l) for a swimming organism becomes instead the greatest length of the organism in the direction of flow.

The Reynolds number is a dimensionless number, i.e. it has no units, and it is a very useful tool for describing the characteristics of flow around a submerged body. It is the ratio of the inertial (velocity × characteristic length × density) to the viscous (dynamic viscosity) forces mentioned earlier. In fact, the Reynolds number can be derived by dividing the formula for momentum (ρSU^2) (inertial force) by that for viscous force (Eq. 1.1).

$$\frac{\rho SU^2}{US\left(\mu/l\right)} = \frac{\rho lU}{\mu} = \mathrm{Re} \quad (1.5)$$

The derivation is done in slightly different ways in Vogel (1981) and in Denny (1993). I have followed Vogel's derivation here. Both books are highly recommended for a more thorough treatment of the forces summarized here.

The useful property of Reynolds number is that you can get a good idea of the physical characteristics of a flow regime with a single number. Low Re (less than 1), such as that experienced by a protist or the moving limb of a swimming crustacean, is dominated by viscous forces. Flow will be laminar. A small swimming crustacean may have Re in the neighborhood of 100–2000 where inertial forces predominate (Torres 1984) but flow is largely laminar. In contrast, a tuna swimming at $10\,\mathrm{m\,s^{-1}}$ with an Re of 30 000 000 (Vogel 1981) is in a highly turbulent flow regime. Most of the species of interest in this book live with Reynolds numbers in the 100s–1000s when moving and feeding.

To get an intuitive sense for the world in which pelagic species live, we need to know what forces they must generate or overcome in order to move and to breathe. Our next topic deals with two of the most important forces acting on any swimming animal: friction drag and pressure drag.

Drag

Recalling the no-slip rule for a solid body in a flow, it follows logically that a swimming individual will be creating a shear as it moves its "no-slip" form through the water. The shear will be resisted by the viscosity of the fluid just as in our example with the flat plate, and the magnitude of that resistance will depend on how much surface is exposed to the flow. The resistance is called the friction drag, or skin friction drag, and since friction is something we have all experienced, it is a pretty easy one to comprehend. Friction drag, being in the province of viscosity, is most important at low Reynolds numbers.

Pressure drag may be less easy to appreciate, but it is very important. If you envision a solid object in a flow (Figure 1.3), e.g. a cylinder with its axis normal to the flow, it is an easy matter to see that the pressure on the object will be highest on its upstream face since it is oriented directly into the flow and takes the brunt of the moving fluid. The shape of the pipe results in an acceleration of the fluid as it passes over and under the pipe. Because of Bernoulli's principle, the increased velocity of the fluid results in a decreased pressure on the downstream side of the cylinder. The differences in pressure from the upstream to the downstream side are rather like pushing from the front and sucking from the rear, and those differences create pressure drag. The expression for dynamic pressure (q) is:

$$q = \frac{\rho U^2}{2} \tag{1.6}$$

which you may recognize as the "fluid equivalent" of the expression for momentum. In fact, drag is the removal of momentum from a flowing fluid.

All swimming species experience drag as they swim. At low Reynolds numbers, usually in smaller species swimming at slower speeds, friction drag is the more important force. At higher Re, pressure drag is more important. Because drag is difficult to measure and more difficult to predict from theory, total drag is most often used to describe the drag force, without discriminating between friction and pressure drag.

Total drag depends on three elements. The first is the dynamic pressure, as expressed earlier, which itself is a function of the velocity of the fluid and its density. The second is the size of the object in the flow since pressure is a force per unit area. The third is the shape of the object in the flow.

Physics and the Physical Environment

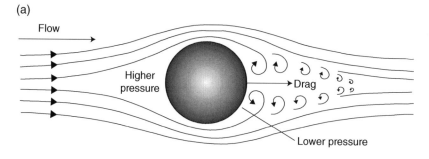

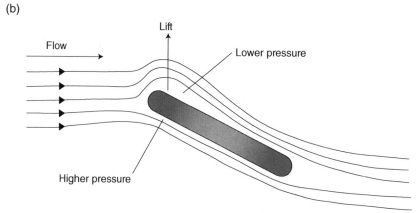

Figure 1.3 Flow-induced differential pressure. (a) Drag. (b) Lift.

Size can take a few different forms. The most common is the frontal, or the maximum cross-sectional area of the object in the flow. It can also be the total area of the object in the flow or the wetted surface area. When giving a value for drag, the way size has been expressed is quite important.

The way drag force behaves with different shapes in a flow regime cannot be predicted from first principles except for the simplest shapes, such as a sphere. Instead, drag force must be measured empirically to create a fudge factor known as the drag coefficient or C_d. The drag coefficient in turn depends on the character of the flow regime itself, which is represented by the Reynolds number. So drag is represented by the equations

$$D = \frac{C_d \rho S U^2}{2} \quad (1.7)$$

and

$$C_d = f(\text{Re}) \quad (1.8)$$

The difficulty of predicting how drag behaves for different shapes in a flow is a real problem for those interested in estimating the cost of overcoming drag to swimming species. You cannot just look up a C_d for your target species unless you happen to be very lucky. Usually you have to use the next best thing, which is a value determined for the closest shape you can find (see e.g. Batchelor 1967).

Temperature

As a factor governing the distribution of open-ocean species, temperature is huge. By far, most pelagic species are ectotherms:

their internal temperature matches that of their environment. All invertebrates and all but a very few species of fishes are ectotherms. Because of their internal biochemistry, ectothermic species have a range of temperatures that is optimal for their survival and that in turn dictates their range, or distribution, within the open-ocean ecosystem. In contrast, marine mammals and seabirds are endotherms: they produce and maintain their own body heat. Endothermy confers some independence from the tyranny of ocean temperature. Many species of marine mammals, particularly the great whales, range over huge distances and experience swings from polar to tropical temperatures each year. A few species of fishes, notably the tunas and mackerel sharks, are regional endotherms or heterotherms: capable of trapping the heat produced in their swimming muscle by using a heat exchanger in their circulatory system. The temperature within the muscle reaches temperatures rivaling that of mammals, allowing regional endotherms to be quite effective swimmers indeed.

Temperature in the ocean varies predictably in both the horizontal (latitude and longitude) and vertical (depth) planes. However, to understand why temperature varies with depth and latitude in the way that it does, we need to know a little about ocean circulation. In turn, to understand ocean circulation, we need to have a clear mental picture of the geography of the ocean basins.

The Oceans and Ocean Basins

Since the year 2000, when the Southern Ocean gained official status from the International Hydrographic Organization, the Earth has had five officially recognized major oceans: The Arctic Ocean, The Atlantic Ocean, The Indian Ocean, The Pacific Ocean, and the Southern Ocean. Vital statistics for the five oceans are given in Table 1.1.

The largest and deepest ocean basin is the Pacific, whose sheer vastness really requires a globe to appreciate. The Atlantic and Indian Oceans, each somewhat less than half the size of the Pacific, follow in size. The two polar oceans, the Southern and Arctic, are the smallest. The Southern Ocean, extending from a latitude of 60° south to the Antarctic Continent, comprises the southernmost portions of the Pacific, Atlantic, and Indian Oceans, which is why it was only recently officially recognized as an Ocean in and of itself. It is worth mentioning here that the Arctic and Antarctic are fundamentally quite different. Both are

Table 1.1 Characteristics of the ocean basins.

	Ocean				
	Atlantic	Arctic	Indian	Pacific	Southern
Ocean area (10^6 km^2)	82.22	14.06	73.48	165.38	20.33
Mean depth (m)	3600	1117	3963	4200	4000–5000
Maximum depth (m)	9560	4440	7725	11034	7235
Location of maximum depth	Puerto Rico Trench	Eurasia Basin	Java Trench	Marianas Trench	South Sandwich Trench

quite cold, of course, and sea ice plays a large role in the ecology of each. However, the Antarctic is a land mass surrounded by ocean, whereas the Arctic is an ocean surrounded by land (Figure 1.4a and b).

The continents define the boundaries of the ocean basins. Within each of the basins, a characteristic circulation transports large quantities of water with all the elements that such bodies of water contain, including plants, animals, gases, salt, and heat. Energy for the water movement is provided by the radiation of the Sun and the rotation of the Earth. The Sun's heat drives circulation within the atmosphere, producing the Earth's prevailing wind patterns that in turn drive surface ocean circulation. Deep ocean circulation, which has a more profound vertical component, is driven by changes in seawater density. Cooler or more saline water will sink below a warmer or less saline body of water. Since the density of seawater is determined by its temperature and salinity, solar radiation ultimately is responsible for deep-ocean circulation as well. Latitudinal gradients in temperature cause heat loss or gain across the ocean–atmosphere interface. Changes in salinity result from evaporation, precipitation, and in polar regions from the freezing and melting of sea ice. An understanding of the influence of the Earth's rotation on ocean circulation is less intuitive; that influence is described in the next section.

Ocean Circulation

For our purposes, ocean currents are of two basic types: surface and deep. Surface currents (Figure 1.5) are responsible for transport of about 10% of the oceanic volume and extend to about 400 m. The Earth's prevailing winds interact with surface waters through friction, literally driving the water before them and forming large current systems, or gyres, in each of the ocean basins. The gyres circulate clockwise in the northern hemisphere and counterclockwise in the southern. The trade winds, persistent surface easterlies between the equator and about 30° latitude, are a major driving force for the central ocean gyres and were critical to past exploration and commerce. Deep-ocean currents (Figure 1.6) transport the remaining 90% of the oceanic volume and are the result of the Earth's thermohaline circulation or the "Great Ocean Conveyor" (Broecker 1992).

Surface Currents: Ocean Gyres and Geostrophic Flow

Four forces combine to produce the surface currents in the global ocean: surface winds, the Sun's radiation, gravity, and the **Coriolis force**. Let us begin with the most

Figure 1.4 The Polar Oceans. (a) Arctic Ocean basin. (b) Southern Ocean surrounding Antarctica. *Source:* NASA. See color plate section for color representation of this figure.

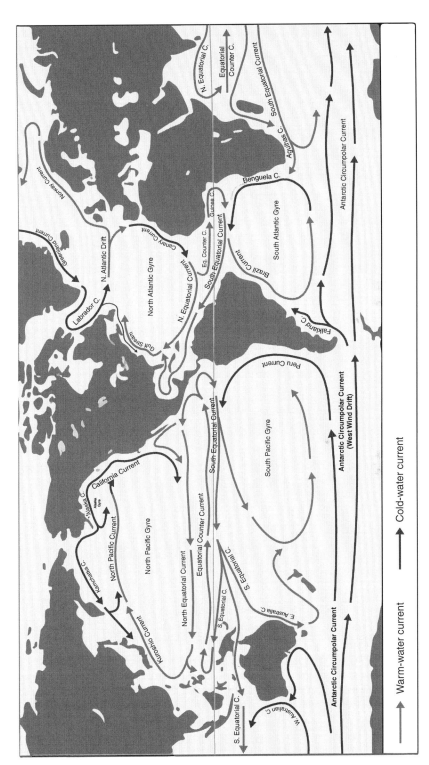

Figure 1.5 Geostrophic (surface) currents. *Source*: NASA. See color plate section for color representation of this figure.

Figure 1.6 Thermohaline (deep) currents. *Source:* NOAA. See color plate section for color representation of this figure.

complicated: the Coriolis force, sometimes known as the Coriolis effect.

Coriolis Force

The Coriolis force, as it applies to water movement in the oceans, is a result of the fact that the vast majority of the ocean's volume is only loosely coupled to the surface of the Earth. The "no slip" condition, discussed in the section on viscosity, applies only to the boundary formed by the ocean bottom. The remaining volume of the ocean moves with the Earth as it rotates, but there is slippage due to the near absence of frictional coupling.

Consider the fact that the Earth is a sphere rotating from west to east. Now imagine that the Earth is sliced in half at the equator and we are looking down at it. It would appear as a disk that rotates through 360° in a 24-hour period. Now imagine taking another slice of the Earth about halfway between the equator and the North pole. This second slice will also appear as a disk but a smaller one than the disk described by the equator (Figure 1.7a). It will also rotate through 360° in a 24-hour period but, because the circumference of this smaller disk is less than the equator's disk, its velocity of rotation is also less. A city on the equator – Bogotá, Colombia, for example – is moving at a velocity of $1668\,km\,h^{-1}$ as the Earth rotates. Meanwhile Ottawa, Canada, at about 45°N, is moving at $1180\,km\,h^{-1}$ (Figure 1.7b). The difference in relative velocities with latitude accounts for a difference of nearly 500 km traveled per hour and is at the heart of the Coriolis effect.

A classic example of the effect of the Coriolis force is to envision a missile or cannonball being fired northward from Bogotá at the equator. Let us stick with the missile. Because it was fired from the equator, the missile has an eastward velocity of $1668\,km\,h^{-1}$ when it leaves the ground, along with its northward velocity. As it speeds north the Earth is rotating beneath it, so the velocity at the Earth's surface is declining with increasing distance from the equator. The result is that the missile, when viewed from the perspective of missile control at the equator, has veered to the right or clockwise (Figure 1.8). Imagine now a similar experiment with the missile being fired from latitude 45°N toward the equator. The missile has an eastward velocity of $1180\,km\,h^{-1}$ when it leaves the launch pad and is heading south toward the equator, which is moving east at $1668\,km\,h^{-1}$. In this case, the Earth is literally moving east more quickly than the missile is moving as it heads south and, once again, the missile appears to veer to the right or clockwise.

This brings us to three general rules about Coriolis force. In the northern hemisphere, the Coriolis force deflects moving bodies, including fluids, in a clockwise direction: to the right. In the southern hemisphere, deflection is counterclockwise, to the left. Third, Coriolis force is nonexistent at the equator and strongest at the poles. Consider also that the influence of the Coriolis force will be very much greater on slow-moving bodies such as parcels of water than on a quickly moving object such as our imaginary missile, which spends only minutes in the air.

The effect of the Coriolis force on ocean circulation is evident in Figure 1.5. In the North Atlantic gyre, the westerlies initially drive the water to the northeast, but the Coriolis force deflects the currents to the right, producing an easterly flowing current that is eventually deflected southward by the continental margins of Europe and Africa. The flow of the current is approximately 45° to the right of the wind direction. Its analogue in the South Atlantic, driven westward by the trade winds, is the North Equatorial Current. The circuit is

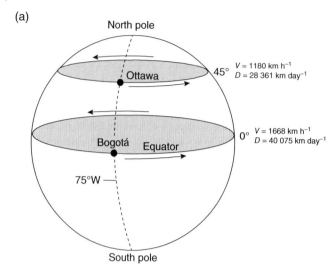

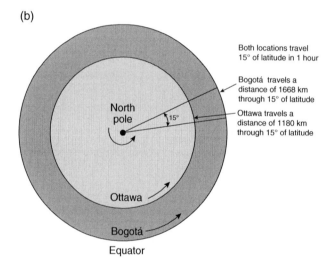

Figure 1.7 Coriolis effect. Differences in velocity of Earth's surface as a function of latitude. (a) Equatorial view. (b) Polar view.

completed by the powerful Gulf Stream on the western limb and the Canary Current on the eastern one.

Ekman Transport

Ekman Transport is another phenomenon important to an understanding of geostrophic currents. Named for the Swedish oceanographer who first described it, and other principles, Ekman Transport is the net spiral motion down a water column created by Coriolis force and drag. Away from the surface and the direct effects of the wind, the current direction veers slightly toward the right or left with increasing depth down to about 100 m. A good way to envision this is to think of the water column as a stack of layers or playing cards, with each layer moving slightly to the right (or left in the southern hemisphere) of the layer above it (Figure 1.9). Due to friction, current speed decreases with increasing depth so that each successively deeper layer moves more slowly than the one above. In theory the resultant net flow over the upper 100 m is at 90° to wind direction, 90° to the right of wind direction in the northern hemisphere, and 90° to the left in the southern. However, the actual net flow of the Ekman Spiral is closer to 45° in both halves of the globe.

Geostrophic currents are the result of a dynamic balance between the driving force of the wind, the turning effects of the Coriolis force, and pressure gradients caused by differences in sea-surface height. Ekman Transport and wind stress act to create a slight hill of water, or topographic high, roughly in the middle of a gyre. Water in the high attempts to flow downhill but is offset by the Coriolis force so that the current in the gyre becomes parallel to the elevated sea surface, flowing clockwise in the northern hemisphere and counterclockwise in the southern.

Ocean Gyres and Geostrophic Flow

Six great circuits are found in the world ocean, four in the southern hemisphere (South Atlantic, South Pacific, Indian, and Antarctic Circumpolar) and two in the northern (North Atlantic, North Pacific). The gyres correspond fairly well to the biogeographic distribution of oceanic species.

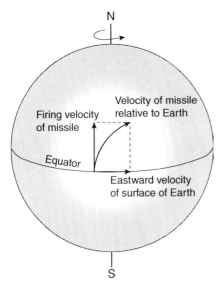

Figure 1.8 Coriolis effect. Apparent curved path of an object not coupled to the Earth's surface, moving in the northern hemisphere. *Source:* Brown et al. (1989), figure 1.2a (p. 7). Reproduced with the permission of Pergamon Press.

Figure 1.9 Ekman transport. Net spiral pattern of wind-driven motion down through a water column due to Coriolis effect and drag. The result is a net flow 90° to the right of the wind direction in the northern hemisphere. See color plate section for color representation of this figure.

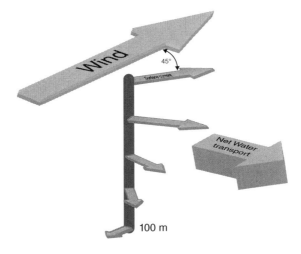

With the exception of the Antarctic Circumpolar Current (ACC), which is not technically a gyre since it flows uninterrupted around the Antarctic continent, all gyres flow around the continental margins (Figure 1.5). The circuits are bounded on the east and west by eastern and western boundary currents and on the north and south by transverse currents. The Gulf Stream in the North Atlantic is an excellent example of a western boundary current and the Canary Current, along west Africa, of an eastern boundary current. Partly because of the convergence of the westerly winds, western boundary currents are far stronger and more clearly defined than their counterparts in the east. Additional contributors to the greater strength of western boundary currents are the fact that the "hill of water" at the heart of the geostrophic circulation is displaced to the west by the eastward rotation of the Earth, creating a steeper pressure gradient and a stronger poleward flow of water. The phenomenon is termed "westward intensification." Table 1.2 gives a comparison of the characteristics of eastern and western boundary currents.

Upwelling

Wind-driven movement of water can also induce vertical circulation, particularly in coastal regions. On the west coasts of land masses in the northern hemisphere, winds out of the north or northwest cause alongshore water movement, which is moved offshore (clockwise) by Ekman transport. In the southern hemisphere, winds from the south will result in similar coastal upwelling. The water moved offshore is replaced by cooler, nutrient-laden water welling up from below (Figure 1.10), resulting in an ideal situation for increased productivity. Upwelling regions are the most fertile oceanic areas of the world, often supporting large fisheries for small coastal pelagic fishes such as sardines and anchovies. As we will see later, areas of high ocean productivity are often associated with zones of low oxygen at mid-depths, resulting from the biological degradation of sinking particulates in a stratified water column. Upwelling can also occur offshore, which it consistently does at the equator where the north and south equatorial currents meet and at the Antarctic divergence. Downwelling, the opposite situation, occurs when winds and Ekman transport cause surface water to converge along a coast.

Deep-Ocean Circulation

Oceanic waters are stratified by density, which is mainly a function of salinity and water temperature. The colder or saltier a parcel of water is, the greater its density. Temperature–salinity plots, or T-S diagrams, help characterize ocean layering and water-column characteristics. Figure 1.11, a T-S curve for an oceanographic station in the Atlantic, is an example. Note that the lines of equal density, or isopycnals, each comprise a variety of temperatures and salinities; the same density can result from many different temperature–salinity combinations.

Vertical structure in the ocean can be divided into three density zones: an upper mixed layer, a layer of changing density, and the deep layer.

The upper mixed layer is a region of fairly uniform density because of the action of wind mixing, waves, and currents. Depending on place and season, it can vary from being very shallow (<30 m) to depths of greater than 200 m and is the only region of the ocean that interacts with the atmosphere. The upper mixed layer contains about 2% of the volume of the ocean.

Beneath the upper mixed layer is the pycnocline, where density increases rapidly with depth and the increasing density acts

Table 1.2 Boundary currents in the Northern Atlantic and Pacific Oceans. Data from Gross (1990), Schwartzlose and Reid (1972), Sverdrup et al. (1942), Zhou et al. (2000).

Location	Current	Speed (cm s^{-1})	Transport (sv^a)	Common features	Special features
Western Atlantic	Gulf Stream	120–140	55	Narrow (100–150 km) and deep (2 km)	Sharp boundary with coastal circulation system; little or no coastal upwelling; waters tend to be depleted in nutrients, unproductive
Western Pacific	Kuroshio Current	89–180	65		
Eastern Atlantic	Canary Current	10–15	16	Broad (~1000 km) and shallow (<500 m)	Diffuse boundaries separating from coastal currents; coastal upwelling common
Eastern Pacific	California Current	12.5–25	10		

^{a} sv = sverdrup (1 sv = 1 million cubic meters per second)

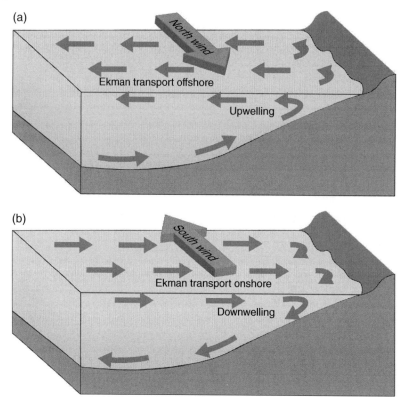

Figure 1.10 Upwelling and downwelling. (a) Ekman transport caused by wind blowing from the north moving surface water offshore, results in deeper water upwelling to the surface in the northern hemisphere. (b) Ekman transport due to winds blowing from the south moves surface water onshore and subsequently down slope. See color plate section for color representation of this figure.

as a barrier between the upper and deep layers. If declining temperatures are mainly responsible for the increasing density, the pycnocline is also a thermocline. If salinity is the major cause, it is a halocline. The pycnocline extends from the bottom of the mixed layer to the cold and stable deep zone. In most regions of the world ocean, the pycnocline ends at about 1000 m of depth and contains about 18% of the world ocean volume.

In the cold and relatively stable deep zone, temperature varies very little with depth and density increases only gradually. The deep zone contains the remaining 80% of the global ocean at depths greater than 1000 m, well away from surface influences.

Water Masses

The global ocean has a variety of different water masses, parcels of ocean identifiable by their temperature, salinity, and density characteristics that determine their place within the vertical structure of the oceanic water column. It is important to keep in mind that certain properties of a water mass are determined during its sojourn at the surface, e.g. temperature and salinity. Those are conservative properties and are changed only when a water mass mixes with another.

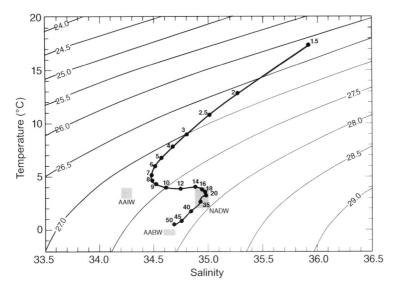

Figure 1.11 T-S diagram. Temperature–salinity plot from an oceanographic station in the Atlantic. The axes represent salinity (X) and temperature (Y). The curved lines represent isopycnals (equal density). AAIW, Antarctic Intermediate Water; NADW, North Atlantic Deep Water; AABW, Antarctic Bottom Water. *Source:* Brown et al. (1989), figure 6.26 (p. 191). Reproduced with the permission of Pergamon Press.

In the deep ocean, water masses can mix only when their densities are roughly equal, otherwise they remain stratified. Therefore vertical movements of water, away from or toward the surface, require a weakly stratified water column, such as is found near the poles. Figure 1.12 is a standard diagram of oceanic temperatures at depth at low, middle, and high latitudes.

Five generic water masses are found at temperate and tropical latitudes. Surface water extends from the surface to about 200 m depth and includes the seasonal thermocline. Central water extends from just below surface water to the bottom of the permanent thermocline, usually at about 1000 m. Intermediate water resides below central water to a depth of about 1500 m, where deep water begins. Deep water is found below intermediate water but is not in contact with the bottom; it is found between 1500 and 4000 m. Deepest of the oceanic layers is bottom water, which is in contact with the seafloor.

Each of the generic water masses has a large number of specific examples that can be identified by their temperature and salinity; they are named for their source region. Upper water masses (surface and central water), shown in Figure 1.13, are numerous and correspond fairly closely with surface currents. Intermediate waters are mapped in Figure 1.14. It is important to note the difference in areal extent between upper and intermediate waters. Antarctic Intermediate Water (AAIW), for example, is widespread at intermediate depths throughout the world ocean. The temperature and salinity characteristics that define AAIW (temperatures of 2–4 °C and salinities of about 34.2) provide a fairly uniform environment for the species that live within it. It is thus easy to understand why deeper-living oceanic species are typically far more

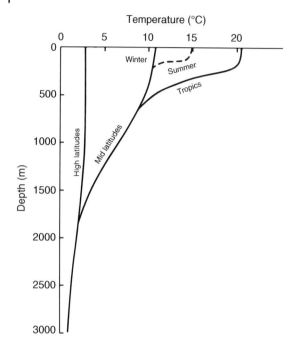

Figure 1.12 Standard depth profiles of temperature at low, middle, and high latitudes.

wide-ranging than those inhabiting surface waters (Briggs 1974).

Deep and bottom water masses are even less numerous and more widespread than intermediate water masses. North Atlantic Deep Water (NADW) forms in the north Atlantic above 60°N when cold, saline waters from the Norwegian and Greenland seas spill over the mid-Atlantic ridge into the depths of the Atlantic (Figure 1.15). Those waters mix with overlying water as well as water flowing south out of the Labrador Sea to create a general southward flow along the west side of the mid-Atlantic ridge. NADW is the most important deep-water mass in the Atlantic, extending well into the southern hemisphere.

Antarctic Bottom Water (AABW) is the most widespread of the water masses, dominating the bottom water in all three ocean basins. It is formed in winter near the Antarctic continent, mainly in the Weddell and Ross Seas, the southernmost portions of the Atlantic and Pacific Ocean basins, respectively. As discussed earlier, when ice crystals form in seawater most of the dissolved salts are excluded as brine, creating very cold and saline water. AABW is the densest water mass in the world ocean; the source water mixes very little with any less dense waters. In many areas of the Antarctic, particularly in the Ross Sea, the temperature from the edge of the continental shelf to the coast is about −2.0°C from surface to bottom.

The Antarctic is the southern end of the line for several water masses and the mixing in the Southern Ocean is complex (Figure 1.16). Warmer water from the north, including NADW, reaches the surface at the Antarctic divergence where it gives up its heat. Cooled water flows northward and sinks at the Antarctic Convergence, creating AAIW. The cycle of heat delivery, cooling, and sinking is the southern counterpart to the same basic process in the North

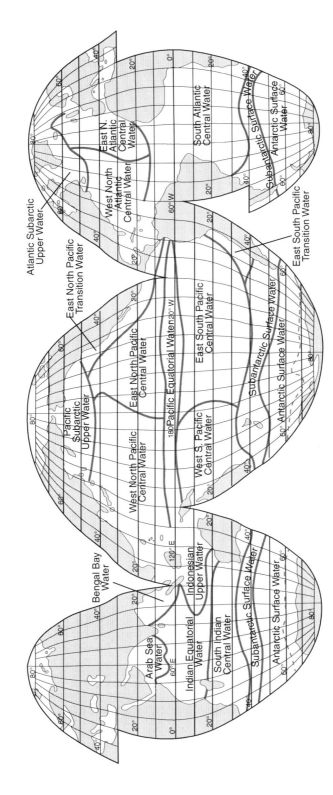

Figure 1.13 Surface and central water masses. *Source*: Lalli and Parsons (1993), figure 2.13 (p. 37). Reproduced with the permission of Pergamon Press.

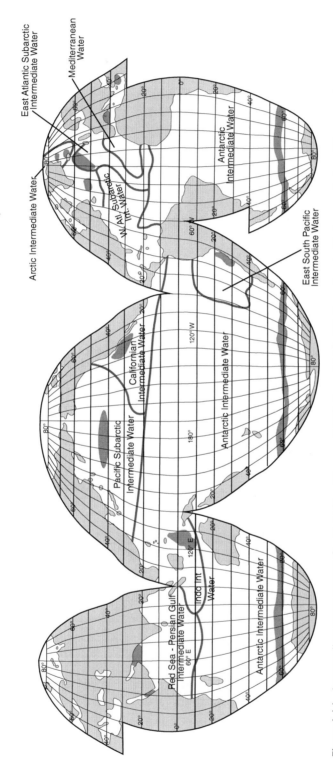

Figure 1.14 Intermediate water masses. *Source:* Lalli and Parsons (1993), figure 2.15 (p. 38). Reproduced with the permission of Pergamon Press.

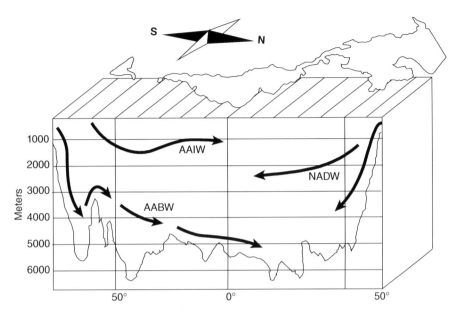

Figure 1.15 Deep and bottom water masses: sources and flow patterns. AAIW, Antarctic Intermediate Water; AABW, Antarctic Bottom Water; NADW, North Atlantic Deep Water. *Source:* Lalli and Parsons (1993), figure 2.17 (p. 39). Reproduced with the permission of Pergamon Press.

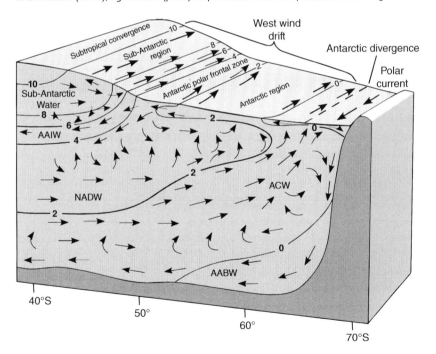

Figure 1.16 Flow patterns and mixing of water masses in the Southern Ocean. AAIW, Antarctic Intermediate Water; NADW, North Atlantic Deep Water; AABW, Antarctic Bottom Water; ACW, Atlantic Central Water. *Source:* Brown et al. (1989), figure 6.20 (p. 184). Reproduced with the permission of Pergamon Press.

Atlantic. The process at both poles is termed the thermohaline circulation or the "Great Ocean Conveyor" (Broecker 1992).

Oxygen

Oxygen is introduced into oceanic waters (and all other waters) by diffusion from the atmosphere, aided by wind-induced turbulence and mixing, and sometimes supplemented to a small degree by photosynthetically produced oxygen. Its solubility in water is an inverse function of salinity and temperature. Once a water mass has left the surface, its dissolved oxygen is consumed through time, mainly by microorganisms but also by larger species such as fishes and crustaceans. As a consequence, oxygen content is an indication of how long the water has been away from the surface. It is a nonconservative property of a water mass that is sometimes used as a tracer. Water masses vary substantially in their time away from the surface. At the extremes, AABW in the Pacific retains its character for 1600 years (Garrison 2002), whereas the residence time for most deep water is 200–300 years. NADW takes about 1000 years to reach the surface after sinking at the northern end of the Great Ocean Conveyor.

Regions of low oxygen concentration are termed oxygen minima, and they vary widely in their severity. If an oceanic region includes a highly productive surface layer with large seasonal algae blooms, the bacterial oxidation of sinking organic matter may remove nearly all the oxygen from the deeper waters. Examples of such areas include the waters off the California coast, the Eastern Tropical Pacific, and the Arabian Sea. Compare the oxygen profiles from the Arabian Sea and the California Current in Figure 1.17 with the profiles for the Antarctic and the Gulf of Mexico. Oxygen concentrations at the surface and at 500 m of depth in the world ocean are shown in Figure 1.18. Most of the world's severest minima are in regions where upwelling is very strong, and algae blooms are episodically very extensive.

Oxygen minima may also occur because of topography. For example, the Black Sea and the Cariaco Basin off the coast of Venezuela have restricted communications with the rest of the open ocean because of a shallow sill restricting circulation of their deeper waters. As a consequence, their deep waters are isolated and become anoxic.

Because deeper waters have always spent some time away from the surface, an oxygen minimum of some degree is always present. However, in most places it is not severe and not limiting to animal life. In the Gulf of Mexico, for example, the oxygen drops to about 50% of surface values at a depth of 600 m, as it does in the Antarctic. In contrast, the waters off southern California drop to about 5% of surface values, and in the Arabian Sea oxygen levels drop to zero below 200 m. Such low values pose severe challenges to animal life. Moreover, oxygen minima in the Cariaco Basin and the Black Sea include the presence of sulfides, which are metabolic poisons. How animals cope (or not) with such low levels of oxygen will be covered in the next chapter.

Pressure

The easiest physical characteristic of the ocean to understand is pressure. Pressure increases by 1 atmosphere (atm) (the barometric pressure at sea level) or 14.7 pounds per square inch (psi) with each increase of 10 m in depth. The metric unit of pressure is the pascal (Pa). One atmosphere is equivalent to 101.3 kPa. Pressure in the

Pressure | 25

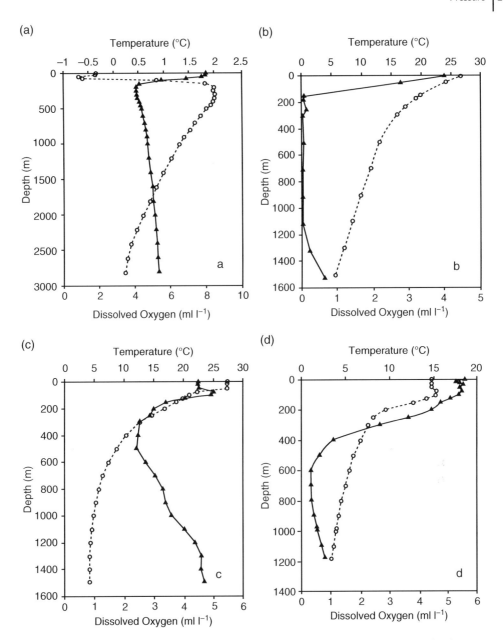

Figure 1.17 Oxygen and temperature profiles from four oceanic regions. (a) Antarctic (Southern Ocean); (b) Arabian Sea; (c) Gulf of Mexico; (d) California Current. *Source:* Torres et al. (2012), figure 1 (p. 1909). Reproduced with the permission of The Company of Biologists.

deepest point in the ocean, the Challenger Deep (depth: 10 916 m), is 16 046 psi or 1.11×10^5 kPa. In contrast, pressure at the average depth of the ocean is 5586 psi or 3.85×10^4 kPa. Pressure can be important in shaping the characteristics of species living in the deep sea and will be discussed in the next chapter.

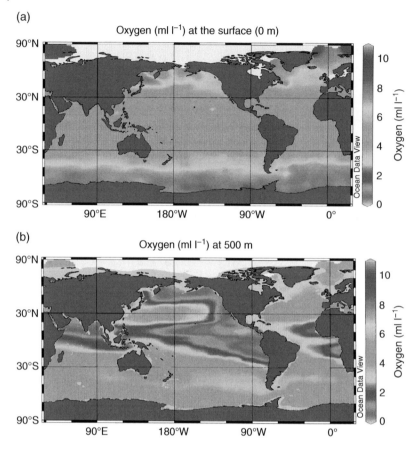

Figure 1.18 Typical oceanic oxygen concentrations. (a) Surface; (b) 500 m. See color plate section for color representation of this figure.

Sound

The ocean is sometimes characterized as a noisy place, though it may seem silent to scuba divers in the open ocean because most of the sounds are not discernible to human ears. Sound levels do vary considerably in the sea in the horizontal and vertical planes, and only a part is generated by the activities of humans. Detection of sound and other vibrations is a sensory modality shared by virtually all oceanic species. Some pelagic species, bottlenose dolphins, for example, use echolocation to locate prey, just as a bat does.

Sound levels do not vary predictably in the ocean except in a general sense. Increasing distance from the crashing waves of a rocky shore will decrease the levels of ambient sound, as will increasing depth and distance from the wind-induced turbulence of surface waters. However, the properties of sound do vary predictably, and a presentation of some basic concepts now will help with later discussions of hearing and mechanoreception in open-ocean fauna. The physics of sound is quite complex; only the most rudimentary aspects will be covered in this book.

"Sound is a longitudinal mechanical wave that propagates in a compressible medium" (Rogers and Cox 1988). A mechanical wave results from the displacement of an elastic medium from its original position, causing it to oscillate about an equilibrium position (Figure 1.19a; Halliday and Resnick 1970). The trick here is to recognize that the disturbance, or mechanical wave, moves through the medium with no resulting movement in the medium itself. A good visualization of the passage of a mechanical wave is to think of a cork bobbing while a surface wave passes underneath it. The cork moves up and down as the wave passes by, transferring some energy to it, but the cork does not follow in the wave's path.

A longitudinal wave is propagated in a back and forth motion along the direction of propagation (Figure 1.19a). This is in contrast to a transverse wave (such as an electromagnetic wave – see the treatment of light as follows), which yields a displacement at right angles to the axis of propagation. Visualize the propagation of sound through a medium by imagining the movement of a rapidly oscillating piston in a tube. As the piston moves back and forth, it creates regions of higher and lower pressure, areas of compression and rarefaction (Figure 1.19b). Tiny volumes of water

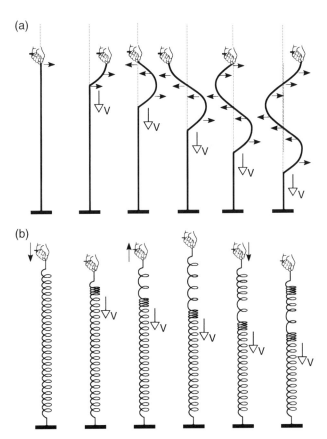

Figure 1.19 Mechanical wave propagation. (a) Transverse wave. Particles displaced perpendicular to the direction the wave travels; (b) Longitudinal wave. Particles displaced parallel to the direction the wave travels. *Source:* Halliday and Resnick (1970), figure 16.1 (p. 301). Reproduced with the permission of John Wiley & Sons.

(water particles) oscillate in place as the areas of compression, or wave fronts, of the sound wave propagate past them. The hair-like sensory elements of open-ocean fauna are designed to detect the water motion, or vibration of sound, as it moves past them.

The speed of sound in a medium is a function of the medium's compressibility: the stiffer the medium, the faster sound will propagate through it. That is why the speed of sound in water is very much faster (4.3 times faster) than it is in air. However, to know if the speed of sound varies with depth in the ocean, we need to know a little more than that. We already know that the density of water does not increase much with increasing pressure. The ratio of the change in pressure on a volume of water (Δp) to the resulting change in volume of that water ($-\Delta V/V$) is known as its bulk modulus of elasticity ("B", Halliday and Resnick 1970, Denny 1993). B is positive because an increase in pressure results in a decrease in volume (or increase in density).

$$B = -V\left(\frac{\Delta p}{\Delta V}\right) \quad (1.9)$$

where Δp is the change in pressure, ΔV is the change in volume, p is the ambient pressure, and V is the volume at the original pressure. Put in a more empirical way, the same equation can be expressed as (Denny 1993):

$$B = \frac{-(p - p_0)}{\left((V - V_0)/V_0\right)} \quad (1.10)$$

where p is the ambient pressure, p_0 is the pressure at 1 atm, V is the volume at pressure p, and V_0 is the volume at 1 atm. The bulk modulus of water is about 2×10^9 Pa depending on the temperature, which is a very considerable pressure. As mentioned earlier, the Challenger Deep at about 11 km of depth would yield a pressure of about 10^9 Pa, not nearly enough to double the density of water.

The speed of sound through water is equal to the square root of the ratio of its bulk modulus to its density (Denny 1993).

$$c = \sqrt{\frac{B}{\rho}} \quad (1.11)$$

where c is the speed of sound, ρ is the density, and B is the bulk modulus.

The declining temperature and increasing pressure with increasing depth in temperate and tropical regions act to change the speed of sound such that there is both a maximum and a minimum velocity in the top 1500 m (Figure 1.20). At the bottom of the mixed layer, the pressure has increased very slightly, increasing the B also very slightly, but the density has remained the same. The result is a maximum sound velocity at the bottom of the mixed layer. The minimum speed results from the influence of declining temperatures over the permanent thermocline. The increase in density with depth in the denominator is not enough to offset the decline in bulk modulus with depth resulting from the declining temperatures, producing a minimum speed near the bottom of the permanent thermocline. Once the bottom of the permanent thermocline is reached, the pressure continues to increase, but the temperature changes little. Pressure then becomes the main factor governing the speed of sound (Eqs. 1.10 and 1.11), which continues to increase with depth. Refraction of sound at the depths of maximum and minimum velocity has importance in submarine warfare, where echoes from pulses of high-energy sound (SONAR) are used to locate enemy submarines.

Before leaving the properties of sound in the ocean, there is one more item of

Figure 1.20 Velocity of sound in seawater as a function of depth. Maximum velocity at the bottom of the mixed layer. Minimum velocity at the base of the permanent thermocline.

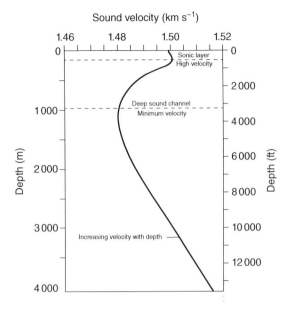

importance: the relationship between speed, wavelength, and frequency. Simply put,

$$c = f\lambda \qquad (1.12)$$

where c is the speed of sound (m s^{-1}), f is the frequency (cycles per second or Hertz (Hz)), and λ is the wavelength (m). An important thing to note here is that the wavelength of sound in water is 4.3 times that of sound in air at equal frequency.

Light

Visible light is electromagnetic radiation, or light energy, with wavelengths between 400 nm (violet) and 700 nm (red), the sensitivity of the human visual system. Plants capitalize on the same range of wavelengths for the process of photosynthesis, so the 400–700 nm range is also known as photosynthetically active radiation (PAR). Visible light is a small segment of the total spectrum of electromagnetic radiation, which ranges from very high-energy, short-wavelength, gamma rays to the longer-wavelength, lower-energy radiation of radio waves (Figure 1.21).

Light behaves as a particle (photons or quanta) and as a wave with a characteristic frequency and wavelength. The waves oscillate at 90° to the axis of movement and are called polarized if the photons have the axis of oscillation in the same plane. For normal light, the axis of oscillation is random between 0° and 180°. Polarized light is created by passing through a polarizing filter, by reflection off a surface (such as the surface of the ocean) or by scattering from small particles (Withers 1992).

The energy of a photon of light is an inverse function of its wavelength, which is expressed as:

$$E = h\upsilon = hc/\lambda \qquad (1.13)$$

where E is the energy of a photon (joules) – joules (J) are the SI unit for energy = 0.24 calories, h is Planck's constant (6.626 × 10^{-34} J s), υ is frequency (s^{-1}), λ is wavelength (m), and c is the speed of light (2.998 × 10^8 m s^{-1})

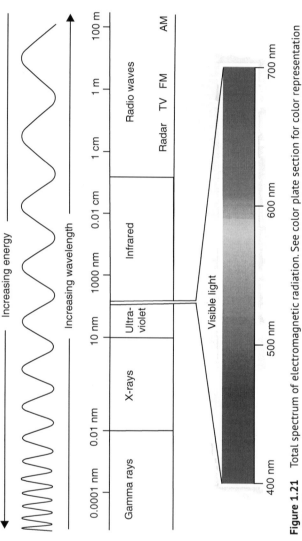

Figure 1.21 Total spectrum of electromagnetic radiation. See color plate section for color representation of this figure.

The equation may be rearranged:

$$E(\text{kJ·mol}^{-1}) = \frac{119660}{\lambda(\text{nm})} \quad (1.14)$$

where the energy of a photon of light varies from about 170 kJ mol^{-1} in red light to 300 kJ mol^{-1} in violet light. A mole is Avogadro's number (6.022×10^{23}) of photons.

Absorption and Scattering

When light impinges upon a molecule of water or a gas (or upon any form of matter), it may be either scattered or absorbed. When absorbed, the energy of the photon is entirely retained, at least for a while. When scattered, the energy of the photon is re-emitted as another electromagnetic wave of the same wavelength and is scattered in all directions. Both processes are highly wavelength-dependent.

The wavelength of light that is maximally absorbed is an inherent property of a substance, a convenient property of matter that is exploited in a variety of ways, e.g. in spectrophotometry. In contrast, scattering of light is proportional to the inverse fourth power of the wavelength (λ^{-4}), which means that blue light is scattered far more readily than red. The blue sky above us is a result of the differential scattering of light by Earth's atmosphere. Similarly, the blue water of the open ocean is partially the result of the same phenomenon, though it is also influenced by the reflection of the sky.

Both absorption and scattering contribute to the attenuation of light in the open ocean. A beam of light traveling over a distance loses part of its energy, whether it be in air or water. The relationship is expressed as Lambert's law:

$$I_x = I_0 e^{-\alpha_\lambda x} \quad (1.15)$$

where I_0 is the intensity of the incident light, I_x is the intensity after traveling through distance x, and $\alpha\lambda$ is the attenuation coefficient, a function of wavelength.

The attenuation coefficients for the different colors of light are quite different in water from those in air (Figure 1.22). Even though blue light is the most prone to scattering, it is attenuated the least with water depth. As a consequence, the character of light changes drastically with depth, a situation that must be reckoned with by the life in the open ocean. When light reaches a depth of 500 m, its wavelength has been narrowed to a very small range on either side of 480 nm in the blue and its intensity has been drastically reduced.

Three vertical zones are recognized in the ocean based on the availability of light. The first one is the euphotic zone, where the intensities and wavelengths of light are sufficient to support growth and reproduction in plant life. The bottom of the euphotic zone is defined by the compensation depth where photosynthesis (energy production)

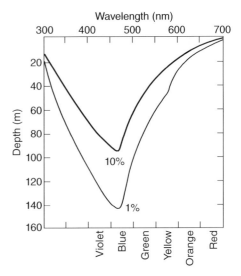

Figure 1.22 Attenuation of different wavelengths of light as a function of depth. *Source:* Lalli and Parsons (1993), figure 2.4 (p. 25). Reproduced with the permission of Pergamon Press.

balances respiration (energy consumption) in plants. Two equations apply (after Lalli and Parsons 1993). The first is a simplification of Eq. (1.15):

$$k = \frac{\ln I_0 - \ln I_D}{\text{depth}(m)} \quad (1.16)$$

where I_0 is the intensity of light at the surface, I_D is the intensity at depth, and k is the extinction coefficient of the seawater at that location (usually for the wavelength 550 nm), where ln denotes the natural logarithm. The second defines the compensation depth itself:

$$D_c = \frac{\ln I_0 - \ln I_c}{k} \quad (1.17)$$

where D_c is the compensation depth, I_0 is the light at the surface, and I_c, the compensation light intensity is an experimentally determined value that ranges between 0.001 and 0.01 cal cm^{-2} min^{-1} depending on the type of dominant algae and how well it gathers light.

The bottom of the euphotic zone is highly variable (Figure 1.23) and is dependent upon water clarity. In clear ocean water, the bottom of the euphotic zone is at about 150 m. Below the euphotic zone is the disphotic zone, where there is still enough light to allow vision but not enough to support plant life. Like the euphotic zone, the depth of the disphotic zone varies with water clarity. In clear open-ocean water, the disphotic zone extends to about 1000 m. The aphotic zone, where light is extinguished, makes up the remainder of the water column. Remembering that only 3% of the ocean's volume is above 1000 m, light is

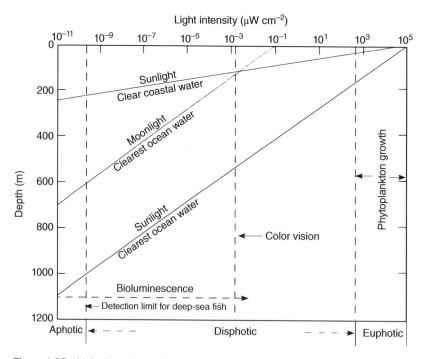

Figure 1.23 Light intensity and photic zonation as a function of depth in coastal and open ocean environments. *Source:* Lalli and Parsons (1993), figure 2.5 (p. 26). Reproduced with the permission of Pergamon Press.

therefore absent from the vast majority of the open sea.

Traditional Depth Zones in the Ocean

Figure 1.24 shows the traditional oceanic depth zones, their characteristics, and representative fauna that reside at those depths. The uppermost layer, the epipelagic zone, extends from the surface to a depth of 200 m and includes the euphotic zone. The next layer down is the mesopelagic zone (200–1000 m), which largely corresponds to the disphotic zone described earlier. Several important changes happen in the mesopelagic zone. Light is gradually extinguished or very nearly so (only in the clearest open-ocean waters does light prevail below 1000 m), and the wavelengths are limited to the blue-green (480 nm). The permanent thermocline results in a temperature change from near-surface values to the more cold and monotonous temperatures characteristic of the very deep-sea and the cold-water masses that comprise it. Temperatures at 1000 m are usually between 4 and 8 °C, declining very gradually to 2 °C in the next vertical stratum, the bathypelagic zone, which extends from 1000 to 6000 m (Herring 2002). The bathypelagic zone includes the depths characteristic of the abyssal plain (4000 m) and the average depth of the ocean, but not the depths characteristic of the ocean trenches that include the ocean's deepest points. Depths below 6000 m are sometimes referred to as hadal or hadopelagic and are associated mainly with oceanic trenches. The three major regions of the oceanic water column are the epipelagic, mesopelagic, and bathypelagic zones, and

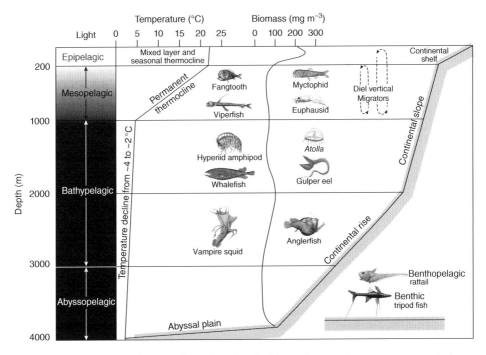

Figure 1.24 Oceanic depth zones. Examples of typical fauna from each zone are represented along with temperature, light, and biomass profiles.

they will be our main concern in this book. The hadal regions are a minor contributor to the seafloor and sea life of the world ocean, (2%, Herring 2002) though those regions are quite important geologically.

Concluding Thoughts

Whether considered in terms of surface area or volume, the open ocean is a vast living space. It covers more than twice the surface area of the terrestrial biome, and in volume it is 99.5% of the habitable space on Earth! Most of that habitable space is in the deep sea, making the deep sea the earth's "average environment." The properties of water shape the characteristics of marine life. Its density confers structural support to marine species, allowing for the delicate constitutions exhibited by gelatinous species such as the jellyfishes and comb jellies and allowing marine plants such as kelp to reach great heights above the bottom without the use of a massive trunk. The negative side of water density is that it takes considerable energy to move through the aqueous medium because of friction drag and pressure drag. Rapid swimmers have a characteristic spindle (fusiform) shape that minimizes drag. The environment experienced by very small open-ocean species is vastly different from that experienced by the larger, quicker swimmers: small marine species live in a much more viscous environment. The environment that a species experiences can be described with the Reynolds number, a dimensionless unit that relates inertial to viscous forces.

Just as important are the internal adjustments that must be made by marine species to the gradients in temperature that vary with latitude and to the changes in temperature, pressure, and oxygen that vary with depth. Those latitudinal and vertical gradients are governed by ocean circulation. Surface currents, from the surface to 400 m of depth, are driven by the wind and the rotation of the Earth. Deep-ocean circulation, which transports about 90% of the sea's volume, is determined by the ocean's density structure and is far more sluggish.

The Earth has five major oceans: the Atlantic, Pacific, Indian, Arctic, and the recently recognized Southern Ocean. The high-latitude Arctic and Southern Oceans are the most homogeneous with respect to temperature, salinity, and oxygen in the horizontal and vertical planes. The Atlantic, Pacific, and Indian Oceans are far larger, extend from tropical to boreal climates, and are far more heterogeneous in their temperature and salinity profiles. In particular, the Pacific and Indian Oceans have large areas in their coastal regions with well-developed oxygen minima that in some cases are limiting to animal life.

As with temperature, the presence of light in the ocean varies predictably with depth, with the longest and shortest wavelengths being largely attenuated in the upper 100 m. The light that penetrates most readily is that in the blue-green range, with a wavelength of approximately 480 nm. Light penetration depends on water clarity, which in turn depends on the presence of particles that absorb and scatter light. In clear open-ocean water, light penetrates to depths of approximately 1000 m. Vision as a sensory modality is therefore most important in the upper 1000 m of the ocean.

Sound propagates better in water than in air, making the detection of sound and vibration important to most oceanic life. Unlike temperature and light, sound does not vary naturally with depth or location in the ocean, except with respect to noise generated by coastal surf and through proximity to man's activities.

Traditionally, the ocean is divided into three main zones with respect to depth. The epipelagic zone, or upper ocean, extends from 0 to 200 m, roughly corresponding to the depths over which plant life can survive. The mesopelagic zone, or middle ocean, extends from 200 to 1000 m and corresponds roughly to both the permanent thermocline and the maximum penetration of light in the ocean. The bathypelagic zone, or deep ocean, extends from 1000 to 6000 m. It includes the average depth of the ocean, ~3700 m, and all depths but those in the deep ocean trenches. It is characterized by the complete absence of light and very low (2 °C) temperatures.

References

Batchelor, G. (1967). *An Introduction to Fluid Dynamics*. Cambridge: Cambridge University Press.

Bearman, G. (1989). *Seawater: Its Composition, Properties, and Behavior*. Oxford: Pergamon Press.

Briggs, J.C. (1974). *Marine Zoogeography*. New York: McGraw-Hill.

Broecker, W.S. (1992). The great ocean conveyor. In: *Global Warming: Physics and Facts* (eds. B.G. Levi, D. Hafemeister and R. Scribner). New York: American Institute of Physics.

Brown, J., Colling, A., Park, D. et al. (1989). *Ocean Circulation*. Tarrytown, NY, USA: Pergamon Press Inc.

Denny, M.W. (1993). *Air and Water: The Biology and Physics of Life's Media*. Princeton University Press: Princeton.

Gage, J.D. and Tyler, P.A. (1991). *Deep-Sea Biology: A Natural History of Organisms at the Deep-Sea Floor*. Cambridge: Cambridge University Press.

Garrison, T. (2002). *Oceanography: An Invitation to Marine Science*. Pacific Grove: Thomson Learning, Inc.

Gross, M.G. (1990). *Oceanography: A View of the Earth*. Englewood Cliffs: Prentice-Hall.

Halliday, D. and Resnick, R. (1970). *Fundamentals of Physics*. New York: Wiley.

Herring, P. (2002). *The Biology of the Deep Ocean*. New York: Oxford University Press.

Lalli, C.M. and Parsons, T.R. (1993). *Biological Oceanography: An Introduction*. Tarrytown, NY, USA: Pergamon Press Inc.

Rogers, P.H. and Cox, M. (1988). Underwater sound as a biological stimulus. In: *Sensory Biology of Aquatic Animals* (eds. J. Atema, R.R. Fay, A.N. Popper and W.N. Tavolga). New York: Springer-Verlag.

Schwartzlose, R.A. and Reid, J.L. (1972). Nearshore circulation in the California Current. *California Marine Research Commission Calcofi Report* 16: 57–65.

Sverdrup, H.U., Johnson, M.W., and Fleming, R.H. (1942). *The Oceans, Their Physics, Chemistry, and General Biology*. Englewood Cliffs: Prentice-Hall.

Torres, J.J. (1984). Relationship of oxygen consumption to swimming speed in *Euphausia pacifica*. II. Drag, efficiency, and a comparison with other swimming organisms. *Marine Biology* 78: 231–237.

Torres, J.J., Grigsby, M.D., and Clarke, M.E. (2012). Aerobic and anaerobic metabolism in oxygen minimum layer fishes: the role of alcohol dehydrogenase. *Journal of Experimental Biology* 215: 1905–1914.

UNESCO (1983). *Algorithms for Computation of Fundamental Properties of Seawater*. Paris: UNESCO.

Vogel, S. (1981). *Life in Moving Fluids*. Princeton University Press: Princeton.

Withers, P.C. (1992). *Comparative Animal Physiology*. Orlando: Saunders.

Zhou, M., Paduan, J.D., and Niiler, P.P. (2000). Surface currents in the Canary Basin from drifter observations. *Journal of Geophysical Research* 105: 21893–21911.

2

Physiological Accommodation to Environmental Challenges

All open-ocean species have a biogeographic range over which they are typically found. That is to say, there are boundaries or limits to a species range in the horizontal and vertical planes. For species inhabiting the epipelagic zone, those boundaries often coincide well with the patterns in the surface oceanic circulation discussed in Chapter 1 and are shown, with euphausiids (shrimp-like zooplankton – Chapter 7) as an example, in Figure 2.1. That concurrence is not surprising since currents define the living space of open-ocean species. Populations and communities in the open ocean are quite literally traveling together!

It takes some mental adjustment to embrace the three dimensionality of the pelagic lifestyle. As Longhurst (1998) observed, for virtually every physical and biological characteristic in the ocean, there is a far more profound change in the vertical plane than in any horizontal excursion, even at oceanic fronts. A 500 m vertical transit from the surface in a tropical region yields a temperature change of >10 °C, a pressure change of 50 atm, and a reduction in light to <1% of that in surface waters with a concomitant change in wavelengths from the entire visible spectrum to entirely blue-green (Chapter 1). If we compare that with a 500 m surface transit into the Gulf Stream from the Sargasso Sea, we will see temperature changes of <5 °C with little change in visible light and no change in pressure. Considering further, if we are not at an oceanographic boundary, a lateral movement of 500 m would not change anything very much, but a 500 m change in depth will always yield substantial changes in temperature at all but polar latitudes and will result in large changes in ambient pressure and light everywhere. Even the "shallow" deep-sea is different enough from the surface environment that its characteristics must be accommodated within the physiology of the species that live there.

Because they can directly limit survival, three physical factors loom large in determining a species' range boundaries in the horizontal and vertical planes. Those factors are temperature, pressure, and oxygen. No less important from an organismal perspective are the sensory mechanisms that must be adjusted in order to survive vision, perception of sound and motion (mechanoreception), and perception of chemicals (olfaction and gustation). In this chapter, we will cover the basics of how animals respond to temperature, and how pressure and low oxygen may be accommodated in the biology of pelagic species. Sensory mechanisms are treated in later chapters dealing with specific taxa.

As observed in Chapter 1, salinity varies from place to place in the ocean but it does

Life in the Open Ocean: The Biology of Pelagic Species, First Edition. Joseph J. Torres and Thomas G. Bailey.
© 2022 John Wiley & Sons Ltd. Published 2022 by John Wiley & Sons Ltd.

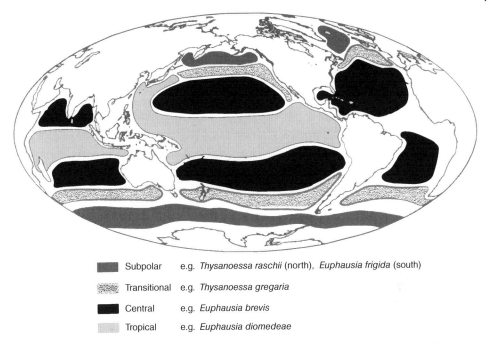

	Subpolar	e.g. *Thysanoessa raschii* (north), *Euphausia frigida* (south)
	Transitional	e.g. *Thysanoessa gregaria*
	Central	e.g. *Euphausia brevis*
	Tropical	e.g. *Euphausia diomedeae*

Figure 2.1 Biogeographic ranges and distributions of four species of epipelagic euphausiids showing their correspondence to surface oceanic circulation patterns (currents and water masses). *Source:* Longhurst (1998), figure 1.2 (p. 13). Reproduced with the permission of Academic Press.

not vary enough to physiologically limit the distributions of oceanic fauna. Nonetheless, it is part of the external milieu that must be dealt with by the physiological systems of oceanic species. We will cover the basics of how oceanic salinity is accommodated by invertebrate and vertebrate fauna with any eye toward developing a basic understanding of osmotic and ionic regulation in both major groups.

Temperature

Active life exists from $-20\,°C$ in arctic insects to $113\,°C$ in some of the Archaea found at hydrothermal vents (Hochachka and Somero 2002). In marine systems, the most challenging habitat is in the intertidal zone, where at low tide, winter cold and summer heat can act directly on an exposed fauna. For pelagic species, the window of life is more narrow and more comfortable, ranging from $-1.86\,°C$ in polar waters to about $40\,°C$ or slightly more in the waters of the Red Sea. With the possible exception of some small species of copepods (e.g. *Oithona similis*), no species is found over the entire range of temperatures in the pelagic zone.

Two basic effects of temperature are our main concern in this book: its effects on survival and its effects on vital rate processes such as metabolism. The literature on temperature and animal life is as fascinating as it is huge. We can only delve into it far enough to appreciate the basics. The best way to start will be with terminology.

Terms

Organisms with a wide range of temperature tolerance are termed eurythermal; "eury" is from the Greek for "wide" or "broad." Eurythermal species have a broad geographic range or live in a habitat subject to wide swings in temperature, or both. Most intertidal species, particularly those that are sessile like barnacles and mussels, are quite eurythermal and tolerate temperature changes of >20 °C in a single tidal cycle. Geographic ranges for intertidal species can be quite broad. On the eastern seaboard of the United States, intertidal species often extend from Cape Cod in Massachusetts to Cape Hatteras in North Carolina and sometimes a great deal further north and south. Eurythermal pelagic species typically have a more modest temperature range. The widely distributed Pacific krill *Euphausia pacifica* has a temperature range of about 10 °C, inhabiting the northern half of the great Pacific Gyre from the Sea of Japan across the northern Pacific and down the US coast to southern California. Pelagic species that vertically migrate from depths >200 m to the surface and back each day can encounter large temperature swings, particularly in the tropics. Vertical migration is an interesting and widespread lifestyle that will be discussed at length later in the book.

The opposite of eurythermal is stenothermal, "steno," the Greek word for "narrow." Stenothermal species have a narrow range of temperature tolerance and are typically found in habitats with small daily and annual temperature deviations. An example of such a habitat for pelagic species is the Antarctic, where the cryopelagic fish *Pagothenia borchgrevinki* has been documented to succumb to heat death at temperatures above 4 °C but can readily tolerate the low temperatures (−1.86 °C) found beneath Antarctic ice.

The next two terms, briefly introduced in Chapter 1, are also opposites: ectotherm and endotherm. They describe organismal body temperature and what determines it. The body temperature of ectotherms is nearly identical to the ambient temperature. Once again, the term is derived from the Greek: "ecto" meaning "outer" or "outside." In layman's terms, the ectotherms are "cold-blooded." Other than mammals and birds and some highly adapted fishes like the tunas, all animals are ectotherms. The particular subjects of this book, the micronekton and macrozooplankton, are all ectotherms.

Endotherms utilize the heat generated by metabolism to maintain a constant body temperature. "Endo" is the Greek root for internal or inside; an endotherm's body temperature results from heat generated within. Mammals and birds modulate the amount of metabolic heat lost to the environment (or gained from it) by a variety of mechanisms, including insulation (fur, feathers, blubber) and adjustment of blood flow to limit or facilitate heat exchange with the external environment. Endotherms, in layman's terms, are "warm-blooded."

The terms endothermy and ectothermy were created to precisely define how a species' internal body temperature comes to be the way it is: by virtue of internally generated heat or by interaction with the external environment.

Another pair of terms formerly used to describe species' body temperature, homeothermy and poikilothermy, are still quite useful though not as widely used as they once were. A homeotherm (from the Greek "alike" or "constant") has a body temperature that is closely regulated around a constant set point. The trick here is that achieving a constant body temperature may be done in a variety of ways. Mammals and birds regulate their internal temperature precisely by controlling loss of metabolic heat. However, a nearly constant body temperature can also be achieved behaviorally, as lizards do by regulating their time spent in sun and shade. In the deep ocean, nearly every species is a homeotherm because temperatures vary little below 1000 m. Some species, e.g. sockeye salmon, have thermal preferenda or optima that they will seek out in a thermal gradient, giving them a nearly constant body temperature as well.

Poikilotherm (from the Greek "poikilo" or "varied") may be considered as the older version of ectotherm. A poikilotherm has a body temperature that changes with the external environment, so it is certainly an ectotherm. However, as we just discussed, ectotherms dwelling at a constant temperature are also homeotherms. So, using the old terminology, a poikilotherm could also be a homeotherm when living at constant temperature, which is confusing at best.

Temperature Effects on Survival: The Tolerance Polygon

It is important to note that the change in species composition at oceanic boundaries such as the Antarctic Polar Front is not due to the short-term lethal effects of temperature change. Instead, a suite of factors is involved, including inefficiencies in reproductive strategies and timing, metabolic inefficiency, absence of preferred prey, and competition from similar species for resources that result in the gradual demise of the replaced species. However, characterizing a species' tolerance to temperature is highly instructive because it introduces two basic rules of physiological response to temperature and to other environmental challenges like salinity. The first rule is that the short-term range of temperature tolerance within a species, population, or individual is not rigid or immutable; animals can adjust their range of tolerance over a period of time in response to changes in external temperature. The second rule is that upper and lower limits exist for all species that cannot be exceeded, even after allowing for biological adjustment.

The internal adjustment process that raises or lowers lethal limits takes time to accomplish and is described by two terms. When the adjustment phenomena take place in the natural habitat (e.g. seasonal temperature change), the process is called acclimatization. When adjustment is induced in the laboratory, the phenomenon is called acclimation.

The best way to define the level of eurythermicity in a species is using an approach that incorporates a species' ability to biologically adjust its temperature range: the thermal tolerance polygon (Figure 2.2a). First introduced in 1952 by the Canadian fish physiologist John R. Brett, the tolerance polygon uses a rigorous experimental protocol to define the upper and lower lethal limits of a species. The lethal T°C was theoretically defined as that temperature at which 50% of a population could withstand for an infinite time. To determine this, a sample of fishes acclimated to a given temperature was subjected to a series of temperatures higher (or lower) levels of which resulted in complete mortality of the sample. The period of tolerance prior to death was termed the resistance time. In each instance, the logarithms of the median resistance time were plotted against temperature and the results formed a straight line (Figure 2.2b). The slope of this line is relatively consistent for most species. In every case, an abrupt change in slope occurred, indicating that mortality due to

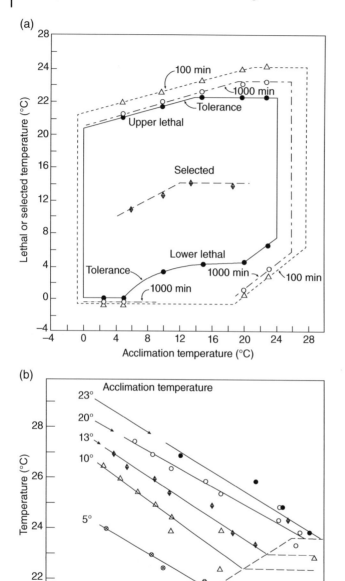

Figure 2.2 Thermal tolerance and lethal limits. (a) Thermal tolerance polygon. Upper and lower lethal limits of the sockeye salmon *Oncorhynhcus nerka* in relation to acclimation temperature. (b) Median resistance times of young chum salmon *Oncorhynchus keta* acclimated to the temperatures indicated. *Sources:* (a) Adapted from Fry and Hochachka (1970), figure 2 (p. 81); (b) Brett (1952), figure 7 (p. 282).

temperature had effectively ceased, marking the change from resistance to tolerance. That point was termed the incipient lethal temperature. Low and high incipient lethal temperatures were determined for each acclimation temperature to form the polygon shown in Figure 2.2a.

The polygon for *Oncorhynchus keta* indicates that it is a fairly eurythermal species. Polygons for Antarctic species would encompass only a small fraction of the lower range, whereas highly eurythermal species such as the brown bullhead catfish (*Ameriurus nebulosus*) would be very much larger.

Studies of temperature tolerance in a variety of different organisms suggest the following.

1) Generally, upper and lower lethal limits can be modified considerably by different acclimation temperatures, e.g. the warmer the temperature of acclimation, the higher the upper lethal limit.
2) There are absolute upper and lower lethal limits beyond which an organism cannot adapt, and these limits can be determined with precision.
3) It takes longer to acclimate to cold temperatures than to warm ones.
4) The tolerance polygon of an organism relates well to habitat and geographic area, as shown in the example below.

Pagothenia borchgrevinkii
(icefish) habitat: Antarctic
[----] tolerance range
  [----[----[----[----[----[----[----[----] °C
−5   0   5 10 15 20 25 30 35
        [-------------------------] tolerance range
Ameiurus nebulosus – brown bullhead catfish, habitat: shallow areas of lakes

In addition to knowing the zones of tolerance or the limits to survival of a species, it is important to understand the physiological responses of an organism to temperatures within its environmental range.

Temperature Effects on Rate Processes – The Q_{10} Approximation

Animals have varying reactions to temperature changes within their zone of tolerance. Reaction to temperature within an animal's environmental range is usually assessed using a rate function, heartbeat for example, or a filtration rate for species such as clams that pump water through their feeding apparatus. Most commonly, metabolic rate is used; metabolism is an excellent index of an animal's rate of energy consumption and is readily measured by monitoring an individual's rate of oxygen consumption. The rate of increase or decrease in reaction rate over a T°C change is standardized by the Q_{10} approximation, which is the factor by which a reaction velocity (e.g. rate of oxygen consumption) is increased for a rise of 10°C.

$$Q_{10} = K(T+10)/K(T) = (K_1/K_2)^{(10/(T_1-T_2))}$$
or $\log Q_{10} = (10(\log K_1 - \log K_2))/(T_1 - T_2)$
(2.1)

where K_1 and K_2 are velocity constants corresponding to temperatures T_1 and T_2. Reaction velocity is generally used instead.

For virtually all rate functions in which we are interested, the biological rate increases by a factor of approximately 2 for each 10°C rise in temperature: that is, $Q_{10} \approx 2$. However, the Q_{10} of an animal's metabolic rate varies slightly over a range of temperatures, being higher in the lower ranges. Therefore, when providing Q_{10} for metabolic rates, it is imperative to also provide the temperature range over which the measurements were taken.

The fact that metabolic rate doubles with a temperature change of 10°C (or halves with a 10°C drop) has a profound effect on ectothermic species. A vertical migrator in the tropical ocean swimming from a depth of 500 m to near surface waters at sunset (a common occurrence, as we shall see) will encounter changes of >10 °C on its way up and again on its way back to depth. As a consequence, it will endure profound changes in its metabolic rate during each leg of its excursion.

In 1914, August Krogh, the father of comparative physiology, first attempted to define a pattern for the change in Q_{10} with temperature by subjecting a narcotized goldfish to temperatures ranging between 0 and 25 °C and measuring its oxygen consumption rate. The curve he derived is called the "Normal Curve." It was popularized considerably in later years when it was found that a similar relationship between metabolism and temperature existed for many species, with the exception of the large Q_{10} in the 0–5 °C temperature interval (e.g. Winberg 1956). Even Krogh stated that his Q_{10} value of 10.9 between 0 and 5 was "obviously erroneous." The general trend was remarkably accurate though, as were the numbers generated above 5 °C. The Q_{10}'s this curve represents is shown below.

T°C intervals	0–5	5–10	10–15
Q_{10}	10.9	3.5	2.9

	15–20	20–25	25–30
	2.5	2.3	2.2

The Q_{10} approximation is a fundamental molecular response to temperature: it applies to chemical reactions taking place in a beaker as well as to rate processes in ectothermic species. However, it is not intuitively obvious why reaction rates should double for every increase in a temperature of 10 °C. The answer is in a concept termed "activation energy," which was pioneered by the Swedish physical chemist Svante Arrhenius, and which earned him a Nobel Prize in 1903.

In the realm of physical chemistry, temperatures are expressed in the absolute temperature scale, in degrees Kelvin. A Kelvin degree is equal to a degree Celsius, but the scale begins at absolute zero, the temperature where all molecular motion ceases: −273 °C. Thus, a temperature of 0 °C is equal to 273 K, and 20 °C is equal to 293 K; by convention, the degree symbol is not used for degrees Kelvin. The temperature range most relevant to the pelagic fauna, −2 to 40 °C (271–313 K), only covers about a 10% change of temperature on the absolute scale. In our range of concern, a change of 10 °C is roughly 3% of the absolute temperature. Why, then, do reaction rates double?

The breakthrough of Arrhenius was his idea that within a population of molecules, only a fraction have sufficient energy to be reactive: those that exceed the activation energy threshold (Figure 2.3). The average thermal energy of the molecules gives us the temperature, but it is not the average that is most important, it is the proportion of molecules that have enough energy to exceed the activation energy threshold and be competent to react. When heat is added to a system, the proportion of molecules that exceeds the activation energy increases more quickly than the average temperature. In fact, an increase in temperature of 10 K results in a doubling of the fraction of molecules exceeding the activation energy. The activation energy concept thus explains the Q_{10}s we observe with biological rates.

Experimentation in the 1940s, 1950s, and 1960s further defined temperature responses as a function of time and acclimation period. Three general time courses were identified.

1) Direct responses of rate functions to changes in temperature persisting for hours: acute measurements

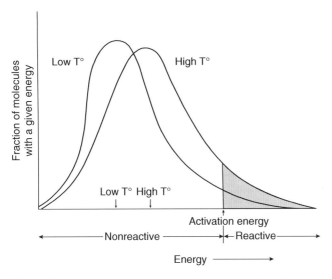

Figure 2.3 Energy distribution curves for a population of molecules at two different temperatures. Only those molecules having energies equal to or greater than the activation energy are reactive. *Source:* Hochachka and Somero (2002), figure 7.1 (p. 296). Reproduced with the permission of Oxford University Press.

Rates measured in this way, with no acclimation period, reflect the short-term flexibility of biological systems. In some cases, acutely measured responses to temperature can show Q_{10} values greatly different from 2. When metabolism is the rate being measured, such a response is termed "metabolic overshoot." It would correspond to a "type 5" response described below. It is the result of a system that is still in the process of adjusting to a new temperature and it can happen in a transition to a warmer or colder temperature.

2) Compensatory acclimation to days or weeks of exposure: the acclimated response

An animal is acclimated only when its rate processes have stabilized to the new temperature. Acclimated animals were utilized in constructing the temperature tolerance polygon shown in Figure 2.2a.

3) Evolutionary adaptation through natural selection: climatic adaptation

Patterns of Thermal Acclimation

Animals moved to a new temperature and kept there for several days often show some compensation in their rate functions, especially metabolic rate. One of the classic treatments of how individuals adjust to a new temperature is that of Precht (1958) who identified five responses (Figure 2.4). Animals were maintained initially at temperature T_2 and moved to T_1 or T_3. Metabolic rates were monitored during the acclimation process.

Type 4 acclimation: no change in rate occurs after time for acclimation ($Q_{10} = 2-3$).
Type 2 acclimation: the animal's rate falls or rises to original rate ($Q_{10} = 1$)
Type 3 acclimation: somewhere in between ($Q_{10} = 1-2$)
Type 1 acclimation: overcompensation – rate lower at higher temperature ($Q_{10} < 0$)
Type 5 acclimation: reverse compensation ($Q_{10} > 3$)

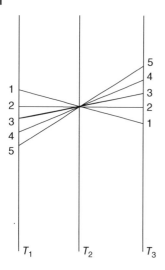

Figure 2.4 Precht's patterns of temperature acclimation. Animals are acclimated at temperature T_2 then transferred to a lower temperature (T_1) or higher temperature (T_3). The five acclimation types are described in the text. *Source:* Prosser (1973), figure 9-13 (p. 375). Reproduced with the permission of Saunders Publishing.

In reality, all five curves are rarely seen. In most cases, the response is type 3 or 4. However, type 2 has been seen in rare cases; it has been documented in some intertidal animals (Newell and Northcroft 1967). The Precht patterns are the simplest way of classifying trends in metabolism vs. temperature, but other useful schemes exist. Prosser (1973) describes species' response to temperature using a family of metabolism vs. temperature (M-T) curves instead of Precht's three points. The advantage of Prosser's scheme is that it can differentiate varied responses to lowered and raised temperature, and the reader is encouraged to examine Prosser's curves.

Climatic Adaptation in Ectotherms

Scholander et al. (1953) first described another important way that ectotherms achieve some degree of freedom from their environment: the phenomenon of metabolic cold adaptation or "MCA." We may think of the changes in lethal limits that are possible with acclimation as the "built-in" flexibility in a species' biochemical machinery that allows it to cope with seasonal change. MCA is a fundamental characteristic of a species' biochemical power plant that allows it to survive in its particular climatic regime. Those adapted characteristics are achieved over the course of evolutionary time. The phenomenon of MCA simply stated is this: "When non cold-adapted ectotherms are introduced to a given low temperature and allowed to acclimate, their metabolism tends to stabilize at a level below that of a species normally adapted to the low temperature" (Wohlschlag 1960). That is, cold-adapted fishes and other cold-adapted taxa have higher metabolic rates at low temperature than would their acclimated tropical counterparts (Figure 2.5). A helpful mental exercise is to extrapolate the metabolism vs. temperature (M-T) curve for temperate fishes shown in Figure 2.5 to a temperature of 0°C. Note that it is well below that of the M-T curves of the polar species.

The rates in this classical figure are similar in the different zoogeographic locations. MCA is very important zoogeographically; it implies an advantage to elevated metabolic rates in cold-adapted species and marked similarity of metabolic rates over the zoogeographic range of fishes (and other taxa). A great deal of thought and experimentation has gone into understanding MCA because it is of paramount importance to ectotherms.

Why should you care about temperature adaptation? The metabolic machinery of ectotherms has evolved to allow metabolic processes to proceed at similar rates over a widely disparate range of temperatures, ameliorating the tyranny of the Arrhenius

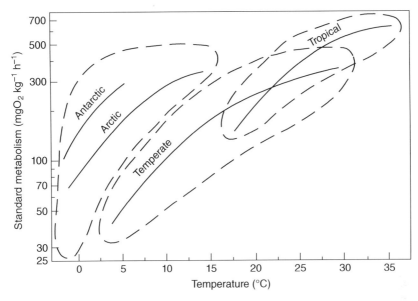

Figure 2.5 Schematic representation of the relation between temperature and standard metabolic rates (log scale) of fish from different climatic zones. Dotted lines indicate the general range of variability within each zone. Adaptive metabolic compensation is shown for the different climatic regimes. *Source:* Brett and Groves (1979), figure 1 (p. 292) with the permission of Academic Press.

concept. Temperature adaptation may be thought of as a biological intervention that allows life to proceed apace over the entire range of temperatures found in the open ocean. It requires modifications of the enzyme systems that underlie all metabolic processes. It is most important to embrace the fact that modifications take place in the long term, on an evolutionary time scale, between species from different regions and in the shorter term (seasonal time scale) within a species as described with the temperature polygon (Figure 2.2a).

Clearly, temperature not only sets boundaries for survival but also governs rate processes within those boundaries. The rate processes are, in turn, governed by enzyme systems: the biological catalysts that make life possible. Alterations in the quality or quantity of enzymes underlie much of the process of temperature adaptation. Thus, it is important to examine temperature adaptation in more depth.

In a comparison of enzymic activities in fishes from different climatic regimes, Kawall et al. (2002) utilized the activities of two important intermediary metabolic enzymes: lactate dehydrogenase (LDH), the terminal enzyme in anaerobic glycolysis, and citrate synthase (CS), the rate-limiting step in the Kreb's Cycle aerobic pathway. To keep the comparison meaningful among fishes of different lifestyles and locomotory abilities, the activity of enzymes within the brains of the fishes was compared at a common temperature (10°C) instead of using enzyme activity in skeletal muscle. Those values were then extrapolated to habitat temperature (0 and 25°C for polar and subtropical species, respectively) using Q_{10} values determined during the study (Figures 2.6 and 2.7). For both enzymes, the activity of the cold-adapted Antarctic species was much higher than that of the subtropical species, showing a considerable degree of temperature compensation in the

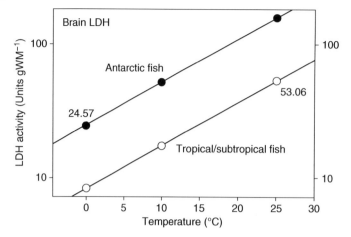

Figure 2.6 Lactate dehydrogenase (LDH) activities (international units gWM^{-1}) in brains of fishes from Antarctic and tropical/subtropical climatic zones in relation to environmental temperature. Curves were generated using the mean LDH activities obtained at 10 °C and adjusted using the experimentally determined Q_{10} of 2.1. Enzyme activities at 0 °C, 10 °C, and 25 °C are denoted by filled circles for Antarctic fish and open circles for tropical species. Activities at the approximate habitat temperatures of the two groups are given numerically next to the relevant symbols. *Source:* Kawall et al. (2002), figure 1 (p. 283). Reproduced with the permission of Springer-Verlag.

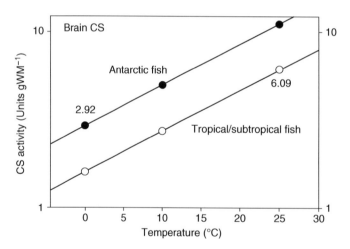

Figure 2.7 Citrate synthase (CS) activities (international units gWM^{-1}) in brains of fishes from Antarctic and tropical/subtropical climatic zones in relation to environmental temperature. Curves generated as in Figure 2.8. *Source:* Kawall et al. (2002), figure 2 (p. 283). Reproduced with the permission of Springer-Verlag.

polar species, and echoing that observed in the metabolic rate determinations of Scholander et al. (1953) shown in Figure 2.5. A careful consideration of enzyme activities at the habitat temperatures within each climatic regime shows that compensation is not complete, i.e. the Antarctic species have an activity at 0 °C that is about 50% of that of the subtropical species at 25 °C. Temperature compensation is not perfect, but it is substantial! If you take the activity at a temperature of 25 °C and apply a Q_{10} of 2 over the 25 °C range, at 0 °C, the rate would be <20% of that at 25 °C.

Temperature Compensation via Changes in Enzyme Concentration: The Quantitative Strategy for Short-term Change

The easiest way to effect a change in rate, as measured by the accumulation of a reaction product, is to alter the concentration of reactants. In the case of an enzymatic reaction, if we assume a constant concentration of substrate and increase the amount of enzyme, the product of the reaction will accumulate more rapidly: an increase in activity. Surprisingly, few studies quantitatively address this issue. Most studies simply assume that short-term changes in enzyme activities are due to enzyme concentration changes. The one study, consistently cited, that does address changes in enzyme activity as a function of enzyme concentration is Sidell et al. (1973). Figure 2.8 shows the activity of cytochrome oxidase, an important enzyme in the electron transport system, from goldfish skeletal muscle in fishes first acclimated to 15 °C and then transferred to either 5 or 25 °C. The enzyme activity per milligram protein in both groups was then monitored for approximately 30 days. Activity was much higher in the fishes transferred to the colder temperature, suggesting that the concentration of enzyme was much higher in the cold-adapted fish.

It should be noted that accelerating or decelerating the activity of an enzymic pathway through differences in enzyme concentrations is best as a short-term solution or for small T°C changes. It is not practical for a long-term (evolutionary) solution to synthesize larger quantities of an inefficient enzyme to deal with the cold.

Compensation via Changes in Enzyme Quality – Isozymes, Allozymes, and Temperature Adaptation

Metabolic adaptation to temperature in eurythermal species may involve a change in the quantity or concentration of enzyme, as observed above in Sidell et al.'s goldfishes, or to actually change the enzyme to one that functions better at the new temperature. For those species that are genetically equipped to do so, the change in enzyme may be the result of an enzyme produced by a different gene locus (a different isozyme) or by a different allele of the same gene locus (a different allozyme).

For example, rainbow trout, *Oncorhynchus mykiss*, shows multiple isozymes of acetyl cholinesterase that can function optimally at different temperatures. The effect of temperature on their apparent Michaelis constant (K_m) is shown in Figure 2.9. The Michaelis constant is the substrate concentration at which the reaction proceeds at 50% of its maximum velocity (V_{max}). The K_m of isozymes in trout acclimated to cold (2 °C) differs considerably from that of trout acclimated to 18 °C (Baldwin and Hochachka 1970; Baldwin 1971). The K_m of the cold-acclimated trout stays within a fairly flat functional range of 0.2–0.5 mM between temperatures of 0 and 14 °C, then above that range the K_m shoots way up. Similarly, the K_m for the warm-acclimated trout at a temperature below 14 °C is very high, then drops to within the functional range between temperatures of 14 and 24 °C, and then shoots up again. Enzymes from the cold- and warm-acclimated trout show different electrophoretic mobilities, meaning that the proteins are structurally different. Thus, within the trout at least, it is possible to have two different versions of the same enzyme, each of which functions optimally only over part of the seasonal

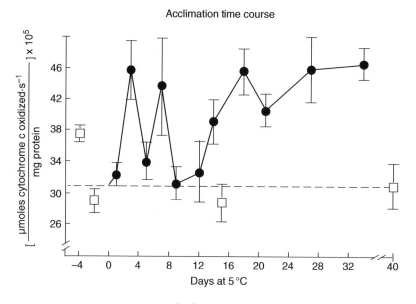

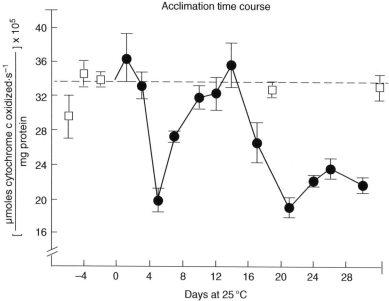

Figure 2.8 Changes in enzyme activity as a function of time after changes in acclimation temperature. Cytochrome oxidase activity (mean ± SE) of goldfish muscle homogenates from fish which were transferred from 15 to 5 °C; Cytochrome oxidase activity (mean ± SE) of goldfish muscle homogenates from fish which were transferred from 15 to 25 °C. Note that values are expressed per milligram protein. *Source:* Sidell et al. (1973), figures 1 and 2 (p. 210). Reproduced with the permission of Springer-Verlag.

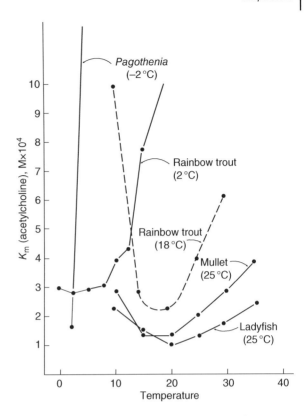

Figure 2.9 The effect of temperature on the K_m of acetylcholine for acetylcholinesterase enzymes of four species of fish acclimated to different environmentally relevant temperatures: rainbow trout, acclimated at 2 and 18 °C; *Pagothenia borchgrevinki*, an Antarctic species; and the mullet (*Mugil cephalus*) and Ladyfish (*Elops hawaiensis*), two tropical species. The approximate temperature to which each fish is adapted or acclimated is given in parentheses below the fish's name. *Source:* Hochachka and Somero (1973), figure 7-14 (p. 231). Reproduced with the permission of Saunders Publishing.

temperature range encountered by the species. In practice, such a multiple isozyme solution to temperature change is a pretty rare occurrence. Trout are tetraploid, having four copies of each chromosome, instead of the diploid structure that we, and most oceanic fauna, have. The trout's larger genome allows for more flexibility in dealing with environmental challenges (cf. Somero 1975; Hochachka and Somero 1984).

Diploid species have two copies of each gene, the two alleles, of which one is normally dominant. A study by Place and Powers (1979) observed the expression of two allozymes for LDH B in the killifish *Fundulus heteroclitus*. LDH B is the type normally found in the heart, as opposed to LDH A normally found in skeletal muscle. *Fundulus* is a widely distributed species along the east coast of the USA, and the Place and Powers study used specimens obtained from Maine to Florida. They found that the two types of LDH B had different efficiencies with respect to temperature, one more efficient in the cold and the other more efficient at warm temperatures. The presence of the different allozymes scaled with the environmental cline in temperature along the eastern US seaboard.

It should be noted that few species have as wide a temperature tolerance as does *Fundulus* or as the brown bullhead mentioned earlier in this chapter. The great majority of species, particularly those in the open ocean, have a more restricted temperature range, and therefore geographic range, over which they are found. The question then remains as to how much temperature change warrants a change in enzyme type (and perhaps species!) in a more typical situation.

A good example of a more typical situation is provided in a study by Graves and Somero (1982) of Pacific barracudas, congeneric pelagic fishes with similar ecologies differing in habitat temperature by only a few degrees: 6–8 °C. The four congeneric species were *Sphyraena argentea* and *S. idiastes*, both cold-temperate species, *S. lucasana*, a warm-temperate species, and *S. ensis*, a tropical-subtropical species. The study used purified enzymes of LDH A to evaluate the effect of temperature on the K_m (Michaelis constant) and the K_{cat} of the four species. K_{cat} is the catalytic efficiency of an enzyme, specifically the rate at which substrate is converted to product per unit time, per active site. Thus, activity of an enzyme $= (K_{cat}) \times$ the concentration of the enzyme.

The electrophoretic properties of the four species were separated into three patterns, with the two cold-temperate species (*S. argentea* and *S. idiastes*) showing identical mobility. *S. lucasana* and *S. ensis* (T-ST) were different from one another, and both were different from *S. argentea* and *S. idiastes*. The electrophoretic study suggested three different enzyme structures, one for the two cold-temperate species, and one each for the warm-temperate and tropical-subtropical species. A look at the kinetic characteristics of the enzymes showed that they differed in those properties as well.

Figure 2.10 is a plot of K_m vs. temperature for the LDHs of three of the four species. Since the plots for the two cold-temperate species, *S. argentea* and *S. idiastes*, were identical, only *S. argentea* is shown. The enzymes from the three species clearly show differences in their K_m vs.

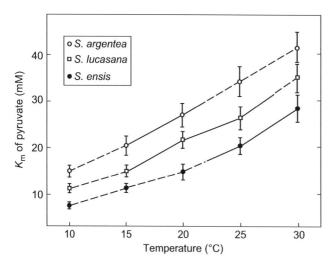

Figure 2.10 The effect of temperature on apparent Michaelis constant (K_m) of pyruvate for the M_4-LDH's of three eastern Pacific barracudas. The K_m values for the southern temperate barracuda, *S. idiastes* (not shown), are statistically indistinguishable from those of the northern temperate species *S. argentea*. Error bars represent standard deviations. Each point represents an average of K_m values determined with at least three different purified LDH preparations from different individuals of each species. Solid portions of the lines connecting the K_m values indicate habitat temperature ranges of the species. *Source:* Graves and Somero (1982), figure 3 (p. 103). Reproduced with the permission of John Wiley & Sons, Inc.

Table 2.1 K_m and K_{cat} values of pyruvate for three congeneric species of barracudas (genus *Sphyraena*) from different habitat temperatures at 25 °C and the temperature mid-range (TM) for each species. *S. argentea* is a cold-temperate species, *S. lucasana* is a warm-temperate species, and *S. ensis* is a tropical-subtropical species. *Source:* Graves and Somero (1982), table 5 (p. 104). Reproduced with the permission of John Wiley & Sons, Inc.

	S. argentea	S. lucasana	S. ensis
K_m of pyruvate at 25 °C	0.34 ± 0.03 mM	0.26 ± 0.02 mM	0.20 ± 0.02 mM
K_{cat} at 25 °C	893 ± 54 s^{-1}	730 ± 37 s^{-1}	658 ± 19 s^{-1}
Temp. midrange (TM)	18 °C	23 °C	26 °C
K_m of pyruvate at TM	0.24 mM	0.24 mM	0.23 mM
K_{cat} at TM	667 s^{-1}	682 s^{-1}	700 s^{-1}

temperature curves, indicating that the enzyme has changed in character among the three locations. When the data are examined in more detail (Table 2.1), it can be seen that the K_m and K_{cat} at the temperature mid-range for each species are virtually identical, but the enzymes show substantial differences when examined at a common temperature. The implications of the data are, first, that a change of <5 °C in a species' temperature range can result in a change in the biochemical makeup of closely related species, and second, that within the barracudas studied here, K_m and K_{cat} both show highly consistent values at environmental temperature.

What Properties of Enzymes Can Be Changed?

The three curves in Figure 2.11 show Michaelis–Menten saturation kinetics, exhibiting three very different relationships between reaction velocity and substrate concentration. They illustrate the balance between the need for efficient catalysis and the need for the cell to be able to regulate its metabolism.

Curve A has a low K_m value; it will always be at or near V_{max}. Any need for an increase in activity to support increased metabolic demand cannot be met; the enzyme is already at maximum. A low K_m is a fine strategy for an enzyme that does not need to be regulated, such as a digestive enzyme, which is best always functioning at maximum velocity. However, for an enzyme involved in metabolism, which varies from a resting to a highly active state, such a curve would be disastrous. Increases in substrate concentration would not affect its activity, and it would be unable to be regulated. Conversely, a high K_m such as that in curve C will have a considerable amount of "reserve capacity" to allow for regulation but will never achieve high velocity and could become a "choke point" for accumulation of metabolites. An enzyme having an intermediate K_m, curve B, not only has a substantial fraction of its V_{max} at cellular concentrations of substrate but also has considerable ability to respond to increases in substrate concentration before it reaches V_{max}. Conclusion: for optimal performance, the enzyme properties and the substrate concentrations available to the enzyme must be complementary.

How variable are physiological substrate concentrations? The answer is, not very: physiological substrate concentrations are

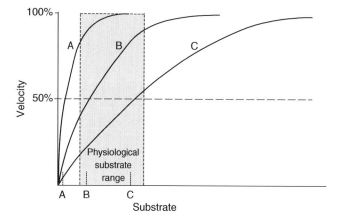

Figure 2.11 Michaelis–Menten saturation kinetics: three types of relationships between reaction velocity and substrate concentration. In (A) the K_m value is low and always at or near V_{max} so that the need for an increase in activity to support increased metabolic demand cannot be met. In (B), the K_m value is intermediate and a substantial fraction of its V_{max} is at cellular concentrations of substrate giving it considerable ability to respond to increases in substrate concentration before it reaches V_{max}. In (C), the K_m is high with a considerable amount of "reserve capacity" to allow for regulation but without the ability to ever achieve high velocity and thus the potential to become a "choke point" for accumulation of metabolites.

highly conserved. Figure 2.12 shows the K_m of pyruvate in LDH from vertebrate species of a wide variety of taxa and habitat temperatures. All K_ms fall between 0.15 and 0.35 mM. Thus, a change in K_m does not appear to be an option in the possibilities available for modification to accommodate temperature change.

What other properties of an enzyme can be modified if K_m is not an option? Consider K_{cat}, the rate at which substrate is converted to product per unit time per active site. If we look at a plot of K_{cat} vs. body temperature in a large suite of differently thermally adapted vertebrate species (Figure 2.13), it is clear that K_{cat} declines considerably with increasing species' habitat temperature. Large differences exist between the K_{cat} of highly cold-adapted species and warm-adapted ectotherms and endotherms. So, unlike K_m, which is highly conserved, we do see large changes with adaptation temperature in K_{cat}, enzyme efficiency.

Let us now consider how enzymes function. The three-dimensional conformation of an enzyme and its overall primary structure are somewhat variable between species. However, the structure of the binding sites within the active site, or catalytic vacuole, is highly conserved (Hochachka and Somero 2002). Further, the actual covalent chemistry that takes place during the enzymatic reaction is fast and likely not the limiting step in the turnover of enzyme. The limiting step is more likely related to conformational changes. How do we reconcile the facts that binding sites in the A_4 – LDHs are invariant from species to species, yet we observe a high degree of variability in K_{cat}s? The answer is in changes of structure outside the catalytic vacuole that confer more conformational flexibility at lower temperature. It is believed that enzymes exist in the cell in a hierarchy or ensemble of conformational states, basically varying between binding-competent

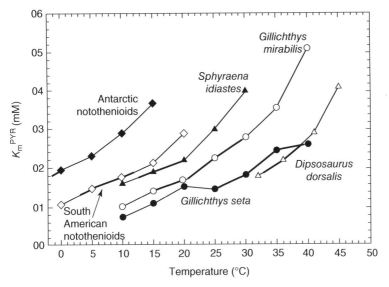

Figure 2.12 Effects of assay temperature on the apparent Michaelis constant of pyruvate (K_m^{PYR}) for A$_4$-LDH orthologs for differently thermally adapted vertebrates: Antarctic and South American notothenioid fishes, a barracuda fish (*Sphyraena ideastes*), to goby fishes (*Gillichthys mirabilis* and *G. seta*), and the desert iguana (*Dipsosaurus dorsalis*). Thick line segments indicate approximate ranges of body temperatures of the species. *Source:* Hochachka and Somero (2002), figure 7.7 (p. 312). Reproduced with the permission of Oxford University Press.

and binding-incompetent (fluctuating) modes. The higher the temperature, the more structurally stable an enzyme has to be to allow the substrate to bind. The trade-off between efficiency and stability is what results in the trends in K_{cat} vs. temperature shown in Figure 2.13.

Lipids and Temperature

We have just been looking at the properties of intermediary metabolic enzymes and observing that changes were needed in enzyme structure to preserve functionality in the face of temperature change. Similarly, the proper environment within the cell with respect to substrates, cofactors, ionic concentrations, and all other properties must be maintained by the cell membrane as temperature changes. The cell membrane itself and its associated proteins govern what crosses it, how much, and the direction of net movement. Ultimately, the cell membrane is the biological barrier that allows the cell its limited autonomy. As a barrier, the membrane not only limits diffusion inward and outward but it also contains embedded transport proteins. The membrane's effectiveness as a barrier and as a center for transport is highly dependent on temperature.

The cell membrane is critical to survival of an organism. If the membrane is breached, its highly regulated internal milieu will be compromised and the cell will die. If enough cells die, wherever they may be located, clearly the whole organism will be in trouble. More importantly, nerve, muscle, and sensory systems are totally dependent on ion transport for their functions. Propagation of signals down a nerve, contraction of muscle, and all sensory

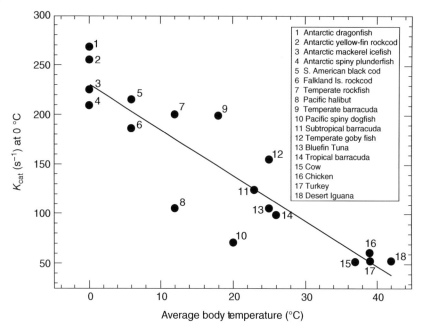

Figure 2.13 The relationship between the catalytic rate constant (K_{cat}) and adaptation temperature for A_4-LDH orthologs from differently thermally-adapted vertebrates. *Source:* Hochachka and Somero (2002), figure 7.3 (p. 302). Reproduced with the permission of Oxford University Press.

mechanisms require an intact, functioning, membrane transport system.

A Membrane Primer

Figure 2.14 gives a schematic representation of a membrane, showing embedded proteins and the lipid bi-layer that makes up the bulk of the functional membrane. Membranes may be thought of as a sheet of phospholipids, individually represented in the figure as a spherical head with two kinked tails that face toward the middle of the lipid bi-layer. The two sets of tails form the inner, hydrophobic, portion of the lipid bi-layer. The backbone of the phospholipid is a glycerol molecule (Figure 2.14). Two fatty acid chains are attached to the glycerol to form the inwardly facing kinky tails, or hydrophobic, portion of the molecule, and a head group is attached to the remaining carbon on the glycerol molecule via a phosphate group (Figure 2.14). The result is a lipid molecule with a hydrophilic head and hydrophobic tails. Phospholipids are named for the head group they contain: one with choline as its head group would be named phosphatidylcholine.

To preserve its function as a protective barrier and transport center, a membrane must strike a balance between being too fluid (sol) and too solid (gel). Too fluid, and it would not be able to act as a barrier or to maintain cellular integrity, and embedded proteins would not have the structural backbone needed to aid in transport across it. Too solid, and transport proteins, which have to change conformations to function, would be constricted, and the lipid membrane itself would be more like a crystalline lattice and more prone to leakage. The phospholipids making up the fabric of the

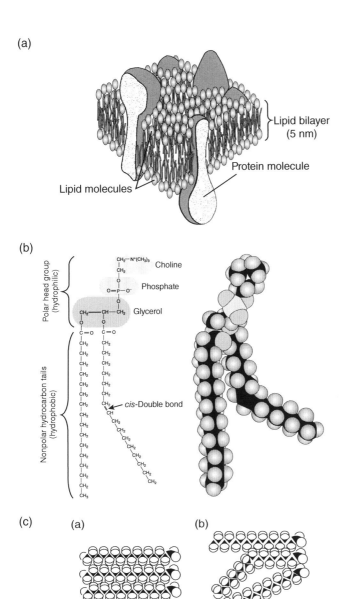

Figure 2.14 Membrane structure. (A) Schematic representation of a membrane showing the lipid bi-layer structure with embedded proteins; (B) chemical structure and a three-dimensional representation of a phospholipid molecule showing the glycerol and phosphate components of the hydrophilic head group and hydrophobic tails that face inward in membranes; (C) saturated vs. unsaturated fatty acids, (a) three molecules of stearic acid (18 carbons, no double bonds, melting temperature 69.6 °C) that pack tightly because of their linear geometries, (b) the addition of a single molecule of oleic acid (18 carbons, one double-bond, melting temperature 56 °C) to a lipid membrane prevent tight packing of the molecules due to the bending of the tail of the unsaturated molecule. See color plate section for color representation of this figure.

membrane are not one single molecular species but are several that, together with aggregate proteins, are organized into domains covering the membrane surface. Structure and function of the membrane differ between domains.

The degree of fluidity of any membrane is determined by the melting point of the fatty acid chains in its phospholipids, and there is a great deal of diversity among them (Table 2.2). You will notice that the fatty acids with the lowest melting points (e.g. linoleic and linolenic) have multiple double bonds. As regards melting points, the "double-bondedness," or saturation state, of a fatty acid is its most critical feature. Why? Because fatty acids with double bonds introduce kinks into the tails of the phospholipids that prevent them from packing tightly (Figure 2.14). The kinks effectively weaken the weak bond interactions between the fatty acid chains, thus lowering the melting point of the lipid. It follows that species living in the cold, that need to maintain membrane fluidity at low temperatures to allow their nerves and muscles to function properly, have lipids rich in double bonds. That is to say, they are unsaturated: the ratio of C to H in the chains is less than it would be if all the bonds between the carbons were single bonds.

The idea that membranes need to maintain an optimum fluidity to allow transport proteins to function properly is the concept of homeoviscosity (Cossins and Bowler 1987; Hochachka and Somero 2002). Mammals and birds have lipids that melt at much higher temperatures than an Antarctic fish. As observed above, the melting point of a membrane lipid correlates inversely with the degree of unsaturation of its fatty acid chains. In fact, a plot of percent-unsaturation vs. adaptation temperature for a group of species from differing thermal environments shows a highly coupled drop in percent-unsaturation of lipids with adaptation temperature (Figure 2.15). Thus, the degree of unsaturation of species' lipids can be added to our growing list of characteristics that change with a species' thermal environment.

Interestingly, just as we see changes in lipid characteristics with a species' habitat temperature, biochemical mechanisms that are usually considered evolutionary adaptation also exist for short-term change in membrane lipids to accommodate more rapid changes in temperature. Figure 2.16 shows a fairly rapid change in lipid classes of trout gill membranes when they were moved acutely from 5 to 20°C and vice versa. In each case, the ratio of phosphatidylcholine (PC) to phosphatidyl ethanolamine (PE) changed profoundly over about five days. You may rightly wonder why changing the head group of a membrane lipid would make much difference. The answer is that PE tends to have fatty acid chains with a greater degree of unsaturation than does PC, affording a greater degree of fluidity at lower temperature (Hochachka and Somero 2002).

Other mechanisms exist for adjusting the fluidity in biomembranes over the short term (hours to days to weeks) in addition to the change in lipid classes just described. Such a capability is particularly important to temperate species that must accommodate changes in temperature associated with seasonal cycles. In most instances, a need for change can be achieved through changes in the biosynthesis of lipids. An example is using enzymes that introduce double bonds into fatty acid chains to make them more suitable for use at cold temperature. Such enzymes are termed desaturases, and they can be up-regulated quickly (Hochachka and Somero 2002).

It is most important to appreciate that not only do species' membrane lipids vary

Table 2.2 Chemical formulas and melting points for a selection of saturated and unsaturated fatty acids.

Carbon atoms	Common name	Empirical formula	Chemical structure	Melting point (°C)
			Saturated fatty acids	
3	Propionic acid	$C_3H_6O_2$	CH_3CH_2COOH	−22
12	Lauric acid	$C_{12}H_{24}O_2$	$CH_3(CH_2)_{10}COOH$	44
14	Myristic acid	$C_{14}H_{25}O_2$	$CH_3(CH_2)_{12}COOH$	54
16	Palmitic acid	$C_{16}H_{32}O_2$	$CH_3(CH_2)_{14}COOH$	63
18	Stearic acid	$C_{18}H_{36}O_2$	$CH_3(CH_2)_{16}COOH$	70
20	Arachidic acid	$C_{20}H_{40}O_2$	$CH_3(CH_2)_{18}COOH$	75
			Unsaturated fatty acids	
16	Palmitoleic acid	$C_{16}H_{30}O_2$	$CH_3(CH_2)_5CH=CH(CH_2)_7COOH$	−0.5
18	Oleic acid	$C_{18}H_{34}O_2$	$CH_3(CH_2)_7CH=CH(CH_2)_7COOH$	13
18	Elaidic acid	$C_{18}H_{34}O_2$	$CH_3(CH_2)_7CH=CH(CH_2)_7COOH$	13
18	Linoleic acid	$C_{18}H_{32}O_2$	$CH_3(CH_2)_4CH=CHCH_2CH=CH(CH_2)_7COOH$	−5
18	Linolenic acid	$C_{18}H_{30}O_2$	$CH_3CH_2CH=CHCH_2CH=CHCH_2CH=CH(CH_2)_7COOH$	−10
20	Arachidonic acid	$C_{20}H_{32}O_2$	$CH_3(CH_2)_4CH=CHCH_2CH=CHCH_2CH=CHCH_2CH=CH(CH_2)_3COOH$	−50

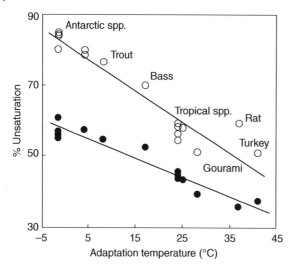

Figure 2.15 The relationship between adaptation temperature and percentage of unsaturated acyl chains in synaptosomal phospholipids of differently adapted vertebrates. Each symbol represents a different species. Open symbols denote phosphatidylethanolamine; filled symbols denote phosphatidylcholine. *Source:* Hochachka and Somero (2002), figure 7.27 (p. 372). Reproduced with the permission of Oxford University Press.

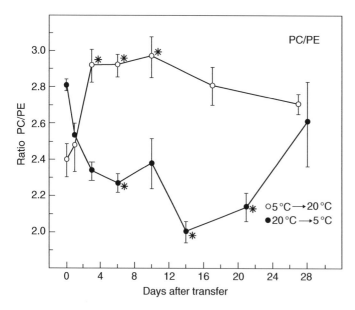

Figure 2.16 Temperature acclimation and phospholipid class. Time course of change in the ratio of phosphatidyl choline (PC) and phosphatidyl ethanolamine (PE) in gill cell membranes of rainbow trout acclimating to the indicated temperatures. *indicates a statistically significant difference ($P<0.05$) compared to the day zero mean. *Source:* Hazel and Carpenter (1985), figure 4 (p. 599). Reproduced with the permission of Springer.

greatly in character with the changes in habitat temperature typical of different zoogeographic regions, but considerable acclimation to temperature change by membrane lipids can also occur within a period of days to weeks. Such short-term change can be considered part of the overall acclimation process that allows a species to adjust its upper and lower lethal limits (see Figure 2.2a, the tolerance polygon).

Pressure

Even though pressure is the most predictable variable in the ocean, increasing by 1 atm with every 10 m increase in depth, pressure is probably the variable most difficult to intuitively understand. Ocean pressure evokes thoughts of dark and forbidding depths, of submarine movies in which the captain and heroic crew must take their craft to depths far greater than she was built to withstand, to there lie on the bottom, evade the enemy, and hope to survive. The great pressure causes the sub to creak and groan, bolts to pop like bullets out of the hull, and leaks to sprout before the ordeal can be successfully ended. However, World War II submarines could not get very deep at all, <300 m, and even modern nuclear subs do not get out of the mesopelagic zone (200–1000 m). Our view on pressure from those movies is one where pressure is acting on gas-filled spaces. A submarine is quite a large gas-filled space and must be immensely strong to withstand even the modest pressure of a dive to 100 m: 11 atm, 162 psi, or 11 143 kPa. In point of fact, most of the species that live under pressure do not have gas-filled spaces, and thus the effects of pressure are far more subtle, especially in the upper 1000 m where much of the ocean's pelagic biomass resides. In our mind's eye though, the pressure associated with even the average depth of the ocean must be a formidable challenge to life.

Pressure at the deepest point (10 916 m) in the ocean, the Challenger Deep in the Mariana Trench near the Philippines, is 16 046 psi (1.11×10^5 kPa). Pressure at the average depth of the ocean of 3800 m (Sverdrup et al. 1942) is 5586 psi (3.85×10^4 kPa). Those are big numbers and are part of the reason we have always held pressure in such high regard as an operator influencing life in the ocean. In fact, respect for pressure has colored some of the history of man's exploration of the sea.

In the 1840s, an eminent British scientist by the name of Edward Forbes was conducting research on the bottom fauna in the coastal waters of the British Isles using an oysterman's dredge. Depths were 600 feet (183 m) and less, and he found the fauna varied and abundant. His reputation as an ocean scientist earned him an invitation to journey to the Mediterranean to do some similar dredging. He found a very sparse bottom fauna in the Mediterranean, but the depths at which he was sampling, 1400 feet (428 m), were about twice those he had sampled off the coast of Great Britain. He assumed that the depth difference between the two regions was responsible for the change in the abundance of the bottom fauna. He decided that if the trend in declining abundance with depth continued as observed, in short order, the bottom fauna would disappear altogether. Based on his results, he declared that below a depth of 1800 feet (550 m), no life would exist and he termed those depths the "azoic zone," the zone without life.

As time went on and exploration of the ocean continued, animals were recovered from greater and greater depths, but rather than accepting the possibility that life existed at the ocean's greatest depths, the azoic zone just kept getting pushed deeper and deeper.

In 1860 just prior to the voyage of the Challenger, the most famous oceanic expedition of all time, a submarine telegraph cable was brought to the surface for repair from a depth of 2000 m in the Mediterranean. The cable was covered with encrusting fauna, animals like barnacles and corals that build calcareous structures. Because those fauna would not have had time to create their structures on the cable during its rapid journey to the surface, the azoic theory was pretty well laid to rest at that point. The theory was further discredited by the discoveries of the Challenger over the next 5 years. Yet, there was still some doubt that life could exist at the very deepest points in the ocean, like the Challenger deep. The challenge to life caused by pressure was at least partially responsible for those doubts, along with the cold, the dark, and the distance from the sun and plant life.

The final demise of the Azoic Theory came in 1960 with the voyage of the submersible Trieste to the deepest point in the ocean. It was one of the great moments in the history of man when Jacques Piccard, son of Auguste Piccard, the submersible's designer, and Lt. Don Walsh of the US Navy descended to the deepest point in the ocean, the Challenger Deep in the southern part of the Mariana Trench. The onboard instrumentation recorded a depth of 11 521 m, which was later revised to 10 916 m. Measurements since then have revised the estimate both up (11 034 m) and down (10 896 m) using different instrumentation, but they are all very close to the original estimate. While at depth, Piccard and Walsh observed swimming shrimps, thus showing that life can exist at the deepest point in the ocean.

Early Work

Unlike temperature or salinity, which can be readily adjusted in the laboratory, conducting experiments with pressure requires specialized instrumentation in the form of pressure vessels. The larger the vessel, the more expensive it is to manufacture, which has limited the amount of research done on the effects of pressure. Despite the difficulties of working with it, enough research has been done on pressure for us to know its basic effects and its main sites of action. In fact, pressure research has a fairly long history, encompassing taxa ranging from bacteria to metazoan invertebrates and fishes. We will cover the points that most directly apply to pelagic fauna.

Regnard (1884, 1891) was the first scientist to study the effects of hydrostatic pressure on invertebrates and fishes. Inspired by the voyages of the Challenger (1872–1876) and the Talisman (1882–1883) and their discovery that life existed at the great depths of the ocean, Regnard decided to bring the environment of the deep sea into the laboratory. He tested the effects of pressure up to 1000 atm on various freshwater and marine animals using a hydraulic pump and rapid pressurization. He found that decapod Crustacea and bony fishes were less resistant to pressure than anemones, echinoderms, mollusks, and annelids. His results on the responses of species to pressure are valid to this day.

Ebbecke (1944) continued the work of Regnard to its logical conclusion also using surface-dwelling and intertidal species. He composed a list of relative pressure tolerance in animal groups going from highest tolerance to lowest: (i) Anemones, (ii) Starfish, (iii) Sea urchins, (iv) Scyphozoan medusae, (v) Gastropods, (vi) Polychaetes, (vii) Shrimp, and (viii) Teleosts.

In addition, he did some behavioral observations of animals exposed to stepped pressures and found that behavioral responses could be divided into three distinct classes.

Phase I. (50 atm) Phase I was characterized by a state of excitement or increased activity as if nervous. Low pressure thus seemed to act as a stimulant.

Phase II. (150 atm) Phase II was characterized by a state of moribundity, almost as if the animal were paralyzed.

Phase III. (200 atm) Phase III caused tetany (maximal contraction of muscles) in shallow invertebrates, and fish were killed immediately. Therefore, surface-dwelling fishes cannot cope with a pressure equivalent to a depth of 2000 m.

Work on tolerance to pressure continued into the late 1960s using similar techniques: shallow-dwelling species exposed to high pressure using small pressure vessels and rapid pressurization. Authors well known for pressure research of this kind were R.J. Menzies, R.Y. George, V. Naroska, C. Schlieper, and H. Flugel (Flugel and Schlieper 1970). An interesting new twist to pressure research was the use of the hydrowinch on an oceangoing research vessel (Menzies and Wilson 1961) as a mechanism for applying pressure. Animals were placed in net-covered jars affixed to the hydrowire and sent to depths ranging from 469 to 3480 m. The lined shore crab *Pachygrapsus crassipes* succumbed to trips below 867 m but survived lesser depths with severe tetany from which they eventually recovered. Mussels (*Mytilus edulis*) were more resistant, surviving trips to 2227 m, but all succumbed to a round trip to 3480 m. It is worth noting that the authors also exposed both species to the lower temperatures they experienced at depth, with no ill effects observed.

At the same time that interest was developing in tolerance to pressure, pressure effects on rate functions were being explored. Regnard initiated this type of research by observing animals through the window of his pressure vessel. The best of the early pressure physiology was by Fontaine (1928) who was the first to study the effects of pressure on the oxygen consumption of plaice (*Pleuronectes platessa*), a European flatfish popular as a menu item. His results mirrored those of Ebbecke in that low and moderate pressures had a stimulatory effect on oxygen consumption rate (VO_2), while high pressures were debilitating. At 25 atm, plaice increased VO_2 by 28%, at 50 atm by 39%, at 100 atm by 58%, and at 100 atm, VO_2 declined.

Workers through the 1930s, 1940s and 1950s continued to test pressure tolerance and the effects of pressure on physiological rates such as rates of ciliary activity in mussels, and the effects on muscle contraction, but all that work was on shallow-living marine or freshwater species. It was not until the early 1960s that experiments began on species that live under pressure. That work was followed by more sophisticated experiments using isolated physiological preparations.

Later Work

One of the most noteworthy studies in the later pressure literature was that of Campenot (1975), who wished to define the neural mechanisms underlying the observed changes in behavior of shallow-dwelling species in response to pressure. The changes in which he was most interested were the excitation caused by pressures <150 atm and the moribundity or depression caused by pressures >150 atm.

Two previous studies provided background relevant to the observed excitability of species at lower pressures. The first noted that hydrostatic pressure dropped the firing threshold for giant axons in squid (Spyropoulos 1957), and the second observed a similar response in frog nerve

(Grundfest 1936). That is, less stimulation was required to elicit neural activity in isolated nerve preparations in both species when they were under low pressure. Their results failed to explain the observed moribundity at high pressures, nor did they give a mechanistic explanation for why neural stimulation occurred at modest pressures.

Campenot used a neuromuscular preparation of the walking legs of two Crustacea to evaluate the effects of pressure. The first preparation was from *Homarus americanus*, the New England lobster dwelling in water 520 m and shallower. The other was of *Chaceon* (formerly *Geryon*) *quinquidens*, a deeper-dwelling red crab found from 300 to 1600 m on the continental slopes of coastal North America. Dr. Campenot's technique was straightforward; he stimulated the excitatory neuron leading to the muscle with one electrode and recorded the response from the muscle with another.

Muscles respond to neural excitation with Excitatory Junction Potentials (EJPs) which then effect a muscle contraction. By examining the amplitude of the EJPs in response to a given fixed stimulus at different hydrostatic pressures, the effect of pressure on the neuromuscular response could be described. Dr. Campenot found that pressure caused an across-the-board depression of EJP amplitude in the lobster (Figure 2.17), but the EJP amplitude was independent of pressure in the deeper-dwelling red crab. In fact, it is now known that independence from pressure effects in

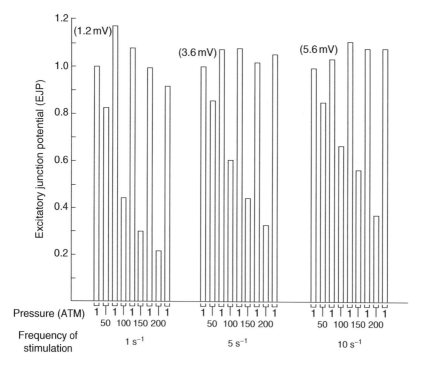

Figure 2.17 The effect of pressure on Excitatory Junction Potentials (EJP) recorded from a lobster muscle fiber. Each bar represents the average of about 20 EJP's. The ordinate is an amplitude index with the first average at 1 atm for each frequency set at 1. Its value in millivolt is given in parenthesis. *Source:* Adapted from Campenot (1975), figure 3 (p. 136). Reproduced with the permission of Elsevier.

species that dwell under pressure is the most common adaptive strategy.

The postulated cause for EJP depression in lobster was a pressure-induced interference of neurotransmitter release at the synapse. At virtually all junctions between nerve and muscle, the neural signal is propagated across the microscopic gap at the neuromuscular junction using a chemical, or neurotransmitter, the best known of which is acetylcholine. Both excitatory and inhibitory neurotransmitters are present at the neuromuscular junction. It was speculated that the observed stimulatory effects of modest pressure were caused by a differential inhibition of transmitter release at inhibitory synapses. In such a situation, excitatory neural activity would then greatly over-ride the depressed inhibitory synapses, resulting in hyperactivity.

Whole Animal Work

The first study of pressure effects on an animal normally living under pressure was that of Napora (1964) who tested pressure effects on the vertically migrating prawn *Systellaspis debilis*. In the western Atlantic, where Napora did his work, *Systellaspis* resides between depths of 500 and 1800 m during the day and 300 and 350 m at night. Napora found that increased pressure resulted in an increased metabolism (measured as oxygen consumption rate) between temperatures of 3 and 20°C and pressures of 0 and 1500 psi. The conclusion from his study was that increases in metabolism as a result of pressure effects offset the decline in metabolism due to the lower temperatures at daytime depth, resulting in a more constant metabolic rate over the diel cycle.

Two additional studies, Teal and Carey (1967) and Teal (1971), improved on Napora's original work, also using species from the northwestern Atlantic. In the first study, the effects of pressure between 0 and 1000 atm were tested on a suite of migrating euphausiids, shrimplike Crustacea 10–25 mm in size found in pelagic waters throughout the world ocean (Chapter 7). The physiological process monitored was once again oxygen consumption rate. Measurements took place at temperatures between 5 and 25 °C, which are typical of the species' vertical range. Oxygen consumption rate (VO_2) was monitored continuously with an oxygen electrode as individuals were rapidly compressed, allowed to remain at pressure for 15–30 minutes, then decompressed. Temperature and pressure were both changed acutely, i.e. without allowing the animal time to acclimate to either variable. Several species of euphausiids were tested in this manner, most of which were epipelagic migrators that came to or near surface waters at night from daytime depths of 200 to 500 m. The rationale for acute measurements was that animals experiencing rapid temperature and pressure changes in the field would be fine with similar treatment in the laboratory, an assumption which was experimentally verified.

The results of the first study (Figure 2.18) showed a different response to pressure in epipelagic migrators (e.g. *Euphausia hemigibba*) than in the mesopelagic migrators (*Thysanopoda monocantha*) that did not go to the upper 50 m of the water column during their nightly migration. Pressure showed no significant effect on metabolism in epipelagic migrators at pressure and temperatures typical of their normal environment. Pressure increased VO_2 only when experimental temperatures and pressures were out of synchrony with those in their normal environment. That is, pressure had no effect unless the pressure–temperature combinations were those that the animals would never normally encounter. In the mesopelagic migratory species *Thysanopoda*

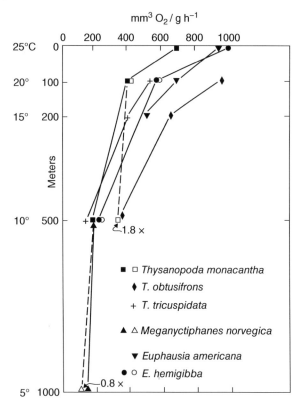

Figure 2.18 Respiration of euphausiids plotted against depth using the indicated depth-temperature distribution, which is typical of summer open-ocean conditions. The solid symbols indicate respiration determined solely by temperature. Where pressure has a significant effect, an open symbol includes the effect of both factors. *Source:* Adapted from Teal and Carey (1967), figure 5 (p. 730). Reproduced with the permission of Elsevier.

monocantha, pressure increased VO_2 enough to offset some of the effects of the lower temperature of their daytime depth such that VO_2 remained fairly constant the animal's normal depth range.

The study's overall conclusions were that

1) Temperature alone determines VO_2 in epipelagic vertical migrators
2) Temperature and pressure working in tandem in mesopelagic vertical migrators make VO_2 more constant over the vertical range of the species.

Teal's 1971 follow-up study dealt with pressure effects on migrating mesopelagic decapod crustaceans, including *Systellaspis debilis*, the experimental species of Napora (1964). His results were similar to those of Napora and to his own 1967 study on euphausiids: pressure increased VO_2 in the vertically migrating mesopelagic species resulting in a more stable VO_2 over their vertical range. Thus, three studies supported conclusion number 2 above. We can add a third conclusion from work on non-migrating mesopelagic species: pressure has little or no effect on the VO_2 of non-migratory mesopelagic species (Meek and Childress 1973).

Molecular Mechanisms of Adaptation to Pressure

The fact that deeper-living species seemed insensitive to pressure piqued the curiosity of biochemical physiologists seeking a molecular basis for the observed data. Using the enzymatic reactions of intermediary metabolism as an experimental system, several studies explored the effects of pressure on deeper-living species. Their a priori reasoning took them back to the basic thermodynamic properties of chemical reactions, which are governed not only by temperature and the concentration of reactants but also by differences in the volume of the reaction system when reactants are changed into products.

1) If a reaction takes place with no volume change, pressure has no effect.
2) If volume increases during the course of the reaction, pressure will inhibit it.
3) If volume decreases, pressure will enhance it.

If reaction rates are expressed incorporating a pressure term, the equation looks like this:

$$K_P = K_D^{-P\Delta V^{\ddagger}/RT} \qquad (2.2)$$

where P is the pressure (atm), T is the absolute temperature (°K), R is the universal gas constant (82 cm^3 atm °K^{-1} mol^{-1}), K_D is the rate constant at 1 atm, K_P is the rate constant at P atm, and $\Delta V^{\ddagger}$ is the activation volume of the reaction; the change in system volume occurring during the rate-limiting step of the reaction, in this case, the activation step.

What is the source of changes in volume in enzymatic reactions? First, consider the total reaction system as the water in which the reactions take place, the enzyme itself, and its substrate or ligand. Those three elements of the system are water–protein–ligand (Hochachka and Somero 1984). Pressure will affect a reaction if there is a volume change in any element of the reaction system during the transition from reactants to products. Though intuitively one might expect that changes in the volume of the enzyme protein during the course of the reaction or production of a higher or lower volume product would be the main source of volume change, most volume changes are actually due to changes in the structure of water. The side-chains of the enzyme protein's amino acids are surrounded by a layer of highly organized water, and this water has a higher density (smaller volume) than the bulk water of the system. Enzymatic reactions by their nature alter the organization of water around the molecule.

The alteration can take several forms (Hochachka and Somero 1984). During ligand binding, the highly organized water may be squeezed out into the bulk phase of the system as the two molecules come together, increasing the total volume of the reaction system. Similarly, when two subunits of an enzyme protein come together to form an aggregate, water can be squeezed out into the bulk phase, increasing system volume. Changes in conformation of the enzyme protein during the reaction can also result in volume change. In one case, exposure to the aqueous medium of hydrophobic residues normally packed within the molecule would increase system volume and would respond negatively to increased pressure. In the opposite case, a normally buried hydrophilic amino acid side-chain exposed to the aqueous medium would allow water to become more densely organized around it, resulting in a decline in volume relative to the bulk water and exhibiting a positive response to increased pressure.

An example from Siebenaller and Somero (1979) of adaptation to pressure in the enzymes of deep-sea fishes is given in Figure 2.19. In their study, the effects of pressure were tested on the LDHs of deep- and shallow-living marine fishes. LDH catalyzes the final reaction in the glycolytic pathway of intermediary metabolism:

Pyruvate + NADH + H$^+$ → Lactate + NAD$^+$

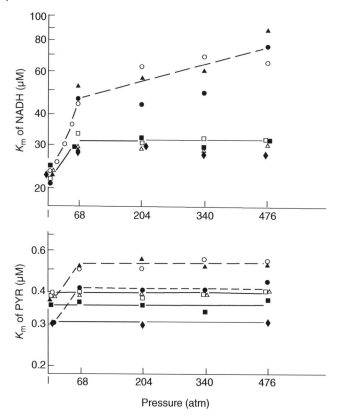

Figure 2.19 The effect of hydrostatic pressure on the apparent Michaelis constant (K_m) NADH (upper) and pyruvate (lower) for M_4-LDH's of four deep-living and three shallow-living species of marine teleost fishes. Shallow-dwelling species indicated by dashed lines in both graphs. *Source:* Siebenaller and Somero (1979), figure 1 (p. 297) Reproduced with the permission of Springer.

K_m values were tested for both NADH and pyruvate at a series of pressures. Not surprisingly, K_ms of deeper living fishes were insensitive to pressure. In contrast, shallow-dwelling species showed a marked elevation of K_ms in response to pressure. As was discussed above, for enzymes to satisfy the dual role of efficient catalysts and regulators of cellular metabolism, their K_ms must fall within a highly conserved range of values (Hochachka and Somero 1984, 2002). The pressure effects noted in Figure 2.20 would be enough to perturb efficient enzyme function in the shallow-dwelling species. Later studies by Siebenaller have found that differences in pressure-sensitive and pressure-insensitive enzymes in closely related species living at different depths can be caused by a change of as little as one amino acid in enzyme structure (Hochachka and Somero 1984).

The final question to consider with respect to enzyme kinetics is that of trade-offs in adaptation to pressure. If only one amino acid separates a pressure-insensitive from a pressure-sensitive enzyme, why would not insensitivity be selected for in most pelagic species? The answer is that pressure-insensitive enzymes are less efficient. Table 2.3 shows the relative velocities

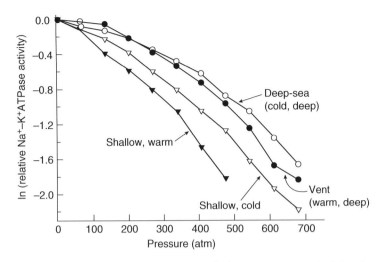

Figure 2.20 Effects of pressure on gill Na$^+$, K$^+$-ATPase activities in fishes from different habitats. The deep-sea fish (cold-deep) was a species living at depths exceeding 2000 m and temperatures of 2–4 °C (*Coryphaenoides armatus* – the grenadier); vent fishes (warm-deep) were two species living near warm hydrothermal vents; shallow-cold was an eastern Pacific fish found at depths less than 2000 m (*Anoplopoma fimbria* – the sablefish); shallow-warm was a species from surface waters near Hawaii (the barracuda *Sphyraena*). *Source:* Gibbs (1997), figure 5 (p. 255). Reproduced with the permission of Academic Press.

Table 2.3 Comparison of lactate dehydrogenase kinetics between two species of deep-sea fishes, three species of shallow-living marine fishes, and a terrestrial species. Velocities are compared at a common temperature and pressure (5 °C and 1 atm.). *Source:* Reprinted by permission from Springer Nature Customer Service Centre GmbH, Nature, Inefficient lactate dehydrogenases of deep-sea fish, Somero and Siebenaller (1979), table 1 (p. 101).

Species (depth, body temp., common name)	ΔH (cal mol^{-1})	ΔS (cal mol^{-1} K^{-1})	ΔG (cal mol^{-1})	Relative velocity
Pagothenia borchgrevinki (surface, −2 °C, ice fish)	10 467	−12.7	14 000	1.00
Sebastolobus alascanus (180–440 m, 4–12 °C, rock fish)	10 515	−12.6	14 009	0.98
Coryphaenoides acrolepis (1460–1840 m, 2–10 °C, rattail fish)	11 813	−8.7	14 222	0.67
Antimora rostrata (1300–2500 m, 2–5 °C, violet cod)	12 557	−6.4	14 343	0.54
Thunnus thynnus (surface to 300 m, 15–30 °C, bluefin tuna)	11 384	−10.0	14 152	0.76
Rabbit (terrestrial, 37 °C)	12 550	−6.4	14 342	0.54

of LDH's from fishes living at different depths measured at a common temperature. Clearly, the reaction velocities of enzymes from deeper-living, pressure-adapted fishes are very much lower than those from shallower dwelling species.

Pressure and Membranes

The sol–gel state of lipids, or their fluidity, has the potential to be profoundly altered by pressure, as it is with temperature. In fact, high pressure and low temperature have similar effects on membrane lipids: both tend to make them more crystalline, i.e. less fluid (Hazel and Williams 1990). Solutions to the problems posed by the ordering effects of hydrostatic pressure and low temperature are solved in a similar manner. In both cases, membrane lipids increase the incidence of double bonds, or their "kinkiness," to increase fluidity.

Evidence supporting the contention that the membranes of deep-sea species are more fluid than those of their shallower dwelling counterparts is more sparse than would be ideal (cf. Hazel and Williams 1990), but it is present, nonetheless. In a benchmark publication from 1984, Cossins and MacDonald found that membrane lipids isolated from the brains of a suite of fishes dwelling between 200 and 4800 m showed significant increases in fluidity with depth consistent with homeoviscous adaptation. Evidence was not conclusive for lipids isolated from other organs, notably liver and kidney, due largely to variability between samples, but trends were similar.

Further evidence supporting membrane adaptation to pressure comes from study of membrane-bound enzymes, notably the ion-pumping enzyme Na-K ATPase, an important player in the osmoregulation of fishes. Gibbs and Somero (1989, 1990) first tested the pressure sensitivity of enzymes from fishes dwelling in a variety of habitats: shallow-warm, shallow-cold, hydrothermal vent (deep-warm), and deep cold (Figure 2.21). As might be expected, they found that the order of pressure sensitivity (highest to lowest) was shallow-warm, shallow-cold, deep-warm, and deep-cold. The investigators then manipulated the membrane environment of the $Na^+ K^+$ ATPase to assess the influence of the membrane fraction on enzyme activity. They found that enzyme from a warm-shallow species (the barracuda *Sphyraena*) placed in a

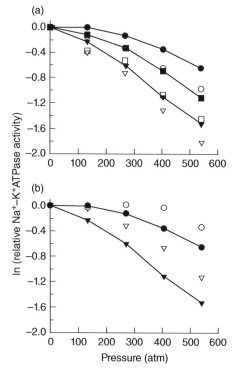

Figure 2.21 Effects of lipid substitution on the pressure responses of Na^+, K^+-ATPase of three species of fish from different depth-temperature habitats. Native membrane lipids were removed and replaced with (a) chicken egg phosphatidylcholine or (b) phospholipids prepared from the gill of the abyssal grenadier *Coryphaenoides armatus*. Filled symbols indicate pressure responses before lipid substitution; open symbols were assays after substitution. Circle symbols represent *Coryphaenoides armatus* (deep sea, cold habitat); square symbols represent the sablefish *Anoplopoma fimbria* (shallow, cold habitat); triangles represent the barracuda *Sphyraena barracuda* (shallow, warm habitat). *Source:* Gibbs (1997), figure 6 (p. 257). Reproduced with the permission of Academic Press.

(a)

(b)

Figure 1.4 The Polar Oceans. (a) Arctic Ocean basin. (b) Southern Ocean surrounding Antarctica. *Source:* NASA.

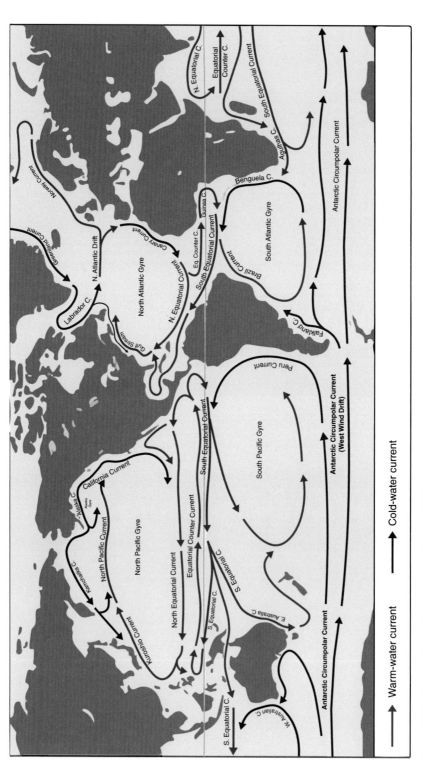

Figure 1.5 Geostrophic (surface) currents. *Source*: NASA.

Figure 1.6 Thermohaline (deep) currents. *Source*: NOAA.

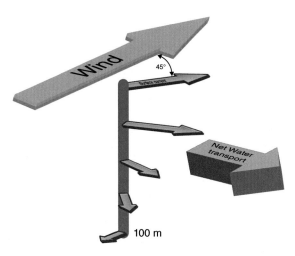

Figure 1.9 Ekman transport. Net spiral pattern of wind-driven motion down through a water column due to Coriolis effect and drag. The result is a net flow 90° to the right of the wind direction in the northern hemisphere.

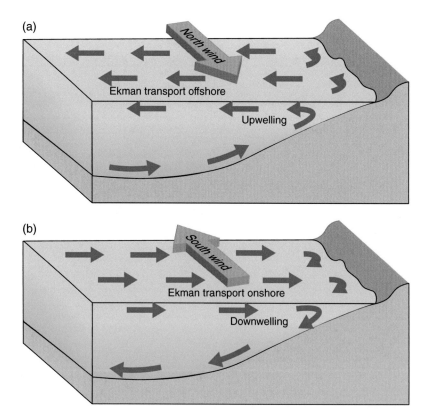

Figure 1.10 Upwelling and downwelling. (a) Ekman transport caused by wind blowing from the north moving surface water offshore, results in deeper water upwelling to the surface in the northern hemisphere. (b) Ekman transport due to winds blowing from the south moves surface water onshore and subsequently down slope.

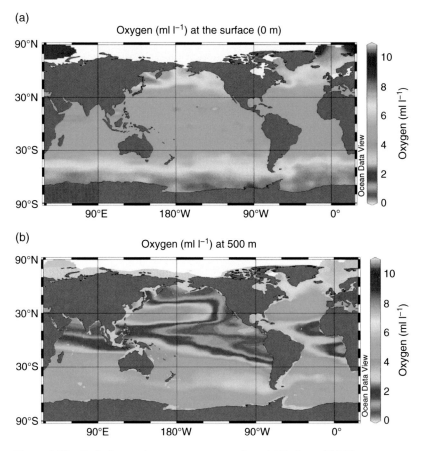

Figure 1.18 Typical oceanic oxygen concentrations. (a) Surface; (b) 500 m.

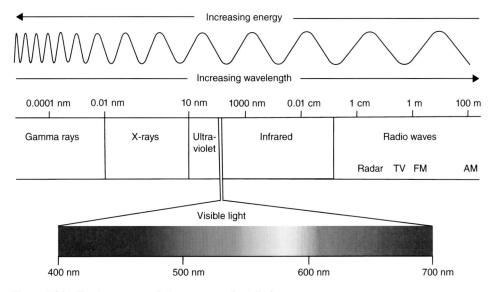

Figure 1.21 Total spectrum of electromagnetic radiation.

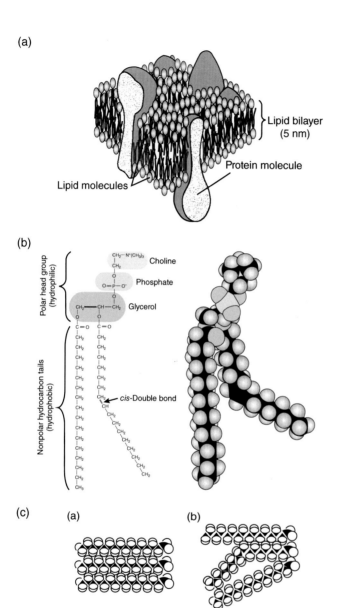

Figure 2.14 Membrane structure. (A) Schematic representation of a membrane showing the lipid bi-layer structure with embedded proteins; (B) chemical structure and a three-dimensional representation of a phospholipid molecule showing the glycerol and phosphate components of the hydrophilic head group and hydrophobic tails that face inward in membranes; (C) saturated vs. unsaturated fatty acids, (a) three molecules of stearic acid (18 carbons, no double bonds, melting temperature 69.6 °C) that pack tightly because of their linear geometries, (b) the addition of a single molecule of oleic acid (18 carbons, one double-bond, melting temperature 56 °C) to a lipid membrane prevent tight packing of the molecules due to the bending of the tail of the unsaturated molecule.

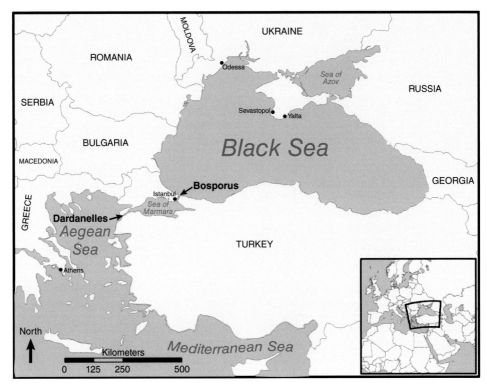

Figure 4.14 The Black Sea region of Eastern Europe showing the two narrow straits, the Bosporus and Dardanelles, that connect the Black Sea, the Sea of Marmara, and the Aegean Sea.

Figure 5.12 The pelagic nemertean *Bathynemertes* species. *Source:* © Dante Fenolio, reproduced with permission.

Figure 6.24 A pelagic tomopterid showing the undulatory sinusoidal shape assumed during swimming. *Source:* © Dante Fenolio, reproduced with permission.

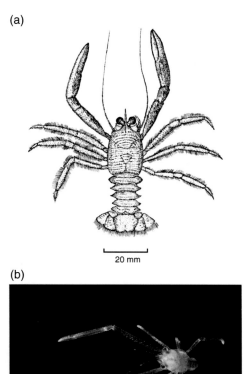

Figure 7.56 Galatheidae: pelagic squat lobsters. (a) The pelagic red crab *Pleuroncodes planipes*; (b) Post-larval grimothea stage of the pelagic squat lobster *Munida* sp. *Sources:* (a) Boyd (1967), figure 1 (p. 395); (b) © Dante Fenolio, reproduced with permission.

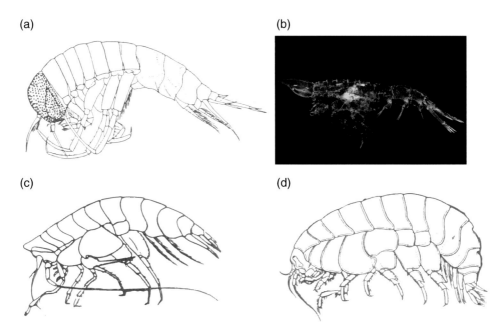

Figure 7.68 A comparison of the forms of two common hyperiid amphipods with two typical gammarids. (a) Female hyperiid *Themisto abyssorum*; (b) The hyperiid *Oxycephalus* sp.; (c) The mesopelagic gammarid *Cyphocaris challengeri*; (d) the bathypelagic gammarid *Eurythenes gryllus*. Sources: (a) Vinogradov et al. (1996), figure 149 (p. 359); (b) © Dante Fenolio, reproduced with permission; (c, d) Vinogradov (1999), figures 4.11, 4.15 (p. 1208).

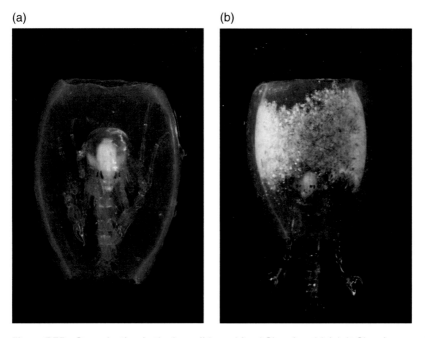

Figure 7.73 Reproduction in the hyperiid amphipod *Phronima*. (a) Adult *Phronima* sp. within a salp barrel; (b) adult female *Phronima* with larvae in a salp barrel. Source: © Dante Fenolio, reproduced with permission.

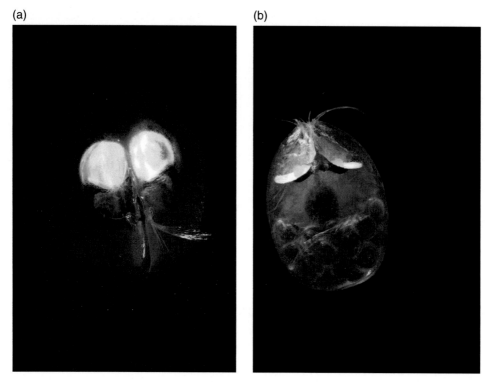

Figure 7.75 The giant ostracod *Gigantocypris*. (a) Frontal view of *Gigantocypris* showing the large naupliar eyes; (b) lateral view of *Gigantocypris* showing the retina and parabolic reflector of the eyes. Note the developing eggs within the carapace. *Source:* © Dante Fenolio, reproduced with permission.

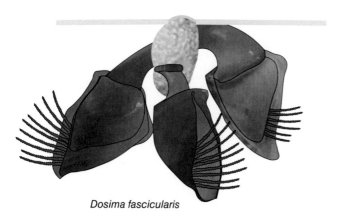

Dosima fascicularis

Figure 7.76 The pleustonic "blue buoy" barnacle *Dosima fascicularis*.

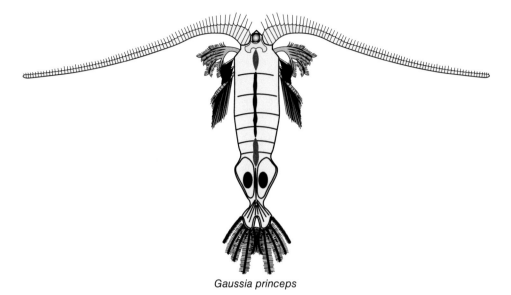

Gaussia princeps

Figure 7.77 The "giant" copepod *Gaussia princeps*. Carapace is deep blue in life.

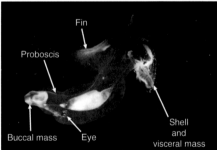

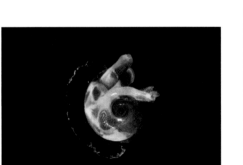

Oxygirus inflatus

Figure 8.33 The atlantid heteropod *Oxygyrus inflatus* with keeled shell. *Source:* © Dante Fenolio, reproduced with permission.

Carinaria lamarcki

Figure 8.34 The carinariid heteropod *Carinaria lamarcki*. *Source:* © Dante Fenolio, reproduced with permission.

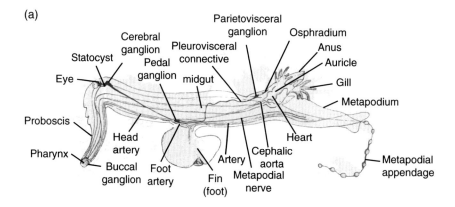
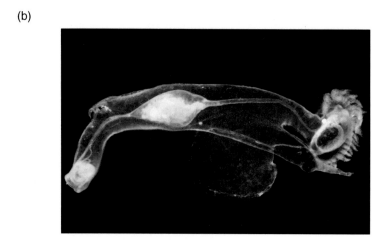

Figure 8.35 The pterotracheid heteropod *Pterotrachea coronata*. (a) Internal anatomy of *P. coronata*. *Source:* From Lang (1900), Figure 10 (p. 8); (b) *P. coronata* in normal swimming orientation; (c) viscera and gills of *P. coronata*. *Source:* (b, c) © Dante Fenolio, reproduced with permission.

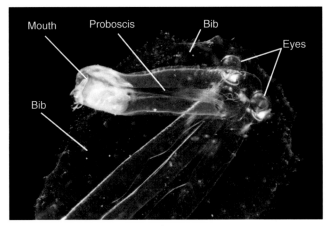

Pterotrachea scutata

Figure 8.36 Ventral view of the anterior end of the pterotracheid heteropod *Pterotrachea scutata* showing the tubular eyes, elongated proboscis, bib, and relatively small mouth. *Source:* © Dante Fenolio, reproduced with permission.

(a) *Cavolinia uncimata*, ventral viewB

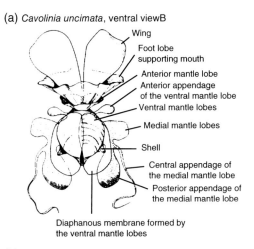

(b) *Cavolinia gibbosa*

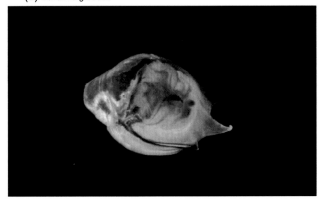

Figure 8.39 Euthecosome pteropods. (a) Anatomy of the euthecosome pteropod *Cavolinia uncimata*. *Source:* From Gilmer and Harbison (1986), Figure 1 (p. 49); (b) the euthecosome pteropod *Cavolinia gibbosa*. *Source:* © Dante Fenolio, reproduced with permission.

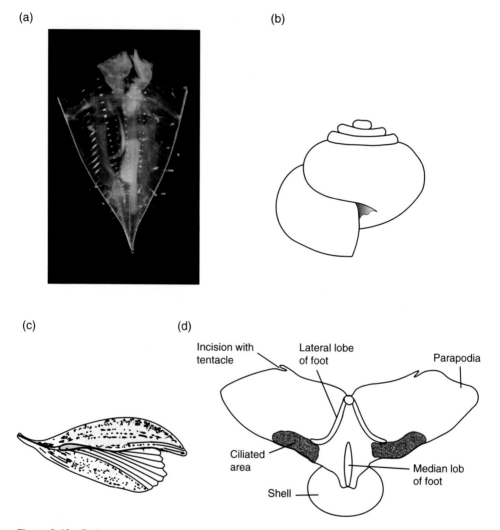

Figure 8.40 Euthecosome pteropods. (a) *Clio* sp. with attached hydroids. *Source:* © Dante Fenolio, reproduced with permission; (b) shell of *Limacina*. *Source:* Hyman (1967), Figure 182 (p. 405); (c) shell of *Cavolina*, lateral view; (d) external anatomy of *Limacina*, ventral view. *Source:* (c, d) From Meisenheimer (1905).

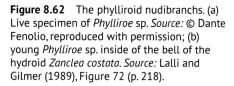

Figure 8.62 The phylliroid nudibranchs. (a) Live specimen of *Phylliroe* sp. *Source:* © Dante Fenolio, reproduced with permission; (b) young *Phylliroe* sp. inside of the bell of the hydroid *Zanclea costata*. *Source:* Lalli and Gilmer (1989), Figure 72 (p. 218).

(a) *Phylliroe* sp.

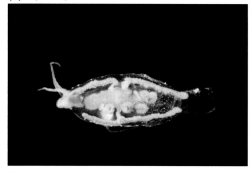

(b) Young *Phylliroe* inside of the medusa of the hydroid *Zanclea costata*

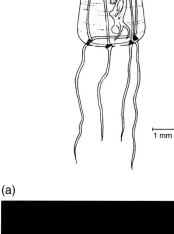

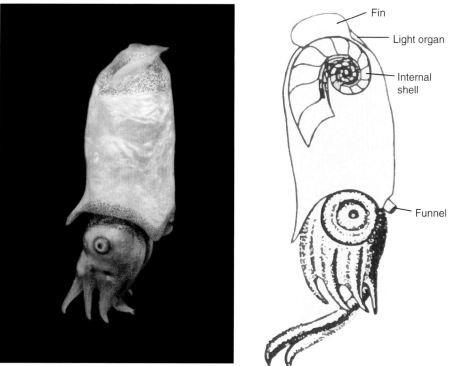

Figure 8.73 The order Spirulida. (a) The Ram's head squid *Spirula spirula*. *Source:* © Dante Fenolio, reproduced with permission; (b) *Spirula spirula* in normal swimming position showing the position of the internal shell. *Source:* Denton and Gilpin-Brown (1973), Figure 7 (p. 216).

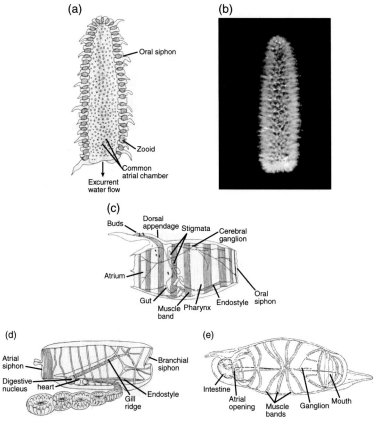

Figure 9.3 Thaliaceans: pyrosomes, doliolids, and salps. (a) Colonial thaliacean *Pyrosoma*. (b) Photo from life of a pyrosome. (c) A solitary thaliacean *Doliolium*. (d) The salp *Cyclosalpa affinis*, solitary form. (e) The salp *Salpa maxima*, aggregate form. *Sources:* (a) Adapted from Brusca and Brusca (2003), figure 23.14 (p. 858); (b) © Dante Fenolio, reproduced with permission; (c) Adapted from Brusca and Brusca (2003), figure 23.14 (p. 858); (d) Adapted from Berrill (1950), figure 104 (p. 288); (e) Berrill (1950), figure 67 (p. 85).

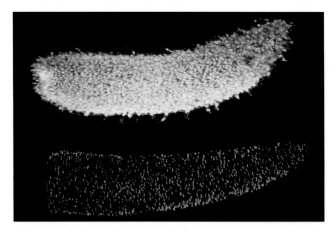

Figure 9.28 Pyrosomes. Photos from life of a pyrosome in natural light (upper) and in the dark after stimulation eliciting bioluminescence (lower). *Source:* © Dante Fenolio, reproduced with permission.

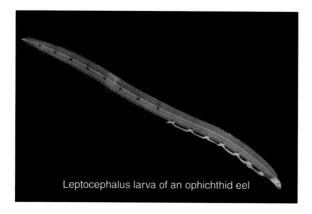

Figure 10.19 Leptocephalus larva of an ophichthid eel. *Source:* © Dante Fenolio, reproduced with permission.

Leptocephalus larva of an ophichthid eel

(a) *Dolichopteryx longipes* (dorsal view)

(b) Head of *D. longipes* (lateral view)

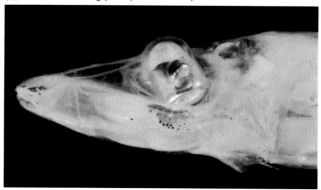

Figure 10.23 Opisthoproctidae. (a) Dorsal view of the spookfish *Dolichopteryx longipes* showing the upward-looking eyes; (b) close-up of the head of *D. longipes* showing the tubular (barrel-shaped) eye with retinal diverticulum. *Source:* © Dante Fenolio, reproduced with permission.

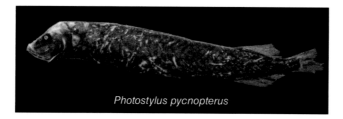

Photostylus pycnopterus

Figure 10.24 The alepocephalid *Photostylus pycnopterus*. *Source:* © Dante Fenolio, reproduced with permission.

(a) *Dolicholagus longirostris*

(b) *D. longirostris*

(c) *Bathylagus pacificus*

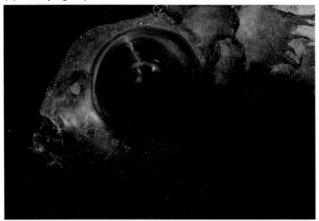

Figure 10.25 Bathylagids. (a) The silvery-colored, upper-mesopelagic bathylagid *Dolicholagus longirostris*; (b) close-up of the head of *D. longirostris* showing the aphakic space in the anterior part of the eye; (c) the dark-colored, deeper-living bathypelagic *Bathylagus pacificus*. Note the relatively larger eye and large scales. *Source:* © Dante Fenolio, reproduced with permission.

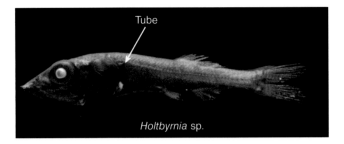

Figure 10.26 The platytroctid *Holtbyrnia* sp. showing the tube that gives the platytroctids the common name "tube shoulder." *Source:* © Dante Fenolio, reproduced with permission.

(a)

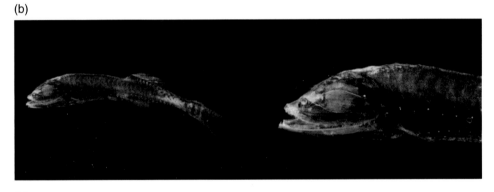

(b)

(c)

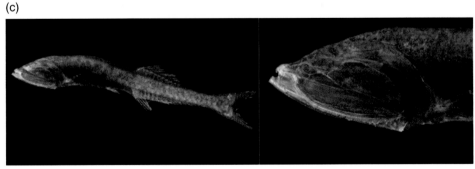

Figure 10.27 A comparison of biological characteristics between three species of *Cyclothone* from different depth zones. Notable differences include coloration, eye size, and photophores. (a) *Cyclothone braueri*: upper mesopelagic, light color, small eyes, and large prominent photophores; (b) *Cyclothone pallida*: lower mesopelagic, dark color, reduced eyes, small photophores; (c) *Cyclothone obscura*: bathypelagic, intermediate dark color, very reduced eyes, no photophores. *Source:* © Dante Fenolio, reproduced with permission.

Sigmops elongatum

Figure 10.28 Lateral view (top) and close-up of the head (bottom) of the large species of gonostomatid, *Sigmops elongatum*. Note the prominent photophore array and dentition. *Source:* © Dante Fenolio, reproduced with permission.

Vinciguerria poweriae

Figure 10.30 The phosichthyid *Vinciguerria poweriae*: whole body (top) and close-up of the head (bottom). Note the well-developed eye with large ventronasal and postorbital photophores and the well-developed ventral photophore array. *Source:* © Dante Fenolio, reproduced with permission.

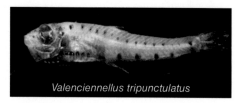

Valenciennellus tripunctulatus

Figure 10.31 The sternoptychid *Valenciennellus tripunctulatus*. Note the well-developed upward-looking eye, the array of large ventral photophores, and the prominent melanophores along the side of the body just above the lateral line. *Source:* © Dante Fenolio, reproduced with permission.

(a)

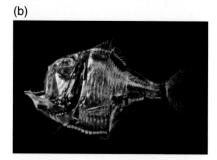

(b)

(c)

Figure 10.32 Sternoptychidae: examples of the three genera of hatchetfishes. (a) *Polyipnus clarus*; (b) *Argyropelecus aculeatus*; (c) *Sternoptyx diaphana*. Note the distinctive tubular photophores along the ventrum and the small pre-orbital photophore that shines directly into the eye, helping to match photophore output with downwelling light intensity (countershading). *Source:* © Dante Fenolio, reproduced with permission.

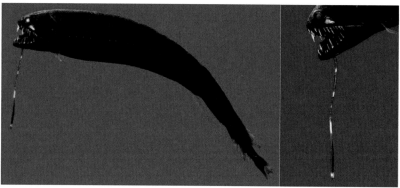

Figure 10.33 Astronesthinae. (a) Lateral view (left) and close-up of the head of *Astronesthes micropogon* showing the long mental barbel, pronounced teeth, and distinctive suborbital photophore. *Source:* © Dante Fenolio, reproduced with permission; (b) day-night vertical distributions of three species of *Astronesthes* from the Gulf of Mexico showing pronounced patterns of vertical migration. *Sources:* (a) © Dante Fenolio, reproduced with permission. (b) adapted from Sutton and Hopkins (1996b), figure 2 (p. 537).

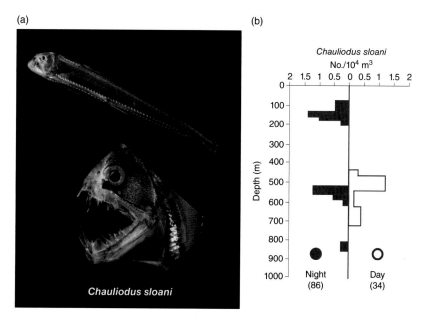

Figure 10.34 Chauliodontinae. (a) Lateral view (top) and close-up of the head of *Chauliodus sloani* showing the two very large mandibular "fangs" and the jaw in the un-hinged position; (b) day-night vertical distributions of *C. sloani* from the Gulf of Mexico showing an asynchronous diel vertical distribution. *Sources:* © Dante Fenolio, reproduced with permission. (b) Adapted from Sutton and Hopkins (1996b), figure 1 (p. 535).

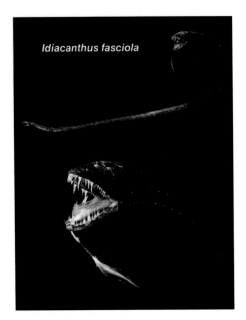

Figure 10.35 Idiacanthinae. Lateral view of the whole body (upper) and a close-up of the head of *Idiacanthus fasciola*. *Source:* © Dante Fenolio, reproduced with permission.

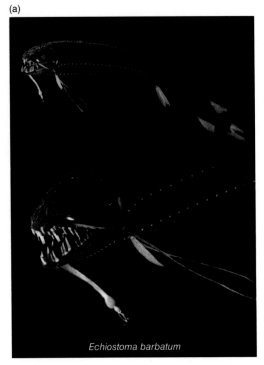

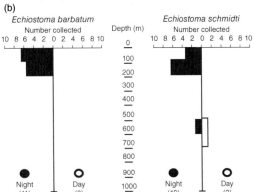

Figure 10.36 Melanostomiinae. (a) Lateral view (upper) and close-up of the head (lower) of *Echiostoma barbatum* (threadfin dragonfish). Note the pectoral fin which is reduced to a single long ray separated from three short rays and the complex esca at the end of the mental barbel; (b) day-night vertical distributions of two species of *Echiostoma* from the Gulf of Mexico showing patterns of diel vertical migration. *Sources:* (a) © Dante Fenolio, reproduced with permission. (b) Adapted from Sutton and Hopkins (1996b), figure 2 (p. 537).

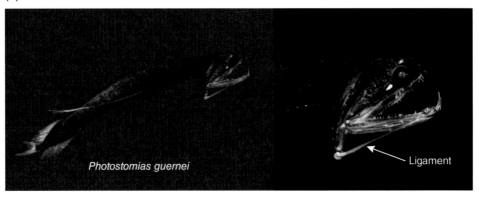

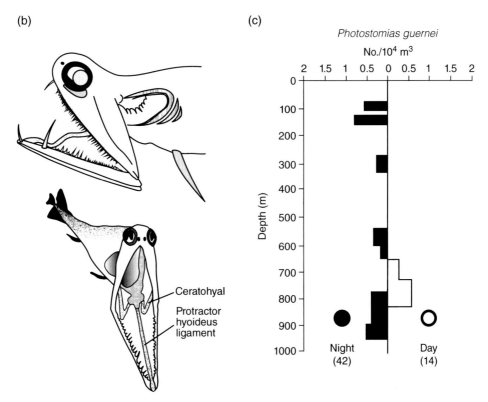

Figure 10.37 Malacosteinae. (a) Lateral view (left) and close-up of the head of *Photostomias guernei*. Note the very long pelvic fins and ligament (protractor hyoideus) extending from the tip of the mandible to the hyoid. *Source:* © Dante Fenolio, reproduced with permission; (b) diagrammatic representation of the lack of ethmoid membrane (floor of the mouth) and showing two of the structures that facilitate the extension of the mandible and expansion of the gape during feeding (the protractor hyoideus ligament and ceratohyal); (c) day-night distributions of *P. guernei* from the Gulf of Mexico showing an asynchronous diel vertical migration pattern. *Sources:* (a) © Dante Fenolio, reproduced with permission. (b) Adapted from Sutton (2005), figure 1 (p. 2067); (c) Adapted from Sutton and Hopkins (1996b), figure 1 (p. 535).

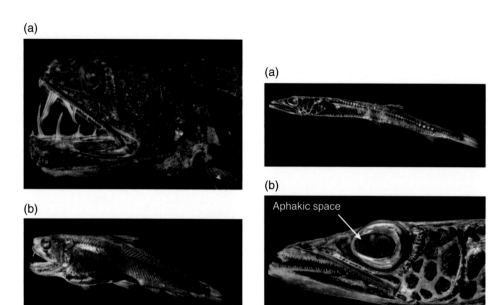

Figure 10.40 Evermanellidae. (a) Close-up of the head of *Odonostomops normalops*; (b) lateral view of *Coccorella atlantica*. Note the formidable fang-like teeth and upward-directed eyes. *Source:* © Dante Fenolio, reproduced with permission.

Figure 10.42 Notosudidae. (a) The notosudid *Scopelosaurus smithi*; (b) close-up of the head of *S. smithi* showing the aphakic space anterior to the lens in the eye. *Source:* © Dante Fenolio, reproduced with permission.

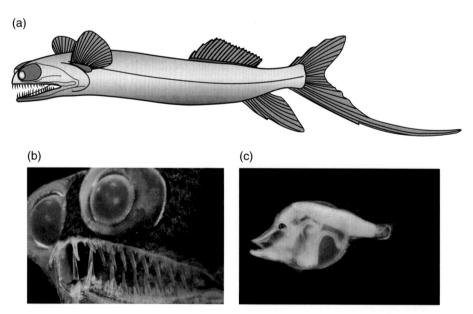

Figure 10.41 Giganturidae. (a) Adult *Gigantura chuni*. (b) Close-up of the head of an adult *Gigantura indica* showing the forward-directed tubular eyes and pronounced dentition; (c) larval giganturid. Note the conventional eye and small mouth. *Sources:* (b, c) © Dante Fenolio, reproduced with permission.

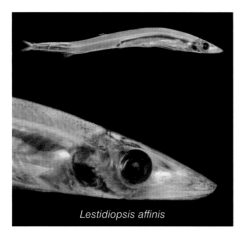

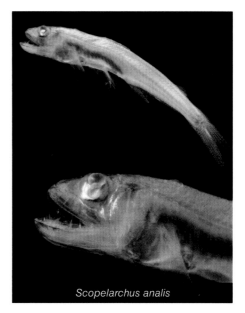

Figure 10.44 Lateral view (top) and close-up of the head of the paralepidid *Lestidiopsis affinis*. Note the darkly pigmented gut that is believed to conceal bioluminescence of prey items such as myctophids. *Source:* © Dante Fenolio, reproduced with permission.

Figure 10.45 Scopelarchidae. Lateral view (top) and close-up of the head of *Scopelarchus analis* showing the upward-looking tubular eye and darkly pigmented gut thought to conceal bioluminescence of prey items. Note the silvery pad beneath the lens of the eye which acts as a light guide allowing light to enter from the side, strike the retina, and thus facilitate detection of lateral point sources of light. *Source:* © Dante Fenolio, reproduced with permission.

(a)

(b)

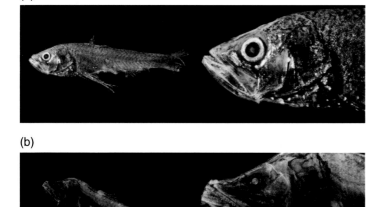

Figure 10.46 Neoscopelidae. External characteristics of shallow and deep living neoscopelids. (a) *Neoscopelus macrolepidotus*, a benthopelagic species found between 200 and 1000 m. Note the well-developed photophores, large eyes and silvery body; (b) *Scopelengys tristis*, a lower mesopelagic species. Note the small eye, dark coloration, and lack of photophores. *Source:* © Dante Fenolio, reproduced with permission.

(a)

Figure 10.48 Two species of myctophids in the subfamily Diaphinae. (a) Lateral view of *Diaphus dumerilii*; (b) close-up of the head of *D. lucidus*. Note the large luminous organs in front and below each eye ("headlights"). Each of these organs consists of a combination of very large dorsonasal, antorbital, and ventronasal photophores, all in contact with each other. The shape and size of these organs vary between species. *Source:* © Dante Fenolio, reproduced with permission.

(b)

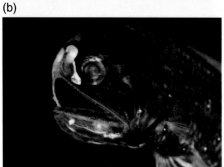

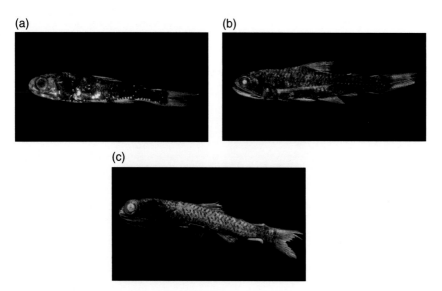

Figure 10.49 A comparison of external characteristics of three species from the subfamily Lampanyctinae from different depth zones. (a) *Ceratoscopelus warmingii*; an active diel migrator with night-time depths in the upper mesopelagic, with silvery body, well-developed photophores, and large eyes; (b) *Lampanyctus alatus*; also an active migrator but considered a lower mesopelagic resident with dark body, smaller photophores and smaller eyes; (c) *Taaningichthys bathyphilus*: a bathypelagic non-migrator with dark body, very small photophores, but a relatively large eye. Note the large subcaudal photophore ("sternchaser") on *Taaningichthys*. *Source:* © Dante Fenolio, reproduced with permission.

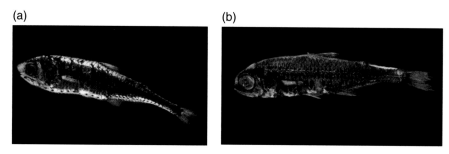

Figure 10.50 Two species of myctophids from the subfamily Myctophinae. (a) *Centrobranchus nigroocellatus*; (b) *Myctophum affine*. Both are active diel migrators appearing in surface waters at night. Note the differences in eye size and photophore development on *Centrobranchus* and the large supracaudal photophore ("sternchaser") on *Myctophum*. *Source:* © Dante Fenolio, reproduced with permission.

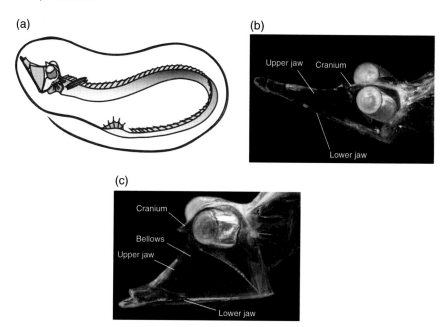

Figure 10.54 The Stylephoriformes. (a) The tube-eye (or thread-tail), *Stylephorus chordatus*. Note the extreme extension of the lower lobe of the caudal fin and the unique configuration of the upper and lower jaws; (b) close-up of the head of *S. chordatus* showing the mouth (upper and lower jaws configured with the bellows in the contracted position); (c) close-up of the head of *S. chordatus* with the bellows in the expanded position. *Source:* (b, c) © Dante Fenolio, reproduced with permission.

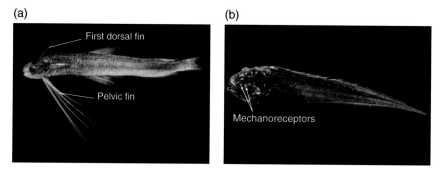

Figure 10.55 The Gadiformes. (a) A codlet, *Bregmaceros* sp. The first dorsal fin is a single long fin ray on the head that folds into a groove when depressed. Note also the very long fin rays of the pelvic fin and its unusual placement on the throat under the operculum; (b) the arrowtail cod, *Melanonus zugmayeri*. The mechanoreceptors (pits) on the head surround the eye and continue along the body (lateral line). *Source:* © Dante Fenolio, reproduced with permission.

(a) Larval *Parataeniophorus brevis* (b) Adult male *Ataxolepis apus*

(c) Post-larval female *Cetostoma regani* (d) Adult female *Cetomimus gilli*

Figure 10.57 Four radically different life stages displayed by the Cetomimidae. (a) Larval form; *Parataeniophorus brevis*. Note the caudal streamer, an extension of the caudal fin rays, that can extend 10–20 times the body length (it is shown incomplete in the figure). (b) Adult male *Ataxolepis apus*. Note the engorged stomach filled with copepods acquired during the larval stage. (c) post-larval female (in transition); *Cetostoma regani*; (d) adult female; *Cetomimus gilli*. *Source:* (d) © Dante Fenolio, reproduced with permission.

(a)

(b)

Anoplogaster cornuta

Figure 10.58 Adult (a) and juvenile (b) *Anoplogaster cornuta*. Note the differences in dentition, relative eye size and lateral line development. *Source:* © Dante Fenolio, reproduced with permission.

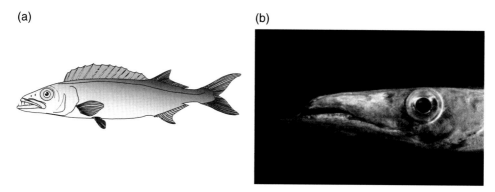

Figure 10.59 The Gempylidae (snake mackerels). (a) The oilfish *Ruvettus pretiosus*. Although edible, the flesh of this species contains a high content of an oil, rich in esters, that is a very effective laxative; (b) close-up of the head of the striped escolar *Diplospinus multistriatus*. This species is an important food fish for local populations. *Source:* (b) © Dante Fenolio, reproduced with permission.

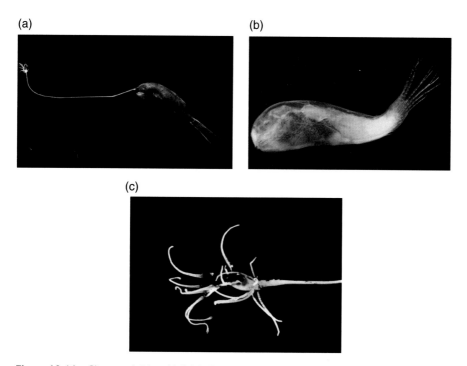

Figure 10.64 Gigantactinidae. (a) Adult female *Gigantactis gargantuai* with long illicium; (b) free-living male *Rhynchactis* sp.; (c) close-up of the esca of *Gigantactis gargantuai*. *Source:* © Dante Fenolio, reproduced with permission.

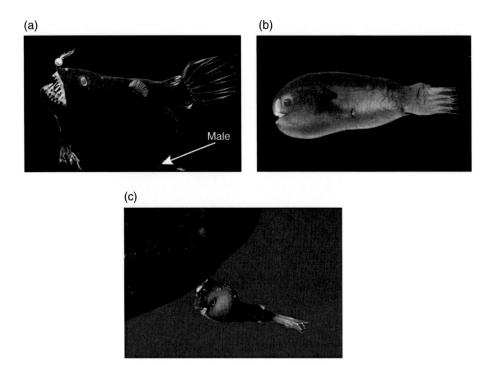

Figure 10.65 Linophrynidae. (a) Adult female *Linophryne* sp. with attached male. Note the large well-developed chin barbel which is unique to the linophrynids; (b) free-living male linophrynid. Note the large well-developed olfactory organs (nostrils) in front of the eyes; (c) close-up of the attached male seen in (a). *Source:* © Dante Fenolio, reproduced with permission.

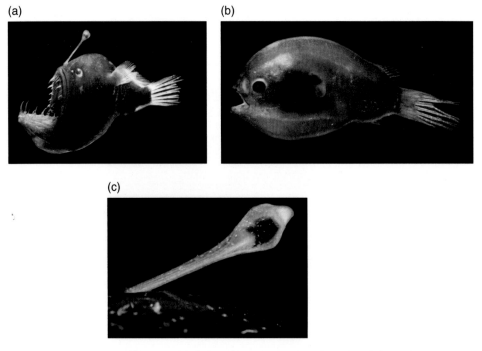

Figure 10.66 Melanocetidae. (a) Adult female *Melanocetus murrayi*; (b) free-living male *Melanocetus* sp. Note the large well-developed olfactory organs (nostrils) in front of the eyes; (c) close-up of the esca of *Melanocetus johnsoni*. *Source:* © Dante Fenolio, reproduced with permission.

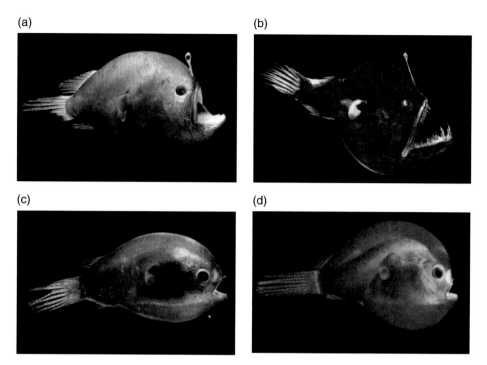

Figure 10.67 Anglerfish life history: metamorphosis. (a) Female *Melanocetus* sp. at metamorphosis; (b) adult female *Melanocetus murrayi*; (c) male *Melanocetus* sp. at metamorphosis; (d) free-living male *Melanocetus* sp. *Source:* © Dante Fenolio, reproduced with permission.

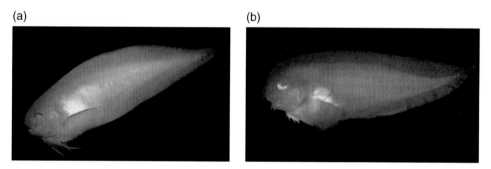

Figure 10.69 Two species of liparids, (a) *Careproctus trachysoma* and (b) *Crystallichthys matsushimae*. *Source:* © Dante Fenolio, reproduced with permission.

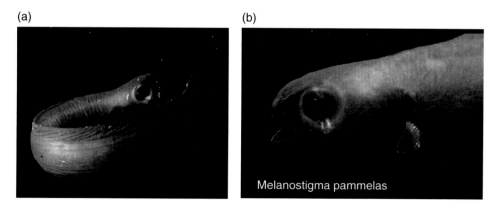

Figure 10.73 Zoarcidae. (a) The eelpout *Melanostigma pammelas*; (b) close-up of the head of *M. pammelas*. *Source:* © Dante Fenolio, reproduced with permission.

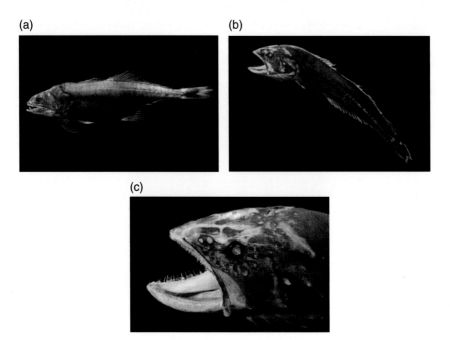

Figure 10.74 Chiasmodontidae. (a) *Chiasmodon niger*, the black swallower. Note the distended belly, evidence of the ability to swallow very large prey including prey larger than their own body; (b) *Dysalotus alcocki*. Note the well-developed lateral line; (c) close-up of the head of *D. alcocki* showing the very large gape with a partially unhinged (protruded) lower jaw and large sensory pits surrounding the eye and continuing into the lateral line. *Source:* © Dante Fenolio, reproduced with permission.

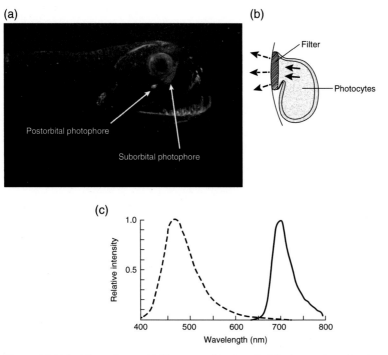

Figure 10.116 Photophores. (a) Close-up of the head of *Pachystomias microdon* showing the blue postorbital and red suborbital photophores; (b) vertical section of the red suborbital photophore of *Malacosteus*. The red light produced by the photocytes is further modified by filter before leaving the photophore; (c) the resulting spectral emission of the suborbital photophore has a maximum at 708 nm (solid line), very different from the typical blue emission (dotted line) of the postorbital photophore. *Sources:* (a) © Dante Fenolio, reproduced with permission. (b) Denton et al. (1985), figure 8 (p. 74); (c) adapted from Widder et al. (1984), figure 2 (p. 512).

membrane environment derived from the cold deep-dwelling fish *Coryphaenoides armatus* was less pressure-sensitive than the enzyme of a cold-shallow species (the sablefish *Anoplopoma fimbria*) that was introduced into a highly ordered lipid environment derived from chicken eggs. In contrast, when enzymes from the three species were introduced into the same lipid environment, one derived from the deep-cold species, *Coryphaenoides armatus*, the order of pressure sensitivity remained as stated above: warm-shallow, cold-shallow, and cold-deep. The results of Gibbs and Somero provide evidence that membrane fluidity influences the function of membrane-bound enzymes and also that changes in enzyme structure play an important role in adaptation to pressure.

Oxygen

At all depths of the ocean, oxygen is removed from the water column as organisms respire and organic matter is biochemically degraded. Wave action, mixing, and photosynthetic processes replenish the lost oxygen in the upper mixed layer, but a zone of minimum oxygen forms at intermediate depth in all the world's oceans. The severity of oxygen depletion in the minimum zones varies considerably, with values of dissolved oxygen ranging from $0\,\mathrm{ml\,l^{-1}}$ (zero) in the Arabian Sea (Hitchcock et al. 1997) to about $4\,\mathrm{ml\,l^{-1}}$ in the Antarctic (Smith et al. 1999) (see Chapter 1 for plots of O_2 vs. depth).

The major factors contributing to the persistence of oxygen minimum zones are global circulation patterns that result in the water at mid-depths being out of contact with the atmosphere for hundreds of years. In areas where the water column is well-stratified and rates of primary production are high, oxygen minima are especially severe. Dissolved oxygen values reach $0\,\mathrm{ml\,l^{-1}}$ and stay near zero for hundreds of meters of water depth. Examples of such regions are the Arabian Sea, Bay of Bengal, Cariaco Basin, Philippine region, the northwest Pacific margin, and the eastern tropical Pacific. Hypoxic conditions in those areas extend from the bottom of the mixed layer to about 1500 m depth.

Although the oxygen concentration in oxygen minimum zones can be $<0.2\,\mathrm{ml\,l^{-1}}$, large populations of organisms can reside there. Those organisms include the decapods, mysids, and copepods and even large populations of midwater (mesopelagic) fishes (Gjøsaeter 1984).

Organisms that inhabit oxygen minima exhibit a range of adaptations for dealing with hypoxic conditions. These adaptations include behavior, such as diel vertical migration out of the layer, in addition to unusual morphological and physiological characteristics (Wishner et al. 2000). Childress and Seibel (1998) have proposed three modes of adaptation to the oxygen minimum: (i) development of mechanisms for efficient removal of oxygen from water; (ii) reduction of metabolic rates; and (iii) use of anaerobic metabolism to compensate for the difference between aerobic metabolism and total metabolic needs. The use of anaerobic metabolism may occur on a sustained basis, during periods of high metabolic demand, or during transient periods spent in the oxygen minimum layer by vertical migrators.

The decreased food energy available in the deep sea (Vinogradov 1970) favors an aerobic existence where there is sufficient oxygen to be utilized. However, an oxygen minimum such as that in the Arabian Sea, where oxygen drops to zero, constrains species to an anaerobic existence while resident within it. Anaerobiosis poses a special problem to vertebrates. With the exception of breath-hold divers such as marine mammals and turtles, most vertebrates, including fish, rely exclusively on aerobic metabolism and only switch to locally active anaerobic pathways during periods of increased activity or ambient low oxygen levels. For example, tuna white muscle is very well adapted for anaerobiosis, which is used during burst swimming (Hochachka 1980).

The main anaerobic pathway used by vertebrates is glycolysis, resulting in lactate production (Withers 1992). However, two major problems arise with use of the glycolytic pathway. First, the energy yield is very low, only 2 ATP per glucose as opposed to 38 ATP per glucose from aerobic metabolism. Second, the accumulation of lactate is potentially harmful to the organism. Build-up of lactate can lead to osmotic imbalance, acidosis, and, ultimately, inhibition of glycolysis (Withers 1992).

Box 2.1   An Aside on Units

Oxygen is expressed in a variety of (nearly) interchangeable ways and those most commonly encountered in the literature have changed over the years. A few basics will help.

1. The total pressure of a gas mixture such as atmospheric air equals the sum of pressures exerted by each constituent gas. This is Dalton's law.

2. Total (standard) atmospheric pressure is expressed as 760 mm Hg (also known as "Torr") or as 101 kilopascals (kPa), which is the SI unit for pressure.

3. The oxygen partial pressure in a standard atmosphere is the product of the mole fraction of oxygen in air × the barometric pressure:

$PO_2 = (0.2095) \times (760 \text{ mm Hg}) = 159.2 \text{ mm Hg}$ (usually rounded up to 160 mm Hg)
$PO_2 = (0.2095) \times (101 \text{ kPa}) = 21.2 \text{ kPa}$ (usually rounded down to 21 kPa)

4. Water in equilibrium with a gas mixture will have the same partial pressures as the gas mixture above it. This is Henry's law. The snag is that the molar concentrations are a function of the partial pressures *and* an additional factor, the solubility coefficient. Thus: $[A] = pA \cdot \alpha_A$
where $[A]$ = mol l^{-1} of A, pA = partial pressure of A (kPa), α_A = solubility coefficient (mol l^{-1} kPa^{-1}).

5. Solubilities of oxygen in seawater decrease with increasing temperature and salinity. They are normally determined empirically (with exacting care!) at air saturation (e.g. Murray and Riley 1969) and compiled in a table for researchers to use. For Oregon waters (Chapter 11), the oxygen concentration at 12 °C and 33‰ was used. It was reported as 6.14 ml l^{-1} in Murray and Riley (1969). Knowing that a mole of oxygen is 22.4 l or 32 g, we can convert 6.14 ml l^{-1} to 8.77 mg l^{-1} (6.14/22.4) × 32. For future reference, to interchange ml l^{-1} with mg l^{-1} simply divide by 0.7. From there, it is a simple matter to convert to millimoles: 8.77/32 = 0.272 millimoles or 272 μmol kg^{-1}, since the mass of a liter is a kg.

6. When comparing P_cs across pelagic systems with widely varying temperatures and oxygen concentrations, it is easiest to use PO_2 because the mole fractions do not change, but relative solubilities do.

Severity of Oxygen Minima, "Dead Zones," and the Intertidal

It is important to appreciate the difference between oceanic oxygen minima, which are persistent, year-round features located predictably in certain regions of the global ocean for millennia, and the seasonally appearing regions of hypoxia or "dead zones" like that at the mouth of the Mississippi River in the Gulf of Mexico (Rabalais et al. 1994) or in the deeper water of some Scandinavian fjords during summer (Diaz and Rosenberg 1995). The persistence and predictability of oceanic oxygen minima have allowed for adaptations in their resident fauna that enable a primarily aerobic existence. In contrast, the seasonally or episodically appearing anoxia of dead zones, as well as their duration and severity, precludes such adaptation. Infaunal species and any others with limited mobility experience 100% mortality in dead zones (Diaz and Rosenberg 1995). Life in both the oceanic oxygen minima and the episodic dead zones differs from the situation in the intertidal, where high intertidal species may experience anoxia twice daily but a return to normoxia is assured with the incoming tide.

At what concentration does oxygen become limiting in oxygen minima? A good indication is a reduction in the biomass or the diversity of the fauna inhabiting them. Childress and Seibel (1998) observed that oxygen has little influence on the biomass or species composition of midwater organisms inhabiting minima down to a level of $0.20\,\mathrm{ml\,l^{-1}}$ or $0.63\,\mathrm{kPa}$. At oxygen concentrations of $0.15\,\mathrm{ml\,l^{-1}}$ and below, such as that in the Eastern Tropical Pacific or Arabian Sea, oxygen minimum zones (OMZ's) exhibit reduced biomass and diversity. In those cases, anaerobiosis while resident in the minimum accompanied by a vertical migration out of the layer to more highly oxygenated waters at night is the most likely strategy (Childress and Seibel 1998).

Intertidal species, including those that dwell in burrows as well as epifaunal species such as bivalves that are exposed for extended periods, are usually highly competent anaerobes (Hochachka 1980). Their situation as intertidal dwellers is fundamentally different from that of oxygen-minimum-layer species. Their anoxia and normoxia are cyclic, varying with tidal exposure. Thus, it is adaptive to be able to extract oxygen efficiently, but only down to the point where a large investment in the systems involved in oxygen uptake and transport is unnecessary. For most species, that point lies in the partial pressure range of $20\text{--}30\,\mathrm{mm\,Hg}$ oxygen or $2.7\text{--}4.0\,\mathrm{kPa}$ (Torres et al. 1979, 1994; Childress and Seibel 1998). For intertidal species, oxygen availability will continue to decline to or near zero oxygen as the tide recedes and anaerobiosis inevitably will become necessary.

In the deep sea, it is the stability of oxygen minima in concert with the limited food resources that have allowed the highly efficient respiratory systems of oxygen-minimum-layer fauna to arise.

Adaptations to Oxygen Minima

The Aerobic Strategy

Species that are able to maintain an aerobic existence in oxygen minima do so by having a highly effective system for removing oxygen from seawater, allowing them to consume oxygen in sufficient quantities to sustain life at very low oxygen concentrations. The ability to regulate oxygen consumption down to very low levels of external oxygen is defined in physiological terms as having a low critical oxygen partial pressure, or P_c. Figure 2.22 shows the

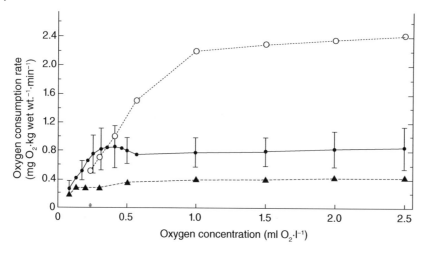

Figure 2.22 Oxygen consumption rate of the lophogastrid *Gnathophausia ingens* as a function of oxygen concentration (in milliliters of oxygen per liter). Open circles represent a single very active animal; closed circles represent the mean of 23 runs with 8 individuals; closed triangles represent a single non-swimming animal. The vertical lines represent plus or minus one standard deviation. *Source:* J. J. Childress, Oxygen minimum layer: vertical distribution and respiration of the mysid *Gnathophausia ingens*, *Science*, 1968, Vol 160, Issue 3833, figure 1 (p. 1242). Reprinted with permission from AAAS.

relationship of oxygen consumption rate and external PO_2 for an oxygen-minimum-layer species, the shrimp-like lophogastrid *Gnathophausia* (now *Neognathophausia*) *ingens*. The species is able to regulate its oxygen consumption (VO_2) down to the lowest oxygen level it experiences in its environment in the oxygen minimum layer off the coast of California (8 mm Hg oxygen or 0.8 kPa). The P_c is the point on the curves where the oxygen consumption declines precipitously toward 0 and it varies with activity level as shown. A study on the relationship of species' critical oxygen partial pressures vs. their minimum environmental PO_2 shows that for most oxygen minimum dwellers, species' P_cs are equivalent to the lowest oxygen concentrations encountered in nature (Childress and Seibel 1998). What is surprising is that all pelagic species living in normoxic waters that have been examined can also regulate their oxygen consumption down to a low PO_2: 4–6 kPa, or 20–30% of air saturation (Childress 1975; Donnelly and Torres 1988; Cowles et al. 1991; Torres et al. 1994). Thus, even at very much higher habitat O_2 levels, pelagic species maintain a P_c near 4 kPa. It is tempting to conclude that 4 kPa reflects a global ocean oxygen minimum that had to be accommodated by most pelagic species at some point during geological history. While possible, evidence does not support it (Childress and Seibel 1998). What is more likely is that the occasional high oxygen demands of increased activity and the respiratory machinery it requires coincidentally equip most species with the ability to take up and transport oxygen down to 4 kPa.

Gnathophausia ingens has been the subject of several detailed studies by Childress and his students. Their studies paint a very complete picture of how the species copes with low oxygen. Rather than any exotic adaptations, such as unusual new respiratory structures, *Gnathophausia* has achieved its ability to thrive in the OMZ by very highly developed gas-exchange and

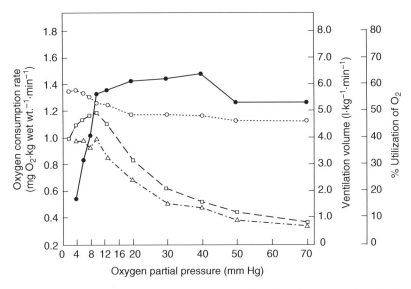

Figure 2.23 Oxygen consumption rate, percent utilization of oxygen, and ventilation volume in *Gnathophausia ingens* as a function of oxygen partial pressure, mean of eight runs. Solid line represents oxygen consumption rate; dotted line represents % utilization; dashed line depicts measured ventilation volume; dot-dash line is calculated ventilation volume. *Source:* Figure 4 from Childress (1971), *Biol. Bull.* 141: 114. Reprinted with permission from the Marine Biological Laboratory, Woods Hole, MA.

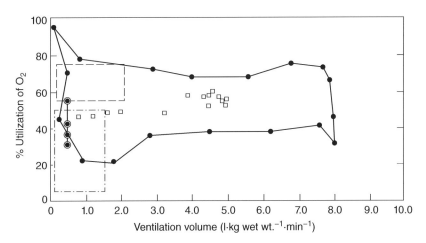

Figure 2.24 Relationship between percent utilization and ventilation volume in *Gnathophausia ingens* utilizing the values given in Figure 2.23. *Source:* Figure 5 from Childress (1971), *Biol. Bull.* 141: 115. Reprinted with permission from the Marine Biological Laboratory, Woods Hole, MA.

circulatory systems. Six major adaptations have been noted. First, *G. ingens* has a very high ventilation volume; that is, it can move a considerable amount of water over its gills per unit time: up to 81 kg min^{-1} (Figures 2.23 and 2.24). Second, it is very efficient at removing oxygen from the ventilatory stream, even when that stream is moving very rapidly over the gills. This property is known as the % utilization of oxygen and in *Gnathophausia* it can be as high as 90% (Figures 2.23 and 2.24). That high % utilization is achieved through its third, fourth, and fifth adaptations: a very high gill surface area to maximize the possibility for exchange (9–14 cm^2 g^{-1} wet mass); a minimal diffusion distance in the gill filaments themselves (1.5–2.5 μm) to minimize the barrier for oxygen diffusion into the blood (very unusual for Crustacea); and a very effective blood pigment (hemocyanin) capable of taking up oxygen at very low concentrations (50% saturated at 0.19 kPa). A well-developed circulatory system rounds out the suite of adaptations with the capability of delivering 225 ml kg^{-1} min^{-1}.

No other oxygen-minimum-layer species has been examined as well as *G. ingens*. Taken together, the several studies on the species' respiratory physiology paint a complete picture of how it is possible to survive at vanishingly low oxygen concentrations. Nonetheless, a few pieces of the puzzle have been collected in other taxa to suggest that elements of the suite have been employed by other species to achieve the same end. The most important characteristic to look for is a P_c at or below the lowest PO$_2$ encountered in nature. That characteristic has been observed in many of the Crustacea living in the oxygen minimum in the California borderland (8 mm Hg, Childress 1975; Childress and Seibel 1998). It has also been seen in at least one crustacean dwelling in the Eastern Tropical Pacific where the oxygen minimum layer is as low as 3 mm Hg O$_2$: the galatheid red crab *Pleuroncodes planipes* with a P_c of 3 mm Hg (Quetin and Childress 1976).

Once an individual reaches its P_c, it responds behaviorally and metabolically. Since metabolism scales positively with activity level, activity is minimized, precipitously dropping metabolic demand for ATP. Any ATP deficit resulting from the inability to meet its needs aerobically must be made up by anaerobic glycolysis. The hypoxia-induced drop in activity resulting in lowered ATP demand is termed metabolic suppression (Seibel 2011; Seibel et al. 2016) and is not confined to pelagic fauna. It is the first weapon any species can wield to reduce the demand for ATP and is exploited by intertidal species, such as bivalves, during low tide exposure (Hochachka and Somero 1984; Hochachka and Guppy 1987).

Pelagic cephalopods dwelling in the California oxygen minimum also exhibit low P_cs (3–7 mm Hg, Seibel et al. 1999). Data were collected from the vampire squid *Vampyroteuthis infernalis*, a fulltime resident of the California borderland's oxygen minimum zone, on two characteristics of the "Gnathophausia suite": gill diffusion capacity and blood pigment efficiency (Table 2.4). The diffusion capacity, DGO$_2$, and oxygen affinity, P_{50}, of *Vampyroteuthis* shown in Table 2.4 are indicative of a highly efficient gas-exchange surface and a blood pigment capable of binding oxygen at extremely low concentrations. Both are quite close to those of *Gnathophausia*, suggesting that, with respect to these two important physiological characteristics, the species are employing a common strategy. Data from the other cephalopod species in Table 2.4 show substantially lower gill diffusion capacity and oxygen affinity in the species inhabiting more normoxic environments.

Vampyroteuthis has a much lower metabolic rate than *Gnathophausia*. On a mass-specific basis, the vampire squid's respiratory rate is a little less than 10% that of *G. ingens*. On one hand, *Vampyroteuthis*' lower rate makes the job of the respiratory system a little easier because the absolute

Table 2.4 Metabolism ($VO_2 = mlO_2$ kg^{-1} min^{-1}), gill diffusion capacity ($DGO_2 = mlO_2$ kg^{-1} kPa^{-1} min^{-1}), blood-water oxygen gradient ($\Delta P_g = VO_2/(DGO_2$; in kPa) and hemocyanin-oxygen affinity ($P_{50} = PO_2$ in kPa at 50% hemocyanin-oxygen saturation) of *Vampyroteuthis ingernalis* in comparison to other cephalopods. Data for the lophogastrid crustacean, *Gnathophausia ingens*, are also shown. Source: Reprinted by permission from Springer Nature Customer Service Centre GmbH: Springer Nature, Experimental Biology Online, Vampire Blood: respiratory physiology of the vampire squid (Cephalopoda: Vampyromorpha) in relation to the oxygen minimum layer, Seibel et al. (1999).

Species	$VO_2{}^a$	DGO_2	ΔP_g	$P_{50}{}^b$	References
Vampyroteuthis infernalis	0.04	2.32	0.02	0.47	Madan and Wells (1996), Seibel et al. (1997)
Nautilus pompilius	0.28	0.38	0.74	2.3	Brix et al. (1989), Wells et al. (1992), Eno (1994)
Octopus vulgaris	0.35	0.45	0.77	2.45	Wells and Wells (1983), Bridges (1994), Eno (1994)
Architeuthis monachis	n.a.	n.a.	n.a.	1.65	Brix et al. (1989)
Gnathophausia ingens	0.56	3.73	0.15	0.19	Belman and Childress (1976), Sanders and Childress (1990)

n.a. = not available.
[a] Normalized to 5°C assuming $Q_{10} = 2$.
[b] Measured at pH 7.4 near environmental temperature.

amount of oxygen required per unit time is less. On the other, because of the physics of diffusion it makes little difference, the oxygen gradient between environment and organism is still a tiny one. What might be expected, *a priori*, in *Vampyroteuthis* would be a lower capability for removing oxygen in quantity. Adaptations such as very high gill surface area and a highly developed circulatory system, important for delivering oxygen in quantity to satisfy tissue demands, would likely be less in evidence. For at least one of those characteristics, gill surface area, this appears to be the case; *Vampyroteuthis* has only a moderate gill surface area relative to other cephalopods (Madan and Wells 1996; Seibel et al. 1999).

The available evidence strongly suggests that most full- and part-time residents of oxygen minima rely primarily on an aerobic strategy to make a living. The suite of adaptations exhibited by *Gnathophausia ingens* are likely found in whole or in part in all oxygen minimum layer residents.

Salinity

We noted in Chapter 1 that salinity in the open ocean does not vary enough to be a major determining factor in the distribution of oceanic species, in contrast to estuaries where large drops in salinity occur over the course of a few kilometers and species compositions change accordingly. In the open ocean, salinity varies little over large oceanic areas, with a low from about 33‰ (in near-surface waters of the Northeastern Pacific for example) to about 38‰ in the Red Sea and Mediterranean.

Marine species exhibit three basic strategies with respect to salinity. The first, typical of the invertebrates, is a strategy of osmoconformity. Virtually all marine invertebrates have a total internal ionic concentration that is nearly identical to that of seawater. As a consequence, they are in osmotic balance with their external environment and the need to regulate internal salinity is minimal. Regulation is usually restricted to ions such as magnesium that have potential for interfering with the function of nerves or muscle.

The second strategy, typified by the bony fishes, is one of hypo-regulation. The prefix hypo is from the Greek meaning "under" or "below" (e.g. hypo-dermic = under the skin). A hypo-regulator keeps its internal salinity well below that of seawater. Bony, or teleost, fishes like tunas, sardines, and swordfishes have internal salinities 45–60% that of seawater. They maintain their low internal salinities using the suite of mechanisms briefly described below. However, it is important to remember that relative to the large difference between a fish's internal salinity and that in the open ocean (e.g. 16 vs. 35‰), the small changes in oceanic salinity (typically <1‰) encountered by a pelagic fish during its lifetime are a trivial matter.

From a physiological perspective, one of a marine fish's greatest problems is the loss of its body water. The basic laws of diffusion are such that ions and water, when left to their own devices, will move across biological membranes until their concentrations are equal on both sides of the membrane. For our purposes here, we can think of a fish's skin as a membrane or barrier that is fighting the natural tendency of the ions in seawater to move into the fish, and the water inside of the fish to move out. Unfortunately for the fish, no skin is perfectly impermeable. Further, in order to breathe, the fishes must extract oxygen from seawater using their gills. In order to function effectively, the physical barrier between blood and seawater at the gill must be very thin indeed. Penultimately, fishes need to open their mouths to feed, which provides yet another chink in the armor. The final insult is that fishes cannot produce a concentrated urine like a kangaroo rat, or for that matter even like a human. The best that they can produce is a urine that is about the same salinity as the blood, so they do not lose much ground with their excretory system, but they do not gain any either.

Fishes replenish the water they lose through their excretory system and the ions they gain through their skin by drinking seawater! Water is extracted as it passes along the fish's gut, along with some ions, but the great majority of the ions are allowed to pass right through. The excess ions that are taken up by the gut and that diffuse inward are excreted at the gills using cells specific for that purpose. Ironically, despite the fact that the blood–seawater barrier is thinnest at the gills, no unwanted ions are taken up there due to the tight-junctions in the cells of the respiratory epithelium. The ion-balancing act of bony fishes is portrayed in Figure 2.25.

The third osmoregulatory strategy exhibited by marine species is shown by the elasmobranchs: the sharks, skates, and rays. It is also employed by the ratfishes (chimaeras) and the ancient lobe-finned fish, the coelacanth. Admittedly, only a very few representatives of those groups are the "small swimmers" that are the focus of this book, but their water-balancing strategy is an important one in the marine system and is included here for completeness.

The third strategy is much like a combination of those used by the invertebrates and the teleosts. The internal salt concentration of the sharks and their kin is about half that of seawater, similar to that of the bony fishes. However, the difference in osmotic concentration between seawater

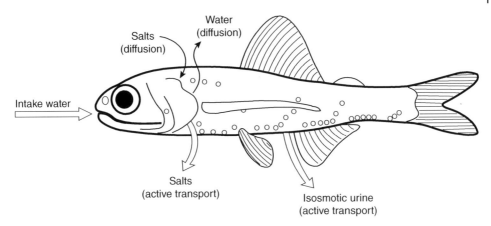

Figure 2.25 Osmoregulation in teleosts. Balance of water and ion concentration in the body of marine fishes through the processes of diffusion and active transport.

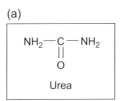

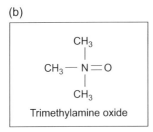

Figure 2.26 Osmoregulation in elasmobranchs. The osmolytes urea (a) and trimethylamine oxide (b) used by cartilaginous fishes to regulate osmotic balance in the blood.

and shark blood is made up by organic osmolytes: urea and trimethyl amine oxide, or TMAO (Figure 2.26). In fact, sharks and kin are slightly hyperosmotic to seawater; that is, they are at a slightly higher ionic concentration than seawater due to the urea and TMAO, which means that water tends to flow in! Urea and TMAO naturally tend to flow out, so the integument of the cartilaginous fishes is highly impermeable to their loss. In addition to using the same mechanisms employed by the bony fishes, i.e. excretion of ions at the gills, elasmobranchs have a rectal gland that can secrete a highly concentrated salt solution to aid in maintaining salt balance.

Depth

By now you may be growing to appreciate the profound changes in the physical environment of the open ocean in the horizontal and vertical planes and their effects on the physiology of open-ocean fauna. Deeper-living species must accommodate the colder temperatures, higher pressures, lower light levels, and, sometimes, lower oxygen levels of the mid-depths within their suite of adapted characters. A fascinating consequence of the changing environment with depth is the metabolic response of many deep-living species to the change: metabolic rate declines precipitously with species' depth of occurrence. It far exceeds that which would be predicted by the changes in the physical environment alone.

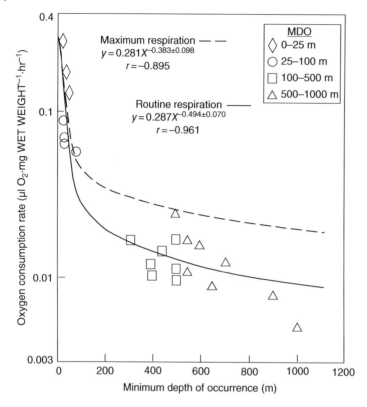

Figure 2.27 Relationship between routine respiration (solid line) and maximum respiration (dashed line) for groups of fishes with different minimum depths of occurrence. *Source:* Adapted from Torres et al. (1979), figure 1 (p. 190). Reproduced with the permission of Elsevier.

Childress (1971) was the first to report an unusually large decline in metabolism with depth in micronektonic species inhabiting the upper 1300 m of the water column off the coast of California, a cold temperate system. He found that species living at depths between 900 and 1300 m had a metabolic rate about 10% of those living in the upper 400 m when measured at the same temperature. The work suggested fundamental differences in the metabolic characteristics of the fauna living in different depth strata.

Nearly forty years later, with investigations spanning the Atlantic, Pacific, Gulf of California, Gulf of Mexico, and Southern Ocean, and using a wide variety of different taxa, the trend has been found to be universal among many taxa. We now know a lot more about the decline in metabolism with depth, and a well-accepted theory of why it occurs has been established.

The first taxa to be studied in detail for trends in metabolism vs. depth were the mesopelagic crustaceans (Childress 1975) and fishes (Torres et al. 1979, Figure 2.27) off the coast of California. In both cases, the difference in metabolism between a species living in surface waters and one living at 1000 m greatly exceeded that which would be caused by temperature alone. Depending on the time of year, the difference in temperature between surface waters (about 15 °C in fall) and those at 1000 m (about 4 °C year-round) would yield an expected change of roughly three-fold, assuming a conventional Q_{10} of 2–3. That is, metabolism at depth would be about one-third of that in surface waters if due only to changes in temperature. Instead the change was about 50-fold in both crustaceans and fishes! A fish dwelling at 1000 m had a metabolic rate about 2% of that of a surface-dwelling species. The difference in metabolism between a surface- and deep-dwelling fish (or crustacean) is huge, akin to the difference in metabolism between an active fish and a jellyfish (Seibel and Drazen 2007).

The fact that both pelagic crustaceans and fishes exhibited profound depth-related declines in metabolism confirmed that the trend was real and not confined to one taxonomic group. The results in turn opened up a Pandora's Box of questions. Why the decline occurs and how it is biologically achieved spring to mind as appropriate queries. In addition, one might wonder how widespread among oceanic taxa the decline is and whether it only occurs among pelagic species or whether it is also observed in bottom-dwelling (benthic) species and species that swim just above the bottom, the benthopelagic species. Enough work has been done to answer many of those questions. It is an instructive journey through the literature to see the questions posed and answered and the explanations for the phenomenon evolve.

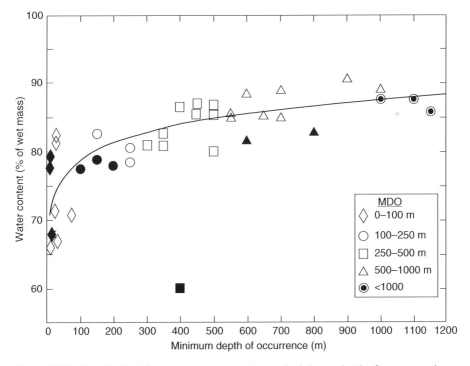

Figure 2.28 The relationship between water content and minimum depth of occurrence in a group of midwater fishes. Filled symbols represent species which have well-developed gas-filled swimbladders. The regression line of water content as a function of depth is for fishes without well-developed gas bladders. *Source:* Adapted from Childress and Nygaard (1973), figure 1 (p. 1098). Reproduced with the permission of Elsevier.

The first question, why the decline occurs, has been answered in two different ways over the years. Initially, we thought that declining metabolism was a response to the lower food availability at depth. The less energy required for the tasks of daily metabolism such as swimming, circulation of blood, and maintaining a constant internal environment, the more energy that could be devoted to growing bigger faster. The "energy limitation" hypothesis was very attractive, made more so by the fact that a large fraction of the metabolic decline was achieved through a reduction in metabolizing tissue: pelagic crustaceans and fishes become more watery with depth (Figure 2.28). The higher the water content of an individual, the lower its protein content, and because muscle is largely protein, it follows that deeper-living species have less muscle. Since muscle commands the lion's share of the energy produced by daily metabolism in most swimming species, watery, deeper-living crustaceans and fishes naturally have a much reduced metabolism. Curiously, the reduction in metabolism cannot be explained by an increased water content alone, it is far too great. Not only is there less muscle, the muscle itself has a greatly reduced metabolic demand.

As more data were collected on metabolism in pelagic species from different locations and from different taxonomic groups, two important trends emerged. The first was that in strong swimmers with good vision, notably the crustaceans, fishes, and cephalopods, metabolism declined profoundly with depth in all areas of the world ocean where they were surveyed, most notably the Pacific off California and Hawaii, the Gulf of Mexico, and in the isothermal waters of the Antarctic (Figure 2.29, Seibel and Drazen 2007). The second trend was that weakly swimming pelagic species with poor vision, such as the arrow worms (chaetognaths) and jellyfishes (hydro- and scyphomedusae) did not exhibit a significant decline in metabolism with depth. Since food availability is lower for all taxa at mesopelagic depths, the energy-limitation hypothesis would have predicted a decline in metabolism for chaetognaths and medusae as well as fishes and crustaceans. Since that was not observed, the hypothesis clearly needed modification.

Let us think our way through the problem. First, lowering daily maintenance energy is always a highly desirable strategy. The less energy that you use, the less food you require, and a greater percentage of the food energy that you do acquire can be used for growth. The fact that surface-dwelling fishes, cephalopods, and crustaceans have a much faster pace of metabolism than deeper-dwelling relatives, means that something about life at depth allows the deeper-living representatives to get away with employing what would seem to be a universally desirable strategy. What is the difference between surface and depth likely to wield the most influence on species' characteristics? The answer, in a word, is light. Metabolic response to the decline in temperature with depth is conventional in pelagic species (Q_{10} = 2 to 3). The lower temperature at mesopelagic depths only explains a fraction of the decline in rate. Salinity does not change enough to make a difference. However, visual predation is commonplace in the epipelagic zone, and, whether predator or prey, a well-developed swimming ability is necessary for survival. A highly developed swimming ability is less important at mesopelagic depths, where the lower light levels mean that visual ranges are greatly reduced. As the need for locomotory ability is relaxed so is the need for investment in musculature, resulting in much of the observed decline in metabolism with depth of occurrence. The arguments

Figure 2.29 Metabolic rates of diverse marine species as a function of minimum habitat depth. (a) Pelagic groups with image-forming eyes, including fish (closed circles), cephalopods (plus signs) and crustaceans (open squares); (b) pelagic taxa lacking image-forming eyes, including chaetognaths (open circles) and medusae (closed circles). *Source:* Seibel and Drazen (2007), figure 1(a) and (b) (p. 3). Republished with the permission of The Royal Society (U.K.), from The Rate of Metabolism in Marine Animals: environmental constraints, ecological demands and energetic opportunities, B.A. Seibel and J. C. Drazen, Philosophical Transactions, Biological Sciences, volume 362, issue 1487, © 2007; permission conveyed through Copyright Clearance Center, Inc.

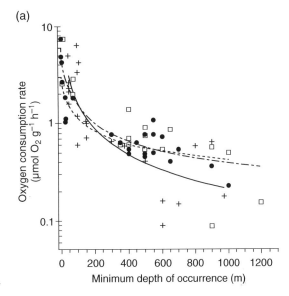

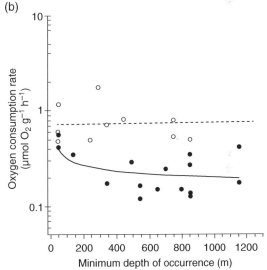

above form the more modern and widely accepted "visual interactions" hypothesis (Childress and Mickel 1985) for the decline in metabolism with depth.

We have already noted that part of the decline in metabolic rate with depth is the result of a more watery structure in deeper-living species, including a lower protein and, by implication, lower muscle content. Further studies revealed that there are also changes in the muscle itself. Activity of the important intermediary metabolic enzymes lactate dehydrogenase and citrate synthase declines with the depth of occurrence in both California and Antarctic pelagic fishes. In fact, the slopes are very similar to that observed in the decline of oxygen consumption rate with depth (Figure 2.30)

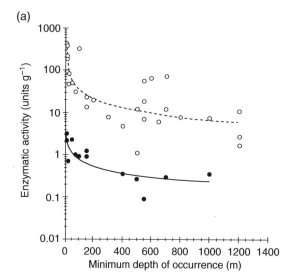
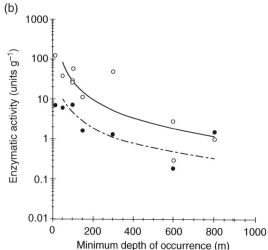

Figure 2.30 Activities of the respiratory enzymes citrate synthase (aerobic; closed symbols) and lactate dehydrogenase (anaerobic; open symbols) in marine animals as a function of minimum depth of occurrence. (a) Pelagic fish; (b) pelagic cephalopods. *Source:* Seibel and Drazen (2007), figure 4 (p. 7). Republished with the permission of The Royal Society (U.K.), from The Rate of Metabolism in Marine Animals: environmental constraints, ecological demands and energetic opportunities, B.A. Seibel and J. C. Drazen, Philosophical Transactions, Biological Sciences, volume 362, issue 1487, © 2007; permission conveyed through Copyright Clearance Center, Inc.

(Childress and Somero 1979; Torres et al. 1979; Torres and Somero 1988; Childress and Thuesen 1995; Seibel and Drazen 2007). Thus, not only is there less muscle in deeper-living species but the muscle that is present is less metabolically active.

Our last question on the decline in metabolism with depth is whether it is confined to pelagic species or whether it can be observed also in benthic and benthopelagic species. Keep in mind that the benthos and the pelagial are fundamentally quite different in their potential food availability. The pelagic realm does not retain the organic matter raining from the sunlit waters above, it's "just passing through." Large near-neutrally-buoyant particles may move through slowly but move through they do. In contrast, the benthos is the final resting place for the organic rain from the surface, however diminutive that rain may be. Even in non-productive waters, the benthos is a predictably richer environment than the deep pelagic realm. Also, due to episodic events like food-falls (those dead whales have to go somewhere), the richness of the environment can vary considerably in the horizontal plane. Thus for mobile species like crabs, shrimps, and fishes, the ability to move well can be very beneficial. Because benthic species like crabs have their weight supported by the sea floor, they do not need to worry about buoyancy. Similarly, though buoyancy is a concern for benthopelagic fishes, changes in their vertical profile are minimal and allow the use of a swim bladder. The advantage of being able to move quickly to a food-fall might make it advantageous to retain a robust musculature.

Does a change in metabolism with depth occur in all open-ocean taxa? Data show that assumption to be both right and wrong, depending upon the taxa of interest. Benthic Crustacea show no change in metabolism with depth of occurrence outside of that predicted by the declining temperature with depth (Childress et al. 1990), a trend very different from that of their pelagic counterparts. Fishes are another matter. Benthic and benthopelagic fishes both show marked declines in metabolism with depth of occurrence (Smith and Brown 1983; Drazen and Seibel 2007), though the slopes for the trend are slightly less than those observed in pelagic species. As in pelagic fishes, the declines in oxygen consumption rate with depth are mirrored by similar declines in enzyme activities (Drazen and Seibel 2007).

The benthic and benthopelagic fishes that have been studied are quite a bit larger than the pelagic species, generally at least 10 times larger and sometimes as much as 100–1000 times (Drazen and Seibel 2007). As adults at least, they are far more likely to be predators than to be prey. Reduced light levels at depths >500 m restrict visual ranges just as profoundly on the bottom as they do in the midwater, so active searching for prey is likely to be a high-cost/low-benefit activity even though hunting is essentially restricted to the horizontal plane. It is thus beneficial for bottom-oriented fishes to cut daily maintenance costs just as the pelagic species do. The tradeoff is a slower journey to the occasional food-fall, but obviously evolution has assured that it is fast enough.

Concluding Thoughts

In most of the world ocean, profound changes in the physical environment occur over very short distances in the ocean's vertical plane. Temperature, pressure, light levels, and sometimes oxygen concentrations vary drastically within a kilometer's journey of the surface. To flourish, open-ocean fauna must accommodate the challenges posed by the environment within their biological characteristics.

Because small swimming species that are the focus of this book are ectotherms, temperature is the environmental variable with the most potential to influence survival, zoogeographic distributions, and biological rate processes such as metabolism and growth. Upper and lower lethal limits of open-ocean species, like those of all ectotherms, are dictated by their physiological and biochemical characteristics. Limited live animal experimentation on species from a variety of open-ocean systems as well as more extensive work with enzymes and membranes strongly suggest that open-ocean pelagic species show the basic responses to temperature described here in detail. That is, we see no exotic or unusual adaptations to temperature. Rate measurements show Q_{10}s in the range of 2–3 and temperature adaptation is observed when ecological analogues in polar and temperate systems are compared.

The increased pressure associated with mesopelagic depths has the potential to influence biochemical and physiological processes ranging from the ion transport necessary for nerve and muscle function to enzyme function in the anaerobic and aerobic pathways of intermediary metabolism. Animals that live at modest pressures (<100 atm) are either insensitive to it, as are the vertically migrating euphausiids, or show a slight acceleratory response as in the deeper-living mesopelagic migrators. In contrast, surface-dwelling species exposed to pressures outside those of their normal environment show excitement at low (50 atm) pressures, moribundity at higher pressures (150 atm), and death due to tetany at high pressures (200 atm). Adaptations to pressure include increases in the fluidity of biological membranes as well as slight changes in the structure of enzymes to confer pressure insensitivity.

Zones of minimum oxygen are present at intermediate depths throughout the world ocean, but in a few locations oxygen reaches levels low enough to limit animal life. Three such locations are coastal California, the eastern tropical Pacific, and the Arabian Sea. When there is oxygen present in sufficient quantities to enable extraction, such as off California, pelagic species have evolved mechanisms to live aerobically despite the vanishingly low oxygen. Such adaptations include a high gill surface area to allow for efficient extraction of oxygen, a well developed circulatory system, and an efficient means of ventilating the gills. Animals that migrate into regions of zero oxygen, such as in the Arabian Sea, use a strategy of minimizing accumulation of toxic end products by changing the end point of their anaerobic metabolism from lactate to ethanol.

Depth itself exerts a profound influence on the metabolic characteristics of pelagic species. In swimming species that are either visual predators or are preyed upon by visual predators, i.e. the crustaceans, squids, and fishes, metabolism declines profoundly with increasing depth of occurrence. A fish living at the surface has a metabolism about 50 times that of a species living at 1000 m. Weaker swimmers where vision plays less of a role, such as jellyfishes and chaetognaths, do not show an equivalent decline in metabolic rate with depth of occurrence. Benthic and benthopelagic fishes show a similar decline with depth of occurrence, but benthic crustaceans do not.

References

Baldwin, J. (1971). Adaptation of enzymes to temperature: acetylcholinesterases in the central nervous systems of fishes. *Comparative Biochemistry and Physiology* 40: 181–187.

Baldwin, J. and Hochachka, P.W. (1970). Functional significance of isoenzymes in thermal acclimatization: acetylcholinesterase from trout brain. *Biochemical Journal* 116: 883–887.

Belman, B.W. and Childress, J.J. (1976). Circulatory adaptations to the oxygen minimum layer in the bathypelagic mysid *Gnathophausia ingens. Biological Bulletin* 150: 15–37.

Brett, J.R. (1952). Temperature tolerance in young *Pacific salmon*, genus *Oncorhynchus*. *Journal of the Fisheries Research Board of Canada* 9: 265–323.

Brett, J.R. and Groves, T.D.D. (1979). Physiological energetics. In: *Fish Physiology*, vol. *8* (eds. W.S. Hoar, D.J. Randall and J.R. Brett). New York: Academic Press.

Bridges, C.R. (1994). Bohr and root effects in cephalopod hemocyanins – paradox or pressure in *Sepia oficinalis*? In: *Physiology of Cephalopod Molluscs: Lifestyle and Performance Adaptations* (eds. H.O. Portner, R.K. O'dor and D.L. Macmillan). New York: Gordon and Breach.

Brix, O., Bardgard, A., Cau, A. et al. (1989). Oxygen binding properties of cephalopod blood with special reference to environmental temperatures and ecological distribution. *Journal of Experimental Zoology* 252: 34–42.

Campenot, R.B. (1975). The effects of high hydrostatic pressure on transmission at the crustacean neuromuscular junction. *Comparative Biochemistry and Physiology* 52B: 133–140.

Childress, J.J. (1968). Oxygen minimum layer: vertical distribution and respiration of the mysid *Gnathophausia ingens*. *Science* 160: 1242–1243.

Childress, J.J. (1971). Respiratory adaptations to the oxygen minimum layer in the bathypelagic mysid *Gnathophausia ingens*. *Biological Bulletin* 141: 109–121.

Childress, J.J. (1975). The respiratory rates of midwater crustaceans as a function of depth of occurrence and relation to the oxygen minimum layer off Southern California. *Comparative Biochemistry and Physiology* 50A: 787–799.

Childress, J.J. and Mickel, T.J. (1985). Metabolic rates of animals from the hydrothermal vents and other deep-sea habitats. *Biological Society of Washington* 6: 249–260.

Childress, J.J. and Nygaard, M.H. (1973). The chemical composition of midwater fishes as a function of depth of occurrence off southern California. *Deep-Sea Research* 20: 1093–1109.

Childress, J.J. and Seibel, B.A. (1998). Life at stable low oxygen levels: adaptations of animals to oceanic oxygen minimum layers. *Journal of Experimental Biology* 201: 1223–1232.

Childress, J.J. and Somero, G.N. (1979). Depth-related enzymic activities in muscle, brain, and heart of deep-living pelagic marine teleosts. *Marine Biology* 52: 273–283.

Childress, J.J. and Thuesen, E.V. (1995). Metabolic potentials of deep-sea fishes: a comparative approach. In: *Biochemistry and Molecular Biology of Fishes* (eds. P.W. Hochachka and T.P. Mommsen). Berlin: Elsevier Science.

Childress, J.J., Cowles, D.L., Favuzzi, J.A., and Mickel, T.J. (1990). Metabolic rates of benthic deep-sea decapod crustaceans decline with increasing depth primarily due to the decline in temperature. *Deep-Sea Research* 37: 929–949.

Cossins, A.R. and Bowler, K. (1987). *Temperature Biology of Animals*. London: Chapman and Hall.

Cossins, A.R. and MacDonald, A.G. (1984). Homeoviscous theory under pressure. 2. The molecular order of membranes from deep-sea fish. *Biochimica et Biophysica Acta* 776: 144–150.

Cowles, D.L., Childress, J.J., and Wells, M.E. (1991). Metabolic rates of midwater crustaceans as a function of depth of occurrence off the Hawaiian Islands: food availability as a selective factor? *Marine Biology* 110: 75–83.

Diaz, R.J. and Rosenberg, R. (1995). Marine benthic hypoxia: a review of its ecological effects and the behavioural responses of benthic macrofauna. *Oceanography and Marine Biology: An Annual Review* 33: 245–303.

Donnelly, J. and Torres, J.J. (1988). Oxygen consumption of midwater fishes and crustaceans from the eastern Gulf of Mexico. *Marine Biology* 97: 483–494.

Drazen, J.C. and Seibel, B.A. (2007). Depth-related trends in metabolism of benthic and benthopelagic deep-sea fishes. *Limnology and Oceanography* 52: 2306–2316.

Ebbecke, U. (1944). Lebensvorgange unter der Einwirking hoher drucke. *Journal of Biological Chemistry* 45: 34–183.

Eno, C.N. (1994). The morphometrics of cephalopod gills. *Journal of the Marine Biological Association of the United Kingdom* 74: 687–706.

Flugel, H. and Schlieper, C. (1970). The effects of pressure on marine invertebrates and fishes. In: *High Pressure Effects on Cellular Processes* (ed. A.M. Zimmerman). New York and London: Academic Press.

Fontaine, M. (1928). Les forte pressions et la consommation d'oxygene de quelques animaux marines. Influences de la taille de l'animal. *Comptes Rendus des Séances et Memoires de la Societe de Biologie* 99: 1789–1790.

Fry, F.E.J. and Hochachka, P.W. (1970). Fish. In: *Comparative Physiology of Thermoregulation* (ed. G.C. Whittow). New York: Academic Press.

Gibbs, A.G. (1997). Biochemistry at depth. In: *Deep-Sea Fishes* (eds. D.J. Randall and A.P. Farrell). San Diego: Academic Press.

Gibbs, A.G. and Somero, G.N. (1989). Pressure adaptation of Na+, K+-ATPase in gills of marine teleosts. *Journal of Experimental Biology* 143: 475–492.

Gibbs, A.G. and Somero, G.N. (1990). Pressure adaptation of teleost gill Na+,K+-adenosine triphosphatase: role of the lipid and protein moieties. *Journal of Comparative Physiology B* 160: 431–439.

Gjøsaeter, J. (1984). Mesopelagic fish: a large potential resource in the Arabian Sea. *Deep-Sea Research* 31: 1019–1035.

Graves, J.E. and Somero, G.N. (1982). Electrophoretic and functional enzymic evolution in four species of eastern *Pacific barracudas* from different thermal environments. *Evolution* 36: 97–106.

Grundfest, H. (1936). Effects of hydrostatic pressures upon the excitability, the recovery, and the potential sequence of frog nerve. *Cold Spring Harbor Symposia on Quantitative Biology* 5: 179–187.

Hazel, J.R. and Carpenter, R. (1985). Rapid changes in the phospholipid composition of gill membranes during thermal acclimation of the rainbow trout, *Salmo gairdneri*. *Journal of Comparative Physiology B* 155: 597–602.

Hazel, J.R. and Williams, E.E. (1990). The role of alterations in membrane lipid composition in enabling physiological adaptation of organisms to their physical environment. *Progress in Lipid Research* 29: 167–227.

Hitchcock, G.L., Wiebinga, C., and Ortner, P.B. (1997). *CTD hydrographic data from the Global Ocean Ecosystem Dynamics (GLOBEC) Indian Ocean Cruises. Technical Report 97-006*. University of Miami.

Hochachka, P.W. (1980). *Living Without Oxygen*. Cambridge: Harvard University Press.

Hochachka, P.W. and Guppy, M. (1987). *Metabolic Arrest and the Control of Biological Time*. Cambridge: Harvard University Press.

Hochachka, P.W. and Somero, G.N. (1973). *Strategies of Biochemical Adaptation*. Philadelphia: Saunders.

Hochachka, P.W. and Somero, G.N. (1984). *Biochemical Adaptation*. Princeton University Press: Princeton.

Hochachka, P.W. and Somero, G.N. (2002). *Biochemical Adaptation: Mechanism and Process in Physiological Evolution*. New York: Oxford University Press.

Kawall, H.G., Torres, J.J., Sidell, B.D., and Somero, G.N. (2002). Metabolic cold adaptation in Antarctic fishes: evidence from enzymatic activities of brain. *Marine Biology* 140: 279–286.

Longhurst, A.R. (1998). *Ecological Geography of the Sea*. London: Academic Press.

Madan, J.J. and Wells, M.J. (1996). Why squid breathe easy. *Nature* 380: 590.

Meek, R.P. and Childress, J.J. (1973). Respiration and the effect of pressure in the pelagic fish *Anoplogaster cornuta* (Beryciformes). *Deep-Sea Research* 20: 1111–1118.

Menzies, R.J. and Wilson, J.B. (1961). Preliminary field experiments on the relative importance of pressure and temperature on the penetration of marine invertebrates into the deep-sea. *Oikos* 12: 302–309.

Murray, C.N. and Riley, J.P. (1969). The solubility of gases in distilled water and sea water-II. Oxygen. *Deep-Sea Research* 16: 311–320.

Napora, T.A. (1964). The effect of hydrostatic pressure on the prawn, *Systellaspis debilis*. *Narragansett Marine Laboratory Occasional Publication* 2: 92–94.

Newell, R.C. and Northcroft, H.R. (1967). Metabolic independence of temperature over limited ranges in poikilotherms. *Journal of Zoology* 151: 277–298.

Place, A.R. and Powers, D.A. (1979). Genetic variation and relative catalytic efficiencies: lactate dehydrogenase B allozymes of *Fundulus heteroclitus*. *Proceedings of the National Academy of Sciences USA* 76: 2354–2358.

Precht, H. (1958). Theory of temperature adaptation in cold-blooded animals. In: *Physiological Adaptation* (ed. C.L. Prosser). Washington, DC: American Physiological Society.

Prosser, C.L. (1973). *Comparative Animal Physiology*. Philadelphia: Saunders.

Quetin, L.B. and Childress, J.J. (1976). Respiratory adaptations of *Pleuroncodes planipes* Stimpson to its environment off Baja California. *Marine Biology* 38: 327–334.

Rabalais, N.N., Wiseman, W.J., and Turner, R.E. (1994). Comparison of continuous records of near-bottom dissolved oxygen from the hypoxia zone along the Louisiana coast. *Estuaries* 17: 850–861.

Regnard, P. (1884). Effect des hautes pressions sur les animaux marins. *Comptes Rendus des Séances et Memoires de la Societe de Biologie* 36: 394–395.

Regnard, P. (1891). *Recherches experimentales sur les conditions physiques de la vie dans les eaux*. Paris: Masson.

Sanders, N.K. and Childress, J.J. (1990). Adaptations to the deep-sea oxygen minimum layer: oxygen binding by the hemocyanin of the bathypelagic mysid, *Gnathophausia ingens* Dohrn. *Biological Bulletin* 178: 286–294.

Scholander, P.F., Flagg, W., Walter, V., and Irving, L. (1953). Climatic adaptation in arctic and tropical poikilotherms. *Physiological Zoology* 26: 67–92.

Seibel, B.A. (2011). Critical oxygen levels and metabolic suppression in oceanic oxygen minimum zones. *Journal of Experimental Biology* 214: 326–336.

Seibel, B.A. and Drazen, J.C. (2007). The rate of metabolism in marine animals: environmental constraints, ecological demands and energetic opportunities. *Philosophical Transactions of the Royal Society B* 362: 1–18.

Seibel, B.A., Thuesen, E.V., Childress, J.J., and Gorodezky, L.A. (1997). Decline in pelagic cephalopod metabolism with habitat depth reflects differences in locomotory efficiency. *Biological Bulletin* 192: 262–278.

Seibel, B.A., Chausson, F., Lallier, F.H. et al. (1999). Vampire blood: respiratory physiology of the vampire squid (Cephalopoda:Vampyromorpha) in relation to the oxygen minimum layer. *Experimental Biology Online* 4: 1–10.

Seibel, B.A., Schneider, J.L., Kaartvedt, S. et al. (2016). Hypoxia tolerance and metabolic suppression in oxygen minimum zone euphausiids: implications for ocean deoxygenation and biogeochemical cycles. *Integrative and Comparative Biology* 56: 510–523.

Sidell, B.D., Wilson, F.R., Hazel, J., and Prosser, C.L. (1973). Time course of thermal acclimation in goldfish. *Journal of Comparative Physiology* 84: 119–127.

Siebenaller, J.F. and Somero, G.N. (1979). Pressure adaptive differences in the binding properties of the muscle-type (M4) lactate dehydrogenases of shallow- and deep-living marine fishes. *Journal of Comparative Physiology* 129: 295–300.

Smith, K.L. and Brown, N.O. (1983). Oxygen consumption of pelagic juveniles and demersal adults of the deep-sea fish *Sebastolobus altivelis*, measured at depth. *Science* 184: 72–73.

Smith, D.A., Hofmann, E.E., Klinck, J.M., and Lascara, C.M. (1999). Hydrography and circulation of the West Antarctic Peninsula continental shelf. *Deep Sea Research, Part I* 46: 925–949.

Somero, G.N. (1975). The role of isozymes in adaptation to varying temperatures. In: *Isozymes II: Physiological Function* (ed. C.L. Markert). New York: Academic Press.

Somero, G.N. and Siebenaller, J.F. (1979). Inefficient lactate dehydrogenases of deep-sea fishes. *Nature* 282: 100–102.

Spyropoulos (1957). Response of single nerve fibers at different hydrostatic pressures. *American Journal of Physiology* 189: 214–218.

Sverdrup, H.U., Johnson, M.W., and Fleming, R.H. (1942). *The Oceans, Their Physics, Chemistry, and General Biology*. Englewood Cliffs: Prentice-Hall.

Teal, J.M. (1971). Pressure effects of the respiration of vertically migrating decapod Crustacea. *American Zoologist* 11: 571–576.

Teal, J.M. and Carey, F.G. (1967). Effects of pressure and temperature on the respiration of euphausiids. *Deep-Sea Research* 14: 725–733.

Torres, J.J. and Somero, G.N. (1988). Metabolism, enzymic activities, and cold adaptation in Antarctic mesopelagic fishes. *Marine Biology* 98: 169–180.

Torres, J.J., Belman, B.W., and Childress, J.J. (1979). Oxygen consumption rates of midwater fishes as a function of depth of occurrence. *Deep-Sea Research* 26: 185–197.

Torres, J.J., Aarset, A.V., Donnelly, J. et al. (1994). Metabolism of Antarctic micronektonic Crustacea as a function of depth of occurrence and season. *Marine Ecology Progress Series* 113: 207–219.

Vinogradov, M.E. (1970). *Vertical Distribution of the Oceanic Zooplankton*. Jerusalem: Israel program for scientific translations.

Wells, M.J. and Wells, J. (1983). The circulatory response to acute hypoxia in Octopus. *Journal of Experimental Biology* 104: 59–71.

Wells, M.J., Wells, J., and O'dor, R.K. (1992). Life at low oxygen tensions: the behavior and physiology of *Nautilus pompilius* and the biology of extinct forms. *Journal of the Marine Biological Association of the United Kingdom* 72: 313–328.

Winberg, G.G. (1956). *Rate of Metabolism and Food Requirements of Fishes*. Dartmouth, Nova Scotia: Fisheries Research Board of Canada.

Wishner, K.F., Gowing, M.M., and Gelfman, C. (2000). Living in suboxia: ecology of an Arabian Sea copepod. *Limnology and Oceanography* 45: 1576–1593.

Withers, P.C. (1992). *Comparative Animal Physiology*. Orlando: Saunders.

Wohlschlag, D.E. (1960). Metabolism of an Antarctic fish and the phenomenon of cold adaptation. *Ecology* 41: 287–292.

3

The Cnidaria

Introduction

The Cnidaria, or stinging jellies, include a bewildering array of groups ranging from aquarium favorites such as anemones to the infamous Portuguese man-o-war and to reef-building corals, deep-dwelling sea pansies, and sea pens. The focus in this book is the cnidarians that are large floaters and weak swimmers: the jellyfishes (the medusae) and the siphonophores. Both are important groups within the polyphyletic assemblage collectively known as the macrozooplankton. In turn, macrozooplanktonic species are important elements of the pelagic community.

The pelagic Cnidaria are particularly confusing because there are two types of medusae: the smaller and less complex hydromedusae and the larger scyphomedusae. Inshore, the scyphomedusae are far more noticeable to the casual observer and are seasonally well represented by species such as the moon jelly *Aurelia* and the scourge of Atlantic beaches, the sea nettle *Chrysaora*. Offshore, in the blue waters of the pelagic realm, the hydromedusae dominate numerically. Adding to the confusion is the fact that many, but not all, of the hydromedusae and scyphomedusae alternate generations from a sedentary, anemone--like, polyp stage to a swimming medusoid form. The alternation of generations is sometimes termed "metagenesis." It was the alternation of generations that captured the imagination of famous natural philosophers such as Cuvier, Lamarck, and especially Ernst Haeckel, who made siphonophores one of his favorite subjects for study.

Siphonophores are unfamiliar to most people not well acquainted with the open ocean because the animals are found predominantly offshore and are very delicate. Therefore, they are difficult to preserve or to view intact after capture. As a consequence, siphonophores have received limited study though they are common predators in the open ocean and especially in the deep sea.

Classification

History

The sedentary nature of many of the Cnidaria (think anemone, not jellyfish) made their position within the animal kingdom difficult to decipher for the early natural philosophers, who considered them something somewhere between an animal and a plant. Aristotle called them Acalephae or Cnidae (from the Greek: akalephe = nettle; cnidos = thread) (Hyman 1940) and placed them with

---

Life in the Open Ocean: The Biology of Pelagic Species, First Edition. Joseph J. Torres and Thomas G. Bailey.
© 2022 John Wiley & Sons Ltd. Published 2022 by John Wiley & Sons Ltd.

a variety of other soft-bodied animals in with the Zoophyta (from the Greek: zoon = animal, phyton = plant). Researchers in the eighteenth century recognized the animal nature of cnidarian polyps, leading the way for natural philosophers of the nineteenth century, Linnaeus, Lamarck, and Cuvier, to place the cnidarians among the animals in their own classification systems, among either the Radiata, recognizing the importance of radial symmetry, or Zoophyta.

Just as happens today, the natural philosophers of the nineteenth century had differences of opinion on the group relationships. Lamarck's system included the medusoid Cnidaria and echinoderms (starfish) in his Radiata, with the polypoid cnidarians simply called the Polyps. In 1829, Sars showed that polyps and medusae were different life stages of the same animal, not separate groups. Not quite 20 years later, Leuckart and Frey (1847) separated the two largest radially symmetrical groups, the echinoderms and the cnidarians, into two groups: the coelenterates and the echinoderms. Leuckart coined the term Coelenterata from the Greek words for body cavity (koilos) and intestine (enteron), noting that the only body cavity in the cnidaria was the intestine (Hyman 1940). Leuckart included the sponges and ctenophores within his Coelenterata. It was up to Hatschek (1888) to separate Leuckart's Coelenterata into the three phyla we recognize today: the Spongiaria (Porifera), the Cnidaria, and the Ctenophora. The term Coelenterata is still extensively used today, most commonly as a synonym for the Cnidaria but sometimes as a way of combining the ctenophores and cnidarians into a single taxon.

Classification Schemes

Classification within the Cnidaria is constantly evolving. While there is consensus on what the different groups within the Cnidaria are, there is considerably less agreement on rank and position within the systematic hierarchy. It is important to recognize that molecular methodologies have opened a new and rapidly evolving way of classifying species at all levels of the taxonomic hierarchy. The half-life of any taxonomic scheme will likely be quite short for some time to come, and the only way to keep up will be with web-based systems such as the world register of marine species (WoRMS) or with taxon-specific sites. The scheme shown below is that of Brusca et al. (2016). Brusca et al. divide the Cnidaria into three subphyla: the Anthozoa, Medusozoa (cf. Bouillon 1999), and the parasitic Myxozoa. The Anthozoa are devoid of the medusa phase, and in the Medusozoa the medusa phase is an important phase of the life cycle. The Anthozoa include one class, the Anthozoa, and two subclasses: the Hexacorallia and Octocorallia. All are benthic forms, with the anemones, black corals, and stony corals included in the subclass Hexacorallia, and the gorgonians, soft corals, sea pens and sea pansies and organ pipe corals included in the subclass Octocorallia.

The Medusozoa include the taxa of main concern, the free-swimming pelagic jellies. Five classes of Medusozoa have been identified, of which three, the Cubozoa (cubomedusae), Scyphozoa (the large jellyfishes), and the Hydrozoa (hydroids and hydromedusae), will garner the most attention in the following pages. Though anthozoans are major players in the benthos of all oceanic systems, our chief interest here is their structural relatedness to the free-swimming jellies. The similarities between polyps and medusae are best visualized by turning a medusa upside down (Figure 3.1). It is easy to see that both polyps and medusae may be characterized as having a central gut (a gastrovascular cavity or stomodeum) surrounded by a row of radially arranged tentacles.

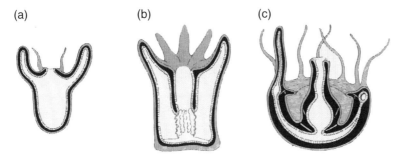

Figure 3.1 Tissue layer homologies in cnidarians. (a) A hydrozoan polyp; (b) an anthozoan polyp; (c) a hydromedusan medusa, shown upside down for similar orientation. The outer tissue layer is ectodermal (epidermis); the inner tissue layer is entodermal (gastrodermis); the middle tissue layer is mesenchyme/mesoglea.

Phylum Cnidaria

Subphylum Medusozoa

Class Hydrozoa
  Subclass Hydroidolina
    Order Athecata (Anthomedusae or Anthoathecata)
      Suborder Capitata (*Porpita, Millepora, Coryne*)
      Suborder Aplanulata (*Tubularia, Corymorpha*)
      Suborder Filifera (*Stylaster, Bouganvillia*)
    Order Thecata (Leptomedusae or Leptothecata) (*Aequorea, Campanularia*)
    Order Siphonophora
      Suborder – Cystonectae (*Physalia*)
      Suborder – Calycophorae (Muggiaea)
      Suborder – Physonectae (Agalma)
  Subclass Trachylina
    Order Limnomedusae
    Order Trachymedusae
    Order Narcomedusae

Class Cubozoa – Sea wasps and box jellies
    Order Chirodropida (*Chyronex*)
    Order Carybdeida (*Carybdea*)

Class Scyphozoa – "true jellyfish"
  Order – Coronatae – coronate medusae, common in deep-sea – *Periphylla*
  Order – Semaeostomae – most typical large jellyfish – e.g. *Aurelia*
  Order – Rhizostomae – cannon-ball jellies

Class Staurozoa – Stalked jellyfish. Small sessile individuals – polypoid (*Halyclistis*)

Class Polypoidozoa – Intracellular parasites of developing fish eggs. Monospecific.

Subphylum Anthozoa

Class Anthozoa
  Subclass Octocorallia – three orders and six suborders, includes soft corals, gorgonians, sea pens, and sea pansies.
  Subclass Hexacorallia – six orders, includes the true sea anemones, stony corals, black corals, and tube-dwelling anemones.

Subphylum Myxozoa

Intracellular parasites of marine and freshwater vertebrate ectotherms, annelids, bryozoa, and sipunculids. Two classes and two orders. About 2200 species.

Open-ocean species are often not as well understood or well described as nearshore species because of insufficient access to specimens. This is particularly true of the Cnidaria. Taxonomic keys and good drawings of species are limited, and many of the best treatments of cnidarian groups are quite old. For decades, the oceanographer's gold standard for identifying medusae resided in the works of two authors, F. S. Russell (*Medusae of the British Isles*, volumes I and II) and P. L. Kramp ("The Hydromedusae of the Atlantic Ocean and Adjacent Waters"; Medusae of the World). Both authors built on the work of luminaries preceding them, most notably A. G. Mayer's three-volume *Medusae of the World*. Recently, the benchmark works of Kramp and Russell have been supplemented with a new, multiauthor, treatment on the South Atlantic Zooplankton (Boltovskoy 1999) that gives a very nice summary of the systematics of the Cnidaria, though as in all things systematic it is not one that is universally accepted. Classification within the Cnidaria is very well treated by Mills et al. (2007) in the new *Light and Smith's Manual* (J. T. Carleton, ed.).

The classification scheme above includes recent consensus views on major groups within the Cnidaria and how they are related. Two groups stand out as unusual within the hydromedusae: the siphonophores and the "by-the-wind-sailors" (sometimes known as the chondrophores). We will treat them separately because their biology is different from the remainder of the hydromedusae, even though they are now considered to be an order (Siphonophorae) and a family (Porpitidae) within the hydromedusae.

If all the classification schemes are considered together, including those of Kramp and Russell, five basic divisions within the medusae of the hydromedusae are apparent: the Anthomedusae, the Leptomedusae, the Limnomedusae, the Trachymedusae, and the Narcomedusae. Species in the first three groups have a polyp stage in most instances and thus show alternation of generations. The second two groups show direct development. We will consider the medusae within the Hydromedusae to have the five basic divisions noted above, which most closely approximate the classical literature. Whether those divisions are orders or suborders is less important than the fact that the hydromedusae can be readily segregated morphologically using those divisions, and they have been for the better portion of 100 years.

The Hydromedusae

Morphology Basics

The Cnidarians are all radially symmetrical (Figure 3.2); their parts are arranged around a central axis much like the spokes of a bicycle wheel are arranged around the hub. Each of the radii has names that are used in classification. The central axis or hub is the oral–aboral axis or gut. Between the "hub" and the "rim," body parts are arranged concentrically (Figures 3.3 and 3.4). In perfect radial symmetry, a plane bisecting the animal in the axis of the hub will result in two perfectly equal halves, no matter where the plane is placed. Most often, and even in the hydromedusae, the morphology is modified to biradial (two equal halves) or tetramerous (four equal quarters) radial symmetry. The radial symmetry of medusae is important to their foraging strategy. Prey are equally accessible from all points of the compass, which, for an ambush predator, is a distinct advantage.

Figure 3.3 shows the basic body form of a hydromedusa. The outer, or exumbrellar, surface and the inner, subumbrellar,

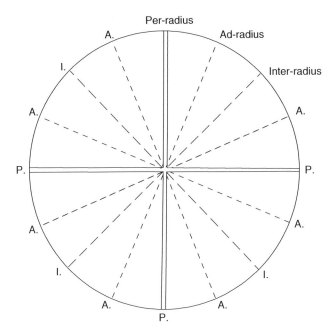

Figure 3.2 Diagram to define the radii of a medusa. *P.* are the per-radii on which the four primary radial canals lie. *I.* are inter-radii and *A.* are ad-radii. *Source:* Adapted from Russell (1954), figure 1 (p. 2).

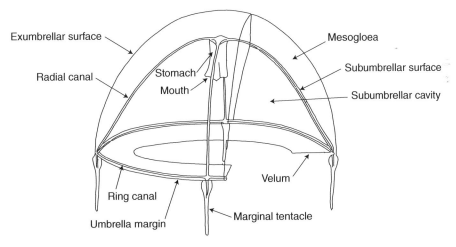

Figure 3.3 Diagram of a medusa with one quadrant removed. *Source:* Adapted from Russell (1954), figure 2 (p. 3).

surface are separated by a deformable gelatinous acellular mesoglea. The mesoglea acts as a primitive skeleton, giving a medusa its characteristic shape and providing a surface against which muscle tissue can act to propel the medusa through the water.

Swimming bells range in shape from tall to spherical to highly flattened in

94 *The Cnidaria*

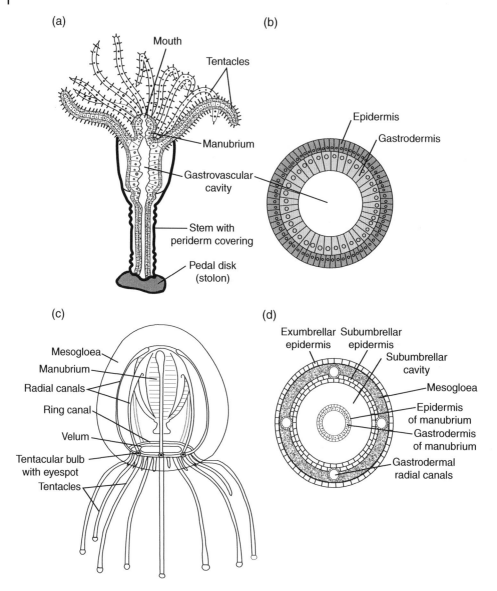

Figure 3.4 Cnidarian structure. (a) Diagram of a hydroid polyp; (b) cross section of a hydroid polyp; (c) diagram of a hydrozoan medusa; (d) cross section of a hydrozoan medusa. *Source:* Adapted from Hyman (1940), figure 106 (p. 368).

appearance (Figure 3.5a, Russell 1954). The stomach is a simple sack that projects downward from the subumbrellar surface and can vary considerably in length and morphology. The opening of the stomach or "lips" may vary from a simple circular opening to a highly crenulated and folded appearance. In some cases, the mouth is surrounded by small oral tentacles (Figure 3.5b).

The radial canals originate at the four corners of the stomach and extend along the

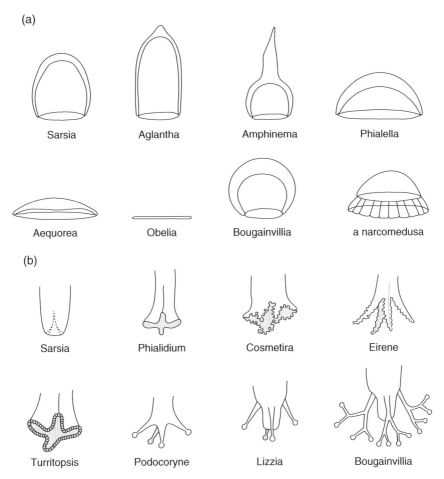

Figure 3.5 Medusae structure. (a) Diagrams of umbrella shapes of different medusae; (b) diagrams of mouths of different medusae. *Sources:* (a) Russell (1954), figure 3 (p. 4); (b) Russell (1954), figure 55 (p. 5).

subumbrellar surface to join the ring canal that runs along the umbrellar margin. The umbrellar margin contains the marginal tentacles that are primarily used for subduing prey. The stomach, radial canal, and ring canal system provide for the distribution of nutrients to the medusa as a whole.

Morphological Detail and Life Histories

A fundamental characteristic in the classification of the hydromedusae is the presence or absence of a polyp stage. Anthomedusae, Leptomedusae, and Limnomedusae all have a fixed polyp stage, whereas the Trachymedusae and Narcomedusae exhibit direct development. Considerable diversity is evident in the polyp stages (Figure 3.6).

Anthomedusae are characterized as usually having a tall bell with gonads mainly on the stomach and manubrium or extending slightly along the radial canals. The hydranths (feeding polyps) of Anthomedusae are athecate, i.e. they lack a surrounding sheath (Figures 3.6b and 3.7), thus giving the group its secondary name

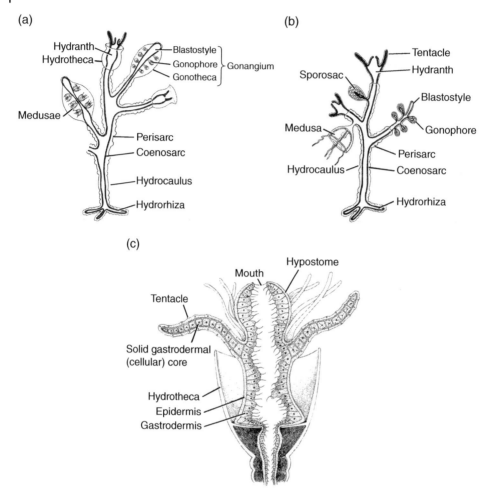

Figure 3.6 Anatomy of hydrozoan polyps. (a) Composite diagram of a thecate hydroid colony showing different reproductive structures. (b) composite diagram of reproductive structures in an athecate hydroid colony; (c) cross section of a thecate gastrozooid (hydranth). *Sources:* (a) Adapted from Bayer and Owre (1968), figure 55 (p. 38); (b) Bayer and Owre (1968), figure 54 (p. 38); (c) Bayer and Owre (1968), figure 53 (p. 38).

of Athecata or Anthoathecata. Common genera include: *Coryne, Bouganvillia, Podocoryne, Cladonema, Amphinema,* and *Leuckartiara.*

Leptomedusae are more dorsoventrally compressed with gonads only on radial canals. Common genera include *Laodicea, Aequorea, Obelia, Eirene, Mitrocoma, Eutima, Phialidium.* Polyps are thecate (have a surrounding sheath, Figure 3.6a and c). The life cycle is shown in Figure 3.8a.

Limnomedusae have some of the characteristics of the Anthomedusae, Leptomedusae, and Trachymedusae. Gonads are either on the stomach wall with continuation along the radial canals or only on the radial canals. A

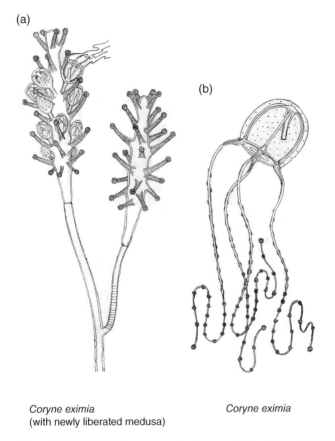

Coryne eximia
(with newly liberated medusa)

Coryne eximia

Figure 3.7 (a) The hydrozoan polyp *Coryne eximia* with newly liberated medusa *Coryne eximia*; (b) a single hydrozoan medusa *Coryne eximia* produced by its polyp stage. Note the capitate (clublike) tentacles on the polyp. *Source:* Redrawn from Mayer (1910), Vol. 1, (p. 56).

sessile hydroid stage with or without tentacles is present but often reduced. The group was created by Kramp (1938), who named it the Limnomedusae because many of the species lived in fresh or brackish water (Russell 1954). Common genera include: *Aglauropsis, Olindias, Cubaia, Vallentinia, Proboscidatyla, Craspedacusta*. The life cycle is illustrated in Figure 3.8b.

Trachymedusae have a hemispherical or deep bell-shaped umbrella with gonads usually confined to the radial canals. The life cycle lacks a polyp stage (Figure 3.8c). Tentacles are solid, and the subumbrellar surface exhibits a heavy musculature suggesting a strong swimming ability (Figure 3.9a). The velum is usually well developed. Common genera include: *Liriope, Rhopalonema, Colobonema, Pantachogon, Crossota, Aglantha, Aglaura,* and *Halicreas*. The trachymedusae are exclusively oceanic.

Narcomedusae are the most bizarre of the hydromedusae (Figure 3.9b). The sides of the umbrella are divided by peronial grooves so that the umbrellar margin may be lobed. They have a broad circular stomach that covers much of the subumbrellar surface, sometimes with

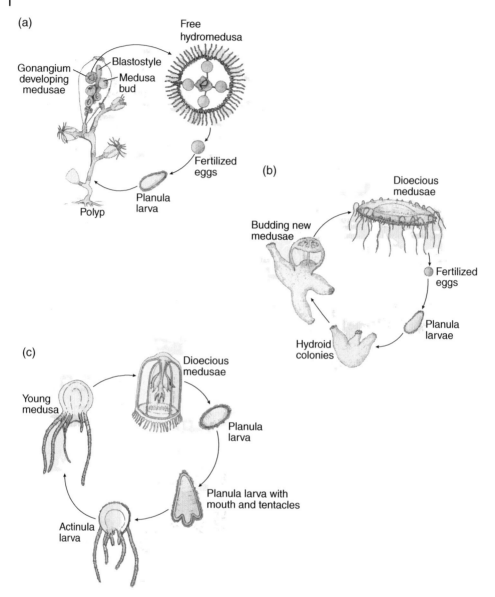

Figure 3.8 Life cycles of hydrozoans. (a) *Obelia*, a thecate hydroid with free medusae; (b) *Limnocnida*, a trachyline hydrozoan with a polypoid stage; (c) *Aglaura*, a trachyline hydrozoan without a polypoid stage. *Sources:* (a) Adapted from Bayer and Owre (1968), figure 153 (p. 101); (b) Bayer and Owre (1968), figure 155 (p. 102); (c) Bayer and Owre (1968), figure 154 (p. 102).

peripheral pouches. Solid tentacles originate above the umbrellar margin. No radial canals are present. Narcomedusae are strictly oceanic. They have no true hydroid stage but may have parasitic larval development (Russell 1954). Common genera include *Solmissus*, *Aegina*, *Aeginura*, and *Solmaris*.

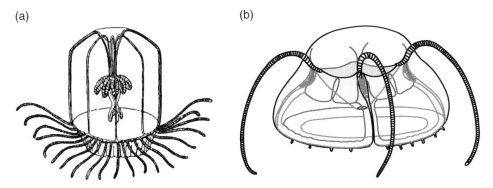

Figure 3.9 Hydrozoan medusae. (a) The trachymedusa *Aglaura hemistoma*. (b) The narcomedusa *Aegina citrea*. *Sources:* (a) Kramp (1959); (b) Mayer (1904), plate IV.

The Scyphomedusae

Basic Characteristics

The Scyphozoa are exclusively marine. They occur from the surface to bathypelagic depths and from polar to tropical oceans. The medusoid stage dominates the life history. When present, the polypoid stage, termed the scyphistoma, is small and sessile. Scyphomedusae range in size from bell diameters of 2 cm to 2 m. About 223 species have been described.

All three orders of Scyphomedusae are pelagic: the Coronatae, Semaeostomae, and Rhizostomae. The coronate medusae are all found in deep water and, because of that, tend to have very wide-ranging distributions. Semaeostome and rhizostome medusae are found primarily in coastal waters. Even so, some species, e.g. the semaeostome *Pelagia noctiluca* and the cannonball jelly *Stomolophus meleagris*, exhibit considerable latitudinal range (Mianzan and Cornelius 1999).

Scyphomedusae differ from Hydromedusae in that Scyphomedusae have no velum, the "skirt" that extends from the umbrellar margin into the subumbrellar space (see Figure 3.3). Additionally, they have gastric filaments in the digestive system, and the embryological origin of the gonads is endodermal (see Chapter 5 for a description of embryological germ layers). Symmetry in the Scyphomedusae is markedly tetramerous in both the polyps and medusae. This characteristic is most obviously manifested by the presence of four oral arms in many species and in the organization of the gastric system (Figure 3.10). The four oral arms lead to the lips of a central mouth that connects to the stomach via a manubrium or oral tube. Cnidoblasts on the oral arms aid in capturing prey and in defense. As in the Hydromedusae, nutrition is conveyed from the stomach to the periphery via radial gastrovascular canals. The position of the canals relative to the axis defined by the oral arms determines their name. Thus, the perradial canals lie in the same axis as the oral arms and the lips of the mouth and are considered the primary axes. In the middle of the quadrants defined by the perradial canals lie the interaradial canals. Finally, the adradial canals bisect the space between the perradial canals and the interradials. The body wall consists of an outer epidermis and inner gastrodermis

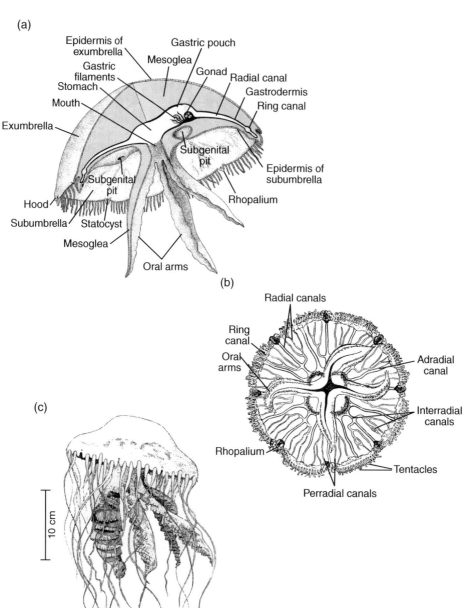

Figure 3.10 Scyphomedusae. (a) General anatomy of a scyphomedusa. (b) Oral view of a scyphomedusa. (c) The scyphomedusan *Chrysaora lactea*. *Sources:* (a) Bayer and Owre (1968), figure 100 (p. 65); (b) Kaestner (1967), figure 5-10 (p. 98); (c) Redrawn from Mayer (1910), figure 368.

separated by mesoglea. The mouth is the only opening to the digestive system; all food, waste, and gametes must move through the same opening.

A further difference between the scyphozoan and hydrozoan medusae is their average size. Though large hydromedusae and young scyphomedusae do overlap in size, scyphomedusae are typically quite a bit larger. Large representatives can reach a meter in diameter (e.g. *Cyanea, Desmonema, Stygiomedusa*) and the largest recorded, *Cyanea arctica*, reaches 2 m (Hyman 1940). Adult hydromedusae range in size from 1–2 mm to 20 cm in diameter.

Morphological Detail and Life Histories

General

Most scyphomedusae are dioecious, though incidences of hermaphroditism have been found. *Chrysaora hysocella* initially produces male gametes and then female (Arai 1997). In most cases, the fertilized egg develops into a planula larva, which settles to the bottom and grows into the scyphozoan polyp stage, known as a scyphistoma. Scyphistomae may reproduce asexually by budding additional scyphistomae or may produce medusae by strobilation. A strobilating polyp develops transverse fissures, which separate from the stalk to form free-swimming ephyrae (larvae), which then rapidly grow into adult medusae. Strobilation takes two forms, monodisk strobilation, where the scyphistoma produces one ephyra at a time, and polydisk strobilation, where ephyrae are stacked up like dinner plates on the scyphistoma and are shed by repetitive transverse fission. Scyphistomae may live for several years. After a period of producing ephyrae, usually during the winter and spring, they resume life as a polyp until the following reproductive season.

Coronatae

The Coronatae are deep-dwelling jellyfish mainly inhabiting mesopelagic to bathypelagic depths, though some genera (*Linuche, Nausithoe*) are found in surface waters. Each has a characteristic deep furrow, the coronal groove, which divides the exumbrella into a central disc and a peripheral zone and gives the order its name (Figure 3.11). Below the coronal groove, the peripheral zone is divided into two subregions. First, the thickened deeply grooved pedalia, and below the pedalia and centered on the inter-pedaliar grooves, the marginal lappets. The pedalia most often have a single, solid tentacle. Coronate medusae are heavily pigmented, usually deep burgundy in color, and most species have stiff noncontractile tentacles that are often held above the bell.

The life histories of coronate scyphomedusae are poorly known, largely because most of them sport a mesopelagic lifestyle. The best-known life cycle is that of *Nausithoe*, which shows a typical scyphozoan pattern similar to that of the semaeostome *Aurelia*. It has a colonial scyphistoma stage that produces ephyrae by polydisk strobilation. The medusae reproduce sexually, producing a planula larva that settles to form a scyphistoma to continue the cycle. It has been speculated that the deeper-dwelling genera such as *Periphylla* and *Atolla* use a strategy of direct development like that of the semeaostome *Pelagia*. In a direct-development strategy, the planula develops directly into an ephyra, bypassing the scyphistoma stage altogether. Evidence in support of this hypothesis is the large size of the eggs in *Periphylla* and *Atolla*, which would facilitate the direct developmental strategy (Larson 1986 in Arai 1997).

Genera include *Atolla, Linuche, Periphylla, Nausithoe, Stephanoscyphus,* and *Tetraplatia*.

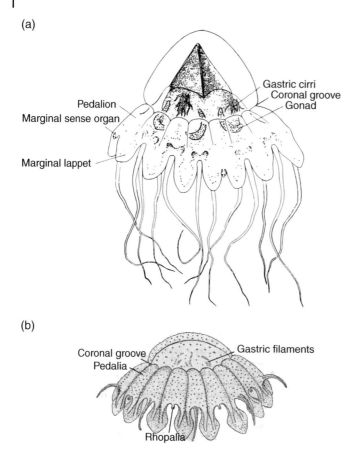

Figure 3.11 Coronatae. (a) The coronate scyphomedusa *Periphylla* (helmet jellyfish). (b) The coronate scyphomedusa *Nausithoe* (crown jellyfish). *Sources:* (a) Adapted from Arai (1997), figure 1.5 (p. 7); (b) Redrawn from Mayer (1910), Vol. III, figure 352.

Semaeostomae

The semaeostomes include the medusae most typical of the class and most familiar to beachgoers. Medusae are large, typically 5–40 cm, with a bell shape ranging from saucer-like (*Aurelia*) to bowl-like (*Chrysaora*) and lack a coronal groove. Tentacles are found along or below the margin of the umbrella, which may be divided into eight or more lappets. Most typical of the semaeostomes are the long frilly oral arms that originate at and form the corners of the mouth (Figure 3.10).

Life histories for many of the semaeostomes are known in detail, owing mainly to their abundance near shore. Two life histories have been noted. The sequence of the first (Figure 3.12a), typified by the moon jelly *Aurelia*, has a planula larva that develops from the fertilized egg and settles to the bottom to form a scyphistoma. The scyphistoma strobilates to form ephyrae that develop into either a mature male or female medusa. In the second life history strategy, seen in the genus *Pelagia*, the planula develops directly into

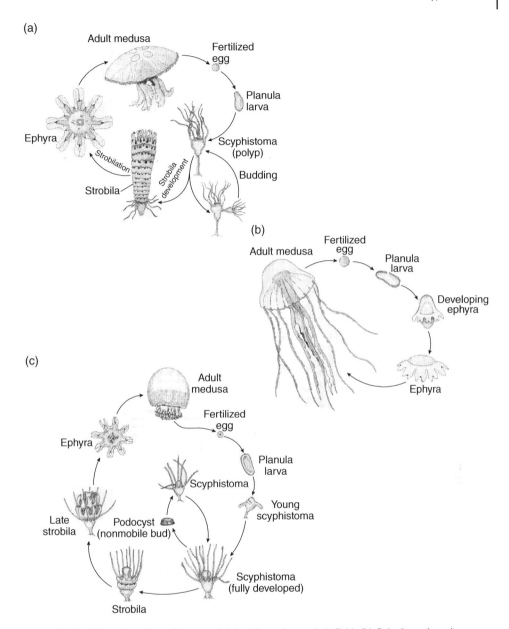

Figure 3.12 Life cycles of scyphozoans. (a) *Aurelia* sp. (moon jellyfish); (b) *Pelagia* sp. (purple-striped jellyfish). (c) *Stomolophus meleagris* (cannonball jellyfish). *Sources:* (a) Bayer and Owre (1968), figure 161 (p. 105); (b) Bayer and Owre (1968), figure 162 (p. 106); (c) Adapted from Calder (1982), figure 4 (p. 156).

an ephyra and then into a mature medusa, bypassing the benthic polyp, or scyphistoma, stage (Figure 3.12b).

Genera include *Aurelia, Chrysaora, Cyanea, Stygiomedusa,* and *Pelagia. Pelagia,* as the name implies, is common in offshore waters

Rhizostomae

Also known as the "cannonball jellies" because the rounded bells of some species (e.g. *Stomolophus meleagris*) resemble an old-fashioned cannonball (dangling an old-fashioned petticoat: Figure 3.12c), the rhizostome medusae lack marginal tentacles and exhibit a proliferation of the oral arms (Figure 3.13). During early development, the lobes of the original four-lobed mouth in the very young medusa grow and bifurcate (Figure 3.13a–c) to form the (usually eight) "mouth arms" typical of the group. As the oral arms develop, the fringed edges close over to form an inner tube or brachial canal that effectively eliminates the central mouth. Once the canal is closed over, the fringed edge forms an outer groove that communicates at intervals with the inner brachial canal through smaller canals that terminate in "suctorial mouths." Most often the oral arms are invested with filamentous appendages that bear the suctorial mouths along with nematocyst batteries and mucous cells that aid in prey capture (Hyman 1940) (Figure 3.13d–f). The brachial canals lead to a central stomach that conveys nutrition to the bell via radial canals.

The life histories of rhizostome medusae are typical of the Scyphozoa (Figure 3.12c) in having a planula larva that settles to the bottom and forms a polypoid scyphistoma. Scyphistomae may strobilate to form ephyrae or may produce other scyphistomae by budding.

The rhizostomes are chiefly a tropical–subtropical group inhabiting shallow waters, though two genera, *Rhizostoma* and *Stomolophus*, are found in temperate climes and may even form blooms. *Stomolophus*

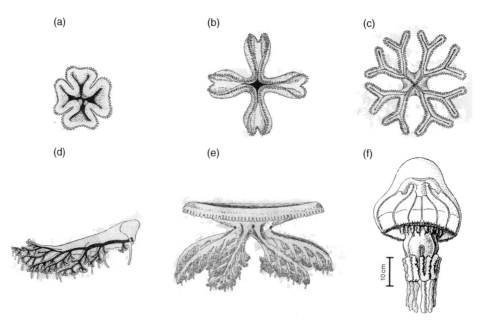

Figure 3.13 Rhizostomeae. Development of the mouth arms of *Mastigias*: (a) early stage, (b) lobe development, and (c) later stage. (d) Mouth arm of *Cassiopeia* showing the gastrovascular canals; (e) adult *Cassiopeia*. (f) Adult *Rhizostoma pulmo*. *Sources:* (a) Adapted from Uchida (1926); (b and c) Kaestner (1967), figure 5-17 (p. 105); (d and e) Hyman (1940), figure 172 (p. 525); (f) Redrawn from Mayer (1910), plate 73.

nomurai forms huge blooms in the Sea of Japan at intervals (Mills 2001). Rhizostomes are generally quite large, with sizes ranging from 4 to 200 cm across the bell.

Genera include: *Rhizostoma, Mastigias, Cassiopeia, Stomolophus, Cephea.*

The Cubomedusae

The Cubomedusae, variously known as the box jellies, sea wasps, or fire medusae, comprise the Cnidarian class Cubozoa. The Cubomedusae were formerly considered to be an order in the Scyphozoa. Now they are considered their own class and comprise two orders, the Carybdeida and the Chirodropida. The group includes about 48 species found in tropical and subtropical latitudes.

Bells of the Cubomedusae are indeed cuboidal, with four flattened sides that are square when viewed from the exumbrellar side or in transverse section (Figure 3.14a). The bell has a simple margin that bends inward to form a velarium (Figure 3.14b). Tentacles arise from the four corners of the bell, either singly or in groups. The base of the tentacle(s), the pedalium, is a blade-like structure that gives rise to a single tentacle in the family Carybdeidae and a palmate structure that gives rise to several tentacles in the family Chirodropidae (Figure 3.14c and d). Below the pedalium, the tentacle is hollow and armed with rings of nematocysts.

A four-sided manubrium leads to a simple, central stomach that is located at the apex of the sub-umbrellar surface. The stomach differentiates into four gastric pockets that occupy the flattened sides of the umbrella.

The life history of the Cubomedusae is much like that of the hydromedusae and scyphomedusae with one important difference: the polypoid stage of the Cubomedusa does not strobilate. Rather, each polyp metamorphoses into an individual medusa. Arneson and Cutress (1976) described the development of *Carybdea alata* in Puerto Rican waters as proceeding from a released blastula stage to a swimming planula in 1 day, settlement of the planula in an additional 4 days, growth and maturation of the polyp for about 60 days, and the metamorphosis culminating in a liberated medusa taking an additional week, for a total of about 75 days for the entire process. Temperature during development was 26–29 °C.

Genera include: *Carybdea, Tripedalia, Tamoya, Chirodropus, Chiropsalmis,* and *Chironex.*

Foraging Strategies

The subject of foraging strategies covers a lot of ground, from diets and prey selectivity to models of encounter rates and predatory behavior (e.g. Gerritsen and Strickler 1977). Our chief concern is to describe what is known about feeding in medusae, including both diet (favorite foods) and elements of the feeding behavior itself. What is important to a weakly swimming, tentaculate, predator?

In most studies of feeding in medusae, hydromedusae are grouped together with scyphomedusae, siphonophores, and sometimes even ctenophores. Clearly, though differing in complexity and size, many elements of hunting will be highly similar between the hydromedusae, scyphomedusae, and cubomedusae because of their similar body shape. The section on foraging strategies will regard the medusae as a whole, though crossing taxonomic boundaries, as will the discussion of locomotion and energetics.

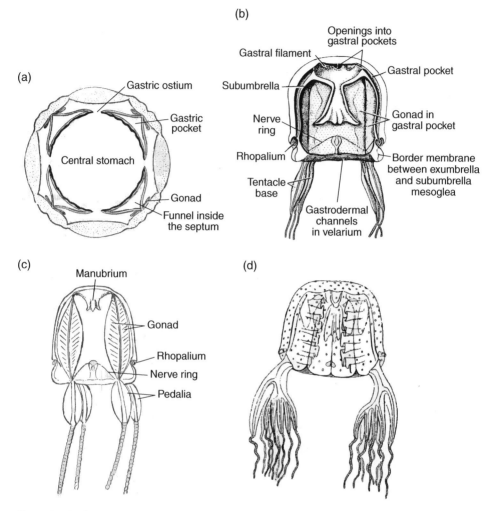

Figure 3.14 Cubomedusae. (a) Cross section through the bell of *Carybdea*; (b) longitudinal section through *Tripedalia*; (c) the Chirodropid medusa *Carybdea*. (d) The Chirodropid medusa *Chiropsalmus*. *Sources:* (a–c) Conant (1898); (d) Redrawn from Mayer (1910), plate 47.

General Considerations

Perspective is important when evaluating the foraging behavior of medusae. Equipped with rudimentary sensory systems and limited locomotory capabilities, they forage in a profoundly three-dimensional environment. Prey are captured on tentacles deployed in a stationary ambush or a slowly moving array as the animal swims forward. Stinging cells (nematocysts) on the tentacles paralyze the prey, which are then conveyed to the mouth and digested. Since both locomotion and the sensory field are quite limited, feeding success of a medusa will be determined by the number of its physical encounters with prey and the effectiveness of its tentacles in subduing the prey item.

In their mathematical model of predator–prey interactions in zooplankton, Gerritsen and Strickler (1977) assumed that the "encounter radius" of a predator was determined by its sensory system. In the case of medusae, it is determined by the volume enclosed within their tentacle array and the likelihood that a prey item once within the "kill zone" will contact a tentacle and be trapped (Madin 1988).

The Cnidae

The stinging organelles, or cnidae, that give the phylum Cnidaria its name are highly complex intracellular structures unique to the phylum. They are formed inside cells called cnidoblasts (Brusca and Brusca 2003), which are formed from interstitial cells in the epidermis and gastrodermis. The mature cnida in its cell is a cnidocyte. The majority of cnidocytes are located on the tentacles in small, blister-like groups called "batteries" or in the epidermis of the oral region.

The cnida in its discharged state consists of a cup-shaped basal capsule, a basal "shaft" or "butt" (Hyman 1940; Arai 1997), and ends in a long hair-like tubule that is often invested with a spiny armor (Figure 3.15). The venom used to subdue prey is either "injected" from a pore at the end of the tubule, much like a syringe works, or is introduced as a coating on the tubule, much like a poisoned dart. The trick to understanding the process of eversion or discharge is to understand how the cnida is coiled in its resting state. If you can imagine poking your finger into that of an empty rubber glove from the wrong side so that it turns "outside in". you will have a good idea of how the cnida is coiled. In its resting state, it is literally turned outside in. When discharged, the pressure from within the cnida's basal capsule causes it to evert, just as the finger on the rubber glove would pop back out to normal if you blew into the bottom of the glove. Cnidae can only be discharged once.

Their older name, nematocysts, is still very much in use, and the term cnidocyte then becomes nematocyte. In the newer terminology, only the stinging cnidae are termed nematocysts, to distinguish them from other types of cnidae that, for example, stick to prey (spirocysts) instead of envenomating them (Brusca and Brusca 2003).

Nematocysts, or cnidae, are considered to be "independent effectors": their discharge is not governed by the nervous system of the medusa but will discharge when stimulated directly by prey contact. The nematocyst has a "lid" or operculum (Figure 3.15) that covers the capsule and acts as a trapdoor. When the cnidocyte discharges, the operculum is flung open. The cnidocil, a bristle located next to the operculum, is believed to be the mechanoreceptor or "trigger" responsible for nematocyst discharge. Though the cnidocytes are considered independent effectors, their sensitivity threshold can be modified by the nutritional state of the medusa. A starved medusa will have a lower threshold for discharge than a well-fed one.

What causes the cnidocyte to discharge? At least three theories purport to explain how a discharge is achieved. The first is the osmotic hypothesis, wherein a high osmotic pressure (high ion concentration) is maintained inside the nematocyst capsule by active ion transport and perhaps by sequestration of small organic osmolytes. Discharge of the cnidae is effected by a change in permeability of the capsule, causing water to rush down its concentration gradient, swell the capsule, and evert the nematocyst tubule. The second explanation is that the cnida formed within the cnidocyte is already in a state of tension owing to

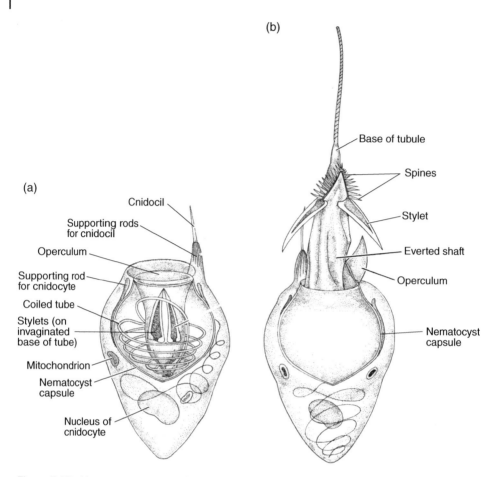

Figure 3.15 Nematocyst structure. (a) Before discharge; (b) after discharge. *Source:* Schultze (1922).

the coiling of its collagen-like structure, much like a jack-in-the-box with the lid closed. The third is that contractile proteins in the cnidocyte squeeze the basal capsule and pop out the tubule. All three explanations are feasible, but the process happens so quickly that it has not been possible to pin one down as the most likely.

Venoms

Venoms are injected when nematocysts discharge. The virulence of the injected toxins varies considerably from species to species, depending partly on the venom itself but also on how much is delivered. For example, the number of tentacles contacted by an unfortunate swimmer brushing up against a Portuguese Man O'War (*Physalia physalis*) will likely be much larger than that in a similar encounter with *Chrysaora*, the sea nettle. The cubomedusae are rightfully most famous for the potency of their venom, particularly the sea wasp, *Chyronex fleckeri*, which can kill a grown man. Within the hydrozoa, the siphonophores are the most potent stingers, with the widely distributed Portuguese Man O'War the best known. It deserves every bit of its nasty reputation as a world-class stinger.

All venoms have a few properties in common, and cnidarian venoms are no exception (Hessinger 1988). First, most venoms act on cell membranes. Membranes are the most accessible part of any cell, always play an important role in cell integrity and function, and can be disrupted in a variety of ways. Second, most venoms are proteinaceous. Of the three major types of biomolecules (proteins, lipids, and carbohydrates), proteins are the most plastic in structure and function. They range from enzymes, which can literally change shape, to blood pigments, to inert structural molecules such as the keratins. They are the most amenable to biological design. Also, most proteins are readily digested, so the predator is not poisoned by its own venom! Third, venoms act rapidly. To be effective, toxins must kill, stun, or paralyze the prey quickly to prevent escape.

When considered on a dose-specific basis ($\mu g\,kg^{-1}$ body mass), cnidarian venoms are among the most deadly known to science, on a par with those of the kraits and the mambas. Purified venoms of the sea wasp and Man O'War show a median lethal dose in mice of less than $50\,\mu g\,kg^{-1}$ body mass. Symptoms in mice of poisoning with purified cnidarian venom include severe respiratory distress, convulsions, and loss of motor control (Hessinger 1988). Because mice are not within the normal prey spectrum of jellies, fiddler crabs have been used as test subjects for effects of cnidarian venom. Their symptoms include violent motor activity, paralysis, and in some cases, autotomy (spontaneous loss) of the walking legs (Hessinger 1988).

Most of the toxic effects associated with cnidarian venoms can be traced to compromised cell membranes. Excitable tissues such as nerve and muscle are particularly susceptible to membrane disruption because they require an intact membrane to function. Cardiovascular distress, loss of motor control, and violent convulsions are all indicative of compromised cell membranes in the neuromuscular system and will nearly always be the most obvious symptoms obtained when an experimental subject is injected with a purified venom.

That said, a world of difference exists between a jellyfish sting and the impacts of purified jellyfish venom, or our beaches would be empty much of the time. Only a very few people die each year from jellyfish stings, and many of those fatalities are due to an allergic response like that seen with bee stings. Most people only feel a minor bit of irritation that goes away in a few hours, even with a venom that is as potent as that of a rattlesnake. Why? The amount delivered by a jellyfish sting is minute compared with the amount delivered by a snake. A large rattler can inject volumes of a milliliter or more of pure venom when it strikes, which makes it very deadly. The larger jellies such as the Man O' War and the Australian sea wasp can deliver large enough quantities of venom to cause severe distress or worse. For small creatures such as fish larvae and many species of zooplankton though, the amount of venom introduced by small jellyfish stings is enough to disable them. The many small "harpoons" of the nematocysts trap them on the tentacle, and they can then be conveyed to the mouth.

Interaction with Prey

As noted above, medusae are radially symmetrical tentaculate predators that rely on contact to entangle their prey. They are not "sight-hunters" or raptors (e.g. Greene 1985) capable of actively seeking a prey item and chasing it down. Yet despite their seeming inability to select prey, virtually all gelatinous predators for which electivity indices have been calculated show some degree of selectivity in their diet (Purcell 1997).

Most pelagic cnidarians feed mainly on copepods, the dominant metazoan zooplankton group in the vast majority of the world ocean (Purcell 1997). This is not surprising; a nonliving bit of "marine flypaper" cut into the shape of a medusa and floating with the ocean current would mainly snag copepods. It is the exceptions to the rule that are intriguing and lead one to wonder about how the selectivity is achieved. For example, Purcell (1997) observed that the hydromedusae *Bougainvillia principis* and *Proboscidactyla* fed mainly on barnacle larvae and molluscan veliger larvae, respectively. Are cnidarian dietary preferences the result of a limited prey field or actual selectivity? We will investigate cnidarian hunting from a theoretical perspective. Interactions with prey have two basic elements: the encounter and the capture (Purcell 1997). Factors influencing the encounter phase have been treated in a number of studies and may be divided into four subcategories: direct interception, encounter zone, water flow and swimming, and attraction between predator and prey.

Direct Interception

Cnidarian tentacles may be considered as a large, loosely configured filter, and the concept of direct interception derives from filtration theory (Rubenstein and Koehl 1977; Purcell 1997). Because the spacing of tentacles in any cnidarian predator is usually much greater than the prey diameter, particularly for small prey, the direct interception of a prey item on a tentacle depends only on the diameter and swimming speed of the prey and the diameter of the tentacle. Further, the theory predicts that larger, faster prey would be selected for by tentaculate predators generally. It applies most directly to ambush predators such as the siphonophores and some of the medusae. Figure 3.16 shows how the ubiquitous medusan predator *Pelagia noctiluca* captures prey on one of its tentacles while swimming. Prey is trapped by nematocyst discharge and conveyed to the mouth with the cooperation of the oral arms as the *Pelagia* continues to swim.

Encounter Zone

Madin (1988) used field observations, videography, and measurements to produce a very useful conceptual model of feeding behavior in medusae, ctenophores, and siphonophores. In his model, the shape of the medusa's bell, its arrangement of tentacles, and its swimming behavior create an encounter zone where probability of prey capture is maximized. Within the encounter zone, the likelihood that various prey types will be caught depends on the interaction of tentacle density, tentacle spacing, prey size, type, and the behavior and properties of nematocysts (Table 3.1).

Figure 3.17 illustrates modes of tentacle deployment for a variety of different medusae and siphonophores. Most of the medusae illustrated are hydromedusae, underscoring the diversity in their morphology. "Type" species represented in Figure 3.17 are listed in Table 3.1 and described briefly below.

Anthomedusae

Calycopsis typa (Figure 3.17a) has a globular bell and thick tentacles held out radially when fishing, creating a discoid volume about three times the diameter of the bell.

Stomotoca pterophylla (Figure 3.17b) is a species with two tentacles that may be 50 times the bell diameter when fully extended. The tentacles occupy only a tiny fraction of

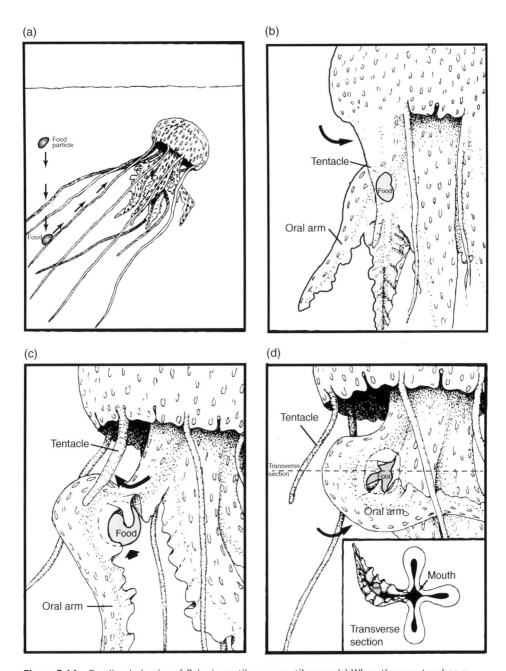

Figure 3.16 Feeding behavior of *Pelagia noctiluca* on motile prey. (a) When the prey touches a marginal tentacle, there is an immediate nematocyst discharge, followed by a tentacle contraction after 2–3 seconds; (b) the stiff tentacle bends toward the nearest oral arm, at the same time the oral arm moves upward, turns slightly, and draws its endodermic layer near the food; (c) the stiff tentacle releases the prey and moves upward, going away from the oral arm; (d) the oral arm grasps the prey completely and starts the peristaltic and mucous movements, which drive the food to the oral arm groove, to the manubrium, and finally to the gastric cavity. The inset figure is a transverse section of "d." *Source:* Rottini-Sandrini and Avian (1989), figure 1 (p. 52). Reprinted by permission from Springer Nature Customer Service Centre GmbH Springer-Verlag, Marine Biology, Feeding mechanism of Pelagia noctiluca (Scyphozoa: Semaeostomeae), Rottini-Sandrini and Avian, 1989.

Table 3.1 Tentacular encounter zones and deployment patterns, tentacle length, volume, and density, and prey types of some medusae and siphonophores. *Source:* Adapted from Madin (1988), table 2 (p. 423).

Species	Encounter zone – shape of space	Tentacle deployment[a] pattern (see Figure 3.17)	Encounter zone – volume (cm³)	Tentacular length (m)	Tentacular volume (cm³)	Tentacle density[b] (ppm)	Prey types
Medusae							
Calycopsis typa	Disc	3.17A	850	12	0.0942	110.20	Large prey – types unknown
Stomotoca pterophylla	Sphere	3.17B	900 000	2.4	0.0033	0.00	Medusae, other gelatinous spp.
Aequorea macrodactyla	Cone	3.17C	1 500 000	200	0.7697	0.53	Salps, ctenophores, pteropods, forams, medusae
Laodicea undulata	Disc	3.17C	175	18	0.0088	50.00	Small crustacea, larval fish
Dichotomia cannoides	Cone	3.17C	50	7.5	0.0019	36.80	Small crustacea types unknown
Liriope tetraphylla	Sphere	3.17B	220 000	3	0.0052	0.02	Heteropods, appendicularia, larval crustacea, juvenile fish
Solmundella bitentaculata	Cylinder	3.17B	2	0.1	0.0118	5000.00	Gelatinous species
Aeginopsis laurentii	Cone	3.17D	100	0.2	0.0471	476.20	No data
Somaris spp.	Cone	3.17D	2	0.3	0.0001	40.00	Small motile species, types unknown
Pelagia noctiluca	Cone	3.17C	8 700 000	24	0.1794	0.21	Salps, ostracods, ctenophores, polychaetes, copepods, fish
Siphonophores							
Sulculeolaria spp.	Cylinder	3.17E,F	257 000	243.2	1.0400	4.10	Copepods
Forskalia spp.	Cylinder	3.17E	434 000	88.2	0.2400	0.56	Copepods, amphipods, chaetognaths, molluscs, fish, fish eggs

[a] See Figure 3.17 for cross-reference to arrangement patterns of deployed tentacles.
[b] The calculated space within the encounter zone occupied by tentacles

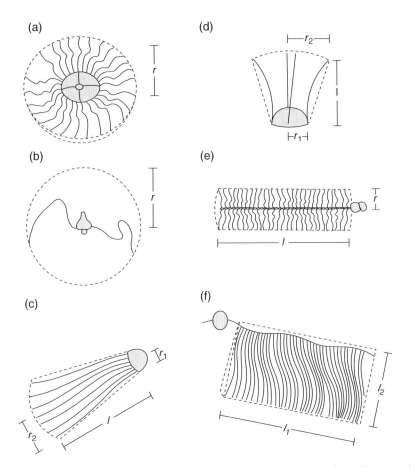

Figure 3.17 Patterns of tentacle deployment seen in medusae, siphonophores, and ctenophores. (a) Tentacles radiate from the body, filling a disk-shaped space; (b) tentacles are somewhere within a sphere around the body; (c) tentacles stream behind the body, filling a truncated cone; (d) tentacles are held ahead of the body in a cylinder or truncated cone; (e) tentacles radiate from a long stem, filling a cylindrical space; (f) tentacles or tentilla form a nearly flat curtain. Some siphonophores may have encounter zones of this shape. *Source:* Madin (1988), figure 1 (p. 416). Reproduced with the permission of the Bulletin of Marine Science.

the spherical volume accessible to them. Prey must swim into the tentacles for successful capture. *Stomotoca* feeds on medusae that are large enough to allow its feeding apparatus to function effectively.

Leptomedusae

Aequorea macrodactyla (Figure 3.17c) are large, lens-shaped, medusae that swim nearly continuously, trailing the tentacles behind the bell to form a conical volume. Tentacles extend to about 50 times the bell diameter. Prey stick to the tentacles of the swimming medusa and are conveyed to the mouth. Like *Stomotoca*, the prey of *Aequorea* are largely gelatinous and include medusae, ctenophores, pteropods, and salps.

Laodicea undulata have a flat shape with large numbers of fine tentacles arranged radially around the bell, giving it an encounter volume like that shown in Figure 3.17a.

It is believed to feed largely on small prey such as copepods.

Dichotomia cannoides has a conical bell with about 50 tentacles arranged around the bell margin. Its tentacles are deployed in a disc (Figure 3.17a) or cone (Figure 3.17c) configuration. It is believed to feed on small prey.

Trachymedusae

Liriope tetraphylla has four tentacles that extend as much as 25 times the bell diameter and an unusually long gastric peduncle with a stomach at the end. Its hunting strategy is most like *Stomotoca* (Figure 3.17b), with a large spherical encounter volume and the tentacles only occupying a small fraction of it.

Narcomedusae

Solmundella bitentaculata (Figure 3.17d) has two tentacles that are held in front of the bell (aboral side) as it swims, forming a cylindrical or conical volume of about three bell diameters in front of the bell.

Aeginopsis laurentii is another D-type (Figure 3.17d) predator but with four tentacles that extend out in front of the bell in a cone. Like *Solmundella*, *Aeginopsis* is a mesopelagic species and little information is available on diet.

Solmaris is a narcomedusan genus with several species living at epi- and mesopelagic depths. It conforms to the narcomedusan D-type tentacle deployment; it has an array of several tentacles forming a larger cone.

Scyphomedusae

Pelagia noctiluca is a semeaostome scyphomedusa. As a scyphozoan, it is larger than its hydrozomedusan brethren with a bell diameter of up to 15cm and tentacles extending outward up 30 times the bell diameter. It swims continuously as a C-type predator feeding on a wide variety of prey types including salps, doliolids, pteropods, forams, amphipods, chaetognaths, and fishes.

The Model

Madin (1988) proposed a general model to describe tentaculate predation incorporating the best elements of previous studies (Gerritsen and Strickler 1977; Mills 1981; Greene 1985; Greene et al. 1986; Larson 1987a, 1987b; Purcell and Mills 1988) and the author's own observations. A successful predatory sequence consists of four parts: encounter, capture/selection, ingestion, and digestion.

Interaction with prey begins with the "encounter" phase, which is determined by the size and type of encounter zone (Figure 3.17a–d) and the swimming behavior of the medusa. It was assumed that sensory mechanisms were not used to target individual prey. Once a prey item is within its encounter zone, how effective a species will be at capturing it will be determined by a suite of characteristics: tentacle density and spacing, prey size and swimming behavior, and effectiveness of the nematocysts and venom.

Prey size is dictated to some degree by the morphology of the predator. Species with fine dense tentacles (A-type in Figure 3.17) may be expected to capture small prey items, e.g. copepods, and those with more widely spaced robust tentacles (B and C-type) to capture larger prey, e.g. ctenophores and other medusae. Many of the B- and C-type predators increase their encounter probability with prey using swimming or hunting behavior. The effectiveness of the nematocysts in paralyzing prey and strength of the tentacles for retaining larger prey are also important in the successful capture of larger prey items. Species that prey on soft-bodied species such as jellies and fishes have one type of nematocyst designed to harpoon and poison the prey, whereas those that prey on crustaceans have four to five types of nematocysts in specialized batteries designed to entangle and hold crustaceans. Clearly, those species (e.g. cubomedusae) that prey on strong swimmers such as fishes and larger shrimp that could potentially tear tentacles must have venom virulent enough to quickly paralyze their prey.

Once a prey item is captured, successful digestion will depend on whether the item can be conveyed to the mouth by the tentacles and whether it can be successfully introduced to the gastric cavity through the mouth and broken down into useful nutrients by the digestive apparatus.

Swimming and Hunting Behavior

Mills (1981) gives an example of four hydromedusan hunting behaviors (Figure 3.18) that correspond well with Madin's general model. The first species, *Proboscidactyla flavicirrata* (Figure 3.18a) corresponds to the A-type predator in Figure 3.18: neutrally buoyant with 40–80 tentacles radiating from a globular bell. *P. flavicirrata* employs a motionless ambush strategy, allowing small zooplankton to enter the "encounter zone" through their own swimming behavior. Its neutral buoyancy makes it an especially effective trap for its plankton prey.

As a congener of *Stomotoca pterophylla*, *Stomotoca atra* is a B-type (see Figure 3.18b) predator. By adding its swimming behavior, we gain a better understanding of how it uses its two long tentacles to hunt. It employs a hop–sink swim cycle to drag its long tentacles up and down through the water column. As it swims and sinks, the tentacles describe a sine curve about the width of the bell and 2 meters from top to bottom. It feeds on large prey (hydromedusae), and it greatly increases the probability of contacting a prey item by its method of interrogating the water column.

The third species, *Phialidium gregarium*, is the primary prey of *Stomotoca atra*. *P. gregarium* also employs a hop–sink feeding strategy but a very different one from that of *S. atra*. It swims upward, bell uppermost, then sinks down with the bell oriented downward and its tentacles trailing behind (Figure 3.18c). As it does so, vortices are created behind the bell that circulate small prey into the tentacles. *P. gregarium* would be classified as a C-type predator in Madin's Figure 3.17 model.

The last example is *Polyorchis pencillatus*, a resident of shallow bays where it spends a great deal of its time on the bottom. Its hunting strategy on the bottom is to perch on its tentacles (Figure 3.18e) and use its manubrium to ingest prey from the surface sediments. At intervals it hops up off the bottom, stirring up the sediments, and then back down. Occasionally it swims up to the surface and drifts downward bell up (Figure 3.18d), becoming an A-type predator in the Figure 3.17 model.

Mills (1981) lists seven factors that contribute to feeding efficiency in medusae: (i) tentacle number and length; (ii) geometry of tentacle posture; (iii) velocity of tentacles moving through water; (iv) swimming pattern of medusa; (v) streamlining effects of the medusa bell on water flow; (vi) diameter of the prey; (vii) swimming pattern and velocity of prey. Together, Mill's observations and Madin's conceptual treatment provide a useful framework for examining the feeding strategies of medusae.

Water Flow and Swimming

The specialized swimming behaviors described above for the hydromedusae increase their hunting efficiency considerably.

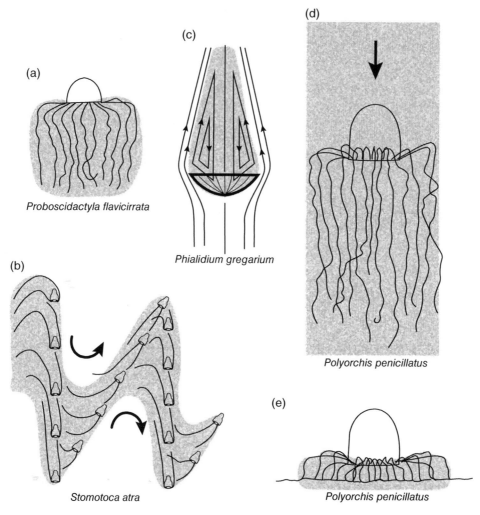

Figure 3.18 Different hunting and feeding behaviors of medusae. Shaded areas indicate effective feeding spaces. (a) Motionless ambush strategy; (b) swimming-sinking search pattern; (c) downward sinking with trailing tentacles creating vortices, which bring food particles within the feeding space; (d and e) two strategies employed by medusae such as *Polyorchis penicillatus;* part-time swimming and fishing in the water column and part-time resting on its tentacles on the bottom and capturing the prey directly with its manubrium. *Source:* Adapted from Mills (1981).

However, even conventional "straight-line" swimming creates flow fields around the swimming medusa that enhance prey capture. Analysis of movement in the disc-like moon jelly *Aurelia aurita* shows that normal swimming motions, particularly the expansion of the swimming bell during the recovery stroke, entrain particles to within easy reach of the tentacles as the water rushes into the subumbrellar space (cf. Costello 1992).

Attraction Between Predator and Prey

Attraction between predators and their prey can take several forms. Siphonophores create "lures" with nematocyst batteries on the fishing tentacles to attract the prey (Purcell 1980). In a different form of attraction, medusae have been observed to swim toward areas of high prey concentration (Arai 1991) in the laboratory, as well as aggregating in areas where prey have recently been, suggesting a chemoreceptive mechanism of attraction resulting in higher densities of medusae.

High concentrations of medusae can be achieved by physical aggregating mechanisms or by rapid reproduction in place to form a true bloom (Graham et al. 2001). Physical cues that have been implicated in high concentrations of medusae include light-mediated migrations such as diurnal vertical migration and aggregations associated with discontinuities in temperature, salinity, and density (pycnoclines) in the vertical plane. The reasons for accumulation of medusae at pycnoclines likely include higher concentrations of prey at the density discontinuities as well as passive mechanisms such as buoyancy at the cline.

Wind, waves, and currents can also act to produce aggregations of medusae. Populations of medusae are often compressed along the shoreline, resulting in rafts of jellies on the beach during periods of onshore winds. Oceanic frontal systems may harbor increased densities of medusae relative to waters outside of the frontal zone, similar to increases in populations observed in fishes and other more mobile species at oceanic fronts. Interestingly, a unique, persistent aggregation of the medusa *Chrysaora fuscescens* may be found in Monterey Bay California, the result of upwelled water entrained by a coastal prominence in the northern part of the bay (Graham et al. 1992).

Diets, Feeding Rates, and Impacts on Prey Populations

Hydrozoan and scyphozoan medusae can have large impacts on local zooplankton prey fields, particularly when prey and predator densities, and therefore encounter probabilities, are high (Table 3.2). Most of the studies on diet and feeding in medusae have taken place in coastal systems with an eye toward describing interactions of medusae with the larvae of commercially important fish species. More general studies on field-caught medusae (Table 3.3) reveal a varied diet that fluctuates with available prey (Mills 1995). It includes copepods, chaetognaths, fish eggs, fish larvae, larvaceans, other medusae, euphausiids, mysids, decapods, and ctenophores.

Impacts of medusae vary considerably and depend largely on predator density. Purcell and Arai (2001) demonstrated that prey-removal rate by the hydromedusa *Aequorea victoria* ranged from 0.1 to 73% of available herring larvae per day from coastal waters off Vancouver Island, British Columbia, depending upon predator concentrations. Clearly, medusan predation can have a profound influence on larval survivorship, particularly when wind and wave or reproductive activity act to concentrate weakly swimming prey and gelatinous predators in one location.

The radial symmetry, stinging tentacles, and gelatinous character of medusae make them highly effective as predators, particularly as ambush predators. However, they also may find themselves as prey in the diets of other medusae. In particular, the semaeostome scyphomedusae often have hydromedusae in their diet when the smaller medusae are available in quantity, e.g. during early spring (Purcell 1991). At this time, no scyphozoan medusa is known to prey exclusively on other medusae, but it may be that the narcomedusae, the slow swimming hydromedusae important in the mesopelagic zone, specialize on other jellies (Purcell and Mills 1988).

Table 3.2 Predation rates, clearance rates, and predation effects from field observations of gelatinous predators feeding on fish eggs and larvae. Prey consumed percentages are estimated consumed daily *in situ*. Source: Adapted from Purcell and Arai (2001).

Species	Size	Prey type (density)	Prey eaten (no. · pred^{-1} · d^{-1})	Clearance rates[a] (no. · pred^{-1} · d^{-1})	Prey consumed[a] (% · d^{-1})	References
Siphonophore						
Physalia physalis	na	Larvae[a] (~0.2 m^{-3})	120	600000	60	Purcell (1984)
Rhizophysa eysenhardti	na	Larvae[a] (28 m^{-3})	9	311	28.3	Purcell (1981a)
Medusae						
Aequorea victoria	33–68 mm	Larvae[c] (<10 m^3)	13 ± 13	5650 ± 6114	18 ± 29	Purcell (1989, 1990), Purcell and Grover (1990)
Aequorea victoria	33–68 mm	Larvae[c] (10–100 m^{-3})	55 ± 48	1357 ± 908	49 ± 35	Purcell (1989, 1990), Purcell and Grover (1990)
Aequorea victoria	33–68 mm	Larvae[c] (<100 m^{-3})	91 ± 47	288 ± 210	33 ± 32	Purcell (1989, 1990), Purcell and Grover (1990)
Aurelia aurita	6–50 mm	Larvae[d] (na)	1.6	na	2.6–4.4	Möller (1980)
Chrysaora quinquecirrha	40–70 mm	Eggs[e] (avg. 164 m^{-3})	343 ± 419	2213 ± 1625	14 ± 4	Purcell et al. (1994)
Chrysaora quinquecirrha	40–70 mm	Larvae[e] (avg. 43 m^{-3})	86 ± 136	1818 ± 1861	29 ± 4	Purcell et al. (1994a)
Stomolophus meleagris	55 mm	Eggs[b] (na)	Na	3120	na	Larson (1991)
Ctenophore						
Mnemiopsis Leidyi	40 ml (live vol.)	Eggs[e] (224 ± 178 m^{-3})	42 ± 33	128 ± 58	9 ± 14	Purcell et al. (1994a)

na = not available
[a] Calculated from data in source.
[b] Mixed species or unidentified.
[c] Pacific herring.
[d] Atlantic herring, *Clupea harengus* Linnaeus.
[e] Bay anchovy.

Table 3.3 Summary of data on stomach contents of field-caught Scyphomedusae. *Source:* Adapted from Arai (1997), table 3.2 (pp. 69–71).

Species (size range if available)	Common prey items (in order of predominance)	Source/references
Aurelia aurita (2.5–150 mm)	Copepods, herring, hydromedusae, crustacea, tintinnids, cladocera, rotifers (in small *Aurelia*)	Matsakis and Conover (1991), Mironov (1967), Möller (1980), Olesen et al. (1994)
Aurelia aurita (80–300 mm)	Copepods, veligers	Hamner et al. (1982), Kerstan (1977)
Chysaora quinquecirrha (<6 mm)	Protozoa and rotifers	Purcell (1992)
Chysaora quinquecirrha (18 to >31 mm)	Copepods, fish eggs	Purcell (1992), Purcell et al. (1994b)
Cyanea capillata (40–700 mm)	Fish larvae, ctenophores, hydromedusae	Plotnikova (1961)
Cyanea capillata	Larvacea, cladocera, fish eggs, fish larvae, copepods	Fancett (1988)
Drymonema dalmatinum	Medusae	Larson (1987b)
Pelagia noctiluca	Fish eggs, copepods, cumacea, chaetognaths	Larson (1987b)
Pelagia noctiluca	Copepods, decapods, cladocerans, chaetognaths	Giorgi et al. (1991)
Periphylla periphylla	Copepods	Fosså (1992)
Phacellophora camtschatica	Fish larvae, larvacea, gelatinous zooplankton, copepods	Purcell (1990)
Pseudorhiza haeckeli	Fish eggs, fish larvae, copepods, larvacea, decapod larvae	Fancett (1988)
Stomolophus meleagris (21–83 mm)	Veligers, copepods, tintinnids	Larson (1991)

Rogue Hydroids: Predatory Polyps in the Midwater

Georges Bank is a shallow (45 m depth) hummock in the Gulf of Maine, USA, made famous by its former bountiful harvests of cod (*Gadus morhua*), sadly now depleted. The area was the subject of an intensive multidisciplinary 1990–2000 oceanographic study as part of the international GLOBEC (*GLOB*al Ocean *EC*osystem Dynamics) program, funded by the USA's National Science Foundation. The mission of GLOBEC was to describe the interaction of physical and biological processes in the life history of important species. In the case of George's Bank, the target species was cod.

A GLOBEC sampling cruise in May of 1994 revealed large numbers of suspended hydroid colonies in the zooplankton over the shallows of Georges Bank (Madin et al. 1996). The colonies were typically fragments of 2–5 polyps each, were widely distributed over the bank, and were the overwhelming dominant component of the net-caught zooplankton, reaching densities of $10000 m^{-3}$ in the water column and $25000 m^{-3}$ nearer the bottom. Colonies were found to be the polypoid life stage of the hydromedusan genus *Clytia*, predominantly *Clytia gracilis*, which normally grows attached to rocks, seaweed, or other available benthic substrate. The polyps suspended in the water column instead of being attached to substrate were not only alive, they were actively hunting. Examination of gut contents and shipboard experiments revealed that the hydroids were catching and digesting larval cod as well as copepod eggs and larvae. Madin et al. (1996) estimated that the hydroids were capable of ingesting half the daily production of copepod eggs and a quarter of the standing stock of copepod larvae per day: an impressive figure for a benthic life stage.

Feeding in the Cubomedusae

The cubomedusae are well known as having a potent sting, particularly the Australian sea wasp *Chyronex fleckeri*, a large jelly (football-sized bell and bigger) that can cause open welts in humans and, with severe stings, even respiratory distress and death. The virulence of cubomedusan venom allows the group to take large prey.

The importance of a strong venom and nematocyst system in cubomedusae is underscored by a description of the feeding behavior of the Caribbean cubomedusa *Carybdea marsupialis* (Larson 1976). *Carybdea* has four robust tentacles that can reach 30 cm in length when extended or about 10 times the height of the bell. Prey are captured on the tentacles by annular nematocyst batteries that paralyze and trap the prey on the tentacle with considerable adhesive force. The "sticking power" of the nematocysts is so strong that fish too large for *Carybdea* to handle must break the tentacle to escape. Once the prey is subdued, it is conveyed into the bell cavity and onward to the digestive system by an inward flexion of the tentacle (Figure 3.19). The outer

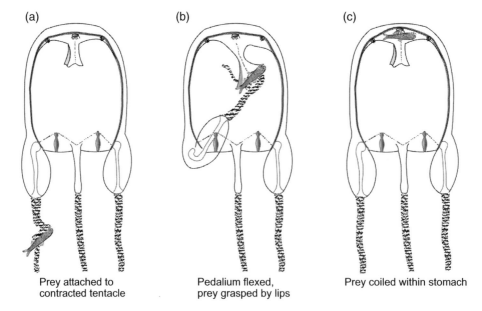

Figure 3.19 Prey capture and ingestion in *Carybdea marsupialis*. (a) Prey capture; (b) Prey transfer to mouth; (c) Prey ingestion. *Source:* Adapted from Larson (1976).

digestive region, or manubrium, is short, and its outer lips are prehensile, so that when a prey item is contacted, it is quickly engulfed.

Though data are limited on the diets of cubomedusae, the little data available suggest a varied menu including polychaetes and small fish as well as the more typical small crustaceans (Table 3.4). Observations reported in Larson (1976) suggested some selectivity for fish by *Carybdea marsupialis*.

Locomotion

Medusae are among the very few aquatic taxa that swim using jet propulsion. Though "jet propulsion" does not necessarily invoke the mental picture of a slowly swimming medusa, the medusae, the squids and octopi, and the salps and doliolids are major marine taxa that swim using jet propulsion. Scallops also use it to escape from starfish predators by rapidly closing their valves to expel a jet of water, allowing brief forays into the water column.

Medusae swim by ejecting the volume enclosed by the bell toward the rear with a forceful contraction, propelling the animal forward (Figure 3.20). Alternate rhythmic contraction of muscles to propel the jet and elastic recoil of mesoglea to refill the jet allow a medusa to make its way forward. The basic locomotory system thus comprises the swimming muscles, the deformable mesoglea that gives the swimming bell its elastic character, and a pacemaker that sets the rhythm and assures that contraction is synchronous over the bell.

The swimming muscles are of two basic types: radial muscles aligned in the same axis as the radial canals, that is, from the center to the periphery of the bell; and the coronal muscles that form a ring parallel to

Table 3.4 Diets of cubomedusae. *Source:* Larson (1976), table 2 (p. 242). Reprinted by permission from Springer Nature Customer Service Centre GmbH, Cubomedusae Feeding, author R. J. Larson, in Coelenterate Ecology and Behavior, G.O. Mackie editor, 1976.

Species	Coelenteron contents	References
Carybdea alata	Polychaetes, mysids, crab megalopae	Larson (unpublished)
Carybdea marsupialis	Polychaetes, misc. crustaceans (copepods, isopods, amphipods, stomatopod larvae, mysids, caridean shrimp and larvae, crab zoeae), chaetognaths, fish	Berger (1900), Larson (unpublished)
Carybdea rastoni	Polychaetes, mysids, fish	Gladfelter (1973), Ishida (1936), Larson (unpublished), Uchida (1929)
Chiropsalmus quadrumanus	Misc. crustaceans (amphipods, cumaceans, stomatopod larvae, *Lucifer* spp., caridean shrimp, crab larvae), fish	Larson (unpublished), Phillips and Burke (1970), Phillips et al. (1969)
Chiropsalmus quadrigatus and *Chironex fleckeri*	Caridean shrimp (*Acetes* spp.), other small crustaceans, fish	Barnes (1966)
Tripedalia cystophora	Copepods (*Oithona* spp.)	Larson (unpublished)

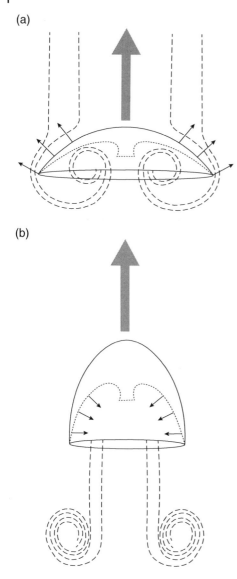

Figure 3.20 Jet propulsion in medusae. (a) Bell expansion and water intake; (b) bell contraction with water expulsion. Direction of motion is indicated by the large arrows.

the margin (Figure 3.21). In the Scyphozoa, swimming is usually initiated by a contraction of the radial muscles, causing the bell to shorten, followed by a contraction of the coronal musculature, which cinches up the margin and forces a jet of water out the bell. In the Hydrozoa, swimming is effected in much the same way with the exception that the presence of a velum, or skirt, gives more direction to the stream of the jet. Direction of the swimming hydromedusa can thus be manipulated by differential contraction of the radial muscles.

Unlike the continuous movement of fishes and shrimp, medusan swimming is intermittent in nature, alternating periods of thrust with the refilling of the jet. We know that in most cases the movement is fairly slow, with velocities of $5\,\text{cm}\,\text{s}^{-1}$ or less, but how efficient is it? The answer is, not very. A few studies have examined the swimming performance of medusae using both modeling (Daniel 1983) and empirical approaches (Daniel 1985). Daniel calculated a Froude efficiency of about 10% for the hydromedusa *Gonionemus*, which means that the work done by the muscle in creating the jet was about 10 times the drag that the jet needed to overcome in order to propel the medusa through the water (Alexander 2003). A medusa expends about 10 times as much effort to move through water as does a fish.

Refilling the medusan bell is accomplished by the deformable and elastic mesoglea, whose recoil to the relaxed state fills the bell with the volume of water that will be forced out in the next jet. Physical properties of the bell were investigated in the hydromedusa *Polyorchis* (DeMont and Gosline 1988) using measurements in the swimming medusa and in isolated blocks of tissue. They found that about 60% of the work done in deforming the mesoglea was returned in its recoil to the relaxed state, a respectable number but not at the high end of elastic mechanisms employed in the locomotion of other species, which can exceed 90% (Alexander 2003).

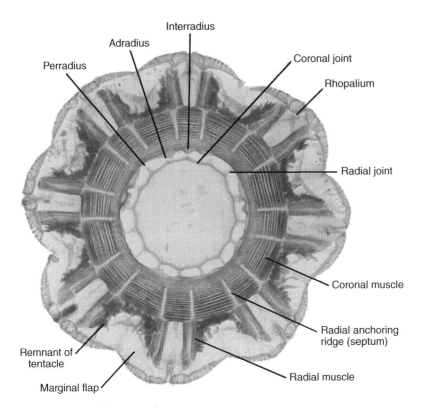

Figure 3.21 Oral surface of the subumbrella of *Cyanea capillata*. The oral arms and tentacles have been removed. *Source:* Gladfelter (1972), figure 1 (p. 151). Reproduced with the permission of Springer-Verlag.

The Mesoglea

The central role of the mesoglea ("or mesenchyme") in the biology of cnidarians is best expressed by the words of Chapman (1966): "the coelenterate has long been regarded as two layers of epithelium stuck to that something which is the mesoglea." Because it is largely acellular, the mesoglea has never been accorded the status of a germ layer, and its character, its importance, and even its name, differ among the cnidarian taxa. It can be regarded as the substance between the inner and outer epithelium. It assumes its greatest importance in the medusae where it acts variously as a source of buoyancy, an anchor for muscle fibers, a primitive skeleton, and even a source of nutrition.

The mesoglea consists of two or three components: fibers, a gel matrix, and where present, cells. Consensus on the chemical composition of the fibers is that they are composed of a collagen-like protein. Evidence supporting the collagen-like nature of the fibers comes from many sources including appearance in the electron microscope, thermal contractility, X-ray diffraction patterns, amino-acid composition, and histochemical staining. Overall, the evidence is quite convincing, and agreement is good among multiple authors (e.g. Arai 1997; Chapman 1966,

1974). The composition of the gel matrix is less well described, but the work of Gross et al. (1958) found mucopolysaccharides (glycosaminoglycans) associated with the collagen fibers, strongly suggesting a mucoprotein (proteoglycan)-based gel. Similar gels are found in the synovial fluid lubricating the joints in vertebrates and are also found along with collagen in the cornea of the vertebrate eye. Proteoglycans and glycosaminoglycans are often highly transparent and quite resilient (Lehninger 1975).

Collagen fibers of varying diameters cross the mesogleal layer of the medusan bell, conferring a memory and resiliency to its shape that is important to the locomotory process described above. Helical elastic fibers of uncertain composition that aid in maintenance of bell shape and elastic recoil have been described in a few species, e.g. *Chrysaora quinquecirrha* (Arai 1997). To put the relative stiffness of jellyfish mesoglea in perspective, in a discussion of the comparative stiffness of a variety of biological materials, Vogel (1988) observed that the mesoglea of a sea anemone was about 500 times more deformable than rubber. Rubber was about 1000 times more deformable than collagen (as animal tendon). Clearly, the ratio of collagen fibers to gel matrix is important in determining the overall resilience of the medusan bell.

Nerve Nets and Nervous Control of Swimming

Swimming in medusae is a rhythmic process that is controlled by neural pacemakers and communicated to the swimming musculature via a neural network known as a nerve net. Nerve nets in the cnidaria are at their most basic in polyps and comprise two grids, one net beneath the outer epidermal layer and one beneath the inner gastrodermal layer. Both nets are located between the outer epithelium and the mesoglea and are considered literally to be a diffuse network, with no polarity in signal propagation. That is, signals propagate equally in all directions.

In the hydromedusae, the inner and outer nerve nets have partially been consolidated into marginal nerve rings located at the inside (subumbrellar) and outside (exumbrellar) of the umbrellar margin (Figure 3.22). The inner nerve ring communicates with the swimming musculature as well as with the marginal sense organs and tentacles and governs the rhythm of the swimming musculature.

With the exception of the coronates, the scyphomedusae do not have the well-defined marginal neural rings of the hydromedusae. Nonetheless most of the action in their nervous system takes place at the umbrellar margin because that is where the sense organs and tentacles are located and, as in the hydromedusae, it is also where the swimming rhythm is generated. The neural networks are a bit more complicated in the scyphomedusae and considerably more is known about them.

The nervous system of scyphomedusae is composed of three parts: the motor nerve net, the diffuse nerve net, and the marginal centers (Arai 1997). The motor nerve net innervates the swimming muscle, the diffuse nerve net conveys sensory information to the marginal centers, and the marginal centers act as pacemakers for the swimming rhythm and integrators for the sensory information provided by the sensory apparatus. Even though only one of the nerve nets is termed "diffuse," both nerve nets are highly complex and diffuse networks that intercalate through different tissues. The neural tissue is difficult to isolate and even more difficult to map. In fact, the two nerve nets are mainly defined by their function, which was described using

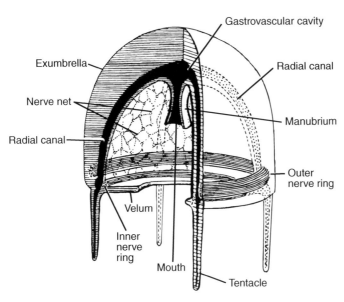

Figure 3.22 Nerve net in a hydromedusa. *Source:* Kaestner (1967), figure 4-20 (p. 61). Reproduced with the permission of John Wiley & Sons.

physiological methods (Anderson and Schwab 1981).

The marginal centers are important junctions for integrating sensory inputs and conveying them as needed to the swimming muscle. They are located directly behind the rhopalia on the bell margin, most easily visualized in the coronates (Figure 3.23a). Their precise location is unknown, but the fact that the centers are located very close to, but not within, the rhopalia itself was determined by methodical and highly localized ablation experiments in the 1980s (Passano 1982). The criterion for determining the location of the marginal center was the presence or absence of a pacemaker signal to the swimming musculature.

Senses and Sensory Mechanisms

The medusae possess at least three sensory modalities: photoreception, equilibrium, or balance – sometimes thought of as gravity reception, and chemoreception. Structures have been described for receptors detecting light and balance but not for waterborne chemicals – equivalent to our senses of taste and smell. The fact that medusae respond to chemicals of various kinds allows us to infer that the sense exists, even if there has been no structure identified to associate with it. Clearly, medusae have well-developed sensory capabilities.

The rhopalia are multifunctional sensory centers, usually possessing a photoreceptor and equilibrium receptor, or statocyst, within the same general structure (Figure 3.23). The photoreceptors vary in complexity from a simple pigment cup without a lens and a limited number of receptors such as that observed in the ocelli of *Aurelia aurita* (Figure 3.23d) to the well-developed ocelli of the cubomedusae that possess a cornea, a lens, and a well-defined retina (Figure 3.23e). Although structures believed to be photoreceptive in nature have been described in medusae since the 1940s, almost no neurophysiological recording has

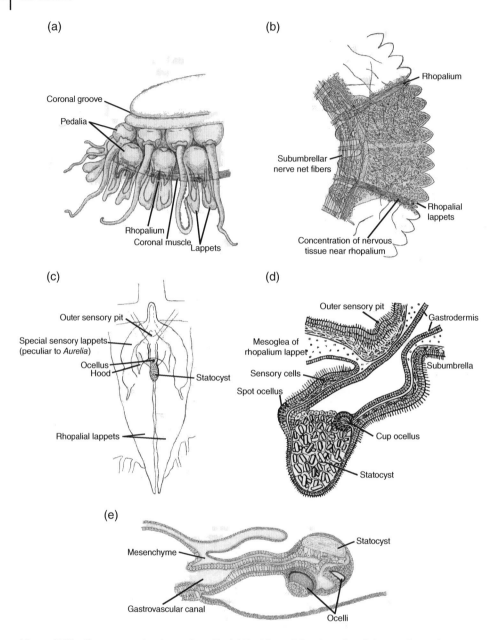

Figure 3.23 Sensory mechanisms: rhopalia. (a) Position of the rhopalia of the scyphomedusa *Atolla*, located between the marginal lappets. (b) Sector of the bell of *Rhizostoma*, showing nerve plexus and gastrovascular network (stippled); (c) rhopalium and surrounding structures in *Aurelia*. (d) Section of the rhopalium of *Aurelia aurita*; diameter at ocelli is 0.4 mm. (e) section of a cubomedusan rhopalium. *Sources:* (a) Redrawn from Maas (1904), plate IV; (b and c) Hyman (1940), figure 163 (p. 504); (d) Kaestner (1967), figure 5-4 (p. 91); (e) Redrawn from Mayer (1910), Vol III, plate 56.

been completed to directly confirm their sensory role. Fortunately, such recordings have been done in other primitive phyla (e.g. flatworms) from highly similar structures, and we may infer their photoreceptive function from those (Land 1990; Withers 1992). In addition, there are a variety of observed behaviors that require a sensitivity to light. Among those are diurnal vertical migration (Hamner 1995) and the orientation of cubomedusae to a point source of light several meters away. The sensory centers of medusae are distributed liberally around the bell margin. Thus even the sensitivity to light and shadow afforded by simple eyes can aid in navigation or alert the individual to the presence of predator or prey.

Statocysts, or equilibrium sensors, are an example of a class of sensory receptors known as mechanoreceptors. At their most sophisticated, mechanoreceptors detect vibration and sound using the same basic principles we observe in organs of equilibrium. The basic structure of a statocyst is depicted in Figure 3.24. In it is a dense body, or statolith, which can be thought of as a stone or concretion secreted by the animal. The sensory epithelium is made up of cells with hair-like projections ("hair cells") that are sensitive to deformation by the statolith. A change in position of the animal will change the position of the statolith on the sensory epithelium, providing information on attitude and equilibrium of the whole animal. Hair cells are present in virtually all types of mechanoreceptors, including those of our own inner ear.

The last type of sensory modality, chemoreception, has been inferred from the behavior of medusae, specifically by the orientation of medusae toward aggregations of prey or even to water that has been conditioned by the presence of prey (Hamner 1995). No specific structures associated with chemoreception have been identified yet, but the fact that cnidarians will show feeding behavior in response to chemical stimuli such as prey homogenates and the tripeptide reduced glutathione has been known for decades.

Sensory mechanisms provide an animal's windows into the physical world. The most telling evidence for the presence or absence of a sensory modality is in a species' behavior. Even if a sense such as touch does not have a discrete, obvious, and easily identified receptor, if a medusa responds to touch, e.g. by suddenly retracting its tentacles, the animal is obviously capable of discriminating touch. Even in more advanced species such as the vertebrates, sensory mechanisms exist that are not easily discriminated, those for heat and cold being two. As different open-ocean dwellers are described in further chapters, we will observe more sophisticated sensory organs in more sophisticated taxa. Sensory mechanisms, and neural processes in general, are exquisitely complex.

The Siphonophores

The Siphonophora comprise an order in the class Hydrozoa, subclass Hydroidolina, as shown in the classification scheme. They are quite distinct, differing from their hydrozoan brethren and the remainder of the pelagic Cnidaria in general morphology and basic organization. Three major biological groups are widely recognized within the Siphonophora, but there are disagreements about their relative taxonomic rank within the order. In this case we are following the World Register of Marine Species (WoRMS) and many decades of history in placing them each in their own suborder. The three suborders are the Cystonectae the Physonectae, and the Calycophorae

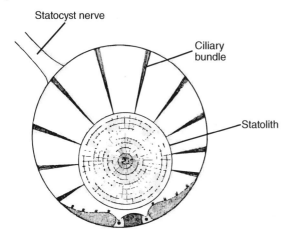

Figure 3.24 Basic statocyst structure showing the calcium carbonate statolith resting on sensitive sensory cells, which respond to changes in position of the statolith. *Source:* Tschachotin (1908), text figure 5 (p. 358).

(Figure 3.25). Altogether, the three suborders contain about 190 species, with the lion's share of them, about 60%, in the suborder Calycophorae. Suborders within the siphonophores are discriminated initially by the presence or absence of a gas-filled float or pneumophore. The suborders Cystonectae and Physonectae (Figure 3.25a–d) both have a gas-filled float, whereas the calycophorans do not (Figure 3.25e and f). The second division for classification is in the presence of swimming bells beneath the float. The physonects have swimming bells, or nectophores, beneath the float; cystonects do not. Table 3.5 is a synopsis of siphonophore classification. The siphonophores are more diverse at the family level than the rest of the Cnidaria.

Siphonophores are possibly the most confusing group in the animal kingdom. A free-floating individual siphonophore is considered to be a colony of various individuals working together to feed, reproduce, and move about, and within each siphonophore colony are individuals with either medusoid or polypoid affinities. Once the concept of a floating colony of individuals working together as a single entity is mastered, a lexicon of terminology replete with historical changes needs to be assimilated before life-history questions can be resolved. How does a colony develop from a single propagule? How do the various bits and pieces work together? Fortunately, some of the great minds in the history of biology have been fascinated by the group: Ernst Haeckel for example. And two very cogent reviews of the group by some of the best talent in gelatinous zooplankton biology (Mackie et al. 1987; Pugh 1999) are invaluable resources.

Terminology and Affinities of Siphonophore "Persons"

As Hyman (1940) notes, siphonophore colonies represent the highest degree of polymorphism in the Cnidaria. Terminology is critical to understanding siphonophores. Individuals within the siphonophore colony are known usually as "zooids" or "persons." To begin, it is best to differentiate between the polypoid and medusoid forms making up the individuals in the siphonophore colony.

Polypoid zooids comprise three basic types: the gastrozooids, the dactylozooids, and the gonozooids (Figure 3.26).

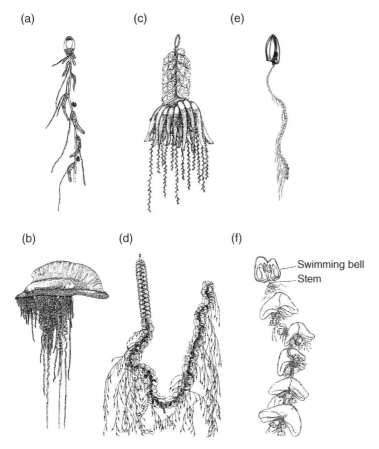

Figure 3.25 Examples of the three suborders of siphonophores. Cystonectae; (a) *Rhizophysa* and (b) *Physalia*, respectively. Physonectae; (c) *Physophora* and (d) *Agalma*, respectively. Calycophorae; (e) *Muggiaea* and (f) *Nectocarmen*, respectively. *Sources:* (a) Pugh (1999), figure 3.2 (p. 495); (b) Pugh (1999), figure 3.1 (p. 495); (c) Pugh (1999), figure 3.16 (p. 497); (d) Kaestner (1967), figure 4-35 (p.75); (f) Adapted from Alvarino (1983), figure 1 (p. 342).

Gastrozooids are the only members of a siphonophore colony that can ingest food and are sometimes called "siphons." The name siphonophore means "siphon-bearer," a siphon in Greek and Latin being a tube or pipe. Gastrozooids have a tubular polyp-like shape but no fringing tentacles at the mouth. Instead, a single long highly contractile tentacle emanates from the base with many side branches or tentilla (Figure 3.26a). The tentilla often terminate in distinctive structures thought to resemble the "prey of the prey" of the siphonophore, as well as in nematocyst batteries.

Dactylozooids resemble gastrozooids without a mouth (Figure 3.26b). They usually have a simple basal tentacle instead of one bearing tentilla and are often called palpons. Dactylozooids may sometimes resemble nothing more than a large, particularly robust tentacle, particularly when they are associated with gonozooids. In those cases they are known as gonopalpons.

Table 3.5 Classification of the order Siphonophora. *Source:* Totton (1965), A Synopsis of the Siphonophora, British Museum of Natural History.

	Family	Genus
Suborder Cystonectae		
	1. Physaliidae	*Physalia*
	2. Rhizophysidae	*Rhizophysa, Bathyphysa, Epibulia*
Suborder Physonectae		
	3. Apolemiidae	*Apolemia*
	4. Agalmidae	*Agalma, Halistemma, Cordagalma*
		Marrus, Moseria, Nanomia, Erenna, Lychnagalma
	5. Pyrostephidae	*Pyrostephos, Bargmannia*
	6. Physophoridae	*Physophora*
	7. Athorybiidae	*Athorybia, Melophysa*
	8. Rhodaliidae	*Rhodalia, Stephalia, Angelopsis, Archangelopsis, Dromalia*
	9. Forskaliidae	*Forskalia*
Suborder Calycophorae		
	10. Prayidae	
	Amphicaryoninae	*Amphicaryon, Maresearsia*
	Prayinae	*Rosacea, Praya, Prayoides, Lilyopsis, Desmophyes, Stephanophyses*
	Nectopyramidinae	*Nectophyramis*
	11. Hippopodiidae	*Hippopodius, Vogtia*
	12. Diphyidae	
	Sulculeolariinae	*Sulculeolaria*
	Diphyinae	*Diphyes, Lensia, Muggiaea, Dimophyes, Chelophyes, Eudoxoides, Eudoxia*
	13. Clausophyidae	*Clausophyes, Chuniphyes, Crystallophyes, Heterophyramis, Thalassophyes*
	14. Sphaeronectidae	*Sphaeronecties*
	15. Abylidae	
	Abylinae	*Ceratocymba, Abyla*
	Abylopsinae	*Abylopsis, Bassia, Enneagonum*

Gonozooids, which are polypoid in origin, are the structures that bear the reproductive gonophores, which are medusoid. They usually are branched stalks called gonodendra (Figure 3.26c)

Medusoid zooids comprise the swimming bells or nectophores (Figure 3.27), the bracts (Figure 3.28), the gonophores, and the gas-filled float or pneumatophore (Figure 3.29).

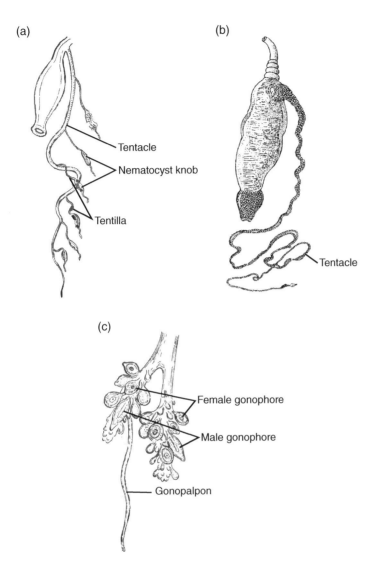

Figure 3.26 Siphonophore zooids. (a) Gastrozooid. (b) Dactylozooid. (c) Gonozooid (gonodendron). *Sources:* (a and c) Adapted from Hyman (1940), figure 148 (p. 470); (b) Bayer and Owre (1968), figure 89 (p. 54).

Nectophores come in a variety of different shapes, ranging from medusa-like to partially flattened rhomboids, to prismatic (Figure 3.27). All are muscular since they are the zooids responsible for the locomotion of the colony and must be able to contract forcefully to provide jet propulsion. Nectophores all have radial canals, usually four, which speaks to their medusoid origins.

Bracts also come in a wide diversity of shapes. Unlike the nectophores, which must retain some internal volume for the water that is ejected to provide jet propulsion for the colony, bracts are often solid chunks of jelly that are believed to serve a

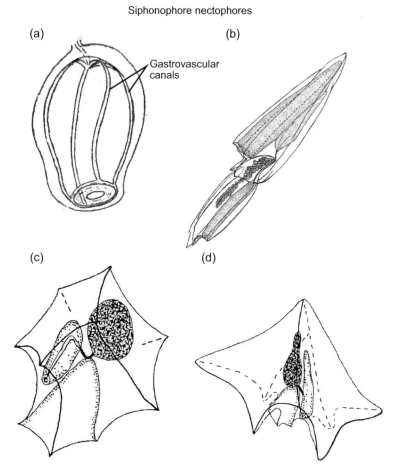

Figure 3.27 Siphonophore nectophores. (a) Basic unmodified nectophore. (b) Calycophoran, family Diphyidae; (c) Calycophoran, *Bassia bassensis*; (d) Calycophoran, *Enneagonum hyalinum*. *Sources:* (a) Hyman (1940), figure 148 (p. 470); (b) Adapted from Pugh (1999), figure 2 (p. 477); (c) Pugh (1999), figure 3.131 (p. 509); (d) Pugh (1999), figure 3.132 (p. 509).

protective function for the colony. They are robust and gelatinous; much of the time they are the only recognizable part of a formerly intact animal after it has been captured in a net. Shapes for bracts have been described as leaf-like, pyramidal, and helmet-shaped (Figure 3.28), and they often have remnant radial canals. They resemble the pieces of an exquisitely complex three-dimensional gelatinous jigsaw puzzle. The oceanographers able to classify siphonophores to species with just the pieces are to be regarded with awe!

Gonophores take many shapes, ranging from what appears to be an intact medusa with the sexual organs taking the place of a manubrium to the more rudimentary sacs observed in Figure 3.26c. Gonophores are dioecious, bearing only male or female gametes. The siphonophore colony as a whole is hermaphroditic; male and female gonophores may be found within the same cluster or may be borne separately.

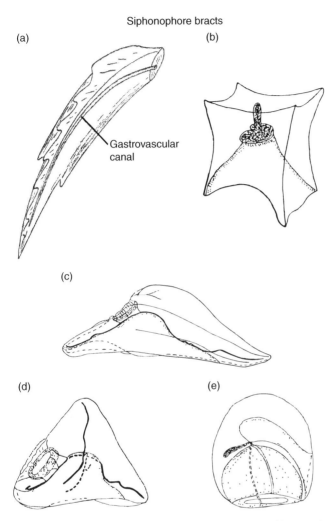

Figure 3.28 Siphonophore bracts. (a) Basic leaf-shaped bract, suborder Physonectae. (b) Calycophoran, *Enneagonum hyalinum*; (c) Calycophoran, *Nectopyramis natans;* (d) Calycophoran, *Nectopyramis thetis*; (e) Calycophoran, *Sphaeronectes gracilis*. *Sources:* (a) Hyman (1940); figure 148 (p. 470); (b) Pugh (1999), figure 3.140 (p. 510); (c–e) Pugh (1999), figures 3.141, 3.142, 3.143 (p. 500).

Pneumatophores are the gas-filled floats of the physonects and cystonects, and like the nectophores, bracts, and gonophores, they also are medusoid in origin. As observed by Hyman (1940), a pneumatophore may be thought of as an inverted medusan bell. The outer wall is termed the pneumatocodon, and the inner subumbrellar wall is termed the pneumatosaccus. The usual mesoglea that would reside between the inner and outer layers of the bell is absent, but the walls of the pneumatocodon and pneumatosaccus retain the usual epidermal and gastrodermal cell layers (Figure 3.29). Inner and outer walls are very robust, as would be expected for a structure that must retain gas under pressure. The air sac is reinforced with a lining of chitin. At

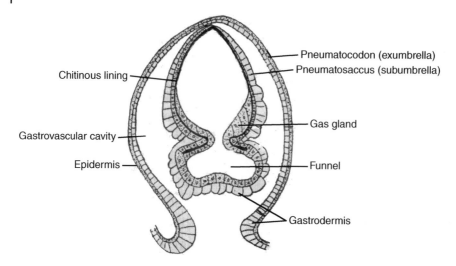

Figure 3.29 Vertical section of the simple type of pneumatophore of *Agalma*. *Source:* Adapted from Woltereck (1905).

the bottom of the air sac is a gas gland that secretes the gas into the pneumatophore. The composition of the gas is 85–95% nitrogen, 1.5% argon, and 7.5–13.5% oxygen or roughly equivalent to air. Figure 3.29 shows the simple pneumatophore of *Agalma*, but they do get more complex, sometimes adding reinforcing septa between the inner and outer walls of the air sac and more highly developed gas glands. In some genera, *Rhyzophysa*, for example, an apical pneumatophore can be utilized to dump the gas within the bladder. Such pores are held tightly shut with a sphincter muscle when present.

Whole Animal Organization

All zooids of the siphonophore colony are budded from and contiguous with a central stem (Figure 3.30). The stem is a robust tube that varies from a couple of centimeters to several meters in length depending on the species and is capable of expanding and contracting. It has a typical cnidarian two-layered structure with a thick mesoglea and a central canal (Figure 3.30a). Canals of the siphonophore persons are all continuous with the central canal of the stem. The stem canal originates at the apex of the colony, either as a somatocyst (Figure 3.30b), a rounded end in the calycophorae, or as a continuation of the gastrovascular canal of the float (Figure 3.30c). Note the continuity of the stem canal with that of the persons in Figure 3.30c. In some species such as *Physalia*, the Portuguese man-o-war, the "stem" is really more of a disk that is contiguous with the wall of the float and the zooids are budded from beneath it (Figure 3.30d).

The stem is divided into two sections, the nectosome and siphosome (formerly siphonosome). In physonect siphonophores (Figure 3.25c and d), the nectosome extends from the base of the float to the bottom of the swimming bells or nectophores. In the calycophorans, lacking a float, the nectosome is apical (Figure 3.25e and f). The cystonects, which lack swimming bells altogether (Figure 3.25a and b), have no nectosome at all.

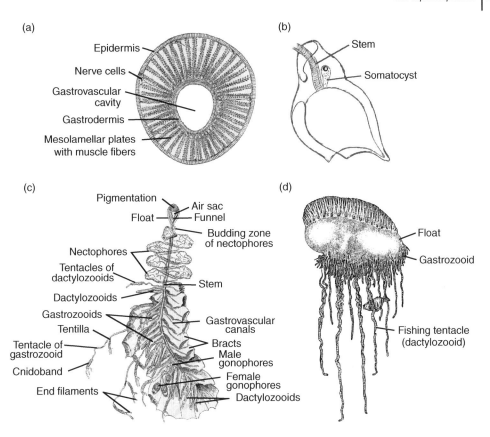

Figure 3.30 Siphonophore colony structure. (a) Cross section of a siphonophore stem; (b) helmet-shaped bract. (c) Colony of the physoncect *Agalma*. (d) Colony of the cystonect *Physalia*. *Sources:* (a) Hyman (1940), figure 151 (p. 476); (b) Hyman (1940), figure 148 (p. 470); (c) Redrawn from Mayer (1900), plate 32; (d) Adapted from Brusca and Brusca (1990), figure 9 (p. 224).

The remaining portion of the siphonophore stem is termed the siphosome. It makes up the majority of the stem in most species of physonects and calycophorans and the entire stem in the cystonects. Nectophores are budded from a zone of proliferation just below the float in the physonects. Similarly, the siphosome proliferates from a budding region just below the nectosome, producing repetitive groups of medusoid and polypoid persons called cormidia (Figure 3.31). Within the Calycophorae, a cormidium often consists of a single gastrozooid, to the base of which is attached a tentacle along with one to several gonophores (Pugh 1999). Overlying the cormidium is a bract that has a characteristic shape and is often useful in identifying species (Figure 3.31c). Cormidia within the physonects are larger and more complex, usually possessing several leaf-like bracts, a gastrozooid, dactylozooids, and clusters of gonophores. The cystonects have simple cormidia consisting of a gastrozooid and clusters of gonophores radiating from a gonodendron or stalk. The cystonects have no bracts.

In the calycophorans, the terminal mature cormidia on the stem break away to form free-swimming "eudoxids." Until their true origins were resolved, eudoxids were thought to be a distinct group within the siphonophores (Hyman 1940). Limited propulsion within the eudoxids is achieved by the gastrozooids pinch-hitting as nectophores. Shedding of cormidia for a brief free-swimming existence is limited to the Calycophorae; neither the physonects nor the cystonects produce eudoxids.

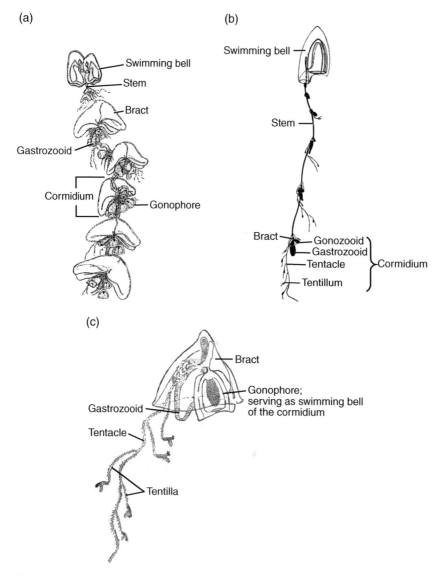

Figure 3.31 Siphonophore cormidia. (a) Calycophoran *Nectocarmen antonioi*. (b) Diphyid *Muggiaea* sp. (c) Enlarged cormidium of *Muggiaea*. *Sources*: (a) Adapted from Alvarino (1983), figure 1 (p. 342); (b) Kaestner (1967), figure 4-34 (p. 74); (c) Adapted from Hyman (1940), figure 150 (p. 174).

Life Histories

Development proceeds from a planula larva and differs between groups. In the physonects, the planula becomes a *siphonula* larva as summarized in Figure 3.32 from Mackie (1986). The two budding zones produce the nectophores and the gastrozooids, gonophores, and bracts as the stem elongates during development. Development in the Calycophorae is somewhat different, with the initial production of a larval nectophore followed by development of the stem and gastrozooid. The larval nectophore is retained in some taxa (e.g. *Sphaeronectes*) and lost in others (e.g. *Hippopodius*). Early development in the cystonects is currently undescribed but defined budding zones beneath the float have been identified in both *Physalia* and *Rhyzophysa*.

The Siphonophore Conundrum

Is a siphonophore an individual or a colony? Siphonophores develop from a single egg, differentiating during development into a complex organism that is made up of parts that at some point in the distant past were probably free-living polyps and medusae. How they came together and whether they are best considered as a colony or individual are questions that have piqued the interest of such luminaries as Ernst Haeckel, T. H. Huxley, E. O. Wilson, and S. J. Gould (cited in Mackie et al. 1987). However, for ease of understanding based on a lifetime of perspective, the observations of G. O. Mackie himself are the most cogent. In Mackie et al. (1987), he observed that "siphonophores . . . are linear in form with little branching, and are polarized, with a distinct anterior end. They are also bilaterally symmetrical. They grow by

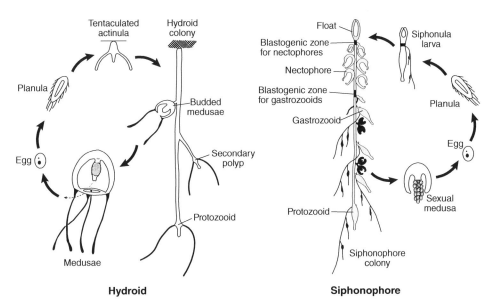

Figure 3.32 Comparison of the life cycle of an agalmid siphonophore with that of an athecate hydroid. *Source:* Reproduced from G. O. Mackie (1986), From Aggregates to Integrates: Physiological Aspects of Modularity in Colonial Animals, *Philosophical Transactions of the Royal Society of London Series B, Biological Sciences*, 1986, Vol 313, No. 1159, page 179, by permission of the Royal Society.

addition of modules at localized growth zones. The result is a high degree of determinacy of form." Earlier (Mackie 1963), he offered this elegant observation.

They have developed colonialism to the point where it has provided them with a means of escaping the diploblastic body plan. The higher animals escaped these limitations by becoming triploblastic using the new layer, the mesoderm, to form organs. The siphonophores have reached the organ grade of construction by a different method – that of converting whole individuals into organs.

It is important to understand the concept that siphonophores function as highly coordinated individual organisms and not as a loosely collected gaggle of different zooids. Natural selection acts on the whole individual.

Feeding

Fishing Behavior

Unlike many of the medusae, siphonophores feed while drifting motionless, creating a "curtain of death" with their extended tentacles. Prey must blunder into the "curtain" for siphonophores to feed. Most calycophorans and physonects spread their tentacles for short bouts of swimming (Figure 3.33) followed by longer periods of drift, usually several minutes in larger species. The frequency of swimming episodes for the purpose of tentacle-reset scales roughly inversely with size, with the larger

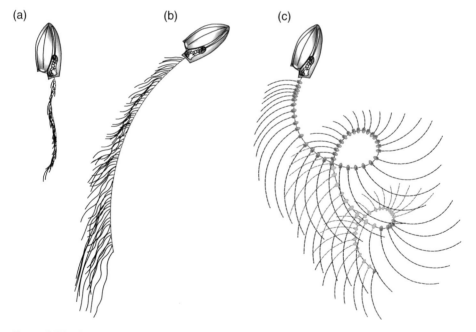

Figure 3.33 Cyclic fishing behavior and optimization of feeding space. (a) The Calycophoran *Muggiaea atlantica* begins a feeding cycle by swimming upward with contracted stem and tentacles; (b) it releases the stem and curtain of tentacles and swims in a circle; (c) swimming stops, with fully extended stem and tentacles; the stem then sinks under its own weight creating a corkscrew shape that slowly moves downward through the water column until the stem is vertical again, where it contracts, and the cycle begins again. This deployment cycle results in the fishing array moving through a maximum volume of water.

physonects swimming every few minutes and smaller calycophorans every minute or less. Species with large siphosomes, e.g. *Apolemia* at 20 m or greater, assume a horizontal position with the tentacles hanging down in a curtain. Swimming is less important for spreading the tentacles in the longer species (Mackie et al. 1987). Changes in the tentacle net happen gradually as structures sort themselves by density, sinking or rising with time. In some physonects, prey are tempted by periodic contractions of tentilla, mimicking the motion of zooplankters. In some cases, structures on the tentacle such as nematocyst batteries resemble "prey of the prey" and bring unwary prey species within the kill zone.

Digestion

When prey contacts a tentacle, only the tentacles attached to the prey contract to bring the prey to the mouth of the gastrozooid. The rest of the colony continues fishing. Gastrozooids work individually or in teams. If the prey is too large to be completely engulfed by a single zooid (e.g. a fish in the case of *Physalia*), more zooids will be recruited to finish the job. Presumably, the additional zooids are joining the job in response to chemical signals released by the captured prey. Digestion of small prey items such as copepods can be quite rapid: 1–2 minutes (Mackie et al. 1987).

As is the case with other cnidarian groups, digestion of prey is achieved using both extracellular and intracellular mechanisms. It takes place primarily in the gastrozooid, where breakdown of food is accompanied by rhythmic contractions. Particles created during the process of extracellular digestion are taken up via phagocytosis by the endodermal cells of the gastrozooid or, in some cases, can be passed along to palpons via the contractions noted above and taken up there. Further processing takes place in the endodermal cells of both structures. Eventually the cells slough off from the endoderm and disintegrate, allowing the nutrients to be absorbed by the rest of the colony via the stem canal. Indigestible parts of prey are ejected from the gastrozooids the same way they entered, out of the mouth with the help of basal contractions. Palpons expel particulates through a pore in their tip.

Diets and Selectivity

Diets of siphonophores correlate roughly with the suborders. Calycophorans mainly consume small copepods, whereas physonects eat larger copepods as well as larger soft and hard-bodied prey such as crustacean larvae, amphipods, ostracods, and pteropods. Cystonects appear to specialize on fish larvae. Even *Physalia*, though capable of taking fish greater than 4 cm in length, feeds primarily on fish larvae 2–20 mm in size (Purcell 1984). Overall, despite *Physalia*'s well-deserved reputation as potent stingers, they appear to favor small weak swimmers as prey.

Some siphonophores appear to capture particular prey species at a greater frequency than would be expected by the abundance of those prey items in the zooplankton community. To explain the apparent selectivity, we need to use the same principles of prey capture that were applied to medusae in the Madin (1988) model described above. Selectivity can be influenced by effectiveness of nematocysts in retaining prey as well as the probability of large vs. small prey encountering a tentacle.

Table 3.6 summarizes ingestion data from a variety of species in a variety of habitats. The take-home lesson is the substantial impact of siphonophores as predators. In sufficiently high concentrations, siphonophores are capable of enormous local impacts on the zooplankton community.

Table 3.6 Ingestion rates of Siphonophores. *Source:* Adapted from Mackie et al. (1987), table 15 (p. 238), with the permission of Academic Press (Elsevier).

Siphonophore	Location	Prey type	Prey abundance	Feeding rate per individual · d^{-1} (carbon content; caloric content)	% of prey items in diet	% of prey population consumed · d^{-1}	References
Physalia physalis	Gulf of Mexico	Fish larvae	0.2 m^{-3}	Avg. 120 prey	94.1	60	Purcell (1984)
Rhizophysa eysenhardtii avg. 8 gastrozooids	Gulf of California	Fish larvae	Avg. 28 m^{-3}	Avg. 8.8 prey (7300 µg C; 107 cal)	100	28	Purcell (1981a)
Sphaeronectes gracilis 38.5 ± 9.6 gastrozooids	Southern California	Copepods	Avg. 250 m^{-3}	8.1–15.5 prey (3.9–6.2 µg C; 0.06–0.09 cal)	100	2–4	Purcell and Kremer (1983)
Muggiaea atlantica avg. 22 gastrozooids	Friday Harbor, WA	Copepods	Avg. 9121 m^{-3}	5.5–10.5 prey (2.6–4.2 µg C; 0.03–0.05 cal)	100	0.1–0.2	Purcell (1982)
Rosacea cymbiformis avg. 40 cormidia/colony	Gulf of California	Copepods	Avg. 1495–1695 m^{-3}	89.4 prey (616–2068 µg C; 9.4–31.5 cal)	75.4	8	Purcell (1981b)

Those studies that have addressed their local influence as predators (Purcell 1982; Purcell and Kremer 1983) concluded that siphonophores were the most important gelatinous predators on copepods in two quite disparate locations: Friday Harbor in Washington and Catalina Island in the California Borderland.

Ecological Importance

Even for gelatinous species, siphonophores are exceptionally difficult to enumerate owing to their delicate colonial structure. Nets tend to reduce them to fragments; those fragments are difficult to quantify as numbers of individuals. The most appropriate techniques for evaluating numbers of siphonophores are mainly visual counts, either from diver-based observations or for deeper-living species, using submersible-based observations either manned or unmanned (Remotely Operated Vehicles – ROV's, or AUV's – Autonomous Underwater Vehicles), Techniques include using diver-powered meter hoops and flowmeters, counting individuals as they passed through the hoops (Purcell 1981a, b) and, in a variation of the same theme, using larger (5 m × 5 m) grids towed behind a slowly moving boat while divers count (Biggs et al. 1981, 1984). Options using submersibles mainly include evaluating nearest-neighbor distances (Mackie and Mills 1983) and mounting a hoop in the front of the submersible.

In the open ocean, siphonophores are found at densities of less than $1/1000 \, m^3$. However, they can number 5–$10 \, m^{-3}$ (Table 3.7) in more productive coastal regions (Mackie et al. 1987). Most often they are outnumbered in the open ocean by ctenophores, but their coastal numbers can exceed those of other gelatinous forms.

Locomotion

Siphonophores differ considerably, by sub-order, in their ability to move about. The cystonects, which have an apical float but no swimming nectophores to aid in propulsion, are limited to contracting and relaxing their stem for movement. Consequently, their swimming ability is weak at best. However, members of the Cystonectae, particularly Physalia, are drifters par-excellence (Totton 1960; Woodcock 1971; Mackie 1974), using their float as a sail to cruise the open sea. Physalia has two basic morphs, a left-hand sailor and right-hand sailor, mirror images of one another that sail 45° to the left and right of the wind direction (Figure 3.34). The crest of the float provides an important part of the sail; its curvature and stiffness may be adjusted to form a characteristic "sailing posture" for most effective movement (Mackie 1974).

Many of the physonects are capable swimmers, combining a float for buoyancy and a battery of nectophores for propulsion. The genus that has received the most attention is *Nanomia,* a capable swimmer often observed from submersibles. Mackie et al. (1987) described three swimming modes:

1) Synchronous forward swimming, usually considered an escape response to stimulation of the siphosome, where all the nectophores contract together for one or two cycles and produce a velocity of 20–$30 \, cm \, s^{-1}$, a respectable velocity for any small swimming species.
2) Asynchronous forward swimming, where the nectophores contract in a less coordinated fashion, producing a forward velocity of 8–$10 \, cm \, s^{-1}$. Sometimes the nectophores on each side of a column contract rhythmically to produce a side-to-side forward movement.
3) Reverse swimming, usually resulting from a stimulation to the float. In this

Table 3.7 Abundance of siphonophores relative to other gelatinous zooplankton, estimated primarily by direct observation. *Source:* Mackie et al. (1987), table 17 (pp. 242–243), with the permission of Academic Press (Elsevier).

Location	Siphonophores			Medusae	Ctenophores	Method	References
	Physonects	*Calycophorans*	*Total*				
Gulf of Mexico (no./10000 m³)							
Spring (4)[a]	2.0±3.4	0		1.7±2.2	10.2±9.9	Grid	Biggs et al. (1984)
Summer (8)	0	0		0	0	Grid	Biggs et al. (1984)
Fall (11)[b]	8.8±7.1	<0.8±1.0		<8.7±20.7	36.6±51.0	Grid	Biggs et al. (1984)
Winter (8)[c]	4.1±2.3	2.6±2.1		3.4±3.5	31.2±53.8	Grid	Biggs et al. (1984)
North Atlantic (no./10000 m³)	*Total*	*Calycophorans*					
Temperate (4)	15.2±24.5	27%[d]		14.2±13.4	38.5±31.9	Grid	Biggs et al. (1981)
Transitional (6)	6.2±10.4	62%[d]		11.5±9.7	24.7±36.1	Grid	Biggs et al. (1981)
Subtropical (7)	8.0±4.5	90%[d]		2.6±3.0	2.1±3.2	Grid	Biggs et al. (1981)
Bahamas (no./10000 m³)			Total 8.3±13.1	9.0±7.8	52.3±59.9	Grid	Biggs et al. (1981)
Southern California (no./m³)	*Polygastric*	*Eudoxids*					
Sphaeronectes gracilis	8.3±4.6	6.3±5.4		4.6±1.7	2.5±1.5	Net	Purcell and Kremer (1983)
Muggiaea atlantica	6.7±3.7	17.3±11.5		4.6±1.7	2.5±1.5	Net	Purcell and Kremer (1983)
Apolemia sp. at 450 m	0.2±0.02	—		0.06	0.05±0.02	Submersible	Alldredge et al. (1984)
Strait of Georgia (no./m³)	*Polygastric*	*Eudoxids*					
Muggiaea atlantica (October)	1.4±0.8	8.0±3.8		0.7±0.5	0.1±0.2	Net	Purcell (1982)
Muggiaea atlantica (November)	2.7±1.8	7.2±4.1		1.4	0.08	Net	Mackie and Mills (1983)
Muggiaea atlantica (November)	2.0±2.6	Common		2.9–10	0.8–5.5	Submersible	Mackie and Mills (1983)

Numbers in parentheses following Gulf of Mexico seasons and North Atlantic zones are the number of SCUBA dives.
[a] *Agalma, Physophora.*
[b] *Forskalia, Agalma, Athorybia, Cordagalma, Rosacea, Stephanophyes.*
[c] *Forskalia, Agalma, Nanomia, Cordagalma.*
[d] *Diphyes, Chelophyes, Rosacea.*

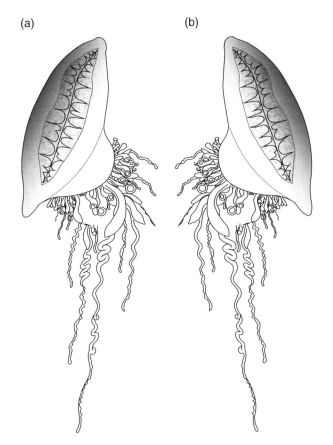

Figure 3.34 Dimorphism of *Physalia*. (a) Right handed (left sailing); (b) left handed (right sailing).

case contraction of muscles at the velar openings directs the jet anteriorly, propelling the siphonophore to the rear.

The calycophorans are the Olympic swimmers of the siphonophores. They lack a pneumophore and are often much smaller than the cystonects and physonects, but are, with a few exceptions, very capable swimmers. In the diphyids particularly, the nectophores have powerful swimming muscles and face directly backward, resembling a miniature spacecraft (Figure 3.27d). Table 3.8 gives the swimming speeds of several siphonophores. The diphyids may cruise using only the smaller posterior nectophore (Bone and Trueman 1982), keeping the larger anterior nectophore in reserve for escape. Keep in mind that for small species such as *Chelophyes*, the escape swimming velocities are many body lengths per second.

Buoyancy

As Mackie (1974) observed, "Cnidarians are fishermen, not hunters." As a group, the siphonophores are the most committed to a fisherman's lifestyle, the best examples being the very large species such as *Apolemia*, which are several meters in length. When observed from a submersible with their feeding tentacles extended, they even resemble a drift net, ambushing their prey in the dimly lit waters of the mesopelagic zone. Flotation is therefore

Table 3.8 Swimming speeds (cm s^{-1}) of siphonophores as estimated in situ by divers. *Source:* Field Studies of Fishing, Feeding, and Digestion in Siphonophores, D. C. Biggs, Marine & Freshwater Behaviour & Physiology, 1977, table 1 (p. 263). Reprinted by permission of the publisher (Taylor & Francis Ltd, http://www.tandfonline.com).

Species	Number measured	Undisturbed ("normal" speed)	Escape speed
Agalma okeni	27	2–5	10–13
Nanomia bijuga	2	—	25
Forskalia spp.	10	1–3	2–5
Stephanophyes superba	3	10–15	—
Rosacea cymbiformis	5	1–3	3
Sulculeolaria monoica	5	2–5	12–16
Chelophyes appendiculata	6	7–16	23
Diphyes dispar	3	1–3	5–10

particularly important to the siphonophores' lifestyle. Neutral or positive buoyancy is achieved in two ways: by the use of floats, as in the cystonects and physonects, and by exclusion of heavy ions, as in the calycophorans.

The cystonects are entirely dependent on their float for buoyancy (Figure 3.35a). For genera such as *Rhysophysa* that remain in the midwater, the ability to adjust their position in the water column may come either from secretion of gas into the float or release of gas from it. An apical pore in the float allows for gas release. In *Physalia*, the large pneumophore is kept inflated as it fishes from the surface. In both species the gas within the float is carbon monoxide, secreted by gas glands intimately associated with the float.

The physonects depend to a varying degree on the float for buoyancy. Streamlined genera such as *Nanomia* that have much of their mass in the nectophores and little gelatinous tissue below them (Figure 3.35b) are apparently quite dependent on the float to maintain their position in the midwater, being negatively buoyant without it. In contrast, genera such as *Agalma* with considerable gelatinous tissue below the nectophores (Figure 3.35c) depend to a much greater degree on the lift provided by the gelatinous tissue.

Calycophorans, having no float, must swim constantly, sink, or rely on their own tissues for lift. Jacobs (1937) first demonstrated that the gelatinous tissues of siphonophores provided lift, and later studies (Robertson 1949; Bidigare and Biggs 1980) provided the mechanism: selective replacement of heavier ions within the tissue, notably the replacement of sulfate with chloride. The lift provided by sulfate exclusion is sufficient in siphonophores and other nearly neutral gelatinous species to enable station-keeping in the water column. It is worth noting that other important pelagic groups, such as the Crustacea (e.g. Sanders and Childress 1988), also exploit sulfate exclusion as a buoyancy mechanism.

Vertical Distribution

As truly open-ocean species, siphonophores are found throughout the water column from the pleustonic surface layers to very great depth (8000 m, Vinogradov 1970). Though vertical distributions for individual

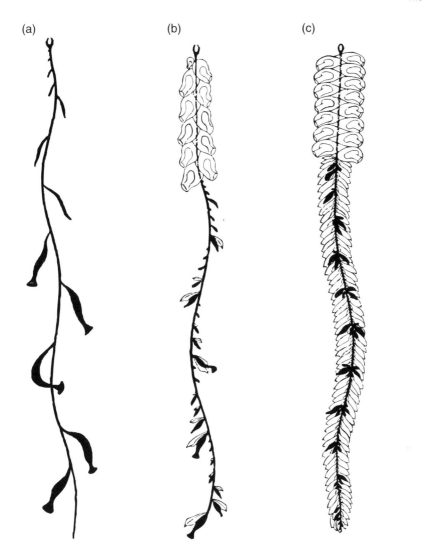

Figure 3.35 Degrees of dependence on gelatinous parts for flotation. (a) *Rhizophysa filiformis* is completely dependent on the apical float for support and lacks gelatinous parts; (b) *Nanomia bijuga* has gelatinous nectophores and bracts but is largely dependent on the float; (c) *Agalma elegans* is buoyed mainly by gelatinous parts with the float only supporting a short anterior portion of the stem. *Source:* Jacobs (1937), figure 7 (p. 594).

species can be extensive, most siphonophores exhibit depth ranges over which they are typically most abundant. Those ranges are the same as seen with most other open-ocean taxa: epipelagic (0–250 m), mesopelagic (250–1000 m), and bathypelagic (>1000 m) (Pugh 1999).

Siphonophores are quite delicate, particularly prone to damage from midwater trawls and even gentle plankton nets. The damage usually involves breaking the animal into its constituent parts, and with particularly delicate forms, the damage may preclude identification altogether. Delicate forms can literally pass through the meshes of a scientific net. Thus, conclusions drawn from a net sampling survey will tend to be biased toward the more robust species. Surveys done primarily with visual observations, by either SCUBA or submersible, will also have bias, usually toward slower moving, larger species. Both types of sampling have their shortcomings, but without question, far more sampling has been done with nets. Trawling techniques have been around longer, trawls sample greater volumes, and they are far cheaper to execute. It is important to recognize the pros and cons of all sampling methodologies and to take away from each what is most useful.

Pugh (1999) observed that calycophophoran species tend to dominate in net samples, but that over 70% of the specimens collected by submersibles are the larger, more delicate, and more highly pigmented physonects. An interesting, though disturbing, corollary to his observations on relative numbers was that about half of the physonects collected by submersibles were new to science. Clearly there remains a lot to learn about the siphonophores.

Data from Pugh (1999) for 93 widely distributed Atlantic siphonophore species are summarized in Table 3.9. Forty-one species are considered to be mainly epipelagic (0–250 m), 17 are epipelagic–upper mesopelagic (<100 m to >250 m), 31 are mesopelagic (200–1000 m), and 4 are bathypelagic (>1000 m). The best information to date suggests that the majority of siphonophores reside in the most productive upper 250 m of the ocean.

Diurnal Vertical Migration

As discussed, swimming ability within the siphonophores varies widely. The idea that siphonophore populations move up to the surface at dusk and back to depth at dawn in response to the waning and waxing illumination in near-surface waters is not a compelling one, especially for the suborders with more limited mobility. Nonetheless, there is a substantial amount of data for many species that suggest precisely that. Moore (1949, 1953) reported vertical excursions of 30–40 m for many species of calycophorans in both the Florida current and in the vicinity of Bermuda on a day–night basis. Similar results were obtained by Musayeva (1976) for calycophorans in the Sulu Sea.

An alternative to a directed vertical migration triggered by sunset and sunrise is a slowly undulating change in vertical profile over a 24-hour period (Pugh 1977). Siphonophores gradually move up and

Table 3.9 Vertical distribution of Siphonophore species in the South Atlantic. *Source:* From the data in Pugh (1999).

	Number of species		
Primary depth range	Order Cystonectae	Order Physonectae	Order Calycophorae
Epipelagic (0–250 m)	3	4	34
Epi-upper Mesopelagic (<100 m to >250 m)	0	5	12
Mesopelagic (200–1000 m)	0	7	24
Bathypelagic (>1000 m)	0	0	4

down with the changing photoperiod. This sinusoidal pattern of migration would explain the unusual depth profiles obtained for some weakly swimming species such as *Hippopodius hippopus*. The theory is still somewhat speculative.

Without question, a changing vertical distribution over the diel cycle is a characteristic of many siphonophore species. However, even among the calycophorans the vertical excursions are usually quite limited in scope (<50 m), a situation to be expected in an order that is morphologically adapted more for ambush predation than long-distance swimming. Because of their float, the physonects are not only good acoustic targets, they face the same problems that fish with swimbladders do when moving vertically: expansion and compression of gas in their flotation system when moving up and down in the water column.

Geographical Distribution

Siphonophores are pan-oceanic; they are found from the equator to the polar oceans in all the major ocean basins. Mackie et al. (1987) concluded that siphonophore distributions coincide reasonably well with the biomes and provinces described in the biogeography literature (e.g. Briggs 1995; Longhurst 1998), which in turn coincide with the earth's major climatic zones. The best data describing faunal relationships with latitude and longitude come from Atlantic waters. Margulis (1976) recognized seven faunal groups among the siphonophores of the Atlantic, which correspond well to the latitudinal zones of Briggs (Figure 3.36): Arctic species, Northern Boreal species (cf. cold temperate), Antarctic species, Bipolar species (found in the Arctic and Antarctic), Tropical species (which include equatorial and warm temperate components), Eurybiotic species (those which live in all biogeographic areas), and Neritic species. An excellent compendium of Atlantic species' latitudinal ranges may be found in Pugh (1999). Most (60 of 96 species) were very wide-ranging, with distributions encompassing 80° of latitude or more, roughly half on each side of the equator.

Mackie et al. (1987) provide a summary of diversity and numbers for 21 common species of calycophoran siphonophores in the North Atlantic. The data show a peak in both numbers and diversity at about 18°N with a gradual decline in species numbers further north. A second peak in abundance is obvious between 40 and 53°N.

Organization
and Sensory Mechanisms

No sensory apparatus has been detected in the siphonophores, i.e. no ocelli, statocysts, or mechanoreceptors such as those observed in the medusae. However, siphonophores are sensitive to touch, light, chemicals, and, in some cases, to waterborne vibration. How? It is likely that the nerves themselves act as receptors although mechanisms effecting the receptor-like responses are undescribed.

Two types of conduction are recognized in the siphonophores, epithelial and neural. Both contribute to coordinated movement and responses. Epithelial conduction is similar to the spread of depolarization in myogenic hearts and is present in the nectophores of physonects and calycophorans. Epithelial conduction was effectively demonstrated in *Nanomia* when its nectophores remained coordinated after severing their nervous connection to the stem (Mackie 1964).

Epithelial conduction as a mechanism for propagating impulses is confusing at best. There is no obvious morphological distinction between epithelia that are capable of conduction and those that are not. Before

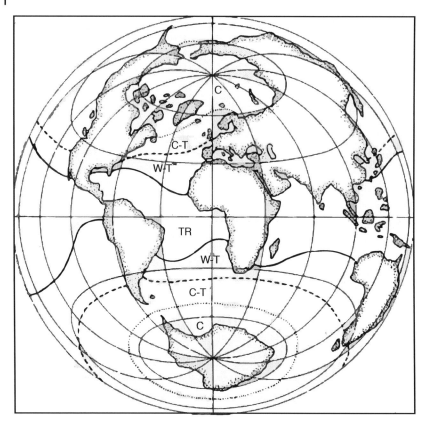

Figure 3.36 The four major temperate zones of the ocean surface. The Tropical Zone (TR) is delimited by the 20 °C isotherm for the coldest month. It is bordered by a Warm-Temperate Zone (W-T), a Cold-Temperate Zone (C-T), and a Cold or Polar Zone (C). *Source:* Adapted from Briggs (1995), figure 55 (p. 209).

the techniques of neurophysiology were available, the presence of epithelial conduction was inferred by the absence of nerves coupled with the presence of coordinated activity (Mackie et al. 1987). Presumably, the epithelial cells themselves possess the membrane channels that are necessary for ion movement and signal propagation. Since all animal cells maintain an ionic disequilibrium with the medium bathing them, whether that medium is seawater or blood, the basic ion transport "equipment" is already in place within the membrane. The signal must spread from cell to cell to function as a conductive pathway. How it happens is still undescribed.

Epithelial Conduction vs. Neural Conduction

Neural pathways exist both in the zooids and in the stem of siphonophores. Communication takes place between them, affording a primitive centralization of coordination. In fact, the neural tissue in the stem of some physonects and calycophores has coalesced to form giant axons that run along the midline of the stem (Mackie et al. 1987). Figure 3.37 is a "wiring diagram" that nicely explains the neural organization of a physonect, including a visualization of the epithelial pathways.

Conductive pathways are more easily defined in the Siphonophora than in the hydromedusae and scyphomedusae. This is

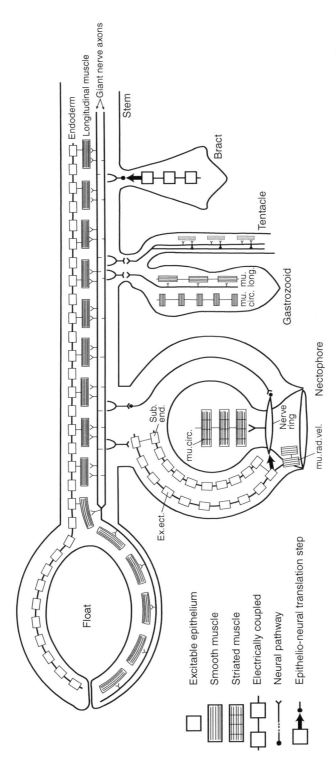

Figure 3.37 Simplified wiring diagram of a physonectid siphonophore. Only ectodermal nerve pathways are included. ex. ect., exumbrellar ectoderm; mu. circ., circular muscle; mu. long., longitudinal muscle; mu. rad. vel., radial muscle of velum ("fibres of Claus"), sub. end., subumbrellar endoderm. *Source:* Adapted from Mackie et al. (1987), figure 35 (p. 187).

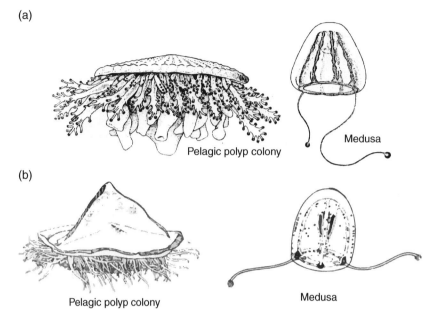

Figure 3.38 Porpitidae. (a) Pelagic polyp colony and medusa of *Porpita porpita*. (b) pelagic polyp colony ("by-the-wind-sailor") and medusa of *Velella velella*. *Sources:* (a) Bouillon (1978); (b) Bouillon (1984).

partially because there has been more neurophysiological research on that group but is also because the neural network in siphonophores is less diffuse, unlike the medusae with their multiple nerve nets.

The Cnidaria Formerly Known as Chondrophora

Classification

The Chondrophora is a legacy taxonomic ranking for a unique group of hydrozoans commonly known as "By-the-Wind Sailors." The name is still widely used to refer to them and will be used here. There are two genera delineating the two morphs within the group: with and without sails (Figure 3.38). Together they form the Porpitidae, a family now classified within the suborder Capitata in the order Athecata and subclass Hydroidolina.

Before any of the present schemes, the chondrophora were considered to be highly specialized members of the Siphonophora. A glance at Figure 3.38 will show you why. They are small colonies of zooids attached to the underside of a chitinous float and are most often encountered in flotillas of varying size. The flotillas form a mini-ecosystem that is exploited by a number of molluscan species taking advantage of the "moving island" as a substrate and source of food. A chance encounter with a raft of *Velella* is one of the real treats of being at sea.

The reason for classification changes over the years is a shift in what was considered the dominant life stage. If the medusa is considered as the primary taxonomic determinant, the "by the wind-sailors" group most closely align with the anthomedusae (Athecata). The best known life stage, the "by the wind-sailors" themselves, is considered to be an aberrant or highly specialized polypoid stage.

Characteristics of the Chondrophoran Medusa

Chondrophoran medusae are very small, 3 mm tall by 2 mm wide, and therefore would be quite easy to miss. They also seem to be rare; they weren't reported in the Atlantic until 1980 (Larson 1980). The chondrophorans are analogous to a floating, single hydranth of an athecate hydroid in structure and development as well as in some behavioral traits. For example, they exhibit whole-organism contractile behavior known as concert behavior.

Velella medusae are brown in color due to high concentrations of zooxanthellae in the subumbrella. The presence of zooxanthellae and the fact that they were first collected in situ by blue-water divers strongly suggest an epipelagic life habit for the medusa stage of *Velella*.

The medusa itself has sensory papillae on the exumbrella, four marginal bulbs, two with pairs of tentacles, one of each pair short, and a conical manubrium. The bell is cylindrical in shape with a flat apex (Larson 1980).

Evolution Within the Chondrophora

There were two schools-of-thought regarding the evolution of the siphonophores themselves. One supported the idea that siphonophores were highly modified medusoid organisms, giving rise by budding from the sub-umbrella to secondary medusae and polyps (Haeckel 1866; Hatschek 1888). The other school-of-thought regarded siphonophores as floating colonies of hydroid polyps, showing division and specialization of labor, and budding off medusae (Leuckart 1848; Vogt 1854; Agassiz 1883). Chondrophora were recognized as a special case by LeLoup (1929) and Garstang (1946), who believed them to be polypoid organisms showing affinities to Tubularia-type hydroids, as they are classified today.

Thus, the new scheme, which may seem to be off-base from the morphological perspective, not only has more substance than was initially obvious but has a long history of argument behind it.

Like other Anthomedusae, *Tubularia* is an athecate hydroid. It does not release free medusae and its hydranth bears numerous gonophores whose eggs develop first into planulae then into actinulae before release (Figure 3.39). In contrast, both *Velella* and *Porpita* have a free medusa stage, and these are presumably dioecious. Either the fertilized egg develops into planula-conaria larva, or planula larvae are released from medusae to become conaria.

Feeding in the Chondrophora

Both Porpita and Velella feed on copepods, both cyclopoid and calanoid, with the size of prey increasing with the size of the predator (Table 3.10). In addition, both genera have zooxanthellae, but it is not clear what contribution the symbionts make to their overall nutrition.

Like their (however distant) relatives, the cystonect siphonophores, the chondrophores are ambush predators that move exclusively where the wind carries them, enacting no active pursuit of prey. It is assumed that prey items that blunder into their stinging tentacles are conveyed to the mouth in a way similar to feeding in the siphonophores and medusae.

Locomotion

The common name for the group, the by-the-wind-sailors, is accurate and descriptive. Like the man-o-war *Physalia*, *Velella* and *Porpita* are part of the pleuston, those organisms floating at the air–water interface. Like *Physalia*, *Velella* occurs in right-hand and left-hand sailors, so named for the direction they will sail in when blown by a wind normal to the long axis

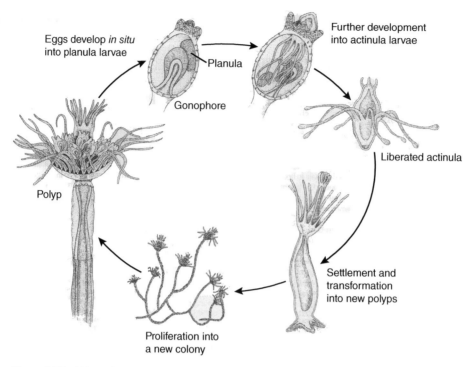

Figure 3.39 Life cycle of *Tubularia*, an athecate hydroid that does not release free medusae. *Source:* Adapted from Allman (1872), plates 20–23.

Table 3.10 Food of *Porpita* collected on Shirahama Beach, Japan. *Source:* Bieri (1970). The food of *Porpita* and niche separation in three neuston coelenterates. Publications of the Seto Marine Biological Laboratory. Vol. 17, table 1 (p. 306). Reproduced with permission of Seto Laboratory. Kyoto University.

Average diameter of *Porpita* (mm)	Number of *Porpita*	Food proportions	Copepod species
1.8	3	1 pontellid 1 unidentified calanid	
18	17	21 pontellids 22 other calanids 1 megalopa	*Labidocera acuta* common *Euchaeta marina* common *E. acuta* one *E. wolfendeni* one *Candacia* sp. one *Pontellopsis villosa* one
24	7	93 calanids including pontellids 1 corycaeid 1 barnacle cyprid	*Euchaeta marina* common *Labidocera acuta* common
31	7	275 calanids including pontellids 25 megalopas 2 corycaeids 1 barnacle cyprid	*Labidocera acula* common *Euchaeta marina* common *Candacia pachydactyla* one *C. truncata* one *Eucalanus crassus* one *Labidocera kroyeri* one

of the sail. *Porpita,* which has no sail, is nonetheless still at the mercy of the wind, which drives all surface water movement.

Interestingly, dispersal of the different morphs of Velella has resulted in dominance of one form or the other in different oceanic regions of the North Pacific. Left-hand sailors dominate in the north and right-hand sailors in the south (Mackie 1974). Though both morphs are found in both regions, one or the other dominates.

References

Agassiz, A. (1883). Exploration of the surface fauna of the gulf stream III. Part I. The Porpitidae and Velellidae. *Memoirs of the Museum of Comparative Zoology, at Harvard College* 8: 1–16.

Alexander, R.M. (2003). *Principles of Animal Locomotion.* Princeton University Press: Princeton.

Alldredge, A.L., Robison, B.H., Fleminger, A. et al. (1984). Direct sampling and in situ observation of a persistent copepod aggregation in the mesopelagic zone of the Santa Barbara Basin. *Marine Biology* 80: 75–81.

Allman, G.J. (1872). *A Monograph of the Gymnoblastic or Tubularian Hydroids.* London: The Ray Society.

Alvarino, A. (1983). Nectocarmen antonoi, a new Prayinae, Calycophorae, Siphonophora from California. *Proceedings of the Biological Society of Washington* 96: 339–348.

Anderson, P.A.V. and Schwab, W.E. (1981). The organization and structure of nerve and muscle in the jellyfish *Cyanea capillata* (Coelenterata: Scyphozoa). *Journal of Morphology* 170: 383–389.

Arai, M.N. (1991). Attraction of *Aurelia* and *Aequorea* to prey. *Hydrobiologia* 216/217: 363–366.

Arai, M.N. (1997). *A Functional Biology of the Scyphozoa.* London: Chapman and Hall.

Arneson, A.C. and Cutress, C.E. (1976). Life history of *Carybdea alata* Reynaud, 1830. In: *Coelenterate Ecology and Behavior* (ed. G.O. Mackie). New York: Plenum Press.

Barnes, J.H. (1966). Studies on three venomous Cubomedusae. *Symposia of the Zoological Society of London* 16: 305–332.

Bayer, F.M. and Owre, H.B. (1968). *The Free-Living Lower Invertebrates.* New York: The Macmillan Company.

Berger, E.W. (1900). Physiology and histology of the Cubomedusae including Dr. F.S. Conant's notes on the physiology. *Memoirs of the Biological Laboratory of Johns Hopkins University* 4: 1–61.

Bidigare, R.R. and Biggs, D.C. (1980). The role of sulfate exclusion in buoyancy maintenance by siphonophores and other gelatinous zooplankton. *Comparative Biochemistry and Physiology* 66A: 467–471.

Bieri, R. (1970). The food of Porpita and niche separation in three neuston coelenterates. *Publications of the Seto Marine Biological Laboratory* 27: 35–37.

Biggs, D.C. (1977). Field studies of fishing, feeding, and digestion in siphonophores. *Marine Behavior and Physiology* 4: 261–274.

Biggs, D.C., Bidigare, R.R., and Johnson, M.A. (1981). Population density of gelatinous macrozooplankton: in situ estimation in oceanic surface waters. *Biological Oceanography* 1: 157–173.

Biggs, D.C., Smith, D.E., Bidigare, R.R., and Johnson, M.A. (1984). In situ estimation of the population density of gelatinous planktivores in Gulf of Mexico surface waters. *Memorial University of Newfoundland Occasional Papers in Biology* 9: 17–34.

Boltovskoy, D. (1999). *South Atlantic Zooplankton,* vol. *1.* Backhuys: Leiden.

Bone, Q. and Trueman, J.R. (1982). Jet propulsion of the calycophoran siphonophores *Chelophyes* and *Abylopsis. Journal of the Marine Biological Association of the United Kingdom* 62: 263–276.

Bouillon, J. (1978). Hydromeduses de la mer de Bismarck (Papouasie, Nouvelle Guinee) II. Limnomedusa, Narcomedusa,

Trachymedusa, et Laingiomedusa (sous-classe nov.). *Cahiers de Biologie Marine* 19: 473–483.

Bouillon, J. (1984). Hydromeduses de la Mer de Bismark. (Papousasie, Nouvelle Guinee). IV. Leptomedusae (Hydromedusae-Cnidaria). *Indo-Malayan Zoology* 1: 25–112.

Bouillon, J. (1999). Hydromedusae. In: *South Atlantic Zooplankton*, vol. *1* (ed. D. Boltovskoy). Leiden: Backhuys Publishers.

Briggs, J.C. (1995). *Global Biogeography Amsterdam*. Elsevier.

Brusca, R.C. and Brusca, G.J. (1990). *Invertebrates*. Sunderland: Sinauer Associates.

Brusca, R.C. and Brusca, G.J. (2003). *Invertebrates*, 2e. Sunderland: Sinauer.

Brusca, R.C., Moore, W., and Shuster, S.M. (2016). *Invertebrates*, 3e. Sunderland: Sinauer Associates.

Calder, D.R. (1982). Life history of the cannonball jellyfish *Stomolophus meleagris* L. Agassiz, 1860 (Scyphozoa, Rhizostomida). *Biological Bulletin* 162: 149–162.

Chapman, G. (1966). The structure and functions of the mesoglea. In: *The Cnidaria and Their Evolution* (ed. W.J. Rees). London: Academic Press.

Chapman, G. (1974). The skeletal system. In: Muscatine, L. & Lenhoff, H. M.(eds.) Coelenterate Biology: reviews and new perspectives. New York: Academic Press.

Conant, F.S. (1898). *The Cubomedusae, Johns Hopkins Univ. Memoirs Biol. Lab.*, vol. 4. Johns Hopkins University, Biological Laboratory.

Costello, J.H. (1992). Foraging mode and energetics of hydrozoan medusae. *Scientia Marina* 56: 185–191.

Daniel, T.L. (1983). Mechanics and energetics of medusan jet propulsion. *Canadian Journal of Zoology* 68: 778–785.

Daniel, T.L. (1985). Cost of locomotion: unsteady medusan swimming. *Journal of Experimental Biology* 119: 149–164.

DeMont, M.E. and Gosline, J.M. (1988). Mechanics of jet propulsion in the hydromedusan jellyfish, *Polyorchis pencillatus*, I-III. *Journal of Experimental Biology* 134: 313–332.

Fancett, M.S. (1988). Diet and prey selectivity of scyphomedusae from Port Phillip Bay, Australia. *Marine Biology* 98: 503–509.

Fosså, J.H. (1992). Mass occurrence of Periphylla periphylla (Scyphozoa, Coronatae). *Sarsia* 77: 237–251.

Garstang, W. (1946). The morphology and relations of the Siphonophora. *Quarterly Journal of Microscopical Science* 87: 103–193.

Gerritsen, J. and Strickler, J.R. (1977). Encounter probabilities and community structure in zooplankton: a mathematical model. *Journal of the Fisheries Research Board of Canada* 34: 73–82.

Giorgi, R., Avian, M., and De Olazabal, S. (1991). Feeding of *Pelagia noctiluca* in open sea. In: *Jellyfish Blooms in the Mediterranean. Proceedings of the II Workshop on Jellyfish in the Mediterranean Sea*. Athens: UNEP.

Gladfelter, W.B. (1972). Structure and function of the locomotory system of the scyphomedusa *Cyanea capillata*. *Marine Biology* 14: 150–160.

Gladfelter, W.B. (1973). A comparative analysis of the locomotory systems of medusoid cnidaria. *Helgoländer Wissenschaftliche Meeresuntersuchungen* 25: 228–272.

Graham, W.M., Field, J.G., and Potts, D.C. (1992). Persistent upwelling shadows and their influence on zooplankton distributions. *Marine Biology* 114: 561–570.

Graham, W.M., Pages, F., and Hamner, W.M. (2001). A physical context for gelatinous zooplankton aggregations: a review. In: *Jellyfish Blooms: Ecological and Societal Importance* (eds. J.E. Purcell, W.M. Graham and H.J. Dumont). Dordrecht: Kluwer Academic Publishers.

Greene, C.H. (1985). Planktivore functional groups and patterns of prey selection in pelagic communities. *Journal of Plankton Research* 7: 35–40.

Greene, C.H., Landry, M.R., and Monger, B.C. (1986). Foraging behavior and prey selection by the ambush entangling predator *Pleurobrachia bachei*. *Ecology* 67: 1493–1501.

Gross, J., Dumsha, B., and Glazer, N. (1958). Comparative biochemistry of collagen. *Biochimica et Biophysica Acta* 30: 293–297.

Hamner, W.M. (1995). Sensory ecology of scyphomedusae. *Marine and Freshwater Behavior and Physiology* 26: 101–118.

Hamner, W.M., Gilmer, R.W., and Hamner, P.P. (1982). The physical, chemical, and biological characteristics of a stratified, saline, sulfide lake in Palau. *Limnology and Oceanography* 27: 896–909.

Haeckel, E. H. P. A. (1866). Generelle Morphologie der Organismen. Allgemeine Grundzuge derOrganischen Formen-Wissenschaft, mechanische Begrundet durch die von CharlesDarwin Descendenz-Theorie. Volume I. Allgemeine Anatomie der Organismen., Berlin,Germany, Georg Reimer.

Hatschek, B. (1888). *Lehrbuch der Zoologie*. Jena: G. Fischer.

Hessinger, D.A. (1988). Nematocyst venoms and toxins. In: *The Biology of Nematocysts* (eds. D.A. Hessinger and H.M. Lenhoff). San Diego: Academic Press.

Hyman, L.H. (1940). *The Invertebrates, Vol. 1, Protozoa Through Ctenophora*. New York: McGraw-Hill.

Ishida, J. (1936). Note on the digestion of *Charybdea rastonii. Annotationes Zoologicae Japonenses* 15: 449–452.

Jacobs, W. (1937). Beobachtungen uber das Schweben der Siphonophoren. *Zeitschrift für Vergleichende Physiologie* 24: 583–601.

Kaestner, A. (1967). *Invertebrate Zoology*, vol. 1. New York: Wiley.

Kerstan, M. (1977). *Untersuchungen zur Nahrungsokologie von Aurelia aurita*. Diploma thesis. Christian-Albrechts-Universitat.

Kramp, P.L. (1938). Die meduse von Ostroumovia inkermanica (Pal.-Ostr) und die systematische Stellung der Olindiiden. *Zoologischer Anzeiger* 122: 103–108.

Kramp, P.L. (1959). The hydromedusae of the Atlantic and adjacent waters. *Dana-Report* 46: 1–283.

Land, M.F. (1990). Optics of the eyes of marine animals. In: *Light and life in the sea* (eds. P.J. Herring, A.K. Campbell, M. Whitfield and L. Maddock). Cambridge: Cambridge University Press.

Larson, R.J. (1976). Cubomedusae: feeding – functional morphology, behavior and phylogenetic position. In: *Coelenterate Ecology and Behavior* (ed. G.O. Mackie). New York: Plenum Press.

Larson, R.J. (1980). The medusa of *Velella velella* (Linnaeus, 1758). *Journal of Plankton Research* 2: 183–186.

Larson, R.J. (1986). Pelagic Scyphomedusae (Scyphozoa: Coronatae and Semaeostomae) of the Southern Ocean. *Antarctic Research Series* 41: 59–165.

Larson, R.J. (1987a). First report on the feeding, growth, and reproduction of the epipelagic scyphomedusa *Dalmonema dalmatinum* in the Caribbean Sea. *Bulletin of Marine Science* 40: 447–454.

Larson, R.J. (1987b). A note on the feeding, growth, and reproduction of the epipelagic scyphomedusa, *Pelagic noctiluca* (Forskal). *Biological Oceanography* 4: 447–454.

Larson, R.J. (1991). Diet, prey selection and daily ration of Stomolophus meleagris, a filter-feeding scyphomedusa from the NE Gulf of Mexico. *Estuarine, Coastal and Shelf Science* 32: 511–525.

Lehninger, A.L. (1975). *Biochemistry*, 2e. Worth: New York.

Leloup, E. (1929). Recherches sur l'anatomie st la Developpement de Velella spirans Forsk. *Archives de biologie, Paris* 39: 397–478.

Leuckart, R.W. (1848). *Über die Morphologie und Verwandtschaftsverhältnisse der wirbellosen Tiere*. Braunschweig: Braunschweig.

Leuckart, R.W. and Frey, H. (1847). *Beltrage zur wirbelloser Tiere*. Braunschweig: Braunschweig.

Longhurst, A. (1998). *Ecological Geography of the Sea*. San Diego: Academic Press.

Maas, O. (1904). Meduses. *Resultats des Campagnes Scientifiques du Prince de Monaco* 28: 1–71.

Mackie, G.O. (1963). Siphonophores, bud colonies and superorganisms. In: *The Lower Metazoa* (ed. E.C. Dougherty). Berkeley: University of California Press.

Mackie, G.O. (1964). Analysis of locomotion in a siphonophore colony. *Proceedings of the Royal Society of London B* 159: 366–391.

Mackie, G.O. (1974). Locomotion, flotation, and dispersal. In: *Coelenterate Biology, Reviews and New Perspectives* (eds. L. Muscatine and H.M. Lenhoff). New York: Academic Press.

Mackie, G.O. (1986). From aggregates to integrates: physiological consequences of modularity in colonial animals. *Philosophical Transactions of the Royal Society of London B* 313: 175–196.

Mackie, G.O. and Mills, C.E. (1983). Use of the Pisces IV submersible for zooplankton studies in coastal waters of British Columbia. *Canadian Journal of Fisheries and Aquatic Sciences* 40: 763–776.

Mackie, G.O., Pugh, P.R., and Purcell, J.E. (1987). Siphonophore biology. *Advances in Marine Biology* 24: 98–263.

Madin, L.P. (1988). Feeding behavior of tentaculate predators: in situ observation and a conceptual model. *Bulletin of Marine Science* 43: 413–429.

Madin, L.P., Bollens, S.M., Horgan, E. et al. (1996). Voracious planktonic hydroids: unexpected predatory impact on a coastal marine ecosystem. *Deep Sea Research, Part II* 43: 1823–1829.

Margulis, R.Y. (1976). On regularities of the distribution of siphonophores in the Atlantic. *Trudy vsesoyuznogo nauchno-issledovatel'skogo institura rybnogo khosyaistva i oceanografii* 110: 1244–1246.

Matsakis, S. and Conover, R.J. (1991). Abundance and feeding of medusae and their potential impact as predators on other zooplankton in Bedford Basin (Nova Scotia, Canada) during spring. *Canadian Journal of Fisheries and Aquatic Sciences* 48: 1419–1430.

Mayer, A.G. (1900). Some medusae from the Tortugas, Florida. *Bulletin of the Museum of Comparative Zoology at Harvard College* 37 (2): 11–82.

Mayer, A.G. (1904). *Medusae of the Bahamas*, Memoirs of Natural Sciences of the Brooklyn Institute of Arts and Sciences, vol. 1. Museum of the Brooklyn Institute of Arts and Sciences.

Mayer, A.G. (1910). *The Medusae of the World*, vol. *1, 2, and 3*. Washington, DC: Carnegie Institution of Washington.

Mianzan, H.W. and Cornelius, P.F.S. (1999). Cubomedusae and Scyphomedusae. In: *South Atlantic Zooplankton*, vol. *1* (ed. D. Boltovskoy). Leiden: Backhuys Publishers.

Mills, C.E. (1981). Diversity of swimming behaviors in hydromedusae as related to feeding and utilization of space. *Marine Biology* 64: 185–189.

Mills, C.E. (1995). Medusae, siphonophores, and ctenophores as planktivorous predators in changing global ecosystems. *ICES Journal of Marine Science* 52: 575–581.

Mills, C.E. (2001). Jellyfish blooms: are populations increasing globally in response to changing ocean conditions. In: *Jellyfish Blooms: Ecological and Societal Importance* (eds. J.E. Purcell, W.M. Graham and H.J. Dumont). Dordrecht: Kluwer.

Mills, C.E., Marques, A.C., Migotto, A.E. et al. (2007). Hydrozoa: polyps, hydromedusae, and Siphonophora. In: *The Light and Smith Manual, Intertidal invertebrates from central California to Oregon* (ed. J.T. Carlton). Berkeley: University of California Press.

Mironov, G.N. (1967). Feeding of planktonic predators III. Food utilization and daily rations of *Aurelia aurita* (L.). In: *Biology and Distribution of Plankton in Southern Seas*. Moscow: Academy Nauka.

Möller, H. (1980). Scyphomedusae as predators and food competitors of larval fish. *Meeresforschung* 28: 90–100.

Moore, H.B. (1949). The zooplankton of the upper waters of the Bermuda area of the North Atlantic. *Bulletin of the Bingham Oceanographic Collection* 12: 1–97.

Moore, H.B. (1953). Plankton of the Florida Current II. Siphonophora. *Bulletin of Marine Science of the Gulf and Caribbean* 2: 559–573.

Musayeva, E.I. (1976). Distribution of siphonophores in the eastern part of the Indian Ocean. *Trudy Institura Oceanologii* 105: 171–197.

Olesen, N.J., Frandsen, K., and Riisgard, H.U. (1994). Population dynamics, growth and energetics of jellyfish *Aurelia aurita* in a shallow fjord. *Marine Ecology Progress Series* 105: 9–18.

Passano, L.M. (1982). Scyphozoa and Cubozoa. In: *Electrical Conduction and Behaviour in "Simple" Invertebrates* (ed. G.A.B. Shelton). Oxford: Clarendon Press.

Phillips, P.J. and Burke, W.D. (1970). The occurrence of sea wasps (Cubomedusae) in Mississippi Sound and the northern Gulf of Mexico. *Bulletin of Marine Science* 20: 853–859.

Phillips, P.J., Burke, W.D., and Keener, E.J. (1969). Observations on the trophic significance of jellyfishes in Mississippi Sound with quantitative data on the associative behavior of small fishes with medusae. *Transactions of the American Fishery Society* 98: 703–712.

Plotnikova, E.D. (1961). On the diet of medusae in the littoral of eastern Murman. In: *Hydrological and Biological Features of the Shore Waters of Murman* (ed. M.M. Kamshilov). Murmansk: Kolski Filial, Akademia Nauk SSSR.

Pugh, P.R. (1977). Some observations on the vertical migration and geographical distribution of siphonophores in the warm waters of the North Atlantic Ocean. In: *Proceedings of the Symposium on Warm Water Zooplankton*. Goa: NIO.

Pugh, P.R. (1999). Siphonophorae. In: *South Atlantic Zooplankton, vol. 1* (ed. D. Boltovskoy). Leiden: Backhuys Publishers.

Purcell, J.E. (1980). Influence of siphonophore behavior upon their natural diets: evidence for aggressive mimicry. *Science* 209: 1045–1047.

Purcell, J.E. (1981a). Feeding ecology of *Rhizophysa eysenhardti*, a siphonophore predator of fish larvae. *Limnology and Oceanography* 26: 424–432.

Purcell, J.E. (1981b). Selective predation and caloric consumption by the siphonophore *Rosacea cymbiformis* in nature. *Marine Biology* 63: 283–294.

Purcell, J.E. (1982). Feeding and growth of the siphonophore *Muggiaea atlantica* (Cunningham 1893). *Journal of Experimental Marine Biology and Ecology* 62: 39–54.

Purcell, J.E. (1984). Predation on fish larvae by *Physalia physalis*, the Portuguese man of war. *Marine Ecology Progress Series* 19: 189–191.

Purcell, J.E. (1989). Predation by the hydromedusa *Aequorea victoria* on fish larvae and eggs at a herring spawning ground in British Columbia. *Canadian Journal of Fisheries and Aquatic Sciences* 46: 1415–1427.

Purcell, J.E. (1990). Soft-bodied zooplankton predators and competitors of larval herring (Clupea harengus pallasi) at herring spawning grounds in British Columbia. *Canadian Journal of Fisheries and Aquatic Sciences* 47: 505–515.

Purcell, J.E. (1991). Predation by *Aequorea victoria* on other species of potentially competing pelagic hydrozoans. *Marine Ecology Progress Series* 72: 255–260.

Purcell, J.E. (1992). Effects of predation by the scyphomedusan *Chrysaora quinquecirrha* on zooplankton populations in Chesapeake Bay, USA. *Marine Ecology Progress Series* 129: 63–70.

Purcell, J.E. (1997). Pelagic cnidarians and ctenophores as predators: selective predation, feeding rates, and effects on prey populations. *Annales de l'Institut Oceanographique* 73: 125–137.

Purcell, J.E. and Arai, M.N. (2001). Interactions of pelagic cnidarians with fish: a review. In: *Jellyfish Blooms: Ecological and Societal Importance* (eds. J.E. Purcell, W.M. Graham and H.J. Dumont). Dordrecht: Kluwer Academic Publishers.

Purcell, J.E. and Grover, J.J. (1990). Predation and food limitation as causes of mortality in larval herring at a spawning ground in British Columbia. *Marine Ecology Progress Series* 59: 55–67.

Purcell, J.E. and Kremer, P. (1983). Feeding and metabolism of the siphonophore *Sphaeronectes gracilis*. *Journal of Plankton Research* 5: 95–106.

Purcell, J.E. and Mills, C.E. (1988). The correlation of nematocyst types to diets in pelagic hydrozoa. In: *The Biology of Nematocysts* (eds. D.A. Hessinger and H.M. Lenhoff). San Diego: Academic Press.

Purcell, J.E., Nemazie, D.A., Dorsey, S.E. et al. (1994b). Predation mortality of bay anchovy (*Anchoa mitchilli*) eggs and larvae due to scyphomedusa and ctenophores in Chesapeake Bay. *Marine Ecology Progress Series* 114: 47–58.

Robertson, J.D. (1949). Ionic regulation in some marine invertebrates. *Journal of Experimental Biology* 26: 182–200.

Rottini-Sandrini, L. and Avian, M. (1989). Feeding mechanisms of *Pelagia noctiluca* (Scyphozoa: Semaeostomae); laboratory and open sea observations. *Marine Biology* 102: 49–55.

Rubenstein, D.I. and Koehl, M.A.R. (1977). The mechanisms of filter feeding: some theoretical considerations. *American Naturalist* 111: 981–994.

Russell, F.S. (1954). *Medusae of the British Isles*, vol. *I*. London: Cambridge University Press.

Sanders, N.K. and Childress, J.J. (1988). Ion replacement as a buoyancy mechanism in a pelagic deep-sea crustacean. *Journal of Experimental Biology* 138: 333–434.

Schultze, P. (1922). Der bau und entladung der penentraten von Hydra attenuata Pallas. *Archiv fur Zellforschung* 16: 383–438.

Totton, A.K. (1960). Studies on *Physalia physalis* (L.) 1. Natural history and morphology. *Discovery Reports* 30: 301–368.

Totton, A.K. (1965). *A Synopsis of the Siphonophora*. London: British Museum.

Tschachotin, S. (1908). Die statocyste der heteropoden. *Zeitschrift für Wissenschaftliche Zoologie* 90: 343–422.

Vinogradov, M.E. (1970). *Vertical Distribution of the Oceanic Zooplankton*. Jerusalem: Israel Program for Scientific Translations.

Uchida, M. (1926). The anatomy and development of a Rhizostome medusa, Mastigias papua L. Agassiz, with observations on the phylogeny of the Rhizostomae. *Tokyo University Faculty Science Journal, Section 4, Zoology* 1: 45–95.

Uchida, T. (1929). Studies on the Stauromedusae and Cubomedusae with special reference to their metamorphosis. *Japanese Journal of Zoology* 2: 103–193.

Vogel, S. (1988). *Life's Devices, the Physical World of Animals and Plants*. Princeton University Press: Princeton.

Vogt, C. (1854). Recherches sur les animaux inferieures de la Mediterranee. 1: Sur les siphonophores de la mer de Nice. *Memoires de l'institut National Genevois* 1: 1–164.

Withers, P.C. (1992). *Comparative Animal Physiology*. Orlando: Saunders.

Woltereck, R. (1905). Entwicklung der Narcomedusen. *Deutsche zoologische gesellschaft. Verhandlungen der deutschen Zoologen* 15.

Woodcock, A.H. (1971). Note concerning *Physalia* behavior at sea. *Limnology and Oceanography* 15: 551–552.

4

The Ctenophora

Introduction

The ctenophores are a beautiful and delicate group of gelatinous animals that may appear inshore during spring and summer months, sometimes in large numbers. Their name comes from the Greek: the "comb-bearers." They are so named for the comb-like rows of tiny hairs, or cilia, that propel them through the water. Each of the combs acts like a row of microscopic oars, and most species have eight rows. Ctenophores are very maneuverable, but not very fast.

When visitors to the ocean see a ctenophore in the water for the first time, they often think that they are looking at a jellyfish without the tentacles. Continued observation, particularly on a sunny day, will reveal the characteristic blue-red shimmer of the beating comb rows as the animal turns or is buffeted by water motion. Common names for ctenophores are "globes" or "gooseberries", and the most common species inshore are about three-quarters of the size of your fist. No need to worry about getting stung by a ctenophore; they have different ways of capturing prey than the jellyfishes.

A remarkable characteristic of the ctenophores is their ability to produce blue-green biological light, or bioluminescence, during the hours of darkness. It is likely that their luminescence is the reason that ctenophores were recognized very early on by scientists classifying animal life, such as Linnaeus and Cuvier. When they are abundant inshore, groups of ctenophores may get trapped in "marine cul-de-sacs," such as between a boat and a dock, or up against a jetty, and produce impressive amounts of light as they are jostled by wave action. If you bump into one in the water at night, it can be bright enough to startle you. Without question, the ctenophores are one of Mother Nature's most beautiful creations.

Most ctenophores are spheroidal in shape, usually with a long axis in the same plane as the gut (Figure 4.1). However, one group is flattened for crawling along the bottom and yet another, the "Venus girdle," takes the form of a thick gelatinous ribbon that can be as long as 1.5 m.

Classification

History

Our knowledge of the ctenophores and the history of their classification parallels that of the Cnidaria. Most ctenophores are large enough to be easily visible from the seashore or from the deck of a ship and all are highly luminescent. They have likely been known since ancient times, but they were

Life in the Open Ocean: The Biology of Pelagic Species, First Edition. Joseph J. Torres and Thomas G. Bailey.
© 2022 John Wiley & Sons Ltd. Published 2022 by John Wiley & Sons Ltd.

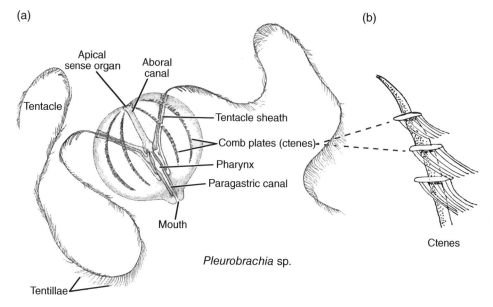

Figure 4.1 Basic ctenophore morphology. (a) The cydippid ctenophore *Pleurobrachia* sp. (b) Close-up of a portion of a comb plate showing the ctenes used as paddles for propulsion. *Sources:* Adapted from (a) Brusca and Brusca (2003), figure 9.2 (p. 272); (b) Mianzan (1999), figure 1 (p. 562).

first recognizably illustrated in 1671 by a ship's doctor who was sailing near the island of Spitsbergen (Hyman 1940). As with the cnidarians, the nineteenth-century natural philosophers had differences of opinion on the position of the ctenophores within the animal kingdom. Jacques Cuvier, famous for his multiple-volume synopsis of animal life, the Règne Animal (1817), grouped them with the cnidarians in the Zoophyta, a taxonomic category in the obsolete class Acalephes. Eschscholtz (1829–1833), working with many ship-collected specimens, created the orders Ctenophorae, Discophorae (medusae), and Siphonophorae within the class Acalepha. The three orders were considered to be intermediate between the zoophytes and echinoderms. The lower animals were further subdivided by Leuckart (1847–1848) who separated the coelenterates and echinoderms but, as discussed in Chapter 3, his Coelenterata retained the sponges with the Cnidaria and ctenophores. The sponges were recognized as their own group and separated from the coelenterates by Vosmaer. In 1888, Hatschek separated the ctenophores from the cnidarians into their own phylum (Hyman 1940), giving them autonomy as a group and the scheme most commonly used today.

Classification Schemes

Two classification schemes have been used for the ctenophores, which at this writing are considered to have between 150 and 200 valid species (Mills n.d.). Most authors agree that the phylum comprises seven orders. The difference between the two schemes is whether the phylum is divided into two classes based on the presence or absence of tentacles. In the older and more classic scheme (e.g. Hyman 1940; Mianzan 1999) two classes were recognized, and they are currently recognized by Mills (n.d.).

Class Tentaculata
    Order Cydippida (7 families)
    Order Lobata (5 families)
    Order Cestida (1 family)
    Order Ganeshida (1 family)
    Order Thalassocalycida (1 family-monotypic)
    Order Platyctenida (1 family)
Class Nuda
    Order Beroida (1 family)

In other treatments of the phylum (Harbison and Madin 1982; Harbison 1985; Podar et al. 2001; Brusca and Brusca 2003; Brusca et al. 2016), the seven orders of ctenophores listed above are given equal weight. As shown below, there is no discrimination between tentaculate and non-tentaculate forms and no initial Class designation within the phylum. Classification within the ctenophores is currently in flux, notably in the number of orders and whether the two classes listed above should be retained (Mills n.d.). Organization within the chapter follows the Harbison, Madin, and Brusca system below. Two additional orders have been reported and are listed in WoRMS: the Cambojiida with one species and the order Cryptoloferida with two. They are based on three publications by Ospovat (1985a, 1985b, 1985c) but have not received much discussion within the ctenophore community since that date (Mills n.d.) and will not factor into the treatment here.

Phylum Ctenophora (classification from Brusca et al. 2016)

Order Cydippida
Order Lobata
Order Cestida
Order Ganeshida
Order Thalassocalycida (1 family-monotypic)
Order Platyctenida
Order Beroida

Ctenophore Basics

The ctenophores pose an interesting problem to systematists. On the one hand, ctenophores resemble cnidarians in that their body structure is organized with respect to a central digestive tract that forms the only internal space and their mesenchyme, located within an external and internal epithelium and constituting the vast majority of the body mass, is gelatinous in character In addition, the absence of true organs places them in the diploblastic or "tissue grade" of animal life. On the other hand, well-defined muscle cells are found within the gelatinous mesenchyme, the digestive tract is loosely compartmentalized, and there is an aboral sensory region (Figure 4.1a) (Hyman 1940). The nervous system of ctenophores, like that of cnidarians, is a subepidermal net but, unlike cnidarians, concentrations of nervous tissue resembling nerves underlie the comb rows. Thus, it is tempting to conclude that they are more advanced than the Cnidaria.

Ctenophores are biradially symmetrical. Two identical halves can be generated by vertically bisecting the animal in two possible ways. The two axes of symmetry shall be described later in this chapter along with enough ctenophore anatomy for them to make sense. Tentacles are widespread within the ctenophores but differ markedly in structure between the orders. In no case do they have stinging cells; instead they have adhesive cells, known as colloblasts, for capturing prey. Also unlike the cnidarians, ctenophores do not have planula larvae. Rather, in all orders for which there is information, there is a cydippid larva that resembles a small cydippid without the tentacles (Figure 4.2). Alternation of generations does not occur within the ctenophores. They are "monomorphic" throughout their lives, and patterns of cleavage in the

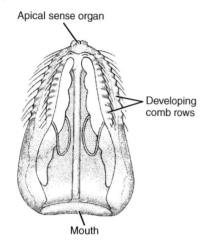

Figure 4.2 Typical young cydippid larva. *Source*: Adapted from Mayer (1912), plates 14, 15.

fertilized egg are very different from those of the Cnidaria.

Virtually all ctenophores are simultaneous hermaphrodites, though at least one exception to this rule has been found (Harbison and Miller 1986). During reproduction, gametes are broadcast into the water where fertilization occurs. The resulting cydippid larvae then differentiate into the adult form. At least two of the Platyctenid genera (*Coeloplana* and *Tjalfiella*) brood their young, though they also have cydippid larvae (Hyman 1940).

Ctenophores have prodigious powers of regeneration. All parts of the individual, including comb rows and statocysts can be regenerated, and animals cut into halves or thirds can produce entire animals from the resulting pieces. If an animal is cut into pieces transversely, regeneration is fastest in the section that retained the original statocyst (Hyman 1940). The regenerative powers of ctenophores have fueled speculation that asexual reproduction is possible in the pelagic forms, but that has never been confirmed. In contrast, the platyctenids are capable of asexual reproduction by fragmenting off small sections as they crawl, with each section developing into a new individual.

Four of the ctenophore orders are quite common: the cydippids, lobates, cestids, and beroids. They are treated in the most detail here. A few lobate and cydippid species are even robust enough to have been manipulated as experimental subjects. However, the vast majority of ctenophores are known only from hand or submersible collections, field observations, or simple behavioral experiments on board research vessels. The ctenophore orders Ganeshida and Thalassocalycida have only two and one species, respectively; knowledge of them is limited and fairly recent. The Platyctenida are also not well described though they have been known since the late 1800s. All Platyctenida are flattened in the oral-aboral plane and some have assumed a benthic mode of life. Ctenophores are considered to be exclusively marine though they are commonly found in estuaries at salinities less than that of full-strength seawater.

Morphology

Cydippida

The cydippids are considered by most authors to be the simplest and most general of the ctenophores, a good choice for addressing basic ctenophore anatomy and the axes of symmetry. We'll use *Pleurobrachia* as our cydippid of choice. Figure 4.1 shows *Pleurobrachia* viewed from the side with the gut oriented straight up and down and the tentacles oriented in the plane of the page. It is nearly spherical in shape. In Figure 4.1, the mouth faces down, comprising the oral pole. Directly opposite, at the end of the (nearly) blind-ending gastric canal, is the aboral or sensory pole. The eight comb rows that provide the animal's locomotion are equally spaced along the surface of *Pleurobrachia*. Each comb plate, or ctene (Figure 4.1b), is composed of fused cilia that act as paddles and each row has hundreds of them. Think of the individual plates as oarsmen on an old Greek trireme and the comb rows as the two sides of the ship (only this is an eight-sided spherical ship). If both sides paddle at the same pace, the ship moves forward in a straight line. If one side paddles more slowly, the ship turns toward the more slowly paddling side. The same principles apply to the ctenophore, but with eight comb rows it is easy to see that a ctenophore would be a highly maneuverable creature.

The tentacles are anchored in two pouches that originate parallel to the gut lumen and bifurcate out to the surface of the comb jelly (Figure 4.3). The tentacles can be completely retracted into their pouches, are highly elastic, and bear side branches known as tentilla that are heavily invested with colloblasts, the sticky cells for trapping prey.

We now know enough ctenophore anatomy to revisit the axes of symmetry. Clearly, the anatomical structures of the ctenophore are radially arranged around the central hub formed by the gut, or oral/aboral axis. The radial structure may be further subdivided into two vertical planes, creating a biradial symmetry most obvious when viewed from the aboral pole (Figure 4.4). One, defined by the plane of the tentacular sheaths or pouches, is called the tentacular plane. It runs from left to right through the middle of Figure 4.4a. The other, the stomodeal plane, is defined by the more flattened axis of the gut, which runs through the middle of the drawings in Figure 4.4 parallel with the long axis of the page. In the cydippids, the flattened axis of the gut is less obvious than the two sensory structures associated with the statocyst known as the "polar fields" (Figure 4.4b and c). The two planes of symmetry are important to understanding the differences between ctenophore orders as they are often elongated or flattened in one plane or the other and are described as such. The stomodeal plane is also termed the median, or sagittal plane. The tentacular plane is also termed the lateral or transverse plane. We will use "tentacular" and "stomodeal" planes in this chapter; the terminology is confusing enough without adding synonyms to the mix.

Let us familiarize ourselves with the remaining high points of basic ctenophore anatomy prior to moving on to the different orders. As with the cnidarians, the gastrovascular canals formed by ramifications of the ctenophore gut as well as the gut itself contribute an important part of the animal's visible structure (Figure 4.3b). Viewed from the side, the mouth leads to a simple tube which expands to a pharynx, or stomodeum, just below the true stomach, or infundibulum. The pharynx has thickened folded walls, presumably to increase surface area and enhance the digestive process. Along the sides of the gut lie the paragastric, or pharyngeal, canals that originate in the stomach and end blindly above the mouth.

Above the stomach and contiguous with it lies the aboral or infundibular canal, which continues to the base of the statocyst. There it gives rise to four small excretory canals, two of which open to the outside as excretory pores (Figure 4.4b). Two large transverse canals branch from the stomach in the tentacular plane to form the tentacular canal, terminating in the tentacular sheath. However, prior to terminating in the tentacular canal each of the transverse canals bifurcates into an interradial canal and then again to form four meridional canals (Figure 4.3b). Each of the meridional

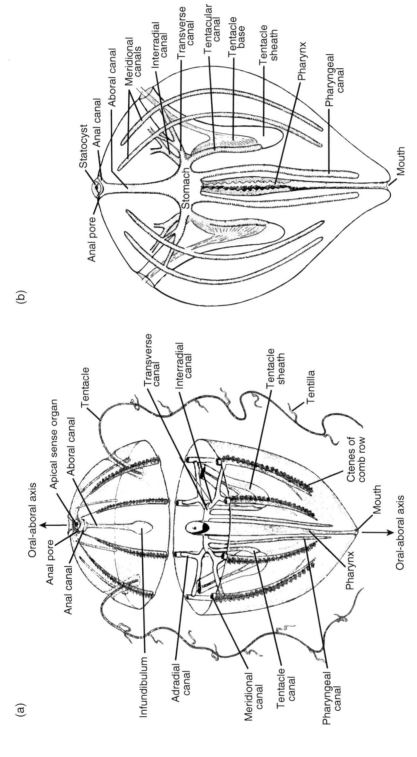

Figure 4.3 Ctenophore anatomy. (a) General anatomy of a cydippid ctenophore, cross-sectional body plan. (b) Diagram of the digestive system of a cydippid ctenophore. *Sources:* (a) Adapted from Bayer and Owre (1968), figure 165 (p.127); (b) Adapted from Chun (1880), plate II.

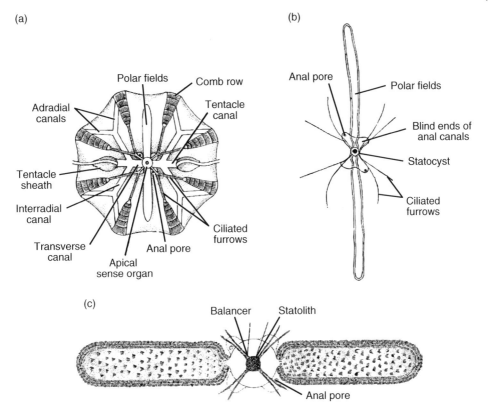

Figure 4.4 Symmetry and polar fields. (a) Aboral view of the cydippid ctenophore *Pleurobrachia*; (b) aboral sense organs and polar fields of *Pleurobrachia*. (c) Diagram of the polar fields and aboral sense organs of a generalized ctenophore. *Sources:* (a, b) Adapted from Hyman (1940), figure 209 (p. 664); (c) Adapted from Chun (1880), plates X, XVI.

canals underlies a comb row, presumably providing it with a source of nutrition.

At the aboral pole lies the statocyst (Figure 4.5) comprising a statolith of calcareous spherules impinging upon four tufts of cilia known as "balancers"; all contained beneath a gelatinous dome or cupule. Each of the four balancers gives rise to a pair of ciliated furrows that connect to the aboral end of each comb row. The statocyst will be discussed in more detail later in this chapter.

Lobata

The lobates differ from the cydippids by having large, muscular, oral lobes, and much-reduced tentacles. Most of the basic structures described for the cydippids are still present but are considerably rearranged (Figure 4.6). Viewed from the oral pole (Figure 4.6c), it is easy to see the extent of the rearrangement. The slit-like mouth is located in the center; the orientation of the slit defines the stomodeal plane. On either side of the mouth lie the small main tentacles defining the tentacular plane. Clearly, the lobates are compressed in the tentacular plane. Structures to notice in the oral view are the four auricles that lie above the mouth in the tentacular plane and the stomodeal comb rows on the two oral lobes. The side view (Figure 4.6b) provides a

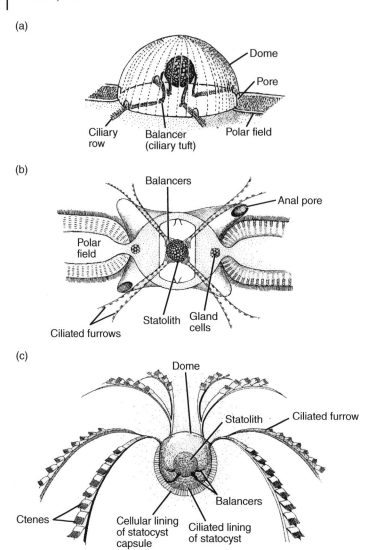

Figure 4.5 Aboral sense organs. (a) Statocyst of a cydippid ctenophore. (b) Apical sense organs of the cydippid *Hormiphora*. (c) Apical sense organ of *Pleurobrachia*. *Sources*: (a) Adapted from Kaestner (1967), figure 6-5 (p. 147); (b) Adapted from Chun (1880), plates XIV, XVI; (c) Adapted from Bayer and Owre (1968), figure 176 (p. 131).

better appreciation of the animal in three dimensions. If you orient to the stomodeal comb rows on the oral lobes, you can see that the oral lobes are large wing-like processes that are reminiscent of a medusan bell, and the oral-aboral canal and auricles are suspended within them much like the manubrium of a medusa.

The shorter comb rows of the tentacular plane have the usual canals underlying them but, unlike in the cydippids, the canals in lobate ctenophores continue past the end of the comb row to underlie the outer margin of the ciliated auricles. The marginal canal of the auricles continues past the "shoulder" of the auricles

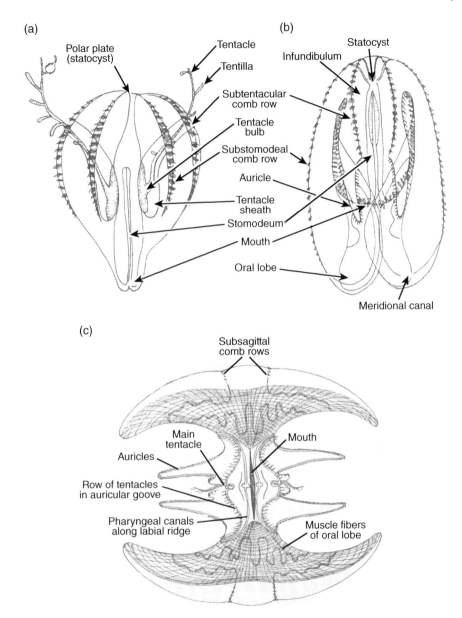

Figure 4.6 Lobate ctenophores. (a and b) Morphological comparison of cydippid and lobate ctenophores, respectively. (c) Oral view of the lobate ctenophore *Mnemiopsis*. *Sources*: (a, b) Adapted from Chun (1880), plates IX, XIV; (c) Adapted from Mayer (1912), plate 7, figure 9.

where they join the main body, now underlying the auricular groove. The auricular groove bears small tentacles, is capable of propelling food particles, and continues to the mouth where it joins its counterpart from the other side at the base of the main tentacle. Confusing? It can be. It can become clearer by following the margin of the auricle in the b and c views of Figure 4.6.

The meridional canals underlying the longer comb rows in the stomodeal plane also continue well past the oral end of the comb rows to form a sinuous loop within the oral lobes themselves. Presumably, these canals provide nutrients to the mass of the oral lobe.

Cestida

The cestids, commonly known as "Venus girdles," take the transverse compression of the lobates a giant step further. The tentacular plane is reduced to a tiny fraction of the stomodeal plane, forming an organism that resembles a large gelatinous belt capable of reaching more than a meter in length (Figure 4.7). The four sub-tentacular comb rows that were reduced considerably in the lobates are tiny in the cestids, and the four sub-stomodeal rows have increased in size to run the length of the animal on the aboral side.

The gut consists of the mouth, pharynx, and stomach with the statocyst residing just above the aboral canal, as shown in the magnification in Figure 4.7. The radial canals arise directly from the stomach, and the sub-stomodeal canals run beneath the sub-stomodeal comb rows on the aboral side of the gelatinous band. Despite the cestids' tiny sub-tentacular comb rows, the four sub-tentacular canals run the entire length of the animal at its oral-aboral midline: the flat side of the belt. The pharyngeal canals run along the oral side of the gelatinous band. The rudimentary main tentacles lie near the mouth.

As shown in Figures 4.6 and 4.7, the basic features of the cydippids are retained in the lobates and cestids even though they have morphed spectacularly from the cydippid blueprint. That same trend continues in the remaining groups.

Beroida

In life, beroids resemble swimming mouths, particularly when their maw is open, and they are about to engulf their prey. They are usually flattened in the tentacular plane (Figure 4.8a), but the amount of compression varies between species, ranging from nearly flat to nearly cylindrical. The mouth and pharynx are greatly expanded (Figure 4.8b), though the stomach is quite small. The statocyst resides above the stomach and aboral canals. It has highly tufted polar fields which presumably have a sensory function (Figure 4.8c).

The canals of beroids are far more straightforward than those of the lobates or cestids. All canals originate in the stomach. Meridional canals run beneath the eight comb rows and fuse to a circumoral ring that runs around the rim of the mouth. Similarly, the pharyngeal canals run down the midline of the flattened side of the animal and fuse to the oral ring. All canals are highly branched, forming an elaborate network of smaller canals that cover most of the surface of the animal (Figure 4.8b).

Within a species, comb rows are all equal in length, but they vary in length from species to species, ranging from about half to nearly full body length. Tentacles are completely absent in Beroe, even as larvae, making it the only order within the Ctenophora without them.

Platyctenida

The platyctenids are another example of a highly modified ctenophore body plan. They are modified for crawling and are highly compressed in the oral-aboral plane. In the genus *Ctenoplana*, most of the important ctenophore characteristics are easily discerned: they just have been flattened out a bit. In some of the other genera

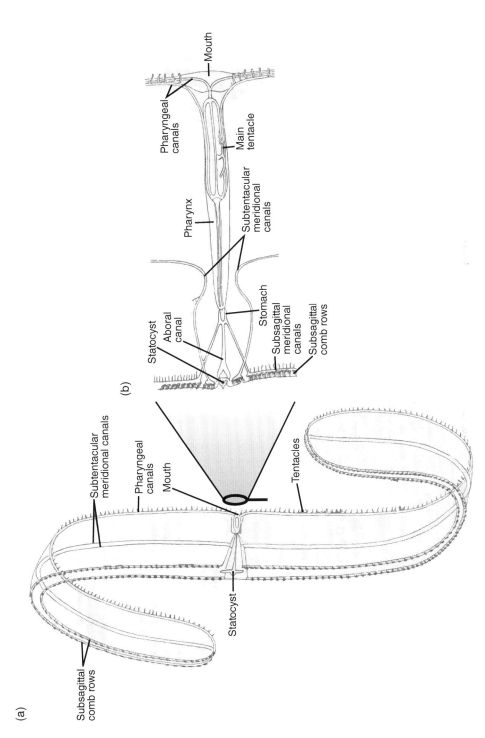

Figure 4.7 Cestid ctenophores. (a) The cestid ctenophore *Velamen*; (b) enlargement of the central region of a cestid ctenophore. *Source:* Adapted from Mayer (1912), plates 13, 14.

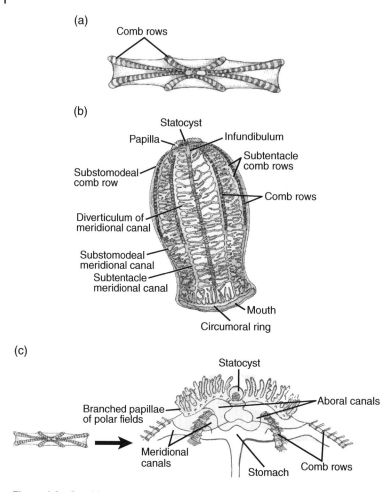

Figure 4.8 Beroid ctenophores. (a) Aboral view of *Beroë* (flattened aspect); (b) *Beroë*, whole animal; (c) enlargement of the aboral view of *Beroë*. *Source:* Adapted from Mayer (1912), plates 15, 16, figure 76.

(e.g. *Coeloplana* and *Tjalfiella*), the comb rows have been lost. Figure 4.9a, a drawing of *Ctenoplana* from Komai (1934), shows both internal and external features. Eight comb rows are present on the aboral surface. Two prominent tentacles with pronounced sheaths and side branches (tentillae) are present, defining the tentacular plane. A system of six gastric canals radiates from each side of the stomach, with four canals associated with the comb plates and two with the tentacles. The canals anastomose into an elaborate peripheral network. Gonads are found within the subtentacular canals, which open to the surface via a duct and a pore.

The oral surface is covered by what is actually the everted pharynx, forming a ciliated "sole" that aids in creeping. The gut within the animal is thus much shortened, essentially the internal portion of the pharynx and the stomach, which leads then to two aboral or anal canals.

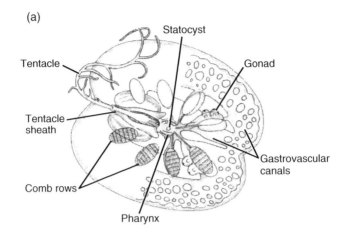

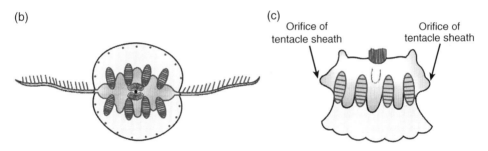

Figure 4.9 Platyctenid ctenophores. (a) The platyctenid *Ctenoplana* sp. showing both internal and external features. (b) *Ctenoplana* sp. configured for creeping on the bottom with tentacles extended; (c) *Ctenoplana* sp. configured for swimming. *Sources:* (a) Adapted from Komai (1934), figure 1 (p. 248); (b, c) Adapted from Willey (1902), figure 13 (p. 720).

Ctenoplana is capable of swimming (Figure 4.9b). A line drawn through the tentacular plane defines two lobes that can be drawn up to form a more typical looking ctenophore. The two lobes can be flapped together to swim, or the animal can use its comb rows for propulsion. Being a small animal, 5–8 mm in length, it can float upside down on the surface film (Hyman 1940).

The platyctenids were discovered in the 1880s near Sumatra and many of the species hail from the Indo-Pacific. They range from about 6 to 60 mm in length. Body forms vary considerably with species, but all possess well-developed tentacles and most species exhibit a similar body form to *Ctenoplana*: a flattened, elongated oval. Unlike most species of ctenophores, the platyctenids are often highly pigmented, making them appear deceptively like large colorful flatworms. Despite their aberrant morphology, the platyctenids comprise about 30% of the total number of species of ctenophores (Mills and Haddock 2007).

Ganeshida

The ganeshids comprise a single family with one genus and two species. They are considered to be intermediate between the cydippids and the lobates in body form (Harbison and Madin 1982), though their mouth resembles that of Beroe. They are diminutive, with adults about a centimeter in height. Ganeshids are found exclusively in Asian seas.

The ganeshid body plan is a straightforward ctenophore design (Figure 4.10a), with nine gastric canals radiating directly from the stomach: one aboral canal, two paragastric canals, two tentacular canals, and four canals that bifurcate to form the eight meridional canals. All but the aboral canal run the length of the animal to meet in a circumoral canal.

Two filamentous tentacles are present, with well-developed sheaths that open to the oral side of the animal. No auricles or lobes are present. Little information is available on the ganeshids other than the initial description (Dawydoff 1946).

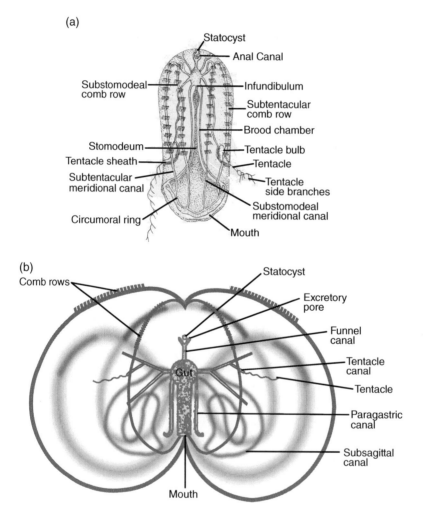

Figure 4.10 Ganeshid and thalassocalycid ctenophores. (a) The ganeshid ctenophore *Ganesha elegans*; (b) the thalassocalycid ctenophore *Thalassocalyce inconstans*. *Source:* (a) Adapted from Harbison and Madin (1982) (p. 710).

Thalassocalycida

The Thalassocalycida is a recently discovered order of ctenophores, containing only one family, genus, and species: *Thalassocalycida inconstans* (Figure 4.10b). It has a greatly expanded body around the mouth, forming a "medusa-like" bell. Tentacles are located near the mouth and lack sheaths, as in the lobates. It has a full complement of canals and, like some of the lobates, the meridional canals form a complex pattern within the bell. It feeds by trapping small crustaceans on the inside of the bell with mucous and conveying them to the mouth with ciliary action (Harbison and Madin 1982).

Foraging Strategies

General Considerations

Like the medusae and siphonophores, ctenophores forage in a three-dimensional environment with very limited sensory input. Their ciliary propulsive system is unique, giving the more active species a high degree of maneuverability but not much forward speed. Ctenophores are exclusively carnivorous and have a greater diversity of feeding mechanisms than the pelagic cnidarians. For example, unlike the medusae, at least one group of ctenophores, the beroids, chase down and engulf their prey.

Four feeding modes have been identified by Haddock (2007): those that use tentacles for feeding, those that use lobes for feeding, those that engulf prey, and feeding specialists. Tentacle feeders among the ctenophores employ a lie-in-wait strategy, similar to the medusae and siphonophores that remain motionless with tentacles spread in a "curtain of death." The difference is that instead of envenomated microharpoons (cnidocytes), the ctenophores have colloblasts, sticky cells that trap, but do not poison, prey.

Tentacles consist of a solid core surrounded by muscle fibers with an outer layer of colloblasts (Figure 4.11a and b), the ctenophore analogue to cnidocytes. Colloblasts have a hemispherical head composed of secretory granules that produce the adhesive used to trap prey. They are anchored to the tentacle core with a contractile spiral filament (Figure 4.11c). Oddly enough, the straight filament running within the spiral filament is a highly modified cell nucleus (Hyman 1940).

Interaction with Prey

The Cydippids

Elements of Madin's (1988) model for feeding in tentaculate predators, discussed at length in the prior chapter, work well for ctenophores that use tentacles as their main feeding tool. The tentilla on the tentacles of cydippids vary considerably in their length and spacing (Figure 4.12). Some form a dense curtain and others, such as in the deep-sea species, are more widely spaced but more adhesive, providing the ability to capture larger prey (Haddock 2007). Species whose tentacles have fewer tentilla (e.g. the genera *Hormiphora* and *Lampea*) seem to target larger prey such as salps and shrimp. Species with a dense curtain of tentilla (e.g. *Callianira* and *Mertensia*) captured smaller prey such as copepods and euphausiids. In all cases, observations disclosed an ambush strategy for hunting, allowing prey to swim into passively deployed tentacles where they were then trapped by the colloblasts. Once a prey item was trapped on a tentacle, the ctenophore contracted the tentacle and rotated its body to bring its mouth into contact with the prey. Apparently, some species contract both tentacles and others only the one that

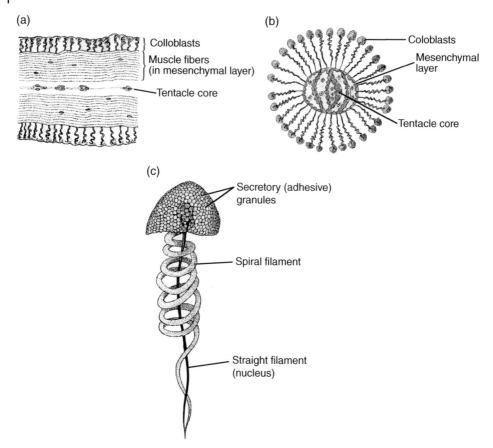

Figure 4.11 Ctenophore tentacles. (a) Longitudinal section through a ctenophore tentacle. (b) Cross section of a ctenophore tentacle. (c) Colloblast structure. *Sources*: (a) Adapted from Hyman (1940), figure 214 (p. 674); (b) Komai (1942), figure 7 (p. 14); (c) Adapted from Chun (1880), plate XVIII.

has ensnared a prey item (Madin 1988). The concept of the "encounter zone" where probability of prey capture is maximized applies as well to tentacle-bearing ctenophores as it does to medusae (Table 4.1).

The Lobates

The basics of prey capture are similar from species to species within the lobates, but there is considerable diversity in the detail. Prey items swim into the space between the lobes or are encouraged into the space by waving auricles and are then trapped in the sticky mucous layer coating the inner surface of the oral lobes and the auricles. When trapped on an auricle, prey items are conveyed to the mouth region via the ciliary activity of the auricle or the tentacles of the auricular groove. When food is trapped in the mucous of an oral lobe, the lobe folds around the prey item and brings it closer to the mouth where it is conveyed within reach of the lips by ciliary action and ingested (Harbison et al. 1978). Lobate tentacles also capture prey. Prey items unlucky enough to contact a tentacle are reeled in and brought to the mouth directly.

Morphology within the lobates varies quite a bit. Some species having fairly short auricles (e.g. *Mnemiopsis* in Figure 4.6c)

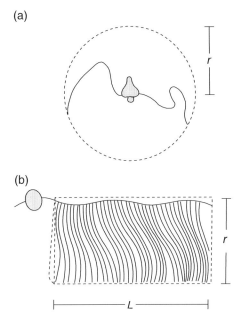

Figure 4.12 Prey encounter zones. (a) Tentacles fish within a volume described by a sphere with radius r, where r = the average tentacle extension and volume = $4/3\pi r^3$. (b) Tentacles form a flat curtain that fishes in a cylindrical volume of radius r around the stem. Radius (r) is tentacle length and cylinder volume = $\pi r^2 L$, where L is the length of the stem. Source: Madin (1988), figure 1 (p. 416). Reproduced with the permission of the Bulletin of Marine Science.

and others like *Leucothea* (Figure 4.13) have long sinuous auricles that beat continuously, creating a current within the volume enclosed by the oral lobes: the "kill zone." Lobates also differ greatly in the size of the lobes, the length of the main tentacles, and in the specifics of how the lobes are used to gather food particles. Most employ the "mucous entrapment strategy" described above. *Ocyropsis,* a robust open-ocean ctenophore, uses its muscular lobes to entrap prey. When contact is made, the lobe folds over the prey trapping it within a tube that is quickly conveyed to the protrusible mouth which "snatches the prey from the lobe" (Harbison et al. 1978). The sequence of entrapment to ingestion is very rapid, a second or less, and enables *Ocyropsis* to capture larger more mobile prey, like euphausiids and pteropods, than can most other ctenophores. *Ocyropsis* is also capable of using its muscular oral lobes to swim, flapping them in a motion resembling the contraction of a medusan bell.

The Cestids

The feeding strategy of the cestids is as unusual as the group's morphology. Recall that four of the eight comb rows are greatly expanded to run the length of the greatly exaggerated stomodeal plane on the aboral side, with the remaining four reduced to a few comb plates, also on the aboral side. The oral tentacles lie on the oral side of the "belt," running the entire length of the oral side of the animal within an oral groove (Figure 4.7).

In the open ocean, the venus girdle hangs nearly motionless in the water or propels itself slowly in an oral direction with the comb rows of the aboral side (Harbison et al. 1978). The oral tentacles extend from the oral groove and flatten against the side of the animal, forming a matrix of tentacles and tentilla that extend the entire width of the "belt," making both sides of the ctenophore a capture zone for small prey like copepods that are unlucky enough to bump into it. Once captured by an oral tentacle or one of its tentilla, the prey item is retracted into the oral groove and moved by cilia within the groove to the mouth.

The Beroids

Beroids feed by engulfing their prey whole or, shark-like, by removing chunks with their enormous mouth and retaining the prey within their sac-like body. They prey mainly on other ctenophores, sometimes forming a predator/prey pair with locally

Table 4.1 Tentacular encounter-zone shape and volume, tentacular length, volume, and density, and prey types. Encounter zone and tentacular data from Madin (1988).

Species	Encounter zone		Tentacular parameters			Prey types
	Shape of space	Volume (cm^3)	Length (m)	Volume (cm^3)	Density[a] (ppb)	
Cydippida						
Callianira bialata	Sphere	87 000	76.1	0.0204	230.00	Hyperiid amphipods, krill, copepods, other
Mertensia ovum	Sphere	113 000	161.2	0.0403	360.00	Primarily large copepods
Horniphora spp.	Sphere	65 500 000	11.0	0.1713	2.60	Heteropods, nudibranchs, fish, other
Lampea pancerina	Sphere	65 500 000	10.5	0.0164	0.20	Specific predator of salps
Lobata						
Leucothea multicornis	Cylinder	2000	—	—	—	Primarily copepods
Cestida						
Cestum veneris	Narrow elongate body	640	—	—	—	Primarily small copepods and nauplii

[a] The calculated space within the encounter zone occupied by tentacles.

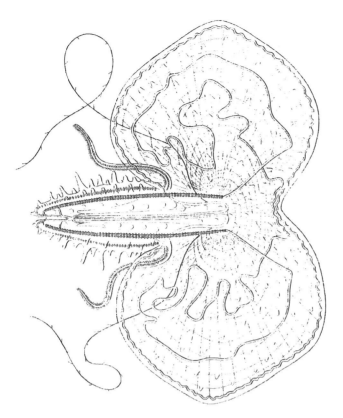

Figure 4.13 The lobate ctenophore *Leucothea ochracea*. *Source:* Redrawn from Mayer (1912), plate 9.

abundant lobates (Swanberg 1974). Swanberg (1974) provides some insight on sensory mechanisms used for detecting prey in a phylum whose only obvious sensory apparatus is the statocyst and its sometimes-associated polar field or, in the case of the beroids, polar tufts. Water conditioned by the presence of prey species caused an increase in swimming by *Beroe ovata* when compared with natural seawater, implying that some level of chemosensory ability is involved in *Beroe's* search behavior.

Once the lips of a Beroe contact prey, they spread over a portion of the prey animal, such as one lobe of a lobate ctenophore. The lips literally crawl up the prey animal, much as a snake engulfs its prey, until the gut of the *Beroe* is full. Macrocilia on the mouth of *Beroe* aid in both the engulfing process and in severing the prey once the *Beroe* is full. Macrocilia are thus quite important in that they allow Beroe to feed on animals larger than themselves (Swanberg 1974).

Though prey specialization has been reported for *Beroe* species (e.g. Mayer 1912; Nelson 1925) more recent evidence suggests that *Beroe* will consume whatever ctenophores are put before them (Swanberg 1974; Harbison et al. 1978). For example, *Beroe ovata* has been observed feeding on the lobates *Bolinopsis vitrea, Ocyropsis chrystallina,* and *Leucothea multicornis,* as well as the cestid *Cestum veneris.*

The Platyctenids

Platyctenids, despite their adaptations for a creeping existence, apparently feed on zooplankton captured with their tentacles (Hyman 1940), presumably in a similar manner to the cydippids: through ambush. Little information is available on the feeding behavior of this unusual group. They have been observed on a variety of substrates, including algae, rock, and floats (Mills and Haddock 2007) and are most common in tropical and subtropical waters (Harbison and Madin 1982).

The Ganeshids

No information is currently available on feeding in the Ganeshids.

The Thalassocalycids

Thalassocalyce has a unique morphology, as described above, but its feeding strategy most closely resembles that of the lobates. Prey items that swim into the cup-shaped bell are trapped by the mucous lining its inner surface. The bell then contracts to bring the prey item closer to the mouth, whereupon it is moved to the mouth by ciliary action and swallowed (Harbison et al. 1978). As observed by Harbison et al. (1978), the animal acts as a passive trap.

Specialists

A few ctenophore species are highly specialized feeders. We will treat the two most unusual examples here, keeping in mind that there are many more (cf. Haddock 2007). The first example is a cydippid, *Lampea pancerina*, that specializes on salps and shows a remarkable degree of adaptability in its feeding. *Lampea* hunts using a conventional ambush strategy, remaining motionless in the water with its tentacles extended. When a salp or salp chain contacts the tentacle, it sticks, and the ctenophore, whether small or large, is pulled along behind the more strongly swimming salp. When the tentacles of the hitchhiking ctenophore contract, it is pulled up to the salp and can begin to devour it. Here is where the story gets unusual. If the *Lampea* is much smaller than the salp, it spreads itself (mouth down!) onto the salp to become the "Gastrodes" form, a flattened form burrowed into the test of the salp that was formerly considered to be a parasite but is now known as a stage in the growth of *Lampea* (Harbison et al. 1978). As *Lampea* grow, they use the same strategy to capture salps, but the larger adults simply devour the salps whole when the mouth contacts them. *Lampea* can ingest salp chains one individual at a time until the entire chain is consumed (Madin 1988).

The second example and probably best-known specialist is the cydippid, *Haeckelia rubra*. In a now-famous study, Mills and Miller (1984) observed *Haeckelia* ingesting the tentacles of the narcomedusa *Aegina citrea* one at a time, bringing to mind a diner inhaling noodles of spaghetti. The ctenophore ingests the undischarged cnidae and incorporates them into its own tentacle, becoming a stinging ctenophore. The process of ingesting and incorporating intact cnidae into the tissues of a predator is known as kleptocnidae. Though very rare, similar behavior has been observed in some pelagic nudibranchs and in some freshwater flatworms (Mills and Miller 1984).

Diet, Feeding Rates, and Impacts on Prey Populations

Most information on the diets of ctenophores and their impact as predators comes from species found nearshore, in particular species of *Mnemiopsis, Pleurobrachia,* and *Beroe*. Less is known about open-ocean genera such as *Leucothea* and *Euramphea*, and the diets and impacts of deep-sea species like *Kiyohimea* remain largely a mystery at this writing. Thus, we have to engage in some cautious extrapolation.

Nearshore species of ctenophores are voracious predators that show decided seasonal peaks in abundance. The most intensively studied genus has been the lobate *Mnemiopsis* due to its abundance in important, easily accessible, estuaries, its role as an invasive species, and its potential impacts on zooplankton populations.

In the temperate areas of its distribution, *Mnemiopsis* exhibits an explosive period of growth in summer (June–September) that either just follows (Narragansett Bay; Kremer and Nixon 1976) or coincides with (Chesapeake Bay; Purcell et al. 2001) the peak in zooplankton biomass for each system (In Narragansett Bay, the overwintering population of *Mnemiopsis* of one or two individuals $10^4 m^{-3}$ increased to abundances of 10 animals m^{-3} in a period of one to two months: an increase of five orders of magnitude (Kremer and Nixon 1976). Such a rapid increase in predator numbers has the potential for very wide-ranging impacts, even if the predator has limited mobility. Common zooplankton prey like copepods and other small Crustacea would be consumed in quantity, but other, highly seasonal, prey such as fish eggs and larvae would also be taken, impacting both higher and lower trophic levels within the system.

Evaluating the impact of any predator requires estimates of its abundance and its feeding rate. Feeding rates can be estimated by examining gut fullness in field-caught specimens captured by repetitive sampling through time. Alternately, experimental observations of feeding rates can be extrapolated to the field (Purcell and Arai 2001; Purcell et al. 2001). Both methods have their positive and negative attributes and both have been widely employed. Estimates of potential removal efficiency by *Mnemiopsis* range from 2% to almost 100% of the local zooplankton standing crop per day, depending on the prey taxa considered. They are particularly lethal to fish eggs and larvae, with most studies obtaining numbers of 20–40% of the eggs and larvae available per day.

Like *Mnemiopsis,* the cydippid *Pleurobrachia pileus* is a neritic species with peaks and valleys in abundance over the course of the year. In Scottish waters, peak abundances typically occurred in the late summer and early fall with the highest numbers in October and November (Fraser 1970). Long-term survey data showed that the differences between the population minimum in February vs. the peak in November were about 14-fold, a substantial change but not as profound as that observed by Kremer and Nixon (1976) in Narragansett Bay. Dietary studies on *Pleurobrachia* revealed a voracious and opportunistic feeder, with the majority of its prey consisting of copepods and crustacean larvae. It was observed that *Pleurobrachia* will eat pretty much whatever it can catch. A consistent observation in the literature describing diet and feeding in ctenophores is that wherever ctenophores are highly abundant, the remainder of the animal plankton is nearly absent (Fraser 1970). Their mettle as carnivores has ramifications up and down the trophic pyramid.

Ctenophores as Invasive Species

A particularly compelling story that underscores the effectiveness of ctenophores as predators as well as their ability to multiply rapidly is the invasion of the Black Sea by *Mnemiopsis*. The ctenophore was introduced to the Black Sea via ballast water in the early 1980s (Purcell et al. 2001) and wreaked havoc there for several years.

The Black Sea (Figure 4.14) is an inland sea, a semi-enclosed body of water that has an unusual hydrography despite its very large size (436 400 km² not including the Sea of Azov). It has a maximum depth of 2212 m. It is connected to the Mediterranean by two very narrow straits that have achieved fame through the many wars that have been fought to control them: the Bosporus, which connects the Black Sea to the Sea of Marmara, and the Dardanelles, which connect the Sea of Marmara to the Aegean Sea. Together the two straits are known as the Turkish Straits. The Bosporus is only 700 m wide, so that connection is indeed restrictive! The Black Sea is fed by several large rivers of which the Don, the Dneiper, and the Danube are the most important. The freshwater input results in a salinity of approximately 18‰ in surface waters. When water flow into the Black Sea is sufficient to produce a positive discharge through the Turkish Straits, there is an exchange flow of higher-salinity Mediterranean Sea water that enters beneath the fresher outflow from the Black Sea, much as in a typical estuary. In some years, there is no exchange with the Mediterranean and the Black Sea is essentially a brackish water lake. Because of the restricted flow into the Black Sea, the shallow depths of the straits, and the large difference in density between the waters above and below the pycnocline, the Black Sea is anoxic below a depth of about 200 m.

Conditions in the Black Sea were nearly perfect for the introduced *Mnemiopsis* to flourish. With summer temperatures in the

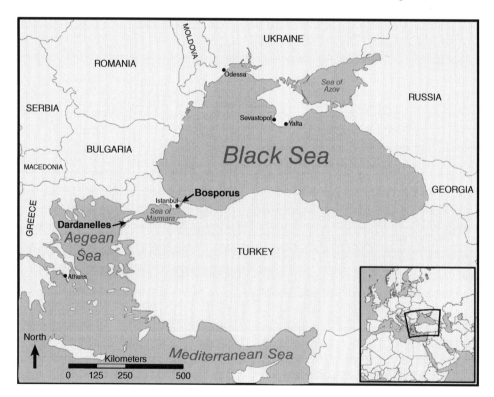

Figure 4.14 The Black Sea region of Eastern Europe showing the two narrow straits, the Bosporus and Dardanelles, that connect the Black Sea, the Sea of Marmara, and the Aegean Sea. See color plate section for color representation of this figure.

mid 20s °C and a surface salinity of 18‰, the physical environment strongly resembled that of its native haunts on the east coast of the United States, Narragansett Bay and Chesapeake Bay. Natural predators were nearly absent. Overfishing of the zooplanktivorous Black Sea Anchovy, Mediterranean horse mackerel, and sprat, had removed potential competitors for zooplankton from the system, leaving *Mnemiopsis* with a clear field of play (Purcell et al. 2001). Over most of the decade spanning the late 1980s to the late 1990s, *Mnemiopsis* was extremely abundant and the remainder of the zooplankton biomass varied inversely with it. To add insult to injury, *Mnemiopsis* coupled removal of the prey of zooplanktivorous fishes with very heavy predation on their eggs and larvae (Figure 4.15). Populations of the zooplanktivorous fishes in the Black Sea suffered a severe decline through the 1990s (Figure 4.15).

Then in 1997, *Beroe ovata*, a natural predator of *Mnemiopsis,* appeared in the Black Sea for the first time, perhaps also introduced via ballast water discharge (Shiganova et al. 2001). It was a fortunate occurrence for the Black Sea system. By 1999, *Mnemiopsis* populations showed a decline with a reciprocal increase in other zooplankton and in numbers of fish eggs (Figure 4.15). It remains to be seen if *Beroe* will aid the Black Sea system in a full recovery to pre-invasion conditions.

Digestion

Digestion in ctenophores is mainly extracellular, which is to say that food breakdown occurs primarily within the lumen of the digestive tract, as it does in most higher animals. In all orders, the mouth leads to a muscular, highly folded, pharynx that has gland cells to produce digestive enzymes to aid in food breakdown. Within the gut, food is moved about by cilia to promote digestion; once digested, it passes into the gastrovascular canals for distribution to the rest of the ctenophore (see Figure 4.3). Digestion is completed within the gastrovascular canal system, nutrients are distributed throughout the body, and absorption takes place. Pores lead from the canals into the mesenchyme to aid in distributing the nutrient soup and that diffusion is enhanced by ciliated cells surrounding the opening (Figure 4.16a). The inner walls of the meridional gastrovascular canals are very much thinner than the outer walls, which contain the gonadal tissue (Figure 4.16b).

Nerves and Sense Organs: Coordination and Conduction

Like the Cnidaria, ctenophores have non-centralized nerve nets that form a diffuse sub-epidermal plexus. Unlike the Cnidaria, the neurons form plexes beneath the comb rows that resemble nerves (Figure 4.17). A similar concentrated plexus surrounds the mouth.

The apical organ at the aboral pole of a ctenophore is the most obvious sensory apparatus, though perhaps "organ" is a misnomer for a phylum considered to be diploblastic. A calcareous concretion, or statolith, within the apical sense organ is supported by four long tufts of cilia known as balancers (Figure 4.5). Each balancer leads to a pair of ciliated furrows, each of which connects to one comb row (Figure 4.17). Thus, each balancer innervates the two comb rows in its quadrant. If a ciliated furrow is cut, the result is an uncoordinated beat in the two comb rows that it is innervating.

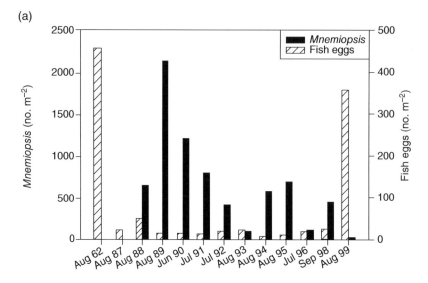
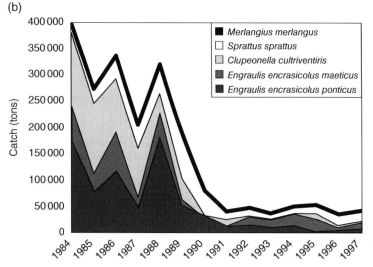

Figure 4.15 Impact of ctenophores on fish populations. (a) Average abundance (no. m^{-2}) of fish eggs and *Mnemiopsis* in June–August in the northern Black Sea. The first bar shows numbers in 1962. The last date is after the arrival of the ctenophore, *Beroe ovata*, which eats *Mnemiopsis*; (b) annual catch of zooplanktivorous fish in the Black Sea and the Sea of Azov in 1984–1997. *Sources:* (a) Purcell et al. (2001), figure 11 (p. 168). (b) Purcell et al. (2001), figure 12 (p. 168).

The cilia beat-frequency changes in response to the weight of the statolith, effecting a change of orientation in the water column or a shift of direction (Figures 4.18 and 4.19) similar to the way a boat turns in response to increased oar-strokes on one side relative to the other.

Locomotion

The meridional comb rows are composed of ctenes (Figure 4.1), transverse bands composed of hundreds of long (3.5 mm) cilia fused together at the base and beating together in unison as a plate Ctenophores

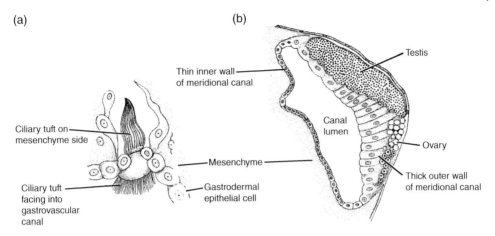

Figure 4.16 Gastrovascular canal structure. (a) A rosette cell from the gastrovascular canal of *Coeloplana*. (b) Meridional canal (in section) of a ctenophore. The gonads are strips of cells in the outer wall of the meridional canal. *Sources:* (a) Adapted from Hyman (1940), figure 215 (p. 676). (b) Adapted from Komai (1934), figure 6 (p. 253).

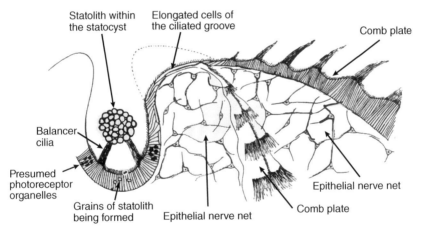

Figure 4.17 The arrangement of the principal components of the apical organ/comb row system. *Source:* Horridge (1974), figure 2 (p. 442). Reproduced with the permission of Academic Press.

are the largest animals to use ciliary-based propulsion as their primary locomotory strategy. Ciliary movement is widely employed by protists (e.g. *Paramecium*) but, other than in the ctenophores, it is not found in the assemblage of metazoan movement.

Some of the lobates (e.g. *Ocyropsis, Bathocyroe*) can clap their muscular oral lobes together to produce an effective jet propulsion quite similar in appearance to the contraction of the bell of a swimming medusa. The mesenchyme of ctenophores performs a similar function to that of the cnidarians, acting as a deformable, elastic skeleton that the musculature can work against to produce movement but has a memory and recoils to its original shape after deformation. Though the musculature of ctenophores clearly varies between

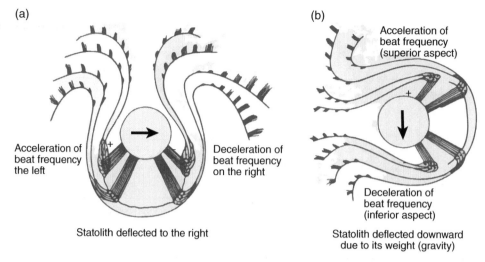

Figure 4.18 Function of balancer cilia in ctenophores. (a) Apical organ pointing upward (animal oriented vertically): a deflection of the statolith to the right causes an increase of ctene beat frequency on the left and a decrease on the right; (b) with the animal oriented horizontally the weight of the statolith deflects it downward causing an increase in beat frequency on the upper aspect and deceleration on the lower. *Source:* Horridge (1974), figure 3 (p. 442). Reproduced with the permission of Academic Press.

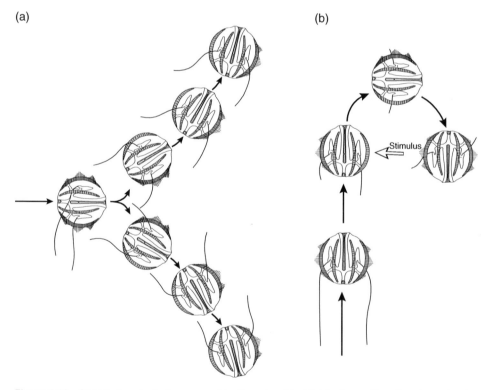

Figure 4.19 Statocyst control of movement in *Pleurobrachia*. (a) A horizontally swimming animal turns upward or downward by increasing the frequency of ciliary waves on the lower or uppermost comb rows, respectively. (b) When an upwardly swimming animal receives a stimulus (indicated by the arrow), tentacles retract and ciliary waves on the stimulated side are temporarily arrested, resulting in a reversal in the direction of swimming. *Source:* Horridge (1974), figure 4 (p. 443). Reproduced with the permission of Academic Press.

orders, the cydippids provide an example. The arrangement of muscle fibers shows an opposing configuration, with radial fibers oriented from the center to the periphery and looped fibers and latitudinal fibers running circumferentially (Figure 4.20). The orientation of the looped fibers tends to retain the undeformed circumference of the animal, while the antagonistic radial fibers tend to elongate it in the oral-aboral plane. The three sets of muscles working in concert allows for a limited repertoire of shape-shifting in an otherwise nearly spherical animal.

Distribution

The ctenophores are a pan-global group, with representatives found from equatorial to polar oceans (Mianzan 1999) and from the surface to bathypelagic depths (Madin and Harbison 1978). They vary considerably in their robustness; deeper-living and open-ocean species are particularly delicate. For that reason, net sampling does not produce a quantitative picture of ctenophore distributions (cf. Harbison et al. 1978). Many species simply pass through the meshes of trawl nets or, if retained, do not remain intact enough to be identified. Thus, meaningful sampling of ctenophores is limited to capture by open-ocean divers and capture or observation using submersibles. Surveys using bluewater diving have been markedly few, and none have been undertaken with manned submersibles.

Harbison et al. (1978) described the abundance and distribution of open-ocean ctenophores collected on 250 dives in the temperate-tropical North Atlantic (Table 4.2). Essentially all dives were far enough from shore to be considered true open-ocean samples. Ctenophores were collected at 188 of the 250 stations, two or more species on 137, three or more on 94, and four or more on 57. On 67 dives, at least one species was considered abundant (i.e. ≥20 specimens observed). Clearly, ctenophores are important predators in the open ocean as they are ubiquitous and often quite abundant, even very far from shore. At 12 stations, ctenophores were estimated to exceed one animal per cubic meter, a density similar to that encountered in the Black Sea during the *Mnemiopsis* blooms but seen in the open ocean!

Table 4.3 gives estimates of three important species of open-ocean ctenophores encountered during dives in the southern Sargasso Sea and Caribbean Sea. The numbers illustrate well the marked patchiness of different species (*Ocyropsis* vs.

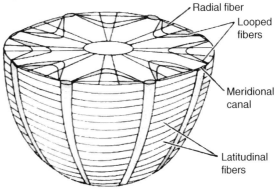

Figure 4.20 Transverse section through the region of the pharynx of Pleurobrachia showing the arrangement of muscle fibers. The gastrovascular system and tentacle sheaths have been omitted for clarity. *Source:* Chapman (1958), figure 3 (p. 356) in Biological Reviews of the Cambridge Philosophical Society, Vol. 33, Issue 3, © John Wiley & Sons. Reproduced with the permission of John Wiley & Sons.

Table 4.2 Abundance and distribution of open-ocean ctenophores collected on 250 dives in the temperate-tropical north Atlantic. Geographic regions are as described in Backus et al. (1977). *Source:* Harbison et al. (1978), table 2 (p. 244). Reproduced with the permission of Pergamon Press/Elsevier.

Region	Province	Number of dives	Dives with ctenophores	% Dives with ctenophores
Temperate	Mediterranean Outflow	1	1	100
	Slope Water	57	38	67
	Slope Water – Western Sargasso	5	4	80
Subtropical	Northern Sargasso	48	35	73
	North African Subtropical	16	8	50
	Southern Sargasso	20	14	70
	Southern Sargasso – Lesser Antillean	10	2	20
Tropical	Lesser Antillean	18	13	72
	Caribbean	10	9	90
	Amazonian	36	36	100
	Equatorial Indian Ocean	29	28	97
	Total all stations	**250**	**188**	**75**

Euramphea and *Cestum*) as well as the peaks and valleys in abundance encountered during open-ocean transits. In the lobates, common pelagic genera included *Ocyropsis*, *Leucothea*, and *Euramphea*. Other common genera encountered were the cydippid *Callianira*, the cestids *Cestum* and *Velamen*, and the beroid *Beroe*.

Ctenophores and Evolution

Because of their similarity in appearance to the cnidarians, the ctenophores have always been considered a closely related group. Both phyla are highly transparent, radially or biradially symmetrical, have a high water content, and have an inner mesoglea that makes up a considerable portion of the body mass as well as acting as a primitive skeleton. Despite their many similarities, in recent years, the affinity between the two groups has been increasingly questioned (Harbison 1985; Podar et al. 2001) as has the classic scheme presented earlier in this chapter for organization within the ctenophores themselves.

Several important items are at issue, the most important of which are the relationships of the ctenophores to other animal phyla and to each other. At a still higher level, the belief that ctenophores may represent the beginnings of bilateral symmetry, an important evolutionary step, is a view that has gained increasing acceptance (Harbison 1985; Nielsen 1995, 2001; Martindale and Henry 1998).

Virtually all treatments of the ctenophores consider them to be a monophyletic group, that is they are derived from a common ancestral species. Which of the current species that ancestor most closely resembled is the source of much of the disagreement, and the absence of a substantive fossil record makes the issue much more difficult to resolve. The discovery of a

Table 4.3 The density of three important open-ocean ctenophores as estimated from dives in the southern Sargasso and Caribbean Seas. Stations 436–443 were the same geographic position in the Sargasso; stations 447–454 were the same position in the Caribbean. *Source:* Harbison et al. (1978), table 4 (p. 249). Reproduced with the permission of Pergamon Press/Elsevier.

Station no.	Total drift (m)	Eurhamphaea vexilligera		Ocyropsis crystallina		Cestum veneris	
		Animals/dive	Animals/1000 m³	Animals/dive	Animals/1000 m³	Animals/dive	Animals/1000 m³
432	60	40	31	0	0	0	0
433	60	30	23	0	0	5	3.9
434	110	20	8.4	0	0	1	0.42
435	290	10	1.6	20	3.2	10	1.6
436	340	0	0	0	0	3	0.41
437	400	0	0	0	0	0	0
438	350	17	2.2	0	0	1	0.13
440	230	10	2.0	0	0	2	0.40
441	230	3	0.60	0	0	2	0.40
442	250	3	0.56	0	0	2	0.37
443	300	1	0.15	0	0	1	0.15
444	170	18	4.9	0	0	1	0.27
445	160	6	1.7	0	0	0	0
446	130	0	0	40	14	4	1.4
447	60	0	0	0	0	15	12
448	60	10	7.7	2	1.5	2	1.5
449	60	0	0	0	0	0	0
450	60	25	19	3	2.3	30	23
451	60	1	0.77	5	3.9	5	3.9
452	60	0	0	0	0	0	0
453	60	7	5.4	5	3.9	0	0
454	100	30	14	30	14	5	2.3

cydippid-like fossil from the Devonian period (Stanley and Sturmer 1983) tilted the scales toward what most people believed to be the most primitive group, the Cydippida (e.g. Hyman 1940). However, later discoveries from the much earlier Cambrian period (Chen et al. 1991; Conway Morris and Collins 1996) more closely resembled the beroids, the only group that lacks tentacles, suggesting a "nude" ancestor. Molecular methods (Podar et al. 2001) point to the cydippids as being the most likely ancestral type, in particular the family Mertensiidae. Further, the evidence strongly supports a monophyletic origin for the ctenophores as a whole.

Within the greater framework of the monophyletic Ctenophora, both morphological

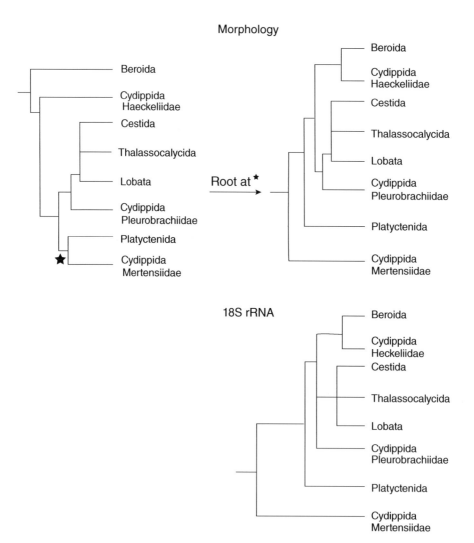

Figure 4.21 Comparison of a hypothesis of ctenophore phylogeny based on morphology with one based on 18S rRNA phylogeny. *Source:* Podar et al. (2001), figure 6 (p. 228). Reproduced with the permission of Elsevier.

(Harbison 1985) and molecular (Podar et al. 2001) evidence supports the idea that the cydippids are themselves not a monophyletic group as was previously believed. The cydippids exhibit enough morphological diversity that they can be split into three subgroups based on whether their tentacles exit toward the mouth or toward the aboral pole, or upon the morphology of their radial canals. Happily, the molecular evidence supports the morphological conclusions, dividing the cydippids into the same three subgroups (Figure 4.21).

Molecular methods have resolved some important questions about the ctenophores, but many remain. Are ctenophores indeed the basal group from which bilateral symmetry arose? Are they more closely related to the flatworms or to the Cnidaria? The morphological and developmental complexity of the Ctenophora suggests they fall between Cnidaria and Platyhelminthes and, for now, it is probably best to accept their traditional position in the metazoan hierarchy.

References

Backus, R.H., Craddock, J.E., Haedrich, R.L., and Robison, B.H. (1977). Atlantic mesopelagic zoogeography. In: *Fishes of the Western North Atlantic*, vol. 7. New Haven: Sears Foundation for Marine Research, Yale University.

Bayer, F.M. and Owre, H.B. (1968). *The Free-Living Lower Invertebrates*. New York: The Macmillan Company.

Brusca, R.C. and Brusca, G.J. (2003). *Invertebrates*, 2e. Sunderland: Sinauer.

Brusca, R.C., Moore, W., and Shuster, S.M. (2016). *Invertebrates*, 3e. Sunderland: Sinauer Associates.

Chapman, G. (1958). The hydrostatic skeleton in the invertebrates. *Biological Reviews of the Cambridge Philosophical Society* 33: 338–371.

Chen, J.Y., Bergstrom, J., Lindstrom, M., and Hou, X.G. (1991). Fossilized soft-bodied fauna. *Research and Exploration* 7: 8–19.

Chun, C. (1880). *Die Ctenophoren des golfes von Neapel und der angrenzenden meeresabschnitte, herausgeben von der zoologishchen station au Neapel. Eine Monographie*. Engelmann: Leipzig.

Conway Morris, S. and Collins, D.H. (1996). Middle Cambrian ctenophores from the Stephen Formation, British Columbia, Canada. *Philosophical Transactions of the Royal Society of London Part B* 351: 279–308.

Cuvier, G. (1817). *La Regne Animal*. Paris: A. Belin.

Dawydoff, C. (1946). Contribution a la connaissance des ctenophores pelagiques des eaux de l'Indochine. *Bulletin Biologique de la France et de la Belgique* 80: 113–170.

Eschscholtz, J.F. (1829). *System der Akalephen*. Berlin: Ferdinand Dummler.

Eschscholtz, J.F. (1829–1833). *Zoologischer Atlas*. Berlin: G. Reimer.

Fraser, H.F. (1970). The ecology of the ctenophore *Pleurobrachia pileus* in Scottish Waters. *Journal du Conseil international pour l'exploration de la mer* 88: 149–168.

Haddock, S.H.D. (2007). Comparative feeding behavior of planktonic ctenophores. *Integrative and Comparative Biology* 47: 847–853.

Harbison, G.R. (1985). On the classification and evolution of the Ctenophora. In: *The Origins and Relationships of Lower Invertebrates* (eds. S.C. Morris, J.D. George, R. Gibson and H.M. Platt). Oxford: Clarendon Press.

Harbison, G.R. and Madin, L.P. (1982). Ctenophora. In: *Synopsis and Classification of Living Organisms*, vol. *1* (ed. S.P. Parker). New York: McGraw-Hill.

Harbison, G.R. and Miller, R.L. (1986). Not all ctenophores are hermaphrodites. Studies on the systematics, distribution, sexuality, and development of two species of *Ocyropsis*. *Marine Biology* 90: 413–424.

Harbison, G.R., Madin, L.P., and Swanberg, N.R. (1978). On the natural history and distribution of the oceanic ctenophores. *Deep-Sea Research* 25: 233–256.

Hatschek, B. (1888). *Lehrbuch der Zoologie*. Jena: Gustav Fischer.

Horridge, G.A. (1974). Recent studies on the Ctenophora. In: *Coelenterate Biology* (eds. L. Muscatine and H.M. Lenhoff). New York: Academic Press.

Hyman, L.H. (1940). *The Invertebrates, Vol. 1, Protozoa through Ctenophora*. New York: McGraw-Hill.

Kaestner, A. (1967). *Invertebrate Zoology*, vol. 1. New York: Wiley.

Komai, T. (1934). On the structure of *Ctenoplana*. *Memoirs of the College of Science, Kyoto Imperial University, Series B* 9: 245–256.

Komai, T. (1942). The structure and development of the sessile ctenophore Lyrocteis imperatoris Komai. *Memoirs of the College of Science, Kyoto Imperial University, Series B* 17: 1–63.

Kremer, P. and Nixon, S. (1976). Distribution and abundance of the ctenophore *Mnemiopsis leidyi* in Narragansett Bay. *Estuarine and Coastal Marine Science* 4: 627–639.

Madin, L.P. (1988). Feeding behavior of tentaculate predators: in situ observation and a conceptual model. *Bulletin of Marine Science* 43: 413–429.

Madin, L.P. and Harbison, G.R. (1978). Thalassocalyce inconstans, new genus and species, an enigmatic ctenophore representing a new family and order. *Bulletin of Marine Science* 28: 680–687.

Martindale, M.Q. and Henry, J.Q. (1998). The development of radial and biradial symmetry: the evolution of bilaterality. *American Zoologist* 38: 672–684.

Mayer, A.G. (1912). *Ctenophores of the Atlantic Coast of North America*. Washington, DC: Carnegie Institute of Washington.

Mianzan, H.W. (1999). Ctenophora. In: *South Atlantic Zooplankton*, vol. 1 (ed. D. Boltovskoy). Leiden: Backhuys Publishers.

Mills, C.E. (n.d.). *Phylum Ctenophora: list of all valid species names*. http://faculty.washington.edu/cemills/Ctenolist.html

Mills, C.E. and Haddock, S.H.D. (2007). Ctenophora. In: *The Light and Smith Manual, Intertidal Invertebrates from Central California to Oregon* (ed. J.T. Carlton). Berkeley: University of California Press.

Mills, C.E. and Miller, R.L. (1984). Ingestion of a medusa (*Aegina citrea*) by the nematocyst-containing ctenophore *Haeckelia rubra* (formerly *Euchlora rubra*): phylogenetic implications. *Marine Biology* 78: 215–221.

Nelson, T.C. (1925). Occurrence and feeding habits of ctenophores in New Jersey coastal waters. *Biological Bulletin* 48: 92–111.

Nielsen, C. (1995). *Animal Evolution: Interrelationships of the Living Phyla*. Oxford: Oxford University Press.

Nielsen, C. (2001). *Animal Evolution: Interrelationships of the Living Phyla*, 2e. Oxford University Press: Oxford.

Ospovat, M.F. (1985a). On relationships between the orders Crytolobiferda, Lobiferida and Thalassocalycida (Ctenophora). *Zoolichesky Zhurnal* 64: 805–812.

Ospovat, M.F. (1985b). On phylogeny and classification of the type Ctenophora. *Zoologichesky Zhurnal* 64: 965–974.

Ospovat, M.F. (1985c). On the origin of symmetry in the Ctenophora. *Zoologichesky Zhurnal* 64: 1125–1132.

Podar, M., Haddock, S.H.D., Sogin, M.L., and Harbison, G.R. (2001). A molecular phylogenetic framework for the phylum Ctenophora using 18S rRNA genes. *Molecular Phylogenetics and Evolution* 21: 218–230.

Purcell, J.E. and Arai, M.N. (2001). Interactions of pelagic cnidarians and ctenophores with fish: a review. In: *Jellyfish Blooms: Ecological and Societal Importance* (eds. J.E. Purcell, W.M. Graham and H.J. Dumont). Dordrecht: Kluwer Academic Publishers.

Purcell, J.E., Shiganova, T.A., Decker, M.B., and Houde, E.D. (2001). The ctenophore Mnemiopsis in native and exotic habitats: U.S. estuaries versus the Black Sea Basin. In: *Jellyfish Blooms: Ecological and Societal Importance* (eds. J.E. Purcell, H.J. Dumont and W.M. Graham). Dordrecht: Kluwer Academic Publishers.

Shiganova, T.A., Bulgakova, Y.V., Volovik, S.P. et al. (2001). The new invader *Beroe ovata* Mayer 1912 and its effect on the ecosystem in the northeastern Black Sea. In: *Jellyfish Blooms: Ecological and Societal Importance* (eds. J.E. Purcell, H.J. Dumont and W.M. Graham). Dordrecht: Kluwer.

Stanley, G.D. and Sturmer, W. (1983). The first fossil ctenophore from the lower Devonian of West Germany. *Nature* 303: 518–520.

Swanberg, N.R. (1974). The feeding behavior of *Beroe ovata*. *Marine Biology* 24: 69–76.

Willey, A. (1902). *Zoological Results Based on Material Collected from New Britain, New Guinea, Loyalty Islands and Elsewhere. Part VI*. Cambridge: Cambridge University Press.

5

The Nemertea

Introduction

The phylum Nemertea (also known as Nemertina or Rhynchocoela), or ribbon worms, are found worldwide in the deep pelagic realm despite being a primarily benthic group (Gibson 1999). A few species are found in freshwater and a very limited number have even succeeded in moist terrestrial environments (Hyman 1951). The vast majority of ribbon worms are predators, with a few acting as endo-commensals and even parasites. From a purely biological standpoint, the body structure of the nemerteans forms an important transition between the medusae and ctenophores and the more advanced and more robust annelids, crustaceans, and molluscs. From an oceanographic perspective, the study of nemerteans is a work in progress and information on their role in the pelagic system is limited. Without question, the ribbon worms are an interesting and unusual group that fits squarely in the realm of the macrozooplankton, with sizes in the 10–100 mm range (Gibson 1999).

Close encounters with ribbon worms usually occur in the intertidal realm with the benthic representatives of the group, and then mainly to biologists in training. Nemerteans are not often in the lexicon of the non-scientist. Intertidal forms can get quite large: the often-quoted maximum length is attributed to the aptly named *Lineus longissimus*, a North Sea species that can reach 30 m, albeit with a diameter of only 9 mm (Kaestner 1967). Benthic species are vermiform (worm-like) in shape. They can be cylindrical like earthworms, though without the obvious segments, or they can be dorso-ventrally compressed (flattened) to be more ribbon-like in appearance (Figure 5.1a), the source of their common name. They have a smooth, almost shiny, skin, are highly extensible, and range in size from a few millimeters to a few meters in length.

The pelagic species are quite a bit shorter than their benthic relatives and are much broader and flatter in appearance (Figure 5.1b). Adaptation to the pelagic realm often includes posterior fin-like projections that further increase their surface area and are likely an aid to buoyancy and swimming.

Since the nemerteans are an important transitional group in our study of open-ocean life, an understanding of their basic biology will lay the groundwork for understanding the more advanced forms of oceanic life discussed in the following chapters. For example, unlike the ctenophores, nemerteans have true organ systems, including innovations such as a one-way

Life in the Open Ocean: The Biology of Pelagic Species, First Edition. Joseph J. Torres and Thomas G. Bailey.
© 2022 John Wiley & Sons Ltd. Published 2022 by John Wiley & Sons Ltd.

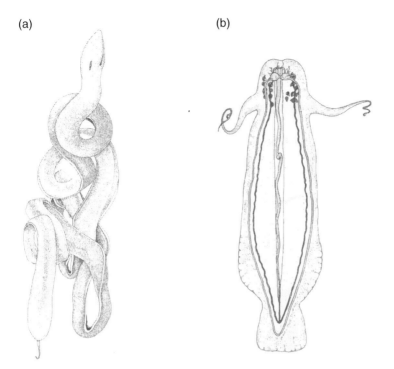

Figure 5.1 Benthic and pelagic nemerteans. (a) *Cerebratulus* sp., a common benthic nemertean (ribbonworm). (b) The pelagic nemertean *Nectonemertes mirabilis*. *Sources:* (a) Adapted from Coe (1943), plate 1; (b) Adapted from Coe (1926), figure 42 (p. 42).

digestive tract. This often-ignored group is much more important than many realize.

Classification

History

Nemerteans were first noted as "sea longworms" in 1758 by Borlase in his book *The Natural History of Cornwall* (Brusca and Brusca 2003), a publication that, surprisingly, is still quite readily available. In 1817, Jacques Cuvier provided two innovations. By describing and naming a specimen of nemertean as Nemertes (a Greek sea nymph one of the Nereids), he provided the most common name for what would later be recognized as a legitimate phylum. Second, he recognized the unique nature of the nemerteans and noted that they should be placed in their own order (Hyman 1951). Yet until 1850, the nemerteans were considered to be a type of flatworm, presumably because like the flatworms they had organ systems but no true body cavity or coelom. Body cavities and their development are important in the evolution of animal life.

Schultze (1851) is credited with the first detailed and accurate description of nemertean anatomy as well as the classification system that is still in use today. Though he considered nemerteans to be a type of flatworm, he called attention to two of their most important attributes: the one-way digestive tract and the eversible proboscis that is their most unique characteristic. He named the group the Nemertina, or Rhynchocoela after the sheath or cavity that houses the proboscis and in many species is the length of the body (Figure 5.1b).

The study of pelagic nemerteans was pioneered by two researchers, Brinkmann (1917a, 1917b) and Coe (1926), in a series of summary monographs and research articles stretching from about 1900 to the mid-1950s. The most famous articles by each author are those just cited. Brinkmann built on Schultze's classification system to better include the pelagic nemerteans and provided most of the families currently used. Coe described many new genera and species and afforded insights into the adaptation of nemerteans to the pelagic life. Widespread acceptance of the nemerteans as a legitimate phylum was cemented by Dr. Hyman's (1951) treatment of the group (Brusca and Brusca 2003).

Most recently, Gibson (1972, 1995, 1999) has contributed a series of scholarly treatments of the nemerteans, including a general description of their biology, an annotated checklist of species and distributions, and a very useful description of pelagic species in the South Atlantic.

Of Germ Layers and Body Cavities

With some exceptions, successful reproduction entails the meeting of sperm and egg (ovum), which initiates embryonic development. After fertilization, the ovum begins to divide in a genetically controlled, species-specific, pattern that eventually results in a functional larva. The process of larval development, or embryogeny, is a highly complex roadmap that has been studied over the last two centuries by biologists interested in both invertebrates and vertebrates. Consequently, some characteristics of early development shared by most animal groups are now well described. An understanding of the landmarks in the evolution of multi-cellular (metazoan) life that occur in the process of early development is critical to a comprehension of how animals relate to one another.

The two phyla of the prior two chapters, the Cnidaria and the Ctenophora, are both diploblastic, or "tissue-grade," groups, so named because as adults they have recognizable tissues but no true organ systems. During early development, their embryos comprise two types of cells, the ectoderm, which becomes the epidermis, and the endoderm, the presumptive gut. The two layers of cells are termed "germ layers" and the Cnidaria and Ctenophora are termed "diploblastic," from the Greek *diplos* (double) and *blastikós* (budding).

Three very important advancements in the developmental sequence of multicellular animal life have been instrumental in producing its present complexity. The first was the evolution of a third germ layer to add to the ectoderm and endoderm characteristic of the diploblastic Cnidaria and Ctenophora. Animal groups with three germ layers are termed triploblastic. The third germ layer, the mesoderm, appears a little later in embryogeny. It usually develops between the ectoderm and endoderm during the embryological process known as gastrulation, as cell numbers increase in the embryo and form multiple layers (Figure 5.2). The mesoderm is the germ layer that develops into muscle; the ectoderm develops into skin and nerve, and the endoderm into the gut. The increased flexibility and raw material afforded by the addition of a third germ layer facilitated development of true organ systems; complexity then had the potential for increasing by leaps and bounds.

The second breakthrough was the development of a coelom, or body cavity, between the gut and the body wall, a cavity usually filled with fluid. Though the presence of a fluid-filled space between the gut and body

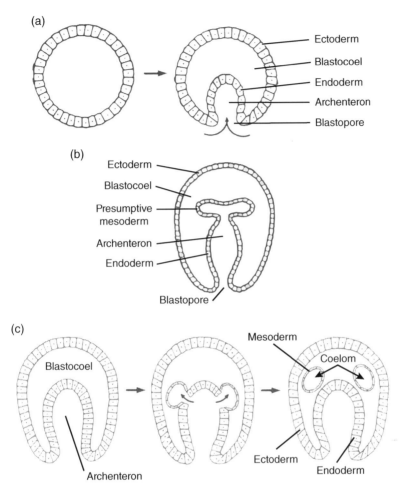

Figure 5.2 Gastrulation and coelom formation. (a) invagination of a coeloblastula to form a coelogastrula; (b) mesoderm formation by archenteric pouching; (c) archenteric pouching and coelom formation. *Source:* Adapted from Brusca and Brusca (2003), (a) figure 4.7 (p. 100); (b) figure 4.8 (p. 101); (c) figure 4.10 (p. 102).

wall may not seem to be of towering importance, it is momentous. The space freed the gut from being attached directly to the body wall except on the ends, a very different arrangement from the layout of the cnidarians and ctenophores where the gut runs through the mesoglea, or that of the flatworms and nemerteans, where it runs through a mesenchyme (internal cellular mass). The cavity provided space for the development of new structures within the body, allowed for a circulating fluid that could bathe those structures, and could act as a hydrostatic skeleton. The coelom greatly expanded the possibilities for locomotion, the very essence of animal life.

Figure 5.3 illustrates the three basic types of body structure. The acoelomate body plan is typical of the flatworms and the nemerteans, although some zoologists consider the proboscis cavity of the nemerteans (Figure 5.4) to be a coelom. The second type

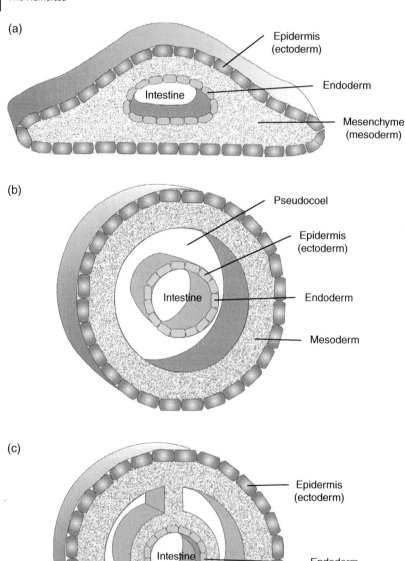

Figure 5.3 Diagrammatic cross sections of grades of structure. (a) acoelomate; (b) pseudocoelomate; (c) eucoelomate (true coelom).

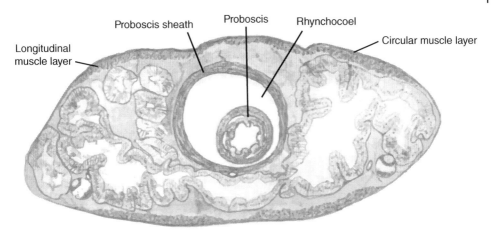

Figure 5.4 Cross section of the pelagic nemertean *Nectonemertes* species. *Source:* Coe and Ball (1920), figure plate IV (p. 483). Reproduced with permission of John Wiley & Sons.

of body plan is the pseudocoelomate, or blastocoelomate, form in which a body cavity is present but is not lined with muscle. The pseudocoelomate body structure is typical of a few phyla (e.g. rotifers and nematodes) not discussed in this book. The pseudo- or blastocoelomate body cavity is in fact an early embryological structure, the blastocoel (see Figure 5.2), that persists through development into the adult. The third type of cavity is the true coelom, typical of the eucoelomate structural grade. Our first eucoelomate phylum will be the annelids, the topic of Chapter 6.

A third advancement was the evolution of metameric segmentation (Clark 1964), or repeated body segments, which we also first see in the annelids. Virtually, all metazoans are segmented at some point in their development. The annelids are the best example of basic or nearly homonomous (equal) segmentation in which limbs, internal structures such as neural nodes or ganglia, excretory organs, and divisions between the segments (septae) are repeated more or less equally down the anterior–posterior axis of the animal. The significance of metameric segmentation is that it allowed modification of individual metameres (segments), the fusing of individual metameres into body "sections," and the modification of those sections to produce many of the important changes in body structure that we see in the diversity of life's forms. Thus, segmentation was a critical step in the evolution of life.

Hydrostatic Skeletons

During our study so far of open-ocean life forms, we have encountered radial symmetry in the Cnidaria and biradial symmetry in the Ctenophora. With the Nemertea, we venture into bilateral symmetry, where there is an easily discerned anterior and posterior as well as a dorsal and ventral surface (Figure 5.5). If we split the animal in half fore and aft, known in biological terms as a median-sagittal section, the two sides are approximate mirror images of each other. Bilateral symmetry is yet another landmark in the evolution of life. It facilitated movement in one direction, thereby making possible creeping and burrowing, as well as the undulatory (side-to-side) swimming of everything from worms to fishes. The concentration of sense organs in

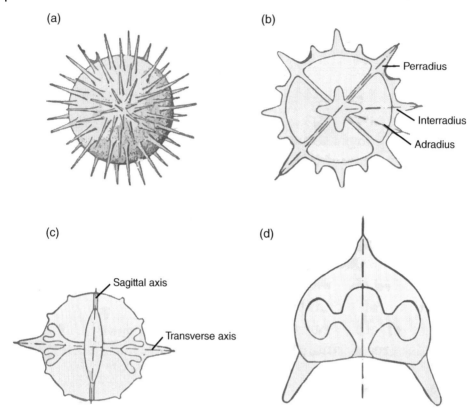

Figure 5.5 Diagrams of symmetry. (a) Spherical; (b) radial (Cnidaria); (c) biradial (Ctenophora); (d) bilateral (Nemertea).

the anterior of the animal to read the environment ahead (cephalization) is yet another important trait that developed with bilateral symmetry.

Figure 5.4 of a pelagic nemertean in cross section shows a well-developed layer of circular muscle just inside the body wall, running the length of the animal much like an unbroken series of elliptical barrel hoops. Just inside of the circular muscle is the longitudinal muscle, which extends like two long muscular cords down the length of the animal. These two opposing muscle groups are important to movement, with the circular muscle allowing for elongation and the longitudinal muscle for shortening in length. The principle of the hydrostatic skeleton is important to understanding movement and the functioning of the nemertean proboscis.

The mesenchyme, or parenchyma, of nemerteans is deformable but, like water, virtually incompressible. It acts much more like a fluid than the more robust gelatinous mesoglea of the jellyfishes and ctenophores, particularly in the benthic nemerteans. Consider Figure 5.6 where, from an initial relaxed state, the circular muscles of the right half of a cylindrical worm contract, elongating the worm much like squeezing a cylindrical balloon filled with water. Since muscles can only contract, the circular muscles alone can only decrease the worm's diameter. It is the

Figure 5.6 Diagrams illustrating the effects of contraction of circular muscles in cylindrical animals. (a) Three possible conditions in an animal with circular muscle bands only; (b) five possible conditions in an animal with both circular and longitudinal muscles. *Source:* Adapted from Chapman (1950), figure 1 (p. 31); figure 2 (p. 32).

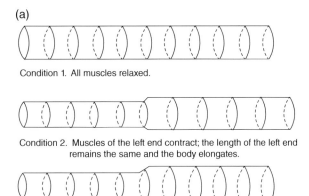

Condition 1. All muscles relaxed.

Condition 2. Muscles of the left end contract; the length of the left end remains the same and the body elongates.

Condition 3. Muscles of the left end contract with no change in length; the muscles of the right end relax and the body thickens.

(b)

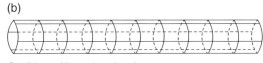

Condition 1. All muscles relaxed.

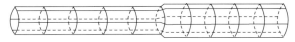

Condition 2. Muscles of left end contract; left end elongates; right end remains unchanged.

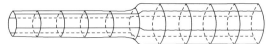

Condition 3. Muscles of left end contract; length of left end unchanged; right end diameter increases.

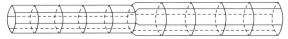

Condition 4. Muscles of left end contract; length of left end unchanged; right end length increases; right end diameter unchanged.

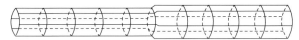

Condition 5. Muscles of left end contract; lengths of both ends increase; diameters of both ends remain unchanged.

presence of the deformable mesenchyme and its need to maintain a constant volume that causes the worm to elongate, just like the water in the balloon. The parenchyma thus gives the circular muscles something to act on, or squeeze, serving as a hydrostatic or fluid skeleton. The proboscis of nemerteans is a particularly creative use of hydrostatic principles, unique in the animal kingdom.

Classification

The basics of nemertean classification were provided by Schultze (1851), Brinkmann (1917a, 1917b), and Coe (1943). The framework presented here is that reported in Brusca et al. (2016). Specifics on pelagic species come from Hyman (1951) and Gibson (1999). Classification in the Nemertea is evolving as molecular insights become available, with many historical groups being retained and others shifted up or down in importance. Nemerteans are classified based on whether or not their proboscis is armed with barbs (stylets), on their larval type, on the musculature present in their body wall, and the position of their mouth relative to the cerebral ganglion (Brusca et al. 2016). Gibson (1995) recognizes 1100 valid species, 97 of which are pelagic. The World Register of Marine Species, www.marinespecies.org (WoRMS), recognizes 1317 total species, with 96 pelagic.

Phylum Nemertea (Rhynchocoela) (after Hyman 1951 and Brusca et al. 2016)

Class Palaeonemertea. Body wall musculature with two or three layers, either circular-longitudinal, or circular-longitudinal-circular. May be lacking ocelli and cerebral organs. Dermis thin and gelatinous. Primarily marine intertidal species (Figure 5.7a).

Class Neonemertea. Proboscis armed or unarmed, with separate or fused mouth and proboscis pores.

Subclass Pilidiophora. Nemerteans with a pilidium larva (Masakova 2010)

 Order Hubrechtida. Unarmed nemerteans with separate mouth and proboscis pores.

 Order Heteronemertea. Body wall musculature with three layers, innermost and outermost layers longitudinal. Ocelli and cerebral organs usually present. Mainly intertidal marine species (Figure 5.7b).

Subclass Hoplonemertea. Mouth anterior to cerebral ganglion, longitudinal nerve cord interior to outer muscle layers and within mesenchyme. Proboscis usually armed and divided into three distinct sections. Mouth and proboscis pores united. Intestine straight with paired lateral diverticulae. Cerebral organs and ocelli usually present. (Figure 5.7c). Marine, freshwater, and terrestrial species.

 Order Monostylifera. Proboscis has one stylet, with two or more sacs containing reserve stylets. Primarily benthic marine species, but with representatives in the terrestrial and freshwater environments.

 Order Polystylifera. Proboscis has multiple stylets present on a basal shield. Primarily benthic marine species, but all pelagic species are in the Polystylifera.

Pelagic species are all within the subclass Hoplonmertea, order Polystylifera. The scheme below is that reported by Gibson (1999) and echoed in WoRMS (n.d.). The number of genera follows each of the 11 families.

Suborder Pelagica
Family Armaueriidae (6)
Family Balaenanemertidae (1)
Family Buergeriellidae (1)
Family Chuniellidae (2)
Family Dinonmertidae (6)
Family Nectonemertidae (1)
Family Pachynemertidae (1)
Family Pelagonemertidae (10)
Family Phallonemertidae (1)
Family Planktonemertidae (7)
Family Protopelagonmertidae (4)

Morphology

Despite their lowly position in the hierarchy of animal life, the nemerteans show a great deal of sophistication, including a one-way

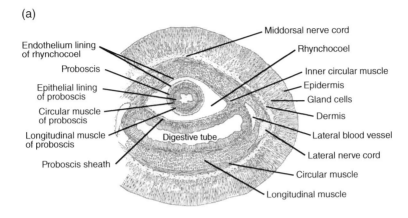
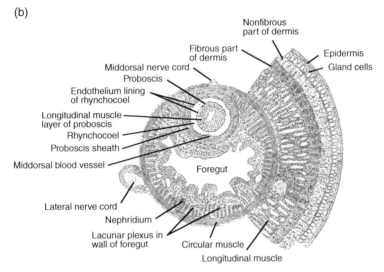
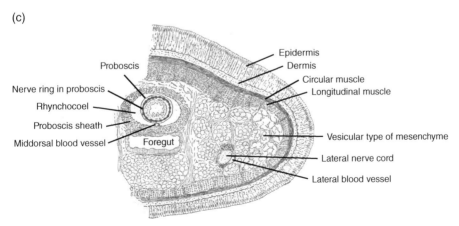

Figure 5.7 Histological structure and anatomy of nemerteans. (a) Section through a palaeonemertine (*Tubulanus*); (b) section through a heteronemertine (*Lineus*); (c) section through a hoplonemertine. *Source:* Adapted from Hyman (1951): (a, b) figure 177 (p. 468); (b) figure 178 (p. 470).

gut, a well-organized body wall musculature, and both circulatory and nervous systems. Our focus will be the pelagic species, the polystyliferan hoplonemerteans. Their defining characteristic is the proboscis (Figures 5.8 and 5.9).

Proboscis Apparatus

The proboscis apparatus (Figure 5.8) begins with the exterior proboscis pore, moveas inward to a short anterior proboscis canal (the rhynchodaeum), and then to the tubular cavity that houses the proboscis, the rhynchocoel. The proboscis itself is an elongated muscular tube that resides free in the rhynchocoel, which is filled with a clear fluid. It is attached at the anterior end. In the hoplonmerteans, the proboscis has three regions (Hyman 1951): an anterior thick-walled tube, a middle bulbous region with the stylet or stylets, and a more slender posterior blind tube (Figure 5.8c). The blind tube is attached to the posterior of the rhynchocoel with a posterior retractor muscle whose main job is to reel in the proboscis

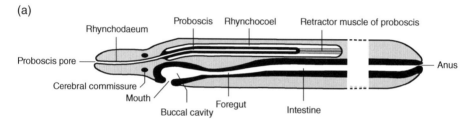

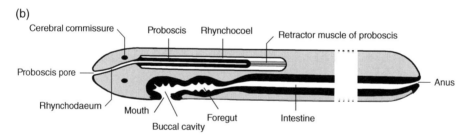

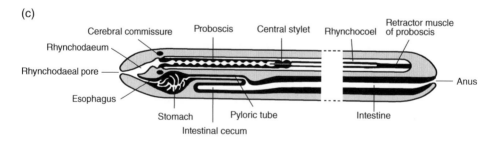

Figure 5.8 Schematic vertical longitudinal sections showing the arrangement of the alimentary canal and proboscis in representatives of three orders of nemerteans. (a) *Cephalothrix bioculata* (Paeonemertini); (b) *Lineus sanguineus* (Heteronemertini); (c) *Amphiporus lactifloreus* (Hoplonemertini). *Source:* Adapted from Jennings and Gibson (1969), (a) figure 1 (p. 406); (b) figure 10 (p. 413); (c) figure 11 (p. 414).

Morphology | 203

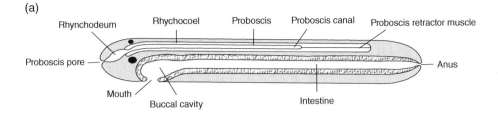

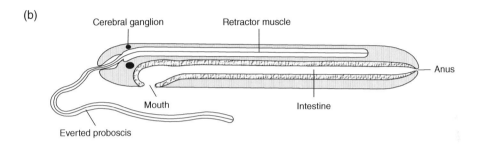

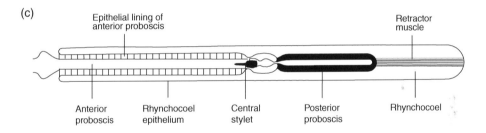

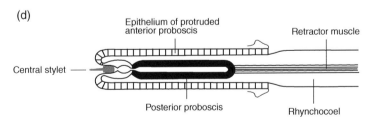

Figure 5.9 Schematic diagrams showing the relationships between the parts of the proboscis in retracted and protruded position. (a and b) Retracted and protruded proboscis of a stylized Hoplonemertean, respectively. (c and d) Retracted and protruded proboscis of the Hoplonemertean *Paranemertes peregrina*, respectively. *Sources*: (a, b) Adapted from Gibson (1982) (p. 831); (c, d) Adapted from Gibson (1972), figure 8 (p. 45).

after it shoots out. Interspersed in the epithelium of the anterior proboscis are gland cells that produce a sticky substance that helps retain prey. The posterior blind tube of the proboscis produces toxins that are introduced into the prey via the stylet wounds, aiding in prey immobilization (Gibson 1972)

Figure 5.9 illustrates the functioning of the nemerteans' characteristic proboscis. At rest, the proboscis is "outside in," like the finger of a rubber glove that has been pushed in from the fingertip side. The inside of the pushed-in finger is a hollow space. If pressure is applied to the hand side of the glove, the finger pops out: it has "everted." Similarly, the nemertean proboscis is everted by muscular contractions exerting pressure on the rhynchocoel fluid, forcing the proboscis inside-out by hydrostatic pressure and exposing the stylets (Figure 5.10) to pierce and hold prey.

The Pelagic Body Form and Locomotion

Pelagic nemerteans tend to be a lot broader and considerably more flattened dorsoventrally than their benthic relatives, often possessing caudal fins (Figure 5.11). The open-ocean species also tend to be more translucent, with a gelatinous parenchyma that helps them to more closely approach neutral buoyancy.

Although pelagic nemerteans have never been observed in situ, movement has been observed in freshly caught specimens. Based on those observations, two modes of locomotion have been postulated. The first, a passive drifting mode, relies on the near-neutral buoyancy afforded them by their gelatinous parenchyma. It allows nemerteans to float passively with the current, swimming only when necessary to hunt prey or to avoid predators. An example of the body form of the first mode is *Planktonemertes* (Figure 5.11), which also exhibits a reduced musculature. The second mode of locomotion is demonstrated by more active swimmers like *Nectonemertes*, with caudal fins (Figure 5.11) and a more highly developed musculature. Swimmers have a longitudinal muscle structure organized into two plates (Coe 1926), one each on the dorsal and ventral sides, to power the caudal fins (see Figure 5.4). Note, however, that the muscle structure in the pelagic species, even in the stronger swimmers, is much reduced relative to the musculature of their benthic relatives.

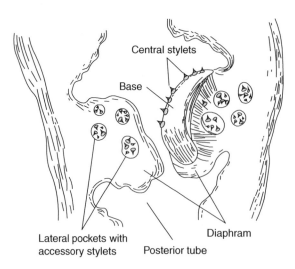

Figure 5.10 Stylet apparatus of *Pelagonemertes*. Source: Adapted from Coe (1926), plate 27, figure 168.

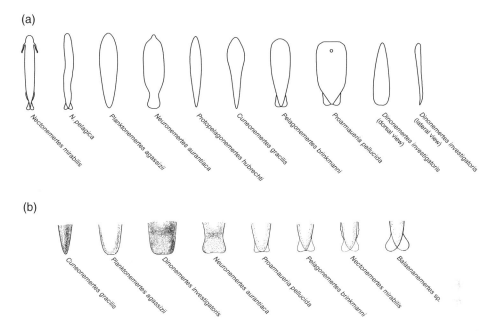

Figure 5.11 Nemertean body and caudal fin shapes. (a) Body outlines of 10 species of bathypelagic nemerteans showing their adaptive configurations for swimming or floating; (b) outlines of the posterior extremities of eight genera of pelagic nemerteans showing the extent of development of the caudal fin. *Sources:* (a) Adapted from Coe (1926), figure 1 (p. 19); (b) Coe (1926), figure 7 (p. 29).

Muscle contractions allow nemerteans to produce strong, fish-like, swimming motions, either an up-and-down motion with the animal's dorsal side up (picture the tail action of a bottlenose dolphin) or a side-to-side fin movement, lateral side up, like the fins of fish. Both types of motions have been seen in freshly caught *Nectonemertes* (Coe 1926). Whether the ribbon worm's attitude is horizontal or vertical, the name for its type of locomotion is "undulatory." Undulatory locomotion is exhibited by several other animal groups, including the subjects of our next chapter, the annelids.

Color

Pelagic nemerteans are brilliantly colored in life, often red, scarlet, orange, or yellow. We captured a vivid orange specimen (Figure 5.12) episodically in our midwater trawls when sampling mesopelagic depths (700 m) off the southern coast of California. As mentioned, pelagic ribbon worms tend to be translucent, with elements of the gut and other internal organs visible through their integument.

Nervous System

The nervous system of nemerteans comprises the brain (cerebral ganglia), a paired set of ganglionated lateral nerve cords that run the length of the body, and a plexus of connectives and commissures that joins them at intervals (Figure 5.13). The brain has four lobes, two dorsal and two ventral. They are joined by a dorsal and ventral commissure, respectively, forming a ring around the proboscis apparatus. Like all hoplonemerteans, the main nerve cords of

206 | The Nemertea

Figure 5.12 The pelagic nemertean *Bathynemertes* species. *Source:* © Dante Fenolio, reproduced with permission. See color plate section for color representation of this figure.

(a)
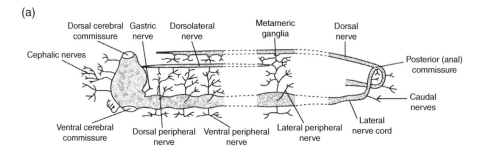

(b)
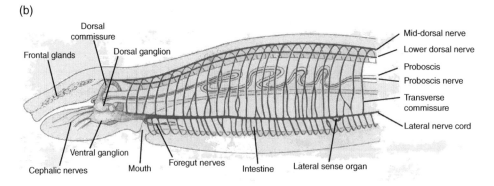

Figure 5.13 Nemertean nervous system. (a) The nervous system of *Neuronemertes aurantiaca*. (b) Internal structure of *Tubulanus* sp. including the nervous system. *Sources:* (a) Adapted from Coe (1926), figure 27 (p. 52); (b) Adapted from Burger (1895).

pelagic nemerteans run interior to the body wall musculature within the mesenchyme.

A dorsal nerve (or dorsomedial nerve) runs longitudinally down the median axis of the body just interior to the circular muscle layer to the posterior commissure joining the two larger lateral nerves. In pelagic species, the dorsal nerve originates just posterior to the dorsal brain commissure. Two smaller dorsolateral nerves run alongside the proboscis sheath. The proboscis is very well innervated, indicating a high degree of neural control.

Sense Organs

The nervous system of nemerteans supports an array of different sensory receptors, few of which are well understood. Most of the sensory organs have been secondarily lost in the pelagic species. However, it is important to note the types of sensory organs present in the Phylum Nemertea, as they are the first sense organs we have encountered on our zoological journey.

Sensory organs are best developed in the littoral species. The receptors are primitive eyes, cephalic grooves or furrows, lateral and frontal sense organs, and sensory epithelial cells (Gibson 1972). The sensory modalities represented include light reception (photoreception), touch perhaps including vibration (mechanoreception), and chemoreception (taste or smell). Figure 5.14a shows the cephalic grooves and ocelli of *Amphiporus bimaculatus*, a hoplonemertean.

Diversity of sensory organs and external structures abounds within the nemerteans. Chemoreception is attributed to a number of organs, including the frontal glands seen in Figure 5.14b and the complex cerebral organs. The cerebral sense organs (Figure 5.14c) have a ciliated central canal that opens to the exterior, usually through a cephalic groove. The central canal circulates the surrounding water into the sensory epithelium of the organ itself, allowing the worm to sample its environment for scent trails from prey.

The pelagic species have lost their ocelli and cerebral organs but do have what appear to be mechanoreceptors within their integument (Figure 5.15; Coe 1926). These integumentary sense organs are reminiscent of the vertebrate hair cells that serve as mechanoreceptors in the lateral line of fishes (Marshall 1971). Sensitivity to water motion is useful in the quiescent waters of the deep sea; it is quite well developed in the bathypelagic fishes (Marshall 1971). Other potential sense organs have been described in the pelagic nemerteans, but at present we can only speculate on their function.

Circulatory System

Blood vascular (circulatory) systems first appear in the nemerteans, and they are a universal characteristic of the phylum. All are considered to be closed systems. They vary in complexity from simple circuits consisting of an unbranched pair of lateral vessels joined at the anterior and posterior ends (Figure 5.16a) to the sophisticated systems of vessels with multiple transverse connections found in some of the littoral species.

Circulatory systems in four species of pelagic nemerteans are shown in Figure 5.16. The pelagic species do not have transverse connectives, rather they have a fairly simple circuit with three longitudinal vessels joined at the anterior and posterior ends by anastomoses (connections or openings). The simpler systems have two types

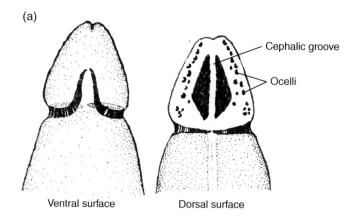
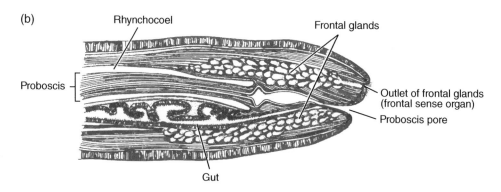
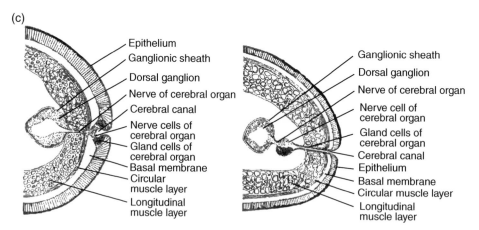

Figure 5.14 Nemertean sense organs. (a) The head of the hoplonemertean *Amphiporus bimaculatus* showing the ocelli and cephalic groove. (b) Frontal sense organs of a hoplonemertean showing clusters of frontal glands. (c) Comparison of the primitive cerebral sense organs of a paleonemertean, *Tubulanus annulatus* (left) and the highly differentiated sense organs of a hoplonemertean *Drepanophorus albolineatus* (right). *Sources:* (a) Coe (1943), figure 27 (p. 182); (b) Burger (1895); (c) Scharrer (1941), figures 4, 5 (pp. 116, 117). Reproduced with permission of John Wiley & Sons.

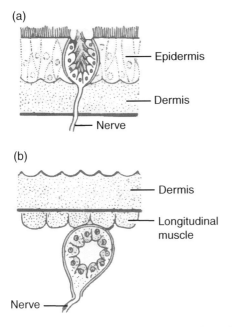

Figure 5.15 Integumentary and subepidermal sense organs in pelagic nemerteans. (a) Integumentary sense organ of *Cuneonemertes gracilis*; (b) subepidermal sense organ of *C. gracilis*. Source: Adapted from Coe (1926), figures 32, 33 (p. 60).

red. Colored blood does not automatically indicate a functional blood pigment aiding in gas exchange; however, it is believed that hemoglobin is present in some with red blood (Hyman 1951). When red is present in nemertean blood, it is in the blood cells or corpuscles, which argues for a respiratory function. Lymphocytes, or phagocytic cells, are also present in the blood of nemerteans, but their function is unclear. Phagocytic cells can perform a variety of roles. When in the blood stream, they may be involved in removing offensive particles like bacteria or in transporting nutrients.

Gas exchange in the nemerteans is accomplished by simple cutaneous diffusion of respiratory gases. Exchange is aided within by the action of the circulatory system and the inward gradient created by a lower internal oxygen concentration. It is aided externally by animal movement to minimize boundary layers and refresh the surrounding medium. Data reported for the pelagic species show a low rate of metabolism (Thuesen and Childress 1993).

Excretory System

Nephridia are found in all nemerteans except the ones of greatest interest to us: the pelagic species. However, benthic hoplonemerteans considered to be most closely related to the pelagic forms, notably *Drepanophorus*, do have nephridial systems and those shall be discussed briefly. The nephridia are blind-ended tubes usually found in close association with blood vessels (Figure 5.18a and b). The blind ends of the tubes are known as flame bulbs. They act as filters, drawing fluid in by the ciliary "flames" that maintain a slightly negative pressure within the tubes themselves. The open ends of the tubes lead to a nephridial tubule that usually joins others

of vessels: lacunae, which are mesenchymal spaces lined with a membrane; and true blood vessels with well-defined walls.

In the more advanced species like the hoplonemerteans, the entire circulatory system consists of true blood vessels. The lateral vessels run alongside the lateral nerves (Figure 5.17a), and the dorsal vessel runs just ventral to the rhynchocoel. In some of the larger species, the lateral vessels show convolutions around the intestinal diverticuli, presumably to better absorb and distribute nutrients from the digestive system. The true blood vessels are of two types: simple; and with a muscular wall that aids in the circulation of blood (Figure 5.17b). Nemerteans lack hearts.

Nemerteans most often have colorless blood, but it can be yellow, green, orange, or

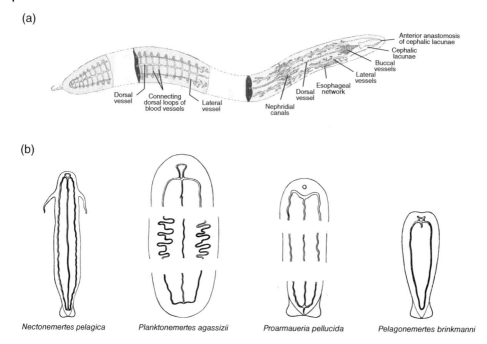

Figure 5.16 Nemertean circulatory system. (a) Circulatory and excretory systems of the heteronemertean *Cerebratulus*. (b) Vascular systems of representative species from four families of pelagic nemerteans. *Sources:* (a) Adapted from Coe (1943), figure 14 (p. 165); (b) Coe (1926), figure 25 (p. 48).

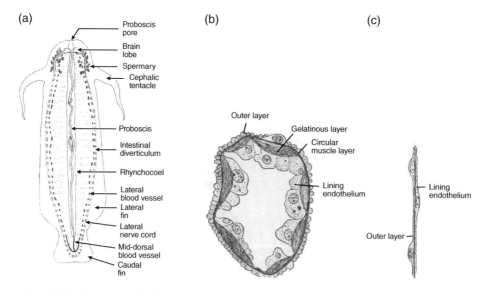

Figure 5.17 Nemertean circulatory system. (a) Dorsal view of the internal structure of the hoplonemertean *Nectonemertes mirabilis* showing dorsal and lateral blood vessels. (b) Cross section of a nemertean contractile blood vessel; (c) section through a non-contractile nemertean blood vessel. *Sources:* (a) Adapted from Coe (1926), plate 16; (b, c) Hyman (1951), figure 480 (p. 489).

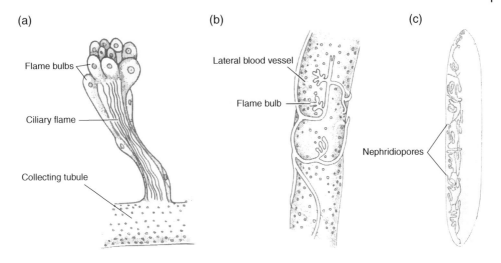

Figure 5.18 Nemertean excretory system. (a) Protonephridial cluster of *Drepanophorus*; (b) nephridial ducts associated with lateral blood vessels in *Amphiporus*; (c) nephridium of *Prostoma* with numerous nephridiopores. *Source:* Adapted from Burger (1895).

to eventually open to the outside via a nephridial pore (Figure 5.18c). The generic name for nephridia like those of the nemerteans is "protonephridia," and they are considered to be the most primitive nephridial type. Protonephridia will appear again in the next chapter: the annelids.

It is believed that, particularly in littoral nemerteans, the nephridial system functions to rid the animal of unwanted solutes or of water from a chance encounter with lower salinity (e.g. after a rain). Pelagic species would have less use for nephridia that function mainly to regulate internal salinity.

Digestive System

The most important characteristic of the alimentary canal in nemerteans is that ingested food gets a one-way trip: it enters at the mouth and leaves at the anus. During its voyage in the canal, food is broken down by the digestive tract, releasing nutrients for absorption. The one-way path allows for regional specialization within the tract and more effective digestion. Contrast this with the digestive systems of the Cnidaria and ctenophores described in Chapters 3 and 4.

Digestive systems in pelagic nemerteans include a mouth and a short esophagus that opens into a well-developed stomach. The stomach terminates in a pylorus, a short tube that in turn empties into the dorsal portion of the intestine, a very capacious organ that runs nearly the whole length of the body, typically possessing paired diverticulae that occupy much of the space between the body walls and contribute significantly to pelagic nemerteans' general appearance. The digestive system is ciliated throughout its length (Gibson 1972). Since many of the pelagic nemerteans are highly transparent, their digestive tracts are readily observed in an intact animal.

Mouth and esophagus. In most pelagic species, the mouth is located terminally, just ventral to the proboscis pore, and leads into a very short, almost non-existent, esophagus. The mouth itself has highly convoluted walls. The esophagus, when distinguishable, is a short tube that leads into the stomach (Figure 5.19a).

Stomach and pylorus. The stomach is highly glandular and often has convoluted walls, implying production of digestive enzymes and lubricating mucous to aid in the movement of a food bolus down the tract. The stomach tapers to a pylorus, a tube that is morphologically distinguishable from the stomach by its lack of glandular cells. The pylorus empties directly into the intestine (Figure 5.19a). The length of the stomach and pylorus varies considerably from species to species.

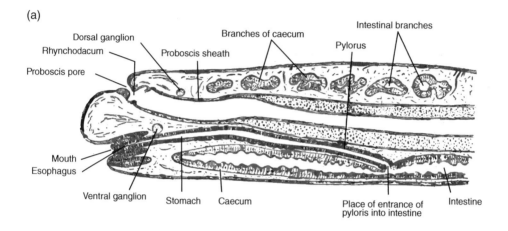

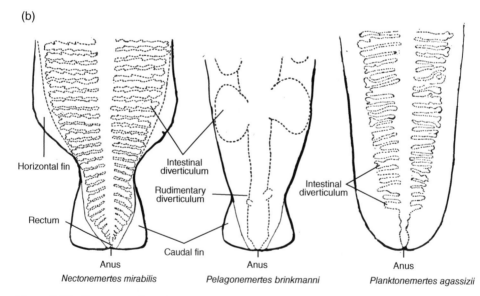

Figure 5.19 Nemertean digestive system. (a) Longitudinal section through the anterior part of *Nectonemertes* showing parts of the digestive system from the mouth to the intestine; (b) posterior parts of three pelagic nemerteans showing the extreme differences in the number and extent of the intestinal diverticula in different forms. *Sources:* (a) Coe (1926), figure 17 (p. 44); (b) Coe (1926), figures 21, 22, 23 (p. 47).

Intestine. The intestine possesses paired lateral diverticulae throughout its length (Figure 5.19b), reaching as many as seventy pairs in *Nectonemertes* (Coe 1926). In some species, such as the *Nectonemertes mirabilis* pictured in Figure 5.19a, there is an anterior portion of intestine termed an intestinal caecum that is ventral to the stomach. Whether or not an intestinal caecum is present depends on where the pylorus enters the intestine. At the posterior end of the intestine, the diverticula taper to a rectum ending in an anus (Figure 5.19b).

Oil droplets are usually present in the epithelium of the intestine. Those can be reddish or yellow in hue and can give a species its characteristic color (Coe 1926, Figure 5.12). The oil droplets may aid pelagic species in achieving neutral buoyancy. The epithelium of the intestine comprises columnar cells of a number of different types, one highly vacuolated that is presumed to have glandular and absorptive function and others that contain the oil droplets. Interspersed among them can be other types of cells, including those with cilia and amoeboid processes important to further breakdown and absorption of nutrients initially processed in the gut lumen.

Digestion. Digestion in nemerteans involves extracellular and intracellular processes. Protein digestion is initiated extracellularly in the lumen of the gut and completed in the cells of the intestinal epithelium. Enzymes for digesting proteins, lipids, and carbohydrates have been identified in nemerteans (Brusca and Brusca 2003). The extensive surface area of the intestine with its multitudinous diverticula and the well-developed circulatory system assure the distribution of nutrients throughout the body. No specialized musculature is associated with the gut wall that might aid in propelling a food bolus down its length. The musculature of the body wall, which is in close proximity to the gut, may aid in food movement in concert with the activity of the ciliated cells of the gut itself.

Reproduction

Nearly, all nemertean species have separate sexes, though a few hermaphroditic forms occur, particularly in the freshwater and terrestrial representatives (Hyman 1951). The pelagic nemerteans are virtually all dioecious. In a very few cases, *Nectonemertes* being an example, males and females are quite different in appearance (Figure 5.20). Littoral nemerteans show prodigious powers of regeneration, which may also be considered a form of reproduction. The favorite experimental subject in this regard is *Lineus socialis*, a paleonemertean. It has been cut into nine pieces, each of which regenerated a whole new worm. The main requirement for successful regeneration was the presence in the worm fragment of a piece of lateral nerve cord (Coe 1943).

Reproductive organs of littoral nemerteans are found serially along the length of the body, usually between the intestinal diverticulae, similar to what is shown for the female *Nectonemertes* in Figure 5.20. Gonads of both sexes are similarly arranged. In contrast, pelagic species are sexually dimorphic, with the male gonads (spermaries) found in the head region and the female ovaries arranged serially. Eggs and sperm are discharged from the gonad via gonoducts that lead to a genital pore (gonopore) on the outside. The character of the ducts, pores, and the gonad itself vary considerably between species, with some of the structures only present near the time of gamete discharge. In some species, gametes are discharged through a rupture in the body wall.

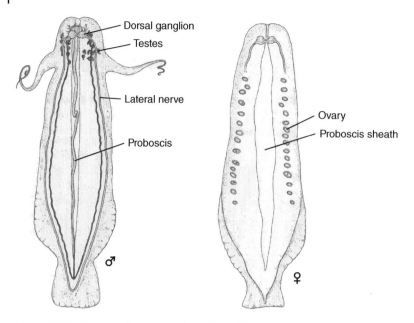

Figure 5.20 Reproductive systems in male and female *Nectonemertes mirabilis*. *Source:* Coe (1943), figure 31 (p. 187).

Breeding behavior in the littoral nemerteans is believed to be chemically cued and shows the diversity typical of the group. In some species, e.g. *Cerebratulus,* the presence of a ripe male or female can elicit gamete release with no physical contact. In others, physical contact is the norm, with a "primitive copulatory behavior" (Coe 1943) involving one individual crawling over the other. Other species, *Lineus* for example, exude a gelatinous mass with several copulating worms contributing gametes that then remain on and develop within the mass.

The situation for the pelagic species is quite different from that of littoral ribbonworms. Individual pelagic nemerteans, never very numerous, are widely separated in an extremely large habitat that extends tens to hundreds of meters in horizontal and vertical space. It is assumed that chemical cues play an important role in mate location among the pelagic species, not only because that is the case with the intertidal species but also because it is the most logical sensory mechanism. Other deep-living pelagic taxa (e.g. anglerfishes) employ a chemotactic mate-locating strategy. Much of what follows is essentially high-quality educated guesswork by luminaries in the field of nemertean biology (Coe, Brinkmann, Gibson, and Hyman), but it works well with the evidence at hand.

First, littoral nemerteans produce very high numbers of gametes. Coe (1926) estimates that a large (up to 2m in length) female *Cerebratulus* produces up to a hundred million eggs in a breeding season. The much smaller (<100mm) pelagic *Nectonemertes* produces only one to two large yolky eggs per ovary, giving a total of 140–300 eggs per individual; the number of testes, or spermaries, are also considerably reduced in number. Second, in the littoral species, ovaries within individual worms show the differences in stage of egg development that are characteristic of multiple spawnings in a breeding season, consonant with what has been observed in intertidal worms. Intertidal nemerteans also experience multiple breeding seasons in a lifetime. In contrast, pelagic species are believed to reach sexual maturity and reproduce only once. The evidence supporting a "once-only" life history comes chiefly from sexually dimorphic species, particularly *Nectonemertes*.

The male of our model pelagic species, *Nectonemertes mirabilis*, has tentacles that only develop when the individual is sexually mature. Coe (1926) hypothesized that the tentacles are used as claspers, to catch and hold females when encountered in the sparsely populated pelagic realm. When ready to spawn, the unusually muscular spermaries of the male *Nectonemertes* can discharge their sperm in close proximity to the oviducts of the female, maximizing the possibility of fertilization. Once fertilization is achieved, the fate of the fertilized eggs is unknown, but the large yolky eggs would yield a more robust larva than the smaller eggs of their intertidal relatives.

A reproductive strategy that includes multiple breeding seasons, such as that observed in the intertidal ribbon worms, is known as interoparity. Its opposite is semelparity: one reproductive event per lifetime, such as what is likely occurring in the pelagic nemerteans. Clearly, each is designed to assure species survival in very different environments. In the pelagic nemerteans, the main problem to be surmounted for reproductive success is mate location. Once a mate is located, reproductive success is best assured with fewer larger eggs and close proximity of the mating worms.

Development

Like all the hoplonemerteans, the newly fertilized eggs of pelagic ribbonworms are assumed to undergo direct development: that is, their embryological development does not include a distinct metamorphosis. Larvae develop within the egg and hatch as young adult worms. In contrast, many of the benthic paleo- and hetero-nemerteans hatch into a free-swimming pilidium larva (Masakova 2010) that undergoes a metamorphic change before descending to the bottom to become a young adult.

Foraging Strategies

Virtually, nothing is known about the foraging strategies and food of the pelagic nemerteans: they are rarely caught with anything in the gut. In one instance, a "small crustacean" was found in the gut of *Nectonemertes mirabilis* (Coe 1926) suggesting a predatory lifestyle, which would be consistent with what is observed in littoral species. Alternately, the prey of pelagic ribbonworms could, like them, be gelatinous in character, making it harder to detect prey items in their guts.

Vertical and Geographic Distributions

Pelagic nemerteans have been captured in all ocean basins including the Southern Ocean (Coe 1950; Gibson 1999). Though commonly referred to as "bathypelagic," catch records suggest that they are frequently found at lower mesopelagic depths (500–1000 m), usually with distributions reaching into the upper bathypelagic zone (1000–1500 m). As we know from Chapter 1, temperatures at depths exceeding 1000 m are uniformly cold (2–8 °C), with temperature and salinity varying little over depths of tens of meters. Examples of the vertical distributions of pelagic nemerteans are given in Figure 5.21 along with the temperature and salinity ranges associated with their vertical profile. *Nectonemertes mirabilis*, the species for which the most data are available, does seem to show a preference for temperatures that do not exceed 6 °C.

An analysis of the geographical distributions of pelagic ribbonworms (Gibson 1999) concluded that the great majority do not have widespread zoogeographic ranges despite the similarities in temperature and salinity found at meso- and bathypelagic depths throughout the global ocean. In most cases (80%), they are found in only one oceanic region. There are a few exceptions. *Pelagonemertes rollestoni* is found in the Atlantic, Indian, and Pacific Oceans. *Nectonmertes mirabilis*, the most common species of pelagic nemertean, is found in both the Atlantic and Pacific. Interestingly, more specimens of *N. mirabilis* have been captured than of any other species.

Though found from the equator to the polar oceans, pelagic nemerteans are always quite rare. It takes a considerable amount of trawling time to catch one, and many species are known from a single specimen only.

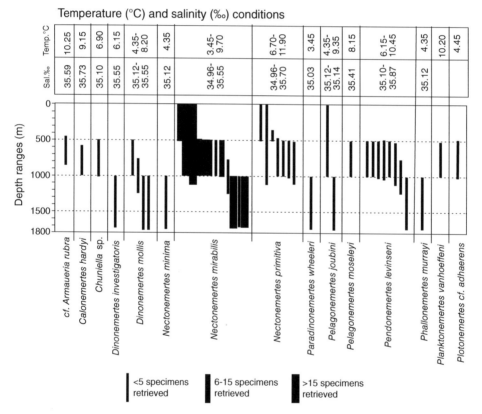

Figure 5.21 Vertical profiles for 15 North Atlantic bathypelagic nemertean species. *Source:* Gibson (1999), figure 5 (p. 584). Reproduced with the permission of Backhuys Publishers.

References

Borlase, W. (1758). *The Natural History of Cornwall*. Oxford: Jackson.

Brinkmann, A. (1917a). Pelagic nemerteans from the Michael Sars North Atlantic Deep-Sea expedition 1910. *Report on the scientific results of the Michael Sars North Atlantic Deep-Sea Expedition 1910* (3): 1–18.

Brinkmann, A. (1917b). Die pelagischen Nemertinen. *Bergens Museum Skrifter* 3: 1–194.

Brusca, R.C. and Brusca, G.J. (2003). *Invertebrates*, 2e. Sunderland: Sinauer.

Brusca, R.C., Moore, W., and Shuster, S.M. (2016). *Invertebrates*, 3e. Sinauer Associates: Sunderland.

Burger, O. (1895). Die nemertinen des Golfes von Neapal und der Angrenzenden Meeres-Abschnitte. *Fauna und Flora des Golfes von Neapel* 22: 1–743.

Chapman, G. (1950). On the movement of worms. *Journal of Experimental Biology* 27: 29–39.

Clark, R.B. (1964). *Dynamics in Metazoan Evolution: The Origin of the Coelom and Segments*. Oxford: Clarendon Press.

Coe, W.R. (1926). The pelagic nemerteans. *Memoirs of the Museum of Comparative Zoology at Harvard College* 49: 1–244.

Coe, W.R. (1943). Biology of the nemerteans of the Atlantic coast of North America. *Transactions. Connecticut Academy of Arts and Sciences* 35: 129–328.

Coe, W.R. (1950). Nemerteans from Antarctica and the Antarctic Ocean. *Journal of the Washington Academy of Sciences* 40: 56–59.

Coe, W.R. and Ball, S.C. (1920). The pelagic nemertean Nectonemertes. *Journal of Morphology* 34: 457–485.

Cuvier, G. (1817). *La Regne Animal*. Gustav Fischer.

Gibson, R. (1972). *Nemerteans*. London: Hutchinson and Co.

Gibson, R. (1982). Nemertea. In: *Synopsis and Classification of Living Organisms*, vol. *I* (ed. S.P. Parker). New York: McGraw-Hill.

Gibson, R. (1995). Nemertean genera and species of the world: an annotated checklist of original names and description citations, synonyms, current taxonomic status, habitats, and recorded zoogeographic distribution. *Journal of Natural History* 29: 271–562.

Gibson, R. (1999). Nemertina. In: *South Atlantic Zooplankton* (ed. D. Boltovskoy). Leiden: Backhuys Publishers.

Hyman, L.H. (1951). *The Acoelomate Bilateria Phylum Rhynchocoela. The invertebrates: Platyhelminthes and Rhynchocoela*. New York: McGraw-Hill Book Company.

Jennings, J.B. and Gibson, R. (1969). Observations on the nutrition of seven species of rhynchocoelan worms. *Biological Bulletin* 136: 405–433.

Kaestner, A. (1967). *Invertebrate Zoology*, vol. *1*. New York: Wiley.

Marshall, N.B. (1971). *Explorations in the Life of Fishes*. Cambridge: Harvard University Press.

Masakova, S.A. (2010). Development to metamorphosis of the nemertean pilidium larva. *Frontiers in Zoology* 7 (30): 1–18.

Scharrer, B. (1941). Neurosecretion. III. The cerebral organ of the nemerteans. *The*

Journal of Comparative Neurology 74: 109–130.

Schultze, M.S. (1851). *Beitrage zur Naturgeschichte der Turbellarien*. Greifswald: C.A. Koch.

Thuesen, E.V. and Childress, J.J. (1993). Metabolic rates, enzyme activities and chemical compositions of some deep-sea pelagic worms, particularly *Nectonemertes mirabiliis* (Nemertea; Hoplonemertinea) and *Poeobius meseres*. *Deep Sea Research, Part I* 40: 937–951.

WoRMS (n.d.). World Register of Marine Species. www.marinespecies.org.

6

The Annelida

Introduction

The annelids, or segmented worms, are a highly diverse group important to both terrestrial and aquatic systems. Over most of the globe, earthworms aerate and mix the soils of terrestrial systems, and a variety of marine annelids perform a similar function in the benthos. Other marine annelid species are predators or sedentary benthic filter-feeders. Annelids are considered one of the major phyla of the animal kingdom (Pettibone 1982) with about 13 000 marine species (WoRMS n.d.) ranging in size from minute (<1 mm) interstitial worms living between sand grains to the giant earthworms of Australia and the palolo worms of the South Pacific that can exceed 3 m in length (Brusca and Brusca 2003). What most annelids have in common is their segmentation, the serially arranged rings, which are usually externally visible, that give the phylum its name (from the Latin *anellus*, "little ring"). The external segmentation is usually accompanied by a serial arrangement of their internal organs.

Traditionally, there were three major groups of annelids: the polychaetes, the earthworms, and the leeches. As mainly freshwater and terrestrial species, earthworms, leeches, and kin are familiar creatures. The polychaetes comprise an impressive array of marine species that crawl, burrow, reside in tubes of their own manufacture, or swim freely in the open ocean.

Classification of the annelids is currently in a state of flux. Newer systems include the Echiura (innkeeper worms or spoonworms) and Sipunculida (peanut worms) in the annelids, while an older invertebrate zoology text would consider the spoonworms and peanut worms as their own separate phyla. Similarly, until recently the pogonophorans, or beardworms, discovered in the early 1900s, and their relatives the vestimentiferan tube worms of the hydrothermal vents described in the 1980s (Jones 1981, 1985), each merited their own phyla. Evidence from cladistic analyses, molecular methodologies, and anatomical study (Jones and Gardiner 1988) has demoted them into the polychaetes at about the family level: Siboglinidae (Figure 6.1) (Rouse and Pleijel 2001; Brusca et al. 2016).

Invertebrate systematics will continue to change rapidly in the next several decades owing to insights provided by the increased use of molecular methods and computational tools such as cladistic analyses. What will change the least will be taxonomy at the level of families and genera, which is where most of our interests in open-ocean life are focused. Rest assured that the

---

Life in the Open Ocean: The Biology of Pelagic Species, First Edition. Joseph J. Torres and Thomas G. Bailey.
© 2022 John Wiley & Sons Ltd. Published 2022 by John Wiley & Sons Ltd.

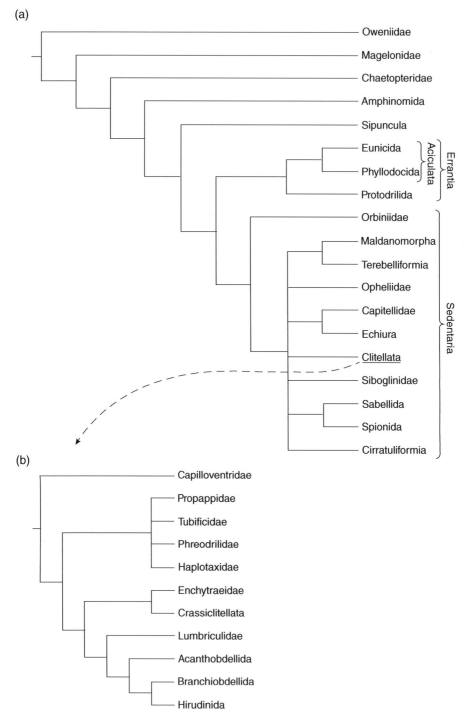

Figure 6.1 Dendrogram showing phylogenetic relationships of annelids (Brusca et al. 2016). Note the position of the Clitellata within the annelids as a whole and the differences between this classification scheme and that in WoRMS. (a) Annelida. (b) Clitellata. *Source:* Brusca et al. (2016), figure 14.41 (p. 598). Reproduced with the permission of Oxford Publishing Limited.

animals themselves will not change at all, but their position in the taxonomic hierarchy very well may.

As is the case with the nemerteans, annelids are a primarily benthic group with only a few representatives in the pelagic fauna. Marine representatives of both nemerteans and annelids are found in the oceans from the equator to the poles. The difference between them is that pelagic nemerteans, though pan-global in their distribution, are nearly always quite rare. The annelids, in contrast, are a conspicuous member of all open-ocean communities that have been analyzed, and they have an ever-present role in the pelagic food web.

Annelids are an elegant example of nature's dabbling, producing an astonishing diversity with the common theme of vermiform shape and (mostly) segmentation. We will briefly treat the basic characteristics of the phylum for background and continue with an emphasis on open-ocean species. The annelids are another very important transitional phylum in the evolution of animal life.

History

The annelids are large enough to be easily seen and ubiquitous enough to probably have been known since the dawn of man. They were included in all volumes of Linnaeus' taxonomic classic, the Systema Naturae, the 10th edition of which, published in 1758, is considered a landmark effort and is the one most often cited as an anchor publication for taxonomic histories. Linnaeus (1758) had only two divisions for the invertebrates, one for the insects (Insecta) that included all the arthropods and one for everything else: the Vermes. The Vermes were subdivided into Intestina, Mollusca, Testacea, Lithophyta, and Zoophyta. Earthworms and leeches were in the Intestina or worms. Linnaeus' "Mollusca" included a host of soft-bodied animals including the marine annelids, cephalopods, anemones, medusae, and some of the echinoderms. Testacea (with a teste or shell) comprised most of the shelled molluscs such as the bivalves and snails but also included the barnacles, a crustacean group. Lithophyta included the stony corals, and the Zoophyta included a potpourri of different groups such as the soft corals and sea pens. Detail and groupings were more accurate in the vertebrate and botanical divisions of the system. It was a good start on biological organization, giving us the genus-species system still in use today, even if groupings were not accurate by modern standards.

Lamarck (1802) first coined the name Annelides for the segmented worms and elaborated further on their relationships in his 1818 Histoire naturelle des animaux sans vertebres, in which he divided the annelids into three major groups based on the presence of antennae and external structures (parapodia, e.g.) and on whether the worms were free-living or tube-dwelling (Fauchald and Rouse 1997). Just prior to Lamarck (1818), Cuvier (1816) had published a classification scheme for the annelids that included the earthworms, polychaetes, leeches, echiurids, and sipunculids. Cuvier grouped the annelids with the arthropods into the Articulata, based on the presence of segmentation in both taxa. The group Articulata was actively in use in some major works as recently as 1969 (e.g. Beklemishev 1969a) and is still a factor considered in studies of higher-order animal systematics. Cuvier did include some misplacements in his annelids, such as the molluscan toothshells, but the annelid classifications of Lamarck and Cuvier showed a great deal of similarity (Fauchald and Rouse 1997).

Classification of the annelids has continued to evolve at a brisk pace, with many benchmark summaries along the way. Highpoints would include, but not be limited to, the following works: Dales (1962), a classic summary of the phylum up to that time including morphology and physiology; Fauchald (1977), a classic in polychaete classification; Pettibone (1982), another definitive work on polychaete classification; Fauchald and Rouse (1997) and Rouse and Fauchald (1997), sister works that summarize the systematics of polychaetes up to that time and use a cladistics approach to further refine the classification scheme; and Rouse and Pleijel (2001), a master summary of the polychaetes including good brief biological descriptions, morphology, and great figures in addition to systematic information. Recent works by Andrade et al. (2015) and Struck et al. (2015) have revolutionized the classification of annelids to the point where many older relationships are no longer considered valid, and Linnaean taxonomic ranks (e.g. Class, Order, and Family) are still being worked out. Though the pelagic representatives are our main focus, a brief overview of this enormous group will help shape our perspective on how much or how little pelagic species differ from their benthic relatives.

Classification

Morphological similarities and patterns in embryological development have been used extensively to deduce animal lineages, forming the basis of the traditional animal phylogenies that persisted from the late 1800s to the 1990s and still form the basic framework for our current understanding of animal relationships. Two new tools for elucidating animal lineages came into play in the latter half of the twentieth century: phylogenetic systematics and molecular phylogenetics. Phylogenetic systematics (cladistics) is a method of analyzing the relatedness of animals based on the observable features of their genotype, which can be morphological, developmental, or the gene sequences themselves (Brusca et al. 2016). Molecular phylogenetics provides the base sequences of the genes themselves, a technology that became practical for phylogenetic analysis in the late 1980s (Field et al. 1988). Together, the two methods have refined the science of animal relationships (systematics) into a more quantitative and predictable field. They have also rearranged traditional animal phylogenies in existence for decades and are continuing to do so.

Based on traditional phylogenetic methodology, annelid classification was fairly stable from the late 1800s into the 1960s, containing three major classes: the polychaetes, oligochaetes, and the leeches (class Hirudinea). Two minor classes were also present: the Myzostomaria, an aberrant group of symbionts associated with crinoids and other echinoderms, and the Archiannelida, a class containing five orders of tiny, interstitial, segmented worms (Dales 1962). In addition, the polychaetes were divided into two main subclasses: the Errantia, for the free-living mobile worms or those that occupied temporary burrows; and the Sedentaria, worms that resided in more permanent tubes or burrows. Those two main subclasses were discarded for many years (e.g. Rouse and Pleijel 2001) but are once again in current use (Brusca et al. 2016; WoRMS n.d.).

Later treatments (Rouse and Fauchald 1997; Rouse and Pleijel 2001, Brusca et al. 2016) put the Archiannelida into the class Polychaeta. The Myzostomaria are now considered to be *incertae sedis*: uncertain placement. In WoRMS, Oligochaeta and Hirudinea have been combined into the annelid class Clitellata and are now subclasses. Brusca et al. (2016) similarly combined the Oligochaeta and Hirudinea using cladistic techniques but did not give them a Linnaean rank. In cladistic analyses, which use both molecular and morphological traits that are ranked in importance and quantitatively compared, relationships within a phylum are not given formal ranks. Classification schemes based on cladistic analyses (e.g. Rouse and Fauchald 1997) yield clades or monophyletic groupings (single common ancestor) that form a dendrogram or branching tree (Figure 6.1). Ancestral and advanced character states within these groups are termed plesiomorphies and apomorphies, respectively (see Box 6.1 for a list of useful terms). Cladistic techniques are complex. A readable summary of the field and its

---

Box 6.1   Useful Terms

Apomorphies More recently derived, evolved, or advanced characters.

Autapomorphies Specialized characters found in only one taxon. Important for that taxon, but not useful in constructing a phylogenetic tree.

Homoplasies Shared independently derived similarities, e.g. parallelisms, convergences, secondary losses, which do not reflect the evolutionary history of a taxon but can appear to

Monophyletic group A group of species that includes all the descendants of a common ancestor: **a clade**. Also known as a **Holophyletic group**.

Outgroup A group closely related to your study group (**ingroup**) that allows you to distinguish the apomorphies of your study group from commonly held symplesiomorphies.

Paraphyletic group A group that contains only descendants of a common ancestor but does not contain the entire list of descendants, i.e. an incomplete list.

Parsimony Ideally when constructing a classification, a taxon can be defined by a number of synapomorphies. However, conflicts often exist. Some characters may point to a closer relationship of group A to B, others group A to C. The principle used to sort out the problem is parsimony – the hypothesis that explains the data in the simplest way. Generally parsimony is achieved using computer programs and large numbers of taxonomic characters.

Plesiomorphies Ancestral, primitive, or generalized characters

Polyphyletic group A group containing descendants from more than one ancestor.

Sister group A monophyletic group that is the closest genealogical relative to your ingroup. The most closely related clades at the nodes of a cladogram.

Symplesiomorphies Shared primitive characters. Not useful for constructing phylogenies because primitive characters may be retained in a wide variety of taxa. Advanced as well as primitive taxa are likely to possess symplesiomorphies.

Synapomorphies Shared derived characters that define monophyletic groups or clades (groups containing an ancestor and all its descendant taxa).

Source: All definitions from Brusca et al. (2016), Wiley and Lieberman (2011), Helfman et al. (2009).

techniques can be found in Wiley and Lieberman (2011). An excellent capsule summary is in Brusca et al. (2016). Although molecular and cladistic methodologies shape our current understanding of animal lineages and relatedness, it is remarkable that the phylogenies produced using the older morphological and developmental methods are still largely intact.

Ultimately, the relatedness of species is defined by similarities in their genetic makeup, but genetic tools robust enough to sort out questions at the higher levels of classification, e.g. at the phylum level and above, have only recently been developed. Molecular phylogenies entered the fray in 1988 (Field et al. 1988) with a sequencing of the 18S ribosomal RNA gene in animal representatives from 10 phyla, founding the field of molecular phylogenetics and giving evolutionary biologists an important new tool for understanding animal relationships.

For everyday reference, the taxonomic system in WoRMS is recommended. It uses the classic taxonomic divisions within the phylum Annelida and provides information for all levels down to species, designated with Linnaean taxonomic ranks. It differs from that presented in Brusca et al. (2016) in three main ways. First, the classes Clitellata and Polychaeta are retained. Second, the peanut worms, Sipuncula, are not included in the phylum Annelida but are retained as their own phylum. Third, the innkeeper worms, Echiura, are a subclass in the class Polychaeta, and in Brusca et al. 2016, the Echiura are roughly at the ordinal or family level.

The Brusca et al. (2016) classification scheme synonymizes the entire phylum Annelida with what was formerly the class Polychaeta. The former class Clitellata, which includes earthworms, leeches, and their kin, is now a clade within the polychaetes (see Figure 6.1a), though their total number of species (about 6000) is almost 35% of the phylum's total numbers (about 17000 species, Chapman 2009). Figure 6.1b provides an expanded look at the Clitellata. In each of the two groups, most clades ending in -ida were formerly orders, those ending in -iformia were suborders, and those ending in -idae were formerly families. Note that of the five uppermost clades shown in Figure 6.1a, four were formerly families and one was a phylum.

Once within the major groupings, (the polychaetes and clitellates), the taxonomic schemes are fairly similar, and within the WoRMS system, there are also large numbers of species with uncertain placement (*incertae sedis*). Until the experts in the field come to a consensus, the nonspecialist must accept the uncertainty. For now, the easiest system to use is that in WoRMS.

Phylum Annelida

The segmented and unsegmented worms; a work in progress. Classic works include Dales (1962), Stuart (1982), Pettibone (1982), Brinkhurst (1982), Fauchald (1977), and Rouse and Pleijel (2001). More recent work includes Siddall et al. (2001), Rousset et al. (2007), Marotta et al. (2008), Andrade et al. (2015), Struck et al. (2015), and Brusca et al. (2016). The taxonomic ranks below roughly follow those in WoRMS with the addition of the sipunculids to the polychaetes. When practical, comparisons are made between WoRMS and Brusca et al. 2016.

Class Polychaeta

The largest (at least 12027 marine species) and oldest group in the phylum Annelida. Most species have a pair of muscular,

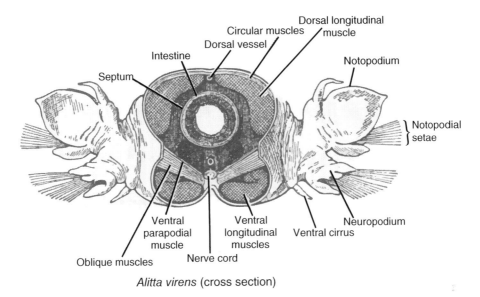

Alitta virens (cross section)

Figure 6.2 Cross section through a segment of the clam worm *Alitta virens* (formerly *Nereis virens*). *Source:* Snodgrass (1938), V97, No. 6, figure 15 (p. 36).

foot-like, parapodia on each segment, usually bearing the bundles of chaetae (setae) that give the class its name (Figure 6.2). The head region consists of two main segments (Figure 6.3). The most anterior is the prostomium, which bears sensory elements (antennae and eyespots) when they are present. The prostomium is followed by the peristomium that bears the mouth. In tube-dwelling forms, the head region can be highly elaborated with feeding and respiratory appendages. The final segment is the pygidium, which contains the anus. The developing polychaete adds its new segments just anterior to the pygidium (teloblastic development).

Polychaetes are highly diverse in form, primarily marine, yet found occasionally in freshwater and even in moist earth. Morphology reflects their life habit: crawling, sedentary, tube-dwelling, or pelagic. Sexes are separate; development is usually indirect with a distinctive larva, the trochophore, that exhibits spiral cleavage in its early embryonic stages.

Subclass Echiura

Plump, highly derived, burrow-dwelling polychaetes with no external segmentation, but with teloblastic development of the nervous system. Since their initial description, they have been in and out of the annelids (Brusca et al. 2016), and their current position within the polychaetes is based on molecular evidence (McHugh 1997). They possess a preoral proboscis that can be extended in some species to many times the body length and their burrows often provide shelter for smaller motile species, such as the scale worm *Harmothoe*, and the tiny pea crab *Scleroplax*, hence the name "innkeeper worms." The best studied species, *Urechis caupo*, feeds by spinning a mucous mesh net within the burrow and pumping water through it, trapping particles, and then

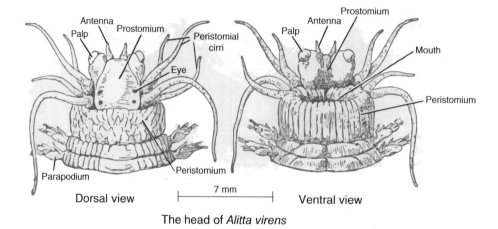

The head of *Alitta virens*

Figure 6.3 Dorsal and ventral views of the head of *Alitta virens*. *Source:* From Snodgrass (1938), V97, No. 6, figure 14 (p. 33).

consuming them net and all. At least 173 marine species (Figure 6.4a).

Subclass Errantia
Mobile worms. Includes the two large orders Eunicida (1410 spp.) and Phyllodocida (4637 spp.) together making up most of the familiar intertidal forms such as the nereids and glycerids as well as the palolo worms. Many of the roughly 150 pelagic species are in the order Phyllodocida. Brusca et al. (2016) include the former Archiannelida in this clade as the Protodrilida (Figure 6.1a), in WoRMS they remain *incertae sedis*.

Subclass Sedentaria (about 5300 species). Includes mainly tubicolous forms such as the featherduster worms and cirratulids as well as the former phyla Pogonophora and Vestimentifera, now in the family Sibloglinidae within the order Sabellida. A small number of pelagic forms, covered below, are in this subclass.

The Sipuncula. Still a phylum in WoRMS, but not in Figure 6.1a, where they form their own clade placed at a fairly basal position within the polychaetes (Brusca et al. 2016). Like the echiuroids, they bear little outward physical resemblance to polychaetes, except in their vermiform shape. They are all benthic and comprise about 157 species. In invertebrate zoology classes fortunate enough to be situated near a fully marine intertidal, they are a stalwart for collection and observation in the lab. Hyman (1959), who erected a phylum for them, describes them as "vermiform animals of cylindroid shape, with the body regionated into a slender anterior part, the introvert, and a plumper posterior trunk." The mouth is located at the terminus of the introvert, usually accompanied by tentacles. The introvert can be fully retracted into the trunk by their robust retractor muscles. When the introvert is withdrawn, the entire worm becomes much like a flexed muscle! They live most often in burrows but are also found in association with kelp holdfasts and coral rock. They feed in a variety of ways including direct deposit feeding and detritivory (Figure 6.4b).

Class Clitellata

Earthworms, leeches, and their relatives. About 6000 total species, 1029 of them are

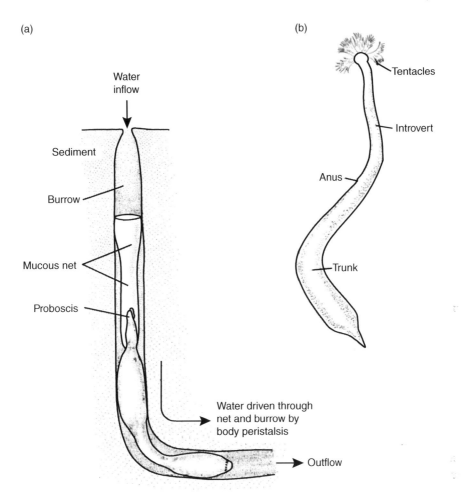

Figure 6.4 (a) The echiuroid *Urechis caupo* in its burrow. Note its body shape and mucous feeding web. Water flow through its U-shaped burrow achieved by peristaltic movements of the body. (b) The sipunculid *Dendrostomum*. Sources: (a) Adapted from Brusca and Brusca (1990), figure 7B (p. 447). (b) Adapted from Hyman (1959), figure 214 (p. 613).

marine. Earthworms and leeches have no parapodia, and chaetae are much reduced in number or absent altogether, hence the name *oligo chaeta* – meaning few chaetae. Oligochaetes are found in freshwater, marine, and moist terrestrial environments. They are mainly hermaphroditic, use copulation during reproduction, and possess a structure called a clitellum that functions in cocoon formation. Development is usually direct.

Subclass Oligochaeta
Earthworms, freshwater annelids (e.g. the familiar tubifex worms), and some marine representatives (about 879 marine spp.). The head region of oligochaetes is simpler than that of polychaetes and lacks sensory structures The oligochaetes are divided into five orders, with the great majority of families in the order Hapolotaxida.

Subclass Hirudinea

The leeches are annelids that possess a fixed number of segments and suckers at one or both ends of the body. The segments are further divided into superficial annuli or rings that are constant within a species and at times also within genera and families. Leeches are primarily found in marine and freshwater systems but are occasionally in moist terrestrial environments. All are parasites or predators.

The Pelagic Polychaetes

Although the overwhelming majority of polychaete species are marine, most are not pelagic. Dales and Peter (1972) list 102 as pelagic, which has been upgraded to about 150 in the intervening years. Thus, only 0.01% of the total number of polychaete species have entered the pelagic realm, a theme we will see repeated in many taxa.

Most of the available information on pelagic polychaetes concerns the species from one order, the Phyllodocida. They include the most commonly captured species in zooplankton and micronekton tows are pan-globally distributed and are an important component of the open-ocean food web. Phyllodocida is a very large order with 28 families, so the pelagic representatives discussed here are only a small part of the total.

Until 2010, only two taxa outside of the Phyllodocida were noted in the literature. The first was the aberrant polychaete *Poeobius meseres* found exclusively in the North Pacific, which in Pettibone (1982) has its own order (Poeobiida) and family (Poeobiidae). WoRMS puts Poeobius in the Order Terebellida and family Flabelligeridae. The second, *Flota,* was classified by Fauchald (1977) as a member of the family Flotidae and Order Fauvelopsida. *Flota* is now also classified within the family Flabelligeridae.

A 2007 expedition to the Celebes Sea, a deep (6200 m) basin to the southeast of the Philippines, revealed a suite of seven new species of pelagic polychaetes (Osborn et al. 2009; Osborn and Rouse 2010). The Celebes Sea is one of a series of basins, including the Sulu, Halmahera, and Banda Seas, in the vicinity of Indonesia that enjoy a relative degree of autonomy because of sill depths that are far shallower than the deepest point in each. The shallow sill depths restrict exchange between the basins and the Pacific Ocean in general, partially isolating the fauna and potentially increasing the chances for speciation. The biological diversity of marine fauna in this region, collectively known as the Indo-West Pacific, is the highest in the world (Briggs 1974). Much of the deep-sea fauna there awaits description.

All seven of the new polychaete species reside in the benthopelagic region of the water column, a transition zone that includes the benthic–water column interface and a few hundred meters of water above it. It is a fairly rich zone within the water column because descending particulates concentrate there.

It is difficult to sample the benthopelagic realm. Midwater trawling gear, designed to sample the water column, gets torn up if it hits the bottom or, if you're "lucky," it fills up with tons of mud. Mud-filled midwater trawls produce hours of misery for the scientific party. To dispose of the mud, the net has to be dismantled while it is still in the water because it could not be winched aboard in that state; the weight of the mud would destroy the net. Gear for bottom trawling is sturdier but tears up delicate animals, rendering it useful only for robust taxa. The new species of worms were discovered using remotely operated vehicles, unmanned submersibles that could record behavior of delicate animals with video and capture specimens for analysis.

The analysis revealed that the new worms, instead of being closely related to the phyllodocid taxa commonly caught in the zooplankton, had closer affinities to tube-dwelling polychaetes and to the aberrant *Poeobius:* a huge discovery. The new species belong to the family Acrocirridae, an unusual, primarily benthic group with a widespread distribution, particularly in the Pacific.

WoRMS places the newly discovered pelagic polychaetes in the Acrocirridae, a family in the order Terebellida and suborder Cirratuliformia (cf. Figure 6.1). Note that the genera *Poeobius* and *Flota* (Flabelligeridae) are also in the Cirratuliformia. Taxonomically, the phyllodocids and the terebellids are quite separate, with most of the phyllodocids being crawlers and most of the terebellids being sedentary or tube-dwellers. At least in *Poeobius* their lineage is reflected in their feeding habits, as discussed in the section "The Suspension Feeders."

One other recent addition to the world of pelagic polychaetes is a chaetopterid, *Chaetopterus pugaporcinus* (Osborn et al. 2007). The chaetopterids are a tube-dwelling family with a morphology well suited to a tubicolous life. Found in the Monterey Canyon, it is unclear whether *C. pugaporcinus* is a neotenic adult form or a usually large and highly modified larva. It appears to feed like its benthic relatives, with a mucous web that traps falling particles, much like that described for *Poeobius*.

Polychaete Subclass Errantia

Individuals in the three families and one tribe listed below, the tomopterids, typhloscolecids, lopadorhynchids, and alciopines (Figure 6.5), are by far the most commonly caught pelagic polychaetes, mainly because they are the largest and best-retained by scientific trawls with typical mesh sizes of 3–6.5 mm. Far more information is available on their biology than on the smaller families or the newly discovered pelagic species, so those four families will be our main focus.

Order Phyllodocida
Suborder Phyllodocida *incertae sedis*
   Family Tomopteridae. Most common genus: *Tomopteris*
   Family Typhloscolecidae. Common genera: *Sagitella, Travisiopsis, Typhloscolex*
Suborder Phyllodociformia
Family Phyllodocida
   Subfamily Eteoninae
   Tribe Alciopini Common genera: *Alciopa, Alicopina, Krohnia, Naiades, Rhynchonereella, Vanadis, Torrea.*
   Family Lopadorrhynchidae. Common genera: *Lopadorrhynchus, Maupasia, Pelagobia.*

The following three families, shown in Figure 6.6, are less common pelagic species considered "minor" by Rouse and Pleijel (2001).
Suborder Phyllodocida
   Family Iospilidae. Genera: *Iospilus, Phalacrophorus.*
   Family Yndolaciidae. Monogeneric family. *Yndolacia*
Suborder Phyllodociformia
Family Pontodoridae. Monogeneric family: *Pontodora*

Polychaete Subclass Sedentaria

The newly discovered pelagic polychaetes are in subclass Sedentaria, as are *Flota* and *Poeobius* (Rouse and Pleijel 2001; Osborn and Rouse 2008; Osborn and Rouse 2010)

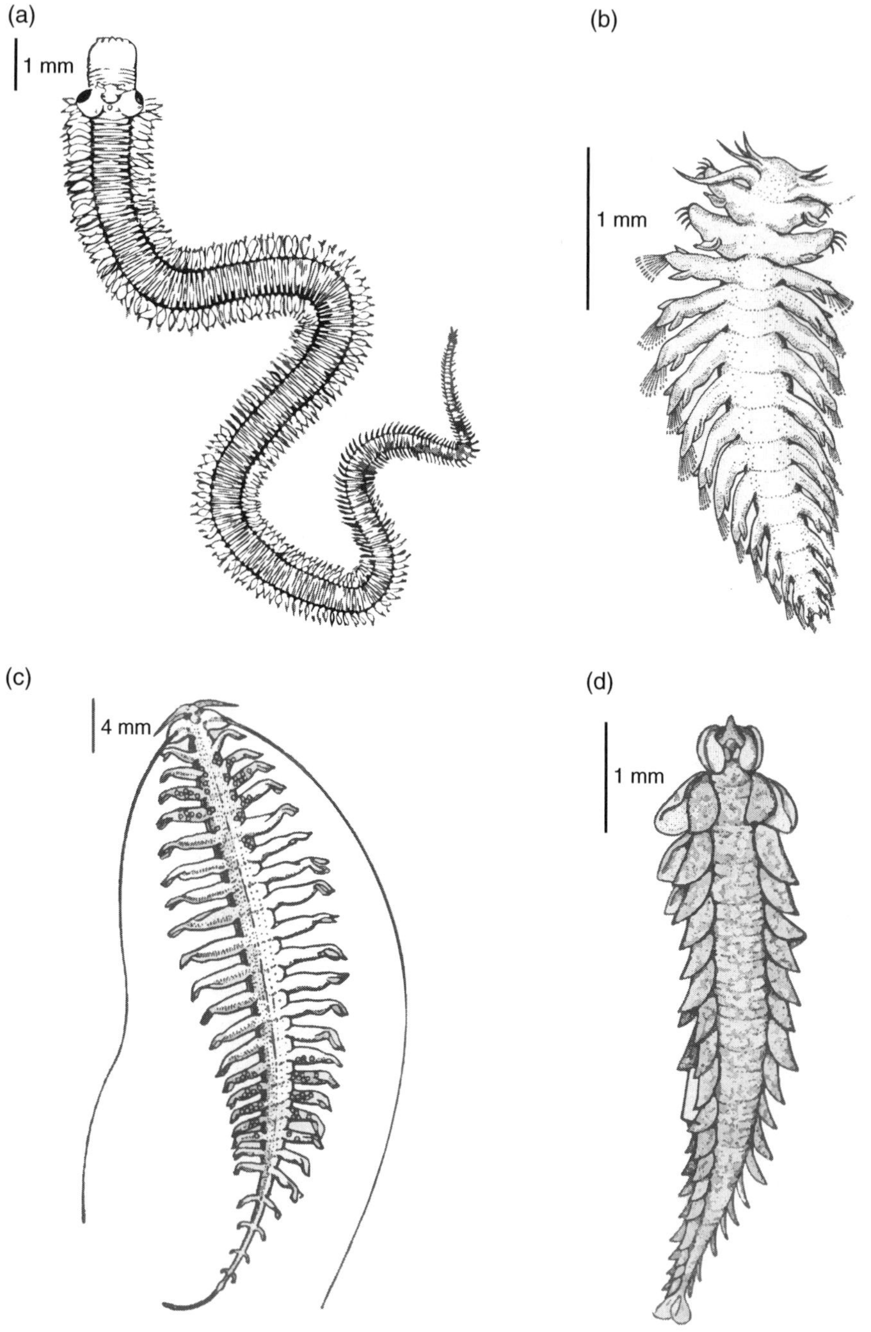

Figure 6.5 Representative examples of the four most commonly caught families of pelagic marine polychaetes. (a) Alciopini: *Rhynchonereella angelini*; (b) Lopadorrhychidae: *Lopadorrhynchus krohnii*; (c) Tomopteridae: *Tomopteris pacifica*; (d) Typhloscolecidae: *Travisiopsis lobifera. Source:* Adapted from Tebble (1962), (a) figure 13 (p. 401); (b) figure 21 (p. 418); (c) figure 6 (p. 386); (D) figure 19 (p. 412).

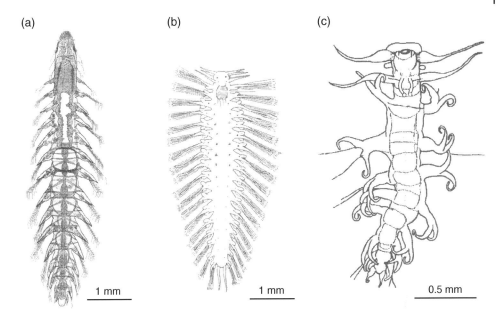

Figure 6.6 Examples of three less common pelagic marine polychaete families. (a) Iospilidae: *Phalacrophorus borealis*. (b) Pontodoridae: *Pontodora pelagica*. (c) Yndolaciidae: *Yndocolaciella polarsterni*. *Sources:* (a) Adapted from Reibisch (1895); (b) Adapted from Uschakov (1972); (c) Buzhinskaja (2004), figure 2 (p. 342). Reproduced with the permission of Taylor and Francis.

Order Terebellida
Suborder Cirratuliformia (Figure 6.7)
    Family Acrocirridae. Two genera thus far comprise the swimming clades: *Swima* and *Teuthidodrilus* (the squid worm)
    Family Flabelligeridae. Genera include *Flota* and *Poeobius*

Fortunately for the study of pelagic annelids, the order Phyllodocida is very well represented in the intertidal. The anatomy, physiology, and ecology of a number of intertidal phyllodocids have received considerable attention, and we can use that information to help fill in the blanks for the more poorly studied pelagic species that are our main focus. In particular, a well-known species in the family Nereididae, *Alitta virens* (recently known as *Nereis virens* and formerly as *Neanthes virens*) provides substantial background. Some of the other polychaete families to be mentioned, particularly with respect to reproductive behavior, are also phyllodocids, e.g. the Syllidae. Another order within the Errantia, the Eunicida, exhibits reproductive swarming behavior and will put in a cameo appearance in the discussion of reproduction. *Glycera*, the blood worm familiar to many as bait, is also in a suborder of the Phyllodocida.

Morphology

General

The annelids possess a true coelom, bilateral symmetry, and, in the polychaetes, a cylindrical body divided into serially repeated segments known as metameres (Figures 6.8 and 6.9). Like the nemerteans, annelids possess a well-developed one-way digestive tract, usually with regional specialization allowing for more efficient

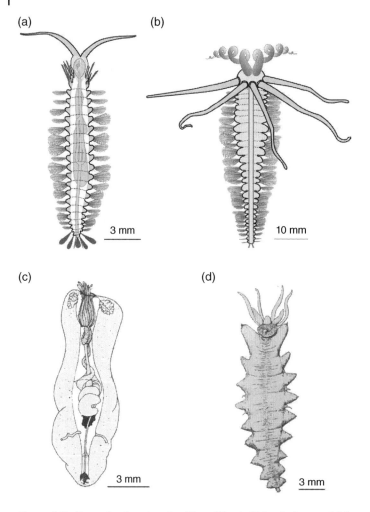

Figure 6.7 Examples from two families of Terebellid polychaetes. (a) Acrocirridae: *Swima bombaviridis*; (b) Acrocirridae: *Teuthidodrilus samae*; (c) Flabelligeridae: *Poeobius meseres*. (d) Flabelligeridae: *Flota flabelligeri*. *Sources:* (c) Robbins (1965), figure 1 (p. 200). Reproduced with permission of John Wiley & Sons; (d) Adapted from Hartman (1967).

initial food breakdown, subsequent digestion, and absorption. A closed circulatory system is normally present and contains intracellular or extracellular blood pigments to aid in gas exchange. The nervous system comprises a dorsal cerebral ganglion or brain, circumesophogeal connectives, and a sometimes-paired ventral nerve cord that is ganglionated within each segment. Excretory needs are met by paired nephridia present in each segment. Polychaetes are most commonly dioecious, but a few sequential hermaphrodites have been described. Fertilization usually occurs through broadcast spawning, particularly in swarming species (e.g. nereids, syllids, and eunicids), but the presence of seminal receptacles in a few species suggests that copulation may occur as well. Annelids are capable of regenerating body parts; however, unlike in the nemerteans, regeneration is not considered an important

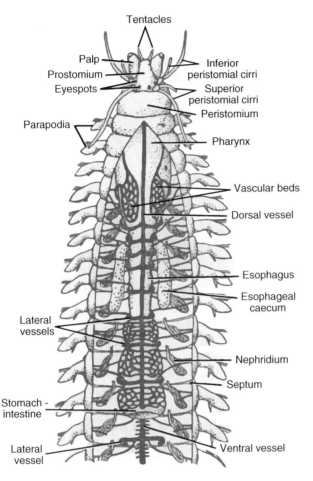

Figure 6.8 Dorsal view of *Alitta virens* showing the general arrangement of internal organs within the metameres. *Source:* Brown (1950), figure 149 (p. 273). Reproduced with the permission of John Wiley & Sons.

reproductive mechanism in annelids. In species that are capable of regeneration, the ability to generate anterior parts of the worm tends to decline on an anterior-to-posterior axis (Kaestner 1967). Thus, the head apparently will always be regenerated, but the number of anterior metameres that are regenerated along with it depends on how far back on the worm the separation occurs. The posterior end will regenerate only the head.

External Anatomy

A simple diagram of polychaete external anatomy, along with some insights on how adult anatomy relates to that of the polychaete larva, the trochophore, is shown in Figure 6.10a. Looking from anterior to posterior, the general regions are the prostomium, peristomium, trunk, and pygidium. The prostomium is a preoral segment that originates from the region anterior to the ciliated prototrochal mid-region of the

(a)

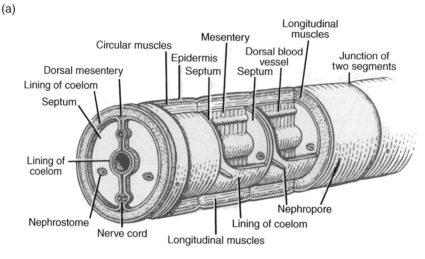

(b)

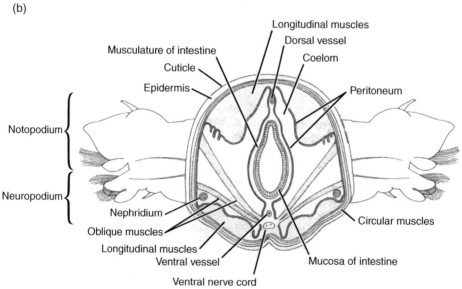

Figure 6.9 Metameric anatomy of annelids. (a) Diagram of the anatomy within several metameres of an annelid with the body wall of two segments removed. (b) cross section of a nereid polychaete metamere. *Sources:* (a) Kaestner (1967), figure 19-2 (p. 456); Brown (1950), figure 150 (p. 274). Reproduced with permission of John Wiley & Sons.

trochophore larva. The peristomium, containing the mouth, derives from the prototrochal zone. Posterior to the peristomium is the trunk region (or metastomium), consisting of a variable number of segments, each bearing parapodia. New segments form in the telotrochal region of the larva as the worm grows and are added just anterior to the pygidium, the terminal segment that surrounds the anus (Figure 6.10b).

The Head Region

The head of *Alitta*, the clam worm (Figure 6.3), provides details on head

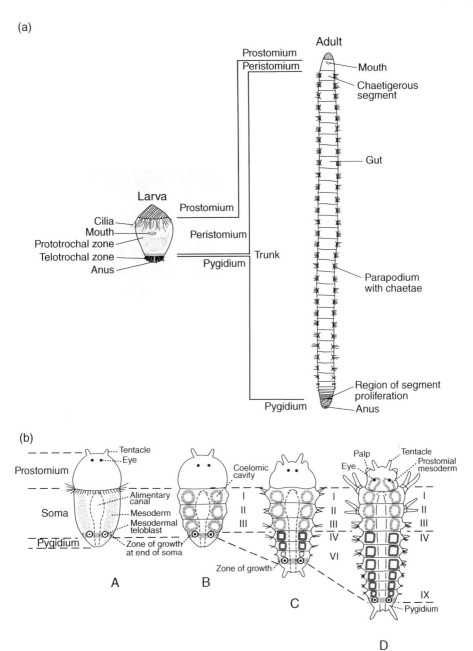

Figure 6.10 External anatomy of polychaetes: regional zonation and growth. (a) Comparison showing regional zonation between a trochophore larva and an adult. (b) larval segmentation and subsequent growth in polychaetes. *Sources:* (a) Pleijel and Dales (1991), figure 1 (p. 3); (b) Snodgrass (1938), V97, No. 6, figure 11 (p. 27)

structure in both dorsal and ventral aspects. *Alitta* is an active, crawling, intertidal predator capable of swimming, and, like most of the pelagic polychaetes, is in the order Phyllodocida. The wide distribution of *Alitta* plus its accessibility and hardiness have made it a favorite subject for study. The fact that it shares an order with the pelagic polychaetes makes it an ideal choice for providing substantial background on the pelagic species.

The prostomium is the first segment in a moving worm to sample the environment ahead and thus carries the main complement of anterior sensory structures. The antennae, believed to be mechanoreceptors sensitive to touch and movement, are usually paired dorsolateral structures, but they can vary in number and size between species. The palps, also paired, originate ventrolaterally and are believed to be chemoreceptors for smell and taste involved in prey detection. One to two pairs of simple cup eyes are located dorsolaterally in those species that have them. In many species, paired structures known as nuchal organs are present on the posterior dorsum of the prostomium. They are believed to have a chemoreceptive function as well.

Just anterior to the prostomium is the peristomium, which bears the mouth and, in many cases, peristomial cirri (Pleijel and Dales 1991). Some controversy exists as to the nomenclature for the cirri present in the head region and on the peristomium itself (Rouse and Pleijel 2001). It is agreed that the peristomium exists as the second segment in the head. However, fusion of one, two, or three segments of the trunk to the peristomium is believed to have occurred as part of development in many species (Figure 6.11a). Since trunk segments all bear parapodial structures, the cirri that appear to be on the peristomium may in fact have an embryological origin in a trunk segment. Rouse and Pleijel consider that the peristomium of *Nereis* is limited to the lips and that the peristomial cirri are part of the first two trunk segments that have fused during development (Figure 6.11b, c). Pleijel and Dales (1991) appreciated the fusion of trunk segments into the peristomium during development, but since the cirri were present on the peristomial segment in the adult, they termed them peristomial cirri.

Even though the proboscis is an internal structure when at rest, it figures largely in the morphology of the head region when it is everted and so will be considered here. Proboscides vary considerably in length and in the presence or absence of chitinous jaws and extra denticle armor. *Olganereis* has a formidable two-hook jaw (Figure 6.11b), but many species lack jaws altogether. Proboscis eversion in polychaetes is quite similar to that in nemerteans (see Figure 5.9c,d). Hydrostatic pressure generated by the anterior body wall musculature, in conjunction with protractor muscles connected to the peristomial wall, forces the pharynx inside out to snare prey. When the body wall muscles relax, the proboscis is withdrawn, along with its prey, aided by retractor muscles.

The Trunk or Metastomial Region

In Figure 6.10a, the segments of the generic polychaete portrayed are quite similar throughout the trunk region, a configuration known as homonomous segmentation. Active crawlers, swimmers, and burrowers, like polychaetes in the subclass Errantia, often exhibit homonomous segmentation. Polychaetes that spend their adult life in tubes, such as the cirratulids and chaetopterids in the subclass Sedentaria, have heteronomous segmentation: regional specialization in the segments along the trunk to aid in the tubicolous life (Figure 6.12).

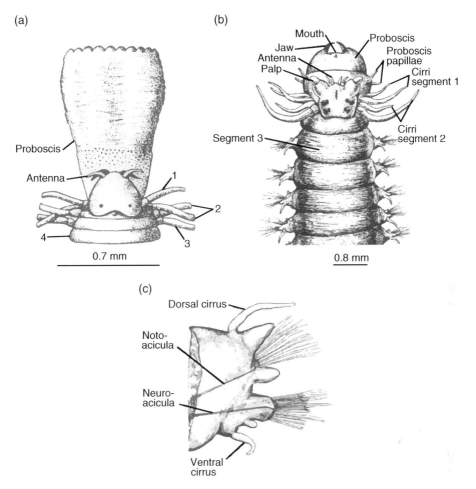

Figure 6.11 Structure of the head and parapodia of marine polychaetes from two families. (a) Phyllodocidae: *Phyllodoce fristedti*. Numbers 1–3 indicate parapodia of the first three segments, number 4 marks the fourth segment. All three segments posterior to the prostomium are free. (b) Nereididae: *Olganereis edmondsi*. Dorsal view of the anterior end with everted proboscis. (c) anterior view of the parapodium from segment 8 of *O. edmondsi*. *Sources:* (a) Bergstrom (1914), plates I–V; (b) Hartman (1954), figure 12 (p. 24); (c) Hartman (1954), figure 14 (p. 24).

Segments, or metameres, are easily discerned externally. Each bears a pair of the fleshy, paddle-like parapodia. The parapodia typically have dorsal and ventral lobes (rami) called the notopodium and neuropodium, respectively (Figure 6.9b). Chitinous rods, or aciculae, stiffen the parapodia by providing internal support. Muscles attached to the proximal side of the aciculae can effect protrusion and retraction of parapodia. Chaetae (setae) project in bundles from epidermal sacs originating deep within the parapodia. Chaetae are a composite of chitin and fibrous protein (glycoprotein) and are replaced as necessary throughout the life of the worm. Though all chaetae are basically rod-shaped, they usually possess a simple shaft leading to a blade that can be flattened, serrated, hooked, or oar-shaped, to name a few variations

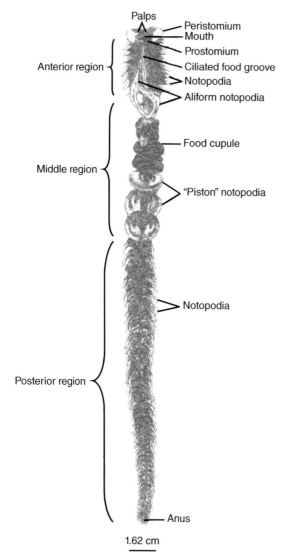

Figure 6.12 The chaetopterid *Chaetopterus variopedatus* showing heteronomous segmentation. *Source:* Enders (1909), plate II.

Chaetopterus variopedatus

(Figure 6.13). When chaetae are bent or jointed, they are considered compound chaetae. The variability in chaetal structure plays an important role in the identification of species.

Chaetae can aid in crawling by providing additional stiffening to the parapodia and providing traction in sediment. Muscles associated with the chaetal sacs move the chaetal bundles. Dorsal and ventral cirri, believed to act as tactile organs, are often present on the parapodia (Figure 6.14a). They are sometimes modified to form rudimentary gills (Figure 6.14b) or an extra paddle. Within the pelagic species, parapodial structure often favors a paddle-like form (Figure 6.14c) as might be expected, with some families also exhibiting reductions in

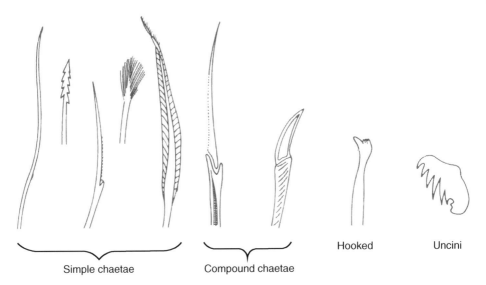

Figure 6.13 Different types of chaetae from various polychaetes. *Source:* Adapted from Brusca and Brusca (2003), figure 13.6 (p. 396). Reproduced with the permission of Oxford University Press.

the number of chaetae (Figure 6.14d) or even their complete absence as in the tomopterids (Figure 6.14e).

Internal Anatomy

A deformable collagenous cuticle is secreted by the epidermis (Richards 1978), serving as a barrier to the exterior but allowing considerable flexibility for sinusoidal motion. Extensibility is less pronounced than in the nemerteans owing to the dual constraints of the annelids' segmental body plan and the cuticle. Beneath the cuticle, the epidermis (or hypodermis) itself is a typical columnar epithelium interspersed with gland cells of various types. A thin layer of connective tissue separates the epithelium from the musculature of the body wall (Figure 6.9).

The outermost section of the body wall comprises a thin layer of circular muscle, followed by a middle oblique layer. The innermost layer is a robust longitudinal musculature divided into four subregions by the mesenteries supporting the dorsal vessel, the ventral nerve cord, and the parapodia (Figure 6.9). Since locomotion in the crawling and swimming worms mainly involves sinusoidal side-to-side or dorsoventral motion, both actions governed by the longitudinal muscles, it is to be expected that the longitudinal musculature would be highly developed.

Just interior to the body-wall muscles is the peritoneal lining of the coelom (body cavity). Note that the peritoneum encloses the coelom completely, isolating it from the gut as well as the body wall and thereby giving it considerable autonomy. That autonomy is refined further in many species by the presence of transverse septa that isolate each segment, or metamere, into a nearly separate, hydrostatically distinct, compartment. The septa are perforated to allow movement of coelomic fluid between segments, but from a hydrostatic perspective remain functionally distinct. In many errant polychaete species, the oblique muscles divide the coelom to form ventrolateral nephridial chambers. The metameres may

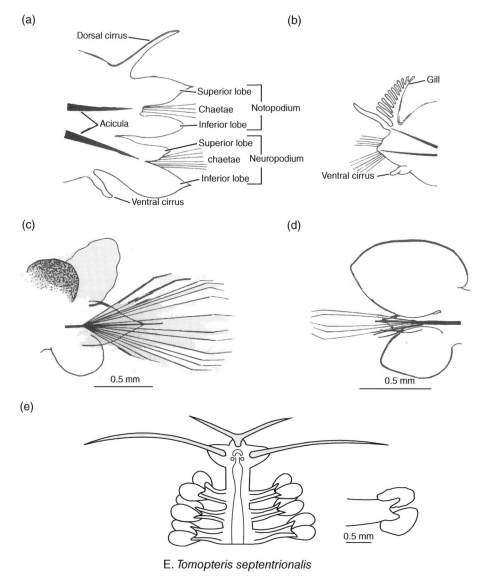

Figure 6.14 Examples of structural differences in polychaete parapodia. (a) general structure of a stylized parapodium; (b) modified parapodium of a eunicid; (c) parapodium of a pelagic species, *Torrea candida*; (d) parapodia of the tomopterid *Tomopteris septentrionalis*. Sources: (a, b) Brusca and Brusca (1990), figure 5 (p. 390); (c) Dales (1957), figure 16 (p. 110); (d) Dales (1957), figure 42 (p. 131).

be separated into left and right halves by the mesenteries that suspend the gut, blood vessels, and nerve cord.

The coelom is filled with a clear fluid, not blood, that acts as a hydrostatic skeleton. Coelomic fluid usually contains amoebocytes, and during the reproductive season, most of the trunk segments contain developing gametes (Pleijel and Dales 1991). Though considerable diversity in form exists among the polychaetes, in the phyllodocids each segment has some form of duct (a gonoduct or coelomoduct) leading from the coelom to the outside to allow for the passage of gametes and has a nephridium for excretion of nitrogenous waste and, where necessary, to allow for osmotic regulation or water balance.

Excretory System

The pelagic families have three types of gonoduct/nephridium structures: protonephridium, the most primitive; protonephromixium; and metanephromixium, also known as metanephridium.

In the protonephridium, found in *Vanadis* (Tribe Alciopini), the gonoduct and nephridium are separate structures (Figure 6.15). The nephridium is a protonephridium, a blind-ended excretory organ like those of the nemerteans, containing flame cells (solenocytes). The gonoduct lies adjacent to the nephridium, a funnel-shaped duct that allows gametes to be discharged to the outside through a pore on the segment immediately posterior.

The second type of gonoduct/nephridium structure, which is found in the phyllodocids, is an integration of the funnel-shaped gonoduct with a protonephridium to form a common duct to the outside (Figure 6.16). All the pelagic polychaetes, including the recently discovered cirratuliform pelagics, are believed to have this combination gonoduct/protonephridium organ, termed a protonephromixium (Goodrich 1945; Rouse and Pleijel 2001).

The third type of combined gonoduct/nephridium is a more complete integration of the two functions into one funnel-bearing structure, the metanephromixium (Figure 6.17). It is seen in our model polychaete, *Alitta virens*. The metanephromixium of nereids allows gametes to make their way to the outside. However, because of the length of the highly coiled tubule leading to the outside nephridiopore, it also provides a mechanism for biological modification of the coelomic fluid within the nephric duct. Important nutrients such as glucose may be retained, and organic wastes voided to the outside. The nephridiopore exits just below the neuropodium on the segment immediately posterior to the location of the nephrostome, the funnel-like anterior end of the metanephromixium.

Names for the combined-function gonoduct/nephric structures of the polychaetes differ from author to author. The mixed function gonoduct/protonephridium of the phyllodocids is termed a protonephromixium by Rouse and Pleijel (2001) and Brusca and Brusca (2003), and that of the nereids a metanephromixium by the same authors. The terminology was originally proposed by Goodrich (1945), in his benchmark monograph on excretory organs, to describe the multifunctional excretory organ of polychaetes. Physiologists interested in the excretory organ itself focus on the nephric structure: protonephridia and metanephridia. The mixed-function organ is more properly termed proto- and metanephromixium, respectively. Protonephridia and metanephridia both exist as "stand-alone" excretory organs.

Virtually all nephridia, from those in nemerteans to those in mammals, are similar in their basic function. Generically they are known as tubular excretory organs. They all function by modifying an initial filtrate along the length of the nephric tubule, either by reabsorbing desirable small molecules such as glucose or by secreting undesirable molecules such as organic acids. The initial filtrate is produced by expressing fluid through a membrane using a hydrostatic pressure differential. In the case of

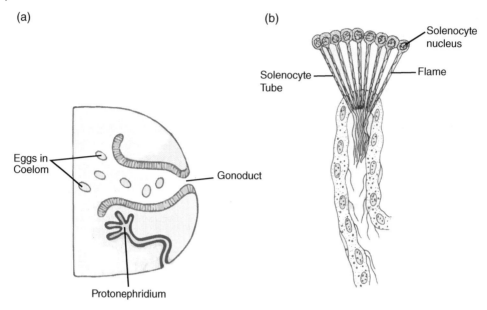

Figure 6.15 Polychaete excretory systems. (a) Hypothetical polychaete excretory system where the gonoduct and protonephridium are separate; (b) the protonephridum of *Phyllodoce* showing solenocyte nuclei, tubes, and ciliary "flames." *Source:* Adapted from Goodrich (1945) (a) text-figure 18 (p. 148); (b) text-figure 23 (p. 154).

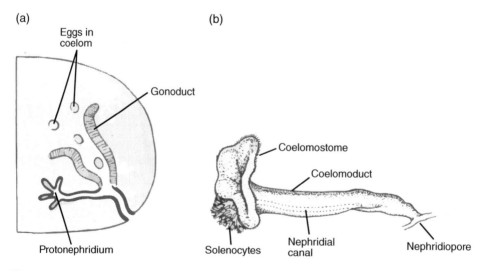

Figure 6.16 Polychaete excretory systems. (a) Hypothetical polychaete excretory system with funnel and protonephridia fused; (b) protonephromixium of *Phyllodoce* with coelomoduct and nephridial canal fused. *Source:* (a) Adapted from Goodrich (1945), text-figure 1 (p. 117); (b) Goodrich (1945), text-figure 24 (p. 156).

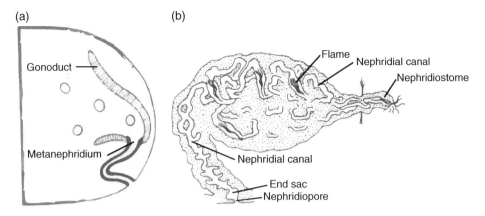

Figure 6.17 Polychaete excretory systems. (a) Hypothetical polychaete excretory system with funnel and metanephridia fused (typical in most polychaetes); (b) metanephridium of the oligochaete annelid *Enchytraeus* with cilia and "flames." *Source:* (a) Adapted from Goodrich (1945), text-figure 3 (p. 117); (b) Goodrich (1945), text-figure 66 (p. 236).

flame cells or solenocytes (Figures 6.15 and 6.16), the filtering membrane is the blind end of the flame cell, and the hydrostatic pressure differential is produced by the ciliary activity inside the nephric tubule, creating a net positive pressure from outside to inside. In the case of the funnel-like metanephridium (Figure 6.17), the initial filters are the capillaries of the circulatory system. Blood pressure within the vascular system causes some fluid to be expressed through capillary walls into the coelomic cavity. In turn, the coelomic fluid is drawn into the metanephridium, through either positive pressure in the coelom or ciliary activity within the nephric tubule, and it is modified during its transit to the nephridiopore.

Nephridia, in whatever form they take, are considered to be segmental organs. In many annelid families, particularly those with homonomous segmentation, segmental organs are serially repeated, one pair to a segment, over the worm's length. In contrast, three important organ systems pass through the segments in a continuous fashion: the nervous system, the circulatory system, and the digestive tract (Figures 6.8 and 6.9).

Pelagic Species

The excretory organs of the pelagic species most closely reflect their systematic positions, which in turn reflect their benthic ancestry. All the segmental organs of pelagic species, phyllodocids and cirratuliforms alike, are considered to be protonephromixia by most authors, although they vary in structure. The work of Bartolomaeus (1997) describes the excretory organs of *Tomopteris* as metanephridia, though in fact they have solenocytes associated with a coelomoduct. The phyllodocid worms for which we have information, which include the tomopterids and alciopini, have serially repeated segmental organs at the base of their parapodia, similar to *Alitta*. The organs then void to the outside via a nephridiopore. In the case of *Tomopteris*, the segmental organs begin well posterior to the head, in the sixth segment. It is likely that the lopadorhynchs and other pelagic families also have serially repeated organs, but there is insufficient information available to confirm that supposition.

In contrast, the benthic acrocirrid polychaetes have a single pair of segmental organs located anteriorly that appear to exit at the first segment (Rouse and Pleijel 2001). A similar arrangement holds in the pelagic cirratuliforms (Osborn and Rouse 2010; Osborn et al. 2011; Robbins 1965), with a single nephridium exiting just behind the head. If you consider the benthic, tube-dwelling ancestry of the cirratuliforms, the concept of excretory organs located so far forward makes sense. Even with little more than the head protruding from a tube, excretory products can still be voided to the environment rather than released inside the burrow.

The Nervous System

The nervous system consists of a brain, or supraesophogeal ganglion, connected to a ventral nerve cord via circumesophogeal connectives. The ventral nerve cord is actually a chain of segmental ganglia beginning in some cases with a large subesophogeal ganglion and continuing with ganglia in each of the metameres along the length of the worm. As an active crawler and swimmer, the system of *Alitta* depicted in Figure 6.18 is an example of an advanced polychaete nervous system. Instead of the single fused ganglion seen in *Alitta*, more primitive nervous systems have paired segmental ganglia joined by lateral connectives and have a paired ventral nerve cord. Note the lateral nerves, or connectives, usually numbering about three per segment (Kaestner 1967), lead from the ventral nerve cord to innervate muscles and accept sensory input (Figure 6.18). Included among those is the parapodial nerve with its pedal ganglion, associated with control of the parapodial muscles.

In many species the ventral nerve cord has a series of giant fibers that are capable of very rapid neural transmission. The giant fibers mediate rapid movement, such as the well-known withdrawal response (Mills 1978) of polychaetes that can move a tube-dwelling worm back into its burrow with lightning speed. The withdrawal response also allows free-living worms to rapidly pull their anterior segments out of harm's way.

Sense Organs

All the primary sensory modalities of mechanoreception, chemoreception, and photoreception are found in the polychaetes, along with a considerable diversity in sensory organ structure. In some cases, such as in the eyes of the alciopini, polychaetes exhibit remarkable sophistication. Structures vary from individual sensory cells widely scattered over the body surface and proboscis to aggregations of cells forming sense organs in the parapodial cirri, antennae, and palps (Mills 1978).

Sensory cells found on the body surface and parapodia are believed to be primarily mechanoreceptors, sensitive to touch, vibration, movement, and position of the parapodia. At least some of the sensory cells on the body surface likely play a role in chemoreception as well. Most of the chemical reception in polychaetes has been attributed to the head region, a reasonable assumption since the head is the first to sample the environment ahead and chemoreception there would be an aid in finding prey. Sensory cells found in the peristomial cirri, antennae, and palps would thus be ideally suited for chemoreception. The paired nuchal organs, located on the dorsal surface of the prostomium, are widely believed to serve as

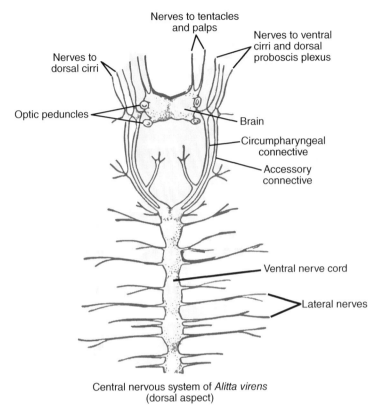

Figure 6.18 Dorsal view of the central nervous system of *Alitta virens*. *Source:* Adapted from Turnbull (1875), plate XLIII, figure 27.

chemoreceptors though direct evidence is lacking. Nuchal organs vary in form, from simple ciliated pits to more elaborate structures that can be everted and retracted. Direct evidence for the chemosensory role of the head region of polychaetes comes by inference from experiments with earthworms (Laverack 1960, 1961), which showed a wider chemical sensitivity in the head when compared to the trunk region of the worm. Unfortunately, the morphology of polychaete sensory cells (Figure 6.19) does not provide the evidence needed to discriminate between sensory functions, between mechanoreception and chemoreception, for example, as can be found in more advanced phyla.

Photoreceptors in polychaetes vary from simple cup eyes with a few receptor cells and no lens to the more complex eyes of *Nereis* with a lens and complex retina to the well-developed vesicular eyes of the pelagic Alciopini (Figure 6.20). Alciopine eyes rival those of the cephalopods and vertebrates in their sophistication. They are widely believed to be capable of resolving images and may even be able to focus near and far, a property termed accommodation (Wald and Rayport 1977). An important finding described in Wald and Rayport was the difference in the spectral sensitivity of the eyes in the surface-dwelling *Torrea candida* versus those of *Vanadis formosa*, a species found primarily below 300 m. Peak

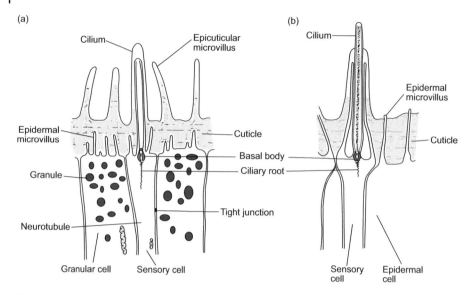

Figure 6.19 Polychaete sense organs. (a) Sensory and supporting granular cells of a nereid. (b) sensory and epidermal cells from a parapodial cirrus of an atokous nereid. *Sources:* (a) Adapted from Dorsett and Hyde (1969), figure 1 (p. 514); (b) Adapted from Boilly-Marer (1972).

sensitivity in *Torrea* was in the blue wavelengths, about 400 nm, which are nearly extinguished by a depth of 100 m. In contrast, the retina of *Vanadis* was most sensitive to wavelengths of about 480 nm, the only wavelength that can penetrate depths greater than 500 m. In both cases, the peak sensitivity was discrete and well-defined.

Yingst et al. (1972) found a flat, broad band of sensitivity over a range of 400–540 nm wavelength in the intertidal polychaete *Nereis mediator*. Peak sensitivity was at 480 nm but, unlike in the Aciopini, sensitivity was nearly equivalent over the entire spectral range.

Circulatory System

The great majority of polychaetes have a well-developed closed circulatory system comprising dorsal and ventral longitudinal vessels that run along the midline of the worm above and below the digestive tract, as well as a complex suite of lateral vessels. The main dorsal vessel bifurcates at the anterior end of the worm, with the two branches proceeding posterolaterally to form the two vascular beds located on the dorsal surface of the pharynx (Figure 6.8). Those vascular reticulae in turn coalesce to form the lateral connectives that join with the main ventral vessel. Circulation is described as a "sluggish flow" (Nicoll 1954) in the principal vessels, characterized by anteriorly moving peristaltic waves in the dorsal vessel accompanied by a posterior flow in the ventral vessel to complete the circuit. The two main vessels are joined at the posterior by a circum-anal ring (Brown 1950).

In *Nereis*, the lateral vessels join a complex system of capillaries in the parapodia (Figure 6.21a) creating a segmental flow that is superimposed on the sluggish axial flow. The segmental flow aids in gas exchange and supply of nutrients to the parapodia.

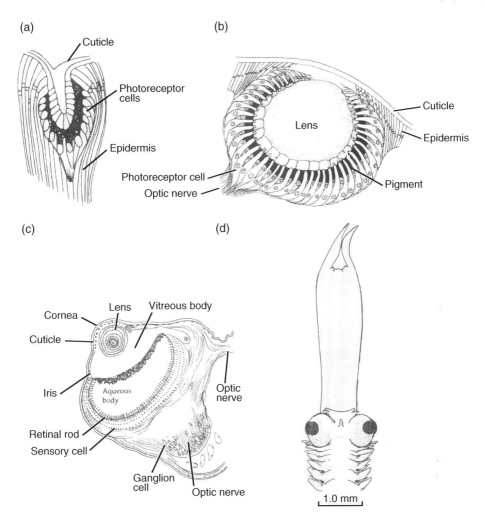

Figure 6.20 Polychaete photoreception (eyes). (a) A simple chaetopterid eye. (b) a complex nereid eye. (c) a vesicular eye of a pelagic alciopid. (d) dorsal view of the head of the alciopine *V. formosa* showing the well-developed complex eyes. *Sources:* (a) Brusca and Brusca (1990), figure 29 (p. 419); (b) Brusca and Brusca (2003), figure 13.29 (p. 423); (c) Hermans and Eakin (1974), figure 2 (p. 248); (d) Dales (1957), figure 21 (p. 115).

Most errant polychaetes have no true hearts. Propulsive force is supplied by contractile musculature in the vessels themselves in addition to the crawling and swimming movements of the worms. Tube-dwelling polychaetes have more specialized pumps to aid in circulation, since their movements are constrained by their sedentary lifestyle. Figure 6.21b shows the circulatory system of a terebellid, a tube-dwelling polychaete in the same order as the cirratuliforms.

It is within the annelids that respiratory pigments first play an important role in gas exchange, and the polychaetes exhibit a high diversity of types. Hemoglobin is the most common pigment. It is usually found in the blood, but in some species it is found

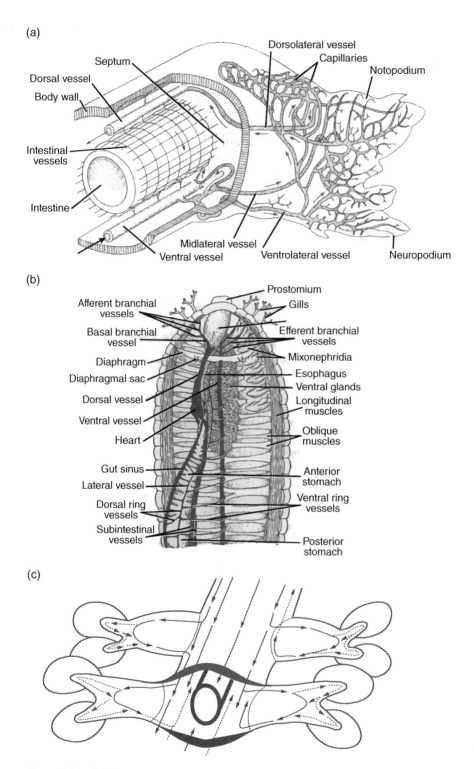

Figure 6.21 Polychaete circulatory systems. (a) Circulation within a segment and parapodium of a nereid. (b) circulatory and digestive systems in a tube-dwelling terebellid. (c) diagram of the flow of coelomic fluid in a tomopterid. *Sources:* (a) Nicoll (1954), figure 3 (p. 74); (b) Brown (1950), figure 158 (p. 292); (c) Beklemishev (1969b), figure 115 (p. 229).

in the coelomic fluid. Red in color, as it is in mammals, the hemoglobin of polychaetes may be confined to cells or be dissolved in solution. An unusual pigment found only in the polychaetes is the green-colored chlorocruorin. It is found mainly in the tube-dwelling worms commonly known as feather-duster worms and in the cirratuliforms, e.g. the acrocirrids and flabelligerids. Respiratory pigments increase the amount of oxygen that may dissolve in the blood, allowing, among other roles, a more active lifestyle by facilitating both oxygen uptake at the respiratory surface and delivery to the tissues.

Pelagic Species

The limited information available on the structure of their circulatory system suggests a considerable reduction in the vascular system of phyllodocid pelagic
polychaetes (Smith and Ruppert 1988; Meyer 1929). The alciopids as a group possess only a dorsal and ventral vessel (Smith and Ruppert 1988; Rouse and Pleijel 2001) with no lateral branching, and the tomopterids possess no circulatory system at all (Meyer 1929). In the tomopterids, the coelomic epithelium is invested with cilia that circulate the coelomic fluid (Figure 6.21c), enabling the transport of oxygen, nutrients, and, during reproductive periods, gametes. The septae that isolate the segments in the nereids are also reduced in the tomopterids, which facilitates the effectiveness of the coelomic fluid as a circulatory medium and allows the movement of gametes to the repositories described in the discussion below on reproduction.

The recently discovered cirratuliform pelagics, like their benthic relatives, apparently have retained a complete circulatory system with a heart (Osborn et al. 2011). The benthic acrocirrids have chlorocruorin as a respiratory pigment, and it is believed that the pelagic acrocirrids do as well (Osborn and Rouse 2010). *Poeobius*, the only pelagic cirratuliform genus whose biology has been described in some detail, has chlorocruorin and a heart body (Robbins 1965).

Gas Exchange

The nearly cylindrical body of polychaetes, their vermiform or worm-like shape, has a very favorable surface-to-volume ratio for gas exchange. Many of the benthic phyllodocids, e.g. *Nereis*, improve on their favorable shape with a vascularized integument and parapodia (Figure 6.21a) to maximize the diffusion of oxygen into the worm. Both the act of crawling and the longitudinal body-pumping motions of burrow-dwellers refresh the seawater around the integument and parapodia, preventing local oxygen depletion and facilitating gas exchange.

Tube-dwelling polychaetes such as the cirratuliforms often possess specialized respiratory structures called branchiae that increase surface area for gas exchange (Figure 6.22). Worms with these structures can remain with part or nearly all of their length in a tube and their branchiae extended outside to take up oxygen.

Pelagic Species

The pelagic phyllodocids have no specialized respiratory structures, but the broad paddle-like parapodia used for swimming in the alciopids, tomopterids, and lopadorhynchids are ideal surfaces for gas exchange. The swimming action refreshes the seawater surrounding the worm, continuously maximizing the outward–inward gradient. Pelagic phyllodocids have particularly high respiratory rates when compared with other pelagic invertebrates (Thuesen

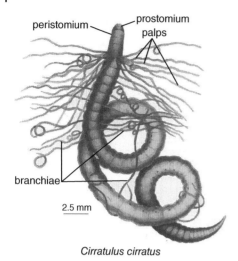

Figure 6.22 Illustration of the tube-dwelling polychaete *Cirratulus cirratus* showing the branchiae (specialized respiratory structures). *Source:* McIntosh (1915), plate XCL, figure 2.

and Childress 1993) so their gas exchange systems are clearly effective.

Pelagic cirratuliforms (Figure 6.7) possess anterior branchiae much like their benthic relatives, and the branchiae are presumably also well vascularized. Information on their gas exchange is not yet available; for now it is logical to assume that they resemble their closest relatives in this regard.

Digestive System

The polychaetes have a fairly simple digestive tract with a ventrally located mouth leading to a pharynx or buccal cavity that is actually part of the proboscis (Figure 6.7). When the proboscis is not everted, the peristomium forms a "buccal ring" surrounding the mouth. When it is everted and the jaws are extended, the mouth is at the business end of the proboscis (Figure 6.11b). Food is snared by the jaws, and the proboscis is then retracted, bringing it into the anterior of the worm. The retracted proboscis thus forms the pharynx or foregut. Both the eversion and retraction of the proboscis are powerful movements aided by specialized muscles. The polychaete *Glycera*, often used as bait and sold as "blood worm," demonstrates the power of that movement. If you have ever seen a "blood worm" evert its proboscis, you were probably surprised at the violence of it. Moreover, *Glycera* has poison glands in its proboscis that can make your hand swell if its jaws puncture your skin. It can be a bit of a shock when your bait fights back!

The pharynx opens into a short esophagus that extends for a few segments before leading into a thin-walled tube that serves as a combination of stomach and intestine. In some carnivorous worms, e.g. *Nereis*, the esophagus has lateral branches, or cecae, that play an important role in initial digestion (Michel and DeVillez 1978). A sphincter muscle usually delineates the end of the esophagus. The wall of the stomach/intestine is invested with a thin layer of longitudinal and circular muscles (Figure 6.9b) that provide only minimal aid in propelling the food bolus down the alimentary canal by peristalsis. Movement of food down the intestine is a cooperative venture involving the body wall and septal muscles in addition to the intestinal musculature.

The lumen of the gut has an internal mucosa that both secretes digestive enzymes and absorbs the nutrients produced by the digestive process. Enzymes for digesting proteins (proteases), lipids (lipases), and carbohydrates have been detected in the intestine as well as proteins that aid in absorption of the resulting nutrients (Michel and DeVillez 1978). Though no clear delineation exists between the midgut and hindgut, there is a rudimentary division of labor in that digestive enzymes are more concentrated in the anterior portions of the intestine, and absorptive functions are more concentrated in the posterior. Data on gut pH are limited; values range from 5.5 to 8.2 with an anterior–posterior gradient of acid to alkaline. The stomach/intestine shows little obvious change in external morphology over its length until reaching the rectum in the final segment. Both the proboscis and rectum have a cuticular lining.

The alimentary canal of polychaetes exhibits the diversity that you would expect in a group with widely different feeding habits that range from predators to suspension feeders to deposit feeders. Sedentary worms have more complex digestive tracts (Figure 6.21b) than the crawling worms and generally have received more attention in the literature. Pelagic phyllodocids are all predators. In the absence of detailed published descriptions, the most conservative course is to assume that the gut morphology of pelagic species is similar to that of the benthic predators, such as *Nereis*, with modifications in the proboscis for better reach in the open sea. The pelagic cirratuliforms have retained a more complex gut with a differentiated esophagus, stomach, and intestine (Robbins 1965).

Reproduction

As a class, the polychaetes are mainly dioecious, with a wide variety of interesting reproductive strategies. All pelagic species studied thus far have separate sexes. Many polychaetes have impressive powers of regeneration, a form of asexual reproduction. In only a very few species is asexual reproduction believed to be a dominant reproductive strategy, and those species are usually quite small, or colonial, or both (Schroeder and Hermans 1975). With a few exceptions, the champion regenerators are sedentary filter-feeders like the featherduster (serpulids and sabellids) and parchment (chaetopterids) worms. Among the pelagic species, *Tomopteris* has been reported as capable of regeneration (Terio 1950).

Hermaphroditic species are found in at least 18 polychaete families, with sequential and simultaneous hermaphroditism represented (Schroeder and Hermans 1975). Both protandrous (male first) and protogynous (female first) sequential hermaphrodites are found; the greatest numbers of representatives are in the filter-feeders. In most cases, hermaphrodites are found in families where the majority of species have separate sexes, e.g. the nereids and syllids. Unlike the earthworms and leeches, where hermaphroditism dominates, polychaetes have retained a more conventional sexual reproductive strategy.

The gonads are not easily recognized, well-defined organs but instead originate as clumps of cells on the wall of the peritoneum that lines the coelomic cavity of the trunk region. In the Errantia, gonads are found in most of the trunk segments. Immature gametic cells are released from the peritoneum into the coelomic cavity to fully mature. Mature sperm and eggs find their way to the outside either via one of the types of gonoduct (Figures 6.15–6.17) or by a rupture of the body wall.

In some species, a form of copulation takes place and fertilization is internal; eggs may be brooded, or they may be released into egg masses where they eventually hatch as free-swimming larvae. Fertilization is external in most errant species, either between individuals, in a small group or in swarms that may vary in size from small to huge. The mating swarms of benthic polychaetes are the stuff of legend; they involve changes in morphology to facilitate swimming, synchronous behaviors cued by the moon, and south Pacific island feast days. Synchronous swarming behavior optimizes the chance for reproductive success in those species that exploit it as a reproductive behavior. A few versions of polychaete swarming behavior exist, both with respect to morphological changes and the degree of synchronicity.

Epitoky

Three families of polychaetes, the nereids, syllids, and eunicids, have representatives that exhibit the phenomenon of epitoky, a marked change in morphology to a sexually active worm called an epitoke or epitokous individual (Barnes 1974; Brusca and Brusca 2003). Strictly speaking, the epitoke is only the gamete-bearing portion of the worm, but the term has become synonymous with the entire sexually transformed individual. Nonreproductive individuals and nonreproductive portions of a sexually active worm are termed atokes. Differences between atokous and epitokous worms are quite profound and happen in two basic ways. Nereids and eunicids transform the body of an existing adult worm into anterior atokous and posterior epitokous sections, appearing much like two different worms joined together (Figure 6.23a–d). In the syllids, an epitoke is formed at the caudal end of an atokous worm (Figure 6.23e). The epitoke then buds off to form a complete new sexually active worm with a new head that is produced either before or after budding. The atoke remains alive to reproduce another day. The syllids can thus be considered to have asexual and sexual elements to their reproductive strategy.

In the nereids and syllids, the parapodia of the sexually active portions transform to more paddle-like structures to facilitate swimming (Figure 6.23b, c). Other morphological changes in the epitokes of nereids include enlarged eyes plus degradation of the gut and some musculature to better accommodate the maximum number of gametes.

Synchronicity

Without question, the most famous example of synchronous swarming occurs in the palolo worms (*Palola viridis*) of Samoa where, on one day of the year, in the third quarter of the moon, epitokes swim to the surface en-masse to release their eggs and sperm. The islanders, who consider the worms a delicacy, can predict the happy day based on the knowledge of countless generations and harvest the swarms for an annual feast. The swarms are created when the palolo worms back out of their burrows and their epitokous posterior sections break off to ascend, headless, to the surface. Once their gametes are released the epitokes die, while the atokous anterior portions of the worms live on.

Nereids and syllids, many of which live in temperate climes, show a few different patterns of synchronicity; none of them are quite as spectacular or as well described as the *Palola* swarms of the south Pacific. All swarming events are tuned to a lunar cycle, and in many species, multiple swarming events occur over the reproductive season. The nights of a new moon, for example, elicit swarming behavior in *Platynereis dumerilii*. Epitokous nereids release their gametes via a rupture of the body wall; once they are released, the entire worm dies, including the atokous anterior section. Epitokes of the syllids, which release their gametes similarly, also expire after gamete release. However, the atokous "parent" in the syllids survives the swarming event because the epitoke has already budded off to form a new and separate worm. Some

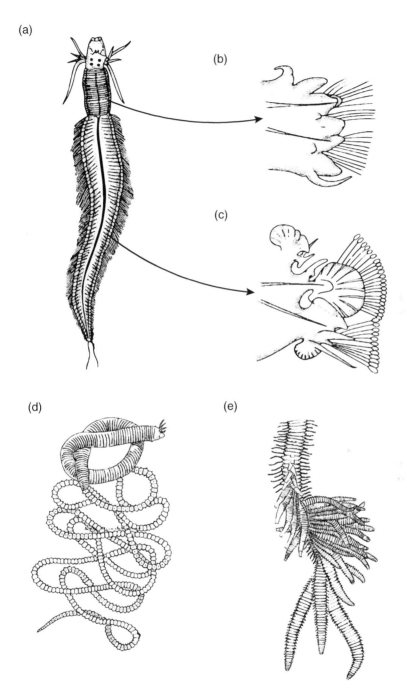

Figure 6.23 (a) An epitokous *Neanthes nubila*; (b and c) the dimorphic parapodia of the epitokous (anterior) end and the atokous (posterior) end, respectively; (d) the epitokous palolo worm *P. viridis* showing marked dimorphism between the epitokous (caudal) and atokous (anterior) ends; (e) multiple epitokes (stolons) on the syllid worm *Trypanosillis asterobia*. *Sources:* (a–c) Barnes (1974), figure 10-41 (p. 275); (d) Brusca and Brusca (1990), figure 32 (p. 423); (e) Adapted from Okada (1933), figure 4 (p. 329).

species of syllids produce several epitokes in the budding process.

The Pelagic Species

Tomopteris

Disappointingly little is known about reproductive habits in the pelagic polychaetes. Åkesson (1962) provides some insights in his detailed monograph on development in *Tomopteris helgolandica*, partially dispelling some older notions but adding his own speculations on the process.

All tomopterids have separate sexes. The earlier literature (Hachfield 1926; Fullarton 1895) made a case for internal fertilization in *Tomopteris*, noting the presence of sperm and eggs in adult females, as well as cleavage stages thought perhaps to be the result of copulation. Åkesson (1962) found no evidence for internal fertilization and observed that the morphology of reproductive organs in *Tomopteris* does not support a case for copulation. He thought instead that a swarming reproductive strategy, similar to that described for the nereids, might also be the case for the tomopterids. However, it would have to be a synchronized broadcast spawn, as epitoky has not been reported in the group. The tomopterids in Åkesson's (1962) study were collected from the Gullmar Fjord on the west coast of Sweden, a place where they are so abundant that they episodically foul the nets of shrimp trawlers. That density certainly could lead to speculation of swarming. Coordinated spawning events in tune with lunar cycles are found in many polychaetes as well as in taxa as disparate as corals and fishes. It is certainly a possibility for tomopterids, though as yet there is no supporting evidence.

Åkesson's 1962 monograph provides additional information on reproduction in tomopterids. He reported that the gametes develop initially in the parapodia and are shed into the coelom, where they mature. Unlike most errant worms, *Tomopteris* has rudimentary segmental septae except in the tail region, so gametes can float freely in the coelom. The eggs apparently are shed to the outside through gonoducts that appear as transverse slits at the base of the fourth and fifth pairs of parapodia in the anterior of the worm. In males, the developing sperm are also released into the coelom. However, the coelomoducts associated with the protonephromixia in the tail of mature males have enlarged to form cavities, or seminal vesicles, and final maturation of the sperm occurs there. What is not clear is whether the vesicles are present in the tail only or in other segments as well.

In *Platynereis megalops* of the Woods Hole region, the mating swarm includes a "mating dance" in which the male wraps around the female and inserts his anus into her mouth, injecting the sperm (Kaestner 1967). The gut of the female has deteriorated during the formation of her epitoke, so the injected sperm can access the eggs of the female; it is internal fertilization without elaborate copulatory organs. Given the reports of the older literature and the swarming hypothesis of Åkesson (1962), it would be quite tempting to meld the two hypotheses and propose a similar situation in the tomopterids. However, again, there is no evidence that it is so.

Alciopini and Lopadorrynchidae

No direct information is available on the mating habits of the alciopines and lopadorrynchs. The closest relatives of the alciopines, the benthic phyllodocids, have been studied in some detail (Olive 1975). *Eulalia viridis* exhibits a swarming behavior with one female being surrounded by multiple males, resulting in the deposition of a benthic egg sac filled with fertilized eggs. The larvae are pelagic.

In contrast, multiple authors (Støp-Bowitz 1948; Dales 1955, Tebble 1960; Schroeder and Hermans 1975) have reported seminal receptacles in the anterior parapodia of alciopine females. Dales (1955) gave a detailed description of sac-like structures on the anterior parapodia of female alciopines that, in mature worms, were filled with sperm. The sperm appeared to be "agglutinated into a kind of spermatophore" (Dales 1955) prior to transferring it to the female. The structure believed responsible for clumping the sperm is a ventral gland on the parapodia of the males. The information in Dales (1955) makes a good case for some type of copulatory, or at least a sperm-transfer, behavior in the alciopines (Schroeder and Hermans 1975).

Development

The embryology of *Tomopteris helgolandicus* was described in detail in Åkesson (1962). Despite the holopelagic lifestyle of *Tomopteris*, its early development is much like that of benthic species, nearly all of which have the pelagic trochophore larvae. *Tomopteris* eggs are fairly large and yolky (315 μ) and after fertilization exhibit spiral cleavage. They develop into a swimming trochophore in ~40 h at 16 °C, the sea-surface temperature at the time of Åkesson's study. in vitro temperature experiments on fertilization success suggested that eggs must be fertilized near the surface. Successful fertilization was greatly diminished, 80% vs 30%, at the colder temperature (e.g. 6 °C) typical of the presumed depth of capture (120 m). In less than two weeks at a temperature of 16 °C, the larvae bore a marked resemblance to adults.

Locomotion

Using cinematography, detailed descriptions have been created on the basics of crawling and swimming in the intertidal polychaetes *Nereis* (Gray 1939) and *Nephthys*, a worm resembling *Nereis* in general body shape (Clark and Clark 1960). Little is known about swimming in pelagic species, so we will extrapolate from what we know about swimming in littoral worms to those that swim full-time. The basic character of the swimming motion is similar in both.

Rapid crawling and swimming are both the result of sinusoidal, undulatory motion. Undulatory, or side-to-side, movement was discussed briefly in Chapter 5 regarding the swimming motions of pelagic nemerteans. It reaches its apex in efficiency and speed in the swimming motion of pelagic fishes and cetaceans, but it is an effective means of swimming in the pelagic worms as well. Unlike the undulatory swimming of smooth-bodied animals such as eels, polychaete swimming exploits both the undulatory motion of the slender body and the paddles formed by the parapodia.

The lateral undulations of swimming polychaetes pass from tail to head; that is, locomotory waves pass anteriorly, opposite to the direction the worm is moving. The metachronal waves are the result of rhythmic contractions in the longitudinal muscles of the worm's body wall, giving the animals a sinusoidal shape as they swim forward. Superimposed on the metachronal waves of the worm's body are the paddle-like parapodia. They can act merely as stiffened paddles that move passively backward as the muscular wave passes under them, creating backward thrust with no directed stroke of their own, or, as in *Nephthys*, they can actively thrust backward. As swimming speed increases in *Nephthys*, the wavelength of the locomotory waves shortens (i.e. fewer segments are involved) and the wave amplitude increases.

In contrast, observations of swimming mechanics in epitokes show a reduced amplitude in the locomotory waves, with the primary thrust being produced by the beating parapodia (Mettam 1967). As shown in Figure 6.23, the parapodia in epitokes are broadened to a more effective paddle, similar to the parapodia in the alciopines and especially to those of the tomopterids. Tomopterids have parapodia that extend a body width from the center of the worm (Figure 6.24), creating effective oars but precluding short-wavelength locomotory waves. Examination of swimming motion in *Tomopteris* shows a reduced-amplitude wave with thrust provided by a strong parapodial beat. The most likely source of muscle power for the rowing parapodial stroke is the oblique muscles shown in Figure 6.25.

Instead of paddle-like parapodia, Cirratuliform pelagic polychaetes have notochaetal fans that perform a similar function (Osborn et al. 2011). Rooted in the fleshy portions of the parapodium, the notochaetal fans provide a paddle-like structure that can be used for swimming. The low Reynolds numbers that typify the flow environment at the surface of the fan limit the flow through it. What would normally be perceived as a rake passing through the water with each stroke functions more as a paddle.

Foraging Strategies

Polychaetes as a group have a wide variety of feeding strategies. In tube-dwellers, feeding habits range from nonselective deposit feeding in some burrowing species (e.g. *Arenicola*) to more sophisticated suspension feeding using the particle-trapping mucous webs seen in *Chaetopterus* to the specialized anterior tentacles in *Amphitrite* that trap small particles and convey them to the mouth via a ciliary tract. Since the most specialized feeding habits and structures are exhibited by tube-dwellers, they are considered to be more evolutionarily advanced than the errant forms. The crawling or errant worms, including all but two of the pelagic families, are more typically carnivores, capturing prey with their proboscis.

The Hunters

Pelagic species are nearly all active hunters, most seizing prey with their proboscides as do their benthic relatives. The tomopterids, alciopines, and lopadorhynchids are all

Figure 6.24 A pelagic tomopterid showing the undulatory sinusoidal shape assumed during swimming. *Source:* © Dante Fenolio, reproduced with permission. See color plate section for color representation of this figure.

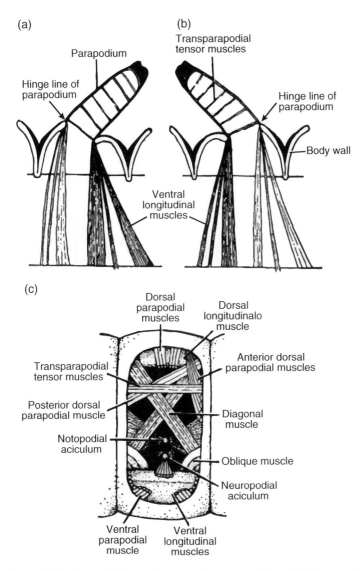

Figure 6.25 Parapodial musculature and movement. (a and b) Movement of a parapodium backward and forward due the contraction of the anterior and posterior ventral longitudinal muscles, respectively; (c) lateral view of the musculature of a segment after removal of a parapodium. *Source:* Mettam (1967), figure 11 (p. 268). Reproduced with the permission of John Wiley & Sons.

reported to be carnivores (Fauchald and Jumars 1979). Though pelagic species do not have the jaws of benthic worms like *Alitta*, it is believed that sticky mucous secretions cause prey to adhere to the everted proboscis (Dales 1955). The proboscides of the alciopines can be quite long, giving a fairly respectable reach (Figure 6.26). The lopadorhynchs have modified antennae and anterior parapodia for grasping prey.

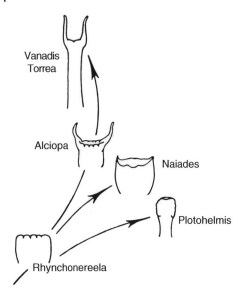

Figure 6.26 Form of the proboscis in alciopines. *Source:* Dales (1955), figure 2 (p. 414). Reproduced with the permission of John Wiley & Sons.

Diets

Alciopini. The remarkable eye of the alciopines (Figure 6.20d) strongly suggests that they are visual predators, tracking their prey by sight and snaring it with a muscular proboscis once it is in range. Smaller representatives of the group, such as *Vanadis minuta,* feed on small crustaceans such as euphausiid larvae and copepods. Larger species feed on gelatinous prey such as doliolids and salps. Diatoms have also been reported in the diet of Rhynchonereella in Antarctic systems (Hopkins 1985b, 1987; Hopkins and Torres 1989).

Lopadorrhynchidae. Unlike the alciopids, which tend to be long and slender, the lopadorhynchids are short, broad, and flattened (Dales 1955, Figure 6.5b). Their proboscis is short and their eyes are reduced, but their prostomium has well-developed antennae and tentacular cirri, suggesting that prey are located primarily by chemo- or mechano-reception. The first two parapodia are enlarged and appear to be modified for grasping; they lack the fleshy paddles of the swimming parapodia and have large, curved chaetae. Diet information is scanty. *Pelagobia longicirrata* has been reported to feed on both diatoms and copepods in Antarctic pelagic systems (Hopkins et al. 1993).

Tomopteridae. Tomopterids have one pair of well-developed eyes with lenses present on most species and pronounced antennae on the prostomium in addition to nuchal organs. Their second segment has a pair of parapodial lobes with whip-like aciculae that may be longer than the body and give the tomopterids their characteristic appearance (Figure 6.24). Their diet includes fish larvae, chaetognaths, tunicates, small siphonophores, and even diatoms (Åkesson 1962; Fauchald and Jumars 1979). The sensory complement of tomopterids includes a full battery of well-developed organs that would allow tracking prey visually as well as with chemo- and mechano-receptive senses. Prey capture and handling have not been described despite the reasonable quantity of information on diet. The tomopterids have a short muscular proboscis that presumably aids in prey capture (Dales 1962).

The Suspension Feeders

The recent discoveries of new and unusual species of pelagic polychaetes (Osborn et al. 2009; Osborn and Rouse 2010), including the addition of several new genera, have revised our concepts of what groups comprise the pelagic polychaete fauna and how they feed. Although diet and foraging strategies have not been described for the newly discovered acrocirrid polychaetes, some information is available on their better-studied relatives.

Poeobiidae, Chaetopteridae

Poeobius meseres, an unusual, if not aberrant, species is more closely related to the newly discovered Acrocirrid genera than it is to the far better-studied phyllodocid hunters. Its morphology is odd (Figure 6.7c), lacking the parapodia and chaetae that are the primary characteristics of the polychaetes. Rather than being a hunter-killer, *Poeobius* is believed to be a suspension feeder, using its near-neutral buoyancy and a mucous web to entangle sinking detritus that it then ingests (Robbins 1965; Uttal and Buck 1996). Its diet, analyzed in detail over the course of a year (Uttal and Buck 1996), comprised mainly copepod fecal pellets, diatoms, and microzooplankton. The diet correlated well with the detrital spectrum available during the different oceanographic seasons in the Monterey Canyon where the study was conducted. Moreover, the diet composition strongly suggested that *Poeobius* was capturing organic aggregates or "marine snow" (Alldredge and Silver 1988) as the particles were sinking to the bottom.

A similar situation is believed to be the case for the aberrant pelagic chaetopterid *Chaetopterus pugaporcinus*. Laboratory observations of captive specimens revealed that the worms produced a diffuse mucous cloud that would be capable of trapping particles (Osborn et al. 2007).

Distributions

Geographical

The pelagic polychaetes enjoy a pan-global distribution (Tables 6.1 and 6.2), with representatives found from the equator to polar seas. The most complete information available is for the phyllodocids, particularly the alciopines, tomopterids, lopadorhynchids, and typhloscolecids, which are the most commonly captured. Several distribution studies have been undertaken, including, but not limited to, the North Pacific (Tebble 1962), south Atlantic (Tebble 1960), Southern Ocean (Monro 1930, 1936), North Atlantic (Støp–Bowitz 1948), Northeastern Pacific (Dales 1957), and in the Atlantic and Indian Oceans bordering South Africa (Day 1967). Information on the cirratuliforms is more limited, with the newly discovered Indo-Pacific species thus far found only in the Celebes Sea (Osborn et al. 2011). Distribution of *Poeobius* in the North Pacific was well described by McGowan (1960).

Tables 6.1 and 6.2 indicate that most pelagic polychaete species have a wide-ranging distribution, with several being truly cosmopolitan. About 30% of the pelagic polychaetes are found in all three major ocean basins. Pelagic species, particularly mesopelagic ones, often exhibit very broad distributions (Briggs 1974) at least partly because of similarities in temperature at depths below 300 m in the global ocean. A few species are cold-water specialists, with three, *Tomopteris carpenteri*, *Rhynchonereella bongraini*, and *Vanadis antarctica*, being endemic to the Antarctic (Tebble 1960).

The horizontal and vertical distributions of the atypical flabelligerid *P. meseres* are explained quite well by its affinity with the Pacific sub-arctic water mass (McGowan 1960); its distribution is limited to the North Pacific and transitional waters. Though most of the pelagic polychaetes do exhibit temperate-subtropical temperature preferenda, their overall distributions are too broad to be explained by affinities with any one water mass. In the North Pacific, Tebble (1962) considers 21 of the 33 species identified in his study to be subtropical, and five, *Tomopteris septentrionalis*, *Typhloscolex mulleri*, *P. longicirrata*, *Maupasia coeca*, and *Phalacrophorus pictus*, to be truly cosmopolitan. Most of the cosmopolitan species are found in the Antarctic.

Table 6.1 Distribution records of pelagic polychaetes with the presence/absence indicated by reference numbers. References are listed at the end of the table.

Family or tribe, genus, species	South Atlantic	North Atlantic	Sub-Arctic Pacific	South Africa/Indian Ocean	Gulf of California	Eastern Tropical Pacific
Alciopini						
Alciopa reynaudii	1, 3			5		
Alciopina parassitica	1			5		
Krohnia lepidota	1, 3	4		5		
Naiades cantraini	1, 3	4		5		
Pseudalciopa modesta	1					
Plotohelmis alata	1			5		
Plotohelmis capitata	1			5	6	
Plotohelmis tenuis	1	4		5		
Torrea candida	1			5		
Watelio gravieri	1					
Rhynchonerella angelini	1, 3		4	5	6	
Rhynchonerella gracilis	1	4		5		
Rhynchonerella longicirrata	1					
Rhynchonerella moebii	1	4		5		
Rhynchonerella petersii	1	4		5	6	
Vanadis longissima	1, 3	4		5		
Vanadis minuta	1	4		5		
Vanadis studeri	1					
Vanadis violacea	1					
Vanadis brevirostris	1					

Vanadis crystallina	1	4	5
Vanadis formosa	1	4	5
Vanadis tagensis	1	4	
Vanadis fuscapunctata			5
Yndolaciidae			
Yndolacia lopadorrhynchoides	1		
Iospilidae			
Iospilus affinis	1		
Iospilus phalacroides	1		
Phalacrophorus pictus	1	4	5	6
Phalacrophorus uniformis	1	4	5	6
Lopadorrhynchidae			
Lopadorrhynchus appendiculatus	1		
Lopadorrhynchus brevis	1,3	4		6
Lopadorrhynchus henseni	1			6
Lopadorrhynchus krohni	1,3	4		
Lopadorrhynchus uncinatus	1,3	4		
Lopadorrhynchus nationalis				6
Maupasia coeca	1,3	4	5	
Pedinosoma curtum	1	4	5	6
Pelagobia longicirrata	1,3	4	5	6

(Continued)

Table 6.1 (Continued)

Family or tribe, genus, species	South Atlantic	North Atlantic	Sub-Arctic Pacific	South Africa/Indian Ocean	Gulf of California	Eastern Tropical Pacific
Flabelligeridae						
Flota flabelligera	1					
Poeobius meseres			2			
Typhloscolecidae						
Sagitella kowalewskii	1	4		5	6	
Typhloscolex muelleri	1		4	5	6	
Travisiopsis coniceps	1			5		
Travisiopsis dubia	1		4	5	6	
Travisiopsis lanceolata	1		4	5		
Travisiopsis levinseni	1		4			
Travisiopsis lobifera	1	4		5		
Travisiopsis lumbricoides	1					
Tomopteridae						
Tomopteris nationalis	1			5	6	7
Tomopteris duccii	1					
Tomopteris elegans	1,3	4		5	6	7
Tomopteris kefersteini	1					
Tomopteris euchaeta	1			5		7
Tomopteris helgolandica	1			5		
Tomopteris kempi	1,3					
Tomopteris ligulata	1	4		5		
Tomopteris septentrionalis	1,3	4		5		7

Species					
Tomopteris onisciformis	1				
Tomopteris planktonis	1,3	4	5	6	7
Tomopteris rolasi	1				7
Tomopteris krampi	1,3		5		7
Tomopteris nisseni	1	4			7
Tomopteris pierantonii	1				
Tomopteris apsteini	3	4			
Tomopteris pacifica		4	5		
Tomopteris dunckeri					7

References
1) Fernandez-Alamo and Thuesen (1999)
2) McGowan (1960)
3) Tebble (1960)
4) Tebble (1962)
5) Day (1967)
6) Fernandez-Alamo (2006)
7) Fernandez-Alamo (2000)

Table 6.2 Distribution records of pelagic polychaetes in the Antarctic and North Atlantic regions. References are listed at the end of the table.

Family or tribe, genus, species	Antarctic/Sub-Antarctic Presence/absence indicated by reference numbers plus depth and temperature if known			North Atlantic Presence/absence indicated by reference numbers plus depth and temperature if known		
	References	Depth (m)	Temperature (°C)	References	Depth (m)	Temperature (°C)
Alciopini						
Alciopa reynaudii				3	100–150	9.5–11.1
Alciopina parassitica				3	25	20.8
Krohnia lepidota				3	50	19.6
Naiades cantraini				3	0–150	17.2–20.8
Torrea candida				3	0–150	16.0–20.6
Watelio longifoliata				3		50
Rhynchonereella angelini				3	50–500	9.1–16.6
Rhynchonereella bongraini	1, 2, 4	0–1000 (50–250)	−0.4 to −1.6			
Rhynchonereella moebii				3	135	
Rhynchonereella petersi				3	0	
Vanadis antarctica	1, 2, 4	0–470 (0–90)	−0.4 to −1.6			
Vanadis longissima	1, 2					
Vanadis crystallina				3	0–150	11.9–19.8
Vanadis formosa				3	0–150	12.1–19.6
Iospilidae						
Phalacrophorus pictus	1					
Lopadorrhynchidae						

Lopadorrhynchus appendiculatus			3	0–400	12.3–14.7
Lopadorrhynchus henseni			3	750	
Lopadorrhynchus uncinatus			3	0–500	11.4–16.6
Lopadorrhynchus nationalis			3	50–150	17.5–17.9
Maupasia coeca	1, 2, 4	160–470	−0.4 to −1.6		
Pelagobia longicirrata	1, 2, 4	50–1000 (100–470)		3	750–1250	
Tomopteridae					
Tomopteris carpenteri	1, 2, 4	0–115	−0.4 to −1.6		
Tomopteris elegans				3	25–1000	16.3–19.3
Tomopteris euchaeta				3	150–300	13.7–17.6
Tomopteris helgolandica				3	50–150	9.5–17.5
Tomopteris ligulata				3	50–1000	6.3–17.9
Tomopteris septentrionalis	1, 2, 4	0–920 (0–150)	−0.4 to −1.6	3	0–1200	3.3–17.8
Tomopteris planktonis	1, 2, 4	40–1000	−0.4 to −1.6	3	150–500	11.1–16.0
Tomopteris krampi				3	0–1000	3.3–13.7
Tomopteris nisseni				3	50–1400	5.9–17.9
Tomopteris apsteini				3	150–300	11.6–16.0
Typhloscolecidae					
Sagitella kowalewskii				3	800–1200	5.1–8.7
Typhloscolex muelleri	1,4	100–1000 (100–470)	−0.4 to −1.6		
Typhloscolex grandis				3	1200	5.1
Travisiopsis coniceps	1, 4	0–200	−0.4 to −1.6		
Travisiopsis dubia				3	10–500	5.7–15.4

(Continued)

Table 6.2 (Continued)

Family or tribe, genus, species	Antarctic/Sub-Antarctic Presence/absence indicated by reference numbers plus depth and temperature if known			North Atlantic Presence/absence indicated by reference numbers plus depth and temperature if known		
	References	Depth (m)	Temperature (°C)	References	Depth (m)	Temperature (°C)
Travisiopsis lanceolata				3	25–1400	4.3–16.0
Travisiopsis levinseni	1, 4	150–920	−0.4 to −1.6	3	500–1200	3.3–7.2
Travisiopsis lobifera				3	0–400	13.5–20.6

References
1 Fernandez-Alamo and Thuesen (1999)
2 Tebble (1960)
3 Stop-Bowitz (1948)
4 Hopkins and Torres (1988)

Vertical

Less information is available on the vertical distributions of polychaetes, particularly any samples obtained with opening-and-closing nets. Classical studies on polychaete distribution, which were remarkably thorough in areal scope (e.g. Støp–Bowitz 1948; Dales 1957; Tebble 1960, 1962), used open nets that were towed horizontally at sampling depth for the great majority of their sampling time, with as rapid a deployment and recovery as the gear would allow. Often, samples were taken with a series of nets towed in tandem at a variety of depths on a single tow cable. The tandem tows were designed to minimize the shortcomings of using open nets by simultaneously covering a series of depths in one deployment. Samples taken with an open net are always subject to question, despite the fact that the net spends most of its time fishing at its intended sampling depth and, in most cases, the catch does give an accurate picture, particularly of a species' overall vertical range and center of distribution. Støp–Bowitz (1948) used the tandem-towing technique just described to collect the data as shown in Table 6.2 on vertical distribution of North Atlantic species and the associated temperature ranges. Hopkins and Torres (1988) used an opening-and-closing net to obtain the vertical profiles of Antarctic species as shown in Table 6.2.

In both the Antarctic and North Atlantic, most species exhibited a broad vertical range with a more restricted center of distribution where most of the population resided. *P. longicirrata,* for example, is found in the Antarctic from 0 to 1000 m, but most of the population resides between 200 and 500 m (Figure 6.27) with little day-night change. Vertical migration, a pronounced change in the center of distribution of a species on a diel, or day-night, basis, was not observed in *Pelagobia*. After considerable searching and trawling on research cruises, I myself have found no evidence to support a case for vertical migration in any polychaete species. What information is available suggests that they occupy a preferred vertical range, but the population center shows no directed change on a diel basis.

Several species were distributed primarily in the upper 150 m: e.g. *Torrea candida, Naiades cantrainii,* and *Tomopteris carpenteri*. Others resided at intermediate depths, e.g. Antarctic *Pelagobia* and North Atlantic *Tomopteris apsteini*, or in the lower mesopelagic realm, such as *Sagitella kowalewskii* and *Travisiopsis levinseni*. The physical characteristics of the water column that co-vary with depth – quantity and quality of light, pressure, and temperature except in the nearly isothermal profile of the Antarctic – are powerful enough to shape the biological characteristics of the species that permanently reside within a given depth range. As we move up the ladder of complexity, we will not only discuss species with well-defined depths of occurrence but also the biological characteristics that are associated with those depths.

Bioluminescence

The ability to produce biological light, bioluminescence is found in a wide variety of benthic and pelagic polychaetes. Within the pelagic taxa, it has been reported in tomopterids (Figure 6.27), alciopines, and in the flabelligerid genus *Flota* (Herring 1978). Primarily benthic families include the tube-dwelling parchment worms, or Chaetopteridae, as well as the cirratulids, terebellids, syllids, and polynoids, or scale worms. In the syllids, luminescence is normally associated with reproductive swarming.

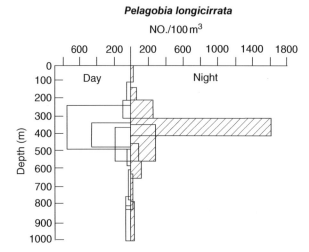

Figure 6.27 Vertical distribution of the Antarctic polychaete *P. longicirrata* showing no evidence of vertical migration. *Source:* Adapted from Hopkins (1985a), figure 4 (p. 165). Reproduced with the permission of Springer.

Sites of luminous tissue vary with the taxon. In tomopterids, luminescence is generated within the parapodia, whereas in *Flota* it comes from small spherical bodies between the parapodial setae (Herring 1978). In *Chaetopterus*, a luminous slime is secreted from several regions on the body (Harvey 1952). Most unusual was the discovery of luminescent "bombs" in the branchial region of the recently described genus *Swima* (Acrocirridae), spheroidal bodies that when released, glowed bright green (Osborn et al. 2009). Hopefully, there will be much more to come from the pelagic sedentaria!

References

Akesson, B. (1962). The embryology of *Tomopteris helgolandica*. Acta Zoologica 43: 135–199.

Alldredge, A.L. and SILVER, M.W. (1988). Characteristics, dynamics, and significance of marine snow. *Progress in Oceanography* 20: 41–82.

ANDRADE, S., NOVO, M., KAWAUCHI, G. et al. (2015). Articulating "archiannelids": Phylogenomics and annelid relationships, with emphasis on meiofaunal taxa. *Molecular Biology and Evolution* 32: 2860–2875.

Barnes, R.D. (1974). *Invertebrate Zoology*. Philadelphia: W.B. Saunders.

Bartolomaeus, T. (1997). Structure and development of the nephridia of Tomopteris helgolandica. *Zoomorphology* 117: 1–11.

Beklemishev, W.N. (1969a). *Principles of Comparative Anatomy of iinvertebrates. Volume 1: Promorphology*. Chicago: University of Chicago Press.

Beklemishev, W.N. (1969b). *Principles of comparative anatomy of invertebrates. Volume 2: Organology*. Chicago: University of Chicago Press.

Bergstrom, E. (1914). Zur systematik der polychaetenfamilie der Phyllodociiden. *Zoologiska Bidrag fran Uppsala* 3: 37–224.

Boilly-Marer, Y. (1972). Etude ultrastructurales des cirres parapodiaux des neridiens atoques (Annelide Polychetes). *Zeitschrift fur zellforschung und mikroskopische anatomie* 131: 309–327.

Briggs, J.C. (1974). *Marine Zoogeography*. New York: McGraw-Hill.

Brinkhurst, R.O. (1982). Oligochaeta. In: *Synopsis and Classification of Living Organisms* (ed. S.P. Parker). New York: McGraw-Hill.

Brown, F.A. (1950). *Selected Invertebrate Types*. New York: Wiley.

Brusca, R.C. and Brusca, G.J. (1990). *Invertebrates*. Sunderland: Sinauer Associates.

Brusca, R.C. and Brusca, G.J. (2003). *Invertebrates*, 2e. Sinauer: Sunderland.

Brusca, R.C., Moore, W., and Shuster, S.M. (2016). *Invertebrates*, 3e. Sinauer Associates: Sunderland.

Buzhinskaja, G.N. (2004). Two new genera of the pelagic family Yndolaciidae from the Arctic Ocean with an addition to the description of *Yndolacia lopadorrhynchoides* Stop-Bowitz. *Sarsia* 89: 338–345.

Chapman, A.D. (2009). *Numbers of living species in Australia and the world (2nd edition)*. Canberra: Australian Biological Resources Study.

Clark, R.B. and Clark, M.E. (1960). The ligamentary system and the segmental musculature of Nephtys. *Quarterly Journal of Microscopical Science* 101: 149–176.

Cuvier, G. (1816). *Le Regne Animal. Les Reptiles, les Poissons, les Mollusques et les Annelides*. Paris: Deterville.

Dales, R.P. (1955). The evolution of the pelagic alciopid and phyllodocid polychaetes. *Proceedings of the Zoological Society of London* 125: 411–420.

Dales, R.P. (1957). Pelagic polychaetes of the Pacific Ocean. *Bulletin of the Scripps Institution of Oceanography* 7: 99–168.

Dales, R.P. (1962). The polychaete stomodeum and the interrelationships of the families of Polychaeta. *Proceedings of the Zoological Society of London* 139: 329–428.

Dales, R.P. and Peter, G. (1972). A synopsis of the pelagic polychaeta. *Journal of Natural History* 6: 55–92.

Day, J.H. (1967). A monograph on the polychaeta of Southern Africa. Part 1. Errantia. *British Museum (Natural History), London, Publication 656*: 1–458.

Dorsett, D.A. and Hyde, R. (1969). The fine structure of the compound sense organs on the cirri of *Nereis diversicolor*. *Zeitscrhift fur Zillforschung und Mikroskopische Anatomie* 97: 512–527.

Enders, H.E. (1909). A study of the life-history and habits of *Chaetopterus variopedatus*. *Journal of Morphology* 20: 479–531.

Fauchald, K. (1977). The polychaete worms: definitions and keys to the orders, families, and genera. *Natural History Museum of Los Angeles County Science Series* 28: 1–190.

Fauchald, K. and Jumars, P.A. (1979). The diet of worms: a study of polychaete feeding guilds. *Oceanography and Marine Biology Annual Reviews* 17: 193–284.

Fauchald, K. and Rouse, G. (1997). Polychaete systematics: past and present. *Zoologica Scripta* 26: 71–138.

Fernandez-Alamo, M.A. (2000). Tomopterids (Annelida-Polychaeta) from the eastern tropical Pacific Ocean. *Bulletin of Marine Science* 67: 45–53.

Fernandez-Alamo, M.A. (2006). Composition, abundance and distribution of holoplanktonic polychaetes from the expedition El Golfo 6311-12 of Scripps Institution of Oceanography. In: *Scientific Advances in Polychaete Research* (eds. R. Sarda, G. San Martin, E. Lopez, et al.). Scientia Marina: Barcelona.

Fernandez-Alamo, M.A. and Thuesen, E.V. (1999). Polychaeta. In: *South Atlantic Zooplankton* (ed. D. Boltovskoy). Backhuys: Leiden.

FIELD, K.G., OLSEN, G.J., GIOVANNONI, S.J. et al. (1988). Molecular Phylogeny of the Animal Kingdom. *Science* 239: 748–753.

Fullarton, J.H. (1895). On the generative organs and products of *Tomopteris onisciformis*. *Eschscholtz Zoologische Jahrbucher* 8: 8.

Goodrich, E.S. (1945). The study of nephridia and genital ducts since 1895. *Quarterly Journal of Microscopic Science* 86: 113–392.

Gray, J. (1939). Studies in animal locomotion. VIII. The kinetics of locomotion of *Neris diversicolor*. *Journal of Experimental Biology* 16: 9–17.

Hachfeld, G. (1926). Beitrage zur Kenntnis der *Tomopteris catharina* Grosse. *Zeitschrift fur wissenschaftliche Zoologie* 128: 133–181.

Hartman, O. (1954). Australian Neridae. *Transactions of the Royal Society of South Australia* 77: 1–41.

Hartman, O. (1967). Polychaetous annelids collected by the USNS Eltanin and Staten Island cruises, chiefly from Antarctic seas. *Allan Hancock Monographs in Marine Biology* 2: 1–387.

Harvey, E.N. (1952). *Bioluminescence*. New York and London: Academic Press.

HELFMAN, G.S., COLLETTE, B.B., FACEY, D.E., and BOWEN, B.B. (2009). *The Diversity of Fishes*. Oxford: Wiley-Blackwell.

Hermans, C.O. and Eakin, R.M. (1974). Fine structure of the eyes of an alciopid polychaete, *Vanadis tagensis* (Annelida). *Zeitschrift fur morphologie und okologie der tiere* 79: 245–267.

Herring, P.J. (ed.) (1978). Bioluminescence of invertebrates other than insects. In: *Bioluminescence in Action*. London and New York: Academic Press.

Hopkins, T.L. (1985a). The zooplankton community of Croker Passage, Antarctic Peninsula. *Polar Biology* 4: 161–170.

Hopkins, T.L. (1985b). Food web of an Antarctic midwater ecosystem. *Marine Biology* 89: 197–212.

Hopkins, T.L. (1987). Midwater food web in McMurdo Sound, Ross Sea, Antarctica. *Marine Biology* 96: 93–106.

Hopkins, T.L. and Torres, J.J. (1988). The zooplankton community in the vicinity of the ice edge, western Weddell Sea, March 1986. *Polar Biology* 9: 79–87.

Hopkins, T.L. and Torres, J.J. (1989). Midwater food web in the vicinity of a marginal ice zone in the western Weddell Sea. *Deep-Sea Research* 36: 543–560.

Hopkins, T.L., Lancraft, T.M., Torres, J.J., and Donnelly, J. (1993). Community structure and trophic ecology of zooplankton in the Scotia Sea marginal ice zone in winter (1988). *Deep-Sea Research* 40: 81–105.

Hyman, L.H. (1959). *The Invertebrates: Smaller Coelomate Groups*. New York: McGraw-Hill.

Jones, M.L. (1981). Riftia pachyptila Jones: observations on the vestimentiferan worm from the Galapagos Rift. *Science* 213: 333–336.

Jones, M.L. (1985). On the Vestimentifera, new phylum: six new species and other taxa, from hydrothermal vents and elsewhere. *Bulletin of the Biological Association of Washington* 6: 117–158.

Jones, M.L. and Gardiner, S.L. (1988). Evidence for a transitory digestive tract in Vestimentifera. *Proceedings of the Biological Society of Washington* 11: 423–433.

Kaestner, A. (1967). *Invertebrate Zoology volume 1*. New York: Wiley.

Lamarck, J.-B. (1802). La nouvelle classes des Annelides. *Bulletin du Museum national d'histoire naturelle*. Paris.

Lamarck, J.-B. (1818). *Histoire Naturelle des Animaux sans Vertibres*. Paris: Deterville and Verdiere.

Laverack, M.S. (1960). Tactile and chemical perception in earthworms. I. Responses to touch, sodium chloride, quinine, and sugars. *Comparative Biochemistry and Physiology* 1: 155–163.

Laverack, M.S. (1961). Tactile and chemical perception in earthworms. II. Responses to acid pH solutions. *Comparative Biochemistry and Physiology* 2: 22–34.

Linnaeus, C. (1758). *Systema naturae per regna tria natura: secundum classes, ordines, genera, species, cum characteribus, differentiis, synonymis, locis.* Holmiae: Laurentii Salvii.

MAROTTA, R., FERRAGUTI, M., ERSEUS, C., and GUSTAVSON, L. (2008). Combined-data analytics and character evolution of Clitellata (Annelida) using 18S rDNA and morphology. *Zoological Journal of the Linnaean Society of London.* 154: 1–26.

McGowan, J.A. (1960). The relationship of the distribution of the planktonic worm *Poeobius meseres* Heath to the water masses of the North Pacific. *Deep-Sea Research* 6: 125–139.

McHugh, D. (1997). Molecular evidence that echiurans and pogonophorans are derived annelids. *Proceedings of the National Academy of Science USA* 94: 8006–8009.

McIntosh, W.C. (1915). *A Monograph of the British Annelids, Vol. 3, Part 1: Polychaeta. Opheliidae to Ammocharidae.* London: Ray Society.

Mettam, C. (1967). Segmental musculature and parapodial movement of *Nereis diversicolor* and *Nephthys hombergi* (Annelida: Polychaeta). *Journal of Zoology* 153: 245–275.

Meyer, A. (1929). On the coelomic cilia and circulation of the body fluid in *Tomopteris helgolandica. Journal of the Marine Biological Association of the United Kingdom* 16: 271–276.

Michel, C. and Devillez, E.J. (1978). Digestion. In: *Physiology of Annelids* (ed. P.J. Mill). London: Academic Press.

Mill, P.J. (ed.) (1978). Sense organs and sensory pathways. In: *Annelids.* London: Academic Press.

Monro, C.C.A. (1930). Polychaete worms. *Discovery Reports, Cambridge* 2: 1–222.

Monro, C.C.A. (1936). Polychaete worms. II. *Discovery Reports, Cambridge* 12: 59–198.

Nicoll, P.A. (1954). Vascular systems in Nereis. *Biological Bulletin* 106: 69–82.

Okada, Y.K. (1933). Two interesting syllids, with remarks on their asexual reproduction. *Memoirs of the College of Science Kyoto Imperial University. Series B* 8 (3): 325–338, Pl. 12.

Olive, P.J.W. (1975). Reproductive biology of *Eulalia viridis* (Muller) (Polychaeta: Phyllodocidae) in the northern U.K. *Journal of the Marine Biological Association of the United Kingdom* 55: 313–326.

Osborn, K.J. and Rouse, G.W. (2008). Multiple origins of pelagicism within Flabelligeridae. *Molecular Phylogenetics and Evolution* 49: 386–392.

Osborn, K.J. and Rouse, G.W. (2010). Phylogenetics of Acrocirridae and Flabelligeridae. *Zoologica Scripta* 40: 204–219.

Osborn, K.J., Rouse, G.W., Goffredi, S.K., and Robison, B.H. (2007). Description and relationships of *Chaetopterus pugaporcinus*, an unusual pelagic polychaete (Annelida, Chaetopteridae). *Biological Bulletin* 212: 40–54.

Osborn, K.J., Haddock, S.H.D., Pleijel, F. et al. (2009). Deep-sea swimming worms with luminescent bombs. *Science* 325: 964.

Osborn, K.J., Madin, L.P., and Rouse, G.W. (2011). The remarkable squidworm is an example of discoveries that await in deep-pelagic habitats. *Biology Letters* 7: 449–453.

Pettibone, M.H. (1982). Annelida. In: *Synopsis and Classification of Living Organisms* (ed. S.P. Parker). New York: McGraw-Hill.

Pleijel, F. and Dales, R.P. (1991). Polychaetes: British Phyllodocoideans, typhloscolecoideans, and tomoperoideans.

Synopses of the British Fauna (New Series) 45: 1–202.

Reibisch, J.G.F. (1895). Die pelagischen Phyllodociden und Typhloscoleciden der Plankton-Expedition. *Ergebnisse des Plankton-expeditions der Humboldt-Stiftung* 2: 1–63.

Richards, K.S. (1978). Epidermis and cuticle. In: *Annelids* (ed. P.J. MILL). London: Academic Press.

Robbins, D.E. (1965). The biology and morphology of the pelagic annelid *Poeobius meseres*. *Journal of Zoology* 146: 197–212.

Rouse, G. and Fauchald, K. (1997). Cladistics and polychaetes. *Zoologica Scripta* 26: 139–204.

Rouse, G.W. and Pleijel, E. (2001). *Polychaetes*. Oxford: Oxford University Press.

Rousset, V., Pleijel, F., Rouse, G. et al. (2007). A molecular phylogeny of annelids. *Cladistics* 23: 41–63.

Schroeder, P.C. and Hermans, C.O. (1975). Annelida: polychaeta. In: *Reproduction of Marine Invertebrates* (eds. A.C. Giese and J.S. Pearse). New York: Academic Press.

Siddall, M. et al. (2001). Validating Livanow: molecular data agree that leeches, branchiobdellidans, and Acanthobdella peledina form a monophyletic group of oligochaetes. *Molecular Phylogeny and Evolution* 21: 351–364.

Smith, P.R. and Ruppert, E.E. (1988). Nephridia. *Microfauna Marina* 4: 231–262.

Snodgrass, R.E. (1938). *Evolution of the Annelida, Onychophora, and Arthropoda*. Washington, DC: Smithsonian Institution.

Stop-Bowitz, C. (1948). Polychaeta from the Michael Sars North Atlantic deep-sea expedition 1910. *Report on the Scientific Results of the Michael Sars North Atlantic Deep-Sea Expedition* 5: 1–191.

Struck, T.H. et al. (2015). The evolution of annelids reveals two adaptive routes to the interstitial realm. *Current Biology* 25: 1993–1999.

Stuart, J. (1982). Hirudinoidea. In: *Synopsis and Classification of Living Organisms* (ed. S.P. Parker). New York: McGraw-Hill.

Tebble, N. (1960). The distribution of pelagic polychaetes in the south Atlantic ocean. *Discovery Reports, Cambridge* 30: 161–299.

Tebble, N. (1962). The distribution of pelagic polychaetes across the North Pacific. *Bulletin of the British Museum (Natural History)* 7: 373–492.

Terio, B. (1950). Autotomia nei Tomopteridi. *Bolletino di zoologia agraria e di bachicoltura* 17: 39–43.

Thuesen, E.V. and Childress, J.J. (1993). Metabolic rates, enzyme activities, and chemical compositions of some deep-sea worms particularly *Nectonemertes mirabilis* (Nemertea;Hoplonemertinea) and *Poeobius meseres* (Annelida;Polychaeta). *Deep-Sea Research* 40: 937–951.

Turnbull, F.M. (1875). On the anatomy and habits of *Nereis virens*. *Transactions. Connecticut Academy of Arts and Sciences* 3: 265–290.

Uschakov, P.V. (1972). Polychaeta I. Polychaetes of the sub-order Phyllodociforma of the Polar Basin and the north-western part of the Pacific (Israel Program for Scientific Translation 1974). *Fauna SSSR* 102: 1–271.

Uttal, L. and Buck, K.R. (1996). Dietary study of the midwater polychaete *Poeobius meseres* in Monterey Bay, California. *Marine Biology* 125: 333–343.

Wald, G. and Rayport, S. (1977). Vision in annelid worms. *Science* 196: 1434–1439.

WILEY, E.O. and LIEBERMAN, B.S. (2011). *Phylogenetic Systematics: Theory and Practice of Phylogenetic Systematics*. Hoboken: Wiley-Blackwell.

WoRMS (n.d.). World Register of Marine Species. https://www.marinespecies.org

Yingst, D.R., Fernandez, H.R., and Bishop, L.G. (1972). The spectral sensitivity of a littoral annelid: *Nereis mediator*. *Journal of Comparative Physiology* 77: 225–232.

7

The Crustacea

Introduction

Our transition into the Crustacea is a major leap, introducing new modes of growth and movement, new body shapes and developmental schemes, and most of all a huge number of relatives and an eye-popping diversity within the group itself. Exciting times are upon us! Added to the diversity is a robust fossil record, which has helped in understanding lineages. Estimates of the numbers of marine crustaceans hover around 52 000 species, with new ones added every year. They comprise about 25% of all marine species in the animal kingdom (201 000) (WoRMS n.d.), including all invertebrate and vertebrate taxa. Table 7.1 puts that number in perspective, showing how the crustaceans fare in relation to numbers in other invertebrate groups. The first thing to notice is the overwhelming superiority of numbers demonstrated by the arthropods, which include the insects, millipedes and centipedes, the chelicerates (spiders and kin), and the crustaceans, together comprising about 82% of all presently known animal species. The second, a corollary of the first, is the contribution of vertebrates to the total, which is about 3.7%. The Crustacea (3.6%) and Vertebrata both weigh in with about the same percent of the total, respectable but not impressively large.

Arthropods fall naturally into two main groups, marine and terrestrial, with the crustaceans overwhelmingly dominating marine systems, and the insects, spiders, and mites dominating the terrestrial. All three groups have representatives in freshwater. Within the marine system, crustaceans are distributed from surface waters to the benthos and from intertidal to deep sea. Pelagic Crustacea, particularly those of open-ocean systems, fall into a few major groups that are distributed pan-globally. That is fortunate. After acquiring a sound background in crustacean basics, we can explore the groups in enough detail to understand how their pelagic representatives make a living.

The strategy in this chapter will be to briefly introduce arthropod classification, arthropod ancestors, universal arthropod characters, and then move more deeply into the crustaceans and the groups most important to the study of the pelagic realm.

Arthropod Classification

History

As noted in Chapter 6, the 10th edition of Linneaeus' Systema Naturae divided all invertebrate groups into two main divisions: the Vermes and Insecta.

Life in the Open Ocean: The Biology of Pelagic Species, First Edition. Joseph J. Torres and Thomas G. Bailey.
© 2022 John Wiley & Sons Ltd. Published 2022 by John Wiley & Sons Ltd.

Table 7.1 Number of species in major animal groups as estimated in the references indicated. The WoRMS list is marine species only. (WoRMS: The World Register of Marine Species. http://www.marinespecies.org).

Taxon	Chapman (2009)	Brusca et al. (2016)	WoRMS
Vertebrates			
Mammals	5 487		137
Birds	9 990		608
Reptiles	8 734		105
Amphibia	6 515		2
Fishes	31 153		19 002
Agnatha	116	Total vertebrates ~58 000	98
Vertebrate total	61 995	58 000	19 952
Invertebrates			
Cephalochordata	33	30	30
Tunicata	2 760	3 000	3 078
Hemichordata	208	135	131
Echinodermata	7 003	7 300	7 446
Arachnida	102 248	110 000	1 594
Pycnogonida	1 340	1 000	1 350
Myriapoda	16 072	16 000	76
Crustacea	47 000	70 000	51 996
Onychophora	165	200	0
Hexapoda	1 009 048	926 990	1 687
Mollusca	85 000	80 000	49 344
Annelida	16 763	20 000	13 332
Nematoda	25 000	25 000	6 282
Acanthocephala	1 150	1 200	506
Platyhelminthes	20 000	26 500	12 657
Porifera	6 000	9 000	9 083
Cnidaria	9 795	13 400	11 966
Ctenophora	200	100	202
Nemertea	1 200	1 300	1 317
Others	11 273	13 247	9 079
Invertebrate total	1 362 258	1 324 402	181 156
Vertebrates + invertebrates	1 424 253	1 382 402	201 108

Interestingly, the insects were divided into several groups based on their wing morphology, e.g. Lepidoptera were, as they are today, the butterflies and moths. The remainder of the arthropods, including the crustaceans, spiders and scorpions, centipedes and millipedes, were all lumped into the group "Aptera" or "without wings" (Winsor 1976). Linnaeus' original system of classification was modified by Lamarck (1802), who separated the arthropods into the Crustacea, Cirripedia, Insecta, and Arachnida (Brusca and Brusca 2003). Cuvier, a foe of Lamarck (and evolution), combined the annelids, crustaceans (decapods, stomatopods, amphipods, isopods, and branchiopods), arachnids, and insects (including the myriapods, or millipedes and centipedes) into the Articulata, a taxonomic category that united animal groups with a segmented body plan when he published his *Regne Animale* in 1817. Arthropods as we know them today were recognized as a separate phylum by Leuckart in 1848 with Haeckel developing an arthropod evolutionary tree a few years later in 1866 (Brusca and Brusca 2003). The Articulata persisted as a taxonomic category, or superphylum, with additions and deletions of animal groups by different zoologists well into the twentieth century (e.g. Beklemishev 1969).

Recent phylogenomic studies have resulted in a new metazoan tree of life that retains some of the classical taxonomic categories as well as establishing new ones (Brusca et al. 2016, Table 7.2). Two major clades incorporate the nemerteans, annelids, molluscs, and arthropods. The Spiralia includes those phyla that exhibit spiral cleavage in their developing embryos (nemerteans, annelids, and molluscs). The Ecdysozoa, united by the need of member phyla to molt at least once in their lifetime, includes the arthropods. The groupings (clades) were initiated by two studies using 18S ribosomal DNA genes (Aguinaldo et al. 1997; Dunn et al. 2008) and are now considered to be fairly well established (Brusca et al. 2016). The Ecdysozoa also includes the nematodes and Nematomorpha, and the smaller phyla Kinorhyncha, Loricifera, and Priapulida. Onychophora, Tardigrada, and Arthropoda are further grouped into a smaller clade called the Panarthropoda which conveniently puts the molecular imprimatur on a group long considered as the arthropods and their closest relatives (cf. Nielsen 2001).

The annelids, formerly considered an evolutionary step on the way to the arthropods, are now, based on molecular evidence, quite separate from them. Annelids and arthropods share a large number of similarities including a segmented body, developmental addition of segments in an identical way (teloblastic development), and a similar basic design in their nervous systems. In this instance, molecular evidence trumps that provided by morphology and embryogeny.

Classification within the arthropods has also evolved with time. The system described below gives the four extant groups of arthropods equal weight as subphyla. Taxonomic ranks of all groups within the arthropod hierarchy have changed many times over the last 50 years and will likely continue to do so. The system presented below is that of Brusca et al. 2016 which has actually simplified classification in the Crustacea by elevating several of the major groups (e.g. Copepoda) to Class status. It differs from that in WoRMS which has retained two superclasses. Below Class level, the two systems

Table 7.2 Classification of the animal kingdom (metazoa). *Source:* Brusca et al. (2016), inside front cover. Reproduced with the permission of Oxford University Press.

Non-bilateria (Diploblasts)	
Phylum Porifera	Lophophorata
Phylum Placozoa	Phylum Phoronida
Phylum Cnidaria	Phylum Bryozoa
Phylum Ctenophora	Phylum Brachiopoda
	Ecdysozoa
Bilateria (Triploblasts)	**Nematoida**
Xenacoelomorpha	Phylum Nematoda
Phylum Xenacoelomorpha	Phylum Nematomorpha
Protostomia	**Scalidophora**
Phylum Chaetognatha	Phylum Kinorhyncha
Spiralia	Phylum Priapula
Phylum Platyhelminthes	Phylum Loricifera
Phylum Gastrotricha	**Panarthropoda**
Phylum Rhombozoa	Phylum Tardigrada
Phylum Orthonectida	Phylum Onychophora
Phylum Nemertea	Phylum Arthropoda
Phylum Mollusca	Subphylum Crustacea[a]
Phylum Annelida	Subphylum Hexapoda
Phylum Entroprocta	Subphylum Myriapoda
Phylum Cycliophora	Subphylum Chelicerata
Gnathifera	**Deuterostomia**
Phylum Gnathostomulida	Phylum Echinodermata
Phylum Micrognathozoa	Phylum Hemichordata
Phylum Rotifera	Phylum Chordata

[a] Paraphyletic group.

are fairly similar. The crustacean class Malacostraca is the one comprising nearly all the micronektonic species and is nearly identical in both systems.

Trilobites were formerly considered a subphylum within the Arthropoda with about 4000 described species in a totally extinct marine group. Their status is currently uncertain and though their relatedness to the arthropods is beyond doubt, how they are related is. We salute the scientists engaged in their study but must move on to the extant groups.

Subphylum Crustacea

The crabs, shrimp, lobsters, and kin. About 52 000 marine species in a highly diverse almost totally aquatic group. Body divided into three tagmata: head, thorax (mid-section), and abdomen, except in the most primitive members of the group, the remipedes, which have only a cephalon (head) and a segmented trunk. All have compound eyes.

Subphylum Hexapoda

The insects and kin. About 1 000 000 species in an almost completely terrestrial group. Body divided into cephalon, thorax, and abdomen. Thorax with three segments and three pairs of uniramous appendages. All have compound eyes.

Subphylum Myriapoda

The millipedes and centipedes. About 16 000 species in a completely terrestrial group. Body divided into a cephalon and a multi-segmented trunk bearing uniramous appendages. Eyes, when present, are simple ocelli.

Subphylum Chelicerata

The spiders, mites, horseshoe crabs, sea spiders, and scorpions. At least 102 200 species in a mainly terrestrial group with a modest presence in the marine environment. Body also divided into two sections, a cephalothorax and abdomen.

The diversity of arthropods has attracted some of the best minds in zoology and spawned numerous theories on origins and relatedness of the major groups. The present consensus is that the arthropods themselves are monophyletic, that is, they arose from a single ancestor. In the later half of the twentieth century, the monophyly of the arthropods was called into question by Tiegs and Manton (1958), Anderson (1973), and Manton (1977) who called for phylum status for the uniramia (insects and myriapods), trilobites, crustaceans, and chelicerates based on limb functional morphology and development (Schram 1978). It is worth noting that though the higher-order status of the major groups listed in the classification scheme above has changed with time, the groups themselves have remained largely intact, even within the Ecdysozoa. What has changed is the evolutionary tree: annelids were moved into a new clade, the Spiralia.

Panarthropoda Phyla

Phylum Onychophora

The velvet worms, with about 200 species (Brusca et al. 2016), were known to many generations of zoology students as the missing link between annelids and arthropods. It is easy to see why. Though all living onychophorans are completely terrestrial, their flexible body and general appearance certainly suggests a "walking worm," a living intermediate between the two phyla.

The onychophorans are an ancient group, having changed very little since their appearance in the Cambrian explosion of 530 mya (**m**illion **y**ears **a**go) (Peck 1982; Brusca and Brusca 2003). Their origins are marine, and one of the most famous fossils of the Burgess Shale, *Aysheia* (Figure 7.1), is believed to be an onychophoran (Gould 1989). Two families of Onychophora make up the phylum: the Peripatopsidae, considered the more primitive, and the Peripatidae. Distribution is southern-hemisphere temperate and wide-ranging tropical, with most found in warmer climes (Peck 1982).

Phylum Tardigrada

The water bears. About 1200 species (Brusca et al. 2016). The water bears are a widely distributed group of tiny (0.3–1.2 mm in length)

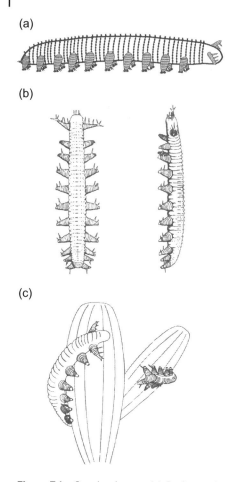

Figure 7.1 Onychophorans. (a) Conjectural reconstruction of *Asheaia*. (b) Another reconstruction of *Asheaia*, dorsal and lateral views; (c) Reconstruction of *Asheaia* feeding on sponges. *Sources:* (b) Adapted from Whittington (1978), figure 86 (p. 187); (c) Whittington (1978), figure 90 (p. 193). Reproduced with permission of The Royal Society.

coelomate aquatic animals, with most representatives below 0.5 mm. Their short body is cylindrical or dorsally convex in shape and is equipped with four pairs of stubby legs (Figure 7.2) that end in a variety of different types of claws or adhesive discs (Morgan 1982). They have no circulatory or respiratory organs (Figure 7.2). Tardigrades are found in all climates, including both poles.

It should be noted that tardigrades are one of only a few taxa that can enter a dormant state at any time during their life. Some protists, rotifers, and nematodes also have a lifetime capability for metabolic arrest, but, for many taxa and even dormancy champions like brine shrimp, dormancy can only occur during early development.

Terminology for the dormant states like those of the tardigrades differs from author to author. Hochachka and Guppy (1987) prefer "anhydrobiosis" or "life without water." Other terms include "anabiosis," from the Greek "return to life," and "cryptobiosis" or "hidden life" (Barnes 1974; Brusca and Brusca 2003). Cryptobiosis and anhydrobiosis both denote the near-absence of detectable metabolic processes. The ability to descend into a dormant state enables tardigrades to survive harsh and unpredictable conditions in addition to prolonging their length of life. Tardigrades in culture live about a year, molting at intervals throughout their lives. However, they can exist in a dormant state for many years, even decades. A rehydrated museum specimen of dried moss that had been in storage for 120 years produced living water bears! (Brusca and Brusca 2003).

Synopsis of Universal Arthropod Characteristics

The true arthropods range in size from mites (chelicerates) and interstitial (living between sand grains) Crustacea of considerably less than a millimeter to a variety of giants, both past and present. The Japanese spider crab *Macrocheira kaempferi* has a leg spread of >3 m, the Tasmanian giant crab (*Pseudocarcinus gigas*) has a carapace width of 46 cm, and New Zealand's giant packhorse lobster *Sagmariasus verreauxi* has a

(a)

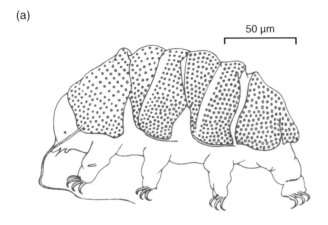

(b)

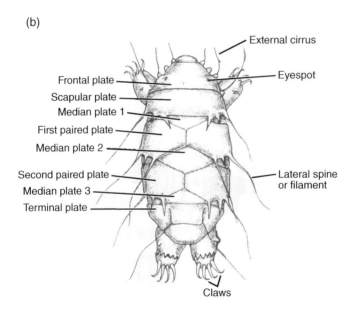

Figure 7.2 Tardigrade general anatomy. (a) A typical "water bear," *Echinicus wendti*; (b) External anatomy of a generalized *Echinicus*. *Sources:* (a) Adapted from Morgan (1982), p. 735; (b) Adapted from Morgan and King (1976).

length of 60 cm. The New England lobster, *Homarus americanus*, can reach weights of 20 kg or more if left to grow old (Kaestner 1968; Brusca and Brusca 2003). Invertebrates almost big enough to ride on!

The fossil record revealed eurypterids (aquatic scorpions) of 1.8 m in length, trilobites of nearly a meter (Kaestner 1968; Barnes 1974), and dragonflies with a wingspread of 65 cm, all of the Paleozoic era (419–541 mya). However, the average size of the two aquatic groups, scorpions and trilobites, was in the neighborhood of a few centimeters; both were primarily benthic.

All arthropods have a chitinous exoskeleton, articulated limbs, and a segmented body arising from teloblastic development, similar to that of the annelids and Onychophora. Depending on the taxon, the first pair of limbs is modified into antennae or mouth parts. The body is further modified into at least a head and trunk region, most often with a thorax and carapace. The primitive condition is for each body segment to bear a pair of appendages. The head or cephalon usually bears compound eyes and simple ocelli.

The main body cavity of arthropods is a hemocoel, or mixocoel, with the true coelomic cavity relegated to the nephridia and gonads. The circulatory system is open, with a tubular neurogenic (beat originating neurally, rather than in the heart muscle itself). The nervous system comprises a brain, or supraesophageal ganglion, with circumesophageal connectives running to a subesophageal ganglion, and a single or double ventral nerve cord with ganglionic swellings in each segment, much like the annelids.

The gut shows considerable regional specialization; the food bolus is contained within a peritrophic membrane as it makes its way down the gut. Muscles are striated and mainly contained within discrete muscle bands rather than as a body wall musculature. Virtually all arthropods are dioecious.

Purely in terms of species numbers, the arthropods are without question the most successful protostomate group. While the cephalopods are the Einsteins of the invertebrates, the arthropods are the most speciose, dominating the marine, freshwater, and terrestrial environments. They exhibit considerable diversity of form, size, complexity, behavior, and locomotion.

Worthy of note is that the arthropods or their ancestors not only dominate the present suite of taxa inhabiting the planet but the fossil record also indicates that they have done so since metazoan beginnings in the Precambrian. Even considering that fossils by their nature are biased toward hard-bodied species, arthropods have enjoyed an astonishing success through evolutionary time. Let us briefly consider why.

First and foremost is the exoskeleton. The exoskeleton not only gives protection but it also provides an anchor for a more efficient musculature, thereby allowing for a great leap forward in diversity of locomotory modes. Combined with diversity in the jointed limbs, the exoskeleton allows for running and walking, swimming, and, with morphological change in the limbs, flying. Arthropods were the first flyers, increasing the suite of habitats and lifestyles available to them. Jointed limbs have also provided the biological raw material for different types of mouthparts, biting, piercing, and sucking among them.

Metameric structure provides a rich template for morphological change during development. Segments can combine into body regions to produce fundamentally different body structures, some of which will succeed, others not, but as the building blocks for diversity of form they are ideal. In sum, the exoskeleton, jointed limbs, and segmented body of arthropods have provided the evolutionary fuel for an astonishing breadth of form.

The Crustacea

Only a few of the major divisions of Crustacea are represented in the pelagic realm. Our strategy is to present the universal characteristics of the Crustacea, briefly touch on classification, describe organ systems common to all pelagic Crustacea, and then move on to the biology of the individual pelagic groups.

Characteristics

A prominent unifying characteristic of all Crustacea is the presence of two pairs of antennae on the head. Overall, the head, or cephalon, contains an anterior acron (anteriormost segment) and five additional segments. Working from anterior to posterior, the segments are the acron, an antennular segment, an antennal segment, a mandibular segment, a maxillulary segment, and a maxillary segment. In appearance, the segments are not obvious and the head appears as one unit with two sets of antennae and three pairs of mouthparts (Figure 7.3). In many cases, a thoracic segment, or thoracomere, is fused to the head, contributing another set of limbs termed maxillipeds.

Posterior to the head is a trunk which usually is composed of a fairly distinct thorax and abdomen. In most crustaceans, a carapace is present. Developmentally, it originates as lateral outgrowths of the posterior head segments, and it may take the form of a flattened head shield or the more familiar shell-like carapace enclosing the entire cephalothorax region as it does in shrimp, lobsters, and krill. The most primitive Crustacea, the recently discovered remipedes (Yager 1981) have a simple, homonomously segmented, trunk that lacks distinct subregions (Figure 7.4).

Crustaceans have a variety of specialized respiratory surfaces for aquatic gas exchange. Advanced taxa exhibit highly modified limbs that serve as gills, often contained in a sophisticated chamber promoting a unidirectional flow of water. More primitive groups may use specialized regions of the body surface. Excretion is achieved with specialized nephridial systems that also vary in complexity and number with the different taxa.

All crustaceans have a well-developed regionalized gut. Compound eyes are usually present, and development spans the gamut of multiple free-living larval stages to direct development in a marsupium, or "brood pouch."

Classification

The classification system below is that of Brusca et al. 2016, which divides the subphylum Crustacea into eleven classes. Two of the three more primitive classes were only recently discovered: the remipedes (Yager 1981) and the cephalocarids (Sanders 1955). Both have a head shield and a

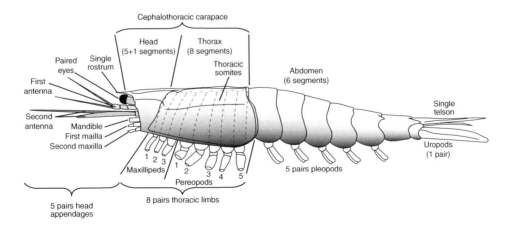

Figure 7.3 Archetypic body architecture of a malacostracan crusteacean.

282 | *The Crustacea*

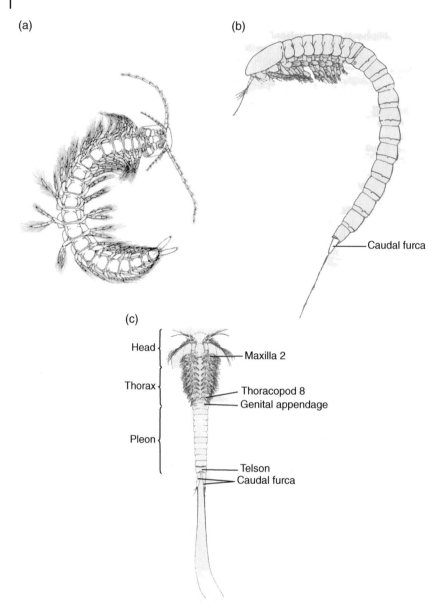

Figure 7.4 Classes Remipedia and Cephalocarida. (a) Stylized drawing of a remipede (ventral view). (b) Lateral view of the cephalocarid *Lightiella incisa*. About 2 mm in length, not including caudal furca. (c) Ventral view of *L. incisa*. About 3 mm in length, not including caudal furca. *Sources:* (a) Adapted from Schram (1986), figure 3-1 (p. 37); (b) Adapted from Gooding (1963), figure 2 (p. 294); (c) Adapted from Sanders (1963), figure 69 (p. 163).

multi-segmented body. In the case of the cephalocarids, a well-defined thorax and abdomen are present but, in the remipedes, the segmentation is homonomous (all segments alike). The third class is the Branchiopoda, the group that includes brine and tadpole shrimp, masters of dormancy. The best-known previous classification systems for Crustacea are Brusca and Brusca (2003), Martin and Davis (2001), Schram (1984), and Bowman and Abele (1982).

The major subdivisions of the Crustacea are listed below to provide deep background for reference. More detail is provided for groups that have pelagic representatives. Our target open-ocean species are in the size range of 1–25 cm: the micronektonic and macrozooplanktonic Crustacea. A summary of the mind-bending diversity of the Crustacea follows. Species numbers are from Brusca et al. (2016) and WoRMS.

Subphylum Crustacea

Class Remipedia

Recently discovered cave dwellers that are considered by many to be the most primitive living Crustacea. Body was divided into a head and trunk section of up to 32 homonomous segments. Laterally directed limbs on each somite and head bear a cephalic shield (Figure 7.4). About 28 species.

Class Cephalocarida

Another primitive crustacean class (Figure 7.4), very small in size (about 3 mm excluding the caudal furcae) and living interstitially. Bears limbs on the thorax but not on the abdomen. Lacks eyes but has a head shield. Found at depths ranging from intertidal to >1000 m. Also has been considered the most primitive crustacean taxon, particularly prior to the discovery of the remipedes. Its limbs are often considered the primitive crustacean archetype. About 12 species.

Class Branchiopoda

A large group (about 900 species, 86 of them marine) containing many familiar species, such as the fairy shrimp ("sea monkeys") and tadpole shrimp. Also considered to be primitive but are far more speciose than either the remipedes or cephalocarids and exhibit much greater diversity of form. Includes the Cladocera or water fleas, an important freshwater zooplankton group.

Class Copepoda

The copepods. An enormously speciose (at least 11 078 marine species) group considered to be the most important primary consumer in most of the world's oceans. They are rivaled by the euphausiid, *Euphausia superba* in the Antarctic. Rarely larger than 4–5 mm in size. The diminutive "cows of the sea." Most free-living, some parasitic groups.

Class Thecostraca

Barnacles and relatives. Important and diverse marine group with 1800 species ranging from the acorn and gooseneck barnacles to the rhizocephalan parasitic castrators of brachyuran crabs. Includes one pelagic genus, *Dosima*, a gooseneck barnacle which hitches a ride on *Velella*, the "by-the-wind-sailor."

Class Tantulocarida

About 38 species. Tiny parasites of deep-sea Crustacea.

Class Mystacocarida

About 13 marine species. Minute (largest is 1.0 mm) interstitial Crustacea most often found in the intertidal.

Class Branchiura

About 45 marine species. Fish lice. Largest species 3 cm.

Class Pentastomida

About 21 species. Tongue worms. Formerly were contained within their own phylum. Parasites of the respiratory systems of reptiles, mammals, and birds.

Class Ostracoda

The ostracods, known in the vernacular as mussel or seed shrimps. Another very important member of the oceanic zooplankton, but the group as a whole is primarily benthic. About 6000 recent species, most of which are about 1 mm in size. The largest *Gigantocypris*, at about 30 mm in size, is a member of the deep-living macrozooplankton from polar to tropical waters throughout the globe. Two superorders, the exclusively marine Myodocopa and the Podocopa. Myodocopans include both pelagic and benthic forms, whereas the podocopans are all tied to the benthos.

Class Malacostraca

Includes the great majority of familiar crustaceans, the krill, shrimps, lobsters, and crabs, as well as many more obscure taxa. More than 32 870 species populate the Malacostraca. Basic anatomy includes a constant eight thoracic segments, or thoracomeres, and six abdominal segments or pleomeres.

Subclass Phyllocarida

Contains one order, the Leptostraca, considered to be an example of the most primitive malacostracan body form (Figure 7.5). About 62 species. All are benthic with the exception of one bathypelagic species, *Nebaliopsis typica*, which is up to 4 cm in length. Most less than a cm in size.

Subclass Hoplocarida

Contains one order, the Stomatopoda or mantis shrimp. About 486 species. Benthic, tube-dwelling predators with a short carapace and long well-developed abdomen (Figure 7.6). Second thoracopod can be used to spear or club prey. Fairly large (15–20 cm). As larvae, stomatopods are perfectly transparent and nearly invisible in a trawl bucket. They bear lethally sharp spines that will spear the unwary (or even the wary, the spines are that invisible!) scientists' hands as they sort through the catch.

Subclass Eumalacostraca

A huge group (32 322 marine species) containing three important superorders, described below. Typically with a carapace, biramous antennae and antennules, abdominal appendages or pleopods, last segment bearing a telson and modified pleopods, called uropods forming a tail fan. Most with compound eyes (*Caridoid facies*, Figure 7.3).

Superorder Eucarida

A species-rich (13 629) superorder containing many of the important pelagic species as well as most of the important commercial ones. All share a well-developed carapace that is fused to the dorsum of all thoracomeres, forming a cephalothorax, and stalked compound eyes. Gills are found only on the thorax.

Order Euphausiacea The euphausiids, or krill. About 90 species of shrimp-like pelagic crustaceans that inhabit the open ocean at all latitudes, sometimes to great depth. Characterized by a carapace that is shortened laterally, exposing the gills. Pereopods are biramous (Figure 7.7).

Order Amphionidacea The longstanding order Amphionidacea is in dispute (DeGrave et al. 2009). The monospecific group may be with the Caridea, most likely with the

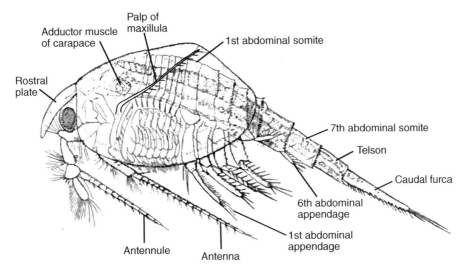

Figure 7.5 The leptostracan crustacean *Nebalia bipes*. *Source:* Calman (1909), figure 87 (p. 152).

(a)

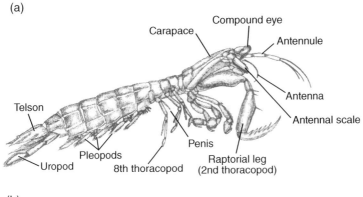

(b)

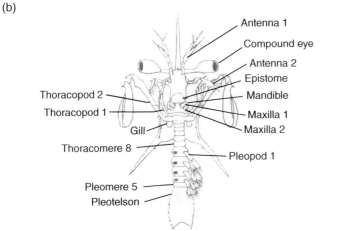

Figure 7.6 Stomatopoda. (a) External anatomy of the stomatopod *Squilla*; (b) First pelagic stage of the pseudozoea of the stomatopod *Squilla*, 4.7 mm. *Sources:* Calman (1909), (a) figure 186, (p. 320); (b) figure 192, (p. 328).

Pandalidae, but its new taxonomy is undecided. It is still recognized by Brusca et al. (2016). At this juncture, it is best to leave it here. The amphionids are pelagic crustaceans (Figure 7.8) with adults primarily found at depths >2000 m. Larvae are found in the upper 30 m. Only three males have been captured. Formerly grouped with the decapods, Amphionidacea were given ordinal status in 1973. They may have now found their way back to the decapods. Found from 35 °N to 35 °S latitude in the Atlantic, Pacific, and Indian Oceans with most records coming from tropical latitudes in the Indian Ocean (Williamson 1973; Schram 1984).

Order Decapoda The shrimp, lobsters, crabs, hermit crabs, and ghost shrimp. About 18 000 species. Current thinking divides the decapods into two suborders

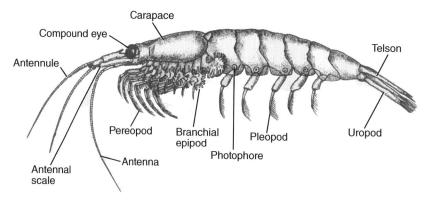

Figure 7.7 General body form and external anatomy of a typical euphausiid *Meganyctiphanes*. *Source:* Adapted from Calman (1909), figure 139 (p. 245).

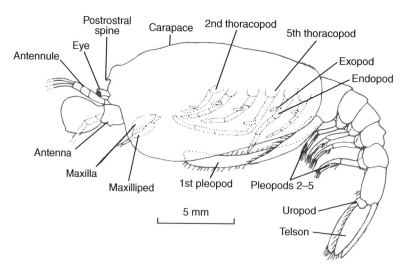

Figure 7.8 External anatomy of the amphionid crustacean *Amphionides reynaudii*. *Source:* Williamson (1973), volume 25, figure 3 (p. 42), Republished with the permission of Brill, from *Amphionides reynaudii, Representative of a Proposed New Order of Eucaridan Malacostraca*.

based on gill morphology. Widely disparate in terms of species numbers.

Suborder Dendrobranchiata The penaeid and sergestid shrimps. Several important open-ocean species as well as many benthic, commercially harvested shrimp. About 450 species altogether. See Figure 7.9.

Suborder Pleocyemata Contains the remainder of the 18 000 species of decapods.

Infraorder Caridea The caridean shrimps. Includes a large suite of important open-ocean forms, commercially harvested shrimp such as the North Atlantic *Pandalus borealis* as well as the snapping shrimp, *Alpheus,* and a variety of other benthic taxa.

Infraorder Achelata The spiny lobsters and kin.

Infraorder Anomura Hermit crabs and relatives. Crabs with an abbreviated or specialized abdominal region. Includes the coconut crabs and sand crabs as well as the pelagic red crabs (pelagic squat lobsters) *Pleuroncodes* and *Munida*.

Infraorder Astacidea The clawed lobsters and crayfish. Includes the New England lobster, *Homarus*.

Infraorder Axiidea The ghost shrimp. Benthic shrimp that excavate impressive networks of tubular burrows. Found mainly in the intertidal.

Infraorder Brachyura The true crabs, including the Atlantic blue crab, western Dungeness crab, fiddler crabs, and ghost crabs.

Infraorder Gebiidea The mud shrimp. Mud-dwelling shrimp that live in tubular burrows. Found mainly in the shallow subtidal.

Infraorder Glypheidea Glypheoid lobsters

Infraorder Polychelida Deep-sea blind lobsters

Infraorder Procarididea Procarididean shrimps.

Infraorder Stenopodidea The stenopodid shrimps. Includes highly specialized reef-dwellers such as cleaner shrimps and sponge-dwellers.

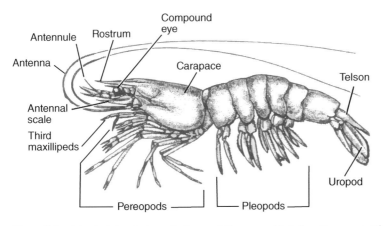

Figure 7.9 A typical dendrobranch decapod; the penaeid shrimp *Penaeus setiferus.*

Superorder Peracarida

The amphipods, isopods, opossum shrimp, and relatives. A very large superorder with about 18 691 marine species. A key characteristic is that all are brooders that have no free-living larvae; most have a marsupium for brooding the developing embryos. Many important open-ocean taxa are peracarids.

Order Mysida The opossum shrimps. About 1166 marine species. Well represented in marine and freshwater, often demersal or associated with aquatic vegetation, nearly always near bottom. Well-developed carapace, pleopods often reduced (Figure 7.10a). Size ranges from 0.2 to 8 cm. Few truly pelagic species.

Order Lophogastrida About 40 species. Formerly lumped with the Mysida as the order Mysidacea, the lophogastrids are an important mesopelagic group found from equatorial to polar seas (Figure 7.10b). Sizes usually range from 1 to 8 cm though occasionally larger; however, the one reported capture of a 35 cm *Gnathophausia ingens*, commonly mentioned as the maximum size for the lophogastrids, has never been repeated.

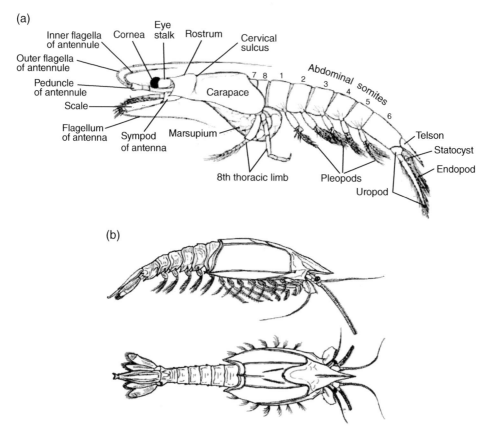

Figure 7.10 Mysid and lophogastrid shrimps. (a) Typical female mysid shrimp. (b) Lateral and dorsal views of a lophogastrid adult female *Gnathophausia ingens*, 15.7 cm. *Sources:* (a) Adapted from Tattersall and Tattersall (1951), diagram 1 (p. 15); (b) Sars (1885), plate II.

Order Isopoda Sea roaches. About 6082 marine species in 14 suborders. Although well represented at all depths in the benthos of marine and freshwater systems, only one genus, *Anuropus*, is considered to be truly pelagic. Even *Anuropus* rides on jellyfish, so its honorary pelagic status is considered marginal by your authors. Isopods have no carapace (Figure 7.11). The only true terrestrial crustaceans, the pillbugs and sowbugs, are isopods and include about 5000 of the 10 000 isopod species. Sizes range from 0.5 mm in the smallest representatives to 50 cm in the

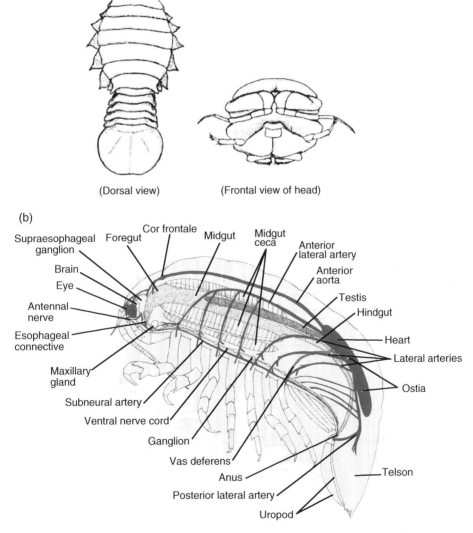

Figure 7.11 Isopoda. (a) The giant bathypelagic isopod *Anuropus bathypelagicus*. (b) Diagrammatic representation of the internal anatomy of a typical isopod. *Sources:* (a) Menzies and Dow (1958), figures 1 and 4 (p. 2); Reproduced with permission of Taylor & Francis. (b) Adapted from McLaughlin (1980), figure 36 (p. 105).

290 | *The Crustacea*

giant isopod *Bathynomus*. Note the position of the heart in Figure 7.11b.

Order Amphipoda Beach hoppers and relatives. About 8100 species in four suborders. The vast majority are benthic, but they are very well represented in the pelagic realm. Like isopods, the amphipods have no carapace (Figure 7.12). The two orders bear a close resemblance, with many features in common.

Suborder Gammaridea The most familiar and most speciose of the amphipods. Though primarily a benthic marine group, gammarids are present in virtually all

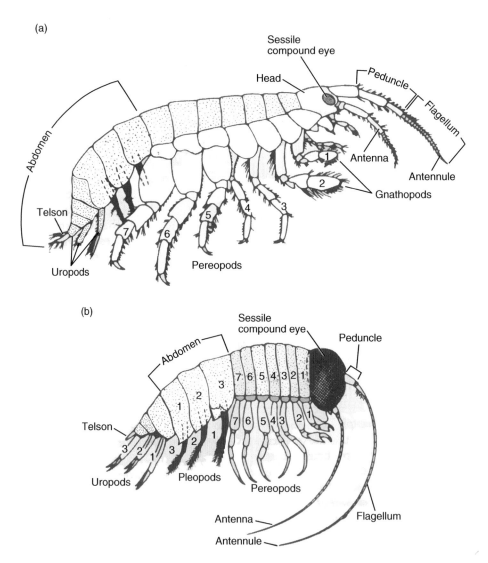

Figure 7.12 Amphipoda. (a) A typical gammarid amphipod. (b) A typical hyperiid amphipod. *Sources:* (a) Adapted from Calman (1909), figure 132 (p. 15); (b) Adapted from Brusca and Brusca (2003), figure 16.15 (p. 538).

aquatic habitats including the intertidal, subterranean, freshwater, and pelagic. About 5000 marine species, but only 150 of those are pelagic.

Suborder Hyperiidea   A totally pelagic group comprising about 260 species, many of which are associated with gelatinous zooplankton either for a "free ride" or as symbionts and ectoparasites. Many, but not all, have very large compound eyes.

Suborder Caprellidea   "Skeleton shrimp." About 300 species adapted for clinging to other organisms. Substrates range from algae to cetaceans. Often seen on hydroids where they exhibit a curious rocking motion.

Suborder Ingolfiellidea   About 40 species. A small elongate suborder found predominantly in caves.

Order Cumacea   Small (0.5–2 cm) benthic crustaceans with a large carapace and distinctive shallow-burrowing behavior. About 1000 species; the great majority are marine.

Order Tanaidacea   Small (0.5–2 cm) benthic marine crustaceans with an abbreviated carapace. Found worldwide and at all ocean depths. About 1500 species.

Order Mictacea   Tiny order (about five species) of small (2–4 mm) crustaceans found in Bermudan caves.

Order Spelaeogriphacea   Another tiny order (four known species) of cave-dwelling freshwater crustaceans.

Order Stygiomysida   Cave-dwelling species formerly grouped with the Mysida (16 species).

Order Thermosbaenacea   First found in African hot springs, but also dwell in freshwater and marine environments as well as anchialine caves (coastal caves formed in limestone or volcanic rock that connect with the ocean). (About 35 species).

Superorder Syncarida
Two orders with about 285 species: the Anaspidacea and Bathynellacea. Primitive Eumalacostra lacking a carapace and inhabiting refugial freshwater habitats, the Bathnellacea are an interstitial subterranean group. First known from fossils, the recent syncarids were discovered in Tasmania and are believed to resemble the ancestral eumalacostracan. Fossil records suggest that they were widespread in the carboniferous period.

Hopefully, the synopsis of the crustacean family tree will give an appreciation for its immense depth and breadth. Clearly, the diversity of form and habitat in the Crustacea is huge. However, as with the primarily benthic nemerteans and polychaetes, the pelagic Crustacea comprise only a tiny fraction of the subphylum as a whole. As it happens, that tiny fraction is still fairly immense. The most important pelagic groups in our target size range are the euphausiids, decapods, lophogastrids, mysids, and amphipods. Cameo appearances are made by the ostracods with the one genus, *Gigantocypris*, the isopods with *Anuropus,* and the copepods with *Gaussia*. We will investigate the major groups in detail and the guest appearances at a level commensurate with the information available.

Within their diversity of form, Crustacea share many basic characteristics that are most easily described in one introductory section. Most notable among those are the integument and limbs, gut, circulation, excretory system, nervous system, sensory mechanisms, and development basics. It would be best to familiarize yourself with the major subdivisions within the Crustacea as the names will reappear many times in the chapter. All of our target species fall within the Class Malacostraca, in the superorders Eucarida and Peracarida.

Crustacean Systems

Integument and Molting

Integument

The exoskeleton of arthropods is their single-most important defining character, a major element in their success as a group. Its basic structure consists of two parts: a thin outer epicuticle and a much thicker three-layer inner cuticle (Figure 7.13a). Skinner (1962) names the three layers of the inner cuticle as the exocuticle, endocuticle, and membranous layer. Beneath the cuticle is the epidermis, which secretes the new cuticle after a molt. The reader should be aware that terminology for the cuticular layers is not uniform between treatments of the subject and particularly between insects and Crustacea. The terminology of

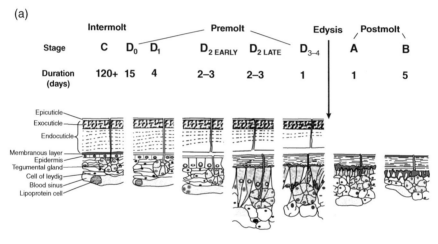

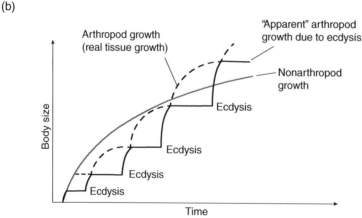

Figure 7.13 (a) Cytology of the integument of the lined shore crab, *Gecarinus*, during each phase of the ecdysis cycle. (b) Growth and molting. Characteristic incremental growth pattern of crustaceans. *Sources:* (a) Adapted from Skinner (1962), figure 9 (p. 644); (b) Adapted from Brusca and Brusca (1990), figure 8 (p. 473).

Dr. Dorothy Skinner, the foremost authority in the field of crustacean molting for several decades, will be used here.

The epicuticle, or outermost region of the arthropod cuticle, usually comprises at least three layers in spite of being only a few microns in thickness. It is chemically inert and nearly impermeable. Moving inward, a consensus description of its three layers includes an outermost lipoprotein cement layer overlying a discrete wax layer, with a hardened protein layer, termed cuticulin in insects, lying just above the exocuticle. The wax layer confers impermeability to water. A thin, highly resistant, membranous layer has also been reported (Hackman 1971). It is refractory to digestion with concentrated acid (!) and is located either between the two outermost layers of the epicuticle described above or is the outermost layer itself. The epicuticle as a whole is the only region of the cuticle that contains no chitin. Without question, the epicuticle is the major barrier to ion and water movement into and out of crustaceans.

Unlike the epicuticle, the inner cuticle of crustaceans consists of closely packed fibrous layers, or laminae, composed of chitin microfibrils surrounded by a protein matrix (Stevenson 1985). The exocuticle (Figure 7.13a) also contains a granular melanin-like pigment that gives the layer a darker color. Below the exocuticle is the endocuticle, which has a similar laminar structure but, particularly in benthic Crustacea, is more highly calcified. Together, the exo- and endocuticular layers make up the lion's share of the inner cuticle and, depending upon the species and its lifestyle, both can be well calcified. Calcium salts are imbedded in the protein matrix, and most often are $CaCO_3$ (calcite) and $Ca_3(PO_4)_2$. The innermost layer of the inner cuticle, the membranous layer, is the smallest in thickness.

Proteins in the crustacean epicuticle and inner cuticle are hardened after their initial deposition by a process called sclerotization or tanning. Protein chains are covalently cross-linked, conferring considerable rigidity to their overall structure. The most highly sclerotized region of the crustacean inner cuticle is the exocuticular region.

The epidermis comprises a single layer of cells underlying the procuticle. It is responsible for secreting the cuticle itself and for the molting fluid that digests part of the cuticle prior to molting. The epidermis also synthesizes the chitin and proteins that form the fibrous lamina of the cuticle. In the past, various functions in cuticle synthesis have also been attributed to the tegmental glands located beneath the epidermis, including secretion of the entire epicuticle or just the outer cement layer. However, the epicuticle is now believed to be secreted by the epidermis, with the tegmental glands possibly contributing enzymes involved in the tanning of epicuticular proteins. The problem with tegmental glands having a large role in crustacean cuticle synthesis is that their distribution is not homogenous enough for them to do the job at all locations within the integument (Stevenson 1985).

Surface setae originate in trichogen cells in the epidermis and remain connected to the epidermis via canals. Setae, when innervated, have a sensory function, primarily mechanoreception, but many are not innervated. Pore canals, minute ducts originating in epidermal cells and traversing the cuticle, may also be present. They play multiple roles in the integument including wound repair.

Molting

The exoskeleton of Crustacea is an effective armor, anchoring an advanced musculature and providing a barrier to the outer world.

It also isolates the crustacean internal milieu, a development that allows for a higher level of internal stability, or homeostasis. On the downside, the integument must be shed and replaced at intervals in order for crustaceans to grow. Growth of crustaceans is thus naturally incremental (Figure 7.13b). Within each step-like increase in size is a hormonally mediated molt cycle that governs the biological pre-molt changes (proecdysis), molting (ecdysis), post-molt hardening (metecdysis), and intermolt interval (anecdysis). Drach (1939) developed a system for staging the molt cycle in Crustacea using the brachyuran crab *Cancer pagurus*. It has been applied with considerable success to molting in a variety of crustacean taxa, including amphipods, isopods, and swimming decapods (Stevenson 1985).

Molt Cycle Stages
Drach (1939) divided the molt cycle into five characteristic stages with E, the final stage, representing ecdysis or the molt itself. Figure 7.13a shows the duration of molt-cycle stages for the lined shore crab, *Gecarcinus lateralis*, the favorite experimental subject of Dorothy Skinner. Following along in Figure 7.13a from left to right, we begin with anecdysis, or intermolt, prior to initiation of the molt sequence. Note the darkened membranous layer above the epidermis in stage C. The epidermis is where the action is. Proecdysis, stage D, begins with the dissolution of the membranous layer, separating the old cuticle from the epidermis. Progress through the D stage is marked by enlargement of the epidermal cells and secretion of the new epicuticle and pigmented layer of the procuticle (D2–D4). At ecdysis (E), the old exoskeleton (exuviae) is shed and the crustacean enters metecdysis, or postmolt. Stage A in metecdysis is characterized by a shrinkage of the epidermal cells back to anecdysial size, and stage B begins the synthesis of the remainder of the procuticle, which continues for an extended period of time. At the end-stage, B, the membranous layer is reformed and the animals again enter anecdysis, the intermolt period. Timing of the different stages differs markedly between taxa. For example, euphausiids can molt every one to two weeks.

Additional metabolic changes accompany the molt process of the cuticle. As the old cuticle is broken down during proecdysis, nutrients and minerals are resorbed and sequestered in the hemolymph. Important compounds such as astaxanthin, an orange carotenoid pigment that cannot be synthesized by Crustacea but is nonetheless present in nearly all crustacean exoskeletons, are stored in the blood. Prior to ecdysis, some muscle atrophy occurs in all the limbs so that they can be smoothly withdrawn from the exoskeleton. After molting, the animal emerges from its old exoskeleton with a pliable, soft shell that swells due to uptake of water into its new larger size before hardening. Muscle is regenerated during metecdysis.

Hormonal Control
The molt cycle is mediated by the crustacean neurohormonal system. Two antagonistic hormones, molting hormone (MH) and molt-inhibiting hormone (MIH), interact to control the cycle. During the intermolt period, production of MH is suppressed by steady-state levels of molt-inhibiting hormone in the blood. When MIH levels drop, either because of a natural molting cycle or the presence of external or internal stimuli, production of MH rises and the molt process begins.

MIH is secreted and stored in the crustacean eyestalk by the X-organ-sinus-gland complex (Figure 7.14). MIH is manufactured by the medulla terminalis component of the X-organ and conveyed to the sinus gland though neurosecretory connections. The sinus gland acts as a reservoir and neurosecretory organ for MIH. It discharges MIH into a blood sinus within the eyestalk and thence to the general circulation.

The other half of the molting equation is the Y-organ, or molting organ, which produces the molting hormone ecdysone. The Y-organ is a small ovoid organ located just below the epidermis in the head region anterior to the eyestalk. Like the X-organs within the eyestalks, Y-organs are paired, one to a side, but because of their size and location, they are considerably more difficult to locate and manipulate experimentally than the X-organs. As MIH levels drop prior to molting, the Y-organs produce α-ecdysone, which is converted peripherally to the active form β-ecdysone. Y-organs increase in size as the molt cycle progresses, reaching their largest at ecdysis and shrinking thereafter. In those Crustacea which have a terminal molt, the Y-organs degenerate after the last ecdysis (Skinner 1985). This last bit of evidence did much to dispel doubts on the role of the Y-organ in molting, which historically had been less certain than that of the easily removed X-organ.

External stimuli that can influence molt frequency include temperature, photoperiod, food availability, and proximity of conspecifics. Internal cues may be growth of soft tissue and a hard-wired seasonal molt cycle. It is best to keep in mind that larval molts must occur over the course of days to weeks and that molting in some taxa, e.g. euphausiids, occurs on a weekly basis. Much of what we know about the molting process in Crustacea has been generated on large, long-lived species that are easily captured alive and maintained in the laboratory. Our basic understanding of the molting process is thus well established over scores of species, but intermolt intervals and hormone peaks and valleys will vary between taxa.

A point of historical interest is the origin of the mysterious names for the X- and Y-organs of the molt cycle. The fact that eyestalk removal induces rapid molting has been known since the publications of Zeleny (1905). Hanström (1939) coined the term "X-organ" to describe the unknown structure in the eyestalks of Crustacea that

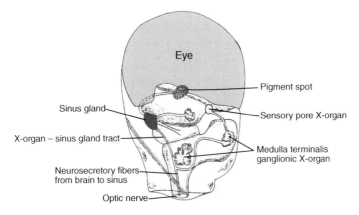

Figure 7.14 The eye and eye stalk of the shrimp, *Leander*, showing the X-organ-sinus-gland complex; the source of molt inhibiting hormone (MIH). *Source:* Barnes (1974), figure 14-77 (p. 611). Reproduced with the permission of W. B. Saunders.

produced MIH, and the name persists today. Gabe (1953) coined the term Y-organ for the organ that produces molting hormone to parallel the "X-organ" label of Hanström.

Joints and Appendages

Joints

The exoskeleton of the arthropods allows a transition away from a system of movement based on hydraulics to one based on levers. The antagonistic action of the polychaete's circular and longitudinal musculature acting on the polychaete's hydrostatic skeleton acts to lengthen and shorten the worm as well as allowing undulatory motion; extrinsic muscles (muscles originating in the body wall and not the appendage itself) move the parapodia. In the onychophorans, the extrinsic muscles are further developed to allow for more sophisticated limb movement, but hydraulics still play a role in the system. In the arthropods, muscles have a firm anchor and intrinsic muscles are present (Figure 7.15), allowing for more efficient movement and for multiple segments, also known as articles, podites, or podomeres.

Crustacean limbs must be hinged to function properly (Figure 7.16). Flexible articular membranes connect each joint within the limb, usually allowing for movement in only one plane. Muscles attach to anchor points termed apodemes, which are actually invaginations of the exoskeleton. Figure 7.16b shows an individual joint in a typical crustacean walking or swimming limb. Two antagonistic muscles operate the limb: the flexor, which moves the limb in its normal plane of motion; and the extensor, which brings it back to status quo. The bearing surface that acts as a fulcrum as the lever-like limb moves back and forth is known as a condyle (Figure 7.16c), and the "lock" is just that – it limits the limb's range of motion. Two items to note are these: first, muscles can only contract, so back-and-forth movement must be achieved with two sets of opposing muscles; second, just as with humans, the larger the cross-sectional

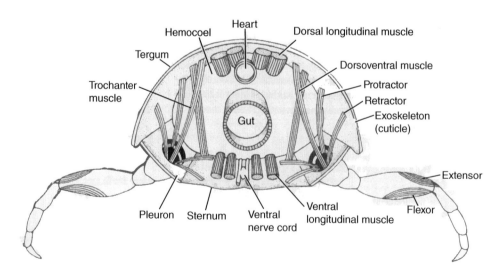

Figure 7.15 Cross section of a body segment of a stylized arthropod showing the typical arrangement of body and limb musculature. *Source:* Barnes (1974), figure 12-1 (p. 435). Reproduced with the permission of W. B. Saunders.

area of a muscle, the stronger it is. In nearly all cases, the flexor muscle is far stronger than the extensor (think of the crushing power of a lobster claw) and is represented that way in the figure.

Figure 7.16d represents a section of a crustacean abdomen showing the articulation and the associated musculature that allow the effective tail-flip escape-response. When encountering a threat, shrimp will often snap their abdomens sharply forward, propelling them rapidly to the rear and taking them out of harm's way. You may have observed the same response when lifting a lobster or shrimp out a tank and received a shower for your troubles.

Appendages

Crustacean appendages are bewilderingly diverse, as might be expected for a taxon with about 70000 species, each bearing multiple limbs with different functions. To minimize confusion, we will cover basic limb morphology and nomenclature here and discuss the limbs of our target taxa as we treat each one.

A generic crustacean limb has two basic sections: the protopod and endopod (also called protopodite and endopodite), which may or may not bear a variety of other structures (Figure 7.17a). The protopod is usually divided into two articles, a coxa and basis; sometimes, a pre-coxa is present though it is unclear if a pre-coxa is part of the limb or the body wall. On its lateral (outer) margin, the protopod may bear epipodites (also called exites) proximally (toward the body) and an exopod distally (away from the body) (Figure 7.17b). On the mesial side (inner, or toward the body center), endites may be present, which constitute the jaws of the mandible in the mouthparts.

The endopod continues the main axis of the limb, and its main articles, or podomeres, are also named. For Crustacea, the pereopods, the walking legs or thoracic appendages of a crab or shrimp, are typically used as examples and we will continue that tradition here. Beginning at the junction of the protopod and endopod and working distally (outward), the podomeres are the pre-ischium, ischium, merus, carpus, propodus, and dactyl (Figure 7.17c). Worthy of note is that the pre-ischium and precoxa are sometimes present but are not common.

Epipodites, or exites, are usually associated with gas exchange and either form the gills themselves or bear more complex gills as in the decapods. Endites, as mentioned above, are usually concerned with food manipulation or mastication. Exopods are usually swimming structures and can play this role on either the thorax or abdomen. On the abdomen, the exopods and endopods are equal in length and well invested with setae that form paddles on the power stroke of the limb.

Excretory System

Unlike the polychaetes, which have multiple tiny nephridia to filter the blood and sometimes as many as one pair per segment, the Crustacea have a single pair of larger excretory organs in the head region that serve the entire organism. In the copepods, barnacles, branchiopods, stomatopods, and isopods, these organs are termed maxillary glands; in the euphausiids, decapods, and amphipods, they are termed antennal glands, named for the segment in which they are located. They are similar in morphology and function in a similar manner, but the antennal gland is considered to be the more advanced of the two structures and considerably more information on its function is available.

Antennal and maxillary glands both function as large individual nephrons, using the

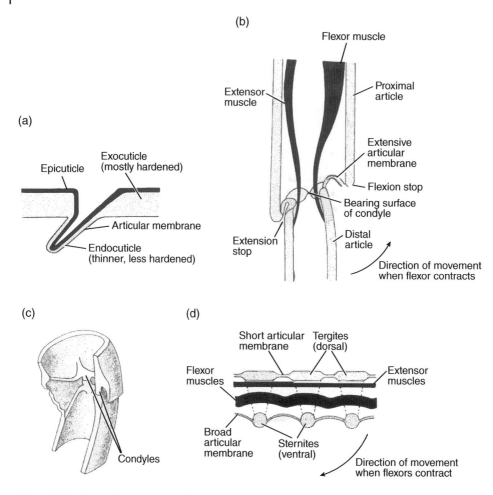

Figure 7.16 Generalized arthropod body and limb joints. (a) Body wall joint with thin articular membrane; (b) longitudinal section of a generalized limb joint showing the arrangement of antagonistic muscles, a condyle, and stops; (c) cut-away view of a simple limb joint in extended condition; (d) longitudinal section of three stylized abdominal joints of a crustacean showing the arrangement of intersegmental muscles and the articular membranes. *Source:* Brusca and Brusca, Invertebrates Second Edition 2003, figure 15.18 (p. 483). © 2003 by Sinauer Associates Inc. Reproduced with the permission of Oxford University Press through PLSclear.

same principles as all tubular excretory organs including the vertebrate kidney. The basics of tubular excretion are the same in all organisms in which they are found, which includes nearly all the taxa we will be covering in this and our remaining chapters.

All tubular excretory organs have four basic functions as illustrated in Figure 7.18.

1. Filtration – formation of an initial urine by passing the hemolymph/coelomic fluid through the pores of a filtering membrane.

2. Secretion – active secretion of waste products such as unwanted ions into the tubular lumen, very important for species like insects that lack a filtration mechanism.

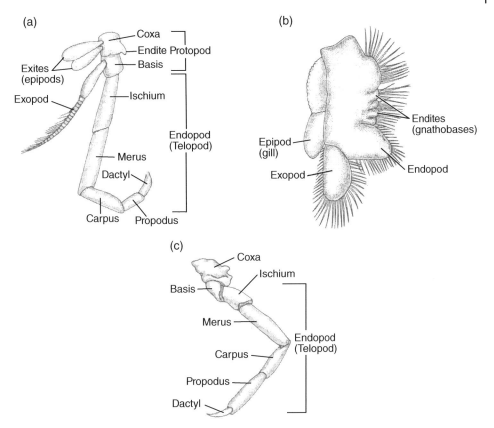

Figure 7.17 Crustacean trunk limbs. (a) Generalized biramous limb. (b) Biramous phyllopodial limb. (c) Uniramous walking leg (stenopod). *Sources:* (a) Adapted from Russell-Hunter (1979), figure 14-1 (p. 241); (b) Adapted from Brusca et al. (2016), figure 20.22 (p. 734); (c) Adapted from Brusca and Brusca (2003), figure 15-17 (p. 480).

3. Reabsorption – since the filtration process is non-selective for smaller solutes, some nutrients are actively resorbed, e.g. glucose and amino acids.

4. Osmoconcentration or dilution – the filtration process produces a urine equal in osmolarity (solute concentration) to the hemolymph/coelomic fluid. Biologically, there are two ways to concentrate it: secretion of additional ions or removal of water. Mammalian and avian species can concentrate their urine by removing water, but they are the only taxa capable of doing that. Many freshwater species, including crayfish, can dilute their urine to lose water by removing ions. However, all marine invertebrate species, including the Crustacea, have a blood osmolarity that is about equal to that of seawater, so there is no need to concentrate or dilute the urine. What differs is the ionic composition, which tends to be lower in magnesium and sulfate than the surrounding seawater (Table 7.3). Thus, while the antennal gland does not concentrate or dilute the initial filtrate in marine Crustacea, it does alter its ionic composition. Marine fishes keep their blood osmolarity about one-third to one-half that of seawater, so water loss is an important problem, and they have a sophisticated excretory system to deal with that problem, as you shall see in Chapter 10.

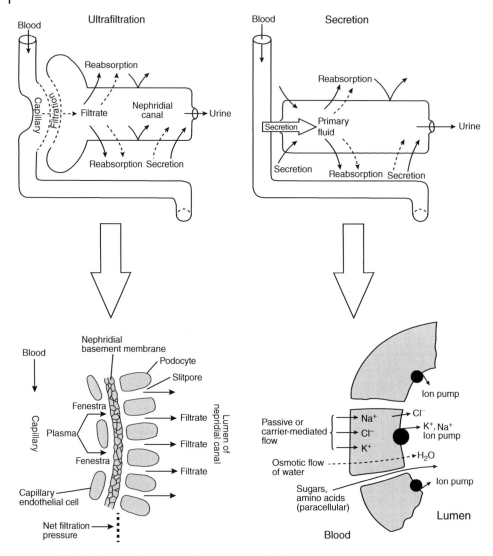

Figure 7.18 Basic principles involved in excretion: ultrafiltration, secretion, reabsorption, and osmoconcentration (or dilution). The upper figures show the basic processes of ultrafiltration and reabsorption (left) and secretion and osmoconcentration (right); the lower figures show the detailed mechanisms of these processes. *Source:* Willmer et al. (2005), figure 5.9 (p. 89). Reproduced with the permission of John Wiley & Sons.

A detailed treatment of the active and passive transport processes that underlie tubular secretion and reabsorption is beyond what we need to cover for a basic understanding of crustacean excretory processes. We will concentrate mainly on the "what happens" and less on the "how."

The mechanism that filters the blood to produce the initial urine is termed "ultrafiltration." Ultrafiltration occurs if the hydrostatic pressure of a solution confined within a semi-permeable membrane exceeds the osmotic pressure. Pure solvent will then be expressed from the solution through the

Table 7.3 Concentrations of ions in the body fluids of marine crustaceans relative to seawater. Values are in mM·l^{-1}. *Source:* Data from Prosser (1973).

Genus	Seawater (SW) reference std.	Na	K	Ca	Mg	Cl	SO$_4$
	SW for 1–3	457.0	9.8	10.1	52.2	535.0	27.5
1. *Palinurus*		545.0	10.3	13.4	16.6	557.0	20.5
2. *Palaemon*		394.0	7.7	12.6	12.6	430.0	2.6
3. *Nephrops*		517.0	8.6	16.2	10.4	527.0	18.7
	SW for 4 and 5	440.0	9.4	9.7	50.2	514.0	26.5
4. *Homarus*		472.0	10.0	15.6	6.8	470.0	—
5. *Maja*		500.0	12.7	13.9	45.2	569.0	14.3
	SW for 6	380.0	7.8	9.3	52.0	493.0	—
6. *Pandalus*		395.0	7.4	12.3	5.8	466.0	—
Composition of average SW		470.2	10.0	10.2	53.6	548.3	28.3

membrane. This is easy to envision. If you fill a porous cloth bag with grapes and squeeze the bag, the juice will be expressed and the seeds and skins will remain behind. On a smaller scale, a cell and its membrane act as a porous sac and an increase in hydrostatic pressure will force water through its pores as well as smaller solutes. Larger molecules will be retained.

The blood pressure generated by the animal's circulatory system provides the hydrostatic pressure needed for ultrafiltration, and the walls of specialized, especially porous, capillaries provide the filter. Once the filtrate enters the nephridial canal, it is modified by transport processes. Desirable small molecules such as amino acids and glucose are reabsorbed by being actively transported out of the nephridial canal. Undesirable ones, such as magnesium and sulfate that have detrimental neuromuscular effects, are actively secreted into the lumen for excretion.

The crustacean antennal gland, or "green gland," has four sections: a coelomic end sac, a labyrinth organ, a nephridial canal, and a bladder (Figure 7.19). The antennal artery directly supplies the antennal gland, joining it at the end-sac and labyrinth where ultrafiltration occurs. Secretion and reabsorption occur in the nephridial canal and the bladder. Marine crustaceans have a very low urine flow.

Extra-renal Mechanisms

The excretory organs are important in maintaining a constant internal environment, but in all, the invertebrate and lower vertebrate phyla homeostasis is much more of a team effort. Contributions from the gut, the body wall, and the gills also play an important role. In purely marine Crustacea, which are our primary focus, the excretory system is concerned with maintaining a status quo with water and ions that is only slightly different from ambient seawater. As noted above, sodium, chloride, potassium, and calcium are approximately equal to, or slightly higher in the hemolymph than in seawater, whereas magnesium and sulfate are lower.

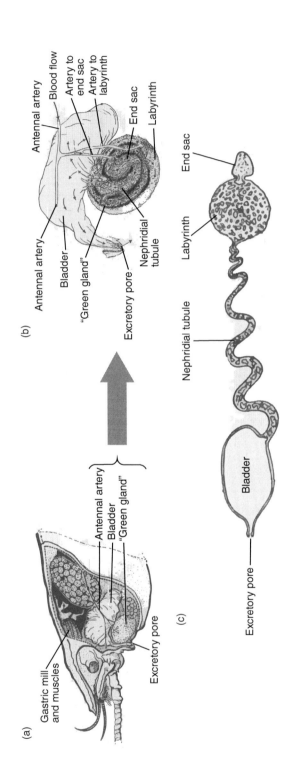

Figure 7.19 The crustacean antennal or "green" gland. (a) Cut-away of the head of the crayfish *Astacus pallipes* showing the position of the antennal gland ("green gland"); (b) enlargement of the antennal gland and renal organs of *A. pallipes*. (c) Expanded representation of the antennal gland of *A. pallipes*. *Sources:* (a, b) Potts and Parry (1964), figure II.7 (p. 69); Reproduced with permission of Pergamon Press. (c) Adapted from Parry (1960), figure 4 (p. 357).

Table 7.4 Urine/blood ratios of ions in marine crustaceans. *Sources:* Data for 1,2,3 from Prosser (1973). Data for 4 from Parry (1954).

	Genus	Na	K	Ca	Mg	Cl	SO$_4$
1)	*Homarus*	0.99	0.91	0.64	1.8	1	1.59
2)	*Nephrops*	0.98	0.83	0.81	1.3	1	1.1
3)	*Maja*	0.99	0.98	0.99	1.1	0.98	2.14
4)	*Palaemon*	0.82	0.86	0.95	6.7	1.06	3.8

How the System Works

Sites of ion and water exchange in crustaceans are the antennal (or maxillary) gland, the gills, the gut, and to a very limited extent the integument or body wall. Even at the low urine flow rates of marine crustaceans (0.2–10 ml kg^{-1} h^{-1}, Withers 1992), the urine generated will be a substantial loss of ions and water that must be replenished. To offset the loss of ions, sodium, calcium, chloride, and potassium are partly reabsorbed by the nephridial canal and bladder of the antennal gland (Table 7.4). The remaining ions and water are replenished by an inward movement at the gills and gut. In the gut, this is achieved with dietary input, or by limited drinking of the medium with an active transport of ions across the gut wall. Water then passively follows the ions into the hemolymph. The gills are a site for active transport of ions in the Crustacea and may aid in ion transport even in the nearly isosmotic marine forms. The gills are actively involved in the inward ion movement of brackish and freshwater Crustacea. While the exoskeleton of crustaceans is a formidable barrier to exchange, the body wall in marine forms is considered to be permeable to water and ions. The exchange that occurs establishes a steady state with the surrounding seawater.

Nitrogen Excretion

The breakdown of dietary proteins produces ammonium, a toxic compound that must be either quickly excreted or metabolically altered to a less toxic form such as urea, or uric acid. Most aquatic animals use ammonium as their primary excretory product, a strategy known as ammonotely. It eliminates the need for synthesizing a more complex excretory molecule such as urea and the need for metabolic expenditure for synthesis. The downside of ammonotely is that ammonium cannot be allowed to accumulate in the blood. Marine crustaceans generally have a blood ammonium concentration of 0.2–1.4 mM (Withers 1992). Ammonium is readily excreted at the gills either by passive diffusion, active transport, or ion exchange.

The Nervous System and Sensory Mechanisms

The Central Nervous System

The central nervous system of Crustacea and other arthropods is quite similar to that of annelids, with a dorsal brain and ventral nerve cord (Figure 7.11). Anteriorly, the system (Figure 7.20) consists of a brain, or supraesophogeal ganglion, circumesophogeal connectives, and a subesophogeal ganglion. The anterior region is followed by a series of segmental ganglia that run the length of the body. Posterior to the brain, the malacostracans generally have 17 pairs of ganglia (Kaestner 1970). The three controlling the mouthparts (mandible and maxillae 1 and 2) and the two controlling the first two maxillipeds comprise the subesophogeal ganglion. Six additional go

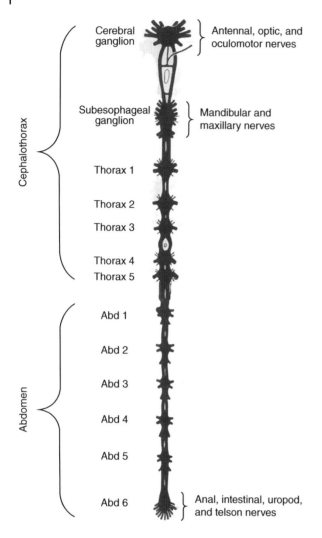

Figure 7.20 Basic structure of the decapod crustacean nervous system. *Source:* Sandeman (1982), figure 1B (p. 4). Reproduced with the permission of Elsevier.

to the remaining thoracomeres, giving a total of eight altogether to control the thoracic appendages. Six supply the limbs of the abdominal somites.

The brain itself has three sub-regions: the protocerebrum, deuterocerebrum, and tritocerebrum. The protocerebrum receives direct input from the eyestalks in the optic nerve. Olfactory and tactile sensory inputs from the antennules go to the deuterocerebrum; motor neurons controlling eye muscles and antennular movement originate there as well. Sensory inputs and motor control of the antennae reside in the tritocerebrum. The sympathetic nervous system controlling the gut, heart, and musculature is the province of the tritocerebrum, subesophogeal ganglion, and thoracic ganglia (Kaestner 1970; Sandeman 1982).

Two giant fibers originating at the junction of the circumesophogeal connectives run the length of the ventral nerve cord. The main role for giant fibers is rapid conduction; in decapod shrimp and euphausiids, they are useful in rapidly communicating the need for a tail-flip escape response. They are homologous to the giant fibers of the annelids that trigger the withdrawal response in tubicolous forms.

Sensory Modalities

Crustaceans have well-developed sensory mechanisms, more advanced than those of the polychaetes and much better described, though the crustacean literature is still not in a league with that available on the vertebrates. We will begin with photoreception, which has the best background information, including some on our target taxa. Arthropod eyes, specifically horseshoe crab eyes, were used as models in groundbreaking early vision research (Hartline et al. 1956), so there is some pride of place in that field. Mechanoreception and chemoreception are present in crustaceans and have received enough attention that the basics are fairly well described.

Photoreception

All micronektonic Crustacea residing in the photic zone (epi- and upper mesopelagic species) have well-developed compound eyes. Compound eyes, or convex eyes, have individual photoreceptor units called ommatidia that radiate from a central point. The simplest compound eyes, apposition eyes (Figure 7.21), have ommatidia that are optically isolated individual units, each with their own lens or cornea, a crystalline cone that acts as a biological light guide, a screening pigment that provides optical isolation for each ommatidium, and light-sensitive photoreceptor cells. The photoceptor cells are termed retinular cells; the light-sensitive regions of several retinular cells are usually fused together to form a photoreceptive unit or rhabdome. Apposition eyes are best suited to well-lit environments since the rhabdome of each ommatidium only receives the light captured by the area of its own lens. Nonetheless, many deeper-living species have apposition eyes adapted to lower light levels (Land 1981a; Herring 2002). The major taxa of concern to us are the gammarid and hyperiid amphipods, both of which have representatives in the mesopelagic zone. Both taxa will be treated in-depth later in this chapter, but their vision is discussed here.

Amphipods, as an order, have sessile (stationary) eyes, i.e. not on eyestalks; apposition eyes are a further characteristic of the order. To solve the problem of gathering sufficient light with an apposition eye, deeper-living hyperiid amphipods such as the very large (and very transparent) *Cystisoma* have an eye with a different morphology than that of their shallow-dwelling relatives. First, as the depth of occurrence of hyperiid species increases, so does the size of their eye, allowing individual ommatidia to become wider. Second, the upper and lower parts of the eye become increasingly different in shape, with the upper part becoming flatter, i.e. declining in curvature and more upwardly directed, and with the ommatidia in the lateral portion of the eye retaining a smaller facet size (Figure 7.22). In Cystisoma, the deepest-living species, the lateral portion of the eye is lost altogether. The combination of a wider ommatidium and a larger portion of the eye directed upward allows more efficient light gathering.

The eye of *Phronima* is unique in possessing an extended crystalline cone (Land 1981b). The interior end of the cone shrinks to a thin fiber that acts as a light

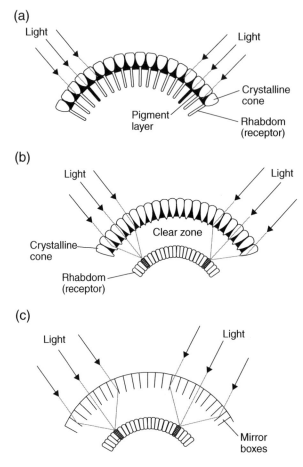

Figure 7.21 Compound eye types. (a) Apposition eye with axial light reception; (b) superposition eye with refracted light reception; (c) superposition eye with reflected light reception. *Source:* Herring (2002), figure 8.9 (p. 179). Reproduced with the permission of Oxford University Press.

guide, transferring the focused image from the cone to the rhabdom. In this way, the eye of *Phronima* can remain nearly transparent, invisible to predators, by obviating the need for screening pigment in the outer eye.

Increased light capture may also be achieved by using a different compound eye design: the superposition eye (Figure 7.21). Superposition eyes separate the light gathering region of the cornea and crystalline cone from the rhabdome with a "clear zone" (Land 1980). The clear zone allows the light from multiple cones to focus on one or more receptors, increasing the sensitivity of the eye considerably. Euphausiids and mysids, two important pelagic groups with a wide vertical range, have superposition eyes with refractive crystalline cones that direct light to the receptors. Both euphausiids and mysids have species with bi-lobed eyes that act in a similar manner to the bi-lobed eyes of the amphipods above, with the caveat that the ommatidia and the rhabdome must be concentrically arranged for the receptors and cones to function properly in both lobes.

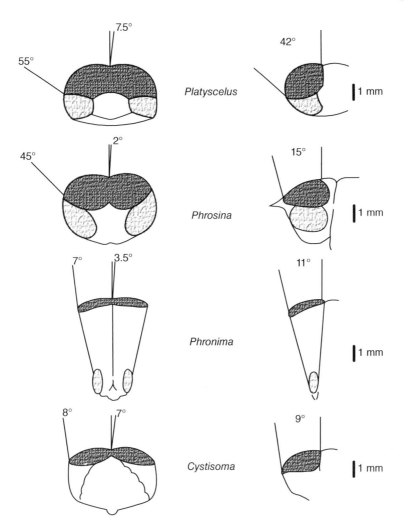

Figure 7.22 Fields of view and binocular overlap of the medial eyes of four hyperiid amphipods viewed from the front (left) and from the left side (right). The four species are arranged in order of their depth of occurrence with the shallowest, *Platyscelus*, at the top and the deepest, *Cystisoma*, at the bottom. Dark stipple represents the medial or upper eyes, light stipple represents the lateral or lower eyes. *Cystisoma* has no lateral eye. *Source:* Land (1981b), figure 6 (p. 215). Reproduced with the permission of Springer.

Most pelagic decapods also have superposition eyes, a few with refracting crystalline cones like euphausiids and mysids (mainly in the suborder Dendrobranchiata), but in most species, the design of the initial light-gathering structures differs. Instead of having crystalline cones, they have "mirror boxes" (Figure 7.21; Land 1980; Herring 2002) that use reflective surfaces composed of guanine crystals to direct light into the rhabdome. The mirror boxes are square in cross section, unlike the refractive superposition eyes which have hexagonal facets. Light that is off-axis is reflected into the rhabdom, obeying the "angle of incidence equals the angle reflection" rule that describes light reflecting off a mirror. Like refractive eyes, the rhabdom of mirror box eyes must be arranged in a concentric manner.

Light attenuates with depth such that at about 1000 m, considered the bottom of the mesopelagic zone, sunlight is effectively extinguished (see Chapter 2). Clear ocean water, like that in mid-ocean gyres or in subtropical systems such as the Gulf of Mexico, may allow light penetration to ~1200 m; in the California Borderland, a deep but more productive coastal system, light only reaches to about 800 m. As the number of available photons diminishes, eyes need to be greatly modified to capture them. In many groups, including the euphausiids, mysids, and decapods, most of the deep-living species do not invest the biological currency needed to develop and maintain eyes capable of capturing photons. Instead, eyes become smaller, less concerned with image resolution than with light detection from bioluminescent sources. Thus, the deepest-living euphausiids, mysids, and decapods have a reduced eye size compared to their shallower-dwelling relatives (Hiller-Adams and Case 1988; Herring 2002). It is interesting that while eyes do become reduced as species' depth of occurrence increases, only very rarely are they lost altogether in the deep ocean as they are in cave dwellers. This is an elegant testament to the importance of bioluminescence in the deep sea.

Mechanoreception

Mechanoreceptors respond to force or displacement (Withers 1992). External mechanoreceptors, our major concern, are sensitive to touch, to vibration (including sound), and to movement, including limb and body position. They are an important sensory modality in any environment but are particularly so in the open ocean, where effective vision is severely limited below 1000 m and chemical and mechanical senses provide a greater fraction of the environmental picture.

In animals without the armor of a crustacean exoskeleton, mechanoreceptors can be as simple as a nerve ending in the skin responding to deformation of the integument. Ion channels in the nerve ending are triggered, and it is perceived as touch. Crustaceans need hollow setae or hairs to penetrate the cuticle and act as micro-levers to communicate motion to the neuronal receptors usually located at the base of the shaft (Figure 7.23). Envision the hollow setae as acting like hair follicles in humans, in which nerve endings provide information on wind presence or absence and its speed and direction by the displacement of hair shafts on the skin.

Crustaceans have a variety of different hair receptors with different shaft morphologies, physiological characteristics, and roles (Bush and Laverack 1982). The basic hair-cell structure of shaft, basal ampulla or cup, and sensory neurons does seem to be universal. Hair sensillae are located everywhere on the crustacean's body, providing information on position and velocity of movement, initiation and cessation of movement, water flow over the crustacean's exterior, and vibration. The hair shaft itself can have a multitude of morphologies including a simple shaft, a feather, a flower bud, and a peg.

External mechanoreceptors are complemented by an internal organ of balance, or statocyst. Statocysts are present in at least some representatives of all the phyla covered thus far, and a fancier model based on the same principles is present in the vertebrates as well. The basic design is a liquid-filled, cuticle-lined sphere with a bed of hair cells that supports a statolith (Figure 7.23). The statolith is a dense body,

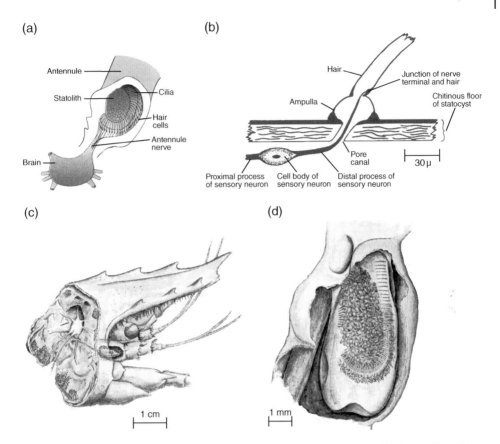

Figure 7.23 Crustacean mechanoreception. (a) Antennal statocyst showing the connection with brain and antennal nerve. (b) Lateral view of statocyst hair receptor. (c) Cut-away view of a statocyst exposed at the base of right antennule; (d) close-up view of an exposed statocyst. *Sources:*
(a) Adapted from Willmer et al. (2005), figure 9.45 (p. 269); (b) Cohen and Dijkgraaf (1961), figure 10 (p. 80); Reproduced with permission of Academic Press. (c, d) Cohen (1955); (c) figure 1 (p. 12); (d) figure 2 (p. 13). Reproduced with permission of John Wiley & Sons.

usually of calcium carbonate, that is free to move as the individual changes position. It responds to gravity as the organism moves, impinging on hair cells that in turn confer information to the brain about the orientation of the animal. Location of the paired statocysts is taxon-dependent within the malacostraca. In mysids, they are located in the uropods. In decapods, as seen in Figure 7.23c, they are found at the base of the antennae.

Early data on statocyst function in decapods resulted from an amusing combination of human ingenuity and a good working knowledge of species' anatomy and natural history (Cohen and Dijkgraaf 1961). In all crustaceans possessing one, the statocyst is lost entirely at molting and must be replaced with a newly generated organ. In some benthic species, e.g. crabs and crayfish, the new statolith is composed of sand grains obtained from the benthos that are cemented together to form the functioning statolith. Early researchers (Kreidl 1893) placed molting experimental subjects in an aquarium with only iron filings available for the new statocyst. The resulting statoliths could then be manipulated with a magnet and the static orientation of the individuals observed, firmly establishing the function of the statocyst.

Sensing flow and vibration in the near field is quite useful for detecting predators and prey in well-lit upper ocean waters as well as in the darkness of the deep sea. The antennae of crustaceans expand the reach of mechanosensors well away from the main body of the animal, forming a "sensory envelope" at the animal's anterior end that aids in detecting incoming particles and nearby swimmers. Copepods, small (1–5 mm) suspension feeders, use the mechanosensory setae on their antennae to help direct the trajectory of incoming algae cells to their mouthparts (Koehl and Strickler 1981). Predatory copepods can also track prey by following the hydrodynamic signature left behind by the prey's swimming motion (Boxshall 1998).

Penaeid shrimp, an important deep-sea group in the Dendrobranchiata, have very long trailing antennae (Figure 7.24) that can detect vibration and flow plus have chemosensing ability (Denton and Gray 1986; Herring 2002), acting in a similar way to the lateral line of fishes. When you observe swimming penaeids, they give the illusion of riding on rails. The antennae are held downward and well away from the body at about a 45° angle and are about double the length of the animal. The length of their sensory field allows them to detect changes in the magnitude of a source vibration over the length of the antennae. Vibrational signals attenuate rapidly, giving the shrimp the ability to localize the point source by virtue of their antennal arrangement, away from the turbulence of their own beating pleopods but long enough to detect change in the signal.

Chemoreception

Chemoreceptors detect chemical compounds in the surrounding environment, providing our senses of taste and smell. In terrestrial systems, the distinction between taste and smell is easily recognized: taste is a contact chemoreceptor, smell a distance receptor. In the marine system, since all water-borne chemicals must impinge on "contact chemoreceptors," taste and smell form a continuum. High concentrations originating near the receptor might be thought of as taste, or gustation. Lower concentrations detected from afar, as smell or olfaction. In most cases though, the receptors themselves are similar but may display differences in sensitivity. Thus, chemoreceptors in the mouthparts of decapods

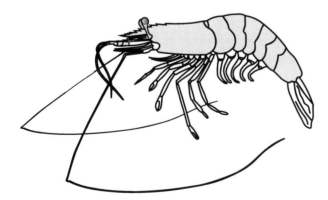

Figure 7.24 Deep-sea penaeid shrimp showing very long trailing antennae.

exhibit a lower sensitivity to chemical stimulation than antennal receptors (Schmitt and Ache 1979). Like mechanoreceptors, chemoreceptors play a larger role in an individual's total sensory picture in the darkness of the deep sea.

Chemoreceptors are useful in locating prey by allowing animals to track a scent trail to its source, but they can also play important roles in mate location, predator avoidance, and mate recognition (Zimmer-Faust 1989; Zimmer and Butman 2000; Herring 2002). In crustaceans, chemoreceptors are located primarily on the anterior limbs (Hindley 1975; Ache 1982), with some sensitivity over the entire body (Figure 7.25a). It should come as no surprise that chemoreceptors take the form of sensory hairs, or aesthetascs, which like mechanoreceptors are basically setae innervated with bipolar neurons (Figure 7.25b). The difference between a chemosensory aesthetasc and setal mechanoreceptor is that the sensory dendrite in a chemoreceptor extends to the end of the setal lumen, which usually ends in a pore. In addition, the hair itself has a "spongy" wall that promotes the inward diffusion of chemical stimuli and the number of dendrites innervating each aesthetasc is very much greater, sometimes as great as 100 (Ache 1982). Mechanoreceptors and chemoreceptors are often located in close proximity to one another, combining perception of flow and scent, which is ideal for homing in on distant prey or localizing a predator (Hamner and Hamner 1977).

Much of the research on crustacean chemoreceptors has been concerned with their sensitivity or threshold, i.e. how low a concentration can they detect? To elicit feeding behavior, stimuli have included clam extracts, quantified using grams of dry tissue per liter, and equimolar amino acid mixtures (Ache 1982). Electrophysiological techniques, recording impulses from antennal nerves or thoracic appendages, have also been used. Both techniques suggest strongly that Crustacea are highly tuned into their chemical environment, with thresholds in the nanomolar to picomolar range (10^{-9}–10^{-12} M) (Figure 7.26) in both shallow subtidal (Cancer) and pelagic species (*Gnathophausia ingens* and *Pleuroncodes planipes*).

Circulatory and Respiratory Systems

All Crustacea have an open circulatory system with a dorsal heart that is either tubular or globular, depending upon the taxon. Most have a copper-based blood pigment, hemocyanin, to facilitate oxygen diffusion inward, to allow more oxygen to be transported by the hemolymph, and to facilitate delivery of oxygen to the tissues. The heart, which is found in the cephalothorax in most of our target taxa (e.g. Figure 7.27a), resides in a pericardial sinus (a pericardium). The heart is anchored within the pericardium by suspensory ligaments, and blood enters from the pericardial sinus through paired one-way valves, called ostia. The heart contracts as a unit, propelling the hemolymph into the arteries and tissue sinuses.

Arterial flow through the large anterior and posterior aortas and the descending, or sternal, artery (terminology differs between authors) directly supplies the limbs, musculature, and organs of the pelagic Crustacea. An accessory heart, the cor frontale, is present on the anterior aorta in many species to aid flow to the head region. Return flow is achieved via a series of sinuses in the abdomen and cephalothorax that channel flow back to the heart. The most important sinus for return flow, and for supplying the gills, is the infrabranchial sinus located in the ventral cephalothorax above the pereiopods (Figure 7.27b). The

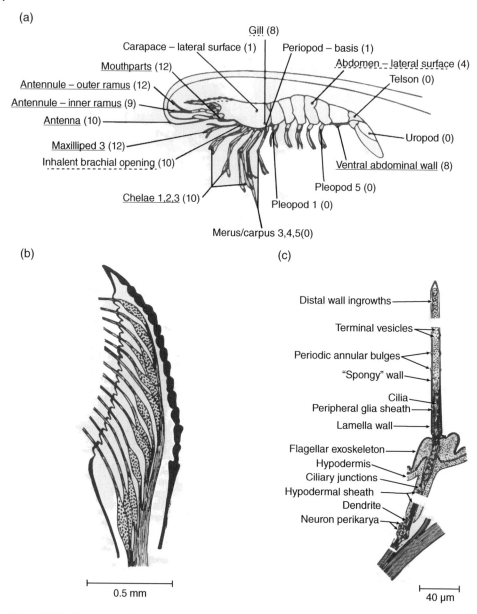

Figure 7.25 Crustacean chemoreception. (a) Distribution of chemosensitivity to meat extract in *Penaeus merguiensis*. Numbers in parentheses represent number of positive responses from 12 trials. Solid underline indicates strong responses were elicited; dashed underlines indicate weak responses. (b) Outer flagellum of *Paragrapsus gaimardii* showing sensory neurons innervating aesthetasc (chemosensory) hairs; (c) innervation of a single aesthetasc hair in *Paragrapsus gaimardii*. *Sources:* (a) Hindley (1975), figure 4 (p. 204); Reproduced with permission of Taylor & Francis. (b, c) Adapted from Ache (1982), figure 2 (p. 373).

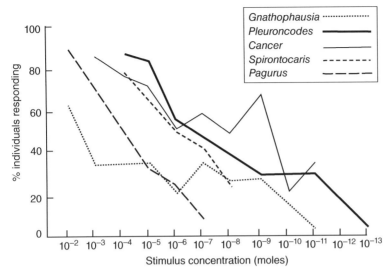

Figure 7.26 Dose response to an equimolar mixture of L-glutamic acid, taurine, and γ-aminobutyric acid. Response was judged by its ability to elicit feeding behavior in the four species of decapods and one lophogastrid tested in the study. *Gnathophausia* and *Pleuroncodes* are open-ocean species, *Cancer* and *Spirontocaris* coastal subtidal and *Pagurus* a hermit crab living in the intertidal or shallow subtidal. *Source:* Adapted from Fuzessery and Childress (1975), table l and figure 3.

infrabranchial sinus is the main collecting point for blood that has bathed the tissues and is returning to the pericardium. The infrabranchial sinus has branches to each of the gills. Branchio-pericardial vessels carry the blood from the bases of the pereiopods, where the gills are located, directly to the pericardial chamber. The oxygenated blood is then pumped once again to the tissues (Belman and Childress 1976).

The trick to understanding flow in an open circulatory system is to remember that water and blood are essentially incompressible at physiological pressures and that no gas-filled spaces exist within the hemocoel. Thus, pressure applied at any point in the blood-filled space of the hemocoel is immediately conveyed throughout. Connectivity and flow within the system are achieved by dint of the fact that the pressure generated by the heart is communicated throughout the hemocoel, assuring continuous flow to expedite return flow despite the presence of sinuses instead of the vessels of a closed circulatory system. Resistance by the tissues themselves does lessen the pressure generated by the heart's contraction, but blood pressure measurements at various internal sites show that sufficient pressure exists even within the infrabranchial sinus to easily propel blood through the gills (McMahon and Wilkens 1983).

The gills of the Eucarida, which include most of our target taxa, arise from the base of the thoracic appendages and, in many decapods, from the wall of the branchial chamber as well. In the euphausiids, the gills branch off the coxa, the initial (most proximal) segment of the pereiopods, and are carried externally (Figure 7.28). In the decapods, the outside wall of the carapace (the branchiostegite) folds over the gills to form a true branchial chamber. The chamber communicates to the outside with

(a)

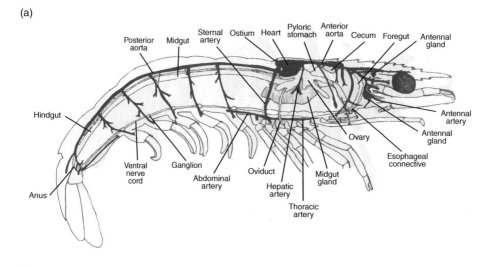

(b)

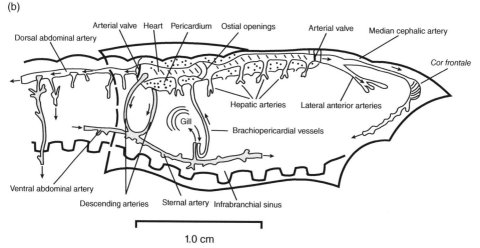

Figure 7.27 Crustacean circulatory systems. (a) Major organ systems of a generalized decapod. (b) Directional blood flow patterns (arrows) in the lophogastrid *Gnathophausia ingens*. Sources: (a) McLaughlin (1983), figure 14 (p. 16); Reproduced with permission of Academic Press. (b) Adapted from Belman and Childress (1976), figure 1 (p. 18).

slit-like openings between the pereiopods and the inner and outer walls of the carapace on the ventral side. Water enters at the base of the pereiopods and is directed anteriorly by a specialized appendage at the anterior end of the gill chamber known as the scaphognathite, or gill bailer, actually the exopod of the second maxilla. This general flow pattern in the branchial chamber holds true for all the pelagic decapods and the lophogastrids (Figure 7.29). Flow past the gills in euphausiids is generated during forward swimming.

The anatomy and number of gills vary between taxa. In fact, gill anatomy is used as a taxonomic character in the decapods. Euphausiids have gills on thoracopods two through eight, originating from a single axis and branching into several branchial filaments. The decapods show a basic pattern

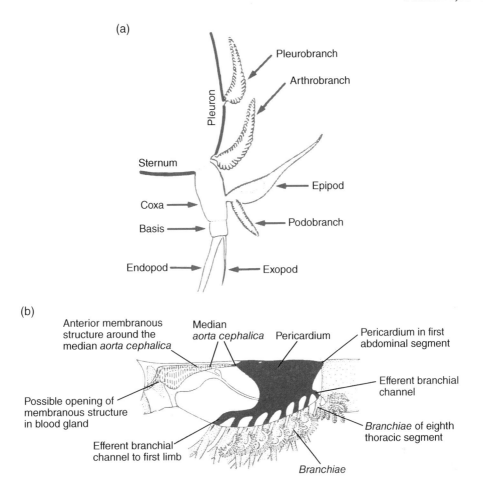

Figure 7.28 Eucarid gills. (a) Positional gill types in decapod crustaceans. The right side of the lateral and ventral body wall (pleuron and sternum) of the cephalothorax is diagrammed, along with the proximal part of a thoracic appendage, to show the attachment points of pleurobranch, arthrobranch, and podobranch gills. (b) Thorax of the euphausiid *Meganyctiphanes norvegica* showing the circulatory anatomy of the gill. *Sources:* (a) Adapted from Bauer (2004), figure 2.12 (p. 32); (b) Mauchline and Fisher (1969), figure 98 (p. 261). Reproduced with permission of Academic Press.

of four gills per appendage with one pleurobranch attached to the internal body wall or pleuron, two arthrobranchs attached to the junction between the body wall and the coxa, and one podobranch originating at the coxa itself (Figure 7.28; the branchiostegite [body wall] that forms the outer wall of the gill chamber is not shown in the figure). Most of the decapods have fewer gills than the full basic complement of 32 gills (4 each on the 3 maxillipeds and 4 each on the 5 pereiopods). Numbers range from a complement of 20 gills in *Vetericaris* (Procarididae) to 4 gills in *Limnocaridina* (Atyidae) (Bauer 2004).

The decapods have three types of gills: dendrobranchiate, phyllobranchiate, and trichobranchiate (Figure 7.30). Two types, phyllobranchiate and dendrobranchiate, are found in pelagic decapods. The

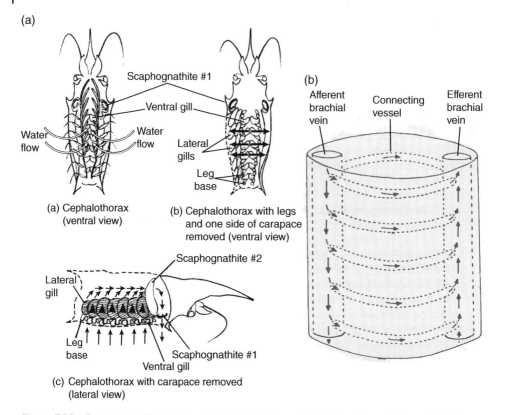

Figure 7.29 Crustacean gills. (a) a,b,c, on left side of figure- Circulation through the cephalothorax and gills of the mysid *Gnathophausia ingens*; (a) water enters the ventral gills (white arrows), (b) water exits the ventral gills and passes through the lateral gills (black arrows), (c) lateral view of the cephalothorax showing water entering the ventral gills and, after passing out of the lateral gills (heavy arrows) being direct anteriorly and out of the body by the scaphognathites. (b) right side of figure - Directional blood flow (arrows) through a part of a gill filament of a euphausiid. *Sources:* (a) Adapted from Childress (1971), figure 3 (p. 113); (b) Mauchline and Fisher (1969), figure 99 (p. 263). Reproduced with permission of Academic Press.

Dendrobranchiata, which include the penaeid and sergestid shrimps, are so-named for their gill structure: dendrobranchiate (branching filaments). The caridean shrimps have phyllobranchiate gills (leaf-like filaments) with flattened lamellae. The third type of gill structure, trichobranchiate (hair-like filaments), is restricted to benthic forms such as the crayfish. Main afferent (flow towards the gill) and efferent (flow away from the gill) vessels are present in each type of gill, and separation of oxygenated from deoxygenated blood as well as countercurrent flow has been demonstrated experimentally in phyllobranchiate gills (Hughes et al. 1969). Flow within the filaments of dendrobranchiate gills has not been described, but efficiency dictates that segregation of deoxygenated from oxygenated blood in the afferent and efferent vessels of the gills is a virtual certainty for these highly active species.

Members of the Peracarida, the other crustacean superorder with many pelagic representatives, are more diverse in the morphology of their respiratory systems and will be treated in the sections devoted

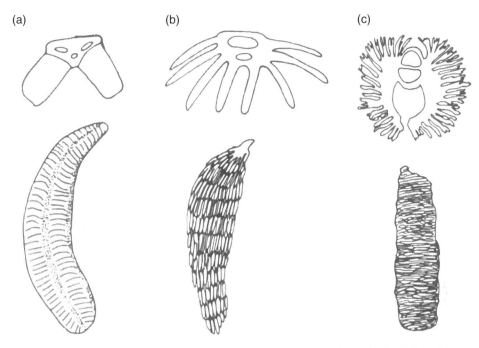

Figure 7.30 Decapod gill types. (a) Phyllobranchiate; (b) trichobranchiate; (c) dendrobranchiate. *Source:* Adapted from Calman (1909), figure 105 (p. 278).

to their biology. The lophogastrids (order Lophogastridae), the most shrimp-like of the peracarids, have been mentioned above and will be treated in more detail later in this chapter.

Circulation and Oxygen Transport in the Blood: Hemocyanin

The primary oxygen-carrier molecule of the pelagic Crustacea is hemocyanin, a copper-containing blood pigment that binds reversibly with oxygen to increase the total solubility of oxygen in the blood. It performs the same function that the more familiar hemoglobin does in the vertebrates and in the many invertebrates where it is found. Hemoglobin is present in some of the lower Crustacea, e.g. the branchiopods and copepods, but it is not present in any of the higher Crustacea (Eucarida or Peracarida) that contain our target taxa.

Hemocyanin gives a modest but important boost to the oxygen-carrying capacity of the blood in marine Crustacea, providing an oxygen content that is usually two to three times that of seawater at the same temperature. In some other invertebrates, such as the molluscs and annelids, the oxygen-carrier molecule affords a jump of 5–10 times that of seawater (Mangum 1983). Oxygen carriers are particularly important to pelagic species that dwell part- or full-time in oxygen-minimum layers where oxygen concentrations can be vanishingly small, 5% of that in surface waters or less (Sanders 1990; Sanders and Childress 1990). Those hemocyanins can bind with oxygen at very low concentrations (2–3% of surface waters), an important part of how aerobic life is possible at naturally occurring very low oxygen concentrations.

Digestive System

Malacostracan crustaceans have a straight gut that consists of three sections, a foregut that includes an esophagus and stomach, a midgut, and a short hindgut (Figure 7.27a). The foregut and hindgut are lined with chitin and have an ectodermal origin, whereas the unlined midgut comes from endoderm. Intimately associated with the stomach and anterior midgut is the digestive gland, known also as the midgut gland or hepatopancreas. Most of the action in the digestive tract occurs in the stomach, anterior midgut, and hepatopancreas.

The stomach has three main tasks: grinding the ingested food, separating small from large particles, and directing those particles to their final destination. Grinding is the province of the cardiac stomach, the heavily-muscled anterior portion of the stomach containing "teeth" to aid in breaking up the food into smaller particles. The cardiac stomach connects directly to the pyloric stomach (Figure 7.31) via a valved constriction that keeps out the largest particles like fish bones and pieces of carapace but allows through the smaller, successfully triturated, particles. Setal filters in the pyloric stomach further separate the particles. Smaller particles are channeled into the midgut gland where they are further broken down enzymatically and absorbed. Coarser particles are directed into the midgut where they are also chemically reduced and absorbed. Material refractory to digestion is eliminated at the anus. The largest particles that do not make it into the pyloric stomach are assumed to exit via the mouth.

Digestive enzymes are synthesized in the hepatopancreas and introduced into the pyloric region of the stomach via its connection to the midgut, eventually reaching the cardiac stomach through the cardiopyloric valve that guards the junction between the cardiac and pyloric stomachs. The

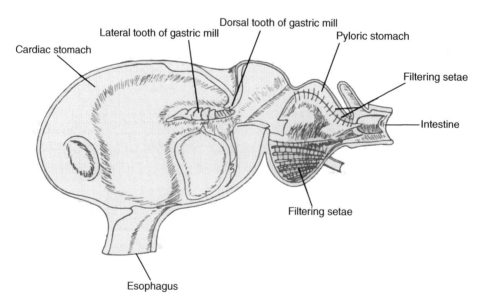

Figure 7.31 Anatomy of the stomach of the crayfish, *Astacus* (lateral view). *Source:* Barnes (1974), figure 14-64 (p. 592). Reproduced with the permission of Elsevier.

hepatopancreas and midgut both have a nutrient absorptive function, but digestive enzymes are only synthesized in the hepatopancreas.

Basic Development

Crustaceans exhibit two basic types of development: anamorphic and epimorphic (Kaestner 1970). Epimorphic development, also known as direct development, is the complete development of the larva within the egg to a full complement of metameres. Direct development is found in the Branchiopoda (most Cladocera), Anaspidacea, all Leptostraca, some Decapoda, and all Peracarida (Kaestner 1970).

In anamorphic development, larvae are free-living. In most instances, anamorphic larvae hatch with three somites as nauplii and then molt repeatedly, adding one or more segments (including appendages) with each molt until reaching the final number typical of the species. Like the polychaetes, additional segments are added by teloblastic budding, which is considered to be a shared primitive character (Brusca and Brusca 2003). Anamorphic development is observed in most of the Crustacea including most or all families in the Branchiopoda, Copepoda, Ostracoda, Cirripedia, Euphausiacea, and Decapoda (Kaestner 1970). Specific larval characters, number of molts, and names of larval stages differ between taxa, but all share the basic teloblastic budding pattern for addition of segments and limbs and progress toward the adult number of metameres with each molt. In some cases, particularly among the decapods and hoplocarids, newly added segments, usually abdominal ones, may develop more quickly than the older thoracic somites. Such "out of sequence" development is sometimes termed metamorphic development (Waterman and Chace 1960; Brusca and Brusca 2003) or irregular anamorphic development (Kaestner 1970).

In species with anamorphic development, it may take as many as ten molts through a series of distinct stages to reach sexual maturity after hatching (Waterman and Chace 1960). Four to five basic stages are commonly recognized, often with different names in different taxa but equivalent progress in development. The nauplius, metanauplius, protozoea, zoea, and postlarva are recognized in the Eucarida, namely in the euphausiids and Dendrobranchiata. In the carideans, larvae hatch at a more advanced stage, either protozoea or zoea. In euphausiids, the protozoeal stage is termed a calyptopis and the zoeal stage is a furcilia. In penaeids, the zoea is termed a mysis stage. A developmental sequence for a penaeid is shown in Figure 7.32. Important behaviors associated with the larvae of each group will be addressed in their respective sections in this chapter.

The Micronektonic Crustacea

A nektonic species is a swimmer capable of purposeful, extended horizontal movement, a very broad definition that includes everything from small crustaceans, squids, and fishes to whales. Our operational definition for micronekton includes all swimming species in the range of 1–25 cm in length. Practically speaking, it includes the range of sizes you will catch in a standard scientific midwater trawl with a mouth area of 4–10 m^2, a mesh size of 3–6 mm, and a tow speed of 2–3 knots (1.0– 1.5 m s^{-1}). Strong swimmers like tuna, mackerel, and most adult squid are not captured by a scientific midwater trawl; they can avoid it too easily.

Micronektonic crustaceans, like their piscine and molluscan counterparts, are

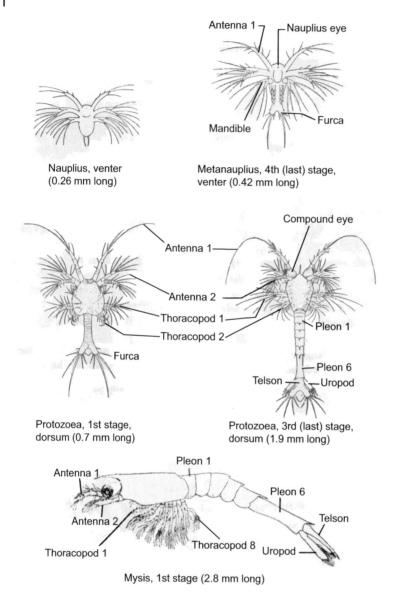

Figure 7.32 Developmental sequence of the penaeid shrimp, *Trachypenaeus constrictus*, showing stages from early nauplius to mysis, first stage. *Source:* Pearson (1939).

considered to be intermediate trophic-level species. Their prey most often consists of primary and secondary consumers. Those species that feed on algae directly (such as copepods) are primary consumers. Those one step up in the trophic pyramid, feeding mainly on algae eaters, are considered secondary consumers. Conceptually, trophic levels are quite discrete, an easily understood way to track the flow of matter and energy in an ecosystem. In practice, most species occupy more than one level as they grow and develop. Additionally, trophic levels for adults in a species often change with variations in food availability.

The main taxa comprising the micronektonic Crustacea are the euphausiids, decapods, and amphipods. Cameo appearances are made by a few other taxa, notably the isopods, ostracods, anomurans, and the cirripedes. The percentage representation by each group changes with latitude, as does the number of species and their total biomass. Decapods are not well represented in polar waters, for example. The primary taxa in open-ocean systems will be addressed first, followed by a brief treatment of the cameo players.

The Pelagic Eucarida

Order Euphausiacea

Introduction

About 85 species of euphausiids are recognized, ranging in adult length from less than a centimeter in some species of *Euphausia* and *Stylocheiron* to greater than 10 cm in *Thysanopoda cornuta* and inhabiting depths from near-surface waters to below 2000 m. Euphausiids are not only found throughout the water column, they are ubiquitous in the world ocean from the tropics to the poles. A few species form large aggregations, two of the most famous being *Meganyctiphanes norvegica* in the northern Atlantic and *Euphausia superba*, the famous Antarctic krill, in the Southern Ocean. Because of their swarming behavior and their importance in the diet of whales, seafarers have been aware of euphausiids for centuries.

Many euphausiid species perform diel vertical migrations, spending their days at mesopelagic depths (200–600 m) and migrating to near surface waters at night. When they are present in sufficient densities, they appear in ships' echograms as the upper portion of a "deep-scattering layer," a community of euphausiids, decapod shrimp, and small fishes that perform a diel vertical migration each day. We will address the deep-scattering layer and vertical migration in greater depth a little further in this chapter.

Euphausiids have piqued the interests of scientists and fishermen alike. Their abundance and general visibility have made them one of the best studied of the open-ocean groups, with several books (Mauchline and Fisher 1969; Mauchline 1980; Tarling 2010), monographs (Boden et al. 1955; Brinton 1962; Lomakina 1978), and journal articles devoted to their biology. The dedicated cadre of champions devoted to euphausiid biology has provided us a solid and innovative background on them.

History

In 1830, the French zoologist Milne-Edwards was the first to describe a euphausiid. He named the genus *Thysanopoda*, species *tricuspide*, now known as *tricuspidata*. He noted the similarity of the biramous swimming legs in the euphausiids to those of the mysids, a similarity which eventually led to the two groups being unified as the Schizopoda. Boas in 1883 was the first to note the differences between the mysids and the euphausiids, recognizing that the euphausiids had more in common with the decapods than the mysids.

Most early work on the group occurred in the North Atlantic and included much confusion on the identity of the dominant species there, *Meganyctiphanes norvegica*. Through the 1800s and as late as the 1930s, the euphausiids were alternately classed as an order by themselves or with the mysids as the order Schizopoda. In 1904, the great carcinologist Calman echoed the view of Boas, recognizing that the mysids and euphausiids were not closely related but rather that the euphausiids were most similar to the penaeid (Dendrobranchiata) decapods. His view eventually prevailed and resulted in the classification scheme we use today, with the euphausiids together with the decapods and amphionidaceans forming the Eucarida.

Classification
Species numbers below are from Mauchline (1980). The euphausiids are divided into two families and eleven genera.

Family Bentheuphausiidae
Monospecific – *Bentheuphausia amblyops*, a deep-living euphausiid with no photophores and reduced eyes

Family Euphausiidae

- Genus *Thysanopoda* (13 spp.)
- Genus *Nyctiphanes* (4 spp.)
- Genus *Pseudeuphausia* (2 spp.)
- Genus *Euphausia* (32 spp.)
- Genus *Thysanoessa* (10 spp.)
- Genus *Nematoscelis* (7 spp.)
- Genus *Nematobrachion* (3 spp.)
- Genus *Stylocheiron* (11 spp.)
- Genus *Meganyctiphanes* (1 spp.)
- Genus *Tessarabrachion* (1 spp.)

Bentheuphausia differs from the remaining genera in the euphausiidae by not only the absence of photophores but also in its deep-red color. Some important taxa are shown in Figure 7.33, illustrating variations in size and general morphology. Note the differences in eye shape between *Thysanoessa macrura* and *Stylocheiron suhmi* and the grasping thoracopod in the *Stylocheiron*.

Euphausiid Anatomy
Euphausiids are shrimp-like in appearance, being laterally compressed and having an abdomen two to three times the length of the cephalothorax (Figure 7.34). Like the decapods, they have a carapace that is fused to the thoracic segments (thoracomeres). Unlike the decapods, the carapace does not extend far enough laterally to form a branchial chamber for the gills, which are exposed. Euphausiids differ from the mysids by the fusion of the carapace to the thoracic segments, or thoracomeres, a eucaridan characteristic.

Euphausiids have eight thoracic limbs, none of which are differentiated into maxillipeds. A typical thoracic limb from a suspension-feeding euphausiid, *Euphausia superba*, is shown in Figure 7.35a. Setae are present on the medial (inner) side of the limb to aid in trapping food particles. The tubular gill branches directly from the coxa as an epipodite on the outer side of the limb. The seventh and eighth pereopods are reduced in size, although they bear the most highly branched gills (Figure 7.34). As noted above, a few genera (e.g. *Nematoscelis, Stylocheiron, Thysanoessa*) have representatives with greatly elongated second or third thoracopods.

Five pairs of biramous pleopods (Figure 7.35b) provide the thrust for swimming. The two rami are fringed with setae that greatly increase their surface area during the power stroke of the paddling limb. In males, the first pleopods are each modified into a copulatory structure known as a petasma, which is important to euphausiid identification. Posterior to the pleopods is a well-developed telson (Figure 7.35c) with uropods, forming an effective tail fan.

Nervous System
Available information suggests a fairly generic crustacean nervous system with long circumesophageal connectives joining the cerebral ganglion to the ventral nerve cord. Ten thoracic ganglia are present, which, depending on the genus, may be separate, partially (*Meganyctiphanes*) or

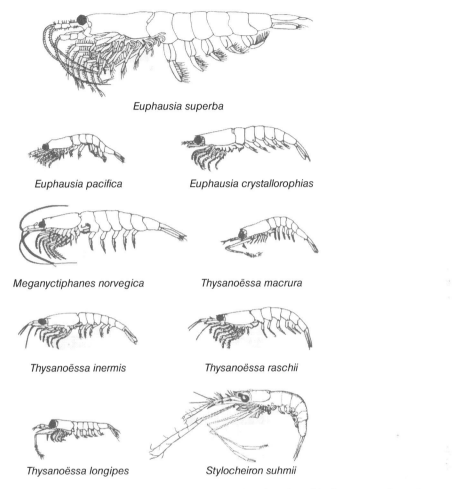

Figure 7.33 Illustrations of nine important species of euphausiids drawn to scale to show differences in size and general morphology, including presence or absence of grasping thoracopods and bi- or single-lobed eyes. *Sources:* Tattersall (1908), plates I–IV for all except *Stylocheiron*. *Stylocheiron suhmii* from Kaestner (1970), figure 13-2 (p. 268).

completely (*Stylocheiron*) fused. Six abdominal ganglia remain separate and well defined (see Figure 7.20). Euphausiids have an X-organ and a sinus gland.

Sensory Mechanisms

Euphausiids have a full complement of sensory hairs and aesthetascs, located mainly on the antennae and mouthparts, which provide information on their mechanical and chemical environment. Unlike the mysids, they do not have a statocyst or organ of balance.

Euphausiids are particularly attuned to light and, with the exception of some deeper-living species such as *Bentheuphausia*, have well-developed, image-forming compound eyes. The eyes of euphausiids are not only important in prey capture and predator avoidance but their vertical distributions are also keyed to particular light levels. The euphausiids all have superposition eyes, but

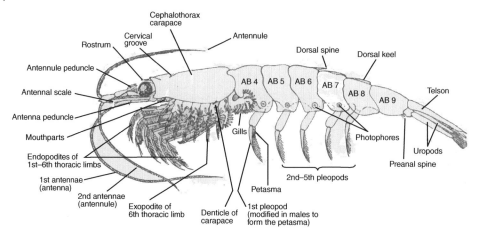

Figure 7.34 External morphology of an euphausiid. *Source:* Mauchline and Fisher (1969), figure 1 (p. 6). Reproduced with the permission of Elsevier.

a fascinating characteristic of the group is that among the different species, eyes have many different shapes: spherical, ovoid, and a variety of bi-lobed eyes. Species that are primarily suspension feeders, like those in the genus Euphausia or Meganyctiphanes, have spherical eyes (Figure 7.33). Predatory species with strongly produced second or third thoracopods, like those in the genera Stylocheiron, Nematoscelis, or Thysanoessa, often have bi-lobed eyes (Figure 7.36). The upper lobe also faces upward, providing a limited field-of-view with high resolution to aid in prey-picking. The lower spherical lobe provides all-around vision with lower resolution (Land 1981a).

Spectral sensitivity of euphausiid eyes (Figure 7.37) peaks in the 460–500 nm wavelength range, similar to that of the ambient light at their daytime depths (470–500 nm; Mauchline and Fisher 1969). Interestingly, many euphausiids also reside at a constant level of irradiance or isolume, moving nearer the surface at nightfall or during a solar eclipse to remain within their isolume as it moves upward, moving down again at sunrise. This behavior, known as diel vertical migration, is widespread in the world ocean, and euphausiids are an important example of the many taxa that exhibit it.

Bioluminescence and Photophores

Light organs, or photophores, are found in all euphausiid species except *Bentheuphausia*. Most species have ten altogether, with one located in each eyestalk (e.g. Figure 7.37), one pair each at the base of the second and seventh thoracic limbs, and one between each of the first four pairs of pleopods on the ventral surface of the abdomen (Mauchline and Fisher 1969, Figure 7.34).

The spectral characteristics of euphausiid bioluminescence are similar to the spectral quality of downwelling light at their daytime depths and to the spectral sensitivity of their eyes. All peak in the blue–green range, with wavelengths at about 480 nm (Figure 7.37). The fact that photophores are directed downward, with an associated musculature that orients them properly even when the euphausiid changes its

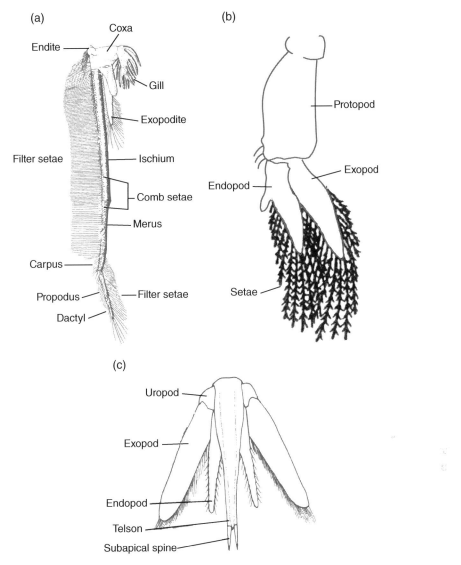

Figure 7.35 Euphausiid appendages. (a) Thoracic limb of *Euphausia superba*. (b) Pleopod of *Bentheuphausia amblyops*. (c) Telson and uropods of a male *Thysanopoda*. *Sources:* (a) Kaestner (1970), figure 13-3 (p. 272); Reproduced with permission of John Wiley & Sons. (b) Adapted from Schram (1986), figure 18-2 (p. 225); (c) Adapted from McLaughlin (1980), figure 39 (p. 119).

attitude relative to the horizontal, strongly suggests that one role for the luminescent organs is to countershade or break up the individual's silhouette when viewed by a predator from below. Euphausiids' photophores allow them to better blend in with downwelling light from the surface, providing a measure of camouflage.

Light organs are basically similar between species and between regions of the body, except that photophores in the eyestalk lack a lens. Each light organ consists of a

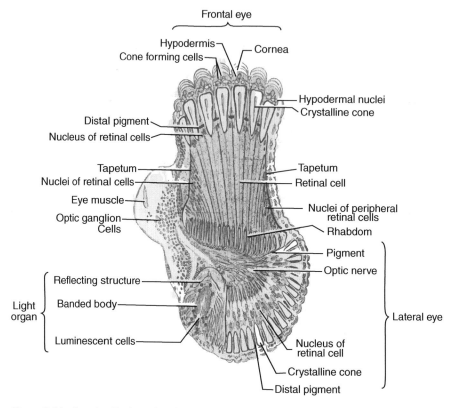

Figure 7.36 Longitudinal section through the bi-lobed eye of *Stylocheiron longicorne*. Source: Mauchline and Fisher (1969), figure 80 (p. 233). Reproduced with the permission of Academic Press.

reflective pigment layer at its base, a layer of light-producing cells, a striated body (also called a rod mass), and a lens to collimate the light and direct it outward (Figure 7.38). The luminescent glow of the photophores is localized in the striated body. It is believed that there are two types of light-producing cells, each manufacturing one of the two necessary biochemicals for the reaction that produces light. Light is generated when the two types of cells secrete their biochemicals into the highly laminated, striated body where they mix, the reaction takes place, and light results. The two biochemicals are luciferin and luciferase, a substrate and enzyme combination found in all taxa that produce biological light.

Circulatory, Respiratory, and Excretory Systems
The euphausiids have a compact globular heart, with two pair of ostia, located in the posterior third of the cephalothorax (see Figure 7.27). Euphausiids exhibit a typical crustacean circulatory pattern with an arterial flow bathing the tissues and collecting in the infrabranchial sinus, where it is oxygenated at the gills and returned to the heart, deoxygenated blood from the infrabranchial sinus and oxygenated blood from the gills are kept separate by the vascular system within the branchiae (see Figures 7.28 and 7.29). Euphausiids have a cor frontale, an accessory heart.

The excretory system of euphausiids comprises paired antennal glands, one at

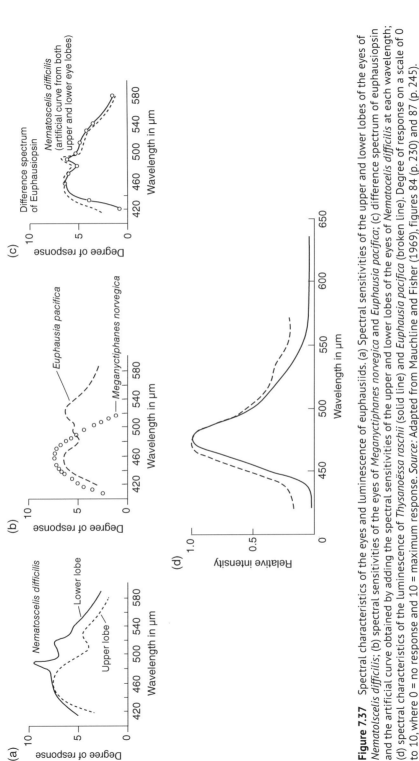

Figure 7.37 Spectral characteristics of the eyes and luminescence of euphausiids. (a) Spectral sensitivities of the upper and lower lobes of the eyes of *Nematolscelis difficilis*; (b) spectral sensitivities of the eyes of *Meganyctiphanes norvegica* and *Euphausia pacifica*; (c) difference spectrum of euphausiopsin and the artificial curve obtained by adding the spectral sensitivities of the upper and lower lobes of the eyes of *Nematocelis difficilis* at each wavelength; (d) spectral characteristics of the luminescence of *Thysanoëssa raschii* (solid line) and *Euphausia pacifica* (broken line). Degree of response on a scale of 0 to 10, where 0 = no response and 10 = maximum response. *Source*: Adapted from Mauchline and Fisher (1969), figures 84 (p. 230) and 87 (p. 245).

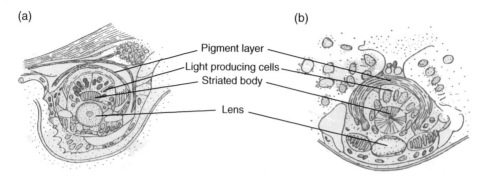

Figure 7.38 Abdominal photophores of (a) *Euphausia* sp. and (b) *Stylocheiron suhmii*. Source: Adapted from Mauchline and Fisher (1969), figure 86 (p. 242). Reproduced with the permission of Academic Press.

the base of each antenna and supplied by the antennal arteries. Anatomy and function of the euphausiid antennal glands do not deviate significantly from the general crustacean model described earlier in this chapter (see Figure 7.19).

Digestive System

The gut of euphausiids has a typical malacostracan design (Figure 7.27) and function, with most of the digestive activity occurring in the midgut region. An anterior cardiac stomach armored with chitin and setae to aid in breaking up food particles is followed by a pyloric stomach that communicates with the midgut gland, or hepatopancreas, where most of the digestion and absorption take place. Some differences in stomach width and armature have been observed between species (Nemoto 1967). Primarily carnivorous species, such as *Bentheuphausia*, exhibit lighter stomach armor than herbivorous species such as *Euphausia superba*, presumably because their prey items are more readily digested. The chitin-lined foregut and hindgut are primarily conductive.

Development

Euphausiids have a long developmental sequence with four stages prior to the post-larval period: nauplius, metanauplius, calyptopis, and furcilia. In older literature, a cyrtopia stage was the final larval stage but that phase of development is now considered to be part of the furcilia sequence. Euphausiids are the only crustaceans with both calyptopis and furcilia larvae. The names originated with Claus (1863), who did not recognize the larvae as euphausiids and described the larval forms as new species. A number of years later Sars (1885) who recognized them as euphausiids, simply kept the species' names of Claus for the appropriate developmental stages (Mauchline and Fisher 1969).

The nauplius is the first stage to emerge from the egg. It molts to a second nauplius and again to a metanauplius (Figure 7.39). All naupliar stages subsist on yolk and swim with the antennae. The metanauplius molts to the first feeding stage, the calyptopis I. Sizes of nauplii during development vary between approximately 0.5 and 1.0 mm.

During the three calyptopis stages, the abdomen develops into a well-defined, segmented region of the body separate from the cephalothorax (Figure 7.39). Adult eyes begin to develop, but swimming is still handled exclusively by the antennae since thoracic and abdominal limbs have not yet developed. Calyptopes do show differences

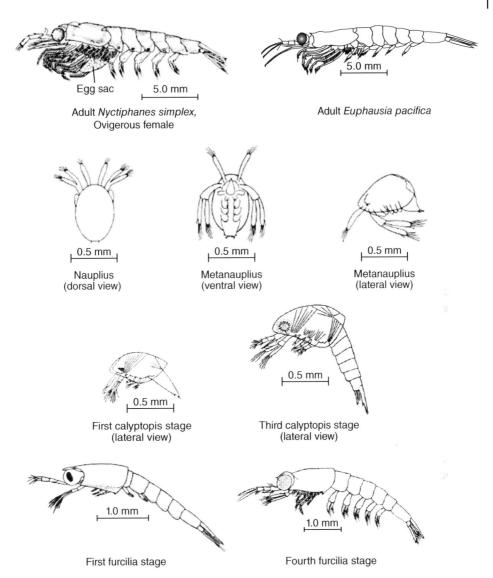

Figure 7.39 Abbreviated sequence of larval development in euphausiids using examples from *Nyctiphanes simplex* and *Euphausia pacifica*. Includes nauplius, metanauplius and first and third calyptopis (*Nyctiphanes*) and first and fourth furcilia stages (*Euphausia*). Note increases in thoracic limbs, abdominal segments and pleopods as well as eye development. *Source:* Adapted from Boden (1950), plate II, and Boden (1951), figures 1, 2, and 5.

in morphology between species, but not usually enough to allow reliable identification. Calyptopes increase in size from about 1 to 3 mm between stages I and III.

Six stages of furcilia complete the larval sequence, with the number of molts needed to complete development varying between species (Figure 7.39). The two major milestones achieved during furcilial development are first the budding and development of fully functional pleopods and, second, the development of the thoracic limbs, abdominal photophores, and telson (Mauchline and Fisher 1969). Once the furcilia molts to an adolescent euphausiid, it is capable of swimming as an adult. Furciliae increase in size from about 3 to 6 mm as they develop.

Gonad Development and Spawning

Gonads are located in the cephalothorax, posterior to the hepatopancreas and beneath the heart and pericardium. In the northern krill *Meganyctiphanes norvegica*, egg and sperm development begins in November with the ovaries and testes maturing in December/January (Mauchline and Fisher 1969). Spawning takes place over a period of six to eight months in *Meganyctiphanes*, which has a protracted spawning season. Other northern species, e.g. *Thysanoessa raschii*, have a spawning season of five months or less (Mauchline and Fisher 1969) and *Euphausia pacifica*, a cold-temperate Pacific species, spawns primarily from April to June (Ross et al. 1982).

Due to its importance as a universal food source in the Antarctic ecosystem, the Antarctic krill *Euphausia superba* has received considerable attention despite its polar location. Gonads in *E. superba* begin developing in September/October (late winter–early spring), and spawning takes place from December to February (Quetin and Ross 1991). Considerable variability is exhibited from year to year in the fraction of the *E. superba* population that reproduces and in the length of the reproductive season, with the variability being dependent upon food availability (Quetin and Ross 2001).

Most species (57 of 85) of euphausiids shed their fertilized eggs directly into the ocean. The remaining 28 species, which are in the genera *Nematobrachion, Nematoscelis, Nyctiphanes, Pseudeuphausia, Stylocheiron,* and *Tessarabrachion,* brood the embryos on their posterior thoracopods until the larvae hatch as nauplii (Mauchline and Fisher 1969). Free spawners produce many more embryos per season than the brooders, which can only accommodate tens of eggs on their thoracopods. Free spawners such as the northern and southern krill produce thousands of embryos in a season.

Euphausiid sperm have no flagella. Mature sperm are sequestered in spermatophores as needed within the male reproductive tract. During copulation, spermatophores are inserted into the thelycum, or sperm receptacle, of the female (Figure 7.40) where they remain attached by a biological adhesive until they are lost at the female's next molt. Figure 7.40 shows the reproductive system of a dendrobranch female, which is quite similar. During mating season, paired fully-formed spermatophores are found in the genital tract of all mature males, with another pair ready to take their place soon after a mating episode. The specialized structures known as petasmae on the first pair of adult male pleopods are responsible for transfer of spermatophores from the male's genital tract to the female thelycum. Once a spermatophore has been successfully inserted, sperms are released from the spermatophore into a cup-like receptacle on the thelycum forming a "sperm plug." Females are receptive to mating just after molting (Cuzin-Roudy 2010).

Eggs are cyclically produced throughout the reproductive season and released in batches (Ross and Quetin 1983; Tarling et al. 2009; Cuzin-Roudy 2010). The entire complement of mature eggs, numbering in the thousands, can be released in a single session lasting for a few hours. As eggs are released, they must pass through a narrow oviduct where they come into close contact with the sperm plug. They are thus fertilized as they are shed, entering the ocean as embryos.

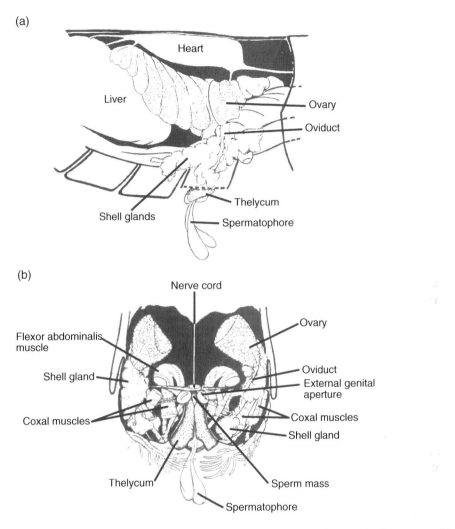

Figure 7.40 Dendrobranchiate reproduction. Dendrobranch female reproductive system with attached spermatophore, (a) lateral view; (b) anterior view. *Source:* (a) Adapted from Bargmann (1937), figure 23 (p. 342); (b) Adapted from Bargmann (1937), figure 25 (p. 344).

Field data suggest that virtually all mature females have a sperm plug during the breeding season, despite the fact that each sperm plug is lost during the female molt and must be replaced (Mauchline and Fisher 1969; Cuzin-Roudy 2010). This male attentiveness is likely the result of pheromones released by the female just prior to molting. Clearly, there is a selective advantage to being the first male to reach a receptive female.

Fully-formed spermatophores are present in the males of many euphausiid species, e.g. *Thysanoessa inermis, T. raschii, and Meganyctiphanes norvegica*, months in advance of mating season (Mauchline and Fisher 1969). More curious is the fact that most mature females of those same species are carrying sperm plugs two months in advance of the reproductive season, a trend also found in *Euphausia superba* and *E. triacantha* (Mauchline and Fisher 1969). Clearly, in many species, there is a time lapse between the presence of sperm in the thelycum of the female and the eventual release of the eggs. An early state of readiness may be advantageous in the event of early favorable conditions for spawning as it allows for a flexible response to an early spring.

Eggs of most free-spawning euphausiid species are negatively buoyant, sinking immediately after spawning with the embryos developing during descent. Sinking rates have been estimated for several species, with *Meganyctiphanes norvegica* showing experimentally determined rates of $132–180\,\text{m}\,\text{d}^{-1}$ at 15° and $96–120\,\text{m}\,\text{d}^{-1}$ at 0 °C (Mauchline and Fisher 1969), and $61–104\,\text{m}\,\text{d}^{-1}$ (Marschall 1983). The fastest sinking rates are shown by *Euphausia superba* at $203–231\,\text{m}\,\text{d}^{-1}$ (Marschall 1983). It is thus a race against time whether the eggs will hatch before reaching the bottom: a trade-off between development time and depth (Tarling 2010). For species that spawn in shelf waters, including fjords or deep shelves within their range, descending eggs have a very real chance of hitting bottom before they hatch.

The best known example of egg-sinking and larval development is that of the Antarctic krill, *Euphausia superba* (Marr 1962). Off the shelf, eggs that are spawned in the upper 100 m of the water column sink to depths of about 850 m prior to then hatching into the motile nauplii after 4.5–6 days (Quetin and Ross 1984, 1991). Nauplii continue to sink for an additional day (Marschall and Hirche 1984) and then make their way to the surface, developing into calyptopis I along the way (Figure 7.41). A fraction of the krill population spawns in shelf waters, where the sinking eggs reach the bottom. Embryos that evade the rich array of deposit and suspension feeders on the Antarctic benthos and survive to the nauplius stage comprise a small anomalous population of early stages found near the surface (Marr 1962).

Food and Feeding

The eleven genera of euphausiids can be roughly divided into three morphotypes based on the morphology of their eyes and thoracic limbs. *Bentheuphausia*, in its own family, exhibits reduced spherical eyes and eight similar pairs of well-developed thoracopods that can be used for grasping (McLaughlin 1980; Torres and Childress 1985). Five of the genera in the family Euphausiidae (*Euphausia, Meganyctiphanes, Nyctiphanes, Pseudeuphausia*, and *Thysanopoda*) have well-developed spherical eyes and a uniform thoracopod structure that allows formation of a "thoracic basket" for suspension feeding. The other five genera (*Nematobrachion, Nematoscelis, Stylocheiron, Tessarabrachion*, and *Thysanoessa*) have bi-lobed eyes (ovoid in some *Thysanoessa*) and a second or third thoracopod that is greatly elongated. The greatest degree of dietary specialization is observed in the five genera with the most highly specialized morphology (Kinsey and Hopkins 1994).

Suspension feeding has been described in detail for *Euphausia superba* (Hamner 1988). It is effected with a "thoracic basket," a basket-like trap for particles formed by the setae of the anterior six pereiopods (Figure 7.42). In the words of Hamner (1988), "As the feeding basket expands laterally, a pressure gradient is created which sucks water and particles into the basket from the front. Water and particles never enter the basket from the sides, below, or

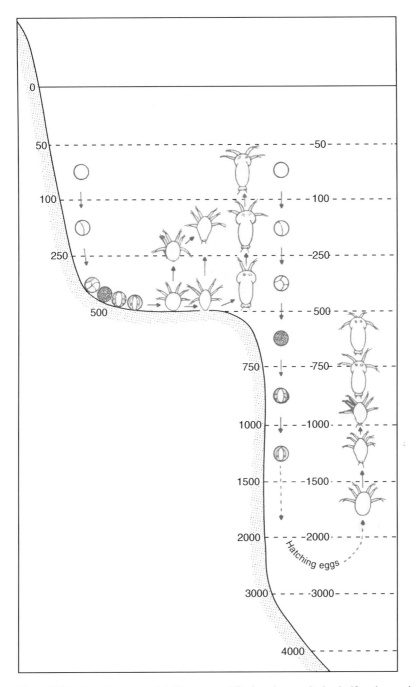

Figure 7.41 Development of sinking eggs of *Euphausia superba* in shelf and oceanic water showing how hatching in the shallower conditions give rise to occurrences of nauplii and metanauplii unusually close to the surface. *Source:* Mauchline and Fisher (1969), figure 51 (p. 135). Reproduced with the permission of Academic Press.

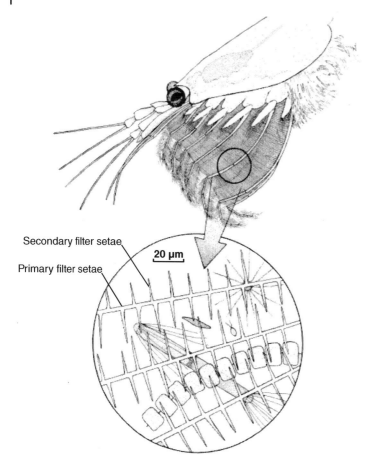

Figure 7.42 The feeding basket of *Euphausia superba*, and the size of common food items in relation to the mesh of the filter. *Source:* Quetin and Ross, Behavioral and Physiological Characteristics of the Antarctic krill, *Euphausia superba*, Integrative and Comparative Biology, volume 31, 1991, figure 2 (p. 53). Reproduced with the permission of Oxford University Press.

behind. Once inside the feeding basket, particles are retained on the filter and water is squeezed out laterally between the setae." Particles are then transferred to the mouthparts where a food bolus is formed prior to ingestion (Hamner et al. 1983).

The biomechanics of suspension feeding described by Hamner (1988) are readily applied to all euphausiids with a similar morphology and diet, which includes most of the swarming species (Mauchline and Fisher 1969). *Euphausia superba* does not feed continuously. It initiates feeding in response to the chemical signature of food particles, which it can localize with the chemosensors on its antennae. The thoracic basket is able to capture small zooplankton like small copepods as well as more passive particles like phytoplankton and microheterotrophs.

As a polar species, *Euphausia superba* encounters sea ice throughout the year, with highest cover during the spring, fall, and winter. Primary production is nearly absent in the winter months, but as sea ice forms in autumn, needle-like frazil ice crystals scavenge algae as they float to the surface, giving rise to a sea ice flora within the ice itself. Krill can use the dactyls (tips) of their thoracopods to scrape algae from the surface of the sea ice into their feeding baskets. This under-ice feeding behavior has been observed in the field (Marschall 1988) and in the laboratory (Hamner 1988).

Observations of feeding behavior are not available for the more carnivorous feeding specialists with bi-lobed eyes and elongate thoracopods. For those species, information on feeding habits must come from stomach contents. Kinsey and Hopkins (1994) described the diets and day-night vertical distributions of 20 species of euphausiids residing in the Gulf of Mexico, including members of the morphologically specialized genera *Nematobrachion, Nematoscelis,* and *Stylocheiron*. They found that species in those three genera also showed the most highly specialized diets, limiting their prey to a smaller number of crustacean taxa or prey sizes.

Geographical Distribution

Euphausiids are found in all major ocean basins and adjacent seas, as summarized by Mauchline and Fisher (1969). A considerable literature exists on euphausiid distributions (Mauchline and Fisher 1969), which will be briefly reviewed here. The vast majority of species are present in more than one ocean basin, with the most cosmopolitan being the bathypelagic and mesopelagic species (Einarsson 1945; Nemoto 1957; Brinton 1962; Ponomareva 1963; Mauchline and Fisher 1969). Species do show thermal preferences, which are reflected in their latitudinal distributions. Species' total geographic ranges are often far greater than the regions in which they spawn.

Overall, Mauchline and Fisher (1969) observed that 62 of the 85 species of euphausiids live in tropical-subtropical regions and 23 have either a northern or southern distribution. Forty-seven, or about half the species, are found in all three ocean basins. Distributional patterns of the euphausiid genera are summarized below from the information in Mauchline and Fisher (1969) and Mauchline (1980).

Bentheuphausia amblyops, the sole member of the monospecific family Bentheuphausiidae, inhabits the lower mesopelagic and bathypelagic regions of all major ocean basins. As such, it is normally found at temperatures of 5 °C or below.

Thysanopoda is an important and widespread genus with thirteen species, most of which are found at mesopelagic or bathypelagic depths. It contains the three largest euphausiids, the giant (total length > 100 mm) bathypelagic species, *egregia, cornuta,* and *spinicaudata*, which are so rare that their distributions are uncertain. Twelve of the thirteen species are believed to inhabit all three ocean basins, the only exception being *microphthalma*, which is not found in the Pacific.

Meganyctiphanes, a monospecific genus that is very important from a biomass perspective, is widely distributed in the North Atlantic, including the Mediterranean, Greenland, and Barents Seas. It has been found as far north as 80 °N in the Greenland Sea. Its temperature range is reported to be between 2 and 18 °C. It is found in coastal, slope, and oceanic waters throughout its range, with its main open ocean presence above 50 °N (Tarling 2010).

Nyctiphanes is a coastal genus containing four species. Two species, *simplex* (western Africa/California) and *australis* (New Zealand), are found in the Pacific. One species, *couchii,* is in the eastern North Atlantic, and one, *capensis*, is found off South Africa.

Peudeuphausia. There are two tropical species, one found coastally in the western Pacific and Indian Oceans (*latifrons*) and the other (*sinica*) in the East China Sea. Some disagreement exists as to whether *sinica* is a true species.

Euphausia, the most speciose of the euphausiid genera with 32 species, is found at all latitudes. Twenty are found exclusively in tropical-subtropical waters, 10 have a circumglobal distribution south of 40°S, including 7 found in polar regions. Several are dominant species, including *Euphausia superba* in the Southern Ocean and *Euphausia pacifica* in the North Pacific.

Tessarabrachion is a monospecific genus (*oculatum*) located in the North Pacific.

Thysanoessa has 10 species, all of which are found in the northern or southern latitudes outside of the tropics. *T. macrura* is a dominant species in Antarctic waters, being found in virtually all habitats there from neritic to open ocean.

Nematoscelis. There are seven species, most with wide-ranging distributions including coastal and open-ocean regions. Three are found in all three major ocean basins. Though distributions for all seven overlap the tropics, at least one (*N. megalops*) shows a behavioral preference for cooler water (Wiebe and Boyd 1978).

Nematobrachion is a tropical-subtropical genus with three species, two of which (*boopis* and *flexipes*) are found in all three ocean basins. The third (*sexspinosum*) is rare; it has only been reported in the Atlantic and Pacific.

Stylocheiron is fairly speciose, ranking third in total species numbers behind *Euphausia* and *Thysanopoda* with 11 species. All 11 have distributions that overlap the tropics, though most are found outside the tropics as well (e.g. *elongatum, longicorne, maximum*). The genus has several mesopelagic representatives. Eight are found in all three major ocean basins.

Vertical Distribution and Vertical Migration

Euphausiids are found at all depths in the open ocean, from near surface waters to bathypelagic depths (Vinogradov 1970). The deepest living species, including *Bentheuphausia amblyops*, and the three giant species of *Thysanopoda: cornuta, egregia,* and *spinicaudata* reside primarily below 1000 m, with only occasional captures in the mesopelagic zone. Most of the remaining species are vertical migrators, residing below 100 m during the day and migrating into near-surface layers at night. A few (e.g. *Stylocheiron elongatum* and *longicorne*) remain at mesopelagic or epipelagic depths day and night (Figure 7.43).

In the early days of sonar, the phenomenon of vertical migration garnered considerable attention in both the scientific community and the popular media. A mysterious "phantom bottom" or "deep-scattering layer" was often observed at mid-depths on ships' echosounders during oceanic transits (Figure 7.44). The layer moved to near-surface waters at night and back down to depths of >200 m during the day. Speculations on its origins included fanciful notions of schools of giant squid, whales, and other large creatures of the deep, but in actuality, the creatures comprising the scattering layer are far smaller than the sea monsters originally envisioned. In 1948, a paper by Hersey and Moore demonstrated that the deep-scattering layer (DSL) was largely the result of sound reflected by siphonophores, euphausiids, decapod shrimp, cephalopods, and small mesopelagic fishes performing their daily migration. Euphausiids are an important member of the DSL community, usually occupying its uppermost layer (Frank and Widder 1997). Penaeid or caridean shrimps are found below the euphausiids, and mesopelagic fishes, primarily lanternfishes, are found below the decapods. Assorted

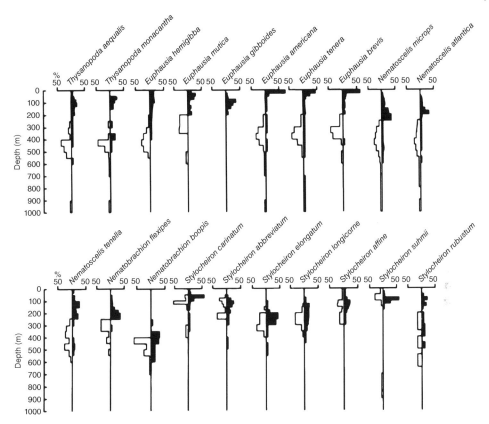

Figure 7.43 Vertical distribution profiles of 20 species of euphausiids in the eastern Gulf of Mexico. Data are percent of total population occurring in each depth zone. *Source:* Adapted from Kinsey and Hopkins (1994), figure 1 and table 2.

predators are found interspersed throughout the DSL including siphonophores, mesopelagic predatory fishes such as the dragonfishes, and medusae. Members of the DSL are also found in the stomachs of more formidable predators such as tunas and billfishes.

The fact that deep-scattering layers could be found at intermediate depths throughout the global ocean and that their faunal composition was similar from place to place raised a number of pressing questions for the ocean science community. Those can be roughly divided into two areas of concern: ecological and adaptive. For our purposes, ecological questions deal with the mechanics of vertical migration such as the physical factors controlling vertical position, effects of temperature, vertical distribution of a species' major food source, and the energy cost of vertical migration. Adaptive questions concern the selective advantage that would result in a migration of several thousand body lengths twice daily, usually toward and away from a food source. In short: the "what" and "why" of vertical migration. Both types of questions have been the subject of numerous field studies and considerable speculation.

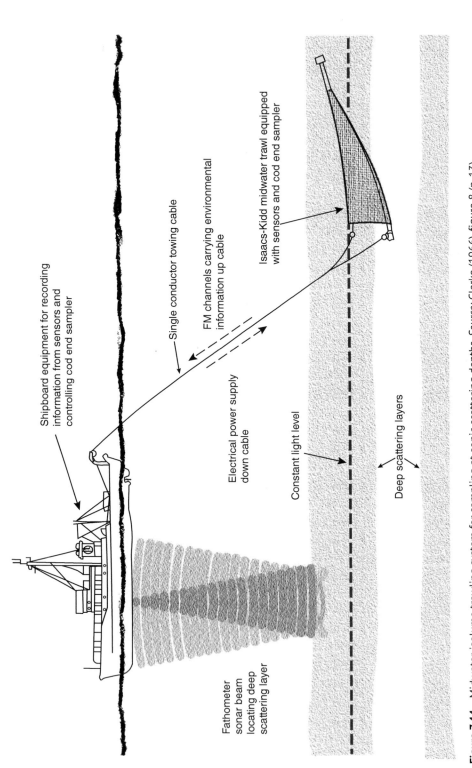

Figure 7.44 Midwater isolume trawling system for sampling at sonic scattering depths. *Source:* Clarke (1966), figure 8 (p. 13).

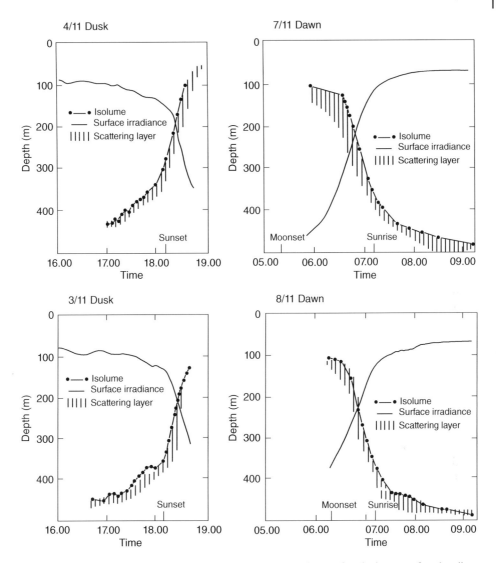

Figure 7.45 Two dawn and dusk examples of the correspondence of an isolume, surface irradiance, and a scattering layer near the Canary Islands. *Source:* Longhurst (1976), figure 6.2 (p. 128). Reproduced with the permission of Elsevier.

Ecological Factors

Light as the Proximal Cue for the Migration The DSL cues its movements to the water column's daily light cycle, making light a natural candidate for controlling the migration. At least three lines of evidence support this notion. First is the close correlation of DSL movement with a particular constant light level, or isolume, shown in Figure 7.45. The scattering layer and isolume depth remain in lockstep throughout the dusk and dawn periods (Boden and Kampa 1967). As the isolume rapidly moves up and down with the setting and rising sun, the DSL keeps pace with it. During the bulk of the day and night periods, the DSL maintains station within its preferred light level (Kampa and Boden 1954).

A second line of evidence supporting light as the factor controlling migration comes from the behavior of the DSL during solar eclipses: an elegant natural experiment. Using a standard ship's echosounder at a frequency of 12 kHz and a submersible photometer, Backus et al. (1965) were one of the first to report a definitive excursion upward by the scattering layer in response to a total eclipse. Later eclipse studies corroborated that result using higher frequency acoustics (Kampa 1975) and with net tows provided more detail on the species responding (Bright et al. 1972). The DSL has been studied in at least five eclipses in widely different locations: the Black Sea (Petipa 1955), the western North Atlantic (Backus et al. 1965; Sherman and Honey 1970; Tont and Wick 1973), the Gulf of Mexico (Bright et al. 1972), and in the vicinity of the Canary Islands in the eastern Atlantic (Kampa 1975). All showed upward vertical excursions by scattering layer communities or by net-captured zooplankton during the daytime darkness of a total eclipse.

Acoustics, Species Composition, Light, and Layer Movement A weakness of scattering layer studies is the fact that reflected sound gives limited information about the species reflecting it. Low-frequency acoustics (3–20 kHz) give the best depth penetration, ideal for seeking the bottom, but low frequencies are best reflected by large animals and those with gas bubbles, such as fishes with swim bladders and siphonophores with gas-filled floats. Smaller species such as euphausiids need to be present in very large numbers to reflect low-frequency sound.

Multiple-frequency echosounders detect a greater spectrum of animal sizes but do not provide information on species composition. Consequently, the actual players in the vertically migrating community associated with a particular light field, or isolume, cannot be identified except by direct observation from a manned submersible or remotely operated vehicle (ROV), by using towed video samplers, or by using high-resolution net tows that have an associated lag time.

The two most likely alterations in the light field cueing a vertical migration are "a threshold change in the absolute light intensity or a relative rate of change in light intensity" (Frank and Widder 1997).

Roe (1983) used a photometer mounted to an opening-and-closing trawl to allow sampling within three pre-determined light levels (cf. Clarke 1966, Figure 7.44), focusing on euphausiids and micronektonic fishes in an open-ocean community of the north Atlantic (42°N 17°W). His findings suggested that virtually, all the species sampled had a depth range that greatly exceeded a single narrow isolume, usually occupying a three-order-of-magnitude change in downwelling irradiance levels. Results thus suggested that affinity of species' populations with light level was looser than suggested by earlier work with DSL movement. It should be noted that studies correlating DSL movement with isolume depth focused mainly on the top of the DSL and not the depth range of individual populations, which are known to be vertically diffuse (cf. Figure 7.43). Roe's results do not change the conclusion that diel vertical migration (DVM) and day-night vertical position within the water column are mediated by ambient light level.

Adaptive Significance: The Environment A lot of creative thinking has been dedicated to explaining why species vertically migrate, but for any theory to be acceptable, it needs to be broadly applicable. DVM is observed from the equator to the poles, so a viable theory would have to apply in systems with widely differing physical characteristics. The taxonomic groups comprising the core of the DVM community are remarkably similar from place to place.

Depth, Light Penetration, Temperature, and Oxygen The distance covered each way by vertical migrators and the water column through which they migrate vary with location, taxon, and size. A typical DSL fauna usually comprises euphausiids, decapods, and micronektonic fishes at a minimum, roughly layering out in that order with increasing depth.

Water-column temperature profiles differ considerably (see Chapters 1 and 2) from the tropics to the poles. In particular, the difference in temperature between daytime and nighttime depths can vary from >10° in the tropics to <2° in polar systems (see Figure 1.12). The day–night temperature differential greatly influences the energy cost of the migration, as well as the total daily energy devoted to metabolism.

Severe oxygen minima may limit the daytime depth of vertical migrators (Childress and Seibel 1998). Oxygen partial pressures at the core of the minimum (about 700 m) in the Eastern Tropical Pacific (3 mm Hg) allow only a few species to live there aerobically, and the anoxic waters in the Arabian Sea (0 mm Hg) and Cariaco Basin (0 mm Hg) preclude aerobic metabolism altogether. Oxygen minima and their influence on vertical migration will be considered in greater depth in Chapter 11.

If the DVMs of three congeneric euphausiids from polar, cold temperate, and subtropical systems are compared, we can appreciate first-hand the differences they experience on a daily basis (Table 7.5). Two of the species, *Euphausia pacifica* and *Euphausia hemgibba*, are primarily herbivorous (Mauchline and Fisher 1969; Kinsey and Hopkins 1994). The Antarctic species, *Euphausia triacantha*, is a carnivore, feeding on euphausiid larvae, pteropods, and foraminifera (Baker 1959). In all three cases, their primary prey items are most numerous in the surface layers. If the approximate centers of distribution are used for daytime and nighttime depths, *Euphausia pacifica* and *Euphausia hemgibba* undergo a vertical migration of 250 and 225 m, respectively, and *E. triacantha* journeys about 375 m to its nightly depth.

The simplest way to calculate the energy cost of the migratory lifestyle is to break down the 24-hour day into three components: daytime residence, nighttime residence, and total time spent migrating. Migration time for the three species can be estimated from published records of isolume movements that approximate the vertical migration of each species (Kampa and Boden 1954; Clarke and Backus 1964) (Figure 7.46). The result is that the migration each way takes about three hours for all species, which is very close to what is seen for isolume movement. Arbitrary times for each part of the day are thus nine hours at daytime depth, nine hours at nighttime depth, and six hours for migrating up and down. In all three species, the migration each way is 10 000 body lengths or more (Table 7.5)

A few data are available on how fast species swim during migration. The best come from Cowles (1994) who determined the swimming speeds of the decapod *Sergestes similis* using a remotely operated vehicle (ROV) in the Monterey Canyon. *S. similis* is a vertically migrating species averaging 43.5 mm in body length. As it happens, *Sergestes* resides just below *Euphausia pacifica* in the scattering layers off the California coast. Routine swimming, the

Table 7.5 Diurnal vertical migration in euphausiids from three oceanic systems. Vertical profiles for *Euphausia triacantha*, *Euphausia pacifica*, and *Euphausia hemigibba* from Baker (1959), Torres and Childress (1983), and Kinsey and Hopkins (1994), respectively. Respiratory rates from Torres et al. (1994), Torres and Childress (1983), and Teal and Carey (1967), respectively. Q_{10}'s taken from literature values (*E. pacifica*), computed from reported rates vs. temperature (*E. hemigibba*), or assumed to be 2.0 (*E. triacantha*). Cost of transport for all species computed from the relationship between routine rate and swim speed (in body-lengths s^{-1}) for *E. pacifica* (Torres and Childress 1983).

	Antarctic	California Borderland	Gulf of Mexico
Genus species	*Euphausia triacantha*	*Euphausia pacifica*	*Euphausia hemigibba*
Size (mm)	35	18	15
Dry mass (DM) (mg)	72	6.9	7
Daytime depth range (m)	250–750	200–400	250–350
Approximate day center of distribution (DCOD) (m)	475	275	300
Temperature at DCOD (°C)	2	8	13
Energy cost for 9 h residence at DCOD – μl O_2 (cal) [joules]	438.7 (20.3) [85.1]	63.3 (2.9) [12.3]	70.6 (3.3) [13.7]
Night depth range (m)	0–500	0–50	0–150
Approximate night center of distribution [NCOD] (m)	125	16	75
Temperature at NCOD (°C)	0	12	23
Energy cost for 9 h residence at NCOD – μl O_2 (cal) [joules]	381.6 (17.7) [74]	97.5 (4.5) [18.9]	199.1 (9.2) [38.6]
Migration distance (m), distance per day (m)	700	508	450
Migration distance (body lengths d^{-1})	20 000	28 222	30 000
Average migration speed (m h^{-1}) (body lengths s^{-1})	117, 0.9	87, 1.3	75, 1.4
Migration cost at average migration temperature, μl O_2 (cal) [joules]	373.2 (17.3) [72.4]	87.4 (4.05) [17.0]	161.7 (7.5) [31.4]
Total daily cost – average speed, μl O_2 (cal) [joules]	1193.5 (55.3) [231.5]	248.2 (11.5) [48.2]	431.4 (20) [83.7]
Total daily cost – 24 h at nighttime depth, μl O_2, (cal) [joules]	1017.8 (47.1) [197.5]	260 (12.0) [50.4]	530.9 (24.6) [103]

normal cruising speed of *S. similis*, was 7.4 cm s^{-1} or 1.7 BL s^{-1} in the field. In a later study on the same species, using a cleverly designed swim tunnel that mimicked day–night temperatures (Cowles 2001), Cowles found routine swimming to be about 1.0 BL s^{-1} both day and night. During simulated migration, he found swim speeds of 6.4 and 5.4 cm s^{-1} (1.4 and 1.2) BL s^{-1} for the upward and downward components,

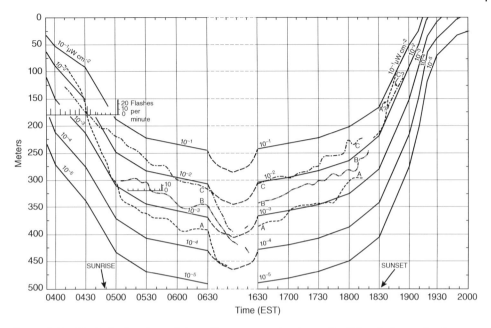

Figure 7.46 Vertical migrations of three discrete scattering layers (A–C) in relation to the descent and ascent of the isolumes of ambient light during 15 August 1959 at a station 150 miles south of Nantucket. The numbers of luminescent flashes at times and depths of rapid migration and slow migration are indicated. *Source:* Clarke (1970), figure 3 (p. 46).

respectively, remarkably similar to the hypothetical average migration speeds for the euphausiids shown in Table 7.5. Observations of downward movement in migrators uniformly report that it is a result of active swimming rather than passive sinking (Cowles 1994).

Water-column temperature profiles typical of each species' range are available from the literature (*E. pacifica*: Torres and Childress 1983; *E. hemigibba*: Torres et al. 2012; *E. triacantha*: Gordon et al. 1982). Similarly, values for oxygen consumption rate vs. temperature are available from direct measurements on each species (*E. pacifica*: Torres and Childress 1983; *E. hemigibba*: Teal and Carey 1967; *E. triacantha*: Torres et al. 1994). Taken together, the migratory profile, temperature profile, and oxygen consumption vs. temperature data provide the basics for estimating migratory costs. A routine-rate multiplier was used to translate the cost of transport from data on *Euphausia pacifica* to the other two species. The relationship between swimming speed, expressed in body lengths per second ($BL\,s^{-1}$), and the routine rate was computed for *E. pacifica* and was then applied to *E. hemigibba* and *E. triacantha* to estimate their cost of swimming. That relationship for *E. pacifica* is as follows:

$$1\,BL \cdot s^{-1} = 1.43 \times \text{routine rate}; 2\,BL\,vs^{-1} = 2.14 \times \text{routine rate}.;$$

Breaking down the energy costs between the three systems shows that the water column temperature profile determines whether energy is saved or lost due to the migration. The polar species loses energy, the cold temperate species roughly breaks even, and the subtropical species saves itself about 20% of what it would cost to remain in warm surface waters for 24 hours.

In a classic study of swimming speeds in a variety of open-ocean fauna, Hardy and Bainbridge (1954) recorded an average value of $120\,m\,h^{-1}$ for the euphausiid *Meganyctiphanes norvegica* swimming vertically in their plankton wheel, a device that allowed monitoring of swimming speeds in untethered animals. A speed of $120\,m\,h^{-1}$ corresponds to a speed of $1\,BL\,s^{-1}$ for an intermediate-size (33 mm) adult. Maximum speed recorded was $215\,m\,h^{-1}$ ($1.8\,BL\,s^{-1}$). Overall, pelagic Crustacea exhibit sustained swimming speeds of $1-2\,BL\,s^{-1}$, with most routine cruising speeds near $1.0\,BL\,s^{-1}$ (Torres and Childress 1983; Cowles and Childress 1988; Cowles 1994, 2001). Shorter periods of higher velocity (burst) swimming have been reported for pelagic species (Torres and Childress 1983; Cowles 2001), ranging from 2.6 to $4.6\,BL\,s^{-1}$.

Adaptive Significance of Vertical Migration Recognizing that the major operational factors involved in diel vertical migration are light, temperature, feeding, and metabolism, a number of theories have been presented to explain why vertical migrations occur. Many of these theories are based on studies of vertical migration of crustacean zooplankton, species smaller than our target size range but more readily captured. However, most of the theories based on these studies apply just as well to micronektonic species as they do to zooplankton. They fall, for the most part, into three general categories: metabolic and energetic models, horizontal displacement advantages, and predator avoidance.

Metabolic models (McLaren 1963, 1974; Enright 1977) argue that vertical migration as a foraging strategy in a thermally stratified water column confers an energetic advantage to individuals that ultimately is expressed as increased reproductive output, a reasonable proxy for increased fitness. The analysis above suggests that any energetic benefits accrued are highly dependent on the water column.

A variant on metabolic models was introduced by invoking the vertical light environment and its relationship to photosynthesis (Kerfoot 1970) as an explanation for vertical migration. Using a technique he called "pathway analysis," he modeled the energy available to vertical migrators based on the primary productivity (and therefore algal abundance) within different isolumes as a function of time of day. In his study, production was maximal in the upper 50 m of the water column, which corresponded to isolumes of 10^4–$10^2\,\mu W\,cm^{-2}$ by day and 10^{-1} and 10^{-2} $\mu W\,cm^2$ by night. This not only allowed individuals residing in the upper 50 m access to high algal biomass on a 24-hour basis but also gave access to individuals residing in the 10^{-1} and 10^{-2} $\mu W\,cm^{-2}$ isolumes (110–150 m) access by night. The constraint on the model was that individuals did not feed at light levels greater or less than an order of magnitude outside of their residence isolume. An interesting analysis foiled by the inconvenient fact that abrupt cessation of feeding outside of a rigid light environment is not observed in marine zooplankton.

Horizontal displacement arguments (Hardy and Gunther 1935; Hardy 1953, 1956; David 1961; Isaacs et al. 1974) suggest that there are advantages conferred due to relative horizontal movement caused by differences in current speed and direction between the surface and daytime depths. Advantages include avoidance of toxic algae at the surface (Hardy and Gunther 1935), improved diversity of the local gene pool (David 1961), potentially fewer predators (Miller 1972), and improved surface feeding

(Hardy 1953, 1956; Miller 1972; Isaacs et al. 1974). However, despite the fact that horizontal displacement may occasionally provide increased feeding opportunities, the one study to experimentally address horizontal movement suggested otherwise (Miller 1972).

The main flaw in energy-based arguments for why migration occurs, whether they focus on energy savings and increased fecundity due to spending days in cooler, deeper waters, or improved nightly feeding grounds resulting from horizontal displacement due to current shear between the surface and daytime depth, is that differential mortality in migrating and non-migrating populations due to visual predation is not considered in the arguments. This was elegantly demonstrated in a detailed study of vertical migration in *Calanus pacificus* by Frost (1988).

Frost (1988) compiled data on seasonal vertical distribution as well as temperature effects on growth and egg production in the dominant calanoid copepod, *Calanus pacificus*, an important member of the north Pacific zooplankton community. *C. pacificus* exhibited variability in its seasonal migratory pattern, showing both migratory and non-migratory years during the spring season. In contrast, *C. pacificus* was invariably a strong vertical migrator in the summer and fall seasons. The spring migratory pattern showed no correlation with either food concentration or degree of thermal stratification, which varied from year to year. However, egg production rates were uniformly high in non-migratory years.

Unlike in spring, a strong vertical migration occurred throughout the summer months in a uniformly well-stratified water column, showing a temperature differential of as much as 8 °C between surface and daytime depth. Food availability was highly variable in the summer samples, as was egg production. Overall, the evidence gathered for *C. pacificus* strongly suggests that vertical migration is not a genetically fixed behavior but varies from year to year and place to place. Moreover, it takes place in a variety of vertical temperature profiles and levels of surface productivity.

The next step in the study was to use existing data on growth and egg production vs. temperature to create a basic life table for *C. pacificus*, comparing three temperatures within the species' normal range with perfect survivorship in each case. Growth rate at surface temperatures (13 °C) was faster than that at either of the two lower temperatures, but adult female size was greater at the lower temperatures, producing more eggs. The question ultimately was whether the larger size and greater number of eggs produced at the lower temperatures was sufficient to offset the disadvantage of a slower growth rate. Differences between migrators and surface dwellers were small enough so that a small difference in mortality rate would likely tip the scales in favor of the migratory habit, and in fact it did.

Like the euphausiids, copepods have multiple life stages, termed copepodites and only the final three (copepodites IV, V, and adult) perform migrations. If mortality in the final three stages of the migrators was reduced by 50% with all else being equal, in 30 days, the migrators would have over twice the population numbers of the surface dwellers, and if reduced by only 12%, their numbers would be equal. In contrast, if migrators and non-migrators were given the same individual mortality rate, the surface dwellers would have a higher population growth rate, due to the fact that the individual mortality rate would have longer to act on the slower growing migrators. Thus, with equivalent mortality rates, the surface population would retain a selective advantage. The analysis of Frost (1988) is particularly cogent because it uses an energetically based argument to support a case for predator avoidance as the selective advantage of vertical migration.

Although metabolic advantages may certainly play a role in perpetuating the phenomenon of vertical migration, and horizontal displacement may occasionally provide increased feeding opportunities, the best evidence to date points to avoidance of visual predation as the major selective advantage to the twice-daily journey.

Field-based and experimental studies provide good support for the predator avoidance hypothesis. In a particularly convincing study that presaged Frost (1988), Ohman et al. (1983) detailed an example of reverse diel vertical migration observed in the copepod *Pseudocalanus* spp. Like the study of Frost (1988), it took place in Dabob Bay, a fjord in Puget Sound, Washington. Briefly, *Pseudocalanus* females descended nocturnally when the abundance of invertebrate predators increased in the upper layers at night, due to the predators' more conventional upward migration. During times of year when the invertebrate predators were scarce, no reverse migration of *Pseudocalanus* was observed. The study is a very direct demonstration of vertical migration as a predator avoidance mechanism.

Robison (2003) used the Antarctic as a model system to argue from first principles that avoidance of visual predators drives vertical migration. The study used micronektonic species, though the major focus happens to be on two vertically migrating midwater fishes: the lanternfish *Electrona antarctica* and the Antarctic silverfish, *Pleuragramma antarctica*, a taxonomic skip ahead for us. Both are strong vertical migrators with centers of distribution of about 600 m during the day and about 200 m at night. During the study, the water column was nearly isothermal at a temperature of 0 °C (Hopkins 1985). Major prey for the fishes was the Antarctic krill, *Euphausia superba*, which resided in the upper 200 m day and night (Lancraft et al. 2004, Figure 7.47).

The isothermal water column removed any potential energetic benefits resulting from spending days at a colder temperature, and krill, the major prey of the fishes, resided in near surface waters, so both fishes were migrating away from their major food source each day at dawn. An impressive array of visual predators known to feed on both species is common in the study area, including penguins, flighted seabirds, and species of phocid and otariid seals (Lancraft et al. 2004). The inescapable conclusion is that the fishes were avoiding visual predators since no energetic benefit could result from a long swim away from their major prey resource in an isothermal sea.

Order Decapoda
Introduction
The decapods display many different body forms, ranging from the "long-tailed" shrimp and lobsters to the hermit crabs and true crabs that sport a far more abbreviated abdomen. The pelagic decapods, the dendrobranchiate and caridean shrimps, conform well to Calman's (1909) "caridoid facies," the ancestral shrimp-like body form (Figure 7.48) that includes the characteristics listed below. Characteristics 8–10 were added by Hessler (1983).

1) A carapace enveloping the entire thoracic region
2) Movable, stalked eyes
3) Biramous first antenna
4) Scale-like exopod on second antenna
5) Oar-like natatory exopods on the thoracic appendages
6) Elongated, ventrally flexible abdomen

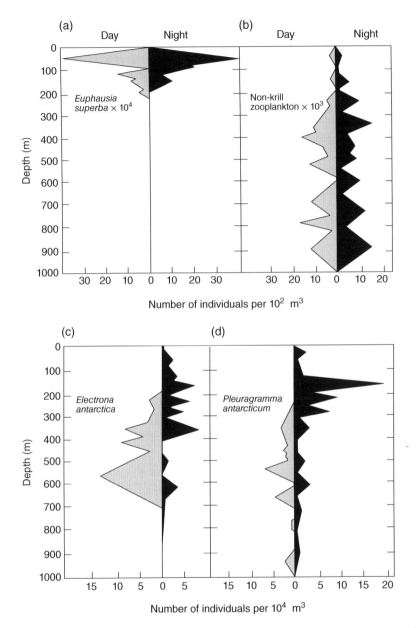

Figure 7.47 Diel vertical distribution patterns in krill, non-krill zooplankton, and midwater fishes in deep water basins in Antarctica. (a) Krill, *Euphausia superba*; (b) non-krill zooplankton; (c) the myctophid fish *Electrona antarctica*; (d) the nototheniid fish *Pleuragramma antarctica*. The area of the triangles indicates the percentage of the total population found within each sampled depth interval. *Source:* Robison (2003), figure 1 (p. 640). Reproduced with the permission of Cambridge University Press.

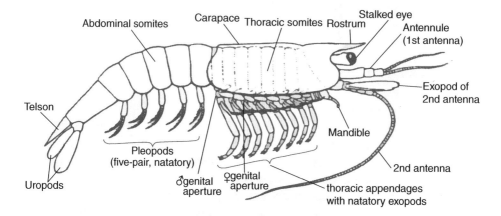

Figure 7.48 Calman's "caridoid facies," the ancestral shrimp form. *Source:* Calman (1909), figure 85 (p. 145).

7) Paddle-like tail fan formed by the uropods on either side of the telson
8) Abdomen characterized by a well-developed musculature
9) Internal organs, excepting the hindgut, absent from abdomen
10) Pleopods 1–5 alike, biramous and natatory

Recent decapods are united in having three pairs of maxillipeds as the first three of their eight pairs of thoracic limbs. The maxillipeds help to manipulate food items and are associated with the mouth. The posterior five pairs of thoracic limbs give the order its name (decapoda = 10 legs). They consist of an endopodite that may be used for grabbing, walking, or swimming, sometimes a swimming exopodite, and one or more respiratory epipodites. The abdomen has five pairs of biramous pleopods for swimming and a sixth pair of limbs, the uropods, which together with the telson forms the tail fan.

Classification within the decapods has seen major changes since the 1970s and is still evolving. As noted earlier, its two major (and very uneven!) subdivisions, the suborders Dendrobranchiata and Pleocyemata, are based on differences in gill morphology. The Dendrobranchiata contains about 531 species (Bauer 2004; WoRMS n.d.) with dendrobranchiate (branching filaments) gills in two superfamilies, the Penaeoidea and Sergestoidea. Pleocyemata contains the remainder of the 13 014 marine decapod species and is further subdivided into seven infraorders, with most pelagic species found in the infraorder Caridea, a group with about 2908 species (WoRMS n.d.). One additional infraorder can be said to have pelagic representatives: the Anomura, with the eastern North Pacific munidid red crab *Pleuroncodes planipes* and the pelagic *Munida gregaria* found off South Africa (Kaestner 1970).

History
The dendrobranchiate and caridean shrimps exhibit similar morphologies and means of locomotion. Up until the 1980s, they were also allied in a single suborder, the Natantia, or swimmers; students of pelagic fauna are likely to encounter taxonomic keys with the older classification

system, e.g. Holthuis (1955). In the old system (Kaestner 1970), the suborder Natantia was divided into two sections: the Penaeidea and Caridea, with the section Penaeidea corresponding closely to the new suborder Dendrobranchiata. In turn, the old section Caridea is nearly identical to the new infraorder Caridea. However, the new infraorder now resides within an entirely new suborder: the Pleocyemata.

The older system split the decapods into two suborders, the swimming Natantia described above and the Reptantia, or crawlers. All the true crabs (Brachyura), burrowing shrimp (Thalassinidea), clawed (Astacura) and spiny (Palinura) lobsters, and hermit crabs (Anomura) fell into the Reptantia. At the family level of classification and below, where most of our interests lie, the old and new systems are very similar. Classification of decapods will continue to evolve, as there has been some disagreement (Martin and Davis 2001) on the validity of the major split based on the gill morphology. The new decapod classification system was introduced by Burkenroad (1963, 1981) and accepted into Bowman and Abele's (1982) highly regarded system of crustacean classification. The Bowman and Abele system was further modified by Martin and Davis (2001) to give the classification system most widely accepted until 2009. With the advent of the 2009 Census of Marine Life and Web-based classification, taxa evolve far more rapidly and the system presented below is from the World Register of Marine Species. It is quite similar to Martin and Davis in most respects. Taxa are presented for the Dendrobranchiata and Caridea only.

 Order Decapoda
  Suborder Dendrobranchiata
   Superfamily Penaeioidea
    Family Aristeidae
    Benthesicymidae
    Penaeidae
    Sicyoniidae
    Solenoceridae
   Superfamily Sergestoidea
    Family **Luciferidae**
    Sergestidae
  Suborder Pleocyemata
   Infraorder Caridea
   Superfamily Alpheoidea
   Superfamily Atyoidea
   Superfamily Bresilioidea
    Family Agostocarididae
    Alvinocarididae
    Bresiliidae
    Disciadidae
    Mirocarididae
   Superfamily Campylonotoidea
   Superfamily Crangonoidea
   Superfamily Nematocarcinoidea
    Family Eugonatonotidae
    Nematocarcinidae
    Rhynchnocinetidae
    Xiphocarididae
   Superfamily Oplophoroidea
    Family **Acanthephyridae**
    Family **Oplophoridae**
   Superfamily Palaeomonoidea
   Superfamily Pandaloidea
    Family **Pandalidae**
    Thalassocarididae
   Superfamily Pasiphaeoidea
    Family **Pasiphaeidae**
   Superfamily Physetocaridoidea
    Family **Physetocaridae**
   Superfamily Processoidea
   Superfamily Psalidopidoidea
   Superfamily Stylodactyloidea

Genera in the Sergestidae have changed considerably since Omori (1974). Judkins and Kensley (2008) split the genus *Sergestes* into six genera: *Allosergestes, Deosergestes, Eusergestes, Neosergestes, Parasergestes, and Sergestes*. The new genus names are not yet common in the literature, and one must be aware when reading about sergestid biology that species in all five genera were once *Sergestes*.

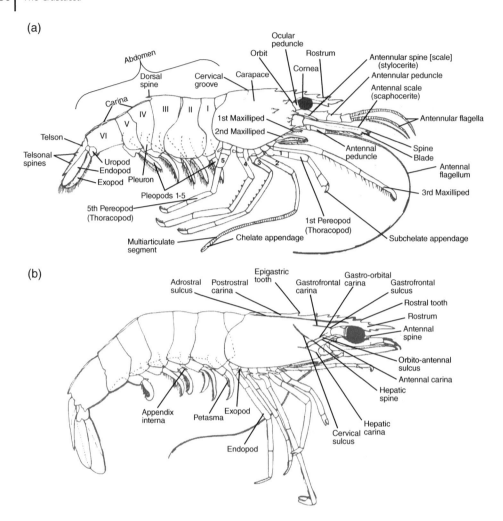

Figure 7.49 Decapod external anatomy. (a) Lateral view of a typical caridean decapod. Note the characteristic "saddle" formed by the second abdominal pleura; (b) lateral view of a typical penaeid (dendrobranch) decapod. *Source:* Adapted from McLaughlin (1980), pages 128 and 135.

Decapod Anatomy

The two main groups of pelagic shrimp, the dendrobranchs and carideans, are similar in anatomy, but not identical. They can easily be separated by eye on the basis of the "caridean saddle": the pleura on the second abdominal segment overlap the first and third (Figure 7.49). In addition, carideans have a "hump"; they usually exhibit a pronounced bend at the level of the third pleomere. The carapace is fused to the thoracic segments and extends far enough laterally so that, unlike in the euphausiids, the gills are enclosed in a branchial chamber.

The posterior five thoracic limbs, or pereopods, also show differences between groups. In the dendrobranchiate shrimps, the first three are almost always chelate (ending in a claw or pincer). In the carideans, never more than two are, and these can be either chelate or subchelate (Figure 7.49). Dendrobranchs often have exopods, whereas the carideans do not. The pelagic pasiphaeids and oplophorids do bear exopods that aid them in swimming, but in most carideans, the exopods described in the caridoid facies are absent or vestigial. Female dendrobranchs possess a thelycum, or sperm receptacle, at the level of the fifth periopod.

The decapod abdomen has six segments, with the first five bearing pleopods and the last bearing a tail fan. The five pairs of biramous pleopods all have fringing setae that fan out during the swimming power stroke, maximizing surface area for thrust, and collapsing inward on the recovery stroke. In the dendrobranchs, the members of a pleopod pair typically beat separately in a metachronal rhythm, as do those of euphausiids. In contrast, pleopods of the carideans usually beat in synchrony. In the carideans, a slender stalk-like structure called the appendix interna, present on the inner ramus (endopod) of each pleopod, links up with that of its neighbor to join the pair together. Tiny coupling hooks (cincinnuli) on the tips of the structures form the link between them, acting much like "crustacean Velcro" (Bauer 2004), allowing the pair of pleopods to beat together.

Secondary sexual characteristics are found on the first two pairs of pleopods. In male dendrobranchs, the inner ramus or endopod of the first pair of pleopods is joined to its neighbor to form a petasma, a male copulatory organ. Because the petasma is species-specific in structure, it is often used as a diagnostic character in taxonomic keys. Instead of a petasma, the second pleopods of males in several caridean families bear a structure called the appendix masculina that has been implicated in spermatophore transfer during copulation (Bauer 2004). Modifications to the first two pleopods of female carideans have been observed during the breeding season as well. In the carideans, the eggs are carried on the pleopods. In both groups, gonopores are located on the coxa of the third pereiopods of females and the fifth of males.

The tail fan in both groups is made up of two broad biramous uropods located adjacent to the telson. Like the pleopods, the telson and uropods are fringed by long setae that greatly increase their surface area and the efficiency of the tail-flip escape response.

Nervous System
The nervous system is similar in dendrobranchs and carideans, conforming to the general decapod plan outlined previously in the Section Crustacean Systems of this chapter. Command and control of the eyes and antennae are the province of the cerebral ganglion, that of the gastric system in the esophageal and stomatogastric ganglia associated with the circumesophogeal commissure, that of the mouthparts in the subesophageal ganglion, and body musculature and limbs in the thoracic, abdominal, and caudal ganglia.

Sensory Mechanisms
Like the euphausiids, dendrobranchs and carideans both have a full spectrum of sensory hairs and aesthetascs for sensing the mechanical and chemical environment. They are located primarily on the antennae and mouthparts as described earlier. Statocysts are found in most decapods, located at the bases of the antennae (Cohen and Dijkgraaf 1961), though little work has been done on the statocysts in pelagic species.

In open-ocean communities, pelagic decapods usually reside more deeply than the vertically migrating euphausiids, sometimes forming a layer below them (Omori 1974; Roe 1974; Omori and Gluck 1979; Roe 1984). Overall, decapods are found in higher abundance at greater depth, with a substantial fraction of their total numbers found below 1000 m (Vinogradov 1970). Evidence suggests that the mechano- and chemo-sensory abilities of decapods are more highly developed than those of euphausiids (e.g. Denton and Gray 1986), perhaps to supplement visual systems in a lower light environment.

Studies on the spectral sensitivity of decapod eyes have revealed a surprising sensitivity to UV wavelengths (Frank and Case 1988a; Frank and Widder 1996). As noted earlier, downwelling irradiance is primarily in the blue–green spectrum (475 nm), which penetrates best in clear ocean waters, but several members of the caridean family Oplophoridae are able to detect light at wavelengths in the near UV (400 nm). The response to UV can be measured electrophysiologically and can be elicited behaviorally in the whole animal (Frank and Widder 1994): quite convincing! At present, there is no single compelling explanation for why the shrimp have the dual sensitivity. It may enable them to resolve predators from conspecifics (individuals of the same species) by slight differences in the quality of their bioluminescence.

Bioluminescence and Photophores

Luminescence is present in most of the pelagic dendrobranchs and carideans, though its character differs considerably among the families and species. There are three basic types: "(i) a luminous secretion from the oral region, (ii) superficial cuticular photophores on the body and limbs, and (iii) internal photophores formed from modified hepatopancreas tubules" (Herring 1985). In some cases, more than one type is present in a single species. Cuticular photophores are smaller than the compound ventral photophores of euphausiids, though they do possess a lens.

Among the dendrobranchs, the most common form of bioluminescence is type 3 and is found mainly in the genus *Sergestes*. The light organs of *Sergestes* are known as the organs of Pesta, and they are best thought of as groups of "ventrally directed luminescent tubules" (Foxton 1972). In practice, they appear as heavily pigmented spherules or regions within the hepatopancreas (Figure 7.50). The organs of Pesta produce a continuously glowing luminescence that matches downwelling light in wavelength and intensity (Warner et al. 1979). Moreover, the organs rotate within the shrimp to face downward whether the shrimp maintains a head-up, head-down, or horizontal attitude (Omori 1974; Warner et al. 1979; Latz and Case 1982). Organs of Pesta are capped dorsally by a screening pigment.

In the deeper-living dendrobranchs, particularly in the genus *Sergia*, cuticular photophores are found on the ventral surface and can be located on the body and appendages (Dennell 1955; Omori 1969; Herring 1978). Other dendrobranchs with cuticular photophores include a few species within the Solenoceridae, Aristeidae, and Penaeidae.

The carideans exhibit a greater diversity in their bioluminescent systems. Herring (1978) has categorized them as follows.

1) A luminous secretion alone (Pasiphaeidae, Oplophoridae, Pandalidae)
2) A secretion and superficial light organs (Oplophoridae)
3) A secretion and hepatic organs (Thalassocarididae)
4) Hepatic organs alone (Pandalidae)

Luminous secretions are produced as a persistent, bright blue, cloud originating in the head region. They can literally light up a catch bucket when viewed in the dark. The luminous material is produced by the

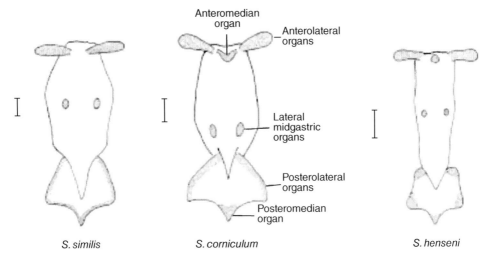

Figure 7.50 Schematic ventral representation of the hepatopancreas and organs of Pesta in three species of *Sergestes*. Scale is 1 mm. Luminescent tissue (organs of Pesta) are shaded. *Source:* adapted from Foxton (1972), figure 1 (p. 184). Reproduced with the permission of Brill.

hepatopancreas and regurgitated from the mouth (Herring 1976b), where it is propelled forward and dispersed by the exhalant respiratory streams from the gill bailers. There is a distinct pulse associated with the production of the cloud that reflects the rhythm of the gill bailers. All the oplophorids are capable of producing a luminescent cloud. The oplophorid genera *Systellaspis* and *Oplophorus* have superficial light organs in addition to being able to generate a luminous secretion. Luminous secretions are also found in members of the Pasiphaeidae and Pandalidae. The pandaloids *Plesionika* (formerly *Parapandalus*) and *Thalassocaris* both possess a caridean equivalent to the hepatic organs of Pesta. However, it is only *Thalassocaris* that has both hepatic organs and the capability of generating a luminescent cloud (Herring and Barnes 1976).

Functions of Bioluminescence

Bioluminescence serves two main functions in the decapods: camouflage and defense. The ventrally directed steady glow produced by the dendrobranch organs of Pesta matches the light that is stimulating the eyes of the shrimp in both spectral quality and intensity. The luminescent hepatic organs of dendrobranchs (and carideans) are thus believed to act as counter-illuminators, eliminating the shrimps' silhouettes when viewed from beneath in much the same way that the silvery undersides of a surface-dwelling fish do in bright sunlight. Counter-illumination is most important in upper mesopelagic species, residing at depths <600 m, where light is dim but still present. Ventral photophores and hepatic luminescent organs thus have a similar role.

Photophore patterns have a species-specific and sometimes sex-specific arrangement in the decapods (Kemp 1910), a trait we have seen in the euphausiids and will also see later in the fishes. It has been assumed that they serve a role in partner recognition, or at the least, to separate friend from foe, but that has not been validated.

Luminous secretions are used in defense. The ejection of a bright luminous cloud in the eyes of a predator could temporarily blind or confuse it long enough for a rapidly swimming shrimp to disappear in the surrounding darkness. Alternately, a bright cloud could serve as a "burglar alarm" (e.g. Fleisher and Case 1995), attracting nearby larger predators to make a meal of the predator stalking the decapod (Widder 2010).

Circulatory, Respiratory, and Excretory Systems

As in the euphausiids, the decapods have a globular heart located in the posterior third of the cephalothorax (see Figure 7.27). The dendrobranchs have three pairs of ostia opening into the pericardium, two pairs dorsally and one pair laterally (McLaughlin 1980). The heart of carideans is nearly identical in location and general structure to that of the dendrobranchs. Most have three pairs of ostia situated as in the dendrobranchs, but at least one freshwater species (*Caridina leavis*) has five pairs: two dorsal, two lateral, and one ventral (Pillai 1965).

Arterial circulation among the decapods is generally as described in the Section Crustacean Systems of the chapter. The anterior aorta provides blood to the eyes and brain; the anterior lateral arteries feed the cephalic appendages, foregut, antennal glands, and musculature (McLaughlin 1983). Flow to the liver is the province of the hepatic artery. The ventral thoracic artery supplies the thoracic appendages. It in turn is fed by the sternal artery, which originates in the heart. Blood flow to the abdominal region is supplied by the posterior aorta via its many segmental arteries.

Respiration and excretion are described in detail in "Crustacean Systems."

Digestive System

Like the euphausiids, the gut of decapods has typical malacostracan form and function (Young 1959; Pillai 1960; Dall 1967; McLaughlin 1980, 1983). Both are described in detail in "Crustacean Systems."

Development

With the exception of the Luciferidae, all Dendrobranchs are broadcast spawners with many small eggs that are released into the plankton to fend for themselves directly after hatching. In *Lucifer*, eggs are brooded on the third pereiopod until hatching (Brooks 1882). The carideans all brood their young, releasing the larvae at a more advanced stage of development. Using the terminology introduced previously, the dendrobranchs exhibit anamorphic development and the carideans, epimorphic.

Dendrobranchs. Names for the larval stages within the dendrobranchs differ from author to author, but the four main stages, the nauplius, protozoea, zoea, and postlarvae, are recognizable in each (Figure 7.51). We will stick with the simplest nomenclature (Omori 1974). The earliest stage posthatch is the nauplius, recognizable by its three pairs of limbs: the antennules, antennae, and mandibles. All three pairs are used for swimming. Naupliar stages do not feed and are normally completed rapidly, generally within 24– 68 hours (Schram 1986). In the next stage, the protozoea, the mouthparts become functional (a diagnostic), but the antennae and antennules are still used for swimming, sometimes along with the first and second maxillipeds. The abdomen develops distinct segments in the protozoeal stage as well as rudimentary limbs and uropods. In the zoeal stage, some or all of the pereiopods and the third maxillipeds are used in swimming. By the end of the zoeal stages, all limbs are present but the

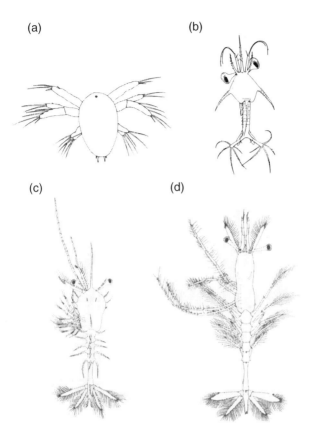

Figure 7.51 Dendrobranch larval stages. (a) First nauplius of *Sergia lucens*; (b) Protozoea of *Sergestes crassus*; (c) Second zoea of *Sergia lucens*; (d) third post-larva of *Sergia lucens*. *Sources:* (a) Omori (1974), figure 12 (p. 274); (b) Gurney (1942), figure 60 (p. 189); (c, d) Omori (1974), figure 15 (p. 278).

pleopods are not yet functional. Zoeal stages swim tail first with their back down.

Schram provides a useful rule-of-thumb for recognizing the different basic stages. "Naupliar and protozoeal stages have antennal locomotion, zoeal stages have thoracopodal locomotion and post-larval stages finally arrive at pleopodal locomotion." In some treatments of dendrobranch development, a metanauplius phase is recognized between the nauplius and protozoeal stages in the Penaeoidea, e.g. Schram (1986). A metanauplius may be distinguished from nauplius stages by the presence of other developing (non-functional) limbs outside of the first three pairs of cephalic appendages (Williamson 1969). Omori (1974) noted that the number of molts per stage varies between species and that, within species, environmental conditions may dictate the number of molts in each. Despite those differences, development in the two dendrobranch superfamilies is considered to be nearly identical and fit the basic nauplius-protozoea-zoea-postlarva pattern (Figure 7.51).

Carideans. Eggs incubated by the female carideans hatch into a late protozoea or zoeal phase larva (Figure 7.52) (Omori 1974). Development during the

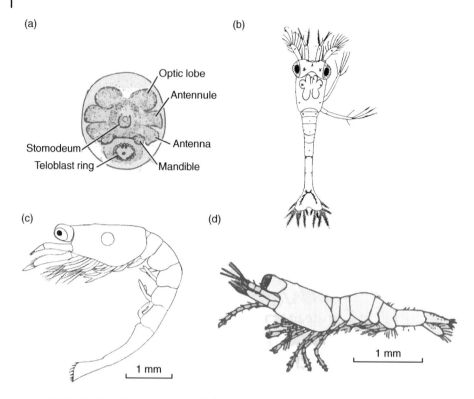

Figure 7.52 Caridean larval stages. (a) Caridean egg after teloblast formation; (b) Stage 1 nauplius of *Processa edulis*; (c) Pasaphaeid zoea; (d) Postlarva of *Thor floridanus*. *Sources:* (a) Oishi (1959), figure 6 (p. 296); Reproduced with permission of John Wiley & Sons. (b) Gurney (1942), figure 77 (p. 213); (c) Adapted from Williamson (1960), figure 1 (p. 333); (d) Adapted from Dobkin (1968), figure 2 (p. 6).

zoeal stages is followed by a post-larval sequence before the shrimp assumes its adult form. The number of molts after hatching varies between species and may be lifestyle-dependent, with deeper-living forms showing fewer zoeal stages prior to achieving the post-larval phase. Mesopelagic genera such as *Hymenodora*, *Oplophorus*, and *Systellaspis* have large, yolky eggs and five zoeal stages, whereas the shallower-dwelling benthic *Pandalus jordani* has smaller eggs and eleven zoeal stages (Modin and Cox 1967; Omori 1974).

The post-larval form is a bit confusing because despite the fact that it is called "post-larval", it is still considered a larval stage and is usually still adding limbs, notably pleopods, prior to assuming a true juvenile stage. Williamson's (1969) definitions for different stages are clearest to your authors, but he combines the protozoeal and zoeal stages. His stages are given below and shown in Figure 7.52.

Nauplius: larva with the first three pairs of cephalic appendages setose and functional, other appendages absent or rudimentary.

Zoea: larva with natatory expopods on some or all of the thoracic appendages with pleopods absent or rudimentary.

Postlarva: (this stage was termed a "megalopa" in the original publication – we are changing it here for clarity) larva with setose natatory pleopods on some or all of the abdominal somites 1–5.

Juvenile: young form, usually small and sexually immature, showing a general resemblance to the adult.

Reproduction

The different early life histories of dendrobranch and caridean shrimp reflect their reproductive habits. The fact that carideans are brooders limits their number of offspring and though considerable overlap occurs, pelagic carideans are usually found at greater depth than the dendrobranchs. Spawning and larval development in carideans must take place at mesopelagic or bathypelagic depths. Once hatched, their larvae must contend with a more limited food supply than that available in surface waters.

Dendrobranchs In dendrobranchs, the reproductive sequence comprises gamete development, parturial molt, mating, and spawning, followed by larval development. The best information on pelagic dendrobranchs is available for the sergestid shrimps, *Sergia lucens* and *Sergestes similis*. As in the euphausiids, the endopods of the first pleopods in male sergestids are modified to form the copulatory structure known as a petasma. Females bear a thelycum, a sperm receptacle, at the level of the third periopod. The testis of male sergestids is located atop the hepatopancreas in the cephalothorax, communicating with the gonopores via the vas deferens. The ovary of females, located in a near-identical position atop the hepatopancreas, communicates with the gonopore via an oviduct.

Copulation takes place after a parturial molt, with the sternum of the males oriented toward that of the females and a body position parallel or perpendicular to the receptive females. A spermatophore is transferred to the open thelycum of the females with the petasma, where it is attached "by means of the plates and hooks of the thelycum, the hook-like wings of the spermatophore, and a secretion produced by the male's terminal ampoule complex," located at the terminus of the vas deferens (Genthe 1969). Spawning takes place a few days after mating. Sperms from the spermatophore are enzymatically released and concentrated near the oviduct opening. Eggs are fertilized directly by the non-motile sperm as they are shed into the sea (Genthe 1969). Time elapsed between spawning and hatching is reported as 24–36 hours in *Sergia lucens* at 22 °C (Omori 1974). *Sergestes similis*' time to hatch varied between 52 hours at 18 °C and 105 hours at 10 °C; median time for development from egg to juvenile in *Sergestes similis* at 14 °C was 109 days (Omori 1979).

Those species that have been studied shed their eggs at night in the upper 50 m, or at the upper end of the adults' vertical range. All studied species are vertical migrators that occupy mesopelagic depths during the day (Omori 1974). Once shed, the negatively buoyant eggs of *Sergestes similis* sank to about 220 m before hatching. The protozoeal and zoeal larval stages were found almost exclusively in the upper 140 m. It was surmised that the most vulnerable but non-feeding naupliar stage developed near the hatching depth. However, the need for food resulted in protozoeal (PZ) and zoeal (Z) stages residing in shallower, more productive depths (20–40 m). Beginning at PZ I, larvae began migrating downward at daybreak with differences in day–night distributions increasing with age Omori (1979).

Seasonality of reproduction varies between species and locations. The cold-temperate *Sergestes similis* shows a peak spawning time between late December and early April off Southern California with some smaller pulses during the summer months (Omori 1979). Both males and females have developing gametes throughout the year (Genthe 1969). The same species is believed to reproduce year-round off Oregon with small pulses in spring and winter (Pearcy and Forss 1969). *Sergestes similis* in Monterey Bay, located between the southern California and Oregon populations, is reported to have two co-existing cohorts with peak reproductive periods in December–January and June–July (Barham 1957). Circulation and seasonality off the northwest coast of the U.S. are complex, with variability in eastern boundary currents and seasonal upwelling influencing temperature and productivity. It is likely that some of the variability observed in *S. similis* reproductive patterns is a result of that complexity.

Reproductive patterns in sergestoids from other systems are more predictable. The boreal *Sergestes arcticus* is reported to spawn in the spring (Matthews and Pinnoi 1973). In the warm-temperate waters off Japan, *Sergia lucens* and *Acetes japonicus* show breeding periods from May to November and May to October, respectively, with strongest pulses in the summer months (Omori 1974). Tropical species such as the *Acetes erythraeus* found off India are believed to reproduce year-round. A recurring, but not universal, theme for the sergestoids is the presence of two overlapping generations in sampled populations (Omori 1974).

Carideans In carideans, the reproductive sequence consists of gamete development, parturial molt, mating, spawning, incubation of eggs, and hatching and release of larvae. Ovaries and testis in carideans are located in the cephalothorax in virtually the same position as those in the dendrobranchs, with the anterior portions lying atop the hepatopancreas and the posterior beneath the heart (Figure 7.27). In males, the vas deferens conducts the mature sperm to the gonopores on the coxae of the fifth pereiopods. The vas deferens terminates in an analogous structure to the ampoule complex of dendrobranchs, a muscular ejaculatory duct coupled with an androgenic gland (Bauer 2004). In females, an oviduct conveys mature eggs to the gonopores located on the third pereopods.

Caridean males may be identified by the presence of a stalk-like appendix masculina on the inner ramus of their second pleopods. Female secondary sexual characters that develop during breeding season include a deepening of the cavity below the abdomen owing to the increased length of the abdominal pleura (side plates of the carapace) coupled with an increase in their fringing setae. The first three pleopods increase in size by lengthening the coxa and basipodites, and their fringing setae also become much longer. The full effect of the lengthening in pleura and pleopods is to form a brood chamber for accepting the eggs during spawning. It is termed "breeding dress" (Bauer 2004) and disappears in the post-parturial molt after the larvae are released.

Copulation occurs shortly (minutes to hours) after the parturial molt. The male shrimp grasps the female and positions his gonopores near hers, against her ventrum. Partners are usually at right angles to one another. At that time, the ejaculatory ducts each squeeze out a spermatophore cord, much like toothpaste from a tube. The twin cords form the dual spermatophore, which adheres to the ventral surface of the female within her brood chamber. Spermatophore transfer normally takes only a few seconds (Bauer 2004).

Soon after mating (again, minutes to hours), the female spawns. She has already broken up the spermatophores through preening, spreading pieces of spermatophore and the sperm they contain throughout the brood chamber. As in the dendrobranchs, the sperm are non-motile so fertilization must take place by direct contact. As eggs leave the oviducts, they are coaxed into the brood chamber by the beating of the first pleopods; the resultant water motion brings them into contact with sperm.

How long eggs are incubated depends largely on environmental temperature. The shortest times, a week or less, are found in tropical species; the longest incubation times, about five months, are in boreal species such as *Pandalus borealis* (Bergstrom 2000; Bauer 2004). *Pasiphaea multidentata*, a boreal pelagic species found in the Gulf of Maine and off the coast of Norway, also as a five-month incubation time (Matthews and Pinnoi 1973). The numbers and sizes of eggs produced are species-specific. However, all other things being equal, the larger the female, the more eggs can be carried (Hopkins et al. 1989). Thus, large size is a reproductive advantage.

Carideans, particularly the smaller species inhabiting warm-temperate and tropical systems, are believed to produce successive broods in a breeding season and in some cases continually until death. To achieve multiple broods, eggs develop in the ovary, while the embryos are incubating in the brood chamber. Each larval release is followed by a fresh parturial molt, copulation, and a new spawning with the eggs that matured in the ovaries during the previous incubation. In cold water species, larval development is slow enough so that only one to two broods are produced per year. *Pasiphaea multidentata* produces two broods per year, one each in the spring and fall with the second showing a lower average number of eggs per female (50 vs. 83: Appollonio 1969).

Deeper-living mesopelagic and bathypelagic species are believed to spawn and incubate their eggs within their normal depth profile; no pronounced seasonal spawning has been reported. Both seasonal and year-round reproductive activity have been reported in pelagic species (Chace 1940; Ziemann 1975; Hopkins et al. 1989). Vertically migrating species that ascend into the upper 400 m at night may show either activity, but the consensus is that lower mesopelagic and bathypelagic species reproduce year-round (Hopkins et al. 1989).

Egg Size and Fecundity The pelagic carideans show a species-specific dichotomy in egg size that is normally termed "small-egg-" versus "large-egg-producing" species. As noted above, larvae of small-egg species take a longer time to develop and have a larger number of stages. For example, the oplophorid *Acanthephyra purpurea*, a small-egg species, has seven or more developmental stages, while its confamilials *Oplophorus gracilirostris* and *Systellaspis debilis* with large yolky eggs have only five (Table 7.6). A shorter larval period confers increased independence on developing young shrimp, giving them better locomotory and feeding capabilities in the food-scarce environment of the deep sea; a useful strategy in highly seasonal polar climes as well. The downside to large eggs is an accompanying sacrifice in fecundity. For example, brood size in *Acanthephyra purpurea* is about 1800 eggs, whereas *Oplophorus gracilirostris* only produces about 27: clearly a trade-off (Table 7.6, Hopkins et al. 1989).

Protandry Sex change has been reported for several species within the Caridea, with the most typical pattern being some form of protandry, that is, a developmental sequence where juveniles initially develop as males and change into females as they increase in size. The advantage to having

Table 7.6 Adult size ranges, egg sizes, and fecundity of pelagic decapods.

Taxon	Carapace length (mm)	Egg size (mm)	Number in brood	Source
suborder DENDROBRANCHIATA				
superfamily Sergestoidea				
family Sergestidae				
Acetes japonicus	3–8	0.25	680–6800	Omori (1974)
Sergia lucens	10–13	0.25–0.27	1700–2300	Omori (1974)
suborder PLEOCYEMATA				
infraorder Caridea				
superfamily Pandaloidea				
family Pandalidae				
Parapandalus richardi	7–9	0.45×0.60	316	Hopkins et al. (1989)
Parapandalus zurstrasseni	7–8	0.79×0.52	320–380	Omori (1974)
superfamily Pasiphaeoidea				
family Pasiphaeidae				
Pasiphaea multidentata	18–30	2.6×2.0	40–120	Omori (1974)
Parapasiphaea sulcatifrons	22–26	5.0×4.0	25	Omori (1974)
superfamily Oplophoroidea				
family Acanthephyridae				
Acanthephyra acanthitelsonis	15–20	0.71×1.0	1856	Hopkins et al. (1989)
A. curtirostris	12–18	0.67×0.83	1748	Hopkins et al. (1989)
A. eximia	41	1.0×0.7	6990	Omori (1974)
A. purpurea	16	0.66×0.96	1820	Hopkins et al. (1989)
A. quadrispinosa	15–17	1.15×0.75	700–1540	Hopkins et al. (1989)
A. stylorostratis	10–11	0.68×0.94	282	Hopkins et al. (1989)
Hymenodora frontalis	14	3.2×2.5	19	Omori (1974)
Meningodora vesca	14	0.70×0.74	254	Hopkins et al. (1989)
Notostomus japonicus	42	0.92×0.72	8014	Omori (1974)
family Oplophoridae				
Oplophorus gracilirostris	15–17	1.64×2.54	27	Hopkins et al. (1989)
Systellaspis brauri	23	4.4×2.8	22	Omori (1974)
S. debilis	14–15	4.6×2.6	16–20	Omori (1974)
S. debilis	12–18	1.74×2.93	14	Hopkins et al. (1989)
S. lanceocaudata	19	3.9×2.8	46	Omori (1974)

females in the larger sizes is the obvious one: they can produce and carry more eggs. Within the Caridea, the best known of the sex-changers are the pandalids (Bauer 2000; Bergstrom 2000).

Sex change has not been confirmed in any of the pelagic species, but differences in size between sexes have been observed (Genthe 1969). Differences in size between males and females can arise from protandry or be just a simple difference in adult size between the sexes. A change in the gonads from testes to ovaries must be observed to confirm protandry. Sex change as a life history tactic is a fascinating subject with multiple levels of complexity. It can be as simple as the advantages offered by the increased fecundity of larger females, as seen in some shrimp, and as complex as the social systems of some reef fishes (Stearns 1992).

Vertical Distribution

The vertical distribution of decapod shrimp is more complex than that of the euphausiids. Decapods have a greater vertical range and behavioral diversity, and in addition to a robust presence at upper and lower mesopelagic depths, they are well represented in the bathypelagic realm.

Omori (1974) categorized the vertical distributions of juvenile and adult pelagic shrimp into seven basic patterns (Figure 7.53) that conform well with the many studies executed before and after his monograph was published. Species are reported here with the original genus-species names given in Omori (1974). In particular, the genus *Sergestes* had not been subdivided into its most recently recognized five genera.

1) Epipelagic. Species that remain in the upper 150 m day and night, including all members of the genera *Acetes*, *Peisos*, and *Lucifer*.

2) Upper mesopelagic. Species that reside in the upper 100 m at night and between 150 and 300 m during the day. (*Sergestes similis* and *S. arcticus*, *Sergia lucens*, *Pasiphaea sivado*).

3) Mesopelagic species. Found between 100 and 300 m at night and 500–700 during the day. *Gennadas incertus*, *Gennadas propinquus*, *Funchalia villosa*, *Sergestes atlanticus*, *S. corniculum*, *S. erectus*, *S. orientalis*, *S. pectinatus*, *S. sargassi*, *Acanthephyra quadrispinosa*, *Systellaspis debilis*, *Parapandalus richardi*.

4) Lower mesopelagic species with a strong migration. Found between 200 and 400 at night and 750–950 m during the day. *Gennadas bouvieri*, *G. parvus*, *G. valens*, *Sergia inequalis*, *S. spendens*, *Acanthephyra purpurea*.

5) Lower mesopelagic species with a weak migration. Found between 550–750 m at night and 800–1000 m during the day. *Gennadas capensis*, *G. elegans*, *Sergia robustus*, *Acanthephyra acanthitelsonis*, *A. pelagica*.

6) Species that straddle the mesopelagic and bathypelagic zones. May or may not migrate. Reside 800–1400 m night and 900–1500 m day. *Bentheogenemma borealis*, *B. intermedius*, *Sergia japonicus*, *Petalidium foliaceum*, *P. obesum*, *Parapasiphae sulcatifrons*, *Acanthephyra prionota*, *Hymenodora frontalis*.

7) Bathypelagic residents. Found below 1400 m. *Acanthephyra stylorostratis*, *Hymenodora glacialis*, *H. gracilis*, *Physetocaris microphthalmis*.

Due to the practical considerations of time and money, only a handful of studies have sampled the bathypelagic realm, particularly in recent years, It takes a long time to deploy a net to depths >1000 m and to then fish it long enough to get a representative sample (6–12 hours or more). Shiptime

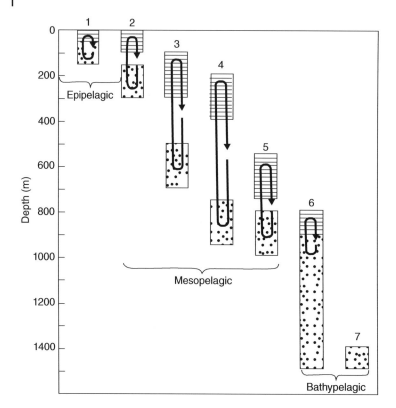

Figure 7.53 Schematic illustration of the diurnal migration of pelagic shrimps. Dotted and hatched areas indicate the depths of the main day and night concentrations: (1) neritic and epipelagic species; (2) upper mesopelagic species; (3) lower mesopelagic species; (4) lower mesopelagic species (strong migrators); (5) lower mesopelagic (weak migrators); (6) lower mesopelagic and upper bathypelagic, some migrators some non-migrators; (7) bathypelagic non-migrators. *Source:* Omori (1974), figure 5 (p. 252). Reproduced with the permission of Academic Press.

is expensive, and multiple research programs are often sharing the vessel: thus, time is at a premium. Even fewer studies have directly compared the fauna of mesopelagic and bathypelagic depths in the same locale using the same sampling gear. Fortunately, just such a set of studies was completed in the eastern Gulf of Mexico (EGOM).

The Gulf of Mexico is a tropical-subtropical system exhibiting high diversity in all elements of its pelagic fauna and particularly in its Crustacea. Its water column has warm surface temperatures throughout the year (20–30 °C) with a shallow mixed layer of 25–75 m and permanent thermocline beginning at about 150 m (Hopkins et al. 1994). Temperatures at 500 and 1000 m are 8–9 °C and 4–5 °C, respectively (Lancraft et al. 1988). Below 1000 m, temperatures remain at about 4 °C to 2500 m and below, and downwelling light is totally absent (Gallaway et al. 2001).

Physical differences between the mesopelagic and bathypelagic environments are minimal at the arbitrary border of 1000 m but are quite different in their overall characteristics. Marked gradients are present in light and temperature within the mesopelagic zone as depth increases from 200 to 1000 m (see Chapter 1). In contrast, the bathypelagic zone is characterized by a uniform absence of sunlight and low constant temperatures that change minimally with increasing depth. From a biotic standpoint, increasing distance from the surface and algal production means that food resources are limited to carnivory or detritivory on sinking particulates. Seasonal change at bathypelagic depths is minimal with the caveat that seasonal peaks and valleys in surface production would be mirrored in the bathypelagic zone by accompanying changes in particulates raining from the surface, though they would be filtered by the entire mesopelagic zone.

The mesopelagic zone of the eastern Gulf of Mexico supports a total micronekton biomass of about $475\,kg\,DM\,km^{-3}$ (Hopkins and Lancraft 1984), with about 250 species of fish (Gartner et al. 1987; Sutton and Hopkins 1996a; Hopkins et al. 1997), at least 47 species of decapods (Heffernan and Hopkins 1981; Flock and Hopkins 1992; Hopkins et al. 1994; Burghart et al. 2007), and at least 43 species of cephalopods (Passarella and Hopkins 1991). Overall, the decapods comprise about 31% of the total biomass of the mesopelagic zone (Hopkins and Lancraft 1984). A summary figure of diel vertical migration patterns for dominant GOM mesopelagic decapods is shown in Figure 7.54. Note the concordance with Omori's patterns described above. In particular, the several *Acanthephyra* species in the lower mesopelagic zone are also important at bathypelagic depths.

A comparison of the mesopelagic and bathypelagic decapod fauna for the Gulf of Mexico using all available data is presented in Table 7.7. A few patterns are evident. There is an overlap in species. The dendrobranchs *Gennadas valens* and *Sergia splendens* ranked in the top ten most abundant in both zones, and four species ranked in the top twenty in biomass in both: *Acanthephyra curtirostris, G. valens, G. capensis,* and *A. purpurea* (Figure 7.54). Of the 59 species of decapods considered here, about half are found at both mesopelagic and bathypelagic depths (Burghart et al. 2007). A similar species overlap is reported in the North Atlantic and Caribbean (Vereshchaka 1994), in the eastern Atlantic in the vicinity of the Canary Islands (Foxton 1970), and off Bermuda (Donaldson 1975).

Though there are considerable similarities in the shrimp fauna in the two zones, meaningful differences are plentiful as well. Shrimps were decidedly less abundant in the bathypelagic zone of the EGOM than in the mesopelagic. Total numbers of decapods in the mesopelagic zone were $2.9 \times 10^6\,km^{-3}$; in the bathypelagic zone, shrimps numbered $1.6 \times 10^5\,km^{-3}$ (Table 7.7, Hopkins et al. 1994; Burghart et al. 2007). Mesopelagic decapod biomass was $179.9\,kg\,DM\,km^{-3}$ versus $48.9\,kg\,DM\,km^{-3}$ for bathypelagic depths. One may conclude from the numbers and biomass that though they were fewer in number, shrimps living in the bathypelagic zone tended to be larger than those in the mesopelagic.

Twelve species occurred only in the bathypelagic zone, ten of which were in the Oplophoridae (Table 7.7), and dominant species within the decapod families changed with depth. *Gennadas valens*, the most important shrimp in the mesopelagic realm, weighing in at about 50% of the total micronektonic crustacean biomass, was replaced by *Bentheogenemma intermedia* as the dominant benthesicymid in the bathypelagic zone. Similarly, in the oplophorids, two of the three most abundant species in the mesopelagic realm, *Systellaspis debilis*

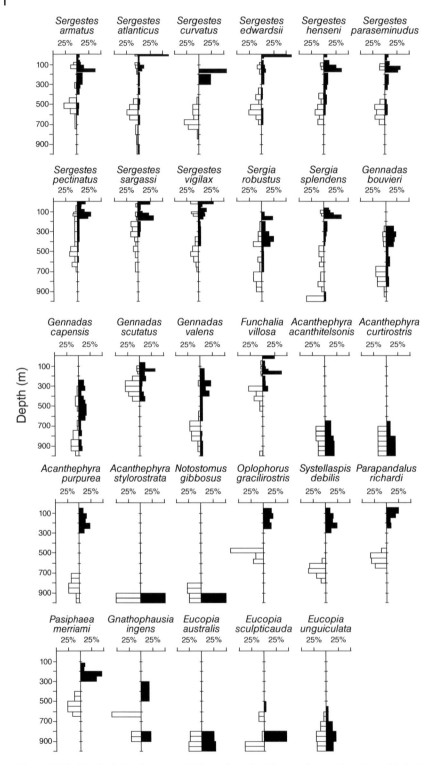

Figure 7.54 Vertical distributions of 29 species of midwater decapods and mysids in the eastern Gulf of Mexico. Open and shaded bars represent day and night, respectively. *Source:* Hopkins et al. (1994), figure 1 (p. 147). Reproduced with the permission of Inter-Research.

Table 7.7 Abundance and biomass of decapods and lophogastrids from mesopelagic and bathypelagic depths in the eastern Gulf of Mexico. *Source*: Data from Flock and Hopkins (1992), Hopkins et al. (1989, 1994), Burghart et al. (2007).

	Mesopelagic zone		Bathypelagic zone	
	Abundance (ind km^{-3})	Biomass (kg DM km^{-3})	Abundance (ind km^{-3})	Biomass (kg DM km^{-3})
Superorder Eucarida				
Order Decapoda				
Suborder Dendrobranchiata				
Superfamily Penaeoidea				
Family Benthesicymidae				
Bentheogenemma intermedia			25 800.00	7.40
Gennadas bouvieri	88 000.00	5.60	600.00	0.10
Gennadas capensis	187 500.00	13.10	8 000.00	0.90
Gennadas scutatus	31 000.00	1.30		
Gennadas valens	1 237 500.00	85.70	19 400.00	2.10
Gennadas spp.			5 100.00	0.30
Family Penaeidae				
Funchalia villosa	8 130.00	2.10		
Superfamily Sergestoidea				
Family Sergestidae				
Sergestes armatus	23 000.00	0.70		
Sergestes atlanticus	97 000.00	1.60		
Sergestes cornutus	31 800.00	0.10		
Sergestes curvatus	5 000.00	0.70		

(*Continued*)

Table 7.7 (Continued)

	Mesopelagic zone		Bathypelagic zone	
	Abundance (ind km^{-3})	Biomass (kg DM km^{-3})	Abundance (ind km^{-3})	Biomass (kg DM km^{-3})
Sergestes edwardsi	36 400.00	0.30		
Sergestes henseni	90 800.00	5.20		
Sergestes paraseminudus	43 600.00	3.80		
Sergestes pectinatus	271 600.00	2.20		
Sergestes sargassi	63 300.00	0.90		
Sergestes vigilax	33 800.00	0.60		
Sergia filictum	14 600.00	3.30		
Sergia grandis	4000.00	1.40	500.00	0.20
Sergia japonicus now S. japonica	2000.00	0.40	800.00	0.30
Sergia regalis			1600.00	0.60
Sergia robustus	68 500.00	3.80		
Sergia splendens	333 400.00	15.30	11 200.00	0.80
Sergia talismani	16 000.00	0.60		
Sergia tenuiremus	4500.00	0.10		
Sergia wolfi			1600.00	0.70
Sergia spp.			2300.00	0.60

Mesopelagic zone

	Abundance (ind km^{-3})	Biomass (kg DM km^{-3})
Suborder Pleocyemata		
Infraorder Caridea		
Superfamily Bresilioidea		
Family Bresilidae		
Lucaya bigelowi		
Superfamily Oplophoroidea		
Family Acanthephyridae		
Acanthephyra acanthitelsonis	6 700.00	3.10
Acanthephyra acutifrons		
Acanthephyra curtirostris	21 800.00	4.10
Acanthephyra purpurea	33 750.00	7.20
Acanthephyra stylorostratis	6 780.00	1.30
Ephyrina benedicti		

Bathypelagic zone

	Abundance (ind km^{-3})	Biomass (kg DM km^{-3})
Suborder Pleocyemata		
Infraorder Caridea		
Superfamily Bresilioidea		
Family Bresilidae		
Lucaya bigelowi	1 600.00	0.10
Superfamily Oplophoroidea		
Family Acanthephyridae		
Acanthephyra acanthitelsonis	900.00	2.50
Acanthephyra acutifrons	700.00	9.30
Acanthephyra curtirostris	8 200.00	2.50
Acanthephyra eximia	100.00	0.10
Acanthephyra gracilipes	3 700.00	0.80
Acanthephyra pelagica	200.00	0.10
Acanthephyra purpurea	2 300	0.80
Acanthephyra quadrispinosa	500.00	0.30
Acanthephyra stylorostratis	24 600.00	3.30
Ephyrina benedicti	600.00	1.50
Ephyrina ombango	1 300.00	1.20

	Abundance (ind km^{-3})	Biomass (kg DM km^{-3})
Family Acanthephyridae		
Hymenodora glacialis	25 800.00	2.50
Hymenodora gracilis	8 000.00	1.00

(*Continued*)

Table 7.7 (Continued)

	Mesopelagic zone			Bathypelagic zone	
	Abundance (ind km^{-3})	Biomass (kg DM km^{-3})		Abundance (ind km^{-3})	Biomass (kg DM km^{-3})
			Meningodora marptocheles	100.00	0.10
			Meningodora miccyla	100.00	0.10
Meningodora mollis			*Meningodora mollis*	1000.00	0.10
Meningodora vesca			*Meningodora vesca*	900.00	0.30
Notostomus gibbosus	1600.00	0.50	*Notostomus gibbosus*	1300.00	6.50
Notostomus westergreni					
Family Oplophoridae			Family Oplophoridae		
Janicella spinicauda			*Janicella spinicauda*	600.00	0.10
Oplophorus gracilirostris	4950.00	0.60			
Systellaspis cristata			*Systellaspis braueri*	300.00	0.30
Systellaspis debilis	70000.00	11.80	*Systellaspis cristata*	100.00	0.10
			Systellaspis debilis	800.00	0.20
			Systellaspis pellucida 300.00	0.10	
Superfamily Pandaloidea					
Family Pandalidae					
ªStylopandalus richardi	36500.00	1.90			
Superfamily Pasiphaeoidea			Superfamily Pasiphaeoidea		
Family Pasiphaeidae			Family Pasiphaeidae		
			Parapasiphaea macrodactyla	500.00	0.10
Parapasiphaea sulcatifrons	7350.00	0.60	*Parapasiphaea sulcatifrons*	1300.00	1.00
Pasiphaea merriami					
Totals	2880860.00	179.90	Totals	162700.00	48.90

Taxon	Value	%	Taxon	Value	%
Superorder Peracarida			Superorder Peracarida		
Order Lophogastrida			Order Lophogastrida		
Family Eucopiidae			Family Eucopiidae		
Eucopia australis	22 900.00	1.00	Eucopia australis	40 200.00	3.10
			Eucopia grimaldi	39 100.00	1.90
Eucopia sculpticauda	91 670.00	3.10	Eucopia sculpticauda	40 500.00	1.60
Eucopia unguiculata	167 600.00	3.60	Eucopia unguiculata	25 300.00	1.20
Family Gnathophausiidae			Family Gnathophausiidae		
			Gnathophausia gigas	900.00	0.40
			Gnathophausia gracilis	800.00	0.20
Gnathophausia ingens	6 346.00	1.10	Gnathophausia ingens	100.00	0.20
			Gnathophausia zoea	600.00	0.10
Family Lophogastridae			Family Lophogastridae		
			Pseudochalaraspidum hanseni	400.00	0.10
Order Mysida			Order Mysida		
Family Mysidae			Family Mysidae		
			Boreomysis spp.	1 700.00	0.10
Totals	288 516.00	8.80	Totals	149 600.00	8.90

[a] Formerly *Parapandalus richa*.

and *Acanthephyra purpurea*, were replaced by *Hymenodora glacialis* and *Acanthephyra stylorostratis*, with *Acanthephyra curtirostris* remaining in the top three in both depth zones.

Perhaps, the most important difference in the mesopelagic and bathypelagic decapod fauna is the prevalence of dendrobranchiate broadcast spawners above 1000 m and of caridean brooders at bathypelagic depths (Burghart et al. 2007). The more fully developed larvae of the carideans at hatching stand a better chance of survival in an environment characterized by a lower biomass of small zooplankton prey and by a complete absence of primary production. In contrast, most of the mesopelagic dendrobranchiate shrimps are strong migrators and can broadcast their eggs in the epipelagic zone, taking advantage of the higher productivity of surface waters. Similar trends have been observed in shrimp fauna of the eastern Pacific (Krygier and Pearcy 1981) and off Hawaii (Walters 1976).

Food and Feeding

Studies of diets in pelagic Crustacea are hampered by the activity of the animals' mandibles and cardiac stomach, which macerate food into tiny particles that require an expert taxonomist to identify. Nonetheless, a respectable amount of information on feeding habits is available, including trends with depth.

Hopkins et al. (1994), working in the mesopelagic zone of the eastern Gulf of Mexico, analyzed the diet composition of 29 micronektonic crustacean species (Table 7.8, Figure 7.55) and then used cluster analysis to identify feeding guilds as well as overlap between species. Overall, the principal food was crustacean, accounting for 46% of the biomass consumed. Euphausiids were the most important items in the diet (28%) followed by copepods (17%).

Six feeding guilds were identified by cluster analysis with guild diet compositions ranging from copepod and euphausiid specialists to euphausiid/copepod mixes, fish specialists, and diverse diets including chaetognaths, ostracods, copepods, fish, euphausiids, and chaetognaths. Cluster analysis of food-size distributions also yielded six clusters, with three of those clusters specializing in crustaceans smaller than 8 mm in length, primarily copepods, ostracods, and small euphausiids. Those clusters included 8 of the 11 sergestids and the eucopiids. The remaining three clusters, including the carideans, benthesicymids, and the remaining sergestids, had a diet with food items larger than 8 mm, including decapods, fishes, larger euphausiids, and chaetognaths.

Principal prey items for the mesopelagic crustacean assemblage shown in Table 7.8 were zooplankton, whose stocks must also feed epipelagic and mesopelagic fishes. Hopkins et al. (1994) estimated the predation impact of the micronektonic crustacean community as approximately 1% of the standing stock per day, assuming a daily ration of 6% of body mass per day. Total micronektonic crustacean biomass was estimated to be 0.18 g dry mass (DM) m^{-2} in the upper 1000 m, and zooplankton biomass was estimated to be 1.2 g DM m^{-2} over the same depth interval. A further exercise in predation impacts used the production-to-biomass ratio, an estimate of new zooplankton biomass produced by the existing standing stock of zooplankton through individual growth and reproduction. For tropical/subtropical systems like the EGOM, it is assumed to be 0.05 : 1 (Shushkina 1973, 1985). Thus, with a standing stock of 1.2 g DM m^{-2}, the daily production of new zooplankton is about 0.06 g DM m^{-2}. Further, if the total biomass of micronektonic Crustacea is 0.18 g DM m^{-2}

Table 7.8 Diet composition of dominant decapods and lophogastrids from the mesopelagic and bathypelagic zones of the eastern Gulf of Mexico. With the exception of the Cnidaria, reported as % occurrence by virtue of the presence of nematocysts only, numbers represent the percent of diet biomass for each taxon. Due to differences in diet composition of meso- and bathypelagic fauna, radiolaria/detritus are reported as % prey biomass for mesopelagic species and % occurrence of detritus in guts for bathypelagic species. Lophogastrids and polychaetes are reported for bathypelagic diets due to their increased relative importance. "ND" = no data. "n"=number of individuals analyzed. *Source*: Data from Hopkins et al. (1994) and Burghart et al. (2010).

Mesopelagic Zone	n	Phytodetritus[a]/debris	Cnidaria[a]/debris	Copepods	Ostracods	Euphausiids	Decapods	Lophogastrids	Polychaetes	Cephalopods	Gastropods	Chaetognaths	Siphono-phores	Radiolaria	Fishes
Superorder Eucarida															
Order Decapoda															
Suborder Dendrobranchiata															
Superfamily Penaeoidea															
Family Benthesicymidae															
Gennadas bouvieri	25	84	40	16.2	1.1	20.4	1.7	ND	ND	1.7	0.2	12	0	14.1	32.3
Gennadas capensis	25	88	48	12.2	1	22.3	8.5	ND	ND	0	0	14.9	0	5.1	36
Gennadas scutatus	25	76	27	14.6	0	38.7	0	ND	ND	0	0	9.4	0	6.3	30.9
Gennadas valens	79	86	29	10.1	0.9	28.3	2.1	ND	ND	0	0.2	16.7	0	13.1	28
Family Penaeidae															
Funchalia villosa	85	55	13	1	0	16.6	0	ND	ND	0	0.3	27.6	0	0	54.5
Superfamily Sergestoidea															
Family Sergestidae															
Sergestes armatus	15	13	0	17.8	0	68	0	ND	ND	0	0	7.2	0	5.8	0
Sergestes atlanticus	20	20	20	38.9	3.5	45.3	0	ND	ND	0	0.3	6.7	0	5.3	0

(*Continued*)

Table 7.8 (Continued)

Mesopelagic Zone	n	Phytodetritus[a]/debris	Cnidaria[a]/debris	Copepods	Ostracods	Euphausiids	Decapods	Lophogastrids	Polychaetes	Cephalopods	Gastropods	Chaetognaths	Siphono-phores	Radiolaria	Fishes
Sergestes curvatus	12	8	33	42.4	3.8	26.5	0	ND	ND	0	0.2	8	0	6.4	12.7
Sergestes edwardsi	10	20	0	10.4	0	77.3	11	ND	ND	0	0	0.2	0	0	0
Sergestes henseni	25	20	32	22.5	4.3	57.4	0	ND	ND	0	0	15	0	0	0
Sergestes paraseminudus	25	16	44	37.8	8.8	34.9	0	ND	ND	0	0	12.6	0	4.9	0
Sergestes pectinatus	24	4	25	90.7	9.3	0	0	ND	ND	0	0	0	0	0	0
Sergestes sargassi	18	22	22	83.8	3.7	12.5	0	ND	ND	0	0	0	0	0	0
Sergestes vigilax	25	20	0	19.4	0	78.5	0.5	ND	ND	0	0	0.6	0	0	0
Sergia robustus	40	48	90	22	20	17.9	0	ND	ND	0	0.1	25.4	0.9	12.7	0
Sergia splendens	80	70	71	15.1	20.4	9.8	9.8	ND	ND	0	2.8	18.7	0.3	18.5	
Suborder Pleocyemata															
Infraorder Caridea															
Superfamily Oplophoroidea															
Family Acanthephyridae															
Acanthephyra acanthitelsonis	49	4	16	8.4	0	3.5	27.2	ND	ND	6.2	2.1	20.4	0.1	0	31.4
Acanthephyra curtirostris	74	30	39	7.6	1	9.8	9	ND	ND	1.1	0.4	23.1	0	0	45.4
Acanthephyra purpurea	53	13	34	3	0.3	31.2	6.2	ND	ND	3.7	0.8	26.3	0	0.3	28

Species															
Acanthephyra stylorostratis	14	21	7.2	3.2	13.6	6.8	ND	ND	0	0.6	11.3	0	0	57.3	
Notostomus gibbosus	13	31	46	1.8	0	16.1	5.4	ND	ND	0	0	46	0	0	25.3
Family Oplophoridae															
Oplophorus gracilirostris	41	0	15	8.7	0	13.1	3.6	ND	ND	10.7	19.1	18.2	0.1	0	26.2
Systellaspis debilis	121	37	77	10.2	2.3	38	4.5			7.1	1.1	18.2	0	0	18.6
Superfamily Pandaloidea															
Family Pandalidae															
Stylopandalus richardi[b]	69	55	68	8.8	0	17.4	0	ND	ND	0	1.3	7.1	19.1	0	46.3
Superfamily Pasiphaeoidea															
Family Pasiphaeidae															
Pasiphaea merriami	22	0	5	1.2	0	62.1	12.1	ND	ND	0	0	0	0	1.9	2.7
Superorder Peracarida															
Order Lophogastrida															
Family Eucopiidae															
Eucopia australis	49	0	0	90.4	0	0	0	ND	ND	0	0	0	0	0	8.7
Eucopia unguiculata	117	0	0	79.6	0.4	0	0	ND	ND	0	0	4.7	0.7	0	14.6
Eucopia sculpticauda	155	0	0	93.4	0	0	0	ND	ND	0	0	6.6	0	0	0
Family Gnathophausidae															
Gnathophausia ingens	45	0	0	7	0.2	17	2.9	ND	ND	0	1.7	6	0	1.4	63.8

(Continued)

Table 7.8 (Continued)

Bathypelagic Zone	n	Phytodetritus[a]/debris	Cnidaria[a]/debris	Copepods	Ostracods	Euphausiids	Decapods	Lophogastrids	Polychaetes	Cephalopods	Gastropods	Chaetognaths	Siphono-phores	Radiolaria	Fishes
Superorder Eucarida															
Order Decapoda															
Suborder Dendrobranchiata															
Superfamily Penaeoidea															
Family Benthesicymidae															
Bentheogennema intermedia	55	65.5	40	11.8	0	0	4.5	0	0.2	0	0	2.5	0	ND	78
Gennadas valens	53	39.6	47.2	11.3	0	0	0	0	0	0	0	2.6	0	ND	85
Suborder Pleocyemata															
Infraorder Caridea															
Superfamily Oplophoroidea															
Family Acanthephyridae															
Acanthephyra acutifrons	8	0	12.5	0.2	0	0	21.6	8.4	0	0	0	0.1	0	ND	69.7
Acanthephyra curtirostris	53	5.7	18.9	4.4	0.3	0	12	2.9	0.2	0	0	2.5	0	ND	77.5
Acanthephyra stylorostratis	59	11.9	25.4	4.2	0.6	0	7.8	1.9	1.3	0	0	1.1	0	ND	82.6
Hymenodora glacialis	83	3.6	3.6	25.3	8.3	0	0	8.8	0.1	0	0	6.4	0	ND	51

Species															
Notostomus gibbosus	13	0	7.7	7.2	0	0.6	0.6	4.7	0	23.5	0	2.1	0	ND	60.7
Superorder Peracarida															
Order Lophogastrida															
Family Eucopiidae															
Eucopia australis	68	0	1.5	89.3	5.6	0	0	0	0	0	0	5	0	ND	0
Eucopia grimaldii	60	0	3.3	93.9	0	0	0	0	0	0	0	5.6	0	ND	0
Eucopia sculpticauda	71	0	2.8	57.6	0	0	0	0	0	0	0	4.3	0	ND	38.1
Eucopia unguiculata	57	0	0	97.9	0	0	0	0	0	0	0	2.1	0	ND	0

[a] Percent occurrence.
[b] Formerly *Parapandalus richardi*.

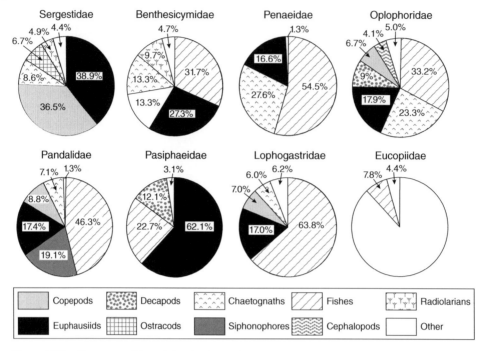

Figure 7.55 Diet composition of the dominant decapods and mysids in the eastern Gulf of Mexico. Percentages of food biomass. *Source:* Hopkins et al. (1994), figure 4 (p. 149). Reproduced with the permission of Inter-Research.

and its daily ration is 6% DM, the daily zooplankton consumption is $0.011\,g\,DM\,m^{-2}$, or about 18% (0.011/0.06) of the daily zooplankton production. In comparison, an important mesopelagic fish group, the lanternfishes, takes only about 8% of zooplankton production. Admittedly, these are very broad-brush numbers with many assumptions attached, but they are useful for generating perspective on the ecosystem as a whole and the role of different taxa within it.

Another important consideration, particularly in a species-rich ecosystem such as the EGOM, is the degree to which species compete with one another for food. A glance at the migration patterns in Figure 7.54 shows a considerable amount of overlap in vertical distributions of the different species. The question is, do species that co-occur in space also share similar diets? Hopkins et al. (1994), using a combined factor (food-taxonomic composition, food size, day and night distributions) species-pairs matrix analysis, found that "no species pair is similar in both diet and space niche parameters." The analysis makes a very strong case for species-specific niche separation based on differences in the use of space and prey resources. Similar conclusions have been reached for the micronektonic Crustacea in other oceanic systems (Omori 1974; Donaldson 1975; Roe 1984; Nishida et al. 1988), but the matrix analysis described here for the EGOM gives an added measure of support. In essence, feeding and migratory behavior by the pelagic species impose biological structure on an environment with very limited physical structure (Hopkins et al. 1994).

Crustacean diets in the bathypelagic realm showed some important differences from those in the mesopelagial (Table 7.9). Fish made up a greater proportion of the diet in bathypelagic decapods. Further, the increase of fish in bathypelagic diets was nearly matched by the loss of euphausiids and chaetognaths that were prominent in the diets of mesopelagic shrimp. Euphausiids and chaetognaths are most important in the upper 500 m, outside the vertical range of bathypelagic individuals.

Though difficult to quantify their biomass contribution, phytodetritus, as well as cnidarian and radioloarian debris, were important dietary components in benthesicymids (Table 7.8, Heffernan and Hopkins 1981; Hopkins et al. 1994) and are also important components of marine snow (Alldredge and Silver 1988). Current thinking is that a substantial fraction of their diet in the mesopelagic zone is marine snow, delicate particulates that are broken up by the turbulence and meshes of the net, but when intact, would be accessible to the feeding appendages of the benthesicymid shrimp (Burghart et al. 2010).

It is likely that the incidence of marine snow at bathypelagic depths is reduced from that in the mesopelagic zone, reducing its importance as a prey item and further increasing the dependence of benthesicymid on fishes in bathypelagic depths.

Resource partitioning in the bathypelagic realm was investigated using only composition and size of prey in the diet, since the vertical distribution information was more limited. Despite the lack of information on space in determining niche overlap, about 50% of species pairs showed no overlap based on diet alone and 30% overlapped in one category only.

Geographic Distribution

As might be expected, meso- and bathypelagic decapods are widely distributed in the world ocean, with most of the EGOM species shown in Table 7.7 found in multiple ocean basins. Only 8 of the 44 species for which zoogeographic data are available are restricted to the Atlantic. Many exhibit tropical/subtropical distributions worldwide, and some are pan-global at higher latitudes as well (Burghart et al. 2007).

Table 7.9 Comparison of the dietary contribution of prey categories between mesopelagic and bathypelagic predators in two families of decapods. Mesopelagic values are taken from Hopkins et al. (1994). *Source:* Burghart et al. (2010), table 4 (p.137), Marine Ecology Progress Series 399:131–149. Reproduced with the permission of Inter-Research.

	Benthesicymidae		Oplophoridae	
	Mesopelagic	Bathypelagic	Mesopelagic	Bathypelagic
Calanoida	13.3	11.6	6.7	5.5
Cephalopoda	0	0	4.1	2.8
Chaetognatha	13.3	2.5	23.3	1.9
Decapoda	0	2.6	0	11.2
Euphausiacea	27.3	0	17.9	0.1
Fish	31.7	81	33.2	72
Ostracoda	0	1.1	0	1
Other	14.4	1.2	5.8	5.2

Infraorder Anomura; Superfamily Galatheoidea; Family Munididae; Genera *Pleuroncodes, Munida,* and *Cervimunida*

The Pelagic Squat Lobsters

Also known as pelagic red crabs or lobster krill (Figure 7.56), 5 species of munidid crabs, out of over 465 munidid species, reach very high densities in the four major eastern Boundary Current systems of the world, which are also the four major oceanic upwelling regions. The regions support major fisheries for sardines and anchovies as well as being home to the large populations of red crabs. *Pleuroncodes planipes* is found in the California Current,

(a)
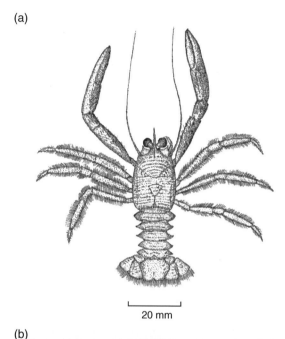
20 mm

(b)
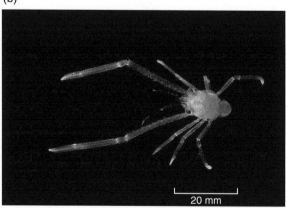
20 mm

Figure 7.56 Galatheidae: pelagic squat lobsters. (a) The pelagic red crab *Pleuroncodes planipes*; (b) Post-larval grimothea stage of the pelagic squat lobster *Munida* sp. *Sources:* (a) Boyd (1967), figure 1 (p. 395); (b) © Dante Fenolio, reproduced with permission. See color plate section for color representation of this figure.

Pleuroncodes monodon and *Cervimunida johni* are found in the Peru Current to the west of Peru and Chile, *Munida rugosa* is found in the Canary Current in the northeastern Atlantic, and *Munida gregaria* is found in the Benguela Current off the west coast of South Africa. *Munida gregaria* is also found in the waters off Patagonia, the Falkland (Malvinas) islands, and in the waters off New Zealand (Matthews 1932). Two species have received the lion's share of study: *Pleuroncodes planipes* and *Pleuroncodes monodon*. Interest in *P. planipes* is due to its large swarms off the California coast during El Niño events, and interest in *P. monodon* is due to its importance to fisheries in Chile and Peru.

Squat lobsters are unusual in that they may be either benthic or pelagic depending upon their age or, in the case of *Pleuroncodes monodon*, on external oxygen concentrations (Boyd 1967; Robinson et al. 2004). *P. planipes* is found year-round in shelf and slope waters off the west coast of Baja California. Older (age 2+ years) individuals of the population are primarily benthic and are found in deeper waters at the edge of the continental shelf including the shallow slope. During their first two years of life, they can live either benthically or pelagically, but the majority are believed to rise to surface waters at night with a daytime residence on the bottom. Nonetheless, swarms of red crabs at densities of 1–100 crabs m^{-2} of sea surface have been observed both day and night (Boyd 1967).

Reproduction occurs during the winter months in those species that have been studied, with the larvae passing through five zoeal stages prior to the final or glaucothoe stage Boyd (1960). The post-larval grimothea stages of *Munida gregaria* and *M. subrugosa* (Figure 7.56b) have been observed in swarms (Matthews 1932).

Squat lobsters are voracious omnivores, capable of grazing on phytoplankton and zooplankton in the water column in addition to benthic polychaetes and crustaceans (Longhurst et al. 1967; Tapella et al. 2002). They, in turn, are important prey for tuna, a variety of benthic fishes, and for whales (Matthews 1932; Boyd 1967).

One of the more interesting attributes of all the pelagic red crabs mentioned above is their affinity for oxygen minima, usually associated with upwelling regions. The minimum of the waters of Baja California in the Eastern Tropical Pacific is particularly severe, with partial pressures of oxygen as low as 3 mm Hg, about 2% of that in well-aerated waters. The ability of *Pleuroncodes planipes* to extract oxygen at very low concentrations was examined by Quetin and Childress (1976), who found that *P. planipes* was capable of living completely aerobically at the lowest oxygen concentrations (<0.1 ml l^{-1}) encountered in their eastern Pacific environment. In contrast, it is believed that the reason *Pleuroncodes monodon* maintains a persistent pelagic existence off Peru and not in Chilean waters is the very low oxygen found on the bottom off Peru (Gutierrez et al. 2008) despite the fact that there is a well-developed oxygen minimum in both regions.

Order Amphionidacea

Brusca et al. (2016) retain the order Amphionidacea, WoRMS relegates the group to the infraorder Caridea with uncertain taxonomic rank.

Introduction and History

The order Amphionidacea has but one species, *Amphionides reynaudii*, a pelagic shrimp with an unusual morphology (see Figure 7.8) and a worldwide distribution between 35°N and 35°S (Schram 1986). It was first described from its larval forms by the famed natural historian Milne-Edwards in 1832. In more recent times, its larval sequence has been described by Gurney (1936), Heegaard (1969), and Williamson (1973). *Amphionides* was formerly classified with the carideans, but its lack of the "caridean saddle" overlapping the first and third pleopods, the presence of a unique brood chamber formed by the greatly elongated first pleopods in the *Amphionides* female, and the absence of a petasma or thelycum showed differences from both the dendrobranchiates and carideans (McLaughlin 1980). Williamson (1973) proposed ordinal status for the Amphionidacea and synonymized all former species names into *Amphionides reynaudii*, honoring the wishes of Milne-Edwards and the observations of Heegaard in doing so.

Development and Natural History

Though the first pleopods of the female are assumed to form a brood chamber, females with eggs have never been captured. Adults in general are quite rare in collections. Most females have been captured at depths exceeding 2000 m (Schram 1986), and only three adult males have ever been captured. Adults have a very thin carapace, which would aid in approaching neutral buoyancy. Adult females exhibit reduced mouthparts and vestigial guts (Schram 1986) suggesting that they are incapable of feeding.

In contrast to adult females, larvae have normally developed digestive tracts. They are found in the upper 30 m and have been the most extensively studied stages of the species. Nine to twelve zoeal stages have been reported for *Amphionides* depending on the author, suggesting some plasticity in developmental sequence such as that observed in the euphausiids. At present, no information is available on the species' feeding habits or other elements of its lifestyle in the pelagic realm.

Superorder Peracarida

The Peracarida form the second major branch of the Eumalacostraca, with nine orders that are briefly described in the classification section earlier in the chapter. The Peracarida include at least 18 700 marine species (WoRMS n.d.). Of major importance to us are the Lophogastrida and Mysida, until recently grouped together as the Mysidacea, and the Amphipoda. The Isopoda, with only one genus, *Anuropus*, found in the pelagic realm is a cameo player and will not be treated in detail. Like many of the amphipods, *Anuropus* are "jelly-riders" (Barham and Pickwell 1969). *Anuropus* is associated with the scyphozoan medusa *Deepstaria enigmatica* and is only encountered episodically in trawl samples.

All peracarids brood their young, most in a ventral brood pouch called the marsupium. A well-developed carapace is present only in the more primitive orders such as the Lophogastrida. Unlike the decapods, the anterior portion of the peracarid carapace is fused to the first three thoracic somites only. Two additional characters that unite the peracarids are a telson without caudal rami and the presence of a lacinia mobilis, a movable toothed article on the chewing side of the mandibles (Kaestner 1970; Brusca and Brusca 2003).

Peracarids are primarily marine, but representatives are found in virtually all aquatic habitats including estuaries, freshwater, caves, and hot-springs. The peracarids also include the most successful terrestrial crustaceans: the pill bugs or "roly-polys," members of the Isopoda.

Orders Lophogastrida and Mysida
Introduction

Until recently, the lophogastrids and mysids were united as two suborders within the order Mysidacea, with the lophogastrids being considered the more primitive group.

Brusca and Brusca (1990) and Martin and Davis (2001) elevated the Lophogastrida and Mysida to ordinal status based on a variety of morphological and molecular characters but kept them within the Peracarida, a venerable group that originated with the pioneering carcinologist W.T. Calman in 1904.

The mysids, with about 1166 marine species, are primarily marine though they also occur in brackish waters and a very few are in freshwater. They are found at all oceanic depths, including the hadal zone (7210 m; Mauchline 1980) though the lion's share of species resides in shallow coastal areas. Their size varies from 2 mm to 8 cm; most are in the 1 to 3 cm range (Kaestner 1970; Brusca and Brusca 2003). Mysids are often found associated with the bottom, either as residents, hovering just above it, or as demersal species, benthic by day and rising into the water column at night. Some are even known to burrow or hide themselves in sand (Kaestner 1970). A few mysids associate closely with coastal plants such as kelp or sea lettuce or are found in seagrass beds. Virtually, all coastal environments including coral reefs have resident mysids. A number of fresh-water cave-dwelling mysids have been found and are primarily included in the families Lepidomysidae and Stygiomysidae.

Mysids are known to form aggregations, sometimes associated with breeding, but other aggregations are persistent enough to be termed schools or swarms similar to those observed in *Euphausia superba* (Mauchline 1980) but on a smaller scale. The aggregations vary in size from a few individuals to huge swarms more than a km in extent (Kaestner 1970; Mauchline 1980) and nearly always comprise a single species. Clutter (1969) observed coordinated behavior in mysid schools reminiscent of that in schooling fishes.

The lophogastrids, with about 53 species, are important contributors to the numbers and biomass of deep-sea Crustacea throughout the global ocean (see e.g. Table 7.7; Vinogradov 1970). The genera *Eucopia* and *Gnathophausia* are exclusively pelagic, inhabiting mesopelagic and bathypelagic depths down to at least 3000 m (Burghart et al. 2007). *Gnathophausia ingens* in particular, because of its hardiness in the laboratory after capture, has added significantly to our knowledge of how deep-sea Crustacea make a living. The lophogastrids are considered to be the ancestral group of the Peracarida because of a number of primitive characteristics, notably a well-developed carapace, an abdomen with seven pleomeres, a heart with several metameric arteries, and the presence of both antennal and maxillary glands (Kaestner 1970).

History
Praunus flexuosus, a common European species, was the first mysid to be recognized in a published description (Muller 1776). It is worth remembering that a great deal of the early history of any taxon was governed by accessibility to specimens. The crustacean groups available to early natural historians were limited to near-shore locations. Thus, seemingly unrelated species were sometimes grouped together. Mysids were originally grouped by Latreille (1802) with the stomatopods and leptostracans. In 1817, Latreille separated the mysids and leptostracans from the stomatopods into a group he termed the Schizopoda. The discovery of the euphausiids in 1830 by Milne-Edwards resulted in their inclusion with the mysids into the Schizopoda, a group that persisted off and on through a number of classification systems with a variety of different members until the definitive formulation by Calman in 1904 created the Eucarida and Peracarida.

A few milestones in mysid classification should be mentioned. The first was the work of G.O. Sars, who provided a suite of reliable keys and figures to the mysids with his publication of the Mysidae of Norway (1872) and of the Mysidae of the Mediterranean (1877). In 1883, Boas showed that the mysids and euphausiids were quite different and considered each to be distinct orders in the Malacostraca. He further divided the mysids into two suborders: the Lophogastrida and the Mysida. His suggestions were adopted by Hansen (1893) but ignored by Sars (1885) and Stebbing (1893). It was not until Calman (1904) that the group Schizopoda was put to rest (Tattersall and Tattersall 1951).

Classification of the mysids continues to evolve at an increasingly rapid pace. At this juncture, it is best to refer the reader to internet-based sources for the latest information. The authors have found the World Register of Marine Species (WoRMS) to be a fairly up-to-date source of ranks in most taxa, and it has a very useful search function. In the case of the mysids and lophogastrids, most of the recent changes have occurred below the ordinal level. As always, the animals themselves have not changed at all and that is the major concern of the book. For general orientation, the classic Martin and Davis (2001) system is presented below.

Classification (Martin and Davis 2001)
    Superorder Peracarida
      Order Lophogastrida
        Family Eucopiidae
          Genus *Eucopia* (9)
        Family Lophogastridae
          Genus *Ceratolepis* (1)
            Chalaraspidum (1)
            Gnathophausia (10–11)
            Lophogaster (20–22)
            Paralophogaster (10)
            Pseudochalaraspidum (1)
      Order Mysida
        Family Lepidomysidae
          Genus *Spelaeomysis* (9)
        Family Mysidae
          169 *Genera*; 1100 spp.
        Family Petalophthalmidae
          Genus *Bacescomysis* (7)
            Ceratomysis (3)
            Hansenomysis (18)
            Parapetalophthalmis (1)
            Petalophthalmis (5–7)
            Pseudopetalophthalmis (2)
        Family Stygiomysidae
          Genus *Stygiomysis* (7)

Anatomy

Mysids and lophogastrids are shrimp-like in appearance, but closer examination reveals differences between them and the euphausiids and decapods. The carapace of the mysids and lophogastrids envelops nearly the entire thorax, but it is fused only to the anteriormost thoracic segments (I–III, rarely IV), whereas in the eucarids, it is fused to all thoracic segments. In the mysids and lophogastrids, one or two of the posterior thoracic segments are usually visible dorsally (Figure 7.57).

The mysids and lophogastrids all have eight thoracic limbs in which at least the first, rarely the second, is modified into a maxilliped (Figure 7.57b). All thoracic limbs are well developed with some diversity exhibited in limb structure between groups. The family Lophogastridae has limbs roughly equal in structure but thoracopods V–VII in the family Eucopiidae are very much longer (Figure 7.58). Well-developed gills with three or four branches are present on thoracopods II–VII in the lophogastrids (Figure 7.59); when present on VIII, they are usually rudimentary (McLaughlin 1980). Exopods are used exclusively in the swimming of most mysid females. During breeding, oostegites are

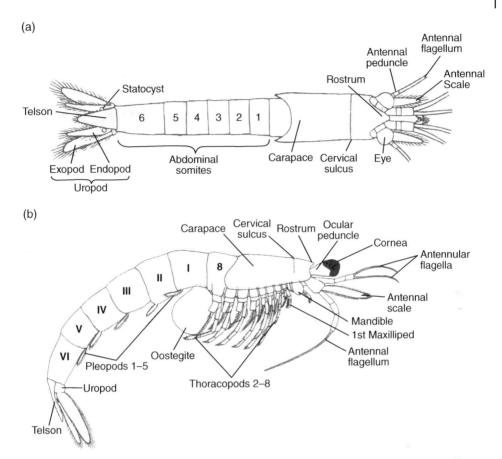

Figure 7.57 Mysid anatomy. (a) Dorsal view of a typical mysid (limbs removed); (b) Lateral view of a typical female mysid. *Sources:* (a) Adapted from Murano (1999), figure 5A (p. 1105); (b) Adapted from McLaughlin (1980), figure 28 (p. 80).

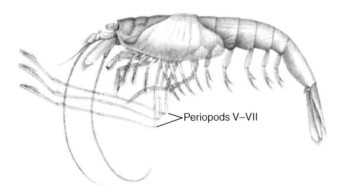

Figure 7.58 Eucopiid anatomy. A young female *Eucopia australis* with elongated thoracopods. The mark of a gentle net is *Eucopia* getting to the surface with its limbs intact! *Source:* Sars (1885), plate X.

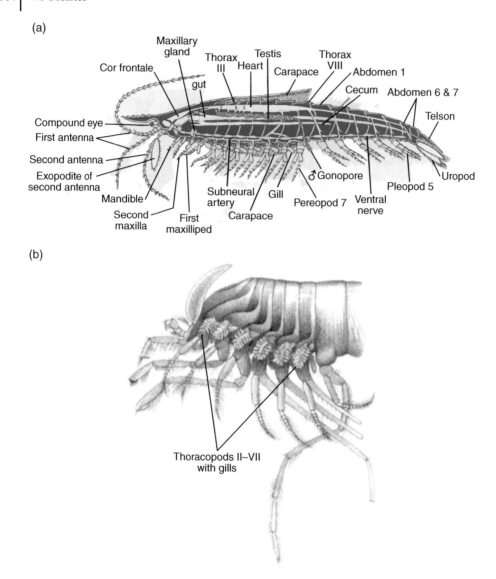

Figure 7.59 Lophogastrid anatomy. (a) Internal anatomy of a male lophogastrid; (b) Thoracic region of a lophogastrid showing well-developed gills on thoracopods II–VII. *Sources:* (a) Adapted from Kaestner (1970), figure 15-3 (p. 374); (b) Sars (1885), plate X.

present as lamellae on thoracopods II–VII of the female lophogastrids and on the posteriormost two to three thoracopods of the mysids.

The abdomen of mysids and lophogastrids has six segments or pleomeres and five pairs of pleopods (Figure 7.59). The sixth segment of the lophogastrids usually has a transverse groove that is sometimes attributed to a seventh segment that fuses during development. Pleopods in the lophogastrids are well-developed biramous appendages that provide the thrust for swimming. In the mysids, pleopods are present in the males

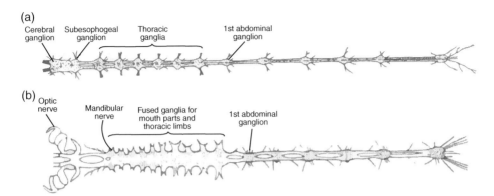

Figure 7.60 Lophogastrid and mysid nervous systems. (a) The nerve cord of the lophogastrid *Gnathophausia longipina*; (b) The nerve cord of the mysid *Mysis relicta*. *Sources:* (a) Mauchline (1980), figure 34 (p. 128); Reproduced with permission of Academic Press. (b) Sars (1867), plate III.

but vestigial in the females. A tail fan similar to those of the euphausiids and decapods is formed by the telson and plate-like uropods.

Nervous System

Mysids and lophogastrids show differences in the anterior portions from the basic crustacean nervous system. The lophogastrids have a distinct subesophogeal ganglion that controls the mouthparts and first maxilliped, followed by six thoracic ganglia (Figure 7.60a). Mysids have no well-defined subesophogeal ganglion; the ganglia of the mouthparts and thoracic limbs are fused into a "long ribbon" that runs the length of the cephalothorax (Figure 7.60b; Kaestner 1970). The number of ganglia comprising the ribbon varies between genera. Mysids and lophogastrids both have six well-defined abdominal ganglia and have X-organs and sinus glands in the eyestalks.

Sensory Mechanisms

Lophogastrids and mysids have a full complement of sensory hairs located mainly on the antennae, mouthparts, and thoracic limbs to provide information on the mechanical and chemical environment (see chemoreception in crustacean systems). Mysids have organs of balance, statocysts, located at the base of their uropods but the lophogastrids do not. The statocysts communicate with the outside via narrow slits (Kaestner 1970).

Eye morphology shows considerable diversity within the mysids, less so in the lophogastrids. Not surprisingly, within each group, the ambient light environment is a major factor shaping the characteristics of the eyes. Mysids living in the well-lit upper layers of the water column have the spherical superposition eyes described earlier, effective for good all-around vision. Some of the mesopelagic mysids (e.g. *Euchaetomera*, Gaten et al. 2002) have bi-lobed eyes similar to those described for the euphausiids (Figure 7.36), with an upward-facing lobe for high resolution in locating prey and a general-purpose lower lobe for general vision. Keep in mind that the eyestalks are mobile; the position of the lobes can be adjusted to improve near-field visual acuity. Kidney and elliptically shaped eyes are also observed among the mysids (Mauchline 1980).

Both the lophogastrids and the mysids show a general reduction in eye size in the bathypelagic forms (as do the dendrobranchs and carideans). Plots of eye size vs. body length in a variety of taxa show a consistent trend of reduced eye size in the deepest-living species over a broad range of pelagic Crustacea (gnathophausiids, sergestids, penaeids, and oplophorids; Hiller-Adams and Case 1988). Interestingly, bathypelagic species with bioluminescent capabilities, using either photophores or a luminescent cloud, have larger eyes than those that do not.

Eye reduction in the deeper-living mysids is most diverse in the subfamily Erythropinae, which also includes the deepest-living mysid species, *Amblyops magna* (7210 m, Belyaev 1966). The Erythropinae are an epibenthic group, living just above the bottom. Eyes vary from plates lacking visual elements to markedly reduced, but normally shaped, stalked eyes containing fewer ommatidia (Mauchline 1980).

Spectral sensitivity was examined in the eyes of *Gnathophausia ingens*, the large mesopelagic (400–900 m) mysid (Frank and Case 1988b). Peak sensitivity was found at a wavelength of 510 nm, but the sensitivity curve was very broad, with responses from roughly 350 to 600 nm. A peak-sensitivity well to the green side (about 480 nm) of the downwelling light, as well as a broad-spectrum sensitivity in what is essentially a monochromatic environment, may yield an increased ability to resolve entire bioluminescent spectra rather than just peak wavelengths. As noted earlier with the decapods, this would confer the ability to discriminate between different sources of bioluminescence (Cronin and Frank 1996; Herring 2002).

Bioluminescence

In the peracarids, bioluminescence has been studied in detail only in the gnathophausiid *Gnathophausia ingens*, but it has been reported in the mysid genera *Siriella* and *Gastrosaccus* with little further elaboration other than its presence (Harvey 1952; Mauchline 1980). It has not been reported since Herring (2002). *Gnathophausia* produces a bioluminescent secretion, another example of the "type 1" luminescent system described earlier (Herring 1978). Those species that produce the luminescent cloud are termed "spewers," and the cloud of *Gnathophausia* is particularly striking. It is a bright blue–green luminescent cloud similar to that of the decapod spewers but is produced using a different mechanism. The luminescent secretion of *Gnathophausia* is produced by a gland in the second maxilla. Glandular cells secrete a luminescent fluid into a muscular reservoir that has an ejection tube exiting at the base of the maxilla's exopod, placing the secretion directly in the path of the exhalant respiratory stream. When *Gnathophausia* is disturbed, the reservoir ejects a luminescent cloud, which rapidly grows and disperses. The cloud is remarkably bright and persists for several seconds. Worthy of note is the fact that one of the necessary chemicals for the bioluminescent reaction, the substrate luciferin, is obtained from its diet (Frank et al. 1984).

Circulatory, Respiratory, and Excretory Systems

While the basics of circulation in lophogastrids and mysids are nearly identical, there are a few differences in the circulatory, respiratory, and excretory systems of the two orders. Circulation in *Gnathophausia ingens*, and the lophogastrids in general, conforms well to the general crustacean model described earlier in the Section Systems. In fact, the circulatory system in *Gnathophausia* is one of the best described in the crustacean literature. The basic design in both orders includes an elongate tubular heart with one to three pairs of ostia that extends from the second thoracomere to the posterior of the thorax (Figures 7.27 and 7.59). An anterior aorta leads to an accessory heart (cor frontale) and then into a series of cephalic arteries supplying the supraesophogeal ganglion, the eyes, and antennal gland. A posterior aorta leaves the heart and splits, part forming a descending artery and part continuing as the posterior aorta that supplies the muscles and pleopods of the abdomen. In the mysids, the descending artery splits three ways. The anteriormost branch becomes the sternal or subneural artery that flows anteriorly to supply the anterior thoracopods; the two posterior branches descend to supply the posterior three pereiopods. In the lophogastrids, the descending artery splits into an anterior-flowing sternal artery and a posterior ventral abdominal artery that runs along the ventral nerve cord.

Additional differences between the lophogastrids and mysids are the many (five) hepatic arteries that supply the gut and hepatopancreas in the lophogastrids (Figures 7.27 and 7.59) versus the two in mysids and the character of the return flow to the heart and the gas exchange apparatus in each. Return flow in the lophogastrids is achieved via the many sinuses that coalesce with the infrabranchial sinus, which in turn branches off to each of the gills. Branchio-pericardial vessels return the blood from the gills to the pericardial chamber. The vessels are discrete channels passing from the bases of each of the pereiopods dorso-laterally to the pericardium and fusing with it directly.

In the mysids, fundamentals of the return flow are the same as in lophogastrids in that all the various body sinuses drain to the infrabranchial sinus, but from there oxygenation of the blood and return to the heart differs. The mysids do not have the foliaceous gills formed by the epipodites of the thoracopods. Instead the interior of the carapace wall has a web of lacunae that allow for oxygenation of the blood (Figure 7.61). Blood passes from the infrabranchial sinus anteriorly into a channel that runs along the edge of the carapace and into a lacunar net that covers the inner respiratory wall of the carapace. From there, it enters the pericardial chamber. Flow of water for gas exchange is maintained in the branchial chamber by the rhythmic beating of the first epipodite of the first maxilliped (gill-bailer) that draws water from the postero-lateral edge of the carapace through the chamber and out the gap between the carapace and the maxillae (Kaestner 1970) (Figure 7.61b), a similar water flow to that observed in *Gnathophausia*.

Lophogastrids are reported to have excretory systems comprising both maxillary and antennal glands, which is fairly unusual. Mysids have antennal glands only. Both function as described earlier.

Digestive System

The alimentary canal of lophogastrids and mysids consists of an esophagus, a stomach with cardiac and pyloric regions, a variety of ceca, a midgut, and a hindgut (Figure 7.62). A short muscular esophagus lined with chitin and armed with posterior-facing spines conducts food items to the cardiac stomach, where they are triturated and passed along to the pyloric region for further reduction and enzymatic breakdown. Both stomach regions are chitin-lined and armored.

Two types of digestive ceca join the gut at the juncture of the pyloric stomach and midgut. The first type is termed a dorsal diverticulum, is usually small in size, and may be single or paired. Dorsal diverticula are believed to secrete the peritrophic

(a)

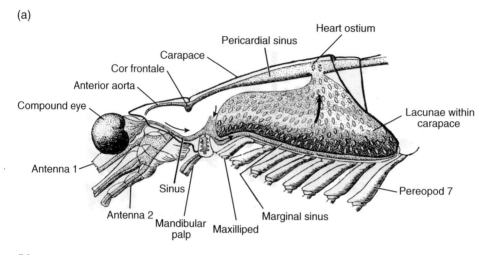

(b)

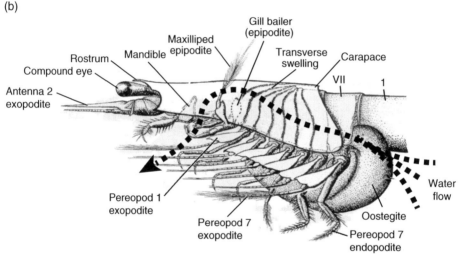

Figure 7.61 Mysid circulation. (a) Principal blood vessels and circulation in the mysids. Arrows indicate direction of blood flow from gills through lacunae of the carapace fold to the pericardium through ostia into the heart. The ostia are seen through the pericardial wall; (b) Water flow enabling gas exchange in mysids. Respiratory stream enters at the posteroventral edge of the carapace propelled by the gill bailer, bathing the exchange surface shown in (a). *Sources:* (a) Adapted from Kaestner (1970), figure 15-4 (p. 376); (b) Adapted from Kaestner (1970), figure 15-2 (p. 373).

membrane, a very thin (0.5 μ) membrane composed of protein and chitin that coats the food bolus for its trip down the remainder of the gut, ultimately forming a fecal pellet. The second type of ceca are the finger-like projections that form the digestive gland in mysids, site of digestive enzyme synthesis and also responsible for a share of the total nutrient absorption. Food particles move freely within the projections of the digestive gland which join the gut ventrally via a common duct located in the posterior region of the pyloric stomach. Differences in the morphology of the

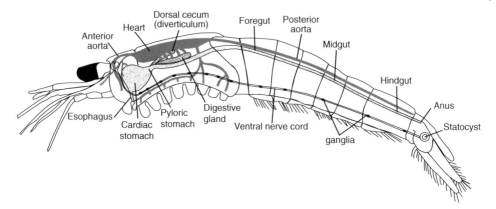

Figure 7.62 Diagrammatic representation of the internal anatomy of a mysid showing the digestive system and associated circulatory and reproductive structures.

digestive gland ceca are observed between the mysids and lophogastrids as well as between species in each. In some cases, the ceca can extend into the abdominal region.

Within the gut itself, nutrient absorption takes place primarily in the long unlined midgut region. Once in the chitin-lined hindgut, fecal pellets are rapidly voided.

Reproduction and Development

Like all peracaridans, larvae of the mysids and lophogastrids exhibit direct development. Fertilized eggs develop within the marsupium until they are released as postlarvae that are capable of feeding. Larvae show characteristic stages within the marsupium.

Stage I – early embryo, egg-like, developing rudiments of antennae and abdomen, egg membrane shed at the end of this stage.

Stage II – larva has hatched, antennae and thoracic appendages develop, eyes become pigmented, stage terminates in a molt.

Stage III – larva now has eyes on stalks, stage terminates with a molt at or shortly after release from marsupium and is then capable of swimming.

Life history strategies differ considerably between the mysids and lophogastrids that have been studied. At least partially due to accessibility, reproductive behavior has been described mainly in temperate coastal mysids, which often have a short life and rapid reproduction.

The mysid *Praunus inermis* starts its reproductive period in April and produces five successive broods before dying in July or August. Successive broods mature and rear their own five before dying in the fall. The last of those broods overwinters to begin the cycle again in the spring. When a brood is released, the female molts, retaining her marsupium to immediately breed again. The mysid *Neomysis intermedia* has a similar life history, with brooding periods of 7–15 days depending on temperature. Generation times vary with season, with winter generations living for 5 months, fall and spring generations for 2 months, and summer for 1.5 months.

Ovaries and testes are located dorsal to the gut and below the pericardium in both orders (Figure 7.59). Mature ova are released into the marsupium via an oviduct that leads to a gonopore on the coxa of the sixth thoracopod. Bundles of non-motile sperm leave the testes via paired vasa deferentia that end in an ejaculatory duct and a genital papilla or penis at the base of the eighth thoracomere.

Males can detect females that have molted within the last 12 hours and have mature ova within the oviduct, presumably by detecting associated pheromones. Sperm is deposited within the marsupium, and eggs are fertilized as they pass from the oviduct into the marsupium. Eggs and developing larvae are irrigated by movements of the lamellae comprising the brood pouch.

Life History of a Lophogastrid

Gnathophausia (now *Neognathophausia*) *ingens* lives at meso- and bathypelagic depths between 30°N and 30°S in all ocean basins. It is particularly abundant in the California borderland where it is found primarily between 400 and 900 m in the deeper coastal basins and off the shelf. Larger individuals are centered at about 700 m during daylight hours. Brooding females are found deeper, between 900 and 1400 m. Its life span is very much longer and its reproductive strategy very different from the smaller mysids (Childress and Price 1978).

Figure 7.63 shows the size-frequency distribution of *Gnathophausia* from the San Clemente Basin, illustrating two important (and very useful) characteristics of its growth pattern. First, size classes are discrete, showing very little overlap between instars. Second, the size or molt classes are consistent from place to place. The molt classes provide a powerful tool because the molt increment (change in length with each molt) is always known. Unlike most crustaceans, animals can be aged once the intermolt interval (time elapsed between molts) is determined for each.

Gnathophausia ingens is very unusual for a deep-sea species in that it can be maintained in the laboratory for about 2.5 years. Intermolt intervals were determined from laboratory observations and from cataloguing the percentages of "soft-shell" individuals captured over multiple cruses to yield the curve in Figure 7.64. From that we can see that *G. ingens* has 13 instars through its life, and lives for about 3000 days, reproducing in one "big bang" at the end of its life. Worthy of note is that eggs are brooded for about 1.5 years, producing large (about 11 mm carapace length) young when released. Females do not feed while brooding.

Life history characteristics are summarized for *Gnathophausia* and the shallow-water mysid, *Metamysidopsis elongata*, in Table 7.10. Overall, the long life, slow growth, and delayed reproduction of *Gnathophausia* reflect a life in a very stable environment with limited available energy and a low risk of mortality. With the exception of the semelparous (reproducing once at the end of life) reproductive strategy, the life history is a very close fit to the classic "k-selected" model (Stearns 1992). In contrast, *Metamysidopsis*, living in a shallow coastal environment with a higher risk of predation but with higher available energy, grows rapidly, reproduces multiple times, and dies in about 200 days, having produced an equivalent number of larvae in its lifetime as does *Gnathophausia*.

A later study describing accumulation of matter and energy throughout the life *of G. ingens* (Childress and Price 1983) confirmed the earlier results of a long life ending in large size and semelparous reproduction, underscoring the very large energy investment in reproduction by the species (Tables 7.10 and 7.11). It was concluded that the production of high numbers of large capable young likely maximized reproductive success per generation in the deep-sea environment.

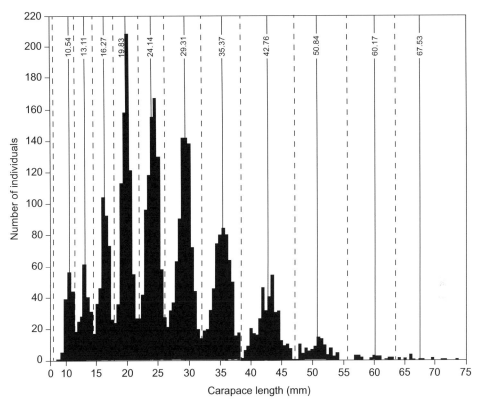

Figure 7.63 Size frequency distribution of 4239 individuals of *Gnathophausia ingens* captured in San Clemente Basin during all seasons between January 1969 and September 1975. Continuous lines and numbers indicate mean sizes for each molt class; dashed lines, boundaries of molt classes; histogram intervals are 0.5 mm ($X + 0.25$ mm to $X + 0.75$ mm and $X + 0.75$ to $X + 1.25$ mm). *Source:* Adapted from Childress and Price (1978), figure 1 (p. 50). Reproduced with the permission of Springer.

Figure 7.64 Growth in size of *Gnathophausia ingens* from egg through final molt to sexual maturity. *Source:* Childress and Price (1978), figure 5 (p. 53). Reproduced with the permission of Springer.

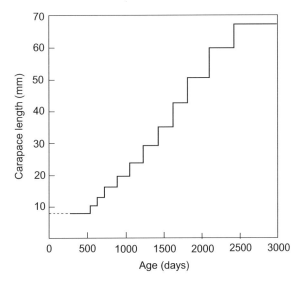

Table 7.10 Comparison of life-history characteristics of the bathypelagic mysid *Gnathophausia ingens* (Childress 1971; Childress and Price 1978) and the shallow-water mysid *Notamysidopsis elongata* (Clutter and Theilacker 1971). *Source:* Childress and Price (1978), table 4 (p. 60). Reprinted by permission from Springer Nature Customer Service Centre GmbH, Springer Nature, Growth rate of the bathypelagic crustacean *Gnathophausia ingens*.

Parameter estimated	G. ingens	N. elongata
Temperature range	3.5°–7.5°	17°–19°
No. of broods	1	>14
Larvae per brood	150–350	5–30
Larvae per lifetime	350	340
Larval development time	530 days	10 days
Juvenile size (earliest free-living stage)	20 mm long	1.2 mm long
Intermolt interval	166–253 days	4–10 days
Molt increment	12–25%	1–31%
Length at onset of reproduction	151 mm	5.3 mm
Age at onset of reproduction	2420 days	53 days
Generation	2950 days	68 days
Feeding of brooding females	No	Yes
Maximum age attained	2950 days	>200 days
Mortality over life span	Low in early stages, high in late stages	Constant
Survivorship curve	Type I[a]	Type II[a]
Growth-curve shape	Exponential	Hyperbolic
O_2 consumption (μl O_2 mg dry weight^{-1} h^{-1})	0.199	4.0

[a] Type I curve has low initial mortality increasing toward the end of life; Type II has a relatively constant mortality over the entire life (Deevey, 1947).

Food and Feeding

The great majority of mysids are found in shallow waters, often with epibenthic, benthic, or demersal habits. Their diets show a mix of detritus, a variety of algae including diatoms, small crustacean zooplankton, and small amphipods (Mauchline 1980). Most are suspension feeders, creating a feeding current with their thoracic exopods that moves phytoplankton or bottom detritus into a central groove that runs from the mouthparts to the end of the thorax. Once in the food groove, particles are moved forward toward setal combs on the second maxillae which form a small filter basket just posterior to the mandibles. Particles can then be sucked in toward the mandibles by outward movements of the maxillae creating a low pressure area between them and thence to the mandibles.

The taxa of most concern to us within the Lophogastrida are the families

Table 7.11 Partitioning of material or energy within production component of the energy budgets of peracarid and euphausiid crustaceans. nd: no data. *Source:* Childress and Price (1978), table 6 (p. 175). Reprinted by permission from Springer Nature Customer Service Centre GmbH, Springer Nature, Growth rate of the bathypelagic crustacean *Gnathophausia ingens*.

	°C	% of production			Source
		Growth	Reproduction	Molts	
Amphipod					
Calliopus laeviusculus	8	44.4	35.2	20.3	Dagg (1976)
Isopods					
Cirolana harfordi	—[a]	78.4	21.6	ND	Johnson (1976)
Ligia dilatata	—[a]	82.4	17.6	ND	Koop and Field (1981)
Euphausiid					
Euphausia pacifica	8	45	40.2	14.8	Ross (1982)
Mysids					
Metamysidopsis elongata	17–19	42	42	15	Clutter and Theilacker (1971)
Mysis relicta	2–10	73.6	26.4	ND	Lasenby and Langford (1972)
Gnathophausia ingens	3–9	33.0	61.3	5.6	Childress and Price (1983)

[a] Field studies in which temperature was not described.

Eucopiidae and Gnathophausiidae. *Lophogaster* is predominantly epibenthic, though it lives at shelf-slope depths. Within the Mysida, the subfamilies Boreomysinae and Erythropinae include most of the pelagic species, including the deep-living ones, with some representatives also in the Gastrosaccinae and Mysinae. Diet information is limited, but some is available for *Boreomysis*, the eucopiids, and *Gnathophausia*. *Boreomysis* is reported to feed on algae including diatoms, and miscellaneous Crustacea (Mauchline 1980). Diatoms, unless associated with sinking marine snow particles, would likely not figure in diets of deeper living *Boreomysis*. Copepods are the preferred dietary item for eucopiids residing at either mesopelagic or bathypelagic depths, with chaetognaths and fish playing a notable role as well. The very long sub-chelate periopods of *Eucopia* (Figure 7.58) are presumed to be useful for plankton picking. *Gnathophausia* clearly feeds mainly on fish (Table 7.8). When feeding, *Gnathophausia* grasps the prey item firmly with its endopods, manipulating it as necessary toward its mandibles as it is consumed. When freshly captured off Southern California and transferred to water at habitat temperature (5.5 °C), many of the *Gnathophausia* in the catch bucket would already be feeding on luckless lanternfish that were captured in the net with them. To transport them back to the home laboratory for long-term study, *Gnathophausia* needed to be isolated by loosely wrapping them up in fine-mesh nylon netting or they would eat each other. A voracious species to be sure.

Locomotion

Normal swimming differs between the mysids and lophogastrids. Pleopods are rudimentary in most female mysids and are responsible for only a fraction of locomotory needs in males. In mysids, the feathery exopods of the thoracic appendages (Figure 7.57b) are used as the main swimming limbs, which may be thought of as high-speed oars. They "hold them towards the sides and rotate them so that their tips describe an oval. They draw water from the sides toward the dorsum so that two strong currents, parallel to the abdomen and some distance from it drive the body forward" (Kaestner 1970) (Figure 7.65a). Rather than the rhythmic motions of the pleopods in euphausiids and decapods, the exopods of mysids move constantly, but slightly out of phase with one another. Mysid movement is very smooth and rapid. They are able to hover, swim backwards and forwards, and up and down.

The lophogastrids, exemplified by *Gnathophausia ingens*, use the pleopods in conjunction with the thoracic exopods (Hessler 1985). Unlike in the mysids, where an elliptical path is described by the thoracic exopods, the power stroke of Gnathophausia's exopods is in the same plane as that of the recovery stroke, a much more straightforward rowing motion (Figure 7.65b). The thoracic exopods are angled slightly downward as they move laterally backward during the power stroke.

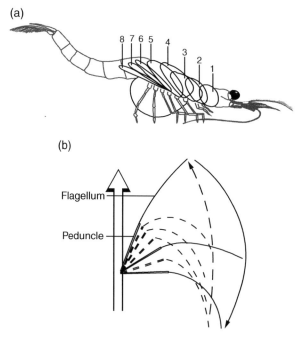

Figure 7.65 Locomotion in lophogastrids and mysids. (a) Lateral view of the mysid *Praunus* showing the paths described by the tips of the exopodites (numbered 1–8 from the front). Note the progression from circular paths of the front limbs to flattened ellipses of the rear limbs; (b) Diagram of the cycle of the swimming movement of a thoracic exopod of *Gnathophausia ingens*, dorsal or ventral view. The large white arrow represents the body moving forward. The solid line and solid arrow represent the exopod and the direction of movement during the power stroke, respectively. The dashed line and arrow depict the exopod and direction of motion during recovery. *Sources:* (a) Laverack et al. (1977), figure 4 (p. 143); Reproduced with permission of The Royal Society. (b) Hessler (1985), figure 8 (p. 118). Reproduced with permission of Cambridge University Press.

"When swimming, the thoracic exopods and pleopods on one side of the animal are unified in a single metachronal beat with a long wavelength such that the first and last limbs are synchronous" (Hessler 1985). The beating limbs on the two sides of an individual are 90° out-of-phase, meaning that movement on one side is initiated when the wave is midway down the opposite side. In euphausiids and dendrobranchs, the pleopodal beat is about 180° out of phase, alternating sides. Keep in mind that the paired pleopods of Gnathophausia, as well as those of the decapods and carideans are "schizopodous" (literally "split-foot") referring to the exopods and endopods of each pleopod that both participate in generating thrust for swimming.

The puzzling question is how the paddle-like limbs of swimming crustaceans can move backward and forward in the same plane and do anything but result in the individual remaining in one place. The secret is in the setae of the pleopods and thoracic exopods (Figure 7.66a) and in the basic physics of flow. Note the basic structure of a

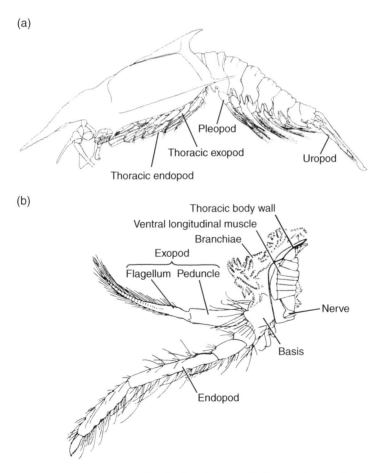

Figure 7.66 Swimming in *Gnathophausia ingens*. (a) Branched appendages of *G. ingens*; (b) Fifth thoracic limb of *G. ingens*. *Source:* R. R. Hessler, Swimming in Crustacea, Earth and Environmental Science, Transactions of The Royal Society of Edinburgh, Vol 76, issue 2–3, Pages 115–116. Reproduced with the permission of Cambridge University Press.

thoracic limb (Figure 7.66b): the exopod has a setose flagellum and a solid peduncle. During the power stroke, the setae on the flagellum spread to a paddle-like shape. During recovery, the setal paddle collapses inward, leaving only the slight drag of trailing setae and the small area of the setal shaft (the flagellum itself) to generate drag during the recovery stroke. The flagellum is flexed during much of the recovery stroke (Figure 7.65b), further reducing drag.

The paddle-like character of the exopod is enhanced by the fact that each seta has setules arranged in a feather-like pattern on the seta itself, conferring further rigidity during the power stroke and enhancing the paddle function by making a smaller-meshed sieve. Reynolds numbers (see Chapter 1) for flow at the level of the setule are very low, (Re = 0.1, Hessler 1985), meaning that viscous forces predominate and the exopod is truly acting as a paddle rather than as a sieve during the power stroke.

The principles described for the swimming limbs of *Gnathophausia* are widely distributed in the swimming Crustacea; virtually all taxa that use the pleopods or thoracic exopodites for swimming exploit collapsing setae on the return stroke in their swimming paddles (Hessler 1985).

Vertical Distributions

Mysids are found in estuaries, caves, freshwater, littoral, coastal, epipelagic, mesopelagic, and bathypelagic environments (Mauchline 1980): virtually every possible aquatic system. Lophogastrids are primarily open-ocean forms, with the families Gnathophausiidae and Eucopiidae predominating in the mesopelagic and bathypelagic environments and the family Lophogastridae predominating in the shelf-slope hyperbenthos with only a few pelagic representatives.

Day–night vertical profiles for common lophogastrids in the mesopelagic zone of the eastern Gulf of Mexico (EGOM) are provided in Figure 7.54. Note that *Gnathophausia ingens* shows evidence of a nightly shallowing in part of its EGOM population and that the eucopiids show little diel change in their vertical profile. Note also that all four species reside below 500 m during the day. Overall, mysids and lophogastrids show less evidence of the pronounced diel vertical migrations that is observed in euphausiids and decapods. They do show diel change, but most data suggest a spreading of the vertical range during nighttime hours and a compression at a central depth during the day.

Lophogastrids and mysids are more important in both numbers and biomass at bathypelagic depths than in the mesopelagic zone of the EGOM. Table 7.7 compares data for decapods, lophogastrids, and mysids from both depth strata. Lophogastrids make up 10% of total numbers and 4.9% of biomass in the mesopelagic zone, whereas between 1000 and 3000 m, they comprise 92% of numbers and 18.2% of biomass. Note that lophogastrids not only make up a greater fraction of the total Crustacea at bathypelagic depths but they are also more speciose (nine spp. versus four spp.) in the bathypelagic zone.

Fewer data are available, but those data we do have show that mysids are also common in the pelagic realm, ranging from epipelagic depths to at least 7210 m (Belyaev 1966). Data on mysid vertical profiles from a variety of regions as well as additional numbers for lophogastrids are presented in Table 7.12.

Geographic Distributions

As was observed in the decapods, many of the mysid and lophogastrid species have a wide-ranging distribution, including multiple ocean basins (Tables 7.12 and 7.13). Temperatures at meso- and bathypelagic depths are similar enough from basin to basin (see Chapter 1) to minimize their influence as a barrier to dispersal (Briggs 1995).

Order Amphipoda
Introduction
The amphipods are a primarily benthic group with about 8100 species (WoRMS n.d.). Martin and Davis (2001) and Brusca et al. (2016) recognize four suborders, two of which are well represented in the pelagic realm: the Gammaridea and the Hyperiidea (Figure 7.12). The principal suborder is the Gammaridea, found primarily in marine, brackish and littoral habitats, less commonly in freshwater and terrestrial. There are about 4100 species, but only about 150 are pelagic (Vinogradov 1999). The 280 species of Hyperiidea are exclusively pelagic and are well-adapted to a pelagic life. Amphipods range in size from the diminutive ingolfiellids to the giant deep-sea gammarid *Alicella gigantea*, which has been recorded at >34 cm in length.

Benthic gammarideans have robust exoskeletons, and unless they are actively swimming, they sink like stones. The pelagic species are more neutrally buoyant, but their anatomy is less than ideal for swimming. In some cases, they may be hitching rides on gelatinous species like their cousins the hyperiids. In others, if there is a substrate, such as ice for polar species, at least a fraction of the population will gravitate to it. This lifestyle, termed "sympagic", is found in amphipods in the Antarctic and Arctic. The relationship of gammarids with the gelatinous zooplankton is not well-explored.

Years of study have led to the conclusion that "most, if not all, hyperiid amphipods are associated with gelatinous zooplankton during some portion of their life histories" (Harbison et al. 1977). The character of the association differs between species of hyperiids. In some cases, it can be parasitic, where the amphipod feeds on the tissues of its host. In others, it feeds on the particulates trapped by its host. In many cases, the symbiosis is believed to be necessary for survival of the amphipod: an obligate relationship.

The taxonomy of amphipods has been undergoing substantive changes. At this juncture, we are at a crossroads, with most of the descriptive literature using the traditional system described in the classification section below and a new system in development. Classification does not change basic biology, but if, for example, you are doing physiological research, you must know what species you are studying.

The two remaining traditional suborders of amphipods are both associated with substrates. The suborder Caprellidea has two basic groups, with about 300 species altogether. The caprellids (Infraorder Caprellida) are small predators (about 1–3 cm in length) that are highly modified for clinging. They can be found clinging to vegetation and sometimes to animals. They are often associated with hydroids (the benthic life-stages of hydrozoa described in Chapter 3) where they prey on polyps, or with starfish where they feed on the detritus that sticks to mucous on the tube feet and spines (Kaestner 1970). The whale lice, in the Infraorder Cyamidae (about 30 species ranging in size up to 1.3 cm), are parasites on the skin of cetaceans and are usually host-specific. The remaining suborder, the Ingolfiellidea, has only one genus: *Ingolfiella*. Ingolfiellids live interstitially in coastal marine sands, coastal groundwater, freshwater caves, and in the fine silts and oozes at great depths. They are mainly from 1 to

Table 7.12 Vertical ranges and sampling locations for a variety of Mysid and Lophogastrid species. Primary depth range contains the majority of the sampled population. *Source:* Adapted from Mauchline (1980). All data from closing trawls with the exception of Pequegnat (1965). WAP = Western Antarctic Peninsula. ND indicates no data.

Species	Total range (m)	Primary day range (m)	Primary night range (m)	Location	Reference
Arachnomysis megalops	200–500	200–500	200–500	Central Pacific	6
Antarctomysis ohlinii	0–500	200–500	0–500	WAP shelf	7
Boreomysis californica	0–1000	500–900	500–900	Northeast Pacific	8
Boreomysis microps	200–1200	800	400	Northwest Atlantic	11
Boreomysis rostrata	550–>1000	>1000	550–>1000	Scotia-Weddell Sea	5
Boreomysis rostrata	0–1000	600	600	Northeast Pacific	8
Caesaromysis hispida	100–3000	300–3000	100–700	Central Pacific	6
Euchaetomera tenuis	200–700	200–500	200–700	Central Pacific	6
Euchaetomera typica	0–300	0–300	0–300	Central Pacific	6
Euchaetomera glyphidophthalmica	60–550	520–550	100–200	Central Pacific	6
Euchaetomeropsis merolepis	100–700	100–300	200–700	Central Pacific	6
Eucopia australis	600–1000	ND	600–1000	Scotia-Weddell Sea	4
Gnathophausia gigas	>1000	>1000	>1000	Scotia-Weddell Sea	5
Euchaetomera zurstrasseni	160–750	500–750	160–710	Weddell Sea	3
Antarctomysis ohlinii	350–820	ND	ND	WAP shelf	2
Eucopia grimaldi	600	ND	ND	Northeast Atlantic	10
Meterythrops picta	600	ND	ND	Northeast Atlantic	10
Katerythrops oceanea	450–600	ND	ND	Northeast Atlantic	10
Eucopia unguiculata	450–600	600	450–600	Northeast Atlantic	10
Eucopia unguiculata	100–1000	700	700	Northeast Pacific	8

Eucopia unguiculata	200–1200	800	400	Northwest Atlantic	11
Gnathophausia gigas	600–4000	2100	2100	Atlantic, Pacific	9
Gnathophausia gracilis	700–3500	1600	1600	Atlantic, Pacific, Indian	9
Gnathophausia ingens	250–2500	1100	500	Atlantic, Pacific, Indian	9
Gnathophausia ingens	150–1500	650–750	400–900	California borderland	1

References:

1) Childress and Price (1978)
2) Hopkins (1985)
3) Hopkins and Torres (1989)
4) Lancraft et al. (1988)
5) Lancraft et al. (1991)
6) Murano (1977)
7) Parker et al. (2011)
8) Pearcy et al. (1977)
9) Pequegnat (1965)
10) Roe (1984)
11) Waterman et al. (1939)

Table 7.13 Oceanic distributions of bathypelagic crustaceans found in the eastern Gulf of Mexico. *Source:* Burghart et al. (2007), table 4 (pp. 325–326), Marine Biology 152, The bathypelagic Decapoda, Lophogastrida, and Mysida of the eastern Gulf of Mexico, © Springer-Verlag 2007. Reproduced with the permission of Springer.

Family; Genus species	Atlantic	Pacific	Indian	Antarctic	Sources
Lophogastridae					
Gnathophausia gigas	X	X	X	X	1
Gnathophausia gracilis	X	X	X		1
Gnathophausia ingens	X	X	X		1
Gnathophausia zoea	X	X	X		1
Pseudochalaraspidum hanseni		X			2
Eucopiidae					
Eucopia australis	X	X	X	X	1
Eucopia grimaldii	X	X	X	X	1
Eucopia sculpticauda	X	X	X	X	1
Eucopia unguiculata	X	X	X	X	1
Benthesicymidae					
Benthogennema intermedia	X	X	X		3
Gennadas bouvieri	X	X	X		3
Gennadas capensis	X	X	X		3
Gennadas scutatus	X	X	X		3
Gennadas talismani	X				3
Gennadas valens	X				3
Sergestidae					
Sergia grandis	X		X		3,4
Sergia japonica	X	X	X		3,4
Sergia regalis	X	X	X		3,4
Sergia splendens	X				3,4
Sergia wolffi	X				3,4
Pasiphaeidae					
Parapasiphaea sulcatifrons	X	X	X		5,6,7
Oplophoridae					
Acanthephyra acanthitelsonis	X				8
Acanthephyra acutifrons	X	X	X		8,9
Acanthephyra curtirostris	X	X	X		8,9
Acanthephyra exima	X	X	X		8,9
Acanthephyra gracilipes	X				8,9
Acanthephyra pelagica	X	X	X	X	7,8,9
Acanthephyra purpurea	X				8
Acanthephyra quadrispinosa	X	X	X		6,8

Table 7.13 (Continued)

Family; Genus species	Atlantic	Pacific	Indian	Antarctic	Sources
Acanthephyra stylorostratis	X	X	X?		8
Ephyrina benedicti		X	X		8,9
Ephyrina ombango	X	X	X		8
Hymenodora glacialis	X	X	X	X	7,8
Hymenodora gracilis	X	X	X	X	7,8
Janicella spinicauda	X	X	X		8
Meningodora marptocheles	X	X			8
Meningodora miccyla	X				8
Meningodora mollis	X	X	X		8,9
Meningodora vesca	X	X	X		8,9
Notostomus gibbosus	X	X	X		8,9
Systellaspis braueri	X	X	X		8,9
Systellaspis cristata	X	X	X		8,9
Systellaspis debilis	X	X	X		6,7,8
Systellaspis pellucida	X	X	X		8

References:

1) Müller (1993)
2) Richter (2003)
3) Pérez Farfante and Kensley (1997)
4) Vereshchaka (1994)
5) Krygier and Pearcy (1981)
6) Iwasaki and Nemoto (1987)
7) Wasmer (1993)
8) Chace (1986)
9) Krygier and Wasmer (1988)

2.5 mm long with one giant representative at 1.4 cm, *Ingolfiella ieleupi*, found in cave waters of central Africa (Kaestner 1970).

History

The order Amphipoda was first created by Latreille in 1816, recognizing the uniqueness of what are today considered the gammarids (Schram 1986). Soon after, the whale lice and caprellids were placed in their own group, the Laemodipoda, equal in rank to the Isopoda and Amphipoda. In 1852, Dana created three of the four traditional suborders described in the introduction above: the caprellideans, the gammarideans, and the hyperiideans. The ingolfiellids were added by Hansen in 1903.

Owing to their many similarities, the isopods and amphipods were united at one time in a group called the Edriophthalma, but they were separated into their own taxa by Calman (1909) in his classification of the peracarids and remain so today. There have been efforts to re-unite them (Schram 1981, 1984), but thus far they remain separated.

The amphipods have had a number of champions with benchmark publications on the classification of what is, without question, a very difficult group. Two that stand out are J. Laurens Barnard (1969) with the gammarids and M.E. Vinogradov et al. (1996) (in English translation) with the hyperiids. Both classic volumes have wonderful drawings and were the gold standard for each group in the later twentieth century. Advances in technology have modified amphipod classification, but the efforts of those two gentlemen set the stage for what has followed.

Classification

Amphipod classification is in the process of being restructured, modifying the widely used traditional groups of Martin and Davis (2001) shown below. However, descriptive treatments of the amphipods up to the present day still use the traditional system, and that is the option your authors have chosen. The system in progress may be found in WoRMS. The best option at this point is to note the important pelagic families and track their eventual position in the new consensus system of classification.

Martin and Davis (2001). Pelagic families from Vinogradov (1999).
Order Amphipoda
Suborder Gammaridea

Traditionally about 125 families in Martin and Davis (2001). Pelagic families only are listed below. No gammaridean infraorders or superfamilies exist in the Martin and Davis system.

    Eusiridae
    Hyperiopsidae
    Iphimediidae
    Lysianassidae
    Pardaliscidae
    Stilipedidae
    Synopiidae
    Vitjazianidae
  Suborder Caprellidea
Infraorder Caprellida
Infraorder Cyamida
  Suborder Hyperiidea
Infraorder Physocephalata
Superfamily Vibilioidea
  Family Cystosomatidae
    Paraphronimidae
    Vibiliidae
Superfamily Phronimoidea
  Family Dairellidae
    Hyperiidae
    Phronimidae
    Phrosinidae
Superfamily Lycaeopsoidea
  Family Lycaeaopsidae
Superfamily Platyceloidea
  Family Anapronoidae
    Lycaeidae
    Oxycephalidae
    Parascelidae
    Playscelidae
    Pronoidae
Infraorder Physosomata
Superfamily Scinoidea
  Family Archaeoscinidae
    Mimonectidae
    Proscinidae
    Scinidae
Superfamily Lanceoloidea
  Family Chuneolidae
    Lanceolidae
    Microphasmatidae
  Suborder Ingolfiellidea

Family Ingolfiellidae
Family Metaingolfiellidae

Anatomy

Morphotypes within the amphipods range from the elongate forms of the caprellids and ingolfiellids to the robust, laterally compressed gammarids. The two of most concern to us are the pelagic forms, the hyperiids and those gammarids that have colonized the pelagic realm.

All amphipods lack a carapace, and the head (cephalon) is fused to one, rarely two, thoracomeres to form the cephalothorax. The abdomen, or pleon, has six segments with the first three pairs of pleopods modified for swimming. The final three abdominal segments are referred to as the urosome, and their three sets of limbs, termed uropods, are often abbreviated, having only one or two articles. In the hyperiids, the last two segments are fused but the urosome still carries the three pairs of uropods. With the exception of some of the caprellids, three pairs of uropods can be considered a diagnostic character of the amphipods (Schram 1986) as a whole. In most gammarids and hyperiids, the first two pereiopods are modified into subchelate gnathopods.

The gammarids (Figure 7.67a) have a tunnel-like ventral channel that is created by coxal plates of the thoracic limbs and continued posteriorly by the pleura (lappets) of the abdominal segments. Much like the crustaceans considered earlier, the epipodites of amphipod thoracic limbs form the gills (Figure 7.67b). The current created by beating pleopods creates a ventilatory flow within the ventral channel as well as providing forward thrust. Pleopods in both the gammarids and hyperiids are biramous.

Figure 7.68 compares two common hyperiids, *Themisto* and *Oxycephalus*, with the two gammarids *Cyphocaris* and *Eurythenes*. *Cyphocaris* is mesopelagic; *Eurythenes*, a large bathypelagic species with a benthopelagic habit, is found near the bottom as well as hundreds of meters above it (Vinogradov 1999). Though hyperiids and gammarids have the same basic body plan, in most cases, hyperiids are more dorsoventrally compressed, a better morphology for swimming. Note the streamlined bodies and well-developed pleopods in *Oxycephalus* and *Themisto* (formerly *Parathemisto*).

The most obvious differences between the two suborders are in the eyes and the cuticle. Most (but not all) hyperiids have very large eyes (Figures 7.12 and 7.68) that often occupy much of the cephalon. The cuticle of hyperiids varies between being well-developed and pigmented as in *Themisto* and *Vibilia* to being transparent and very thin as in *Oxycephalus* and *Cystisoma*. All the pelagic gammarids have a well-developed and pigmented exoskeleton, though not one as robust as the benthic gammarids. The eyes of gammarids are smaller than those in the hyperiids, but they are nonetheless fully functional. The exoskeletons of all pelagic amphipod taxa, whether hyperiid or gammarid, show no calcification. They remain flexible in the interest of approaching neutral buoyancy.

Nervous System

Amphipods have a conventional malacostracan nervous system (Figure 7.69) with a supraesophogeal ganglion (brain) receiving information from the eyes and antennae, a circumesophogeal commissure, and a subesophogeal ganglion controlling the mouthparts and maxillipeds. Each thoracic and abdominal segment up to the final three has an individual ganglion that is

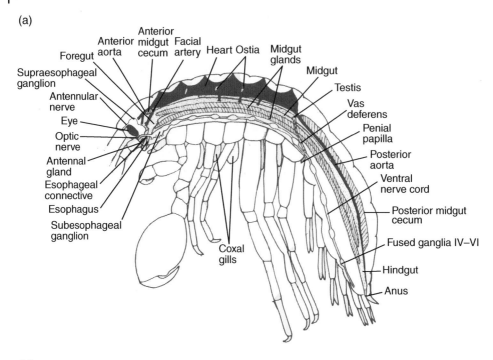

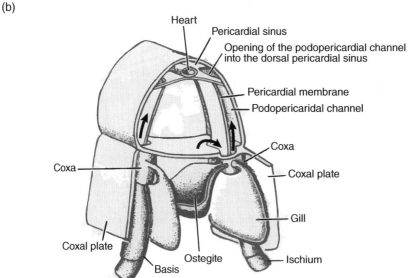

Figure 7.67 Gammarid amphipod structure. (a) Lateral view of the internal anatomy of a typical gammarid amphipod; (b) Posterior view of a cross-section of a thoracic segment of a female gammarid amphipod. Arrows indicate blood flow. *Sources:* (a) McLaughlin (1983), figure 12 (p. 14); Reproduced with permission of Academic Press. (b) Kaestner (1970), figure 18-2 (p. 472). Reproduced with permission of John Wiley & Sons.

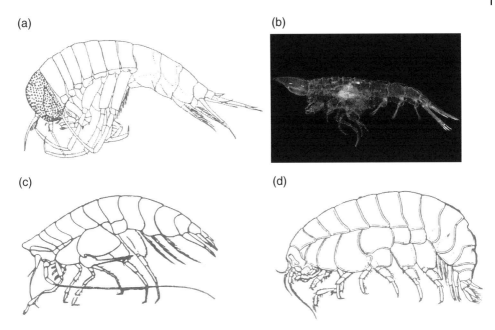

Figure 7.68 A comparison of the forms of two common hyperiid amphipods with two typical gammarids. (a) Female hyperiid *Themisto abyssorum*; (b) The hyperiid *Oxycephalus* sp.; (c) The mesopelagic gammarid *Cyphocaris challengeri*; (d) the bathypelagic gammarid *Eurythenes gryllus*. Sources: (a) Vinogradov et al. (1996), figure 149 (p. 359); (b) © Dante Fenolio, reproduced with permission; (c, d) Vinogradov (1999), figures 4.11, 4.15 (p. 1208). See color plate section for color representation of this figure.

separated from the next by connectives. In gammarids, the ganglia of the last three somites (the urosome) are fused (Figure 7.67a).

Sensory Mechanisms

Amphipods all have sessile (unstalked) compound eyes, called apposition eyes, which have been developed to a remarkable degree in some hyperiid species. The eyes were discussed in detail in the Section Crustacean Systems of this chapter.

Amphipods have a full suite of sensory hairs and aesthetascs located on the antennae, mouthparts, and legs. Statocysts, when present, are located dorsally in the head, innervated by the supraesophogeal ganglion.

Bioluminescence

Bioluminescence has been reported in a number of pelagic amphipod taxa. The ubiquitous gammarid *Cyphocaris* and its fellow lysianassoids *Thoriella* and *Danaella*, the hyperiids *Scina* and *Ancanthoscina*, and the pronoid *Parapronoe* all show bioluminescence at a variety of locations on the body (Figure 7.70). The amphipods lack a true photophore, instead relying on large luminescent cells called photocytes. The photocytes show neural properties, responding to electrical stimulation (Herring 1978).

Circulatory, Respiratory, and Excretory Systems

The tubular heart of gammarids and hyperiids extends from the first to the sixth thoracic

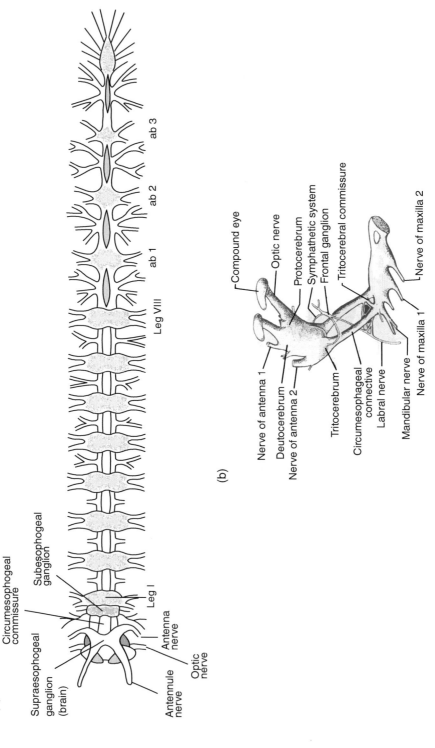

Figure 7.69 Amphipod nervous system. (a) The ventral nerve cord of *Gammarus neglectus*; (b) Lateral view of the anterior part of an amphipod nervous system. *Sources:* (a) Sars (1867), plate VI; (b) Kaestner (1970), figure 4-10 (p. 48).

segments, virtually the entire pereon (Figure 7.67a). Two main arteries leave the heart anteriorly, the anterior aorta and the facial artery. Three pairs of lateral arteries supply the midgut region, and a posterior artery leaves the heart and proceeds posteriorly beneath the posterior gut cecum, ending in the final abdominal segment. Three pairs of ostia are located in the second through fourth pereomeres, one pair per segment.

Blood flow in the gammarids and hyperiids is remarkably well described, at least partly due to the fact that many are sufficiently large and resilient enough to allow experimental manipulation. Like most crustaceans, the heart is located within a pericardial sinus. In amphipods, the sinus is defined on the ventral side by a muscular membranous septum (Figure 7.67) that facilitates blood flow into the heart when its muscles contract by enlarging the pericardial sinus. Laterally, the pericardial membrane extends ventrally, forming a space between the membrane itself and the body wall termed the podopericardial channel. The membrane extends into the pereiopod allowing for an afferent and efferent flow within the leg. Afferent blood originates in the ventral sinus where it has collected after bathing the tissues. From there, it passes from the sinus into the afferent podopericardial channel and its leaf-like gill, exiting the gill into the efferent podopericardial channel to complete the circuit by bringing oxygenated blood to the heart (Figure 7.67b).

Amphipods have paired antennal glands located at the bases of their second antennae that function as described earlier.

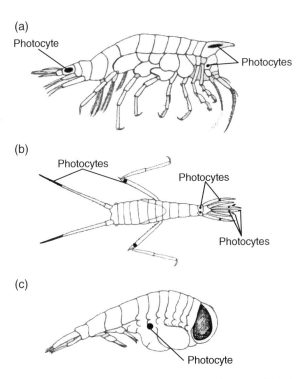

Figure 7.70 Luminescence sites (black) in three amphipod species. (a) The gammarid *Cyphocaris* spp.; (b) the hyperiid *Scina* spp.; (c) the hyperiid *Parapronoë crustulum*. *Source:* Herring (1978), figure 7.7 (p. 224). Reproduced with the permission of Academic Press.

Digestive System

The gut of gammaridean amphipods conforms to the standard malacostracan design, including a foregut with esophagus and stomach followed by an absorptive and conductive midgut region and ending in a short hindgut with a ventrally directed anus (Figure 7.67a). The stomach contains both grinding and pyloric regions. Two to four tubular digestive cecae diverge from the anterior midgut and extend nearly to the urosome. Within the urosome, the midgut gives rise to another pair of anteriorly directed cecae. The cecae associated with the anterior midgut are believed to have secretory and absorptive functions similar to those of the mysids and lophogastrids or to the hepatopancreas of euphausiids and decapods.

Less information is available on the gut of hyperiid amphipods. Brusca (1981) described the digestive tract of *Cystisoma*, a very large and delicate hyperiid that was introduced earlier in the Section Crustacean Systems of the chapter because of its unusually well-developed eyes. The genus is completely transparent, and some species can exceed 100mm in length. Its gut (Figure 7.71) is quite simple and differs considerably from the gammaridean model described above. The esophagus leads into a muscular pouch that was termed the cecum in its original description (Willemoes-Suhm 1874) instead of the stomach, and Brusca continued with that terminology in his 1981 description. The cecum exits into a cecal duct that leads to a thin-walled transparent intestine. The first portion of the intestine is a flattened, highly folded, sac termed the anterior intestine that occupies much of the thoracic region. The anterior intestine leads to a transparent cylindrical intestine that continues on to the thicker-walled rectum or hindgut. Interestingly, transverse muscles attach the hindgut wall to the body wall, allowing it to be expanded and facilitating evacuation. The anatomy of *Phronima's* digestive tract is nearly identical to that of *Cystisoma* (Diebel 1988).

Reproduction and Development

The paired tubular gonads of male and female amphipods are located in the thorax below the heart and adjacent to the midgut, above the midgut cecae. The testes are fairly elongate, extending from the third to the sixth pereomere. The testes continue posteriorly with vasa deferentia that terminate in ejaculatory ducts at the tips of penis papillae, located mid-ventrally at the end of the thorax. Ovaries are located in the same position in females as are the testes in males, but they are not quite as long. Oviducts end in vaginae within the marsupium at the fifth pereomere. Hyperiid and Gammaridean females have four pairs of oostegites attached to the coxae of the second to fifth pereopods (Figure 7.67b) forming the marsupium.

Mating has not been directly observed in either the hyperiids or the pelagic gammarideans. In benthic gammarideans, males detect females approaching the parturial molt and mount them, sometimes for as long as 5–7 days until the molt occurs. After molting, the male and female copulate venter to venter with the male depositing sperm in long strings within the marsupium. Movement of the male pleopods propels the sperm toward the gonopores as the female deposits eggs into the marsupium and it is there that they are fertilized. Young develop within the egg membrane and hatch as miniature adults.

Pelagic male gammarideans exploit their chemosensory abilities to detect females approaching the parturial molt, much like the pelagic decapods and mysids. Because sinking would be a problem, an extended pre-nuptial mounting is unlikely. Copulation

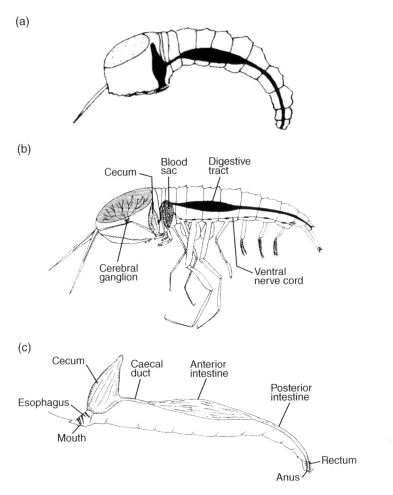

Figure 7.71 Digestive system of the hyperiid amphipod *Cystisoma*. (a) Position of the digestive system within the body; (b) lateral view of the digestive system of a generalized female *Cystisoma*; (c) lateral view of an isolated digestive system of *Cystisome*. *Source:* (a, c) Brusca (1981), figure 2 (p. 360); (b) Brusca (1981), figure 1 (p. 359).

is likely to be similar to that of benthic species but with a more abbreviated coupling such as that observed in decapods and mysids.

Reproduction in the hyperiids presents a very different situation. In many species, the hyperiids must select a gelatinous host for their larvae, which will be incapable of swimming well enough to find one. The use of a host for larval development deviates from the normal reproductive pattern in the Peracarida characterized by the direct development of larvae in a marsupium.

Laval (1980) describes the process of hyperiids' larval deposition as "demarsupiation." Once a suitable host is selected, larvae are coaxed out of the marsupium and deposited in a favorable location in the host. This may be within the gonad of a medusa or inside the barrel-shaped test of a salp, a free-swimming pelagic tunicate described in a later chapter. Hyperiid larvae do not

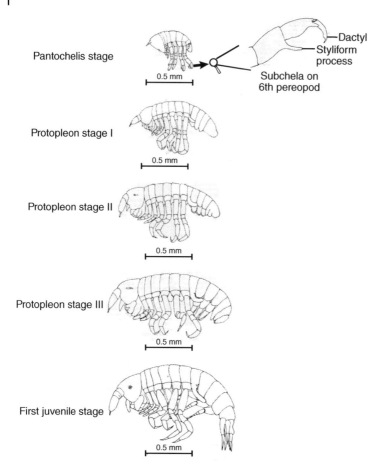

Figure 7.72 Larval developmental stages of the hyperiid amphipod *Vibilia armata* from pantochelis larva through three protopleon stages to the first juvenile stage. Note the modified subchela of the sixth pereopod of the pantochelis larva (magnified on right); this character is lost at the molt into the first protopleon stage. *Source:* Laval (1980), adapted from pages 30, 32, 33.

resemble miniature adults when deposited, rather they have a "precocious hatching" (Laval 1980) and specialized early developmental stages for clinging to the host. The larval development of *Vibilia armata* shown in Figure 7.72 is an example. The hatching stage, the pantochelis, has a subchela on the sixth pereopod used to secure the larva when it is first transferred to the host. The subchela disappears at the next molt. Larvae continue to develop within the host through a series of stages until a final molt produces a juvenile resembling a miniature adult, equivalent to the hatching stage of gammarids.

An interesting correlation to the larval deposition used by hyperiids is that often females will deposit only a few larvae per host. As brooders, the number of larvae available for deposition in each clutch is limited and the females apparently exercise a "bet-hedging" strategy for larval survival by depositing on multiple hosts.

Sheader (1977) was able to observe a successful laboratory mating in *Themisto gaudichaudii*. *Themisto* is a hyperiid,

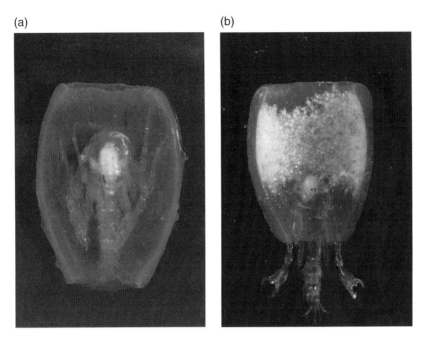

Figure 7.73 Reproduction in the hyperiid amphipod *Phronima*. (a) Adult *Phronima* sp. within a salp barrel; (b) adult female *Phronima* with larvae in a salp barrel. *Source:* © Dante Fenolio, reproduced with permission. See color plate section for color representation of this figure.

occasionally very numerous in Antarctic waters, that spends part of its life as a free-living species. *Themisto* uses a host, which can be a hydromedusa, salp, or scyphomedusa, as a mating platform. How receptive females and males find one another in open waters is unknown, but, as with all other pelagic Crustacea, it is assumed that females close to the parturial molt can be detected by mature males. Diebel (1988) demonstrated a chemosensory basis for detecting preferred hosts by a variety of hyperiids, giving strong support for the importance of chemosensory mechanisms in detecting mates as well. When the female completes her parturial molt aboard the mating platform, the male deposits a sperm bundle into her marsupium. There is no pre-copulatory behavior as in the gammarids. In *Themisto*, the eggs are carried to term and the young hatch as juveniles, a rare occurrence in the hyperiids (Laval 1980).

Phronima sedentaria, perhaps the best known hyperiid, deposits her entire brood pouch as a large cluster within the characteristic barrel that she fashions from the tunic of a salp (Figure 7.73). The larvae coexist together within the barrel, remaining in a group that behaviorally retains its coherence even though it is constantly moving. The mother feeds her larvae with prey she brings back to the barrel and actively keeps her brood within it by "combing" them back from its edge with her gnathopods.

Food and Feeding

Information on amphipod diets comes from three different types of studies: field-based observations, laboratory-based studies, and examination of gut contents. Field studies (Harbison et al. 1977, 1978; Madin and Harbison 1977) use SCUBA for observations in the upper 30 m of open-ocean systems to describe the associations of hyperiids and host organisms in situ. Bluewater diving allows first-hand observations of feeding behavior in a natural situation and, in many cases, gentle capture of host species with their symbionts still on board. Captured individuals can be then observed in shipboard laboratories in near-perfect condition. Host organisms that can be observed and captured in this way are salps, medusae, small siphonophores, and ctenophores.

Laboratory-based studies can use aquaria to observe animals captured by hand or in gentle plankton nets with very short tows and large catch buckets to avoid crowding. Home laboratories with deep water nearby, such as the laboratories in Naples, Villefranche-sur-Mer, or Monterey Bay, allow for longer-term observations of behavior. They can be used for studies of larval development, some feeding behaviors, and some physiological measurements.

Examination of gut contents, a tried and true technique in pelagic animal study, has its advantages and disadvantages when applied to the amphipods. One of its chief advantages is that any item in the gut has definitely been ingested. One of its chief disadvantages, particularly with the hyperiids, is in not knowing where the item originated. Did it come from the food captured by the host, or from the amphipod's own hunting activity? With species that live exclusively below 30 m, examination of gut contents is really the only option for feeding information. Gut contents can be evaluated in light of what is known about species' behavior, morphology, and natural history to sketch a picture of its possible feeding behavior in-situ.

Diets of Antarctic gammarids and hyperiids taken primarily in the Austral fall are presented in Table 7.14. Gammarids showed a higher diet diversity overall, with both groups having a large amount of soft-bodied prey. Since neither pelagic gammarids nor hyperiids have the mouthparts or thoracic appendages for microphagous (suspension- or filter-) feeding, appearance of phytoplankton in the gut suggests that it may be derived from host or prey food capture as has been reported for *Vibilia* (Madin and Harbison 1977). Species in both groups are clearly capable of doing their own hunting, with live krill, copepods and polychaetes appearing in their diets.

The relationship between gammarids and gelatinous zooplankton hosts, if any, remains largely unexplored. Sea ice in both the Arctic and Antarctic is used by gammarids as a platform or "inverted benthos" for feeding and hunting excursions (Carey 1985; Aarset and Torres 1989), a lifestyle called "sympagic". Historically, the Arctic has had large areas of multi-year ice, generally poleward of 80 °N, as well as the seasonal sea ice formed further south (Maykut 1985). Sea ice in the Arctic is thus available as a predictable site for feeding and rest for a large part of the year, and the sympagic lifestyle is widespread. Antarctic sea ice is highly seasonal, forming in the fall and melting in the spring. It predictably covers the Antarctic pelagial but is only available for part the year. The degree to which it is exploited by the amphipod fauna is unknown.

Table 7.15 lists hyperiid families and their observed hosts. Most of the information was obtained using blue-water diving observations, a technique that is near-perfect for this type of research. Its only drawback is that it is depth-limited, so lifestyles in deeper-living species must be addressed with other methods such as gut contents

Table 7.14 Diets of Antarctic amphipods. All diet taxa expressed as percent frequency-of-occurrence in guts. n = number of guts with identifiable remains. The [a]s indicate Hyperiid hosts not confirmed; taken from confamilials (Laval 1980), or correlated capture data (*Cyllopus*: Donnelly et al. 2006; *H. dilatata*: McClintock 2012). Phytoplankton includes all phytoplankton taxa and debris; krill debris is mainly molts. Krill live-prey includes *Euphausia superba*, *E. crystallorophias* and all larval stages of each. In "Other" columns, C = *Clione*, a pteropod, L = *Limacina*, a pteropod, M = Mysid, A = amphipod.

Species (n)	Host	Phytoplankton	Krill debris	Krill live prey	Salps	Cnidaria	Polychaets	Calanoids	Oithona/ Oncaea	Other copepods	Ostracods	Chaetognaths	Fish debris	Other	Location	Season	References
Hyperiids																	
Cyllopus lucasii (3)	Salps[a]	100	0	0	0	0	0	0	0	0	0	0	0	0	1	F	1
Cyllopus lucasii (36)	Salps[a]	39	0	8	50	28	31	6	11	11	0	3	0	0	2	W	3
Cyllopus lucasii (21)	Salps[a]	0	0	0	38	57	5	0	0	0	0	0	0	0	3	W	4
Hyperia macrocephala (10)	Medusae[a]	10	0	0	0	90	0	0	0	0	0	0	0	0	4	S	2
Hyperiella dilatata (23)	Pteropods, medusae[a]	4	13	0	0	87	22	0	0	0	0	0	0	0	1	F	1
Hyperiella dilatata (30)	Pteropods, medusae[a]	63	0	0	0	97	10	0	3	3	0	0	0	7 C	2	F	3
Hyperiella macronyx (5)	Medusae[a]	40	20	40	0	40	40	20	20	40	0	0	0	0	1	F	1
Thermisto gaudichaudii (46)	Free-living, salps, medusae	57	43	28	52	22	13	9	11	20	0	9	20	0	1	F	1
Primno macropa (38)	Free-living, hosts unknown	5	34	47	0	0	8	21	8	8	0	3	0	0	1	F	1
Vibilia stebbingi (2)	Salps	0	100	0	0	0	0	0	0	0	0	0	0	0	1	F	1
Gammarids																	
Cyphocaris faueri (2)	N/A	0	100	0	0	0	0	0	0	0	0	0	0	0	1	F	1

(*Continued*)

Table 7.14 (Continued)

										Diet items						
Species (n)	Host	Phytoplankton	Krill debris	Krill live prey	Salps	Cnidaria	Polychaets	Calanoids	Oithona/ Oncaea	Other copepods	Ostracods	Chaetognaths	Fish debris	Other	Location	Season	References
Cyphocaris richardi (98)	N/A	2	97	1	5	2	4	9	1	7	0	0	0	2 M	1	F	1
Epimeriella macronyx (22)	N/A	36	27	0	5	86	55	41	14	0	0	0	0	0	1	F	1
Eusirus antarcticus (8)	N/A	0	0	63	0	0	0	0	0	38	0	0	0	0	2	F	3
Eusirus microps (16)	N/A	13	69	0	13	6	19	19	0	6	0	6	38	0	1	F	1
Eusirus perdentatus (2)	N/A	100	100	0	0	0	0	0	0	0	0	0	0	0	1	F	1
Eusirus propeperdentatus (62)	N/A	6	55	3	0	6	16	66	2	27	16	5	0	11 M,A	1	F	1
Eusirus tridentatus (18)	N/A	67	0	67	0	22	72	0	0	0	0	0	0	39 L	4	S	2
Eusirus tridentatus (16)	N/A	69	0	25	0	19	44	13	6	0	6	0	0	13 L	4	S	2

Species																
Eurythenes gryllus (8) N/A	13	0	0	0	0	0	0	0	0	0	0	0		1	F	1
Eurythenes obesus (1) N/A	0	0	0	100	0	0	0	0	0	0	0		1	F	1
Orchomene plebs (50) N/A	16	30	18	28	16	36	30	30	30	10	58	0	1	F	1
Orchomene rossi (22) N/A	32	50	27	23	0	41	32	32	41	56	55	0	1	F	1
Parandania boecki N/A (16)	13	0	6	13	100	6	32	0	0	6	13	0	1	F	1

Location:
1) Croker Passage; Western Antarctic Peninsula;
2) Weddell Sea;
3) Scotia/Weddell Seas;
4) McMurdo Sound.

Season:
F, Fall;
S, Summer;
W, Winter.

References:
1) Hopkins (1985)
2) Hopkins (1987)
3) Hopkins and Torres (1989)
4) Lancraft et al. (1991).

and correlative captures (Donnelly et al. 2006). Nearly, all the host taxa in Table 7.15 provide a good platform for rest, reproduction, larval development, or hunting excursions. At the species level, many amphipod taxa have a preferred host species or type of host, e.g. salp, or medusa. At the family level, host specificity is less obvious but a few of the families clearly prefer salps or siphonophores. Immunity to cnidarian stings and digestive enzymes has been observed in a few hyperiid genera, but it is not a universal trait (Harbison et al. 1977; Laval 1980).

Amphipods and Gelatinous Zooplankton

The character of the relationship between hyperiids and gelatinous zooplankton varies considerably. Several genera, e.g. *Vibilia, Lycaea, Hyperia,* and *Lestrigonus* (Hyperiidae), place their young in host organisms. The young feed and develop on host tissues until they are mobile enough to colonize a host of their own. *Vibilia* and *Lycaea* remain highly dependent on salps throughout their lives. In contrast, Themisto's mating behavior and early life is spent in association with salps or hydromedusae (Madin and Harbison 1977; Laval 1980) but is believed to be largely free-living in later life. Overall, the jelly-hyperiid association shows a continuum from near-total dependence for food and transportation to an intermittent home base for hunting, mating, and the occasional meal (cf. Laval 1980)

It is believed that gammarid amphipods also utilize gelatinous organisms as islands and food sources in the three-dimensional environment of the open sea, though the relationships are poorly described (Vinogradov 1999). Note that the diets of the gammarids in Table 7.15 partially resemble those of the free-living hyperiids: both have Cnidaria and polychaetes among their common prey items as well as a variety of copepods. Gammarids also show evidence of preying on fish, as does the hyperiid *Themisto*. Though all are very capable swimmers, gammarid morphology is essentially a benthic design, well suited to clinging and moving about on a substrate and ideal for using jellies and other taxa as a way to invade the pelagic realm.

Gelatinous species are ubiquitous in the open ocean and in many areas of the world are present year-round. Whether SCUBA diving in the upper 30 m or in a submersible in the upper 1000 m, jellies will usually dominate your visual field. To the author, it is quite natural that they have acquired hitchhikers from among the many negatively buoyant species in the open ocean.

Geographic Distributions

Pelagic amphipods inhabit "all depths and latitudes of the world ocean" (Vinogradov 1999). They typically comprise a small yet consistent fraction of the total numbers of midwater crustaceans in the upper 1000 m of the water column. In the Antarctic, amphipods made up 1–2% of the total numbers of crustaceans captured in the Scotia-Weddell Sea region but were 30–50% of the total crustacean species (Lancraft et al. 1989, 1991). In the more productive waters of the Western Antarctic Peninsula shelf, amphipods were 12 and 6% of the total numbers of Crustacea captured in the fall seasons of 2001 and 2002, respectively, but made up 78% of the total crustacean species (Parker et al. 2011). A similar pattern was observed in the Northeastern Atlantic, with amphipods contributing 6% of the total crustaceans captured but comprising 60% of the total number of crustacean species (Roe 1984; Roe et al. 1984). In sum, pelagic amphipods are a substantial fraction of the total number of crustacean species but only a modest contributor to total numbers of the Crustacea.

Table 7.15 Common hosts and diets in hyperiid amphipod families. Host taxa and diets confirmed by field observations in the cited references. Host taxa of *Cyllopus* unconfirmed by direct observation; deduced from net captures and diet analysis. Additional detail at the genus-species level may be found in cited references.

Suborder: Hyperiidea	Common genera	Common host taxa	Diet	References
Infraorder Physocephalata				
Superfamily Vibiliodea				
Family				
Cyllopodidae	*Cyllopus*	Salps, medusae, free-living[a]	Host tissue	Hopkins and Torres (1989)
Paraphronimidae	*Paraphronima*	Siphonophores	Host tissue	Harbison et al. (1977)
Vibiliidae	*Vibilia*	Salps	Host-captured food	Madin and Harbison (1977)
Superfamily Lycaeopsiodea				
Family Lycaeopsidae	*Lycaeopsis*	Siphonophores	Host tissue	Harbison et al. (1977)
Superfamily Phronimoidea				
Family				
Hyperiidae	*Themisto, Hyperoche, Hyperiella*	Medusae, radiolarians, salps, ctenophores, free-living	Host tissue, hunting	Harbison et al. (1977)
Phronimidae	*Phronima*	Salps	Host tissue	Harbison et al. (1977)
Phrosinidae	*Anchylomera, Primno*	Siphonophores, salps	Unknown	Harbison et al. (1977)
Superfamily Platysceloidea				
Family				
Brachyscelidae	*Brachyscelus, Euthamneus*	Medusae, salps	Host tissue	Harbison et al. (1977)
Lycaeidae	*Lycaea*	Medusae, salps	Host tissue	Madin and Harbison (1977)
Oxycephalidae	*Oxycephalus, Glossocephalus*	Ctenophores, salps, radiolarians, pteropods	Host tissue	Harbison et al. (1978)

(*Continued*)

Table 7.15 (Continued)

Suborder: Hyperiidea	Common genera	Common host taxa	Diet	References
Parascelidae	*Schizocelus, Thyropus*	Siphonophores	Host tissue	Harbison et al. (1977)
Platyscelidae	*Amphithyrus, Hemityphis*	Siphonophores	Unknown	Harbison et al. (1977)
Pronoidae	*Paralycaea, Sympronoe, Eupronoe*	Siphonophores, free-living	Host tissue	Harbison et al. (1977)
Infraorder Physosomata				
Superfamily Scinoidea				
Family Scinidae	*Scina*	Siphonophores	—	Laval (1980)
Superfamily Lanceoloidea				
Family Lanceolidae	*Lanceola*	Medusae	—	Laval (1980)

Vertical Distributions

Hyperiids and gammarids are found throughout the water column in all regions that have been sampled (Vinogradov 1970). Within the hyperiids, the two infraorders Physocephalata and Physomata show different trends with depth. The Physocephalata are most diverse in the upper 500 m, with far fewer species, e.g. those within the Cystosomatidae, occurring more deeply. The more primitive group, the Physosomata, are well represented down to 4000 m, tapering off below that (Table 7.16, Figure 7.74).

Many of the physosomatous hyperiids are eurybathic. Vinogradov noted that about 25% of the total species captured were found from 200 to 4000 m, and some, e.g. *Lanceola clausi*, from 200 m to >5000 m. Clearly, eurybathic hyperiids must have a host with an equally prodigious depth distribution or they must be able to switch hosts with depth as needed. When hyperiids do show vertical zonation, they inhabit one of three zones: 200–1000 m, 100–3000 m, and >3000 m.

The gammarids show a different relationship with depth, including a component that leaves the benthos for forays into the water column. Vinogradov (1970) observed three general trends in the vertical distribution of pelagic gammarids. First, many species (11 of the 26 captured in his study) found in the 200–500 m depth stratum do not occur below 500 m, making the 500 m isobath an important zonal boundary. Second, the gammarid fauna between 500 and 3000 m is fairly uniform, inhabited by a suite of eurybathic species. Few species

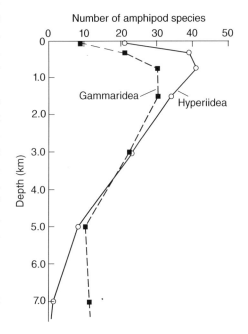

Figure 7.74 Number of species of amphipods as a function of depth. *Source:* Vinogradov (1970), figure 72 (p. 207).

Table 7.16 Depth distribution of pelagic amphipod species. *Source:* Vinogradov (1970), table 48 (p.20).

	Depth (m)						
	0–200	200–500	500–1000	1000–2000	2000–4000	4000–6000	6000–8000
Suborder Gammaridea	9	21	30	30	22	10	12
Suborder Hyperiidea							
Infraorder Physosomata	21	39	41	34	23	8	2

within the 500–3000 m group are found below that depth. Third, the pelagic gammaridean fauna at depths below 6000 m is quite rich, with at least 12 species being found there. Deep-living pelagic genera (200–4000 m) often show an increased number of species per genus, a phenomenon called polytypy. For example, *Hyperiopsis*, a genus that does not occur above 1000 m, has seven species.

A number of benthic amphipods that are commonly seen at food-falls and baited traps on the ocean floor venture hundreds of meters into the water column at typical open-ocean bottom depths (4000 m) and at the hadal depths (>6000 m) found in oceanic trenches. This phenomenon was well documented by Smith and Baldwin (1984) who trapped the gammarid *Eurythenes gryllus* with baited minnow traps up to 1400 m off the ocean floor. Vinogradov (1999) considers benthopelagic scavengers like *Eurythenes* to be the predecessors to the purely pelagic gammarids like *Cyphocaris*. The benthopelagic gammarids form a more important part of the pelagic fauna at the hadal depths within oceanic trenches than over the abyssal plain.

Cameo Players

A few major crustacean taxa have entered the micronektonic pelagic realm with a very few key players. We will deal briefly with four of those. The first is the pelagic isopod *Anuropus*. It is rarely encountered in conventional sampling, and like most of its cousins, the pelagic amphipods, it is associated with a jellyfish, the giant scyphozoan medusa *Deepstaria*. The second is the giant ostracod *Gigantocypris*, the largest member of the ostracods, an enormously important zooplankton group. The third is *Dosima fascicularis*, the "blue buoy" pelagic barnacle. The fourth is the very large and highly luminescent copepod *Gaussia princeps*.

Anuropus, the pelagic isopod
Superorder Peracarida
Order Isopoda
Suborder Cymothoida
Superfamily Cymothoidea
Family Anuropidae
Anuropus Bathypelagicus

The isopods comprise the most speciose order within the Peracarida, with about 8200 species. The group is nearly 100% benthic and ranges in size from a few mm to the giant *Bathynomus giganteus*, which can reach 35 cm. The first specimen of *Anuropus* was collected from a depth of 1070 fathoms by the Challenger expedition in 1875 but it was not described until 11 years later (Beddard 1886). *Anuropus* was initially assumed to be benthic, like most of its brethren. A number of subsequent captures in midwater trawls suggested a pelagic lifestyle (Menzies and Dow 1958) for the genus, later confirmed beyond doubt by observations from a submersible (Barham and Pickwell 1969).

Anuropus bathypelagicus (see Figure 7.11) is a large species, reaching at least 8.0 cm, and is believed to be an ectoparasite on the large (>70 cm diameter) mesopelagic medusa, *Deepstaria enigmatica*. Barham and Pickwell's observations of the animal in situ suggest that the species lives in the bell cavity of the medusa, clinging with the sharp dactyls on its thoracic appendages and feeding on its tissues. Menzies and Dow (1958) examined the gut contents of Anuropus and found mainly cnidarian debris, including nematocysts. *Anuropus* was observed to be neutrally buoyant, obtaining its lift from sequestered lipids (Barham and Pickwell 1969).

Anuropus has no eyes and apparently varies somewhat in its coloration. Barham and Pickwell reported the species to have the deep-red coloration typical of mesopelagic Crustacea. A purple specimen of

Anuropus was maintained in the laboratory for several months during the author's tenure as a graduate student at the University of California Santa Barbara. Both variants, purple and deep red, would be well camouflaged at mesopelagic depths.

Gigantocypris, the Giant Ostracod
Superclass Oligostraca
Class Ostracoda
Subclass Myodocopa
Order Myodocopida
Suborder Myodocopina
Family Cypridinidae
Subfamily Cypridininae
Tribe Gigantocypridinini
Gigantocypris agassizii

Gigantocypris is the largest of the ostracods, or "mussel shrimps," so named because "the limbs and trunk are completely enclosed within a bivalved carapace that resembles a mussel shell" (Kaestner 1970). The ostracods are prominent members of the marine zooplankton and are found in abundance in freshwater as well. Currently (WoRMS n.d.) there are about 6000 species. Ostracods, particularly *Cypridina*, are important in the diets of many bioluminescent species as a source of luciferin (Widder 2010). Nearly, all the ostracods are diminutive in size, with a great many <1 mm. *Gigantocypris* is the only representative that exceeds 10 mm, and it reaches about 3.2 cm, or about the size of a large "shooter" marble. We have not discussed other members of the class, despite their importance in the zooplankton, because *Gigantocypris* is the only representative that is even close to our target size range (2–20 cm).

The genus *Gigantocypris* is found in all major ocean basins at depths >500 m and is distributed from the equator to both poles. It is a ubiquitous member of the micronekton/macrozooplankton fauna, never abundant but nearly always present. The author has captured it at 65 °S in the Atlantic Sector of the Southern Ocean and at 69 °S in George VI Sound at the southern end of Marguerite Bay on the Western Antarctic Peninsula.

Gigantocypris is very unusual in appearance (Figure 7.75), with its large naupliar eyes being its most remarkable feature. Its anterior perspective (Figure 7.75a) shows the paired eyes appearing much like the headlights of a car, a particularly good analogy because its eyes and a car's headlamp are constructed in a similar way. The rear of each eye is composed of a mirror-like parabolic reflector that is visible in Figure 7.75b. The retina, or sensory element, is the orange, sausage-shaped structure in the center of the eye. Light reflected toward the center of the eye by the shape of the parabolic reflector is captured by the retina. *Gigantocypris* eyes are extremely effective at detecting light but are not believed capable of forming images (Land 1981a). At the depths they normally inhabit (>700 m), most of the available light would be produced by bioluminescence. They are believed to prey on copepods, and their unusual visual system likely helps in prey location.

Gigantocypris swims with its paired second antennae by extending them between the valves of the carapace and beating them in a motion that appears much like the breaststroke. The second antennae are partially extended in Figure 7.75; with a little imagination you can create the motion in your mind's eye. Videos of swimming *Gigantocypris* are available on the internet.

***Dosima*, the blue buoy barnacle.**
Superclass Multicrustacea
Subclass Thecostraca
Infraclass Cirripedia
Superorder Thoracica
Order Lepadiformes
Suborder Lepadomorpha

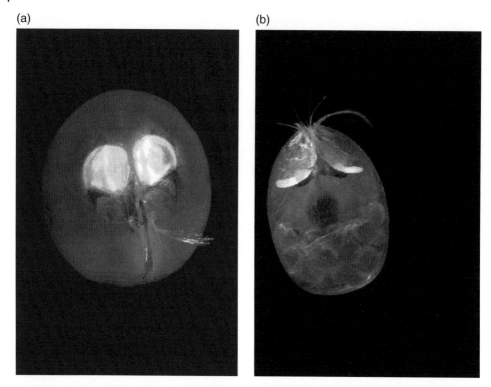

Figure 7.75 The giant ostracod *Gigantocypris*. (a) Frontal view of *Gigantocypris* showing the large naupliar eyes; (b) lateral view of *Gigantocypris* showing the retina and parabolic reflector of the eyes. Note the developing eggs within the carapace. *Source:* © Dante Fenolio, reproduced with permission. See color plate section for color representation of this figure.

Family Lepadidae
Dosima fascicularis

Dosima fascicularis, the "blue buoy" barnacle, is a member of the pleuston, the blue-water oceanic surface community that includes *Velella*, the "by-the-wind-sailor" and *Physalia*, the "Portuguese Man-o-War" siphonophore, described in Chapter 3. *D. fascicularis* (Figure 7.76) is a close relative of the goose-neck barnacles found in the intertidal but is highly specialized for a pelagic life (Cheng and Lewin 1976). Its shell is almost entirely uncalcified, and its body is "open-ocean" blue, the characteristic aqua blue color found in *Physalia*, *Velella*, and other members of the pleuston community.

Dosima may be found attached to *Velella*, bird feathers, tar balls, sea-grass blades, or terrestrial debris (Cheng and Lewin 1976). It also can manufacture its own float and may be found floating singly or in small colonies that share a common float of their own manufacture. The thin layer of cement that it initially uses for adhering to a floating object is expanded as the animal grows, morphing into a porous, gas-filled float that resembles a ball of large-celled urethane foam. The cells of the cement foam are sealed to the outside and are resistant to lowered pressure and water intrusion.

The proteinaceous cement is produced by glands in the animal's stalk and conveyed to the base of the stalk by specialized cement

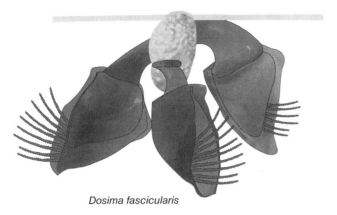

Dosima fascicularis

Figure 7.76 The pleustonic "blue buoy" barnacle *Dosima fascicularis*. See color plate section for color representation of this figure.

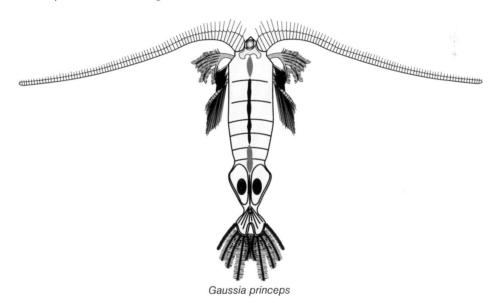

Gaussia princeps

Figure 7.77 The "giant" copepod *Gaussia princeps*. Carapace is deep blue in life. See color plate section for color representation of this figure.

canals. The gas within the cells is believed to be CO_2 produced by the animal's metabolism (Zheden et al. 2012).

This author's one brief experience with *Dosima* was impressive enough to linger in the memory banks for over 40 years because the contrast between *Dosima* behavior and that of *Balanus*, a common intertidal barnacle, was remarkable. *Balanus* is extremely light-sensitive. If you pass your hand over a group of *Balanus* that are feeding with the typical rhythmic sweeping motions of their feeding limbs, or cirri, the cirri will be retracted into the *Balanus* shells with blinding speed. I passed my hand over a *Dosima*, and it tried to grab my finger. A very different feeding strategy, to be sure.

***Gaussia princeps* the "black knight" copepod.**

Superclass Multicrustacea

Subclass Copepoda
Infraclass Neocopepoda
Superorder Gymnoplea
Order Calanoida
Family Metridinidae
Gaussia princeps

Gaussia princeps, with a size exceeding 10 mm, is an honorary member of the micronekton (Figure 7.77). It is found worldwide at tropical and temperate latitudes. In the cool waters of the California borderland, it may be captured alive and maintained for extended periods (Barnes and Case 1972; Childress 1977).

Gaussia is best known for its bioluminescent displays and as the ultimate source of a commercially available luciferase, the enzyme which when combined with its substrate luciferin produces bioluminescent light. *Gaussia princeps* is one of the few species known to be metabolically sensitive to pressure within its normal bathymetric range (Childress 1977). In the California borderland, *Gaussia* spends its days in the oxygen-minimum layer below 400 m and its nights above it at 200–300 m.

References

Aarset, A.V. and Torres, J.J. (1989). Cold resistance and metabolic responses to salinity variations in the amphipod *Eusirus antarcticus* and the krill *Euphausia superba*. *Polar Biology* 9: 491–497.

Ache, B.W. (1982). Chemoreception and thermoreception. In: *Biology of Crustacea, Volume 3, Neurobiology: Structure and Function* (eds. H.L. Atwood and D.C. Sandeman). New York and London: Academic Press.

Aguinaldo, A.M.A., Turbeville, J.M., Linford, L.S. et al. (1997). Evidence for a clade of nematodes, arthropods, and other moulting animals. *Nature* 387: 489–493.

Alldredge, A.L. and Silver, M.W. (1988). Characteristics, dynamics, and significance of marine snow. *Progress in Oceanography* 20: 41–82.

Anderson, D.T. (1973). *Embryology and Phylogeny in Annelids and Arthropods*. New York: Pergamon Press.

Appollonio, S. (1969). Breeding and fecundity of the glass shrimp, *Pasiphaea multidentata*. *Journal of the Fisheries Research Board of Canada* 26: 1969–1983.

Backus, R.H., Clark, R.C., and Wing, A.S. (1965). Behaviour of certain marine organisms during the solar eclipse of July 20, 1963. *Nature* 205: 989–991.

Baker, A.D.C. (1959). The distribution and life history of *Euphausia triacantha* Holt and Tattersall. *Discovery Reports* 29: 309–340.

Bargmann, H.E. (1937). The reproductive system of *Euphausia superba*. *Discovery Reports* 14: 325–350.

Barham, E.G. (1957). The ecology of sonic scattering layers in the Monterey Bay area, California. *Hopkins Marine Station, Stanford University, Technical Report* 1: 1–182.

Barham, E.G. and Pickwell, G.V. (1969). The giant isopod Anuropus: a scyphozoan symbiont. *Deep-Sea Research* 16: 525–529.

Barnard, J.L. (1969). The families and genera of marine gammaridean Amphipoda. *U.S. National Museum Bulletin* 271: 1–535.

Barnes, A.T. and Case, J.F. (1972). Bioluminescence in the mesopelagic copepod, *Gaussia princeps* (T. Scott). *Journal of Experimental Biology and Ecology* 8: 53–71.

Barnes, R.D. (1974). *Invertebrate Zoology*. Philadelphia: W.B. Saunders.

Bauer, R.T. (2000). Simultaneous Hermaphroditism in caridean shrimps: a unique and puzzling sexual system in the Decapoda. *Journal of Crustacean Biology* 20 (special number 2): 116–128.

Bauer, R.T. (2004). *Remarkable Shrimps, Adaptations and Natural History of the*

Carideans. Norman: University of Oklahoma Press.

Beddard, F.E. (1886). Preliminary notice of the Isopoda collected during the voyage of H.M.S. Challenger, Part III. *Proceedings. Zoological Society of London* 1: 97–122.

Beklemishev, W.N. (1969). *Principles of Comparative Anatomy of Invertebrates*. Chicago: University of Chicago Press.

Belman, B.W. and Childress, J.J. (1976). Circulatory adaptations to the oxygen minimum layer in the bathypelagic mysid *Gnathophausia ingens*. *Biological Bulletin* 150: 15–37.

Belyaev, G.M. (1966). *Hadal Bottom Fauna of the World Ocean*. Jerusalem: Israel Program for Scientific Translations.

Bergstrom, B. (2000). The biology of Pandalus. In: *Advances in Marine Biology*, vol. *38*. San Diego: Academic Press.

Boas, J.E.V. (1883). Studien uber die Verwandtschaftsbeziehungen der Malakostraken. *Morphologie Jb.* 8: 485–579.

Boden, B.P. (1950). The post-naupliar stages of the crustacean *Euphausia pacifica*. *Transactions of the American Microscopical Society* 69: 373–386.

Boden, B.P. (1951). The egg and larval stages of Nyctiphanes simplex, a euphausiid crustacean from California. *Proceedings. Zoological Society of London* 121: 515–527.

Boden, B.P., Johnson, M.W., and Brinton, E. (1955). The Euphausiacea (Crustacea) of the North Pacific. *Bulletin of the Scripps Institution of Oceanography* 6: 287–400.

Boden, B.P. and Kampa, E.M. (1967). The influence of natural light on the vertical migration of an animal community in the sea. *Symposia of the Zoological Society of London* 19: 15–26.

Bowman, T.E. and Abele, L.G. (1982). Classification of the recent Crustacea. In: *Biology of Crustacea, Volume 1, Systematics, the Fossil Record, and Biogeography* (ed. L.G. Abele). New York: Academic Press.

Boxshall, G.A. (1998). Mating biology of copepod crustaceans. *Philosophical Transactions of the Royal Society of London B* 353: 669–815.

Boyd, C.M. (1960). The larval stages of *Pleuroncodes planipes* Stimpson (Crustacea, Decapoda, Galatheidae). *Biological Bulletin* 118: 17–30.

Boyd, C.M. (1967). The benthic and pelagic habits of the red crab, *Pleuroncodes planipes*. *Pacific Science* 21: 394–403.

Briggs, J.C. (1995). *Global Biogeography*. Amsterdam: Elsevier.

Bright, T., Ferrarri, F., Martin, D., and Franceschini, G.A. (1972). Effects of a total solar eclipse on the vertical distribution of certain oceanic zooplankters. *Limnology and Oceanography* 17: 296–301.

Brinton, E. (1962). The distribution of Pacific Euphausiids. *Bulletin of the Scripps Institution of Oceanography* 8: 51–270.

Brooks, W.K. (1882). Lucifer: a study in morphology. *Philosophical Transactions of the Royal Society of London* 173: 57–137.

Brusca, G.J. (1981). On the anatomy of Cystisoma. *Journal of Crustacean Biology* 1: 358–375.

Brusca, R.C. and Brusca, G.J. (1990). *Invertebrates*. Sunderland: Sinauer Associates.

Brusca, R.C. and Brusca, G.J. (2003). *Invertebrates*, 2e. Sinauer: Sunderland.

Brusca, R.C., Moore, W., and Shuster, S.M. (2016). *Invertebrates*, 3e. Sinauer Associates: Sunderland.

Burghart, S.E., Hopkins, T.L., and Torres, J.J. (2007). The bathypelagic Decapoda, Lophogastrida, and Mysida of the eastern Gulf of Mexico. *Marine Biology* 152: 315–327.

Burghart, S.E., Hopkins, T.L., and Torres, J.J. (2010). Partitioning of food resources in bathypelagic micronekton in the eastern Gulf of Mexico. *Marine Ecology Progress Series* 399: 131–140.

Burkenroad, M.D. (1963). The evolution of the Eucarida (Crustacea, Eumalacostraca) in relation to the fossil record. *Tulane Studies in Geology* 2: 3–16.

Burkenroad, M.D. (1981). The higher taxonomy and evolution of Decapoda (Crustacea). *Transactions of the San Diego Society of Natural History* 19: 251–268.

Bush, B.M.H. and Laverack, M.S. (1982). Mechanoreception. In: *Biology of Crustacea, Volume 3, Neurobiology: Structure and Function* (eds. H.L. Atwood and D.C. Sandeman). New York and London: Academic Press.

Calman, W.T. (1904). On the classification of the Crustacea Malacostraca. *Annals and Magazine of Natural History, Series 7* 13: 144–158.

Calman, W.T. (1909). *Crustacea*. London: Adam and Charles Black.

Carey, A.G. (1985). Marine Ice Fauna: Arctic. In: *Sea Ice Biota* (ed. R.A. Horner). Boca Raton: CRC Press.

Chace, F.A. (1940). Plankton of the Bermuda Oceanographic Expeditions. IX.: the bathypelagic caridean Crustacea. *Zoologica* 25: 117–209.

Chace, F.A. (1986). The caridean shrimps (Crustacea: Decapoda) of the Albatross Philippine expedition, 1907–1910, part 4: Families Oplophoridae and Nematocarcinidae. *Smithsonian Contributions to Zoology* 432: 1–82.

Chapman, A.D. (2009). *Numbers of Living Species in Australia and the World*, 2e. Australian Biological Resources Study: Canberra.

Cheng, I. and Lewin, R.A. (1976). Goose barnacles (Cirripedia, Thoracica) on flotsam beached at La Jolla, California. *Fishery Bulletin U.S.* 74: 212–217.

Childress, J.J. (1971). Respiratory adaptations to the oxygen minimum layer in the bathypelagic mysid *Gnathophausia ingens*. *Biological Bulletin* 141: 109–121.

Childress, J.J. (1977). Effects of pressure, temperature, and oxygen on the oxygen-consumption rate of the Midwater copepod *Gaussia princeps*. *Marine Biology* 39: 19–24.

Childress, J.J. and Price, M.H. (1978). Growth rate of the bathypelagic crustacean *Gnathophausia ingens* (Mysidacea: Lophogastridae) I. Dimensional growth and population structure. *Marine Biology* 50: 47–62.

Childress, J.J. and Price, M.H. (1983). Growth rate of the bathypelagic crustacean *Gnathophausia ingens* (Mysidacea: Lophogastridae) II. Accumulation of material and energy. *Marine Biology* 76: 165–177.

Childress, J.J. and Seibel, B.A. (1998). Life at stable low oxygen levels: adaptations of animals to oceanic oxygen minimum layers. *Journal of Experimental Biology* 201: 1223–1232.

Clarke, G.L. and Backus, R.H. (1964). Interrelations between vertical migrations of the deep-scattering layers, bioluminescence, and changes in daylight in the sea. *Bulletin. Institut Océanographique (Monaco)* 64: 1–36.

Clarke, W.D. (1966). *Bathyphotometric Studies of the Light Regime of Organisms of the Deep Scattering Layers*. Santa Barbara: Sea Operations Division, General Motors Corp.

Clarke, W.D. (1970). Comparison of different investigative techniques for studying the deep scattering layer. In: *Proceedings of an International Symposium on Biological Sound Scattering in the Ocean* (ed. G.B. Farquhar). Washington, D.C.: Maury Center of Ocean Science, Department of the Navy.

Claus, C. (1863). Uber einige Schizopoden und niedere Malakostraken Messinas. *Zeitschrift für Wissenschaftliche Zoologie* 13: 422–454.

Clutter, R.I. (1969). The microdistribution and social behavior of some pelagic mysid shrimps. *Journal of Experimental Marine Biology and Ecology* 3: 125–155.

Clutter, R.L. and Theilacker, G.H. (1971). Ecological efficiency of a pelagic mysid shrimp; estimates from growth, energy budget and mortality studies. *Fishery Bulletin U.S.* 69: 93–117.

Cohen, M.J. (1955). The function of receptors in the statocyst of the lobster *Homarus americanus*. *Journal of Physiology (London)* 130: 9–34.

Cohen, M.J. and Dijkgraaf, S. (1961). Mechanoreception. In: *Physiology of Crustacea*, vol. *II* (ed. T.H. Waterman). New York: Academic Press.

Cowles, D.L. (1994). Swimming dynamics of the mesopelagic vertically migrating penaeid shrimp *Sergestes similis*: modes and speeds of swimming. *Journal of Crustacean Biology* 14: 247–257.

Cowles, D.L. (2001). Swimming speed and metabolic rate during routine swimming and simulated diel vertical migration of *Sergestes similis* in the laboratory. *Pacific Science* 55: 215–226.

Cowles, D.L. and Childress, J.J. (1988). Swimming speed and oxygen consumption in the bathypelagic mysid *Gnathophausia ingens*. *Biological Bulletin* 175: 111–121.

Cronin, T.W. and Frank, T.M. (1996). A short-wavelength photoreceptor in a deep-sea shrimp. *Proceedings of the Royal Society of London B* 163: 861–865.

Cuvier, G. 1817. *La Regne Animal*.

Cuzin-Roudy, J. (2010). Reproduction in Northern Krill (*Meganyctiphanes norvegica* Sars). In: *Advances in Marine Biology, Volume 57, The Biology of Northern Krill* (ed. G.A. Tarling). Amsterdam: Academic Press.

Dagg, M.J. (1976). Complete carbon and nitrogen budgets for the carnivorous amphipod Calliopius laevisculus. *Internationale Revue der Gesamten Hydrobiologie* 61: 297–357.

Dall, W. (1967). The functional anatomy of the digestive tract of a shrimp *Metapenaus bennettae* Racek and Dall (Crustacea: Decapoda: Penaeidae). *Australian Journal of Zoology* 15: 699–714.

David, P.M. (1961). The influence of vertical migration on speciation in the oceanic plankton. *Systematic Zoology* 10: 10–16.

Deevey, E.S.J. (1947). Life tables for natural populations of animals. *Quarterly Review of Biology* 22: 28–314.

DeGrave, S., Pentcheff, N.D., Ahyong, S.T. et al. (2009). A classification of living and fossil genera of decapod crustaceans. *The Raffles Bulletin of Zoology* Supplement No. 21: 1–109.

Dennell, R. (1955). Observations of bathypelagic Crustacea Decapoda of the Bermuda area. *Journal of the Linnean Society of London, Zoology* 42: 393–406.

Denton, E.J. and Gray, J.A.B. (1986). Lateral-line-like antennae of certain of the Penaidea (Crustacea, Decapoda, Natantia). *Proceedings of the Royal Society of London B* 226: 249–261.

Diebel, C. E. 1988. The sensory mediation of symbiosis between hyperiid amphipods and salps. PhD, Massachusetts Institute of Technology.

Dobkin, S. (1968). The larval development of a species of Thor (Decapoda: Caridea) from south Florida. *Crustaceana suppl. 2*: 1–18.

Donaldson, D.M. (1975). Vertical distribution and feeding of sergestid shrimps (Decapoda: Natantia) collected near Bermuda. *Marine Biology* 31: 37–50.

Donnelly, J., Sutton, T.T., and Torres, J.J. (2006). Distribution and abundance of micronekton and macrozooplankton in the NW Weddell Sea: relation to a spring ice edge bloom. *Polar Biology* 29: 280–293.

Drach, P. (1939). Mue et cycle d'intermue chez les crustaces decapodes. *Annales de l'Institut Oceanographique de Monaco* 19: 103–391.

Dunn, C.W., Hejnol, A., Matus, D.Q. et al. (2008). Broad phylogenomic sampling improves resolution of the animal tree of life. *Nature* 452: 745–749.

Einarsson, H. (1945). Euphausiacea I. North Atlantic species. *Dana-Report* 27: 1–85.

Enright, J.T. (1977). Diurnal vertical migration: adaptive significance and timing. Part 1. Selective advantage: a metabolic model. *Limnology and Oceanography* 22: 856–872.

Fleisher, K.J. and Case, J.F. (1995). Cephalopod predation facilitated by dinoflagellate bioluminescence. *Biological Bulletin* 189: 263–271.

Flock, M.E. and Hopkins, T.L. (1992). Species composition, vertical distribution, and food habits of the sergestid shrimp assemblage in the eastern Gulf of Mexico. *Journal of Crustacean Biology* 12: 210–223.

Foxton, P. (1970). The vertical distribution of pelagic decapods (Crustacea: Natantia) collected on the SOND cruise 1965. I. The caridea. *Journal of the Marine Biological Association of the United Kingdom* 50: 939–960.

Foxton, P. (1972). Further evidence of the taxonomic importance of the organs of Pesta in the genus *Sergestes* (Natantia, Penaeidiea). *Crustaceana* 22: 181–189.

Frank, T.M. and Case, J.F. (1988a). Visual spectral sensitivities of bioluminescent deep-sea crustaceans. *Biological Bulletin* 175: 261–273.

Frank, T.M. and Case, J.F. (1988b). Visual spectral sensitivity of the bioluminescent deep-sea mysid *Gnathophausia ingens*. *Biological Bulletin* 175: 274–283.

Frank, T.M. and Widder, E.A. (1994). Comparative study of behavioral sensitivity thresholds to near-UV and blue-green light in deep-sea crustaceans. *Marine Biology* 121: 229–235.

Frank, T.M. and Widder, E.A. (1996). UV light in the deep-sea: in situ measurements of downwelling irradiance in relation to the visual threshold sensitivity of UV-sensitive crustaceans. *Marine and Freshwater Behaviour and Physiology* 27: 189–197.

Frank, T.M. and Widder, E.A. (1997). The correlation of downwelling irradiance and staggered vertical migration patterns of zooplankton in Wilkinson Basin, Gulf of Maine. *Journal of Plankton Research* 19: 1975–1991.

Frank, T.M., Widder, E.A., Latz, M.I., and Case, J.F. (1984). Dietary maintenance of bioluminescence in a deep-sea mysid. *Journal of Experimental Biology* 109: 385–389.

Frost, B.W. (1988). Variability and possible adaptive significance of diel vertical migration in *Calanus pacificus*, a planktonic marine copepod. *Bulletin of Marine Science* 43: 675–694.

Fuzessery, Z.M. and Childress, J.J. (1975). Comparative chemosensitivity to amino acids and their role in the feeding activity of bathypelagic and littoral crustaceans. *Biological Bulletin* 149: 522–538.

Gabe, M. (1953). Sur l'existence, chez quelques Crustaces Malacostraces, d'un organe comparable a la glande de la mue des Insectes. *Comptes Rendus. Académie des Sciences* 237: 111–113.

Gallaway, B.J., Cole, J.G., and Martin, L.R. (2001). *The Deep-Sea Gulf of Mexico: An Overview and Guide*. New Orleans, LA: U.S. Department of the Interior, Minerals Management Service, Gulf of Mexico Region. OCS STUDY MMS 2001-165. New Orleans, LA.

Gartner, J.V., Hopkins, T.L., Baird, R.C., and Milliken, D.M. (1987). Ecology of the lanternfishes (Pisces: Myctophidae) of the eastern Gulf of Mexico. *Fishery Bulletin U.S.* 85: 81–98.

Gaten, E., Herring, P.J., and Shelton, P.M.J. (2002). Eye morphology and optics of the double-eyed mysid *Euchaetomera typica*. *Acta Zoologica* 83: 221–230.

Genthe, H.C. (1969). The reproductive biology of *Sergestes similis* (Decapoda, Natantia). *Marine Biology* 2: 203–217.

Gooding, R.U. (1963). Lightiella incisa sp. nov. (Cephalocarida) from the West Indies. *Crustaceana* 5: 293–314.

Gordon, A.L., Molinelli, E.J., and Baker, T.N. (1982). *Southern Ocean Atlas: Thermohaline Chemical Distributions and the Atlas Data Set*. New York: Columbia University Press.

Gould, S.J. (1989). *Wonderful Life: The Burgess Shale and the Nature of History*. New York: W.W. Norton.

Gurney, R. (1936). Larvae of decapod crustacea, 2. Amphionidae. *Discovery Reports* 12: 392–399.

Gurney, R. (1942). *Larvae of Decapod Crustacea*. London: The Ray Society.

Gutierrez, M., Ramirez, A., Bertrand, S. et al. (2008). Ecological niches and areas of overlap of the squat lobster "Munida" (*Pleuroncodes monodon*) and anchoveta (*Engraulis ringens*) off Peru. *Progress in Oceanography* 79: 256–263.

Hackman, R.H. (1971). The integument of arthropods. In: *Chemical Zoology, Volume VI, Arthropoda Part B* (eds. M. Florkin and B.T. Scheer). New York: Academic Press.

Haeckel, E.H.P.A. (1866). *Generelle Morphologie der Organismen. Allgemeine Grundzuge der Organischen Formen-Wissenschaft, mechanische Begrundet durch die von Charles Darwin Descendenz-Theorie. Volume I. Allgemeine Anatomie der Organismen*. Berlin, Germany: Georg Reimer.

Hamner, P. and Hamner, W.M. (1977). Chemosensory tracking of scent trails by the planktonic shrimp *Acetes sibogae australis*. *Science* 195: 886–888.

Hamner, W.M. (1988). Biomechanics of filter feeding in the Antarctic krill Euphausia superba: review of past work and new observations. *Journal of Crustacean Biology* 8: 149–163.

Hamner, W.M., Hamner, P.P., Strand, S.W., and Gilmer, R.W. (1983). Behavior of Antarctic krill, *Euphausia superba*: chemoreception, feeding, schooling, and molting. *Science* 220: 433–435.

Hansen, H.J. (1893). Zur morphologie der Gleidmassen und Mundteile bei Crustacen und Insecten. *Zoologischer Anzeiger* 16 (193-198): 201–212.

Hansen, H.J. (1903). The Ingolfiellidae, a new type of Amphipoda. *Zoological Journal of the Linnaean Society of London* 28: 117–132.

Hanström, B. (1939). *Hormones in Invertebrates*. London and New York: Oxford University Press.

Harbison, G.R., Biggs, D.C., and Madin, L.P. (1977). The associations of Amphipoda Hyperiidea with gelatinous zooplankton – II. Associations with Cnidaria, Ctenophora and Radiolaria. *Deep-Sea Research* 24: 465–488.

Harbison, G.R., Madin, L.P., and Swanberg, N.R. (1978). On the natural history and distribution of the oceanic ctenophores. *Deep-Sea Research* 25: 233–256.

Hardy, A.C. (1953). Some problems in pelagic life. In: *Essays in Marine Biology*. London: Oliver and Boyd.

Hardy, A.C. (1956). *The Open Sea*. London: Collins.

Hardy, A.C. and Bainbridge, E.R. (1954). Experimental observations on the vertical migrations of plankton animals. *Journal of the Marine Biological Association of the United Kingdom* 33: 409–448.

Hardy, A.C. and Gunther, E.R. (1935). The plankton of the South Georgia whaling grounds and adjacent waters. *Discovery Reports* 11: 1–456.

Hartline, H.K., Wagner, H.G., and Ratliff, F. (1956). Inhibition in the eye of Limulus. *Journal of General Physiology* 39: 651–671.

Harvey, E.N. (1952). *Bioluminescence*. New York and London: Academic Press.

Heegaard, P. (1969). Larvae of decapod Crustacea. The Amphionidae. *Dana-Report* 77: 1–82.

Heffernan, J.J. and Hopkins, T.L. (1981). Vertical distribution and feeding of the shrimp genera *Gennadas* and *Bentheogennema* (Decapoda: Penaeidea) in the eastern Gulf of Mexico. *Journal of Crustacean Biology* 1: 461–473.

Herring, P.J. (1976b). Bioluminescence in decapod Crustacea. *Journal of the Marine Biological Association of the United Kingdom* 56: 1029–1047.

Herring, P.J. (1978). Bioluminescence of invertebrates other than insects. In: *Bioluminescence in Action* (ed. P.J. Herring). London and New York: Academic Press.

Herring, P.J. (1985). Bioluminescence in the Crustacea. *Journal of Crustacean Biology* 5: 557–573.

Herring, P.J. (2002). *The Biology of the Deep Ocean*. New York: Oxford University Press.

Herring, P.J. and Barnes, A.T. (1976). Light-stimulated bioluminescence of *Thalassocaris crinita* (Dana) (Decapoda, Caridea). *Crustaceana* 31: 107–110.

Hersey, J.B. and Moore, H.B. (1948). Progress report on scattering layer observations in the Atlantic Ocean. *Transactions of the American Geophysical Union* 29: 341–354.

Hessler, R.R. (1983). A defense of the caridoid facies: wherein the early evolution of the Eumalacostraca is discussed. In: *Crustacean Issues 1. Crustacean Phylogeny* (ed. F.R. Schram). Rotterdam: A. A. Balkema.

Hessler, R.R. (1985). Swimming in Crustacea. *Transactions of the Royal Society of Edinburgh* 76: 115–122.

Hiller-Adams, C.P. and Case, J.F. (1988). Eye size of pelagic crustaceans as a function of habitat depth and possession of photophores. *Vision Research* 28: 667–680.

Hindley, J. (1975). The detection, location and recognition of food by juvenile banana prawns, *Penaeus merguiensis*. *Marine Behaviour and Physiology* 3: 193–210.

Hochachka, P.W. and Guppy, M. (1987). *Metabolic Arrest and the Control of Biological Time*. Cambridge: Harvard University Press.

Holthuis, L.B. (1955). The recent genera of the caridean and stenopodidean shrimps (Class Crustacea: Order Decapoda: Supersection Natantia) with keys for their identification. *Zoologische Verhandelingen* 26: 1–157.

Hopkins, T.L. (1985). The zooplankton community of Croker Passage, Antarctic Peninsula. *Polar Biology* 4: 161–170.

Hopkins, T.L. (1987). Midwater food web in McMurdo Sound, Ross Sea, Antarctica. *Marine Biology*, 96: 93–106.

Hopkins, T.L., Flock, M.E., Gartner, J.V., and Torres, J.J. (1994). Structure and trophic ecology of a low latitude midwater decapod and mysid assemblage. *Marine Ecology Progress Series* 109: 143–156.

Hopkins, T.L., Gartner, J.V., and Flock, M.E. (1989). The caridean shrimp (Decapoda:Natantia) assemblage in the mesopelagic zone of the eastern Gulf of Mexico. *Bulletin of Marine Science* 45: 1–14.

Hopkins, T.L. and Lancraft, T.M. (1984). The composition and standing stock of mesopelagic micronekton at 27 N 86 W in the eastern Gulf of Mexico. *Contributions in Marine Science* 27: 145–158.

Hopkins, T.L., Sutton, T.T., and Lancraft, T.M. (1997). The trophic structure and predation impact of a low latitude midwater fish assemblage. *Progress in Oceanography* 38: 205–239.

Hopkins, T.L. and Torres, J.J. (1989). Midwater food web in the vicinity of a marginal ice zone in the western Weddell Sea. *Deep-Sea Research* 36: 543–560.

Hughes, G.M., Knights, B., and Scammel, C.A. (1969). The distribution of PO_2 and hydrostatic pressure changes within the branchial chambers of the shore crab, *Carcinus maenas*. *Journal of Experimental Biology* 51: 203–220.

Isaacs, J.D., Tont, S.A., and Wick, G.L. (1974). Deep scattering layers: vertical migration as a tactic for finding food. *Deep-Sea Research* 21: 651–656.

Iwasaki, N. and Nemoto, T. (1987). Pelagic shrimps (Crustacea: Decapoda) from the Southern Ocean between 150 E and 115 E. *Memoirs of National Institute of Polar Research* 38: 1–40.

Johnson, W.S. (1976). Population energetics of the intertidal isopod *Cirolana harfordi*. *Marine Biology* 36: 351–357.

Judkins, D.C. and Kensley, B. (2008). New genera in the family Sergestidae (Crustacea: Decapoda: Penaeidea). *Proceedings of the Biological Society of Washington* 121: 72–84.

Kaestner, A. (1968). *Invertebrate Zoology*, vol. 2. New York: Wiley.

Kaestner, A. (1970). *Invertebrate Zoology*, vol. 3. New York and London: Wiley.

Kampa, E.M. (1975). Observations of a sonic scattering layer during the total solar eclipse, 30 June 1973. *Deep-Sea Research* 22: 417–423.

Kampa, E.M. and Boden, B.P. (1954). Submarine illumination and the twilight movements of a sonic scattering layer. *Nature* 174: 867–873.

Kemp, S. (1910). Notes on the photophores of decapod Crustacea. *Proceedings. Zoological Society of London* 2: 639–651.

Kerfoot, W.B. (1970). Bioenergetics of vertical migration. *American Naturalist* 104: 529–546.

Kinsey, S.T. and Hopkins, T.L. (1994). Trophic strategies of euphausiids in a low-latitude ecosystem. *Marine Biology* 118: 651–661.

Koehl, M.A.R. and Strickler, J.R. (1981). Copepod feeding currents: food capture at low Reynolds number. *Limnology and Oceanography* 26: 1062–1073.

Koop, K. and Field, J.G. (1981). Energy transformation by the supralittoral isopod *Ligia dilatata* Brandt. *Journal of Experimental Marine Biology and Ecology* 53: 221–233.

Kreidl, A. (1893). Weitere beitrage zur physiologie des ohrlabyrinthes (II. Mitth.) Versuche an Krebsen. *Sitzungberichte Akademie der Wissenschaften, Mathemmatische-Naturwissenschaftliche Klasse. Abteilung III* 102: 149–174.

Krygier, E.E. and Pearcy, W.G. (1981). Vertical distribution and biology of pelagic decapod crustaceans off Oregon. *Journal of Crustacean Biology* 1: 70–95.

Krygier, E.E. and Wasmer, R.A. (1988). Zoogeography of the pelagic shrimps (Natantia: Penaidea and Caridea) in the North Pacific Ocean (with synopses and keysto the species of the subarctic transitional zones). *Bulletin of the Ocean Research Institute. University of Tokyo* 26: 43–98.

Lamarck, J.-B. (1802). *La nouvelle classes des Annelides,* 27 Floreal, Disc. d'ouverture.

Lancraft, T.M., Hopkins, T.L., and Torres, J.J. (1988). Aspects of the ecology of the mesopelagic fish *Gonostoma elongatum* (Gonostomatidae, Stomiiformes) in the eastern Gulf of Mexico. *Marine Ecology Progress Series* 49: 27–40.

Lancraft, T.M., Hopkins, T.L., Torres, J.J., and Donnelly, J. (1991). Oceanic micronektonic/macrozooplanktonic community structure and feeding under ice covered Antarctic waters during the winter (Ameriez 1988). *Polar Biology* 11: 157–167.

Lancraft, T.M., Reisenbichler, K.R., Hopkins, T.L. et al. (2004). A krill-dominated micronekton and macrozooplankton community in Croker Passage, Antarctica with an estimate of fish predation. *Deep Sea Research, Part II* 51: 2247–2260.

Lancraft, T.M., Torres, J.J., and Hopkins, T.L. (1989). Micronekton and macrozooplankton in the open waters near Antarctic ice edge zones (Ameriez 1983 and 1986). *Polar Biology* 9: 225–233.

Land, M.F. (1980). Compound eyes: old and new mechanisms. *Nature* 287: 681–685.

Land, M.F. (1981a). Optics and vision in invertebrates. In: *Handbook of Sensory Physiology, Volume VII/6B. Comparative Physiology and Evolution of Vision in Invertebrates* (ed. H. Autrum). Berlin: Springer-Verlag.

Land, M.F. (1981b). Optics of the eyes of Phronima and other deep-sea amphipods. *Journal of Comparative Physiology A* 145: 209–226.

Lasenby, D.C. and Langford, R.R. (1972). Growth, life history and respiration of *Mysis relicta* in an arctic and temperate lake. *Journal of the Fisheries Research Board of Canada* 29: 1701–1708.

Latreille, P.A. (1802). *Histoire naturelle, generale et particuliere des Crustaces et des Insectes*. Paris: F. Dufart.

Latreille, P.A. (1817). Les Crustaces. In: *Regne Animale* (ed. G. Cuvier). Gustav Fischer.

Latz, M.I. and Case, J.F. (1982). Light organ and eyestalk compensation to body tilt in the luminescent midwater shrimp, *Sergestes similis*. *Journal of Experimental Biology* 98: 83–104.

Laval, P. (1980). Hyperiid amphipods as crustacean parasitoids associated with gelatinous zooplankton. *Oceanography and Marine Biology. Annual Review* 18: 11–56.

Laverack, M.S., Neil, D.M., and Robertson, R.M. (1977). Metachronal expodite beating in the mysid *Praunus flexuosus*: a quantitative analysis. *Proceedings of the Royal Society of London B* 198: 139–154.

Leuckart, R.W. (1848). Über die Morphologie und Verwandtschaftsverhältnisse der wirbellosen Tiere. Braunschweig: Braunschweig.

Lomakina, N.B. (1978). *Euphausiids of the World Oceans*. Leningrad: Academy of Sciences of the Ussr.

Longhurst, A.R. (1976). Vertical migration. In: *The Ecology of the Seas* (eds. D.H. Cushing and J.J. Walsh). Philadelphia: W.B. Saunders.

Longhurst, A.R., Lorenzen, C.J., and Thomas, W.H. (1967). The role of pelagic crabs in the grazing of phytoplankton off Baja California. *Ecology* 48: 190–200.

Madin, L.P. and Harbison, G.R. (1977). The associations of Amphipoda Hyperiidea with gelatinous zooplankton – I. Associations with Salpidae. *Deep-Sea Research* 24: 449–463.

Mangum, C.P. (1983). Oxygen transport in the blood. In: *Biology of Crustacea, Volume 5, Internal Anatomy and Physiological Regulation* (ed. L.H. Mantel). New York and London: Academic Press.

Manton, S.M. (1977). *The Arthropoda*. Oxford: Oxford University Press.

Marr, J.W. (1962). The natural history and geography of the Antarctic krill (*Euphausia superba* Dana). *Discovery Reports* 32: 33–464.

Marschall, H.-P. (1983). Sinking speed, density and size of euphausiid eggs. *Meeresforschung* 30: 1–9.

Marschall, H.-P. (1988). The overwintering strategy of the Antarctic krill under the pack-ice of the Weddell Sea. *Polar Biology* 9: 129–135.

Marschall, H.-P. and Hirche, H.-J. (1984). Development of eggs and nauplii of *Euphausia superba*. *Polar Biology* 2: 245–250.

Martin, J.W. and Davis, G.E. (2001). An updated classification of the recent Crustacea. *Natural History Museum of Los Angeles County Science Series* 39: 1–124.

Matthews, H. (1932). Lobster krill, Anomuran Crustacea that are the food of whales. *Discovery Reports* 5: 467–484.

Matthews, J.B.L. and Pinnoi, S. (1973). Ecological studies on the deep-water pelagic community of Korsfjorden, western Norway. The species of *Pasiphaea* and *Sergestes* (Crustacea Decapoda) recorded in 1968 and 1969. *Sarsia* 52: 123–144.

Mauchline, J. (1980). *Advances in Marine Biology, Volume 18, The Biology of Mysids and Euphausiids*. London: Academic Press.

Mauchline, J. and Fisher, L.R. (1969). *Advances in Marine Biology, Volume 7, The Biology of Euphausiids*. London: Academic Press.

Maykut, F.A. (1985). The ice environment. In: *Sea Ice Biota* (ed. R.A. Horner). Boca Raton: CRC Press.

Mcclintock, J.B. (2012). *Lost Antarctica*. New York: Palgrave Macmillan.

Mclaren, I.A. (1963). Effects of temperature on growth of zooplankton, and the adaptive value of vertical migration. *Journal of the Fisheries Research Board of Canada* 20: 685–727.

Mclaren, I.A. (1974). Demographic strategy of vertical migration by a marine copepod. *American Naturalist* 108: 91–102.

Mclaughlin, P.A. (1980). *Comparative Morphology of Recent Crustacea*. San Francisco: W.H. Freeman and Company.

Mclaughlin, P.A. (1983). Internal anatomy. In: *Biology of Crustacea, Volume 5, Internal Anatomy and Physiological Regulation* (ed. L.H. Mantel). New York: Academic Press.

Mcmahon, B.R. and Wilkens, J.L. (1983). Ventilation, perfusion, and oxygen uptake. In: *Biology of Crustacea, Volume 5, Internal Anatomy and Physiological Regulation* (ed. L.H. Mantel). New York and London: Academic Press.

Menzies, R.J. and Dow, T. (1958). The largest known bathypelagic isopod, *Anuropus bathypelagicus* n. sp. *Annals and Magazine of Natural History (13)* 1: 1–6.

Miller, C.B. (1972). Some environmental consequence of vertical migration in marine zooplankton. *Limnology and Oceanography* 15: 727–741.

Milne-Edwards, H. (1830). Memoire sur une disposition particuliere de l'appareile branchial chez quelques Crustaces. *Annales des Sciences Naturelles* 19: 452–460.

Modin, J.C. and Cox, K.W. (1967). Post-embryonic development of laboratory-reared ocean shrimp, *Pandalus jordani* Rathbun. *Crustaceana* 13: 197–219.

Morgan, C.I. (1982). Tardigrada. In: *Synopsis and Classification of Living Organisms*, vol. 2 (ed. S.P. Parker). New York: McGraw-Hill.

Morgan, C.I. and King, P.E. (1976). *British Tardigrades: Keys and Notes for the Identification of the Species*. New York: Academic Press.

Müller, H.-G. (1993). *World Catalogue and Bibliography of the Recent Mysidacea. Laboratory for Tropical Ecosystems Research and Information Service* 491. Wetzlar.

Muller, O.F. (1776). *Zoologica Danicae Prodromus*. Havniae.

Murano, M. (1977). Mysidacea from the central and western Pacific. IV. Genera *Euchaetomera, Euchaetomeropsis, Arachnomysis, Caesaromysis, Echinomysides, Meterythrops* and *Nipponerythrops* (Tribe Erythropini). *Publications of the Seto Marine Biological Laboratory* 24: 141–192.

Murano, M. (1999). Mysidacea. In: *South Atlantic Zooplankton* (ed. D. Boltovskoy). Leiden, The Netherlands: Backhuys Publishers.

Nemoto, T. (1957). Food of baleen whales in the northern Pacific. *Scientific Reports of the Whales Research Institute, Tokyo* 14: 149–290.

Nemoto, T. (1967). Feeding pattern of euphausiids and differentiations in their body characters. *Information Bulletin on*

Planktology in Japan, 61st Annual Number: 143–160.

Nielsen, C. (2001). *Animal Evolution: Interrelationships of the Living Phyla*, 2e. Oxford University Press: Oxford.

Nishida, S., Pearcy, W.G., and Nemoto, T. (1988). Feeding habits of mesopelagic shrimps collected off Oregon. *Bulletin of the Ocean Research Institute, Tokyo University.* 26: 99–108.

Ohman, M.D., Frost, B.W., and Cohen, E.B. (1983). Reverse diel vertical migration: an escape from invertebrate predators. *Science* 220: 1404–1407.

Oishi, S. (1959). Studies on the teloblasts in the decapod embryo. 1. Origin of teloblasts in *Hepatocarpus rectrirostris*. *Embryologia* 4: 283–309.

Omori, M. (1969). The biology of a sergestid shrimp *Sergestes lucens* Hansen. *Bull. Ocean Res. Inst., Tokyo Univ.* 4: 1–83.

Omori, M. (1974). The biology of pelagic shrimps in the ocean. In: *Advances in Marine Biology*. New York: Academic Press.

Omori, M. (1979). Growth, feeding, and mortality of larval and early postlarval stages of the oceanic shrimp *Sergestes similis* Hansen. *Limnology and Oceanography* 24: 273–288.

Omori, M. and Gluck, D.L. (1979). Life history and vertical migration of the pelagic shrimp *Sergestes similis* Hansen off the southern California coast. *Fishery Bulletin U.S.* 77: 183–198.

Pérez Farfante, I. & Kensley, B. 1997. *Penaeoid and Sergestoid Shrimps of the World,* Paris, Editions du Museum National D'Histoire Naturelle.

Parker, M.L., Donnelly, J., and Torres, J.J. (2011). Micronekton and macrozooplankton in the Marguerite Bay region of the Western Antarctic Peninsula. *Deep Sea Research, Part II* 58: 1580–1598.

Parry, G. (1954). Ionic regulation in the prawn Palaemon. *Journal of Experimental Biology* 31: 601–613.

Parry, G. (1960). Excretion. In: *Physiology of Crustacea*, vol. *I* (ed. T.H. Waterman). New York: Academic Press.

Passarella, K.C. and Hopkins, T.L. (1991). Species composition and food habits of the micronektonic cephalopod assemblage in the eastern Gulf of Mexico. *Bulletin of Marine Science* 49: 638–659.

Pearcy, W.G. and Forss, C.A. (1969). Depth distribution of oceanic shrimps (Decapoda: Natantia) off Oregon. *Journal of the Fisheries Research Board of Canada* 23: 1135–1143.

Pearcy, W.G., Krygier, E.E., Mesecar, R., and Ramsey, F. (1977). Vertical distribution and migration of oceanic micronekton off Oregon. *Deep-Sea Research* 24: 223–245.

Pearson, J.C. (1939). The early life histories of some American Penaeidae. *U.S. Department of Commerce Bulletin of the Bureau of Fisheries* 49: 1–73.

Peck, S.B. (1982). Onychophora. In: *Synopsis and Classification of Living Organisms*, vol. 2 (ed. S.P. Parker). New York: McGraw-Hill.

Pequegnat, L.H. (1965). The bathypelagic mysid *Gnathophausia* (Crustacea) in the eastern Pacific Ocean. *Pacific Science* 19: 399–421.

Petipa, T.S. (1955). Observation of the behaviour of zooplankton during a solar eclipse (in Russian). *Doklady Akademii Nauk SSSR* 104: 323–325.

Pillai, R.S. (1960). Studies on the shrimp *Caridina laevis*. I. The digestive system. *Journal of the Marine Biological Association of India* 2: 57–74.

Pillai, R.S. (1965). The circulatory system of *Caridina laevis* Heller. *Crustaceana* 8: 66–74.

Ponomareva, L.A. (1963). *The euphausiids of the North Pacific, their distribution and ecology*, vol. 1966. Israel Program for Scientific Translation.

Potts, W.T.W. and Parry, G. (1964). *Osmotic and Ionic Regulation in Animals.* Oxford: Pergamon Press Ltd.

Prosser, C.L. (1973). *Comparative Animal Physiology.* Philadelphia: Saunders.

Quetin, L.B. and Childress, J.J. (1976). Respiratory adaptations of *Pleuroncodes planipes* Stimpson to its environment off Baja California. *Marine Biology* 38: 327–334.

Quetin, L.B. and Ross, R.M. (1984). Depth distribution of developing *Euphausia superba* embryos, predicted from sinking rates. *Marine Biology* 79: 47–53.

Quetin, L.B. and Ross, R.M. (1991). Behavioral and physiological characteristics of the Antarctic krill, *Euphausia superba*. *American Zoologist* 31: 49–63.

Quetin, L.B. and Ross, R.M. (2001). Environmental variability and its impact on the reproductive cycle of Antarctic krill. *American Zoologist* 41: 74–89.

Richter, S. (2003). The mouthparts of two lophogastrids, *Chalaraspidium alatum* and *Pseudochalaraspidium hanseni* (Lophogastrida, Peracarida, Malacostraca), including some remarks on the monophyly of the Lophogastrida. *Journal of Natural History* 37: 2773–2786.

Robinson, C.J., Anislado, V., and Lopez, A. (2004). The pelagic red crab (Pleuroncodes planipes) related to active upwelling sites in the California Current off the west coast of Baja California. *Deep Sea Research, Part II* 51: 753–766.

Robison, B.H. (2003). What drives the diel vertical migrations of Antarctic midwater fish. *Journal of the Marine Biological Association of the UK* 83: 639–642.

Roe, H.S.J. (1974). Observations on the diurnal vertical migrations of an oceanic animal community. *Marine Biology* 28: 99–113.

Roe, H.S.J. (1983). Vertical distributions of euphausiids and fish in relation to light intensity in the Northeastern Atlantic. *Marine Biology* 77: 287–298.

Roe, H.S.J. (1984). The diel migrations and distributions within a mesopelagic community in the North East Atlantic. 2. Vertical migrations and feeding of mysids and decapod Crustacea. *Progress in Oceanography* 13: 269–318.

Roe, H.S.J., James, P.T., and Thurston, M.H. (1984). The diel migrations and distributions within a mesopelagic community in the North East Atlantic. 6. Medusae, Ctenophores, Amphipods and Euphausiids. *Progress in Oceanography* 13: 425–460.

Ross, R.M. (1982). Energetics of *Euphausia pacifica* II. Complete carbon and nitrogen budgets at 8 and 12 C throughout the life span. *Marine Biology* 68: 15–23.

Ross, R.M., Daly, K.L., and English, T.S. (1982). Reproductive cycle and fecundity of *Euphausia pacifica* in Puget Sound, Washington. *Limnology and Oceanography* 27: 304–314.

Ross, R.M. and Quetin, L.B. (1983). Spawning frequency and fecundity of the Antarctic krill, *Euphausia superba*. *Marine Biology* 77: 201–205.

Russell-Hunter, W.D. (1979). *A Life of Invertebrates.* New York: Macmillan Publishing Co., Inc.

Sandeman, D.C. (1982). Organization of the central nervous system. In: *Biology of Crustacea, Volume 3, Neurobiology: Structure and Function* (eds. H.L. Atwood and D.C. Sandeman). New York and London: Academic Press.

Sanders, H.L. (1955). The Cephalocarida, a new subclass of Crustacea from Long Island Sound. *Proceedings of the National Academy of Sciences of the USA* 41: 61–66.

Sanders, H.L. (1963). Significance of Cephalocarida. In: *Phylogeny and Evolution* of *Crustacea* (eds. H.B. Whittington and W.D.I. Rolfe). Cambridge: Museum of Comparative Zoology.

Sanders, N.K. (1990). A comparison of the respiratory function of the haemocyanins of vertically migrating and non-migrating pelagic, deep-sea oplophorid shrimps. *Journal of Experimental Biology* 152: 167–187.

Sanders, N.K. and Childress, J.J. (1990). Adaptations to the deep-sea oxygen minimum layer: oxygen binding by the hemocyanin of the bathypelagic mysid, *Gnathophausia ingens* Dohrn. *Biological Bulletin* 178: 286–294.

Sars, G.O. (1867). *Histoire Naturelle des Crustaces d'eau douce de Norvege 1. les Malacostraces*. Johnsen. Christiana.

Sars, G.O. (1885). Report of the Schizopoda collected by H.M.S. Challenger during the years 1873–1876. *Challenger Reports, Zoology* 13: 1–228.

Schmitt, B. and Ache, B.W. (1979). Olfaction: Responses of a decapod crustacean are enhanced by flicking. *Science* 205: 204–206.

Schram, F.R. (1978). Arthropods: a convergent phenomenon. *Fieldiana: Geology* 39: 61–108.

Schram, F.R. (1981). On the classification of the Eumalacostraca. *Journal of Crustacean Biology* 1: 1–10.

Schram, F.R. (1984). Relationships within Eumalacostracan Crustacea. *Transactions of the San Diego Society of Natural History* 20: 301–312.

Schram, F.R. (1986). *Crustacea*. New York: Oxford University Press.

Sheader, M. (1977). Breeding and marsupial development in laboratory-maintained *Parathemisto gaudichaudi* (Amphipoda). *Journal of the Marine Biological Association of the United Kingdom* 57: 943–954.

Sherman, K. and Honey, K.A. (1970). Vertical movements of zooplankton during a solar eclipse. *Nature* 227: 1156–1158.

Shushkina, E.A. (1973). Evaluation of the production of tropical zooplankton. In: *Life Activity of Pelagic Communities in the Ocean Tropics* (ed. M.E. Vinogradov). Moscow: Akad. Nauk. SSSR.

Shushkina, E.A. (1985). Production of principal ecological groups of plankton in the epipelagic zone of the ocean. *Oceanology* 25: 653–658.

Skinner, D.M. (1962). The structure and metabolism of a crustacean integumentary tissue during a molt cycle. *Biological Bulletin* 123: 635–647.

Skinner, D.M. (1985). Molting and regeneration. In: *Biology of Crustacea, Volume 9, Integument, Pigments, and Hormonal Processes* (eds. D.E. Bliss and L.H. Mantel). New York: Academic Press.

Smith, K.L. and Baldwin, R.J. (1984). Vertical distribution of the necrophagous amphipod, *Eurythenes gryllus*, in the North Pacific: spatial and temporal variation. *Deep-Sea Research* 31: 1179–1196.

Stearns, S.C. (1992). *The Evolution of Life Histories*. New York: Oxford University Press.

Stebbing, T.R.R. (1893). *A history of Crustacea*. London: D. Appleton and Co.

Stevenson, J.R. (1985). Dynamics of the integument. In: *Biology of Crustacea, Volume 9, Integument, Pigments and Hormonal Processes* (eds. D.E. Bliss and L.H. Mantel). New York: Academic Press.

Sutton, T.T. and Hopkins, T.L. (1996a). The species composition, abundance, and vertical distribution of the stomiid (Pisces: Stomiiformes) fish assemblage of the Gulf of Mexico. *Bulletin of Marine Science* 59: 530–542.

Tapella, T., Romero, M.C., Lovrich, G.A., and Chizzini, A. (2002). Life History of the Galatheid Crab Munida subrugosa in Subantarctic Waters of the Beagle Channel, Argentina. In: *Crabs in Cold Water Regions: Biology, Management, and Economics* (eds.

A.J. Paul, E.G. Dawe, R. Elner, et al.). Fairbanks: University of Alaska Sea Grant.

Tarling, G.A. (2010). Population dynamics of northern krill (*Meganyctiphanes norvegica* Sars). In: *Advances in Marine Biology, Volume 57, The Biology of Northern Krill* (ed. G.A. Tarling). Amsterdam: Academic Press.

Tarling, G.A., Cuzin-Rowdy, J., Wooton, K., and Johnson, M.L. (2009). Egg release behavior in Antarctic krill. *Polar Biology* 32: 1187–1194.

Tattersall, W.M. (1908). Crustacea. VII. Schizopoda. National Antarctic Expeditions 1901-1904. *Natural History (Zoology)* 4: 1–42.

Tattersall, W.M. and Tattersall, O.W. (1951). *The British Mysidacea*. London: Ray Society.

Teal, J.M. and Carey, F.G. (1967). Effects of pressure and temperature on the respiration of euphausiids. *Deep-Sea Research* 14: 725–733.

Tiegs, O.W. and Manton, S.M. (1958). The evolution of the Arthropoda. *Biological Reviews of the Cambridge Philosophical Society* 33: 255–337.

Tont, S.A. and Wick, G.L. (1973). Response of a deep-scattering layer to the 1972 total eclipse. *Deep-Sea Research* 20: 769–771.

Torres, J.J., Aarset, A.V., Donnelly, J. et al. (1994). Metabolism of Antarctic micronektonic Crustacea as a function of depth of occurrence and season. *Marine Ecology Progress Series* 113: 207–219.

Torres, J.J. and Childress, J.J. (1983). Relationship of oxygen consumption to swimming speed in *Euphausia pacifica* 1. Effects of temperature and pressure. *Marine Biology* 74: 79–86.

Torres, J.J. and Childress, J.J. (1985). Respiration and chemical composition of the bathypelagic euphausiid, *Bentheuphausia amblyoops*. *Marine Biology* 87: 267–272.

Torres, J.J., Grigsby, M.D., and Clarke, M.E. (2012). Aerobic and anaerobic metabolism in oxygen minimum layer fishes: the role of alcohol dehydrogenase. *Journal of Experimental Biology* 215: 1905–1914.

Vereshchaka, A.L. (1994). North Atlantic and Caribbean species of Sergia (Crustacea, Decapoda, Sergestidae) and their horizontal and vertical distribution. *Steenstrupia* 20: 73–95.

Vinogradov, G.M. (1999). Amphipoda. In: *South Atlantic Zooplankton*, vol. *2* (ed. D. Boltovskoy). Backhuys: Leiden.

Vinogradov, M.E. (1970). *Vertical Distribution of the Oceanic Zooplankton*. Jerusalem: Israel program for scientific translations.

Vinogradov, M.E., Volkov, A.F., and Semenova, T.N. (1996). *Hyperiid Amphipods (Amphipoda, Hyperiidea) of the World Oceans*. Lebanon: N.H., Science Publishers, Inc.

Walters, J.F. (1976). Ecology of Hawaiian sergestid shrimps (Penaeidea: Sergestidae). *Fishery Bulletin U.S.* 74: 799–836.

Warner, J.A., Latz, M.I., and Case, J.F. (1979). Cryptic bioluminescence in a midwater shrimp. *Science* 203: 1109–1110.

Wasmer, R.A. (1993). Pelagic shrimps (Crustacea: Decapoda) from six USNS Eltanin cruises in the southeastern Indian Ocean, Tasman Sea, and southwestern Pacific Ocean to the Ross Sea. In: *Biology of the Antarctic Seas XXII* (ed. S. Cairns). Washington, D.C.: American Geophysical Union.

Waterman, T.H. and Chace, F.A. (1960). General crustacean biology. In: *The Physiology of Crustacea, Volume 1, Metabolism and Growth* (ed. T.H. Waterman). New York: Academic Press.

Waterman, T.H., Nunnemacher, R.F., Chace, F.A., and Clarke, G.L. (1939). Diurnal vertical migrations of deep-water plankton. *Biological Bulletin* 76: 256–279.

Whittington, H.B. (1978). The lobopod animal Aysheia pedunculata Walcott, Middle Cambrian, Burgess shale, British Columbia. *Philosophical Transactions of the Royal Society of London B* 284: 165–197.

Widder, E.A. (2010). Bioluminescence in the ocean: origins of biological, chemical, and ecological diversity. *Science* 328: 704–708.

Wiebe, P.H. and Boyd, S.H. (1978). Limits of *Nematoscelis megalops* in the Northwestern Atlantic in relation to Gulf Stream cold core rings. I. Horizontal and vertical distributions. *Journal of Marine Research* 36: 119–142.

Willemoes-Suhm, R.V. (1874). On a new genus of amphipod crustaceans. *Philosophical Transactions of the Royal Society of London* 163: 629–638.

Williamson, D.I. (1960). Larval stages of *Pasiphaea sivado* and some other Pasiphaeidae. *Crustaceana* 1: 331–341.

Williamson, D.I. (1969). Names of larvae in the Decapoda and Euphausiacea. *Crustaceana* 16: 210–213.

Williamson, D.I. (1973). Amphionides reynaudii (H. Milne-Edwards), representative of a proposed new order of Eucaridan Malacostraca. *Crustaceana* 25: 35–50.

Willmer, P., Stone, G., and Johnston, I. (2005). *Environmental Physiology of Animals*. Suffolk: Blackwell.

Winsor, M. (1976). The development of Linnaean Insect Classification. *Taxon* 25: 57–67.

Withers, P.C. (1992). *Comparative Animal Physiology*. Orlando: Saunders.

WoRMS. World Register of Marine Species (n.d.). https://www.marinespecies.org.

Yager, J. (1981). Remipedia, a new class of Crustacea from a marine cave in the Bahamas. *Journal of Crustacean Biology* 1: 328–333.

Young, J.H. (1959). Morphology of the white shrimp, *Penaeus setiferus* (Linnaeus 1758). *Fishery Bulletin U.S.* 59: 1–168.

Zeleny, C. (1905). Compensatory regulation. *Journal of Experimental Zoology* 2: 1–102.

Zheden, V., Von Byern, J., Kerbl, A. et al. (2012). Morphology of the cement apparatus and the cement of the buoy barnacle *Dosima fascicularis* (Crustacea, Cirripedia, Thoracica, Lepadidae). *Biological Bulletin* 223: 192–204.

Ziemann, D.A. (1975). Patterns of vertical distribution, vertical migration, and reproduction in the Hawaiian mesopelagic shrimps of the family Oplophoridae. PhD, University of Hawaii.

Zimmer, R.K. and Butman, C.A. (2000). Chemical signaling processes in the marine environment. *Biological Bulletin* 198: 168–187.

Zimmer-Faust, R.K. (1989). The relationship between chemoreception and foraging behavior in crustaceans. *Limnology and Oceanography* 34: 1367–1374.

8

The Mollusca

Introduction

The Mollusca comprise a highly successful and morphologically diverse animal phylum that is well represented in terrestrial, freshwater, and marine systems. Without question, they are one of the most familiar invertebrate groups in a vertebrate-oriented world. Clams, snails, squid, and octopus have been used as food since prehistoric times, and the exquisite shells of the gastropods have engendered a huge and devoted following among collectors.

Molluscs have been of great service to science and medicine (besides feeding the scientists) by providing the laboratory preparations that allowed pioneering neurophysiologists to elucidate how nerves function. The squid giant axon was used to decipher how nerves generate electrical impulses in addition to providing important basic information on the biochemistry of membranes and active transport of ions. The sea hare, *Aplysia,* is still a model system for studying the neural basis of behavior. Neuroscientists can use its large, pigmented ganglia (they resemble clusters of grapes!) as reliable, replicable sites for neural recording from subject to subject, allowing the mapping of neural circuitry.

The molluscs, along with the arthropods and chordates, are one of the three most successful animal phyla both in terms of species numbers and in the different habitats that they have successfully invaded. Current estimates of species numbers vary from 80000 to 85000 (Chapman 2009; Brusca et al. 2016, respectively) in a phylum with eight classes. The great majority of molluscs are benthic, but representatives of two classes, the gastropods and cephalopods, have invaded the pelagic realm. The pelagic snails include the gelatinous "sea butterflies" that are episodically abundant in the open sea. The cephalopods include "extreme nekton" such as the giant squid, reaching a body length of greater than 6.6 m and a total length including tentacles of at least 18 m. A large suite of slower moving deep-sea forms populate mesopelagic and bathypelagic depths.

Despite their morphological diversity (consider clams and squid), the molluscs share a similar basic body plan (Hyman 1967; Brusca et al. 2016). As a phylum, they are coelomate and bilaterally symmetrical, though the coelom has been reduced to the pericardial cavity and the lumen of the nephridia and gonoducts. They are soft-bodied and usually protected by a calcareous shell that is secreted by the mantle, a fold in the body wall forming a cavity surrounding the viscera. It may be sheet-like, lining the shell as in bivalves, or robust and muscular as in cephalopods, and it is unique to the phylum. Molluscs have a large, ventrally located, muscular foot that may be differentiated into tentacles or, as in the gastropods, flattened for creeping along the bottom. The pharynx has a toothed band called the radula, a second feature unique to the phylum. Gills, a myogenic heart, and metanephridia (see "Excretory System" section in Chapter 6 for detailed definition) are uniformly present as well. Circulatory systems are open except in the cephalopods, which exhibit a largely closed and efficient circulation. With the exception of the Monoplacophora, molluscs show little evidence of metamerism (segmentation).

Classification

History

The history of molluscan classification dates back to the time of Aristotle, who divided them into two groups, the Malachia for the cephalopods and the Ostrachodermata for the shelled forms, noting differences between univalves and bivalves (Hyman 1967). The name "Mollusca" was introduced by Jonston (1650). He used it as a grouping for the cephalopods and barnacles, an association that persisted on and off until it was finally put to rest in the 1830s by Thompson (1830) and Burmeister (1834) who observed that barnacles produced crustacean-type larvae.

Linnaeus used the term Mollusca in his Systema Naturae as one of the five groups within his "Vermes" classification, which included all invertebrates except the insects. Linnaeus' mollusca included cephalopods, slugs, pteropods, tunicates, medusae, echinoderms, anemones, and polychaetes. The shelled forms were placed in the "Testacea" which included chitons, bivalves, the nautiloid cephalopods, univalves, and the serpulid polychaetes with their calcareous tubes. Cuvier in 1795 revised Linnaeus' scheme, approximating more modern views, but much of the ninteenth century went by before the molluscs were purged of all extraneous groups and all present groups were included. The scaphopods, or tooth shells, were definitively added in a series of works spanning the 1800s and culminating in the work of Keferstein (1863–1866) who gave them their name. Tunicates were eliminated from the Mollusca by Kowalevsky (1866), who observed the chordate features in their larvae.

Two of the main molluscan groups were discovered fairly recently, the Aplacophora in 1841 and the Monoplacophora in 1952. Aplacophora (Figure 8.1a) was formerly a taxonomic class with the Caudofoveata and Solenogastres as subclasses within it (e.g. Brusca and Brusca 2003), but each is now a class in its own right. The aplacophorans are also known in the older literature simply as solenogasters. Both classes are considered primitive. The aplacophorans were definitively placed in the Mollusca by Graff (1875). Thereafter, a popular classification scheme grouped the chitons and aplacophorans together as the subphylum Amphineura based on the similarity of their nervous systems. Though that scheme has not been accepted since the 1960s, it is possible to still see reference to it in older texts.

The monoplacophorans are found only in the deep sea. Prior to their discovery in 1952, they were known only from the fossil record and were thought to have been extinct since the Devonian period [419 million years ago (mya)]. The first living specimens were found by the science party of the Danish research vessel Galathea, who were dredging off the coast of Costa Rica in 1952 at a depth of 3570 m. The value of the collections was not appreciated until

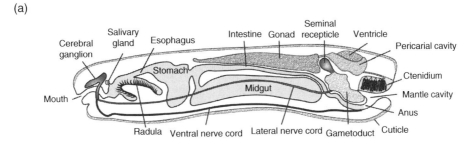

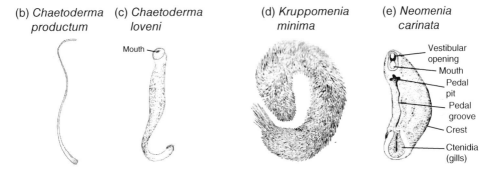

Figure 8.1 Molluscan classes Caudofoveata, and Solenogastres, formerly subclasses in the now defunct Class Aplacophora. (a) Internal anatomy of a stylized mollusc in the Class Caudofoveata; (b and c) two examples of molluscs in the Class Caudofoveata; (d, e) two examples of molluscs in the Class Solengastres. *Sources:* (b) From Wiren (1892); (c) From Nierstrasz (1902); (d) From Nierstrasz (1905); (e) From Hansen (1888).

1956 when the specimens were first examined and recognized as living fossils by Lemche (1957), who named them *Neopilina galathea*. They are the only segmented molluscs, and their extinct members had been identified by the presence of paired, serially repeated, muscle scars on the inside of their fossil shells. Since then a number of expeditions dredging in deep-ocean trenches have recovered several more monoplacophoran species (Clarke and Menzies 1959; Menzies and Layton 1962).

The Pelagic Molluscs

Two molluscan classes have successfully invaded the pelagic realm, the gastropods and the cephalopods. The gastropods are represented by four quite different taxa: the heteropods, pteropods, janthinid snails, and a tiny group of nudibranchs. The heteropods are an unusual group of gelatinous hunters that prey largely on other pelagic gastropods; the pteropods, or sea butterflies, comprise a gelatinous group that includes particulate feeders as well as predators. Both have morphologies that are very atypical of the gastropods. The rafting janthinid snails have a more conventional appearance including a shell, whereas the pelagic nudibranchs have a small, highly adapted, group of gelatinous species and a small group of species that differ little from their benthic relatives. Altogether the pelagic gastropods number about 179 species in a class that has about 38 000 marine species (WoRMS - World

Register of Marine Species. http://www.marinespecies.org) and about 70 000 altogether (Brusca et al. 2016).

The cephalopods include the nautiloids, octopods, and squids, numbering about 810 species altogether. The octopods have about 295 species, the majority of which are benthic, and the nautiloids have five (WoRMS). The remainder are squids of various types, which we will consider 100% pelagic for a total of about 511 pelagic cephalopod species including the octopod *Vampyroteuthis*.

There are about 49 400 species of marine molluscs, of which about 690 are pelagic gastropods and cephalopods, or 1.4% of the total. A fairly respectable showing for molluscan pelagic invaders! The strategy in this chapter will be to briefly introduce the major classes of molluscs for zoological background followed by a more detailed treatment of classes with pelagic members and the pelagic groups themselves.

Phylum Mollusca

Class Caudofoveata

About 142 species of burrowing, marine worm-like molluscs with a cylindrical body (Figure 8.1a–c), lacking a head, eyes, foot, shell, and nephridia; with a chitinous cuticle bearing calcareous spicules, with paired ctenidia, a radula and small posterior mantle cavity; without a ventral groove. Size usually ranges from 0.5 to 3 cm in length, but at least one species, *Chaetoderma productum*, may reach 14 cm (Hyman 1967).

Class Solenogastres

About 290 species of benthic marine, worm-like molluscs with a spiny cuticle and ventral groove, lacking a head, eyes, shell, ctenidia, and nephridia. They have a small vestibular cavity located anterior to the mouth and a posterior mantle cavity, often with respiratory folds (Figure 8.1d, e). Size ranges from 1 to 5 cm with at least one species, *Epimenia babai*, reaching 30 cm; often hermaphroditic.

Class Monoplacophora

About 30 species of living marine fossils with a univalve (limpet-like) shell and bilaterally symmetrical segmentation. They have serially repeated external gills, nephridia, auricles, ventricles, foot retractor muscles, and nerve branches as well as a median terminal anus (Hyman 1967). Possess a well-developed foot and distinct head, but no eyes. A benthic group found at depths exceeding 2000 m (Figure 8.2) and reaching at least 3.7 cm in length. Believed to be a deposit feeder.

Class Polyplacophora

A benthic group with about 1000 species. The chitons are a familiar sight in the Pacific rocky intertidal. The group as a whole ranges from the intertidal to the deep sea. The largest species is *Criptochiton stelleri*, the gumboot chiton, reaching 33 cm and found on both sides of the Pacific. The chitons are dorsoventrally flattened molluscs (Figure 8.3) with a distinct head lacking eyes and tentacles, a prominent foot with a flattened creeping sole, and a shell comprised of eight, longitudinally arranged, overlapping valves. The mantle forms a girdle surrounding the shell valves on the dorsal side. It is usually toughened by a heavy cuticle including calcareous scales or spicules. The mantle forms the outer edge of the ventral surface. Between the foot and outer edge of the mantle is the pallial groove, actually the mantle cavity, which contains a varying number of bipectinate gills. The chitons have a well-developed radula, a simple ladder-like nervous system, and a pair of nephridia.

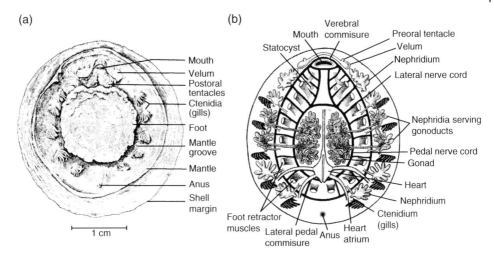

Figure 8.2 Class Monoplacophora. (a) External anatomy of *Neopolina galathea*, ventral view, a monoplacophoran mollusc; (b) internal anatomy of *N. galathea*, ventral view. *Sources:* (a) Barnes (1974), figure 11-37 (p. 368); (b) Kaestner (1967), figure 13-4 (p. 295).

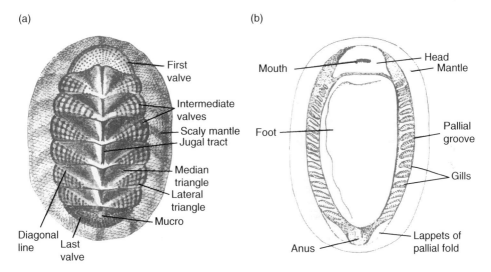

Figure 8.3 Class Polyplacophora (chitons). (a) Stylized external anatomy of a mollusc in the Class Polyplacophora; (b) ventral view of the chiton *Trachydemon cinereus*, *Sources:* From (a) Hescheler 1900; (b) From Pelseneer (1899).

Class Scaphopoda

The tusk or tooth shells. About 576 marine species with the largest about 12 cm and many about 3–4 cm. Individuals occupy a tapering, tooth-like conical shell lined entirely with the mantle that secretes it, forming a tube-like mantle cavity. They possess a rudimentary proboscis-like head with no eyes. The anterior end has a well-developed foot that may be extended with hydrostatic pressure to loosen up the sediment at the organism's anterior end, allowing for movement and feeding. Slender club-shaped tentacles termed captaculae originate near the mouth and are deployed into the loosened sediment where they capture small prey such as diatoms and foraminifera and convey them to the mouth. Scaphopods have a strong radula. No specialized respiratory structures are present; gas exchange is effected by the water bathing the mantle wall as it circulates in and out of the posterior shell. Typically, scaphopods are found head-down in the sand with the posterior end of their shells protruding into the water above (Figure 8.4).

Class Bivalvia

The mussels, clams, oysters, and kin. About 9200 marine and freshwater species (Brusca et al. 2016). Distributed from the intertidal to abyssal depths. Almost exclusively benthic filter feeders, though some species may bore into rock and the shipworms (Teredinidae) into wood. The largest of the bivalves is the giant clam, *Tridacna gigas*, with a shell length of up to 1.35 m and a mass of 200 kg (Kaestner 1967).

Bivalves are bilaterally symmetrical, laterally compressed, molluscs with a body

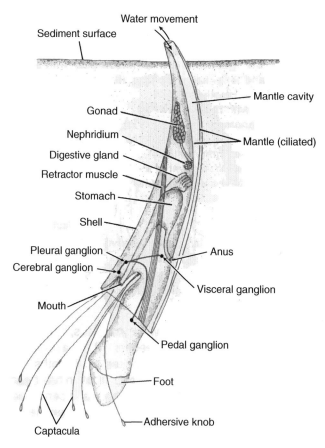

Figure 8.4 Class Scaphopoda (tusk or tooth shells). General anatomy of a stylized scaphopod. *Source:* From Brusca et al. (2016), figure 13.9 (p. 470). Reproduced with the permission of Oxford University Press.

completely enclosed by the two lobes of the mantle that in turn line the hinged, bivalved shell (Figure 8.5). They are without a distinct head, pharynx, radula, or tentacles, and most are without eyes. They have nephridia and paired leaf-like gills (ctenidia). Water movement through the mantle cavity is largely the result of ciliary activity.

Class Gastropoda

Snails, limpets, nudibranchs, sea hares, and kin with about 70 000 species represented in marine, terrestrial, and freshwater systems (Brusca et al. 2016) and about 38 000 marine species (WoRMS). Gastropoda are asymmetrical Mollusca, usually with a univalve spirally coiled shell. They have a distinct head, bearing tentacles and eyes, and a well-developed muscular foot that in most groups is used as a creeping sole (Figure 8.6) but in pelagic forms may be modified to be a swimming fin. With a radula, nephridium, and usually with one or two ctenidia (gills). A signature characteristic of the class is their torsion: during early development, the visceral mass is rotated 90–180° on the foot (Figure 8.7), so that the anus faces anteriorly (Figure 8.7c) or to the right (Figure 8.7b, d). Note that the torsion of gastropods is what makes them asymmetrical. Gastropods that have secondarily lost their torsion (e.g. the heteropods and pteropods) are called "detorted."

Class Cephalopoda

Nautilus, cuttlefish, octopus, and squids. A marine group with about 820 species. Bilaterally symmetrical molluscs with an elongated dorsoventral axis, a well-developed head and large eyes rivaling those of the vertebrates in their complexity. Cephalopods have jaws, a radula, and a foot that has been modified into a circlet of arms surrounding the mouth. With the exception of the nautiloids, the arms are equipped with suckers. Nautiloids have an external spirally chambered shell. Other cephalopods may have a reduced shell imbedded in the dorsal mantle cavity; some have no shell at all. All have a closed circulatory system with one to two pairs of bipectinate gills, heart with two to four auricles, sometimes with branchial hearts and with two to four nephridia. Robust, muscular mantle forming a large ventral cavity that exits through a muscular siphon. Contraction of the mantle musculature provides jet propulsion; direction is determined by the position of the siphon. Dioecious.

Body Organization

Generic representatives of the different classes of molluscs are shown in Figure 8.8. Note especially the comparative positions and structures of the foot, position of the anus relative to the head, and the character of the mantle. The radula, unique to the molluscs and present in all classes but the bivalves, is located in the anterior buccal cavity.

The Gastropoda

Classification

Molluscan classification has changed considerably since the early twenty-first century, particularly within the gastropods. The previous system originated with Milne-Edwards (1848) and persisted well into the 2000s. Milne-Edwards based his system on branchial position and structure, dividing the gastropods into the prosobranchs (anterior-facing gill), opisthobranchs (posterior-facing gill), and pulmonates (air breathers with a

(a) A typical eulamellibranch (cross section)

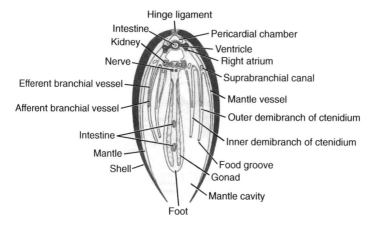

(b) The mussel, *Mytilus*, seen from the right side after removal of the right shell and the mantle.

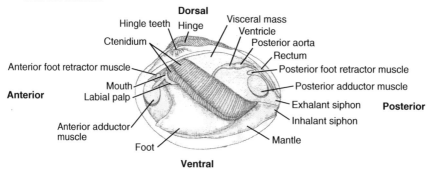

(c) Internal anatomy of *Mercenaria*

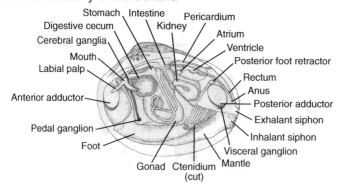

Figure 8.5 Class Bivalvia. (a) Internal anatomy of a stylized eulamellibranch bivalve; (b) internal anatomy of the mussel *Mytilus*; (c) internal anatomy of the clam *Mercenaria*. *Sources:* (a) Brusca et al. (2016), figure 20.8 (p.713) (b and c) Adapted from Sherman and Sherman (1970), figure 8.6 (pp. 294–295).

(a) A typical coiled-shell gastropod (female)

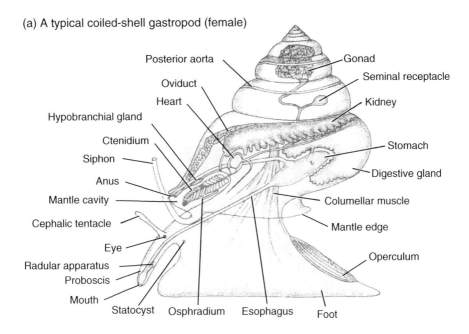

(b) The periwinkle, *Littorina*, removed from its shell (anterior view)

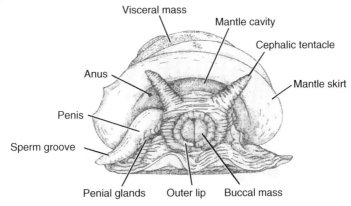

Figure 8.6 Class Gastropoda. (a) Internal anatomy of a stylized coil-shelled gastropod; (b) external anatomy of the periwinkle *Littorina*. *Sources:* (a) Brusca and Brusca (2003), figure 20.6 (p.709); (b) Fretter and Graham (1962), figure 3 (p. 17).

primitive "lung" in the mantle cavity not shown) (Figure 8.7). The currently evolving system of classification includes phylogenetics as well as the morphology of both organism and shell. As a result, classification within the Gastropoda is unsettled. Below the infraclass level, many Linneaean taxonomic rankings have dissolved and final placement of traditional groupings has yet to be determined. The system presented below is an amalgam of the phylogenetically based subdivisions in Brusca et al. (2016) down to infraclass, and the more traditional divisions below the cohort level. It is less a systematic

448 | *The Mollusca*

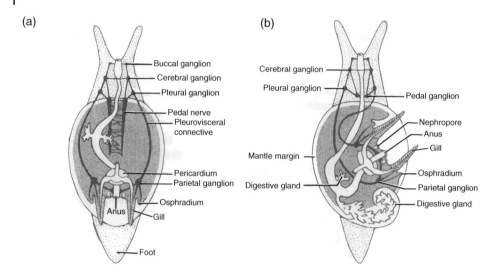

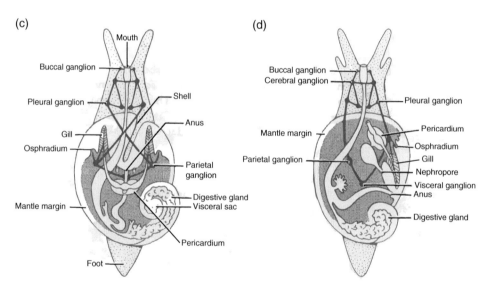

Figure 8.7 Torsion in gastropod molluscs. (a) Hypothetical ancestral snail with mantle cavity posterior of the visceral sac; (b) prosobranch snail, with 90° torsion of the visceral sac, right midgut gland degenerates; (c) primitive prosobranch snail, *Zeugobranchia*, torsion completed; (d) opisthobranch snail, the mantle cavity is turned to the right side. *Source:* Permission from Springer Nature Customer Service Center: Springer Nature, *Zoologie im Grundriss*, by Stempell (1926). © 1926, Figure 314.

statement than a way to keep track of the pelagic groups. Information from WoRMS on species numbers and classification below cohort is included.

Because most of what is known about gastropods is couched in the old Milne-Edwards system, a very brief "rosetta stone" is provided for target taxa.

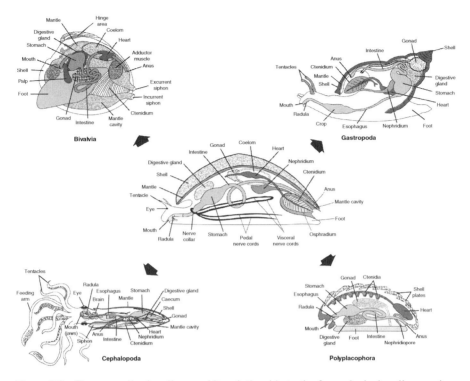

Figure 8.8 The generalized mollusc and its relationship to the four principal molluscan classes. *Source:* Adapted from Sherman and Sherman (1970), figure 8.1 (p. 217).

Class Gastropoda

Classification Below Subclass Given for Pelagic Species Only

Total marine species (only) given in parentheses for each taxon.

SUBCLASS PATELLOGASTROPODA – (360) the true limpets.
SUBCLASS VETIGASTROPODA – (4000) Abalones and topshells.
SUBCLASS NERITIMORPHA – (240) Marine, freshwater, and terrestrial neritid snails. Includes *Helix*, the edible terrestrial snail, as well as many common aquarium snails.
SUBCLASS NEOMPHALIONES – (112) Hydrothermal vent snails
SUBCLASS CAENOGASTROPODA – Marine, terrestrial, and freshwater snails with about 25000 marine species, making it the most speciose of the gastropod subclasses. Includes many of the most familiar snails: periwinkles, conchs, and cowries, as well as the heteropods, one of the two most important groups of swimming snails. The Heteropoda was previously a formal taxonomic designation (suborder or superfamily, depending on the time period) but is now an informal (but quite useful) term used to describe the superfamily Pterotracheoidea. Informal taxonomic names like the heteropods are known as "legacy names" (Brusca et al. 2016).
INFRACLASS SORBEOCONCHA
  COHORT HYPSOGASTROPODA
    SUPERORDER LITTORINIMORPHA (6750) Diverse superorder containing the conchs, periwinkles, slipper shells, vermetids and

heteropods, among others. Fourteen additional superfamilies in addition to the Pterotracheoidea.
SUPERFAMILY PTEROTRACHEIOIDEA (38)
    Family Atlantidae (24)
    Family Carinariidae (9)
    Family Pterotracheidae (5)
ORDER CAENOGASTROPODA – (unassigned taxa) (3317)
  SUPERFAMILY EPITONIOIDEA (793)
  Formerly Family Janthininidae (9), now in Family Epitoniidae
SUBCLASS HETEROBRANCHIA – (8630) A highly diverse and speciose group including the sea hares, nudibranchs, pulmonates, and pelagic pteropods. Two of the traditional gastropod subclasses, the Opisthobranchia and Pulmonata, are found in this new subclass.
INFRACLASS EUTHYNEURA (8205)
  COHORT EUOPISTHOBRANCHIA
    ORDER PTEROPODA (146)
      SUBORDER GYMNOSOMATA (54)
        SUPERFAMILY CLIONOIDEA (52)
        SUPERFAMILY HYOMYLOIDEA (2)
      SUBORDER EUTHECOSOMATA (69)
        SUPERFAMILY CAVOLINOIDEA (61)
      SUPERFAMILY LIMACINOIDEA (8)
      SUBORDER PSEUDOTHECOSOMATA (23)
        SUPERFAMILY CYMBULOIDEA (23)
  COHORT NUDIPLEURA
    ORDER NUDIBRANCHIA (2310)
      SUBORDER CLADOBRANCHIA (1015)
        SUPERFAMILY AEOLIDOIDEA (284)
          Family Glaucidae (5)
        SUPERFAMILY FIONOIDEA (313)
          Family Fionidae (4)
        SUPERFAMILY DENDRONOTOIDEA (187)
          Family Phylliroidae (3)

Traditional Classification of the Pelagic Gastropoda (From Hyman 1967; Barnes 1974; Lalli and Gilmer 1989; Brusca and Brusca 2003)

CLASS GASTROPODA

SUBCLASS PROSOBRANCHIA. Torted gastropods with the pallial complex (mantle cavity, gills, anus, nephridiopore, and gonopore) anteriorly directed. Usually with a spirally coiled shell.

  ORDER ARCHAEOGASTROPODA. Most primitive of the prosobranchs. Includes the limpets, abalones, and turban shells. Twenty-six families.

  ORDER MESOGASTROPODA. Prosobranchs usually with a siphon and operculum. One gill, one auricle, and one nephridium. Includes the periwinkles, conchs, heteropods, and janthinid snails. Many of the caenogastropods were originally classified as Mesogastropoda. About 100 families.
  SUBORDER HETEROPODA (Pterotracheoidea), the pelagic heteropods.
  SUBORDER PTENOGLOSSA
    Family Janthinidae, the violet, or janthinid snails.
  ORDER NEOGASTROPODA. Prosobranchs with short to very long siphonal canal, one auricle, one nephridium. Dioecious. Includes the whelks, muricids, and cone shells. About 24 families of marine snails.

SUBCLASS OPISTHOBRANCHIA. Torted or detorted gastropods. Shell usually without ornamentation, often reduced or lost. Sea hares, nudibranchs, and pteropods. Nine to 13 traditional orders and over 100 families. Hyman (1967) retains the original order Pteropoda which is then subdivided into thecosomes and gymnosomes, whereas older editions of Barnes (1974), Lalli and Gilmer (1989), and Brusca and Brusca (2003) give the thecosomes and gymnosomes ordinal status.

ORDER THECOSOMATA – The shelled pteropods

ORDER GYMNOSOMATA – The naked pteropods

ORDER NUDIBRANCHIA – the pelagic nudibranchs, e.g. *Fiona* and *Phylliroe*.

SUBCLASS PULMONATA – The air-breathing snails and slugs. Detorted snails with or without a shell, mainly terrestrial and freshwater in habit. Mantle cavity altered into a lung with a contractile opening, the pneumostome.

Gastropod Systems and Structures

Figure 8.6 gives a cutaway view of a typical torted gastropod.

The Digestive Tract

The approximately 70 000 species of gastropods have a diversity of dietary habits ranging from microphagy (suspension feeding) to herbivory and carnivory. Our target taxa occupy the two ends of the spectrum: they are either suspension feeders (the thecosomes) or carnivores (the janthinids, nudibranchs, heteropods, and gymnosomes). Clearly, the diversity in their diets is reflected in their digestive systems, but they all have a similar basic framework.

The gastropod digestive tract includes an initial buccal region, usually lined with cuticle and further stiffened with a pair of jaws of varying shapes and sizes on its lateral walls. The buccal region is followed by a muscular pharynx with ducts leading from one or two pairs of salivary glands. The pharynx in turn leads to a long esophagus and a stomach that receives the ducts of a large midgut gland, similar in function to the hepatopancreas of the Crustacea. A long intestine follows the stomach, ending in a hindgut or rectum. The hindgut may pass directly through the pericardium, particularly in torted species. The stomach, midgut gland, intestine, and hypobranchial gland are parts of what is collectively called the visceral mass.

The pharynx is dominated by the ventral mass comprising the radular apparatus (Figure 8.9). The radula is a highly complex, muscular structure that is variously used for rasping in herbivores and prey capture in carnivores. It is essentially a biological conveyor belt with imbedded chitinous teeth that moves back and forth over the tip of a fulcrum, or bolster termed the odontophore, composed of muscle and cartilage. Teeth of the radula are arranged in a series of transverse rows on the chitinous radular membrane. The radular membrane, or ribbon, is moved back and forth by retractor and protractor muscles. In turn, the entire radular apparatus may be moved anteriorly and posteriorly by the odontophore protractor and retractor muscles, much like a vertebrate tongue. Radular teeth are continuously secreted in the posterior of the radular sac, gradually moving forward to replace the teeth broken off at the other end of the ribbon. Note that the teeth are posteriorly directed, helping the inward stroke of the radula to move the prey item (or piece thereof) into the pharynx. The number and type of radular teeth are taxon-specific, a useful tool in gastropod taxonomy. The tooth pattern of most caenogastropods (e.g. the heteropods) is called taenioglossate ("ribbon tongue," Figure 8.10a) which bears seven teeth per row: a central tooth, with two median teeth, and two lateral teeth per side. Altogether there are seven patterns, only two of which are important to our study of pelagic gastropods: the taenioglossate of the heteropods and the ctenoglossate pattern of janthinid snails.

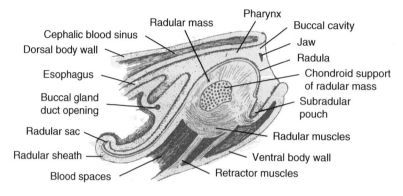

Figure 8.9 Gastropod pharynx. Sagittal section of the head of a prosobranch, without proboscis. *Source:* From Hescheler (1900).

The radula of heteropods is designed to capture and hold their pelagic prey, usually salps or other pelagic gastropods, and swallow it whole or in pieces. The teeth on their taenioglossate radula are highly modified and hook-like to accomplish this (Figure 8.10b). As the radular membrane is pulled over the tip of the odontophore, the teeth are spread; continued movement forward impales the prey, allowing it to be conveyed back to the pharynx and then to the esophagus (Figure 8.10c–e). Radular teeth allow for removal of large chunks of food (Hamner et al. 1975).

The radula of janthinids has a ptenoglossate pattern ("wing tongue"), lacking a central tooth and with a radular membrane that is split in two. The radular teeth themselves are also long and hook-like, allowing them to function much like the teeth of the heteropods described above. The janthinids are pelagic, preying on species such as *Velella* and *Physalia* from the rafts they create for themselves in the warm surface waters of the tropical and subtropical ocean pleuston. Their close relatives, the epitoniids, are benthic, feeding on anthozoa. The buccal region of janthinids includes a pair of jaws that allow them to tear off pieces of the gelatinous species they feed on, aiding digestion.

Movement of food items past the pharyngeal region and into the esophagus is facilitated by the initial reductive activity of the radula, the lubricatory secretions of the salivary glands, and, especially in the Caenogastropoda, ciliary activity in the foregut. The gut of caenogastropods usually does not have the circular and longitudinal musculature that allows for peristalsis. It is assumed that the absence of a muscular gut wall may be a holdover from a microphagous ancestor for the group (Morton 1958b). In any event, that absence increases the importance of chemical food breakdown, at least partially explaining the universally well-developed midgut gland in gastropods. Figure 8.11 shows the digestive tract of *Murex*, a carnivorous genus, with all associated glands. Note the presence of a hypobranchial gland which is found in some gastropods (see also Figure 8.6) and is presumed to aid in chemical breakdown of food. Many caenogastropods have a crystalline style, a rod composed of protein and digestive enzymes

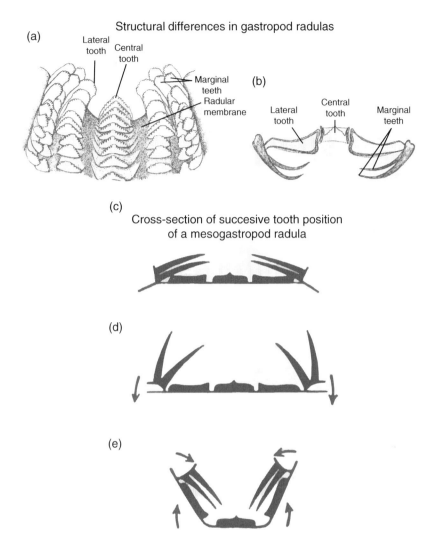

Figure 8.10 Structure of gastropod radulas. (a) Radula of the gastropod *Viviparus*; (b) radula of the heteropod *Pterotrachea*; (c) caudal view of a cross section of the odontophore of a mesogastropod radula; (d) view on the edge of a cross section of the odontophore of a mesogastropod; (e) view in the longitudinal groove of the anterior side of a cross section of the odontophore of a mesogastropod. *Sources:* (a and b) Hyman (1967), figure 105 (p. 239); (c) Kaestner (1967), figure 13-33 (p. 323);

(mainly for carbohydrates) which is located in a "style sac" in the stomach proper. Cilia rotate the rod against a hardened gastric shield in the stomach to release its carbohydrases. The rod also aids in winding food particles into the stomach and moving them through to the intestine. In contrast to the caenogastropods, the heterobranchs (pteropods and nudibranchs) usually have a more muscular gut capable of peristalsis, sometimes including a crop or gizzard to aid in food breakdown.

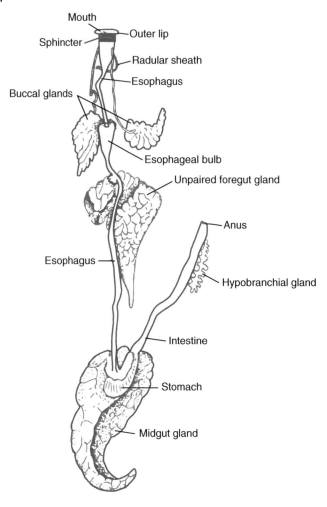

Digestive system of *Murex*

Figure 8.11 Digestive system of the carnivorous marine snail *Murex*. *Source:* Adapted from Haller (1888), plate III.

Food breakdown is the province of the stomach and glands, with most absorption of nutrients taking place in the glands as well as the walls of the intestine. Intracellular digestion plays a far more important role in the molluscs via phagocytic cells in the stomach and midgut gland than in the annelids or arthropods, particularly in the caenogastropods.

Circulation

Circulation in all classes of molluscs follows a similar basic pattern, reaching its greatest level of sophistication in the cephalopods. The number of auricles in the heart and the number of gills, or ctenidia, vary within and between classes, usually with the number of auricles and number of gills being equal. Figure 8.12 is a representation of the

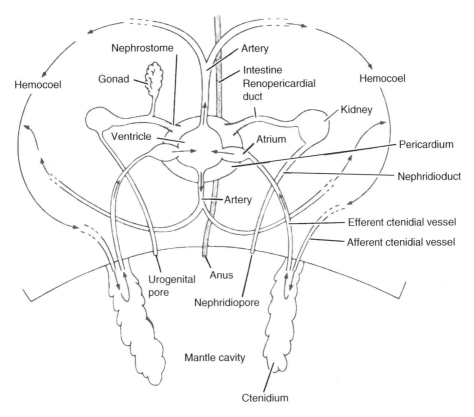

Hemolymph flow in a typical mollusc

Figure 8.12 Circulatory system and circulation pattern in a typical mollusc. *Source:* Brusca et al. (2016), figure 13.34 (p. 503). Reproduced with the permission of Oxford University Press.

pattern in the "archetypical" or generic mollusc, which includes two ctenidia. The archetypical mollusc is the reincarnation of what used to be known as the ancestral mollusc (Hypothetical Ancestral Mollusc or "HAM") and serves the same purpose: portrayal of basic molluscan patterns.

Our model organism for gastropod circulation is the periwinkle *Littorina littorea*, a fairly close relative to our caenogastropod target taxa and a good example of an advanced caenogastropod (Figure 8.13). All molluscs have a systemic heart, meaning that, like in the Crustacea, it receives oxygenated blood as returned-flow from the gills and pumps it to the organs. The molluscan heart is located in a well-defined pericardial cavity bounded by a pericardial membrane or wall, which, along with the lumen of the gonads, is the only remaining part of the true coelom. A system of arteries, including a large anterior, posterior, and esophageal artery, branch repeatedly to distribute blood directly to the visceral organs and muscles of the gastropods. After bathing the organs and muscle, the blood is collected in a variety of sinuses for the venous return to the heart (Figure 8.13b, c). The main blood sinuses are the buccal sinus, radular sinus, cephalopedal sinus, the

(a) *Littorina littorea*: animal removed from shell (left side view)

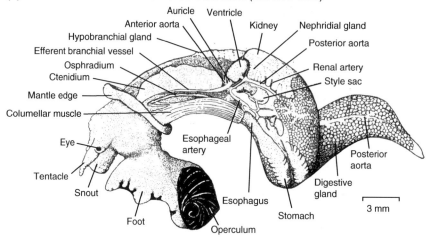

(b) *Littorina littorea*: arterial system (left side view)

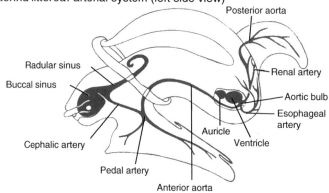

(c) *Littorina littorea*: venous system (left side view)

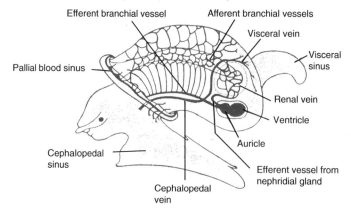

Figure 8.13 Anatomy of the periwinkle *Littorina littorea*. (a) General anatomy; (b) arterial system; (c) venous system. *Sources:* (a) Fretter and Graham (1962), figure 2 (p. 16); (b and c) Fretter and Graham (1962), figure 18 (p. 36).

circumferential pallial sinus at the edge of the mantle, and the visceral sinus. Note in particular the pallial (mantle) sinus, which supplies much of the blood flow to the afferent branchial vessels that feed the ctenidium. After passing through the gill, the oxygenated blood collects in the efferent branchial vein for return to the heart. The second major return pathway to the heart is through the kidney (Fretter and Graham 1962).

Respiration

Gills and gill morphology vary considerably within the pelagic species as well as within the gastropods as a whole. The janthinids are typical of a fairly advanced torted gastropod in having a monopectinate or comblike gill (also known as a pectinibranch), with filaments or plates on one side only (Figure 8.14). Water is circulated through the mantle cavity by ciliary activity on the gill plates (Figure 8.14b). Blood has a one-way flow through the gill filaments that effectively segregates oxygenated from deoxygenated blood. The one-way flow also allows for an effective countercurrent exchange as water and blood flow in opposite directions while crossing the gill. Gills are normally anchored to the mantle wall (Figure 8.14c).

Gill structure in most of the heteropods is quite different from that of the janthinids. The exception is the atlantid heteropods (Figure 8.15a), which have a full, though thin-walled, shell and mantle cavity. They retain a monopectinate gill and presumably have a water circulation through the mantle cavity similar to that of the janthinids. Atlantids are quite small, usually with a shell diameter of 10 mm or less. Heteropods in the genus *Carinaria* have a much-reduced shell covering only the externally located visceral mass with the gill filaments (ctenidia) protruding beneath the shell along the right side (Figure 8.16). The filaments are highly contractile, which would allow them to be moved under the shell when threatened, and they have a ciliated epithelium to facilitate gas exchange. Their external position on the visceral mass would allow for effective gas exchange, particularly during swimming. The most advanced of the heteropods, the pterotracheids (Figure 8.15b, c), have no shell and a very small visceral mass. Their gill filaments remain exposed. The filaments are highly contractile and invested with cilia to expedite water movement over their surface. The genus *Firoloida* has no gills (Hyman 1967).

A variety of gas exchange structures may be found within the pteropods and nudibranchs. A mantle cavity enclosing the visceral mass is present in the thecosomatous pteropods, lining the shell when present or containing the internal pseudoconch of the more advanced pseudothecosomes (Hyman 1967; Lalli and Gilmer 1989). A gill is present in the mantle of the thecosome family Cavoliniidae and pseudothecosome family Peraclidae, but unlike the pectinibranch gills of the janthinids it is a "plicate" (folded lengthwise) gill (Figure 8.17a) originating during development as a fold in the mantle and serving the same function as a more conventional molluscan gill. In the case of *Cavolinia*, the gill itself extends in a long crescent roughly following the contour of the mantle roof (Figure 8.17b).

The naked pteropods (sea angels) have no mantle cavity, but specialized regions with thinner integument girdling the lower trunk or forming the posterior tip of the body are believed to act as gills (Lalli and Gilmer 1989) (Figure 8.18). Both are normally associated with a ciliary band, presumably enhancing flow over the respiratory surface. The posterior tip is well invested with small blood sinuses.

458 | *The Mollusca*

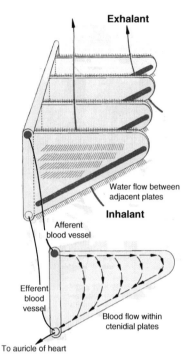
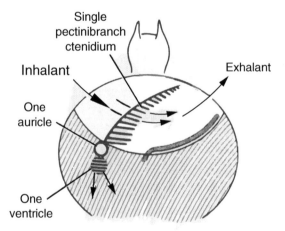

(a) Monopectinate ctenidium

(b) Water current and blood flow in a monopectinate ctenidium, ventral inhalant to dorsal exhalant

(c) Gastropod mantle cavity, dorsal view

Figure 8.14 Monopectinate ctenidia (gills). (a) Single gill filament; (b) water and blood flow patterns in a monopectinate gill; (c) water flow pattern through the mantle cavity of a mollusc with a monopectinate gill. *Sources:* (a and b) Adapted from Russell-Hunter (1979), figure 19-4 (p. 353); (c) Russell-Hunter (1979), figure 20-6 (p. 371).

(a) Heteropod *Atlanta*, showing crested shell

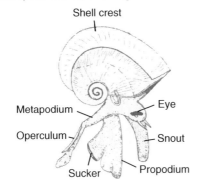

(b) Heteropod *Pterotrachea*, without shell

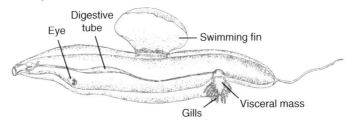

(c) Visceral mass of the heteropod *Pterotrachea*, showing gills

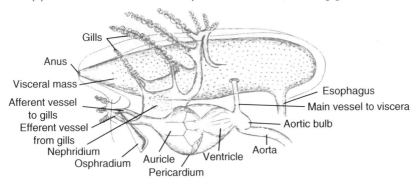

Figure 8.15 Heteropod respiration. (a) External anatomy of the atlantid heteropod *Atlantis* which possess a monopectinate gill and full shell; (b) external anatomy of the carinariid heteropod *Pterotrachea* with reduced shell; (c) visceral mass of *Pterotrachea* showing gill filaments protruding beneath the shell. *Sources:* (a) From Souleyet (1852), plate 23; (b) Souleyet 1852, plate 22; (c) Hyman (1967), figure 85 (p. 206).

Gas exchange within the nudibranchs is assumed to be the province of the cerata when present. Cerata are projections of the body wall that can be arranged in dorsal or lateral tracts or clusters (Figure 8.19). Cerata contain branches of the midgut gland in addition to blood sinuses just beneath the epithelium. At the tip of each ceras is a cnidosac that contains "live" nematocysts that have been ingested by the

(a) *Carinaria mediterranea*, male

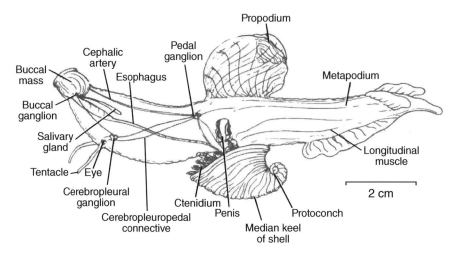

(b) Visceral mass of male *C. mediterranea*

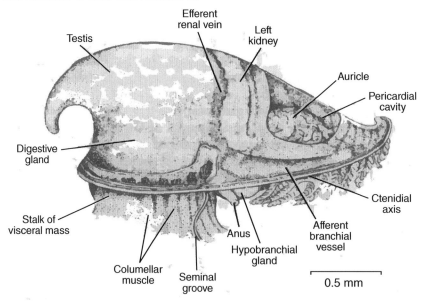

Figure 8.16 Heteropod respiration. (a) Anatomy of *Carinaria mediterranea* showing the gills extending beyond the edge of the shell; (b) visceral mass of *C. mediterranea*. *Sources:* (a) Fretter and Graham (1962), figure 300 (p. 564); (b) Fretter and Graham (1962), figure 301 (p. 565).

glaucids, usually from *Physalia* but also from other Cnidaria. The cnidosac ends in a pore at the distal end of the ceras (Hyman 1967). The most highly adapted pelagic nudibranchs, the family Phylliroidae, are very laterally compressed (Figure 8.20) and have no obvious respiratory structures. Their large surface-to-volume ratio likely

(a) Plicate gill in the mantle cavity of an opisthobranch

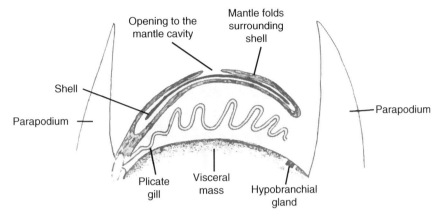

(b) Dorsal view of plicate gill in *Cavolinia tridentata*

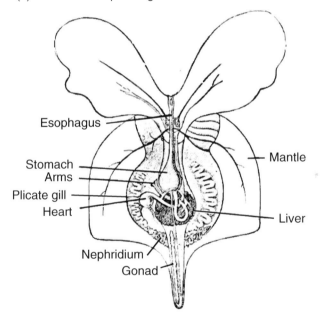

Figure 8.17 Plicate gills. (a) Typical opisthobranch mantle cavity; (b) mantle cavity of the pteropod *Cavolinia tridentata*. *Sources:* (a) Hyman (1967), figure 169 (p.420); (b) From Hertwig (1905), figure 376.

allows simple diffusion to satisfy their respiratory needs.

Excretion

All molluscs have a similar strategy for ridding the blood of metabolic waste and other impurities. The kidney, or nephridium, is a tubular excretory organ using the same principles discussed in Chapter 6: ultrafiltration followed by tubular secretion and reabsorption. In the gastropods, the ultrafilter producing the primary urine is the wall

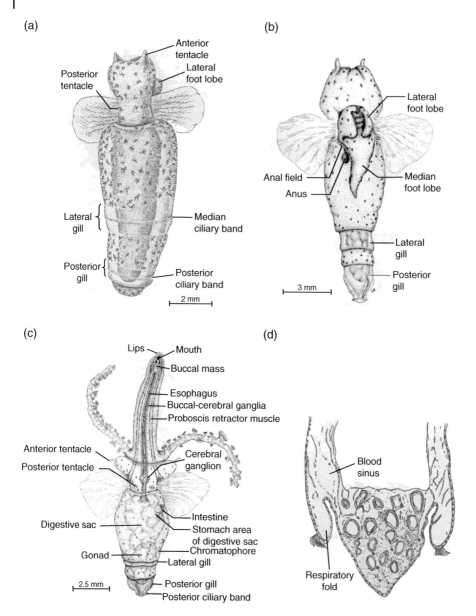

Figure 8.18 Pneumodermatid (sea angels) respiratory structures. (a) Lateral and posterior gills with accompanying ciliary bands in a juvenile *Spongiobranchaea* sp.; (b) lateral and posterior gills in *Pneumodermopsis* sp.; (c) gills and ciliary band in *Pneumodermopsis* sp. in feeding position; (d) section through the posterior tip of *Spongiobranchaea* sp. showing the respiratory fold and blood sinus. *Sources:* (a) Lalli and Gilmer (1989), figure 52 (p. 52); (b) Lalli (1970a), figure 2 (p. 4); (c) Lalli 1970a, figure 1 (p. 3); (d) Hyman (1967), figure 170 (p. 423).

of the auricle, which produces the hydrostatic pressure needed for ultrafiltration by the force of its contraction, expelling the ultrafiltrate into the pericardium (Andrews and Taylor 1988). The auricle is thus considered to be "two pumps in parallel, one with

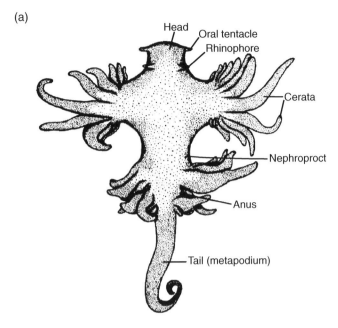
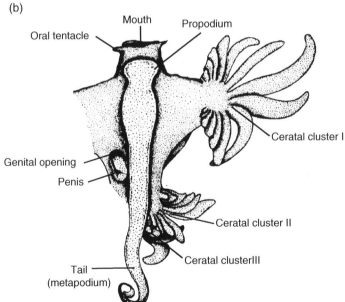

Figure 8.19 Nudibranch respiration. (a) External anatomy (dorsal view) of the pelagic nudibranch *Glaucus*, showing the cerata; (b) ventral view of the external anatomy of *Glaucus* showing a ceratal cluster. *Sources:* (a) Thompson and McFarlane (1967), figure 1 (p. 108); (b) Thompson and McFarlane (1967), figure 2 (p. 109).

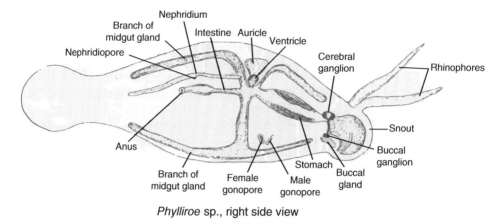

Figure 8.20 Internal anatomy of the Phylliroid nudibranch *Phylliroe* sp. with no apparent respiratory structures. *Source:* Adapted from Hescheler (1900), figure 21 (p. 14).

a leaky patch for producing primary urine from its wall, the other for refilling the ventricle" (Andrews and Taylor 1988). The ultrafiltrate thus becomes the pericardial fluid which drains to the kidney via the ciliated renopericardial canal. Once within the kidney lumen, the pericardial fluid is modified by the processes of secretion and reabsorption to produce a final urine that is voided to the mantle cavity for excretion.

The structure of the gastropod kidney (Figure 8.21) makes it difficult to identify the specific sites for secretion and reabsorption during the modification of primary urine to its final state. The internal kidney sac is clearly modified to increase surface area by extensive infolding. It is believed that the nephridial gland plays a role in reabsorption of organic solutes (Andrews and Taylor 1988). Much of the available information on the excretory process has been deduced from use of tracers such as inulin, a low-molecular-weight polysaccharide that is entirely ultrafiltered but not secreted or reabsorbed, or phlorizin, an inhibitor of glucose reabsorption. Experiments using inulin initially identified the auricular wall as the system ultrafilter. Similarly, appearance of glucose in the final urine after administering phlorizin confirmed that reabsorption was taking place in the kidney lumen (Martin 1983). Introduction of tracers into the blood coupled with analysis of the final urine yields a good picture of basic excretory system function and effectiveness.

Shell Formation

When a shell is present in our target taxa, it is reduced to its minimal mass to keep individuals at near-neutral buoyancy. The shell of cephalopods can actually be used as a buoyancy mechanism! Shells are present in the janthinid snails, the atlantid and carinariid heteropods, and the thecosomatous pteropods. The structure of a typical molluscan shell is depicted in Figure 8.22. It comprises three layers, a very thin protein outer layer or periostracum, an intermediate prismatic layer, and an inner nacreous layer. Overall, the shell structure may be thought of as a protein matrix imbedded with crystalline calcium carbonate. The protein of the periostracum and the shell matrix is conchiolin, a tanned protein unique to the Mollusca. Two forms of calcium carbonate make up

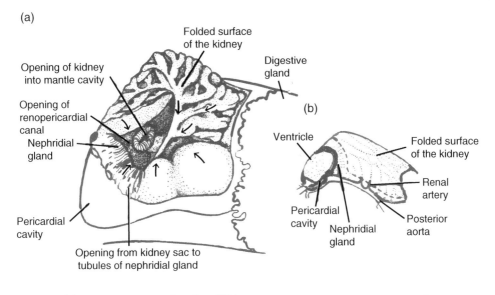
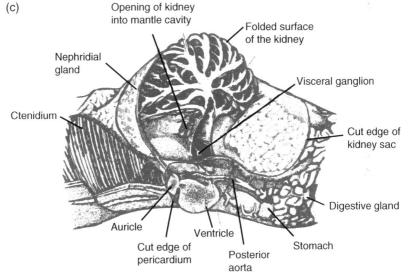

Figure 8.21 Kidney structure in the periwinkle *Littorina littorea*. (a) View of kidney structures with the kidney sac opened along the dashed line shown in (b). The arrows show the direction of ciliary currents; (b) kidney and heart of *L. littorea*; (c) blood vessels in relation to the kidney. The pericardial cavity and kidney sac have been opened to show the relationship between the kidney, heart, and gills within the visceral mass. *Sources:* (a and b) Fretter and Graham (1962), figure 16 (p. 33); (c) Fretter and Graham (1962), figure 148 (p. 274).

the inner shell layers. The prismatic layer is composed of calcite crystals arranged normal to the shell surface, whereas the nacreous layer is made up of aragonite crystals arranged parallel to it (Fretter and Graham 1962). Mass reduction in the shells of pelagic snails is achieved by minimizing shell mineralization (Lalli and Gilmer 1989).

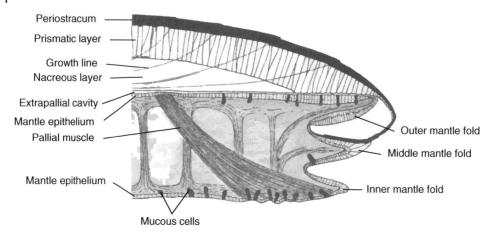

Figure 8.22 Cross section through the mantle and shell of a bivalve. The periostracum thickness and the size of the mantle folds are exaggerated in the figure. *Source:* Barnes (1974), figure 11-46C (p. 376). Reproduced with the permission of W. B. Saunders Company.

New shell is secreted by folds at the very edge of the mantle skirt (Figure 8.22) beginning very early in life after the formation of the protoconch, or larval shell. Most of the deposition at the shell edge is believed to be the responsibility of the outer mantle fold. Note that the mantle edge is quite small (about 1 mm) as are the folds, and the figure exaggerates their size. In addition, the muscles associated with the mantle edge can adjust its position as needed. As the snail grows, the protein conchiolin is deposited at the shell margin followed by deposition of calcium carbonate. Calcium carbonate sequestration and deposition is the province of the mantle epithelium and it occurs in the extrapallial space between the mantle and shell. In benthic forms, new shell grows at the margin and the inner layer of shell already formed is thickened for strength as the animal grows larger. In pelagic forms, the shell remains quite thin; it provides a defensive surface but must remain as close to neutrally buoyant possible. Examples of heteropod and pteropod shells are shown in Figure 8.23.

The Nervous System
The basic pattern of the molluscan nervous system, best represented by the chitons, is a circumesophageal ring that gives rise to two, paired, longitudinal nerve cords running posteriorly, one dorsally toward the viscera and one ventrally within the foot. In chitons, the ventral, or pedal nerve cord pair show a ladder-like series of commissures reminiscent of that seen in the primitive polychaetes. The paired dorsal cords, known as the pallial or pleural nerves, do not. No ganglia are present (Fretter and Graham 1962).

Gastropods show a more advanced nervous system, based on the primitive molluscan plan but including some centralization in the form of a suite of ganglia and complicated in many species by the effects of torsion. Figure 8.24a shows the basic nervous system design in a generic, torted gastropod similar in anatomy to *Janthina*. The head region contains a circumesophageal ring with three pairs of important ganglia: the dorsally located cerebral ganglion followed ventrally by the pleural and pedal ganglia.

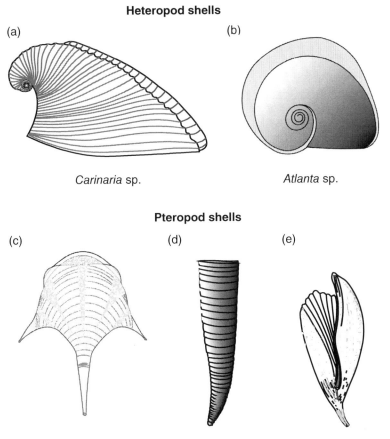

Figure 8.23 Heteropod and pteropod shells. (a) *Carinaria* sp.; (b) *Atlanta* sp.; (c) *Diacria* sp.; (d) *Creseis* sp.; (e) *Cavolina* sp.

A buccal ganglion is located anterior to the circumesophageal ring with connectives to the cerebral ganglion. Connectives from the pleural ganglia form the "visceral loop," servicing the alimentary canal, mantle, and other posterior organs. The visceral loop comprises the supra- and sub-esophageal ganglia, the visceral ganglion, and the connectives between them.

The cerebral ganglion innervates the tentacles, optic nerve, and statocysts as well as the musculature of the head region. The radula and its associated musculature are controlled by the buccal ganglion as is the buccal region in general, including the lips, salivary glands, and esophageal glands. The pleural ganglion innervates the visceral loop and its associated ganglia, ultimately connecting the mantle, ctenidia, alimentary system, and its associated glands. The pedal ganglion innervates the foot and shell (columnar) muscles. Note the effects of torsion on the path of the visceral loop, which must pass under and over the gut.

Compare the generic neural layout of Figure 8.24a with the actual anatomy of *Janthina*'s nervous system in Figure 8.24b. In addition to the actual length of the connectives and commissures when shown in planar fashion, a few other details are

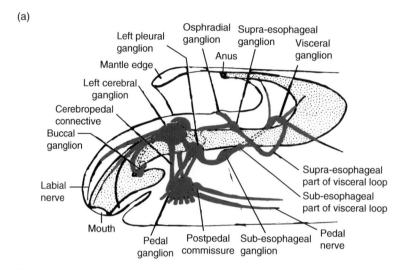

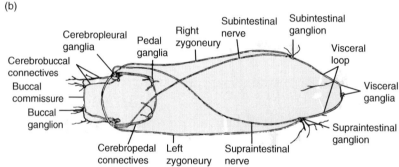

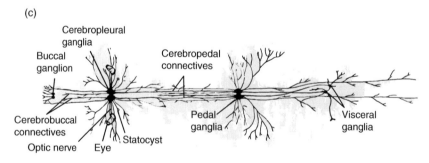

Figure 8.24 Molluscan nervous system. (a) Nervous system pattern in a generalized tortedgastropod. (b) nervous system of the torted gastropod *Janthina* sp. (c) nervous system of the detorted gastropod *Pterotrachea* sp. *Sources:* (a) Adapted from Fretter and Graham (1962), adapted from figure 162 (p. 311); (b) Hescheler (1900); (c) Hyman (1967), figure 117 (p. 260).

worthy of attention. First, the cerebral and pleural ganglia are fused into cerebropleural ganglia, following the evolutionary trend within the caenogastropods of increasing condensation of the main ganglia in more advanced taxa. The cerebropleural ganglia are also connected to the sub- and supraintestinal ganglia (different terminology for the sub- and supraesophageal ganglia) on the same side of the body by connectives termed "zygoneuries," giving a more complete neural circuit to the viscera.

Of particular interest is the central nervous system (CNS) of the heteropod *Pterotrachea*, depicted in Figure 8.24c. *Pterotrachea*, like its relatives *Carinaria* and *Firoloida*, has a detorted elongate body and correspondingly long connectives between ganglia. A wide separation exists between the buccal and cerebral ganglia, and the pedal ganglion has moved posteriorly to the base of the foot, or swimming fin. The visceral loop is discernible through the cerebral, pedal, and visceral connectives. The pleural ganglia are completely incorporated into the cerebral ganglia. Note, too, the well-developed optic nerve servicing the advanced heteropod eye.

Pteropod nervous systems are far more concentrated than those of either the heteropods or the janthinids. In general, the heterobranchs show a much more condensed CNS than the caenogastropods. Figure 8.25a is an enlargement of the CNS of *Gastropteron*, a heterobranch in the order Cephalaspida and a sea slug with large parapodia used for swimming like those in the pteropods. Figure 8.25b shows the CNS in perspective with the remainder of the body and its large swimming parapodia, illustrating the central condensation of all major ganglia. In the CNS of the shelled pteropod *Limacina*, all major ganglia are condensed into what is essentially a single mass, the commissures mainly having been reduced to slightly narrowed necks between ganglia (Figure 8.25c). The entire CNS of *Limacina* lies on the ventral surface of the anterior esophagus (Pelseneer 1888b) with a commissure that connects the two cerebral ganglia running across the dorsal surface of the esophagus.

The gymnosomes, or naked pteropods, exhibit a less condensed nervous system than their shelled relatives, reflecting their more elongated body plan. All major ganglia and commissures of the CNS of *Pneumoderma* are shown in Figure 8.26. The complex feeding structures of gymnosomes, the buccal cones, are innervated by the cerebral ganglia. The radular apparatus is controlled by the buccal ganglia.

Sensory Mechanisms

Gastropods are equipped with a variety of mechanoreceptors, chemoreceptors, and photoreceptors that vary considerably in form and complexity, particularly within our target taxa. With a few exceptions, and despite the groundbreaking work of Eric Kandel on memory and learning in sea hares, much of the information concerning sensory mechanisms in gastropods consists of behavioral observations rather than detailed descriptions of the sensors themselves. The observations do provide a reasonable picture of the gastropod sensory repertoire, however.

The entire exposed surface of "conventional" shelled gastropods such as *Janthina* is sensitive to touch, eliciting a withdrawal response when mechanically stimulated. Chemoreceptors have been localized in various regions of the epidermis, most notably in the oral region, by positive behavioral responses to food extracts. Ability to discriminate tastes has been shown in the terrestrial pulmonate, *Helix* (Kaestner 1967). Similarly, escape responses can be elicited in the intertidal snail *Nassarius* by stimulating the dorsum of its posterior foot with "predator extracts."

Surface projections, such as the cephalic tentacles (e.g. Figure 8.13a) are presumed to be sensory, containing neurosensory cells with bristles (Figure 8.27a) that are usually most numerous at the tip. The cephalic tentacles themselves can have papillae (Figure 8.27b), smaller projections invested

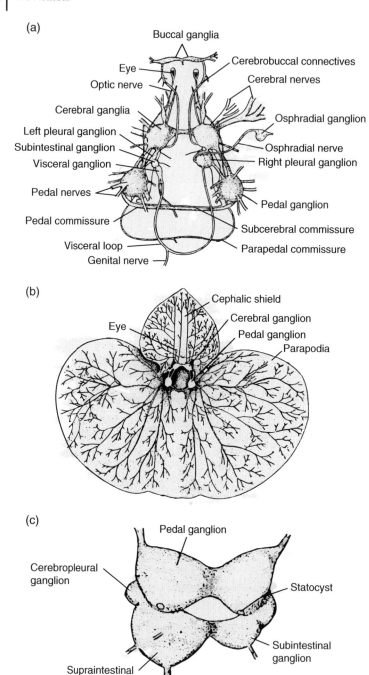

Figure 8.25 Heterobranchia nervous system. (a) Anterior portion of the central nervous system of the sea slug *Gastropteron*; (b) innervation of the parapodia of *Gastropteron*; (c) condensed configuration of the major ganglia in the CNS of the shelled pteropod *Limacina*. *Sources:* (a and b) Hyman (1967), figure 188 (p. 458); (c) From Pelseneer (1888b), plate I.

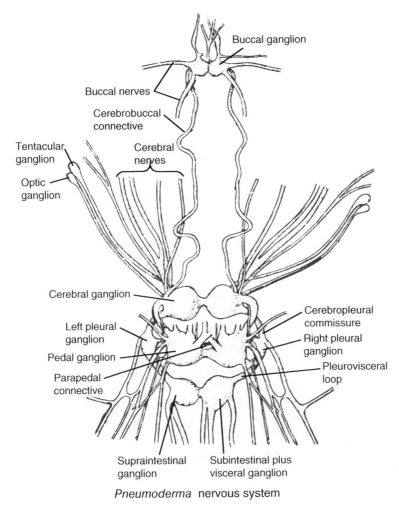

Figure 8.26 Anterior portion of the CNS of the naked pteropod *Pneumoderma*. *Source:* From Meisenheimer (1905).

with neurosensory cells that increase surface area and potentially enhance sensitivity. Cephalic tentacles of the heteropod *Carinaria* have papillae. The janthinid snails have large forked cephalic tentacles that may serve a variety of sensory roles. Neurosensory bristle cells probably have a range of functions including mechanoreception and chemoreception, similar to the functioning of crustacean hair cells.

The main chemoreceptive organ in gastropods (and molluscs in general) is the osphradium, a unique organ that in some gastropods resembles a small gill, in others a tubular structure (Figure 8.13a), oval, or series of bumps. It is always located adjacent to the gill(s) in the respiratory current. The fact that it is in the path of the inhalant stream means that it samples the surrounding medium as the snail breathes, monitoring for evidence of prey or predators. Within our target taxa, the shelled gastropod *Janthina* and the heteropods possess osphradia.

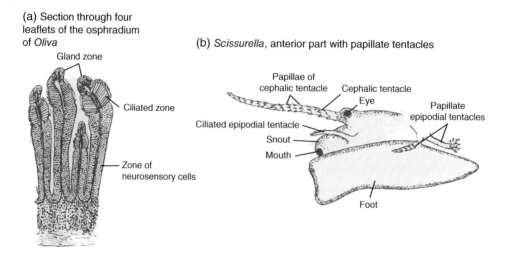

Figure 8.27 Sensory organs. (a) Neurosensory cells in the olfactory organ (osphradium) of the olive snail *Oliva*. (b) anterior portion of the little slit snail *Sciissurella* showing cephalic tentacles with neurosensory cell bearing papillae. *Sources:* (a) Küttler (1913); (b) Vayssiere (1894).

Effective vision in the pelagic Caenogastropoda is limited to the heteropods, which, as it happens, have the most highly developed eyes within the gastropods as a whole. The janthinid snails are reported as either having no eyes (Hyman 1967) or minute eyes at the base of the cephalic tentacles (Fretter and Graham 1962). They do exhibit behavioral photosensitivity (Bayer 1963; Lalli and Gilmer 1989). The heteropods have large image-forming eyes complete with a cornea, lens, and a stalk that allows for binocular vision and eye mobility (Figure 8.28a–c) (Land 1981). Behavioral observations reported in Hamner et al. (1975) suggest that heteropods are visual predators, able to detect their gelatinous prey from a distance of at least 60 cm.

A pair of statocysts, variously described as gravity detectors, organs of equilibrium, or motion detectors (see Chapter 7) are present in most of the caenogastropods, usually in close proximity to the pedal ganglia. However, they are innervated by the cerebral ganglia. Statocysts are particularly well developed in the heteropods (Figure 8.28c, d), easily visible though their transparent bodies. As in the Crustacea, the statocysts are fluid-filled vesicles lined with sensory hairs, each of which contains a calcium carbonate statolith. Positional information is conveyed when the statolith impinges on the sensory hairs. The CNS of a pelagic visual hunter must integrate detailed information on its own position and orientation relative to its prey. Well-developed statocysts would presumably aid in its hunting success.

The pteropods and pelagic nudibranchs (Heterobranchia) exhibit many of the same basic sensory mechanisms as the caenogastropods despite their differences in morphology. A general surface sensitivity to touch and chemical stimulation similar to that observed in the caenogastropods is

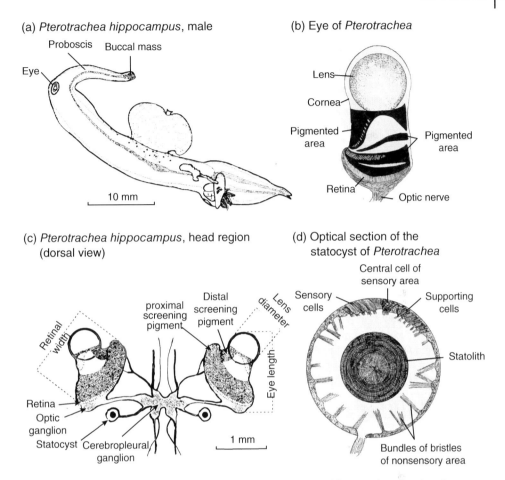

Figure 8.28 Heteropod vision and equilibrium. (a) The heteropod *Pterotrachea*, showing the position of the well-developed eyes; (b) structure of the eye of *Pterotrachea*; (c) dorsal view of the hippocampus, head region of *Pterotrachea* showing the relation of the eyes and statocysts to the central ganglia; (d) optical structure of the statolith of *Pterotrachea*. *Sources:* (a) Seapy (1985), figure 1 (p. 126); (b) Hyman (1967), figure 122 (p. 272); (c) Seapy 1985, figure 2 (p. 127); (d) Hyman 1967, figure 120 (p. 268).

assumed for the heterobranchs, though the presence of epidermal neurosensory cells have only been demonstrated in a few species (Hyman 1967). Tentacles are also assumed to have a sensory function; neurosensory cells are present in those species that have been examined. Both the naked and shelled pteropods bear cephalic tentacles that may possess a rudimentary eye in addition to neurosensory cells. Pseudothecosomes may also possess parapodial tentacles (*Desmopterus*) and a tail tentacle (*Cymbulia*) that have a sensory function. Statocysts are present in all the heterobranchs, usually located in the vicinity of the pedal ganglia.

The rudimentary eyes of pteropods are located in the tips of their cephalic tentacles. The eyes are simple vesicles with a fluid-filled space containing retinal cells

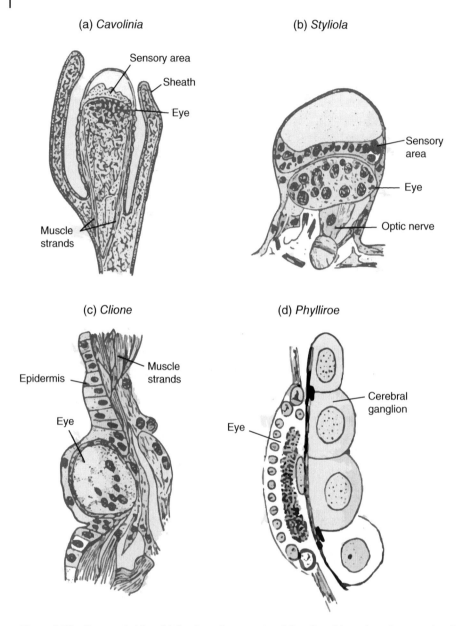

Figure 8.29 Pteropod vision. (a) Section of a tentacle of *Cavolina*; (b) section of a tentacle of *Styliola*; (c) rudimentary eye of *Clione*; (d) flattened rudimentary eye of *Phylliroe*. *Sources:* (a–c) From Meisenheimer (1905); (d) From Born (1910).

but lacking any lens or pigment (Figure 8.29a–c). The flat eyes of the pelagic nudibranch *Phylliroe* (Figure 8.29d) are even more primitive (Hyman 1967). In both cases, the eyes have no image-forming capability but would be sensitive to ambient light and shadow. Osphradia are present in both the naked and shelled pteropods.

The Pelagic Gastropods: Anatomy and Habits

The Janthinid Snails

The family Epitoniidae comprises two pelagic genera: *Janthina* with five species and *Recluzia* with two. What little is known about the habits of *Recluzia* suggests that they have a similar lifestyle to the janthinids. Janthinid (known in the older literature as Ianthinid) snails are normally found in the large wind-driven rafts of the cnidarian "by-the-wind-sailors" *Velella* and *Porpita*, often in *Physalia* as well. Janthinid's external appearance is that of a typical snail with a globular shell, a well-developed foot, and cephalic tentacles (Figure 8.30a). Their shell is the violet-blue color typical of many pelagic species, including *Velella* and *Physalia*. When upside down, the dark underside of the janthinid foot faces upward and the lighter colored shell faces down, providing a measure of countershading.

Janthinids are considered to be truly pelagic. They inhabit the pleuston, the top few centimeters of the ocean, by remaining attached to floats that they can generate themselves, similar to the pelagic barnacle *Dosima*. They are endemic to the subtropical and tropical ocean but during wind events are sometimes blown into temperate coastal waters, where they wash ashore. Shell size of janthinids ranges from 9 mm in *J. umbilicata* to 39.5 mm in *J. janthina* (Lalli and Gilmer 1989).

Janthina is incapable of floating on its own. Its normal attitude is to be attached upside down to its own bubble raft by the posterior portion of its foot (epipodium), allowing the anterior third of the foot (propodium) to range freely (Figure 8.30a). The bubble raft is created by the propodium, which traps a bubble of air by reaching though the sea surface film much like "a hand about to grasp" (Fretter and Graham 1962) and snares a bubble of air within the spoon-like shape created by the foot muscles folding in the edges of the foot. The bubble is conveyed below the surface and coated with mucous, which hardens quickly into an elastic skin. The bubble is added to the growing raft float by being pressed against it. The whole process takes 10–20 seconds (Bayer 1963). When complete, the raft is "firm, elastic and dry to the touch" (Lalli and Gilmer 1989). If a snail has abandoned its float to feed on the polyps at the underside of its main prey item, *Velella*, it will construct a new bubble raft from the edge of its prey, which conveniently provides flotation as well as a meal.

Food and Feeding

Janthinids feed on pleustonic Cnidaria, including *Velella, Porpita,* and *Physalia*. In addition, they are known to feed on other janthinids (Graham 1965), leading one to conclude that they will feed on "anything within their reach." When feeding on *Velella*, they graze the underside clean of tentacles, leaving only the sail and horny skeleton behind (Fretter and Graham 1962). They have a protrusible radula with hook-like teeth that may be used to seize tissue and pull it into the buccal cavity (Figure 8.30b, c). Note that the radular membrane is split into two wings (ptenoglossan). The genus *Janthina* is apparently unaffected by the venom of nematocysts, but *Recluzia* may not be immune (Lalli and Gilmer 1989).

A purple secretion, produced by a hypobranchial gland present in the mantle cavity, has been observed in association with feeding in the janthinids. Anecdotal observations (Hardy 1956) suggest that the secretion may anesthetize the cnidarian prey, which do act lifeless when the secretion is present, but the observations have not been confirmed by others (Lalli and Gilmer 1989).

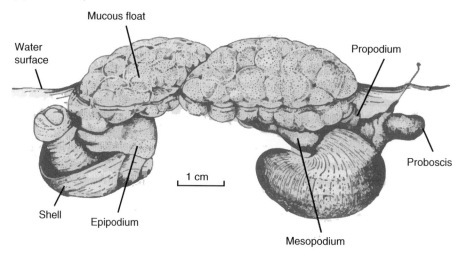

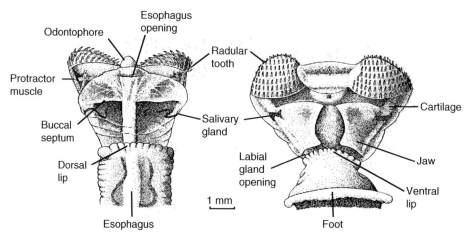

Figure 8.30 Janthinid snails. (a) Two janthinid snails with bubble floats; (b and c) the protrusible radula of a janthinid snail, dorsal and ventral views, respectively. *Sources:* (a) Fretter and Graham (1962), figure 299 (p. 561); (b and c) Graham (1965), figure 5 (p. 330).

Reproduction and Development

Janthinids are variously described as sequential (protandrous) hermaphrodites, where young snails are initially males that change sex to females at a larger size, and simultaneous hermaphrodites, that are male and female at the same time. Large adult males have also been reported (Laursen 1953). Most authors describe them as protandrous hermaphrodites, which is likely the most common mating system. However, the mating systems of many hermaphroditic species can be altered by environment or population structure (Stearns 1992) and that of the janthinids may be as well. It has been suggested that sex may change more than once in the janthinids or that some of the population may remain as males throughout their lives (Fretter and Graham 1962).

Reproduction in the janthinids is unusual in a few additional ways. Unlike most of the caenogastropods, there are no copulatory organs: no penis when in the male phase or bursa copulatrix and receptaculum when female (Fretter and Graham 1962). Thus, there is no mechanism for directly inseminating the female. Yet, all fertilization of eggs takes place in the ovary. To solve the problem, males produce two types of sperm: conventional (eupyrene) sperm with a full haploid complement of chromosomes that are fully capable of fertilizing eggs and "oligopyrene" sperm that are very much larger than the tiny eupyrene sperm. Oligopyrene sperm retain typical sperm morphology, with a flattened head and slender tail, but do not have a complete haploid set of chromosomes and thus cannot fertilize eggs. Instead, a very large number of the tiny eupyrene sperm aggregate on the tail of the larger oligopyrene sperm giving it the appearance of having a "hairy" tail. Together, the two types of sperm have enough mobility, or at least enough buoyancy, to reach the female genital duct and fertilize the eggs. The head of the oligopyrene sperm is capable of undulatory movement. The specialized combination of oligopyrene and eupyrene sperm is called a spermatozeugma (Figure 8.31a). Apparently, the close relatives of the janthinids, the epitoniids, or wentletrap snails also lack copulatory organs and use the same reproductive strategy.

Lacking the organs for physical copulation, mating snails must nonetheless be in very close proximity for gamete exchange to be successful, because the swimming prowess of the spermatozeugma is marginal. Release of the spermatozeugma by the male must be close enough for them to be drawn into the female's mantle cavity and thence into her oviduct via her respiratory current (Robertson 1983; Lalli and Gilmer 1989).

In all but one species of *Janthina*, the female packages the eggs into capsules (Figure 8.31b) within the oviduct once they are fertilized and they are then cemented to the underside of the float. Egg capsules vary in size from 2 to 8 mm and in number of eggs from 80 to 7300 per capsule (Lalli and Gilmer 1989). Larvae develop within the capsules and hatch as veligers that are capable of swimming and feeding. *Janthina janthina* does not form egg capsules but rather broods its young within its ovary and genital duct until they are released as veligers. It is the only janthinid species known to do so (Graham 1954).

The only gastropods to have a free-swimming trochophore larva (Figure 8.32a) are those whose eggs are shed directly into the sea like the eggs of the limpets (Patella: Patellogastropoda) or abalones (Haliotis: Vetigastropoda). The molluscan trochophore larva greatly resembles that of the polychaetes. All the pelagic gastropods, when they have a free-swimming larval form, release theirs as veligers (Figure 8.32b–e), which are further along in development than a trochophore.

The Heteropods

The heteropods (a legacy name for the superfamily Pterotracheoidea) have three families, differing considerably in size and external anatomy but sharing a transparent body and carnivorous lifestyle. The first of the three families, the atlantidae, comprises three genera and 24 species at latest count (WoRMS), making it the most speciose of the three. Twenty-one of the species are in the genus *Atlanta*, with two species in the genus *Proatlanta* and one *Oxygyrus*. The atlantids are considered to be the most

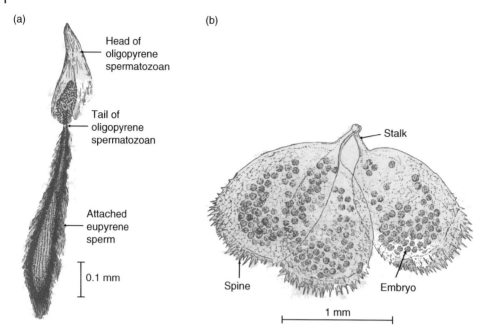

Figure 8.31 Janthinid reproductive structures. (a) Oligopyrene spermatozoan with attached eupyrene spermatozoa (spermatozeugma) of *Janthina pallida*; (b) egg capsules of *Janthina umbilicata*. *Sources:* (a) Lalli and Gilmer (1989), figure 4 (p. 17); (b) Lalli and Gilmer (1989), figure 5 (p. 19).

primitive of the heteropods, retaining the coiled shell typical of the caenogastropods but with a much reduced level of calcification. They are quite small, typically with shell diameters of 10 mm or less. The atlantid shell is flattened in profile, or "planospiral" (Thiriot-Quievreux 1973) and the outer whorl is topped by a keel (Figure 8.33), or carena, that acts as a stabilizer during swimming (Lalli and Gilmer 1989). The body of atlantids can be withdrawn completely into the shell and sealed in by their chitinous operculum. As is characteristic of the heteropods, atlantids have well-developed image-forming eyes, both movable and pigmented. The red of myoglobin in the muscles of the radula and foot sucker is easily visible through the translucent shell typical of *Oxygyrus* (Figure 8.33).

The carinariid heteropods include three genera: *Cardiapoda* with two species, *Carinaria* with six species, and *Pterosoma* with one. Shell size in the carinariids is greatly reduced relative to the body size. The shell is external in *Carinaria* and *Pterosoma*, covering only the visceral mass and allowing the pectinate gill to protrude beneath the shell (Figure 8.34). In *Cardiapoda*, the shell is greatly reduced and often buried in tissue, so that only a small fraction of the viscera is protected. Sizes vary from 2 cm in *Cardiapoda richardi* to at least 46 cm in *Carinaria cristata* (Thiriot-Quievreux 1973; Lalli and Gilmer 1989), the largest of the pelagic gastropods. The head includes a well-developed muscular proboscis. Note the movable image-forming eyes and, again, the red coloration caused by myoglobin in the radular muscle and sucker portion of the fin. The fin, or modified foot, is located mid-ventrally, but since heteropods swim "upside down" their normal attitude is facing upward despite the fact that they are often portrayed with the viscera

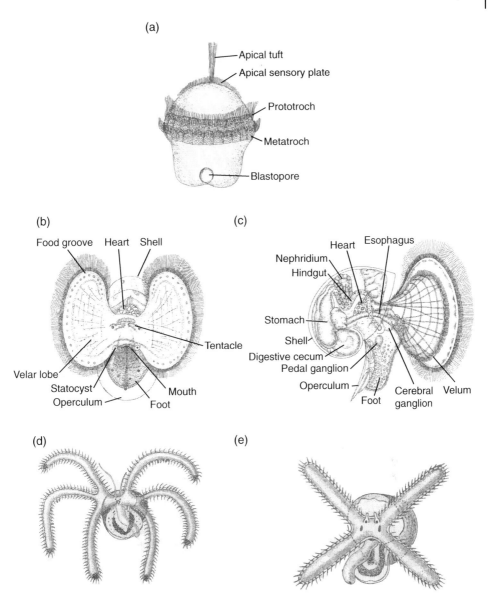

Figure 8.32 Gastropod larvae. (a) Trochophore larva of the limpet *Patella* sp.; (b and c) velliger larvae of the marine snail *Crepidula*, frontal and lateral views, respectively; (d) velliger larva of the heteropod *Carinaria lamarcki*; (e) velliger larva of the heteropod *Firoloida desmaresti*.
Sources: (a) From Patten (1886); (b and c) Werner (1955); (d and e) Thiriot-Quievreux (1973), figure 8 (p. 255).

pointing upward and the fin facing down. The sucker portion of the fin, used in feeding, is considered by some to be a remnant of the typical snail's creeping sole (Hyman 1967). The body ends in a robust tail, which is considered by some to be the metapodium, or posterior portion, of the foot. Pigmentation in the form of

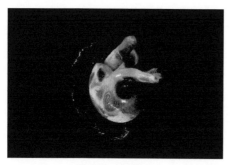

Oxygirus inflatus

Figure 8.33 The atlantid heteropod *Oxygyrus inflatus* with keeled shell. *Source:* © Dante Fenolio, reproduced with permission. See color plate section for color representation of this figure.

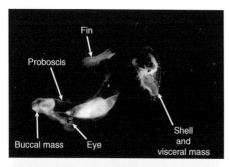

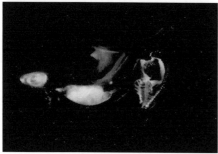

Carinaria lamarcki

Figure 8.34 The carinariid heteropod *Carinaria lamarcki*. *Source:* © Dante Fenolio, reproduced with permission. See color plate section for color representation of this figure.

chromatophores is present in many carinariids, particularly *Pterosoma*. Like the chromatophores of cephalopods, the pigment spots of carinariids can rapidly change in size (Lalli and Gilmer 1989); it is likely they are under neural control.

Pterotracheids are considered the most advanced of the heteropods. There are five species in two genera. They have no shell, so their viscera and retractable gills are permanently exposed (Figure 8.35). The proboscis is elongate, and the mouth is reduced in size in comparison to *Carinaria*. Species in the genus *Pterotrachea* may have a tail filament, as pictured in Figure 8.35a. Maximum sizes range from about 4 cm in *Firoloida desmaresti* to at least 33 cm in *Pterotrachea coronata* (Lalli and Gilmer 1989), with some reports up to 53 cm (Kaestner 1967). The complex tubular eyes with their spherical lens and ribbon-like retina are visible in the ventral view of *Pterotrachea scutata* shown in Figure 8.36.

Locomotion

The two body types within the heteropods, the spiral-shelled atlantids with a complete foot and operculum and the detorted elongate carinariids and pterotracheids, nevertheless swim using the same mechanism: a sculling motion of the median fin, or foot (Figure 8.37). As noted above, the shell, when present, and the visceral mass are oriented downward when swimming, so that the fin faces up, close to vertical even in the atlantids. In the atlantids, the crest or keel on the planospiral shell acts as a "cut-water" (Morton 1964), stabilizing the snail during sculling motions of the fin. Swimming atlantids have their bodies fully extended from the shell and stretched nearly horizontal, so that they resemble swimming carinariids and pterotracheids. Laboratory observations of swimming (Land 1982) revealed a "hop-sink" behavior in atlantids: short bouts (one to two seconds) of upward swimming alternated with 10-second periods of rest and sinking. During the sinking phase, the atlantids' movable eyes scan

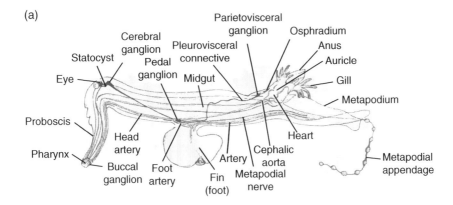

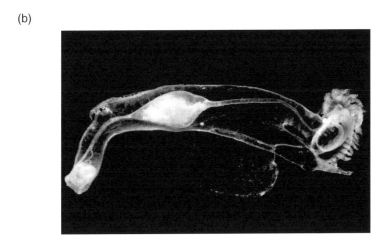

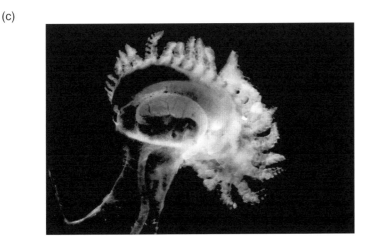

Figure 8.35 The pterotracheid heteropod *Pterotrachea coronata*. (a) Internal anatomy of *P. coronata*; (b) *P. coronata* in normal swimming orientation; (c) viscera and gills of *P. coronata*. *Sources:* (a) Lang (1900), figure 10 (p. 8); (b and c) © Dante Fenolio, reproduced with permission. See color plate section for color representation of this figure.

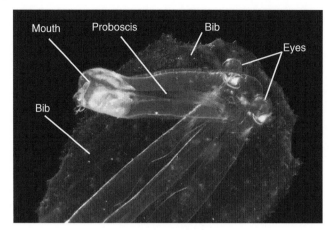

Pterotrachea scutata

Figure 8.36 Ventral view of the anterior end of the pterotracheid heteropod *Pterotrachea scutata* showing the tubular eyes, elongated proboscis, bib, and relatively small mouth. *Source:* © Dante Fenolio, reproduced with permission. See color plate section for color representation of this figure.

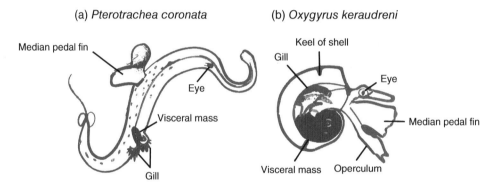

Figure 8.37 Heteropod swimming behavior. (a) The pterotracheid heteropod *Pterotrachea coronata* in normal swimming position with the visceral mass oriented downward and fin upward; (b) the Atlantic heteropod *Oxygyrus keraudreni* in swimming position with the shell held vertically and swimming fin extended. *Sources:* (a) Morton (1964), figure 18, p. 407; (b) Yonge (1942), figure 2 (p. 199).

downward and then back to horizontal, presumably searching for prey or predators. Keep in mind that their tubular eyes have a narrow visual field and that their retina is ribbon-like. Regular scanning allows for effective monitoring over a wider environmental swath.

Even with their thin, predominantly conchiolin shells, atlantids are negatively buoyant, but they are also quite small, so that their orientation in the water column greatly influences their sinking rate. Lalli and Gilmer (1989) report that when disturbed and fully withdrawn into their shells, atlantid snails naturally assume a horizontal attitude, retarding their sinking rate to $1.4 \, cm \, s^{-1}$. At night, divers have seen atlantids suspended from "mucous strings that emanate from the foot" with no detectable sinking. When disturbed, the animals release the strings and sink at a rate of up to 10 cm s^{-1}, with the keel down and fin up for stability (Lalli and Gilmer 1989). Atlantids are capable of routinely swimming at 2–3 cm s^{-1} and can burst to speeds of at least three times that figure (Lalli and Gilmer 1989). In short, they are quite capable swimmers.

The cylindrical body shape of carinariids and pterotracheids gives them superior hydrodynamics when compared to the shell-bearing atlantids, though the basics of their movement are the same. An anteroposterior wave is propagated down the muscular fin (Morton 1964) (Figure 8.37a) resulting in a "corkscrew-like undulation" that acts as a sculling oar, propelling the heteropod forward. A reversal of the wave, i.e. posteroanterior, reverses the direction of movement. The entire body can also be undulated when changing direction or when a burst of rapid swimming is needed, such as when fleeing a threat. Speeds of up to 50 cm s^{-1} have been observed in the carinariid *Cardiapoda placenta* (Lalli and Gilmer 1989).

The pterotracheids, and probably the carinariids as well, are at near-neutral buoyancy (Denton 1974), which they achieve using a strategy of ion replacement. In ion replacement, the heavy sulfate ions normally found in seawater are partially excluded from the body fluids of the heteropod, affording them a measure of static lift. Heavy ion exclusion and increased body lipid content are two common mechanisms for increasing individual buoyancy (Childress and Nygaard 1974; Sanders and Childress 1988; Withers 1992). Another is to have a high water content, or watery composition (Childress and Nygaard 1973). To achieve neutral or positive buoyancy, the density of an individual must be equal to, or less than, that of seawater. A commonly used standard value for seawater density is 1.026 g ml^{-1} (Alexander 1982), which will serve our purposes here. The density of muscle is about 1.060 g ml^{-1}, which is greater than seawater and thus negatively buoyant. The density of lipid is 0.930 g ml^{-1}, positively buoyant. A solution of sodium sulfate that is isosmotic with seawater is 1.040 g ml^{-1}, and that of sodium chloride is 1.018 g ml^{-1}, so excluding the sulfate ion will reduce the density of body fluids relative to seawater, yielding an increase in buoyancy (all values from Alexander 1982).

Let us consider how much low-density tissue must be added to a robust marine species to achieve neutral buoyancy. A typical surface-dwelling squid with a robust musculature has a density of about 1.070 g ml^{-1}, which means that it is decidedly negatively buoyant. The question remains as to how much lipid it must add to achieve neutral buoyancy. Using the equation (Withers 1992):

$$\frac{V_{ld}}{V_a} = \frac{(\rho_a - \rho_w)}{(\rho_w - \rho_{ld})} \quad (8.1)$$

where

V_{ld}/V_a = volume of low-density material that needs to be added to achieve neutral buoyancy
ρ_a = animal density
ρ_w = density of seawater
ρ_{ld} = density of lipid
$V_{ld}/V_a = (1.070 - 1.026)/(1.026 - 0.930)$
$= (0.044)/(0.096)$
$= 46\%$

Therefore, the squid would have to add an additional 46% body volume of lipid to become neutrally buoyant in seawater, clearly a difficult proposition. The conclusion here is that you need to add a considerable volume of low-density tissue to achieve neutral buoyancy in a robust animal, something that will be revisited later in our discussion of the cephalopods. The least-cost tactic for getting closer to neutral buoyancy is with a watery, low-density, body composition.

Food and Feeding

Heteropods are visual predators, using their well-developed eyes to locate prey and their swimming abilities for pursuit and capture. The main mechanism of food capture is with their radula, sometimes aided by the sucker on the foot to help hold the prey. Two types of feeding data are available: descriptions of feeding behavior obtained by direct observation and gut contents analysis. In both cases, enough data are available to suggest general patterns within the group. Based on laboratory observations of feeding behavior, Richter (1968) classified the atlantids chiefly as "scrapers or tearers" and the carinariids and pterotracheids primarily as "swallowers" (Thiriot-Quievreux 1973). However, gut contents on field-captured specimens suggest that either feeding method may be used as needed in all three families (Lalli and Gilmer 1989). Atlantids show a preference for pelagic gastropods in their diets, followed by crustaceans, chaetognaths, and foraminifera (Richter 1982).

In the atlantids, "prey is brought to the mouth by the movement of the foot, the *Atlanta* holds the shell of the captive animal with the sucker while the *Atlanta* inserts its buccal mass into the aperture of the shell. Due to the movement of the protracted radula, the body of the prey is torn and ingested" (Thiriot-Quievreux 1973). The transparency of the atlantid body allows the interested observer to track the progress of prey tissue as it passes through the gut. Movement from the esophagus to the crop is rapid. Food resides in the crop until it is ready for passage to the stomach, achieved via muscular contractions, and thence to the digestive gland. The entire process takes 24 hours to complete (Thiriot-Quievreux 1973).

On five separate occasions Hamner et al. (1975) observed the carinariid *Cardiapoda placenta* feeding in situ on the salp *Salpa cylindrica*. Once a salp was located, the heteropod attacked by swimming rapidly to it, capturing it with its highly protrusible radula. The entire buccal mass of carinariids is protrusible owing to a band of circular muscles surrounding the terminal mouth. The radula itself is armed with "powerful lateral and marginal teeth" well suited to prey capture (Fretter and Graham 1962). Smaller individuals were ingested whole by pulling them entirely into the gut with the radula, aided by muscular contractions of the crop. Larger salps were pulled in as far as possible and the remainder severed with the radular teeth (Hamner et al. 1975). In at least one case observed in the field, prey was visually detected by *Cardiapoda placenta* from a distance of 60 cm. Though it may first seem that visual detection from a little more than half a meter is not all that impressive for species with well-developed image-forming eyes, keep in mind that salps are nearly 100% transparent, making them nearly invisible in the water column.

Additional data on feeding in the carinariids are available from Seapy (1980), who examined feeding in the *Carinaria japonica* of the California borderland, and Rast (1990), who studied distribution and feeding of carinariids and pterotracheids in the eastern Gulf of Mexico (GOM). *Carinaria japonica* is found in abundance off the California coast during the summer months (Seapy 1974), feeding largely on doliolids, small thaliaceans that are close relatives of salps, and on a host of other organisms including copepods, chaetognaths, polychaetes, euphausiids, and siphonophores. Diets of the GOM carinariids *Carinaria lamarcki* and *Cardiapoda placenta* included heteropods, siphonophores, copepods, polychaetes, euphausiids, and pteropods (Table 8.1).

Heteropods are clearly opportunistic feeders with a diet that is composed mainly of soft-bodied zooplankton. When the diets of the four species shown in Table 8.1 are compared with the taxonomic composition of the prey available, they show distinct preferences as noted by the Ivlev (1961) electivity index. The electivity index is defined as follows:

$$E = \frac{(r_i - p_i)}{(r_i + p_i)} \quad (8.2)$$

r_1 = the percentage of a particular taxon in the diet
p_1 = the percentage of that taxon in the zooplankton community

Values range from −1 to 1, with negative values indicating avoidance and positive values indicating selection. Soft-bodied prey including polychaetes, heteropods, and chaetognaths were selected for in all four species examined and copepods were selected against in all four (Table 8.1). Despite the negative electivity for copepods, they were still a substantial fraction of the diet (16–19%) in three of the four species shown in Table 8.1. The three GOM heteropods showed active selection for pteropods. Siphonophores were present in the diets of all four species, but no electivity for them could be determined in either Pacific or GOM species due to the lack of abundance data on siphonophores. Overall, the diets of heteropods differ markedly from those of Crustacea (Chapter 7), which showed a dominance of crustacean prey.

Fewer data are available on pterotracheid diets. GOM *Pterotrachea hippocampus*, not shown in Table 8.1, had a high percentage of copepods in its diet (33%) with siphonophores (29%), euphausiids, and chaetognaths also present (Rast 1990). Lalli and Gilmer (1989) also reported field and lab observations of *P. hippocampus* feeding on copepods ensnared in mucous produced by its bib, a dermal thickening of the region between the fin and proboscis, presented as ventral folds (bib) or laterally as a coronalike disc (Figure 8.36). Divers usually encounter pterotracheids in areas with high concentrations of siphonophores, where the heteropods were observed feeding on bracts (Hamner et al. 1975; Lalli and Gilmer 1989). Field observations thus track well with the laboratory-based diet information available.

The common-sense observation that the available prey spectrum has a marked influence on the prey ingested is worth keeping in mind when reviewing information on diet. The prey field influences estimates of electivity as well. For example, in a system like the GOM where copepods overwhelmingly dominate the zooplankton field, opportunistic species that take a variety of prey will exhibit negative values of electivity for copepods despite their presence in the diet. The diet data from two different oceanic systems described here do suggest strongly that soft-bodied prey are preferred by heteropods but also that they are opportunistic feeders. The low incidence of thaliaceans in the diets of GOM heteropods relative to the high numbers observed by Seapy for *Carinaria japonica* underscores the importance of the prey field in any diet study. Thaliaceans are far less abundant in the GOM than in the California borderland (Hopkins and Lancraft 1984).

Feeding Chronology

Day–night comparisons of feeding activity are also quite scanty. Seapy (1980) reported

Table 8.1 Diet composition of *Carinaria japonica* from the California borderland, and of *Cardiopoda placenta*, *Carinaria lamarcki* and *Pterotrachea scutata* from the eastern Gulf of Mexico. Diet percentages for *Carinaria japonica* were computed using the global average for each taxonomic category in all four stations of the data set. *Sources*: Data for *C. japonica* from Seapy (1980), all others from Rast (1990).

Prey taxa	Carinaria japonica		Cardiopoda placenta		Carinaria lamarcki		Pterotrachea scutata	
	Mean % of diet	Electivity	Mean % of diet	Electivity	Mean % of diet	Electivity	Mean % of diet	Electivity
Thaliacea	48	0.35	6	−0.53	0	−1.0	3	0.33
Siphonophora	5	ND	29	ND	13	ND	25	ND
Polychaeta	3	0.41	10	0.97	3	0.90	5	0.96
Pteropods (all taxa)	0	ND	16	0.84	17	0.8	8	0.71
Heteropods (all taxa)	3	0.76	10	0.99	23	0.99	2	0.99
Copepods	16	−0.68	19	−0.52	19	−0.59	3	−0.82
Other Crustacea	8	−0.23	5	0.86	7	0.88	11	0.97
Chaetognaths	15	0.55	6	0.67	2	0.05	3	0.65
Fish	0	ND	8	0.97	2	0.90	2	−1.00

feeding activity at all times of the day and night in *Carinaria japonica* with peak gut fullness observed in the early evening, suggesting that peak feeding occurred during daylight hours. Rast (1990), lacking daytime feeding data for comparison, nonetheless observed high feeding activity in the GOM carinariids throughout the night, as well as in *Pterotracha scutata*. At this point, generalizations cannot be made on feeding periodicities, nor can opportunistic feeding throughout the diel cycle be ruled out.

Vertical Distributions and Migrations

Heteropods of all three families are found primarily in the upper 300 m of the water column, with evidence for limited diel vertical migrations in some of the larger species (Rast 1990; Seapy 1990a, b; Michel and Michel 1991). For the atlantids, the most consistent pattern of vertical distribution observed between studies was a modest shoaling within the epipelagic zone (upper 200 m) at night (Figure 8.38a). Carinariids and pterotracheids in the GOM exhibited a more pronounced movement, with daytime centers of distribution for *Carinaria lamarcki, Pterotrachea coronata* and *P. scutata* occurring between 200 and 300 m and nighttime peaks in the upper 200 m (Rast 1990, Figure 8.38b). *P. hippocampus* from the GOM showed a more modest shift, being centered between 100 and 200 m during the day and the upper 100 m at night. Atlantids consistently exhibit the shallowest distributions of the three families.

Seapy's (1990a) work used shallow discrete-depth tows in the waters off Hawaii, evaluating the vertical profiles of individual atlantid species as well as their numerical abundance in a series of depth strata from 0 to 300 m. Atlantids consistently exhibited their highest densities in the upper 140 m of the water column, an observation underscored by comparing densities found in tows limited to the upper 50 m with those in oblique tows from 0 to 300 m (shown in Table 8.2). Densities in the upper 50 m were an order of magnitude higher than those from the entire upper 300 m (Seapy 1990b). In the carinariidae and pterotracheidae, vertical ranges usually include depths below 200 m (Rast 1990; Seapy 1990b; Michel and Michel 1991).

Abundance and Geographic Distribution

Heteropods are normally present in "moderately low to low abundances" (Richter and Seapy 1999), a consistent observation in studies from different ocean systems. Using a variety of different sampling technique, estimates of heteropod density acquired from tows in the upper 300 m of the water column vary from less than 10 to over 100 individuals per $10\,000\,m^{-3}$ (Table 8.2). When information is available for all three families, atlantids are found in highest abundance, followed by pterotracheids and carinariids (Richter and Seapy 1999).

Distribution data from the GOM (Hopkins and Lancraft 1984) using the same sampling gear as Rast (1990) allow comparisons between heteropods and other invertebrate taxa within the micronekton and macrozooplankton fauna (Table 8.3). Each of the major groups of crustaceans is at least four times as abundant as the heteropods, with the dendrobranchiates more than an order of magnitude higher in numbers. Heteropods are similar in numbers to the more robust Cnidaria and the larger tunicates. Thus, as Rast (1990) concludes, they are not a dominant taxon, but neither are they a rare one.

Data are scanty on seasonal variation in heteropod abundance. Seapy (1974) found highest abundances of *Carinaria japonica* in August in the California borderland and Guadalupe basin (Table 8.2). In an earlier study encompassing the waters off Oregon

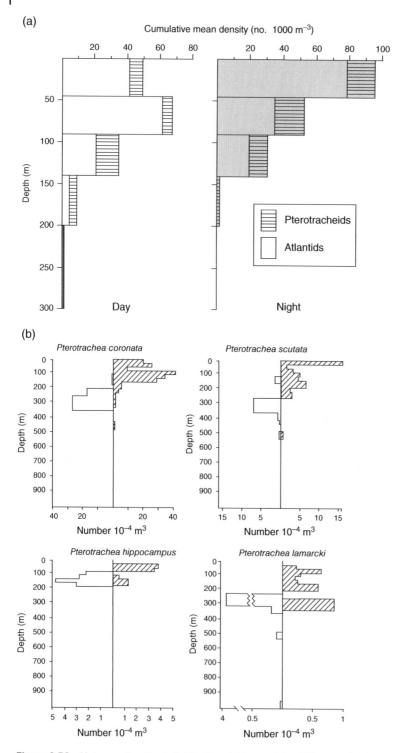

Figure 8.38 Heteropod vertical distribution. (a) Day–night differences in density as a function of depth in pterotracheid and atlantid heteropods off Hawaii; (b) day–night distributions of four species of *Pterotrachea* from the Gulf of Maine. *Sources:* (a) Seapy (1990b), figure 16 (p. 244); (b) From Rast (1990), figure 2 (p. 21).

Table 8.2 Heteropod abundances from four different oceanic regions. Species abundances were all normalized to number of individuals per 10 000 m³. Note the difference in abundance between the California and Florida systems. *Sources*: Data from Seapy (1974); Seapy (1990a); Rast (1990).

Region	Species or families	Abundance (individuals 10 000 m⁻³)					Sampling gear	Depths sampled	References
		Min	Max	Mean	Std. dev.	n			
So. California/ Guadalupe Island	*Carinaria japonica*, summer	2	821	129	282	55	10 m² IKMT (12 mm mesh)	0–100 m	Seapy (1974)
	C. japonica, (fall, winter, spring)	0	28	7	9	68			
Florida Straits	Atlantidae	0	13	2.7	2.9	33	10 m² MOCNESS (3 mm mesh)	0–400 m	Michel and Michel (1991)
	Carinariidae	0	4	0.9	0.9	33			
	Pterotracheidae	0	16	2.2	3.4	33			
E. Gulf of Mexico	Carinariidae			3		418	2.6 m² and 5.3 m² Tucker trawls (4 mm mesh)	0–250 m	Rast (1990)
	Pterotracheidae			13.4					
	Carinaria lamarcki			1.14					
	C. placenta			1.89					
	Pterotrachea coronata			2.95					
	P. scutata			0.79					
	P. hippocampus			4.52					
	Filaroides desmaresti			5.12					
Hawaii	Atlantidae (0–300 m)			73			0.4 m² Bongo and 4 m² Ring nets (0.5 mm mesh)	0–300 m	Seapy (1990a)
	Atlantidae (0–50 m)			627				0–50 m	

Table 8.3 Abundance of invertebrate micronekton in the Gulf of Mexico. *Sources:* Data from Hopkins and Lancraft (1984) and Rast (1990).

Taxon	Abundance (individuals·10 000 m^3)
Heteropods	0.59
Carideans	2.19
Dendrobranchiates	11.42
Euphausiids	3.16
Mysids and Lophogastrids	2.15
Cnidaria exclusive of siphonophores	0.44
Tunicates	0.1

and California, Dales (1953) also found peak abundances of *Carinaria* and *Pterotrachea* in the August–September months. In the East China Sea, Xu and Li (2005) found a pronounced autumn peak in the abundance of atlantid heteropods, with a mean value of 21.03 ind.·100 m^{-3}. A secondary peak was observed in summer (4.89 ind.·100 m^{-3}). A strong positive correlation was observed in the East China Sea between temperature and heteropod abundance as a function of both location and season. The values for heteropod abundance reported by Xu and Li (2005) are the highest the authors have seen in the literature.

Heteropods are found pan-globally between 40°N and 40°S latitude (Richter and Seapy 1999), and most have a very cosmopolitan distribution. Of the 39 currently recognized species of heteropods, at least 22 are found in the Atlantic, Pacific, and Indian Ocean basins, 8 in 2 of the 3, and 2 are found in 1 basin only. All five species of pterotracheids are found in all three ocean basins (Richter and Seapy 1999). A recent survey in the mid-Atlantic between 46°N and 46°S (Burridge et al. 2016) confirmed the extensive latitudinal range of heteropods in the Atlantic Basin. In a separate study, the range of *Atlanta gaudichaudii* was extended into northern sub-Antarctic waters in the Pacific (Howard et al. 2011).

Reproduction and Development

Unlike the janthinids, heteropods are believed to be entirely dioecious (Fretter and Graham 1962), though sex ratios in the field may differ from 1 : 1 (Lalli and Gilmer 1989). Males have a conspicuous penis (see e.g. Figure 8.16) in all three families. In the pterotracheids, only the male has a sucker on the fin, whereas in the atlantids and carinariids suckers are present in both sexes. The sucker faces anteriorly in the atlantids and has been observed to help in the capture and consumption of prey (Thiriot-Quievreux 1973). In the carinariids and pterotracheids, the fin sucker faces posteriorly and would be of little use in feeding. Thus, another function postulated for the fin sucker is to act as a "clasper" for holding onto a mate during copulation, which, though likely, is as yet unconfirmed. Fertilization is internal and sperm transfer to the female is accomplished using spermatophores, almost certainly via the penis though the mechanism of transfer remains undescribed. Spermatophore production has been confirmed in all three families of heteropods (Lalli and Gilmer 1989). In the atlantids, fertilized eggs may be laid singly as in *Atlanta peroni* and *A. lesueri*, or in gelatinous strings of varying length (Thiriot-Quievreux 1973). The carinariids and pterotracheids produce egg strings exclusively, in many cases far longer than the length of the female producing them (Owre 1964). Owre considered the egg strings to be analogues of the egg cases produced by benthic species. Each egg string extruded by the female was considered to be a single spawning event. Egg strings typically contain thousands of eggs (Lalli and Gilmer 1989). In Florida waters, swimming veliger larvae hatched within 48 hours in *Pterotrachea hippocampus*. A time of 72 hours was quoted for hatching in *Carinaria* embryos in the cooler waters of the north Atlantic (Fretter and Graham 1962).

The morphology of the larval shell (number of whorls) and the velum (number of lobes) differ between heteropod veligers and may be used to identify them in the plankton (Figure 8.32d, e). The velum may have two, four, or six lobes (Thiriot-Quievreux 1973; Lalli and Gilmer 1989) and is the main mechanism of veliger propulsion and food gathering. As the veliger grows, it develops the internal organ systems and foot modifications of the adult. At metamorphosis, having transformed into a miniature adult, it sheds and consumes the velum (Owre 1964; Thiriot-Quievreux 1973). The duration of the planktonic veliger stage is unknown.

The seasonality of reproduction is not well studied in the heteropods. The two studies (Dales 1953; Seapy 1974) off the west coast of the United States that examined seasonal heteropod abundance both found peak abundances in the months of August and September. Dales (1953), who also recorded size classes, noted a main spawning period in May with a possible secondary event in late September. Small sizes (<10 mm) were found throughout the year, peaking in June and July.

The Pteropods

The order Pteropoda (wing-feet), has gone back and forth as a legacy name and a taxonomic order and is now back to being an order, including the three suborders Euthecosomata, Pseudothecosomata, and Gymnosomata, the shelled and naked pteropods. The euthecosomes and pseudothecosomes are differentiated by the presence of either a calcareous shell or a pseudoconch. The order Pteropoda is found within the subclass Heterobranchia and infraclass Euthyneura (Brusca et al. 2016; WoRMS). Far more information is available on the pteropods than on the heteropods because of their greater abundance and their role in the oceanic carbonate cycle.

Mariners have been aware of the pteropods for centuries. One of the first written records of pteropods dates back to a 1675 travelog describing voyages to Spitsbergen and Greenland (Martens 1675). They were recorded and illustrated in the 1700s by Forsskal, a student of Linnaeus, on a Danish voyage to explore Arabia. Forsskal died of malaria on the way home, but his work was published in a monograph by his colleague Niebuhr (1772). In 1817, Cuvier assigned the pteropods as a class within the Mollusca in his Regne Animal. De Blainville (1824) later placed them among the Gastropoda, rejecting Cuvier's classification. At the end of the nineteenth century, Pelseneer's (1887) Challenger reports on the pteropods firmly established them within the opisthobranch molluscs. Their common name, sea butterflies, was initially given them by French fishermen in the 1700s, who observed them from their vessels in coastal waters (Lalli and Gilmer 1989). The French have been important contributors to the study of both heteropods and pteropods from the earliest literature until the present day.

The euthecosomes and pseudothecosomes, together known as the shelled pteropods or thecosomes, and the gymnosomes, or naked pteropods differ considerably in morphology and foraging strategies. The thecosomes feed on small particulates and the gymnosomes feed mainly on thecosomes. Current thinking puts thecosome species numbers at 92 and gymnosomes at 54 (WoRMS), more than the heteropods but not in a league with crustacean taxa. Classification to the family level is presented below with species numbers for each family given in parentheses.

Order Pteropoda
SUBORDER EUTHECOSOMATA
    Superfamily Cavolinoidea (61)
        Family Cavoliniidae (44)
        Family Cliidae (11)
        Family Creseidae (5)
        Family Hyalocylidae (1)
    Superfamily Limacinoidea (8)
        Family Heliconoididae (1)
        Family Limacinidae (6)
        Family Thieleidae (1)
SUBORDER PSEUDOTHECOSOMATA
    Superfamily Cymbuloidea (23)
        Family Cymbulidae (11)
        Family Desmopteridae (4)
        Family Peraclidae (8)
SUBORDER GYMNOSOMATA
    Superfamily Clionoidea (52)
        Family Clionidae (16)
        Family Cliopsidae (3)
        Family Notobranchaeidae (8)
        Family Pneumodermatidae (25)
    Superfamily Hydromyloidea (2)
        Family Hydromylidae (1)
        Family Laginiopsidae (1)

Euthecosomata
Anatomy

As pelagic species living in a three-dimensional environment, pteropods are not constrained by the same demands as their benthic relatives. The need for a robust musculature to support a thick protective shell has been replaced by the needs for near-neutral buoyancy and finding food in the open sea. To serve those needs, the euthecosome mantle has become a very elaborate system of structures that aid the animal in maintaining its vertical position but can only be observed in its entirety by divers (Figure 8.39a). Once captured and confined, particularly if captured in a net, the mantle (and usually the entire animal) will be fully or partially contracted and hard to resolve in its entirety (Figure 8.39b) though basic, more robust, structures such as the wings remain obvious. Visual observations in situ have contributed much to our present understanding of thecosome biology.

The euthecosomes all have thin aragonitic shells that vary in thickness from 6 to 100 µm (Lalli and Gilmer 1989). Their fragility is best appreciated in photographs of recently captured specimens in which their appearance is reminiscent of finely blown glass (Figure 8.40a). Note the hydroids on the shell. The most primitive superfamily, the Limacinoidea, have spirally coiled shells (Figure 8.40b). The Cavolinioids have lost the spiral coiling; they have detorted (Figure 8.40c). The euthecosomes can all withdraw into their shells for protection, but only the limacinids have an operculum to cover the opening. The two wings of the euthecosomes are always separate, lying dorsally and laterally to the footlobes (Figure 8.40d).

The Pseudothecosomata differ from the euthecosomes in having "the wings fused into a single plate, by having two cephalic tentacles that are symmetrical in size, and by a proboscis formed by the elongation and modification of part of the footlobes" (Lalli and Gilmer 1989). Advanced pseudothecosomes have a gelatinous pseudoconch that protects the viscera in a similar manner to the shell of euthecosomes.

(a) *Cavolinia uncimata*, ventral view

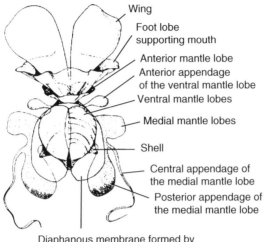

- Wing
- Foot lobe supporting mouth
- Anterior mantle lobe
- Anterior appendage of the ventral mantle lobe
- Ventral mantle lobes
- Medial mantle lobes
- Shell
- Central appendage of the medial mantle lobe
- Posterior appendage of the medial mantle lobe
- Diaphanous membrane formed by the ventral mantle lobes

(b) *Cavolinia gibbosa*

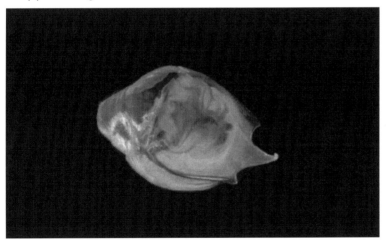

Figure 8.39 Euthecosome pteropods. (a) Anatomy of the euthecosome pteropod *Cavolinia uncimata*; (b) the euthecosome pteropod *Cavolinia gibbosa*. *Sources:* (a) Gilmer and Harbison (1986), figure 1 (p. 49); (b) © Dante Fenolio, reproduced with permission. See color plate section for color representation of this figure.

The Peraclidae is a monogeneric (genus *Peracle*) family that is considered to be the most primitive of the three pseudothecosome families. They resemble *Limacina* in having an external sinistrally coiled calcareous shell but differ in the presence of a single wingplate (Figure 8.41).

The Cymbuliidae (genera: *Cymbulia, Corolla, Gleba*) are considered to be the most advanced of the pseudothecosomes. Instead of a shell, they have an internal, translucent, gelatinous pseudoconch secreted under the integument that provides some protection and support for the

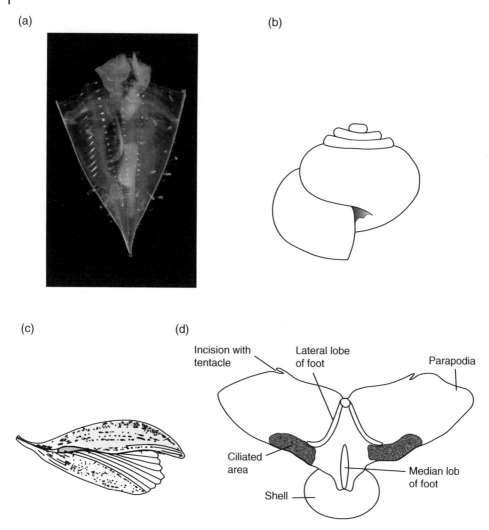

Figure 8.40 Euthecosome pteropods. (a) *Clio* sp. with attached hydroids; (b) shell of *Limacina*; (c) shell of *Cavolina*, lateral view; (d) external anatomy of *Limacina*, ventral view. *Sources:* (a) © Dante Fenolio, reproduced with permission; (b) Hyman (1967), figure 182 (p. 405); (c and d) Meisenheimer (1905). See color plate section for color representation of this figure.

visceral mass. In all three genera, the pseudoconch has numerous small denticles or tubercles (Figure 8.42a) and a species-specific shape that may be used for identification. The proboscis is well developed in the cymbuliids, reaching its greatest length and range of movement in *Gleba* (Figure 8.42b–d). Neither *Corolla* nor *Gleba* have a radula (Hyman 1967).

The third family in the Cymbuloidea, the monogeneric Desmopteridae (genus *Desmopterus*), is very different in morphology from the other two. It lacks a shell and mantle cavity altogether. A small cylindrical body adjoins a head with a terminal mouth equipped with jaws and a radula. Two ciliated wing tentacles protrude from incisions on the posterior margins of each

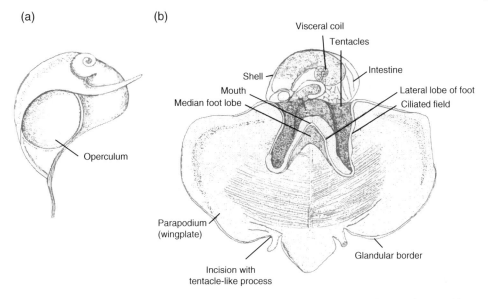

Figure 8.41 The Peraclidae. (a) The shell of *Peracle*; (b) external anatomy of *Peracle*. *Sources:* (a) Meisenheimer (1905); (b) Hyman (1967), figure 212 (p. 512).

side of the wing plate (Figure 8.43). Desmopterids are quite small (wingspan ~6mm). They are reported to reach very high abundances in coastal waters of the Indian Ocean (Lalli and Gilmer 1989). Little is known about their biology.

Like the cephalopods, the cymbuliids possess chromatophores on the wings and body that are neurally controlled, enabling them to change appearance rapidly (Lalli and Gilmer 1989).

Locomotion

The muscular parapodia, or "wings," of pteropods make them very capable swimmers. If you have been fortunate enough to observe their swimming behavior in nature, you would likely agree with the French fishermen that gave them the common name "sea butterfly," as their swimming behavior does resemble a butterfly in flight. The basics of parapodial swimming in the euthecosomes and pseudothecosomes have been described by Morton (1954a, 1964).

Limacinids, the smallest and most primitive of the euthecosomes, have retained a typical gastropod torsion along with their spiral shell. Though their shell is thin, and its mass greatly reduced relative to that of a benthic snail, it still is enough of a dead weight to make limacinids negatively buoyant. When swimming, the aperture of the shell faces upward, and the parapodia ("wings") are long and narrow at the base (Figure 8.44). A downward power stroke with the wings outstretched propels them upward. The wings are "feathered" for recovery and another downward stroke propels them further upward. Upward movement usually describes a spiral course. When the wings are held parallel, the weight of the shell causes a rapid descent. Observations by divers (Gilmer and Harbison 1986) suggest that sinking is greatly retarded when the wings are held outstretched perpendicular to the shell and that upward swimming bouts alternate with periods of slow sinking and neutral

496 | *The Mollusca*

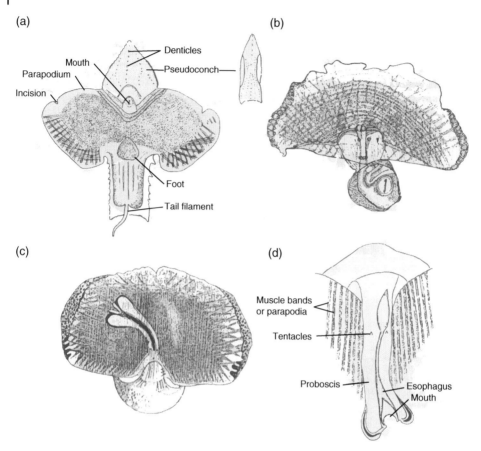

Figure 8.42 The Cymbuliidae. (a) Ventral view of *Cymbulia* with a lateral view of the pseudoconch; (b) *Corolla cupula*; (c) *Gleba cordata*; (d) anterior end of *Gleba* sp. showing buccal structures. *Sources:* Tesch (1904); (b and c) Van Der Spoel and Dadon (1999), figure 6.58 (p. 700), figure 6.62 (p. 701); (d) Hyman (1967), figure 172 (p. 428).

buoyancy. Neutral buoyancy is achieved by generation of the mucous web used in feeding. Keep in mind that *Limacina* is only a few mm in size at most; its mass is quite small and its wingspread substantial.

Recent research on locomotion in *Limacina*, using sophisticated analysis of wing movement and of the flow fields generated during swimming, has demonstrated a remarkable similarity between the underwater "flight" of Limacina and that of small, winged insects (Murphy et al. 2016). Interestingly, the downward transverse movement of the shell during each power stroke as the shell slides beneath the flapping wings, termed "body pitching," is a major factor determining the efficiency of the wings during underwater flight. Though difficult to tease out from visual observations, the wing tips describe a "figure-of-eight" pattern similar to that of the gymnosomes when movement of the shell is accounted for, optimizing generation of lift.

The cavolinioids have a variety of shell shapes (Figure 8.45). The elongated cone of the creseids is considered to be the most primitive and the more compressed shells

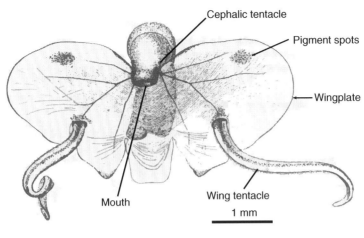

Desmopterus papilio

Figure 8.43 The desmopterid pteropod *Desmopterus papilio*. Source: From Chun (1889).

of Clio and Cavolinia with their laterally elongated apertures, to be the more advanced. The creseids are well suited to up and down movement, which is beneficial for their feeding strategy but not for horizontal movement. The detorting of the cavolinoids resulted in the mantle cavity moving from the dorsal to the ventral side of the body below the visceral mass, resulting in a center of gravity at the structural center of the animal when oriented horizontally, with the wings above (Morton 1964). This change allowed for better maneuverability, postural stability, and horizontal movement in addition to retaining upward swimming ability.

The pseudothecosomes, at a size range of 25–50 mm, tend to be larger than the euthecosomes, but their locomotory basics are similar. The animals are enclosed within the pseudoconch, which is open above "like a slipper or narrow boat" (Morton 1964) with the wingplate spread above it (Figure 8.46). In *Cymbulia*, the pseudoconch thus gives the appearance of a boat's prow as the animal swims forward (Figure 8.46b, c). As in the euthecosomes, the parapodia of pseudothecosomes act as paddles, able to row the animals upward in addition to maintaining stability. Forward thrust for horizontal movement is the result of an anteroposterior wave attending the downward wingbeat. Diver observations (Lalli and Gilmer 1989) reveal that normal swimming in pseudothecosomes conforms to the pattern shown in Figure 8.46d; the pseudoconch is below the wingplate (the "backstroke"). However, in more rapid escape-swimming, the animals flip over with the pseudoconch behind and the anterior edge of the wingplate forward. Swim speeds during escape can reach 40–45 cm s^{-1} in the pseudothecosomes *Corolla* and *Gleba*, respectively. Average swim speeds for euthecosomes range from 7 to 14 cm s^{-1} in the cavolinioids. Escape swimming in *Limacina* can reach 8–12 cm s^{-1}, quite a feat for an animal with a shell diameter of 10 mm or less (Lalli and Gilmer 1989).

Neutral buoyancy in thecosomes was reported in the older literature (e.g. Rang and Souleyet 1852) and has been observed repeatedly by divers (Gilmer 1974; Gilmer and Harbison 1986). Vertical station-keeping

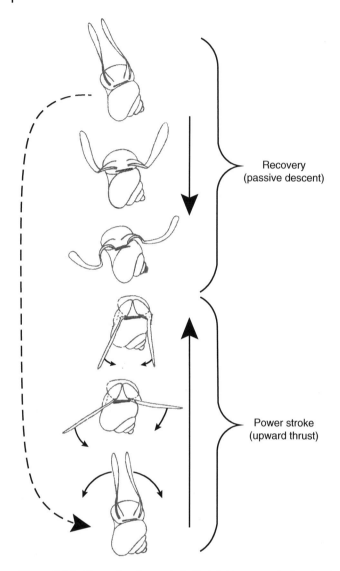

Figure 8.44 The swimming cycle in limicinid pteropods showing both power and recovery strokes with corresponding ascent and descent periods.

in the euthecosomes is achieved with a number of mechanisms. The first is the proliferation of mantle surface area (Figure 8.39), providing a physical brake to rapid sinking via friction drag. A second, observed only in *Cavolinia*, is the production of a neutrally buoyant temporary pseudoconch that resides within the folds of the mantle outside the shell (Gilmer and Harbison 1986), increasing overall buoyancy. A third is the production of the mucous web used in feeding, which acts as a buoy to retard sinking. The pseudoconch of pseudothecosomes, being neutrally buoyant, conveys a measure of buoyancy and the same mechanisms observed in the

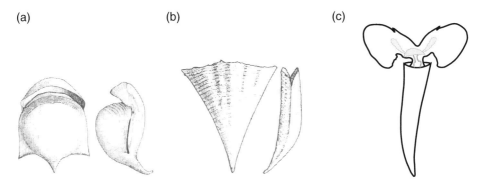

Figure 8.45 Shell shapes of cavoliniod pteropods. (a) *Cavolinia uncinata*; (b) *Clio chaptali*; (c) *Creseis* sp. with parapodia. *Sources:* (a) Van Der Spoel and Dadon (1999), figure 6.16 (p. 693); (b) Van Der Spoel and Dadon (1999), figure 6.35 (p. 696).

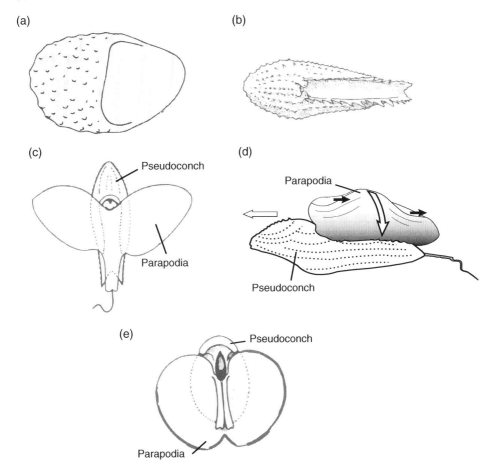

Figure 8.46 Pseudoconch shape and swimming in pseudothecosomes. (a) Pseudoconch of *Corolla ovata*; (b) pseudoconch of *Cymbulia peroni*; (c) *Cymbulia peroni* showing the relationship of the pseudoconch and parapodia; (d) lateral view of *Cymulia peroni*, the small black arrows indicate the undulatory movement of the parapodia used for horizontal movement, the smaller of the two white arrows indicates the horizontal direction of motion, the larger white arrow indicates the wingbeat of the parapodia; (e) *Gleba neapolitana* showing the relationship of the pseudoconch and parapodia. *Sources:* (a) Van Der Spoel (1976); (b) Van Der Spoel and Boltovskoy (1981); (c and d) Morton (1964), figure 16 (p. 406). Reproduced with permission of Elsevier.

euthecosomes apply to and work equally well for pseudothecosome species. Yet to be investigated is the potential role of the ion exclusion described earlier for heteropods. Gilmer (1974) and Gilmer and Harbison (1986) have measured sinking rates in quiescent euthecosomes at $0.5–1.0\,cm\,s^{-1}$ and in pseudothecosomes at less than $0.5\,cm\,s^{-1}$. In contrast and though observations are few, captured individuals in aquaria have been described as sinking rapidly (Morton 1964). Clearly, our understanding of the behavior of pelagic gelatinous species has been refined and improved considerably by diver observations.

Food and Feeding

Perhaps the best new insights into gelatinous zooplankton behavior afforded by in situ observations are in the realm of feeding by pteropods. Early detailed descriptions of pteropods feeding (Yonge 1926; Morton 1954a) were necessarily done in laboratory aquaria. Yonge, noting the activity of ciliary fields on the wings of Cavolinoids and Cymbulioids, concluded that ciliary-induced flows conducted particles to the mouth where they could be ingested (Figure 8.47a). In contrast, Morton (1954) observed a through-mantle flow of water in *Limacina retroversa* that allowed them to trap particles in mucous produced by the large pallial gland, resulting in food strings (Figure 8.47b). The food strings were then conveyed to the mouth by the exhalant ciliary currents of the mantle cavity. Morton (1954a) noted the inconsistency of the well-developed mucous-producing pallial glands found in all the thecosomes with the different feeding modalities he observed in *Limacina* and Yonge (1926) observed in the cavolinoids.

Observations by Gilmer (1972, 1974) of thecosome (cavoliniid) feeding in situ showed them ingesting mucous strands, leading him to initially conclude that Morton's (1954a) description of thecosome feeding was basically correct. Later work (Gilmer and Harbison 1986) revealed that thecosomes all deploy a delicate mucous web many times their body size to entrap a variety of protists, phytoplankton cells, and zooplankton, which are then reeled in and ingested (Figure 8.48). The web not only ensnares larger quantities of prey than a cilia-generated flow through a small mantle cavity, it also is able to trap faster-moving prey such as copepods and crustacean larvae by tangling them in its mucous strands. The main source of the web is believed to be the large pallial gland found in the mantle cavity throughout the Thecosomata as well as in the glands on the wing margins of the pseudothecosomes. The mucous webs can be deployed quickly: 5 seconds in the limacinids and 2.5 minutes in *Cavolinia* (Lalli and Gilmer 1989). Further, the webs can be ingested nearly as quickly as they are deployed, about 20 seconds in limacinids and 1–3 minutes in *Cavolinia*. The diaphanous character of the webs means that they can be manipulated during ingestion by the ciliary currents generated in the mantle cavity and foot lobes. It also means that near-neutral buoyancy is an important factor; wing movement would disrupt their effectiveness. Once the webs with prey are ingested, any hard parts (e.g. diatom frustules, crustacean cuticles) can be broken up by the muscular gizzard found in all pteropod digestive systems (Lalli and Gilmer 1989).

Gut contents of thecosomes appear to scale with size, with the limacinids taking predominantly diatoms, dinoflagellates, tintinnids, foraminifera, and radiolaria. Cavoliniid guts contain all the taxa observed in limacinids as well as coccolithophores, copepods, and occasionally even young thecosomes and atlantid heteropods. Metazoan taxa are better represented in the guts of

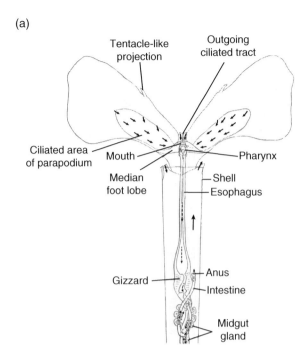

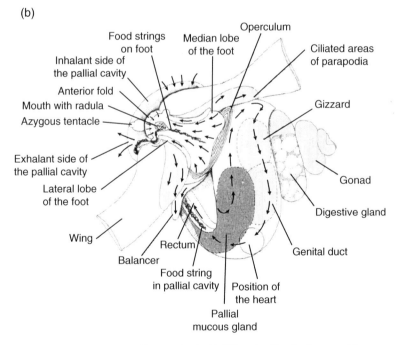

Figure 8.47 Pteropod feeding methods. (a) Ciliary feeding in the creseid pteropod *Creseis* sp., the arrows indicate the direction of flow of ciliary fields on the wings that move food particles to the mouth; (b) suspension feeding in the limacinid pteropod *Limacina retroversa*, arrows indicate the direction of flow of water through the mantle allowing food particles to be trapped in mucous strings, which are then conveyed to the mouth. *Sources:* (a) Yonge (1926), figure 2 (p. 420); (b) Morton (1954a), figure 2 (p. 300). Reproduced with permission of Cambridge University Press.

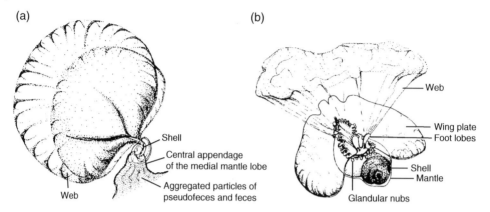

Figure 8.48 Pteropod feeding webs. (a) Feeding web of *Cavolinia uncinata*; (b) feeding web of *Peraclis reticulata*. *Sources:* Gilmer and Harbison (1986), figure 4 (p. 52); (b) Gilmer and Harbison (1986), figure 7 (p. 54).

pseudothecosomes. Gastropod and crustacean larvae are taken, as are copepods, young pteropods and heteropods, in addition to the diatoms, foraminifera, and radiolaria common in the cavoliniids and limacinids (Lalli and Gilmer 1989).

Reproduction and Development

Thecosomes are protandrous hermaphrodites, initially maturing as males and changing sex to females in later life. An intermediate hermaphroditic stage exists during the transition (Hsiao 1939b). Like the heteropods, most species release egg masses that produce veliger-stage larvae upon hatching. The most complete information is available for *Limacina* (Hsiao 1939a, b; Lalli and Wells 1978; Lalli and Gilmer 1989), which we'll use as a "model system" for the thecosomes as a whole.

The reproductive system of *Limacina* is depicted in Figure 8.49 (Lalli and Wells 1978). It is a combination figure that displays together the anatomy of both mature males and females for instructional purposes; the two do not coexist in the same animal in a mature state in nature. In functional males, sperm is generated in the gonad (ovotestis). When mature, the sperm is stored in the bulbous area of the hermaphrodite duct, which acts as a seminal vesicle. The duct leads to two female accessory glands (which remain undeveloped until sex change occurs) and then to the common genital pore which empties into the mantle cavity. Eggs and sperm are both released through the common genital pore. An external ciliated sperm- groove connects the genital pore with the male pore.

In most species, a large penis is located in the head region (Figure 8.49). It resides in a penial sheath and opens externally through the male pore on the right side of the head (Lalli and Wells 1978). The base of the penis is connected to a prostate gland; two tentacle-like extensions, termed the accessory copulatory organ, maintain contact during copulation. Visual observations and capture of specimens *in copulo* indicate that copulation occurs between males or immature females (hermaphrodites). When sperm is exchanged via spermatophores, e.g. in *Limacina helicina* (Lalli and Gilmer 1989), each of the copulating individuals deposit a spermatophore on the underside of the mate's right wing. Spermatophores may then be stored until

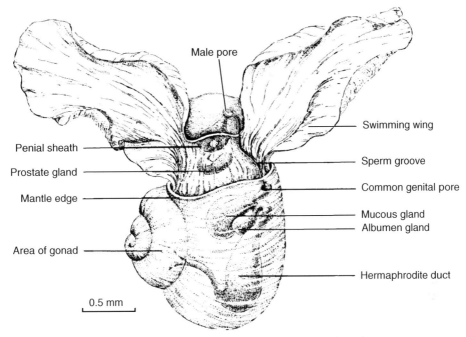

Composite reproductive anatomy of the genus *Limacina*

Figure 8.49 Pteropod reproduction. Composite reproductive anatomy of the hermaphroditic pteropod *Limacina* sp.; both male and female structures are depicted in the same individual, a condition which does not occur in nature. *Source:* From Lalli and Wells (1978), figure 1 (p. 99). Reproduced with the permission of John Wiley & Sons.

sex change is complete. Spermatophores have only been reported in a few thecosome species; it is unclear how widespread they are as a reproductive mechanism within the group. They do allow for a quick transfer of large numbers of sperm, which would presumably be an advantage in the pelagial. Spermatophores have been observed in *Limacina helicina* and *L. inflata* (Lalli and Gilmer 1989).

In mature females, the penis and prostate glands have been resorbed and the gonad is filled with oocytes. Mature oocytes move into the hermaphrodite duct and it is there that fertilization takes place. Whether sperm is stored in a seminal vesicle within the hermaphrodite duct after copulation or remains in a spermatophore in the vicinity of the mantle cavity and is later released into the hermaphrodite duct is likely a species-specific trait. What is clear in the great majority of thecosome species is that fertilized eggs pass through the albumen gland where the egg capsule is formed and then move into the mucous gland where they are embedded into the egg mass before being released from the common genital pore. Egg masses contain a large number of eggs, varying between 150 and 10 000 per spawn depending on the species, their location, and, most importantly, their size (Lalli and Wells 1978; Lalli and Gilmer 1989).

A few species, usually those inhabiting mesopelagic depths, are ovoviviparous or viviparous and brood a smaller number of young within their mantle cavity (Lalli and Wells 1973, 1978). Species exhibiting brooding behavior include *Limacina helicoides*,

Limacina inflata, and two species of *Clio*, *C. chaptalii* and *C. recurva*. *L. inflata* is an epipelagic species that broods its young; because of its accessibility to researchers, it's brooding is the best described. Larvae of various stages are present, ranging from early developmental stages (gastrulae) found in the reproductive tract near the common genital pore to shelled veligers in the anterior mantle cavity (Figure 8.50a). The average number of brooding larvae seen within the mantle cavity is 45 (Lalli and Wells 1973). *L. inflatus* is considered an ovoviviparous species because its young are released as shelled veligers (Figure 8.50b). The two species of *Clio* and *L. helicoides* release their young as miniature adults and are thus designated as viviparous (Tesch 1946; van der Spoel 1970; Lalli and Wells 1973).

The basics of pteropod reproduction described here for Limacinids, including reproductive anatomy, copulatory behavior, and a protandrous reproductive strategy, are mirrored in the cavolinoids and the pseudothecosomes. Both the direct exchange of sperm (*Creseis, Cavolinia*) and the exchange of spermatophores (*Diacria*) have been observed (Lalli and Gilmer 1989). Asexual reproduction in *Clio* has been reported (van der Spoel 1962, 1973a; Pafort-van Iersel and van der Spoel 1986), but that view is not yet widely accepted.

Seasonality of reproduction varies with species and location. In the tropical waters of the West Indies, *Limacina inflata* breeds continuously (Lalli and Wells 1973). *Limacina retroversa*, a boreal-temperate species, appears to reproduce in the spring and summer months in the Gulf of Maine (Hsiao 1939b) and year round in English waters (Morton 1954a). *Limacina bulimoides* found in the Benguela Current off South Africa has one peak period of reproduction in the fall (Morton 1954b). In the arctic waters of Svalbard's Kongsfjorden, *Limacina helicina*, the dominant Arctic thecosome, has a one-year life cycle with a peak spawn in August. The offspring overwinter as veligers, metamorphosing to juveniles and males in the spring and early summer. In July and August, they mature into females to begin the annual cycle once again (Gannefors et al. 2005). An earlier seasonal study on *L. helicina* in the central Arctic waters of the Canadian Basin

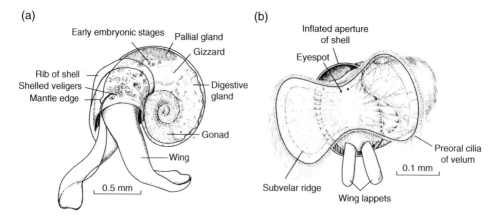

Figure 8.50 Pteropod reproduction. (a) The ovoviviporous pteropod *Limacina inflata* with early stage embryos and shelled veliger larvae within the mantle cavity; (b) shelled veliger of *Limacina inflata*. *Sources:* (a) Lalli and Wells (1973), figure 1 (p. 935); (b) Lalli and Wells (1973), figure 2 (p. 936). Reproduced with permission of Bulletin of Marine Science.

(Kobayashi 1974) yielded a nearly identical cycle, with the caveat that low-level reproduction was also observed in the winter months. The reproductive cycle of the southernmost euthecosome, *Limacina antarctica*, is presently undescribed but likely is similar to that of its arctic congener.

Limacina retroversa found off the coast of Argentina in the cold-temperate waters of the Argentine Sea (Latitude 40–50°S) produce two generations per year (Dadon and de Cidre 1992). Their first peak in reproduction occurs in the early spring, coinciding with the annual spring bloom. Offspring from the spring peak, taking advantage of the abundant phytoplankton, mature rapidly over the spring and summer to produce another generation in the fall. The fall generation then overwinters to reproduce in the spring. Unlike the study of Kobayashi (1974) in the Canadian Basin, no reproductive activity was observed in the overwintering Argentine Sea population. Populations of *Limacina retroversa* in the Argentine Sea during fall can reach as high as 67 000 ind. $1000 \, m^{-3}$.

Pteropods, pH, and the Carbonate Cycle

As is the case with most of the gastropods, the shell of euthecosomes is composed of aragonite, a crystalline form of calcium carbonate, embedded in a lattice of the protein conchiolin. Aragonite is one of the two major crystalline forms of calcium carbonate found in the ocean, the other of which is calcite. The two forms differ in their crystalline structure, and, of the two, aragonite is the more easily dissolved. Each of the two crystals has a characteristic depth in the ocean at which it dissolves, which is termed its compensation depth. The compensation depth for aragonite is typically between 2 and 3 km and for calcite it is between 4 and 5 km, depending upon location. Why is this important? Because the shells are negatively buoyant they sink to the bottom and, above their compensation depth, they accumulate to form a record that can be used to decipher what the ocean above them was like in the past. In the Atlantic, pteropod oozes are generally found in the vicinity of oceanic islands and other elevated areas of the ocean floor where depths are less than 2800 m (Berner 1977). With the higher acidity at depth in the Pacific and Indian Oceans, pteropod oozes are only found at depths of 500 m or less (Lalli and Gilmer 1989).

The community composition of pteropods in the sediments can be used to evaluate water column temperatures through geologic time, as can the isotopic composition of their shells. Since thecosome shells have a species-specific morphology and thecosomes have a characteristic temperature range over which they live, their past community composition can be described based on the species composition of the shells in the sediment and thus provide a good idea of what the surface ocean was like at the time they fell to the bottom. Alternately, a purely chemical strategy can be employed, one based on the stable isotope ratios of oxygen in the calcium carbonate ($CaCO_3$) of the shells. The stable isotope used as a paleotemperature indicator is the naturally occurring one with the heaviest atomic mass, ^{18}O, a very small percentage of the total. By far the most abundant isotope is ^{16}O. The ratio of the two isotopes, known as the $\delta^{18}O$, is a reliable indicator of ocean temperature when a shell was formed; ^{18}O is found at lower concentrations in shells formed in warm water. Community species composition and stable isotope ratios thus provide two complementary pathways to the same goal. Moreover, the same techniques can be used for other types of organisms that accumulate in the sediments such as planktonic foraminifera.

Since the early 1800s, the industrial revolution and human population growth have brought a marked increase in atmospheric CO_2 that is reducing the pH of the ocean, making it more acidic. CO_2 dissolves in water to form carbonic acid, which breaks down to form hydronium ions and bicarbonate:

$$CO_2 + H_2O \rightarrow H_2CO_3 \rightarrow H^+ + HCO_3^-$$

The H^+ ions produced when CO_2 dissolves cause the decline in pH.

A consensus number for the reduction is about 0.1 pH unit, from a previous value of 8.2 to a value of 8.1. The surface oceans are still slightly alkaline, i.e. above a neutral pH of 7.0, but the seemingly slight reduction of 0.1 unit is a 25% increase in acidity owing to the fact that pH is a logarithmic number. The ramifications of increasing acidity in the oceans are enormous, and shell-bearing taxa are among the most vulnerable. Calcium deposition in the shell is the result of biomineralization, a process whereby calcium is actively concentrated in the microenvironment where the shell is being formed, allowing it to precipitate as $CaCO_3$ that in pteropods takes the form of aragonite. That is a remarkable feat considering that calcium constitutes about 1% of the total ion concentration in normal seawater. Historically, the aragonite shell, once formed, was stable for the life of the organism in the surface ocean. However, experiments using computer-predicted pH changes in limacinids suggested that by 2050 the habitat for shell-bearing species will be in serious decline, particularly at high latitudes (Orr et al. 2005). Now there is strong evidence that etching and weakening of pteropod shells is already occurring in the California borderland (Bednarsek et al. 2014).

Gymnosomata
Anatomy and Feeding

Gymnosomes (gymno = naked; some = body) have no shell and therefore no mantle cavity. They vary in appearance from the fairly streamlined *Clione* to the more sac-like *Thliptodon* (Clionidae) and *Pneumodermopsis* (Pneumodermatidae) (Figure 8.51). They vary in size from about 2 mm in *Paedoclione* to as much as 85 mm in subarctic *Clione limacina* (Lalli and Gilmer 1989) with many species in the range of 5–20 mm (van der Spoel and Dadon 1999). The body usually comprises an anterior or cephalic section with two pairs of tentacles, a pair of eyes, a reduced foot with lateral and median lobes, and the swimming parapodia or "wings." The posterior, or trunk, section of gymnosomes often has an elevated and unpigmented region about two-thirds of the way down the trunk believed to function as a lateral gill, as well as a posterior gill at the end of the trunk (Figures 8.18b and 8.52a). In both gills, the epidermis is thin and the areas are well invested with blood spaces (Hyman 1967; Lalli 1970a), presumably to aid in gas exchange. The remainder of the body is covered with "an opaque and glandular integument bearing stellate chromatophores" (Lalli 1970b). The chromatophores are under muscular control, but they do not respond to changes in light level with the rapidity of those of cephalopods.

The most remarkable features of the naked pteropods are the prey-catching elements of their buccal apparatus, sometimes including an eversible proboscis and usually either buccal cones or the suckered buccal tentacles shown in Figure 8.52a for *Pneumodermopsis macrochira* (formerly *Crucibranchia*). In *Pneumodermopsis*, the buccal mass itself is situated at the end of the proboscis and contains paired hook sacs, a radula, and jaws (Figure 8.52b)

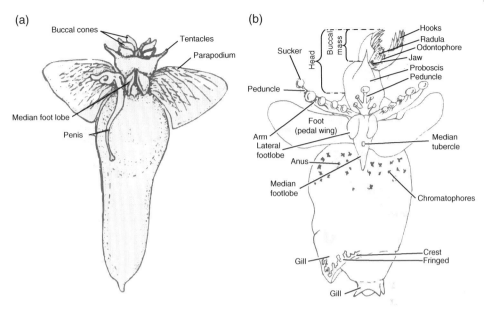

Figure 8.51 Gymnosome pteropods. (a) External anatomy of *Clione limacina*; (b) external anatomy of *Pneumodermopsis* sp. *Sources:* (a) Hyman (1967), figure 212 (p. 512); (b) Van Der Spoel and Dadon (1999), figure 1 (p. 652).

(Lalli 1970a). The alimentary canal of gymnosomes is simple, with a muscular ciliated esophagus leading from the buccal mass to a large digestive sac within the digestive gland. The stomach in gymnosomes is reduced to a small ciliated region in the digestive gland. A short intestine leads to the anus, which opens to a long ciliated groove termed the "anal field" below the right parapodium. Currents generated by wing motion propel fecal material away from the body.

All gymnosomes are predators and though there is considerable diversity in the number and types of structures making up the buccal apparatus (cf. Lalli and Gilmer 1989), observations of prey capture and ingestion in different species show a similar pattern (Lalli 1970b, 1972; Conover and Lalli 1972). Once a thecosome prey species is detected, the faster-swimming and more maneuverable gymnosome captures it with either buccal cones or buccal tentacles (Figure 8.53a).

In the example of *Clione* feeding on *Limacina*, the prey is manipulated quickly by the buccal cones until the shell aperture faces the mouth of the gymnosome and chitinous hooks are everted from their sacs on either side of the radula (Figure 8.53b, c) to grab onto the prey. Once secured, the prey is pulled from its shell and swallowed whole (Lalli 1970b). Eversion of the hooks and of the proboscis or buccal cones is effected through hydrostatic pressure, similar to that described in Chapters 5 and 6 for the proboscides of nemerteans and polychaetes. Circular muscle in the wall of each structure contracts around a hemocoel, forcing the hooks out or the proboscis to elongate. In *Clione*, a diaphragm separating the anterior cephalic region from the general body hemocoel is believed to regulate the amount of fluid in the cephalic region, confining it so that the pressure required for protrusion of the buccal cones and hooks can develop (Lalli 1970b). In *Pneumodermopsis*, the

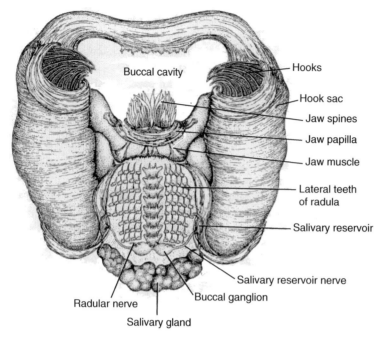

Pneumodemopsis macrochira
(buccal mass, dorsal view)

Figure 8.52 The buccal mass of *Pneumodermopsis macrochira*. *Source:* From Lalli, Morphology of *Crucibranchaea Macrochira* (Meisenheimer), A Gymnosomatous Pteropod, Journal of Molluscan Studies, (1970a), volume 5, Issue 1, pp 1–14, figure 3. Reprinted by permission of Oxford University Press on behalf of The Malacological Society of London.

same general effect is achieved by a contraction of the trunk region when the proboscis and tentacles are everted. Retraction of the various buccal structures, including the hooks, cones, and proboscides when present, is achieved by shortening retractor muscles in each case. Prey is drawn into the buccal cavity by the retracting hooks and radula working together. Once detected, the timing of prey capture and ingestion varies from 2 to 45 minutes (Conover and Lalli 1972).

Gymnosomes appear to be selective feeders, both in their choice of prey species and prey size. Given choices of four thecosome species in two genera, *Cavolinia* and *Creseis*, the gymnosome *Pneumodermopsis paucidens* selected only *Creseis* (Sentz-Braconnot 1965; Lalli 1970b). Likewise, based on visual observations and gut contents, *Clione limacina* appears to feed only on *Limacina helicina* and *Limacina retroversa* (Lalli 1970b; Conover and Lalli 1972). The considerable diversity in feeding structures within the gymnosomes do support a case for a high degree of prey specialization in the group as a whole. Lalli and Gilmer (1989) identify 19 different predator–prey combinations in the Gymnosomata, an order with only 50 species (WoRMS).

Conover and Lalli (1972) examined size preferences of *Clione limacina* with a series of experiments offering a range of prey (*Limacina retroversa*) sizes to *Clione* of four

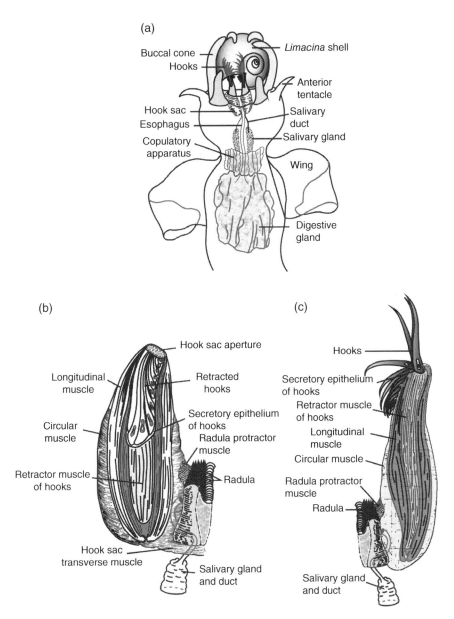

Figure 8.53 Gymnosome feeding. (a) *Clione limacina* feeding on a *Limacina helicina* using both buccal cones and hooks to secure the prey; (b) hook sac of *Clione limacina* with hooks retracted; (c) hook sac of *Clione limacina* with hooks everted. *Sources:* Lalli (1970b), figure 1 (p. 103); (b and c) Adapted from Lalli (1970b), figure 4 (p. 108). Reproduced with permission of Elsevier.

different sizes. They found that larger *Clione* actively selected larger prey, leaving smaller individuals uneaten. Anatomical, descriptive, and experimental evidence all clearly suggest that the gymnosomes are highly selective predators; how they successfully choose size and species with their limited sensory array is less clear.

Locomotion

Gymnosomes swim more rapidly than their thecosome brethren and are far more maneuverable, owing partly to their more streamlined shell-free bodies, their wing structure and location, and their swimming technique (Figure 8.54). Their wings are shorter and more rounded than those of the thecosomes, acting more as "two-way sculling organs" than as oars (Morton 1964). Morton (1958) described the sculling motion of the wings like this: "The wings move synchronously dorsally and ventrally, that is toward and away from the sides of the body. At its narrow attachment to the body, the wing makes a small twist after each stroke so that the leading anterior edge is directed more strongly downward in a down-stroke and upward on the return. The tips of the wings roughly describe a figure-eight (Figure 8.54b). Their shorter wings allow gymnosomes to provide thrust on both upward and downward strokes, increasing efficiency, and maneuverability.

A suite of more recent papers (Satterlie and Spencer 1985; Satterlie et al. 1985; Satterlie 1993; Borrell et al. 2005; Rosenthal et al. 2009; Szymik and Satterlie 2011) analyze the swimming of *Clione* in far greater detail, using high-speed cinematography and video to better evaluate the attitude (kinematics) of the wings during up and down strokes and the neuromuscular organization of the swimming system. The above description by Morton (1958) is fundamentally correct for the basic swimming pattern (Satterlie et al. 1985). Later work shows the presence of a "two-speed" system in *Clione limacina*: both fast and slow swimming and the neuromuscular underpinnings (Satterlie 1993; Rosenthal et al. 2009; Szymik and Satterlie 2011). The fast and slow speeds are useful for chase and capture of prey and during manipulation of captured prey, respectively (Szymik and Satterlie 2011). Gymnosomes will continue to be important subjects for study in the field of animal locomotion because they are among the smallest marine animals to use flapping wings for movement (Borrell et al. 2005) and are found over a wide range of temperatures, enabling researchers to study temperature effects and temperature adaptation in a particularly creative way (Rosenthal et al. 2009).

Reproduction

Based on their reproductive anatomy and its similarities to that of thecosomes, fundamentals of reproduction in gymnosomes are believed to be similar, including a protandrous reproductive strategy. Limited observations of copulating gymnosomes as well as sex-change with size (Morton 1958; Lalli and Conover 1973) make a strong case for protandry, though there is some doubt about particulars. Notable among those are whether male mating structures persist in adult females and how widespread spermatophores are in the order as a whole. *Pneumodermopsis macrochira* utilizes them and other gymnosomes may as well (Lalli and Gilmer 1989).

As in the thecosomes, gametogenesis in the gymnosome reproductive system is the province of the ovotestis, a conical sac located beneath the other viscera at the base of the trunk (Figure 8.55). A narrow hermaphrodite duct exits the ovotestis, leading first to the albumen gland, which is believed to serve the same function as the albumen gland in the thecosomes (Figure 8.49) (Lalli and Wells 1978), covering eggs with a thin nutritive coating before they enter the larger mucous gland where in *Clione* they are coated with mucous to form an egg mass, or ribbon (cf. Lebour 1931), about 1.2 mm across containing 30–50 eggs. The egg mass corresponds well in size to the entire contents of the mucous gland

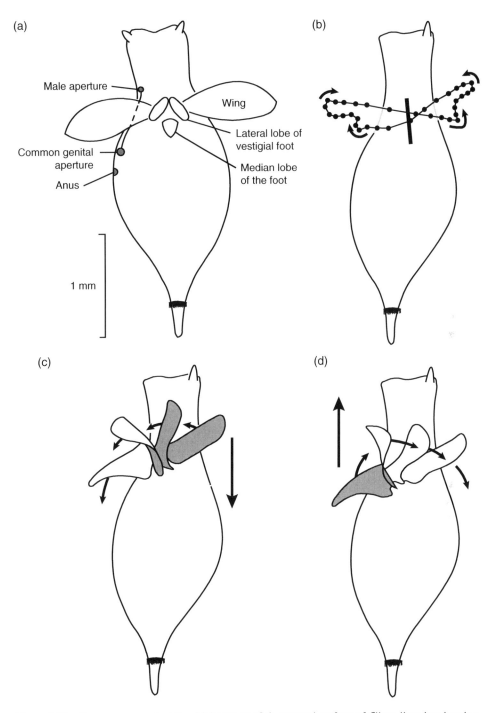

Figure 8.54 Gymnosome swimming. (a) Anatomy of the ventral surface of *Clione limacina* showing the relative positions of the wings, genital pores, anus and foot; (b) lateral view of *C. limacina* showing wing tip positions during hovering; (c) wing positions during downward swimming strokes in *C. limacina*; (d) wing positions during upward swimming strokes in *C. limacina*. *Sources:* (a, c, and d) Adapted from Morton (1958), figure 1 (p. 288) Reproduced with the permission of Cambridge University Press; (b) Satterlie et al. (1985), figure 10 (p. 200).

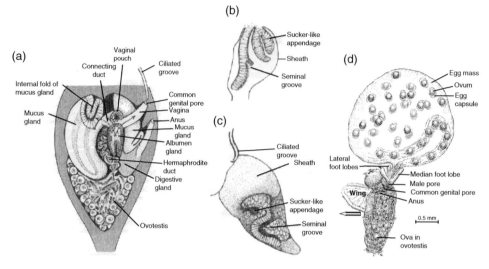

Figure 8.55 Gymnosome reproduction. (a) Reproductive structures of *Clione limacina*; (b) sheathed penis of *C. limacina*; (c) everted penis of *C. limacina*; (d) spawning egg mass of *Paedoclione doliiformis*. *Sources:* (a–c) Adapted from Morton (1958), figure 3 (p. 293). (d) From Lalli and Conover (1973), figure 2 (p. 15). Reproduced with permission of Springer Nature.

(Morton 1958a). Once through the mucous gland, the egg mass passes through to the common genital pore (termed "vagina" by Morton 1958a) and is discharged to the outside (Lebour 1931).

In male *Clione*, sperm formation occurs in the central region of the ovotestis. Sperm exit the ovotestis through the hermaphrodite duct and must pass through the albumen and mucous glands prior to reaching the common genital pore. As in the thecosomes, mature sperm may be stored in the hermaphrodite duct. A penis, branched copulatory organ, and prostate gland are present in the head region below the male pore, through which the penis is everted during copulation. A ciliated groove leads from the common genital pore to the male pore. The penis is deeply furrowed and ciliated (Figure 8.55b, c), and it is assumed that when the penis is everted during copulation the penial groove becomes continuous with the ciliated groove leading from the common genital pore, allowing sperm to make their way via the ciliated groove to the penial groove and thence on to the partner.

Copulation in *Clione* has now been observed in situ by scuba divers and by researchers in submersibles (Lalli and Gilmer 1989), and those observations agree with the earliest descriptions of mating taken from specimens preserved *en copulo* (Boas 1886). Mating occurs venter to venter with the heads facing in the same direction and the penes crossing to reach the partner's common genital pore. Partners are held in close proximity by their copulatory organs. Copulation is believed to take a long time; a timing of three to four hours is reported by Lalli and Gilmer (1989).

The common genital pore, or "vagina", is the terminus of the duct leading from the mucous gland (Figure 8.55a), and there is a pouch at its proximal end believed to store incoming sperm. In this way, eggs can be fertilized as they exit during spawning. It is presently unclear whether male mating structures eventually degenerate in mature females as they do in thecosomes or if they persist to allow simultaneous hermaphroditism. Mating has only been observed in hermaphrodites: mature females bearing fully functional male sexual characteristics as well (Lalli and Gilmer 1989).

Data collected on *Paedoclione doliiformis* and *Clione limacina* suggest that spawning occurs in a summer–fall time frame when their prey, *Limacina helicina*, is at its peak abundance. Both species spawn floating egg masses (Figure 8.55d) containing from 30 to 165 eggs, with most in the 30–80 range (Lebour 1931; Lalli and Conover 1973). The number of eggs in each egg mass scales positively with the size of the spawning female, the temperature, and recent feeding success. Successive egg masses may be continually produced over a period of days resulting in hundreds to thousands of eggs produced. Larvae hatch as veligers in 3–4 days in temperatures ranging from 16 to 19 °C and begin feeding on phytoplankton.

Gymnosome veligers hatch with a cup-shaped shell (Figure 8.56a), which in *Clione* and *Paedoclione* is lost after 11–14 days in the plankton at temperatures ranging from 16 to 19 °C. Once the shell is lost the larva, now a "polytroch," sheds its velum, elongates, and gradually morphs into its adult form. *Paedoclione* is considered to be neotenous due to the fact that it never loses its three sets of larval ciliary rings (Figure 8.56b–d), even when sexually mature. Compare *Paedoclione* with *Pneumodermopsis* or *Clione* and note the absence of anterior and middle ciliary bands in the adults of the latter two genera (Figure 8.51). Polytroch larvae of *Clione* and *Paedoclione* feed initially on limacinid veligers, presumably moving on to larger prey as they mature.

Superfamily Hydromyloidea

Only two species populate the superfamily Hydromyloidea: *Hydromyles globulosus* and *Laginiopsis trilobata*, each in its own family (WoRMS). *Hydromyles* is found in the eastern Pacific and Indian Oceans and has been observed to reach very high densities in the tropical Indo-Pacific (Tesch 1950), with up to 3000 specimens per plankton tow (Lalli and Gilmer 1989). It has a unique appearance, with very long tentacles and a distinct head and trunk (Figure 8.57a). Although it possesses a radula, it lacks any buccal cones or tentacles, proboscis, or hook sacs. Nonetheless, thecosome debris has been found in its gut. *Hydromyles* are unique among the gymnosomes in that they brood their young, which emerge as juveniles. In other respects, their reproductive system differs little from the general gymnosome model described above. A gland located adjacent to the viscera is apparently capable of producing a "pteropod ink" which can be released through the anal opening when the animal is disturbed (Lalli and Gilmer 1989).

The other species in the Hydromyloidea, *Laginiopsis trilobata*, is known from only a single specimen. It has a long, nonretractable, proboscis that ends in three lobes attached to a distinct head equipped with two pairs of tentacles (Figure 8.57b). It lacks a buccal mass altogether (Pruvot-Fol 1926; Lalli and Gilmer 1989).

Pteropod Distributions

Surprisingly, pteropod biogeography is one of the better-known aspects of pteropod biology, which is largely due to the efforts of a few dedicated scientists. More recent treatments of pteropod distributions (van der Spoel 1967, 1976; Be and Gilmer 1977; van der Spoel and Dadon 1999; Hunt et al. 2008; Burridge et al. 2016) were able to build on the work of pioneers (Pelseneer 1887, 1888a, b;

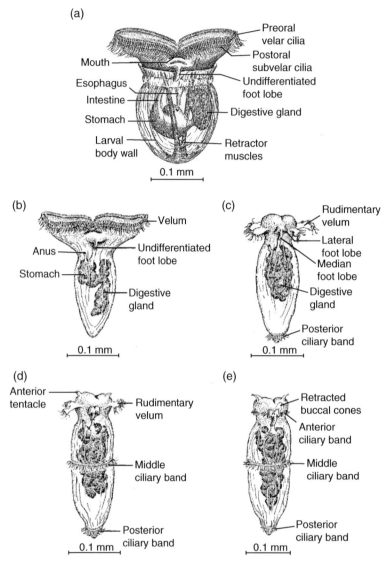

Figure 8.56 Gymnosome larval development. (a) Anatomy of the veliger of *Paedoclione doliiformis*; (b–d) stages of larval development; (b) phase one: shell-less veliger, shells cast off 11 days after hatching; (c) phase two: start of velum loss and development of first ciliary band, generally less than 12 hours from time of shell loss; (d) phase three: development of second ciliary band; (e) phase four: polytrochous larva, development of third ciliary band. *Sources:* (a) From Lalli and Conover (1973), figure 4 (p. 18). (b-e) Lalli and Conover (1973), figure 6 (p. 19). Reproduced with the permission of Springer Nature.

Tesch 1904, 1946, 1948, 1950) to provide a good framework of pteropod distributions worldwide.

Highest diversities in south Atlantic pteropods were found in tropical–subtropical waters, declining in species numbers with declining ambient temperatures (Van der Spoel and Dadon 1999), a trend mirrored by more recent work (Burridge et al. 2016) in the Atlantic. Only a small fraction of

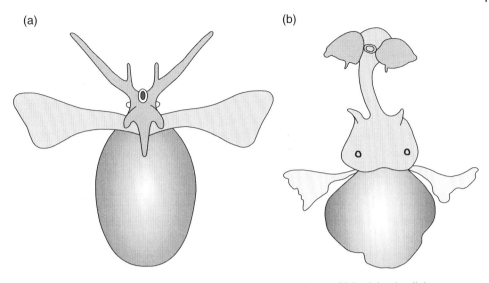

Figure 8.57 Superfamily Hydromyloidea. (a) *Hydromyles globulosus*; (b) *Laginiopsis trilobata*.

pteropod species reach polar latitudes (van der Spoel and Dadon 1999) but when present in cold waters, they can reach high abundances. *Limacina helicina antarctica* has been reported at levels of 10000 ind. $1000\,m^{-3}$ (Chen 1968a) near the Antarctic Convergence. For comparison, Wormuth (1981), working in the Sargasso Sea, found the mean abundance of *Helicoides inflatus* (formerly *Limacina inflata*), the most abundant species, to be 97.2 ind. $1000\,m^{-3}$. Circumglobal distributions were exhibited in 29–61% of south Atlantic pteropod species depending on suborder (29% in pseudothecosomes, 53% in euthecosomes, and 61% in gymnosomes) with the gymnosomes being the most cosmopolitan (van der Spoel and Dadon 1999).

Distributions by depth show that the great majority of euthecosomes reside in the epipelagic zone (0–200 m), with a very limited degree of overlap with deeper layers. As a consequence, many euthecosomes can also be found on the shelf in neritic waters, unlike their relatives. In contrast, the pseudothecosomes show a marked preference for mesopelagic (200–1000 m) life, particularly within the shell-bearing Peraclidae. Numbers of bathypelagic species are limited, with the best representation in the Cliidae (e.g. *Clio andreae* and *Clio chaptali*) and the best known, the ubiquitous *Limacina helicoides* (van der Spoel and Dadon 1999). Gymnosomes are well represented in all three depth strata, but epipelagic species predominate.

Vertical Migrations

Diel vertical migrations have been reported in a variety of euthecosomes (Wormelle 1962; van der Spoel 1973a, b; Kobayashi 1974; Wormuth 1981, 1985) though overall, data on vertical distributions are fairly limited (van der Spoel and Dadon 1999). Using a closing net and stratified tows in the Sargasso Sea, Wormuth (1981) found that the most abundant euthecosomes fell into migratory and nonmigratory groups, with nonmigrators predominating. The primary migratory pattern was a daytime center of distribution between 200 and 300 m and a nighttime center in the upper 100 m

(Table 8.4). Data from the nearby Florida Current (Wormelle 1962) taken using similar trawling gear are presented for comparison. Diel vertical distributions for the same species were similar in many cases but not identical. In both systems, a diel shift in center of distribution was observed in several species, though a nighttime excursion toward the surface from daytime depth was more consistent in the Sargasso Sea data.

It is worth noting that two of the four Limacinids, with shell diameters of about 10 mm, migrated roughly 200 m on a diel basis in the Sargasso Sea. The journey entailed swimming 20 000 body lengths twice daily, an impressive feat.

The Nudibranchs

The nudibranchs (naked gill) are a highly diverse and often strikingly beautiful group that is mainly confined to the benthos. They have "no shell, mantle cavity, osphradium or internal gill and the mantle, visceral mass and foot are incorporated into one body, externally bilaterally symmetrical and often bearing cerata or other outgrowths" (Hyman 1967). Their common name, sea slugs, belies their visual appeal. Few resemble terrestrial slugs, though both groups, like the shelled gastropods, primarily move about by creeping on the sole of their well-developed foot. Swimming for most benthic nudibranchs is an escape from predators and only occurs for short periods. In some species, escape involves violent side-to-side movements of the entire body, while in others it is accomplished by dorsoventral flexion or an undulatory wave of the mantle margin (Morton 1964). Flapping movements of the dorsal cerata, resembling the motion of bird wings in flight, are used to propel some aeolid nudibranchs off the seabed (Thompson 1976).

Only a very few nudibranchs are considered to be pelagic and only two genera are capable swimmers: *Phylliroe* and *Cephalopyge*, both in the family Phylliroidae. The remaining pelagic nudibranch families, the Glaucidae and Fionidae, have a lifestyle that is reminiscent of the janthinid snails, remaining in the pleustonic community comprised of rafts of *Vellela* and *Porpita*. The nudibranch *Scyllaea pelagica* is associated with the sargassum community commonly seen in the GOM and Caribbean, but its biology is poorly described and won't be treated here.

Family Glaucidae

The glaucids are Aeolidoid nudibranchs, famous for their habit of feeding on cnidarians, ingesting their nematocysts, and deploying the nematocysts as weapons in their own cerata (Figure 8.58) (Thompson 1976). Like the janthinids, Glaucids are pleustonic in habit, feeding mainly on *Physalia*, *Velella*, and *Porpita*. They can sting quite effectively, causing pain in humans for up to two hours. The digestive process is believed to select for the particularly virulent nematocysts of *Physalia* for use in the ceratal defense. How this would be achieved is unclear, though size selectivity is possible. Whether there is active digestive sorting or whether the presence of *Physalia* nematocysts merely reflects a predominance in the diet is a matter of conjecture, but the virulence of the nematocysts is consistent enough for *Glaucus* to be considered potentially dangerous for humans to handle.

Gut contents of *Glaucus* usually contain an abundance of nematocysts from *Velella* and *Porpita* as well as from *Physalia*. Up to 50% of those are already discharged. Undischarged *Physalia* nematocysts are translocated through the digestive gland to the cnidosacs, where they are ingested by cells within the cnidophore (Figure 8.58b, c). There they are nourished and stored

Table 8.4 Pteropod day–night vertical distributions in the Sargasso Sea and Florida Current. MD (mean depth) is the depth at which 50% of the population was reached when integrating the total population numbers from the surface to the deepest sampling depth. Sp (Spread) refers to distance (depth range) between which 25 and 75% of the total population was obtained. *Sources*: Sargasso Sea data are from Wormuth (1981); Florida Current data from Wormelle (1962).

Family/species	Abundance (ind·1000 m³)	Sargasso Sea			Florida Current					
		Day range (m)	Night range (m)		10 mile stations			40 mile stations		
				Day MD (m)	Day Sp (m)	Night MD (m)	Day MD (m)	Day Sp (m)	Night MD (m)
Limacinidae										
Helicoides inflatus	97.2	200–300	0–100	236	131	232	218	332	263
Limacina bulimoides	15.5	100–200	0–100						
Limacina leseuri	7.4	200–300	0–100	103		85	289		398
Limacina trachiformis	2.9	0–100	0–100	165		99	88		120
Cavolniidae										
Styliola subula	12.2	200–300	0–100	234	136	81	594	318	171
Creseidae										
Creseis acicula	20.8	0–100	0–100	157	139	52	185	200	200
Creseis virgula	8.4	0–100	0–100	206		98	319		167
Cliidae										
Clio pyramidata	12.1	200–500	0–150						
Clio cuspidata	2.6	300–800	300–800						

until needed. When disturbed, the nematocysts discharge through the central pore of the cnidosac. Nematocysts are replaced in the next feeding (Thompson 1976) and any cerata that have been nipped off are rapidly regenerated.

Though differing little in external morphology from a benthic Aeolidoid, glaucids do show a behavioral modification that allows them to more easily remain near the surface. They are able to gulp air and trap it in their stomach to provide positive buoyancy, enabling them to float ventral side up within rafts of *Velella*, navigating by creeping under the surface film. Like the janthinids, they exhibit countershading with an "open-ocean blue" ventrum and silvery dorsal surface. Thompson (1976) reports that well-developed sphincter muscles (Figure 8.58a) guard the exits from the stomach into the cerata and intestine to keep the buoyancy bubble from escaping.

Like the pteropods, the nudibranchs are hermaphroditic and much of the reproductive anatomy is the same. Unlike pteropods, they are simultaneous hermaphrodites, with sexually mature individuals fully functional as both male and female. The reproductive system of simultaneous hermaphrodites is more complex than the protrandrous system of the pteropods, mainly in the conductive system for sperm and ova (Figure 8.59a, b). The gametes follow very different paths. The common hermaphrodite duct leading from the ovotestis bifurcates into a male and female channel, kept separate by a muscular valve. The valve stops the movement of ova into the male system, particularly during copulation. Just past the valve in the male channel is a constriction or bottleneck (Figure 8.59b, c) presumed to block the further passage of "stray" ova into the vas deferens. The vas deferens connects directly to the penis. Mature sperm are stored in the vesicula seminalis prior to copulation.

During copulation, the penis of each partner is inserted in the vagina of the other and sperm transfer is effected by contractions of the vesicula seminalis and vas deferens. Introduced sperm travel up the vagina to the receptaculum seminis where they can be stored for a limited time until oviposition. Movement of ova from the ovotestis into the oviduct is accomplished by ciliary activity within the hermaphrodite duct and vesicular seminalis. The oviduct is similarly lined with cilia. Self-fertilization is avoided by compartmentalization of the autosperm (i.e. the individual's own sperm) within the vesicula seminalis such that the ova do not encounter mature autosperm on their journey from the ovotestis to the oviduct. Fertilization is accomplished with the allosperm ("foreign" sperm from a copulatory partner) that travel from the receptaculum seminalis down the allosperm duct to the oviduct, and thence to the albumen gland (Figure 8.59b). Eggs are fertilized within the albumen gland, an egg capsule is formed around them, and they proceed via ciliary activity in the oviduct to the mucous gland, where they are coated with mucous and formed into the egg string or ribbon that is extruded from the genital opening. Egg strings are small, usually 20 mm or less and contain 10–36 eggs. Eggs hatch into free-swimming veligers in about 48 hours at temperatures of 25 °C (Lalli and Gilmer 1989).

Family Fionidae
Fiona pinnata, though no longer the only species in the family Fionidae, remains the one with the greatest body of research describing its biology. It is another nudibranch commonly found in the circumtropical pleustonic community (Figure 8.60). It reaches a maximum length of 50–60 mm though is most commonly found at sizes between 20 and 30 mm. Its cerata have been reported with cnidosacs (Hyman 1967) and

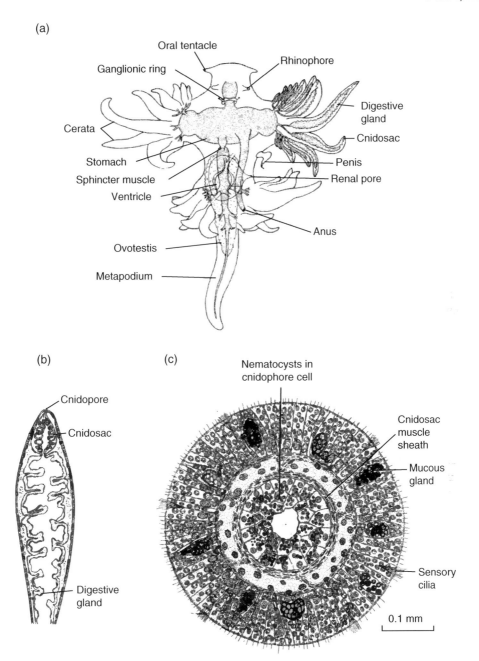

Figure 8.58 Aeolid nudibranchs. (a) Anatomy of the glaucid nudibranch *Glaucus atlanticus*; (b) longitudinal section through a ceras of an aeolid nudibranch; (c) transverse section through a ceras of an aeolid nudibranch. *Sources:* (a) Thompson (1976) (p. 27); (b and c) Thompson (1976), figure 21 (p. 51).

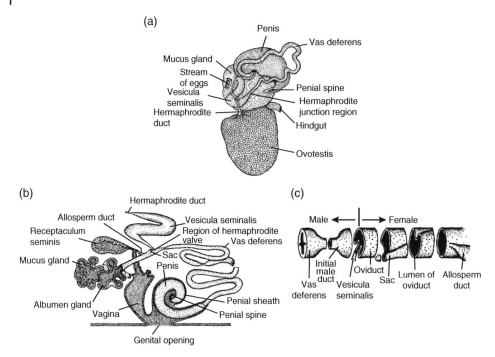

Figure 8.59 Nudibranch reproduction. (a) Anatomy of the anterior genital mass of *Glaucus* (dorsal view); (b) reproductive system of *Glaucus*; (c) Hermaphrodite valve position in relation to neighboring parts of the reproductive system. *Sources:* (a) Thompson and McFarlane (1967), figure 8 (p. 116); (b and c) Thompson and McFarlane (1967), figure 9 (p. 117).

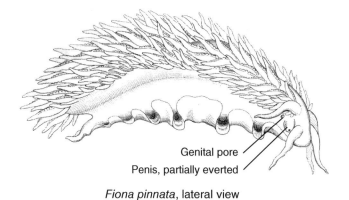

Fiona pinnata, lateral view

Figure 8.60 The fionid nudibranch *Fiona pinnata*. *Source:* From Alder and Hancock (1851), plate IX.

without (Lalli and Gilmer 1989). The original species description (Alder and Hancock 1851) does mention ramifications of the midgut that "communicate with the glands of the cerata" suggesting that cnidosacs are possible from an anatomical point of view, but their presence is unconfirmed. Based upon laboratory behavioral studies, gut contents, and the observations of divers, Fiona feeds preferentially on

Velella. However, it also preys readily on *Lepas anatiferas* (Bayer 1963; Bieri 1966; Lalli and Gilmer 1989), one of the two pelagic gooseneck barnacles commonly seen in the pleustonic community, the other being *Dosima fascicularis*. *Lepas* does not secrete its own float; it is dependent on floating debris or the buoyancy of other species (e.g. *Velella*, turtles), and it may be found well outside the pleustonic community.

The anatomy of *Fiona*'s reproductive tract is very similar to that of *Glaucus*, with a hermaphrodite duct that bifurcates into a (male) vas deferens and (female) oviduct, allowing for simultaneous hermaphroditism. The genital pore and penis are located on the right side of the head region (Figure 8.60). Mating occurs head to head with the right sides of the two partners touching, allowing reciprocal fertilization (Holleman 1972). Holleman (1972) reported that egg masses were extruded 12–48 hours after copulation in the cool temperate waters of Bodega Bay, CA, hatching into veligers in 5 days. Time-to-hatch was 46 hours in the subtropical waters off Miami (Bayer 1963). Time to metamorphosis in the veliger larvae was not available, but it may depend on proximity to available substrate. Clearly, for a species that has little control over its buoyancy and cannot swim, delaying metamorphosis until suitable substrate is near would be advantageous (Lalli and Gilmer 1989).

Growth in *Fiona* is astonishingly rapid. Individuals that were fed *Lepas ad libitum* for 24 days grew from an average length of 3.4–30 mm and gained in average mass of from 1.3 to 448.8 mg in the 24-day period (Holleman 1972). Similarly, Bayer (1963) reported four *Fiona* fed *ad libitum* on *Velella* grew in length from 8 to 32 mm in 5 days, whereupon they began copulating and producing multiple egg masses. Temperatures were not reported in either study, but summer temperatures for Bodega Bay (Holleman) are about 13 °C and those for Miami Beach in March (Bayer) are about 24 °C (NOAA, NCEI). The entire life cycle of *Fiona* can be completed in about 30 days (Lalli and Gilmer 1989).

Fiona has no reported behaviors that aid its pelagic lifestyle such as ingesting bubbles of air, as *Glaucus* does, to make it positively buoyant. It is dependent on the buoyancy of *Velella* or floating debris like driftwood for a free ride and is found on a variety of substrates. Kropp (1931) reports that *Fiona* can move about by creeping upside down along the surface film as has been observed in *Glaucus*. With a well-developed radula and strong paired jaws, *Fiona* is well suited for feeding on Cnidaria and the barnacle *Lepas*. *Lepas* is attacked where the peduncle (gooseneck) meets the body proper just below the plates (Figure 8.61). There it rasps its way inside the body until the barnacle relaxes its protective plates, which begin to gape. The nudibranch then moves to the front of the barnacle, enters through the gap between the plates and begins to feed, leaving only the tough integument of the peduncle and the plates behind (Kropp 1931). The color of *Fiona* depends on its diet, with "open-ocean blue" predominating in nudibranchs that have been feeding on *Velella* (Kropp 1931) but pink being its primary color after feeding on *Lepas anatifera* (Bayer 1963).

Family Phylliroidae

Three species in two genera comprise the Phylliroidae. *Phylliroe bucephala* and *P. lichtensteinii* make up the genus *Phylliroe* and *Cephalopyge trematoides* is the sole member of its monospecific genus. The three species share a similar morphology (Figures 8.20 and 8.62a) with laterally compressed transparent bodies, no cerata, and prominent rhinophores. Their leaf-like shape, including a pronounced tail in *Phylliroe*, allows effective undulatory (fish-like) swimming. They are reported to achieve speeds of 15 cm s^{-1} (Lalli and Gilmer 1989), which is a good clip for animals with a body length of less than 55 mm. The foot has been reduced to a pedal gland in *Phylliroe* and *Cephalopyge* (Figure 8.63) which, though fairly diminutive, may be used to attach itself to prey.

In its early life, *Phylliroe* is associated with the hydromedusa *Zanclea* (Figure 8.62b). It attaches itself to the medusa, living within the bell and consuming its prey as it grows, using its jaws and small radula for rasping and a sucking action by its muscular pharynx to ingest the prey's soft tissues (Lalli and Gilmer 1989). Once it reaches adult size, *Phylliroe* is capable of capturing and ingesting whole *Zanclea* as well as other prey, and divers have observed them feeding on larvaceans (Martin and Brinckmann 1963; Lalli and Gilmer 1989). Comprehensive data on their natural diet are not available. Zooxanthellae have been observed in the midgut gland of *Phylliroe* but their function, if any, is a mystery.

The ovotestis of *Phylliroe* is divided into multiple round lobes, easily visible through the transparent body (Figure 8.62a). These join to form a hermaphrodite duct that bifurcates into male and female channels as in *Glaucus* and *Fiona*. Penis and genital pores are located on the anterior right side of *Phylliroe* (Figure 8.20) and, as in *Fiona*, reciprocal mating occurs head to head, facing in opposite directions. After fertilization, up to 20 multiple-egg strings are produced with about 12 eggs per string (Martin and Brinckmann 1963). Fecundity of *Phylliroe* is considered to be quite low, particularly in comparison to other gastropods. Veligers hatched from the eggs in 7–9 days at a temperature of 13 °C (Martin 1966). It is thought that veligers of *Phylliroe* do not metamorphose unless they are in close proximity to *Zanclea*. Though likely, the behavior is unconfirmed.

Cephalopyge bears a strong anatomical resemblance to *Phylliroe* (Figure 8.63) but is more slender in its dorsoventral aspect, making it more eel-like in appearance. Its maximum size is about 25 mm and its buccal mass includes a small radula and chitinous jaws. Like *Phylliroe*, it has an undulatory swimming mode and is capable of swimming at least 12 cm s^{-1} (Lalli and Gilmer 1989).

Cephalopyge feeds on the siphonophore *Nanomia bijuga*, attaching by its pedal gland (Figure 8.63b) to the stolon and gradually consuming the colony. When offered other siphonophores in laboratory experiments, it rejected all but *Nanomia* (Sentz-Braconnot and Carre 1966) and is thus considered to be a feeding specialist. The pedal gland of *Cephalopyge* is larger and more capable of expansion than that of *Phylliroe*.

The reproductive tract of *Cephalopyge* is much like that of *Phylliroe*, with

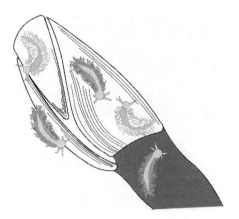

Fiona pinnata feeding on *Lepas* sp.

Figure 8.61 *Fiona pinnata* feeding on the gooseneck barnacle *Lepas* sp.

(a) *Phylliroe* sp.

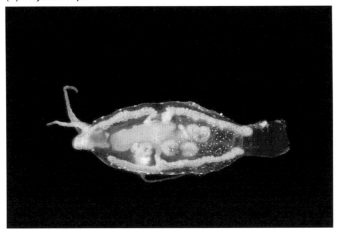

(b) Young *Phylliroe* inside of the medusa of the hydroid *Zanclea costata*

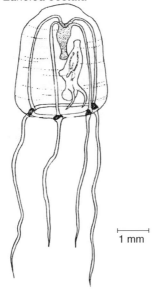

1 mm

Figure 8.62 The phylliroid nudibranchs. (a) Live specimen of *Phylliroe* sp. (b) young *Phylliroe* sp. inside of the bell of the hydroid *Zanclea costata*. *Sources:* © Dante Fenolio, reproduced with permission; (b) Lalli and Gilmer (1989), figure 72 (p. 218). See color plate section for color representation of this figure.

multiple globular ovotestes joined by a common duct. *Cephalopyge* has been observed by divers to mate in swarms. Copulation is brief, lasting about a minute, and individuals have been observed to mate more than once. Eggs are produced in floating filaments over a number of days, with individuals producing more than 3000 eggs. *Cephalopyge* is thus far more fecund than its relative *Phylliroe*

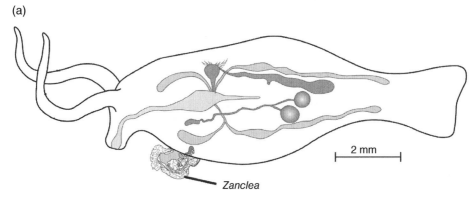

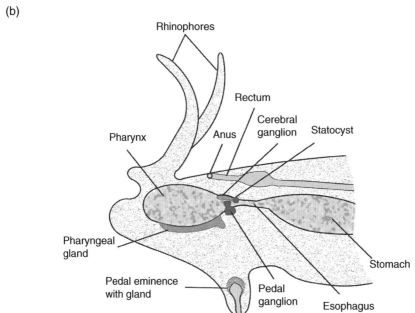

Figure 8.63 Feeding in the phylliroid nudibranchs. (a) *Phylliroe bucephala* with the bell of *Zanclea* sp. attached to its pedal gland; (b) anterior end of *Cephalopyge trematoides* showing the pedal gland and associated feeding structures. *Sources:* Adapted from Lalli and Gilmer (1989), Figure 72 (p. 218); (b) Adapted from Hyman (1967), figure 165 (p. 410).

(Sentz-Braconnot and Carre 1966). Eggs hatch into bi-lobed veligers.

Phylliroe is strongly bioluminescent; it "scintillates from points all over the body and tentacles" (Figure 8.64) (Herring 1978). The photocytes have not been described or even definitively identified. No luminescence has been elicited from *Cephalopyge*. In the case of *Phylliroe*, luminescence is most likely to be defensive, with a goal of blinding or confusing predators.

The Phylliroidae are found in warm temperate to subtropical waters with most reports coming from the Gulf of Mexico, Mediterranean, Tasman Sea, and the NW Pacific in the vicinity of Japan (WoRMS; Lalli and Gilmer 1989).

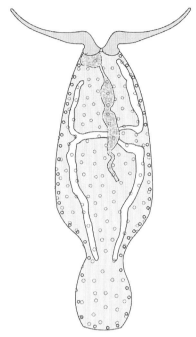

Phylliroe bucephalum

Figure 8.64 Dorsal aspect of *Phylliroe bucephalum* showing the widely distributed photocytes responsible for bioluminescence.

The Cephalopoda

Cephalopods have enough remarkable characteristics to make them the royalty, or "zenith," of invertebrate life forms (cf. Russell-Hunter 1979). They include the largest of the invertebrates, the giant squid *Architeuthis* with a mantle up to 6.6 m in length, a total length up to 20 m, and a weight of 500–1000 kg (Roper et al. 1984). Not to be forgotten are other sizable species including *Mesonychoteuthis*, the Colossal Squid, with a total length up to 14 m and mass of 350 kg, and *Dosidicus gigas*, the Humboldt Squid, with a length of 4 m. The giant of the octopods is *Octopus dofleini*, with a length of 3 m and weight of 50 kg. Adult cephalopods range in size from 2 cm in the Australian pygmy squid *Idiosepius notoides* to the 20-m giant squid.

Cephalopods all have exceptionally well-developed eyes, a closed and highly efficient circulatory system, and a sophisticated nervous system with a large brain. Their jet propulsion and diverse buoyancy systems are unique and highly effective. They are all marine and are all dioecious. They are nearly all carnivorous, with pelagic forms preying largely on shrimps, fishes, crabs, and other cephalopods. Benthic octopuses add clams to their diet as well.

The first cephalopods appeared in the fossil record in the late (or upper) Cambrian, about 540 mya, with the most famous fossil, *Plectronoceras cambria*, found in northeast China and first described by Walcott (1905). It is classified as a nautiloid (Schrock and Twenhofel 1953), but with a horn-like shell instead of the more familiar planispiral coiled shell (Figure 8.65). Like the living nautiloids, the shell of *Plectronoceras* contains a phragmocone, a series of chambers within the shell filled with gas in the living animal and conferring buoyancy to the shell and the mollusc within. The buoyancy system of the nautiloids is nearly identical and uses the same principles. Neutral buoyancy emancipated the early cephalopods from the bottom despite their shell, allowing them to enter the pelagic realm. The departure from the seabed was a singularly important step in the evolution of the group as a whole (Boyle and Rodhouse 2005). The immediate ancestor of the early cephalopods was believed to be a monoplacophoran, perhaps resembling *Knightoconia*, a monoplacophoran that had the right morphological characteristics to be a forebear, including a chambered shell (Yochelson et al. 1973; Webers and Yochelson 1989).

Cephalopods with an external shell, or ectocochleates, have two main lineages important in the fossil record and are

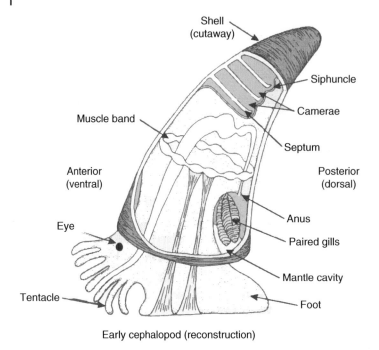

Figure 8.65 Reconstruction of the ancestral cephalopod *Plectronoceras cambria*. *Source:* From Yochelson et al. (1973), figure 3 (p. 282). Reproduced with the permission of John Wiley & Sons.

presently considered to be subclasses: the nautiloids and the ammonoids (Figure 8.66). The nautiloid lineage is fairly clear, arising in the later Cambrian and continuing to the present day. The ammonoids arose later, in the upper Silurian (Morton 1958), and were extinct by the end of the Cretaceous 65 mya, another victim of the great K-T extinction that wiped out the non-avian dinosaurs. Like the present-day nautiloids, ammonoids had chambered shells and are believed to have achieved neutral buoyancy in a similar way.

The coleoids form the third important cephalopod lineage, comprising the squids, the octopods, and the belemnoids, an extinct coleoid group with an internal shell. Though the coleoids clearly dominate the present-day cephalopod assemblage, their current species numbers pale in comparison to the 10 000 fossil ammonoid and nautiloid species known to science (Schrock and Twenhofel 1953), most of which lived in the Mesozoic era (225–65 mya). The first coleoid fossil known is the belemnoid *Eobelemnites,* which dates back to the early Carboniferous period (Mississipian epoch: 345 mya). Since *Eobelemnites* was already fairly advanced, it is likely that the group arose in the late Devonian era (Schrock and Twenhofel 1953).

The belemnoids (Figure 8.67) are considered the ancestral group of the coleoid line. They are believed to have arisen from a straight-shelled ectocochleate ancestor with the mantle gradually enveloping the external shell over evolutionary time. The belemnoids had an internal shell with three distinct regions: a heavily calcified posterior rostrum (or guard) of varying length, a phragmocone for buoyancy, and a proostracum that gave rigidity to the mantle cavity

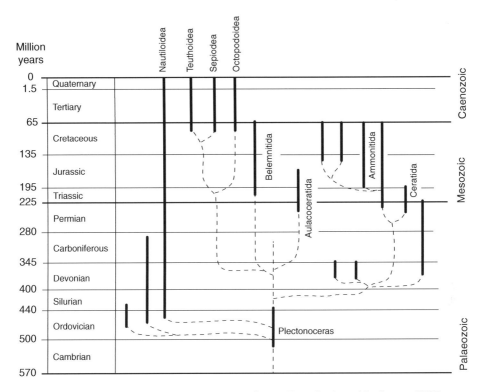

Figure 8.66 Cephalopod evolution and lineages. *Source:* From Boyle and Rodhouse (2005), Figure 3.3 (p. 41). Reproduced with the permission of John Wiley & Sons.

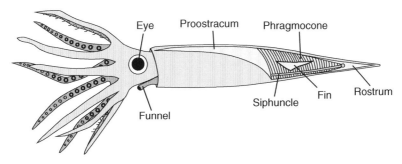

Figure 8.67 Reconstruction of an upper cretaceous belemnoid cephalopod.

and a site for mantle attachment. It is believed that the belemnoids were rapid horizontal swimmers much like modern squid, with the cigar-shaped rostrum conferring rigidity and balancing the phragmocone that was providing buoyancy at the middle of the body. The guard or rostrum is usually the only part of the belemnoid shell that survives fossilization. Belemnoids were modest in size with most being 30 cm or less, though the largest reached more than 2 m (Morton 1958). It is believed that most had 10 arms of equal length (Boyle and Rodhouse 2005). Like the ammonoids, the belemnoids died out at the end of the Cretaceous. One theory for their demise at the K-T boundary was that both the ammonoids and belemnoids produced small (1–2 mm) planktivorous larvae. The K-T extinction has been associated with a comet or asteroid impact at the end of the Cretaceous, resulting in an "impact winter" that would have severely restricted photosynthesis and made the survival of small planktivorous larvae very difficult. The cephalopod taxa that survived the K-T extinction event all produce large yolky eggs and well-developed young, substantially improving their chances for survival (Boyle and Rodhouse 2005).

Though the coleoiods are an ancient lineage, most of the radiation that produced our modern cephalopods occurred in the Tertiary period (65–1.5 mya). The chief characteristic of the modern cephalopods is the reduction or loss of the internal shell. The loss was accompanied by an immense increase in swimming ability, power, and activity (Boyle and Rodhouse 2005).

Classification

Cephalopod classification has not changed as profoundly as that of the gastropods and many of the crustacean groups, even with newly added phylogenetic information. Some orders and suborders have moved about, but the basic framework is much like the system of Naef (1921). We will consider living cephalopods only. The classification system below is from the World Register of Marine Species. Species numbers are given in parentheses for each group. The system below is similar but not identical to that of Brusca et al. (2016). Where taxonomic position is currently uncertain, groups are given a designation of "unassigned," such as for the Bathyteuthoidea and Idiosepiidae below. Those families designated with an "A" are considered ammoniacal, using sequestered ammonium as a buoyancy aid (Voight et al. 1994), which will be discussed in the "Locomotion and Buoyancy" section.

CLASS CEPHALOPODA (811)
  SUBCLASS COLEOIDEA (806)
    SUPERORDER DECAPODIFORMES (509)
      Order **Unassigned** Decapodiformes (14)
        Superfamily Bathyteuthoidea
          Family Bathyteuthidae (3) (A)
          Family Chenopterygidae (3)
          Family Idiosepiidae (8)
      Order Myopsida (49)
        Family Australiteuthidae (1)
        Family Loliginidae (48)
      Order Oegopsida (250)
        Family Ancistrocheiridae (1) (A)
        Family Architeuthidae (1) (A)
        Family Batoteuthidae (1) (A)
        Family Brachioteuthidae (7)
        Family Chiroteuthidae (19) (A)
        Family Cranchiidae (35) (A)
        Family Cycloteuthidae (4) (A)
        Family Enoploteuthidae (43) (A)
        Family Gonatidae (19)
        Family Histioteuthidae (18) (A)
        Family Joubiniteuthidae (1) (A)
        Family Lepidoteuthidae (1) (A)
        Family Lycoteuthidae (1)
        Family Magnapinnidae (3)
        Family Mastigoteuthidae (16) (A)
        Family Neoteuthidae (4) (A)
        Family Octopoteuthidae (8) (A)
        Family Ommastrephidae (22)
        Family Onychoteuthidae (27) (A)
        Family Pholidoteuthidae (2)
        Family Promachoteuthidae (3) (A)
        Family Psychroteuthidae (1)
        Family Pyroteuthidae (7)
        Family Thysanoteuthidae (1)
      Order Sepiida (195)
        Family Sepiadariidae (8)

Family Sepiidae (115)
Family Sepiolidae (72)
Order Spirulida (1)
Family Spirulidae (1)

SUPERORDER OCTOPODIFORMES (297)
Order Octopoda (296)
Suborder Cirrata (47)
Family Cirroctopodidae (4)
Family Cirroteuthidae (4)
Family Opisthoteuthidae (37)
Family Stauroteuthidae (2)
Suborder Incirrata (249)
Superfamily Argonautidae (10)
Family Alloposidae (1)
Family Argonautidae (4)
Family Ocythoidae (1)
Family Tremoctopodidae (4)
Superfamily Octopodoidea (239)
Family Amphitretidae (9)
Family Bathypolypodidae (7)
Family Eledonidae (7)
Family Enteroctopodidae (35)
Family Idioctopodidae (1)
Family Megaleledonidae (39)
Family Octopodidae (141)
Order Vampyromorpha (1)
Family Vampyroteuthidae

SUBCLASS NAUTILOIDEA
Order Nautilida (5)
Family Nautilidae (5)

Basic Anatomy of the Major Cephalopod Groups

The four major groups of living cephalopods are the nautilids, sepiids, squids, and octopods (Figure 8.68). We will briefly examine the major anatomical features of each, including unusual forms such as the spirulids and argonautids.

General

Cephalopods are bilaterally symmetrical, but their anteroposterior axis has shifted considerably from the "hypothetical ancestral mollusc" with its anterior head, dorsal external shell and visceral mass, and posterior anus. Unlike many of the gastropods, which exhibit a twist in the visceral mass relative to the foot known as torsion (see Figure 8.7), the cephalopods have shifted the orientation of the head and anterior foot into line with the visceral mass, with the mantle cavity and posterior foot bending around into a "U" (Figure 8.69). The posterior foot forms the siphon, and the relationship between anterio-posterior and dorsal-ventral is no longer intuitive (Figure 8.70). For example, squid normally swim backward. Russell-Hunter (1979) lists three basic differences between the cephalopods and the remainder of the molluscs. The first is the shifting of the anteroposterior axis illustrated in Figure 8.70. The second is the reduced importance of cilia in circulating seawater through the mantle cavity for respiration. The pumping activity of the muscular mantle wall ventilates the entire mantle cavity and provides the jet propulsion characteristic of the group. "Thirdly, in all modern cephalopods but Nautilus, the shell is either an internal structure of limited skeletal significance or, as in the octopods, almost entirely absent" (Russell-Hunter 1979).

The Nautilida

The nautilids (Figure 8.71) have more than 10 circumoral appendages, usually between 63 and 94 of them, and a chambered, coiled (planispiral), external shell. Despite the ancient lineage of the nautilids, the shell exhibits a typical three-layered molluscan structure with an outer proteinaceous periostracum, an intermediate layer of calcium carbonate (calcite) crystals embedded in a conchiolin matrix, and an innermost aragonitic nacreous layer. The animal lives in the outermost chamber with the body firmly

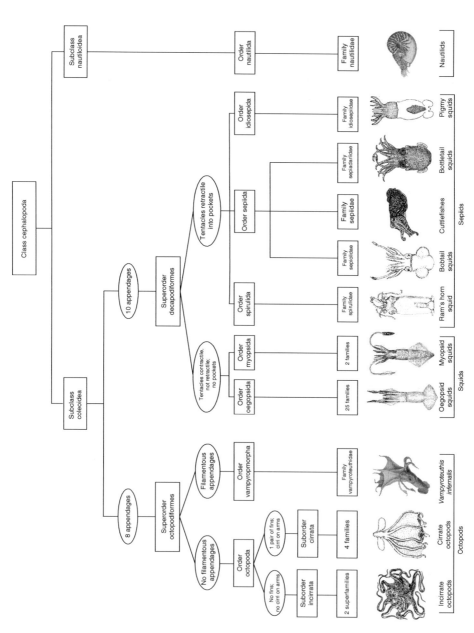

Figure 8.68 Classification scheme for the class Cephalopoda. *Source:* Food and Agricultural Organization of the United Nations (2005), P. Jereb and C.F.E. Roper, Cephalopods of the World Volume 1. Reproduced with permission.

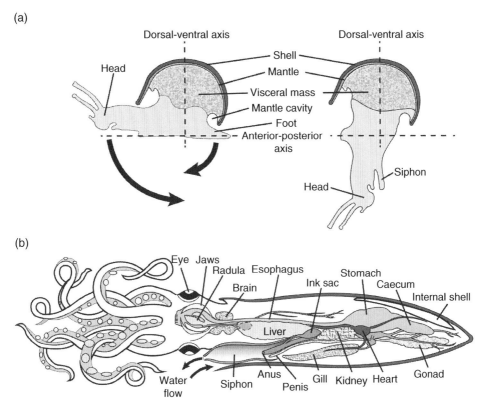

Figure 8.69 Basic cephalopod anatomy. (a) Reorganization of the major axes of an archetypic mollusc to that of a cephalopod; (b) basic anatomy of a stylized modern cephalopod.

attached to the outermost septum. A vascularized siphuncle, or siphon tube, connects the body with the gas-filled chambers behind it. The digestive tract includes a beak, radula, and muscular crop. The nautilid eye is unique, functioning analogously to a pinhole camera. Nautilids have two pairs of gills. Their buoyancy and vision will be treated along with other major organ systems in the "Cephalopod Systems" section.

The Sepiida and Spirulida

Until recently, the sepiids and spirulids formed a single order, the Sepioidea, which also included the pygmy squids, the Idiosepiidae. Present classification has split them into three orders. The order Sepiida includes the family Sepiidae (cuttlefishes), the family Sepiadariadae (the bottletail squids), and the family Sepiolidae (the bobtail squids) (Figure 8.72). The order Spirulida (ram's horn squids) has one species, *Spirula spirula*, in the family Spirulidae (Figure 8.73). Pygmy squids (Figure 8.74) have their own order, the Idiosepida.

The Sepiida and Spirulida all have ten sucker-bearing circumoral appendages (eight arms and two longer tentacles), sac-like bodies with lateral fins, and lack an external shell. Tentacles are contractile (able to be shortened) and can be partially retracted into pockets on the ventrolateral sides of the head. Arms have suckers over

532 | The Mollusca

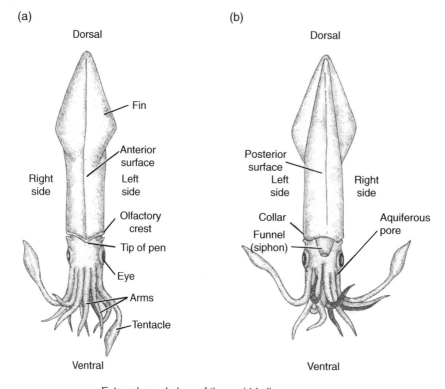

Figure 8.70 External morphology of the squid *Loligo*. (a) Anterior view; (b) posterior view. *Source:* From Brusca and Brusca (2003), figure 20.11 (p. 715). Reproduced with the permission of Oxford University Press.

their entire length; tentacles have a cluster of suckers only at their distal ends, forming a tentacular club (Figure 8.75) (Roper et al. 1984). Individual suckers are mounted on short stalks and are equipped with a ring of chitinous teeth around the periphery (Figure 8.75). The family Sepiidae have a laminate calcareous internal shell, or cuttlebone, that provides buoyancy, whereas the internal shell of the Sepiolidae is a chitinous gladius or pen, and the Sepiaridae lack an internal shell altogether. The spirulids have an internal chambered shell (Figure 8.73) resembling that of the nautilids and conferring buoyancy in a similar way. Sepiid internal anatomy (Figure 8.76) includes a beak, radula, muscular digestive tract, one pair of gills, and a heart with two auricles. Circulation through each gill is aided by a branchial heart. Note the position of the internal shell. Like most of their cephalopod brethren, the sepiids, spirulids, and idiosepiids have exceptionally well-developed eyes.

The Myopsida and Oegopsida

The orders Myopsida and Oegopsida comprise the true squids. They were formerly united as suborders in the now-defunct order Teuthoidea (Roper et al. 1984). Like the sepiids and spirulids, the myopsids and oegopsids have 10 circumoral appendages, eight arms plus two tentacles, with stalked suckers (myopsids) or stalked suckers and

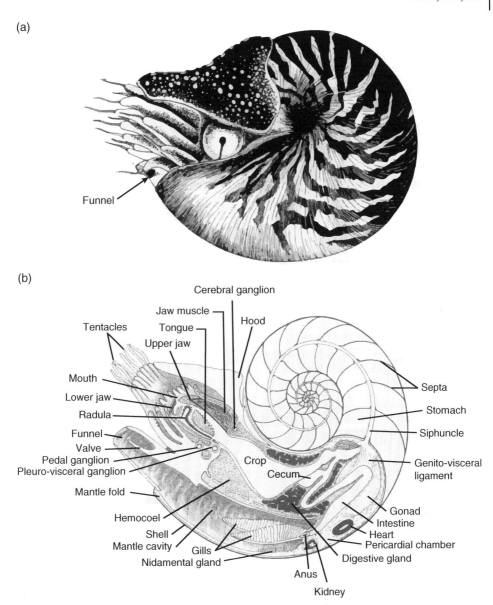

Figure 8.71 Anatomy of the nautilids. (a) External view of *Nautilus* sp. (b) internal anatomy of *Nautilus* sp. – stomach hidden by shell. *Sources:* (a) Roper et al. (1984), figure 33 (p. 15); (b) Barnes (1974), figure 11-79 (p. 412).

hooks (oegopsids) (Figure 8.75d, e). The chief anatomical differences between them and the cuttlefishes are in the tentacles and the internal shell. The tentacles of squids are contractile, but there are no pockets on the side of the head for them to retract into and their internal shell (gladius or pen) is always chitinous and always straight and simple, with either a feather-like or rod-like shape.

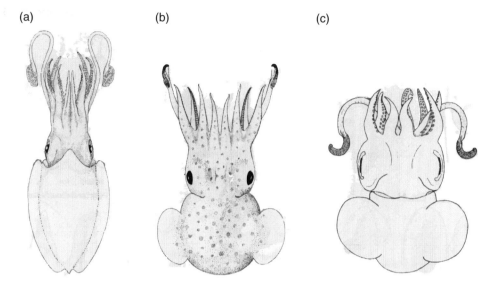

Figure 8.72 The order Sepiida. (a) The cuttlefish *Sepia* (family Sepiidae); (b) the bottletail squid *Sepiadarium* (family Sepiadariidae); (c) the bobtail squid *Rossia* (family Sepiolidae). *Sources:* (a) Roper et al. (1984), figure 35; (b) Roper et al. (1984), figure 37; (c) Roper et al. (1984), figure 36. Reproduced with permission of The Food and Agriculture Organization.

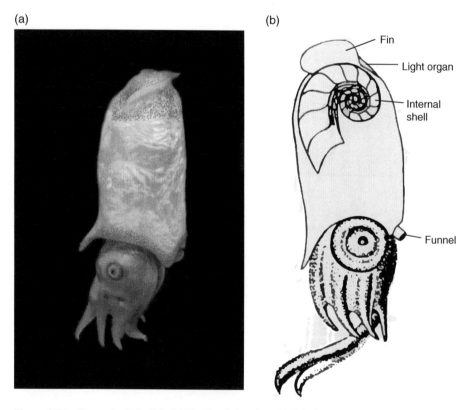

Figure 8.73 The order Spirulida. (a) The Ram's head squid *Spirula spirula*; (b) *Spirula spirula* in normal swimming position showing the position of the internal shell. *Sources:* (a) © Dante Fenolio, reproduced with permission; (b) Denton and Gilpin-Brown (1973), figure 7 (p. 216). Reproduced with permission of Elsevier. See color plate section for color representation of this figure.

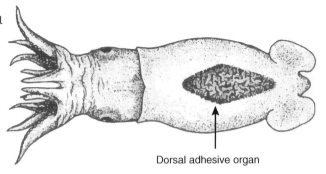

Figure 8.74 The pigmy squid *Idiosepius* sp. showing the dorsal adhesive organ. *Source:* From Boyle and Rodhouse (2005), figure A17 (p. 357). Reproduced with the permission of John Wiley & Sons.

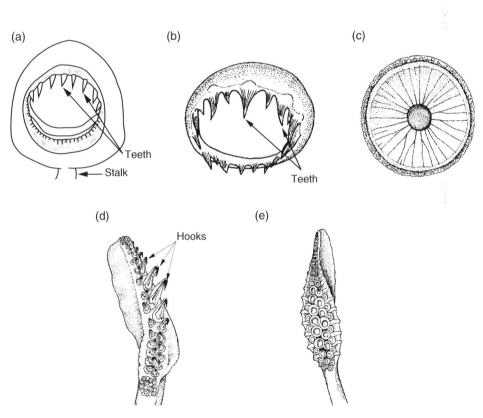

Figure 8.75 Cephalopod tentacular structure. (a) Stalked tentacular sucker with chitinous teeth; (b) close up of teeth; (c) octopod sucker; (d) tentacular club with hooks and stalked suckers; (e) tentacular club with stalked sucker and no hooks. *Sources:* (a) Roper et al. (1984), figure 29 (p. 13); (b) Roper et al. (1984), figure 30 (p. 13); (c) Jereb et al. (2016), figure 29 (p. 31). Reproduced with permission of The Food and Agriculture Organization; (d) Roper et al. (1984), figure 44 (p. 17). Reproduced with permission of The Food and Agriculture Organization; (e) Jereb and Roper (2005), figure 52 (p. 39). Reproduced with permission of The Food and Agriculture Organization.

Their internal anatomy (Figure 8.76) is similar to that of the cuttlefishes, with a beak, radula, muscular gut, single pair of gills, and a heart with two auricles. The chief difference between the myopsid and oegopsid squids is that the eyes of the myopsids, such as *Loligo*, have a corneal membrane and the oegopsids do not. The oegopsid eye is in direct contact with seawater.

The Octopodiformes

The superorder Octopodiformes includes all the cephalopods with eight circumoral appendages; they lack tentacles. Two orders comprise the Octopodiformes: the Octopoda, with 2 suborders and 296 species, and the Vampyromorpha, a monotypic order with one widely distributed deep-sea species, *Vampyroteuthis infernalis*. With the exception of *Vampyroteuthis*, the octopods have suckers without stalks or chitinous teeth (Figure 8.75c). The suckers of *Vampyroteuthis* are on stalks.

The internal anatomy of the octopods (Figure 8.77) has the same basic layout as that of the squids and cuttlefishes. A muscular gut with a crop and multiple **cecae** to aid in food breakdown and absorption ends in an anus that discharges to the mantle cavity. A closed circulatory system is powered by a systemic heart with two auricles (termed diotocardiac) aided by a branchial heart for each of the two gills. A sophisticated nervous system, robust mantle musculature, and advanced eyes provide the basis for a highly mobile visual predator. The predominantly benthic lifestyle of the octopods is evident in their globular visceral hump, a very different design from that of

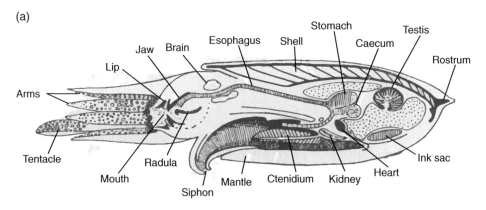

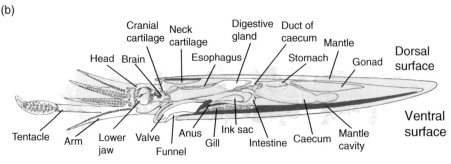

Figure 8.76 Sepiida anatomy. (a) Generalized internal anatomy of *Sepia* sp.; (b) internal anatomy of the myopsid squid *Loligo* sp.

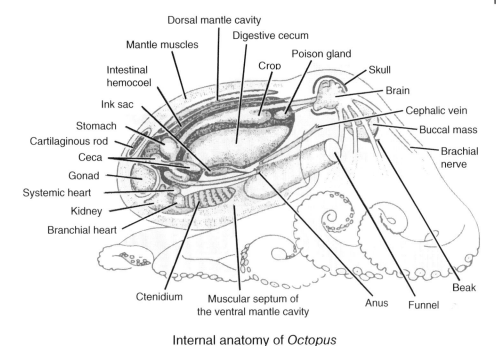

Internal anatomy of *Octopus*

Figure 8.77 Internal anatomy of *Octopus* sp. *Source:* From Brusca and Brusca (2003), figure 20.12 (p. 715). Reproduced with the permission of Oxford University Press.

the torpedo-like squids. All that remains of the internal shell in octopods is a pair of small cartilaginous rods.

Two suborders, Cirrata and Incirrata, make up the Octopoda. The Cirrata are distinguished by the presence of cirri (slender, elongate, fleshy protuberances) on their arms and paddle-shaped fins, both of which are lacking in the Incirrata (Figure 8.78) (Roper et al. 1984). In addition, many of the Cirrata have pronounced webbing between the arms to aid in locomotion. Both groups have pelagic representatives, with the Cirrata being better represented in the deep sea.

Argonauta, the paper nautilus, is a true octopod with an unusual morphology and some unusual habits. Those include the manufacture of a paper-thin external shell in females, usually termed an egg case but persists as a shelter. The male *Argonauta* is diminutive, less than 10% of the size of the female, and has no shell. Female *Argonauta* (Figure 8.79) reside with their visceral hump inside the shell and their arms turned backward, exposing their suckers and beak. The most dorsal (first) pair of arms bear an extensive web, a very thin, membranous, flattened extension of the dorsal side of each arm (Young 1959). The web of each dorsal arm covers its side of the shell and is responsible for its secretion. Figure 8.79a is a drawing of *Argonauta* with the web of the left arm covering the shell and that of the right spread out on the side of its aquarium. The drawing is meant to show the character of the web but at first glance is quite confusing. When swimming, each side of the animal is symmetrical, with the webs of the first pair of arms covering their respective

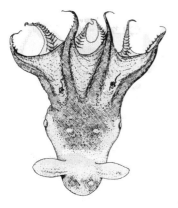
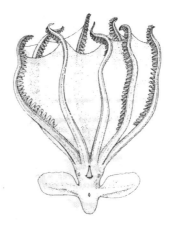

(a) Order Vampyromorpha,
Family Vampyroteuthidae,
Vampyroteuthis infernalis
(dorsal view)

(b) Order Octopoda,
Suborder Cirrata,
Family Cirroteuthidae,
Cirrothauma species
(ventral view)

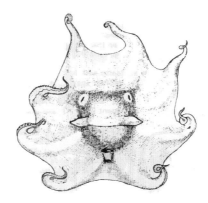

(c) Order Octopoda,
Suborder Incirrata,
Family Alloposidae,
Haliphron atlanticus
(ventral view)

(d) Order Octopoda,
Suborder Cirrata,
Family Opisthoteuthidae,
Opisthoteuthis species
(top view)

Figure 8.78 Representatives of the orders and suborders of the Octopodiformes. (a) Order Vampyromorpha: *Vampyroteuthis infernalis*; (b) order Octopoda, suborder Cirrata: *Cirrothauma* sp.; (c) order Octopoda, suborder Incirrata: *Haliphron atlanticus*; (d) order Octopoda, suborder Cirrata: *Opisthoteuthis* sp. *Sources:* (a) Roper et al. (1984), figure 67; (b) Roper et al. (1984), figure 68; (c) Roper et al. (1984), figure 70; (d) Roper et al. (1984), figure 69. Reproduced with permission of The Food and Agriculture Organization.

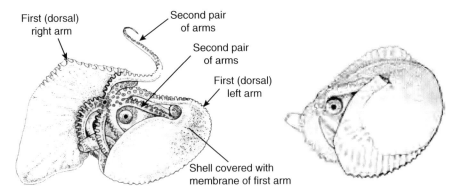

(a) Argonautidae, *Argonauta* sp., female, lateral view

(b) *Argonauta* sp., female, swimming position

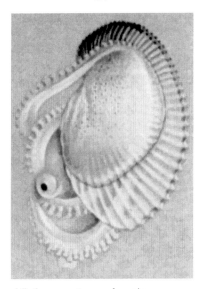

(c) *Argonauta* sp., female, web being pulled across shell by arm

(d) *Argonauta* sp., female, further spreading of the web

Figure 8.79 The paper nautilus *Argonauta*. (a) Lateral view of a female *Argonauta* showing the arrangement of the first and second pairs of arms; (b) lateral view of a female *Argonauta* in swimming position; (c) lateral view of a female *Argonauta* showing the web being pulled across the shell; (d) further spreading of the web across the shell. *Sources:* Roper et al. (1984), figure 76 (p. 24). Reproduced with permission of The Food and Agriculture Organization; (b) Jereb et al. (2016) (p. 230); (c and d) Young (1959), plate 2. Reproduced with permission of The Food and Agriculture Organization.

sides of the shell or tucked inside the shell (Figure 8.79b). Figures 8.79c and d show the web being deployed across the shell, pulled by the suckers on the arms (Young 1959). The shell of the paper nautilus is unchambered and, unlike its distant relative the pearly nautilus, the animal is not attached to it. It may leave as needed. The web covers the shell at intervals, often for long periods, and is iridescent white,

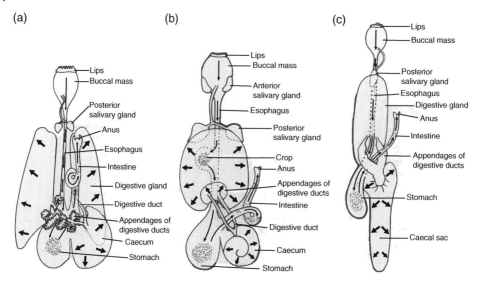

Figure 8.80 Familial differences in cephalopod digestive systems. Thin arrows show the course of particles within the digestive tract; thick arrows show the sites of absorption. (a) Family Sepiidae; *Sepia officinalis;* (b) Family Octopodidae, *Octopus vulgaris;* (c) Family Loliginidae; *Loligo vulgaris* (ink sac not represented). *Source:* From Boucaud-Camou and Boucher-Rodoni (1983), figure 4 (p. 165). Reproduced with the permission of Academic Press.

presumably to blend in with the shimmering surface waters in the epipelagic zone where it resides.

Cephalopod Systems

Feeding and Digestion

Cephalopods are active visual predators feeding mainly on crustaceans and fishes and, depending upon habitat and lifestyle, on other molluscs as well. Most field-based observations of feeding behavior have been obtained on coastal and benthic species, e.g. octopuses, cuttlefishes, and *Loligo* (Boucaud-Camou and Boucher-Rodoni 1983), with additional information provided by laboratory-based studies (Boletzky and Hanlon 1983, Rodhouse and Nigmatullin 1996). Squid and cuttlefish both capture prey by maneuvering to within striking range and snagging the prey with a rapid extension of the tentacles, usually accompanied by a rapid jet forward. The tentacles convey the prey to the mouth region where it is manipulated by the arms; the beak dispatches the prey into bite-sized chunks that are swallowed whole (Messenger 1968; Kier and Van Leeuwen 1997; Boyle and Rodhouse 2005).

Benthic octopods capture prey with a similar strategy. Prey is located visually while moving over the bottom, approached to within striking distance, and trapped by a jet-propelled leap (rush and grab) onto the prey, trapping it within the webbing between the arms (interbrachial webbing) and holding it there until it is poisoned, rendered, and consumed (Wells 1978). Cirrate octopods have been observed floating just above the benthos at depths exceeding 2500 m (Roper and Brundage 1972) with cirri erect and their legs extended parallel to the bottom, their interbrachial webbing giving them a decidedly umbrella-like appearance. The cirri were believed to act as mechanoreceptors, an aid in detecting the small-scale turbulence of prey in the quiet waters of the deep ocean. Similarly, the subtropical epipelagic *Argonauta* has a well-developed tactile sense, providing a strong feeding response to prey touching the web that overhangs its shell. It would also seize nearby prey detected visually or chemically (or both). In hunting mode, *Nautilus* spreads a group of its many arms to form a "cone of search" while swimming forward near the bottom. When a food item is encountered, it is captured and conveyed to the mouth by another set of arms (Bidder 1962). Haven (1972) noted that an aquarium specimen of *Nautilus pompilius* used its "cone of search" to explore the bottom with the tips of its arms rapidly touching the sediment. Food introduced to the aquarium was quickly located, presumably via chemoreception, but the food item was not seized until it was directly touched, underscoring the importance of the tactile senses to feeding in *Nautilus*. Haven (1972) concluded that *Nautilus* is primarily a bottom feeder.

Cephalopods all kill or paralyze their prey prior to ingesting it. Squids kill their prey outright. Those feeding on fish may sever the spinal cord with their beak; crustaceans are dispatched by simple dismemberment. In contrast, Sepiida and most of the octopods poison their prey. It has been experimentally demonstrated using gland extracts that the posterior salivary glands are actually poison glands, containing a variety of bioactive compounds including neurotransmitters and other amines, enzymes, and toxins. Cephalopod toxins include a class of compounds termed cephalotoxins, glycoproteins that inhibit neurotransmission thereby paralyzing and eventually killing their prey (Boucaud-Camou and Boucher-Rodoni 1983). Perhaps most famous for its poison is the blue-ringed octopus, which kills with tetrodotoxin, a particularly virulent class of neurotoxin that attacks nerve cells directly. Tetrodotoxin's mechanism of action is the same as that of the famous Japanese puffer fish used in Fugu, and like the flesh of a poorly prepared Fugu fish, the bite of the blue-ringed octopus can be fatal to humans.

The Digestive Tract

The alimentary canal of cephalopods comprises the buccal mass, the esophagus and crop (if present; not in decapodiforms), the stomach, digestive gland, caecum, and intestine. The cephalopods show a few embellishments on the typical molluscan design (Figure 8.80). The buccal mass, located within the circle of arms, is quite mobile. It can be rotated and protruded to make the most out of the powerful cephalopod beak (Boucaud-Camou and Boucher-Rodoni 1983). The prominent beak of cephalopods defines the internal cavity of the buccal mass, which in nearly all cases includes a strong radula. The buccal cavity receives the secretion of a variety of glands including the anterior and posterior salivary glands. The glands' secretions produce enzymes involved in the initial stages of digestion, mucous for lubricating the food's trip down the digestive tract, and toxins for poisoning captured prey in the sepiids and octopods. The chitinous beak is responsible for biting during food capture and, in tandem with the arms, for tearing the prey into digestible pieces.

The esophagus, with the food fragments rendered by the beak, exits the buccal mass dorsally and descends into what is essentially a U-shaped digestive tract (Boucaud-Camou and Boucher-Rodoni 1983) with descending and ascending limbs. In octopods and nautilids, the esophagus is swollen into a crop (Figure 8.80b) used for storage of food prior to its arrival at the stomach. In the coleoids and nautilids, food fragments are moved by the esophagus to the stomach via peristalsis. Stomach and esophagus are both lined with a soft chitinous cuticle.

In all cephalopods, the stomach and caecum open side by side into the bend of the "U," with the short intestine then continuing (ascending) into the mantle cavity and terminating in an anus near the funnel for excretion (Figure 8.80). The large digestive gland is anterior to the caecum and connected to it by the digestive ducts. Thus, in all the coleoids, the stomach, digestive gland, caecum, and intestine are connected at the bend of the "U." The connection takes place in the anterior portion of the caecum, which is actually an elaborate switching system. Each of the components (the stomach, digestive gland, caecum, and intestine) may be connected to one another or may be isolated using a system of biological valves. The valves are in the form of sphincters and flow is directed by grooves in the anterior, spiral, section of the caecum. To further complicate matters, flow between components may proceed in two directions.

Food reaching the stomach is broken down by a combination of the churning action of the muscular stomach and a suite of digestive enzymes manufactured in the digestive gland. Enzymes synthesized in the digestive gland reach the stomach through the paired digestive ducts connecting it to the caecum. The enzymes are then delivered to the stomach via the digestive groove in the anterior caecum, part of the switching system mentioned above. The role of the stomach is to break down the chunks of food received by the esophagus and separate useful food particles from larger indigestible fragments. The useful food particles are passed along into the caecum for further breakdown in the digestive portions of the organ. The larger indigestible portions of the food are passed at intervals into the anterior caecum where they are twirled into skeins of mucous and moved into the intestine for excretion by a specialized region of the caecum called the ciliated organ.

The anterior section of the caecum contains the ciliated organ as well as the digestive groove, openings from the intestine and digestive gland, and their host of complex structures. The posterior is a simple sac that is concerned only with digestion and absorption. The entire organ is muscular and lined with a ciliated epithelium which in the ciliated organ is interspersed with mucous cells (Bidder 1950). Cilia in the posterior sac keep the digestive fluid stirred to maximize contact and absorption. In the squids, it is believed that much of the final breakdown and absorption occurs in the large posterior caecum and a little in the intestine. In the sepiids and octopods, those functions are also attributed in large measure to the digestive gland (Boucaud-Camou and Boucher-Rodoni 1983). More recent work by Semmens (2002) found evidence of absorption in the digestive gland of *Sepioteuthis*, a loliginid squid. The consensus is that the digestive gland plays less of a role in absorption in the squids than it does in the other two groups of coleoids.

The digestive gland, originally described as the liver and pancreas, comprises two quite different parts connected in series (Boucaud-Camou and Boucher-Rodoni 1983). The anterior portion, known historically as the "liver," is a large brown gland located anterior to the caecum and covered by a "delicate, transparent muscular sheath" (Bidder 1950). It is connected to the caecum by a pair of ducts that fuse into one just before entering the anterior caecum. The "pancreas" is a spongy outgrowth of the paired digestive ducts and its secretions, like those of the liver, pass into the caecum via the digestive ducts. Bidder (1976) renamed the "pancreas" the "digestive ducts appendages" (Figure 8.80), a more accurate name that is now the accepted term for the gland. Both portions of the digestive gland are basically a system of branching tubules draining secretory cells and opening into the digestive ducts (Bidder 1950).

Proteases, carbohydrases, and lipases have all been identified in the alimentary canal of squids, octopods, and cuttlefish. The major site of synthesis in all cases is believed to be the digestive gland, with small supplementary contributions provided by the salivary glands (Boucaud-Camou and Boucher-Rodoni 1983).

The short intestine has a substantial layer of circular muscles and a thinner layer of longitudinal muscles in its wall. It is well vascularized, supporting peristalsis and providing an absorptive function as well. The intestinal lumen is highly ciliated with an abundance of mucous-producing cells that presumably protect the intestinal lining from any sharp edges in the indigestible material voided by the stomach. The intestine is essentially continuous with the outflow from the ciliated organ of the caecum. Cilia help to move the food along to the anus as well as keeping absorptive surfaces free from mucous. The ink gland connects to the intestine just before the anus.

Movement of Food

Food is killed, either by physical trauma or poison from the salivary glands, bitten into chunks, and swallowed with the help of the radula. Peristalsis in the esophagus aided by lubrication from the buccal gland moves the fragments to the crop and eventually the stomach in octopods and sepiids, or directly to the stomach in squids. In the stomach, the food is broken down into a slurry of liquid and fine particles by its churning action and by the lytic activity of digestive enzymes originating in the digestive gland. The partially digested food then passes at intervals into the caecum where it is conveyed to the posterior sac for further digestion and absorption. In sepiids and octopods, partially digested food passes through the digestive ducts into the digestive gland as well, which performs similar breakdown and absorption duties to the posterior sac of the squids. Muscular movements and ciliary activity in the posterior sac facilitate breakdown and absorption, aided by enzymes from the digestive gland. Solid particles are moved from the caecum to the mucous groove of the intestine by the action of the ciliated organ. The intestine, which produces mucous and is itself heavily ciliated, allows final absorption before egesta are excreted as mucous strings, exiting the mantle cavity through the funnel. Large, indigestible fragments are passed directly from stomach to intestine through the anterior caecum where they are wrapped in mucous and excreted. Large particles do not enter the posterior caecum (Bidder 1950, 1966; Boucaud-Camou and Boucher-Rodoni 1983).

Total time for digestion varies considerably between taxa. In sepiids and octopods, it ranges from 15 to 20 hours; in the loliginid squids, it is approximately 4 hours. Numbers are unavailable for deeper-living oceanic squids. The rapid digestion time in the loliginid squids is attributed to the fact that most of their final digestion and absorption takes place in the posterior caecum and intestine rather than in the digestive gland, which is dedicated to producing digestive enzymes. The digestive gland of octopods and sepiids must serve three functions: enzyme production as well as breakdown and absorption of food. Enzyme secretion and transport to the stomach for initial breakdown cannot take place at the same time as food breakdown and absorption in the gland itself, which therefore increases total digestion time.

Assimilation, that portion of digestible food translated into growth and metabolism, is high in all cephalopods. Growth rates tend to be fast and life spans short (Boyle and Rodhouse 2005). Part of the reason for rapid cephalopod growth is their carnivorous diet and hunting prowess. Another reason is the efficiency of their digestive tract, specifically the early separation of digestible from indigestible fractions by the stomach, maximizing the efficiency of final digestion and absorption (Bidder 1966; Boucaud-Camou and Boucher-Rodoni 1983).

Circulation

Circulation in the cephalopods is in a league with the vertebrates in terms of both structure and efficiency. The basic layout of the cephalopod circulatory system resembles that of their gastropod relatives, with a systemic heart receiving oxygenated blood from the gills and pumping it throughout the body, followed by an eventual venous return through the gills (Figure 8.81). In the nautilids, a suite of venous sinuses drains the organ and limb systems prior to the blood's return. In the coleoids, the arterial circulatory system is entirely closed; venous sinuses are well defined and fewer in number but nonetheless present. Return to the heart through the single pair of gills is facilitated by a pair of auxiliary pumps, the branchial hearts, absent in the nautilids (Wells 1983).

All cephalopods have three main arteries leaving the heart (Figure 8.81b). The largest is the dorsal aorta, which supplies all organs and structures of the digestive tract and cephalic region including the buccal mass, arms, and funnel. The posterior aorta supplies the mantle musculature and fins, a substantial job in the active squids, as well as the hearts themselves. The third, or gonadal artery, directly supplies the gonad.

In squids, anterior venous return to the heart includes a large buccal sinus enveloping the buccal mass, followed by a series of three sinuses directly ventral to it, the optic, cephalic, and salivary sinuses, respectively (Figure 8.81c). All drain into the large anterior vena cava. Blood returns from the mantle and fins via the posterior vena cava. Anterior and posterior vena cavae feed into the lateral vena cavae, which in turn supply the branchial hearts and gills.

The blood pigment of cephalopods, like that of gastropods and crustaceans, is hemocyanin, a copper-containing pigment with a high molecular weight: 800 000 to several million daltons. Hemocyanin increases the amount of oxygen dissolved in cephalopod blood, its "carrying capacity," from about 0.54 Vol. % (for seawater at 20 °C) to about 3.5 Vol. %: a 6.5-fold change. The unit for carrying capacity, Vol. %, comes from the medical literature. It is shorthand for ml O_2 100 ml^{-1} of blood. The increase in carrying capacity provided by the hemocyanin of cephalopods is modest relative to the hemoglobin of fishes (4.9–19.7 ml O_2 100 ml^{-1}) or mammals (14–32 ml O_2 100 ml^{-1}) (Withers 1992). Nonetheless, it greatly increases the amount of oxygen that can be taken up by the gills and delivered to the tissues, supporting a high rate of metabolism and an active lifestyle.

Gas Exchange

Ventilation of the gills in the cuttlefish and squid is a straightforward tidal process. Expansion of the mantle by contraction of its longitudinal muscle layer draws water into the mantle cavity between the head and its free anterior edge, or collar

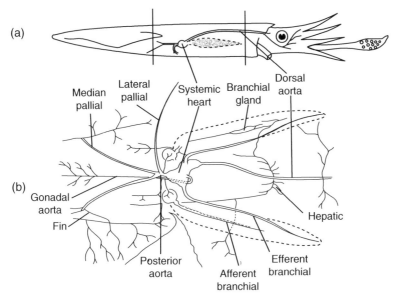

Arterial system of *Loligo* (a): side view, (b): detail, ventral view

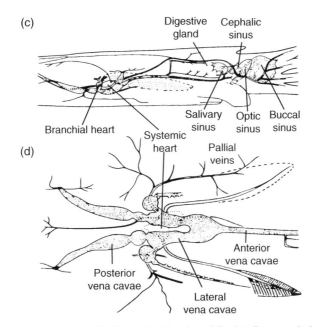

Venous system of *Loligo*. (c): side view, (d): detail, ventral view

Figure 8.81 Cephalopod circulation. (a) Lateral view of the basic arterial circulatory system of *Loligo*; (b) detailed ventral view of the arterial circulatory system of *Loligo*; (c) lateral view of the venous circulatory system of *Loligo*; (d) ventral view of the venous circulatory system of *Loligo*. *Sources:* From Wells (1983), (a and b) figure 2 (p. 243); (c and d) figure 4 (p. 245).

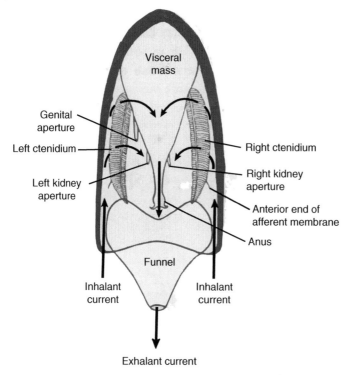

Sepia officinalis. Mantle cavity showing ctenidia and respiratory currents

Figure 8.82 Cephalopod respiration. Respiratory system of the cuttlefish *Sepia officinalis*. Arrows show the direction of flow of inhalant and exhalant currents and the pathway through the ctenidia (gills). *Source:* Yonge (1947), figure 35 (p. 502). Reproduced with permission of The Royal Society.

(Figure 8.76b). Contraction of the mantle's circular muscle layer decreases the size of the cavity and draws the mantle tightly around the head, forcing the water out the funnel. Water flows through the gill and into the posterior expanded mantle cavity, or post-branchial chamber, during the inspiratory phase. It is expelled through a "channel below and between the gills and then out through the funnel" during the expiratory phase (Wells 1983) (Figure 8.82).

Direction of the ventilatory flow opposes that of the blood flow through the ctenidia, creating a countercurrent exchange for efficient oxygen extraction (Figure 8.83). However, despite well-developed gills and a countercurrent flow, oxygen extraction is highly variable within and between species (Table 8.5). The main reason for the variability is that the water flow ventilating the gills is also used for jet propulsion. For benthically oriented species with a more sedentary lifestyle, such as the cuttlefishes and octopods, the residence time of the tidal flow within the mantle chamber is longer, time for diffusion is greater, and the percentage utilization of oxygen is far higher (33–80%) than in the pelagic species which are constantly swimming (5–10%). Within species, the efficiency of oxygen extraction depends on how vigorously the animals are jetting, with the lowest values

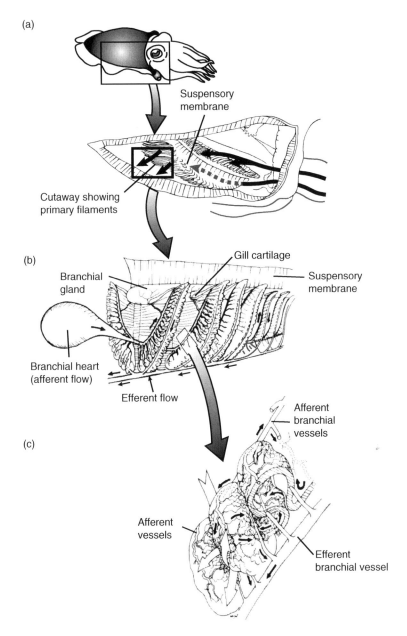

Figure 8.83 Countercurrent exchange in cephalopod respiration. (a) General position of the gills in *Sepia*. Cutout of the area indicated by the box showing the gills and heart. Arrows show direction of water flow through the gills; (b) close-up of the area indicated by the box in (a) showing the relationship of the primary gill lamellae and circulatory structures. Black arrows show the direction of blood flow; (c) close-up of the secondary gill lamellae. White arrows show direction of water flow, black arrows show direction of blood flow opposite to that of the water resulting in countercurrent exchange of respiratory gases. Note that both sides of the secondary lamellar folds are active in gas exchange. *Source:* Adapted from Wells (1983), figure 10 (pp. 256, 257). Reproduced with the permission of Academic Press.

Table 8.5 Efficiency of oxygen extraction in cephalopods.

Order	Family	Species	Percentage utilization of O_2	References
Sepiida	Sepiidae (cuttlefish)	Sepia officinalis	51	Wells and Wells (1982)
		Sepia officinalis	53	Wells and Wells (1991)
		Sepia officinalis	80	Melzner et al. (2006)
Myopsida (squid)	Loliginidae	Lolliguncula brevis	<10	Wells et al. (1988)
Oegopsida (squid)	Ommastrephidae	Illex illecebrosus	<10	Melzner et al. (2006)
Nautilida	Nautilidae	Nautilus pompilius	<10	Wells and Wells (1985a)
Octopoda (suborder Incirrata)	Octopodidae	Octopus vulgaris	>60	Wells and Wells (1982)
		Octopus vulgaris	46	Wells and Wells (1985b)
		Octopus vulgaris	50–80	Hazelhof (1939)
		Octopus vulgaris	70	Winterstein (1909)

accompanying the most rapid swimming (Wells and Wells 1982).

Excretion

The fundamentals of excretion in cephalopods are similar to those of the gastropods, and the annelids and crustaceans as well. A primary urine is formed by ultrafiltration of the blood that is subsequently modified by secretion and reabsorption prior to excretion. However, the excretory system of the cephalopods is the most complex we have encountered thus far due to the many elements composing it.

Structures important to the production of the final urine are the anterior, posterior, and lateral vena cavae (Figure 8.81d), the branchial hearts and branchial heart appendages, the renal appendages, and the paired renal sacs. The paired kidney sacs of the squids and cuttlefishes are connected to a third median dorsal pouch, also connected to the pericardial cavity and considered to be part of the nephric system, but its function in urine production is unclear. The best information available on form and function of the excretory system is from the cuttlefishes and octopods, largely because they are more readily captured and maintained than the more highly active squids. For the present discussion, our model organism is the cuttlefish, *Sepia officinalis*.

Figure 8.84a shows the internal anatomy of *Sepia* including the large median coelomic cavity that is retained through adulthood in the squids and cuttlefishes. In gastropods and other molluscs, the pericardium remains as the last vestige of the coelomic cavity, but in *Nautilus*, squids, and cuttlefishes the pericaridium joins via a constriction with another coelomic space that contains the gonads: a gonocoel (Kaestner 1967).

Excretion in the cephalopods is best understood by following the path of blood returning to the systemic heart. A good diagram is presented in Figure 8.85, showing the entire excretory system of *Sepia* in posterior aspect (Schipp and von Boletzky 1975). Blood returns from the cephalic region along the body midline via the large anterior vena cava, which bifurcates into the lateral vena cavae just prior to reaching the renal (or kidney) sacs. The lateral vena cavae enter and pass through the corresponding kidney sacs. Within the sacs, the vena cavae are invested with the renal appendages, globular protrusions of the vein that have a secretory function. Blood entering the renal appendages passes through progressively smaller channels until reaching a thin epithelium (Martin 1983). It is believed that much of the ammonia excreted in the final urine originates by diffusion from the blood via the renal appendages (Boucher-Rodoni and Mangold 1994).

Once past the renal sacs the lateral vena cavae join the branchial hearts, as do the posterior vena cavae (shown in Figure 8.81) bringing venous blood from the posterior mantle. The ultrafilters that form the initial urine for cephalopods are the pericardial glands, also identified as the branchial heart appendages. The hydrostatic pressure required for ultrafiltration is in turn provided by each of the branchial hearts. The pericardial glands drain to the pericardial coelom which itself drains to the kidney sacs via the renopericardial ducts. The kidney sacs void at intervals to the mantle cavity via the nephropores.

Ultrafiltration, reabsorption, and secretion can be experimentally demonstrated by introducing tracers into the blood and monitoring the relative concentrations of the tracers in the urine and blood over time: the urine to blood or U/B ratio. Ultrafiltration has been demonstrated experimentally by noting the rapid appearance of inulin in the

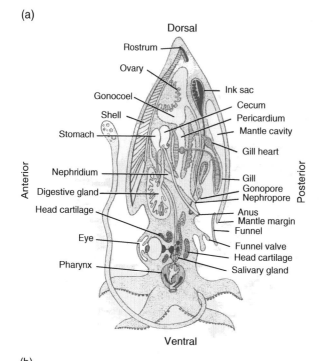
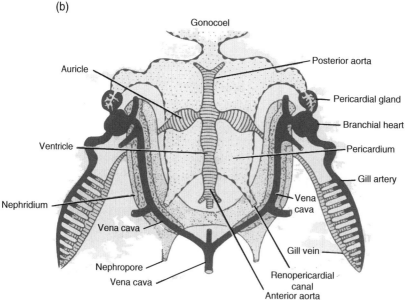

Figure 8.84 Cephalopod excretory systems. (a) Internal anatomy of the cuttlefish *Sepia* with eye turned 90° (longitudinal section); (b) excretory and respiratory organs of *Sepia* (dorsal view). *Sources:* Kaestner (1967), figure 15-12 (p. 404); (b) Kaestner (1967), figure 15-14 (p. 405). Reproduced with permission of John Wiley & Sons.

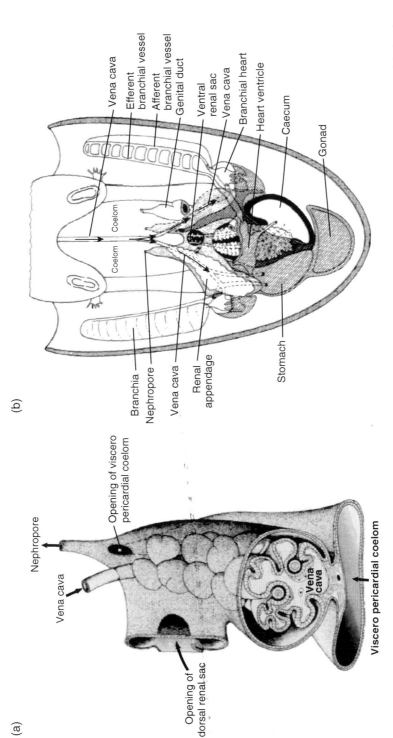

Figure 8.85 Cephalopod excretion. (a) Renal sac structure in the cuttlefish *Sepia*; (b) excretory system of *Sepia* showing the relationship with the visceral organs and circulatory system. Arrows show blood flow through the excretory organs where filtration and secretion occur. *Sources*: From Schipp and von Boletzy (1975), figure 10 (p. 100); (b) Schipp and von Boletzy (1975), figure 1 (p. 90).

urine. Inulin is a tracer compound that is 100% ultrafiltered but neither secreted nor reabsorbed (Potts 1967). It disappears quickly into the urine once introduced into the blood. Glucose is actively reabsorbed by the cephalopod nephridium, and the indicator compound para-aminohippuric acid (PAH) is actively secreted. The metabolic inhibitor phlorizin when introduced into the blood effectively prevents glucose reabsorption, causing glucose to appear in the urine when administered. In contrast, if the secretory inhibitor 2,4-dinitrophenol (DNP) is introduced into the blood, PAH concentrations in the urine rapidly decline.

Main sites of secretion and reabsorption in cephalopods are the kidney sacs themselves, which exhibit considerable infolding to increase surface area. Glucose reabsorption in the renopericardial canal has been demonstrated as well (Potts 1967; Martin 1983). Extra-renal excretion of ammonia takes place in the gills of cephalopods, equaling or exceeding the kidneys in importance as a site of nitrogen excretion (Potts 1967).

Nervous System and Sensory Mechanisms

The cephalopod nervous system is equipped with a large, sophisticated brain, a giant fiber system facilitating rapid communication with the mantle musculature, and a highly developed sensory system. Without question, the cephalopod nervous system is unmatched within the invertebrates, enabling a complex suite of behaviors not seen outside of the vertebrates.

The Brain

The circumesophageal nerve ring and ganglia typical of the gastropods have consolidated and greatly increased in complexity to become a true multi-lobed brain in the cephalopods (Figure 8.86) (Wells 1966b). The best-studied experimental subject has been *Octopus vulgaris*, largely due to the twin facts that it is readily maintained in the laboratory and is amenable to experimental manipulation. *Octopus* will be our model organism; squids and cuttlefish will play a cameo role.

The brain of all cephalopods is protected by a suite of cephalic cartilages forming a cranium around the brain, analogous in function to the vertebrate skull and underscoring the importance of the brain to cephalopod survival. A few regions of the cephalopod brain correspond roughly to the basic molluscan pattern previously observed in the gastropods. The supraesophageal region is responsible for the functions of the gastropod cerebral and buccal ganglia and the anterior subesophageal region incorporates the role of the pedal ganglia. The posterior subesophageal region corresponds to the pleural and visceral ganglia (Russell-Hunter 1979).

There are about 15 paired lobes in the *Octopus* brain that may be categorized in 3 hierarchical levels of organization (Table 8.6). The first level is direct motor control, where the anatomical regions affected by direct stimulation of a brain lobe produce reflexive or isolated responses, showing a definite connection but little coordination. Those lobes are found primarily in the subesophageal region of the brain (Figure 8.86). The second level in the response hierarchy is a coordinated behavioral response to electrical stimulation of a brain lobe such as rapid-escape behavior or a prey-capture response. Those intermediate level lobes are found in the supraesophageal region of the brain. The third and highest level of organization is associated with large lobes of the brain, including the superior frontal, vertical, and subvertical lobes, which do not produce an obvious response when stimulated but have been

(a) The brain of *Octopus*, viewed from above, (broken line is the position of the esophagus)

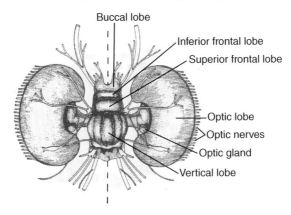

(b) The brain of *Octopus*, lateral view from right side (broken line is the position of the esophagus)

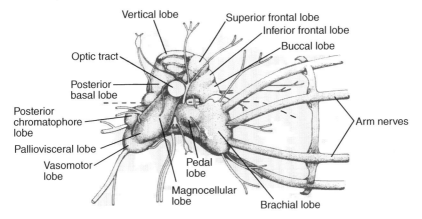

(c) The brain of *Octopus*, supraesophageal lobes, viewed from the right side

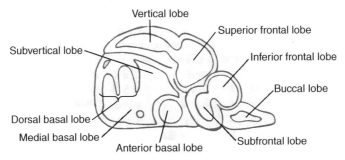

Figure 8.86 Cephalopod nervous systems. (a) Superior view of the brain of *Octopus* showing the basic lobed structure; (b) lateral view of the brain of *Octopus* showing more detailed lobed structure; (c) lateral view of the supraesophageal lobes of the brain of *Octopus*. *Sources:* (a) From Young (1961), figure 1 (p. 36); (b) Wells (1966b), Figure 1 (p. 550); (c) Adapted from Wells (1961), figure 1 (p. 812).

Table 8.6 Levels of organization and responses in the brain of *Octopus*.

Brain region	Organizational level	Anatomical influence
Subesophageal lobes		
Brachial	Direct motor control	Arms
Pedal	Direct motor control	Siphon ink gland
Palliovisceral	Direct motor control	Viscera and mantle
Vasomotor	Direct motor control	Mantle
Chromatophore	Direct motor control	Chromatophore muscles
Supraesophageal lobes I		
Anterior basal	Integration and coordination	Head and arms (prey capture)
Posterior basal	Integration and coordination	Swimming and escape
Magnocellular	Integration and coordination	Swimming and escape
Lateral basal	Integration and coordination	Chromatophore patterns
Buccal	Integration and coordination	Beak and radula
Supraesophageal lobes II		
Inferior frontal	Sensory integration	Tactile centers in arms
Subfrontal		
Optic	Visual integration, hormonal	Eyes and vision, reproduction
Olfactory lobes (on optic stalks)	Sensory integration	Chemoreceptors
Supraesophageal lobes III		
Superior frontal	Memory and learning	
Vertical	Memory and learning	
Subvertical	Memory and learning	

demonstrated experimentally to be important in memory and learning (Wells 1966b; Russell-Hunter 1979).

Differences between the brain of octopods and that of decapods lie mainly in differences in anatomy or behavior. For example, the brain of decapods includes a lobe controlling the fins, a nonexistent structure in octopods. In turn, the exceptional tactile development of *Octopus* is reflected in its brain structure (Wells 1966b). Another difference is the relative size of different lobes. In pelagic decapods like *Loligo* and *Architeuthis*, the optic and basal lobes are very large relative to those of the benthic octopods. In turn, the inferior frontal and subfrontal lobes of benthic octopods are more highly developed than those of their pelagic counterparts. Overall, evidence strongly suggests that homologous brain lobes control the same functions throughout the cephalopods (Wells 1966b).

Octopuses have been trained to visually discriminate between shapes and tactilely discriminate between textures. Moreover, they are capable of solving simple problems (Wells 1966b). Years of research by Drs. J.Z. Young, M.J. Wells, and B.B. Boycott describing cephalopod brain function have resulted in major advances in our understanding of brain and behavior in cephalopods and on the function of brains in general. Review publications include Young (1961, 1964, 1967, 1971, 1991), Wells (1966b, 1978), and Boycott (1961).

In the field, cephalopods demonstrate a curiosity that sets them apart not only from all other invertebrates but from most vertebrates as well. Sharks, pinnipeds, and cetaceans will often check out divers and submersibles, either as a potential meal or just a large new object in their sensory field. Squids will as well. On multiple occasions, a small school of *Loligo opalescens* (20–30 cm) followed the Deep Rover submersible for greater than 100 m while it was descending into the Monterey Canyon, matching its rate of descent as it dropped into the darkness below. There were a number of stimuli the squid might have been responding to, e.g. the large object emitting light and sound, but they also effected an "ink and escape" in response to sudden motion within the acrylic bubble of the submersible. Their awareness and curiosity were impressive (Torres, personal observation).

The Giant Fiber System

The giant axons of squid were used by A.L. Hodgkin and A.F. Huxley (1952), a pair of highly talented physiologists, skilled in engineering and biophysics, to elucidate how all nerves work. Their seminal papers and those of their colleagues not only described neural function, they ushered in decades of important studies on membrane ion transport and helped pave the way to our present understanding of muscle function. The chief characteristic of squid axons that made them highly desirable experimental subjects was their diameter, which is about 1 mm: huge by neural standards. Large-diameter axons conduct neural signals far faster than narrow ones, which is why giant fibers are usually associated with life-or-death responses that require extreme speed. In tube-dwelling annelids, they mediate the withdrawal response into the tube, in crustaceans the tail-flip response, and in squids the rapid-escape response.

A rapid escape in squid requires the circular muscles of the mantle to contract simultaneously (Figure 8.87a), forcefully ejecting water out the funnel and allowing the squid to jet away. In turn, simultaneous contraction requires a neural system that invests the entire mantle and conveys the signal from one end to the other with extreme rapidity. Figure 8.87b is a simplified diagram of *Loligo*'s giant fiber system, which has three parts and begins in the brain.

The response originates in the paired magnocellular lobes of the brain (Figures 8.86b and 8.87b) with a corresponding pair of short first-order giant fibers, one on each side. Each of the first-order fibers crosses to the opposite side of the brain, joining its twin in a neural chiasma, or crossing, and assuring that both sides function in synchrony. The first-order fibers terminate in the palliovisceral lobes on the contralateral side of the brain, where they connect with seven second-order giant axons innervating the funnel, head-retractor muscles, and the stellate ganglia (Boyle and Rodhouse 2005). The stellate ganglia house the giant cell bodies for the third-order giant axons radiating out into the mantle's circular muscle layer, causing its powerful contraction. Synchrony in muscle contraction results from a gradient in diameter among the axons of the mantle musculature: those with the shortest distance to traverse are narrowest. Since speed of conduction is directly proportional to axonal diameter, the command to contract arrives simultaneously throughout the mantle musculature. The largest diameter third-order giant axons were the nerves

556 | The Mollusca

(a) Muscle effectors of jetting in cephalopods

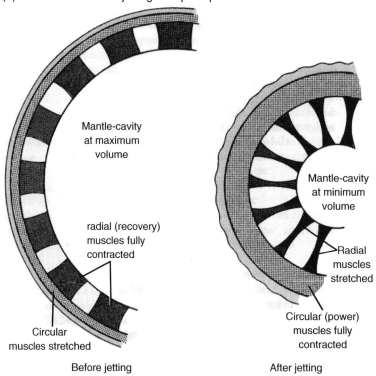

(b) Giant axon complex in the squid (stylized)

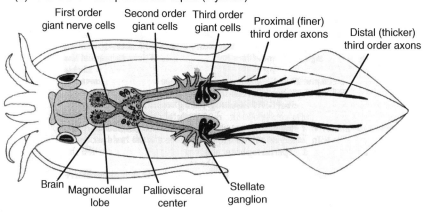

Figure 8.87 Cephalopod rapid escape response. (a) Arrangement of circular and radial muscle fibers in squids before and after jetting; (b) stylized giant axon system in squids. *Sources:* Adapted from Russell-Hunter (1979), (a) figure 24-4 (p. 438); (b) Russell-Hunter (1979), figure 24-7 (p. 442).

used in the experiments of Hodgkin and Huxley (1952).

Sensory Mechanisms

Cephalopods have a full suite of highly developed senses as well as other, more unusual, talents like the ability to rapidly change color. Their eyes are the most highly advanced of the invertebrates, rivaling the vertebrates in their complexity.

Touch, Taste, and Smell

Like the gastropods, all exposed surfaces of cephalopods are highly sensitive to touch, but their suckers and arms are exquisitely so. Blind octopuses can be trained to discriminate between different textures using only touch. Using a combination of touch and chemical cues, they can use their arms to distinguish between inedible and edible objects such as a stone versus a bivalve, or between a living and dead bivalve of the same species (Wells 1966a). Going even further, blind octopuses can readily discriminate between porous, otherwise identical, objects that have been soaked in different chemical solutions (e.g. quinine and sucrose), even when the solutions are fairly dilute (10^{-2} mM). They discriminate tastes better with their arms than humans do with their tongues (Wells 1963)!

Though sensory cells for touch and taste abound in the skin of cephalopods, the cells are particularly concentrated in their lips (see Figure 8.80) and in the rims of their suckers, where they may number in the hundreds mm^{-2} (Wells 1966a; Boyle and Rodhouse 2005). Chemoreceptors and mechanoreceptors have both been identified based on their structural resemblance to those classes of cells in other taxa (Wells 1966a). Chemoreceptors are also present in a specialized organ located between the eye and the upper margin or collar of the mantle in octopods and decapods. It is known by a variety of names including the olfactory pit and olfactory crest or tubercle, and it is believed to sample the chemical environment of the inhalant respiratory stream in an analogous fashion to the osphradium of the gastropods. Good behavioral and electrophysiological evidence is available to support an olfactory function for the organ (Gilly and Lucero 1992; Lucero et al. 1992), and there is considerable evidence that cephalopods respond to waterborne chemicals (Boyle 1986a). A similar organ, the rhinophore, is found in *Nautilus* (Boyle and Rodhouse 2005).

Mechanoreception: Equilibrium and Acceleration

Equilibrium, acceleration, and hearing all fall within the province of mechanoreception. Though seemingly unrelated, all are measurements of force and displacement, including vibration. The basic equilibrium detection system is composed of two structural elements: a mass, whose position depends on forces applied (e.g. gravity), and biological sensors that are mechanically affected by the position of the mass. The mass usually takes the form of a single calcium carbonate statolith or multiple smaller statoliths, termed statoconia, that serve the same function. Sensors usually take the form of hair cells, which are similar in basic structure from phylum to phylum (Budelmann 1988). Hair cells detect motion by the displacement of cilia located on their apical surface, creating a neural impulse that is sent to the brain. The agent displacing the cilia is the statolith or the fluid bathing the hair cell. In invertebrates, the basic equilibrium receptor system is the statocyst (cf. Chapter 7).

For all cephalopods, gravity as well as linear and angular acceleration are detected by a pair of statocysts embedded in the cartilaginous cranium protecting the brain. In squids, statocysts are located in the cephalic cartilage below the palliovisceral lobe of the brain (Pierce 1967). Cephalopod statocysts are "the most sophisticated equilibrium receptor organs of all those found in the invertebrates" (Budelmann 1988). They may be roughly divided into two morphological types, the octopod morphology typified by the statocysts of *Octopus* (Figure 8.88a–c), and a decapod

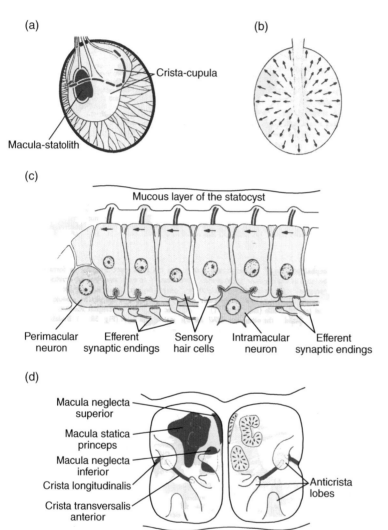

Figure 8.88 Cephalopod mechanoreception. (a) Stylized "octopod" statocyst; (b) polarization pattern of macula hair cells in an "octopod" statocyst; (c) cellular organization of the macula in an "octopod" statocyst. Arrows in the distal ends of hair cells indicate the direction of the cell's polarization; (d) anterior view of the left and right decapod-type statocyst of the cuttlefish *Sepia*. In the left-hand statocyst, the three macula of the gravity detecting system are depicted. In the right-hand statocyst, the polarization pattern of the macula hair cells is shown. *Sources:* (a-c) From Budelmann (1988), figure 30.9 (p. 769); (d) Budelmann (1988), figure 30.10 (p. 770). Reprinted by permission from Springer Nature Customer Service Centre GmbH, Morphological Diversity of Equilibrium Receptor Systems in Aquatic Invertebrates by Budelmann, in Sensory Biology of Aquatic Animals, 1988.

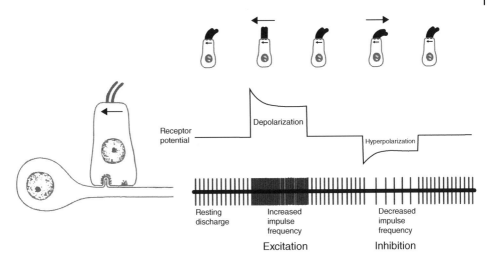

Figure 8.89 Gravity detection in cephalopods: polarized hair cells. The relationship between direction of displacement of cilia on the apical surface of hair cells (arrows above the cells) in, or opposite to, the direction of the cell's morphological polarization (arrows within the cells). Below the cells are recordings of the receptor potential and nerve impulse frequency that correspond with the deflections show in the cells. *Source:* From Budelmann (1988), figure 30.3 (p. 762). Reprinted by permission from Springer Nature Customer Service Centre GmbH, Morphological Diversity of Equilibrium Receptor Systems in Aquatic Invertebrates by Budelmann, in Sensory Biology of Aquatic Animals, 1988.

morphology exhibited by both cuttlefishes and squids (Figure 8.88d). Generally, size and sensitivity of statocysts correlate with an organism's speed of movement. Octopods, mainly benthic and slower moving, have larger and more sensitive statocysts than their decapod relations.

Within each statocyst, there are two separate systems, one for detecting gravity, or equilibrium, and the other for detecting acceleration. Both have "polarized" hair cells that are not only sensitive to displacement but also to the direction of that displacement, and they communicate the direction with a change in the type of neural signal sent to the brain (Figure 8.89). The basic gravity receptor system is a statolith resting on a bed of polarized hair cells termed a macula. A change in body orientation moves the statolith, which in turn displaces the cilia on the hair cells, alerting the brain to the change. In octopod statocysts, there is one macula for gravity reception; in decapods, there are three in three orthogonal planes, an arrangement similar to that of the semicircular canals in the vertebrate ear.

Angular acceleration receptors differ from gravity receptors in two basic ways. First, there is no statolith associated with them; acceleration is detected by movement of the fluid, the endolymph, in the statocyst cavity. Second, though the sensory epithelium is composed of hair cells, it is not a discrete bed of hair cells in the form of a macula. Instead, the sensory epithelium is arranged in a series of sensory ridges termed cristae that wind around the statocyst cavity to form a sensory field arranged in three orthogonal (x–y–z) planes. The ridges or cristae are divided into nine sections in the octopods and four in the decapods. In each section of the crista, the ciliary detectors of many hair cells are joined together to form a cupula (Figure 8.90), a delicate flap-like structure that protrudes freely into the fluid-filled statocyst cavity and acts "as a swinging door" (Budelmann 1988). Because of their high surface area, the cupulas are quite sensitive, bending in response to the fluid motion within the statocyst resulting from angular acceleration. The fact that the cristae are arranged in the x–y–z planes gives the brain a picture of angular acceleration in three-dimensional space (Budelmann 1996). Note that the statocysts of squid have protrusions into the statocyst cavity termed "anticristae." The anticristae serve to damp fluid flow in the statocyst cavity as a whole, making them less sensitive to slow movement as well as forming channels that improve detection of angular acceleration in the faster moving decapods.

The "lateral line" system of squids and cuttlefishes provides an additional mechanoreceptive sensory field that is sensitive to external water motion. Considered analogous to the lateral line of fishes, it consists of 8–10 lines of epidermal hair cells that run in an anterior-posterior direction along the head and arms. They are equivalent in sensitivity to the hair cells of fish lateral lines and are believed to aid in prey detection and predator avoidance (Budelmann 1996).

Hearing

Receptors in the statocysts and lateral-line system of cephalopods clearly can detect vibrational stimuli. Whether they can truly "hear" has been argued back and forth (see Moynihan 1985; Hanlon and Budelmann 1987). Behavioral responses (e.g. flight) to sound have been observed in the laboratory and the field (Hanlon and Budelmann 1987), and electrophysiological recordings have been made from in vitro preparations of octopus statocysts stimulated by low-frequency sound (Williamson 1988). However, systematic studies of cephalopod behavior in response to sound stimuli have not yet been performed, nor have neural recordings been attempted that include the most likely sensory receptors for sound vibrations: the maculae in the statocysts and the lateral-line receptors. At this point, there is no doubt that cephalopods respond to vibration, but there is a lot more to learn

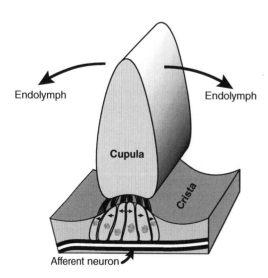

Figure 8.90 Cross section of a stylized octopod crista with attached cupula showing the polarized hair cells with afferent neurons underlying the cupula.

about how sensitive their "hearing" is. In the light-limited deep-sea particularly, sensitivity to vibration is an important asset in detecting prey and avoiding predators.

Vision

The cephalopod eye is without question the most sophisticated invertebrate photoreceptor, as well as the largest. Giant squid eyes as large as 40 cm in diameter have been recovered from the stomachs of sperm whales (Kaestner 1967). There are two types of Cephalopod eyes, the pinhole eye of nautiloids and the lens eyes of the coleoids. The pinhole eye of *Nautilus* (Figure 8.91) is the only example of such an eye in the animal kingdom (Land 1981).

The pinhole eye of *Nautilus* functions using the same principle as a pinhole camera: light passes through a tiny aperture in the eye producing a clear inverted image on the retina due to the rectilinear properties of light. In effect, the pinhole acts as a lens. In *Nautilus*, the aperture may vary between 0.4 and 2.4 mm (Hurley et al. 1978) depending upon ambient light. Despite its primitive

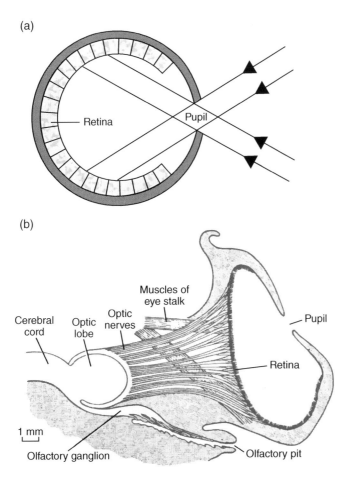

Figure 8.91 The pinhole eye of *Nautilus*. (a) Diagram of the basic pinhole type of eye; (b) transverse section of the pinhole eye of *Nautilus*. *Sources:* (a) Land (1990), figure 9.1B (p.151); (b) Land (1981), figure 18 (p. 511).

character, the eye of *Nautilus* is a highly functional image-forming eye that has been of prime importance to the success of the nautiloids through evolutionary time (several hundred million years: Land 1981). It is similar in resolving power to the average insect eye, but its lack of a lens is a severe drawback in dim light. The quantity of light reaching the retina of an octopus eye of the same size (about 1 cm) as the eye of a *Nautilus* would be 6400 times greater (Land 1981), assuming an 8-mm pupil in the octopus and 0.1 mm in the *Nautilus* producing ideal resolution in both cases. *Nautilus* is normally found in shallow, well-lit environments.

Eyes of the coleioids are a step up in complexity from those of the nautiloids. They are similar in many respects to those of vertebrates and are commonly hailed as an excellent example of convergent evolution. Eyes of octopods and decapods are nearly identical, differing only in the number of extrinsic muscles (Packard 1972) and in the absence of a corneal membrane over the eye in oegopsid squids, which are the majority of decapod coleioids.

Gross anatomy of the octopus eye comprises a thin cornea, an iris diaphragm, a spherical lens and a hemispherical retina (Figure 8.92a). The teleost eye in Figure 8.92b shows the same basic internal structure. In fact, the most obvious difference between the two is in the number of axons exiting the rear of the eye. Unlike in terrestrial vertebrates, the cornea of cephalopods (when present) and of fishes does not aid in focusing the perceived image. Just as light is refracted as it passes from air through the sea surface, it is similarly bent as it transits from air to the fluid within the vertebrate eye. The natural refraction in moving from air to liquid is adjusted by the vertebrate cornea to help the lens focus the image on the retina, in effect acting as an auxiliary lens. Since cephalopods and fishes dwell in an aqueous medium, there is no refraction as the light enters the eye and the preliminary focusing function of the cornea is lost. All focusing must be done by the lens of the eye itself, and the ideal shape for such a lens is a sphere (Land 1981).

Four properties of spherical lenses are remarkable and certainly worth noting. First, they have a very short focal length. Objects are in focus at a distance of 2.55 times the lens radius (known as Mathiessen's ratio), making them ideal for service in a photoreceptor that obviously must be limited in size. Second, because the lens is a sphere, objects are equally in focus in all parts of the concentric or nearly concentric retina and there is no one spot on the retina, as there is on the fovea of terrestrial vertebrates, that receives the highest quality image. Third, to achieve a sharp image, the spherical lenses of cephalopods and fishes have a gradient of refractive index from the center to the periphery with highest values (1.53) at the center and lowest (1.33, near seawater) at the periphery (Packard 1972; Land 1981). The gradient allows light entering the lens at either the center or periphery to be focused on the same point (Figure 8.92c) and is achieved through a gradient in the concentration of crystalline proteins in the lens. Fourth, the depth of field, i.e. the distance in front of the eye over which an object remains in focus, extends from infinity to a few centimeters (Packard 1972). Thus, the need to "accommodate" the eye by moving the lens forward or back to achieve a focused image on the retina is minimal. Nonetheless, a well-developed musculature for moving the lens back and forth and "fine-tuning" the focus is present in both cephalopods and fishes.

The amount of light admitted to the cephalopod eye is regulated by the size of the pupil, which is in turn controlled by the

Figure 8.92 Cephalopod vision. (a) The eye of *Octopus*; (b) teleost fish eye showing differences and similarities to the octopod eye; (c) paths of light rays from a distant source passing through an aplanatic spherical lens showing how objects are focused equally well regardless of the incident angle of entry of the light. *Sources:* (a) Wells (1966a), figure 5 (p. 534). Reproduced with permission of Elsevier; (b) Helfman et al. (2009), figure 6.8 (p. 85). Reproduced with permission of John Wiley & Sons; (c) From Packard (1972), figure 22 (p. 271).

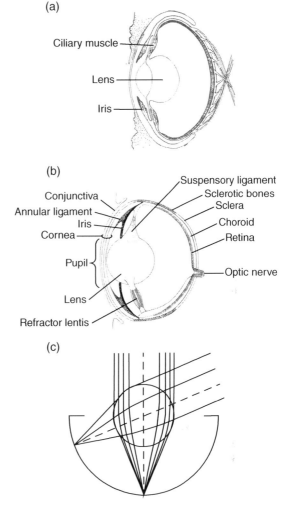

muscles of the iris. Pupils vary in shape from the horizontal slits of coastal octopods and cuttlefishes to the more conventionally shaped round pupils of oceanic squids. In all cases, a change in light intensity elicits a quick response in the iris, whose sphincter-like musculature can allow the pupil to contract or expand rapidly (Wells 1966a). The iris of teleosts is less mobile but performs a similar function.

The retinal structure of cephalopods is quite different from that of fishes: the layering of cells within the retina is exactly in reverse. Tall light receptor cells, termed retinula cells, face inward toward the lens in cephalopods, comprising the first layer (Figure 8.93a). The retinula cells are supported on a basement membrane, and supporting cells containing screening pigment interleave between them. The retinula cells are oval in cross section and bear regions of countless microvilli, termed rhabdomeres, on both sides of the oval that contain light-sensitive visual pigment. Retinula cells are

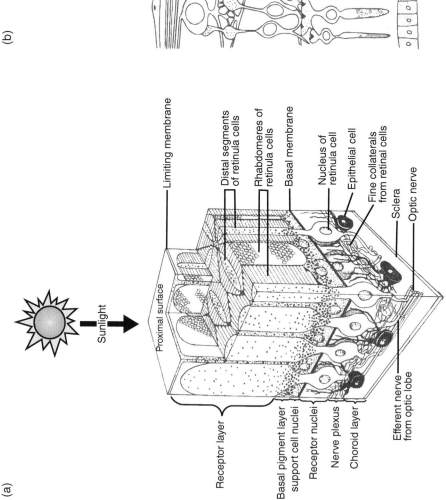

Figure 8.93 Cephalopod vision. (a) The converse retina of *Octopus*; (b) the inverse retina of vertebrates. *Sources*: (a) Adapted from Wells (1966a), figure 6 (p. 536); (b) Adapted from Withers (1992), figure 7-44 (p. 305).

organized into groups of four at right angles to one another, forming a square sensory unit called a rhabdom. Below the basement membrane are the retinula cell nuclei and below them a neural plexus containing efferent (probably inhibitory) nerves from the optic lobe and collateral nerves from the retinula cells themselves. Epithelial cells and a choroid layer of connective tissue, blood vessels, and muscle finish the retinal layering. A cartilaginous sclera bounds the rear of the eye, penetrated only by blood vessels and the nerve bundles entering and leaving the retina. The most important thing to note is that the first, inward-facing layer of the retina contains the light receptor cells (Young 1966); it is a converse retina.

Fish and other vertebrates have an inverse retina (Figure 8.93b). Light entering the eye must pass through several layers of nervous tissue before reaching the photoreceptors located in the outermost layer of the retina. A considerable amount of sensory integration occurs in the vertebrate retina. The number of rods and cones greatly exceeds the number of ganglion cells, indicating that much signal processing takes place in the retina prior to any visual information leaving for the brain. In effect, the retina has assumed an interpretative as well as a sensory role in the visual process (Withers 1992). In cephalopods, sensory integration takes place in the highly developed optic lobes of the brain (Wells 1966a; Boyle and Rodhouse 2005).

Receptor density in cephalopod and fish retinas is roughly equivalent, with cephalopods ranging from 50 000 to 70 000 mm^{-2} (*Loligo* and *Octopus*, respectively; Young 1966) and fishes (small sharks 24 000–75 000 mm^{-2}, *Torpedo* 21 600 mm^{-2}; Packard 1972). Many species of cephalopods and fishes have areas within the retina with increased concentrations of receptors, giving increased sensitivity analogous to the fovea of higher vertebrates. Within the cephalopods, the length of photoreceptive rhabdoms as well as their density can be increased in such areas to greatly improve visual sensitivity. In the ventral retina of the bathypelagic squid *Bathyteuthis*, photoreceptors reach concentrations of 250 000 mm^{-2}, forming a true fovea (Packard 1972). The eyes of deep-sea fishes show similar adaptations in the retina as well as profound changes in the overall shape of the eye.

The retinula cells of cephalopods and the all-rod retinas of deep-sea fishes have a single visual pigment that absorbs maximally in the blue-green region of the visible light spectrum ($\lambda = 475$–500 nm), the wavelength that penetrates seawater most readily. However, shallower-dwelling fishes also have cone cell photoreceptors with multiple visual pigments, allowing them to see color and, in some cases, conferring visual sensitivity into the UV spectrum (Bone and Moore 2008). Extensive behavioral and physiological work indicates that cephalopods do not detect color (Boyle and Rodhouse 2005).

Color

All the coleoids are capable of rapidly changing color, altering their appearance to mimic their background, entice their mates, or frighten away rivals and predators. The skin of coleoids is best thought of as a sophisticated three-dimensional system for changing color and body pattern (Messenger 2001). Color change is achieved with neurally controlled chromatophores found within the multi-tiered dermal skin layer located just below the transparent epidermis (Messenger 2001). In addition to chromatophores, the dermis is populated with connective tissue fibers as well as reflective iridophores and leucophores. The five colors of individual chromatophores are black, brown, red, orange, and yellow. They layer-out by color in the dermis, e.g. in *Octopus vulgaris* black chromatophores are the most superficial, followed by red and then yellow (Messenger 2001). Below the chromatophores are the crystalline iridophores. Though colorless themselves, the iridophores produce colors, often blue-green, by reflection and constructive interference, (Messenger 2001; Boyle and Rodhouse 2005). Below the iridophores are the leucophores, which act as broadband reflectors and when exposed can produce a white background. The multiple layers of innervated chromatophores, reflective elements, and a centralized system of command and control represent a considerable investment of biological currency. We must assume that color change and body patterning is of great importance to survival.

Chromatophores themselves consist of an "elastic sacculus containing pigment granules, to which is attached a set of radial muscles. Color is produced when the muscles contract, expanding the chromatophore; when they relax, the energy stored in the elastic sacculus retracts it" (Messenger 2001) (Figure 8.94). Chromatophores are controlled by the brain. Specific, presumably pre-programmed, body color patterns are initiated in response to visual cues, including mimicking a sandy bottom or countershading in the open ocean with a blue dorsum and a silvery ventrum. Due to the neural control, the color patterns can be achieved nearly instantaneously. Though the possible combinations of color and pattern in nature are nearly endless, those cephalopod species that have been studied display a limited number of basic color patterns.

Recent thinking suggests that basic cephalopod body patterns are part of a pre-programmed library of possibilities rather than being composed extemporaneously based on visual cues (Messenger 2001). One reason for this line of thought is the limited number of basic body patterns observed in behavioral studies. Another is the fact that cephalopods can't see color, and patterns therefore must be based on tonal matching. A third is that a system of visual feedback, or situational adjustment, has not been detected for the brain's control of body patterning. It is assumed that patterns of camouflage have been selected for over evolutionary time by their past effectiveness

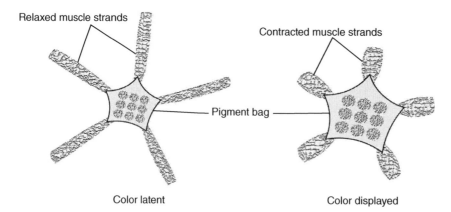

Figure 8.94 Dermal chromatophores. Contraction of muscle bands causes expansion of the pigment bag and display of color.

against visual predators, notably fishes (and early on by ichthyosaurs!). Successful patterns have been retained as part of a neural library (Messenger 2001). This fascinating subject can be further explored in Packard (1972), Hanlon and Messenger (1996), and Messenger (2001).

The main function of color change and body patterning is for camouflage or concealment. Other functions are for communication, either in courtship displays, as a warning or attempt to frighten away predators, or as a distraction to make prey more vulnerable. The well-known zebra display of male *Sepia oficinalis* is a way of signaling readiness and strength to females during courtship and a way of letting other males know that they are males, and that they are willing to fight. Similar black and white banding displays during courtship have been observed in loliginid squids and octopods (Messenger 2001). Deimatic or "scare" displays to ward off visual predators are similar in all three major groups of cephalopods. The displays usually include dark rings around the eyes, dilating the pupil, whitening the central part of the body and darkening the periphery, and creating large black eyespots with the chromatophores. Where possible, as in cuttlefish and octopods, the body is spread and flattened to make it appear larger. Prey items, e.g. shrimp or crabs, can be distracted with the "Passing Cloud" display, which involves dark waves moving across the body prior to ejecting tentacles or pouncing on the luckless crustacean (Messenger 2001).

Bioluminescence

Greater than 60% of squid and cuttlefish genera contain luminescent species (Herring 1978) whose light organs vary considerably in location, design, and complexity (Figure 8.95). Bioluminescence in octopods is very rare, with confirmed luminescence in only three deep-sea genera, the amphitretids *Japetella* and *Eledonella* (Robison and Young 1981) and the stauroteuthid *Stauroteuthis syrtensis* (Johnsen et al. 1999).

The myopsid squids including *Loligo,* and the Sepiida, have paired luminous organs located next to the ink sac which in some cases are invested with luminous bacterial symbionts and in others (subfamily Heteroteuthidinae) produce a luminous secretion originating in the organ itself (Dilly and Herring 1978; Herring 1978). The deep pelagic bobtail squids, e.g. *Heteroteuthis*, are capable of ejecting a cloud of luminous particles when disturbed.

Spirula, a deep-living decapodiform best known for its internal nautilus-like shell (Figure 8.73), and Vampyroteuthis, a deep-living Octopodiform, both occupy their own monotypic orders in the current classification scheme. Both are luminescent, with *Spirula* bearing a single light organ between its fins and *Vampyroteuthis* showing a pair at the base of its fins, another pair just posterior to its eyes and a sprinkling of organs on its surface (Herring 1978). The oegopsid squids have not only the most species of bioluminescent molluscs but also the greatest diversity of luminescent organs and displays, rivalling those of the fishes. The organs vary in complexity from the simple patches of luminous tissue seen in *Sthenoteuthis* (formerly *Symplectoteuthis*) and other ommastrephids to the lensed and filtered photophores of the enoploteuthid, *Abraliopsis* (Young and Bennett 1988). Luminous organs of all types are most commonly found on the ventral surface where they can act as counterilluminators (e.g. Young and Roper 1977), allowing the squid to blend in with downwelling light when viewed from beneath. Photophore sites (Figure 8.95) include the ventral surface of the eyeball, the ventral arms, along the

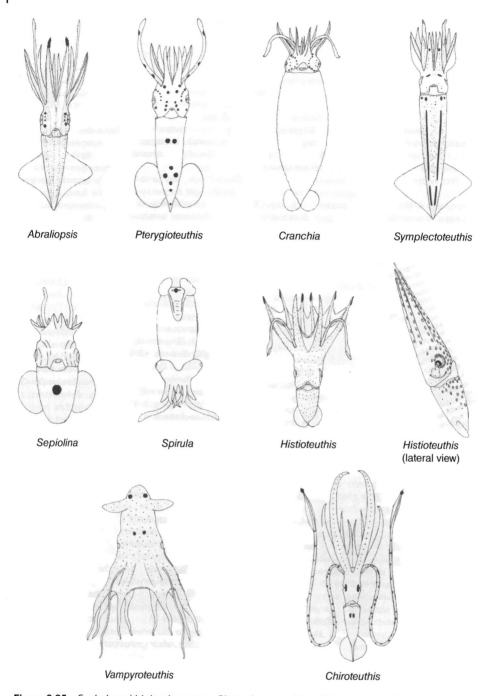

Figure 8.95 Cephalopod bioluminescence. Photophore positions (black areas or dots on bodies and tentacles) in a selection of cephalopods. *Source:* From Herring (1978), figure 7.3 (p. 209). Reproduced with the permission of Academic Press.

tentacles and in the tentacle tips (Herring 1978). Light organs are generally absent in shallow-dwelling squids, present and well developed on the ventral surfaces of mesopelagic species, and variable in true bathypelagic species. Photophore patterns are species-specific and incidences of sexual dimorphism are rare (Herring 1978).

Photophores show varying levels of sophistication within and between species. In a 1988 study on photophore anatomy in the enoploteuthids, Young and Bennett showed three different levels of complexity in the photophores of the integument alone: simple, lensed, and filtered. In each, the centrally located photogenic tissue is backed by a parabolic reflector and its generated light is further directed outward and diffused by a light-guide system. In lensed and filtered photophores, outgoing biological light may be altered in either quantity or quality to match downwelling irradiance.

Counterillumination has been directly demonstrated in several species of cephalopods (Young and Roper 1976, 1977; Young 1977, 1978). For a counterillumination system to be effective, it requires several elements. First and most obvious is the ability to produce light. Second, and more complicated, is the ability to match the light produced by the photophores with downwelling light in both intensity and spectral quality. This absolutely requires monitoring both, either visually or with other photoreceptors, and making the necessary adjustments. Clearly implied is the need for neural control and feedback, and there is good evidence for both. Adjustments of luminous output to match ambient light have been observed behaviorally in several species (Young and Roper 1976, 1977), as has synaptic input to photocytes (Arnold and Young 1974). Photosensitive vesicles, extraocular photoreceptive organs located near the brain in the cephalic cartilage, are believed to aid the eyes in the monitoring and matching requirements of countershading in cephalopods (Young 1972a, 1978).

Locomotion and Buoyancy

Four types of locomotion are employed by the cephalopods. Common to all is the jet propulsion achieved by ejecting water from the contracting mantle cavity though the small aperture of the ventral funnel. A second type of jet propulsion is more akin to that of cnidarian medusae; it is observed to varying degrees in all octopods and in the deep-living squid *Histioteuthis*. In medusiform propulsion, the circlet of arms and the interbrachial webbing between them is contracted like the bell of a medusa, propelling the animal backward. A well-developed interbrachial web is present in most of the deep-living octopods including the opisthoteuthids and the monotypic Vampyromorpha (see Figure 8.78a). A third type of locomotion is typified by the benthic octopods that move rapidly along the bottom by pulling themselves forward with their powerful arms in a scrambling motion. Lastly, the fins of squids and cuttlefishes can be used for hovering or to slowly propel themselves by a parallel undulating wave running down the paired fins. Cuttlefish can move slowly backward and forward using only the undulatory motion of their fins (Trueman 1983), providing fine-scale maneuverability. Fins can also be used as hydrofoils to provide lift (Alexander 1982) and to stabilize the cylindrical body of squids during rapid swimming and maneuvering (Boyle and Rodhouse 2005).

Octopods and decapods are all capable jetters but, as a propulsive system, jet propulsion reaches its apex in the fast-swimming coastal and oceanic squids. Benthic octopods use it in conjunction with crawling for escaping predators and pouncing on prey: whenever rapid movement is required. The more sluggish pelagic forms rely on medusiform locomotion and near-neutral buoyancy for station-keeping and jetting when necessary for prey capture and predator avoidance.

The basics of jet propulsion in squid begin with its nearly cylindrical mantle cavity. The wall of the mantle is essentially a sheet of muscle consisting of circular and radial muscles anchored in an outer and inner integument with a tough collagen matrix (Figure 8.96). Contraction of the mantle's circular muscle results in a 30% reduction in volume (Figure 8.87a), forcing a jet of water through the funnel aperture to the outside (Trueman 1980). The mantle cavity is refilled by a contraction of the radial muscles, causing the mantle wall to expand outward, drawing in the inhalant stream. Water is drawn in through the free "collar" area of the mantle between the head and the funnel. Extensions of the funnel sidewalls into the mantle cavity act as a one-way valve, preventing the escape of water through the inhalant region when the collar tightens around the head during the power stroke. A muscular valve within the funnel itself prevents water from entering the funnel during the inhalant, or recovery, stroke thus creating a truly one-way flow in and out of the mantle cavity during the propulsive cycle.

The jet propulsion of squid is less efficient than the undulatory swimming of fishes, whether the two are compared using first principles, i.e. hydrodynamic

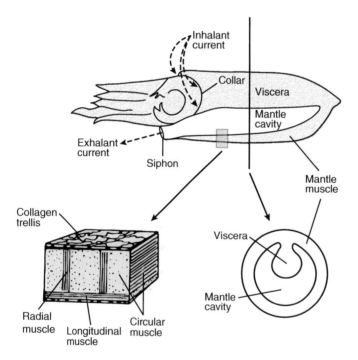

Figure 8.96 Schematic lateral view of a squid showing the relative positions of viscera, mantle cavity, and mantle musculature. Arrows show the direction of inhalant and exhalant currents. Below on the right is a cross-sectional representation of the body taken at the location indicated by the vertical line on the upper figure. Below on the left is a detailed diagram of the structure of the mantle musculature. *Source:* From Trueman (1980), figure 2 (p. 95). Reproduced with the permission of Cambridge University Press.

theory, or direct measurements of energy consumption in animals of similar size swimming at similar speeds. Alexander (2003) estimated the Froude efficiency (useful power/useful power + power lost to the fluid) of a swimming squid (*Illex illecebrosus*; Weber and O'Dor 1986) to be about 0.38, approximately half that of a swimming fish (0.75) of the same size. Direct comparisons of the net cost of transport for *Illex illecebrosus* and *Oncorhynchus nerka*, the sockeye salmon, measured under nearly identical conditions with animals of nearly identical size, revealed that it costs *Illex* 3.2 times the energy to swim the same distance as the salmon: net cost of transport of $5.4 \, J \, kg^{-1} \, m^{-1}$ for *Illex* and $1.7 \, J \, kg^{-1} \, m^{-1}$ *Onchorynchus nerka*. The net cost of transport is the metabolic energy cost of swimming a distance, in this case a meter, minus the standard (basal) metabolism. The cost of transport incorporates the metabolic costs of moving locomotory muscle as well as the hydrodynamic costs incurred by the propulsive system itself (e.g. efficiency and drag), which are the total costs to the animal. Overall, the metabolic efficiency of squid swimming is about a third of that in fish. Despite the low efficiency, peak swimming speeds in squids are as high as $6.7–11 \, m \, s^{-1}$, impressively fast (Boyle and Rodhouse 2005).

Buoyancy

Cephalopods employ three mechanisms of buoyancy control. The nautilids, spirulids, and cuttlefishes use gas-filled spaces within their shells to provide flotation. "Ammoniacal" squid retain low-density solutions of inorganic ions to provide static lift. Many of the deep-sea squid have high water contents, nearing the density of seawater by lowering the density of their tissues. We'll begin with the nautilids, spirulids, and cuttlefish. The principles are the same in all three groups, although the shell structure differs. Studies on shell strength in all three groups confirm that they are easily capable of withstanding the external pressures in their depth range (Denton 1974).

The best-studied buoyancy system and the easiest to understand is that of the chambered nautilus. *Nautilus* has a vertical range down to 540 m (Denton 1974), to a pressure of 54 atm, and is believed able to maintain neutral buoyancy to that depth. In the living nautilus, all but the newest chambers are empty of liquid (Figure 8.97), filled instead with a gas at 80–90% of atmospheric pressure despite the prodigious external pressures, with the youngest chambers at 40–50%. The gas within the chambers is clearly derived from atmospheric air, with a partial pressure of 0.8 atm nitrogen and about 0.1 atm oxygen with a trace of argon (air is at 0.8 N_2 and 0.2 O_2). The question is how are the chambers drained of fluid and replaced with gas?

The answer is using the principles of osmotic movement. All chambers, including the oldest, are connected to the hemocoel of the living nautilus via an extended tube of tissue, the siphuncle. The siphuncle has an active ion-transporting epithelium, enabling it to remove the fluid within the new chambers by osmotic withdrawal. The initial fluid within the chamber, like the hemolymph of *Nautilus*, is isosmotic with seawater and is termed "cameral fluid." The siphuncle's epithelium actively transports salts from the cameral fluid into its own lumen, raising the salt concentration within the lumen and greatly diluting that of the cameral fluid. The dilute cameral fluid diffuses down the osmotic gradient into the lumen of the siphuncle, leaving a nearly empty chamber. Gas diffuses out of the siphuncle into the empty chamber until it equals the dissolved gas concentrations

572 | The Mollusca

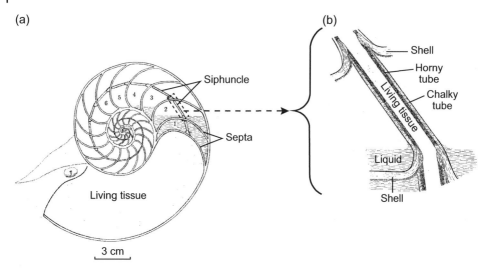

Figure 8.97 Buoyancy control in *Nautilus*. (a) Section of *Nautilus* showing the position of the animal within the shell; (b) detail of the siphuncle marked by the rectangle in (a). *Source:* Adapted from Denton (1974) (p. 288). Reproduced with the permission.

within the hemocoel of the nautilus, a little less than 1 atm (Denton 1974) with nearly the same gas composition as atmospheric air. The buoyancy mechanism of *Nautilus* thus elegantly exploits the basic physics of water and air.

The internal spiral shell of *Spirula* and the cuttlebone of cuttlefish are both charged with gas using identical principles, though the shell structure and the relation between the siphuncular tissue and shell are different. *Spirula*, the only species in the order Spirulida, is a true mesopelagic resident, found down to 1200 m and not associated with the bottom as are the nautiliids and cuttlefishes. Figure 8.98 shows the position of the siphuncle in *Spirula*, which runs along the inner coil of the spiral shell instead of through the middle of the chambers as in *Nautilus* but extracts cameral fluid in the same way.

The cuttlebone of sepiids differs greatly from the spiral shells of *Nautilus* and *Spirula*. It also is chambered, but the chambers are more numerous, more elongated, and smaller in volume. The cuttlebone itself is oriented along the dorsum (Figure 8.99). A siphuncular membrane that is capable of pumping water out of the cuttlebone via osmotic movement underlies its posterior section.

Buoyancy adjustment is a more active process in the cuttlefish than in the nautilids and *Spirula*. The cuttlebone of *Sepia* can actively adjust its buoyancy by pumping liquid into or out of its shell. When hunting over the bottom at night, it maintains itself at neutral buoyancy. *Sepia oficinalis*, the European cuttlefish and best-studied cephalopod, is found to a maximum depth of 150 m and normally rests on the bottom during the day.

Samples of liquid taken from *Sepia* cuttlebone show that, unlike in *Nautilus* and *Spirula*, the oldest chambers are always partially full of liquid. Since they are also the most posterior chambers, the extra liquid acts as ballast, allowing them to maintain a horizontal orientation when swimming. That orientation is in contrast to *Spirula*,

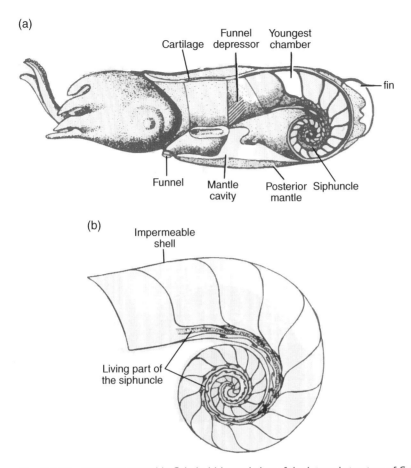

Figure 8.98 Buoyancy control in *Spirula*. (a) Lateral view of the internal structure of *Spirula spirula* with mantle and shell partly removed; (b) the shell of *S. spirula* showing the living part of the siphuncle. *Sources:* (a) Kaestner (1967), figure 15-21 (p. 412). Reproduced with permission of John Wiley & Sons; (b) Denton and Gilpin-Brown (1973), figure 7 (p. 216). Reproduced with permission of Elsevier.

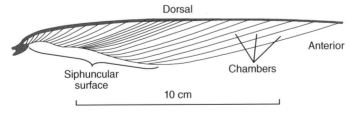

Figure 8.99 Diagrammatic longitudinal section through the cuttlebone of an adult *Sepia officinalis*; it would have about 100 chambers. *Source:* From Denton and Gilpin-Brown (1973), figure 8 (p. 217). Reproduced with the permission of Academic Press.

in which the gas-filled shell chambers maintain a vertical attitude. Experiments with *Sepia* at various pressures show that the higher the pressure, the more dilute the liquid in the anterior cuttlebone chambers involved with maintaining buoyancy. Since hydrostatic pressure forcing liquid into the cuttlebone must be countered by the diffusion pressure forcing water out, as the hydrostatic pressure increases so must the osmotic gradient between the fluid within the cuttlebone and that of the hemocoel to maintain neutral buoyancy, hence the lower ion content in the cuttlebone fluid.

The last method of achieving, or approaching, neutral buoyancy is by decreasing the density of body tissues themselves (cf. Childress and Nygaard 1973, 1974), either by adding positively buoyant compounds like lipids, excluding heavy ions like sulfate, or simply by having a watery composition, i.e. becoming more "gelatinous." This strategy was discussed earlier in the chapter, including relevant equations, in "The Heteropods" section.

Decreased tissue density can be achieved in a variety of ways. In many cephalopods, a variant of heavy ion exclusion is employed by using ammonium ion for lift. Those cephalopods are called "ammoniacal squid" and their buoyancy mechanism is truly unique. Sixteen squid families are known to exploit ammonium as a buoyancy aid (Voight et al. 1994, see table in "Classification"), but the best described and most unusual are the cranchiids.

Cranchiid squids, when brought on board alive and undamaged, will hang motionless in seawater. They are truly neutrally buoyant. The source of their buoyancy is a large thinly walled coelom that fills most of the mantle cavity and is itself filled with a fluid that is isosmotic with seawater (Denton and Gilpin-Brown 1973). Although isosmotic, the fluid within the coelom has a very different ionic composition than seawater: it is a solution of nearly pure ammonium chloride (86%) and it confers enough static lift to make the squids almost perfectly neutral. It has a specific gravity of 1.010, considerably less dense than seawater's specific gravity of 1.026. When the coelomic cavity is drained of fluid, the squids sink rapidly because their density has jumped to 1.046 (Denton and Gilpin-Brown 1973).

The energy-saving buoyancy conferred by the coelomic cavity of cranchiids is not without its costs. The mass of the intact coelomic cavity (coelom + fluid) is responsible for about 65% of the mass of the animal in air. It is a very large structure that takes up much of the mantle cavity, giving the cranchiids a balloon-like appearance and certainly limiting their potential for rapid swimming. Another consideration is the need to sequester and retain ammonium, which, though it is an end product of protein metabolism, is toxic as NH_3, and is usually voided soon after its formation. To sequester enough ammonium for creating their coelomic fluids, the cranchiids must retain about 40% of the ammonium they produce during their lifetime (Denton and Gilpin-Brown 1973). The secret to their ability for retaining ammonia in the coelomic cavity at concentrations very much greater than those ever reached in their blood is believed to be in the characteristics of the coelomic cavity itself. The pH of the fluid in the coelomic cavity is quite acid, about 5.0, which greatly favors the ammonium ion (NH_4^+) over the ammonia molecule (NH_3) and results in a lower concentration of NH_3 in the coelom. Due to its lower concentration in the coelom, any free NH_3 in the blood would tend to move into the coelom where it would rapidly be converted into NH_4^+ and trapped in the cavity. Free ammonia readily diffuses through membranes because it is small and has no charge. Once ionized to NH_4^+, its mobility is greatly diminished.

The muscular loliginid and ommastrephid squids are negatively buoyant and must swim continuously to maintain their vertical position, either using the inherent maneuverability of their funnel-based swimming or by using their fins as hydrofoils for lift or stationary hovering. Even some of the otherwise muscular families, such as the Neoteuthidae, the Enoploteuthidae, and the Onychoteuthidae, have representatives using ammoniacal buoyancy (Voight et al. 1994).

Instead of the single ammonium-filled coelomic cavity of the cranchiids, most of the other families have their ammonium sequestered in smaller fluid-filled compartments peppered through a variety of tissues. Examples include a vacuolar network within the mantle (e.g. Chiroteuthidae), mantle and arms (Architeuthidae and Histioteuthidae), gelatinous outer layer or inner connective tissues (Octopoteuthidae, Cycloteuthidae, and Bathyteuthidae), and a variety of other combinations (Voight et al. 1994). All in all, the cephalopods exhibit a remarkable degree of evolutionary creativity in their quest for buoyancy in the open sea.

Life Histories

Life histories differ markedly between the nautiloids and the coleoids. The nautiloids take 6–15 years to reach sexual maturity and their natural longevity is unknown (Martin et al. 1978; Saunders 1983; Ward 1987). They reproduce multiple times after reaching maturity. In contrast, the great majority of coleoids live for one to two years reproducing once at the end of their lives. A good outline of basic life history traits is that of Stearns (1992); they are listed below.

The principal life history traits:

- Size at birth
- Growth pattern
- Age at maturity
- Size at maturity
- Number, size, and sex ratio of offspring
- Age- and size-specific reproductive investments
- Age- and size-specific mortality schedules
- Length of life

For many cephalopod species, information is available for only a few of the life history traits. Complete data sets are usually limited to commercial species, mainly in the interests of fishery management. In particular, information is scanty for the deep-sea species and for the very large species that are difficult to capture, e.g. *Architeuthis*, *Mesonychoteuthis*.

Nautiloids

The five currently recognized species of Nautilus (WoRMS) are all found in the tropical Pacific including, but not limited to, the Philippines, Palau, Australia's Great Barrier Reef, Fiji, Samoa, and Tonga (Boyle and Rodhouse 2005). Even living at tropical temperatures, they take many years to reach sexual maturity and their length of life in the wild is unknown. Keep in mind that their vertical range is between the surface and 500 m, and that even in tropical seas, temperatures decline with increasing depth. Nautilids do reproduce in captivity, laying one to two large eggs per month throughout the year (Norman 2000). Embryos take up to a year to develop and hatch (Boyle and Rodhouse 2005). If we refer to Stearns' list of traits, we get a picture of a slow steady growth pattern, with late maturity, low fecundity, large fully functional offspring, and a long life.

Coleoids

Since the coleoids include the diverse array of cuttlefishes, octopods, and squids and their many habitats, one might expect an equally large diversity in their life history characteristics. Such is not the case, and the life history traits that have been successful for cephalopods are not common in large-bodied species. The great majority of coleoids have a short life span (one to two years) with rapid growth and a single breeding season, followed by death. In many species, the breeding season is a protracted one with multiple spawnings, such as in the cirrate deep-sea octopods and the pygmy squid *Idiosepius pygmaeus* (Villanueva 1992; Jackson 1993). However, in all cases observed thus far, once the breeding season has ended gametogenesis cannot be restarted and death eventually ensues (Boyle and Rodhouse 2005). Yet there is often enough variability in the timing of breeding and spawning within a single location, sometimes with separate breeding groups, to result in multiple cohorts, increasing resilience. The monocyclic life history of cephalopods is termed "semelparous" and is observed in other invertebrate and vertebrate groups, including some species of deep-sea Crustacea, e.g. the lophogastrid *Gnathophausia ingens*, as well as in many species of mesopelagic lanternfishes (Childress and Price 1978, 1983; Gartner 1991).

A cephalopod life history often used as a paradigm case is that of *Loligo opalescens*, the Pacific coast edible squid (Fields 1965). As is typical in the loliginids, the species deposits egg masses attached to the bottom in coastal shallows. The preferred spawning grounds are usually sheltered embayments at depths ranging from 3 to 40 m. In Monterey Bay, CA, peak spawning occurs from April to July. Because females will preferentially fasten their newly laid egg capsules to an existing cluster, pillars of squid eggs almost two meters high can form on the bottom during peak spawning and be so closely grouped that divers "have difficulty thrusting their way through them" (Fields 1965). At 16 °C, the eggs take 3–4 weeks to hatch into tiny (3-mm mantle length), fully functional, young squid. Adults of the species have a mantle length of 130–160 mm (Fields 1965).

Death typically occurs after spawning in *Loligo opalescens*; the bottom of spawning grounds become littered with squid carcasses. During the process of spawning, squids transform from robust healthy adults with fully charged reproductive organs to completely spent, weakly swimming, husks. A 151-mm female (mantle length) weighing 77 g when entering the spawning ground weighs 36 g after spawning: a loss of 53% of its total weight. Its reproductive system goes from 27.9 to 4.3 g for an 85% loss. A similarly sized (149 mm, 76 g) male squid incurs a loss of 32% of its total weight and 40% of its reproductive weight (Fields 1965).

All of the loliginid squids spawn in shallow coastal waters. Although not all have large spawning aggregations like *L. opalescens*, it appears that most have an annual cycle (Boyle and Rodhouse 2005). Overlying a straightforward annual cycle for many loliginid populations, including *L. opalescens*, is a layer of natural variability that includes mature animals and low-level spawning found throughout the year, multiple pulses of juvenile recruitment, and large, temperature-related, differences in hatch times for the developing embryos. The observed variability gives enough plasticity to the life cycle (Boyle et al. 1995) to minimize the risk of local extinctions in the event of a bad year.

An analogous study on the coastal octopod *Octopus bimaculatus* (Ambrose 1988), also found off the California coast, revealed a similar life history pattern. However, after spawning, *O. bimaculatus* females persisted long enough to brood their eggs to hatching before they too expired as spent animals. Thus, spawning was a terminal event in the life history of the species, confirming a semelparous life history. In addition, the study by Ambrose demonstrated how variability in hatching dates during a single breeding season coupled with seasonal temperature change can result in a population with multiple cohorts.

Spawning in *O. bimaculatus* occurred from April to August and hatching from June to September. After hatching, larvae spent one to several months in the plankton prior to settling out as juveniles at 5-mm mantle length in the nearshore kelp forest, where they used small shelters and kelp holdfasts as refugia until reaching a size of 50-mm mantle length, when they were considered adults. The period from initial settlement to a size of 50 mm was estimated at 8–10 months. In turn, the length of the adult stage was estimated at 11–12 months for a total life span of 19–22 months based on field data alone (Ambrose 1988).

Two life history models were proposed (Figure 8.100). The first, a better fit for growth rates at the temperatures observed in the field, was an "alternating years" model whereby individuals in a given year class skip the first reproductive season after settling and spawn in the following breeding season. In the "alternating years" model, late hatchlings would simply spawn later during the breeding season in their second year of life. In the second model, or "alternating generations" model, early hatchlings would spend less time in the plankton, settling out early enough to take advantage of the higher late-summer and fall temperatures for rapid growth, enabling them to spawn in their first breeding season 13 months after settlement. The "alternating generations" model doesn't match the field data as well as the "alternating years" model, but growth data from a sibling species (*O. bimaculoides*) and variability in the field data made Ambrose (1988) propose it as a theoretical possibility. The "alternating generations" life history has been proposed

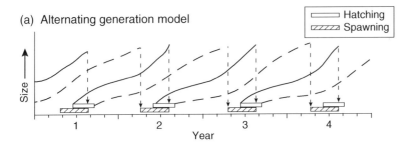

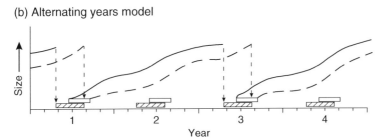

Figure 8.100 Life history models. (a) Alternating generation model: young of the year grow and mature rapidly and spawn in their first breeding season; (b) alternating years model: individuals in a given year class skip the first breeding season. *Source:* From Boyle and Rodhouse (2005), figure 6.9 (p. 97). Reproduced with the permission of John Wiley & Sons.

multiple times for other cephalopod species (Mesnil 1977; Boletzky 1983; Boyle 1983). In *Octopus bimaculatus*, as in *Loligo opalescens*, the life history of the bulk of the population is supplemented by a low level of spawning and hatching throughout the year (Ambrose 1988).

We have now seen examples from two coastal species, both of which are well described, providing good theoretical framework for the more pelagic cephalopods. The ommastrephid squids such as *Illex illecibrosus* complete their entire life cycle seaward of the continental shelf, meaning that they must also have pelagic eggs, a new twist on the life histories previously described. *Illex illecibrosus* is found from the Florida Straits to the Labrador Sea, generally following the axis of the Gulf Stream (O'Dor and Dawe 1998). Peak spawning is believed to occur off the Florida coast in late winter (Figure 8.101). Spherical gelatinous egg masses produced by the female swell rapidly after release, becoming near-neutrally buoyant "egg balloons" (O'Dor and Dawe 1998). Egg balloons are close to neutral buoyancy and are believed to drift in the midwater in association with pycnoclines or currents, but none have been captured using conventional sampling gear. Eggs hatch in a matter of days (6–16 days at 26 °C and 13 °C respectively), and larvae ascend to the surface to feed on the plentiful zooplankton typical of the spring bloom. The young squid continue to grow as they move northward, eventually moving inshore in late summer and fall where they enter fisheries in Newfoundland, Nova Scotia, and northern Maine. Most adults are believed to migrate south to the spawning area in early winter to begin the annual cycle again. A secondary smaller peak in spawning occurs in summer, most of which are believed to remain closer to the spawning grounds. Summer-spawned *Illex* also provide a "fail-safe" segment of the population, spreading out the reproductive effort and acting as a buffer in the event of unfavorable conditions for the winter-spawned animals. Life span of *Illex illecibrosus* is believed to be one year (O'Dor and Dawe 1998).

Reproduction and Development

All cephalopods are dioecious, and none exhibit sex change. Reproductive anatomy is similar enough in the octopods and decapods to allow a focus on squid to serve as a general description with *Loligo* as our model system (Figure 8.102). *Loligo* females have four major components to their reproductive system (Arnold and Williams-Arnold 1977): a large median ovary, paired nidamental glands, paired accessory nidamental glands, and the oviduct including the oviducal gland. *Loligo* females also have a pouch-like seminal receptacle, located on the arm web close to their buccal membranes, which receives male spermatophores during copulation (Boyle and Rodhouse 2005). Seminal receptacles are not found in all squids, with the ommastrephids being an important example of those that don't.

Mature eggs are shed from the ovary into the proximal thinly walled portion of the oviduct, where they accumulate until copulation. Note that Figure 8.102a shows only the distal portion of the oviduct. When the laying process begins, eggs pass into the oviducal gland where the individual eggs receive a coating of egg jelly. As they exit the oviduct, the eggs are formed into a spiral and coated with mucous from both pairs of nidamental glands to form the egg capsules for deposition. The oegopsid squids lack accessory nidamental glands and have paired oviducts; the remainder of the reproductive anatomy is the same (Roper et al. 1984; Boyle and Rodhouse 2005). Egg capsules of *Loligo* each contain 90–120 eggs.

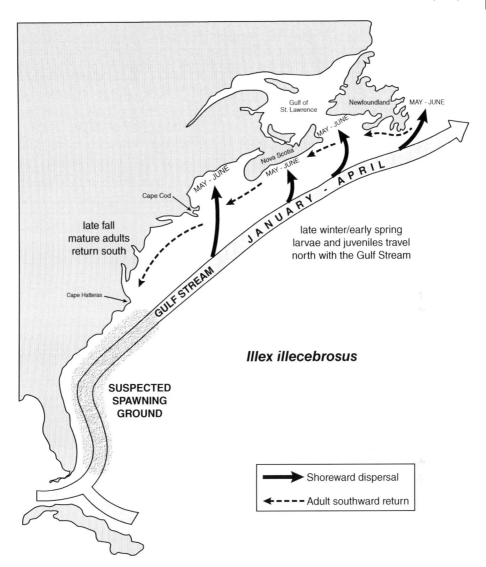

Figure 8.101 Life cycle of *Illex illecebrosus*. Larger solid arrows show northward movement and shoreward dispersal of larvae and juveniles; smaller dashed arrows show southward migration and dispersal of adults to spawning areas. *Sources:* Data from O'Dor and Dawe (1998); Black et al. (1987).

Male reproductive anatomy comprises five major components: the testis, the proximal vas deferens, the spermatophoric organ, Needham's sac (also called the spermatophoric sac), and the distal vas deferens or penis. A mature testis is pure white and, like the ovary, is located medially in the apex of the mantle. Mature sperm are shed into the ciliated opening of the proximal vas deferens which leads to the spermatophoric organ, a complex system of glandular tissue and lumina that empty into a central canal. The spermatophoric organ forms a quantity of sperm into a spiral mass and coats them with the membranes and tunics that form a mature spermatophore. Spermatophores

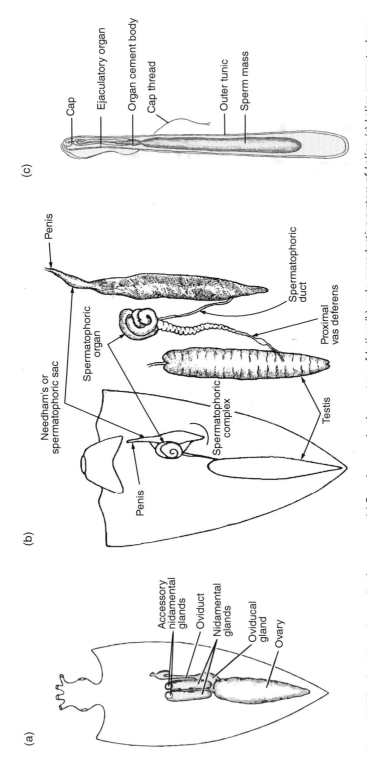

Figure 8.102 Cephalopod reproductive systems. (a) Female reproductive system of *Loligo*; (b) male reproductive system of *Loligo*; (c) *Loligo* spermatophore. *Sources*: (a) Brusca et al. (2016), figure 13.49 (p. 517); (b) Jereb and Roper (2005), figure 29. Reproduced with permission of The Food and Agriculture Organization; (c) Barnes (1974), figure 11-92 (p. 429).

are stored in Needham's sac like arrows in a quiver until needed during copulation. During copulation, spermatophores are ejected in small bunches by the muscular distal vas deferens (penis). Each spermatophore contains 7 200 000–9 600 000 sperm. (Williams 1909; Austin et al. 1964; Arnold and Williams-Arnold 1977).

Copulation in *Loligo* has been observed in the laboratory (Arnold and Williams-Arnold 1977) and in the field (Fields 1965), and though different species have different courtship behaviors, the basics are similar. Mating in *Loligo pealii* can be initiated in the laboratory by introducing a previously spawned egg mass into a tank holding a small school of mature *L. pealii*. Males become amorous when the egg mass is introduced and quickly seek a mate. Once a female is selected, the male attempts to sequester her, defending her with aggressive displays toward other suitors. If he is successful, the mated pair swims parallel until the male grasps the female, positioning himself parallel to her and near the anterior margin of her mantle. When in position, he passes his hectocotylus into his own mantle cavity and picks up a group of spermatophores that have been ejected by the penis and rapidly transfers them to the mantle cavity of the female, where he holds them until they ejaculate, usually a few seconds (Arnold and Williams-Arnold 1977). The entire process of spermatophore transfer takes about 10 seconds. It is astonishingly rapid, almost too fast for the eye to follow. Field observations of mating *Loligo* confirm that partners frequently switch mates after a period of copulation (Fields 1965).

Worthy of note is the role of the seminal receptacle of *Loligo*. Filling of the buccal pouch requires head-to-head copulation (Arnold and Williams-Arnold 1977) which is rarely observed during mass-mating events in the field or when staged in the lab. It is believed that head-to-head mating occurs offshore, perhaps even before sexual maturity is reached, filling the seminal receptacles to insure fertilization of the eggs. Females captured prior to mass-mating events inshore usually have full sperm pouches (Arnold and Williams-Arnold 1977). Sperm is stored in the seminal receptacle and can be used as needed.

Fecundity varies greatly between species, depending on taxon, reproductive strategy, and organismal size. Within the sepiids, fecundity ranges from 35 to 500 eggs; in the loliginids, from about 4 000 to 53 000; in the ommastrephids from about 100 000 to 650 000. The octopods also show a very large range of fecundities with numbers varying from 200 to 500 000 in the genus *Octopus*, 100 to 54 000 in the genus *Eledone*, and 20 to 80 in the species *Bathypolypnus arcticus* (Boyle and Rodhouse 2005). Fecundity within any taxon scales roughly, but not absolutely, with size.

Size and types of egg capsules, and where they are deposited, vary with species and lifestyle. An example of a coastal spawner (*Loligo opalescens*) and offshore spawner (*Illex illecibrosus*) have been described earlier. Information on reproduction in deep-sea species is scanty. The deep-sea Gonatids are believed to produce gelatinous egg masses that are carried within the arms and brooded until hatching (Okutani et al. 1995; Bjorke et al. 1997; Seibel et al. 2000a). Because of the low temperatures (near 3 °C) at the meso- and bathypelagic depths where *Gonatus onyx* was found brooding her eggs, the brooding period is thought to be as long as 9 months. (Seibel et al. 2000a). Many of the shallower-dwelling oegopsid squids found in the open ocean produce gelatinous pelagic egg masses (Young et al. 1985) that are believed to reside in the epipelagic zone until hatching. Squids of the

Enoploteuthidae, lacking the nidamental glands to produce the jelly for egg masses, lay their pelagic eggs singly, where they develop in near-surface waters.

Virtually, all cephalopods exhibit direct development, producing a well-developed hatchling that resembles a miniature adult, capable of swimming and hunting. Energy for development is provided by the yolky cephalopod egg, and an internal portion of the yolk sac may persist to help fuel the period immediately after hatching (Boletzky 1983). Hatch times vary considerably between species, with temperature and egg size being the main determining factors influencing incubation time. Cold water octopods with large eggs have incubation times of a year or more, cold-temperate squids with large eggs have incubation times of 3–4 weeks, and tropical squids have incubations of 2–10 days (Fields 1965; Sakurai et al. 1995; Boyle and Rodhouse 2005).

Terminology describing the early life history of cephalopods is unique to the group. Growth in young squid proceeds without the obvious transitions observed in crustacean larvae, yet they normally have a period in the plankton with a lifestyle greatly different from that of the adult, and the term "paralarva" was introduced as a distinct life stage (Young and Harman 1988). The paralarval period covers the post-hatch period when the young squid is "pelagic in near-surface waters during the day and has a distinctively different mode of life from that of older conspecific individuals." A cephalopod was defined as a "subadult" when "all diagnostic morphological features used to define the species other than those relating to sex and size" were attained. Features such as photophore and chromatophore patterns, or the shape and sucker patterns of tentacular clubs, are good examples, the same morphological characters used in a taxonomic key. An adult has all features of a subadult but is sexually mature. The stage in between the paralarva and subadult, if present, is the "juvenile" stage (Young and Harman 1988).

Vertical Distribution and Migration

Cephalopods are found from the intertidal zone to abyssopelagic depths, occupying both the benthic and pelagic habitats. Octopods such as *Grimpoteuthis*, the "dumbo squid," are found near-bottom at depths to 7000 m, and pelagic forms such as the infamous giant squid *Architeuthis* and the vampire squid *Vampyroteuthis* may be found to 3000 m.

Life histories typical of coastal (*Loligo opalescens*) and outer shelf/slope (*Illex illecibrosus*) species have been briefly described, as well as a more speculative one for a widely distributed mesopelagic genus (*Gonatus*). Each of the three environments has its own suite of characteristics that shape the life histories of the cephalopods dwelling there and select for the morphotypes that will be most successful. For pelagic species, three characteristics of each habitat are especially important as selective pressures: the photic environment, the level of annual productivity (food availability), and proximity to the bottom (depth of occurrence). Photic environment and annual production are also important to benthic species, with a third major operator being available cover and concealment.

The coastal environment, typified by maximum depths of about 200 m at the shelf break for continental shelves outside the Antarctic (see Chapter 1), is a well-lit, physically dynamic system with high levels of annual production. Nutrients are provided by riverine inputs, resuspended sediments and coastal run-off, fueling algal growth and a diverse food web in the water column and benthos. Estuarine systems, a potentially rich source of prey, are rarely exploited by cephalopods due to the lower salinities in those environments. However, there are exceptions. *Lolliguncula brevis*, a common species in the GOM, is found at reduced salinities of 22 ⁰/₀₀ in Florida estuaries (Dragovich and Kelly 1967). And some juvenile octopods (e.g. *Octopus bimaculoides* and *Octopus rubescens*) can be found in the California intertidal zone where chance encounters with low salinity would be unavoidable.

The pelagic coastal system is largely the province of the loliginids, strong swimmers that take a variety of benthic and pelagic prey, mainly crustaceans and small fishes but sometimes small cephalopods as well. Benthic and demersal cephalopods inhabiting the coastal region include the octopod genera *Octopus* and *Eledone* and cuttlefishes in the families Sepiolidae and Sepiidae. Octopods and cuttlefishes both prey on crabs, shrimps, and fishes; shelled molluscs form a large part of the octopod diet as well. Octopods use bottom structure for shelter but can also partially bury themselves while mimicking their background with their chromatophores, as can the cuttlefishes. Even the more pelagic loliginids remain close to the bottom during the day, dispersing upward at night (Boyle and Rodhouse 2005). All three coleoid groups use the ocean bottom to deposit their egg masses.

The well-lit water column and diverse prey spectrum of the coastal habitat are ideal for strongly swimming visual predators such as the cephalopods. Likewise, large visual predators including jacks (Carangidae) and groupers (Serranidae) as well as weasel and dogfish sharks (Hemigaleus and Squalus, respectively) prey on loliginids, octopods, and cuttlefishes (Smale 1996). One may conclude that a robust locomotory musculature and well-developed eyes are absolutely required for cephalopod survival in the coastal zone, both for success in hunting capable swimmers and for avoiding their own predators.

Moving further offshore to the shelf/slope break and greater depths results in a more stable oceanic habitat and a different cephalopod fauna. Increased depth dampens seasonal temperature swings and episodic storm turbulence, making for a slightly more predictable but still dynamic environment. Shelf/slope breaks throughout the globe are the sites of the world's major boundary currents (see Chapter 1), highly productive regions that form the gateways to the open ocean and are home to many of the world's major fisheries. Boundary currents are the eastern and western limbs of the world's oceanic gyre systems. Interaction of the boundary currents with coastal currents and a shoaling bottom topography generates a variety of physical effects. Those may include topographic upwelling as well as meanders and rings, all of which impact the resident fauna. Most of all, the regions combine a well-lit water column with a rich food supply, making them ideal for pelagic cephalopods. The pelagial of the boundary current regions is the province of the ommastrephid squids, large powerful squids that along with the loliginids are the champion swimmers of the cephalopod fauna.

Cephalopod fisheries associated with boundary currents are listed below (Boyle and Rodhouse 2005).

Western boundary currents	Eastern boundary currents
Kuroshio Current: *Todarodes pacificus*	Peru Current: *Dosidicus gigas*
Brazil Current: *Illex argentinus*	California Current: *Loligo opalescens*
Gulf Stream: *Illex illecebrosus*	Canaries Current: *Octopus vulgaris*
Arabian Sea: *Sthenoteuthis oualaniensis*	

The Antarctic Circumpolar Current, which circumnavigates the Antarctic continent and, in some areas, such as the Antarctic Peninsula, acts as a boundary current, supports a fishery for three ommastrephid squid: *Notodarus sloani, N. gouldi*, and *Martialia hyadesi*. Note that all harvested species except *Loligo opalescens* and *Octopus vulgaris* are ommastrephids.

We have already been introduced to the reproductive habits of *Illex illecebrosus*, including the use of floating egg masses as a means of pelagic egg development. *Illex argentinus* (Arkihipkin 1993; Rodhouse et al. 1995) and *Todarodes pacificus* exhibit a similar reproductive strategy (Okutani 1983; Sakurai et al. 2000) including pelagic egg masses, a strategy now believed to be widespread among the ommastrephid squids (Boyle and Rodhouse 2005). Ommastrephids are mobile enough to use the productive waters of the outer shelf, shelf break, and nearby oceanic waters for hunting, exploiting the boundary currents as part of their migratory life histories.

Shelf-break ommastrephids take a variety of pelagic and demersal fauna, primarily crustaceans and fishes, as prey. Euphausiids, small pelagic fishes, and even polychaetes have been reported as prey items (Nicol and O'Dor 1985; Tanaka 1993; Rodhouse and Nigmatullin 1996). In particular, lanternfishes (family Myctophidae – see Chapter 10) are an important part of the diet of both shelf-break and oceanic ommastrephids (Rodhouse and Nigmatullin 1996).

Several different families of dogfish sharks (Squaliform) feed on cephalopods inhabiting the outer shelf environment. They include the dogfish sharks themselves (Squalidae), as well as the gulper sharks (Centrophoridae: genus *Deania*), lantern sharks (Etmopteridae: genus *Centroscyllium*), and sleeper sharks (Somniosidae: genus *Centroscymnus*), with a significant fraction of their prey being ommastrephids (Smale 1996).

The productive coastal and shelf environments continue out to the shelf-break and boundary current regions supporting a rich prey field for the cephalopods and a large suite of rapidly swimming visual predators. Once in the open ocean, the well-lit productive layer extends only to the bottom of the epipelagic zone, usually considered to be 200 m (see Chapter 1). Below that, declining light levels increasingly limit visual predation, which is at once a boon and a bane to the open-ocean cephalopod community: more difficult to be hunted, but more difficult to hunt.

Vertical distributions of major cephalopod groups and some notable species are summarized in Figure 8.103. The 24 families of oegopsid squids dominate the midwater fauna of the open ocean with many species reaching depths below 2000 m (Vecchione and Pohle 2002; Young 1978). Nautilids are found at depths up to 500 m on and above the continental slope at several locations within the tropical Pacific, including the Philippines and Australia's Great Barrier Reef. The argonautid octopods *Argonauta* and *Tremoctopus* inhabit the epipelagic zone, with the "paper nautilus" *Argonauta argo* found pan-globally in tropical and warm-temperate seas (Roper et al. 1984) and *Tremoctopus violaceus* in the Atlantic between 40°N and 36°S as well as in the Mediterranean. The Ram's horn squid, *Spirula*, is found at mesopelagic depths panglobally, wherever the temperature at 400 m is 10 °C or warmer. It is believed to vertically migrate between 300 and 600 m on a diel basis (Reid 2005). The "vampire squid," *Vampyroteuthis infernalis* is found between 600 and 1200 m pan-globally, never abundant but always present. In reality an Octopodiform, *V. infernalis*, is a monotypic species in its own order.

Benthic and benthopelagic species of cephalopods occur from the shelf break to abyssal depths. In the north Atlantic, the

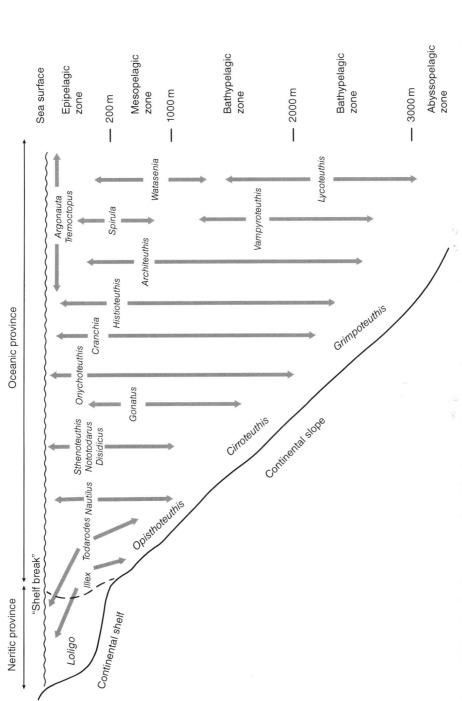

Figure 8.103 Distribution as a function of depth and habitat for cephalopod genera characteristic of the open-ocean and deep-sea realms. *Source:* From Boyle and Rodhouse (2005), figure 12.2 (p. 182). Reproduced with the permission of John Wiley & Sons.

sepiolids (bobtail squid) dominate the shallow and upper slopes (150–500 m) with at least one species (*Neorossia caroli*) found from 400 to 1535 m (Collins et al. 2001). Within the Octopoda, the deep-sea incirrates tend to be found at shallower depths than the cirrates, ranging from 200 to 4000 m but with most found to 500 m (Voss, G.L. 1988a). The deepest living family within the incirrates is the Megaleledonidae containing the genera *Graneledone*, *Thaumeledone*, and *Bentheledone*, all living between 1000 m and depths of 3500 m or greater. Deepest of the octopods are the cirrates, or finned octopods, gelatinous octopods with well-developed brachial webbing used to help them hover above the bottom. Deepest among the cirrates are the Opisthoteuthids, specifically the genus *Grimpoteuthis* which has been trawled up from as deep as 7279 m (Voss 1988a).

Selective pressures on pelagic squid change with increasing depth. Threat of visual predation remains high in the upper 200 m. Below that, increasing distance from near-surface primary and secondary production limits the prey field and available energy (Vinogradov 1970). Thus, the threat of visual predation declines with increasing depth, but so does the available energy. Based on what we have seen in other taxa, at least three possible foraging strategies could be employed by oceanic squids to deal with the highly stratified open-ocean environment. The first would be simply to remain in the upper 200 m for the entire day–night cycle. Since cover and concealment are essentially absent in the well-lit surface layers of the open ocean during the day, remaining in the euphotic zone would require a highly capable locomotory musculature, relying on swim speed both to hunt and to avoid predation in the productive upper layers. The second strategy would be to retain a strong locomotory capability and reside in poorly lit deeper layers during the day, taking advantage of the cover and concealment provided by the constant darkness at depth. Nightly excursions to hunt in near-surface waters would allow them to remain constantly in dimly lit waters while also allowing them to hunt in a more productive environment for about half of the day. The third strategy would be to permanently reside at mesopelagic depths using an ambush foraging strategy and limited movement, a low-energy lifestyle that exploits the darkness at mid-depths for camouflage and accommodates the lower energy available in the mesopelagic realm. Good evidence exists for all three strategies.

Ommastrephid squids in the genera *Sthenoteuthis*, *Ommastrephis*, *Martialia*, *Dosidicus*, and *Nototodarus* all are very strong swimmers and are theoretically quite capable of remaining in the epipelagic zone over the entire day–night cycle, even with the threat of large, fast-moving, pelagic predators like the tunas and swordfish (Smale 1996). It is difficult to obtain data on diel vertical distributions of the ommastrephids using nets, they are simply too fast for quantitative sampling. We have to let their predators do the sampling for us, and the evidence supports a case for at least part-time epipelagic residence in the ommastrephids. They are present in the diets of epipelagic hunters such as tunas and swordfishes (Smale 1996), as well as those of flighted seabirds such as the albatrosses (Rodhouse et al. 1990). Evidence is strongest for daytime residence in the region of the Antarctic Polar Front (Rodhouse and White 1995), where it is believed that the cephalopods occupy a similar niche to that of the pelagic fishes in warmer systems.

Good evidence supports a case for vertical migration as well. Nakamura (1993) directly demonstrated diel vertical migration in the ommastrephid squid *Ommastrephis bartrami* using a tagged individual. Using both closing trawls and a more rapidly towed net off Hawaii to catch faster moving individuals, Young (1978) made a convincing case for vertical migration in most of the pelagic cephalopods there. About 60% of the species and 80% of the individuals vertically migrated from depths of 500–800 m during the day into the upper 250 m at night. The most abundant vertical migrators were enoploteuthids in the genera *Pterygioteuthis, Abralia, Abraliopsis,* and *Pyroteuthis*, all considered to be strong swimmers. Ommastrephids were not captured during the study. A similar breakdown of the cephalopod fauna was reported by Passarella and Hopkins (1991) for the GOM, with most of the cephalopods occurring shallower than 200 m at night and centered between 100 and 400 m during the day.

The available literature strongly suggests a part- or full-time epipelagic presence for ommastrephids. Absence of data makes it difficult to categorize them as full-time residents of the epipelagic zone or obligate vertical migrators, itself implying a measure of behavioral independence for the group. Available data suggest that lanternfishes are important in the diet of several species (Rodhouse and Nigmatullin 1996). Since most lanternfishes are strong migrators, their presence in the diet means that ommastrephids are hunting in the epipelagic zone at night or at mesopelagic depths during the day.

Mesopelagic cephalopods exhibiting a low-energy lifestyle have a few characteristics in common. The first is an ammoniacal buoyancy mechanism such as the NH_4Cl-filled coelomic cavity of cranchiids or the NH_4Cl-filled vesicles intercalated through a variety of tissues in the histioteuthids, architeuthids, chirotuthids, and 13 other oegopsid squid families (Voight et al. 1994). The second is the presence of photophores to aid in cover and concealment (Herring 1977, 1978). The third is a lowered rate of metabolism (Seibel et al. 1997), particularly relative to the muscular swimmers in the Ommastrephidae and Loliginidae. Generally, the metabolism of oceanic cephalopods showed a profound decline with increasing minimum depth of occurrence (MDO: the depth below which 90% of the population lives, Childress and Nygaard 1973). Deeper living species showing a mass-specific rate less than 1% of that of strong swimmers migrating into or residing in epipelagic waters. Lowest metabolic rates recorded in the study were for the vampire squid, *Vampyroteuthis infernalis* ($0.07\,\mu l\ O_2\ g^{-1}\ h^{-1}$), residing below 600 m, and highest were for *Gonatus onyx* ($8.79\,\mu l\ O_2\ g^{-1}\ h^{-1}$), a vertical migrator reaching 100 m during its nightly excursions. Intermediate values were recorded for *Histioteuthis heteropsis* ($1.02\,\mu l\ O_2\ g^{-1}\ h^{-1}$), an ammoniacal squid that nonetheless vertically migrates between 500 and 150 m. Lower values, based as much on a sluggish lifestyle as depth of occurrence, were recorded for the neutrally buoyant cranchiids. The range was from $0.29\,\mu l\ O_2\ g^{-1}\ h^{-1}$ for *Cranchia scabra* (MDO of 10 m) to $0.97\,\mu l\ O_2\ g^{-1}\ h^{-1}$ for *Heliocranchia* (MDO of 300 m) (Seibel et al. 1997). Similar declines have been reported for a broad spectrum of mesopelagic Crustacea and fishes (Childress 1975; Torres et al. 1979). The lower rates at mesopelagic depths are due to a reduction in locomotory musculature and its associated metabolic costs, the end result of a reduced light field, and a diminished effectiveness of visual predation as both hunter and hunted.

A remarkable fauna is found in the benthopelagic realm at lower (>500 m) mesopelagic and bathypelagic depths. The umbrella-like cirratid octopods float above the bottom by sculling with their fins, using their brachial webbing for medusoid locomotion and their flaccid bodies for near-neutral buoyancy. It is believed that they feed on benthic crustaceans such as amphipods, perhaps by descending on them from above (Boyle and Rodhouse 2005).

The large-finned deep-sea squids in the families Mastigoteuthidae and Octopoteuthidae as well as the recently described big-fin squid in the family Magnapinnidae (Vecchione and Young 1997; Vecchione et al. 2002) are remarkable in appearance and behavior. Submersible observations revealed a benthopelagic habit for individuals in each of the families, floating just above the bottom at depths ranging from 600 to 850 m. Posture was maintained by sculling with their large fins. *Mastigoteuthis* has exceptionally long tentacles that it trails along the bottom with the tips of its tentacular clubs touching, looking as if it is "trolling through the epibenthic soup of small organisms and organic material" for prey (Vecchione et al. 2002). Most spectacular of the three types of squid observed was the magnapinnid, which has extraordinarily long arms and tentacles that dangle to the bottom from a 90° bend or "elbow" a short distance from the head (Vecchione et al. 2002). Total length of the squid with its arms fully extended was estimated at about 7 m, nearly half the length of a giant squid!

The powerfully swimming open-ocean fish taxa, notably the tunas and billfishes, are important predators on oceanic squids (Smale 1996), as are the deeper-living lancetfishes (Alepisauridae). Flighted and diving seabirds also prey on cephalopods, with the best information coming from work in temperate and polar regions. Squids make up a varying portion of the diet of albatrosses and petrels in sub-Antarctic waters. In some species, squids can be as important to their diets as fishes and crustaceans. Subantarctic and Antarctic penguins prey heavily on squids as well, chiefly on the families *Onychoteutidae, Gonatidae*, and *Psychroteuthidae* (Croxall and Prince 1996). It is estimated that squids make up between 1 and 54% of the food intake of seabirds depending on species and location (Boyle and Rodhouse 2005). The emperor penguin *Aptenodytes forsteri* feeds heavily on squid, which predominate in its diet in certain locations and seasons.

Seals and toothed whales both include cephalopods in their diet. Virtually, all seal species in the northern and southern hemispheres take cephalopods occasionally, and in a few they dominate the diet (Klages 1996). Southern elephant seals (*Mirounga lenonina*) are considered squid specialists, as is the rarely observed Ross Seal, a member of the Antarctic pack-ice seal community. Over 80% of the toothed whales regularly consume cephalopods, including the sperm whales (Physeteridae), beaked whales (Ziphiidae), and the coastal and oceanic dolphins (Phocenidae and Delphinidae). The great majority of squids in the diets are the ommastrephids, histioteuthids, and cranchiids in the open ocean, with the loliginids being most important in shelf waters. Onychoteuthids and gonatids are most important diet items in polar regions and in the North Pacific (Clarke 1996). The battles between sperm whales and giant squid, whether real or imagined, are the stuff of legend and hopefully will be the inspiration for sea stories for many years to come.

Geographic Distribution

Cephalopods occupy the entire range of marine habitats ranging from the intertidal to great depths and from the equator to the poles. Detailed information on the systematics and biogeography of cephalopods is available in a series of excellent publications that provide illustrations and distributional maps for the very great majority of described species (Roper et al. 1984; Voss N.A. 1988b; Sweeney et al. 1992; Nesis 1997; Voss et al. 1998a, b; Nesis 1999; Jereb and Roper 2005; Jereb and Roper 2013; Jereb et al. 2016). A single broad synthesis of cephalopod biogeographic patterns is not available, but the work on individual species and regional patterns suggests that cephalopods conform to the same patterns that are observed in pelagic fishes and Crustacea (e.g. Briggs 1974, 1995; Backus et al. 1977).

References

Alder, J. and Hancock, A. (1851). Descriptions of two new species of nudibranchiate Mollusca, one of them forming the type of a new genus. *Annals and Magazine of Natural History* 8 (46): 290–302.

Alexander, R.M. (1982). *Locomotion of Animals*. Glasgow: Blackie.

Alexander, R.M. (2003). *Principles of Animal Locomotion*. Princeton University Press.

Ambrose, R.F. (1988). Population dynamics of *Octopus bimaculatus*: influence of life history patterns, synchronous reproduction and recruitment. *Malacologia* 29: 23–39.

Andrews, E.B. and Taylor, P.M. (1988). Fine structure, mechanism of heart function and haemodynamics in the prosobranch gastropod mollusc *Littorina littorea* (L.). *Journal of Comparative Physiology B* 158: 247–262.

Arkihipkin, A. (1993). Age, growth, stock structure and migratory rate of prespawning short-finned squid, Illex argentinus, based on statolith ageing investigations. *Fisheries Research* 16: 313–338.

Arnold, J.M. and Wiliams-Arnold, L.D. (1977). Cephalopoda, decapoda. In: *Reproduction of Marine Invertebrates* (eds. A.C. Giese and J.S. Pearse), 243–290. London: Academic Press.

Arnold, J.M. and Young, R.E. (1974). Ultrastructure of a cephalopod photophore. I. Structure odf the photogenic tissue. *Biological Bulletin* 147: 507–521.

Austin, C.R., Lutwak-Mann, C., and Mann, T. (1964). Spermatophores and spermatozoa of the squid *Loligo pealii*. *Proceedings of the Royal Society of London Ser. B* 161: 143–152.

Backus, R.H., Craddock, J.E., Haedrich, R.L., and Robison, B.H. (1977). Atlantic mesopelagic zoogeography. In: *Fishes of the Western North Atlantic, vol. 7* (eds. R.H. Gibbs, F.H. Berry, E.B. Bohlke, et al.). New Haven: Sears Foundation for Marine Research, Yale University.

Barnes, R.D. (1974). *Invertebrate Zoology*. Philadelphia: W.B. Saunders.

Bayer, F.M. (1963). Observations on pelagic mollusks associated with the siphonophores Velella and Physalia. *Bulletin of Marine Science of the Gulf and Caribbean* 13: 454–466.

Be, A.W.H. and Gilmer, R.W. (1977). *A Zoogeographic and Taxonomic Review of Euthecosomatous Pteropoda Oceanic Micropaleontology*. London: Academic Press.

Bednarsek, N., Feely, R.A., Reum, J.C.P. et al. (2014). Limacina helicina shell dissolution as an indicator of declining habitat suitability owing to ocean acidification in the California Current ecosystem. *Proceedings of the Royal Society B* 281: 20140123.

Berner, R.A. (1977). Sedimentation and dissolution of pteropods in the ocean. In: *The Fate of Fossil Fuel CO2 in the Oceans* (eds. N.R. Andersen and A. Malahoff). New York: Plenum Press.

Bidder, A.M. (1950). The digestive mechanism of the European squids *Loligo forbesi, Alloteuthis media* and *Alloteuthis subulata. Quarterly Journal of Microscopic Science* 91: 1–43.

Bidder, A. (1962). The use of the tentacles, swimming and buoyancy control in the pearly nautilus. *Nature* 192: 925–926.

Bidder, A.M. (1966). Feeding and digestion of cephalopods. In: *Physiology of Mollusca* (eds. K.M. Wilbur and C.M. Yonge). New York: Academic Press.

Bidder, A.M. (1976). New names for old: the cephalopod midgut gland. *Journal of Zoology* 180: 441–443.

Bieri, R. (1966). Feeding preferences and rates of the snail Ianthina prolongata, the barnacle *Lepas anserifera*, the nudibranchs *Glaucus atlanticus* and *Fiona pinnata*, and the food web in the marine nesuton. *Publications of the Seto Marine Biological Laboratory* 14: 161–170.

Bjorke, H., Hansen, H., and Sundt, R.C. (1997). Egg masses of the squid *Gonatus fabricii* (Cephalopoda, Gonatidae) caught with a pelagic trawl off northern Norway. *Sarsia* 82: 149–152.

Black, G.A.P., Rowell, T.W., and Dawe, E.G. (1987). Atlas of the biology and distribution of the squids *Illex illicebrosus* and *Loligo pealei* in the northwest Atlantic. *Canadian Special Publication Fisheries Aquatic Science* 100: 1–62.

Boas, J.E.V. (1886). Spolia Atlantica. Bidrag til Pteropdernes. Morfologi og Systematic samt til Kundskaben om deres geografiski Udbederelse. *Kongelige Danske videnskabernes selsklabs skrifter. Naturvidenskabelig og mathematisk afdeling* 4: 1–231.

Boletzky, S.V. (1983). Sepia officinalis. In: *Cephalopod Life Cycles. Species Accounts* (ed. P.R. Boyle). London and New York: Academic Press.

Boletzky, S.V. and Hanlon, R.V. (1983). A review of the laboratory maintenance, rearing and culture of cephalopod molluscs. *Memoirs of the National Museum, Victoria* 44: 147–187.

Bone, Q. and Moore, R.H. (2008). *The Biology of Fishes*, 3e. Taylor & Francis: New York.

Born, E. (1910). Anatomie der Phyllirhoe. *Zeitschrift fur wissenschaftliche Zoologie* 97.

Borrell, B.J., Goldbogen, J.A., and Dudley, R. (2005). Aquatic wing flapping at low Reynolds numbers: swimming kinematics of the Antarctic pteropod *Clione antarctica*. *Journal of Experimental Biology* 208: 2939–2949.

Boucaud-Camou, E. and Boucher-Rodoni, R. (1983). Feeding and digestion in the cephalopods. In: *The Mollusca* (eds. A.S.M. Saleuddin and K.M. Wilbur). New York: Academic Press.

Boucher-Rodoni, R. and Mangold, K. (1994). Ammonia production in cephalopods, physiological and evolutionary aspects. *Marine and Freshwater Behavior and Physiology* 25: 53–60.

Boycott, B.B. (1961). The functional organization of the brain of the cuttlefish, *Sepia officinalis. Proceedings of the Royal Society of London B* B153: 503–534.

Boyle, P.R. (ed.) (1983). Eledone cirrhosa. In: *Cephalopod Life Cycles. Species Accounts*. London and New York: Academic Press.

Boyle, P.R. (1986a). Responses to water-borne chemicals by the octopus *Eledone cirrhosa*. *Journal of Experimental Marine Biology and Ecology* 69: 129–136.

Boyle, P.R. and Rodhouse, P. (2005). *Cephalopods, Ecology and Fisheries*. Oxford: Blackwell.

Boyle, P.R., Pierce, G.J., and Hastie, L.C. (1995). Flexible reproductive strategies in the squid *Loligo forbesi*. *Marine Biology* 121: 501–508.

Briggs, J.C. (1974). *Marine Zoogeography*. New York: McGraw-Hill.

Briggs, J.C. (1995). *Global Biogeography*. Amsterdam: Elsevier.

Brusca, R.C. and Brusca, G.J. (2003). *Invertebrates*, 2e. Sunderland: Sinauer.

Brusca, R.C., Moore, W., and Shuster, S.M. (2016). *Invertebrates*, 3e. Sunderland: Sinauer Associates.

Budelmann, B.U. (1988). Morphological diversity of equilibrium receptor systems in aquatic invertebrates. In: *Sensory Biology of Aquatic Animals* (eds. J. Atema, R.R. Fay, A.N. Popper and W.N. Tavolga). New York: Springer-Verlag.

Budelmann, B.U. (1996). Active marine predators: the sensory world of cephalopods. In: *Zooplankton Sensory Ecology and Physiology* (eds. P.H. Lenz, D.K. Hartline, J.E. Purcell and D.L. Macmillan). Amsterdam: Gordon and Breach.

Burmeister, C. (1834). *Beitrage zur Naturgeschichte der Rankenfusser*. Berlin: G. Reimer.

Burridge, A.K., Goetze, E., Wall-Palmer, D. et al. (2016). Diversity and abundance of pteropods and heteropods along a latitudinal gradient across the Atlantic Ocean. *Progress in Oceanography* 158: 213–223.

Chapman, A.D. (2009). *Numbers of Living Species in Australia and the World*, 2e. Canberra: Australian Biological Resources Study.

Chen, C. (1968a). The distribution of thecosomatous pteropods in relation to the Antarctic convergence. *Antarctic Journal, U.S.* 3: 155–157.

Childress, J.J. (1975). The respiratory rates of midwater crustaceans as a function of depth of occurrence and relation to the oxygen minimum layer off Southern California. *Comparative Biochemistry and Physiology* 50A: 787–799.

Childress, J.J. and Nygaard, M.H. (1973). The chemical composition of midwater fishes as a function of depth of occurrence off southern California. *Deep-Sea Research* 20: 1093–1109.

Childress, J.J. and Nygaard, M.H. (1974). Chemical composition and buoyancy of midwater crustaceans as a function of depth of occurrence off southern California. *Marine Biology* 27: 225–238.

Childress, J.J. and Price, M.H. (1978). Growth rate of the bathypelagic crustacean *Gnathophausia ingens* (Mysidacea: Lophogastridae) I. Dimensional growth and population structure. *Marine Biology* 50: 47–62.

Childress, J.J. and Price, M.H. (1983). Growth rate of the bathypelagic crustacean *Gnathophausia ingens* (Mysidacea: Lophogastridae) II. Accumulation of material and energy. *Marine Biology* 76: 165–177.

Chun, C. (1889). Bericht uber eine nach den Canarischen Inseln im Winter 1887/1888 ausgefuhrte Reise. *Sitzungsber Preuss Akad Wiss* 2: 519–553.

Clarke, M.R. (1996). Cephalopods as prey. III. Cetaceans. *Philosophical Transactions of the Royal Society of London, Series B, Biological Sciences* 351: 1053–1065.

Clarke, A.H. and Menzies, R.J. (1959). Neopilina (Vema) ewingi, a second living species of the Paleozoic class Monoplacophora. *Science* 129: 1026–1027.

Collins, M.A., Yau, C., Allcock, L., and Thurston, M.H. (2001). Distribution of deep-water benthic and bentho-pelagic cephalopods from the north-east Atlantic. *Journal of the Marine Biological Association of the United Kingdom* 81: 105–117.

Conover, R.J. and Lalli, C.M. (1972). Feeding and growth in *Clione limacina* (Phipps) a pteropod mollusc. *Journal of Experimental Marine Biology and Ecology* 9: 279–302.

Croxall, J.P. and Prince, P.A. (1996). Cephalopods as prey. I. Seabirds.

Philosophical Transactions of the Royal Society of London, B 351: 1023–1043.

Dadon, J.R. and De Cidre, L.L. (1992). The reproductive cycle of the thecosomatous pteropod *Limacina retroversa* in the western South Atlantic. *Marine Biology* 114: 439–442.

Dales, R.P. (1953). The distribution of some heteropod molluscs off the Pacific coast of North America. *Proceedings of the Zoological Society of London* 122: 1007–1015.

De Blainville, H.M.D. (1824). *Dictionnaire des Sciences Naturelles*, vol. 32. Paris: Levrault.

Denton, E.J. (1974). On buoyancy and the lives of modern and fossil cephalopods. *Proceedings of the Royal Society of London* B 185: 273–299.

Denton, E.J. and Gilpin-Brown, J.B. (1973). Floatation mechanisms in modern and fossil cephalopods. *Advances in Marine Biology* 11: 197–268.

Dilly, P.N. and Herring, P.J. (1978). The light organ and ink sac of Heteroteuthis dispar. *Journal of Zoology* 186: 47–59.

Dragovich, A. and Kelly, J.A. (1967). Occurrence of the squid, *Lolliguncula brevis*, in some coastal waters of Western Florida. *Bulletin of Marine Science* 17: 840–844.

Fields, W.G. (1965). The structure, development, food relations, reproduction, and life history of Loligo opalescens Berry. *California Fish and Game, Fishery Bulletin* 131: 1–108.

Fretter, V. and Graham, A. (1962). *British Prosobranch Molluscs*. London: The Ray Society.

Gannefors, C., Boer, M., Kattner, G. et al. (2005). The Arctic sea butterfly *Limacina helicina*: lipids and life strategy. *Marine Biology* 147: 169–177.

Gartner, J.V. (1991). Life histories of three species of lanternfishes (Pisces: Myctophidae) from the eastern Gulf of Mexico. II. Age and growth patterns. *Marine Biology* 111: 21–27.

Gilly, W.F. and Lucero, M.T. (1992). Behavioural responses to chemical stimulation of the olfactory organ in the squid *Loligo opalescens*. *Journal of Experimental Biology* 162: 209–229.

Gilmer, R.W. (1972). Free floating mucous webs: a novel feeding adaptation for the open ocean. *Science* 176: 1239–1240.

Gilmer, R.W. (1974). Some aspects of feeding in the thecosomatous pteropod molluscs. *Journal of Experimental Marine Biology and Ecology* 15: 127–144.

Gilmer, R.W. and Harbison, G.R. (1986). Morphology and field behavior of pteropod molluscs: feeding methods in the families Cavoliniidae, Limacinidae and Peraclididae (Gastropoda: Thecosomata). *Marine Biology* 91: 47–57.

Graff, L.V. (1875). Anatomie des Chaetoderma. *Zeitschrift fur Wissenschaften Zoologie* 26.

Graham, A. (1954). Some observations on the reproductive tract of *Ianthina janthina*. *Proceedings of the Malacological Society of London* 31: 1–6.

Graham, A. (1965). The buccal mass of Ianthinid prosobranchs. *Proceedings of the Malacological Society of London* 36: 323–338.

Haller, B. (1888). Die morphologie der prosobranchier. *Morphologisches Jahrbuch* 14: 54–169, pl. 3-9.

Hamner, W.M., Madin, L.P., Alldredge, A.L. et al. (1975). Underwater observations of gelatinous zooplankton: sampling problems, feeding biology and behavior. *Limnology and Oceanography* 20: 907–917.

Hanlon, R.T. and Budelmann, B.U. (1987). Why cephalopods are probably not deaf. *The American Naturalist* 129: 312–317.

Hanlon, R.T. and Messenger, J.B. (1996). *Cephalopod Behavior*. Cambridge: Cambridge University Press.

Hansen, G. (1888). Neomenia, Proneomenia und Chaetoderma. *Bergen Mus. Aarsbrecht*.

Hardy, A.C. (1956). *The Open Sea*. London: Collins.

Haven, N. (1972). The ecology and behavior of *Nautilus pompilius* in the Philippines. *Veliger* 15: 75–81.

Hazelhof, E.H. (1939). Uber die Ausnutzung des Sauerstoffs bei verschniedenen Wassertierren. *Zeitschrift fur Vergleichende Physiologie* 26: 306–327.

Helfman, G.S., Collette, B.B., Facey, D.E., and Bowen, B.B. (2009). *The Diversity of Fishes*. Oxford: Wiley-Blackwell.

Herring, P.J. (1977). Luminescence in cephalopods and fish. *Symposium of the Zoological Society, London* 38: 127–159.

Herring, P.J. (1978). Bioluminescence of invertebrates other than insects. In: *Bioluminescence in Action* (ed. P.J. Herring). London and New York: Academic Press.

Hertwig, R. (1905). *A Manual of Zoology*. New York: Henry Holt and Company.

Hescheler, K. (1900). Mollusca. In: *Lehrbuch der vergleichenden anatomie der wirbellosen*, 3e (ed. A. Lang): Lief. 1.

Hodgkin, A.L. and Huxley, A.F. (1952). A quantitative description of membrane current and its application to conduction and excitation in nerve. *Journal of Physiology* 117: 500–544.

Holleman, J.J. (1972). Observations on growth, feeding, reproduction, and development in the opisthobranch, *Fiona pinnata*. *Veliger* 15: 142–146.

Hopkins, T.L. and Lancraft, T.M. (1984). The composition and standing stock of mesopelagic micronekton at 27 N 86 W in the eastern Gulf of Mexico. *Contributions in Marine Science* 27: 145–158.

Howard, W.R., Roberts, D., Moy, A.D. et al. (2011). Distribution, abundance and seasonal flux of pteropods in the Sub-Antarctic zone. *Deep-Sea Research II* 58: 2293–2300.

Hsiao, S.C.T. (1939a). The reproductive system and spermatogenesis of Limacina (Spiratella) retroversa (Flem.). *Biological Bulletin* 76: 7–25.

Hsiao, S.C.T. (1939b). The reproduction of *Limacina retroversa* (Flem.). *Biological Bulletin* 76: 280–303.

Hunt, B.P.V., Pakhomov, E.A., Hosie, G.W. et al. (2008). Pteropods in Southern Ocean ecosystems. *Progress in Oceanography* 78: 193–221.

Hurley, A.C., Lange, G.D., and Hartline, P.H. (1978). The adjustable "pin-hole camera" eye of Nautilus. *Journal of Experimental Biology* 205: 37–44.

Hyman, L.H. (1967). *The Invertebrates, Vol. 6, Mollusca I*. New York: McGraw-Hill.

Ivlev, V.S. (1961). *Experimental Ecology of the Feeding of Fishes*. New Haven: Yale University Press.

Jackson, G.D. (1993). Seasonal variation in reproductive investment in the tropical squid Loligo chinensis and the small tropical sepioid Idiosepius pygmaeus. *Fishery Bulletin U.S.* 91: 260–270.

Jereb, P. and Roper, C.F.E. (2005). *Cephalopods of the World. An Annotated and Illustrated Catalogue of Species Known to Date. Volume 1. Chambered Nautiluses and Sepioids (Nautilidae, Sepiidae, Sepioloidae, Sepiadariidae, Idiosepiidae and Spirulidae)*, FAO Species Catalogue for Fishery Purposes. No. 4, vol. 1. Rome: FAO.

Jereb, P. and Roper, C.F.E. (2013). *Cephalopods of the World. An Annotated and Illustrated Catalogue of Species Known to Date. Volume 2. Myopsid and Oegopsid Squids*, FAO Species Catalogue for Fishery Purposes. No. 4, vol. 2. Rome: FAO.

Jereb, P., Roper, C.F.E., Norman, M.D., and Finn, J.K. (2016). *Cephalopods of the World. An Annotated and Illustrated Catalogue of Cephalopod Species Known to Date. Volume*

3. Octopods and Vampire Squids, FAO Species Catalogue for Fishery Purposes. No. 4, vol. 3. Rome: FAO.

Johnsen, S., Balser, E.J., Fisher, E., and Widder, E.A. (1999). Bioluminescence in the deep-sea cirrate octopod *Stauroteuthis syrtensis* Verrill (Mollusca: Cephalopoda). *Biological Bulletin* 197: 26–39.

Jonston, J. (1650). *Historia naturalis de examgubus. Libri IV*. Amsterdam: J.J. Schipperi.

Kaestner, A. (1967). *Invertebrate Zoology Volume 1*. New York: Wiley.

Keferstein, W. (1863–1866). Kopftragende Weichthiere (Malacozoa, Cephalophora). In: *Klassen und Ordnungen der Weichthiere (Malacozoa), Volume III, Abt 2* (ed. H.G. Bronn). C.F. Winter'sche Verlagshandlung.

Kier, W.M. and Van Leeuwen, J.L. (1997). A kinematic analysis of tentacle extension in the squid *Loligo pealei*. *Journal of Experimental Biology* 200: 41–53.

Klages, N.T.W. (1996). Cephalopods as prey. II. Seals. *Philosophical Transactions of the Royal Society of London, Series B, Biological Sciences* 351: 1045–1052.

Kobayashi, H.A. (1974). Growth cycle and related vertical distribution of the thecosomatous pteropod *Spiratella* ("*Limacina*") *helicina* in the central Arctic Ocean. *Marine Biology* 26: 295–301.

Kowalevsky, A. (1866). Entwicklungsgesichte der einfachen Ascidien. *Memoires de l'Academie imperial des sciences de St. Petersburg* 7: 10.

Kropp, B. (1931). The pigment of *Velella spirans* and *Fiona marina*. *Biological Bulletin* 60: 120–123.

Lalli, C.M. (1970a). Morphology of *Crucibranchea macrochira* (Meisenheimer), a gymnosomatous pteropod. *Proceedings of the Malacological Society of London* 39: 1–14.

Lalli, C.M. (1970b). Structure and function of the buccal apparatus of *Clione limacina* (Phipps) with a review of feeding in gymnosomatous pteropods. *Journal of Experimental Marine Biology and Ecology* 4: 101–118.

Lalli, C.M. (1972). Food and feeding of *Paedoclione doliiformis* Danforth, a neotenous gymnosomatous pteropod. *Biological Bulletin* 143: 392–402.

Lalli, C.M. and Conover, R.J. (1973). Reproduction and development of *Paedoclione doliiformis* and a comparison with *Clione limacina* (Opisthobranchia: Gymnosomata). *Marine Biology* 19: 13–22.

Lalli, C.M. and Gilmer, R.W. (1989). *Pelagic Snails: The Biology of Holoplanktonic Molluscs*. Stanford: Stanford University Press.

Lalli, C.M. and Wells, F.E. (1973). Brood protection in an epipelagic thecosomatous pteropod, *Spiratella* ("*Limacina*") *inflata* (D'Orbigny). *Bulletin of Marine Science* 23: 933–941.

Lalli, C.M. and Wells, F.E. (1978). Reproduction in the genus Limacina (Opisthobranchia: Thecosomata). *Journal of Zoology, London* 186: 95–108.

Land, M.F. (1981). Optics and vision in invertebrates. In: *Handbook of Sensory Physiology, Volume VII/6B. Comparative Physiology and Evolution of Vision in Invertebrates* (ed. H. Autrum). Springer-Verlag.

Land, M.F. (1982). Scanning eye movements in a heteropod mollusc. *Journal of Experimental Biology* 96: 427–430.

Land, M.F. (1990). Optics of the eyes of marine animals. In: *Light and Life in the Sea* (eds. P.J. Herring, A.K. Campbell, M. Whitfield and L. Maddock). Cambridge: Cambridge University Press.

Lang, A. (1900). *Lehrbuch der Vergleichenden Anatomie der Wirbellosen thiere: Mollusca*. Jena: Verlag von Gustav Fischer.

Laursen, D. (1953). The genus Ianthina: a monograph. *Dana Reports* 38: 40.

Lebour, M.V. (1931). *Clione limacina* in Plymouth waters. *Journal of the Marine Biological Association of The United Kingdom* 17: 785–795.

Lemche, H. (1957). A new living deep-sea mollusc of the Cambro-Devonian class Monoplacophora. *Nature* 179: 413–416.

Lucero, M.T., Horrigan, F.T., and Gilly, W.F. (1992). Electrical responses to chemical stimulation of squid olfactory receptor cells. *Journal of Experimental Biology* 162: 231–249.

Martens, F. (1675). *Spitzbergische oder gronlandische Reise Beschreibung gethan im Jahr 1671*. Schultzen: Hamburg.

Martin, R. (1966). An attempt to infect in vitro medusae of *Zanclea costata* (Anthomedusae) with the veliger of *Phyllirrhoe bucephala* (Opisthobranchia). *Pubblicazioni di Stazione Zoologica di Napoli* 35: 130–131.

Martin, A.W. (1983). Excretion. In: *The Mollusca Volume 5* (eds. A.S.M. Saleuddin and K.M. Wilbur). New York and London: Academic Press.

Martin, A.W. (1983). Excretion. In: *The Mollusca* (eds. A.S.M. Saleuddin and K.M. Wilbur). New York: Academic Press.

Martin, R. and Brinckmann, A. (1963). Zum Brutparasitismus von Phlliroe bucephala Per. & Les. (Gastropoda, Nudibranchia) auf der Meduse Zanclea costata Gegenb. (Hydrozoa, Anthomedusae). *Pubblicazioni di Stazione Zoologica di Napoli* 33: 206–223.

Martin, A.W., Catala-Stucki, I., and Ward, P.D. (1978). The growth rate and reproductive behaviour of *Nautilus macromphalus*. *Neues Jahrbuch Geologie und Palaontologie* 156: 207–225.

Meisenheimer, J. (1905). Pteropoda. *Wissensch. Ergebnisse Dtsch. Tiefsee Exped. Valdivia*.

Melzner, F., Bock, C., and Portner, H.O. (2006). Temperature-dependent oxygen extraction from the ventilatory current and the costs of ventilation in the cephalopod *Sepia oficinalis*. *Journal of Comparative Physiology B* 176: 607–621.

Menzies, R.J. and Layton, W. (1962). A new species of monoplacophoran mollusc, Neopilina (Neopilina) veleronis from the slope of the Cedros Trench, Mexico. *Annals and Magazine of Natural History* 13: 401–406.

Mesnil, B. (1977). Growth and life cycle of the squid, *Loligos pealei* and *Illex illecebrosus* from the north-west Atlantic. *ICNAF Selecte Papers* 2: 55–69.

Messenger, J.B. (1968). The visual attack of the cuttlefish, *Sepia officinalis*. *Animal Behaviour* 16: 342–357.

Messenger, J.B. (2001). Cephalopod chromatophores, neurobiology and natural history. *Biological Reviews* 76: 473–528.

Michel, H.B. and Michel, J.F. (1991). Heteropod and thecosome (Mollusca: Gastropoda) macroplankton in the Florida Straits. *Bulletin of Marine Science* 49: 562–574.

Milne-Edwards, H. (1848). Note sur la classification naturelle chez Mollusques Gasteropodes. *Annales des Sciences Naturalles Series 3* (9): 102–112.

Morton, J.E. (1954a). The biology of *Limacina retroversa*. *Journal of the Marine Biological Association of The United Kingdom* 33: 297–312.

Morton, J.E. (1954b). The pelagic mollusca of the Benguela Current. Part I. First survey, R.R.S. "William Scoresby," March 1950, with an account of the reproductive system and sexual succession of *Limacina bulimoides*. *Discovery Reports* 27: 163–169.

Morton, J.E. (1958a). Observations on the gymnosomatous pteropod *Clione limacina*. *Journal of the Marine Biological Association of The United Kingdom* 37: 287–297.

Morton, J.E. (1958b). *Molluscs*. London: Hutchinson and Co.

Morton, J.E. (1964). Locomotion. In: *Physiology of Mollusca*, vol. *1* (eds. A.S.M. Saleuddin and K.M. Wilbur). New York: Academic Press.

Moynihan, M. (1985). Why are cephalopods deaf? *American Naturalist* 125: 465–469.

Murphy, D.W., Adhikari, D., Webster, D.R., and Yen, J. (2016). Underwater flight by the planktonic sea butterfly. *Journal of Experimental Biology* 219: 535–543.

Naef, A. (1921). Cephalopoda, Jerusalem, 1972, Israel program for scientific translation.

Nakamura, Y. (1993). Vertical and horizontal movements of mature females of *Ommastrephes bartrami* observed by ultrasonic telemetry. In: *Recent Advances in Cephalopod Fisheries Biology* (eds. T. Okutani, R.K. O'Dor and T. Kubodera). Tokyo: Tokai University Press.

Nesis, K.N. (1997). *Cephalopods of the World*. New York: T.N.P. Publications.

Nesis, K.N. (1999). Cephalopoda. In: *South Atlantic Zooplankton* (ed. D. Boltovskoy). Leiden: Backhuys Publishers.

Nicol, S. and O'Dor, R.K. (1985). Predatory behaviour of squid (*Illex illecebrosus*) feeding on surface swarms of euphausiids. *Canadian Journal of Zoology* 63: 8–14.

Niebuhr, C. (1772). *Beschreibung von Arabien*. Copenhagen: Moller.

Nierstrasz, H. (1902). *The Solenogastres of the Siboga Expedition*, Siboga Expeditie Monogr., vol. 47. E.J. Brill.

Nierstrasz, H. (1905). Kruppamenia und die radula der solenogastren. *Zoologishe Jahrbucher Abteilung Anatomie* 18.

Norman, M. (2000). *Cephalopods, A World Guide*. Hackenheim: ConchBooks.

O'Dor, R.K. and Dawe, E.G. (1998). Illex illecebrosus. In: *Squid Recruitment Dynamics*, *FAO Fisheries Technical Paper*, vol. *376* (eds. P.G. Rodhouse, E.G. Dawe and R.K. O'Dor). Rome: FAO.

Okutani, T. (1983). Todarodes pacificus. In: *Cephalopod Life Cycles. Species Accounts* (ed. P.R. Boyle). London: Academic Press.

Okutani, T., Nakamura, I., and Katsunori, S. (1995). An unusual egg-brooding behaviour of an oceanic squid in the Okhotsk Sea. *Venus* 54: 237–239.

Orr, J.C., Fabry, V.J., Aumont, O., and AL, E. (2005). Anthropogenic ocean acidification over the twenty-first century and its impact on calcifying organisms. *Nature* 437: 681–686.

Owre, H.B. (1964). Observations on development of the heteropod molluscs *Pterotrachea hippocampus* and *Firoloida desmaresti*. *Bulletin of Marine Science* 14: 529–538.

Packard, A. (1972). Cephalopods and fish: the limits of convergence. *Biological Review* 47: 241–307.

Pafort-Van Iersel, T. and Van Der Spoel, S. (1986). Schizogamy in the planktonic opistrhobranch Clio - a previously undescribed mode of reproduction in the Mollusca. *International Journal of Invertebrate Reproduction and Development* 10: 43–50.

Passarella, K.C. and Hopkins, T.L. (1991). Species composition and food habits of the micronektonic cephalopod assemblage in the eastern Gulf of Mexico. *Bulletin of Marine Science* 49: 638–659.

Patten, W. (1886). Embryology of Patella. *Arbeiten aus dem Zoologischen Instituten der Universität Wien* 6.

Pelseneer, P. (1887). Report on the Pteropoda collected by H.M.S. Challenger during the years 1873-1876. I. The Gymnosomata. *Reports on the Scientific Results of the Voyage of H.M.S. Challenger During the Years 1873–1876* 23: 1–74.

Pelseneer, P. 1888a. Report on the Pteropoda collected by H.M.S. Challenger during the years 1873-1876. II. The Thecosomata.

Scientific Reports of the "Challenger," Zoology, 23, 1-132.

Pelseneer, P. (1888b). Report on the Pteropoda collected by H.M.S. Challenger during the years 1873-1876. III. Anatomy. *Scientific Reports of the "Challenger,"* Zoology 23: 1–97.

Pelseneer, P. (1899). Recherches morphologiques et phylogenetiques sur les mollusques archaiques. *Mem. Savants Etrang. Acad. Roy. Soc. Lettres Beaux-Artes Belgique* 57.

Pierce, M. (1967). Loligo pealii. In: *Selected Invertebrate Types* (ed. F.A. Brown). New York: Wiley.

Potts, W.T.W. (1967). Excretion in the molluscs. *Biological Reviews* 42: 1–41.

Pruvot-Fol, A. (1926). *Mollusques Pteropodes Gymnosomes Provenant des Campagnes du Prince Albert I de Monaco, Résultats des campagnes scientifiques accomplies sur son yacht par Albert Ier, Prince souverain de Monaco*, vol. 70, 1–60. Monaco: Impr. de Monaco.

Rang, P.C.A.L. and Souleyet, L.F.A. (1852). *Histoire Naturelle des Mollusques Pteropodes. Monographie Comprenant la Description de Toutes le Especes de ce Groupe de Mollusques.* Paris: J.B. Bailliere.

Rast, J.P. (1990). *The Species Composition, Vertical Distribution and Feeding Ecology of Heteropods in the Families Carinariidae and Pterotracheidae at a Station in the Eastern Gulf of Mexico.* M.S., University of South Florida.

Reid, A. (2005). Family Spirulidae. In: *Cephalopods of the World. An Annotated and Illustrated Catalogue of Species Known to Date*, Chambered Nautiliuses and Sepioids (Nautilidae, Sepiidae, Sepiolidae, Sepiadariidae, Idiosepiidae and Spirulidae). FAO species catalogue for fishery purposes. No. 4, vol.1, vol. 1 (eds. P. Jereb and C.F.E. Roper). Rome: FAO.

Richter, G. (1968). Heteropoden und heteropodenlarven im Oberflachenplankton des Golfs von Neapel. *Publ. Staz. zool. Napoli* 36: 346–400.

Richter, G. (1982). Mageninhaltsuntersuchungen an Oxygyrus keraudreni (Lesueur) (Atlantidae, Heteropoda). Beispiel ener Nahrungskette im tropischen Pelagial. *Senckenberg. marit.* 14: 47–77.

Richter, G. and Seapy, R.R. (1999). Heteropoda. In: *South Atlantic Zooplankton* (ed. D. Boltovskoy). Leiden: Backhuys.

Robertson, R. (1983). *Observations on the Life History of the Wentletrap Epitonium Albidum in the West Indies*, American Malacological Bulletin, vol. 1, 1–11. Hattiesburg: American Malacological Union.

Robison, B.H. and Young, R.E. (1981). Bioluminescence in pelagic octopods. *Pacific Science* 35: 39–44.

Rodhouse, P.G. and White, M.G. (1995). Cephalopods occupy the ecological niche of epipelagic fish in the Antarctic Polar Frontal Zone. *Biological Bulletin* 189: 77–80.

Rodhouse, P. and Nigmatullin, C.M. (1996). Role as consumers. *Philosophical Transactions of the Royal Society of London, Series B, Biological Sciences* 351: 1003–1022.

Rodhouse, P.G., Prince, P.A., Clarke, M.R., and Murray, A.W.A. (1990). Cephalopod prey of the grey-headed albatross Diomodea chrysostoma. *Marine Biology* 104: 353–362.

Rodhouse, P.G., Barton, J., Hatfield, E.M.C., and Symon, C. (1995). Illex argentinus, life cycle, population structure, and fishery. *ICES Marine Science Symposium* 199: 425–432.

Roper, C.F.E. and Brundage, W.L.J. (1972). Cirrate octopods with associated deep-sea organisms, new biological data based on deep benthic photographs (Cephalopoda). *Smithsonian Contributions to Zoology* 121: 1–46.

Roper, C.F.E., Sweeney, M.J., and Nauen, F.E. (1984). Cephalopods of the world. An annotated and illustrated catalogue of species of interest to fisheries. *FAO Fisheries Synopsis* 125: 1–277.

Rosenthal, J. J. C., Seibel, B. A., Dymowska, A. & Bezanilla, F. 2009. Trade-off between aerobic capacity and locomotor capability in an Antarctic pteropod. *Proceedings of the National Academy of Sciences of the United States of America*, 106(15): 6192-6196.

Russell-Hunter, W.D. (1979). *A Life of Invertebrates*. New York: Macmillan Publishing Co., Inc.

Sakurai, Y., Young, R.E., Hirota, J. et al. (1995). Artificial fertilization and development through hatching in the oceanic squids *Ommastrephes bartramii* and *Sthenoteuthis oualaniensis*. *The Veliger* 38: 185–191.

Sakurai, Y., Kiuofuji, H., Satoh, S. et al. (2000). Changes in inferred spawning areas of *Todarodes pacificus* (Cephalopoda: Ommastrephidae) due to changing environmental conditions. *ICES Journal of Marine Science* 57: 24–30.

Sanders, N.K. and Childress, J.J. (1988). Ion replacement as a buoyancy mechanism in a pelagic deep-sea crustacean. *Journal of Experimental Biology* 138: 333–434.

Satterlie, R.A. (1993). Neuromuscular organization in the swimming system of the pteropod mollusc *Clione limacina*. *Journal of Experimental Biology* 181: 119–140.

Satterlie, R.A. and Spencer, A.N. (1985). Swimming in the pteropod mollusc, *Clione limacina*. II. Physiology. *Journal of Experimental Biology* 116: 205–222.

Satterlie, R.A., Labarbera, M., and Spencer, A.N. (1985). Swimming in the pteropod mollusc, *Clione limacina*. I. Behavior and morphology. *Journal of Experimental Biology* 116: 189–204.

Saunders, W.B. (1983). Natural rates of growth and longevity of *Nautilus belauensis*. *Paleobiology* 9: 280–288.

Schipp, R. and Von Boletzky, S. (1975). Morphology and function of the excretory organs in dibranchiate cephalopods. *Fortschritte der Zoologie* 23: 89–110.

Schrock, R.R. and Twenhofel, W.J. (1953). *Principles of Invertebrate Paleontology*. New York: McGraw-Hill.

Seapy, R.R. (1974). Distribution and abundance of the epipelagic mollusk *Carinaria japonica* in the waters off Southern California. *Marine Biology* 24: 243–250.

Seapy, R.R. (1980). Predation by the epipelagic heteropod mollusk *Carinaria cristata* forma Japonica. *Marine Biology* 60: 137–146.

Seapy, R.R. (1985). The pelagic genus Pterotrachea (Gastropoda: Heteropoda) from Hawaiian waters: a taxonomic review. *Malacologia* 26: 125–135.

Seapy, R.R. (1990a). The pelagic family Atlantidae (Gastropoda: Heteropoda) from Hawaiian waters: a faunistic survey. *Malacologia* 32: 107–130.

Seapy, R.R. (1990b). Patterns of vertical distribution in epipelagic heteropod molluscs off Hawaii. *Marine Ecology Progress Series* 60: 235–246.

Seibel, B.A., Thuesen, E.V., Childress, J.J., and Gorodezky, L.A. (1997). Decline in pelagic cephalopod metabolism with depth reflects differences in locomotory efficiency. *Biological Bulletin* 192: 262–278.

Seibel, B.A., Hochberg, F.G., and Carlini, D.B. (2000a). Life history of *Gonatus onyx* (Cephalopoda: Teuthoidea), deep-sea spawning and post-spawning egg care. *Marine Biology* 137: 519–526.

Semmens, J.M. (2002). Changes in the digestive gland of the loliginid squid

Sepioteuthis lessoniana. Journal of Experimental Marine Biology and Ecology 274: 19–39.

Sentz-Braconnot, E. (1965). Sur la capture des proies par le pteropode gymnosome Pneumodermopsis pauciden (Boas). *Cahiers de Biologie Marine* 6: 191–194.

Sentz-Braconnot, E. and Carre, C. (1966). Sur la biologie du nudibranche pelagique *Cephalopyge trematoides*. Parasitisme sur le Siphonophore Nanomia bijuga, nutrition, developpement. *Cahiers du biologie Marine* 7: 31–38.

Sherman, I.W. and Sherman, V.G. (1970). *The Invertebrates: Function and Form*. London: The Macmillan Company.

Smale, M.J. (1996). Cephalopods as prey. IV. Fishes. *Philosophical Transactions of the Royal Society of London* 351: 1067–1081.

Souleyet, L. (1852). *Mollusques*.

Stearns, S.C. (1992). *The Evolution of Life Histories*. New York: Oxford University Press.

Stempell, W. (1926). *Zoologie im Grundriss*. Berlin: Verlag von Gebruder Borntraeger.

Sweeney, M.J., Roper, C.F.E., Mangold, K.M. et al. (1992). *Larval and Juvenile Cephalopods, a Manual for Their Identification, Smithsonian Contributions to Zoology*, vol. 513. Washington, D.C.: Smithsonian Institution Press.

Szymik, B.G. and Satterlie, R.A. (2011). Changes in wingstroke kinematics associated with a change in swimming speed in a pteropod mollusk, *Clione limacina. Journal of Experimental Biology* 214: 3935–3947.

Tanaka, Y. (1993). Japanese common squid Todarodes pacificus preys on benthic polychaete (*Nereis pelagica*). In: *Recent Advances in Cephalopod Fisheries Biology* (eds. T. Okutani, R.K. O'Dor and T. Kubodera). Tokyo: Tokai University Press.

Tesch, J. (1904). *Thecosomata and Gymnosomata of the Siboga Expedition*, Siboga Expedition Monographs, vol. 52. Leyden: E.J. Brill.

Tesch, J.J. (1946). The thecosomatous pteropods. I. The Atlantic. *Dana Reports* 28, 82pp.

Tesch, J.J. (1948). The thecosomatous pteropods. II. The Indo-Pacific. *Dana Reports* 30: 45.

Tesch, J.J. (1950). The Gymnosomata II. *Dana Reports* 36: 55.

Thiriot-Quievreux, C. (1973). Heteropoda. *Oceanography and Marine Biology Annual Review* 11: 237–261.

Thompson, J. (1830). *Zoological Researches. Memoir 4*.

Thompson, T.E. (1976). *Biology of Opisthobranch Molluscs*. London: Ray Society.

Thompson, T.E. and Mcfarlane, I.D. (1967). Observations on a collection of Glaucus from the Gulf of Aden with a critical review of published records of Glaucidae (Gastropoda, Opisthobranchia). *Proceedings of the Linnaean Society of London* 178: 107–123.

Torres, J.J., Belman, B.W., and Childress, J.J. (1979). Oxygen consumption rates of midwater fishes as a function of depth of occurrence. *Deep-Sea Research* 26: 185–197.

Trueman, E.R. (1980). Swimming by jet propulsion. In: *Aspects of Animal Movement* (eds. H.Y. Elder and E.R. Trueman). New York: Cambridge University Press.

Trueman, E.R. (1983). Locomotion in molluscs. In: *The Mollusca, Volume 5* (eds. A.S.M. Saleuddin and K.M. Wilbur). New York: Academic Press.

Van Der Spoel, S. (1962). Aberrant forms of the genus Clio Linnaeus, 1767, with a review of the genus Proclio Hubendick, 1951 (Gastropoda: Pteropoda). *Beaufortia* 9: 173–200.

Van Der Spoel, S. (1967). *Euthecosomata, A Group with Remarkable Developmental Stages (Gastropoda, Pteropoda), Gorinchem.* J. Noorduijn.

Van Der Spoel, S. (1970). The pelagic mollusca from the "Atlantide" and "Galathea" expeditions collected in the East Atlantic. *Atlantide Report* 11: 99–139.

Van Der Spoel, S. (1973a). Growth, reproduction and vertical migration in *Clio pyramidata* Linne, 1767 forma lanceolata (Lesueur, 1813) with notes on some other Cavoliniidae (Mollusca, Pteropda). *Beaufortia* 21: 117–134.

Van Der Spoel, S. (1973b). Strobilation in a mollusc; the development of aberrant stages in *Clio pyramidata* Linnaeus, 1767 (Gastropoda, Pteropoda). *Bijdragen tot de Dierkunde* 43: 202–214.

Van Der Spoel, S. (1976). *Pseudothecosomata, Gymnosomata, and Heteropoda (Gastropoda)*. Utrecht, Bohn: Scheltema and Holkema.

Van Der Spoel, S. and Boltovskoy, D. (1981). Pteropoda. In: *Atlas del zooplancton del Atlantico Sudoccidental y metodos de trabjajo con el zooplancton marino* (ed. D. Boltovskoy). Publ. Esp. Inst. Nac. Inv. Desarollo Pesq., Mar del Plata. Instituto nacional de Investigacion y Desarrollo Pesquero, (INIDEP) Argentina.

Van Der Spoel, S. and Dadon, J.R. (1999). Pteropoda. In: *South Atlantic Zooplankton* (ed. D. Boltovskoy). Leiden: Backhuys Publishers.

Vayssiere, A. (1894). Etude de la Scissurella. *Journal of Conchyliologie* 42.

Vecchione, M. and Pohle, G. (2002). Midwater cephalopods in the Western North Atlantic Ocean off Nova Scotia. *Bulletin of Marine Science* 71: 883–892.

Vecchione, M. and Young, R.E. (1997). Aspects of the functional morphology of cirrate octopods, locomotion and feeding. *Vie et Milieu* 47: 101–110.

Vecchione, M., Roper, C.F.E., Widder, E.A., and Frank, T.M. (2002). In situ observations on three species of large-finned deep-sea squids. *Bulletin of Marine Science* 71: 893–301.

Villanueva, R. (1992). Continuous spawning in the cirrate octopods *Opisthoteuthis agassizii* and *O. vossi*, features of sexual maturation defining a reproductive strategy in cephalopods. *Marine Biology* 114: 265–275.

Vinogradov, M.E. (1970). *Vertical distribution of the oceanic zooplankton*. Jerusalem: Israel Program for Scientific Translations.

Voight, J.R., Portner, H.O., and O'Dor, R.K. (1994). A review of ammonia-mediated buoyancy in squids (Cephalopoda: Teuthoidea). *Marine and Freshwater Behavior and Physiology* 25: 193–203.

Voss, G.L. (1988a). The biogeogrpaphy of the deep-sea cephalopoda. *Malacologia* 29: 295–307.

Voss, N.A. (1988b). Systematics and zoogeography of cephalopods. *Malacologia* 29: 209–214.

Voss, N.A., Vecchione, M., Toll, R.B., and Sweeney, M.J. (1998a). *Systematics and Biogeography of Cephalopods, Volume I, Smithsonian Contributions to Zoology*, vol. 586. Washington, D.C.: Smithsonian Institution Press.

Voss, N.A., Vecchione, M., Toll, R.B., and Sweeney, M.J. (1998b). *Systematics and Biogeography of Cephalopods, Volume II, Smithsonian Contributions to Zoology*, vol. 586. Washington, D.C.: Smithsonian Institution Press.

Walcott, C.D. (1905). Cambrian faunas of China. *Proceedings of the United States National Museum* 29: 1–106.

Ward, P.D. (1987). *The Natural History of Nautilus*. London: Allen and Unwin.

Webber, D.M. and O'Dor, R.K. (1986). Monitoring the metabolic rate and activity of free-swimming squid with telemetered jet pressure. *Journal of Experimental Biology* 26: 205–224.

Webers, G.F. and Yochelson, E.L. (1989). Late Cambrian molluscan faunas and the origin of the Cephalopoda. *Geological Society of London, Special Publications* 47: 29–42.

Wells, M.J. (1961). Centres for tactile and visual learning in the brain of octopus. *Journal of Experimental Biology* 38: 811–826.

Wells, M.J. (1963). Taste by touch: some experiments with Octopus. *Journal of Experimental Biology* 40: 187–193.

Wells, M.J. (1966a). Cephalopod sense organs. In: *Physiology of Mollusca, Vol. II* (eds. K.M. Wilbur and C.M. Yonge). New York and London: Academic Press.

Wells, M.J. (1966b). The brain and behavior of cephalopods. In: *Physiology of Mollusca, Vol. II* (eds. K.M. Wilbur and C.M. Yonge). New York and London: Academic Press.

Wells, M.J. (1978). *Octopus: Physiology and Behavior of an Advanced Invertebrate*. London: Chapman and Hall.

Wells, M.J. (1983). Circulation in cephalopods. In: *The Mollusca* (eds. A.S.M. Saleuddin and K.M. Wilbur). New York: Academic Press.

Wells, M.J. and Wells, J. (1982). Ventilatory currents in the mantle of cephalopods. *Journal of Experimental Biology* 99: 315–330.

Wells, M.J. and Wells, J. (1985a). Ventilation frequencies and stroke volumes in acute hypoxia in Octopus. *Journal of Experimental Biology* 118: 445–448.

Wells, M.J. and Wells, J. (1985b). Ventilation and oxygen uptake in Nautilus. *Journal of Experimental Biology* 118: 297–312.

Wells, M.J. and Wells, J. (1991). Is Sepia really an octopus? In: *La Seiche, First International Symposium on the Cuttlefish Sepia* (ed. E. Boucaud-Camou). Caen: Centre de Publications, Universite de Caen.

Wells, M.J., Hanlon, R.T., Lee, P.G., and Dimarco, F.P. (1988). Respiratory and cardiac performance in *Lolliguncula brevis* (Cephalopoda, Myopsida): the effects of activity, temperature and hypoxia. *Journal of Experimental Biology* 138: 17–36.

Werner, B. (1955). Anatomie, Entwicklung und Biologie des Veligers und der Veliconchia von Crepidula. *Helgolander Wissensch. Meeresuntersuchungen* 5: 169–217.

Williams, L.W. (1909). *Anatomy of Loligo pealii*. Leiden: Brill.

Williamson, R. (1988). Vibration sensitivity in the statocyst of the northern octopus, *Eledone cirrosa*. *Journal of Experimental Biology* 134: 451–454.

Winterstein, H. (1909). Zur kenntnis der blutgase wirbelloser seetiere. *Biochemie Zhurnal* 19: 384–424.

Wiren, A. (1892). Studien uber die solenogastres. *Kong. Svenske Vetensk. Akad. Handl.*, new ser. 25, no. 6.

Withers, P.C. (1992). *Comparative Animal Physiology*. Orlando: Saunders.

Wormelle, R.L. (1962). A survey of the standing crop of plankton of the Florida Current. VI. A study of the pteropods of the Florida Current. *Bulletin of Marine Science* 12: 95–136.

WoRMS. World Register of Marine Species. www.marinespecies.org.

Wormuth, J.H. (1981). Vertical distributions and diel migrations of Euthecosomata in the northwest Sargasso Sea. *Deep-Sea Research* 28A: 1493–1515.

Wormuth, J.H. (1985). The role of cold-core Gulf Stream rings in the temporal and spatial patterns of euthecosomatous pteropods. *Deep-Sea Research* 32: 773–788.

Xu, Z.-L. and Li, C. (2005). Horizontal distribution and dominant species of heteropods in the East China Sea. *Journal of Plankton Research* 27: 373–382.

Yochelson, E.L., Flower, R.H., and Webers, G.F. (1973). The bearing of the new Late Cambrian monoplacophoran genus Knightoconus upon the origin of the Cephalopoda. *Lethaia* 6: 275–310.

Yonge, M. (1926). Ciliary feeding mechanisms in the thecosomatous pteropods. *Journal of the Linnean Society of London, Zoology* 36: 417–429.

Yonge, C.M. (1942). Ciliary currents in the mantle cavity of the Atlantidae. *Quarterly Journal of Microscopic Science* 83: 197–203.

Yonge, C.M. (1947). The pallial organs in the aspidobranch Gastropoda and their evolution throughout the Mollusca. *Philosophical Transactions of the Royal Society of London. Series B, Biological Sciences* 232: 443–518.

Young, J.Z. (1959). Observations on Argonauta and especially its method of feeding. *Proceedings of the Zoological Society of London* 133: 471–479.

Young, J.Z. (1961). Learning and form discrimination by Octopus. *Biological Reviews of the Cambridge Philosophical Society* 36: 32–96.

Young, J.Z. (1964). *A Model of the Brain*. London and New York: Oxford University Press.

Young, J.Z. (1966). *The Memory System of the Brain*. Oxford: Oxford University Press.

Young, J.Z. (1967). Some comparisons between the nervous system of cephalopods and mammals. In: *Invertebrate Nervous Systems* (ed. C.A.G. Wiersma). Chicago: University of Chicago Press.

Young, J.Z. (1971). *The Anatomy of the Nervous System of Octopus vulgaris*. Oxford: Clarendon Press.

Young, R.E. (1972a). Function of extra-ocular photoreceptors in bathypelagic cephalopods. *Deep-Sea Research* 19: 651–660.

Young, R.E. (1977). Ventral bioluminescent countershading in midwater cephalopods. In: *The Biology of Cephalopods* (eds. M. Nixon and J.B. Messenger). London: Academic Press.

Young, R.E. (1978). Vertical distribution and photosensitive vesicles of pelagic cephalopods from Hawaiian waters. *Fishery Bulletin U.S.* 76: 583–615.

Young, J.Z. (1991). Computation in the learning system of cephalopods. *Biological Bulletin* 180: 200–208.

Young, R.E. and Bennett, T.M. (1988). Photophore structure and evolution within the Enoploteuthinae (Cephalopoda). In: *The Mollusca* (ed. K.M. Wilbur). New York: Academic Press.

Young, R.E. and Harman, R.F. (1988). Larva, paralarva, and subadult in cephalopod terminology. *Malacologia* 29: 201–207.

Young, R.E. and Roper, C.F.E. (1976). Bioluminescent countershading in midwater animals: evidence from living squid. *Science* 191: 1046–1048.

Young, R.E. and Roper, C.F.E. (1977). Intensity regulation of bioluminescence during countershading in living midwater animals. *Fishery Bulletin U.S.* 75: 239–252.

Young, R.E., Harman, R.F., and Mangold, K.M. (1985). The common occurrence of oegopsid squid eggs in near-surface oceanic waters. *Pacific Science* 39: 359–366.

9

The Chordata

Introduction

The phylum Chordata includes our closest invertebrate relatives as well as our next of kin: the mammals, birds, reptiles, amphibians, and fishes. As a phylum, we are moderately successful in terms of species numbers and highly successful in the number of habitats we have invaded, being well represented in all terrestrial and aquatic systems. The chordates include about 65 000 species (Chapman 2009) with the great majority being vertebrates. The phylum is named for the dorsal notochord, a flexible rod providing structural support that is always present at some stage in development. Chordates are bilaterally symmetrical and coelomate with deuterostomate development. At some stage during development, all have pharyngeal gill slits, a dorsal hollow nerve chord, and a muscular post-anal tail. A complete gut and a circulatory system with a ventral pump, either a heart or contractile blood vessel, are also present. Chordates may be monoecious or gonochoristic, but all have a tadpole stage during their life history. Unless specifically referenced, all species numbers are from the World Register of Marine Species (WoRMS – World Register of Marine Species. www.marinespecies.org).

The phylum Chordata comprises three subphyla.

- Cephalochordata (the lancelets) – 30 species
- Tunicata (the sea-squirts, pyrosomes, salps, doliolids, and larvaceans) – 3080 species
- Vertebrata (mammals, birds, amphibians, reptiles, fishes, and agnathans) – 61 995 species

The cephalochordates and tunicates are the invertebrate members of the phylum, with the pelagic tunicates being our major focus. The cephalochordates, or lancelets, are a small (<5 cm) but important group of burrowing "fish-like" animals that possess gills and a laterally compressed body that resembles a flattened lance (Figure 9.1). They are considered an important link between the invertebrates and vertebrates and are a mainstay in early zoology instruction. Tunicates are united by the presence of an exterior "tunic" composed of tunicin, a cellulose-like polysaccharide. The vast majority of tunicates are the sessile ascidians or sea-squirts (Figure 9.2) (2935 spp.), so named for the squirts of water produced by intertidal ascidians when prodded. They are found from the intertidal to depths below 5000 m (Berrill 1950). Pelagic tunicates are composed of two major groups,

Life in the Open Ocean: The Biology of Pelagic Species, First Edition. Joseph J. Torres and Thomas G. Bailey.
© 2022 John Wiley & Sons Ltd. Published 2022 by John Wiley & Sons Ltd.

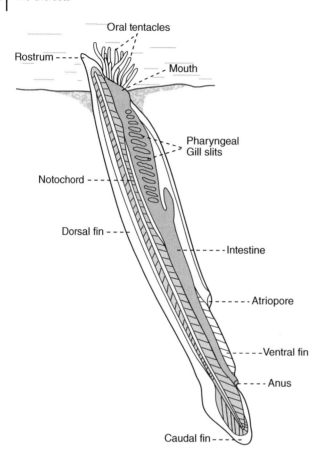

Figure 9.1 The lancelet, *Amphioxus* species.

the Appendicularia (Larvacea), with 68 species and the Thaliacea with 77 species. Thaliaceans include the pyrosomes, salps, and doliolids (Figure 9.3). All tunicates are marine, and representatives of benthic and pelagic taxa are found pan-globally.

Historical milestones for the study of pelagic tunicates date back to the 1700s (Bone 1998a). Salps were first described by Browne (1756) and Forsskaal (1775); a description of pyrosomes followed a few years later (Peron 1804). The affinity of the two groups was recognized by Cuvier (1804) who united them with the ascideans as the Tunicata, a class of Mollusca. Tunicates were later accorded separate taxonomic status (Savigny 1816). Appendicularia (Chamisso and Eysenhardt 1821) and doliolids (Quoy and Gaimard 1827) were described in the years following. Huxley (1851) first recognized the affinity of Appendicularia for the rest of the tunicates despite a very different morphology, and Bronn (1862) gave them equal taxonomic rank with the pyrosomes, salps, and doliolids (Bone 1998a). The four major groups remain today, with the Appendicularia occupying its own class.

Deuterostomes and the Phylogenetic Toolkit

Our first foray into the chordates takes us into a new developmental scheme and the other major branch of the metazoan diphyletic tree: the deuterostomate phyla. In deuterostomate development, the blastopore of the developing embryo becomes the anus,

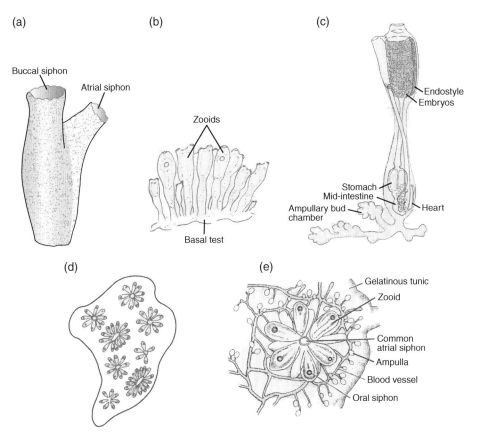

Figure 9.2 Solitary and colonial ascidians. (a) The solitary ascidian *Ciona intestinalis*. (b) A colony of the social ascidian *Clavelina lepadiformis*; (c) an individual zooid of *C. lepadiformis*; (d) a colony of the compound ascidian *Botryllus schlosseri*. (e) Zooid cluster of *Botryllus*. *Sources:* (b, c) Adapted from Berrill (1950), figure 14 (p. 71); (d) Adapted from Berrill (1950), figure 74 (p. 218); (e) Adapted from Brusca and Brusca (1990) (p. 859).

whereas in protostomes the blastopore becomes the mouth. Current thinking places three phyla within the Deuterostomia: the Chordata, the Hemichordata (acorn worms), and the Echinodermata (starfish, sea cucumbers, and sea urchins). Traditionally, the deuterostomes also included the Chaetognatha, a highly abundant, pan-global member of the oceanic zooplankton, and the three diminutive lophophorate phyla: the Brachiopoda, Bryozoa, and Phoronida (Brusca et al. 2016).

Morphological similarities and patterns in embryological development have been used extensively to deduce animal lineages, forming the basis of the traditional animal phylogenies that persisted from the late 1800s to the 1990s and still form the basic framework for our current understanding of animal relationships. Two new tools for elucidating animal lineages came into play in the latter half of the twentieth century, phylogenetic systematics and molecular phylogenetics. Phylogenetic systematics (cladistics) is a method of analyzing the relatedness of animals based on the observable features of their genotype, which can be morphological, developmental, or the gene sequences themselves (Brusca et al. 2016). Molecular phylogenetics provides the base sequences of the genes

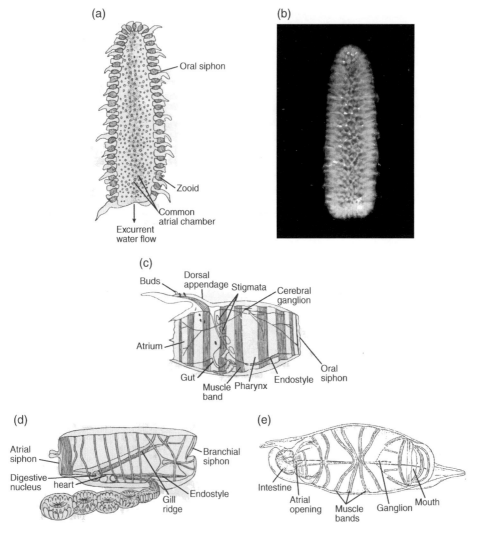

Figure 9.3 Thaliaceans: pyrosomes, doliolids, and salps. (a) Colonial thaliacean *Pyrosoma*. (b) Photo from life of a pyrosome. (c) A solitary thaliacean *Doliolum*. (d) The salp *Cyclosalpa affinis*, solitary form. (e) The salp *Salpa maxima*, aggregate form. *Sources*: (a) Adapted from Brusca and Brusca (2003), figure 23.14 (p. 858); (b) © Dante Fenolio, reproduced with permission; (c) Adapted from Brusca and Brusca (2003), figure 23.14 (p. 858); (d) Adapted from Berrill (1950), figure 104 (p. 288); (e) Berrill (1950), figure 67 (p. 85). See color plate section for color representation of this figure.

themselves, a technology that became practical for phylogenetic analysis in the late 1980s (Field et al. 1988). Together, the two methods have refined the science of animal relationships (systematics) into a more quantitative and predictable field. They have also rearranged traditional animal phylogenies in existence for decades, and they are continuing to do so. Figure 9.4 is an example of a modern animal phylogeny.

Ultimately, the relatedness of species is defined by similarities in their genetic makeup, but genetic tools robust enough to sort out questions at the higher levels of classification, e.g. at the phylum level and above, have been developed only recently. Molecular phylogenies entered the fray in 1988 (Field et al. 1988) with a sequencing of the 18S ribosomal RNA gene in animal representatives from 10 phyla, founding the field of molecular phylogenetics and giving evolutionary biologists an important new tool for understanding animal relationships.

Prior to the advent of molecular phylogenetics, use of phylogenetic systematics, or cladistics, (Hennig 1979) was becoming more widespread. The goal of cladistics is to understand the genealogies of animal taxa by identifying monophyletic groups (single common ancestor) and ancestral and advanced character states within those groups, known as plesiomorphies and apomorphies, respectively. The techniques of phylogenetic systematics produce dendrograms, or branching trees, that ideally can order a group of species within a monophyletic group or clade from most primitive or ancestral to the most highly derived or advanced. The techniques are complex. A readable summary of the field and its techniques can be found in Wiley and Lieberman (2011). An excellent capsule summary is in Brusca et al. (2016). Molecular and cladistic methodologies shape our current understanding of animal lineages and relatedness. It is remarkable that the phylogenies produced using the older morphological and developmental methods are still largely intact.

Classification

Three well-established classes comprise the Tunicata, along with a fourth whose taxonomic position is less concrete. The Appendicularia, Ascidiacea, and Thaliacea have been recognized since the 1800s. Individuals of a fourth class, the Sorberacea, were first widely recognized as unusual members of the hydrothermal vent fauna and given class status in the early 2000s (Brusca and Brusca 2003). The sorberaceans are a carnivorous benthic group, most of whom are found at great depth. While they lack the branchial chamber that is characteristic of most of the tunicates, they have a dorsal nerve chord that persists into adulthood. Their current taxonomic position is uncertain. Brusca et al. (2016) accept them as a class within the subphylum Tunicata, but WoRMS does not, positing at least one family of them, the Octacnemidae, in the order Phlebobranchia within the class Ascidiacea. For the purposes of this book, we will accept them as a group within the ascidians. The classification presented here is that of WoRMS. Species numbers are given in parentheses.

Subphylum Tunicata

Class Ascidiacea (2935)

Order Aplousobranchia (1558)
Order Phlebobranchia (338)
Order Stolidobranchia (1039)

Class Appendicularia (68)

Order Copelata (68)
   Family Fritillariidae (30)
      Subfamily Fritillariinae (28)
      Subfamily Appendiculariinae (2)
   Family Kowalevskiidae (2)
   Family Oikopleuridae (36)
      Subfamily Oikopleurinae (31)
      Subfamily Bathychordaeinae (5)

Class Thaliacea (78)

Order Doliolida (24)
   Suborder Doliolidina (19)
      Family Doliolidae (16)
      Family Doliopsoididae (3)

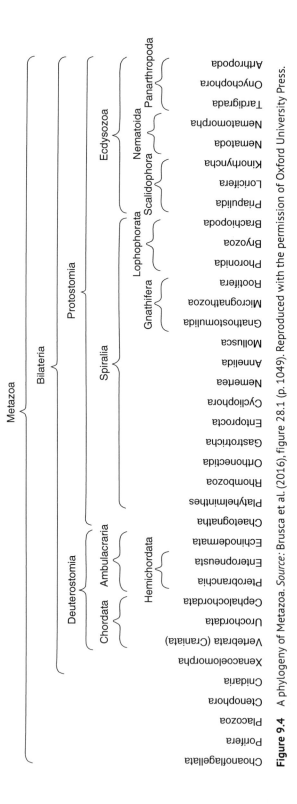

Figure 9.4 A phylogeny of Metazoa. *Source*: Brusca et al. (2016), figure 28.1 (p. 1049). Reproduced with the permission of Oxford University Press.

> **Box 9.1 Useful terms.**
>
> **Apomorphies**. More recently derived, evolved, or advanced characters.
>
> **Autapomorphies**. Specialized characters found in only one taxon. Important for that taxon, but not useful in constructing a phylogenetic tree.
>
> **Homoplasies**. Shared independently derived similarities, e.g. parallelisms, convergences, secondary losses, that do not reflect the evolutionary history of a taxon but can appear to.
>
> **Monophyletic group**. A group of species that includes all the descendants of a common ancestor: **a clade**.
>
> **Outgroup**. A group closely related to your study group (**ingroup**) that allows you to distinguish the apomorphies of your study group from commonly held symplesiomorphies.
>
> **Paraphyletic group**. A group that contains only descendants of a common ancestor but does not contain the entire list of descendants, i.e. an incomplete list.
>
> **Parsimony**. Ideally, a taxon can be defined by a number of synapomorphies when constructing a classification. However, conflicts often exist. Some characters may point to a closer relationship of groups A–B, other groups A–C. The principle used to sort out the problem is parsimony, the hypothesis that explains the data in the simplest way. Parsimony is usually achieved using computer programs and large numbers of taxonomic characters.
>
> **Plesiomorphies**. Ancestral, primitive, or generalized characters
>
> **Polyphyletic group**. A group containing descendants from more than one ancestor.
>
> **Sister group**. A monophyletic group that is the closest genealogical relative to your ingroup. The most closely related clades at the nodes of a cladogram.
>
> **Symplesiomorphies**. Shared primitive characters. Not useful for constructing phylogenies because primitive characters may be retained in a wide variety of taxa. Advanced taxa as well as primitive taxa are likely to possess symplesiomorphies.
>
> **Synapomorphies**. Shared derived characters that define monophyletic groups or clades (groups containing an ancestor and all its descendant taxa).
>
> *Source:* All definitions are from Brusca et al. (2016), Wiley and Lieberman (2011), and Helfman et al. (2009).

Suborder Doliopsidina (5)
   Family Doliolunidae ((1)
   Family Doliopsidae (3)
   Family Paradoliopsidae (1)
Order Pyrosomatida (8)
   Family Pyrosomatidae (8)
   Subfamily Pyrosomatinae (6)
   Subfamily Pyrostremmatinae (2)
Order Salpida (45)
   Family Salpidae (45)
   Subfamily Cyclosalpinae (14)
   Subfamily Salpinae (31)

Basic Anatomy and Life History

The Ascidians

The ascidians have the most straightforward life history of the tunicates, hatching as a larva that metamorphoses into a settled benthic existence. The salps, pyrosomes, and doliolids alternate generations so that anatomy changes with life history stage, necessarily tying the two areas of study together. Pyrosomes are considered the most primitive of the thaliaceans with the most easily understood life history. They will be treated first after a brief grounding with the ascidians as a reference model.

All ascidians are benthic and favor hard substrates to which they become firmly attached. There are three basic types (Figure 9.2): solitary forms that live singly and tend to have larger zooids (e.g. *Molgula, Ciona*); social forms that live in clumps and are vascularly attached at their bases (e.g. *Clavelina*); and compound forms that have multiple individual zooids living together in a gelatinous matrix (e.g. *Botryllus*). Internal anatomy, flow, and feeding are best visualized in the solitary ascidians and will help clarify the feeding strategies of the pelagic forms discussed later in this chapter. Virtually all tunicates, whether benthic or pelagic, are suspension feeders, trapping particles with a directed flow generated by the animal using either ciliary activity or contractions of the body wall, or both.

In the ascidians, water and any suspended particulates enter the mouth (buccal siphon) and pass into the pharyngeal basket (Figure 9.5). A large number of cilia present on the gill bars lining the basket generate the flow of water into the basket. Flow continues through the gill slits and out into the atrial cavity, exiting via the atrial siphon. A mucous film lining the internal surface of the basket is generated by the endostyle and spread across the basket by ciliary activity. The mucous film traps particulates as the water passes through the gill slits, also known as "stigmata." Opposite the endostyle is the dorsal lamina, which may be thought of as the opposite end of the "mucous-film conveyor." Film generated at the endostyle makes its way across both sides of the pharyngeal basket and is wound into a cord by ciliary activity at the dorsal lamina. The cord and its trapped particles are conveyed into the esophagus, located at the base of the pharyngeal basket (Figure 9.5a).

Ascidians can reproduce sexually, producing a tadpole larva that settles and metamorphoses into the adult form (Figure 9.6), or they can reproduce asexually by budding. Considerable diversity is found within the class for each type of reproduction. Tunicate buds are usually termed blastozooids and originate in different parts of the body in different ascidians. The simplest type is that of the colonial *Clavelina*, which buds from its basal stolon (Figure 9.2b and c), Like all tunicates, the ascidians are hermaphroditic with the simplest reproductive system consisting of an ovary and testis located at the loop of the intestine (Figure 9.5a). Some species are broadcast spawners; eggs and sperm are shed into the sea, develop into tadpole larvae, and settle on the bottom using adhesive papillae. Colonial species with smaller zooids may produce one or a few embryos that are fertilized internally and brooded until the tadpole larvae are fully developed (Barnes 1974; Berrill 1950).

The Pyrosomes

Pyrosomes are the pelagic analogue to compound ascidians. They inhabit warm and temperate seas throughout the globe, most commonly in epipelagic to upper mesopelagic depths (<500 m). Colonies vary in size from a few cm (*Pyrosoma aherniosum*) to over 20 m in length with a diameter of 1.5 m (*Pyrostremma spinosum*) (Esnal 1999b). Pyrosomes comprise a colony of many tiny, ascidian-like zooids embedded in a gelatinous matrix (Figure 9.7a and b). The oral (buccal) siphon faces outward and the atrial siphon, instead of exiting adjacent to the buccal siphon as in ascidians, is at the opposite end of the

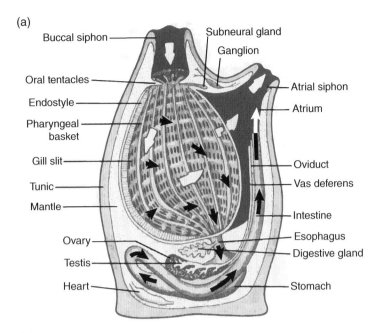

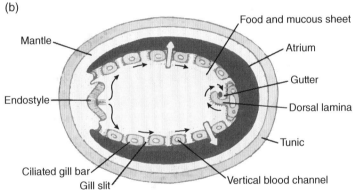

Figure 9.5 Diagrammatic lateral view (a) and cross section (b) of a tunicate showing major internal organs. Large arrows (white) represent the course of the current of water; small arrows (dark), that of food and mucous sheet. Stomach, intestine, and other visceral organs are embedded in the mantle. *Source:* Barnes (1974), figure 20-8 (p. 809). Reproduced with the permission of W. B. Saunders.

zooid and empties into a common cloaca (atrial chamber). The ciliary flow generated by the many zooids produces an excurrent flow out of the chamber that is the source of the colony's propulsion. Feeding and generation of the ciliary flow in a pyrosome's individual zooids are believed to be nearly identical to that of benthic ascidians except that the flow is straight through the zooid instead of describing a "U."

Reproduction and alternation of generations within the pyrosomes are abbreviated relative to the doliolids and salps. The zooids imbedded within the gelatinous matrix are

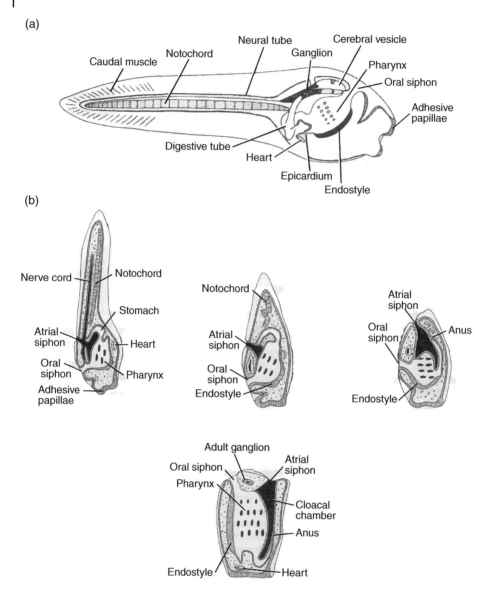

Figure 9.6 Ascidian larvae and metamorphosis. (a) An ascidian tadpole larva. (b) Metamorphosis of a settled tadpole larva showing tail resorption followed by reorientation of the body to bring the siphons to the adult positions. *Source:* Brusca and Brusca (2003), figure 23.17 (p. 864). Reproduced with the permission of Oxford University Press.

considered blastozooids but in the pyrosomes are known as ascidiozooids, a legacy name dating back to early descriptions in the 1800s (Esnal 1999b). The equivalent to the oozooid is the cyathozooid and is very short lived.

Pyrosomes are hermaphrodites, with an ovary and testis located posterior to the branchial basket in each of the ascidiozooids (Figure 9.7c). The reproductive process is unusual as it combines a sexual and asexual component, and the sexual portion of the

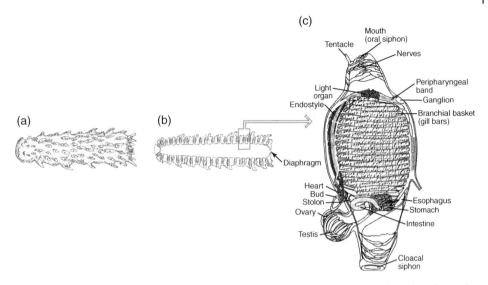

Figure 9.7 Pyrosomes. (a) A colony of *Pyrosoma atlanticum*; (b) longitudinal section of a colony of *P. atlanticum*; (c) a single ascidiozooid of *P. atlanticum*. Sources: (a, b) Metcalf and Hopkins (1919) (p. 199); (c) Metcalf and Hopkins (1919) (p. 270).

cycle is highly compressed. In *Pyrosoma atlanticum*, the ovary produces a single yolky egg that is internally fertilized. The source of the sperm is unclear, though the ovary itself is near the cloacal siphon, and its oviduct opens into the cloacal cavity of the zooid (Metcalf and Hopkins 1919). Sperm thus could gain entry into the cloacal siphon and ovary of the zooid through the common cloaca of the colony, though the mechanism for that is undescribed. Self-fertilization from sperm generated by zooids elsewhere within the colony is certainly possible. In fact, there is a gradient within the colony with the oldest zooids (near the "closed" end) showing protrandric development and the youngest a protgynous one (Godeaux et al. 1998), assuring a ready supply of sperm for mature eggs within the colony.

Once fertilized, cleavage begins, and the embryo (cyathozooid) develops as a cap on the yolk. However, development is never fully completed in *Pyrosoma atlanticum*: the embryo never completely matures. Organs are not fully formed before a series of four blastozooid buds are produced (Figure 9.8a) that develop into a tetrazooid colony (Figure 9.8b). When the four buds are fully mature, the tetrazooid colony is released into the common cloaca and makes its way out to sea in the excurrent flow of the colony. The tetrazooid colony forms the closed end of the pyrosome, developing into the larger colony by active budding of the individual zooids. Pyrosomes are considered ovoviviparous with a very short-lived (cyathozooid) stage.

The Salps

Like the pyrosomes, the salps alternate generations (a "metagenetic" life history: Godeaux 1990; Esnal and DaPonte 1999b). Two distinct morphs are present. The first is an asexual solitary oozooid form (Figure 9.9a) that reproduces prodigiously by budding large numbers of the second morph: the aggregate, or blastozooid, form (Figure 9.9b). Blastozooids reproduce sexually. Unlike the pyrosomes, the oozooid or solitary form in salps matures completely and persists from

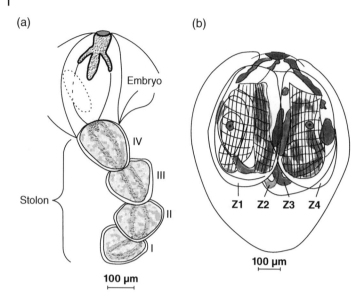

Figure 9.8 Developmental stages of *Pyrosoma atlanticum*. (a) Early embryo and its stolon divided into the four primary blastozooids (labeled I–IV); (b) a later young tetrazooid colony stage with the first four blastozooids (labeled Z1–Z4). *Source:* Adapted from Godeaux et al. (1998).

weeks to months in the open ocean (Foxton 1961, 1966). Salps range in length from 1.5 to 19 cm with most species falling in the middle portion of the range.

Solitary salp zooids live as individuals; the asexually produced aggregates persist as chains or whorls of connected zooids that act as a pseudo-colony. The shape of the aggregate chains and how individuals are attached differ between species (Figure 9.10). Turbulence or predation will often break up the chains, but the individual blastozooids are quite capable of surviving on their own.

Externally, the oozooid morph is the more symmetrical in appearance (Figures 9.9a and 9.11a).

The aggregate forms are adapted for swimming in chains or whorls as a de-facto colony, and their external symmetry reflects the need for swimming as an attached group (Figure 9.11b).

Basic internal anatomy of the two morphs is quite similar, the major differences being the number of muscle bands (Figure 9.9) and the reproductive structures (Figure 9.12). Each form has a pouch-like mucous feeding web that fills the entire pharyngeal cavity and is anchored dorsally by the gill bar and ventrally by the endostyle. Anteriorly, the hoop-like peripharyngeal band forms the mouth of the pouch. The web tapers posteriorly to a close where it enters the esophagus. All water pumped through the salp by its swimming activity passes through the meshes of the feeding web, which have a size of 4–5 μm by 0.3–0.7 μm (Bone et al. 1991; Madin and Deibel 1998). The esophagus leads to the stomach and intestine, with the whole gut forming a conspicuous-colored spheroid referred to in the older literature as the "nucleus" (Foxton 1966). Water passes through the filter and into the atrial, or cloacal, cavity and out through the atrial opening, or siphon. Both the oral and atrial openings are valved (see Section "Tunicate Systems").

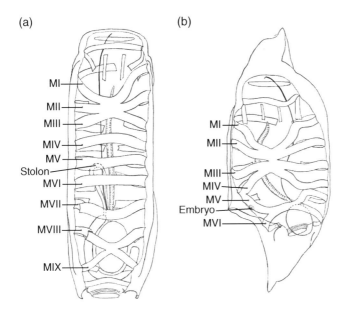

Figure 9.9 Illustrations of the two distinct morphs of salps. (a) Solitary form of *Salpa thompsonii* showing the stolon and muscle bands (MI–MIX); (b) the aggregate form of *S. thompsonii* showing the embryo and muscle bands (MI–MVI). *Source:* Adapted from Foxton (1966), text-figure 1.

Salps are essentially tubes with thick transparent walls forming a test that houses the internal organ systems (Figure 9.12). The body wall also provides the elastic recoil against which the muscle bands work to drive the animal's jet propulsion. Each compression of the circumferential muscle bands forces water out the atrial opening to drive the salp forward; the body wall's recoil refills the salp for the next pulse. Circulation is driven by a ventrally located heart that propels blood through a system of lacunae in the test to nourish and provide oxygen to the tissues. A dorsally located brain, often termed a ganglion, is the neural center for salps. Nerves radiate from it to supply the muscle bands, viscera, and sense receptors (Bone 1998c). A simple eye located atop the brain provides a measure of photoreception.

A brief summary of salp life history is depicted in Figure 9.13 starting with mature solitary asexual individuals (oozooids). During their development, oozooids produce a stolon which is functional at the time they are released into the water column. The stolon has a complete complement of tissue types: ectoderm, mesoderm, and endoderm. It emerges from behind the endostyle of the mature oozooid (Figure 9.12a). During reproduction, the stolon swells and forms buds of blastozooids in distinct blocks or groups, with each group ending in a weak link (Figure 9.14b). The groups break off at the weak link to form a free chain of aggregates. All segments are genetically identical. In the well-studied species *Salpa fusiformis*, the aggregate chains are 100–150 individuals long. New chains of similar length can be liberated every 2 days (Braconnot et al. 1988). In culture at 15 °C, *S. fusiformis* can produce up to five chains, but in a favorable oceanic environment, e.g. in a spring algae bloom, it is likely that many more could be produced (Godeaux et al. 1998).

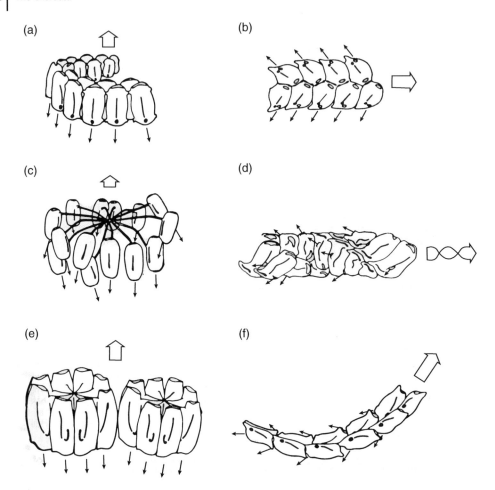

Figure 9.10 Patterns of arrangement of aggregate salps. Representative genera are included for each pattern. (a) Transverse chain; *Pegea*. (b) Cluster; *Cyclosalpa*. (c) Whorl; *Cyclosalpa*. (d) Helical; *Helicosalpa*. (e) Oblique; *Brooksia*. (f) Linear; *Salpa*. *Source:* Godeaux et al. (1998), figure 1.29 (p. 23). Reproduced with the permission of Oxford University Press.

Salp blastozooids are protogynous: the older individuals in a blastozooid chain produce sperm for fertilizing the eggs in the younger, more recently budded, female zooids. Each blastozooid has an ovary located in the postero-dorsal portion of the cloacal cavity, visible in Figure 9.12b as a tiny sperm-like structure. The "tail" of the ovary is actually the oviduct leading to the ovarian follicle itself. In most salp species, each blastozooid ovary contains a single egg that is present from the time the blastozooid chain is released by the oozooid. Sperm broadcast from the older zooids mainly enter the salp through the oral opening as it "inhales" water for locomotion and feeding. From there sperm must ride the feeding current, or actively swim, through the feeding filter until they reach the atrial cavity, where they can enter the oviduct and fertilize the egg. Sperm head size reported in Holland and Miller (1994) is theoretically

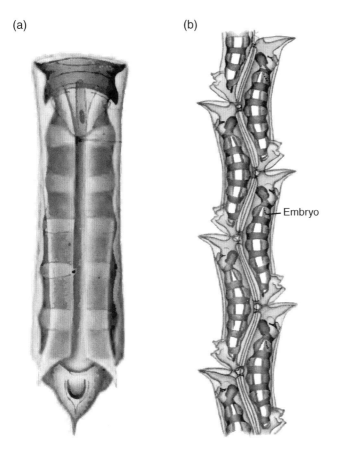

Figure 9.11 The two morphs of the salp *Soestia zonaria*. (a) Solitary form; (b) aggregate form. *Source:* Brooks (1893), plate IV.

small enough to pass through the typical mesh size of a salp feeding filter (Madin and Deibel 1998), and the sperm are apparently strong swimmers (Holland and Miller 1994). A fraction of the water "inhaled" by the salp to refill its body for jet propulsion enters via the atrial opening (about 18%: Bone 1998b) so some sperm may enter the cloacal cavity directly.

Once fertilized, the developing embryo grows within the ovarian follicle, nourished by the blood of the maternal blastozooid. The ovary with embryo is located just beneath the epithelium of the atrial cavity, appearing as a "blister" on the wall of the atrial cavity (Figure 9.9b). As the embryo grows it protrudes into the atrial cavity, enveloped by a fold of the atrial epithelium that forms a pouch or uterine sac. During development, the cells of the ovarian follicle form a large placenta (Figures 9.14a and 9.15) that interacts with the maternal circulation to provide oxygen and nourishment to the embryo. As the embryo continues to grow, a mass of embryonic mesodermal cells develops on the posterior portion of the developing embryo, forming a large elaeoblast. The elaeoblast is believed to supply cells important to muscle and blood cell formation. When the mature

(a)

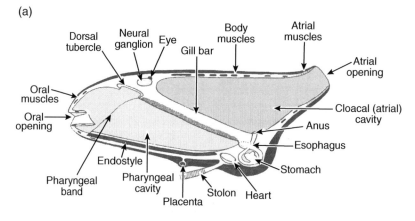

(b)

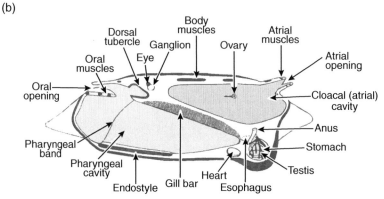

Figure 9.12 Morphology of salps. (a) Morphology of an oozooid (solitary zooid) of a generic salp; (b) morphology of a blastozooid (aggregate zooid) of a generic salp. *Source:* Esnal and Daponte (1999b) (p. 1428). Reproduced with the permission of Backhuys Publishers.

embryo breaks away from the blastozooid, the regressing elaeoblast is believed to supply nutrition to aid the first-feeding young oozooid.

In *Salpa thompsoni*, the Antarctic salp pictured in Figures 9.14 and 9.15, embryos are usually 3–5 mm in length when released into the atrial cavity and then out the atrial siphon to fend for themselves in the open sea. Maximum size of the aggregate stage in *S. thompsoni* is 45–55 mm (Foxton 1966).

When mature, the blastozooid testis is very much larger than the young ovary and is located near the stomach and intestine. Once the testis ripens, its tubular vas deferens conducts sperm into the cloacal cavity, where they are discharged into the sea through the atrial opening. Typically, a blastozooid with ripe testis has already released its oozooid progeny and is classified as "spent."

The Doliolids

The doliolids (from the Latin doliolum or "little cask") resemble miniature salps in general appearance, though they are very much smaller, usually between 1 and 3 mm in length (Bone et al. 1997b). They tend to be

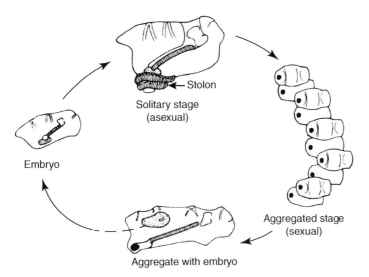

Figure 9.13 Three basic stages in the salp life cycle: (i) solitary asexual individuals (oozooids); (ii) aggregates (sexual) in chains (blastozooids) budded from oozooids. These free-swimming pseudo-colonial units have individuals that are initially female and then transform into males after release of eggs. Sperm from males fertilize the eggs; (iii) embryos in the water column develop into solitary asexual individuals (oozooids). *Source:* Alldredge and Madin (1982), Pelagic Tunicates: Unique Herbivores in the Marine Plankton, figure 2A, BioScience, 1982, volume 32, issue 8, p. 659. By permission of Oxford University Press.

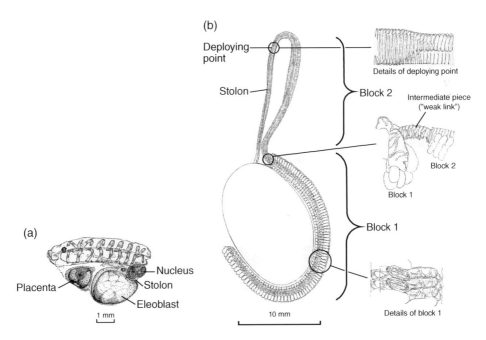

Figure 9.14 Salp reproduction. (a) Lateral view of a solitary embryo (oozooid); (b) ventral view of a stolon releasing blocks of blastozooids. *Source:* (a) Adapted from Foxton (1966), text-figure 11 (p. 30); (b) Adapted from Foxton (1966), text-figure 9.

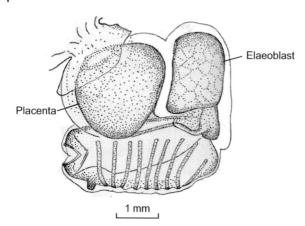

Figure 9.15 The developed oozooid of *Salpa thompsonii*. Source: Adapted from Foxton (1966), text-figure 28.

found further inshore, usually in epipelagic waters over the continental shelf. Their internal design and feeding mechanisms are similar to those of salps. Their life history is exquisitely complex, with five different types of zooids and an independent larval stage. Like salps, doliolids basically alternate an asexual oozooid generation with a sexual blastozooid one, but the process in doliolids involves more steps and additional morphs.

Basic anatomy is best illustrated with the young oozooid (Figure 9.16a). Doliolids have a highly transparent test with a pharyngeal feeding filter and organ systems arranged in a similar manner to salps (Figure 9.16b and c). All doliolids have nine muscle bands that are completely circumferential, unlike the muscle bands that do not entirely girdle the animal in most species of salps (see e.g. Figure 9.9). As in salps, the pharyngeal and atrial cavities are divided by the gill septum. The branchial and atrial siphons, or apertures, are each guarded by a muscle band and series of flap-like valves and that work together to effectively close the anterior or posterior end of the animal if the need for jetting arises. Normal locomotion and feeding use only the cilia present on the gill septum. The beating cilia produce a continuous feeding current that results in a slow forward motion. When disturbed, the doliolid can use its muscle bands to produce a powerful jet thrust forward or backward for escape.

A feeding filter is secreted by the endostyle and deployed by the ciliated peripharyngeal band that encircles the opening to the pharyngeal cavity. The filter itself terminates in the entry to the esophagus and is digested by the stomach and intestine.

As the young oozooid ages and grows larger, its structure changes considerably. In some species, the interior muscle bands (II–VIII) thicken and grow together. In all species, the gut, gills, and endostyle eventually disappear, leaving only the stolon, heart, and brain (Godeaux et al. 1998). When the oozooid reaches this stage, it can no longer feed and is called an "old nurse." This point is where the most unusual portion of doliolid life history unfolds.

The process begins with an "early nurse" stage in which the stolon lengthens and forms a series of buds that, remarkable as it may seem, migrate from the stolon to the dorsal appendix where they attach themselves in discrete rows (Braconnot 1971a; Godeaux et al. 1998; Holland 2016) (Figure 9.17a). The two lateral rows of buds on the appendix develop into gastrozooids (or trophozooids) (Figure 9.17b). The central

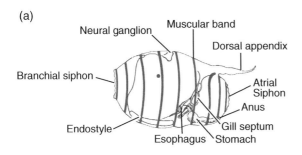

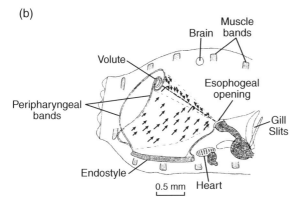

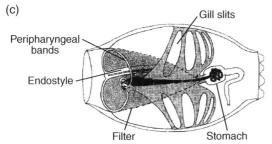

Figure 9.16 Doliolid anatomy. (a) Basic anatomy of a young oozooid of *Doliolletta gegenbauri*. (b) Anatomy of a feeding oozooid of *Doliolum nationalis*, arrows indicate the paths of algal cells trapped on the filter. (c) Components of a doliolid feeding filter (dorsal view). *Sources*: (a) Neumann (1913) (p.2); (b) Bone et al. (1997b), figure 3 (p. 185); Reproduced with permission of Elsevier. (c) Adapted from Werner and Werner (1954).

row of buds develops into phorozooids (Figure 9.17c). Each type of zooid has a specific task. The dorsal appendix, also known as a cladophore (Paffenhöfer and Koster 2011), caudal peduncle (Godeaux et al. 1998), or dorsal process (Godeaux et al. 1998), expands greatly in length to accommodate the developing zooids.

Gastrozooid buds begin the migration from the stolon to the dorsal appendix. They are the feeding specialists responsible for nourishing both the developing zooids and the old nurse using canals that run the length of the dorsal appendix. They connect to the canals via their stalk (Figure 9.18a) and are presumed to exchange nutrients

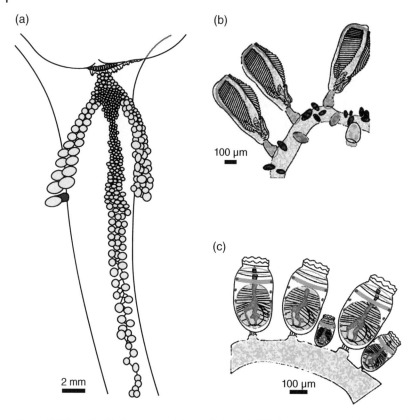

Figure 9.17 Doliolid development: "nurse" stage. (a) "Early nurse" stage with rows of zooids on the dorsal appendix; (b) three gastrozooids and several small phorozooid buds; (c) phorozooid buds.

with the nurse and young phorozooids with a shared circulation effected by the canals. Gastrozooids trap and digest phytoplankton using a mucous net and ciliary feeding current in a manner similar to oozooids but with a greatly enlarged gill area and buccal opening. They remain attached to the appendix. Once the number of gastrozooids sufficient to feed the nurse have reached the appendix, the "young nurse" loses its feeding apparatus and becomes an "old nurse" (Paffenhöfer and Koster 2011).

Phorozooids are the next step in the reproductive cycle of doliolids. They are attached to the dorsal appendix by a ventral peduncle (Figure 9.17c). As they mature, they assume a form similar to a young oozooid but with eight muscle bands (Figure 9.18b). Phorozooids are capable of feeding themselves and eventually detach from the dorsal appendix to become free-swimming independent entities. The newly independent phorozooids begin budding gonozooids on their ventral peduncle, forming clusters that are initially supported by the phorozooid but begin feeding on their own while still attached. In *Dolioletta gegenbauri*, gonozooids are released at a size of 1–2 mm over a period of several days (Paffenhöfer and Gibson 1999).

Gonozooids closely resemble the phorozooids, but they are protogynous hermaphrodites that develop an ovary and testis over the course of several days after release from

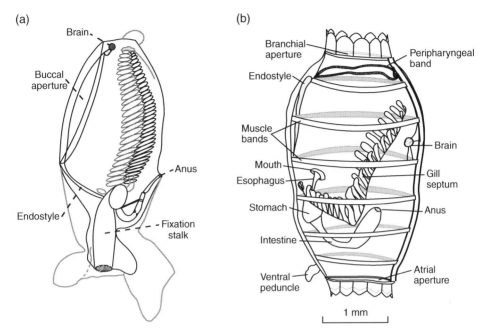

Figure 9.18 Doliolid zooid morphology. (a) A gastrozooid of *Dolioletta gegenbauri*. (b) A phorozooid of *Doliolum nationalis*. Sources: (a) Adapted from Godeaux (1998), figure 17.9 (p. 281); (b) Adapted from Madin and Deibel (1998), figure 5.1 (p. 82).

the phorozooid. The ovary matures first, producing up to three eggs (Godeaux et al. 1998). As in salps, eggs are fertilized in the ovary, but once fertilized they are shed into the sea to produce free-living larvae instead of being incubated within the gonozooid. After the eggs are released, the testis matures to produce sperm for the younger gonozooids, again a strategy like that of the salps.

The fertilized egg develops into a tailed swimming larva in most species (Figure 9.19a and b), though the embryo of *Dolioletta gegenbauri* is contained within a spherical envelope (Figure 9.19c). Doliolid larvae resemble the tadpole larvae of ascidians but have a simpler structure: an anterior trunk, or cephalenteron, and a rudimentary tail used only for locomotion. All organs develop in the cephalenteron. The tail, used for locomotion during the larval stage only, consists of a notochord for support and a simple lateral musculature for generating side-to-side motion. As the larva develops, the tail is resorbed and the mature oozooid escapes the follicular envelope as a free-swimming form, beginning the cycle once more (Godeaux 2003). From egg to oozooid takes about 2 days at 18 °C. Figure 9.20 shows a summary life history of the Doliolida.

The asexual portions of the doliolids life history allow them to reach enormous densities in favorable conditions (90–7200 ind m^{-3}, Paffenhöfer and Gibson 1999), a characteristic they share chiefly with the salps but occasionally with the pyrosomes and appendicularians. The oozooid old-nurse portion of the reproductive cycle, in particular, promotes a rapid population response to plentiful food. Old nurses do not expire when they release their first batch of phorozooids; they continue to produce phorozooids for an indefinite period, shedding them

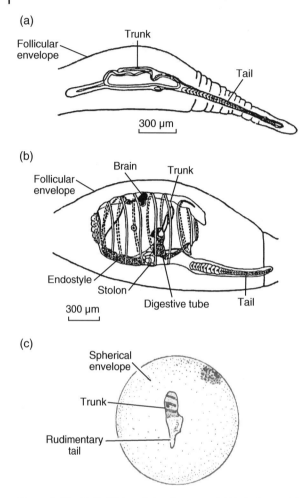

Figure 9.19 Doliolid larval development. (a) Early larva of *Doliolum denticulatum*; (b) late (metamorphosing) larva of *D. denticulatum*. (c) Larva of *Dolioletta gegenbauri*. *Sources:* (a, b) Adapted from Godeaux (2003), figure 4 (p. 197). (c) Adapted from Esnal and Daponte (1999a) (p. 1410).

from the distal end of the dorsal appendix and recruiting new buds at the continuously growing proximal end. Nurses released 36 phorozooids day^{-1} in culture. Average nurse longevity in culture was 20.2 days in *Dolioletta gegenbauri* with a range of 16–27 days. Phorozooid production scaled with appendix length and phytoplankton concentration in the study (Paffenhöfer and Gibson 1999). Based upon appendix lengths in field-caught specimens, it was estimated that nurses in the field would be capable of releasing more than 70 phorozooids a day.

Phorozooids budded off the old nurse also contribute substantially to the asexual portions of the doliolid reproductive cycle. In favorable conditions, phorozooids release 9–14 gonozooids per day over a period of 8–18 days (Paffenhöfer and Koster 2011), averaging 11 gonozooids day^{-1} over 11.2 days (Paffenhöfer and Gibson 1999). This tallies to a lifetime production of 83–163 gonozooids

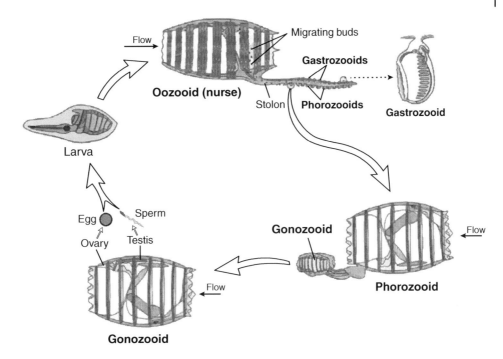

Figure 9.20 Doliolid life cycle. Larvae develop into a non-feeding oozooid (nurse), which then produces a stolon that bears two types of zooids, gastrozooids, and phorozooids. Gastrozooids only feed and stay attached to the stolon, whereas phorozooids break off the stolon and then bud off gonozooids, each of which develops an ovary and a testis. Eggs, released and fertilized in the surrounding seawater, develop into larvae.

for each phorozooid, with an average of 121. If one couples the prodigious reproductive capacity of the nurses with that of the phorozooids, the results are impressive. A single oozooid/nurse producing 50 phorozooids day^{-1} for 20 days results in 1000 phorozooids. They in turn produce about 120 gonozooids for a grand total of 120 000 individuals produced by a single oozooid! It is easy to see how doliolids can rapidly dominate an oceanic area.

The Appendicularia

The Appendicularia, or larvaceans, are highly unusual, even for a group as full of surprises as the tunicates. They occupy their own class in the subphylum Tunicata and are a considered a textbook example of neoteny, the retention of juvenile features in the adult animal: hence their legacy name, Larvacea. The changes that occur in Appendicularia during development (embryogeny) are fertile ground for bringing about changes in adult morphology, a complex and fascinating topic treated well in Steven J. Gould's 1977 book *Ontogeny and Phylogeny*. We will thoroughly explore the characteristics of the Appendicularia, noting similarities to and differences from their tunicate relatives.

Larvaceans differ from the rest of the tunicates in two basic ways: first in their adult morphology, which superficially resembles the tadpole larva of ascidians and doliolids; and second in their use of a gelatinous "house" for feeding and near full-time residence. Adult and house morphologies differ between the three families of Appendicularia (Figure 9.21) as well as between species within each family, but the basics of feeding and development are similar. Without question the most information is available for the Oikopleuridae, which will serve as a model for the class and be bolstered by examples from the Fritillariidae and Kowalevskiidae.

The great majority of appendicularians are diminutive in size, reaching only a few millimeter in length. However, the giant deep-living forms in the subfamily Bathychordaeinae can achieve lengths of 8–9 cm.

The tail of all three families is similar in basic structure (Figure 9.21). At the central axis of the tail is the notochord, a flexible fluid-filled tube that confers enough rigidity to act as a skeletal-like structure for the muscles but is flexible enough to allow the tail to be folded nearly double when entering a new house. Tail flexure and its eel-like, rhythmic, undulatory motion are provided by the flattened band of striated muscle on each side of the notochord in all three families. The flattened fin-like projections that increase the tail's surface area are composed of epithelium.

The tail is under direct neural control. A ganglionated nerve cord originating in the trunk's caudal ganglion at the base of the tail runs alongside the notochord. Interestingly, the appendicularian tail has been twisted 90° so that it oscillates in the sagittal plane (dolphin-like), different from the tail of doliolid and ascidian larvae that, like fishes, move their tails laterally back and forth. The tail is nourished by blood sinuses that run its length on either side of the notochord (Fenaux 1998). The subchordal cells seen in Figure 9.21 are present in many of the oikopleurids, but their function is currently uncertain.

It is important to appreciate that the trunk of the larvacean contains all its major organ systems and that the tail is used only for locomotion or to generate the feeding currents within its house; its only function is to move water. The anatomy of the larvacean trunk resembles that of a salp or doliolid without a posterior atrial cavity. The digestive system begins with a mouth that leads to a pronounced pharyngeal cavity equipped with a ventral endostyle and peripharyngeal bands forming a ciliated hoop encircling the entrance of the pharyngeal cavity. As in the salps and doliolids, the endostyle secretes a conical mucous filter whose anterior border is defined by the peripharyngeal bands, entering the esophagus at its posterior. Water flow through the pharyngeal cavity is driven by ciliary beating in the spiracular rings, creating a flow into the mouth, through the filter cone, and out the spiracles. Food particles are trapped by the meshes of the mucous filter and are digested along with the filter itself. The filter is wound into a strand at the mouth of the esophagus, similar to feeding in the doliolids and salps.

Early digestion and some assimilation are thought to be the province of the stomach lobes, with the remaining assimilation of nutrients attributed to the intestine (Fenaux 1998). The esophagus joins the left lobe of the stomach, which leads to a right stomach lobe and a well-developed intestine with upper and lower sections, followed by a rectum. The gut terminates in an anus located between the spiracles (Figure 9.21). Most food particles ingested by appendicularians are already quite small (1–20 μm, Deibel 1998a) so there is little need for heavy grinding by the stomach lobes. The anatomy of the digestive tract differs enough between species to allow its use in classification of the Oikopleuridae (Fenaux 1998).

Movement of hemolymph takes place between the ectoderm and internal organs, propelled by a two-chambered heart and

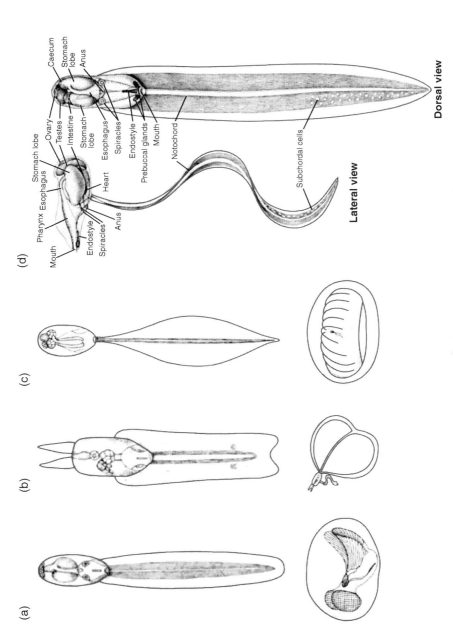

Figure 9.21 Appendicularian anatomy. (a–c) are comparisons among the three families of appendicularians to show differences in morphology of the animals and structure of their houses. (a) Oikopleuridae; (b) Fritillaridae; (c) Kowalevskiidae; (d) lateral and dorsal views of *Oikopleura albicans* showing prominent external and internal anatomical structures. *Source:* (a–c) Adapted from Alldredge (1976c) (p. 97); (d) Adapted from Alldredge (1976c) (p. 96).

the rhythmic movements of the muscular tail. The hemocoel is composed of colloidal jelly with sinuses and lacunae that act as vessels to direct flow to the major organ systems, much as in the salps and doliolids. Overall, circulation in Appendicularia is poorly described (Fenaux 1998).

Trunk morphology in the Fritillariidae and Kowalevskiidae differs from that in the Oikopleuridae (Figure 9.22). The trunk in the Fritillariidae is less compact than the trunk in the Oikopleuridae, tending to be more slender and more flattened dorsoventrally. Major organs are arranged much like those in the Oikopleurids. An endostyle and peripharyngeal bands are located at the entrance to the pharyngeal cavity, which leads to an esophagus and then on to a stomach and intestine in the midtrunk region. The gonads are located posteriorly.

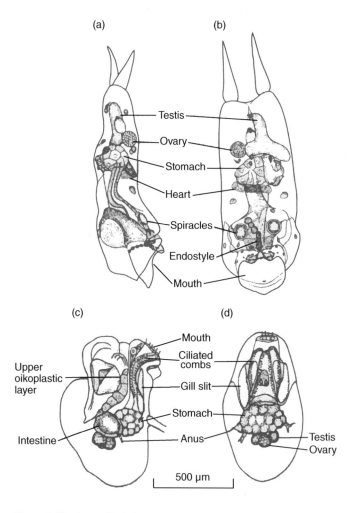

Figure 9.22 Appendicularian anatomy. Anatomy of *Fritillaria pellucida*: (a) lateral view (left side); (b) ventral view. Anatomy of *Kowalewskia oceanica*: (c) lateral view (right side); (d) ventral view. *Sources:* (a) Adapted from Fenaux (1998), figure 2.9 (p. 31); (b) Adapted from Fenaux (1998), figure 2.12 (p. 33).

A mucous feeding net is generated by the endostyle, and water is drawn through the pharyngeal cavity by ciliary beating in the spiracles. The anatomy of the tail is much like that of the oikopleurids, as are the roles of the stomach and intestine (Fenaux 1998).

The kowalevskiids are the smallest and most poorly known of the larvaceans. The trunk is globular in appearance and has a pharyngeal cavity, esophagus, stomach, and intestine with a pattern of water flow like that of the other two families of Appendicularia. The big difference between the kowalevskiids, oikopleurids, and fritillariids is in the absence of an endostyle and peripharyngeal bands in the kowalevskiids and a narrower pharyngeal cavity. Instead of a conical feeding web, the kowalevskiids have two rows of cilia that run longitudinally on each of the dorsal and ventral portions of the cavity, forming three compartments within (Figure 9.22c and d). The absence of a feeding web suggests that the animals obtain sufficient particulate food from the food concentrating mechanisms of their house.

Appendicularians are all protandric hermaphrodites, with one exception: *Oikopleura dioica* has separate sexes. Gonads are located in the posterior trunk, in the oikopleurids usually comprising a median ovary and lateral testes (Fenaux 1998). Unlike the thaliaceans, reproduction in appendicularians is straightforward; it is also semelparous. The testes mature first, and sperm are shed through a tiny spermiduct that leads to the trunk's exterior. Oocytes mature later but are only released when the walls of the ovaries and the trunk rupture, shedding the eggs into the sea and killing the animal. Fertilization takes place externally, a reproductive strategy known as broadcast spawning. It is most effective when spawning takes place in groups of mature larvaceans, maximizing the probability of fertilization. Within 24–48 hours after fertilization, the embryos have developed into juveniles fully capable of creating a mucous house and feeding on their own (Alldredge 1976b, 1976c). Asexual reproduction is not found in the Appendicularia. Figure 9.23 illustrates a summary of the larvacean life cycle.

The Appendicularian House

The gelatinous house used by larvaceans for feeding and maintaining their position in the open sea is secreted by a mosaic of epithelial glandular cells, the oikoblastic epithelium, that cover much of the trunk's anterior and mid-regions. Each section of the mosaic is responsible for secreting a specific portion of the house. Figure 9.24 shows, for example, the regions responsible for secreting the intake and feeding filters

The house originates in collapsed form as a rudiment secreted by the oikoblastic epithelium. Fully formed house rudiments overlay the oikoblastic epithelium as a kind of second skin until needed. In many instances, multiple rudiments are present as additional layers that are secreted below the uppermost one. Each rudiment is bound by a limiting membrane that keeps it in its collapsed form until needed. When a house needs replacing, the larvacean will break through its existing house with vigorous tail movements and escape (Alldredge 1976b, 1976c). Once free, the animal expands the uppermost house rudiment and enters it in an orchestrated series of behaviors (Figure 9.25). First, the rudiment swells and lifts away from the trunk, bursting through its limiting membrane. Vigorous nodding and tumbling movements by the tadpole force water into the rudiment until it has expanded enough to allow the tail to enter base-first into the house. Once in, the tail's sinusoidal movement draws water in through the intake filters, expanding the house until it has reached

630 | The Chordata

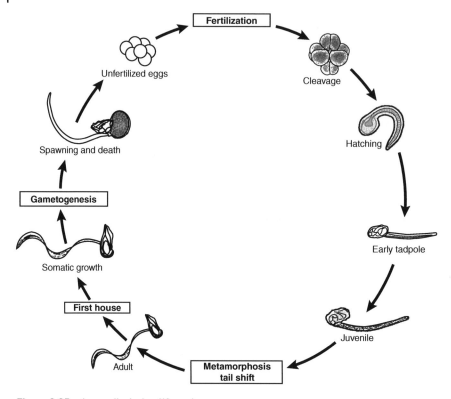

Figure 9.23 Appendicularian life cycle.

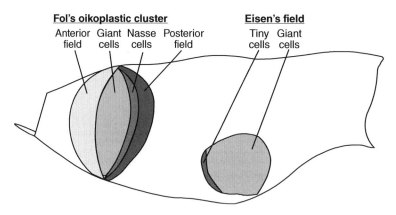

Figure 9.24 Appendicularian house production. Diagrammatic lateral view of the appendicularian *Oikopleura albicans* showing the regions on the body with mosaics of epithelial cells responsible for producing the feeding and intake filters of the house. Fol's oikoplastic cluster is responsible for secreting the feeding filters and Eisen's field is responsible for the intake filters.

its full size and normal feeding begins. The diameter of a full-size oikopleurid house is about twice the total length of the animal (Alldredge 1976b, 1976c; Flood and Deibel 1998). The process of house expansion has been reported to take from one to several minutes (Alldredge 1976b, 1976c; Flood and Deibel 1998).

At present, there is no consensus as to what cue initiates house replacement. Filter-clogging and damage to the house have been suggested (Lohmann 1909a; Alldredge 1976b, 1976c) as have food availability and rudiment-secretion rate (Fenaux 1985; Flood and Deibel 1998). Since none of those cues are mutually exclusive, it is most likely that house renewal happens for a variety of reasons.

A couple of facts about the larvacean house are worth stressing. First, the animal cannot feed without it; from a functional perspective the house is as much a part of the anatomy of a larvacean as are the trunk and tail. Second, all appendicularia abandon and replace their houses repeatedly. Depending upon temperature, replacement rates range from 2 to 5 houses per day, with smaller species replacing their houses more often than larger ones (Flood and Deibel 1998).

In oikopleurids, the best-studied group, the house is a transparent, hollow spheroid that surrounds the animal and is equipped with two sets of filters: one set of inlet filters

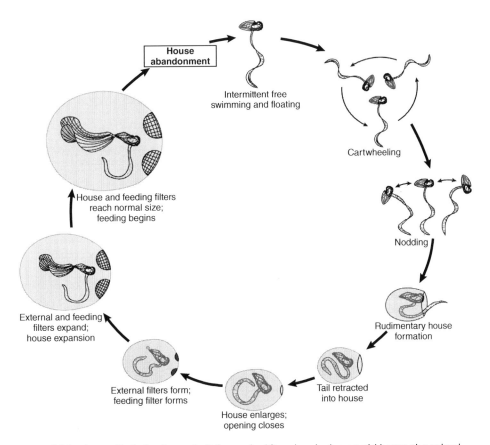

Figure 9.25 Appendicularian house building cycle. After abandoning an old house, the animal spends a short time floating free (30–120 seconds). A series of violent movements, including cartwheeling and head nodding, enlarges a rudimentary house created just before abandoning the old house. The animal enters the rudimentary house after which tail undulations further enlarge the house as the animal proceeds to make the intake and feeding filters.

that limit the size of incoming particles, and one set of feeding filters that greatly improve feeding efficiency by concentrating the incoming particles by as much as 1000-fold (Flood 1991) before the animal ingests them. The house is composed of a gelatinous mucopolysaccharide (Korner 1952; Alldredge 1977). Water flow through the house is generated by the tail beat of the appendicularian, which acts as a pump.

Water flow and chamber design differ substantially between the three families (Figure 9.21), but the following basic principles remain the same.

1) Water enters through inlet channels guarded by a coarsely meshed inlet filter, with mesh sizes (length:width) varying from $13 \times 13\,\mu m$ to $100 \times 300\,\mu m$ (Flood and Deibel 1998).
2) Water and smaller particulates are pumped into a finely meshed filter ($1.4 \times 0.24\,\mu m$ to $0.15 \times 0.61\,\mu m$) that significantly concentrates the food particles.
3) The particles are conveyed to the mouth by a food-collecting duct.
4) Particles are drawn into the mouth and trapped in the appendicularian's pharyngeal basket, which is wound into a strand at the esophagus and ingested in a strategy similar to that of doliolids.

The best way to understand house design and function is to focus initially on what happens and then progress to how it happens, particularly with respect to the food-concentrating filter. The house is an astonishing feat of bio-engineering.

Figure 9.26 is a schematic showing the basics of oikopleurid house design. The animal is situated at the midline of the house in a two-part form-fitting chamber that accommodates its trunk in one portion and its tail in the other. The volume of the tail

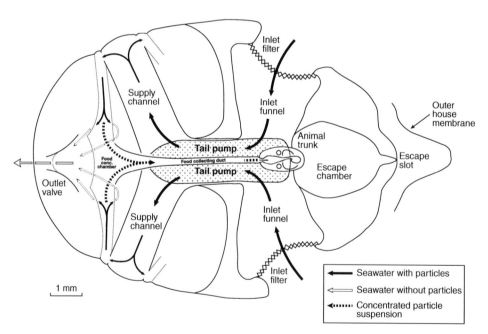

Figure 9.26 Water circulation through an oikopleurid house. *Source:* Flood and Deibel (1998), figure 6.1 (p. 107). Reproduced with the permission of Oxford University Press.

chamber is just enough to allow the tail to beat sinusoidally, creating an impeller pump that draws water into the house through the two cylindrical inlet channels. The inlets join the animal chamber near the base of the tail where it meets the trunk (Alldredge 1976b, 1976c). Near the tip of the tail, the animal chamber bifurcates into two supply channels that direct the water, with its included particulates, into two wing-like feeding filters. Particulates are concentrated up to 1000-fold in the filter (Flood 1991) and are conveyed to the animal's mouth through a food-collecting duct. At intervals, the animal aspirates the particulates into its pharyngeal basket using its internal feeding current and digestion proceeds. Water from which particulate matter has been removed exits the house through a pressure-actuated outlet valve.

The functions of the filter and movement of the particulates from the filter to the feeding duct are complex operations. As described, water enters through the coarsely filtered inlet channels, goes through the tail chamber into the supply channels, and then to the food filters. Note the position of the wing-like filters and how they join at the center of the house. The scalloped outer edges of the filter wings are nearly flush with the internal house wall and, though it is not obvious, the edges open obliquely inward, facing the supply channels. Water enters the filters along their entire outer margins and through lateral openings at their base (Figure 9.27) (Deibel 1986). During normal feeding, all water drawn into the house passes through the feeding filters (Alldredge 1977). For simplicity, it is easiest to think of the feeding filters as a series of about 25 long, parallel, tubes set with their ends facing outward in an inverted U-shape, creating the scalloped appearance of the filter edge. The inner edges of the filter tubes empty into the buccal, or feeding, tube. The recurved length of the tubes between their inner and outer edges forms the corrugated "wings."

The filters themselves (Figure 9.27c) are tubular in configuration but are not truly tubes because they are not physically isolated from one another; water can pass freely though the wide meshes of the suspensory screens that define and strengthen the tubes. The upper and lower surfaces of the filter tubes are the actual food-concentrating filters. Their mesh sizes vary between 0.15 and 0.24 μm (Flood and Deibel 1998). Once within the filter, water entering at the filter edges is subjected to a slightly elevated hydrostatic pressure due to the propulsive activity of the animal tail and the confined volume of the filter itself. Water is expressed through the upper and lower filters, leaving its particulates behind within the filter itself (Flood 1991). Water devoid of its particulates collects in upper and lower posterior chambers of the house and exits through the posterior exit valve (Figure 9.26).

Particulates concentrate within the filters, trapped by the tiny meshes. Every few seconds, the larvacean's tail-beat ceases, causing a slight reflux of water within the house as the exit valve closes. The tubes of the delicate food-concentrating filters partially collapse, freeing particles from the filter meshes and, in some cases, agglutinating them into larger particles. When tail-beating resumes, the re-suspended particulates are more readily moved toward their end destination in the feeding tube. The particles are carried into the feeding tube as a concentrated suspension in a bulk flow of the water remaining in the filter. The feeding filter thus functions as a true filter; it concentrates particles by allowing the majority of the water drawn into the house to simply pass through, leaving behind a concentrated suspension of particles to be harvested (Deibel 1986).

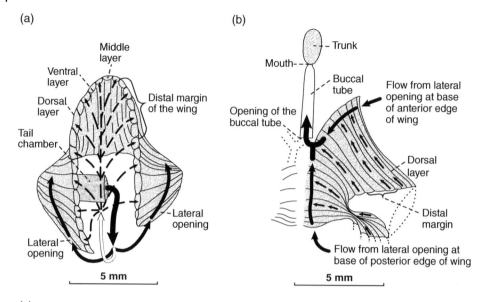

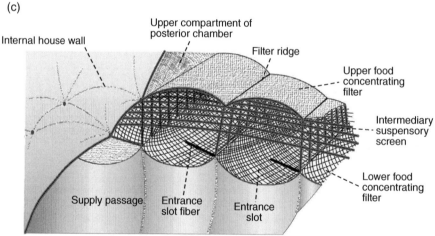

Figure 9.27 Feeding filter dynamic. (a) Lateral view into the sinus at the open end of one wing of a feeding filter of *Oikopleura vanhoeffeni* showing flow patterns. Solid and dashed arrows depict flow streamlines; (b) dorsal view of one wing of a feeding filter of *O. vanhoeffeni*. Arrows depict flow streamlines; (c) Illustration of the structural organization of the triple-layered food-concentrating trap and its junction to the supply passage in the house. *Sources:* (a, b) Deibel (1986), figure 4 (p. 433). Reproduced with permission of Springer Nature. (c) Flood (1991), figure 6 (p. 102). Reproduced with permission of Springer Nature.

Tunicate Systems

Locomotion and Buoyancy

Two basic types of locomotion are employed by the pelagic tunicates. The Thaliacea use continuous or intermittent jet propulsion, whereas the Appendicularia use their tail in undulatory fashion to propel themselves forward (Bone 1998b). Basics of locomotion were briefly covered earlier for each group; more detail will be added here.

Pyrosomes

Feeding currents generated by all zooids within the pyrosome colony accumulate in the cloacal cavity, producing an excurrent flow from the open end of the colony that moves it gently forward. Feeding currents are generated by cilia on the gill bars of each individual, which are under neural control and cease when the colony is touched. Touching the colony will elicit a brilliant bioluminescent response (Figure 9.28).

The excurrent opening of a pyrosome colony is just that; it has no muscular sphincter to direct flow. Consequently, the colony has no way to change direction in any way, including even up or down, using jet propulsion. Yet, substantial vertical migrations have been documented in pyrosomes (Andersen and Sardou 1994; Bone 1998b). Bone (1998b) has speculated that pyrosome colonies may rely on adjustment of sulfate-ion concentrations to allow vertical movement based solely on sulfate exclusion for the upward trip and re-entry for the downward one: an interesting concept. Sulfate exclusion is believed responsible for the near-neutral buoyancy of salps (Denton and Shaw 1961), and its exclusion contributes to static lift in the decapod crustacean *Notostomus* (Sanders and Childress 1988). At present, ion adjustment is the best explanation available for the vertical migrations observed in *Pyrosoma atlanticum*, but at present it is still just an educated guess.

Salps

Locomotion in salps is unusual in that the water used for jet propulsion has a one-way flow, entering through the oral aperture and exiting through the atrial opening. Salps are capable swimmers, though not particularly speedy ones: mean free-swimming velocities for five species of salp oozooids were all about one body length per second ($BL\,s^{-1}$; Madin 1990; Nishikawa and Terazaki 1995; Bone 1998b) with maximum speeds falling between 1.5 and $2\,BL\,s^{-1}$ (Madin 1974; Bone 1998b). To be fair, most fishes cruise at $1\,BL\,s^{-1}$ as well; for many species that speed coincides with their minimum cost of transport (Webb 1975), i.e. the speed that gives them the most distance traveled with the least energy expended. Fishes do exhibit higher maximum sustained speeds ($4\,BL\,s^{-1}$ Brett and Groves 1979). Besides normal forward swimming, salps can stop, reverse, accelerate, decelerate, and change their swimming direction (Bone and Trueman 1983). They are rightfully considered to be agile swimmers.

From a locomotory standpoint, blastozooids and oozooids are structurally similar; the

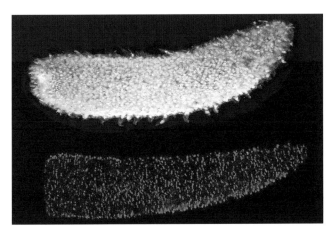

Figure 9.28 Pyrosomes. Photos from life of a pyrosome in natural light (upper) and in the dark after stimulation eliciting bioluminescence (lower). *Source:* © Dante Fenolio, reproduced with permission. See color plate section for color representation of this figure.

major differences are the numbers of muscle bands and length of the propulsive tube (Figure 9.29). In addition, there are fewer muscle bands (5 vs. 13) in the exhalant aperture of the blastozooid (Bone and Trueman 1983). The aperture muscle bands of both zooids are capable of sealing the posterior aperture to force the propulsive jet out the anterior opening for reversing direction. Muscle bands in the posterior aperture also confer the ability to direct the posterior jet for changing direction, which presumably would be more important to the solitary form. Blastozooid aggregates, or chains, do vary from species to species in their swimming capabilities depending on the way the zooids are attached to one another. Neither *Pegea confederata* (Figure 9.10a), attached laterally, nor the whorls of *Cyclosalpa affinis* (Figure 9.10c) swim as effectively as the streamlined linear chain of *Salpa cylindrica* (Figure 9.10f), where the thrust is aligned with the colony axis, the frontal area is minimized, and the zooids attached to form a fairly smooth cylindrical surface as the colony moves through the water (Madin 1990). Overall, salp aggregates tend to swim more slowly than solitaries of the same species (mean: 3.9 vs. 4.9 cm s^{-1}, Madin 1990).

Locomotion is achieved by cyclical compression of the body wall musculature; forward speed can be altered, for example during "escape swimming," by both rate and degree of compression in the muscle bands. Mean pulse rates range from 0.7 to 2.7 per second (Madin 1990). A normal cycle begins by sealing the oral aperture with the front valve or "lips" followed by a

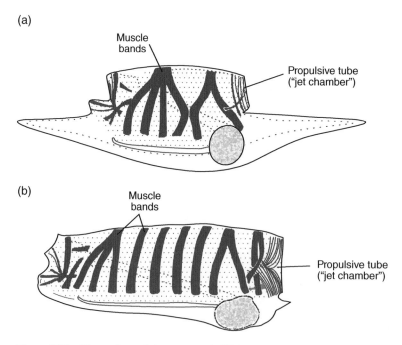

Figure 9.29 Illustrations of the structural differences between the blastozooid (a) and oozooid (b) of *Salpa fusiformis*. Note the greater complexity of the rear aperture muscle bands and longer propulsive tube of the oozooid. Volume of propulsive tube in each indicated by stippling. *Source:* Bone and Trueman (1983), figure 3 (p. 41). Reproduced with the permission of John Wiley & Sons.

near-simultaneous contraction of the body wall muscles to produce the jet. When the front of the animal is stimulated to produce rearward swimming as an escape response, the cycle works in reverse. During normal forward swimming, about 40% of the chamber volume is expelled through the atrial aperture during each cycle. Volume is refilled through both openings, with 33% entering through the oral aperture and 7% through the atrial (Bone 1998b).

Efficiency of salp locomotion is similar to that of other taxa that use jet propulsion (Table 9.1). The measure used here is the efficiency of transfer of momentum from the animal's jet to the water, resulting in animal forward velocity. It is more economical to eject a large-volume jet at low velocity than a low-volume jet at high velocity (Table 9.1). Salp propulsion compares favorably to that of siphonophores and squid. As noted earlier, living salps have been shown to achieve near-neutral buoyancy using sulfate exclusion from the fluid within the test (Robertson 1957; Denton and Shaw 1961; Bidigare and Biggs 1980). When salps die, they sink (Wiebe et al. 1979).

Doliolids

Doliolids normally swim using only the feeding currents generated by the cilia on their gill bar: as they feed, they move forward. However,

Table 9.1 Relative mechanical efficiency, E, of jet-propulsion in *Salpa fusiformis* versus a squid and two siphonophore species. *Source*: Bone and Trueman (1983), table V (p. 503). Reproduced with the permission of John Wiley & Sons.

	Chamber pressure (Pa)	Mean jet velocity, $\bar{u}$ (cm s^{-1})	Mean forward velocity, $\bar{U}$ (cm s^{-1})	$E = \dfrac{2\bar{U}}{(2\bar{U}+\bar{u})}$
Salpa fusiformis				
Small blastozooid	100	22.8	3.8	0.25
Large blastozooid	80	29	4.8	0.25
Small oozooid	45	18.5	3.8	0.29
Large oozooid	60	27.9	6.6	0.32
Blastozooid chain of 14, one active		22.8	2	0.15
Same blastozooid chain, all active[a]		22.8	6.5 (est)	0.36
Siphonophores				
Chelophyes, anterior nectophore[b]	400	71	16	0.31
Abylopsis, posterior nectophore[b]	30	22	3	0.21
Squid				
Alloteuthis[c]	20 000	625	80	0.20

[a] Assumes all zooids at maximum performance; if not, $\bar{u}$ will be lower and E will be greater.
[b] From Bone and Trueman (1982).
[c] Trueman, unpublished.

when stimulated on either end they can move quite rapidly in response. Sensory cells present in the scalloped valves of their anterior and posterior apertures (Figure 9.18b) cause the muscle band furthest from the point of stimulation to contract. Almost immediately after (5–7 milliseconds), the remaining bands contract simultaneously to produce the escape jet. When it contracts along with the rest, the muscle band encircling the stimulated aperture also narrows the jet, increasing its velocity. Doliolids generate higher chamber pressures and higher instantaneous velocities (50 BL s^{-1}) than salps (Bone and Trueman 1984). About 60% of their chamber volume is lost during the jet pulse. As in salps, the contraction of the body musculature of doliolids can be graded to vary the strength of the jet (Bone 1998b). Though instantaneous velocities are impressive, the diminutive size of doliolids means that the distance traveled with each pulse is small. For example, a 0.4 cm doliolid can move forward about 3 cm with a single pulse (Bone 1998b), sufficient to take the tiny tunicate out of immediate harm's way in its Lilliputian world.

Phorozooid, gonozooid, and oozooid stages typically do not use pulsed jet propulsion unless they need to escape, but the larger old nurse stages (e.g. a *Doliolletta gegenbauri* old nurse at 5 cm) contract rhythmically. The pulsing body musculature aids the heart in circulating nutrients between the gastrozooids, phorozooids, and the old nurse itself.

Doliolid zooids are slightly negatively buoyant. They maintain their position in the water column by orienting with their anterior end tilted obliquely up, presumably as a result of their viscera (and therefore most of their body mass) being located in the posterior of the animal (Fedele 1923; Bone 1998b). As a consequence, their feeding current points slightly downward, giving them a small but steady upward thrust and enabling them to maintain vertical station. Occasional jet pulses supplement the continuous feeding current. Old nurses that have lost their gill bar, and therefore their feeding current, can maintain their vertical station with their rhythmic pulsing, a second benefit in addition to helping their circulation. Doliolid eggs and larvae are positively buoyant, believed to be the result of ammonium sequestration in their follicle cells and larvae (Bone 1998b).

Appendicularia

Like fishes, appendicularians are undulatory swimmers, but they do not exist outside of their houses except for brief periods of frenetic swimming activity after exiting their old house and before expanding their new one. Apparently, their repertoire of tail movement patterns is also quite limited, consisting of three "programs" in the oikopleurids that have been studied. The first program is a series of regular, low-frequency (2–3 Hz) short bursts that mimic the tail-beat pattern associated with feeding in the house. The second mimics the more prolonged tail beating at higher frequency (4–5 Hz) that characterizes the nodding movements used when expanding a new house. The third and highest-frequency tail beating (5–6 Hz) is most typical of free-swimming animals that are maintaining their position in the water column. Oikopleurids are negatively buoyant and will sink rapidly when outside their house and not actively swimming (Bone 1998b).

Nervous Systems and Sensory Mechanisms

The Thaliacea all have a tiny, centralized brain, present in most figures as a "ganglion," with nerves that radiate out to the musculature, viscera, and gills (Bone 1998c). The brain of Appendicularia (Oikopleuridae) is located in the antero-dorsal region of the trunk (Fenaux 1998). It shows a prominent connection with the caudal ganglion and nerves of the tail as well as multiple connections with sensory cells located in the mouth region (Bone 1998c). Sensory modalities that have been identified in the pelagic tunicates include photoreception, mechanoreception in the form of cupular organs, hair-cell-like receptors located in the epithelium that are believed to be sensitive to touch and water motion, and statocysts. Pelagic tunicates, particularly the salps and appendicularia, exhibit a well-developed system of epithelial conduction that acts in concert with the nervous system to propagate sensory information and, in the salps, to control swimming in blastozooid chains (Mackie and Bone 1977; Bone 1998c).

Pyrosomes

Each zooid in the pyrosome colony has a small central brain and associated rudimentary eye (Figure 9.30a). Nerves run from the brain to the oral siphon as well as into the pharyngeal cavity, where they innervate the gill bars (Figure 9.30b). Mechanoreceptors on the oral sphincter muscle are highly sensitive to touch (Mackie and Bone 1978): when stimulated they close the sphincter.

Work on nervous system function has been limited to basic studies on ciliary control in the branchial cavity and bioluminescent responses. The feeding currents of pyrosome zooids result from coordinated ciliary activity on the gill bars propelling water through the branchial cavity. Mechanical stimulation of individual zooids by vibration or touch results in an abrupt cessation of activity in all gill cilia of an individual's branchial basket, producing a neural signal termed a ciliary arrest potential (CAP).

Pyrosomes are brilliantly bioluminescent (Figure 9.28), and colonies can be stimulated to flash by mechanical stimulation as well as by photic stimulation, i.e. they will luminesce in response to a flash of light, much like many fireflies. Luminescent flashes are always accompanied by ciliary arrest and CAP's (Mackie and Bone 1978), yet the large light organ (Figures 9.7c and 9.30b) within each zooid is not directly innervated. At present, a mechanistic explanation for how the two phenomena are connected is lacking, but epithelial conduction has been proposed (Mackie and Bone 1978). Interestingly, pyrosomes respond most strongly to light stimuli of the same wavelength as their own luminescent flashes (490nm, Bone 1998c), and their luminescent response increases with increasing intensity of the stimulating light (Bowlby et al. 1990). The pyrosome light organ thus behaves as if it were under direct neural control although its light is believed to be produced by a culture of bacteria within the organ. Its rapid response and scaling with stimulus intensity are truly unique for a light organ of that type (cf. Herring 1978).

Doliolids

Doliolids have a small brain (ganglion or cerebral ganglion) located midway between the buccal and atrial siphons with nerves that radiate to the locomotory muscle bands, lip muscles of the two apertures, and to the viscera, endostyle, gill bar, and a variety of sensory receptors (Figure 9.31a). It is believed that brain function is mainly concerned with locomotory muscle control, particularly escape movements. The brain and nervous system organization of doliolids has received only a modest amount of study (Bone 1998c).

Doliolid sensory receptors are present in a number of sites including the lips of the oral and atrial siphons and the external epithelium. In the oozooid stage only, a statocyst is present just anterior to the fourth muscle band (Figure 9.31a). Photoreceptors are not found in any doliolid stage. Morphology of most receptors is the typical

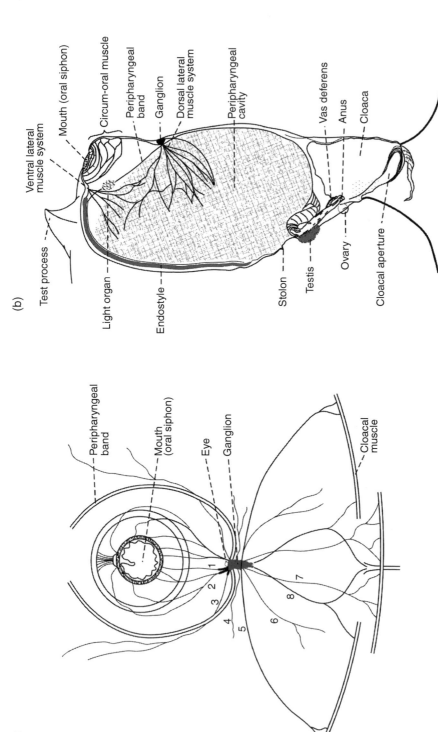

Figure 9.30 Pyrosome nervous system. (a) Nervous system in the mouth region of the pyrosome *Pyrosomella verticillata* showing the central ganglion from which the major nerves (numbered) originate. Note the proximity of the small eye to the ganglion; (b) nervous system of *Pyrostemma spinosum* showing nerves radiating from a central ganglion (brain) to the oral siphon and into the pharyngeal cavity. Note that the light organ is not directly innervated. *Sources*: (a) Metcalf and Hopkins (1919), plate 22; (b) Metcalf and Hopkins (1919), plate 19.

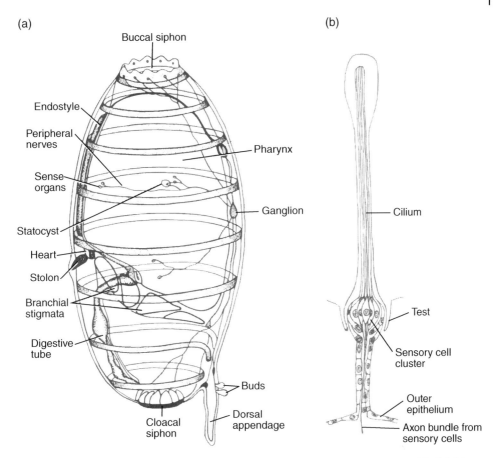

Figure 9.31 The nervous system of thaliaceans. (a) Nervous system of the oozooid of *Dolioloides rarum*. (b) Sensory cell (cupular organ) of a tunicate. *Sources*: (a) Neumann (1913) (p. 9); (b) Bolles-Lee (1891), plate X.

ciliated tunicate "hair-cell" type (Figure 9.31b). Sensory modalities that have been identified are mechanoreception, including sensitivity to vibration and touch, chemoreception at the front lips, and presumably, gravity and orientation in the oozooids via their statocyst (Bone 1998c).

Doliolids exhibit a rhythmic cessation in the beating of their gill cilia, similar to the rhythmic cessation of tail beating in oikopleurids, that is believed to be triggered by neural pacemakers in the gill bars themselves. As in the pyrosomes, the cessation of beating results in a CAP that can be recorded readily with electrodes. Gill cilia will cease beating temporarily in response to vibration as well (Bone and Mackie 1977).

Salps

Salps exhibit a major leap forward in neural complexity. A sophisticated epithelial conduction system (see Chapter 3) interacts with the nervous system proper to greatly increase the effectiveness of sensory communication and motor control in solitary salps and blastozooid colonies. The interaction between neural and epithelial systems is known only in salps and appendicularians and may be unique in the animal kingdom (Anderson et al. 1979; Bone 1998c).

As in the doliolids, the cerebral ganglion of salps is located dorsally (Figure 9.32a), comprising a central core of nerve fibers of various types (a neuropil) and an outer region (cortex or rind) of nerve cells (Figure 9.32b) (Bullock and Horridge 1965). Nerves leave the brain to supply the muscle bands and viscera. The locomotory rhythm of salps

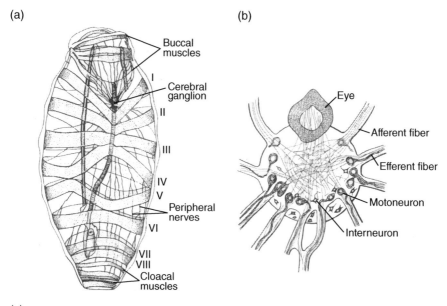

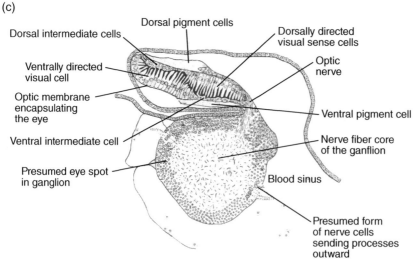

Figure 9.32 The nervous system and eye of thaliaceans. (a) The nervous system of an oozooid of *Ihlea racovitzai*. (b) Section of the ganglion of *Salpa*. (c) Eye and ganglion of *Cyclosalpa*. Sources: (a) Adapted from Brien (1948); (b) Fedele (1938); (c) Metcalf and Lentz-Johnson (1905), plate VIII.

originates in the nerve cells of the brain's cortex (Anderson et al. 1979) but can be modified by sensory input that is processed in the neuropil of the brain. Early studies by the pioneer Fedele (1933a, b) revealed that ablating the core of the brain did not destroy the normal locomotory rhythm of the salp but did prevent it from being modified by sensory input (Bone 1998b).

Some motor axons from the brain connect directly to adjacent epithelial cells capable of eliciting epithelial action potentials (impulses) that spread over the outer epithelium as outer skin potentials or "OSPs": epithelial conduction (Anderson et al. 1979; Bone 1998c). OSPs and the epithelial conductive system are important in salp locomotory behavior in both solitary zooids and blastozooid chains. Posterior stimulation of a salp or salp chain results in propagated OSPs and an escape response of the individual or entire chain by accelerating forward. An analogous response may be obtained by stimulating anteriorly, in which case the oozooid or chain swim backwards. The importance of coupling epithelial and neural conductive systems is huge. In the current example, stimulation of the excitable outer epithelium stimulates the nearest mechanoreceptor, which neurally conveys the location of the stimulus to the brain, which can then coordinate the proper locomotory response. The locomotory response initiated by the brain can then be propagated neurally and epithelially to maximum effect (Bone 1998c).

The most conspicuous sensory receptor of the salp is the eye, located just above the brain (Figure 9.32c). Eye morphology differs between species and between oozooids and blastozooids of the same species (Metcalf 1893; Metcalf and Lentz-Johnson 1905). Photoreceptor cells are not equipped with lenses, so any sense of directionality would be limited, but it has been experimentally demonstrated that they do respond to light (Gorman et al. 1971).

Salps also have sensory receptors termed cupular or tentacular sense organs (Figure 9.31b), generally resembling arthropod hair cells, that are imbedded in the test and connected via axons to the brain. They are believed to be mechanoreceptors, sensitive to deformation, but may play alternate roles including chemoreception (Bone 1959). Salps lack statocysts in both life stages.

Appendicularia

The appendicularians are not only diminutive in size, their nervous systems have been significantly miniaturized so that the total cell numbers of the brain and associated nerves as well as the numbers of sensory receptors are remarkably small (Bone 1998c), even for animals of their size. For example, the brain of oikopleurids contains only 70 cells (Martini 1909a, b; Bone 1998c). The small number of cells in the central and peripheral nervous systems of oikopleurids has allowed mapping (Martini 1909a, b; Bollner et al. 1986; Olsson et al. 1990) of neural pathways in the brain and anterior trunk relevant to appendicularian feeding behavior. Brain circuitry and behavioral studies suggest that input from sensory cells located beneath the lower lip of oikopleurids and fritillariids (Bone et al. 1979; Bone 1998c) helps to govern the direction of water flow in the pharyngeal cavity of feeding larvaceans via the spiracles, also termed ciliary rings (Figure 9.21). Reversal of normal inward flow allows the tadpole to reject undesirable particles, or to prevent entry of excessive numbers of particles into the pharynx.

Oikopleurids have a statocyst located on the left side of the brain, but it is unknown whether it is sensitive to vibration, acceleration, or orientation (gravity).

Nerves from the brain connect directly to the caudal ganglion that in turn is connected to the ganglionated nerve cord of the tail. At present, the origin of the tail's rhythmic patterns is unclear. The caudal ganglion also connects directly to the paired Langerhans receptor cells located on either side of the posterior trunk (Figure 9.33). The Langerhans receptor cells each bear a bristle that is long enough to touch the inside of the house. When stimulated, a bout of rapid escape swimming ensues, and it is presumed that the receptors are involved in emergency escapes from the house.

The Langerhans cells and escape swimming can be stimulated by direct contact or

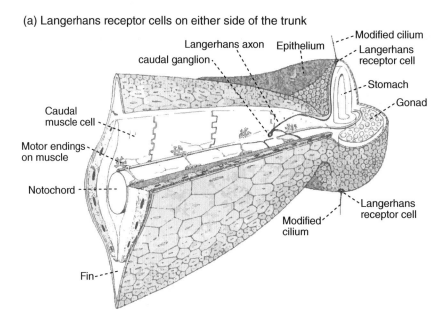

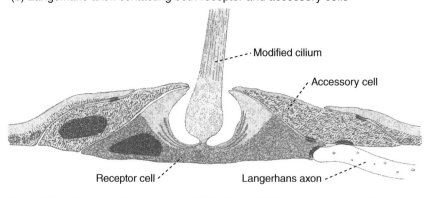

Figure 9.33 The Langerhans receptor of *Oikopleura*. (a) Stereogram showing the position of receptor cells on either side of the trunk. (b) Schematic section across a receptor showing a Langerhans axon connecting both receptor and accessory cells. *Sources*: (a) Adapted from Bone and Mackie (1975), figure 2 (p. 269); (b) Adapted from Bone (1998c), figure 4.31 (p. 78).

by stimulation of the surrounding outer epithelial cells, which propagate epithelial action potentials (OSPs) over the tail and posterior trunk that trigger the Langerhans receptors (Bone 1998c). The epithelial cells surrounding the Langerhans receptors are connected to them by electrical junctions (Bone and Mackie 1975), another example of interacting epithelial and neural conductive systems similar to that observed in salps. In addition, there is direct axonal innervation of an adjacent epithelial cell by the caudal ganglion (Figure 9.33b), assuring the epithelial-neural connection. In effect, the excitable outer epithelium acts to greatly widen the receptor field of the Langerhans cells (Bone 1998c).

Gas Exchange, Circulation, and Excretion

Anatomy and feeding strategies of the tunicates, particularly the thaliaceans, provide an ideal environment for diffusive gas exchange. The pharyngeal baskets of pyrosomes and gill bars of salps and doliolids all present a large surface area to a continuous water flow, facilitating inward diffusion of oxygen. Diffusion at the gills is likely supplemented by diffusion through the thin-walled tests into the general circulation of salps and doliolids. Perhaps as a result of the ideal diffusive environment, no detailed information is available for characteristics such as ventilation volume or percent utilization that would describe the effectiveness of thaliaceans gas-exchange systems. In normoxic waters, it is unlikely that oxygen is a limiting variable. Salps do vertically migrate at times (Wiebe et al. 1979; Lancraft et al. 1989), but migrations into severe oxygen minima have not been reported. The giant appendicularians of the Bathychordaeinae are found at hypoxic depths in the California borderland (Steinberg et al. 1994), but they, and the appendicularia in general, await detailed physiological study.

Basic information describing circulation is available for the salps and doliolids (Heron 1975a, b; Bone et al. 1997a). In salps, colorless blood is propelled through an elaborate system of lacunae that permeate the test, acting as blood vessels by providing definite paths for blood flow. A variety of particles and blood cell types are present, at least some of which are assumed to be nutritive (Heron 1975a). The tubular heart (Figure 9.12) of salps has a rapid peristaltic beat, with a contraction that passes from one end to the other, pushing the blood ahead of it. Like all tunicates, the heart regularly reverses itself with the peristaltic contraction simply moving in the opposite direction. Normal direction of flow propels the blood from the anterior end of the heart to the endostyle and up the sides of the body around the muscle bands. It collects in dorsal lacunae near the brain before proceeding down the gill bar to the viscera and posteriorly to the posterior muscle bands before returning to the heart (Figure 9.34a) (Heron 1975b; Godeaux et al. 1998). Periodic heart reversal propels the blood through the same pathway in the opposite direction. It was postulated by Heron (1975b) that heart reversal allowed the blood to bathe the viscera more effectively, perhaps aiding in the pick-up and distribution of nutritive particles.

Circulation in the doliolids is quite similar to that in salps with the caveat that the lacunar pathways are absent. The hemolymph circulates in a hemocoel that is a "single open space in which the internal organs including the heart are found" (Bone et al. 1997a). Normal flow traces a similar path to that seen in salps, moving anteriorly to the endostyle up the sides of the body to the brain and then posteriorly to the gill bar and posterior muscle bands before

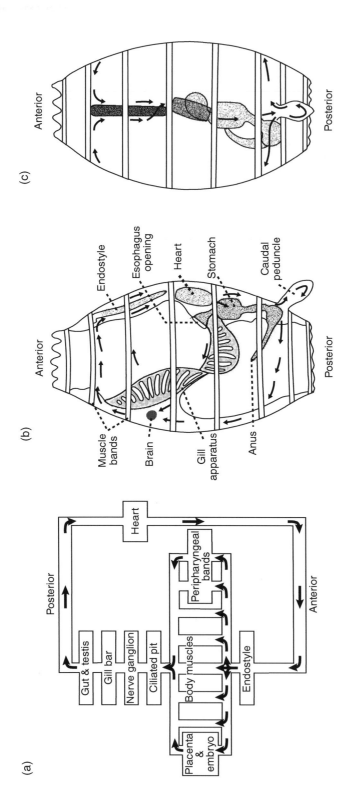

Figure 9.34 Thaliacian circulatory systems. (a) Blood flow in *Thalia democratica*. Direction of flow is shown when the heart is beating forward. (b) Lateral view of *Doliolum nationalis* showing the arrangement of the major organs. Arrows indicate the direction of flow when the heart is beating in reverse. (c) Ventral view of *D. nationalis* showing circulation paths along the ventral region. *Sources*: (a) Heron (1975b), figure 1 (p. 960); Reproduced with permission of Cambridge University Press. (b) Adapted from Bone et al. (1997a), figure 1 (p. 190).

returning to the heart (Figure 9.34b and c). Note that arrows showing the path of doliolid circulation in Figure 9.34 are depicting the pattern during heart reversal, not during normal flow. In comparison to salps, fewer cells can be observed in the hemolymph of doliolids, and those are amoeboid in form.

Little is known about circulation in the Appendicularia. A ventral heart and a system of lacunae are present in the trunk, and they are believed to direct blood flow in a pattern similar to that observed in salps and doliolids, proceeding anteriorly from the heart to the endostyle and then dorsally, following the peripharyngeal bands. Flow continues posteriorly along the dorsum, bathing the gut and gonads before proceeding down and up the tail and returning to the trunk and heart (Fol 1872; Fenaux 1998).

A tube-or sac-shaped excretory organ located posterior to the hindgut has been described in the salp *Thalia democratica* by Heron (1976). It discharges excretory blood cells filled with uric acid crystals from the bloodstream to the outside. Similar structures have also been reported in *Salpa fusiformis* Godeaux et al. (1998) and *Pegea confederata* (Heron 1976). Origin of the uric acid is believed to be from nucleic acid breakdown. The salps' sessile tunicate relatives, the ascidians, are known to accumulate uric acid and guanine crystals, which they sequester over the course of their lives without discharging them (Goodbody 1974; Heron 1976). Since such crystals are highly reflective but metabolically benign, Heron believed that discharging them would preserve the transparency and relative invisibility of salps, an advantage in the open ocean but not in the benthos. It is believed that ammonia resulting from protein breakdown is readily excreted by diffusion alone (Heron 1976).

Trophic Role

Pelagic tunicates are important in open-ocean ecology for a variety of reasons. Three of the most important are their high filtration rates, the size range of particles that they are capable of ingesting, and their ability to rapidly form huge blooms thus completely dominating the zooplankton community. Their unique system of particle filtration allows them to rapidly filter very small plankton, limited only by the mesh size of their pharyngeal feeding webs. Their efficient feeding strategy coupled with prodigious population growth rates resulting from asexual reproduction means that they can respond quite rapidly to phytoplankton blooms with swarms of their own. Swarms of salps have been recorded in the literature ranging in numbers from 0.15 to 1000 ind m^{-3} (Table 9.2) (Andersen 1998).

Looking at matters from the opposite perspective, prodigious ingestion leads to equally large numbers of egested particles in the form of rapidly sinking fecal pellets. Those in turn create a vertical flux of organic and inorganic nutrients: nutrition for deeper-dwelling microbes and microplankton as well as export of trace elements to the benthos (Andersen 1998).

Table 9.3 (Alldredge and Madin 1982) provides a brief summary of tunicate filtration rates, including a comparison with other taxa on a per-animal and per-unit-mass basis. Tunicates clearly exhibit the highest rates of filtration, ranging from 2 to 5 times those of crustaceans when considered on a mass-specific basis, and up to 100 times the crustacean rates on a per-animal basis. The tunicates are quite effective at ingesting phytoplankton, particularly the salps.

Table 9.4 (Gorsky and Fenaux 1998) gives the size spectrum of the smallest-sized particles ingested by a variety of taxa, including

Table 9.2 Abundance data for salp swarms reported for locations throughout the world's oceans. *Source*: Andersen (1998), table 7.1. Reproduced with the permission of Oxford University Press.

Species	Location	Depth (m)	Concentration (no·m^{-3})	References
Salpa aspera	NW Atlantic	100	65	Wiebe et al. (1979)
Salpa fusiformis	Gulf of Guinea	Surface	1000	Roger (1982)
	Gulf of Guinea	Surface	165	LeBorgne (1983)
	Off west coast of Ireland	1–100	700	Bathmann (1988)
	NW Mediterranean	0–200	19	Nival et al. (1985)
	NW Mediterranean	100	7.5	Morris et al. (1988)
Salpa thompsoni	Scotia Sea	0–1000	0.15	Lancraft et al. (1989)
	Scotia Sea	400–600	0.40	Lancraft et al. (1989)
Thalia democratica	Off east coast of Florida	0–30	1000	Paffenhöfer and Lee (1987)
	Off east coast of Florida	0–12	60	Atkinson et al. (1978)
	Off southern California	0–70	275	Berner (1967)
	Off Ivory Coast	Surface	100	Binet (1976)
	SE Australian shelf	50	44	Heron and Benham (1984)

several copepods. Nearly all are in the range of 1–5 μm. Of particular interest is the tunicate taxon missing from the table: the Appendicularia. The appendicularians not only exploit particles in the nanoplankton (2–20 μm) size range, they are capable of taking submicron-sized particles in the range known as picoplankton (0.2–2.0 μm) (Deibel and Lee 1992). Picoplankton include tiny photosynthetic bacteria: cyanobacteria (blue-green algae) and prochlorophytes, as well as other bacterioplankton that are of paramount importance in the primary production of oligotrophic systems. Though extremely abundant (10^3–10^7 cells ml^{-1}), picoplankton are not accessible to most of the filter/suspension-feeding zooplankton community. They are too small to be effectively trapped by the feeding appendages of copepods and most pass through the meshes of the thaliaceans. By ingesting picoplankton and incorporating them as tissue into their body mass, larvaceans make picoplankton energy available as food to larger organisms. They do this in three ways: in the bodies of the tadpoles themselves, as part of the debris that remains in their discarded houses, and in the houses themselves.

Particles of submicron size, such as bacterioplankton and small organic aggregates, are part of an important oceanic micro-community known as the microbial loop (Figure 9.35) (Azam et al. 1983; Fenchel 1988). The microbial loop is best thought of as an ecosystem in miniature. Primary production is the province of the cyanobacteria and prochlorophytes, which

Table 9.3 Size, filtering rate, and weight-specific filtering rates of protozoan, crustacean, and tunicate herbivores. *Source:* Alldredge and Madin (1982) Pelagic Tunicates: Unique Herbivores in the Marine Plankton, table 1, BioScience, 1982, volume 32, issue 8 (p. 657). By permission of Oxford University Press.

Species	Length (mm)	Weight (µgC)	Filtering rate[a] (ml animal^{-1} h^{-1})	Weight specific filtering rate (ml µgC^{-1} h^{-1})	References
Protozoa					
Tintinopsis acumata	—	0.008	0.0015	1.9	Heinbokel (1978)
Copepoda					
Calanus helgolandicus	—	10	7.9	0.8	Paffenhöfer (1971)
Temora longicornis	—	10	4.7	0.5	Harris and Paffenhöfer (1976)
Pseudocalanus elongatus	—	10	5.0	0.5	Paffenhöfer and Harris (1976)
Calanus pacificus	—	68	4.3	0.1	Frost (1972)
				$\bar{x} = 0.5$	
Euphausiacea					
Euphausia pacifica	11.0	18.06	38.0	0.02	Lasker (1966)
Appendicularia					
Oikopleura dioica	1.25	10	5.8	0.6	King et al. (1980)
Oikopleura dioica	1.25	10	18.3	1.8	Paffenhöfer (1975)
Oikopleura dioica	1.25	10	13.6	1.4	Alldredge (1981)
Stegasoma magnum	1.36	10	59.0	5.9	Alldredge (1981)
				$\bar{x} = 2.4$	
Doliolida					
Dolioletta gegenbauri (gonozooid)	—	10	5.3	0.5	Deibel (1980)
Dolioletta gegenbauri (oozooid)	3.4	10.5	12.3	1.2	Deibel (1980)

(*Continued*)

Table 9.3 (Continued)

Species	Length (mm)	Weight (μgC)	Filtering rate[a] (ml animal^{-1}·h^{-1})	Weight specific filtering rate (ml μgC^{-1} h^{-1})	References
Salpida					
Thalia democratia (aggregate)	3.3	10	3	0.3	Deibel (1980) and Madin et al. (1981)
Cyclosalpa floridana (solitary)	20	278	308	1.1	Madin et al. (1981) and Harbison and McAlister (1979)
Cyclosalpa floridana (aggregate)	20	139	129	0.9	Madin et al. (1981) and Harbison and McAlister (1979)
Cyclosalpa affinis (solitary)	50	4512	2306	0.5	Madin et al. (1981) and Harbison and McAlister (1979)
Cyclosalpa affinis (aggregate)	50	1819	448	0.2	Madin et al. (1981) and Harbison and McAlister (1979)
Pegea confoederata (solitary)	50	1591	1554	1.0	Madin et al. (1981) and Harbison and Gilmer (1976)
Pegea confoederata (aggregate)	50	1340	3230	2.4	Madin et al. (1981) and Harbison and Gilmer (1976)
Salpa maxima (aggregate)	50	3231	2522	0.8	Madin et al. (1981) and Harbison and Gilmer (1976)
Salpa cylindrica (solitary)	50	1265	5131	4.1	Madin et al. (1981) and Madin and Cetta (1980)

$$\bar{x} = 1.3$$

Food concentrations ranged from 20 to 200 μgCl^{-1} except for Harbison and Gilmer (1976) and Harbison and McAlister (1979), which ranged from 200 to 1400 μgCl^{-1}. Temperature ranged from 12 to 24 °C.
[a] Filtering rates for salps are "Fmax" values.

Table 9.4 Minimum size of food particles ingested by several taxa that are potential competitors of appendicularians for food resources. *Source*: Fenaux (1998), table 10.1 (p. 163). Reproduced with the permission of Oxford University Press.

Taxon	Food size (μm)	References
Salps		
Cyclosalpa polae	2.1	Harbison and McAlister (1979)
Cyclosalpa affinis	3.4	Harbison and McAlister (1979)
Cyclosalpa floridana	1.5	Harbison and McAlister (1979)
Pegea confederata	3.3	Bone et al. (1991)
Pegea confederata	2.8	Harbison and Gilmer (1976)
Salpa maxima	2.6	Harbison and Gilmer (1976)
Salpa fusiformis	<5	Silver and Bruland (1981)
Thalia democratica	1.2	Mullin (1983)
Doliolids		
Dolioletta gegenbauri	0.2–5	Crocker et al. (1991)
Pteropods		
Cavolinia tridentata	1–5	Gilmer (1974)
Euphausiids		
Euphausia pacifica	5	Parsons et al. (1967)
Euphausia superba	3	Schnack (1985)
Copepods		
Acartia clausi	2.5–8	Nival and Nival (1976)
Calanus pacificus	8	Parsons et al. (1967)
Neocalanus plumchrus	2.5	Frost et al. (1983)
Neocalanus cristus	3	Frost et al. (1983)
Rhincalanus gigas	3	Schnack (1985)
Cladocerans		
Penilia avirostris	2	Turner et al. (1988)
Tintinnids		
Stenosemella ventricosa	>1.3	Rassoulzadegan and Etienne (1981)

are preyed upon by larger single-celled heterotrophic flagellates and ciliates. Flagellates and ciliates are within the size range that can be taken by crustacean zooplankton as well as the salps, doliolids, and pyrosomes, thus entering the more traditional marine food web. Bacterioplankton play an important role by breaking down submicron colloids, tiny organic aggregates that result from the decomposition of phytoplankton cells, and other cellular debris. The filtering abilities of Appendicularia enable them to "short-circuit" the microbial loop, providing an energy pathway from picoplankton to organisms as large as fish.

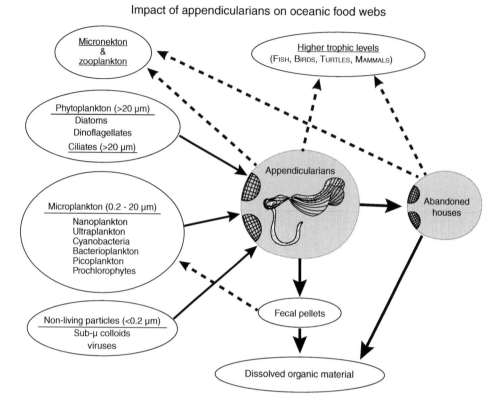

Figure 9.35 Impact of appendicularians on oceanic food webs and their involvement in the microbial loop. Large solid arrows represent energy flow into the appendicularians. Dashed arrows represent appendicularian production and input into the ecosystem. Smaller solid arrows represent flow of energy that is lost to the microbial loop and system.

Discarded appendicularian houses themselves are a unique energy source in the oceanic water column. House turnover ranges from two to greater than five times per day (Flood and Deibel 1998) depending on species and temperature, so productivity is substantial. When discarded, the houses contain noningested phytoplankton cells still trapped in their filters as well as fecal pellets, colloidal material, and other particulates concentrated by the filtering activity of the larvaceans (Gorsky and Fenaux 1998). The houses are thus food-rich morsels that are utilized by copepods, chaetognaths, fish larvae, and adult fish (Alldredge 1976a; Davoll and Silver 1986; Ohtsuka and Kubo 1991; Hamner and Robison 1992). Bacteria flourish in the rich organic environment of the discarded houses, rapidly creating a microbial ecosystem that further increases their food value to larger species (Bedo et al. 1993). Larvaceans and their houses are thus a major vehicle for repackaging the production of micron-sized particles into larger packets accessible to higher trophic levels (Gorsky and Fenaux 1998).

Bioluminescence

Bioluminescence is well known and well described in two groups of the pelagic tunicates: the pyrosomes and the appendicularians. The luminescent display of pyrosomes gives them their name (pyro = fire, soma = body). The glow from a freshly captured pyrosome is brilliant (Figure 9.28); pyrosome swarms are capable of producing enough light to easily allow "reading a book of small print" at night (Bennett 1834; Bone 1998c). As noted earlier, pyrosomes respond to photic as well as mechanical stimulation.

Bioluminescence has been confirmed in many, but not all, species of Appendicularia. The deep-sea genera of the Bathychordaeinae have been confirmed as luminescent (Fenaux and Youngbluth 1990; Hamner and Robison 1992; Haddock et al. 2010) as have about half the species of Oikopleurids (Galt and Flood 1998). All luminescent species have oral glands (or prebuccal glands) and subchordal cells (Figures 9.21 and 9.36); their presence is considered to be an indicator of a luminescent species (Galt and Flood 1998).

Luminescence in the tadpole itself is restricted to the trunk region, in particular, the house rudiments. It originates in "inclusion bodies," bean- or kidney-shaped structures that form species-specific patterns in the house rudiment present on the trunk of the tadpole and later in the expanded house itself (Figure 9.36) (Galt and Flood 1998). Within each of the inclusion bodies are tiny granular luminescent sites termed "lumisomes" that are the source of the luminescence. It is important to note that inclusion bodies are secreted structures that are not an interactive part of the animal's tissues as are the photophores of Crustacea or cephalopods. At this point, it is believed that the oral glands are involved in the secretory process, but a complete picture still awaits further research (Galt and Flood 1998). Once incorporated into the house rudiment, inclusion bodies function autonomously, responding to mechanical stimulation with a luminescent flash. Apparently, each lumisome is only capable of flashing once before disintegrating, giving the inclusion bodies

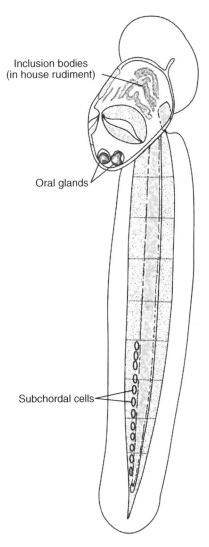

Figure 9.36 Oikopleurid appendicularian illustrating the location of inclusion bodies, oral glands, and subchordal cells. *Source:* Galt and Flood (1998), figure 13.1 (p.216). Reproduced with the permission of Oxford University Press.

and their luminescent capabilities, like the larvacean house itself, a limited life (Galt and Flood 1998).

Interestingly, the luminescence of discarded Appendicularian houses attracts consumer species, adding to their value in the trophic pyramid. The luminescence comes from the houses themselves, non-ingested phytoplankton cells left in the filters, and whatever biological light is produced by the microbial communities that grow on the dissolved and particulate organic matter left behind.

Salps and doliolids have been historically considered nonluminescent members of the Thaliacea (Haddock et al. 2010; Widder 2010) and this is still the case for the salps. However, recent in-situ and laboratory-based studies on a unique mesopelagic doliolid from the Monterey Canyon, *Doliolula equus*, revealed a bioluminescent glow originating in the tunic of the species (Robison et al. 2005), confirming luminescence in the Doliolida. Further observations have revealed additional luminescent doliolids in the mesopelagic depths of the Monterey Canyon (Haddock et al. 2010).

Predators, Parasites and Other Interactions

Tunicates occupy a unique niche in the marine macrozooplankton. Under the right conditions, they are capable of forming enormous blooms, dominating the zooplankton community, and yet as prey items they are of limited food value, with a watery composition (95% water) and a mucoid, gelatinous body. It is tempting to assume that much of their biomass sinks to the bottom, but in fact they contribute to the diet of a large number of pelagic species representing virtually all marine animal taxa from the Cnidaria to the vertebrates. In addition, the larger thaliacean taxa, particularly the salps, serve as hosts to a variety of commensals and parasites that use them for a free ride and a free meal (Harbison 1998). A comprehensive list of species that exploit thaliaceans for food or shelter is reported in Harbison (1998) and metazoan taxa are briefly summarized in Table 9.5. In many cases, Thaliacea are occasional prey that do not form a large part of the diet, but a number of taxa specialize in them.

At least ten protozoan species are associated with salps and are considered to be parasitic, with most found in the gut. Classification in the protozoa is currently in a state of flux, but up until quite recently, the taxa would have been classified as dinoflagellates, ciliates, and sporozoans. Colonial radiolarians (now radiozoans) have been observed to prey upon the salp *Thalia democratica* (Swanberg and Anderson 1985).

A broad array of species exploits thaliaceans as prey (Table 9.5), but as Harbison (1998) observed, the reported interactions are likely only a small fraction of those that actually occur. During bloom conditions, when salps overwhelmingly dominate the local marine fauna, some species will temporarily alter their diet to exploit the available resource. For example, during a salp bloom encountered in the Scotia-Weddell Sea region of the Southern Ocean, several flighted seabirds that normally prey on krill and lanternfishes were observed to feed on salp stomachs as well (Ainley et al. 1992). Further, the dominant Antarctic salp, *Salpa thompsoni*, which blooms often in the coastal Antarctic, is noteworthy in the diets of several Antarctic endemic fishes (Nototheniodei) (Harbison 1998), most of which are benthic in habit, so at least some of the sinking salps directly benefit the benthic and benthopelagic communities!

Appendicularia are also preyed upon by a variety of taxa, as are their discarded houses (Fenaux et al. 1990; Gorsky and Fenaux 1998). Invertebrates known to feed on them include

Table 9.5 Species numbers for taxa confirmed to prey on thaliaceans. Species are considered specialists when they feed heavily on gelatinous zooplankton including thaliaceans. Data from Harbison (1998). Numbers for seabird predators include data from Ainley et al. (1992).

Taxon	Interaction	Prey	Number of species	Number of specialists
Phylum Cnidaria				
Class Hydrozoa	Predation	Salps and doliolids	13	
Class Scyphozoa	Predation	Salps and doliolids	1	
Phylum Ctenophora				
Class Tentaculata	Predation	Salps	4	
Class Nuda	Predation	Salps	1	
Phylum Mollusca				
Class Gastropoda	Predation	Salps	5	
Class Cephalopoda	Predation	Salps, pyrosomes, doliolids	2	
Phylum Arthropoda				
Subphylum Crustacea				
Subclass Copepoda	Parasitism and predation	Salps and doliolids	10	
Class Malacostraca	Parasitism and predation	Salps	11 genera	
Phylum Chordata				
Subphylum Vertebrata				
Superclass Gnathostomata				
Superclass Pisces				
Class Elasmobranchii	Predation	Salps, pyrosomes	3	
Class Holocephali	Predation	Salps, pyrosomes	3	2
Class Actinopterygii	Predation	Salps, pyrosomes, doliolids	202	71
Superclass Tetrapoda				
Class Reptilia	Predation	Salps, pyrosomes	7	3
Class Aves	Predation	Salps	10	

the Foraminifera, Chaetognatha, hydro- and scyphomedusae, Ctenophora, and Copepoda. They are important in the diets of several fish species as well, particularly during their larval stages (Ikewaki and Tanaka 1993). Families reported to feed on larvaceans include the Scombridae (mackerels and tunas), Clupeidae (sardines and herrings), Gadidae (cods), Sparidae (porgies and sea breams), Argentinidae (herring smelts), Pleuronectidae (right-eye flounders), Salmonidae (salmon and trout), Bathylagidae (deep-sea smelts), and Myctophidae (lanternfishes) (Gorsky and Fenaux 1998). Because of the fact that appendicularia are quite small and, like the rest of the pelagic tunicates, can form dense blooms, they are particularly beneficial to fish larvae.

Geographic and Vertical Distributions

Zoogeography of the pelagic tunicates has received the benefit of a large number of studies, particularly those summarized in Van Soest (1975, 1979, 1998), Van der Spoel and Heyman (1983), Fenaux et al. (1990, 1998), Esnal (1999a, b), and Esnal and Daponte (1999a, b). Though all but the pyrosomes have polar representatives, both the Thaliacea and Appendicularia show the highest number of species in warm temperate and tropical regions, with many considered to be warm-water cosmopolitans (Tables 9.6 and 9.7). The group that is best represented at polar latitudes is the Appendicularia with 26 species followed by the salps with six. Though salps are not as speciose as the larvaceans in polar waters, *Salpa thompsoni* often dominates the Southern Ocean macrozooplankton community with massive spring blooms (e.g. Lancraft et al. 1989).

Salps are the most capable swimmers of the pelagic tunicates and exhibit the greatest ability to vertically migrate. Those considered capable of strong migrations are all in the genus *Salpa* (Andersen 1998) and include *Salpa aspera*, *Salpa thompsoni*, and *Salpa fusiformis*. Most typical of the weak migrators are *Cyclosalpa bakeri*, *Ihlea punctata*, *Thalia democratica*, *Pegea confoederata*, *Salpa cylindrical*, and *S. maxima* (Andersen 1998). The most impressive example of salp migration was reported in Wiebe et al. (1979): *Salpa aspera* exhibited a diel migration of 800 m (800 m to surface waters) off the coast of the U.S. in the slope waters of the western North Atlantic.

It is worth noting that salps may not behave uniformly from place to place or even from study to study in the same place. For example, in a study by Franqueville (1971) in the northwestern Mediterranean, *Salpa fusiformis* exhibited strong diel migratory behavior, rising from 300 to 800 m during the day to surface waters at night. In the same general location, a later study showed part of the *S. fusiformis* population residing in epipelagic waters (0–200 m) and part at greater depth (400–600 m) with no directed diel movement (Laval et al. 1992; Andersen 1998). In a very different location, a developing bloom of *Salpa thompsoni* in the Scotia-Weddell Sea region of the Antarctic (vic 60°S 40°W) began with the bulk of the population in epipelagic waters, spreading throughout the upper 1000 m during the course of the study and showing little migratory behavior. A follow-up study in nonbloom conditions at the same longitude, but further south in the Weddell Sea, showed evidence of weak migratory behavior: peak population density at 50–115 m during the day vs. 0–50 at night (Lancraft et al. 1989). In contrast, Casareto and Nemoto (1986) observed a more pronounced diel migratory pattern in *S. thompsoni* (200–300 m during the day vs. 0–75 m at night) at a different location in the Southern Ocean.

It is tempting to conclude that migratory behavior for at least some species of salps is dictated by conditions in the field. The salp bloom described in Lancraft et al. (1989) coincided with a *Phaeocystis* bloom, a marine alga that forms massive colonies accompanied by the production of a polysaccharide gel that is quite capable of clogging the pharyngeal baskets of salps (cf. Harbison et al. 1986), disrupting their ability to swim and feed and perhaps forcing them deeper in

Table 9.6 Regional distribution of Thaliacean species enumerated by genus. Total number of species in each genus given in parentheses. The only Arctic species listed is *Brooksia rostrata*; the other polar species are Antarctic. Species are counted wherever they appear: cosmopolitan species like *Salpa fusiformis*, *Pyrosoma atlanticum*, and *Doliolina krohni* are counted in all three ocean basins as both temperate and tropical. Data from Van Soest (1998).

Taxa	Polar[a]	Atlantic Ocean basin		Indian Ocean basin		Pacific Ocean basin		Red Sea	Mediterranean Sea
		Temperate[b]	Tropical[c]	Temperate[b]	Tropical[c]	Temperate[b]	Tropical[c]		
Salpida									
Salpa (7)	2	5	5	4	5	5	4	2	2
Weelia (1)		1	1	1	1	1	1	1	
Ritteriella (3)		2	1	3	3	2	3		
Metcalfina (1)		1		1	1	1	1		
Ihlea (3)	2	2	1	2	1	2	1		
Thetys (1)		1	1	1	1	1	1		1
Iasis (1)	1	1	1	1	1	1	1	1	1
Pegea (3)		3	1	2	2	2		1	2
Traustetedtia (1)		1	1	1	1	1	1		
Thalia (7)		4	3	4	4	3	6	4	2
Brooksia (2)	1	2	1	1	2	1	2	1	1
Cyclosalpa (11)		6	6	5	8	8	9		3
Helicosalpa (3)		1	1	1	3	2	3		1
Total species: 44	6	30	23	27	33	30	33	10	13
Pyrosomatida									
Pyrosoma (4)		3	2	4	3	2	2		1
Pyrosomella (2)				1	2	1	2		
Pyrostremma (2)		2	1	1	2	2	2		
Total species: 8	0	5	3	6	7	5	6	0	1

(*Continued*)

Table 9.6 (Continued)

Taxa	Polar[a]	Atlantic Ocean basin		Indian Ocean basin		Pacific Ocean basin		Red Sea	Mediterranean Sea
		Temperate[b]	Tropical[c]	Temperate[b]	Tropical[c]	Temperate[b]	Tropical[c]		
Doliolida									
Doliolum (2)		2	2	1	2	1	2	2	2
Dolioletta (5)		1	1	3	5	1	2	2	1
Doliolina (9)	2	3	3	4	4		5	2	3
Doliopsis (1)		1							1
Doliolopsoides (3)			2		1		1		1
Total species: 20	2	7	8	8	12	2	10	6	8

Regions
[a] Polar = north of 75°N or south of 50°S; circumglobal.
[b] Temperate = 20–50°N, or 20–50°S, or both.
[c] Tropical = 20°N–20°S.

Table 9.7 Regional distribution of Appendicularian species enumerated by family. Total number of species in each family in parentheses. Polar regions are considered to be circumglobal with Arctic species being those found north of 75°N and Antarctic species south of 50°S. Data from Fenaux et al. (1998).

Taxa	Atlantic Ocean	Pacific Ocean	Indian Ocean and Red Sea	Mediterranean Sea	Arctic Ocean	Antarctic Ocean
Oikopleuridae (37)	26	23	17	17	3	9
Fritillariidae (30)	19	19	22	21	2	11
Kowalevskiidae (2)	2	1	1	2	0	1
Total species in region	47	43	40	39	5	21

the water column. Multiple studies on *Salpa fusiformis* showing a suite of different day-night distribution patterns (Andersen 1998) also support the idea of a flexible response to differing field conditions, even in species known to vertically migrate.

Pyrosoma atlanticum, though seemingly incapable of doing so with its limited locomotory ability, is a well-documented vertical migrator, ranging in migration distance from 515 to 650 m in the waters of the NW Mediterranean and NE Atlantic on a diel basis (Franqueville 1971; Angel 1989; Andersen et al. 1992; Andersen 1998). Since its locomotory mechanism provides no way of adjusting its swimming direction, it has been assumed that it achieves its up and down movement by adjusting its buoyancy (see Section "Locomotion and Buoyancy").

Mesoscale (tens to hundreds of km) studies of doliolid distributions suggest that they are most numerous in the upper 100 m of neritic environments. They are capable of forming massive swarms, at times reaching densities of 1000–4500 m^{-3} and covering hundreds of square km (Deibel 1998b). However, their small size and prodigious reproductive abilities can also result in very localized areas (patches) of high population density on the horizontal and vertical scale, detectable only with manned submersibles or optical samplers such as the VPR (Video Plankton Recorder) (Davis et al. 1992). Paffenhöfer et al. (1991) detected changes in abundance of several hundred m^{-3} in a vertical distance of only 2 m (Deibel 1998b). High concentrations were associated with the pycnocline at the base of the upper mixed layer.

Appendicularia, because of their diminutive size and rapid reproductive abilities, share some distributional characteristics with the doliolids. On the mesoscale, they are found in both neritic and open-ocean environments but are most abundant in the neritic (Esnal 1999a). Globally, they are conspicuous members of the zooplankton from the equator to the poles in all ocean basins (Table 9.7). Spectacular blooms of larvaceans have been reported, with levels of *Oikopleura dioica* reaching 25 600 ind m^{-3} in the upper few meters of the water column in Saanich Inlet, British Columbia, and extending for several km (Seki 1973). Uye and Ichino (1995) reported still higher levels (53 200 ind m^{-3}) in an inlet of the Sea of Japan during July of 1987.

Though appendicularians do not reproduce asexually, they have a very short generation time: as little as 1 day^{-1} for *Oikopleura dioica* at high (27–29 °C) temperatures (Esnal and Castro 1985; Hopcroft and Roff 1995; Uye and Ichino 1995). Even at temperatures more typical of a temperate system, 10 and 15 °C, generation times for *Oikopleura dioica* were estimated to be 15.8–7.8 days, respectively (Uye and Ichino 1995). Thus, they can respond rapidly to favorable conditions with explosive growth (Fenaux et al. 1998).

Appendicularia reach their greatest densities in the epipelagic zone. In a vertically stratified (0–500 m) sampling series collected monthly over two years with closing nets, Fenaux (1963, 1968a, b) reported that 70% of specimens collected were captured in the upper 80 m. In contrast, the number of species collected showed a maximum in all three families at depths of 75–150 m (Fenaux et al. 1998). The general trend was fewer species with greater individual abundance at shallow depths and higher diversity with lower individual abundance at mid-depths (Fenaux et al. 1998). Like doliolids, larvaceans can form layers of high abundance 1–2 m thick in the fine scale (meters to tens of meters, Haury et al. 1978) that have been observed by divers but would be very difficult to detect using nets.

Submersible observations at depth (Barham 1979; Fenaux and Youngbluth 1990; Hamner and Robison 1992) introduced the mesopelagic species of Larvacea to the zooplankton community. Now larvaceans have been reported at mesopelagic and bathypelagic depths from several locations in the global ocean, sometimes in high densities (Fenaux et al. 1998).

References

Ainley, D.G., Ribic, C.A., and Fraser, W.R. (1992). Does prey preference affect habitat choice in Antarctic seabirds? *Marine Ecology Progress Series* 90: 207–221.

Alldredge, A.L. (1976a). Discarded appendicularian houses as sources of food, surface habitats and particulate organic matter in planktonic environments. *Limnology and Oceanography* 21: 14–23.

Alldredge, A.L. (1976b). Field behavior and adaptive strategies of appendicularians (Chordata: Tunicata). *Marine Biology* 38: 29–39.

Alldredge, A.L. (1976c). Appendicularians. *Scientific American* 235: 94–100.

Alldredge, A.L. (1977). House morphology and mechanisms of feeding in the Oikopleuridae (Tunicata, Appendicularia). *Journal of Zoology (London)* 181: 175–188.

Alldredge, A.L. (1981). The impact of appendicularian grazing on natural food concentrations in-situ. *Limnology and Oceanography* 26: 247–257.

Alldredge, A.L. and Madin, L.P. (1982). Pelagic tunicates, unique herbivores in the marine plankton. *Bioscience* 32: 655–663.

Andersen, V. (1998). Salp and pyrosomid blooms and their importance to biogeochemical cycles. In: *The Biology of Pelagic Tunicates* (ed. Q. Bone). Oxford: Oxford University Press.

Andersen, V. and Sardou, J. (1994). *Pyrosoma atlanticum* (Tunicata, Thaliacea): diel migration and vertical distribution as a function of colony size. *Journal of Plankton Research* 16: 337–349.

Andersen, V., Sardou, J., and Nival, P. (1992). The diel migrations and vertical distributions of zooplankton and micronekton in the Northwestern Mediterranean Sea. 2. Siphonophores, hydromedusae and pyrosomids. *Journal of Plankton Research* 14: 1155–1169.

Anderson, P.A.V., Bone, Q., Mackie, G.O., and Singla, C.L. (1979). Epithelial conduction in salps. II. The role of nervous and non-nervous conduction system interactions in the control of locomotion. *Journal of Experimental Biology* 80: 241–250.

Angel, M.V. (1989). Vertical profiles of pelagic communities in the vicinity of the Azores Front and their implications to deep ocean ecology. *Progress in Oceanography* 22: 1–46.

Atkinson, L.P., Paffenhöfer, G.-A., and Dunstan, W.M. (1978). The chemical and biological effect of a Gulf Stream intrusion off St. Augustine, Florida. *Bulletin of Marine Science* 28: 667–679.

Azam, F., Fenchel, T., Field, J.G. et al. (1983). The ecological role of microbes in the sea. *Marine Ecology Progress Series* 10: 257–263.

Barham, E.G. (1979). Giant larvacean houses: observations from deep submersibles. *Science* 205: 1129–1131.

Barnes, R.D. (1974). *Invertebrate Zoology*. Philadelphia: W.B. Saunders.

Bathmann, U.V. (1988). Mass occurrence of Salpa fusiformis in the spring of 1984 off Ireland: implications for sediment processes. *Marine Biology* 97: 127–135.

Bedo, A.W., Acuna, J.-L., Robins, D., and Harris, R.P. (1993). Grazing in the micronic and sub-micronic particle size range: the case of Oikopleura dioica (Appendicularia). *Bulletin of Marine Science* 53: 2–14.

Bennett, G. (1834). *Wanderings in New South Wales, Batavia, Pedir Coast, Singapore, and China Being the Journal of a Naturalist in Those Countries, During 1832, 1833, and 1834*. London: R. Bentley.

Berner, L.D. (1967). Distribution atlas of Thaliacea in the California Current region. *California Cooperative Oceanic Fisheries Investigations Atlas* 8: 1–322.

Berrill, N.J. (1950). *The Tunicata, with an Account of the British Species*. London: The Ray Society.

Bidigare, R.R. and Biggs, D.C. (1980). The role of sulfate exclusion in buoyancy maintenance by siphonophores and other gelatinous zooplankton. *Comparative Biochemistry and Physiology* 66A: 467–471.

Binet, D. (1976). Contribution a l'ecologie de quelques taxons du zooplancton de Cote d'Ivoire II. Colioles, salpes, appendiculaires. *Documents scientifiques du centre de recherches Oceanographiques d'Abidjan* 7: 45–61.

Bolles Lee, A. (1891). On a little known sense organ in salps. *Quarterly Journal of Microscopical Science* 32: 89–96.

Bollner, T., Holmberg, K., and Olsson, R. (1986). A rostral sensory mechanism in Oikopleura dioica (Appendicularia). *Acta Zoologica (Stockholm)* 67: 235–241.

Bone, Q. (1959). Observations upon the role of the nervous systems of pelagic tunicates. *Quarterly Journal of Microscopical Science* 100: 167–181.

Bone, Q. (1998a). Introduction. In: *The Biology of Pelagic Tunicates* (ed. Q. Bone). Oxford: Oxford University Press.

Bone, Q. (1998b). Locomotion, locomotor muscles, and buoyancy. In: *The Biology of Pelagic Tunicates* (ed. Q. Bone). Oxford: Oxford University Press.

Bone, Q. (1998c). Nervous system, sense organs, and excitable epithelia. In: *The Biology of Pelagic Tunicates* (ed. Q. Bone). Oxford: Oxford University Press.

Bone, Q. and Mackie, G.O. (1975). Skin impulses and locomotion in Oikopleura (Tunicata: Larvacea). *Biological Bulletin* 149: 267–286.

Bone, Q. and Mackie, G.O. (1977). Ciliary arrest potentials, locomotion and skin impulses in Doliolum (Tunicata: Thaliacea). *Rivista di biologia normale e patologica* 3: 181–191.

Bone, Q. and Trueman, J.R. (1982). Jet propulsion of the calycophora siphonophores Chelophyes and Abylopsis. *Journal of the Marine Biological Association of the United Kingdom* 62: 263–276.

Bone, Q. and Trueman, E.R. (1983). Jet propulsion in salps. *Journal of Zoology (London)* 201: 481–506.

Bone, Q. and Trueman, E.R. (1984). Jet propulsion in Doliolum. *Journal of Experimental Marine Biology and Ecology* 76: 105–118.

Bone, Q., Gorsky, G., and Pulsford, A.L. (1979). On the structure and behavior of Fritillaria. *Journal of the Marine Biological Association of the United Kingdom* 59: 399–411.

Bone, Q., Braconnot, J.-C., and Ryan, K.P. (1991). On the pharyngeal feeding filter of the salp *Pegea confoederata* (Tunicata: Thaliacea). *Acta Zoologica (Stockholm)* 72: 55–60.

Bone, Q., Braconnot, J.-C., and Carre, C. (1997a). On the heart and circulation in Doliolum (Tunicata, Thaliacea). *Scientia Marina* 61: 189–194.

Bone, Q., Braconnot, J.-C., and Carre, C. (1997b). On the filter-feeding of Doliolum (Tunicata, Thaliacea). *Journal of Experimental Marine Biology and Ecology* 214: 179–193.

Bowlby, M.R., Widder, E.A., and Case, J.F. (1990). Patterns of stimulated bioluminescence in two pyrosomes (Tunicata: Pyrosomatidae). *Biological Bulletin* 179: 340–350.

Braconnot, J.-C. (1971a). Contribution a l'etude des stades successifs dans le cycle des Tunicieres pelagiques Dolioloides. II. Les stades phorozoide et gonozoide des Doliolides. *Archives de Zoologie Experimentale et Generale* 112: 5–31.

Braconnot, J.-C., Choe, S.-M., and Nival, P. (1988). La croissance et le developpement de Salpa fusiformis Cuvier (Tunicata, Thaliacea). *Annales de l'Institut Oceanographique Paris* 64: 101–114.

Brett, J.R. and Groves, T.D.D. (1979). Physiological energetics. In: *Fish Physiology*, vol. *8* (eds. W.S. Hoar, D.J. Randall and J.R. Brett). New York: Academic Press.

Brien, P. (1948). Potentialites totales, potentialites restreintes et correlatives d'une meme ebauche, a propos de la regeneration de l'epicarde chez les tuniciers. *Bulletins de l'Academie Belgique de la Classe des Sciences* 34 (5): 724–734.

Bronn, H.G. (1862). *Die Klassen und Ordnungen der Weichtiere (Malacozoa)*. Leipzig, Heidelberg: C.F. Winter.

Brooks, W.K. (1893). *The Genus Salpa*. Memoirs from the Biological Laboratory of the Johns Hopkins University II.

Browne, R. (1756). *A Civil and Natural History of Jamaica*. White.

Brusca, R.C. and Brusca, G.J. (1990). *Invertebrates*. Sunderland: Sinauer Associates.

Brusca, R.C. and Brusca, G.J. (2003). *Invertebrates*, 2e. Sinauer: Sunderland.

Brusca, R.C., Moore, W., and Shuster, S.M. (2016). *Invertebrates*, 3e. Sinauer Associates: Sunderland.

Bullock, T.H. and Horridge, G.A. (1965). *Structure and Function in the Nervous Systems of Invertebrates*, vol. *II*. San Francisco: W.H. Freeman and Company.

Casareto, B.E. and Nemoto, T. (1986). Salps of the Southern Ocean (Australian Sector) during the 1983–1984 summer, with special reference to the species Salpa thompsoni Foxton 1961. *Proceedings of the 7th Symposium on Polar Biology, National Institute for Polar Biology* 40: 221–239.

Chamisso, A.D. and Eysenhardt, C.G. (1821). De animalibus quibusdamde classe vermium linneana, in circumnavigatione terrae, auspicante comite N. Romanzoff, duce Ottone de Kotzebue, annis 1815-1818, peracta, observatis. *Nova Acta Physico-medica Academiae Caesareae Leopoldino-Carolinae Naturae Curiosorum* 10: 343–374.

Chapman, A.D. (2009). *Numbers of lving species in Australia and the world*, 2e. Canberra: Australian Biological Resources Study.

Crocker, K.M., Alldredge, A.L., and Steinberg, D.K. (1991). Feeding rates of the doliolid, *Dolioletta gegenbauri*, on diatoms and bacteria. *Journal of Plankton Research* 13: 77–82.

Cuvier, G. (1804). Memoire sur les Thalides et sur les Biphores. *Annales de la Musee d'Histoire Naturelle de Paris* 4: 360–382.

Davis, C.S., Gallager, S.M., and Solow, A.R. (1992). Microaggregations of oceanic plankton observed by towed video microscopy. *Science* 257: 230–232.

Davoll, P.J. and Silver, M.W. (1986). Marine snow aggregates: life history sequence and microbial community of abandoned larvacean houses from Monterey Bay, California. *Marine Ecology Progress Series* 33: 111–120.

Deibel, D. (1980). Feeding, growth and swarm dynamics of neritic tunicates from the Georgia Bight. PhD, University of Georgia.

Deibel, D. (1986). Feeding mechanism and house of the appendicularian *Oikopleura vanhoeffeni*. *Marine Biology* 93: 429–436.

Deibel, D. (1998a). Feeding and metabolism of Appendicularia. In: *The Biology of Pelagic Tunicates* (ed. Q. Bone). Oxford: Oxford University Press.

Deibel, D. (1998b). The abundance, distribution, and ecological impact of doliolids. In: *The Biology of Pelagic Tunicates* (ed. Q. Bone). Oxford: Oxford University Press.

Deibel, D. and Lee, S.H. (1992). Retention efficiency of sub-micrometer particles by the pharyngeal filter of the pelagic tunicate *Oikopleura vanhoeffeni*. *Marine Ecology Progress Series* 81: 25–30.

Denton, E.J. and Shaw, T.I. (1961). The buoyancy of gelatinous marine animals. *Journal of Physiology (London)* 161: 14–15P.

Esnal, G.B. (1999a). Appendicularia. In: *South Atlantic Zooplankton* (ed. D. Boltovskoy). Leiden: Backhuys Publishers.

Esnal, G.B. (1999b). Pyrosomatida. In: *South Atlantic Zooplankton* (ed. D. Boltovskoy). Leiden: Backhuys Publishers.

Esnal, G.B. and Castro, R.J. (1985). *Oikopleura albicans* (Leuckart 1853), un estudio biometrico (Tunicata: Appendicularia). *Neotropica, La Plata* 31: 111–117.

Esnal, G.B. and Daponte, M.C. (1999a). Doliolida. In: *South Atlantic Zooplankton* (ed. D. Boltovskoy). Leiden: Backhuys Publishers.

Esnal, G.B. and Daponte, M.C. (1999b). Salpida. In: *South Atlantic Zooplankton* (ed. D. Boltovskoy). Leiden: Backhuys Publishers.

Fedele, M. (1923). Le attivita dinamiche ed i rapporti nervosi nella vita de Doliolidae. *Pubblicazione della Stazione Zoologica di Napoli* 4: 129–240.

Fedele, M. (1933a). Sul complesso della funzioni che intervengono nel meccanismo ingestivo dei Salpidae. *Rendiconti del Accademia dei Lincei* 27: 241–245.

Fedele, M. (1933b). Sul ritmo muscolare somatico delle salpe. *Bolletino della Societa Italiano di Biologia Sperimentale* 8: 475–478.

Fedele, M. (1938). Sistema nervoso degli "Ascidiacea" nel piano di organizzazione dei cordati. *Rediconti del Accademia dei Lincei* 27 (6): 370–376.

Fenaux, R. (1963). Ecologie et biologie des Appendiculaires mediterraneans (Villefranche-sur-mer). *Vie et Milieu* supplement 16: 1–142.

Fenaux, R. (1968a). Appendiculaires. (Campagne de la Calypso au large des cotes atlantiques de l'Amerique du Sud (1961-1962) (premiere partie)). *Annales de l'Institut Oceanographique Paris* 45: 33–46.

Fenaux, R. (1968b). Quelques aspects de la distribution verticale chez les Appendiculaires en Mediterranee. *Cahiers de Biologie Marine* 9: 23–29.

Fenaux, R. (1985). Rhythm of secretion of oikopleurid's houses. *Bulletin of Marine Science* 37: 498–503.

Fenaux, R. (1998). Anatomy and functional morphology of the Appendicularia. In: *The Biology of Pelagic Tunicates* (ed. Q. Bone). Oxford: Oxford University Press.

Fenaux, R. and Youngbluth, M.J. (1990). A new mesopelagic appendicularian, *Mesochordaeus bahamasi*, gen. nov., sp. nov. *Journal of the Marine Biological Association of the United Kingdom* 70: 755–760.

Fenaux, R., Galt, C.P., and Carpine-Lancre, J. (1990). Bibiographie des Appendiculaires. *Memoires de l'institut Oceanographique, Monaco* 15: 1–129.

Fenaux, R., Bone, Q., and Deibel, D. (1998). Appendicularian distribution and zoogeography. In: *The Biology of the Pelagic Tunicates* (ed. Q. Bone). Oxford: Oxford University Press.

Fenchel, T. (1988). Marine plankton food chains. *Annual Review of Ecology and Systematics* 19: 19–38.

Field, K.G., Olsen, G.J., Lane, D.J. et al. (1988). Molecular phylogeny of the animal kingdom. *Science* 239: 748–753.

Flood, P.R. (1991). Architecture of, and water circulation and flow-rate in, the house of the planktonic tunicate *Oikopleura labradoriensis*. *Marine Biology* 111: 95–111.

Flood, P.R. and Deibel, D. (1998). The appendicularian house. In: *The Biology of Pelagic Tunicates* (ed. Q. Bone). Oxford: Oxford University Press.

Fol, H. (1872). Etudes sur les appendiculaires du Detroit de Messine. *Memoires de la Societe de Physique et d'Histoire naturelle de Geneve* 21: 445–499.

Forsskaal, P. (1775). *Descriptiones Animalium, quae in itinere orientali observavit. Post mortem auctoris edidit C. Niebuhr.* Moller Haauniae.

Foxton, P. (1961). Salpa fusiformis Cuvier and related species. *Discovery Reports* 32: 1–32.

Foxton, P. (1966). The distribution and life history of Salpa thompsoni-Foxton with observations on a related species, Salpa gerlachei-Foxton.... *Discovery Reports* 34: 1–116.

Franqueville, C. (1971). Macroplancton profond (invertebres) de la Mediterranee nord-occidentale. *Tethys* 3: 11–56.

Frost, B.W. (1972). Effects of size and concentration of food particles on the feeding behavior of the marine planktonic copepod *Calanus pacificus*. *Limnology and Oceanography* 17: 805–815.

Frost, B.W., Landry, M.R., and Hassett, R.P. (1983). Feeding behavior of the large calanoid copepods *Neocalanus cristatus* and *N. plumchrus* from the subantarctic Pacific Ocean. *Deep-Sea Research* 30: 1–13.

Galt, C.P. and Flood, P.R. (1998). Bioluminescence in the Appendicularia. In: *The Biology of Pelagic Tunicates* (ed. Q. Bone). Oxford: Oxford University Press.

Gilmer, R.W. (1974). Some aspects of feeding in the thecosomatous pteropod molluscs. *Journal of Experimental Marine Biology and Ecology* 15: 127–144.

Godeaux, J. (1990). Urochordata-Thaliacea. In: *Reproductive Biology of Invertebrates* (eds. K.G. Adiyodi and R.G. Adiyodi). New Delhi: Oxford and IBH Publishing.

Godeaux, J. (1998). The relationships and systematics of the Thaliacea, with keys for identification In: *The Biology of Pelagic Tunicates* (ed. Q. Bone). Oxford: Oxford University Press.

Godeaux, J. (2003). History and revised classification of the order Cyclomyaria (Tunicata, Thaliacea, Doliolida). *Bulletin de l'institut royal des sciences naturelles de Belgique, Biologie* 73: 191–222.

Godeaux, J., Bone, Q., and Braconnot, J.-C. (1998). Anatomy of Thaliacea. In: *The Biology of Pelagic Tunicates* (ed. Q. Bone). Oxford: Oxford University Press.

Goodbody, I. (1974). The physiology of ascidians. *Advances in Marine Biology* 12: 1–149.

Gorman, A.L.F., Mcreynolds, J.S., and Barnes, S.N. (1971). Photoreceptors in primitive chordates: fine structure, hyperpolarizing potentials and evolution. *Science* 172: 1052–1054.

Gorsky, G. and Fenaux, R. (1998). The role of Appendicularia in marine food webs. In: *The Biology of Pelagic Tunicates* (ed. Q. Bone). Oxford: Oxford University Press.

Gould, S.J. (1977). *Ontogeny and Phylogeny*. Cambridge: Harvard University Press.

Haddock, S.H.D., Moline, M.A., and Case, J.F. (2010). Bioluminescence in the Sea. *Annual Review of Marine Science* 2: 443–493.

Hamner, W.M. and Robison, B.H. (1992). In situ observations of giant appendicularians in Monterey Bay. *Deep-Sea Research* 39: 1299–1313.

Harbison, G.R. (1998). The parasites and predators of Thaliacea. In: *The Biology of the Pelagic Tunicates* (ed. Q. Bone). Oxford: Oxford University Press.

Harbison, G.R. and Gilmer, R.W. (1976). The feeding rates of the pelagic tunicate *Pegea confoederata* and two other salps. *Limnology and Oceanography* 21: 517–528.

Harbison, G.R. and McAlister, V.L. (1979). The filter-feeding rates and particle retention efficiencies of three species of Cyclosalpa (Tunicata: Thaliacea). *Limnology and Oceanography* 24: 875–892.

Harbison, G.R., McAlister, V.L., and Gilmer, R.W. (1986). The response of the salp, *Pegea confoederata*, to high levels of particulate material: starvation in the midst of plenty. *Limnology and Oceanography* 31: 371–382.

Harris, R.P. and Paffenhöfer, G.-A. (1976). Feeding, growth and reproduction of the marine planktonic copepod *Temora longicornis* Muller. *Journal of the Marine Biological Association of the United Kingdom* 56: 675–690.

Haury, L.R., Mcgowan, J.A., and Wiebe, P.W. (1978). Patterns and processes in the time-space scales of plankton distributions. In: *Spatial Pattern in Plankton Communities* (ed. J.H. Steele). New York: Plenum Press.

Heinbokel, J.F. (1978). Studies on the functional role of tintinnids in the Southern California Bight I. Grazing and growth rates in laboratory cultures. *Marine Biology* 47: 177–189.

Helfman, G.S., Collette, B.B., Facey, D.E., and Bowen, B.B. (2009). *The Diversity of Fishes*. Oxford: Wiley-Blackwel.

Hennig, W. (1979). *Phylogenetic systematics*, 3e. Urbana: University of Illinois Press.

Heron, A.C. (1975a). A new type of heart mechanism in the invertebrates. *Journal of the Marine Biological Association of the United Kingdom* 53: 425–428.

Heron, A.C. (1975b). Advantages of heart reversal in pelagic tunicates. *Journal of the Marine Biological Association of the United Kingdom* 55: 959–963.

Heron, A.C. (1976). A new type of excretory mechanism in the tunicates. *Marine Biology* 36: 191–197.

Heron, A.C. and Benham, E.E. (1984). Individual growth rates of salps in three populations. *Journal of Plankton Research* 6: 811–828.

Herring, P.J. (1978). Bioluminescence of invertebrates other than insects. In: *Bioluminescence in Action* (ed. P.J. Herring). London and New York: Academic Press.

Holland, L.Z. (2016). Tunicates. *Current Biology* 26: R146–R152.

Holland, L.Z. and Miller, R.L. (1994). Mechanism of internal fertilization in *Pegea socia* (Tunicata: Thaliacea), a salp with a solid oviduct. *Journal of Morphology* 219: 257–267.

Hopcroft, R.R. and Roff, J.C. (1995). Zooplankton growth rates: extraordinary production by the larvacean *Oikopleura dioica* in tropical waters. *Journal of Plankton Research* 17: 205–220.

Huxley, T.H. (1851). Observations upon the anatomy and physiology of Salpa and Pyrosoma. *Philosophical Transactions of the Royal Society of London* 141: 567–594.

Ikewaki, Y. and Tanaka, M. (1993). Feeding habits of the Japanese flounder (Paralichthys olivaceus) in the western part of Wakasa Bay, the Japan Sea. *Nippon Suisan Gakkaishi* 59: 951–956.

King, K.R., Hollibaugh, J.T., and azam, F. (1980). Predator-prey interactions between the larvacean Oikopleura dioica and bacterioplankton in enclosed water columns. *Marine Biology* 56: 49–57.

Korner, W.F. (1952). Untersuchungen uber die Gehausebildung bei Appendicularien (*Oikopleura dioica* Fol). *Zeitschrift für Wissenschaftliche Zoologie* 51: 613–628.

Lancraft, T.M., Torres, J.J., and Hopkins, T.L. (1989). Micronekton and macrozooplankton in the open waters near Antarctic ice edge zones (Ameriez 1983 and 1986). *Polar Biology* 9: 225–233.

Lasker, R. (1966). Feeding, growth, respiration and carbon utilization of a *Euphausiid crustacean*. *Journal of the Fisheries Research Board of Canada* 23: 1291–1317.

Laval, P., Braconnot, J.-C., and Linds DA Silva, N. (1992). Deep planktonic filter-feeders found in the aphotic zone with the Cyana submersible in the Ligurian Sea (NW Mediterranean). *Marine Ecology Progress Series* 79: 235–241.

LeBorgne and R. (1983). Note sur les proliferations de Thaliaces dans le Golfe de Guinee. *Oceanographie Tropicale* 18: 49–54.

Lohmann, H. (1909a). Die Gehause und Gallertblasen der Appendicularien und ihre Bedeutung fur die Eforschung des lebens en Meer. *Verhandlung Deutsche Zoologiche Gesellschaft* 19: 200–239.

Mackie, G.O. and Bone, Q. (1977). Locomotion and propagated skin impulses in salps (Tunicata: Thaliacea). *Biological Bulletin* 153: 180–197.

Mackie, G.O. and Bone, Q. (1978). Luminescence and associated effector activity in Pyrosoma (Tunicata: Pyrosomida). *Proceedings of the Royal Society of London B* 202: 483–495.

Madin, L.P. (1974). Field observations on the feeding behavior of salps (Tunicata: Thaliacea). *Marine Biology* 25: 143–147.

Madin, L.P. (1990). Aspects of jet propulsion in salps. *Canadian Journal of Zoology* 68: 765–777.

Madin, L.P. and Cetta, C.M. (1980). Estimating in situ grazing rates of salps from plant pigments in the gut. *Abstract, Aslo Third winter meeting, Seattle, WA*.

Madin, L.P. and Deibel, D. (1998). Feeding and energetics of Thaliacea. In: *The Biology of Pelagic Tunicates* (ed. Q. Bone). Oxford: Oxford University Press.

Madin, L.P., Cetta, C.M., and McAlister, V.L. (1981). Elemental and biochemical composition of salps (Tunicata: Thaliacea). *Marine Biology* 63: 217–226.

Martini, E. (1909a). Studien uber die Konstanz histologischer Elemente. I. *Oikopleura longicauda*. *Zeitschrift für Wissenschaftliche Zoologie* 92: 563–626.

Martini, E. (1909b). Studien uber die Konstanz histologischer Elemente. II. *Fritillaria pellucida*. *Zeitschrift für Wissenschaftliche Zoologie* 94: 81–170.

Metcalf, M.M. (1893). The eyes and subneural glands of Salpa. In: *The Genus Salpa* (ed. W.K. Brooks). Memoirs of the Biological Laboratory Johns Hopkins University.

Metcalf, M.M. and Hopkins, M.S. (1919). Pyrosoma. A taxonomic study based upon the collection of the U.S. Bureau of Fisheries and the U.S. National Museum. *Bulletin of the U.S. National Museum* 100: 195–276.

Metcalf, M.M. and Lentz-Johnson, M.E.G. (1905). The anatomy of the eyes and neural glands in the aggregated forms of *Cyclosalpa dolichosoma*-virgula and *Salpa punctata*. *Biological Bulletin* 9: 195–212.

Morris, R.J., Bone, Q., Head, R. et al. (1988). Role of salps in the flux of organic matter to the bottom of the Ligurian Sea. *Marine Biology* 97: 237–241.

Mullin, M.M. (1983). In situ measurement of filtering rates of the salp, Thalia democratica, on phytoplankton and bacteria. *Journal of Plankton Research* 5: 279–288.

Neumann, G. (1913). Salpae II: Cyclomyaria et Pyrosomida. *Das Tierreich* 40: 1–38.

Nishikawa, J. and Terazaki, M. (1995). Measurement of swimming speeds and pulse rates in salps using video equipment. *Bulletin of the Plankton Society of Japan* 41: 170–173.

Nival, P. and Nival, S. (1976). Particle retention efficiency of an herbivorous copepod, *Acartia clausi* (adult and copepodite stages) effects on grazing. *Limnology and Oceanography* 21: 24–38.

Nival, P., Braconnot, J.-C., Andersen, V. et al. (1985). Estimation de l'impact des salpes sur le phytoplancton en mer Ligure. *Rapports et Proces-verbaux des Reunions – Commission Internationale pour l'Exploration scientifique de la Mer Mediterranee* 29: 283–286.

Ohtsuka, S. and Kubo, N. (1991). Appendicularians and their houses as important food for some pelagic copepods. *Proceedings of the 4th International Conference on Copepoda, Bulletin of the Plankton Society Special Volume*: 535–551.

Olsson, R., Holmberg, K., and Lilliemarck, Y. (1990). Fine structure of the brain and brain nerves of *Oikopleura dioica* (Urochordata, Appendicularia). *Zoomorphology* 110: 1–7.

Paffenhöfer, G.-A. (1971). Grazing and ingestion rates of nauplii, copepodids, and adults of the marine planktonic copepod *Calanus helgolandicus*. *Marine Biology* 11: 286–298.

Paffenhöfer, G.-A. (1975). On the biology of Appendicularia of the southern North Sea. In: *Proceedings of the Tenth European Symposium on Marine Biology 2* (eds. G. Persoone and E. Jaspers). Wetteren, Belgium: Universa Press.

Paffenhöfer, G.-A. and Gibson, D.M. (1999). Determination of generation time and asexual fecundity of doliolids (Tunicata, Thaliacea). *Journal of Plankton Research* 21: 1183–1189.

Paffenhöfer, G.-A. and Harris, R.P. (1976). Feeding, growth and reproduction of the marine planktonic copepod *Pseudocalanus elongatus* Boeck. *Journal of the Marine Biological Association of the United Kingdom* 56: 327–344.

Paffenhöfer, G.-A. and Koster, M. (2011). From one to many: on the life cycle of *Dolioletta gegenbauri* Uljanin (Tunicata, Thaliacea). *Journal of Plankton Research* 33: 1139–1145.

Paffenhöfer, G.-A. and Lee, T.N. (1987). Development and persistence of patches of Thaliacea. *South African Journal of Marine Science* 5: 305–318.

Paffenhöfer, G.-A., Stewart, T.B., Youngbluth, M.J., and Bailey, T.G. (1991). High resolution vertical profiles of pelagic tunicates. *Journal of Plankton Research* 13: 971–981.

Parsons, T.R., Lebrasseur, R.J., and Fulton, J.D. (1967). Some observations on the dependence of zooplankton grazing on the cell size and concentration of phytoplankton blooms. *Journal of the Oceanographical Society of Japan* 23: 10–17.

Peron, F. (1804). Memoire sur le nouveau genre Pyrosoma. *Annales de la Musee d'Histoire Naturelle de Paris* 4: 437–446.

Quoy, J.R.C. and Gaimard, J.-P. (1827). Observations zoologiques faites a bord d l'Astrolabe, en Mai, 1826, dans le detroit de Gibraltar. *Annales des Sciences Naturelles* 10: 1–21.

Rassoulzadegan, F. and Etienne, M. (1981). Grazing rate of the tintinnid *Stenosemella ventricosa* (Clap. & Lachm.) Jorg. on the spectrum of the naturally occurring particulate matter from a Mediterranean neritic area. *Limnology and Oceanography* 26: 258–270.

Robertson, J.D. (1957). Osmotic and ionic regulation in aquatic invertebrates. In: *Recent Advances in Invertebrate Physiology* (ed. B.T. Scheer). Eugene: Oregon University Press.

Robison, B.H., Raskoff, K.A., and Sherlock, R.E. (2005). Ecological substrate in midwater: *Doliolula equus*, a new mesopelagic tunicate. *Journal of the Marine Biological Association of the United Kingdom* 85: 655–663.

Roger, C. (1982). Macroplancton et micronecton de l'Atlantique tropical. I. Biomasses et composition taxonomique. *Oceanographie Tropicale* 17: 85–96.

Sanders, N.K. and Childress, J.J. (1988). Ion replacement as a buoyancy mechanism in a pelagic deep-sea crustacean. *Journal of Experimental Biology* 138: 333–434.

Savigny, J.C. (1816). *Memoires sur les Animaux sans Vertebres*. Paris: Deterville.

Schnack, S.B. (1985). Feeding by *Euphausia superba* and copepod species in response to varying concentrations of phytoplankton. In: *Antarctic Nutrient Cycles and Food Webs* (eds. W.R. Siegfried, P.R. Condy and R.M. Laws). Berlins: Springer-Verlag.

Seki, H. (1973). Red tide of *Oikopleura* in Saanich Inlet. *La Mer (Bulletin de la Societe Franco-Japonaise d'Oceanographie)* 11: 673–711.

Silver, M.W. and Bruland, K.W. (1981). Differential feeding and fecal pellet composition of salps and pteropods, and the possible large-scale transport of particulate organic matter to the deep sea. *Marine Biology* 53: 249–255.

Steinberg, K.D., Silver, M.W., Pilskaln, H.C. et al. (1994). Midwater communities on pelagic detritus (giant larvacean houses) in Monterey Bay, California. *Limnology and Oceanography* 39: 1606–1620.

Swanberg, N.R. and Anderson, O.R. (1985). The nutrition of radioloarians: trophic activity of some Spumellaria. *Limnology and Oceanography* 30: 646–652.

Turner, J.T., Tester, P.A., and Fergusson, R.L. (1988). The marine cladooceran *Penilia avirostris* and the "microbial loop" of pelagic food webs. *Limnology and Oceanography* 33: 245–255.

Uye, S.I. and Ichino, S. (1995). Seasonal variations in abundance, size composition, biomass and production rate of *Oikopleura dioica* (Fol) (Tunicata: Appendicularia) in a temperate eutrophic inlet. *Journal of Experimental Marine Biology and Ecology* 189: 1–11.

Van Der Spoel, S. and Heyman, R.P. (1983). *A comparative atlas of zooplankton. Biological Patterns in the Oceans*. Utrecht: Bunge.

Van Soest, R.M.W. (1975). Zoogeography and speciation in the Salpidae (Tunicata: Thaliacea). *Beaufortia* 23: 181–215.

Van Soest, R.M.W. (1979). North-South diversity. In: *Zoogeography and Diversity of Plankton* (eds. S. Van Der Spoel and A.C. Pierrot-Bults). Utrecht: Bunge.

Van Soest, R.M.W. (1998). The cladistic biogeography of salps and pyrosomas. In: *The Biology of the Pelagic Tunicates* (ed. Q. Bone). Oxford: Oxford University Press.

Webb, P.W. (1975). Hydrodynamics and energetics of fish propulsion. *Bulletin of the Fisheries Research Board of Canada* 190: 1–158.

Werner, E. and Werner, B. (1954). Uben den mechanismus der nahrungserwerbs der tunicaten, speciell der ascidien *Helgolander Wissensch. Meeresuntersuchungen* 5: 57–92.

Widder, E.A. (2010). Bioluminescence in the ocean: origins of biological, chemical, and ecological diversity. *Science* 328: 704–708.

Wiebe, P.W., Madin, L.P., Haury, L.R. et al. (1979). Diel vertical migration by *Salpa aspera*: potential for large-scale particulate organic matter transport to the deep sea. *Marine Biology* 53: 249–255.

Wiley, E.O. and Lieberman, B.S. (2011). *Phylogenetic Systematics: Theory and Practice of Phylogenetic Systematics*. Hoboken: Wiley-Blackwell.

10

The Fishes

Introduction

Our second foray into the chordates brings us to the vertebrates and the remarkable world of fishes. Total numbers of fish species range from 32 000 (Nelson et al. 2016) to 33 700 (fishbase.org); the literature describing them is vast. They are highly diverse in morphology, behavior, and habitat, and many consider them to be quite elegant as well. Much of the focus of this chapter will be on the open-ocean pelagics, the micronekton, with emphasis on their special characteristics as members of the deep-sea fauna. Deep-living species are particularly important in the open sea, where the darker waters at mesopelagic depths serve as a refugium for smaller fishes. The basic biology of fishes, including their physiological systems, will be covered here. For further detail, we recommend an ichthyology text; Bone and Moore (2008) and Helfman et al. (2009) both provide excellent coverage.

Until very recently, the cephalochordates, the Lancelets, (Figure 10.1) were considered unarguably the closest relatives of the vertebrates (e.g. Bone and Moore 2008; Helfman et al. 2009) based mainly on their morphology (Holland et al. 2015). Their basic characteristics include a notochord that extends to the anterior end of the animal, i.e. in front of the brain – hence the name. They have no cranium, no vertebrae, and no cartilage or bone, but they do have v-shaped myotomes similar to those of fishes. They spend most of their time buried in the sand where feeding occurs by straining minute organisms from the water that is constantly drawn through the mouth.

Recent molecular phylogenetic evidence places the tunicates as the sister group of the vertebrates (Holland 2006; Holland et al. 2015), a clade called the Olfactores (Brusca et al. 2016), moving the cephalochordates down a step in the chordate line. Both the lancelet and tunicate camps have support in the ichthyology community, and there is no doubt that the morphology of the lancelet is decidedly more vertebrate-like than the sea squirt, though the ascidian's tadpole larva is also similar. Hopefully, time will produce a consensus.

The higher-level taxa of the vertebrates are also a work in progress. Many systems (e.g. WoRMS) initially divide the vertebrates into two major groups: the jawless (Agnatha: hagfish and lampreys) and jawed (Gnathostomata) vertebrates. In WoRMS, each occupies its own superclass. The jawed vertebrates include the fishes and tetrapods (amphibians, reptiles, birds, and

Life in the Open Ocean: The Biology of Pelagic Species, First Edition. Joseph J. Torres and Thomas G. Bailey.
© 2022 John Wiley & Sons Ltd. Published 2022 by John Wiley & Sons Ltd.

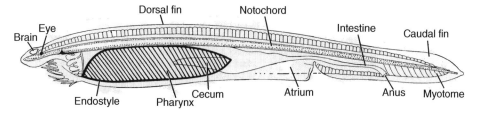

General anatomy of the cephalochordate *Branchiostoma*

Figure 10.1 Internal anatomy of the cephalochordate *Branchiostoma*.

mammals). The fishes have slightly more species (33 700) than the combined numbers of tetrapods (31 000, Chapman 2009).

The following terms will be useful for understanding the many formal taxonomic categories now used in the world of systematics (Nelson et al. 2016). The terms provide a hierarchy for understanding animal relationships, and each category is intended to be monophyletic: a single clade (Nelson et al. 2016). In descending order, the terms are phylum, subphylum, superclass, grade, class, subclass, infraclass, division, subdivision, superorder, series, subseries, infraseries, order, suborder, infraorder, superfamily, family, subfamily, tribe, genus, subgenus, and species. The entire suite of categories is not used for every taxon. For those of us that were weaned on the basic phylum, class, order, family, genus, and species hierarchy of yesteryear, the sheer number of new categories is a bit daunting. However, you will likely encounter many of these new categories in the literature.

Besides the taxonomic hierarchy described above, there is another terminology that is useful to know and that is the concept of stem and crown groups (Figure 10.2). Nelson et al. (2016) define them like this: "For any given taxon with extant members, the crown group is all those species descended from the last common ancestor (LCA) of all the extant members. The stem group taxa are all those extinct taxa known by fossils that are more closely related to a particular crown group than they are to any other extant clade. The total group is the sum of the stem group and the crown group." For the fishes, this is best exemplified by the diagram in Figure 10.2b, showing the phylogenetic relationship between the major groups of teleosts.

For a classification of the fishes, see Appendix A.

The epipelagic, mesopelagic, and bathypelagic realms of the Earth's oceans are described in Chapters 1 and 2 and illustrated in Figure 1.24. For the pelagial, the open ocean, two additional terms are often used for the deeper zones: the abyssopelagic for depths from 3000 to 6000 m and the hadopelagic for all depths >6000 m. Similar designations are used for benthic environments; Priede's (2017) terms are similar to the pelagic ones and will be used here for our occasional treatment of bottom-dwelling fishes. The bathyal zone includes depths from 200 to 3000 m, incorporating the continental slope (200–1000 m) and much of the continental rise (2000–4000 m). The abyssal region ranges from 3000 to 6000 m, and the hadal includes all depths >6000 m. Two details useful to remember are the average depth of the ocean, 3800 m, often rounded up to 4000 m (Sverdrup et al. 1942), and the fact that ocean depths >6000 m are a very small fraction of the total: about 2% (Priede 2017). Other

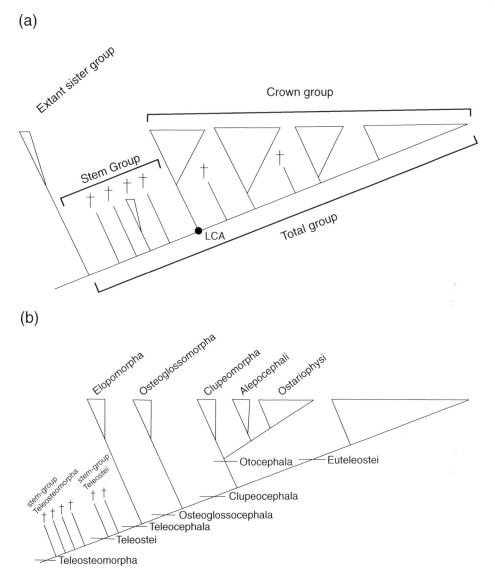

Figure 10.2 Taxonomic concepts and teleostean phylogeny. (a) Taxonomic concepts of stem-group and crown group. LCA denotes the last common ancestor; daggers indicate extinct groups; (b) phylogenetic relationships of the major groups of Teleostei. *Sources:* (a) Nelson et al. (2016), page 7; (b) Nelson et al. (2016), page 129. Reproduced with permission of John Wiley & Sons.

depth-designation schemes may be found in Haedrich and Merrett (1988) and UNESCO (2009).

Three additional definitions from Drazen and Sutton (2017) will be employed with respect to bottom and near-bottom dwellers.

<u>Benthic</u> fishes live mainly on the seafloor, including fishes such as the flatfishes and searobins.

<u>Benthopelagic</u> fishes live in association with the seafloor but spend little time in contact with it. Examples would be the

grenadiers (Macrouridae), slickheads (Alepocephalidae), and most rockfishes (Scorpaenidae).

Demersal fishes live in association with the seafloor. The term applies to benthic and benthopelagic species.

The Deep-Sea Groups

A quick survey of the many orders of fishes in Appendix A may have surprised you with the large number of deep-sea representatives: about 4000 species of agnathans, elasmobranchs, and teleosts out of a total of 19000 marine species (WoRMS) are deep-sea dwellers. A cut-off of 200 m (Priede 2017) for a deep-sea designation roughly coincides with the shelf-slope break at the seaward edge of an average continental shelf, as well as with the bottom of the epipelagic zone in open-ocean waters. No serious metabolic adjustments are required to accommodate a depth of 200 m; a surface dweller's metabolism will be adequate to about 500 m (Hochachka and Somero 1984). Physical and ecological changes will be more substantial (see Chapters 1 and 2). Open-ocean species will be below the seasonal thermocline and will experience a drop in zooplankton biomass as well as changes in the quality and quantity of ambient light. Benthic and benthopelagic species will experience similar physical changes in the water column above them along with a change in the benthic prey spectrum and a decrease in the organic rain from surface waters.

A survey of the many orders of fishes in Appendix A shows that a few orders consist entirely, or nearly entirely, of deep-sea species, but even more orders have just a few deep representatives. Andriashev (1953) noted that discrepancy and also the fact that many of the deep-water families have "peculiarities in structure" that enhance survival in the deep-sea environment. He concluded that there were two types of deep-sea fishes: ancient deep-water forms and secondarily deep-water forms. Ancient forms exhibit a very wide, nearly pan-global distribution in addition to their unusual adaptations in structure and life history. They also belong to phylogenetically ancient, less advanced orders of fishes. Secondarily, deep-water forms belong to phylogenetically younger groups of fishes, often perciforms, and have more restricted distributions. They are found locally on continental slopes and sometimes on adjacent abyssal plains but do not show the elaborate modifications of structure observed in the ancient fishes. He considered them to be "modified shore forms."

The history of deep-sea colonization by fishes is now thought to be more complex than an early colonization by ancient groups and their modification through time, followed by more recent invaders that have had less time to be sculpted by natural selection. Increased knowledge of ocean history and development of molecular clocks to age fish lineages have made the story more complicated and more interesting.

From the explosion of life in the Cambrian Period 500 mybp (million years before present), the geological history of the oceans includes five mass extinctions and seven periods of widespread anoxia known as Oceanic Anoxic Events or "OAE's" (Figure 10.3) (Takashima et al. 2006; McClain and Hardy 2010; Priede 2017). Temperature and oxygen in the deep ocean have fluctuated strongly through geologic time, as has the character of the oceanic fauna (Figures 10.3–10.5). Records obtained by the Deep-Sea Drilling Program (DSDP) have been responsible for much of our present knowledge of the

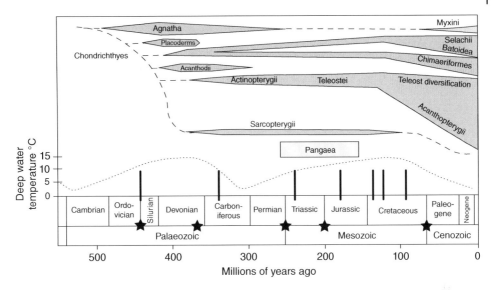

Figure 10.3 Fish evolution and ocean paleohistory. Stars indicate the five global mass extinction events. Vertical black bars indicate Oceanic Anoxic Events (OAEs) (Takashima et al. 2006). Deep water temperature from Schopf (1980). The box defines the period when all land masses coalesced into one supercontinent: Pangaea. *Source:* From Priede (2017), figure 2.7 (p. 73). Reproduced with the permission of Cambridge University Press.

ocean's paleohistory, including temperature and oxygen records that help us interpret the fossil record. The timeline in Figure 10.3 shows that the deep ocean has had many periods when it was inimical to life and that colonization by fishes and other taxa likely has happened repetitively (McClain and Hardy 2010; Priede and Froese 2013). Also important is the fact that although anoxic sediments effectively integrate and preserve the fallout from the water column above them, pockets of normoxic water may have been present to act as refugia (Guinot et al. 2013) but were not reflected in the sedimentary records analyzed.

Two recent hypotheses have been advanced to explain the modern distribution of fishes in the deep sea (Priede 2017):

1) *Repetitive Colonizations.* The deep-sea has been repeatedly invaded since the Cambrian Period whenever conditions would allow it. Those ancient faunas were episodically culled or wiped out by extinction events and OAE's; they were recolonized from the shallows. The modern deep-sea fauna is the result of recolonization since the Cretaceous (Priede and Froese 2013).

2) *Persistence of lineages via refugia* (Guinot et al. 2013). Examples of persistence may be found in two important fish lineages, the now-extinct cladodontomorph sharks and the coelacanths, or lobe-finned fishes. Both had large gaps in their fossil records (120 and 70 mybp, respectively), and both persisted through major extinction events, the Permian–Triassic at 252 mybp and the Cretaceous–Paleogene at 66 mybp (Figure 10.3). Both are believed to have survived by occupying refuge environments in deep water (Guinot et al. 2013). For the cladodontomorphs, persistence is documented by new fossil

evidence; for the coelacanths, it was confirmed by the exciting discovery in 1938 of living specimens.

The two hypotheses are not mutually exclusive. Because there is good evidence in support of both, it is likely that both mechanisms have been in play to produce the modern deep-sea fish fauna.

A Brief History of Fishes

The Jawless Fishes

"Fishes have an ancestry that goes back at least 500 million years. Some fossil groups can be linked with extant taxa, some extant taxa lack obvious fossil antecedents, and numerous groups arose, prospered, and were extinguished" (Helfman et al. 2009). The oldest fossil fish enjoying a wide consensus of that designation is *Myllokunmingia fengjiaoa*, found in the Chengjiang fossil beds of China's Yunnan province and dated at 530–540 mya, in the early Cambrian (Nelson et al. 2016).

The extinct group Conodonta is important to the history of fishes and to present-day paleontologists who use the cone-shaped teeth (conodonta = cone-teeth; Figure 10.4a) to identify the age of fossil beds (Aldridge and Briggs 1989). Their very long history ended in the Triassic period. Further along in geological time, four additional superclasses of jawless fishes appeared in the fossil record (Figure 10.5). Collectively, they are known as ostracoderms (shell-skinned), so named for the bony shield covering their head and thorax. The superclass Thelodontomorphi is primarily known from the widely deposited denticles or scales that, like the teeth of conodonts, are used in biostatigraphy.

The last major group of jawless fishes is the Osteostracomorphi, widely considered to be the sister group to the gnathostomes, the jawed fishes. All have complex head shields, paired pectoral fins, a body triangular in cross section and dorsoventrally compressed, and a heterocercal tail with a pronounced upper lobe (i.e. "epicercal") (Figure 10.4b). All have an ossified endoskeleton (Nelson et al. 2016). The Osteostracomorphs date from the early Silurian to the end of the Devonian.

(a) Conodont, reconstructed

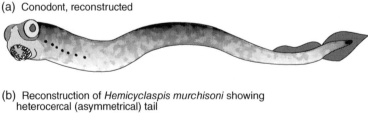

(b) Reconstruction of *Hemicyclaspis murchisoni* showing heterocercal (asymmetrical) tail

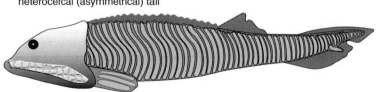

Figure 10.4 Ancestral fish groups. (a) Class Conodonta; (b) superclass Osteostracomorphi; note the heterocercal tail with pronounced upper lobe.

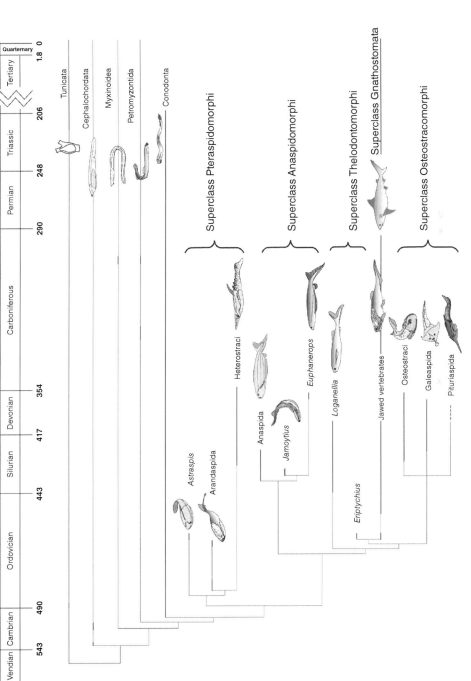

Figure 10.5 A proposed chordate phylogeny (gray) with fossil presence (stratigraphic ranges) shown in black. Numbers are in millions of years before present. *Source:* Adapted from Donoghue et al. (2000).

The Jawed Fishes

Teleosts

Jawed fishes first appeared during the early Silurian period, so they coexisted for millions of years with the jawless fishes, most of which vanished at the end of the Devonian. Fish jaws developed from gill arches; the development of jaws is considered the "greatest of all advances in vertebrate history" (Romer and Parsons 1986). Jaws greatly expanded the prey types available to fishes, resulting in the rapid increase in fish diversity that is reflected in the fossil record and persists to this day. As yet no fish ancestor has been found to connect the jawless and jawed fishes, but the search goes on. A second milestone in fish anatomy that significantly enhanced the success of jawed fishes was the development of paired pectoral and pelvic fins (Pough et al. 1989; Helfman et al. 2009). Paired fins greatly increased the agility of fishes, particularly in the area of precision swimming like that seen in reef fish species. Many species use the pectoral fins as their primary means of propulsion.

Two extant fish groups are quite ancient and are considered the ancestors of the tetrapods: the Coelacanthi, the lobe-finned fishes, and the Dipneusti, the lungfishes. The lungfishes are also lobe-fins, but they are set apart from the coelacanths by their ability to breathe air. In many classification systems, the two classes are grouped together as the Sarcopterygii (lobe fins). The coelacanths appeared in the middle Devonian but reached their highest diversity in the Triassic. Though not nearly as diverse now as they were in Palaeozoic and Mesozoic times when 83 species of coelacanths cruised the paleo-ocean, two species of coelacanths remain today, along with six species of lungfishes.

Living lungfishes, Class Dipneusti, commonly referred to as Dipnoi (double-breathers), are all in the order Ceratodontiformes, but they have many extinct relatives important to the evolution of the tetrapods. The lungfishes are more closely related to the tetrapods than to other fishes (Nelson et al. 2016).

A hypothetical lineage leading to the teleosts, or bony fishes, is presented in Figure 10.6. The lineage begins with an

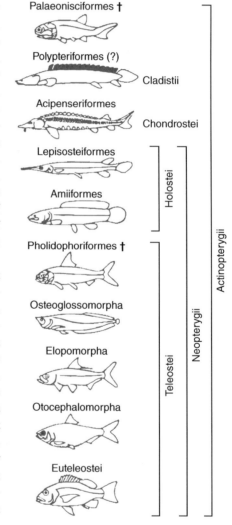

Figure 10.6 Phylogenetic relationships among Actinopterygian fishes. Daggers indicate extinct groups. *Source:* Adapted from Helfman et al. (2009).

extinct assemblage of fishes known as the Paleonisciformes, primitive actinopterygians that were found from the late Silurian to the Cretaceous. Trends within the fossil record show a gradual reduction in heavy scales, increased ossification of the vertebral column, a more highly developed internal musculature and development of a robust homocercal (crescent-shaped) tail, all suggesting an increased reliance on locomotion for prey capture (Helfman et al. 2009).

The lineage includes all groups making up the Actinopterygii, or ray-finned fishes. In our adopted system from the Cal Academy (Appendix A), the Bichirs (Polypteriformes) have their own class, the Cladistii, and the Actinopterygii include sturgeons (Acipenseriformes), gars (Lepisosteiformes), and bowfins (Amiiformes) as well as the true bony fishes, or Euteleosts. The sturgeons are usually given their own infraclass, the Chondrostei, as are the gars and bowfins which are included in the infraclass Holostei (Nelson et al. 2016). The Neopterygii, or "new fins", comprise an unranked clade that is believed to be monophyletic (Nelson et al. 2016); it includes the gars, bowfins, and euteleosts.

A second extinct group dating from the late Triassic, the Pholidophoriformes, is an important stem group leading us into the crown-group teleosts. The crown teleosts comprise four distinct groups: the cohort Osteoglossomorpha, or bonytongues; the cohort Elopomorpha, eels, tarpons, and ladyfish; the cohort Otocephala (ear-head), herrings and catfish; and the cohort Euteleostei, a huge group that includes 96% of all living fishes (Bone and Moore 2008) and the remainder of the bony fishes (see Appendix A). There is some disagreement about where to draw the line between the Euteleostei and its sister groups. Its position in our classification system agrees with Nelson et al. (2016) and with the older literature (Greenwood et al. 1966).

Elasmobranchs

The elasmobranchs (plate gills) are often grouped with the chimaeras, or Holocephali (whole heads), in the class Chondrichthyes, or cartilaginous fishes (Helfman et al. 2009; Nelson et al. 2016). The Cal Academy system and WoRMS confer class status on both the Elasmobranchii and Holocephali. However, Chondrichthyes is a widely used category with a long history in the literature.

The paleohistory of the elasmobranchs is not as rich as that of the teleosts because their cartilaginous skeletons do not preserve as well as do the bony ones of the teleosts. Several fossil groups are considered likely stem elasmobranchs. Three particularly well-known examples are the Cladoselachimorpha, Ctenocanthomorpha, and Xenacanthomorpha pictured in Figure 10.7. An additional fossil group, the Hybodonta (Figure 10.8a), is considered the closest extinct sister group to the elasmobranchs. They lived from the Mississipian epoch (359 mybp) to the Cretaceous period (145 mybp). Note the transition from a terminal to a ventral mouth through time (Figures 10.7a–d and 10.8a, b). The increase in size of the rostral region accompanied a developing olfactory system (Figure 10.8b). Like the teleosts, modern sharks have a protrusible upper jaw, allowing bigger bites and the generation of suction (Helfman et al. 2009). "Most orders of living elasmobranchs had appeared in the upper Jurassic and all orders had appeared by the end of the Cretaceous" (Helfman et al. 2009).

Holocephali

The holocephalans, or chimaeras, are an interesting benthopelagic group with a long paleohistory. Almost all are found in deep water. The small number of species extant

678 | *The Fishes*

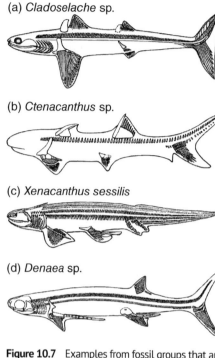

Figure 10.7 Examples from fossil groups that are possible stem ancestors of the elasmobranch. (a) Cladoselachimorpha; (b) Ctenacanthamorpha; (c) Xenacanthomorpha; (d) Denaeamorpha. *Sources:* (a) Adapted from Schaefer and Williams (1977), figure 5 (p. 299); (b) Adapted from Schaefer and Williams (1977), figure 7 (p. 300); (c) Adapted from Schaefer and Williams (1977), figure 3 (p. 27); (d) Adapted from Schaefer and Williams (1977), figure 4 (p. 298).

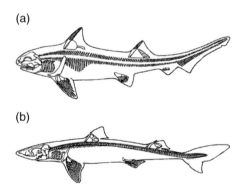

Figure 10.8 Ancestral and modern sharks. (a) The ancestral shark *Hybodus* sp. which lived from the end of the Permian to late Cretaceous; (b) the modern shark *Squalus* sp. (spiny dogfish). *Sources:* (a) Adapted from Schaefer and Williams (1977), figure 6 (p. 300); (b) Adapted from Schaefer and Williams (1977), figure 9 (p. 301).

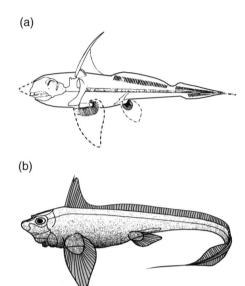

Figure 10.9 Holocephali (chimaeras). (a) Restoration of the ancestral holocephalan *Ischyodus schuebleri* from the upper Jurassic; (b) the modern plownose chimaera *Hydrolagus affinis*. *Sources:* Adapted from Patterson (1965), figure 2 (p. 111); (b) Adapted from Jordan and Evermann (1900), plate XIX.

today are the remnants of a large and diverse group that dates back to the late Devonian. Though the great majority of that group are now extinct, the three living families have fossil records dating back to the Jurassic and Cretaceous periods. The plownose chimaeras are an example. Figure 10.9 shows *Ischyodus*, a callorhinchid of the upper Jurassic, and *Hydrolagus affinis* (formerly *Chimaera affinis*), a modern fish of the same family.

The Classes of Living Fishes

Class Myxini

The jawless fishes, the agnathans, first appear in the fossil record in the early Cambrian, about 530 mybp. Because they appear with well-developed tissues and fins

(Helfman et al. 2009), it is likely they date to pre-Cambrian times. Five classes of extinct jawless fishes were important members of the fish fauna, persisting until the late Devonian period, 400 mybp (Figure 10.5). The two surviving groups of agnathans are the hagfishes, Class Myxini, and the lampreys, Class Petromyzonti. The hagfishes, also called slime hags or slime eels (Figure 10.10), as presently known originated in the Carboniferous period. Current thinking suggests that they represent a successful derived form that lost much of the complexity of its predecessors (Heimberg et al. 2010).

Found in temperate waters from shelf depths (<200 m) to about 3000 m (Bone and Moore 2008; Yeh and Drazen 2009; Priede 2017), hagfishes are primarily scavengers, feeding on dead and dying fish and invertebrates. They are quite capable however of active predation as well (Zintzen et al. 2011, 2013). Prey is located by olfaction; their dual eyespots are capable only of distinguishing light from dark (Jensen 1966). When they encounter prey, hagfishes attack with their rasping tongue, burrowing in and tearing out pieces of flesh. They can tie their bodies into a knot to gain leverage for tearing out flesh by pressing the knot against the prey.

Hagfishes are unique in the fish world in having a cartilaginous notochord located just below their spinal cord instead of a vertebral column. The notochord provides the suppleness needed to tie themselves into a knot. They can produce prodigious amounts of slime from the tiny pores that line their body. The slime is used to deter predators by clogging up their gills, forcing them to rid themselves of the slime instead of dining on the hagfish. In a similar manner, the slime can be used to eliminate the competition for a prey item by choking up the competitor. Bone and Moore (2008) state that a single hagfish can solidify a bucket of water with its slime.

The hagfishes have another important distinguishing characteristic: they are the only fishes with a blood ionic concentration that is equivalent to seawater. Their blood ionic composition is slightly different, but their blood ionic concentration is more similar to that of an invertebrate than that of a fish or tetrapod.

Hagfishes are believed to reproduce multiple times throughout their lives; however, the details of their reproductive behavior, e.g. seasonality and age at maturity, are undescribed. They have a single large gonad that runs the length of the body. The gonad appears to be differentiated into an anterior ovary and posterior testis in immature hagfishes, but mature fishes are either male or female. The female produces large sausage-shaped eggs with filaments that anchor the eggs to each other and to the sea bottom. Fertilization is believed to be external, and development is direct; the eggs hatch as miniature adults.

Class Petromyzonti

The lampreys (Figure 10.11) like the hagfishes originated in the upper Carboniferous period. As yet, no definitive connection has been made between either the lampreys or the hagfishes and the jawless fishes of the Devonian and earlier periods. The lampreys more closely resemble their jawed relatives in having a second semicircular canal, a lateral line, radial muscles in the fins, functional eyes, and some cartilaginous vertebral elements around the large notochord (Bone and Moore 2008). In addition, like the jawed fishes, the lampreys have a blood ionic concentration about one-third that of seawater. The large difference between the internal and external ionic environments must be accommodated by the integument, kidney,

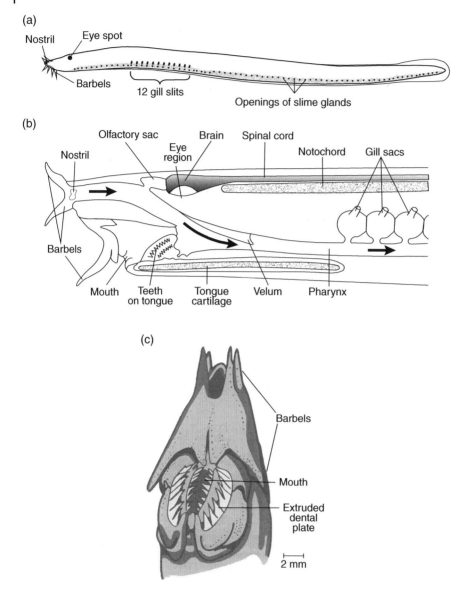

Figure 10.10 Agnatha (hagfishes). (a) Lateral view of a stylized modern hagfish; (b) internal anatomy of a hagfish. Arrows indicate the direction of water flow from the single nostril through the olfactory sac to the gills; (c) the rasping apparatus of the hagfish. *Sources:* (a) Adapted from Jensen (1966), page 84; (b) Jensen (1966), page 87; (c) Bone and Moore (2008), figure 1.12 (p. 18).

gut, and gills working together. When lampreys are in freshwater, they are hypertonic to their external environment, so the balancing systems must, in effect, work in reverse.

The 55 species of lampreys include nonparasitic as well as the better-known parasitic ones. Without question, the best-known species is the sea lamprey, *Petromyzon marinus*, which can reach up to 1 m in

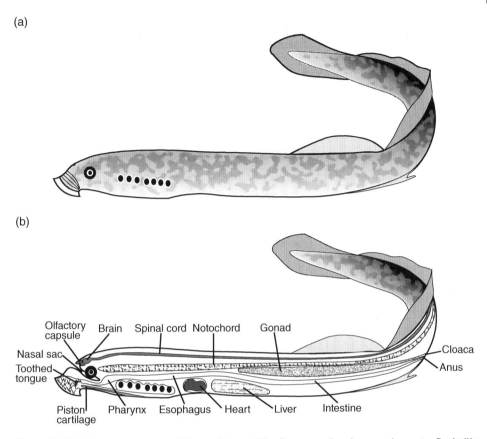

Figure 10.11 Lamprey anatomy. (a) Lateral view of the European river lamprey *Lampetra fluviatilis*; (b) internal anatomy of *L. fluviatilis*.

length. Their normal life cycle is to hatch in a stream, spend several years there as larvae, migrate to the ocean to spend their relatively short adult life, then migrate back to a stream to spawn and die (Figure 10.12). Lampreys are found in rivers north of 30°N and south of 30°S (Bone and Moore 2008).

Unfortunately, sea lampreys have now invaded the great lakes, first Lake Ontario, then Huron and Michigan in the 1930s via man-made canals and devastated the trout fisheries there. Once established, instead of downstream migration to the ocean to spend their adult life, they migrated to the lakes to feed on the abundant lake trout.

The spawning and early life of the sea lamprey are particularly interesting. After arriving in a suitable stream in early spring, the ripe male creates a nest, or redd, by thrashing out a pit with his tail and encircling the downstream side with cobbles. Females are attracted by pheromones and in some species help construct the nest. Cobbles are picked up and transported using their sucker-like mouths. Once the temperature is warm enough, about 10°C for sea lampreys, spawning begins. The female extrudes a small batch of eggs that are immediately fertilized by the male. The fertilized eggs are washed downstream

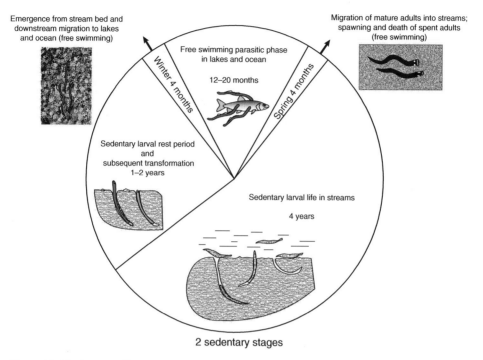

Figure 10.12 Lamprey life cycle. Segments of the pie chart represent the different stages of the life cycle and are proportional to the time spent in each stage. Note that most of the lamprey's life is as a larval stage.

where they lodge in the nooks and crannies of the nest cobbles and are then covered with sand. This process is repeated over a one- to three-day period with the female eventually producing thousands of eggs; an average number for the sea lamprey is 61 500 eggs. Once spawning is complete, both the male and female die, sometimes within a few hours (Applegate and Moffett 1955).

The larvae hatch in 10–12 days, leaving the nest at 20 days as tiny (6.4 mm) needle-like ammocoetes. They drift downstream toward quiet waters and dive for the bottom where they create a burrow for themselves. There the larvae filter-feed on microplankton and detritus for four years (Figure 10.12). In their fifth year, they metamorphose into young adults with the characteristic horny teeth and file-like tongue of the species and assume their parasitic life. The young adults use their sucker-like mouths to attach to prey, rasp a hole in the fish host, and feed on its blood and tissues. After one or two years, they mature and seek out streams for spawning. Their sex organs expand, and their digestive tract becomes non-functional (Applegate and Moffett 1955).

Class Elasmobranchii

The elasmobranchs represent an alternative evolutionary strategy to that of the bony fishes. All possess an ossified backbone with an otherwise cartilaginous skeleton (Figure 10.13) and have internal

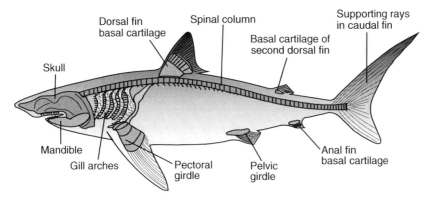

Figure 10.13 Shark skeletal structure.

fertilization. Internal fertilization allows for a variety of reproductive strategies including the production of egg cases (skates) and bearing of live young (requiem sharks) (Bone and Moore 2008). Except for a very few landlocked species, all elasmobranchs are marine.

Sharks and rays are noteworthy in that all use urea as an osmolyte to help them maintain osmotic balance in the marine environment. Like the bony fishes, they have an internal ionic concentration approximately a third that of seawater. Unlike the bony fishes, urea makes up the remaining two-thirds in the elasmobranchs so that their internal osmolarity is approximately equal to that of seawater. Their osmotic strategy minimizes the outward movement of water and inward movement of ions.

Relative to the bony fishes, elasmobranchs are large. They range in size from the dwarf lantern shark (*Etmopterus perryi*) at 17 cm long and 15 g to the 10 m-long, 4000 kg basking shark (*Cetorhinus maximus*) and the 12-m long, 12 000 kg whale shark (*Rhincodon typus*). The well-documented maximum sizes for basking and whale sharks noted above are exceeded by reliable reports of whale sharks of 18 m in length and basking sharks at 12–15 m. The discrepancy may be attributed to the difficulty in measuring large animals in the field (Helfman et al. 2009).

Quite a bit of controversy has attended the maximum length of white sharks, particularly considering their close relationship to the extinct giant predator *Carcharodon megalodon*. A widely accepted length for the largest white shark (*Carcharodon carcarias*) is 5.94 m, a measurement from off Australia in 1984 (Mollet et al. 1996). However, bite-marks on whale carcasses suggest lengths up to 7 m. The fossilized teeth of *C. megalodon* suggest a maximum length of 16 m and a maximum weight of 18 000 kg (Compagno et al. 1993). About 90% of sharks have a body length >30 cm, 50% exceed 1 m, and 20% exceed 2 m (Springer and Gold 1989). With increased size comes increased cruising speed and overall mobility, favorable attributes for a predator.

Sharks can be classified into coastal and oceanic species based on observed habitat preferences and home ranges (Table 10.1). The bull (*Carcharinus leucas*, 643 km) and nurse sharks (*Ginglymostoma cirratum*, 541 km) are considered local, with home ranges of a few hundred square kilometers in the nearshore. Coastal sharks often range from nearshore nursery grounds out to the offshore regions of the shelf and slope, with home ranges of 1000 km and up. Examples are the sandbar (*Carcharhinus plumbeus*, 3776 km), silky (*Carcharhinus falciformis*, 1340 km), and dusky shark (*Carcharhinus obscurus*, 3800 km, Kohler and Turner 2001). Of primary interest to us are the oceanic species, which range over entire ocean basins, sometimes crossing them repeatedly. Oceanic species include the blue shark (*Prionace glauca*, 7871 km), mako (*Isurus oxyrhynchus*, 4543 km), oceanic whitetip (*Carcharhinus longimanus*, 2811 km), tiger shark (*Galeocerda cuvieri*, 6747 km), whale shark (*Rhincodon typus*, 8025 km), and white shark (*Carcharodon carcharias*, 22 000 km) (Kohler and Turner 2001; Bonfil 2005).

Growth in sharks is slow, sexual maturity is later in life, and they typically live longer than most bony fishes (Table 10.2) (Cailliet and Goldman 2004). The sharks listed in Table 10.2 are all livebearers. Pups are miniature versions of the adult fish, varying in size from 2.9 to 33% of adult maximum size. Gestation periods range from 3–4 months in the bonnethead shark to 3.5 years in the Basking Shark, with an overall average of 9–12 months (Pratt and Casey 1990; Helfman et al. 2009). The combination of slow growth rate, late age at sexual maturity, long gestation periods, and alternate years of reproduction make sharks particularly susceptible to overfishing. It has been estimated that shark populations have declined by 50–80% overall since the 1970s (Helfman et al. 2009). Some species suffered even more profound declines between the 1950s and 1990s (Baum and Myers 2004; Baum et al. 2005). Declines have been estimated to be as much as 90 and 99% in the silky shark and the oceanic whitetip, respectively, though these numbers have been disputed as being too pessimistic (Burgess et al. 2005).

The early life history strategies of sharks do vary, but a few generalities can be made. Shark pups are most susceptible to predation at sizes <100 cm, and predation is often from other sharks. Once pups reach 100 cm, their size and their increased swim-speed allow them to evade most predators. They are also at this point more effective predators themselves (Branstetter 1990). Most species in Table 10.2 have pups that are large relative to the size of the mother, 25% or greater of her maximum size, and grow rapidly to a length of 100 cm in their second year. Many of the pelagic sharks have pups of 100 cm or greater at birth. Since the number and size of pups are limited by the size of the mother and her ovaries, the larger the pups, the smaller the size of the litter.

Coastal species like bull sharks and lemon sharks may use protected nursery grounds such as bays and estuaries or coral reef flats to minimize predation on their young. Species using protected nurseries exhibit slower growth than their relatives born in areas subject to higher predation (Branstetter 1990). An exception to this rule is the oceanic whitetip. Pups of the oceanic whitetip shark take 2–3 years to reach 100 cm despite the fact that they are vulnerable to predation in their open-ocean environment. The slow growth of the pups is offset by larger litter sizes to assure cohort survival (Branstetter 1990).

Pups of the blue shark, another large open-ocean species, show rapid growth with even larger litter sizes (40–80), as do those of the tiger shark (30–70), a wide-ranging coastal species with a presence in the open ocean as well. Presumably, predation is heavy enough that very large litters are needed to assure cohort survival.

Sharks are outnumbered by the second major grouping of Elasmobranchs, the Batoidea, or rays (537 spp. vs 668 spp.) All rays are dorsoventrally compressed, with

Table 10.1 Elements of shark early life history. *Source:* Data from Hamady et al. (2014), Eckert and Stewart (2001), Kohler and Turner (2001), Compagno (1984), and fishbase.org. Home ranges are based on the tag-recapture data in Kohler and Turner, most of which are results from thousands of samples. ND = no data.

Family		Habitat		
Genus species	Common name	Home range (km)	Max. size (cm)	Sex
		Coastal		
Carcharhinidae	Requiem sharks			
Carcharhinus acronotus	Blacknose shark	315	165	Both
Carcharhinus brevipinna	Spinner shark	1665	240	Both
Carcharhinus isodon	Finetooth shark	6	160	Female
Carcharhinus leucas	Bull shark	643	300	Both
Carcharhinus limbatus	Blacktip shark	2146	200	Both
Carcharhinus plumbeus	Sandbar shark	3776	250	Both
Carcharhinus porosus	Smalltail shark	37	134	Both
Galeocerdo cuvieri	Tiger shark	6747	450	Both
Negaprion brevirostris	Lemon shark	426	300	Female
Rhizoprionodon terraenovae	Atlantic sharpnose shark	1037	94	Both
Ginglymostomatidae	Carpet sharks			
Ginglymosotoma cirratum	Nurse shark	541	430	Female
Hexanchidae	Cow sharks			
Notorynchus cepedianus	Sevenfill shark	539	300	Female

(*Continued*)

Table 10.1 (Continued)

Family		Habitat		
Genus species	Common name	Home range (km)	Max. size (cm)	Sex
Odontaspididae	**Sand tiger shark**			
Carcharias taurus	Sand tiger shark	1 897	300	Female
Sphyrnidae	**Hammerhead sharks**			
Sphyrna lewini	Scalloped hammerhead	1 671	310	Female
Sphyrna makarran	Giant hammerhead	1 180	560	ND
		Pelagic		
Alopidae	**Thresher sharks**			
Alopias superciliosus	Bigeye thresher shark	2 767	400	Female
Alopias vulpinus	Common thresher shark	1 556	650	Female
Carcharhinidae	**Requiem sharks**			
Carcharhinus falciformis	Silky shark	1 340	310	Female
Carcharhinus longimanus	Oceanic whitetip shark	2 811	275	Both
Carcharhinus obscurus	Dusky shark	3 800	360	Female
Charcharhinus signatus	Night shark	2 669	275	Both
Prionace glauca	Blue shark	7 871	350	Female
Ginglymostomatidae	**Carpet sharks**			
Rhincodon typus	Whale shark	up to 13 000	2000	Female

Lamnidae	**Mackerel sharks**		
Carcharodon carcharias	White shark	3800–22000	Female
Cetorhinus maximus	Basking shark	3400	Female
Isurus oxyrinchus	Shortfin mako shark	4543	Female
Isurus paucus	Longfin mako shark	3430	Both
Lamna nasus	Porbeagle	4260	Female
Sphyrnidae	**Hammerhead sharks**		
Sphyrna zygaena	Smooth hammerhead	1122	Both

		650	
		980	
		375	
		420	
		270	
		390	

Table 10.2 Elements of shark early life history. *Source:* Data from Hamady et al. (2014), Santana et al. (2009), Cailliet and Goldman (2004), Branstetter (1990), Pratt and Casey (1990), Gruber and Brown (1988), Compagno (1984), and fishbase.org. Home ranges are based on the tag-recapture data in Kohler and Turner, most of which are results from thousands of samples. See Table 10.1 for data on sex. ND = no data. TL = total length.

Family Genus species	Common name	Age at sexual maturity	Max. age	Birth size L (%Lmax)	Number of young	Year 1 growth (%BL)	Gestation (months)
Carcharhinidae	Requiem sharks						
Carcharhinus acronotus	Blacknose shark	4.5 y	10–16 y	45 cm (27%)	4–6	17 cm (38%)	12
Carcharhinus brevipinna	Spinner shark	8–10 y	17–19 y	65 cm (27%)	6–10	40 cm (62%)	12
Carcharhinus isodon	Finetooth shark	3.9 y	8.0 y	50 cm (31%)	6–8	25 cm (50%)	12
Carcharhinus leucas	Bull shark	20–21 y	29–32 y	70 cm (23%)	6–10	15 cm (22%)	11
Carcharhinus limbatus	Blacktip shark	165 cm TL	12 y	55 cm (27%)	6–10	35 cm (65%)	11
Carcharhinus plumbeus	Sandbar shark	15–16 y	18–25 y	65 cm (26%)	6–18	15 cm (23%)	12
Carcharhinus porosus	Smalltail shark	6 y	12 y	30 cm (22%)	6	ND	ND
Galeocerdo cuvieri	Tiger shark	7 y	23 y	70 cm (16%)	30–70	70 cm (100%)	12
Negaprion brevirostris	Lemon shark	13 y	25 y	65 cm (22%)	6–18	15 cm (23%)	12
Rhizoprionodon terraenovae	Atlantic sharpnose shark	1.3–3 y	10 y	32 cm (29%)	6–8	22 cm (69%)	11
Ginglymostomatidae	Carpet sharks						
Ginglymostoma cirratum	Nurse shark	25 y	35 y	ND	21–28	ND	ND
Hexanchidae	Cow sharks						
Notorynchus cepedianus	Sevengill shark	11–21 y	49 y	40–53 cm (18%)	82	ND	ND
Odontaspididae	Sand tiger shark						
Carcharias taurus	Sand tiger shark	9–10 y	17 y	100 cm (33%)	2	30 cm (30%)	ND
Sphyrnidae	Hammerhead sharks						
Sphyrna lewini	Scalloped hammerhead	8.8 y	19 y	45 cm (15%)	30–40	17 cm (38%)	12
Sphyrna makarran	Giant hammerhead	ND	ND	70 cm (13%)	30–40	ND	12

Family / Species	Common name	Col1	Col2	Col3	Col4	Col5	Col6
Alopidae	**Thresher sharks**						
Alopias superciliosus	Bigeye thresher shark	9–9.2 y	16 y	100 cm (25%)	2	40 cm (55%)	ND
Alopias vulpinus	Common thresher shark	5.8 y	22 y	140 cm (22%)	2	40 cm (28%)	9
Carcharhinidae	**Requiem sharks**						
Carcharhinus falciformis	Silky shark	12 y	22 y	70 cm (23%)	10–15	45 cm (65%)	12
Carcharhinus longimanus	Oceanic whitetip shark	6–7 y	13–17 y	65 cm (24%)	10–15	20 cm (30%)	12
Carcharhinus obscurus	Dusky shark	21 y	33 y	90 cm (25%)	10–12	15 cm (16%)	12
Charcharhinus signatus	Night shark	10 y	17 y	65 cm (26%)	10–18	32 cm (50%)	ND
Prionace glauca	Blue shark	5–6 y	15–16 y	45 cm (13%)	40–80	40 cm (90%)	12
Ginglymostomatidae	**Carpet sharks**						
Rhincodon typus	Whale shark	22 y	38–80 y	58–64 cm (2.9–3.2%)	300	ND	ND
Lamnidae	**Mackerel sharks**						
Carcharodon carcharias	White shark	8–13 y	43 y	140 cm (22%)	8–10	40 cm (30%)	ND
Cetorhinus maximus	Basking shark	5 y	ND	150 cm (15%)	6	ND	18
Isurus oxyrinchus	Shortfin mako shark	18 y	32 y	70 cm (20%)	8–10	40 cm (70%)	12
Isurus paucus	Longfin mako shark	205–228 cm TL	ND	110 cm (26%)	2	40 cm (40%)	ND
Lamna nasus	Porbeagle	13 y	24 y	70 cm (26%)	2	35 cm (45%)	8
Sphyrnidae	**Hammerhead sharks**						
Sphyrna zygaena	Smooth hammerhead	265 cm TL	21 y	50 cm (13%)	20–40	ND	ND

most having large pectoral fins that are fused to the head (Bone and Moore 2008). Gill openings are ventrally located. Body forms vary (Figure 10.14). The shark-like guitarfishes have a robust tail used in swimming, while the stingrays have well-developed pectoral fins and a barbed whip-like tail of little use in locomotion. Most rays are tied to the bottom, but a few such as the pelagic stingray (*Pteroplatytrygon violacea*) and manta rays (e.g. *Manta birostris*) have a pelagic lifestyle. The pelagic stingray is an active hunter in the midwater, feeding mainly on small fish and squid. Manta rays can reach an astonishingly large size, up to 7 m in width.

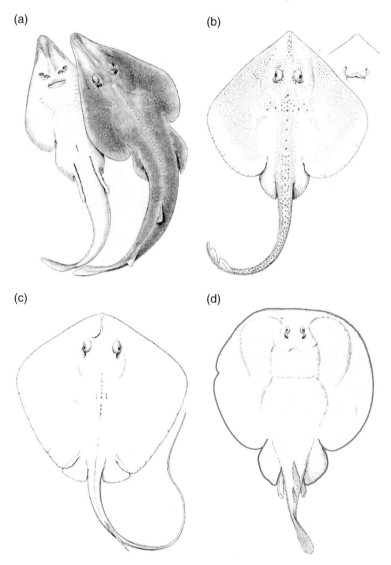

Figure 10.14 The Batoidea. (a) The guitarfish *Rhinbatos lentiginosus*; (b) the skate *Raja bathyphila*; (c) the stingray *Dasyatis say*; (d) the electric ray *Torpedo nobiliana*. *Sources:* (a) Bigelow and Schroeder (1953), figure 14 (p. 61); (b) Bigelow and Schroeder (1953), figure 30 (p. 160); (c) Bigelow and Schroeder (1953), figure 90 (p. 379); (d) Bigelow and Schroeder (1953), figure 22 (p. 97).

Like many large marine species, including the whale shark, basking shark, and megamouth shark, manta rays filter-feed on zooplankton like euphausiids, copepods, and schools of small fish. Filtration is achieved by swimming forward with a wide-open mouth (ram feeding), ingesting the zooplankton with the incoming water and trapping it on branchial sieves as the water passes through gill slits (Figure 10.15b). The esophagus can be constricted so that water flows out only through the branchial chamber. The resulting "plankton concentrate" can then be digested. The manta rays' cephalic fins, or "horns," aid in the filtration process by dropping down like doors at the sides of the mouth, increasing the size of the "scoop." When swimming without feeding, the cephalic fins are furled and face forward like horns.

The filters in mantas, basking sharks, megamouth sharks, and whale sharks are similar in principle but different in structure. All are associated with water flow over the gill arches, and all use a biological sieve to strain the particles out of that flow, concentrating the tiny prey so they can be more efficiently digested. In the manta, the sieves are very fine "filter pads" located on the gill arch (Figure 10.15c, d). In the basking shark (Figure 10.15e), the sieves are closely spaced, extended, gill rakers (Figure 10.15f, g).

Like most sharks, mantas are live bearers. Live-bearing, or viviparity, in elasmobranchs is achieved using one of three mechanisms. The first is commonly termed ovoviviparity: the embryo is retained in the ovary and nourished with its own yolk sac until its birth as a miniature adult. In the second, termed oophagy, the developing embryo feeds until birth on new eggs produced by the mother within the ovary after its own yolk sac is exhausted. In the third process, placental viviparity, the spent yolk sac of the developing embryo attaches to the uterine wall and forms an umbilical cord that nourishes the embryo and carries away its metabolic wastes. In some cases, a "uterine milk" is produced that can be absorbed either through processes on the umbilical or through the skin, mouth, or developing gills of the embryo. Manta embryos and those of other rays in the order Myliobatiformes develop solely by absorbing uterine milk without any placental connection (Helfman et al. 2009).

Class Holocephali

The holocephalans (whole head), or chimaeras, are a benthopelagic group that share many characteristics with the elasmobranchs. Three families with a total of 55 species make up the single order, Chimaeriformes: the shortnose chimaeras, the *Chimaeridae* (Figure 10.16a); the longnose chimaeras, the *Rhinochimaeridae*, (Figure 10.18b); and the plownose chimaeras, the *Callorhynchidae* (Figure 10.16c). Mucous-filled canals on the head are sensory detectors (Figure 10.16d). The three extant families are remnants of a once very diverse group dating back to the Devonian.

Chimaeras are also known as ratfish because of their tails or as rabbitfish because of their toothplates (Figure 10.16d). They dwell from shallow shelf depths down to at least 3000 m (Priede 2017), and many species (e.g. *Harriotta raleighana*) enjoy a very large zoogeographic range (Bigelow and Schroeder 1953; Priede 2017). Chimaeras share a cartilaginous skeleton with the elasmobranchs as well as their dermal denticles. Reproductive systems and brain structure are also quite similar, as are the presence of claspers on the pelvic fins of the male. Their many similarities, particularly the presence of cartilaginous skeletons in

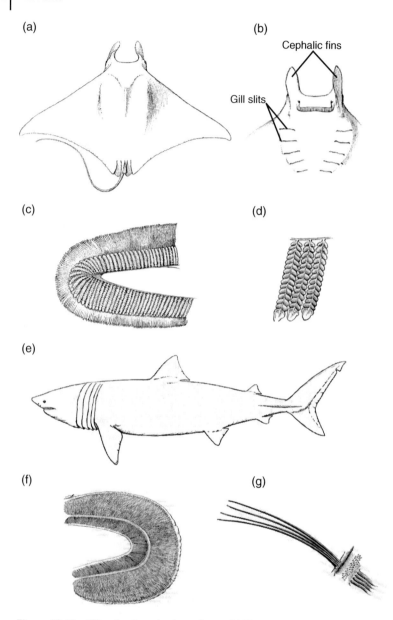

Figure 10.15 Filter-feeding sharks and rays. (a) The manta ray *Mobula hypostoma*; (b) ventral view of the head of *M. hypostoma* showing the cephalic fins and gill slits; (c) manta ray branchial arch showing the arrangement of gill plates and gill folds; (d) margins of manta ray gill plates showing the arrangement of the lateral lobes; (e) the basking shark *Cetorhinus maximus*; (f) gill arch from *C. maximus* showing the gill folds and gill rakers; (g) four gill rakers from a gill arch of *C. maximus*, with bases of the gill folds. *Sources*: (a, b) Adapted from Bigelow and Schroeder (1953), figure 113, (p. 489); (c, d) Adapted from Bigelow and Schroeder (1953), figure 110 (p. 481); (e) Adapted from Bigelow and Schroeder (1948), figure 23 (p. 148), (f, g) Adapted from Bigelow and Schroeder (1948), figure 24 (p. 149).

The Classes of Living Fishes | 693

and Moore 2008; Helfman et al. 2009; Nelson et al. 2016)

Fertilization in chimaeras is internal, effected by the male claspers, but the group as a whole is oviparous, producing large horny egg cases for embryonic development. The egg cases are large relative to the body of the female, and two are laid at a time, one from each ovary. Incubation times range from 9 to 12 months (Bigelow and Schroeder 1953), similar to the gestation times in sharks (Table 10.2).

Chimaeras swim with undulatory motions of their tail and posterior body and with flapping movements of their substantial pectorals (Figure 10.17). The pectorals can also be used in fine-scale movement through wavelike undulations along the fin margin, resembling those in reef fishes (Dean 1906). Pelvic fins aid in balance but largely remain stationary. Water for respiration enters through the nares, flows through branchial chambers with four sets of gills, and exits through a single flap on each side, unlike the elasmobranchs with their five to seven gill slits.

Chimaeras feed on benthic invertebrates, including gastropods and crustaceans, as well as small fishes. Examination of stomach contents suggests that they can subdue and ingest prey seemingly too large for their small mouths (Dean 1906).

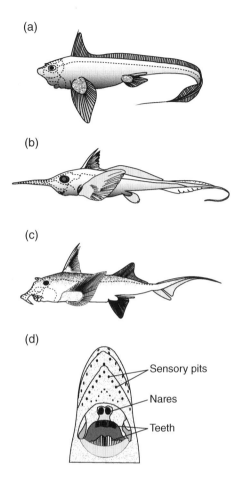

Figure 10.16 The Holocephali (chimaeras). (a) Short-nose chimaera *Hydrolagus affinis*; (b) long-nose chimaera *Harriotta* sp.; (c) plow-nose chimaera *Callorhynchus* sp.; (d) ventral view of the head of a long-nose chimaera showing the teeth, nares, and sensory pits. *Sources:* (a) Adapted from Jordan and Evermann (1900), plate XIX; (b) Adapted from Bigelow and Schroeder (1953), figure 124 (p. 553); (c) Adapted from Smith and Heemstra (1991), figure 34.1 (p. 148).

Class Coelacanthi

Coelacanths, so named for the hollow rays in their tailfin (coel = space, acanth = spine, Bone and Moore 2008), are a well-known fossil group dating back to mid-Devonian times. Their fossils had been studied for about 100 years prior to the discovery of a living specimen in 1938. They were "rediscovered" by Marjorie Courtenay-Latimer, a scientist working at the East London Museum in South Africa who,

both groups, have traditionally resulted in their being grouped with the elasmobranchs as the *Chondrichthyes*, or cartilaginous fishes. Our system has no such formal designation for the two groups, but Chondrichthyes as a taxonomic rank, usually a Class, is still used extensively (Bone

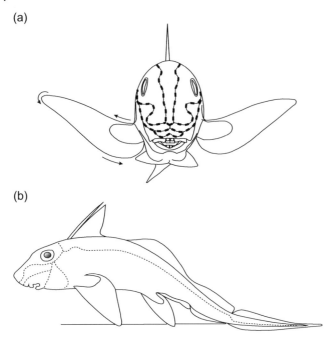

Figure 10.17 *Hydrolagus colliei*. (a) Anterior view of *H. colliei*. Arrows show the direction of undulations of the pectoral fins when swimming. Undulations begin at the proximal-anterior edge of the fin and progress out to the tip and then back to the proximal-posterior edge; (b) *H. colliei* in resting position with pectoral and pelvic fins and a small area of the ventral surface between the vent and caudal fin in contact with the substrate. *Source:* Dean (1906), figure 4 (p. 16).

when looking over the catch of the *Nerine*, a local trawler that often provided specimens for the museum, spotted an unusual fish in the catch (Courtenay-Latimer 1979; Helfman et al. 2009). Ms. Courtenay-Latimer encountered many difficulties transporting and preserving the large fish (1.5 m in length), but she persisted and with the help of a local taxidermist was able to save the very tough skin and get the fish mounted for viewing. Dr. J.L.B. Smith of the South African Museum examined the fish and pronounced it a Coelacanth, a fish that supposedly had been extinct for about 70 Ma (Figure 10.18a). Dr. Smith named the fish *Latimeria chalumnae* after Ms. Courtenay-Latimer and the Chaluma River, near the capture site. The discovery captured the world's imagination: who knew how many other living fossils were out there waiting to be discovered?

Many further specimens of *L. chalumnae* were captured off southern Africa around the islands of the Comoro Archipelago near Madagascar. Later, a new species of *Latimeria, L. menadoensis,* was discovered in the vicinity of Sulawesi (or Celebes). The two species are quite similar in appearance but have been discriminated using molecular methods (Bone and Moore 2008).

The coelacanths have a cartilaginous skeleton (Figure 10.18b) with ossified fin lobes supporting their lobular pectoral, pelvic, anal, and second dorsal fins and a large, completely unossified, notochord running from the skull completely down the midline of the body. The pectoral and pelvic fins are

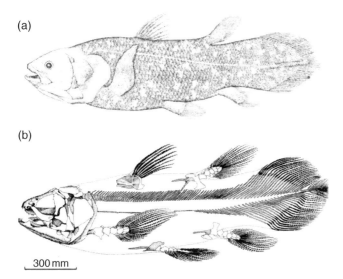

Figure 10.18 Coelacanths. (a) Adult male coelacanth *Latimeria chalumnae*; (b) skeletal structure of *L. chalumnae*. *Sources:* (a) Adapted from Smith and Heemstra (1991), figure 35.1 (p. 152); (b) Adapted from Forey (1998), figure 8.1 (p. 210).

used for sculling to maintain vertical station and orientation in the current flow. A lipid-filled swimbladder aids in maintaining neutral buoyancy. Like the elasmobranchs, coelacanths employ urea as an osmolyte, balancing the osmotic concentration of the blood with that of seawater.

Coelacanths are ovoviviparous, producing litters of pups that have absorbed their yolk sacs in the safety of the mother's ovaries. Like sharks, coelacanths are long-lived (20–40 years), mature late in life (15 years), and have a very long gestation time, perhaps as long as three years per litter (Helfman et al. 2009). Coupled with their small geographic range, their life histories leave them quite vulnerable to over-exploitation.

Class Dipneusti

All extant lungfishes live in riverine systems, but their fossil ancestors were largely marine. The lungfishes are important members of the lineage leading to the tetrapods. Like the coelacanths, all lungfishes have an unrestricted notochord and cartilaginous skeletons.

Class Cladistii

The cladistians, or bichirs, are a diminutive group of ~16 freshwater species restricted to tropical Africa. The importance of the Cladistii is that they are considered the primitive sister group to the Actinopterygii (Nelson et al. 2016). Though their fossil record is sketchy, several of their basic traits, including their scales, which are heavy and rhomboidal in shape (ganoid), the structure and arrangement of their teeth, their low chromosome count, HOX gene sequences, and skull structure all place them at the base of the bony fish lineage (Helfman et al. 2009).

Class Actinopterygii

The world of the ray-finned fishes is very large (17 500 marine species) and is subdivided into several major groups that are mentioned frequently throughout the fish literature.

Subclass Chondrostei

The most primitive of the ray-finned fishes are the Chondrostei, or "cartilage bone," the sturgeons and paddlefishes. As the name suggests, their skeleton is indeed cartilaginous, with a free notochord that runs the length of the body. Some sturgeons, e.g. the lake sturgeon *Acipenser fulvescens*, spend their lives entirely in freshwater. Others like the Atlantic sturgeon (*Acipenser oxyrhynchus*) are mainly oceanic but return to rivers to spawn, a lifestyle termed anadromous.

Sturgeons are long-lived, surviving over 100 years in some cases (Beverton and Holt 1959), and slow to mature. For example, the Atlantic sturgeon takes from 5 to 30 years to reach maturity and may only reproduce every 3–5 years thereafter. When they do reproduce, they produce very large numbers of eggs that have been prized as a delicacy for centuries, most notably those of the Beluga. Their long life, slow growth, late maturity, and high fecundity worked as a life-history strategy for millions of years, but the world of today is defeating the sturgeon. Many natal rivers are compromised by pollution or damming, large adults are harvested, and young sturgeon do not enter the reproducing population quickly enough to offset the damage. It will be very difficult to prevent extirpation of the most sought-after species.

Subclass Holostei

The Holostei or "whole bone," the next subclass in the Actinopterygii, is named for the bony skeleton that characterizes the group. They are the most primitive group in the all-inclusive but unranked clade Neopterygii ("new fin") and are the most primitive fishes that have well-developed vertebrae and an ossified axial skeleton. The group includes the gars (Lepisosteiformes – seven species) and the bowfins (Amiiformes – one species), both primarily resident in the sluggish freshwater of slow-moving rivers and marshes. Gars also venture into brackish water.

The next subclass, the Teleostei or "bony fishes," comprises about 96% of all living fishes (Bone and Moore 2008).

Subclass Teleostei

All classification systems to date agree that there have been four major radiations within the teleosts (Greenwood et al. 1966; Bone and Moore 2008; Helfman et al. 2009; Nelson et al. 2016) which are identified as cohorts in the classification system in Appendix A. These are the four: the *Elopomorpha*, the tarpon, bonefish and eels; the *Osteoglossomorpha* or bonytongues, an entirely freshwater group; the *Otocephala*, including the sardines and anchovies; and the *Euteleostei*, by far the largest and including all other fishes. Of the four, three are well-represented in the pelagic realm: the Elopomorpha, Otocephala, and Euteleostei. Fishes will be treated from most primitive to most advanced.

Cohort Elopomorpha

The elopomorphs are united in possessing a leptocephalus (slim head) larva, a transparent, laterally compressed, larva (Figure 10.19) varying in length from 7 cm to over a meter, with most of them >50 mm (Smith 1989). They are a significant contributor to micronekton numbers and biomass in the upper 1000 m of the Gulf of Mexico (Hopkins and Lancraft 1984), migrating into epipelagic waters at night. Residence times in the water column vary from 30 days to as much as a year (Castonguay 1987; Schmidt 1925; Crabtree et al. 1992). In fact,

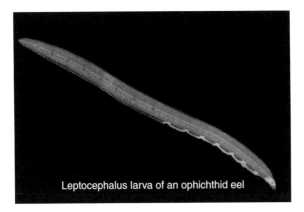

Figure 10.19 Leptocephalus larva of an ophichthid eel. *Source:* © Dante Fenolio, reproduced with permission. See color plate section for color representation of this figure.

the onset of metamorphosis to the juvenile phase may be delayed until appropriate physical conditions are met (Tseng 1990).

Leptocephali are fascinating. Their developmental strategy is quite different from that of the larvae of most fishes (Pfeiler 1986). In most larval fishes, increases in somatic mass are accomplished by the accumulation of protein in the form of muscle (Balbontin et al. 1973; Ehrlich 1974a, b; Cetta and Capuzzo 1982). Little energy is stored; all energy exceeding metabolic requirements is devoted to growth (e.g. Brightman et al. 1996). The larval period typically ranges from a week to a month and may be dictated by the growth rate (Werner and Gilliam 1984). Overall, the strategy of a typical larva is to reduce vulnerability to predation by rapidly increasing in size and minimizing duration of the larval phase.

In leptocephali, development proceeds in two main phases (Pfeiler 1986). In phase I, the larvae grow until they reach a maximum typical of the species. During phase I, energy reserves are accumulated within the leptocephalus as lipid, in addition to an acellular mass contained within a mucinous pouch that runs the length of the body. The mucinous pouch contains proteoglycans, compounds made up of a conjugated peptide and glycosaminoglycan carbohydrates, most familiar as mucous and cartilage. The accumulated proteoglycans act as a "skeleton" that the muscles can work against during swimming as the larva attains its maximum size. Leptocephali exhibit well-developed backward and forward anguilliform (eel-like) locomotion, yet, except for their teeth and otoliths, they are completely unossified. The glycosaminoglycans act as a firm gelatinous skeleton for the musculature to work against, conferring an exceptional swimming ability without a bony skeleton (Bishop and Torres 1999). Phase II of leptocephalus development consists of a shrinkage in size and a profound change in shape to the juvenile morph, fueled by combustion of the accumulated energy reserves in the form of glycosaminoglycans and lipids (Pfeiler 1996; Bishop and Torres 1999, 2001; Bishop et al. 2000).

One of the great mysteries of the sea has been what leptocephali used for food. Nothing was ever visible in the guts of captured specimens, yet they were able to grow rapidly (greater than a mm day^{-1}), and sometimes to substantial size, particularly for a larva (Bishop et al. 2000). Speculation on their diet ranged from uptake of dissolved organic carbon (DOC) to marine snow (small organic aggregates, the sinking residue of surface productivity) to larvacean houses (Chapter 9) (Otake and Maruyama 1993; Mochioka and Iwamizu 1996). Mochioka and Iwamizu's (1996) study of leptocephalus gut contents made the convincing case that larvacean houses, which leptocephali grabbed with their long teeth, are the food source.

The elopomorphs include five orders of fishes with very different adult lifestyles. The tarpons (Figure 10.20a) and ladyfish are both large (1 m) coastal midwater predators They both spawn at sea, and their leptocephali migrate inshore to metamorphose into their juvenile stage. Bonefishes (Figure 10.20b) are typically inhabitants of

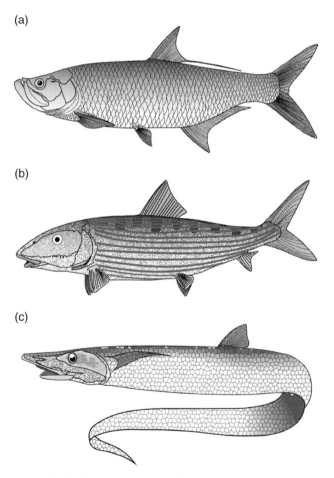

Figure 10.20 The elopomorphs. (a) The tarpon *Megalops atlanticus*; (b) the bonefish *Albula vulpes*; (c) the halosaur (spiny eel) *Halosauropsis macrochir*.

estuaries and inshore flats, feeding on benthic invertebrates with their ventrally directed mouth. Their deeper-living relatives, the pterothrissins, have a similar lifestyle with a deeper distribution. All the albulids spawn at sea. Note that the leptocephali of the albulids and elopiforms have an obvious tail fin.

The notacanthiformes (halosaurs and spiny eels) are a deep benthopelagic group with representatives in bathyal (200–3000 m) and abyssal (3000–6000 m) depths of the ocean. Eel-like in appearance (Figure 10.20c) they are a truly deep-living group with no shallow-water representatives (Priede 2017).

The most important elopomorphs to students of open-ocean fauna are the orders *Anguilliformes* and *Saccopharyngiformes*. Though the family *Derichthyidae* (longneck eels) inhabits mesopelagic and bathypelagic waters worldwide (Priede 2017), they are rarely captured in midwater trawls. In contrast, the snipe eels (*Nemeichthyidae*) and sawtooth eels (*Serrivomeridae*) (Figure 10.21) are commonly caught at mesopelagic depths in waters of the Atlantic, Pacific, and Gulf of Mexico. Though many of the remaining families have deep-water representatives, the Anguilliformes are predominantly benthic.

The Saccopharyngidae include the famous swallowers (*Saccopharynx*) (Figure 10.21) and umbrellamouth gulpers (*Eurypharynx*), both resident at lower mesopelagic and bathypelagic depths. Less well known are the bobtail snipe eels, resembling a small version of the Nemeichthyids with a more typical fish-like tail. The saccopharyngiforms are rarely captured in midwater trawls. Though enjoying a panglobal distribution (Briggs 1974), they are sparsely distributed at bathypelagic depths. *Saccopharynx* is capable of taking fishes larger than itself, which it digests in its

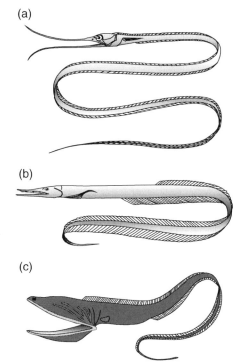

Figure 10.21 The Anguilliformes and Sachopharyngiformes. (a) The snipe eel *Nemichthys scolopaceus*; (b) the sawtooth eel *Serrivomer beanii*; (c) the gulper eel *Eurypharynx pelecanoides*.

highly distensible stomach. Though its mouth is much larger than that of *Saccopharynx*, *Eurypharynx* feeds mainly on smaller prey such as shrimp and juvenile cephalopods. It opens its mouth wide to engulf its prey and squeezes the water out through its gill rakers, retaining the prey in its gullet (Priede 2017). Its feeding strategy is much like the basking shark or baleen whale, in miniature.

The one-jaw gulpers, or Monognathidae, are very rare residents of the bathypelagic zone. Their common name comes from the fact that they have no true upper jaw. The lower jaw closes directly on the neurocranium (Priede 2017). Little is known about their biology, but their gut contents suggest a diet of shrimp.

Cohort Otocephala

The cohort Otocephala (ear-head) comprises two superorders: the Clupeomorpha (sardines and herrings) and the Ostariophysi (milkfishes, carps, characins, knifefishes, and catfishes). The sardines and herrings are important coastal pelagics that reach their greatest abundances in the highly productive regions associated with either upwelling or riverine inputs (Figure 10.22) (Checkley 2009). The Ostariophysi are a large and interesting group living primarily in freshwater. Marine representatives include the milkfishes (Gonorhynchiformes) and the sea catfishes of the family Ariidae in the order Siluriformes.

The Otocephala are labeled "ear-head" because their swimbladders are connected to the auditory structures in the head, greatly improving their hearing. Their gas-filled swimbladders act as superb resonators. As discussed further in "Mechanoreception," a variety of different mechanisms are used to effect the connection so that sound resonating in the swimbladder is communicated to the inner ear.

Cohort Euteleostei

The euteleosts comprise the most advanced of the four cohorts making up the bony fishes. The cohort includes about 14 600 marine species with body morphs ranging from pikes to parrotfish. They are divided into eight superorders, of which six have pelagic representatives. Moving from most primitive to most advanced, the six are the Osmeromorpha, the Cyclosquamata, the Scopelomorpha, the Lamprimorpha, the Paracanthopterygii, and the Acanthopterygii.

The Acanthopterygii, or spiny-finned fishes, include many of the important epipelagic, mesopelagic, and bathypelagic taxa. The orders Beryciformes (Bigscales), Trachichthyiformes (Fangtooths), Scombriformes (Tunas and Mackerels), Istiophoriformes (Billfishes), Beloniformes (Flying fishes and Sauries), and Lophiiformes (Anglerfishes) are notable groups with multiple pelagic families. Others such as the Tetraodontiformes have important individual species, in this case the Mola Mola or Ocean Sunfish.

The amount of information on the different pelagic taxa varies considerably, ranging from simple numerical abundance to detailed information on life history and zoogeography. Our knowledge is predicated on the results of several dedicated sampling programs whose main mission was to elucidate the character of the open-ocean biota. Those programs include, but are not limited to, the following: the SOND cruise (1965) of the Institute of Oceanographic Sciences, UK; the Gulf of Mexico sampling program of T.L Hopkins (1970–2000); the tropical Pacific (Hawaiian) sampling program of T.A. Clarke (1970–2000); the subarctic Pacific sampling program of W.G. Pearcy at Oregon State University; the physiologically oriented program of J.J. Childress based in the California Borderland; the 1967–1969 Bermuda Ocean Acre program in the northern Sargasso Sea (R.H.Gibbs and C.F.E. Roper); the observational program based at MBARI led by B.H. Robison; the wide-ranging sampling of R.H. Backus and P. H. Wiebe at WHOI; the recent Gulf of Mexico DEEPEND program led by T.T. Sutton; and the Antarctic micronekton program led by J.J. Torres.

One important characteristic of the midwater fish communities from boreal to tropical waters is that the dominant taxa are quite consistent from place to place. In all communities examined, the two overwhelmingly dominant fish families are the

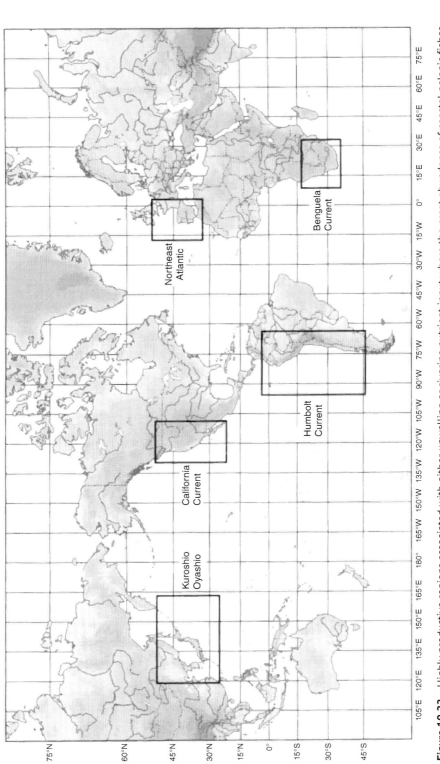

Figure 10.22 Highly productive regions associated with either upwelling or riverine nutrient input where the great abundance of coastal pelagic fishes supports major fisheries. *Source*: Checkley (2009), figure 3.1 (p. 13). Reproduced with the permission of Cambridge University Press.

Gonostomatidae, or bristlemouths, and the Myctophidae, or lanternfishes. In many cases, the two families comprise 90% of the mesopelagic fishes captured (e.g. Gibbs and Roper 1970; Hopkins and Lancraft 1984). Other universal dominants are the Sternoptychidae, or hatchetfishes, and the Stomiidae, or dragonfishes. Families that are consistently present in smaller numbers include the Melamphaidae (bigscales) and the Bathylagidae (deep-sea smelts).

It is important to embrace the fact that the way we envision the pelagic environment is, in large measure, a function of how we sample it. Almost all sampling of pelagic fauna relies on scientific trawls, the three main types being the IKMT (Isaacs-Kidd Midwater Trawl) (Figure 7.60), the MOCNESS (Multiple Opening and Closing Net and Environmental Sampling System) (Figure 7.44), and the Tucker trawl. Scientific trawls are towed slowly, at 2–3 knots, thus selecting for animals that are either slow swimmers or are slow to react to the approaching net, e.g., *Serrivomer* sp. (slow to react) and some species of *Cyclothone* (too slow). Fortunately for studies of mesopelagic and bathypelagic fishes, most are weak swimmers, since in dimly lit zones, the need for strong locomotory ability is greatly reduced (e.g. Childress et al. 1990).

Superorder Osmeromorpha
Order Osmeriformes The Osmeromorpha comprise two orders with several important mesopelagic families, the orders Osmeriformes and the Stomiiformes. The most commonly encountered osmeriform families in midwater trawls are the Bathylagidae (deep-sea smelts), the alepocephalids (slickheads), and the platytroctids (tubeshoulders). The slickheads and tubeshoulders are never numerous, and when captured in the midwater, they are taken from lower mesopelagic to bathypelagic depths (500–1500 m). The spookfishes (Opisthoproctidae) are also captured in midwater trawls sampling in lower mesopelagic depths, but only infrequently. They are unusual in appearance, with either a slender, elongated body (spookfish) or a stubby, short one (barreleyes), and are equipped with upward or forward-looking tubular eyes. In at least one species, *Macropinna microstoma*, a resident of the California borderland, the fish can rotate its eyes from looking upward to looking forward, presumably to help track prey (Robison and Reisenbichler 2008). Some spookfishes are highly transparent (Figure 10.23), which gives them their common name. The species shown here, *Dolichopteryx longipes*, has a retinal diverticulum protruding from the ventrolateral aspect of each eye, which aids in gathering light from below and to the side. Light from below is reflected into it off the silvery sides of the eye, giving an indication of a point-source of light such as that produced by a bioluminescent animal (Herring 2002). The main axis of the tubular eye faces upward, providing the best chance of seeing an animal silhouetted in the light welling down from above.

The Alepocephalids (Figure 10.24) are primarily benthopelagic (Crabtree and Sulak 1986; Crabtree 1995; Priede 2017). They are present in the bathyal with some species reaching abyssal depths. Alepocephalids are taken as bycatch, sometimes in large numbers, by deep commercial bottom trawls. Reproduction in the Alepocephalids is species-dependent, with some having pelagic eggs and others depositing their eggs on the seafloor (Priede 2017). Drazen and Sutton (2017) classify the larger genera (e.g. *Alepocephalus*) as demersal

(a) *Dolichopteryx longipes* (dorsal view)

(b) Head of *D. longipes* (lateral view)

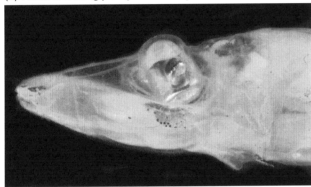

Figure 10.23 Opisthoproctidae. (a) Dorsal view of the spookfish *Dolichopteryx longipes* showing the upward-looking eyes; (b) close-up of the head of *D. longipes* showing the tubular (barrel-shaped) eye with retinal diverticulum. *Source:* © Dante Fenolio, reproduced with permission. See color plate section for color representation of this figure.

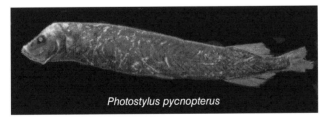

Figure 10.24 The alepocephalid *Photostylus pycnopterus*. *Source:* © Dante Fenolio, reproduced with permission. See color plate section for color representation of this figure.

micronektivores, feeding on small midwater fishes, a variety of crustaceans, and cephalopods. Smaller genera (e.g. *Conocara*) are considered to be gelativores, primarily feeding on cnidarians, ctenophores, and pyrosomes.

The Bathylagids (Figure 10.25) are a mesopelagic-upper bathypelagic family with two basic colorations, dark with big scales in the deeper-living species (e.g. *Bathlagus pacificus*) and silvery in the upper mesopelagic species (e.g.

(a) *Dolicholagus longirostris*

(b) *D. longirostris*

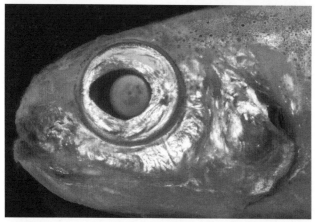

(c) *Bathylagus pacificus*

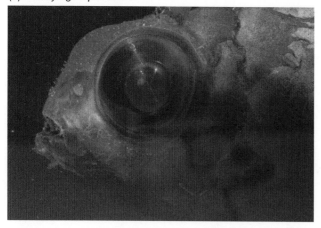

Figure 10.25 Bathylagids. (a) The silvery-colored, upper-mesopelagic bathylagid *Dolicholagus longirostris*; (b) close-up of the head of *D. longirostris* showing the aphakic space in the anterior part of the eye; (c) the dark-colored, deeper-living bathypelagic *Bathylagus pacificus*. Note the relatively larger eye and large scales. *Source:* © Dante Fenolio, reproduced with permission. See color plate section for color representation of this figure.

Dolicholagus longirostris). Commensurate with their deeper-living habits, the species in the lower mesopelagic have very large eyes for gathering ambient light. Bathylagids vary considerably in their abundance. They are more numerous in the waters of the California borderland, particularly the deep inshore basins, in the Pacific Northwest, and in the Antarctic than they are in the Gulf of Mexico. They are zooplanktivores, taking copepods and euphausiids as well as cnidarians and salps. Bathylagids are broadcast spawners with many species having highly evolved, stalked-eyed larvae.

The Platytroctids, or tubeshoulders, are another family with species distributed in the lower mesopelagic and upper bathypelagic zones (Figure 10.26). The tube that gives the group its common name produces a bright, blue–green, luminescent fluid described by Nicol (1958) as "myriads of blue-green sparks," presumably as a defense mechanism. The displays are impressive and can light up the whole catch bucket from a trawl. Each of the tubes connects to a large sac on either side of the fish that manufactures the fluid. Platytroctids are found in tropical and temperate waters worldwide (Priede 2017).

Order Stomiiformes The order stomiiformes includes four of the most important midwater families. With the exception of polar regions, the gonostomatids are always in the top two mesopelagic families in terms of numerical abundance, usually closely followed by the sternoptychids (Gibbs and Roper 1970; Badcock 1970; Pearcy et al. 1977; Hopkins and Lancraft 1984; Lancraft et al. 1989, 1991). As predators, the stomiids are further down the ladder in abundance, but in temperate and particularly tropical systems, they are often the most diverse (Sutton and Hopkins 1996a).

The Gonostomatidae, or bristlemouths, are a fabulously abundant group considered by many to be the most abundant vertebrate on the planet. Most important of the eight genera comprising the family is the genus *Cyclothone* (Figure 10.27), with a total of seven species. The genus is found panglobally, as are several of its species, occupying depths from the upper mesopelagic to the upper bathpelagic zone. They are diminutive in size, typically <70mm, and dark, half-dark, or pale white in color. Their common name comes from their numerous bristle-like teeth. Pale *Cyclothone* like the *Cyclothone braueri* shown in Figure 10.27 inhabit the upper mesopelagic zone (Miya and Nishida 1996). The silvery stomach is thought to trap the light emitted by its zooplankton prey. *Cyclothone pallida* and *C. obscura* are deeper-living species residing in the lower mesopelagic and upper bathpelagic zones, respectively. Photophores are absent in *C. obscura*, and eye-size is reduced.

The Gonostomatids come in four general body morphs. The genus *Cyclothone* is

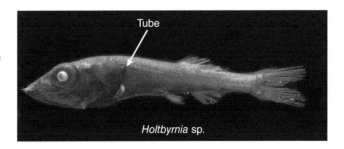

Figure 10.26 The platytroctid *Holtbyrnia* sp. showing the tube that gives the platytroctids the common name "tube shoulder." *Source:* © Dante Fenolio, reproduced with permission. See color plate section for color representation of this figure.

(a)

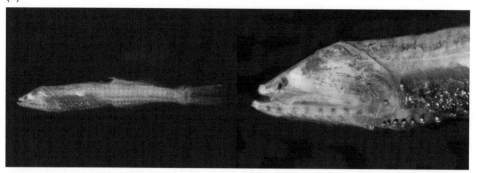

(b)

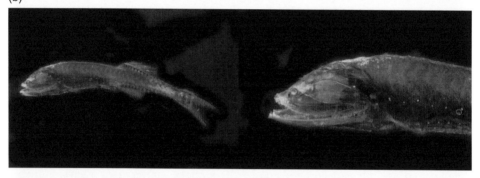

(c)

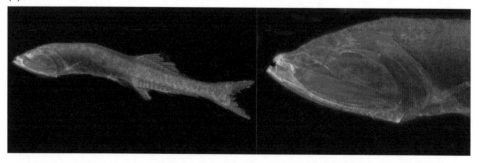

Figure 10.27 A comparison of biological characteristics between three species of *Cyclothone* from different depth zones. Notable differences include coloration, eye size, and photophores. (a) *Cyclothone braueri*: upper mesopelagic, light color, small eyes, and large prominent photophores; (b) *Cyclothone pallida*: lower mesopelagic, dark color, reduced eyes, small photophores; (c) *Cyclothone obscura*: bathypelagic, intermediate dark color, very reduced eyes, no photophores. *Source:* © Dante Fenolio, reproduced with permission. See color plate section for color representation of this figure.

small, with a reduced eye and, in most species, an array of small photophores distributed ventrally and ventrolaterally. The genus *Margrethia* is deeper-bodied with silvery sides, large well-developed photophores, and a pronounced eye. The genus *Bonapartia* also has silvery sides, large photophores, and a prominent eye but is not as deeply bodied as *Margrethia*. Largest of the Gonostomatids are the two

Figure 10.28 Lateral view (top) and close-up of the head (bottom) of the large species of gonostomatid, *Sigmops elongatum*. Note the prominent photophore array and dentition. *Source:* © Dante Fenolio, reproduced with permission. See color plate section for color representation of this figure.

genera *Gonostoma* and *Sigmops* (Figure 10.28), whose members may reach >175 mm in standard length (Lancraft et al. 1988). They are dark with silvery sides, well-developed photophores, and small eyes. *Cyclothone, Gonostoma,* and *Sigmops* lead the gonostomatids in numerical abundance, often dominating the mesopelagic fish fauna.

Several studies (e.g. Miya and Nemoto 1991; Miya and Nishida 1996) suggest that *Cyclothone* species do not vertically migrate but instead remain within their preferred depth strata on a diel basis (Figure 10.29). Further, their preferred strata are consistent from location to location. In contrast, *Gonostoma* and *Sigmops* tend to be strong vertical migrators (e.g. Lancraft et al. 1988). In the Gulf of Mexico, *Sigmops elongatum* (formerly *Gonostoma elongatum*) migrates from daytime depths between 500 and 700 m to the upper 250 m at night.

Cyclothone are zooplanktivores, primarily consuming copepods and ostracods in the mesopelagic (Hopkins et al. 1996) and bathypelagic (Burghart et al. 2010) zones of the Gulf of Mexico. *S. elongatum* is planktivorous, primarily taking euphausiids, copepods, and ostracods (Lancraft et al. 1988).

Sex change is evident in at least two Gonostomatid species: *Cyclothone atraria* (Miya and Nemoto 1985) and *Gonostoma gracile* (Kawaguchi and Marumo 1967). In both cases, the change is from male to female, or protandry, and, in both species, all larger fish are female. Since the number of eggs produced increases with increasing size, larger females produce more eggs, increasing their likelihood of reproductive success in any given reproductive season.

Miya and Nemoto (1991) provide an interesting comparison of the life histories of *Cyclothone* species residing at different depths in Sagami Bay, Japan. Three species were compared. The following parentheses designate size at first female reproduction, depth range, and center of distribution: *C. alba* (21 mm, 300–500 m, 350 m), *C. pseudopallida* (31 mm, 400–600 m, 450 m), and *C. atraria* (40 mm, 400–1000 m, 600 m). In *C. alba*, the smallest species of the three, females exhibited an early age at first reproduction (two years), semelparity (reproducing once at the end of its life), and low fecundity (255 eggs). *C. pseudopallida*, intermediate in size, reproduces at least twice beginning in its third year and has an intermediate level of fecundity (590 eggs). *C. atraria*, the largest species, reproduces at least twice beginning in its fourth year with a high level of fecundity (990 eggs). Egg size is virtually identical in all three species (0.53–0.54 mm), and all eggs are positively buoyant. Thus, larvae for all three species develop in near-surface waters. Larvae must make their way to their resident depth, potentially exposing themselves to predators during the ontogenetic journey.

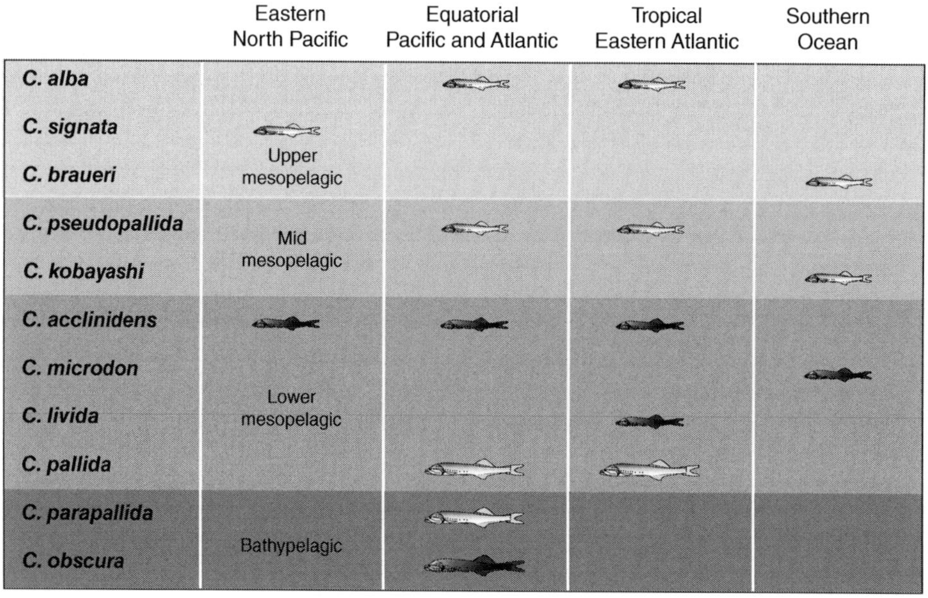

	Eastern North Pacific	Equatorial Pacific and Atlantic	Tropical Eastern Atlantic	Southern Ocean
C. alba		🐟	🐟	
C. signata	🐟 (Upper mesopelagic)			
C. braueri				🐟
C. pseudopallida		🐟	🐟	
C. kobayashi	(Mid mesopelagic)			🐟
C. acclinidens	🐟	🐟	🐟	
C. microdon				🐟
C. livida	(Lower mesopelagic)		🐟	
C. pallida		🐟	🐟	
C. parapallida		🐟		
C. obscura	(Bathypelagic)	🐟		

Figure 10.29 Combinations of co-dominant species of *Cyclothone* from the Pacific, Atlantic, and Southern Ocean. Vertical sequences of fishes within each depth category do not imply actual depth stratification. *Source:* Adapted from Miya and Nishida (1996), page 391.

Cyclothone alba has the shortest journey, so its metamorphosing larvae face the least exposure. *C. alba* dwells in the well-lit upper mesopelagic zone, and adults are likely subject to higher levels of visual predation than those of its deeper-living congeners. It conforms reasonably well to a classical r-selected species, reproducing early in life with one "big bang" reproductive event and then dying. The deeper-living *C. pseudopallida* delays initial reproduction to its third year and reproduces at least once more after that, suggesting a lower probability of adult mortality. Its females are larger than those of *C. alba*, and individual lifetime fecundity is considerably higher. The deepest living species, *C. atraria*, achieves sexual maturity as a male in its third year and, after changing sex, as a larger female in its fourth year, reproducing at least once more after that. Its females are much larger than either of its congeners, and its annual and lifetime individual fecundities are considerably higher. Growth in size proceeds at an identical rate in *C. alba* and *C. pseudopallida* and at an only slightly slower rate in *C. atraria* (Miya and Nemoto 1991). The differences in adult size and annual egg production are solely a result of greater longevity in the deeper living species.

Two environmental factors important to survival covary with depth. Light declines exponentially over the upper 1000 m and is extinguished, or nearly so, at 1000 m depth. The availability of zooplankton prey also declines exponentially with increasing depth (e.g. Vinogradov 1970; Hopkins 1982). The size-increase with increasing depth reported by Miya and Nemoto confers faster swim-speed and a larger gape, effectively increasing the prey spectrum available and survivability at depth. Miya and Nemoto concluded that the main reason for the increased fecundity in the deeper-living species was the longer journey their developing larvae must make to reach their resident depth. That is an important factor but others, such as the lower availability of zooplankton prey to the younger stages of deeper-living species, are important as well. Comparative data sets like those described in Miya and Nemoto (1991) are rare in the literature, and theirs is particularly well executed. The most important conclusion is that each of the three species is well-adapted to its depth of occurrence.

The lightfishes, the family Phosichthyidae, are conspicuous members of the midwater fish fauna with two widely distributed genera, *Vinciguerria* (Figure 10.30) and *Ichthyococcus*. They are often present in midwater trawls sweeping the upper 1000 m of the water column but are never abundant. Species in the genus *Vinciguerria* are vertical migrators with a daytime distribution in the middle mesopelagic and a

nighttime distribution in the upper mesopelagic/epipelagic. For example, *Vinciguerria nimbaria* is found from 200 to 500 m during the day and in the upper 100 m at night in the seas off Southern Japan (Ozawa et al. 1977). *V. nimbaria* is zooplanktivorous, feeding at night primarily on small copepods in a variety of genera and on euphausiid larvae, small amphipods, decapod larvae, and pteropods. There are five additional genera and a total of 24 species in the family (Priede 2017).

The family Sternoptychidae comprises two subfamilies, the Maurolycinae, or pearlsides, and the Sternoptychinae, or hatchetfishes. Of the two, the hatchetfishes are the most abundant, often occupying the "number three" spot in abundance after the gonostomatids and myctophids (e.g. Gibbs and Roper 1970; Hopkins and Lancraft 1984).

The pearlsides include 30 species in seven genera. They are found in temperate and tropical latitudes worldwide, with either benthopelagic or mesopelagic distributions (Priede 2017). The most abundant and widely distributed pelagic species is the Constellationfish, *Valencienellus tripunctulatus*, found worldwide between 67°N and 34°S (Figure 10.31). Hopkins and Baird (1981) and Baird and Hopkins (1981a, b)

Figure 10.30 The phosichthyid *Vinciguerria poweriae*: whole body (top) and close-up of the head (bottom). Note the well-developed eye with large ventronasal and postorbital photophores and the well-developed ventral photophore array. *Source:* © Dante Fenolio, reproduced with permission. See color plate section for color representation of this figure.

Figure 10.31 The sternoptychid *Valenciennellus tripunctulatus*. Note the well-developed upward-looking eye, the array of large ventral photophores, and the prominent melanophores along the side of the body just above the lateral line. *Source:* © Dante Fenolio, reproduced with permission. See color plate section for color representation of this figure.

describe the vertical distribution, diet, feeding chronology and strategy, and energetics of *Valenciennellus*.

V. tripunctulatus is a non-migrator residing between 250 and 550 m. In the Gulf of Mexico, highest densities during the day were between 290 and 460 m and at night between 180 and 500 m. Feeding occurs mainly between 1200 and 2200 hours, with copepods and ostracods as the primary prey. *Valenciennellus* appears to be actively selecting the copepod genus *Pleuromamma* because it appears in the stomachs at a greater frequency than would be expected based on its numbers in the plankton. Growth based on daily ration was modeled, with the most likely scenario yielding an adult fish of 30 mm (max) length being 1 year old.

The subfamily Sternoptychinae (Figure 10.32) comprises three genera: *Argyropelecus* with 7 species, *Sternoptyx* with 4 species, and *Polyipnus* with 32 species (Priede 2017). All three genera are quite small in size, with standard lengths typically in the 3–5 cm range. However, *A. aculeatus* can reach 83 mm (Howell and Krueger 1987), and *A. gigas* can reach 11.5 cm (Badcock 1970). All three genera are strikingly silver, laterally compressed, and deep-bodied: hatchet-like in shape. They are all equipped with a suite of ventral photophores that are among the most sophisticated in the animal kingdom.

Argyropelecus and *Sternopyx* are prominent members of the mesopelagic community from boreal to tropical latitudes (e.g. Badcock 1970; Hopkins and Baird 1973; Badcock and Merrett 1976; Pearcy et al. 1977; Howell and Krueger 1987; Hopkins and Lancraft 1984). They are absent from south polar waters, in particular the Scotia-Weddell Sea region of the Atlantic sector (Lancraft et al. 1989, 1991). In any given ocean system, the genus *Argyropelecus* tends to be distributed above

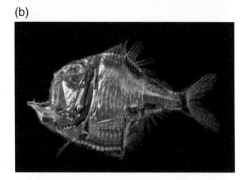

Figure 10.32 Sternoptychidae: examples of the three genera of hatchetfishes. (a) *Polyipnus clarus*; (b) *Argyropelecus aculeatus*; (c) *Sternoptyx diaphana*. Note the distinctive tubular photophores along the ventrum and the small pre-orbital photophore that shines directly into the eye, helping to match photophore output with downwelling light intensity (countershading). *Source:* © Dante Fenolio, reproduced with permission. See color plate section for color representation of this figure.

Sternoptyx, though there is considerable overlap (e.g. Badcock and Merrett 1976; Howell and Krueger 1987). The genus

Polyipnus is more speciose than its two confamilials, and many of its members are considered benthopelagic (Priede 2017), often captured in bottom trawls (Harold et al. 2002). The fish pictured in Figure 10.32, *Polyipnus clarus*, may be found over the continental shelf; though in the open-ocean system of the Gulf of Mexico, it is occasionally captured in upper mesopelagic (300–500 m) depths well off the shelf.

Argyropelecus aculeatus (Figure 10.32) provides a good example of the genus. It is common at temperate and tropical latitudes in the Atlantic, Indian, and Pacific Ocean basins. Depth distributions differ slightly between regions of study. A typical day-night depth profile found by Badcock and Merrett (1976) in the vicinity of the Canary Islands (30°N 23°W) showed the fish residing between 300 and 500 m during the day and spreading out to 100–500 m at night with a nighttime center of distribution between 200 and 300 m.

In the Bermuda Ocean Acre study (32°N 64°W, Gibbs and Krueger 1987; Howell and Krueger 1987), *Argyropelecus aculeatus* was captured primarily (86% of total numbers captured) between 301 and 500 m during the day and primarily (91%) between 201 and 500 m at night, suggesting a slight upward displacement at night. Analysis of developmental stages during the study suggested that the species breeds continuously throughout the year with a peak during the summer months. Changes in population structure during the year suggested a two-year life history with most adults spawning in the summer or fall months and then dying, but with a few surviving to spawn a second time. "Postlarvae produced during those months develop into juveniles by the following winter, and into subadults by the following late summer or fall. At that time, some reproduce, but most spend a second year as subadults, maturing and spawning near the end of their second year" (Howell and Krueger 1987). In its early life stages, *A. aculeatus* feeds mostly on copepods and ostracods, moving to euphausiids, pteropods, and fish in its later and larger life sizes (Hopkins and Baird 1977).

The family Stomiidae comprises the barbeled dragonfishes, the most diverse and spectacular looking family of midwater predators in the open ocean, with 6 subfamilies and 287 total species (Figures 10.33–10.38). Despite their ferocious appearance, their average maximum length is only 18 cm (Priede 2017) and, like many deep-living species, their bodies tend to be somewhat flaccid due to water contents above 80% of wet mass (Childress and Nygaard 1973). Nonetheless, ichthyologists specializing in the stomiids consider them to be the "Kings of the Sea." Your authors agree. Stomiids are the top resident predators in the world's largest living space. The tunas and billfishes are certainly more formidable predators, but they are only visitors to the mesopelagic zone.

Stomiids are identified by their barbel morphology, which can be problematic; the barbels are easily lost in the net during the fish's journey to the surface. The great majority of stomiid species have a mental barbel on their lower jaw (Figure 10.35) which is believed to act as a lure, enticing smaller fishes to within range of the dragonfish's formidable maw. The prominent suborbital photophores of the stomiids are believed to act as a sighting mechanism, generating bioluminescence to illuminate prey. Observations of stomiids suggest that the fishes have muscular control over their subocular photophores, enabling them to direct the light downward or forward and in some cases to shield with a pigment screen, effectively "turning it off" (Nicol 1960). Most of the observations suggest that a blue light is emitted by the subocular photophores. The luminescence is generated not by bacteria but by the fish's own photosystem (Nicol 1960).

Stomiids are important members of the mesopelagic community worldwide, reaching their highest diversity at tropical/subtropical latitudes (e.g. Clarke 1974; Sutton and Hopkins 1996a). Because they are predators, they never reach the numerical abundances of the zooplanktivores but the stomiid community is often very diverse. There are 83 species of stomiids in the Gulf of Mexico. They are present, but neither as abundant nor as diverse at higher latitudes. The astronesthid, *Borostomias antarcticus*, has been captured as far south as 66°S in the Atlantic sector of the Southern Ocean (Lancraft et al. 1991) and is found at 54°S in the Pacific sector vic. Macquarie Island (Gaskett et al. 2001). Several stomiid species are found in the subarctic Pacific at latitudes of 44–51° (Pearcy et al. 1977; Frost and McCrone 1979; Beamish et al. 1999).

The Astronesthinae or "snaggletooths" (Figure 10.33) are a diverse subfamily with six genera and 59 species worldwide (Nelson et al. 2016). They are represented in all ocean systems. Data on vertical distribution and diet are available from the Gulf of Mexico (GOM) (Sutton and Hopkins 1996a, b) and the waters off Hawaii (Clarke 1974, 1982). The two areas show similar patterns. In the subtropical Pacific (22°N 158°W), eight common Astronesthinid species exhibited a pronounced vertical migration, residing between 500 and 800 m during the day and between 30 and 200 m at night (Clarke 1974). In the GOM, sufficient data were obtained to describe the vertical migration patterns for four of the eight species of Astronesthines. Three of the four were migrators, including *Astronesthes micropogon*, resided between 400 and 700 m during the day, ascending to between 0 and 200 m at night (Figure 10.33a). *Astronesthes niger*, the remaining species, showed a discontinuous distribution. Daytime residence for the entire population was between 400 and 700 m, but the nighttime distribution was split, with most of the population migrating to between 0 and 200 m and the remainder staying at daytime depth. The three dominant stomiids in the Gulf also exhibited discontinuous distributions. In the GOM, the Astronesthines comprised about 12% of the stomiid assemblage in numerical abundance.

Astronesthines in the GOM were classified as generalized zooplankton/micronekton feeders with euphausiids as their primary prey in both abundance and biomass ("Feeding Guild 2": Sutton and Hopkins 1996b). Myctophids and caridean shrimp ranked second in abundance and biomass, with copepods a distant third place. Clarke (1982) observed copepods and ostracods in the guts of smaller (<60mm). Astronesthinids, with euphausiids and fishes dominating in the larger sizes.

The subfamily Chauliodontinae, or "viperfishes", contains one of the three dominant stomiids in the GOM: *Chauliodus sloani* (Figure 10.34), comprising about 20% of the total stomiid catch (Sutton and Hopkins 1996a). Their long fang-like teeth and ferocious appearance give them their name. Though important in numerical abundance and biomass, the chauliodontinids only comprised two species in the gulf, *Chauliodus sloani* and *danae*, with *danae* only being represented by two specimens. *C. sloani* was also important in the Pacific but was not a dominant (Clarke 1974).

C. sloani showed an asynchronous diel vertical distribution (Figure 10.34), with about 70% of the population migrating from a depth of 400–700 m during the day to a depth above 200 m at night and the rest remaining at daytime depth. Clarke (1974)

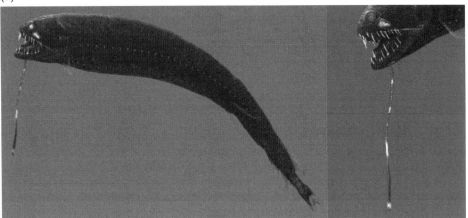

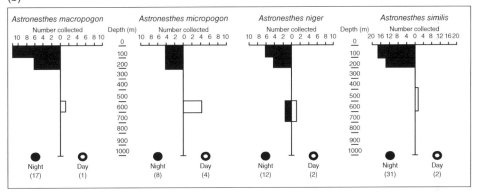

Figure 10.33 Astronesthinae. (a) Lateral view (left) and close-up of the head of *Astronesthes micropogon* showing the long mental barbel, pronounced teeth, and distinctive suborbital photophore; (b) day–night vertical distributions of three species of *Astronesthes* from the Gulf of Mexico showing pronounced patterns of vertical migration. *Sources:* (a) © Dante Fenolio, reproduced with permission. (b) Adapted from Sutton and Hopkins (1996b), figure 2 (p. 537). See color plate section for color representation of this figure.

observed vertical migration in the Pacific as well, with daytime depths of 450–825 m and nighttime depths of 45–225 m. In the GOM, using opening and closing nets, no size differences were observed between the individuals remaining at depth and those migrating to near surface waters. That observation is both interesting and important. It means that the "divided nighttime distributions are ethological in nature and not the result of ontogenetic descent, such as been observed in other mesopelagic species" (Sutton and Hopkins 1996a). Such being the case, it may be that degree of satiation overrides the drive to vertically migrate: only hungry individuals participate (Sutton and Hopkins 1996a).

Chauliodus sloani preys almost exclusively on myctophids (lanternfishes), with minor amounts of dendrobranchiate shrimps (Sutton and Hopkins 1996b). Oddly enough, the feeding chronology of the

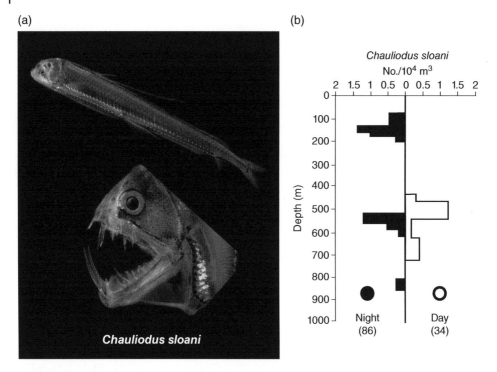

Figure 10.34 Chauliodontinae. (a) Lateral view (top) and close-up of the head of *Chauliodus sloani* showing the two very large mandibular "fangs" and the jaw in the un-hinged position; (b) day-night vertical distributions of *C. sloani* from the Gulf of Mexico showing an asynchronous diel vertical distribution. *Sources:* (a) © Dante Fenolio, reproduced with permission. (b) Adapted from Sutton and Hopkins (1996b), figure 1 (p. 535). See color plate section for color representation of this figure.

species showed no diel periodicity, suggesting that they either fed at different times of day or fed once and digested their prey over more than one day.

In the Pacific, Clarke (1974) observed differences in the size composition of *Chauliodus sloani* over the course of the year, with smaller individuals appearing only in the summer and early fall. He concluded that spawning took place during the spring or early summer. Data on size composition vs sexual maturity suggested that it took several years to become reproductively active, since only the largest fish (225–250 mm) were mature.

The subfamily Idiacanthinae (black dragonfishes or blackdragons) are the most eel-like of the six stomiid subfamilies, possessing a greatly elongated dorsal fin that runs greater than half the length of its slender body (Figure 10.35). One genus, *Idiacanthus*, and three species comprise the subfamily (Nelson et al. 2016). *Idiacanthus* is found circumglobally, with *I. fasciola* in the Atlantic, Indian, and Pacific oceans and *I. atlanticus* in the southern hemisphere (25–60°S). *I. antrostomus* replaces *I. fasciola* in the eastern Pacific (Priede 2017); it is the only idiacanthine in the California borderland and off Oregon (Pearcy et al. 1977).

Idiacanthus fasciola is a piscivore, feeding mainly on myctophids (Clarke 1982; Sutton and Hopkins 1996b). Its diel vertical distribution is unclear, mainly because of its

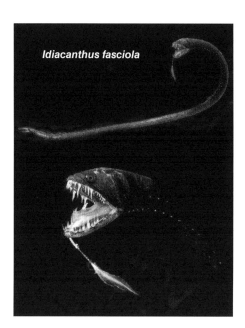

Figure 10.35 Idiacanthinae. Lateral view of the whole body (upper) and a close-up of the head of *Idiacanthus fasciola*. *Source:* © Dante Fenolio, reproduced with permission. See color plate section for color representation of this figure.

rarity. In both the Pacific near Hawaii and the GOM, it has been captured in the upper 300 m at night. The synopsis by Clarke (1974) suggests an asynchronous vertical migration because the species has been captured during the day-depth range (400–800 m) both day and night.

Idiacanthus has an unusual life history, with the females growing to a much larger maximum size than the males: 48 cm for the females and 7 cm for the males (Priede 2017). Females do not mature until they reach a size of 250 mm (Clarke 1974). The males are found in the bathypelagic realm. They have a rudimentary morphology, lacking teeth and paired fins as well as the characteristic mental barbel. The gut is similarly poorly developed, but the gonads fill the body cavity. It is believed that spawning takes place in the bathypelagial during winter and that the males are short-lived (Clarke 1974; Priede 2017). Females are longer-lived, vertically migrating predators.

The Melanostomiinae (scaleless black dragonfishes: Figure 10.36) are the most speciose of the dragonfish clan, with about 191 species in at least 15 genera (Nelson et al. 2016). By far, the genus with the most species is *Eustomias* (Sutton and Hartel 2004), with 115 total. The many species of Eustomias are best discriminated by the morphology of their chin barbels, a structure that is easily lost during capture. As a group, the Melanostomiinae are wide-ranging, with many species exhibiting a circumglobal distribution (Sutton and Hopkins 1996a; Priede 2017) at temperate to tropical latitudes.

What information is available on diel vertical habits for the melanostomiines suggests that most vertically migrate, either synchronously or asynchronously (Clarke 1974; Sutton and Hopkins 1996a), most from depths of 600–800 m to the upper 300 m at night. The species shown in Figure 10.36, *Echiostoma barbatum*, is believed to migrate from depths below 1000 m into the upper 200 m at night, an example of a bathypelagic to epipelagic migration. Most of the melanostomiines are piscivores, feeding mainly on myctophids. Species found in both the GOM and Hawaiian waters showed similar vertical distributions and diet. A common melanostomiine species in Hawaiian waters, *Thysanactis dentex*, showed some ontogenetic changes in diet with smaller fishes taking euphausiids and large copepods and moving up to sergestid shrimps and lanternfishes as they grew larger.

The loosejaws (Malacosteinae) are named for their ability to create a very large gape using a jaw structure that exceeds the size of the skull. The species shown in Figure 10.37a, *Photostomias guernei*, is the dominant stomiid in the GOM, exhibiting

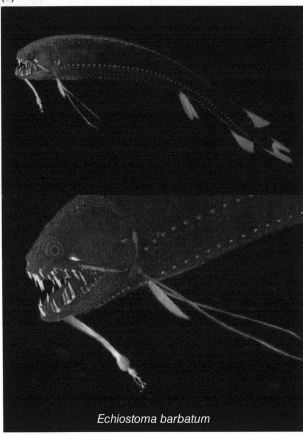

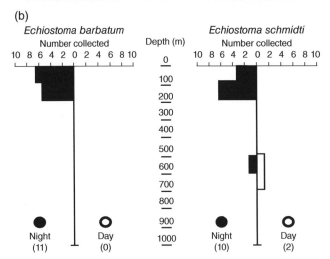

Figure 10.36 Melanostomiinae. (a) Lateral view (upper) and close-up of the head (lower) of *Echiostoma barbatum* (threadfin dragonfish). Note the pectoral fin which is reduced to a single long ray separated from three short rays and the complex esca at the end of the mental barbel; (b) day-night vertical distributions of two species of *Echiostoma* from the Gulf of Mexico showing patterns of diel vertical migration. *Sources:* (a) © Dante Fenolio, reproduced with permission. (b) Adapted from Sutton and Hopkins (1996b), figure 2 (p. 537). See color plate section for color representation of this figure.

(a)

(b)

(c)

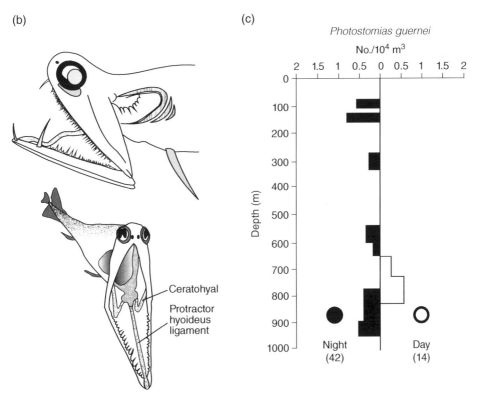

Figure 10.37 Malacosteinae. (a) Lateral view (left) and close-up of the head of *Photostomias guernei*. Note the very long pelvic fins and ligament (protractor hyoideus) extending from the tip of the mandible to the hyoid; (b) diagrammatic representation of the lack of ethmoid membrane (floor of the mouth) and showing two of the structures that facilitate the extension of the mandible and expansion of the gape during feeding (the protractor hyoideus ligament and ceratohyal); (c) day-night distributions of *P. guernei* from the Gulf of Mexico showing an asynchronous diel vertical migration pattern. *Sources:* (a) © Dante Fenolio, reproduced with permission. (b) Adapted from Sutton (2005), figure 1 (p. 2067); (c) Adapted from Sutton and Hopkins (1996b), figure 1 (p. 535). See color plate section for color representation of this figure.

(a)

Stomias boa ferox

(b)

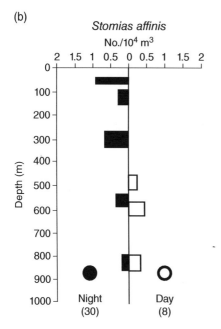

Figure 10.38 Stomiinae. (a) The stomiid *Stomias boa ferox*; (b) day-night distributions of *Stomias affinis* from the Gulf of Mexico showing an asynchronous diel vertical migration pattern. *Sources:* (a) Adapted from Morrow (1964), figure 75 (p. 297); (b) Adapted from Sutton and Hopkins (1996b), figure 1 (p. 535).

an asynchronous vertical migration with about 50% of the population migrating to the upper 200 m at night (Figure 10.37c, Sutton and Hopkins 1996a).

The jaws (Figure 10.37a, b) appear to have a "double hinge." The head is rocked back when the fish is swallowing prey, and the muscle/ligament structure pulls the lower jaw wide to accept the prey item (Figure 10.37b). Soft vertebrae in the spine just behind the head allow it to rock quite far back. The lower jaw does not have the ethmoid membrane that forms the "floor" of a typical fish jaw; it is open to the sea.

Most of the malacosteines in the GOM feed on lanternfishes; however, *Photostomias guernei* feeds primarily on dendrobranchiate shrimp (Sutton and Hopkins 1996b). Curiously, the namesake of the malacosteines, *Malacosteus niger*, feeds primarily on large calanoid copepods despite having a jaw structure that is clearly adapted for larger prey. It was concluded by Sutton (2005) that *Malacosteus* may be taking copepods as a stopgap measure in between meals of larger prey, but none of the specimens examined had large items in their gut.

Figure 10.39 The alepisaurid *Alepisaurus ferox*.

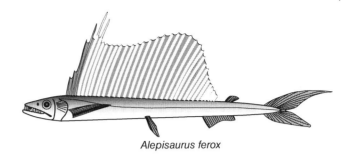

Alepisaurus ferox

Hawaiian malacosteines exhibited similar feeding habits and migratory patterns to those in the GOM (Clarke 1974, 1982). *Malacosteus niger* had a copepod diet, and *Photostomias* fed exclusively on larger crustaceans, mostly sergestid shrimps. Examination of seasonal size composition suggested that *Photostomias* was a year-round spawner.

The subfamily Stomiinae, or scaly dragonfishes (Figure 10.38a), includes the third most abundant stomiid in the GOM: *Stomias affinis*. The subfamily is well-represented in all non-polar ocean systems, with one genus and 11 species altogether. Four species show a circumglobal distribution, including *Stomias affinis*, which is found worldwide at temperate to tropical latitudes 35°N to 39°S (Priede 2017). Both *Stomias affinis* in the GOM and *Stomias danae* in Hawaiian waters were piscivores (Clarke 1974; Sutton and Hopkins 1996b), feeding primarily on myctophids.

Stomias affinis showed an asynchronous vertical distribution (Figure 10.38b) with daytime residence between 450 and 850 m. About 60% of the population vertically migrated to above 300 m at night, presumably to intersect with their preferred prey.

Superorder Ateleopodimorpha The superorder Ateleopodimorpha (Nelson et al. 2016) contains one order, the Ateleopodiformes, with one family containing 13 species. Their common name is the jellynose fishes, named for their gelatinous snout. They are a benthopelagic group associated with the continental slopes at tropical latitudes. Their long tail and pronounced head give them a superficial resemblance to macrourids or rattails (Priede 2017).

Superorder Cyclosquamata The superorder Cyclosquamata contains one order, the Aulopiformes, or Lizardfishes, with 16 families and 269 species. Most of the families are benthic or demersal, but a few inhabit the pelagial. They are all formidable predators, but, like the Stomiiformes, are never abundant in scientific trawls (cf. Sutton et al. 2010). Several of their numbers include unique adaptations to mesopelagic life, particularly in their eyes. The Aulopiformes are all synchronous hermaphrodites (Davis and Fielitz 2010; Priede 2017), an unusual and highly useful adaptation to deep-sea life.

Family Alepisauridae: The Lancetfishes (Figure 10.39). *Alepisaurus* is the only genus in the family Alepisauridae, and it contains two species: *ferox* and *brevirostris*, both of which are found circumglobally and have very wide latitudinal distributions. *A. ferox* is found between 84°N and 57°S (Priede 2017). Both are voracious predators, ingesting other members of their genus, as well as pteropods, heteropods, cephalopods, crustaceans, polychaetes, and a variety of midwater fishes (Haedrich 1964). *A. ferox* reaches over 2 m in length, and it is often caught on longlines and shallow trolling lines. Though it has been captured as deep as 1830 m (Priede 2017), its morphology suggests that, like the tunas and billfishes, it is probably more of a visitor than a resident at depth.

Family Evermannelidae: The Sabertooth fishes (Figure 10.40). The evermanellids are a widely distributed family of mesopelagic predators with representatives at temperate to tropical latitudes throughout the global ocean. They are only captured occasionally in scientific trawls. Their characteristic fang-like teeth, blunt square-shaped head, tubular upward-directed eyes, and pronounced lateral line make them well-equipped for their predatory role in the dim light of the mesopelagial. Their diet consists mainly of cephalopods and fishes (Hopkins et al. 1996)

Family Giganturidae. The unusual morphology of the giganturids, the telescope fishes, (Figure 10.41) makes them one of the quintessential deep-sea families. The two giganturid species, *Giganturus indica* and *G. chuni*, have a circumglobal distribution but are nonetheless rare in midwater trawls. They are considered to be bathypelagic but do occur in the upper 1000 m (Sutton et al. 2010). The transformation from larva to juvenile brings about enormous change (Figure 10.41) with the eye morphing from a conventional structure to the characteristic forwardly directed tubular eyes of the adult. Like most of the Aulopiformes, the giganturids have a highly distensible stomach, allowing them to take large prey.

Family Notosudidae. The Waryfishes (Figure 10.42) comprise 17 species in three genera, *Ahliesaurus* (2 spp.), *Luciosudis* (1), and *Scopelosaurus* (14) (Nelson et al. 2016). Waryfishes are found at lower (500–800 m) mesopelagic depths worldwide, primarily inhabiting temperate and tropical latitudes. They are long slender fishes with a large gape and well-developed eyes. *Scopelosaurus smithi*, pictured in Figure 10.42, has an eye with a textbook-quality aphakic gap (cf. Herring 2002). The anterior portion of the eye has a large pear-shaped gap between the lens and the iris, allowing incoming light to reach the surface of the entire lens from the front, giving the fish an excellent forward sight-line and binocular vision. As a result of the aphakic gap, the lens focuses the light from the front on a well-defined region or pit within the retina, the fovea, that has a very high density of receptor cells (Munk 1977; Locket 1985). In the case of *Scopelosaurus*, the fovea is made up largely of cone cells, usually only present in

Figure 10.40 Evermanellidae. (a) Close-up of the head of *Odonostomops normalops*; (b) lateral view of *Coccorella atlantica*. Note the formidable fang-like teeth and upward-directed eyes. *Source:* © Dante Fenolio, reproduced with permission. See color plate section for color representation of this figure.

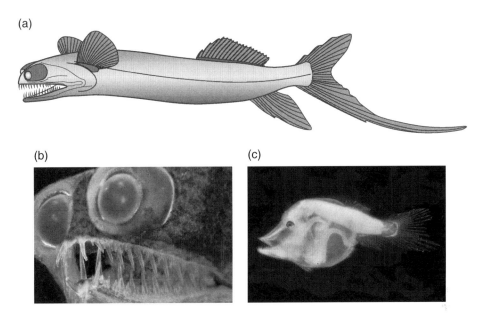

Figure 10.41 Giganturidae. (a) Adult *Gigantura chuni*. (b) Close-up of the head of an adult *Gigantura indica* showing the forward-directed tubular eyes and pronounced dentition; (c) larval giganturid. Note the conventional eye and small mouth. *Sources:* (b, c) © Dante Fenolio, reproduced with permission. See color plate section for color representation of this figure.

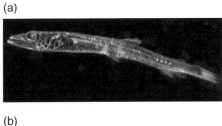

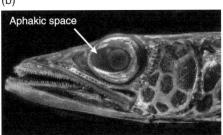

Figure 10.42 Notosudidae. (a) The notosudid *Scopelosaurus smithi*; (b) close-up of the head of *S. smithi* showing the aphakic space anterior to the lens in the eye. *Source:* © Dante Fenolio, reproduced with permission. See color plate section for color representation of this figure.

shallow-dwelling species because they are less sensitive to light than rods. The reason for their presence in the foveae of many aulopiforms is unclear (Herring 2002). Several other aulopiform species have eyes with similar structure, including those in the Chlorophtalmidae, Ipnopidae, and Omosudidae (Munk 1977).

An additional characteristic of the notosudid eye is that outside of the cone-rich fovea, the retina is equipped with rods for receiving the lower levels of light impinging on areas outside the foveal region. Further, there is a well-developed *tapetum lucidum*, or reflective layer, behind the retina. The tapetum reflects back light that has already passed through the eye, increasing the probability of photons colliding with receptor cells, greatly increasing the effectiveness of the retina.

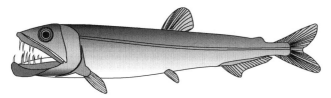

Omosudis lowei

Figure 10.43 The omosudid *Omosudis lowei*.

Family Omosudidae. The "Hammerjaws" (Figure 10.43) comprise a monotypic family, with the one species, *Omosudis lowei*, found at mesopelagic depths in temperate and tropical waters worldwide. Like many of its aulopiform relatives, *Omosudis* is a mesopelagic predator, feeding primarily on cephalopods (68% of diet) and fishes (32%) (Hopkins et al. 1996). Classification of *Omosudis* has varied between placing it in its own family as we have it here (Appendix A), following the Catalog of Fishes and WoRMS, or including it with the Alepisauridae (Nelson et al. 2016). *Omosudis* is reported to reach 20 cm maximum size and is found at both meso- and bathypelagic depths (Hopkins et al. 1996; Sutton et al. 2010).

Family Paralepididae. The Barracudinas (Figure 10.44) are the most speciose of the pelagic aulopiforms, with about 60 species in 12 genera. The Barracudinas are closely related to the Daggertooths (Anotopteridae), and Nelson et al. (2016) group them together with the paralepidids. The barracudinas are truly pan-global, being found in all oceans including polar latitudes. A southern hemisphere species, *Notolepis coatsi*, is commonly captured as a juvenile in scientific trawls below 60°S in the Scotia Sea (South Atlantic) (Lancraft et al. 1989), usually residing between 200 and 750 m. Adults were never captured during that study, and it may be assumed that they avoided the sampling gear.

Paralepidids are long slender fishes with a large gape and a conventional eye (Figure 10.44) and are assumed to be agile swimmers (Priede 2017). Some species of

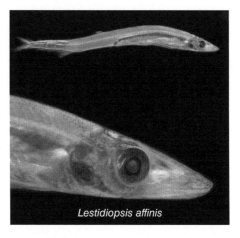

Lestidiopsis affinis

Figure 10.44 Lateral view (top) and close-up of the head of the paralepidid *Lestidiopsis affinis*. Note the darkly pigmented gut that is believed to conceal bioluminescence of prey items such as myctophids. *Source:* © Dante Fenolio, reproduced with permission. See color plate section for color representation of this figure.

barracudinas can reach over a meter in length (Nelson et al. 2016), but mean maximum size is 26 cm (Priede 2017). Several species have luminous tissue along their ventrum, presumably to provide countershading (Graae 1967).

Ninety-nine percent of the diet biomass of eight paralepidid species from the GOM was fish. Their prey were myctophids, sternoptychids, and bremacerotids, as well as other paralepidids (Hopkins et al. 1996).

Family Scopelarchidae. The Pearleyes are named for their unusual eye morphology (Figure 10.45). They have an upward-looking tubular eye with accessory light the binocular vision provided by the tubular eyes (Herring 2002).

The scopelarchids have five genera, and 18 species with representatives found from the tropics to polar waters. A southern hemisphere species, *Benthalbella elongata*, is one of the very few mesopelagic predators reaching Antarctic waters. Adults are located primarily between 500 and 1000 m (Nelson et al. 2016). In the GOM, scopelarchids feed primarily on myctophids and sternoptychids, continuing the trend of piscivory observed in other aulopiforms (Hopkins et al. 1996).

Superorder Scopelomorpha The superorder Scopelomorpha contains one order, the Myctophiformes, with two families: the Myctophidae, or lanternfishes, and the Neoscopelidae or blackchins. In all deep-ocean systems, the myctophids comprise one of the three most important families in numerical abundance and biomass and in many, they are the lead group. The blackchins are considerably less abundant; individuals of the genus *Neoscopelus* inhabit shelf and slope waters in tropical and subtropical regions worldwide, and *Scopelengys* is an occasional capture in midwater trawls reaching the lower (500–1000 m) mesopelagic and upper bathypelagic zone.

Family Neoscopelidae. The blackchins comprise six species in three genera: *Neoscopelus* with three species, *Scopelengys* with two species, and *Solivomer*, found only in the Philippines, with one. Little is known of *Solivomer*, but the other two genera are well-described, and they differ considerably in life habits and morphology.

Neoscopelus (Figure 10.46) has a large eye, a half-silvered body with abundant photophores on its ventrum (and its tongue!), and a firm musculature (Nafpaktitis 1977). They are considered to be benthopelagic, but they are occasionally

Figure 10.45 Scopelarchidae. Lateral view (top) and close-up of the head of *Scopelarchus analis* showing the upward-looking tubular eye and darkly pigmented gut thought to conceal bioluminescence of prey items. Note the silvery pad beneath the lens of the eye which acts as a light guide allowing light to enter from the side, strike the retina, and thus facilitate detection of lateral point sources of light. *Source:* © Dante Fenolio, reproduced with permission. See color plate section for color representation of this figure.

collectors that form a pad located beneath the lens giving them their name. The pad acts as a light guide, allowing light to enter from the side and strike the retina, alerting the fish to lateral point sources of light. It is assumed that no focused image would be produced but flashes of bioluminescence from prey items would be detected (Herring 2002). It is worth noting that accessory light collectors will work well only in poorly lit environments such as the lower mesopelagic zone. In well-lit waters nearer the surface, the ambient light passing through the biological light guides in the "pearl" would flood the retina and ruin

(a)

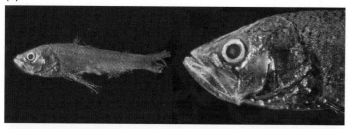

(b)

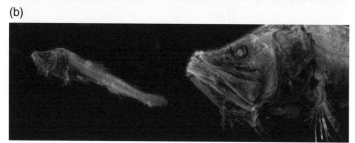

Figure 10.46 Neoscopelidae. External characteristics of shallow and deep living neoscopelids. (a) *Neoscopelus macrolepidotus*, a benthopelagic species found between 200 and 1000 m. Note the well-developed photophores, large eyes and silvery body; (b) *Scopelengys tristis*, a lower mesopelagic species. Note the small eye, dark coloration, and lack of photophores. *Source:* © Dante Fenolio, reproduced with permission. See color plate section for color representation of this figure.

captured in midwater trawls. They are typically distributed between 200 and 1000 m (Priede 2017).

Scopelengys is black/brown with a small eye, no photophores, and a flaccid body. It is captured fairly routinely in midwater trawls working the lower mesopelagic zone of the California borderland, though it is never abundant. It is a truly pelagic genus, whose coloration, eye morphology, metabolic rate (Torres et al. 1979), and compositional attributes (Childress and Nygaard 1973) reflect its distribution in the lower mesopelagic zone.

Based on its morphology, Nafpaktitis (1977) speculated that *Neoscopelus* may have been ancestral to *Scopelyngys*, with *Solivomer* as an intermediate, but recent molecular data places *Solivomer* as the sister group to both *Neoscopelus* and *Scopelengys* (Poulsen et al. 2013).

Further, the same study finds that the neoscopelids are the sister group to the Myctophidae.

Family Myctophidae. The myctophids are considered the sardines and anchovies of the open ocean, occupying the intermediate trophic levels of all deep-sea systems from the equator to the poles, sometimes in very large numbers (Gjosaeter and Kawaguchi 1980). Systematic surveys of the mesopelagial in several areas of the global ocean (e.g. Badcock 1970; Clarke 1973; Merrett and Roe 1974; Badcock and Merrett 1976; Pearcy et al. 1977; Gjosaeter and Kawaguchi 1980; Hopkins and Lancraft 1984; Gibbs and Krueger 1987; Lancraft et al. 1989, 1991; Beamish et al. 1999; Ross et al. 2010) consistently place the myctophids in the top three families in numerical abundance. They comprise about 252 species in 33 genera, making them the most speciose family of mesopelagic fishes. Depending on the authority, myctophids have been divided into two sub-families and seven tribes (Priede 2017; Paxton et al. 1984), two sub-families and six tribes (Poulsen et al. 2013) or two sub-families only (Nelson et al. 2016; WoRMS n.d.). Eschemeyer's Catalog of Fishes recognizes five subfamilies: *Myctophinae, Lampanyctinae, Diaphinae, Gymnoscopelinae,* and *Notolychninae*.

Myctophids are identified to species by their photophore patterns (e.g. Nafpaktitis et al. 1977, Figure 10.47). Subdivisions within the family are based on similarities in osteological characteristics (e.g. jaw morphology), larval stages (Moser and Ahlstrom 1974), photophore patterns and, most recently, by mitochondrial DNA gene sequences (Poulsen et al. 2013). In virtually all studies on their biology, myctophids are treated at the family level, and that will be our strategy here.

Ecology. Myctophids are clearly adapted to the mesopelagic zone, with large well-developed eyes and prominent photophore arrays (Figures 10.48–10.50). They have limited representation in the bathypelagic realm. They are small fishes, with adult standard lengths ranging from about 22 mm (*Notolychnus valdiviae*) to 162 mm

Figure 10.48 Two species of myctophids in the subfamily Diaphinae. (a) Lateral view of *Diaphus dumerilii*; (b) close-up of the head of *D. lucidus*. Note the large luminous organs in front and below each eye ("headlights"). Each of these organs consists of a combination of very large dorsonasal, antorbital, and ventronasal photophores, all in contact with each other. The shape and size of these organs vary between species. *Source:* © Dante Fenolio, reproduced with permission. See color plate section for color representation of this figure.

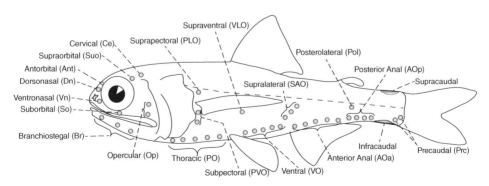

Figure 10.47 Names of individual photophores and photophore groups in the Myctophidae. Presence or absence of individual photophores and numbers of photophores within groups vary between species and are the primary means of identification. No species possesses all of the photophores shown in the figure.

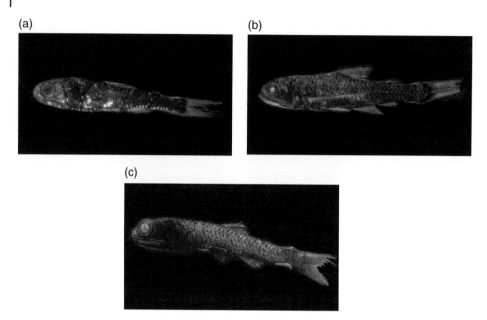

Figure 10.49 A comparison of external characteristics of three species from the subfamily Lampanyctinae from different depth zones. (a) *Ceratoscopelus warmingii*; an active diel migrator with night-time depths in the upper mesopelagic, with silvery body, well-developed photophores, and large eyes; (b) *Lampanyctus alatus*; also an active migrator but considered a lower mesopelagic resident with dark body, smaller photophores, and smaller eyes; (c) *Taaningichthys bathyphilus*: a bathypelagic non-migrator with dark body, very small photophores, but a relatively large eye. Note the large subcaudal photophore ("sternchaser") on *Taaningichthys*. Source: © Dante Fenolio, reproduced with permission. See color plate section for color representation of this figure.

(*Gymnoscopelus opisthopterus*). Adult sizes depend on the oceanic region. For example, species in tropical-subtropical systems are typically in the 30–65 mm range (Clarke 1973; Gartner et al. 1987), whereas those in the polar climes of the Southern Ocean range from 60 to 160 mm (Hully 1990).

Virtually all myctophids exhibit a diel vertical migration from mesopelagic depths into the upper 200 m (Badcock 1970; Clarke 1973; Merrett and Roe 1974; Badcock and Merrett 1976; Pearcy et al. 1977; Gjosaeter and Kawaguchi 1980; Hopkins and Lancraft 1984; Gibbs and Krueger 1987; Lancraft et al. 1989, 1991; Beamish et al. 1999; Ross et al. 2010). Lower mesopelagic/bathypelagic residents such as *Taaningichthys bathyphilus* (Figure 10.49) (Gartner et al. 1987) tend to remain at depth.

Forty-nine species in 17 genera reside in the GOM (Gartner et al. 1987). They rank second to the Gonostomatidae in numerical abundance but are first in biomass (Hopkins and Lancraft 1984). Diel vertical distributions for four of the seven most abundant myctophids in the GOM (Figures 10.48–10.50) are shown in Figure 10.51. All four are strong vertical migrators, with *Ceratoscopelus warmingii*, the most abundant species in the GOM, having the largest diel depth range. A fraction of *Ceratoscopelus*' population remains at depth during nighttime hours. Those individuals tend to be small juveniles, a trend observed in other myctophids in the GOM as well as

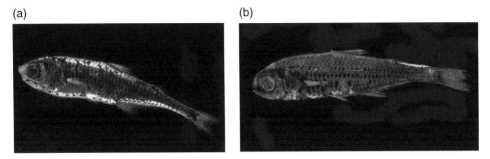

Figure 10.50 Two species of myctophids from the subfamily Myctophinae. (a) *Centrobranchus nigroocellatus*; (b) *Myctophum affine*. Both are active diel migrators appearing in surface waters at night. Note the differences in eye size and photophore development on *Centrobranchus* and the large supracaudal photophore ("sternchaser") on *Myctophum*. Source: © Dante Fenolio, reproduced with permission. See color plate section for color representation of this figure.

in other oceanic systems (Clarke 1973; Badcock and Merrett 1976; Willis and Pearcy 1980; Karnella 1987).

Myctophids are zooplanktivorous, feeding largely on copepods, ostracods, and euphausiids (Clarke 1978; Hopkins and Gartner 1992; Hopkins et al. 1996). In the GOM and the waters off Hawaii, and presumably in most other systems, they feed at night in the upper 200 m of the water column, overlapping with the peak abundance of the zooplankton community (Clarke 1978; Hopkins 1982).

The suite of photophores along the ventral and lateral aspects of the myctophid body is presumed to serve two functions: countershading and species recognition, respectively. Countershading has been demonstrated in lanternfishes (Case et al. 1977) using a little technical wizardry in the laboratory very soon after capture, by stimulating them with light from above and recording their response from below. The lanternfishes matched the downwelling light in intensity and spectral quality, making a convincing case for counter-illumination. The case for species/sexual recognition is unsupported by data other than the obvious: the lanternfish species each have unique photophore patterns.

Further, the caudal photophores known as "stern-chasers" that are present in some species (see Figures 10.49 and 10.50) are dimorphic in the males and females, potentially aiding in sexual recognition (Nafkaptitis and Nafpaktitis 1969; Herring 2002). Herring (2000) notes that nearest neighbor distances between conspecifics in the mesopelagic zone are most likely to be 10 m or greater, even in abundant species. However, by using sensory modalities in concert, notably chemoreception in the far-field and photoreception to detect photophore patterns in the near-field, the chance of successful encounters is greatly increased. The pronounced photophores on the head of the genus *Diaphus* are a defining character of the genus. Those and the subopercular photophores of the stomiids are both believed to illuminate prey items for the hunting fish.

The vertical distributions of lanternfishes as a whole cover the entire mesopelagic zone and the upper bathpelagic. Figure 10.49 highlights the difference in the coloration and photophore size between the lower mesopelagic-upper bathypelagic species *Taaningichthys bathyphilus* and two species that migrate to the surface at night, *Myctophum affine* and *Centrobranchus*

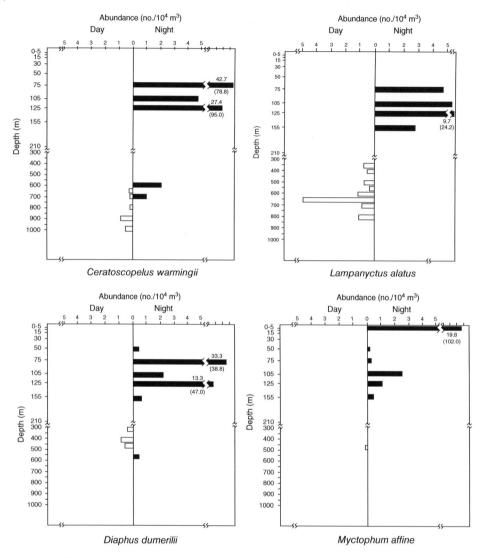

Figure 10.51 Diel vertical profiles of four dominant myctophid species in the eastern Gulf of Mexico. Numbers not in parenthesis indicate average abundance at depth; numbers in parenthesis indicate maximum abundance at depth. *Source:* Adapted from Gartner et al. (1987), figures 3 and 4.

nigroocellatus. The photophores are prominent in both migrators, providing counter-illumination, and their sides are half-silvered, making them less visible when viewed from above or below in downwelling light. In contrast, *Taaningichthys*, living where light is absent or nearly so, has nearly invisible photophores and dark coloration.

Myctophids are the only deep-living pelagic fish species on which there is confirmed information on age and growth, data which are not only intrinsically valuable but also allow evaluation of the concept that the cold, high pressures, and lack of light in the deep-sea environment results in a slowing of life processes. Age and growth data in three species of myctophids from the GOM were collected using the primary growth increments, or "daily rings," in their sagittal otoliths (Gartner 1991a, b). Otoliths, components of the auditory system in all fishes, act in a similar manner to the statoliths in the equilibrium receptors of invertebrates (statocysts). In the GOM myctophids, and in all other fish that have been tested, primary growth rings are deposited daily. Myctophids' growth rings are tiny, often too small to be determined by light microscopy but can resolved using electron microscopy (e.g. Greely et al. 1999). Gartner (1991a) monitored the deposition of primary growth rings over the course of the diel cycle in three of the GOM dominant species, *Benthosema suborbitale, Diaphus dumerilii,* and *Lepidophanes guentheri,* confirming that indeed growth of the rings was completed on a daily basis. Data on *B. suborbitale* showed a lifespan of slightly less than a year, *Lepidophanes* a lifespan of one year, and *D. dumerilii* a lifespan of about two years (Gartner 1991b). Later data on growth in *Ceratocscopelus* also showed a lifespan of less than a year (Gartner 1993). Growth rates in the GOM myctophids were quite similar to those of congeners and conspecifics from the Gulf of Oman, Gulf of Aden, and Arabian Sea that had also been aged using daily rings (Gjosaeter 1987).

A comparable study (Greely et al. 1999) was done on the lanternfish residing in the frigid (−2 to 4 °C) waters of the Southern Ocean, *Electrona antarctica,* the end-member species in the continuum of vertically migrating myctophids that extends from the equator to the polar circle (Nafpaktitis et al. 1977; McGinnis 1982). Daily rings were enumerated using Scanning Electron Microscopy. *E. antarctica* has a maximum age of about four years, making it the longest-lived myctophid for which data are available (Table 10.3). It grows to a length of about 100 mm. The growth curve generated was similar to that of lanternfishes from warmer climes.

Figure 10.52 compares the growth of *Electrona antarctica* to that of the Pacific sardine, *Sardinops caerulea,* and the anchovy *Engraulis mordax*. The sardine and anchovy are considered cold-temperate species, living at about 8–12 °C. While the first two years of growth in *E. antarctica* are quite a bit slower than those of the sardine and

Table 10.3 Maximum-age estimates for six tropical-subtropical and two subantarctic myctophid species based on primary growth increments (*SL* standard length). *Source:* Reprinted by permission from Springer Nature Customer Service Centre GmbH, Springer Nature, Age and Growth of Electrona antarctica. Greeley et al. (1999), table 2 (p. 156).

Species	Max. SL (mm)	Max. age (months)	References
Benthosema suborbitale	33	10–11	Gartner (1991a)
Diaphus dumerilli	63	18–24	Gartner (1991a)
Lepidophanes guentheri	65	12–15	Gartner (1991a)
Benthosema fibulatum	100	12	Gjøsaeter (1987)
Benthosema pterotum	50	9–12	Gjøsaeter (1987)
Lampanyctodes hectoris	73	36	Young et al. (1988)
Electrona risso	72	18	Linkowski (1987)
Electrona ventralis	110	24	Linkowski (1987)

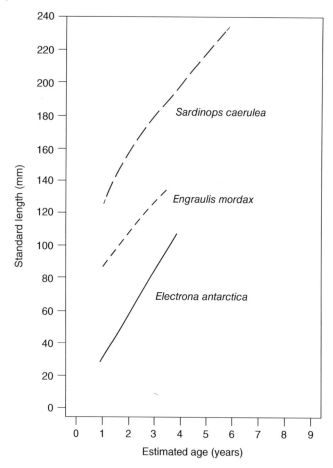

Figure 10.52 A comparison of growth rates for two cold-temperate species, the sardine *Sardinops caerulea* and the anchovy, *Engraulis mordax*, with that of the polar myctophid, *Electrona antarctica*. *Source:* Greely et al. (1999), figure 8 (p. 156). Reproduced with the permission of Springer.

anchovy, by year 2, its growth (0.07 mm day^{-1}) is equivalent to that of the sardine.

The main conclusion to be drawn from the multiple studies on growth in myctophids (Linkowski 1985, 1987; Gjosaeter 1987; Young et al. 1988; Gartner 1991b; Greely et al. 1999) is that their growth is neither much faster nor much slower than other small pelagic fishes with similar feeding habits living at similar temperatures. The concept that "life is slower in the deep sea" may apply below the photic zone, but data on the Myctophidae argue against it for the mesopelagial.

Reproduction. Myctophids in the GOM exhibited two main reproductive strategies. In the first, spawning occurred throughout the year in "protracted spawning periods of 4–6 months duration, with individuals spawning every 1 to 4 days" (Gartner 1993), otherwise known as "continuous wave" spawning (cf. Lisovenko and Prut'ko 1987). This pattern was exhibited by *Benthosema suborbitale*, *Lampanyctus* *alatus, Lepidophanes guentheri,* and *Notolychnus valdiviae.* In the second strategy, exhibited by *Ceratoscopelus warmingii,* spawning was confined to two distinct periods in the winter and spring-summer months. During those spawning periods, individuals spawned every seven days and the spawning period lasted about three months in total. However, because *Ceratoscopelus* has a lifespan of less than a year, the two spawning seasons were spread over two different generations.

Repetitive batch-spawns allow lanternfishes to maximize their lifetime reproductive output while also meeting their daily energy requirements. The numbers of eggs released per batch varied with species' size, with a low in the diminutive *Notolychnus valdiviae* (40–130 eggs, fish size 18–22 mm) to a high in *Ceratoscopelus warmingii* (3269–12626 eggs, fish size 53–70 mm). *Benthosema suborbitale* (101–427 eggs, fish size 24–32 mm) and *Lepidophanes guentheri* (494–2294 eggs, fish size 37–62 mm) were between the two extremes. Integrating batch numbers over their reproductive lives yields lifetime fecundities of about 19600 eggs for *B. suborbitale*, 102500 for *L. guentheri*, and 166300 eggs for *Ceratoscopelus warmingii*: respectable numbers for fish 3–7 cm in length and living less than a year (Gartner 1993).

It is important to note that most of the dominant species in the GOM are found at tropical-subtropical latitudes worldwide and that congeners and conspecifics exhibit regional differences in diel and seasonal reproductive characteristics. For example, in the Gulf of Oman, *Benthosema pterotum* spawned at depth during evening hours (1700–2000, Gjosaeter and Tilseth 1988), whereas in the GOM, its congener *B. suborbitale* spawned after midnight in the epipelagic zone (Gartner 1993). After spawning at depth, the positively buoyant eggs of *B. pterotum* ascend to the upper 50 m overnight and hatch the next day (Dalpadado 1988).

In Hawaiian waters, *Benthosema suborbitale* showed a spring-summer pattern of reproduction (Clarke 1973) contrasting with its pattern in the GOM where the species reproduced evenly throughout the year (Gartner 1993). Conversely, the seasonal pattern for *Ceratoscopelus warmingii* in the GOM described above contrasts with that of *C. warmingii* in Hawaiian waters, where it has a prolonged spawning season over first six months of the year (Clarke 1973). Available data suggest that widely distributed species such as *C. warmingii* and *B. suborbitale* have enough plasticity in their reproductive habits to suit the demands of their local environment.

Superorder Lamprimorpha The Superorder Lamprimorpha (Nelson et al. 2016) contains only one order, the Lampriformes: the opahs, oarfishes, and ribbonfishes, with at least 22 species in 6 families. Though they are too mobile to be captured in scientific trawls except as larvae, they have two important claims to fame. The first is that the oarfish, *Regalecus glesne* (Figure 10.53a), is the largest living teleost, achieving lengths up to at least 11 m (Priede 2017). The second is that the Opah, *Lampris guttatus* (Figure 10.53b), has achieved limited whole-body endothermy by using a circulatory heat-exchange system similar to the tunas.

The elongated body of the Oarfish and large disc-like body of the Opah represent the two body morphs of the Lampriformes. Most of the lampriforms (Lophotidae, Radiicephalidae, and Trachipteridae) have an elongated body similar to that of the Oarfish. Like the Oarfish, Opahs are very large, with those off the coast of California reaching at least 1.4 m in length and 73 kg in

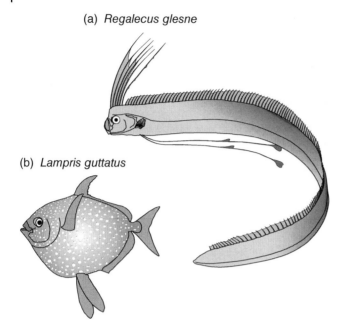

Figure 10.53 The Lampriformes. (a) the oarfish, *Regalecus glesne*; (b) the opah, *Lampris guttatus*.

weight (Fitch and Lavenberg 1968). However, they are reported to reach a maximum mass as high as 250 kg (Priede 2017). Opahs are voracious feeders, feeding on fishes, squids, and crustaceans. In contrast, Oarfish feed mainly on euphausiids (Fitch and Lavenberg 1968). None of the lampriforms are common, and Oarfish in particular are known mainly from large specimens swimming feebly at the surface or stranded on shore. All families in the order except the Veliferidae are considered epipelagic or mesopelagic in habit (Priede 2017).

Superorder Paracanthopterygii The Paracanthopterygii includes 5 orders and 674 species. Members of one order, the Percopsiformes, inhabit freshwater. Families in three of the remaining four orders: Zeiformes (dories), Gadiformes (cods, grenadiers and hakes), and Polymixiiformes (beardfishes) are nearly all benthopelagic (Priede 2017). Exceptions are the rarely captured Stylephiformes (tube-eyes), inhabiting the lower mesopelagic zone, and three families within the Gadiformes, the Bregmacerotidae (codlets), Melanonidae (pelagic cods), and Merlucciidae (merluccid hakes). *Steindachneria argentea*, the luminous hake, is the only species in its subfamily (Steindachneriinae) within the Merlucciidae. The grenadiers (Macrouridae) or rattails are a highly important benthopelagic family with 411 species found in all deep-sea habitats. They are sometimes captured in midwater trawls as juveniles, but as adults they are mainly bottom-dwellers (Priede 2017).

The order Stylephoriformes contains only one species, *Stylephorus chordatus*, a very unusual fish with tubular eyes, a bellows-like feeding apparatus, and a long, threadlike, lower lobe to its tail (Figure 10.54a). *Stylephorus* is a zooplanktivore, feeding mainly on copepods. The tubular mouth leads into a bellows-like membranous vestibule defined by the long

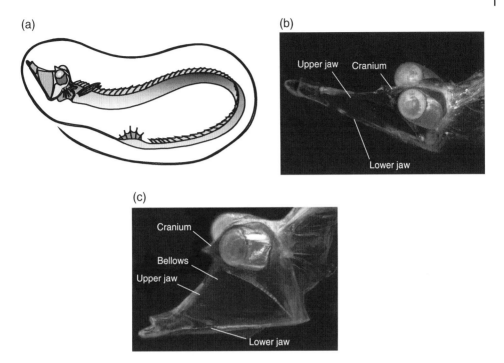

Figure 10.54 The Stylephoriformes. (a) The tube-eye (or thread-tail), *Stylephorus chordatus*. Note the extreme extension of the lower lobe of the caudal fin and the unique configuration of the upper and lower jaws; (b) close-up of the head of *S. chordatus* showing the mouth (upper and lower jaws configured with the bellows in the contracted position); (c) close-up of the head of *S. chordatus* with the bellows in the expanded position. *Source:* (b, c) © Dante Fenolio, reproduced with permission. See color plate section for color representation of this figure.

lower jaw and the cranium (Figure 10.54b, c). When a prey item is within range, *Stylephorus* tilts back its head, dropping the lower jaw and greatly expanding the volume of the membranous vestibule behind the tubular mouth, sucking the prey into the buccal cavity. The tubular mouth acts as a Venturi tube, greatly increasing the velocity of fluid flow into the membranous vestibule as it expands. It is a brilliant biological application of the principle of continuity in fluid flow. Pietsch (1978) calculated suction velocities as high as $3.25\,\mathrm{m\,sec^{-1}}$ for feeding *Stylephorus*.

Much of the literature describes *Stylephorus* with the eyes facing upward during feeding. However, *Stylephorus* may be able to feed without directing its eyes away from its prey.

The eyes are still facing forward in Figure 10.54c despite the dropped lower jaw and the difference in the volume of the membranous vestibule between (b) and (c) in Figure 10.54. The two photos of *Stylephorus* in that Figure are of freshly captured specimens in excellent shape and taken in a gentle net. It may be that the eyes-upward feeding position reported in the literature is a result of the head being rocked completely back in the net during capture and that enough suction velocity can be generated for most feeding encounters without the head being rocked back enough to tilt the eyes upward and away from the prey.

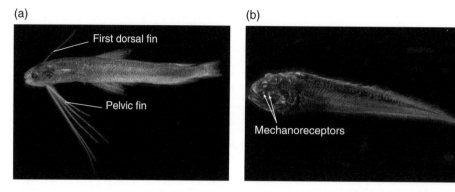

Figure 10.55 The Gadiformes. (a) a codlet, *Bregmaceros* sp. The first dorsal fin is a single long fin ray on the head that folds into a groove when depressed. Note also the very long fin rays of the pelvic fin and its unusual placement on the throat under the operculum; (b) the arrowtail cod, *Melanonus zugmayeri*. The mechanoreceptors (pits) on the head surround the eye and continue along the body (lateral line). *Source:* © Dante Fenolio, reproduced with permission. See color plate section for color representation of this figure.

Recent research on Stylephorid eyes (Musilova et al. 2019) has revealed that they not only have a visual field attuned to the bioluminescent spectrum, roughly 420–520 nm in the blue–green wavelengths, but they also have the ability to further discriminate distinct wavelengths within that limited range. Since bioluminescent displays of prey and predator species often have signature wavelengths (Widder 2010), this heretofore undescribed visual acuity may allow them to discriminate between food and danger in the dimly lit deep-sea environment.

The order Gadiformes is an extremely important one on a variety of levels. Cods (Gadidae) fed the western world long before the time of the nineteenth-century tall ships. Adventurous Basques and Spaniards caught them on the Grand Banks as early as the fifteenth century, preserved them by salting, and brought them home as bacalao, a staple in the Spanish diet. Prior to the collapse of the Grand Banks fishery, cod were responsible for the wealth of many families in the New England states. In terms of historical influence, no fish can compare to it.

In the deeper waters of the open ocean, the Gadiformes are represented by the benthopelagic grenadiers (Macrouridae) and deep-sea cods (Moridae) inhabiting benthic environments ranging from seamounts to the abyssal plain. The hakes (Merlucciidae), like their famous relatives the gadids, are mainly found at shelf depths (Priede 2017).

Of the three truly pelagic gadid families, the Bregmacerotidae, or codlets (Figure 10.55a), is the most ubiquitous, being found at mesopelagic depths in tropical-subtropical waters worldwide. They are small fishes, typically <8 cm, and are a small but consistent fraction of the midwater fauna in tropical-subtropical systems (e.g. Hopkins and Lancraft 1984; Sutton et al. 2010). However in one particularly interesting location, the Cariaco Basin, they are a dominant species.

The Cariaco Basin is a 1400-m-deep depression on the continental margin off Venezuela that is openly connected to the surface Atlantic Ocean above a shallow (c. 140 m) sill. Basin waters turn over slowly (Astor et al. 2003), and decomposition of sinking material leads to permanent anoxia below about 275 m.

Pugh (1972) described the fish fauna of the Cariaco Basin itself and from a station immediately north of it. He found a typical tropical-subtropical midwater fish assemblage in the waters north of the basin and a less speciose but otherwise unremarkable assemblage within it. Of particular interest was the dominance of the codlet *Bregmaceros*, lanternfishes in the genus *Diaphus*, and the luminous hake, *Steindachneria*. Mead (1963) reported a catch of 223 *Bregmaceros* and virtually nothing else in a single tow to 400–600 m in the Cariaco Basin, well within the anoxic zone. About 10 years later, Wilson (1972), using a 12 kHz echosounder, observed a scattering layer within the Cariaco Basin that exhibited unusual behavior. At the top of the oxygen minimum, the scattering layer split into two components. One component remained above the anoxic zone, but the other continued downward to depths >400 m. Clearly, the implication was that *Bregmaceros* comprised the deeper portion of the scattering layer.

The suspicion that *Bregmaceros* was the deeper migrator was confirmed by later discrete-depth tows (Baird et al. 1973) from which the codlet was virtually the only species to be captured within the anoxic zone. Two species comprised the upper layer: the lanternfish *Diaphus taaningi* and the luminous hake *Steindachneria argentea* (Baird et al. 1974; Love et al. 2003). The lanternfish genus *Diaphus* is widespread throughout the world ocean, particularly in tropical-subtropical waters (Smith and Heemstra 1991). It is well-represented in areas with severe oxygen minima such as the Arabian Sea (Gjosaeter 1984; Kinzer et al. 1993) as well as in regions with more benign oxygen profiles such as the Gulf of Mexico (Gartner 1993). However, *Bregmaceros* is not generally found in the numbers reported for the Cariaco Basin, even in regions very nearby. It is logical to assume that *Bregmaceros* has physiological, and perhaps morphological, attributes that allow it to flourish in an environment that is deadly to other pelagic fishes.

The remaining two pelagic gadiforms are the Melanonids (Figure 10.55b), with their own family, and the Steindachneriinae, or luminous hakes, that comprise a monospecific subfamily in the hake family Merluciidae. The family Melanonidae has only two species, *Melanonus zugmayeri* and *M. gracilis*. *Melanonus zugmayeri* is always a rare catch, but it is found in all three major ocean basins between 60°N and 49°S (Priede 2017). Its congener has a circumpolar distribution in waters south of about 40°S (Priede 2017). Note the mechanoreceptors on the head of *Melanonus*. *Steindachneria* is found in the tropical western Atlantic and is so named for the photophores on its ventrum.

Superorder Acanthopterygii The superorder Acanthopterygii, spiny-rayed fishes, comprises the most, and the most advanced, bony fishes. It should be noted that the systematics of spiny-rayed fishes is a very dynamic field, and classification systems differ on what species are the most advanced.

About 13 600 marine species make up the superorder. It includes the most highly derived species, such as the parrotfishes common to coral reef environments, as well as the puffers and the enormous ocean sunfishes, the *Mola mola*s. Only a handful of the Acanthopterygii are open-ocean fishes, and still fewer are captured in scientific trawls. We will briefly discuss those groups most commonly found in pelagic systems worldwide.

Order Beryciformes The order Beryciformes includes several important families that are widely distributed and commonly appear in open-ocean trawls. Families in its two suborders, Berycoidei and Stephanoberycoidei, differ significantly in habits and appearance. The melamphaids or bigscales (Berycoidei) typically inhabit mesopelagic depths; the whalefishes (Stephanoberycoidei) live in the bathypelagial.

Family Melamphaidae. The suborder Berycoidei includes the Berycidae, or alfonsinos, which are typically benthopelagic and associated with seamounts and upper slope environments (Priede 2017), and the Melamphaidae, or bigscales (Figure 10.56). As adults, melamphaids range in length from about 22 to 140 mm. They are found in the catch from virtually every trawl that sweeps the upper 1000 m of the water column, and there will usually be more than one species (Ebeling 1962; Ebeling and Weed 1963). Though they never reach the abundances of the gonostomatids and myctophids, melamphaids are ever-present. In the Bermuda Ocean Acre study (Keene et al. 1987), they ranked fourth in abundance, with 15 species in 4 genera; in most other survey sites, they are less common (Pearcy 1964; Badcock 1970; Clarke and Wagner 1976; Hopkins and Lancraft 1984; Sutton et al. 2010).

Much of the data on vertical distributions in melamphaids are from tropical/subtropical systems, in particular Hawaii (Clarke and Wagner 1976) and Bermuda (Keene et al. 1987). In both systems, the majority were vertical migrators, with a daytime residence between 500 and 1000 m and nightly excursions into the upper 300 m (Clarke and Wagner 1976; Keene et al. 1987). However, in both the Hawaii and Bermuda locations, some remained at lower mesopelagic depths, below 600 m, on a diel basis. Thus, melamphaid species' diel distributions may differ from place to place, migrating in one location and remaining at depth in another (Ebeling and Weed 1963; Clarke and Wagner 1976).

Melamphaids are zooplanktivorous, feeding on a wide variety of zooplankton, including copepods, ostracods, euphausiids, amphipods, polychaetes, and gelatinous animals like salps and pteropods (Clarke 1978; Hopkins et al. 1996). Deeper-living species such as *Poromitra gibbsi* showed a preference for gelatinous prey. Data on feeding chronology are limited, but in the GOM, shallow-living copepods dominated the diet of at least one vertically migrating species (*Melamphaes simus*) suggesting that feeding took place mainly at night. In Hawaiian waters, the vertical migrator *Melamphaes danae* showed

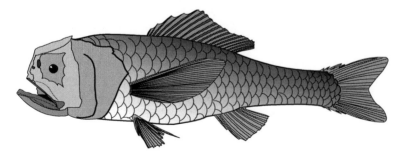

Melamphaes suborbitalis

Figure 10.56 The melamphaid *Melamphaes suborbitalis*.

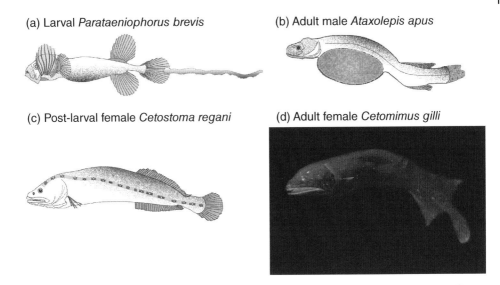

Figure 10.57 Four radically different life stages displayed by the Cetomimidae. (a) Larval form; *Parataeniophorus brevis*. Note the caudal streamer, an extension of the caudal fin rays, that can extend 10–20 times the body length (it is shown incomplete in the figure). (b) Adult male *Ataxolepis apus*. Note the engorged stomach filled with copepods acquired during the larval stage. (c) post-larval female (in transition); *Cetostoma regani*; (d) adult female; *Cetomimus gilli*. Source: (d) © Dante Fenolio, reproduced with permission. See color plate section for color representation of this figure.

increased stomach fullness at night suggesting a peak in feeding in the shallow waters of its nighttime distribution.

Data on age and reproductive habits suggest a variety of life histories within the group. In the Bermuda Ocean Acre study, the small (24 mm maximum size) dominant species *Melamphaes pumilus* had a lifespan of one year and reproduced in the spring and summer. Its larger congener *M. typhlops* (73 mm, 2-years life-span) reproduced year-round. The two largest species *Melamphaes ebelingi* (137 mm, 3–4 years life-span) and *Poromitra capito* (99 mm, 3 years life-span), both reproduced in the fall-winter months. Ages were determined by following modal size-classes (Keene et al. 1987.)

Suborder Stephanoberycoidei. The most important Stephanoberycoidei in the pelagial are the three families Barbourisiidae, the velvet whalefishes, Cetomimidae, the flabby whalefishes, and the Rondeletiidae, or redmouth whalefishes. Fishes in the remaining three families of the suborder are rarely captured, and little is known about their biology. By far, the cetomimids are the most speciose and the most likely to appear in a midwater trawl. Whalefishes are major components of the bathypelagic realm, increasing in importance below 1800 m (Paxton 1989; Priede 2017), but they are captured in the lower mesopelagic zone as well (Clarke and Wagner 1976). Most genera have a very wide latitudinal range that covers temperate and tropical waters and sometimes polar as well (Priede 2017). Several genera, *Gyrinomimus* being one example, are panglobal in distribution. All three families of whalefishes have a characteristically large gape and a body shape that, using imagination, generally resembles a whale. However, they are commonly 20 cm or less in length; the largest species, *Barbourisia rufa*, reaches only 39 cm. Limited data suggest an upward displacement at night in *Rondeletia* from lower to upper mesopelagic depths (Clarke and Wagner 1976).

Recent groundbreaking information from Johnson et al. (2009) has given us a completely new understanding of development in the cetomimids. There is marked sexual dimorphism in the group (Figure 10.57). The males are very much smaller than the females and have highly developed nasal organs with complementary olfactory regions in the brain. As larvae and juveniles, males engorge themselves with copepods and acquire a very large food bolus (Figure 10.57b) in the stomach at metamorphosis. The copepod bolus fuels the development of the very large liver that sustains them through their adult life. As adults, males' stomach and esophagus deteriorate and they no longer feed. Gonads are large, filling their peritoneal (midgut) region. The chemoreceptive abilities of the males are believed to help them find mates. Females are also believed to feed heavily on copepods as larvae and juveniles prior to their metamorphosis into adults, but they retain their digestive system as they develop into the large-gaped adult (Figure 10.57d).

One of the most interesting findings from Johnson et al. (2009) is that the larvae and juveniles of the cetomimids had been classified as two separate families within the order Stephanoberyciformes: the Mirapinnidae (tapetails) and the Megamycteridae (bignose fishes), respectively (Nelson 2006; Nelson et al. 2016). Differences in morphology between the extremely elongated larvae (up to 816 mm total length), the juveniles, and the adults are quite pronounced; it is easy to understand why the early life stages were thought to be two additional families, particularly in a group where specimens were scarce. The mystery was resolved by research partnerships between systematists, molecular biologists, and field samplers.

Particularly noticeable in all three families of whalefishes is their well-developed lateral line, suggesting that mechanosensory ability is important to survival in the bathypelagic realm. In turn, the cetomimids and barbourisiids have reduced eyes, supporting the idea that vision plays a reduced role in the disphotic and aphotic environment of the lower mesopelagic and bathypelagic zones. Interestingly, *Rondeletia*, with a well-developed eye, was found by Clarke and Wagner (1976) to move into the upper mesopelagic zone at night.

Order Trachichthyiformes The order Trachichthyiformes contains five interesting and important families, only one of which is often captured in scientific trawls: the Anoplogastridae, or fangtooths (Figure 10.58). *Anoplogaster cornuta* is the iconic mesopelagic fish. Its fearsome visage adorns book covers and, when speakers have access to good images, *Anoplogaster* always gets a gasp from the audience. Despite its ferocious appearance, *A. cornuta* only reaches a maximum size of 15–18 cm and is most often captured at 15 cm and less. Like the dragonfishes, it is a lot of teeth in a small package. It is found worldwide at latitudes between 65°N and 46°S, most often at lower mesopelagic and upper bathypelagic depths (500–1500 m) (Smith and Heemstra 1991; fishbase.org).

Anoplogaster cornuta juveniles feed primarily on Crustacea; the adults are piscivorous. The juveniles are silvery in appearance and have a less pronounced lateral line (Figure 10.58b).

Note the difference in eye size relative to body size between the adult and juvenile. Information on the congener of *Anoplogaster cornuta, A. brachycera,* is quite

Figure 10.58 Adult (a) and juvenile (b) *Anoplogaster cornuta*. Note the differences in dentition, relative eye size and lateral line development. *Source:* © Dante Fenolio, reproduced with permission. See color plate section for color representation of this figure.

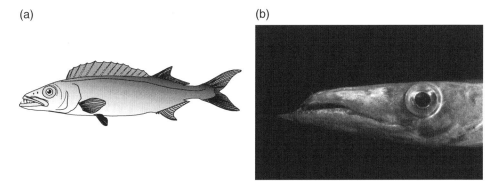

Figure 10.59 The Gempylidae (snake mackerels). (a) The oilfish *Ruvettus pretiosus*. Although edible, the flesh of this species contains a high content of an oil, rich in esters, that is a very effective laxative; (b) close-up of the head of the striped escolar *Diplospinus multistriatus*. This species is an important food fish for local populations. *Source:* (b) © Dante Fenolio, reproduced with permission. See color plate section for color representation of this figure.

scarce, but it appears to be restricted to tropical/subtropical waters.

The three remaining families in the Trachichthyiformes are the Diretmidae or spinyfins, the Trachichthyidae or roughies, and the shallow-dwelling non-pelagic Monocentridae, the pinecone fishes.

The spinyfins are considered mesopelagic during at least part of their life. The family has four species in three genera. Two of them, *Diretmus argenteus* and *Diretmichthys parinus,* are distributed worldwide at temperate and tropical latitudes; *Diretmus* reaches boreal latitudes. All the spinyfins are believed to be benthopelagic in their larger sizes (fishbase.org). Recent findings on vision in *Diretmus argenteus* (Musilova et al. 2019) echo those found for *Stylephorus*: they probably have the ability to discriminate color despite the monochromatic world in which they live.

The roughies, or slimeheads, are the most diverse of the trachichthyiforms with about 49 species. The most well known is the orange roughy, *Hoplostethus atlanticus,* a widely distributed species at latitudes from 65°N to 46°S in all three major ocean basins. It was the target of major trawl fisheries in several locations but owing to the fish's late maturity (20 years), extremely long life (>100 years), and synchronous spawning aggregations, roughies were highly vulnerable to over-fishing, and the fisheries in most locations rapidly crashed. There is still a limited harvest off New Zealand. The orange roughy is a benthopelagic species that reaches 75 cm in length. Spawning aggregations occur on continental slopes and seamounts. They forage on pelagic crustaceans, squid, and fishes (Priede 2017).

The remaining trachichthyiforms are primarily benthopelagic, living from shallow to deep water. Four species of "luminous roughies" in the genus *Aulotrachichthys* inhabit the upper mesopelagic zones of the tropical Atlantic, Pacific, and Indian Oceans. They are so named for the small, donut-shaped, luminous organs populated by luminous bacteria that are found near the anus (Herring and Morin 1978).

Order Scombriformes The order Scombriformes is divided into three suborders. The Scombroidei includes the highly mobile mackerels and tunas with their fusiform bodies and advanced locomotory systems as well as two deep-water families, the Gempylidae, or snake mackerels, and the Trichiuridae or cutlassfishes. The suborder Scombrolabracoidei has only one rare species, Scombrolabrax heterolepis, found at upper mesopelagic depths in tropical and subtropical systems, usually associated with bottom features such as the continental shelf or underwater rises (fishbase.org). The Stromateoidei comprises six families of pelagic fishes, two of which, the Amarsipidae and Stromateidae, are shallow coastal groups. The other four families, discussed below, are pelagic.

Suborder Scombroidei

Family Gempylidae. The snake mackerels (Figure 10.59) are the only scombriform occasionally taken in scientific trawls, and then it is usually the smaller representatives that are captured. As a family, the gempylids have a global distribution. One species, *Paradiplopinus antarcticus,* is found in the frigid waters of the Southern Ocean. Gempylids have two body morphs: a tuna-like shape capable of rapid swimming and a more elongated, slowly swimming, form. Nakamura and Parin (1993) consider them to be intermediate between the highly fusiform, powerfully swimming scombrids, and the elongated, benthopelagic trichiurids. The family is found in both mesopelagic and benthopelagic habitats. Gempylids feed on pelagic crustaceans and small fishes (Priede 2017), and some, e.g. *Diplospinus* *multistriatus,* vertically migrate from lower mesopelagic depths into the epipelagic zone (Nakamura and Parin 1993).

Family Trichiuridae. The Trichiurids, or cutlassfish (Figure 10.60), are a benthopelagic group found worldwide at temperate and tropical latitudes. Their common name comes from their laterally compressed yet elongated body form and silvery sides. Trichiurids range from 80 to 200 cm in length and are found at depths ranging from 200 to 1000 m. They feed on fishes, cephalopods, and Crustacea.

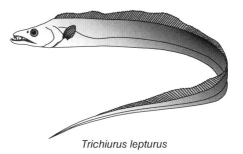

Trichiurus lepturus

Figure 10.60 The trichiurid (cutlassfish) *Trichiurus lepturus.*

Family Scombridae. Fast and elegant, the mackerels and tunas include some of the most highly capable swimmers of the fish world. They are epipelagic hunters, often visiting mesopelagic depths to supplement their diet with midwater fishes and so are doubly important members of the open-ocean community. Unlike most of the fishes covered here, which support fisheries only as prey items, the scombrids support important commercial and recreational fisheries worldwide to the point that many species are now endangered or threatened. There are about 51 species of mackerels and tunas in 14 genera.

Scombrids are divided into two subfamilies, the monospecific Gasterochismatinae and the Scombrinae, which includes all the remaining species. *Gasterochisma melampus* is circumglobal in temperate waters south of the equator. It differs from the rest of the tunas in its large cycloid scales and the enormous pelvic fins of its juvenile stage. In some classification systems (Collette and Nauen 1983; Nelson et al. 2016), the Scombrinae are divided into four tribes (Figure 10.61): the Scombrini or mackerels, the Scomberomorini or Spanish mackerels, the Sardini or bonitos, and the Thunnini or tunas. The four tribes have changed little in composition since 1982 and are quite useful for discriminating between differences in habits of the 50 species of Scombrinae. All scombrids are dioecious, and all are broadcast spawners.

The most primitive of the Scombrinae, the mackerels, are zooplanktivorous, straining copepods and other zooplankton out of the water with their gill-rakers as they swim. The remaining three tribes feed on larger crustaceans, small fishes, and squids, including mesopelagic taxa such as the myctophids. The scombrids are all capable of deep-diving, but most of the information on vertical movement has been recorded from the largest species, in the tribe Thunnini. Atlantic Bluefin have been recorded at depths of as great as 700 m following vertically migrating prey items. Vertical migratory patterns that coincide with the movements of the deep-scattering layer have been observed in all three species of Bluefin tunas (*Thunnus thynnus, T. orientalis, T. maccoyi*: Atlantic, Pacific and Southern Bluefin respectively) as well as in Bigeye tunas (*Thunnus obesus*) (Gunn and Block 2001). Yellowfin (*T. albacares*) tends to remain in the mixed layer.

Thirteen species within the Thunnini are heterothermic, a physiological term that means they are partially endothermic, or warm-blooded. Heat generated in the muscle by swimming activity is retained in the

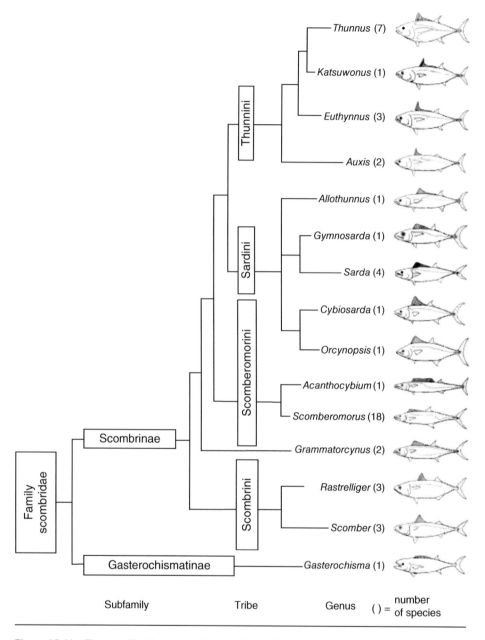

Figure 10.61 The classification system for the Scombridae showing the four-tribe division of the subfamily Scombrinae. *Source:* Adapted from Collette and Nauen (1983), figure 1.

muscle via a circulatory heat-exchange system, allowing the tunas to maintain a core body temperature significantly above ambient and in some cases to truly thermoregulate (Carey and Lawson 1973; Graham and Dickson 2001; Katz 2002). Without the circulatory heat exchanger, heat generated within the muscle would be lost at the gills. Just like an automobile radiator, the very high surface area needed by the gills for gas-exchange would rapidly dissipate all heat generated within the muscle. The heterothermy of tunas significantly improves their muscle function and thereby their swimming ability. Bluefins, *Thunnus thynnus*, can cross the Atlantic in 50 days (Bone and Moore 2008).

Archival tagging studies have considerably improved our knowledge of diel vertical distributions in several species of tunas as well their impressive abilities to control their core body temperature in the face of external temperatures many degrees colder, and to do so for long periods of time. The studies have also given us what knowledge we have on the seasonal and annual movement of individuals, the best of which is on the Atlantic bluefin (Block et al. 2001). Bluefins tagged in the western Atlantic off North Carolina exhibited three types of behaviors: "western residency with no visitations to spawning areas, western residency with Gulf of Mexico breeding, and trans-Atlantic migrations to the eastern Atlantic or Mediterranean Sea" (Gunn and Block 2001).

Suborder Stromateoidei. The stromateoids comprise four open-ocean families, the Ariommatidae or ariomids, the Centrolophidae or medusafishes, the Nomiidae or driftfishes, and the Tetragonuridae, or squaretails. All have wide-ranging distributions at temperate and tropical latitudes, and all have representatives that live at mesopelagic depths. Species in three of the four families have some association with gelatinous zooplankton, either for food or shelter. Juveniles in the Centrolophidae associate with jellyfish and other drifting objects and feed on salps in addition to small crustaceans and fishes. The nomeids associate with floating objects and feed on salps (Priede 2017). The Tetragonuridae have the unusual lifestyle of living within the tests of pelagic tunicates (Janssen and Harbison 1981). Younger, smaller tetragonurids (40–60 mm) are epipelagic and live within the body cavities of salps, consuming their commensal amphipods and stomachs. Larger squaretails (up to 70 mm) inhabit mesopelagic depths and are believed to live in the tests of pyrosomes or larger, deeper-living salps like *Thetys vagina*. Tetragonurids have an elongated body, dorsal, and pectoral fins that can be lowered into grooves, and the ability to swim forward and backward equally well (Janssen and Harbison 1981).

Order Istiophoriformes The billfishes. The billfishes comprise 12 species in 6 genera (Figure 10.62). They are found in the upper 200 m of temperate and tropical latitudes (45°N to 45°S) worldwide (Nakamura 1985). Like the tunas, they are the subject of fisheries throughout their ranges and, where data are available, IUCN (International Union for the Conservation of Nature) species assessments vary from Vulnerable (e.g. *Kajikia albida, Makaira nigricans*) to Endangered (*Xiphias gladius*), a testament to the quality of their flesh.

Billfishes vary in maximum length from 184 cm in the roundscale spearfish (*Tetrapturus georgii*) to 500 cm in the Atlantic blue marlin (*Makaira nigricans*) and in mass from 21.5 kg in *T. georgii* to 750 kg in *Istiompax indica* (Black marlin) (fishbase.org). Most species prefer warmer water and remain above the thermocline, but the swordfish *Xiphias gladius* ventures into colder waters and the mesopelagial, as deep as 600 m. To cope with the much colder waters at depth, swordfishes have developed a brain heater (Carey 1982). It consists of brown adipose tissue (BAT) that generates heat, and a heat exchange system, the rete mirabile, that warms the blood

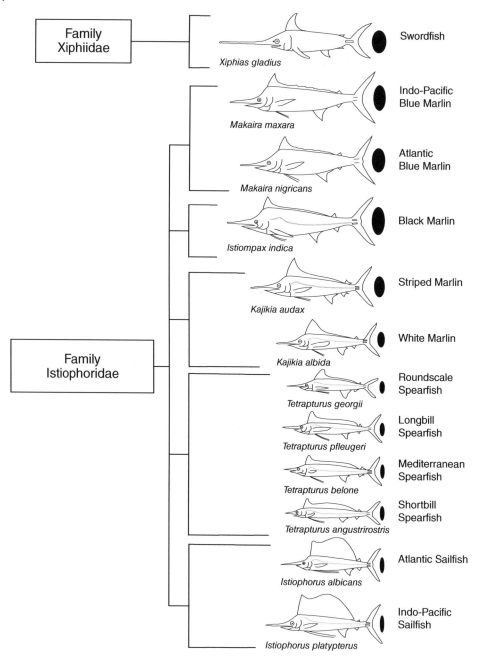

Figure 10.62 Classification system for the suborder Xiphioidei (the billfishes). Sizes of drawings correspond roughly to maximum lengths of species. The black ovals to the right of each drawing show cross sections of bodies at the level of the pectoral fin bases. *Source:* Adapted from Nakamura (1985), figure 1.

(a) *Hyporhamphus unifasciatus* (silverstripe halfbeak)

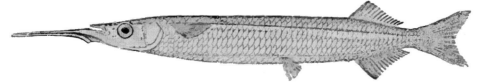

(b) *Fodiator acutus* (sharpchin flyingfish)

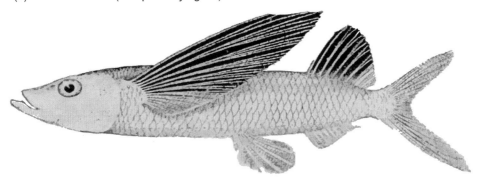

Figure 10.63 The Beloniformes. (a) The hemiramphid *Hyporamphus unifasciatus* (silverstripe halfbeak); (b) the exocoetid *Fodiator acutus* (sharpchin flyingfish). *Sources:* (a) Jordan and Evermann (1900), plate CXVI; (b) Jordan and Evermann (1900), plate CXVII.

going to the brain and eyes of the fish. The brain heaters are believed to facilitate effective vision and hunting in the cold waters at depth, offsetting the large temperature changes encountered during its vertical excursions. Swordfishes are opportunistic feeders, feeding on epipelagic species including tunas, Dolphin fishes, flying fishes, mesopelagic species like myctophids and sternoptychids, and Ommastrephid squids to name but a few. The bill of swordfishes and their billfish relatives is used to slash, spear, and disable prey. Billfishes are not continuous swimmers like the tunas; they are "stalkers and sprinters" (Carey 1982).

Billfishes are all dioecious broadcast spawners, with most species spawning in tropical and subtropical waters. Eggs hatch into pelagic larvae and develop in the epipelagic zone. Virtually, all billfishes show a seasonal migration into temperate or cold waters for feeding and back to subtropical or tropical water for overwintering and spawning (Nakamura 1985).

Order Beloniformes Three families within the Beloniformes are important epipelagic forage fishes: the halfbeaks (Hemiramphidae), the sauries (Scomberesocidae), and the flying fishes (Exocoetidae) (Figure 10.63). Peak species' diversities are in tropical/subtropical waters. All are dioecious broadcast spawners. They are opportunistic feeders with diets ranging from zooplankton to small fishes.

Habitats vary between families. Most of the halfbeaks, like the common halfbeak *Hyporhamphus unifasciatus*, are found schooling nearshore. Others, e.g. *Hyporhamphus roberti*, range out to pelagic waters. Sauries are schooling fishes that are primarily oceanic though sometimes shoaling close to shore.

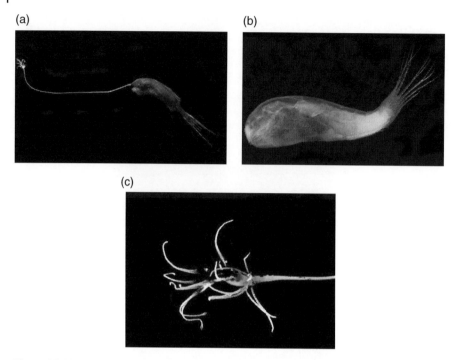

Figure 10.64 Gigantactinidae. (a) Adult female *Gigantactis gargantuai* with long illicium; (b) free-living male *Rhynchactis* sp.; (c) close-up of the esca of *Gigantactis gargantuai*. *Source:* © Dante Fenolio, reproduced with permission. See color plate section for color representation of this figure.

The flying fishes are oceanic schooling species, most with very wide-ranging distributions. Some, e.g. the two-wing flying fish *Exocoetus volitans*, range into coastal pelagic waters, but most species dwell exclusively further offshore. They are all capable of propelling themselves through the sea surface and gliding just above it, usually for 50 m or less. The distance record is 400 m, and the record for flight-time is 45 seconds (Ross Piper 2007). Flights are usually induced by the presence of predators. Flying fishes are the prey of all the billfishes, especially the marlins and swordfish, as well as tunas, dolphinfish, and a variety of porpoises and seals. A flotilla of flying fish seen above the sea surface indicates predators beneath them, often a school of tunas. If a predator succeeds in tracking them through their initial flight, they can remain aloft by sculling with the long ventral lobe of their tailfin. That sometimes does the trick and allows escape.

Order Lophiiformes (Anglerfishes and Frogfishes)
Eleven of the eighteen families of lophiiforms have a deep pelagic distribution as adults, ranging from the lower mesopelagic through the entire bathypelagic zone but reaching their highest densities between 2000 and 3000 m (Bertelsen 1951; Pietsch 2009). The remainder of the lophiiform families are primarily benthic, benthopelagic, or demersal. Pietsch (2009) and Nelson et al. (2016) place the eleven pelagic lophiiforms in their own suborder, the Ceratioidei, whereas the Cal Academy system does not subdivide the order. It should be noted that the different classification authorities also differ in their placement of the Lophiiformes in the hierarchy

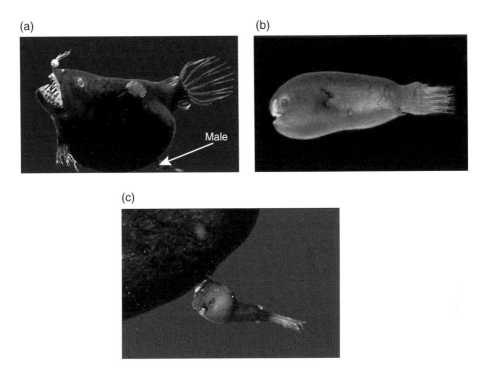

Figure 10.65 Linophrynidae. (a) Adult female *Linophryne* sp. with attached male. Note the large well-developed chin barbel which is unique to the linophrynids; (b) free-living male linophrynid. Note the large well-developed olfactory organs (nostrils) in front of the eyes; (c) close-up of the attached male seen in (a). *Source:* © Dante Fenolio, reproduced with permission. See color plate section for color representation of this figure.

of fish advancement. Without question, they are a highly derived group. Ceratioid families are designated with an asterisk in Appendix A.

All the deep-sea anglers have unusual morphologies and reproductive habits. Body shapes vary among families, with the Gigantactinids (Figure 10.64a), Centrophrynids, Ceratiids, and Himantolophids having a more elongated fish-like body than their more globular relatives, the Linophryinids (Figure 10.65a), Melanocetids (Figure 10.66a), and Oneirodids. Female anglerfishes range in size from that of a kumquat (*Melanocetus johnsoni*, Figure 10.66a) to that of a football, e.g. *Linophryne* (Figure 10.65a). The maximum size of adult deep-sea anglers is typically between 100 and 250 mm (Pietsch 2009). The Gigantactinidae, or whipnoses, combine a moderate standard length with a very long lure. Pietsch (2009) reports a *Gigantactis macronema* with a standard length of 232 mm and a lure 1 m in length.

Sexual dimorphism is highly pronounced in the deep-sea anglers. Females are all free-living and have a large maw, an elaborate bioluminescent lure derived from the first two spines of the dorsal fin, reduced eyes, and a posterior triplet of dorsal, caudal, and anal fins. Many are globular in shape (e.g. Figures 10.65a and 10.67a). Males are diminutive, ranging in length from 10 to 40 mm, and have a very different morphology. Not only are they very small compared to the females but also the males have a

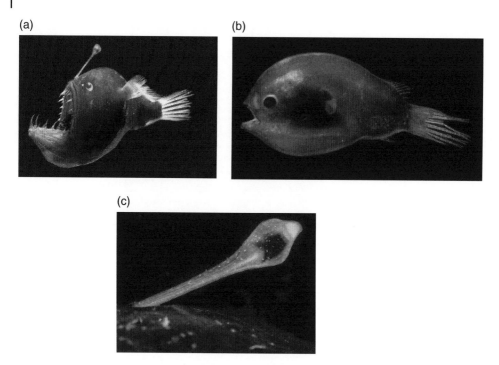

Figure 10.66 Melanocetidae. (a) Adult female *Melanocetus murrayi*; (b) free-living male *Melanocetus* sp. Note the large well-developed olfactory organs (nostrils) in front of the eyes; (c) close-up of the esca of *Melanocetus johnsoni*. *Source:* © Dante Fenolio, reproduced with permission. See color plate section for color representation of this figure.

conventional fish-shaped body, well-developed eyes and nares, and a smaller mouth (Figures 10.64b–10.66b).

There are at least two reproductive strategies in the ceratioids: one with free-living males, the other with parasitic males. Consider the difficulties the males face finding a mate in total darkness, in a water column at least 2000 m deep and many thousands of kilometers in horizontal extent, after having spent their early life in epipelagic waters. In those families of ceratioids that use sexual parasitism, males are believed to use a highly developed olfactory sense in tandem with their visual and mechanoreceptive abilities to home in on females of their own species by following their pheromone trail (Marshall 1979). Males destined to be parasitic are incapable of feeding, subsisting instead on energy acquired during their larval stage and stored in their enlarged livers, similar to the whalefishes. Once a female is located, the male uses his needle-like teeth to permanently clamp onto her body (Figure 10.65c). The head of the male fuses to the female, initiating a continuity in circulation that nourishes the now parasitic male. After fusion, the male increases several-fold in size and the pair act as a single self-fertilizing hermaphrodite for multiple broods during the life of the female. The parasitic male lives as long as its host. In many taxa, the limit is one male per female; in others, multiple males may be attached to a single female. Sexual parasitism has been confirmed in the Ceratiidae, Caulophrynidae, Neoceratiidae, Linophrynidae, and some members of the Oneirodidae (Pietsch 2009).

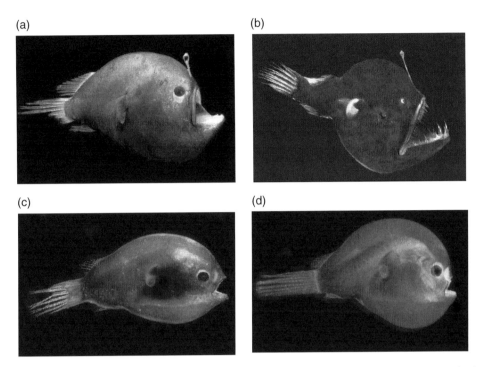

Figure 10.67 Anglerfish life history: metamorphosis. (a) Female *Melanocetus* sp. at metamorphosis; (b) adult female *Melanocetus murrayi*; (c) male *Melanocetus* sp. at metamorphosis; (d) free-living male *Melanocetus* sp. *Source:* © Dante Fenolio, reproduced with permission. See color plate section for color representation of this figure.

In a few of the ceratioid families, there is no evidence of sexual parasitism. Those families include the Gigantactinidae, Melanocetidae, Diceratiidae, Himantolophidae, and some members of the Oneirodidae. It is believed that the tracking strategy of the males in those families is similar to that of the parasitic males for finding a mate, but that sexual attachment ends after spawning and fertilization. Nonparasitic males are capable of feeding on their own. There is insufficient information on the Centrophrynidae and Thaumatichthyidae to judge whether they employ sexual parasitism as a reproductive strategy (Pietsch 2009).

Spawning of ceratioids is believed to take place at bathypelagic depths, with the fertilized eggs rising to near-surface waters to develop and hatch. The larvae of *Himantolophus groenlandica*, a common species in the North Atlantic, are about 2.5 mm at hatching and remain in surface waters until the onset of metamorphosis, which occurs in about two months (Bertelsen 1951). Metamorphosing larvae sink to bathypelagic depths where the females develop into subadults over a period of 6 months, attaining a length of about 25 mm (Figure 10.67a, b). Similarly, males sink to the bathypelagic zone during metamorphosis but only have a free-living longevity of approximately six months, when their liver-based energy depots are expended. Parasitic males do not mature sexually until they have attached to a female, which in turn stimulates her ovarian development. In contrast, the more capable free-living males of *Melanocetus* (Figure 10.67c, d) *Gigantactis* and

Himantolophus have well-developed testes in their late larval and metamorphic stages and can feed and grow after metamorphosis. Females take more than one year, and perhaps several years, to reach sexual maturity (Bertelsen 1951; Marshall 1979).

The diets of ceratioids vary with species and size. In the Gulf of Mexico, *Melanocetus johnsoni* is piscivorous, mainly consuming the dominant gonostomatid *Cyclothone* spp. at lower mesopelagic and upper bathypelagic depths (Hopkins et al. 1996; Burghart et al. 2010). Larger size classes of the Triplewart Seadevil, *Cryptopsaras couesii*, primarily consumed small cephalopods, whereas small size classes of its confamilial *Ceratias* fed on copepods. Those findings agree well with other data on ceratioid diets, which are dominated by fish, squids, and crustaceans (Pietsch 2009).

Female anglers are "gape-and-suck" feeders: a large negative pressure is created by rapid expansion of the oral and opercular cavities, drawing prey into their jaws (Marshall 1971; Pietsch 2009). Once inside the fish's maw, prey is trapped by the hinged inwardly facing teeth of the jaw and moved along to the stomach by the pharyngeal teeth on the roof of the pharyngeal cavity. Anglerfishes are capable of taking large prey, including prey larger than themselves, and retaining it in their highly distensible stomachs. Depending on the size of the prey, this ability can prove fatal. Since the entire suite of feeding structures, the large jaw, inward-facing mandibular teeth, pharyngeal teeth, and large buccal/opercular cavity, are engineered for prey retention, prey cannot be released once taken. A prey item so large that it occludes the pharynx results in the death of both prey and predator. There are no data on how often this scenario occurs, but it does happen. Dead ceratioids have occasionally been found at the surface with prey protruding from their mouths (Pietsch 2009). Their feeding strategy must, however, work quite well the great majority of the time, as with 11 families and 173 species, the Ceratioidei are the most diverse group of bathypelagic fishes.

Due to their unusual morphology, ceratioids are theoretically capable of three types of locomotion. The first is pectoral-fin propulsion: their large pectoral fins beat with a coordinated breast-stroke-like motion to propel the fish forward or backward. Pectoral-fin swimmers like the wrasses and butterflyfishes can also execute precision maneuvers such as plucking a coral polyp out of its skeleton. The pectoral fins of ceratioids are set too far back on the body to be used like those of the reef-fishes but are believed to aid in turning, including quick "turn and bite" movements described by Marshall (1979) for female anglers. The second mode is undulatory propulsion, characterized by side-to-side movements of the tail and posterior body. Undulatory propulsion is used in the locomotion of males and females. The third is jet-propulsion.

The opercular chamber in anglerfishes ends behind the pectoral fins in a restricted tube-like opening instead of a large flap-like gill cover The musculature that allows the fishes to rapidly expand their large opercular chamber in the "gape and suck" feeding strategy also controls the velocity of the outgoing (respiratory) flow. By a vigorous contraction of the opercular chamber, anglers can expel water rapidly through the small opercular openings, generating a forward thrust. Though never directly observed, the ceratioid morphology and the fact that the closely related family Antennariidae with similar tube-like opercular openings has been documented to employ jet propulsion in such a way strongly suggest that jet propulsion is quite possible for anglers (Bertelsen 1951; Pietsch 2009).

Luck and Pietsch (2008) observed a swimming female *Oneiroides sp.* in the Monterey Canyon off the California coast, a globular species that, based on their observations, may be considered a "sit-and-wait" predator (cf. Gartner et al. 1987). During the course of their observations, the fish drifted passively 73% of the time. About 25% of the time she moved with low-level sculling of her pectoral fins and tail, a pattern echoed in the laboratory observations of *Melanocetus johnsoni* by Cowles and Childress (1995). The pectoral fins oscillated in synchrony with each other and with the undulatory motions of the tail, unlike most coral reef fishes that beat their pectoral fins with a "breast-stroke" motion. The fish was also capable of short rapid bursts of swimming as it would need to escape or for prey capture.

The luminescent lure that gives the ceratioids their common name of sea devils is composed of two parts, the illicium, or "fishing rod," and the bulbous luminescent esca, or lure, at the tip. The illicium is under muscular control and in many species can be dangled in front of the mouth when fishing or laid back in a groove when swimming (Luck and Pietsch 2008). The morphology of the esca is a species-specific characteristic, ranging from a simple globe as on *Melanocetus* (Figure 10.66c) to elaborate mimicry (Figure 10.64c). Regardless of its ornamentation, the esca is basically a cup-shaped light gland containing luminescent bacteria, kept in culture by the fish.

The central cavity, or cup, of the esca holds the luminous bacteria that are nourished by a surrounding layer of glandular tissue. The glandular tissue is in turn surrounded by an outer silvery reflective cup that confines the light produced by the bacteria and directs it anteriorly. A pigment layer surrounds the entire core of luminous tissue which is visible in Figure 10.66c as the dark region in the center of the esca. The inner structure of the esca acts as a light guide, its silvery sides bringing the luminescence generated in the central cavity toward the tip. At the tip of the esca, a window allows the light to escape, producing the glow that lights the lure. In the photo (Figure 10.66c), the reflective tissue is pigmented white. Lures with accessory appendages such as those shown in Figures 10.64c and 10.65a use silvery internal reflective layers to act as light guides, conveying the luminescence produced in the central cavity to the ends of the appendages.

Escal luminescence is normally a steady glow with a color in the blue–green range, though other colors including white, yellow–green, and orange have been reported (Herring and Morin 1978; Pietsch 2009). Hansen and Herring (1977) observed a pulsing of bright turquoise light from the esca of *Linophryne brevibarbata,* but the mechanism(s) producing the pulse remain unresolved. It is present in other species. The esca is very scantily innervated and even more sparingly invested with muscle, but it may be enough to produce the observed pulses, which peaked for about six seconds and trailed off for an additional six seconds before peaking again. In *L. brevibarbata*, there are rings of smooth muscle fibers located at the escal "windows" that may play a role in producing the observed pulses. Other possibilities are a movement of the pigment layer to occlude the "windows" in the esca (Pietsch and Van Duzer 1980) and a restriction of blood flow to the esca (Bertelsen 1951), reducing the oxygen available for the production of luminescence.

The luminescence of the large chin barbel found only in the genus *Linophryne* (Figure 10.65a) is self-generated rather than bacterial in origin. It is unique among the ceratioids and constitutes an entirely separate bioluminescent system.

Order Tetraodontiformes (Puffers and Molas) The tetraodontiform fishes of most interest to studies of the open ocean are the molas, or ocean sunfishes, the family Molidae in the suborder Tetraodontoidei. The five species of molas are distributed pan-globally in temperate to tropical waters and are found from the surface to mesopelagic depths (Harbison and Janssen 1987). They are quite large, with *Mola mola, Mola alexandrine*, and *Masturus lanceolatus* reaching at least 3.3 m in length and weighing up to 2.3 tons (fishbase.org). Ocean sunfishes are a familiar sight at the surface in the California borderland and offshore in the GOM, basking in the sun and flapping their dorsal fins from side to side like a metronome. Molas (Figure 10.68) have a reduced caudal fin; they propel themselves forward by sculling motions of their opposed dorsal and anal fins (Marshall 1965). They feed on gelatinous zooplankton (medusae, siphonophores, ctenophores, and salps) and are frequent visitors to mesopelagic depths to seek their prey (Harbison and Janssen 1987; Potter and Howell 2011). Observations by Potter and Howell (2011) of tagged fish revealed that the daily depth profile of *Mola mola* changed with location and season. They resided primarily in the upper 200 m in the summer months when in the northeastern Atlantic but had a diel vertical migration from 400 to 800 m during the day to near-surface waters at night when in the Gulf Stream.

Order Scorpaeniformes The scorpaeniforms comprise about 1570 species in 6 suborders and 35 families: an enormous group. The order includes benthic, demersal, and benthopelagic species. One family, the Liparidae or snailfishes (Figure 10.69), has members that occasionally foray into

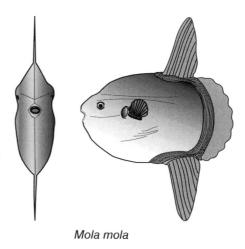

Mola mola

Figure 10.68 Frontal and side views of the Ocean Sunfish *Mola Mola*.

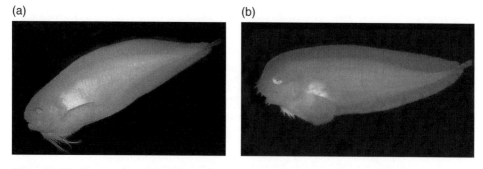

Figure 10.69 Two species of liparids, (a) *Careproctus trachysoma* and (b) *Crystallichthys matsushimae*. Source: © Dante Fenolio, reproduced with permission. See color plate section for color representation of this figure.

pelagic waters and are captured in midwater trawls, but the great majority of the 406 species in the family lead a benthic or benthopelagic existence. They range from pole to pole and the intertidal to the very deep-sea, including hadal depths. Hadal liparids such as the Mariana Snailfish (*Pseudoliparis swirei*) are now considered to be "trench endemics," with the same species occupying trenches thousands of km apart (Linley et al. 2016; Gerringer et al. 2018). Most importantly, it now appears that the liparids are the dominant hadal fish family, living at depths of >6000 m in five different trenches: the Japan, Kermadec, Peru-Chile, Kurile-Kamchatka, and Mariana Trenches (Gerringer et al. 2018). The deepest-living fish species for which reliable depth information is available is *Pseudoliparis swirei* with a confirmed depth of 8078 m. Another liparid (the "ethereal snailfish") was observed at 8145 m (Linley et al. 2016) but was not captured and so could not be described and named. Yancey et al. (2014) have proposed an inherent biochemical depth limit of 8200 m for teleost fishes. Thus far, direct observations have supported their supposition, with both the "ethereal snailfish" and *Pseudoliparis swirei* observed very close to that limit.

Order Perciformes The order Perciformes comprises about 5757 species, 11 suborders, and 130 families, by far the largest of the teleost orders. It includes most of the coral reef fishes, the seabasses, the croakers, the snappers, the coastal Antarctic fishes, and a wide variety of other, mainly coastal fishes as well as a host of freshwater species. Only those families most likely to be observed directly in the open ocean, brought up in a midwater trawl, or with interesting or unusual characteristics are discussed below.

Suborder Percoidei. **Family Carangidae.** The carangids, or jacks, are a large family with about 30 genera and 150 species, ranging from well inshore all the way out to open-ocean waters. The family includes the pompanos, jack mackerels, runners, and scads as well as the amberjack, *Seriola dumerili*, the largest in the family, reaching 190 m and 81 kg in size. Most carangids reach between 25 and 100 cm as adults. They are active predators and excellent swimmers, with most species forming small schools (Honebrink 2000; McEachran and Fechhelm 2005) in coastal surface waters. Habitats include coral reefs, bays, and continental shelf waters out to the shelf break and beyond. A few, e.g. the black jack (or black trevally: *Caranx lugubris*), prefer clear oceanic waters.

Body form in the jacks is highly diverse (Figure 10.70), with the familiar, laterally compressed, form of the Crevalle Jack (*Caranx hippos*) intermediate between that of the fusiform Mackerel Scad (*Decapterus macarellus*) and the highly compressed moon-like shape of the Lookdown (*Selene vomer*). Feeding habits among the jacks are equally diverse, ranging from zooplanktivory in the decapterids, which feed on a variety of crustacean zooplankton, fish larvae and small fishes to the piscivory of the larger black jack and amberjack. Adult giant trevally, *Caranx ignobilis* (up to 170 cm), benthopelagic in habit, are known to feed on a variety of fishes, cephalopods, and crustaceans including large lobsters. The pilotfish (*Naucrates ductor*), well known for a commensal relationship with sharks and other large fishes, feeds on pieces of its benefactor's food as well as small fishes and invertebrates it encounters along the way (Honebrink 2000).

Cooperative hunting has been observed in small schools of yellowtail amberjack (*Seriola lalandi*) and young giant trevally

(Major 1978; Schmitt 1982). In both cases, the predators penetrated a large school of smaller fishes, separating out smaller groups of them before encircling and attacking.

Jacks are all gonochoristic, and all are broadcast spawners. Spawning locations are species-specific, with some spawning pelagically and others inshore or nearer the bottom. Species from Hawaiian waters that have been studied have a protracted spawning season, usually lasting from spring to fall with a peak in the summer months. Multiple broods are produced, with spawning periodicity varying with the species but ranging from every three days to once a month Honebrink (2000). Mating aggregations of 100 fishes or more have been

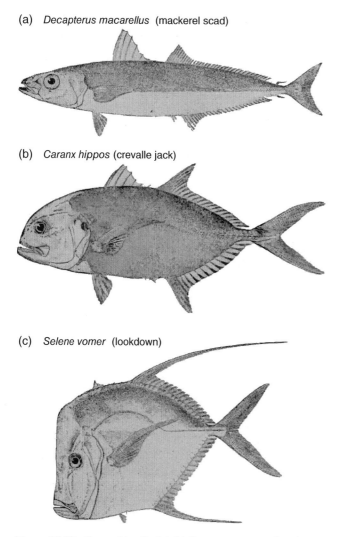

Figure 10.70 Carangidae (jacks). (a) *Decapterus macarellus*, the mackerel scad; (b) *Caranx hippos*, the crevalle jack; (c) *Selene vomer*, the lookdown. *Sources:* (a) Jordan and Evermann (1900), plate CXL; (b) Jordan and Evermann (1900), plate CXLI; (c) Jordan and Evermann (1900), plate CXLV.

described for giant trevally, which spawn close to shore. The aggregations break up into smaller groups and then into male–female pairs that sink to near the bottom, where spawning occurs (von Westernhagen 1974).

Jacks are panglobally distributed in temperate and tropical waters, and all have coastal affinities. Even the black jack, considered an oceanic species, is not found much past the shelf break, but it is associated with the landmasses in all three major ocean basins between 30°N and 30°S (fishbase.org).

Nelson et al. (2016) place the jacks in their own order, the Carangiformes, along with the Nemastiidae (roosterfishes), Coryphaenidae (dolphinfishes), Rachycentridae (cobias), Echeneidae (remoras), and Menidae (moonfishes). While Eschemeyer's catalog of fishes does not, they are grouped together within the Perciformes, indicating a close relationship. The six families in the Carangiformes of Nelson et al. (2016) were grouped together based on scale morphology (small cycloid scales), fin placement, and molecular evidence. The most important family to studies of pelagic fauna is the Coryphaenidae, which are commonly observed in open-ocean waters worldwide.

Family Caristiidae. Multiple species of veilfins have been observed in association with oceanic siphonophores at mesopelagic depths (Priede 2017), strongly suggesting a general affinity of the family with the siphonophores. They are laterally compressed with a high "forehead," large manelike dorsal fins, and long pectoral fins (Figure 10.71) that provide excellent maneuverability (Fitch and Lavenberg 1968). Janssen et al. (1989) described the association of *Caristius* sp. with the siphonophore *Bathyphysa conifera* as viewed from the Johnson Sea-Link submersible. *Caristius* apparently steals food from its nominal host including pieces of midwater fishes, shrimps, and zooids of the siphonophore itself. The fish also used the siphonophore as a form of protection, maneuvering to keep the siphonophore between itself and the submersible. It was speculated that the fish might, in turn,

Figure 10.71 Caristiidae (manefish). The manefish *Caristius macropus*.

provide some protection from the pelagic amphipods that prey on siphonophores.

Family Coryphaenidae. The common dolphinfish (Figure 10.72) *Coryphaena hippurus* is one of two species in the family, the other being *C. equiselis*, the pompano dolphinfish. Both have panglobal distributions at subtropical to tropical latitudes, living in near-surface waters across all major ocean basins, a true open-ocean species. Both form schools and can be found in coastal waters as well as offshore. *C. hippurus* reaches a maximum length of 2.1 m, and *C. equiselis* reaches 1.4 m (fishbase.org). They are dioecious with pelagic spawning and larval development. They grow quickly, maturing at four to five months. Dolphinfish are voracious predators, feeding mainly on fishes, but taking squid and crustaceans as well. They are sometimes found in association with rafts of sargassum or floating ocean debris. Without question, they are one of the most beautiful fish in the ocean, with metallic blue and golden coloration, a streamlined body, and deeply forked tail.

***Suborder Zoarcoidei.* Family Zoarcidae:** eelpouts. The eelpouts (Figure 10.73) have about 294 species in 59 genera, the great majority of which are benthic, demersal, or benthopelagic. Widely distributed, zoarcids are found from the intertidal to depths typical of the lower

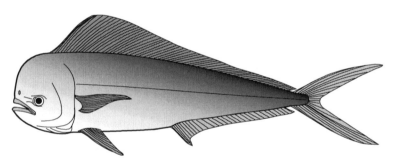

Coryphaena hippurus

Figure 10.72 Coryphaenidae (dolphinfish or mahi mahi). The dolphinfish *Coryphaena hippurus*.

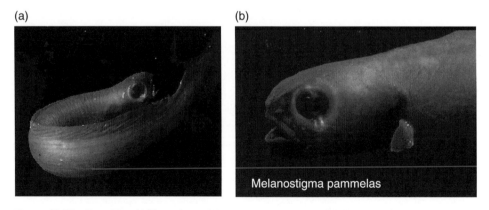

Figure 10.73 Zoarcidae. (a) The eelpout *Melanostigma pammelas*; (b) close-up of the head of *M. pammelas*. *Source:* © Dante Fenolio, reproduced with permission. See color plate section for color representation of this figure.

continental slope (2000–3000 m) and from pole to pole. Several species of zoarcids have been found in association with hydrothermal vents either as endemics (e.g. *Pachycara thermophilum*) or facultatively (*P. gymnominium*) (Priede 2017). A very few have invaded pelagic waters: *Melanostigma gelatinosum* and *Seleneolychus laevifasciatus* inhabit the deep trenches crossing the Antarctic coastal zone (Donnelly and Torres 2008), and the well-studied midwater eelpout, *Melanostigma pammelas*, is in the California borderland up to the waters off Oregon and Washington (fishbase.org). *M. pammelas* can be maintained for up to 18.5 months in the laboratory and makes a great experimental subject (Belman and Anderson 1979; Belman and Gordon 1979). It feeds on small fishes and crustaceans and is generally found below 500 m.

***Suborder Trachinoidei*. Family Chiasmodontidae:** the swallowers. The chiasmodontids (Figure 10.74) are an ancient deep-water group and one of the few deep pelagic families within the advanced fishes of the Perciformes. There are about 32 species in the four genera: *Chiasmodon*, *Dasyotus*, *Kali*, and *Pseudoscopelus*. Able to swallow fishes larger than themselves, Chiasmodontids inhabit lower mesopelagic (750 m) and bathypelagic depths as adults; younger ones are found shallower in the water column. As a family, Chiasmodontids have a

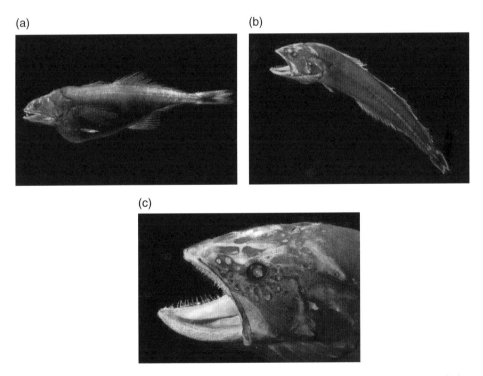

Figure 10.74 Chiasmodontidae. (a) *Chiasmodon niger*, the black swallower. Note the distended belly, evidence of the ability to swallow very large prey including prey larger than their own body; (b) *Dysalotus alcocki*. Note the well-developed lateral line; (c) close-up of the head of *D. alcocki* showing the very large gape with a partially unhinged (protruded) lower jaw and large sensory pits surrounding the eye and continuing into the lateral line. *Source:* © Dante Fenolio, reproduced with permission. See color plate section for color representation of this figure.

worldwide distribution ranging from tropical to subpolar waters, though some species reside in only one ocean basin, e.g. *Chiasmodon niger* in the Atlantic (Priede 2017). Despite their ubiquitous presence, they are rarely taken in midwater trawls (e.g. Sutton et al. 2010).

Fish Systems

Basic Anatomy

External Features and Terms

There are four main types of taxonomic characteristics.

- Meristic characteristics are any countable structures, e.g. the numbers of fin rays or lateral-line scales, though the term originally referred to features corresponding to body segments, like the number of vertebrae.
- Morphometric characteristics are measurable features such as body lengths, head lengths, eye diameters, or ratios between such measurements like head-length as a percentage of standard length. Morphometric characters can vary between juvenile and adult stages of the same species.
- Anatomical characteristics are features of the skeleton or soft anatomy, like the shape of sensory canals on the head or the position of light organs in myctophids. Anatomical characteristics are the most variable and subjective of the three main types (Cailliet et al. 1986; Helfman et al. 2009).
- Molecular characteristics are a work in progress. Their utility depends upon the taxonomic level of discrimination needed. Molecular data currently play a major role in determining higher level systematic relationships (e.g. Nelson et al. 2016). At the species level, molecular "barcoding" shows great promise and species libraries are being assembled by a collaborative of scientists active in sampling and molecular research. There will always be debate over which technique is more relevant, morphological or molecular, which is as it should be.

A composite teleost is shown in Figure 10.75 and a composite elasmobranch in Figure 10.76. Though fin names are the same for both teleosts and sharks, the fins of teleosts are far more maneuverable. One major external difference between sharks and bony fishes is the presence of spiracles in the shark, which are important to their gas exchange system (Figure 10.76).

Skull and Skeleton

Figure 10.76 illustrates the skull and skeletal anatomy of the bull shark, *Carcharhinus leucas*. Of particular importance for comparative purposes is the anterior third of the shark. Noteworthy features in the anterior skeleton are the brain case or chondrocranium, the pterygoquadrate or upper jaw, and the lower jaw, or Meckel's cartilage. The chondrocranium has an anterior fontanelle, or "hole in the head," as well as areas of the brain responsible for the olfactory, optical, and auditory functions. Moving posteriorly, cartilaginous skeletal elements include the hyoid arch suspending the jaw, the gill arches, the pectoral girdle, and the vertebral column (Lagler et al. 1962).

Figure 10.77 illustrates the skull and skeleton of a perciform teleost. Of particular importance is the suspensorium which is the series of bones connecting the jaw apparatus to the neurocranium (ethmoid region, supraorbital and infraorbital bridges, and the cranial vault). There are two suspensoria, one on each side of the head. The suspensorium is just that: it

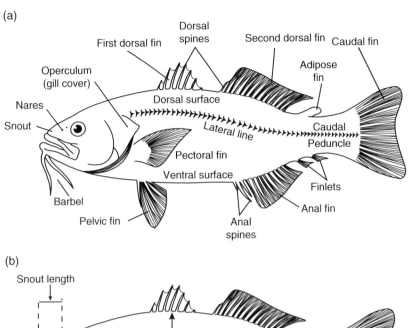
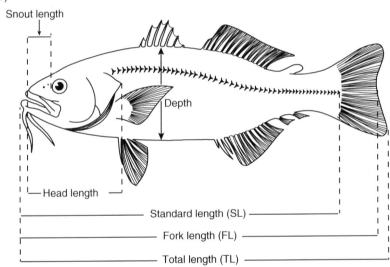

Figure 10.75 External anatomy and morphometrics. (a) A stylized teleost showing most of the common external features; (b) the common morphometric measurements used with teleosts.

suspends the jaw from two points on the neurocranium, the ethmoid region and the cranial vault. The suspensoria are particularly important because they serve as an attachment site for the opercular bones as well as for the muscles and ligaments controlling the jaw. They effectively form the walls and roof of the orobranchial cavity. The suspensoria are able to be abducted (moved outward) and adducted (inward), changing the volume of the orobranchial cavity. As such, they play an important role in the suction feeding of fish. The plate-like opercular bones (opercles and interopercles) just posterior to the suspensorium form the bony operculum, covering the gill

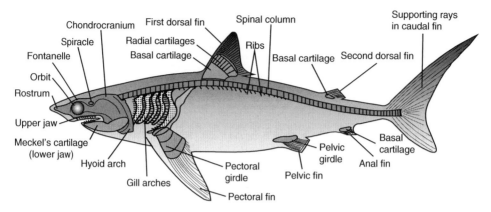

Figure 10.76 The cartilaginous skeletal elements of a shark.

arches and providing structural support for the flap-like motions that help ventilate the branchial cavity (Stiassny 2000).

Elasmobranch jaws are bit more straightforward than those of the teleosts, as is their skull structure (Figure 10.76). A fused chondrocranium provides the suspension site for a jaw apparatus consisting of the hyoid arch, the upper jaw or palatoquadrate and the lower jaw, or Meckel's cartilage. Despite its basic simplicity, the elasmobranch jaw allows a large repertoire of feeding strategies including suction feeding, (nurse sharks) ram feeding (oceanic whitetip sharks), biting (great white sharks), and filter feeding (manta rays, basking, and whale sharks) (Motta and Wilga 2001).

The axial skeleton of teleosts (Figure 10.77b) comprises a series of rigid vertebrae with flexible intervertebral joints. It has a variety of functions along its length, acting as a rigid site for muscle attachment, support for the fins and tail, and a firm, jointed attachment to the skull (Stiassny 2000). The pectoral and pelvic fins are supported by the pectoral and pelvic girdles, respectively. The pectoral girdle consists of the cleithral bones (Posttemporal, Supracleithrum and Cleithrum) that attach directly to the cranial vault, providing a robust anchor for the bones directly supporting the pectoral fin (Coracoid and Scapula). In the spiny-rayed fishes, the pelvic girdle or pelvic bones (Basipterygia) are located directly beneath the pectoral girdle. In less advanced species, e.g. the salmonids, the pelvic fins and girdle are located more posteriorly. Note the modifications of the vertebrae associated with the dorsal and caudal fins and the support for the rib cage. Precaudal vertebrae have transverse processes supporting the ribs; caudal vertebrae do not (Figure 10.77b).

The vertebral column of elasmobranchs (Figure 10.76) is similar to that of the bony fishes (Figure 10.77b). Like the teleosts, the column is modified along its length, first to provide a firm attachment to the cranium, then for support of the fins and tail. As in the teleosts, the shark vertebral column also has lateral processes supporting its ribs.

Feeding and Digestion

Food Acquisition, the Three Dominant Modes: Ramming, Sucking, and Biting

Ram Feeding
Flow through the mouth and gills generated by forward movement is known as

"ram" flow, and it is used to great effect in both feeding and gas exchange. In its simplest form, the gill rakers are used as a mechanical sieve to trap particles from the water flowing through the branchial arches (Figure 10.78a). This is known as a dead-end sieve (Sanderson et al. 2001; Brainerd 2001): particles too large to fit through the sieve are trapped and retained, eventually to be moved to the esophagus for consumption. The clupeids, basking sharks, whale sharks, and manta rays were all believed to use this type of feeding based on the densely arrayed, filamentous structure of their gill rakers (Figure 10.15), and that still may be the case. However, fluid flow through the branchial chamber is complex, and it is now believed that instead of a

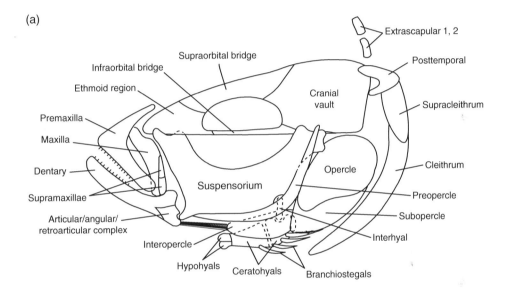

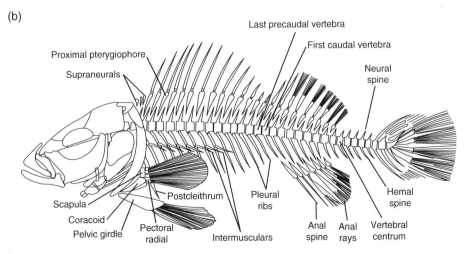

Figure 10.77 The teleost skeleton. (a) The bones of the skull; (b) the axial skeleton. *Sources:* (a) Stiassny (2000), figure 6.2 (p. 111); (b) Stiassny (2000), figure 6.3 (p. 115). Reproduced with permission of Elsevier.

(a) Dead-end filtration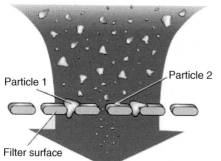

(b) Crossflow filtration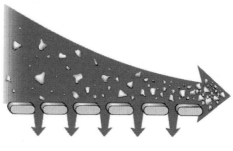

Figure 10.78 Dead-end and crossflow filtration. (a) Dead-end filtration with the filter acting as a simple mechanical sieve. Particles are retained when they exceed pore size or stick to the filter itself; (b) crossflow filtration where flow parallels the filter sieve and, due to a higher pressure on the inside of the filter caused by ram flow, the filtrate passes through leaving a more concentrated suspension for ingestion.

dead-end sieve, many fishes use the principle of cross-flow filtration, wherein the particle-containing fluid ingested by the fish flows parallel to the gill-raker sieve (tangential flow) instead of impacting it directly (Figure 10.78b). In cross-flow filtration, the filtrate (or permeate) passes out through the gill-raker sieve leaving a more concentrated suspension of particles (the retentate) behind to continue toward the esophagus. Most important to the cross-flow process is the trans-sieve pressure: it must be higher on the inside than on the outside, and that pressure is provided by ram flow. The principle has been elegantly demonstrated by Sanderson et al. (2001), and it likely applies to many of the large suspension-feeding fishes such as basking sharks as well as to the gizzard shad and goldfish that were the experimental subjects for the study.

Ram feeding is not limited to passive suspension feeders swimming forward with an open gape through concentrations of zooplankton; forward locomotion is used to aid in suction feeding and ram capture. Ram capture is typical of many sharks as well as the barracudas (Motta and Wilga 2001; Porter and Motta 2004). "During ram capture the shark swims over its more slowly moving prey, either engulfing it whole or seizing it in its jaws and using manipulation bites to reduce it in size" (Motta and Wilga 2001). The food is then conveyed to the esophagus using hydraulic suction generated within the pharyngeal cavity. Several species of lamnid and carcharinid sharks use ram capture as their dominant feeding strategy. The oceanic whitetip (*Carcharhinus longimanus*) has been seen engulfing prey as it swims slowly through schools of tuna with its mouth open (Compagno 1984; Motta and Wilga 2001).

Great white sharks (*Carcharodon carcharias*) also use a type of ram feeding, swimming rapidly to their pinniped prey, removing a bite and releasing it, allowing it to die or go into shock before beginning serious feeding (Tricas and McCosker 1984). Alternately, Klimley (1994) has found that the sharks drag their prey below the surface after initially attacking, sometimes removing a bite, then allowing the prey to float to the surface and expire before feeding: a similar strategy.

Suction Feeding

Suction feeding is the dominant feeding strategy of teleosts (Liem 1980) and is believed by many to be the ancestral mode of feeding, beginning with the jawless fishes (Bone and Moore 2008). Inertial suction is achieved by rapidly expanding the size of the buccopharyngeal cavity, generating a large negative pressure, and drawing prey into the open mouth. The small pelagic fishes most abundant in temperate and tropical open-ocean systems, notably the gonostomatids, myctophids, and sternoptychids, are all suction feeders: "plankton pickers" that feed on copepods and other small Crustacea. Some ram feeders like the clupeids and whale shark use suction feeding with or instead of the more passive ram filter-feeding, particularly when prey is less concentrated (Motta and Wilga 2001; Bone and Moore 2008).

Though the shark jaw is not as mobile as that of teleosts, it is protrusible and the musculature of the buccopharyngeal cavity allows for capable suction feeding in some elasmobranch species. It is most highly developed in bottom feeders such as the nurse sharks (Orectolobiformes) that use suction exclusively. It is used in conjunction with ram capture in others, e.g. the dogfish (Squaliformes), that use a rapid approach (ram) toward their prey and then suction to engulf it (Wilga et al. 2007). Generally, sharks relying on suction for their prey-capture have a smaller, more laterally enclosed, mouth than their ram-feeding brethren, providing more efficient suction (Motta and Wilga 2001).

Biting

Biting is the removal of a portion of the prey organism by the oral jaws of the fish, followed by ingestion. Any prey too large to be engulfed must be torn apart and ingested in smaller pieces. Sharks, tunas, and billfishes use their teeth or "sword" to tear prey into manageable bits. The maximum protrusion of the shark's upper jaw occurs during the compressive phase of prey capture, facilitating the bite, unlike that of the teleosts where the maximum jaw protrusion occurs early in the expansive phase of prey capture.

The cookie-cutter shark *Isistius brasiliensis* is a classic example of a biting predatory strategy when the prey is much larger than the predator. It attaches to prey using suction created by its buccopharyngeal musculature in concert with a seal formed by its lips. It then bites into the quarry and forcefully spins its body along its longitudinal axis, tearing out a plug of flesh that is as neat as a Christmas cookie. Prey includes tunas, dolphinfish, swordfish, and elephant seals (Jones 1971; Motta and Wilga 2001).

In most cases, biters have a large gape and a jaw with cutting teeth. In some cases, particularly in fishes like the porgies (Sparidae) feeding on crustaceans and bivalves, molar-like teeth in the jaw can be used for mastication of prey.

Biting is the last part of the ram-feeding sequence in Carcharinid and Lamnid sharks, but it is also used by smaller fish that feed on a variety of prey ranging from benthic polychaetes to coral polyps or even clam siphons (Horn 1998). In sharks, the maximum protrusion of the upper jaw occurs when the mouth has closed (compressive phase), facilitating the bite. In teleosts, the maximum jaw protrusion occurs early in the prey capture sequence.

The mouth cavity of teleosts is often equipped with teeth outside those of the jaw proper, including the vomer (roof of the mouth), glossohyal (floor of mouth), and other locations depending on the species.

The "extra" teeth may aid in retaining prey after capture as well as moving it rearward to the pharyngeal jaws.

Food Sorting: The "Pharyngeal Jaws"

The pharyngeal jaws are not true jaws at all but are modifications of the posterior gill arches (Figure 10.79) that guard the entrance to the esophagus. They differ considerably between species. The upper pharyngeal jaw is composed of the posterior two pairs of pharyngobranchial bones: the dorsal-most bones of gill arches 3 and 4. Those are modified in many fishes to form toothplates in the upper pharyngeal jaws. The lower pharyngeal jaws are formed by the paired ceratobranchial bones of the relict fifth gill arch which also may form robust toothplates. The two jaws are under muscular control, allowing them to bite against each other (Figure 10.79b, c). In some species, they are powerful enough to be used for grinding and crushing.

The most highly developed pharyngeal jaws are found in fishes that feed on heavily armored prey such as crabs and bivalves or prey that is accompanied by debris when it is ingested and the digestible items need to be sorted out of the mix, like the surfperches. Open-ocean pelagic fishes feed mainly on other fishes and pelagic crustaceans like krill, decapod shrimp, and the smaller crustacean zooplankton, all of which are minimally armored and readily digested.

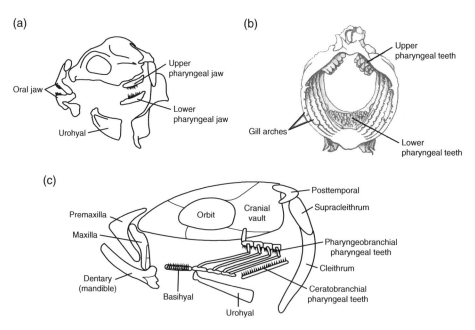

Figure 10.79 Pharyngeal jaws and teeth. (a) Schematic lateral view of the head of a surfperch showing the relative positions of the oral and pharyngeal jaws; (b) anterior view of the gill arches and pharyngeal teeth of a surfperch (Embiotocidae); (c) lateral view of a stylized teleost skull with lateral bones removed to show the branchial arches with upper and lower pharyngeal jaws. Note the basihyal (the tongue) also has teeth which are not part of the oral or pharyngeal jaws. *Sources:* (a) Adapted from Horn (1998), figure 3 (p. 51); (b) Adapted from Bond et al. (1996), figure 25-3 (p. 421); (c) Adapted from Stiassny (2000), figure 6.2 (p. 112).

Digestion: The Alimentary Canal

Once a food item exits, the pharyngeal cavity it enters the digestive tract proper (Figure 10.80), where it begins further mechanical and chemical processing. The digestive tract of fishes is divided into four basic sections: the foregut, including the esophagus and stomach, the midgut, including the intestine, and in some species, a hindgut and rectum (Bone and Moore 2008). Layers of smooth and longitudinal muscle in the walls generate peristalsis which moves material through the digestive tract. The esophagus in most fishes is a short, highly distensible, muscular tube. It is mainly a conductive region that leads directly to the stomach, secreting only mucous for lubricating the food bolus as it begins its travel through the gut. It does not secrete digestive enzymes or mechanically break down the food. The stomach is the second element in the foregut of both the sharks and teleosts. Its primary functions are mastication, mixing, and chemical breakdown of food items. Muscularity in fish stomachs varies on a continuum with diet: the more roughage (e.g. crustacean shells or algae with bottom debris) the more gizzard-like the stomach. Stomachs of carnivores and zooplanktivores, with the chemical help of an acidic environment and digestive enzymes, are robust enough to reduce muscle, bone, and chitinous tissue to a semi-solid slurry. Exit from the stomach to the upper intestinal region is regulated by the pyloric valve.

Species that feed on tiny prey are termed microphagous feeders, and many of them lack a stomach altogether. Well-known examples of microphagous feeders are the

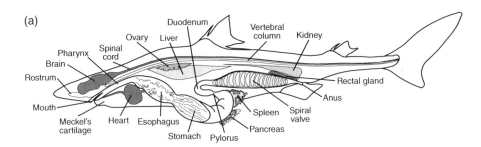

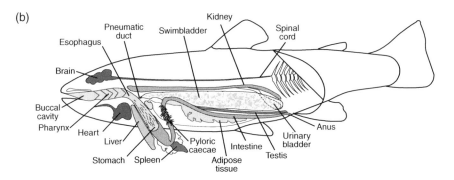

Figure 10.80 Digestive systems. (a) Internal anatomy of a shark (spiny dogfish, *Squalus acanthias*); (b) Internal anatomy of a teleost (brook trout, *Salvelinus fontinalis*).

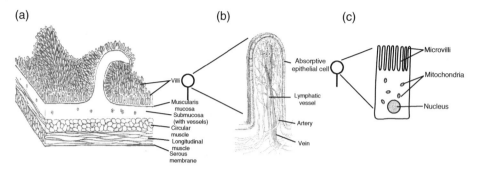

Figure 10.81 Intestine anatomy. (a) Three-dimensional representation of the small intestine posterior to the pyloric valve showing the major tissue layers of the gut wall; the mucosa with villi which increase the surface area of the gut, the muscularis mucosa (smooth muscle layer at the base of the mucosa), submucosa with blood vessels, circular and longitudinal muscles layers of the gut wall, and serous membrane (lining of the body wall); (b) enlargement of a single villus showing the absorptive epithelial cells as well as the lymphatic and blood vessels; (c) enlargement of an absorptive epithelial cell showing the microvilli which further increase the surface area of the gut. *Sources:* (a, c) Florey (1966), figure 11-17 (p. 244); (b) Adapted from Moog (1981), p. 156.

seahorses and pipefishes, sucking in tiny copepods through their highly specialized tubular mouth. Microphagous feeders need to feed nearly continuously to survive, producing a constant throughput transiting the alimentary canal. None of the open-ocean fish species exhibits microphagous feeding.

Absorption of digested nutrients occurs in the midgut and, in some cases, the hindgut. Absorption across the intestinal wall is a three-step process:

- Movement of nutrients from the lumen to the gut wall
- Transport across cell membranes into the epithelial cell cytoplasm
- Transport out of the epithelial cell into the extra-cellular body fluids (blood and lymph).

The acidic semi-solid bolus of food leaving the stomach is quickly neutralized when it enters the upper intestine by the secretion of bicarbonate from the pancreas. The entire suite of intestinal enzymes functions optimally in a neutral to alkaline environment, including the carbohydrases and lipases as well as the proteases outside of pepsin. In teleosts, the region of the intestine just past the pyloric valve is usually invested with pyloric caecae, diverticuli that increase the surface area for food breakdown and absorption (Buddington and Diamond 1986). Numbers of pyloric caecae are highly variable, ranging from 0 to 1000 depending on the species (Bone and Moore 2008). Absorption is further facilitated by maximizing the surface area for exchange in the intestinal wall with tiny finger-like projections on the mucosal surface called villi (Figure 10.81). Each villous is invested with a dense network of blood and lymph vessels that receive the nutrients leaving the epithelial cells. Sharks maximize the surface area for absorption with a tortuous path though the intestine termed the spiral valve (Figure 10.80a).

The length of the intestine is diet-dependent, with herbivorous fishes usually having a longer gut. Carnivores, such as the trout and dogfish pictured in Figure 10.80, have a shorter midgut. Feeding habits of open-ocean species range from zooplanktivory in gonostomatids, myctophids, and sternoptychids to piscivory in the stomiids,

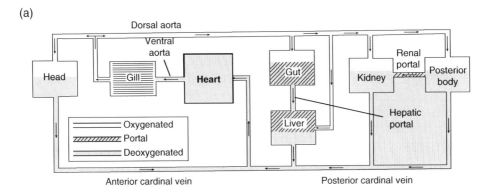

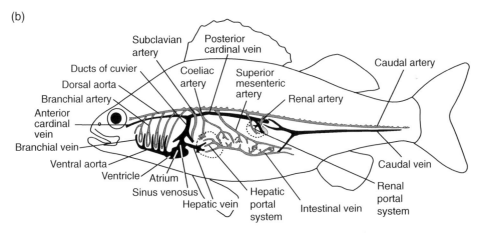

Figure 10.82 Circulatory system. (a) Schematic diagram of the circulatory system of a teleost; (b) anatomical diagram of the circulatory system of a teleost. *Sources:* (a) Adapted from Willmer et al. (2005), page 159; (b) Lagler et al. (1962), figure 3.17 (p. 92).

tunas, and billfishes: all carnivores, but over a large prey size-spectrum. The midgut region is the longest section of the alimentary canal. In some herbivorous species, the posterior intestine is differentiated into a true hindgut and rectum, but in carnivores, the intestine ends in a terminal anus.

Circulation, Respiration, and Excretion

Circulation

The teleosts and elasmobranchs both have a single muscular heart with a closed vascular system, comprising the most advanced circulatory system discussed thus far. The basic circulatory path is shown schematically in Figure 10.82a and anatomically in Figure 10.82b. Blood leaving the heart flows first through the gills, with the oxygenated blood then bathing the head region through the carotid arteries and the remainder of the body through tributaries of the dorsal aorta. The gills account for 30–57% of the total resistance to blood flow depending on the gill surface area of the fish (Bone and Moore 2008). Return to the heart from the posterior body is through the renal portal system of the kidney and thence to the posterior cardinal vein. Blood from the head region returns via the anterior cardinal vein, joining the posterior cardinal vein before connecting with the ducts of Cuvier and entering the sinus venosus of the heart. The sinus venosus also receives blood from the liver via the hepatic veins.

Elasmobranch and Teleost Hearts

Elasmobranch and teleost hearts differ slightly in structure (Figure 10.83), but both have four chambers in series. With the exception of the elastic teleost bulbus arteriosus, all chambers are contractile. Shark and teleost hearts both reside in a pericardial cavity below the gills (Figure 10.83) but differ in the nature of the pericardium. In sharks, the pericardium is rigid and in the teleosts, it is membranous and deformable, creating a difference between the two groups in the type of blood flow into and out of the heart chambers.

The rigid pericardium surrounding the shark heart allows it to act as a suction and pressure pump. Contraction of the ventricle (systole) causes a negative pressure within the pericardium, drawing blood into the atrium as it expands to fill the void (suction phase). The sino-atrial valve is then closed as the atrium contracts to fill the ventricle. The ventricle in turn pumps its volume into the contractile conus arteriosus and on to the ventral aorta (pressure phase). Puncturing the pericardial wall reduces the volume pumped by the heart, providing experimental verification for a suction-pressure pumping function (Eckert et al. 1988). Though rare among the teleosts, tunas also have a rigid pericardium, allowing them to utilize the suction-pressure pumping strategy for maximizing cardiac output in support of their highly active lifestyle (Bone and Moore 2008).

Most of the teleosts have a membranous pericardium, meaning that the heart acts only as a pressure pump, with sequential filling and contraction of the atrium and ventricle. The ventricle has a rapid synchronous contraction, forcing its blood into the highly elastic bulbus arteriosus, which expands to accept the entire volume. A valve between the ventricle and the bulbus arteriosus prevents backflow (Figure 10.83b), so the elastic bulbus pushes the blood toward the gills with a continuous steady flow as it relaxes, greatly reducing the pulsatile character of the blood flow as it leaves the heart. Since gas exchange at the gills is most effective with a continuous steady-state flow, the teleost heart design improves the acquisition of oxygen. Tunas have the effective suction-pressure pumping system of a rigid pericardium and the smoothing effects of a highly elastic bulbus ateriosus: the best of both worlds.

Arteries, Veins, and Continuity

Branching and coalescence of the vascular system are shown schematically in Figure 10.84. As the arteries branch into arterioles and capillaries, the cross-sectional area of the tubular system increases by about 800 times and the total surface area available for exchange diffusion is far greater: over 2300 times (Waterman et al. 1971; Withers 1992). Since the vascular system is a closed system of pipes, the hydrodynamic principle of continuity applies: when a given volume of fluid enters one end, an equal volume must exit the other end. Not only must the volumes be equal but entry and exit must occur at precisely the same time so the volumes per unit time are equal. The principle of continuity has several implications for blood flow. As the arteries decline in size to capillaries, the blood flow slows greatly due to the huge increase in total cross-sectional area (Figure 10.84). The slow pace of blood as it flows through the capillaries maximizes the time for diffusional exchange. It has been estimated that a milliliter of blood would

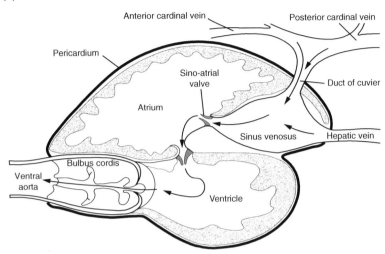

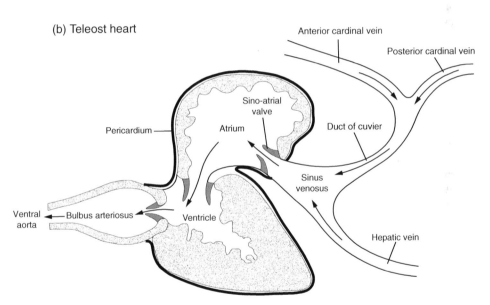

Figure 10.83 Elasmobranch and teleost heart structure. (a) Elasmobranch heart; (b) teleost heart. Arrows show direction of blood flow. *Sources:* (a) Adapted from Waterman et al. (1971), figure 8-37 (p. 335).

take 14 years to pass through a single capillary (Waterman et al. 1971). In addition, the tiny diameter (8 μ) of capillaries maximizes the surface-to-volume ratio, further facilitating exchange. Capillaries are just wide enough to allow one blood cell through them at a time. Finally, capillary walls are extremely thin, consisting of a single flat layer of endothelial cells, minimizing diffusion distance.

A secondary circulation system has been identified in fish that roughly parallels the

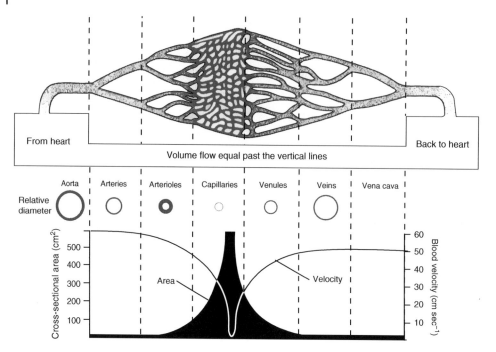

Figure 10.84 Schematic diagram of the vertebrate vascular system showing the branching and coalescence of vessels between the aorta and vena cava. The dashed lines divide the figure into different size categories within which flow is equal. The relative diameters and wall thickness of vessels within each category are shown below the top figure. The bottom section shows the changes in total cross-sectional area and flow velocity as the vessels branch and then coalesce.

structure of the primary system (Figure 10.85). Its function is a bit of a mystery. This secondary circulation was initially believed to be a true lymphatic system like that of the higher vertebrates, comprising a thin-walled array of vessels with the main function of scavenging fluid and plasma proteins that had leaked into the interstitial spaces from the capillaries and returning them to the general circulation. Though flow through a lymphatic system is very slow and at a very low pressure, its total volume can equal or exceed that of the primary system. The secondary circulation found in fish has a volume of approximately 1.5 times that of the primary circulation (Olson 1996). However, while the secondary circulation system of fish does have many features of a lymphatic circulation in that it is a low-flow system with thin-walled vessels, the fish system connects with arterial flow at several junctures through tortuous interarterial anastomoses which prevent the transfer of red blood cells. Those arterial connections conflict with one of the basic characteristics of a lymphatic system: scavenged fluids and proteins are returned to sites in the venous circulation with the lowest pressures and never connect with arterial flow. In teleosts, the secondary system forms its own capillary beds in the fins, gills, and skin and joins the primary circulation in caudal and cutaneous veins (Olson 1996; Bone and Moore 2008). It now appears that the secondary circulation in fish performs some of the functions of a lymphatic system but has additional functions yet to be determined.

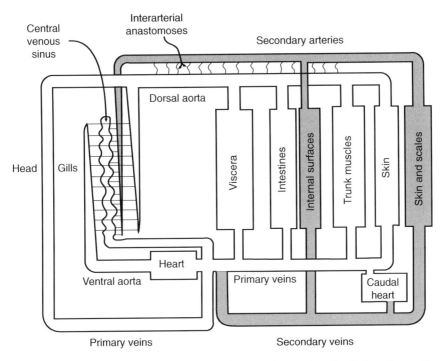

Figure 10.85 Block diagram of the teleost primary and secondary circulations. *Source:* Bone and Moore (2008), figure 5.21 (p. 147). Reproduced with the permission of Taylor and Francis.

Gas-Exchange in the Teleosts and Elasmobranchs

Respiratory gas exchange in fishes occurs primarily at the gill surfaces by the process of diffusion. Factors that play a role in the rates and efficiency of the process include the following:

1) the relative concentrations of the exchange gases (oxygen and carbon dioxide) in the seawater and blood;
2) the distance over which diffusion takes place, i.e., the thickness of the gill membranes;
3) the surface area of the surface across which diffusion occurs, i.e., gill surface area;
4) speed of the water passing over the gills;
5) the chemical environment, especially the pH of blood and tissue fluid.

The gill of fishes is composed of three subunits: the arch, the filament, and the lamellae (Figure 10.86). Elasmobranchs generally have five gill arches (sometimes six or seven) and a spiracle anterior to the arches (Figures 10.76 and 10.87A). Each gill has a septum, obvious as a flap, joining the filaments over most of their length and giving the group its name: elasmobranch = plate-gill. The gill arches provide skeletal support for the gills and vasculature. Teleosts typically have four gill arches (Figure 10.87A), but their gill septa are shorter, allowing more freedom of movement in the gill filaments. Filaments in both sharks and teleosts are supported by skeletal gill rays (Figure 10.87B).

In both sharks and bony fishes, the filaments extend from each arch in the direction of water flow. Gills have multiple tasks.

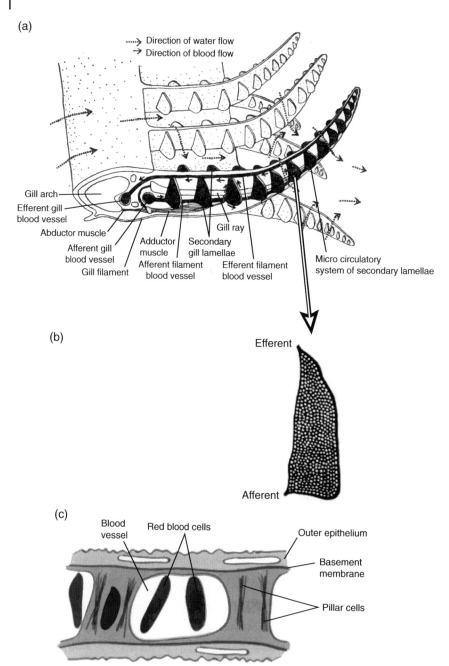

Figure 10.86 Teleost gill structure. (a) Arrangement of gill filaments with secondary lamellae supplied by afferent and efferent vessels. Blood flow from afferent to efferent vessels is opposite (countercurrent) to that of water flow. The bony gill arch and gill rays are linked by intrinsic muscles which can change the position of the gill filaments; (b) cast of vascular spaces in secondary lamella, white dots are where pillar cells interrupt the space; (c) section across a secondary lamella showing pillar cells, blood vessels with red blood cells, and the basement membrane separating the pillar cells from the outer epithelium. *Sources:* (a) Adapted from Hughes and Morgan (1973), page 422, (b) Adapted from Hughes and Morgan (1973), page 431; (c) Adapted from Bone and Moore (2008), figure 5.6 (p. 131).

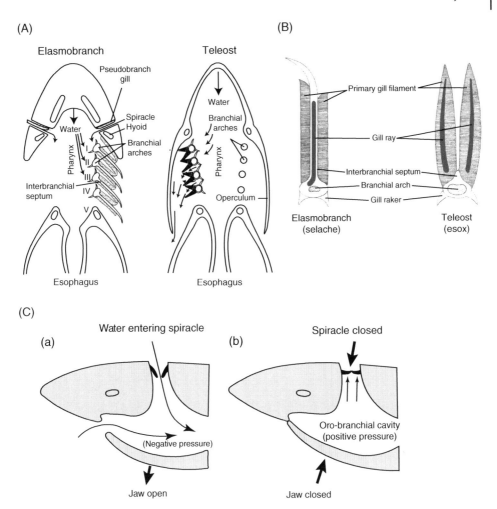

Figure 10.87 Gill structure in elasmobranch and teleost fishes. (A) Horizontal sections through the heads of an elasmobranch and teleost, respectively, showing the position of the gill arches and the water flow patterns; (B) gill structure of an elasmobranch (*Selache*) and a teleost (*Esox*), respectively; (C) one-way respiratory flow through the buccal regions of a dogfish (*Squalus*). (a) illustrates the inspiratory phase and the accompanying negative pressure in the oral region; in (b), the spiracle closes and the floor of the mouth is raised, increasing the pressure in the oral cavity and forcing the respiratory current into the gills. *Sources*: (A) Adapted from Waterman et al. (1971), figure 8-12 (p. 294); (B) Waterman et al. (1971), figure 8-13 (p. 295); (C) Waterman et al. (1971), figure 8-15 (p. 297).

In the teleosts, they are the primary site for ion balance, nitrogen excretion, and gas exchange. In the elasmobranchs, they are most important in gas exchange with a lesser role in ion balance and nitrogen metabolism. In both groups, gas exchange occurs at the secondary lamellae (Figure 10.86a).

Each of the secondary lamellae is a small sinus intercalated with pillar cells that act to break up the blood flow much as if they were capillary beds (Figure 10.86b, c). They are arranged on the gill filament such that they maximize countercurrent flow. In teleosts, the attitude of the gill filaments relative to water flow can be adjusted by muscles connecting the gill rays and the gill arch. In sharks, the long septum keeps the gill lamellae fairly stationary.

Unidirectional Flow and Countercurrent Exchange: Maximizing the Concentration Gradient

Blood circulating through the secondary lamellae flows in the opposite direction to the water crossing its surface. In this way, the gradient between the blood and seawater oxygen concentrations is always maximized, greatly facilitating diffusion. The best that could be hoped for if blood and water were flowing in the same direction (co-current exchange) would be 50% removal. Oxygen concentration in seawater is always low even when at air-saturation because of its basic physical properties. That fact makes unidirectional flow an imperative for active aquatic species. It is achieved in fishes using two strategies: buccal pumping and ram ventilation.

Buccal pumping (Figure 10.87) uses much of the same musculature that is employed in suction feeding, with the caveat that the timing and strength of contraction is more leisurely. Water is drawn into the mouth by expansion of the buccal and opercular cavities with the operculae sealed. The mouth is then closed, and water is forced through the resistance of the gills by sequential compression of the buccal and opercular chambers, ending with the opening of the operculae and exit of the inspired water. In effect, the buccal cavity acts as a pressure pump and the operculae act as suction pumps. Sharks use the same basic strategy with additional participation by the spiracular openings during inspiration. Water is inspired through the spiracles as well as the mouth. The mouth and spiracular valves then close, the orobranchial cavity is compressed, and water is forced through the gills and out the gill slits.

In faster-swimming fishes such as the tunas and billfishes, branchial pumping ceases when speed is sufficient to ventilate the gills by just keeping the mouth open. Some scombroids, such as *Katsuwonus*, the skipjack tuna, can only respire by this method. Ram ventilation is much less expensive energetically; it has been estimated that branchial pumping costs about 15% of a fish's energy budget, whereas ram ventilation costs only 9% (Bone and Moore 2008).

The structure of gills in fast-swimming ram ventilators is quite different from that of fishes relying on branchial pumping (Figure 10.88). The ends of the filaments are fused together, and they have bridges binding them together along their length. The additional strength prevents deformation of the delicate respiratory surface at high swimming speeds and gives protection from foreign objects accidentally inhaled with the high-speed respiratory stream.

Two other elements of gill structure are highly correlated with lifestyle: the total gill surface area and the diffusion distance, which translates as the thickness of the gill epithelium. If several species with different life habits and levels of activity are compared (Table 10.4), the clear trend is for species with the highest gill surface areas (GSA) to be actively swimming pelagic species. The remaining species, both teleosts and elasmobranchs, ranged from 1.9 to 3.7 $cm^2\ g^{-1}$, less efficient for oxygen uptake but clearly adequate for those active predators. Diffusion distance is clearly lowest in the most active species, *Katsuwonus* and *Trachurus*, again maximizing their capability for oxygen uptake when compared with the other teleosts and the elasmobranchs. The limited data available suggest that elasmobranchs have higher diffusion distances than the teleosts.

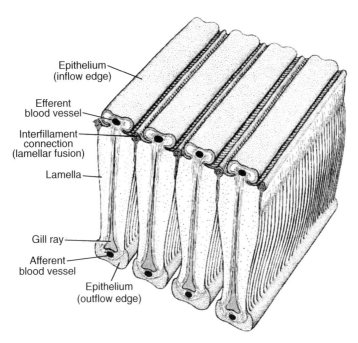

Figure 10.88 Diagram of a cross section through four filaments of the gill of *Katsuwonus pelamis* (skipjack tuna). Water flows through the interlamellar spaces from top to bottom of diagram. Blood flows through lamellae in opposite direction. *Source:* Muir and Kendall (1968), figure 1 (p. 390). Used with the permission of BioOne, from Structural Modifications in the Gills of Tunas and Some Other Oceanic Fishes, Muir and Kendall, *Copeia,* 1968 No. 2; permission conveyed through Copyright Clearance Center, Inc.

The study of Hughes and Morgan (1973) agrees well with previous studies on GSA in fishes such as Gray (1954) who obtained similar numbers for GSA and came to the same overall conclusions. It works for crabs as well (Gray 1957). More active species require more oxygen and have larger GSA's to procure it. Numbers on diffusion distances are comparatively rare in the literature but lead to the same conclusions.

Blood and Oxygen at the Respiratory Surface
Blood Basics
Oxygen has a very low solubility in water; in distilled water at 15 °C, the dissolved oxygen (DO) at air saturation is $7.05\,\text{ml}\,\text{l}^{-1}$ and in seawater at 35 ‰, it is $5.72\,\text{ml}\,\text{l}^{-1}$ (Riley and Chester 1971). The amount of oxygen dissolved in water or blood at air saturation (21.8 kPa, or 160 mm Hg) is called its carrying capacity and is expressed in ml O_2 $100\,\text{ml}^{-1}$ (ml deciliter^{-1}), or Vol. %, a standard term that comes from the medical profession. Thus, the value for DO in seawater at 15 °C is 0.57 Vol. %.

Dissolved oxygen in fish blood varies from 1.0 to 1.2 Vol. % in the hagfish *Myxine* and the lamprey *Petromyzon* to 19.7 Vol. % in *Electrophorus*, the electric eel: an air-breathing fish. Values for the coastal sharks *Squalus* and *Scyliorhinus* are intermediate at 4.5 and 4.4 Vol. %, respectively, while the mackerel *Scomber*, an active pelagic species, has a carrying capacity of 15.7 Vol. % (Table 10.5) Two things to notice about the data in Table 10.5 are first, that among

Table 10.4 Total number of gill filaments, secondary lamellae (one side only), gill surface area, and diffusion distance (thickness of gill epithelium) in fishes with very different activity levels. Note the large differences in gill surface area and diffusion distance between the very active skipjack and yellowfin tunas and the other species. *Source:* Data from Hughes and Morgan (1973) and Withers (1992).

Class/Species	Mass (g)	Lifestyle	Total number of filaments	Number of lamellae (mm)	Gill surface area (cm² g⁻¹)	Diffusion distance (µ)	References
Elasmobranchii							
Scyliorhinus canicula (catshark)	520	Benthopelagic moderately active	749	11.25	2.1	11.3	Hughes and Morgan (1973)
Squalus acanthias (spiny dogfish)	1000	Benthopelagic moderately active	1000	7	3.7	10.1	Hughes and Morgan (1973)
Teleostei							
Katsuwonus pelamis (skipjack tuna)	3258	Pelagic predator very active	6066	32	13.5	0.6	Hughes and Morgan (1973)
Trachurus trachurus (horse mackerel)	26	Pelagic predator active	1665	38.5	7.83	2.2	Hughes and Morgan (1973)
Oncorhynchus mykiss (rainbow trout)	394	FW predator moderately active	1606	19	2	6.4	Hughes and Morgan (1973)
Opsanus tau (oyster toadfish)	251	Benthic sluggish	660	11	1.9	5	Hughes and Morgan (1973)
Thunnus albacares (yellowfin tuna)	1000–2000	Pelagic predator very active			14.4	0.533	Brill and Bushnell (2001)
Katsuwonus pelamis (skipjack tuna)	1000–2000	Pelagic predator very active			18.4	0.596	Brill and Bushnell (2001)
Oncorhynchus mykiss (rainbow trout)	900–1500	FW predator moderately active			2	6.37	Brill and Bushnell (2001)

Table 10.5 Oxygen carrying capacity of the blood of fishes with different lifestyles and habitats.

Species	Lifestyle	Blood carrying capacity (ml dl^{-1})	References
Superclass Agnatha			
Myxine (hagfish)	Scavenger	1	Prosser (1973)
Petromyzon (lamprey)	Predator/parasite	1.2	Prosser (1973)
Class Elasmobranchii			
Scyliorhinus (catshark)	Benthopelagic	4.4	Prosser (1973)
Squalus (spiny dogfish)	Benthopelagic	4.5	Prosser (1973)
Class Actinopterygii			
Electrophorus (electric eel)	FW air-breather	19.7	Prosser (1973)
Symbranchus (swamp eel)	FW air-breather	14.7	Prosser (1973)
Oncorhynchus mykiss (rainbow trout)	FW midwater predator	10.5	Brill and Bushnell (2001)
Scomber (mackerel)	Pelagic predator	15.7	Prosser (1973)
Thunnus albacares (yellowfin tuna)	Pelagic predator	12.6	Brill and Bushnell (2001)
Katsuwonus pelamis (skipjack tuna)	Pelagic predator	14	Brill and Bushnell (2001)
Seriola (yellowtail)	Pelagic predator	11.2	Brill and Bushnell (2001)

aquatic breathers the most active fish have the highest carrying capacities, and second, that the quantity of oxygen in the blood of fish is one to two orders of magnitude higher than that dissolved in seawater. The difference in O_2 between blood and seawater is carried by the fish's respiratory pigment: hemoglobin. Hemoglobin (Hb) combines reversibly with O_2, acquiring it at the respiratory surface and releasing it at the tissues.

The basic unit of hemoglobin is a ferrous (Fe^{++}) iron enclosed in a porphyrin ring, called the heme, that is further enclosed in a folded protein chain, termed a globin. In fishes (and most other vertebrates), the basic unit of 1 globin to 1 heme has a molecular weight (MW) of about 16 000 Da, an example of which is the monomeric (single subunit) myoglobin (Myb) contained in muscle. Hemoglobin (Hb) found in the blood has four such subunits, forming a tetrameric molecule. Each of the four subunits resembles myoglobin in structure and is capable of combining with a molecule of oxygen. The quaternary structure of Hb is responsible for the shape of its oxygen dissociation curve (Figure 10.89).

Two general types of curves are seen, hyperbolic and sigmoidal (Figure 10.90). Hyperbolic curves imply the presence of a monomeric form acting alone. Such a curve is seen in myoglobin. Sigmoidal curves imply the interaction of subunits, "cooperativity"; this type of curve is seen in the tetrameric form of Hb seen in fish blood and indicates subunit interaction, i.e. each additional O_2 added changes the affinity of other sites for O_2. The general measure of the O_2 affinity of a pigment is the P_{50} or pO_2

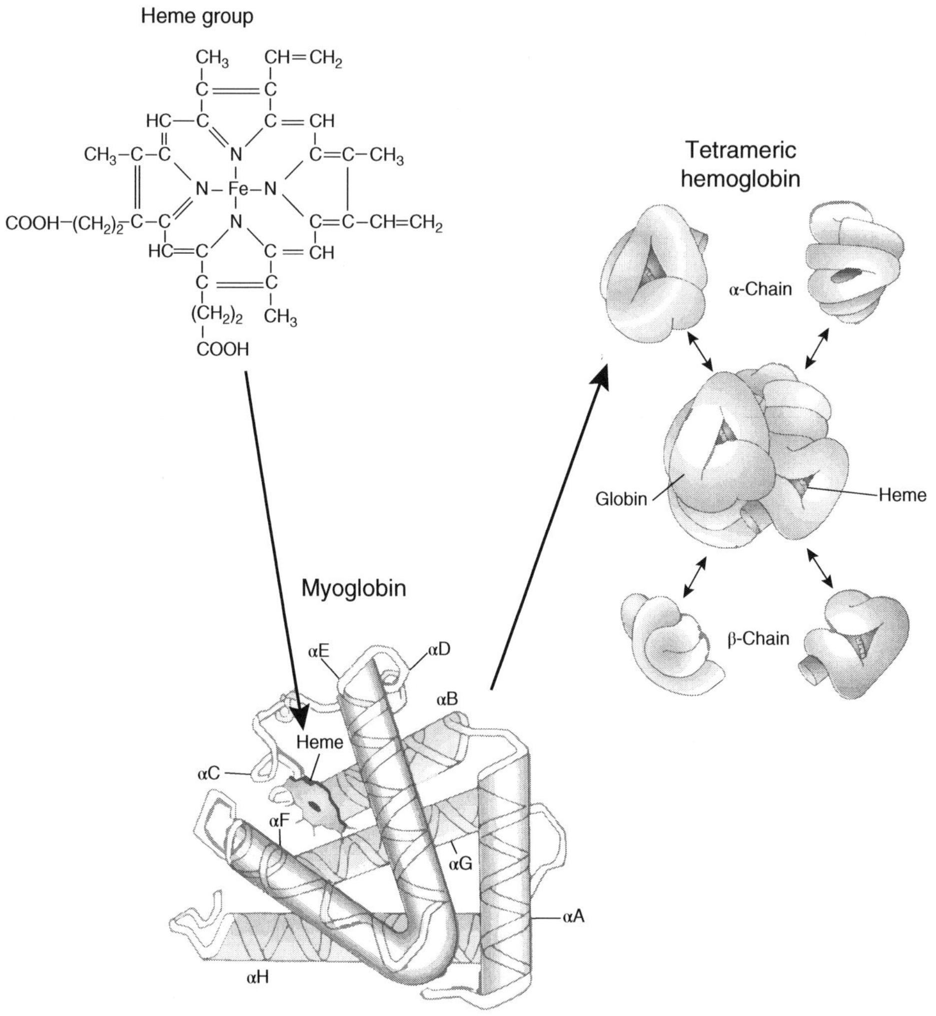

Figure 10.89 The structure of heme (porphyrin ring plus a ferrous iron), of myoglobin, and of tetrameric hemoglobin. *Source:* Willmer et al. (2005), figure 7.21, (p. 163). Reproduced with the permission of Blackwell Science Ltd.

at which the pigment is ½ saturated, and it varies considerably from species to species. It is a function of the species' lifestyle and the ambient O_2 in its habitat. The range for teleost fishes is from 1.5 mm Hg in the African FW catfish *Bagrus*, to 23 mm Hg in the mackerel tuna (*Euthynnus affinis*), with most in the range of 10–20 mm Hg (Prosser 1973; Lowe et al. 2000). The sharks *Squalus* and *Scyliorhinus* have P_{50}s of 17 and 12 mm Hg, respectively. A higher P_{50} indicates a *lower* affinity for oxygen in the pigment.

Lower pH or higher PCO_2 in the blood, such as that caused by exercise or just general tissue metabolism, results in a lower affinity for oxygen in the circulating Hb. The drop is known as the Bohr effect. It is a shift to the right in the equilibrium curve: a higher P_{50} at lower pH. The Bohr effect

Figure 10.90 Oxygen equilibrium curves for myoglobin and hemoglobin.

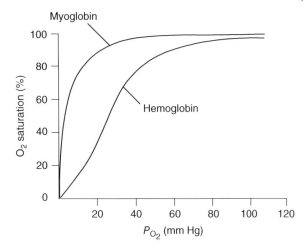

facilitates dumping of O_2 at the tissues (Figure 10.91a, b). The Root effect, seen in fish blood and that of several other vertebrates (not humans!), is similar to the Bohr effect but more drastic (Figure 10.91c). It is a decrease in the maximum saturation of Hb, (or its P_{100}) due to lowered pH, making the Hb unable to saturate, thereby lowering its carrying capacity, even at high pO_2. Both the Root effect and Bohr effect are intimately involved in the function of the swimbladder.

Fishes show a great variety of patterns in their O_2 equilibrium curves as related to their mode of life. In general, sluggish fishes have a higher affinity (lower P_{50}) pigment; this is also seen in animals living at very low pO_2s. For example, P_{50} for the sluggish freshwater catfish the brown bullhead is 6 mm Hg; that for the active pelagic mackerel is 16 mm Hg (Prosser 1973). High affinity pigments are not optimally effective at O_2 delivery to the tissues, but they are useful down to lower O_2 partial pressures. More active animals have a lower affinity pigment that drops its O_2 to the tissues at higher oxygen concentrations, thereby keeping the entire system running at a higher pO_2. The main disadvantage of a high-affinity pigment is that it holds onto its oxygen tightly, forcing the tissues to operate at a lower pO_2. In turn, if the fish must acquire its O_2 at low ambient concentrations, a high-affinity pigment is a distinct advantage.

Red Cell Modifications

A physiologically important characteristic of Hb is that the O_2 equilibrium curve of Hb enclosed in red cells (erythrocytes) is different from the equilibrium curve of the same Hb free in solution. Thus, the oxygen-containing properties of a fish's blood can be modified quickly by changing the medium inside the erythrocytes without the necessity of replacing the original Hb molecule with a different one. In fishes, ATP combines reversibly with deoxy Hb, decreasing the affinity of the Hb for O_2 and shifting the equilibrium curve to the right. Thus, the equilibrium curve of blood can be altered by modifying the ATP levels within the erythrocyte. Whole blood (containing erythrocytes) normally has a lower affinity than Hb alone in solution, termed "stripped Hb."

A classic study using the brown bullhead catfish (*Ameiurus nebulosus*), a favorite

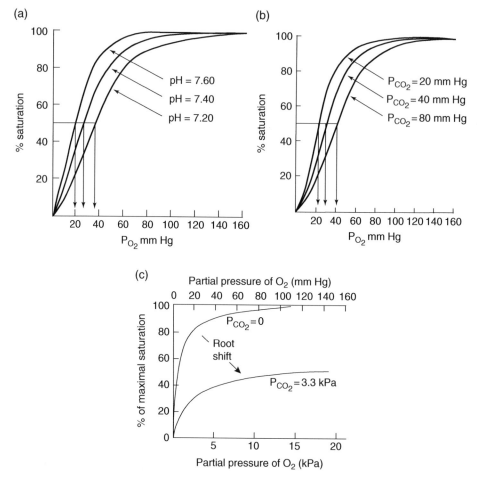

Figure 10.91 The Bohr and Root Effects. Bohr Effect: shifts in the oxygen equilibrium curves for hemoglobin as functions of pH (a) and CO_2 concentration (b), respectively; (c) Root Effect: right shift in the oxygen equilibrium curve for fish hemoglobin due to increases in PCO_2 or pH, with a decrease in O_2-binding capacity. *Sources:* (a, b) Keyes (1985), figure 3.8 (p. 57); (c) Willmer et al. (2005), figure 7.27 (p. 167).

experimental subject for fish physiologists, elegantly demonstrates the importance of red cell modifications in the equilibrium curve of fish blood. The whole blood of fish that had been acclimatized to 24 °C had a higher affinity for O_2 than did whole blood of fish acclimatized to 9 °C, a useful change because of the lower ambient O_2 levels of water at 24 °C. In contrast, the affinity of a stripped Hb solution remained the same. This was done by measuring O_2 affinity in fishes captured at different times of year (Grigg 1969).

CO_2 Transport–Far Different from O_2

The basic difference is in the solubility of the two gases. Not only can CO_2 physically dissolve but it also combines chemically with H_2O. CO_2 is carried in blood in four forms:

Physically dissolved CO_2 – CO_2 is 22× more soluble than O_2 in water at the same T °C. In mammals – physically dissolved CO_2 accounts for ~8% of the CO_2 transported from tissues to the respiratory surface.

Carbonic acid

$$CO_2 + H_2O \rightarrow H_2CO_3 \rightarrow H^+ + HCO_3^-$$

There is approximately 1 molecule of H_2CO_3 for every 340 molecules of CO_2. Thus, it accounts for <1% of CO_2 transported. This reaction proceeds very slowly except in the presence of carbonic anhydrase (CA). CA is located in the red cell, and concentrations usually found are sufficient to increase the rate of interconversion by 100 to 1000-fold. CA is found in oxygen secreting tissues such as the swimbladder of fishes, as well as kidneys, muscle and retina.

Bicarbonate. The majority of CO_2 transported is in the form of HCO_3^-. The equilibrium for interconversion shown below is far to the right, greatly favoring the dissociation of H_2CO_3 into HCO_3^-. About 81% of the CO_2 to be excreted is in the form of HCO_3^-.

$$H_2CO_3 \rightarrow H^+ + HCO_3^-$$

Carbamino compounds. CO_2 also combines chemically with certain amino groups on plasma proteins and on the Hb itself (at the N-terminal end of the four globin chains). The vast majority is on the Hb; contributions by plasma proteins are negligible.

$$Hb-NH_2 + CO_2 \rightarrow Hb-NHCO_2^- + H^+$$

This reaction proceeds quickly and requires no enzyme. The imidazole groups of the amino acid Histidine combine with H^+, effecting H^+ transport (not shown) and the N-terminal ends of the globins combine with CO_2. Approximately 11% of the total CO_2 transported from tissues to the respiratory surface is in the form of carbamino compounds. Oxygenation of deoxy-Hb drives the above reaction to the left. The loss of oxygen at the tissues promotes the combination of CO_2 and Hb, driving the reaction to the right. Note that the CO_2 combines at a different site on the Hb, on the globins rather than the heme.

The Bohr and Haldane Effects
At the tissues, CO_2 shifts the oxygen equilibrium curve to the right, increasing the P_{50} and reducing the affinity of Hb for O_2. There are three components to the effect of added CO_2 to blood. First, CO_2 binds with Hb to form carbamino Hb. Second, CO_2 is hydrated to form H_2CO_3 which dissociates to form $H^+ + HCO_3^-$. Third, H^+ combines with NH and NH_2 groups on Hb, shifting the equilibrium curve further to the right. This is the Bohr effect (Figure 10.92a).

At the respiratory surface, when O_2 is added to the blood, the total CO_2 content is reduced. There are two components here. First, when O_2 is added to blood at a *constant* PCO_2, there is a decrease in carbamino Hb because of a decreased affinity of Hb for CO_2, probably due to changes in tetramer conformation. This accounts for about 70% of the reduction in Hb affinity for CO_2. The remaining 30% is due to the release of H^+ ions from Hb, which is a strong acid when oxygenated. The H^+ ions combine with HCO_3^- to form H_2CO_3, which drives the dissociation reaction to the left, in turn driving the CO_2 off (see equation above). The reduced affinity of oxy-Hb for CO_2 that occurs at the respiratory surface is called the Haldane effect (Figure 10.92b).

A summary of values for the fate of CO_2 in human gas exchange follows (Eckert et al. 1988).

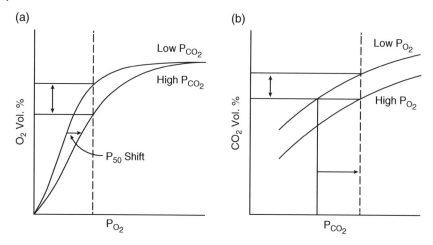

Figure 10.92 A comparison of the Bohr and Haldane Effects. (a) Bohr Effect: a shift to the right in the oxygen equilibrium curve with an increase in the P_{50} when CO_2 concentration is increased; (b) Haldane Effect: The change in CO_2 content (vertical arrow) when O_2 is added to reduced hemoglobin at a constant PCO_2. *Source:* Keyes (1974). Reproduced with the permission of Elsevier.

- <u>48.2 Vol. % CO_2 in arterial blood</u>
- 2.4 dissolved (near sat)
- 2.2 carbamino
- 43.7 HCO_3^-
- <u>3.8 more Vol. % added in tissues*</u>
- 0.3 in solution 8%
- 1 in carbamino 27%
- 2.4 from HCO_3^- 65%
- These are excreted*

Though the values for human blood are different from those of fishes, the relative importance of the three different mechanisms for CO_2 transfer is not.

Secretion of Gases into the Swimbladder of Fishes

Most bony fishes have gas-filled swimbladders to help maintain neutral buoyancy, though some deep-sea teleosts do not, and elasmobranchs do not (Figures 10.80 and 10.93a). Fishes that do not have a gas bladder utilize other methods to achieve neutral buoyancy. There are two types of swimbladders in fishes that do have them: physostomous (open bladder) and physoclistous (closed bladder). Typically, physostomous swimbladders are found in the soft-rayed fishes, which in the hierarchy of fishes include the superorders Clupeomorpha (herrings and kin) Ostariophysi (carps and catfishes), salmonids, and other more primitive fishes (Marshall 1965). The physostomous swimbladder joins the esophagus via a pneumatic duct (Figure 10.80b), which can be closed via a sphincter to retain gas. The physoclistous bladder has lost the connection to the esophagus but has retained the same basic location within the fish's anatomy (Figure 10.93a). To fill the physostomous swimbladder, air is gulped at the surface and forced down the duct, sometimes with the aid of a pneumatic bulb, a muscular structure on the duct that pumps air into the sac. More advanced fishes with functional swimbladders have physoclistous swimbladders, an elegant example of bioengineering using the basic properties of blood in tandem with a highly developed blood circuitry.

Fishes need to be as close to neutrally buoyant as possible to conserve energy. The specific gravity of fishes is about 1.07, so in freshwater their swimbladders must comprise 7% of body volume to achieve neutrality and in seawater about 5% of body volume, aided by the greater density of seawater. In addition, the pressure of the gas in the swimbladder must equal the ambient pressure. The swimbladder itself is a soft-walled structure, so deeper-living fishes must have gas in the bladder at tremendous pressures. For example, fishes with gas-filled swimbladders living at 4000 m must have a bladder at 400 atm pressure.

Volume and pressure are controlled in physoclistous swimbladders using the structure of the swimbladder and the chemical properties of the blood. Structurally, there is a gas-secreting complex consisting of a gas gland and the rete mirabile or "wonderful net." The gas gland is located on the wall of the swimbladder and is supplied by an artery that branches directly off the dorsal aorta (Figure 10.93b). The artery supplying the gas gland divides into a mesh of many fine capillaries, creating an enormous surface area. The venous return is also finely subdivided, with the venous capillaries interleaving with the arterial capillaries and oriented so that flows are in opposite directions. This arrangement results in countercurrent exchange between arterial and venous systems. This capillary network constitutes the rete. Krogh (1929) found a total of 100 000 arterial capillaries in the rete of the common eel *Anguilla* with an equivalent number of venous capillaries, producing a total length of 400 m for each kind, and a surface area of more than 10 cm^2 in a structure the size of a drop of water (Schmidt-Nielsen 1990).

To understand how gas is maintained in the swimbladder of a fish at a depth of 1000 m, or at 100 atm pressure, let us assume the swimbladder already has gas at 100 atm pressure. The blood in the arterial supply to the rete is in equilibrium with the outside partial pressure, which for our purposes here we will assume to be at air saturation or 0.20 atm O_2 and 0.80 atm N_2. Normal circulation to the gas gland would suffice only to remove gases. This problem is solved by the rete mirabile, which serves as a gas trap. The arteries contain blood O_2 at outside levels, and when they pass through the gas gland, they pick up gases at high pressures. However, the arterial and venous capillaries show countercurrent exchange (Figure 10.93c). The venous blood leaving the rete has a much higher gas concentration than the incoming arterial blood. That gas is exchanged along the length of the capillaries in the rete. All the gas picked up by the venous capillaries at the gas gland diffuses across into the incoming arterial capillaries until equilibrium is reached. Gas is thus conserved.

It is now a small step to understanding how gas is added to the swimbladder if the fish swims a little deeper. Consider the O_2 content of blood. For our example, the arterial blood is at 10 ml O_2 100 ml^{-1} and the venous blood is slightly lower at 9 ml O_2 100 ml^{-1}. Therefore, the amount that is lost must have been deposited in the swimbladder. Now assume that the gas gland produces lactic acid (it does!). Fish blood exhibits the root effect, meaning that in acidic conditions, the Hb cannot saturate even at very high pO_2. Thus, the presence of lactic acid causes Hb to dump its O_2, raising the pO_2 in the blood because it cannot recombine chemically with Hb and must remain as a gas. This happens in the venous capillaries. The O_2 diffuses across to the arterial side because of the very high pO_2s (Figure 10.93c) in the loop around the gas gland. The amount of O_2 that finally leaves the rete is less by 1 Vol. %, but partial pressures of O_2 at the gas gland and in the venous capillaries remain very high due to the countercurrent multiplier effect and the inability of blood to recombine with O_2.

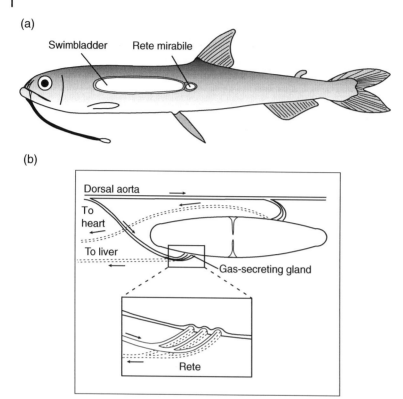
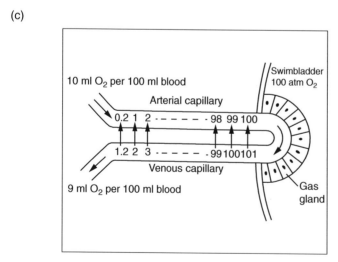

Figure 10.93 The swimbladder. (a) Position of the swimbladder and rete mirabile in a deep-sea astronesthid; (b) diagram of the blood circulation to and from the swimbladder. Note that there are two sources: an artery to the rete mirabile and a vessel to the posterior portion where a fine capillary bed spreads over the walls. These capillaries can serve to absorb gases rapidly. When gas is being secreted, the gas-absorbing vessels are closed and carry no blood; (c) diagram of the countercurrent multiplier system of the swimbladder. *Sources:* (a) Adapted from Lagler et al. (1962), figure 8.14 (p. 255); (b) Schmidt-Nielsen (1990), figure 10.32 (p. 445); (c) Schmidt-Nielsen (1990), figure 10.33 (p. 447).

For vertical migrators, the need to continually adjust pressure with depth changes is metabolically too costly and most species that have retained the bladder have filled it with lipid. These lipid-invested bladders are a major factor in achieving neutral buoyancy.

Nitrogen Excretion

Excretion may be defined as the elimination of waste products resulting directly from food breakdown as well as from the turnover of materials in the body. Major wastes are CO_2, resulting from the cellular combustion of carbohydrates and fats that are excreted at the respiratory surface, and nitrogen compounds originating from protein and nucleic acid breakdown. About 90% of the excreted nitrogen comes from the amino acids of ingested proteins, with the remaining 10% from a variety of sources including the breakdown of nucleic acids (Hochachka and Somero 1973).

Waste nitrogen can be excreted in three major forms: ammonia/ammonium (NH_3/NH_4^+), urea [$CO(NH_2)_2$], and uric acid ($C_5H_4N_4O_3$). The two used by aquatic species are the two simplest, ammonia and urea. Ammonia is the primary excretory product of the bony fishes and urea is the primary product of the elasmobranchs.

Ammonia Excretion

Species that excrete ammonia nitrogen are termed ammonotelic, which includes the teleosts as well as virtually all the aquatic invertebrates. At physiological pH's, which in fishes are near neutral, ammonia is in the form of ammonium. Ammonium is by far the most toxic of the nitrogenous excretory products, so it cannot be allowed to accumulate in the blood. It can be lethal at levels as low as $0.5-5.0$ mmol-N-l^{-1} (Withers 1992) depending on the species. In teleosts, values reported for total ammonia in the blood plasma ranged from 0.04 to 1.30 mmol-N-l^{-1} with most in the range of 0.1–0.5 (Wood 1993). Ammonium is energetically the least expensive nitrogenous waste product to create, but it is only used by aquatic organisms because of the need to excrete it as rapidly as it is formed. The constant excretion of NH_3/NH_4^+ requires an aqueous environment, which is why ammonotelism is pretty much restricted to aquatic forms.

In teleost fishes, about 85% of the total nitrogen excreted is excreted at the gills using the processes of diffusion and possibly carrier-mediated transport (Wood 1993; Walsh 1998). A small amount of ammonia is processed by the kidneys of teleosts, but the primary function of the kidneys in teleosts is in salt and water balance.

Urea

Urea is the major excretory product of the elasmobranchs. It is an uncharged molecule with a high aqueous solubility and low toxicity, allowing it to be accumulated and retained at high concentrations in the blood. Urea is accumulated in the blood of sharks at levels of 441 milliosmoles (mOsm) (Table 10.6) and is a major part of the elasmobranch osmoregulatory strategy. The few teleosts that use urea as an excretory product include species that live in high stress environments such as lungfish and fishes living in highly alkaline lakes. All fishes examined that primarily excrete urea, excrete it through the gills (Wood 1993; Walsh 1998).

Osmotic and Ionic Regulation

Introduction

The physiological repertoire of fishes includes a much tighter regulation of their internal milieu when compared to the invertebrates, a process known as homeostatic control. Not only is the fish internal

Table 10.6 Ionic concentrations (mOmol l^{-1}) of plasma and rectal gland fluid of various marine fishes.

	Na	Cl	K	Mg	Ca	SO$_4$	Urea	TMAO	Total
Seawater	439	513	9.3	50	9.6	26	0	0	1050
Myxine glutinosa (Hagfish)	486	508	8.2	12	5.1	3.0	—	—	1035
Petromyzon marinus (Lamprey)	156	159	32	7.0	3.5	—	—	—	333
Scyliorhinus canicula (Spotted Catshark)	255	241	6.0*	3.0*	5.0*	0.5*	441	72	1118
Squalus acanthis (spiny dogfish) Rectal gland fluid	540	533	7.1	>1	>1	—	14.5	—	1082
Lophius piscatorius (Monkfish)	180	196	5.1	2.5	2.8	2.7	—	—	452
Pleuronectes flesus (Flounder)	142	168	3.4	—	3.3	—	—	—	297

Note: Plasma data for *Squalus acanthis*, denoted with a *, are included with the data for *Scyliorhinus* and are from Forster et al. (1972), other plasma data are from Evans (1979). Rectal gland fluid data for *Squalus acanthius* are from Burger and Hess (1960).

milieu more tightly regulated but its ionic concentration is also quite different from that of its surrounding marine environment. The elasmobranchs and teleosts employ very different strategies for maintaining a constant internal ionic environment, both of which are highly successful. Understanding the complex processes involved requires a basic knowledge of osmosis and diffusion.

Osmosis and Diffusion

Basic properties of ion and water movement are usually couched in terms of their movement through a semi-permeable membrane. A semi-permeable membrane is one that is permeable to water molecules but impermeable to solute particles. Semi-permeable membranes do not really exist in nature; their closest equivalent in biological systems is the cell membrane.

"When two aqueous solutions of different concentration are separated by a semipermeable membrane, water will pass through the membrane until the molal concentrations on the two sides are equal" (Potts and Parry 1964). This movement of water is termed osmosis. Similarly, when two aqueous solutions are separated by a barrier that is permeable to both water and solutes, water and solutes will pass through it until their concentrations on both sides are equal. The movement of a solute through a membrane is termed diffusion. Those two basic properties of ion and water movement are what biological systems must counteract to achieve a constant internal ionic environment. The term osmotic pressure, which you may find curious, derives from its definition: it is the amount of hydrostatic pressure needed to prevent any movement of water from pure water through a semipermeable membrane into a solution. The basic unit of measure is the osmole, or 1 mole of solute particles, and concentrations are expressed in $mOsm\,kg^{-1}$ or $mOsm\,l^{-1}$ (milliosmolal or milliosmolar, respectively). The osmolarity of seawater is about $1000\,mOsm\,l^{-1}$.

Terms.

1) Hyposmotic regulator: maintaining an internal osmotic pressure less than that of the external environment. The resulting standing gradients continually induce water to flow out and ions to flow in. Marine teleosts are hypoosmotic regulators, typically with an internal osmotic pressure (OP) of $450\,mOsm\,l^{-1}$.
2) Hyperosmotic regulator: maintaining an internal osmotic pressure (OP) greater than that of the external environment. Standing gradients induce water to flow in and ions to flow out. Marine elasmobranchs and freshwater teleosts are hyperosmotic regulators.

Elasmobranchs are substantially hyperosmotic to seawater due to the combination of an ionic composition roughly equal to that of teleosts plus a large retention of urea and trimethylamine oxide (TMAO): by-products of nitrogen metabolism. Consequently, they must regulate the movement of water into the body along with the loss of electrolytes to the external environment. Thus, they are termed hyperosmotic regulators. Virtually, all freshwater fishes are hyperosmotic regulators.

In teleosts, the body wall, the kidney, the gills, and the gut work together to maintain a constant internal environment. In elasmobranchs, the important systems utilized to achieve osmotic balance include the kidney, the rectal gland, and the gill secretory system. Of these three systems, the kidney and the rectal gland are the most important.

The skin and scales of fishes are the major barrier defending their internal milieu from

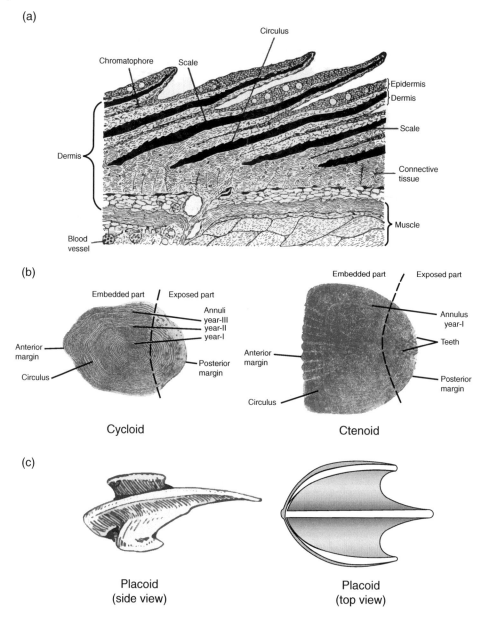

Figure 10.94 Skin and scales. (a) Section through the skin of a teleost showing scales embedded in the dermis; (b) the two most common types of teleost scales: cycloid (left, soft-rayed fishes) and ctenoid (right, spiny-rayed fishes); (c) side- and top-view of typical placoid scales (elasmobranchs). *Source:* Adapted from Lagler et al. (1962), pp. 109, 113, 115, 116.

loss of water and intrusion of ions. Teleost skin is composed of two layers: the epidermis and dermis, shown in Figure 10.94a. Their overlapping scales are embedded in the dermis (Figure 10.94b). Two basic types are of most concern to us: the cycloid scales of the soft-rayed fishes and the ctenoid scales typical of spiny-rayed (acanthopterygian) fishes. Note the embedded vs exposed portions of the scales and the annuli resulting from seasonal growth patterns in the fish. The elasmobranchs have placoid scales, composed of a hard enamel-like outer layer and a pulp layer beneath. "Each scale has a disc-like basal plate in the dermis with a cusp projecting outward from it through the dermis" (Lagler et al. 1962).

Data on the relative contribution of the body wall as a barrier to ion and water flux are scanty in both teleosts and elasmobranchs. It is assumed that the body wall blocks most water leaving the fish and the inward movement of most ions. That assumption has been supported by studies using radioactive tracers to show the balancing of ion flux (Figure 10.95; Potts 1976).

Teleost and elasmobranch kidneys (see Figure 10.80 for location) are both more sophisticated than those of the crustaceans, but the principles of ultrafiltration discussed for the crustaceans remain the same. They are all tubular excretion systems. In marine teleosts, the kidney functions primarily as an excretory device for Mg^{++} and $SO_4^=$ ions. There are two basic structural types of kidneys in teleosts: glomerular and aglomerular, with glomerular kidneys being the more typical case. Aglomerular kidneys eliminate the ultrafiltration step, relying solely on secretory processes to rid the fish of unwanted ions. The kidney is the only pathway for excretion of Mg^{++} and $SO_4^=$; other ions (Na^+, Cl^-) are excreted by the gill and gut.

Since all marine teleosts are hypoosmotic regulators, they are faced with the dual problems of passive water loss to the more saline marine environment and the passive intrusion of electrolytes down their concentration gradient into the fish. The kidneys are not capable of producing a concentrated urine; the urine produced is slightly hypoosmotic to the blood. The hypoosmotic urine thus involves the loss of H_2O, which is counterproductive to their regulation of osmotic balance. Water lost in urine production and through the gills and integument is replaced by drinking seawater, usually in a quantity of 0.2–2.3% of the body mass per hour. As seawater passes through the intestine, 60–85% of the water is absorbed. Monovalent ions are also absorbed through the intestinal wall. Divalent ions remain behind. Most of the monovalent ions absorbed through the intestine are excreted extra-renally at the gills.

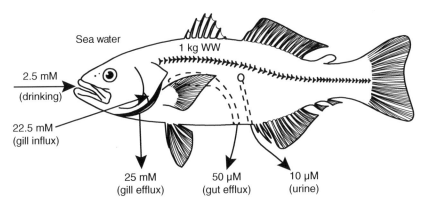

Figure 10.95 Salt balance in a marine teleost showing influx through the mouth and gills and efflux through gills, gut and kidneys (urine).

Because the sharks, skates, and rays are hyperosmotic in a marine environment, they must rid themselves not only of Mg^{++} and $SO_4^=$ ions but also of excess water. Water enters the body through the gills and integument by osmosis. Electrolytes also enter by diffusion because the concentrations of Na^+ and Cl^- inside are very much less than outside.

The high internal osmotic pressure passively supplies the water needed for formation of urine without the need for imbibing seawater as in marine teleosts. Indeed, elasmobranchs do not appear to drink their medium. However, they still must contend with an influx of unwanted electrolytes. The system used by elasmobranchs to achieve osmotic balance is termed the hydromineral balancing system, which includes the kidney, rectal gland, and the gill secretory system. Of the three systems, the kidney and rectal gland are the most important.

Urine formation by elasmobranch kidneys differs from that in teleosts in that both urea and TMAO are actively reabsorbed from the filtrate such that very little appears in the urine. Generally, about 70–85% of the water in the filtrate is reabsorbed. Na^+ and Cl^- are also reabsorbed but not as completely as in teleosts. In fact, Na^+ and Cl^- are the dominant solutes in elasmobranch urine.

The rectal gland produces a colorless fluid isosmotic to the plasma, containing no urea or TMAO but containing Na^+ and Cl^- at two times its plasma concentration. Experimentally increased Na^+ and Cl^- in blood, "salt loading," results in increased excretion of Na^+ and Cl^- by the rectal gland with the urine osmolality remaining constant. Elasmobranchs in general are more tolerant of changes in blood osmolality than are teleosts.

The gills of elasmobranchs have a minor secretory function, with most ionic regulation done by the rectal gland and kidney. However, as in the teleosts, the gills are a major site for entry of electrolytes.

The gills

The gills of teleosts are a very complex tissue, participating in ion regulation and nitrogen excretion in addition to their important role in gas exchange. Within the gill, there are three types of cells important to ion balance and gas exchange:

1) Those typical of the branchial epithelium, which are simple and found in the lamellae: respiratory cells and involved primarily in gas exchange.
2) Mucous cells found mainly in the filament. The mucous, which provides a protective barrier to abrasive substances, is also involved in ion regulation, particularly sodium and chloride ions.
3) The mitochondria-rich cells called chloride cells – found exclusively on the filament, containing an apical pit. These specialized cells play a major role in osmoregulation.

The respiratory cells of the gill lamellae are characterized by "tight junctions" between cells and are considered to be relatively impermeable to ions and water movement. Thus, they play no active role in ionic regulation. The chloride cells are gathered around the afferent blood flow in the filament (the flow *to* the filament from the heart), and in marine teleosts, they are characterized by leaky junctions. The chloride cells are the major driver in Na^+ and Cl^- excretion by marine teleosts.

Morphology of a chloride cell in a marine teleost is shown in Figure 10.96a. Chloride cells, now mainly known as MR (mitochondria-rich) cells, occur in groups in the gill filament, extending from the basal lamina (blood-side) to the exterior environment of the fish. Each is accompanied by an accessory cell (Figure 10.96a), which together with the chloride cell form a salt-secreting unit (Marshall and Grosell 2006).

The basal or blood side of the chloride cell has a crenulated appearance due to the infolding of an extensive tubular system (also known as the smooth tubular system or STS) that extends from the base almost to the apical pit of the cell. Mitochondria are densely packed within the cell cytoplasm in close proximity to the STS. A high concentration of energy-requiring ion pumps in the form of Na^+-K^+ ATPases are located within the STS on the basolateral side of the MR cells.

The actual mechanics of MR cell function are complex, involving active and passive transport processes (Figure 10.96b). The system is driven by the Na^+-K^+ ATPases on the basolateral side of the MR cell, which deplete the Na^+ within the cell, creating a large concentration gradient between the cell cytoplasm (about 20 mM) and the blood plasma bathing the base of the cell (about 180 mM). Na^+-K^+ ATPase activity results in a high cytoplasmic K^+ concentration which facilitates its passive diffusion back into the blood via dedicated channels in the basal membrane.

Cl^- is transported into the cell by an epithelial cotransporter (Figure 10.96b) whereby a membrane bound protein facilitates the diffusion of one Na^+, one K^+, and two Cl^- through the basolateral membrane. This is a passive process because the low cytoplasmic Na^+ concentration generated by the Na^+-K^+ ATPase activity facilitates Na^+ diffusion inward. Activity of the co-transporter causes Cl^- to accumulate in the cell above its electrochemical equilibrium; it then diffuses out passively through specialized anion channels in the apical membrane.

Sodium is secreted outward through what is termed a "paracellular pathway," comprising the lateral intercellular spaces and leaky junctions between the MR cells and their accessory cells (Marshall and Grosell 2006). The leaky junctions where the accessory cells and chloride cells meet are cation selective, only allowing positive ions to pass to the outside. The question now is why Na^+ diffuses out passively against the substantial concentration gradient between the blood plasma and seawater. The answer is in the transgill electrical potential. The chloride cell itself is electronegative to seawater as a result of its low Na+ and high Cl^-. This results in a microenvironment whereby the blood plasma at its base is substantially electropositive to seawater, and Na^+ diffuses down the charge gradient to the outside. The salt-secretory system of marine teleosts is one of the most effective ion-transporting systems in the animal kingdom (Bone and Moore 2008).

The gut. The gut is the final operator in marine teleost osmoregulation. As noted above, water is lost at the kidney as well as through the gills and integument. That water is replaced by drinking seawater at a constant continuous rate, ranging from 1 to 5 ml kg^{-1} h^{-1}. It is important that SW be ingested continuously; it allows a slow progression over the gut epithelium and successive steps of treatment during its transit down the alimentary canal. The efficiency of the gut is high: most monovalent ions and H_2O are absorbed, while most divalent ions remain in the lumen and are voided with feces. Absorbed monovalent ions are excreted at the gills. During its transit down the alimentary canal, ingested seawater changes considerably in its ionic composition and 60–85% of the water has been removed.

Esophagus
In marine teleosts, the esophagus is very permeable to Na^+ and Cl^-, but nearly

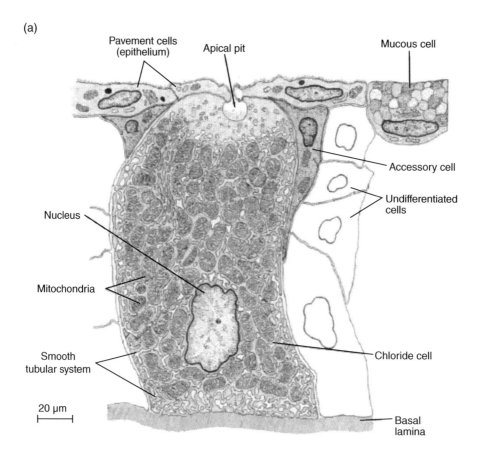
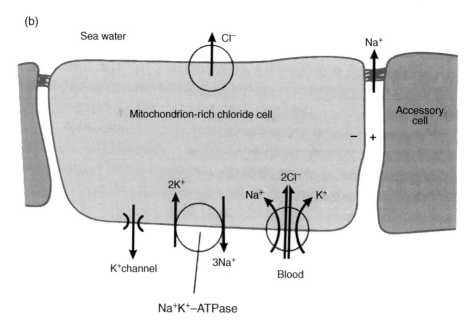

Figure 10.96 (a) Schematic representation of a chloride cell from the opercular epithelium of a seawater-adapted *Fundulus heteroclitus*. Note the apical pit (apical crypt), accessory cell, and the abundant mitochondria; (b) Model for NaCl extrusion in the chloride cells of the marine teleost gill showing location of ion pumps and paracellular pathways. *Sources:* (a) Degnan et al. (1977), text-figure 1 (p. 162); (b) Bone and Moore (2008), figure 6.8 (p. 170).

impermeable to water. Fifty to seventy percent of the Na$^+$ and Cl$^-$ ions ingested move passively outward from the luminal fluid to the blood plasma down their electrochemical gradient or are actively transported outward with Na$^+$-K$^+$ ATPases. Concentrations of Cl$^-$ at the end of the esophagus are 36–67% of their seawater value. The end result is a "desalinated" luminal fluid at a little below the internal OP of the fish.

Stomach
The stomach has little or no osmoregulatory function.

Intestine
Most water absorption occurs in the intestine, and it is once-again driven largely by ion pumps (Na$^+$–K$^+$ ATPases). In the case of intestinal cells, the apical side is facing toward the intestinal lumen and the basolateral side toward the circulating blood. Between the cells is a microenvironment of extracellular fluid that extends to the apex of the cell as a paracellular pathway. The basolateral membranes of intestinal epithelial cells are heavily invested with Na$^+$–K$^+$ ATPases, which deplete the Na$^+$ inside the cell, providing the electrochemical gradient driving the passive Na$^+$, K$^+$, and Cl$^-$ uptake from the intestinal luminal fluid.

All three major ions, Na$^+$, K$^+$, and Cl$^-$, exit into the extracellular fluid via either active transport (Na$^+$) or co-transporters (K$^+$ and Cl$^-$), creating a standing hyperosmotic environment in the extracellular fluid that draws water inward through the paracellular pathways between the cells. The end result is a rectal fluid that is roughly isosmosmotic to the blood but greatly reduced in volume from the imbibed seawater, and with much higher concentrations of Mg^{2+} and SO$_4^{2-}$. Just as the fishes drink continuously, so do they excrete rectal fluid continuously, but at a rate 15–40% of the drinking rate. Thus, 60–85% of the volume of ingested seawater is absorbed as water (Marshall and Grosell 2006).

Locomotion

Open-ocean fishes span the gamut of the five basic fish locomotory types. Those forms were elucidated by Breder (1926) and are still the standard today; anguilliform (eel-like), thuniform (tuna-like), carangiform (jack-like), labriform (bass-like, or hatchetfish-like), and tetraodontiform (puffer-fish-like or Mola-Mola-like) (Figure 10.97).

Musculature
Muscles and their arrangement in fishes are at once unique in the vertebrate world and highly complex. Their contractions must produce the axial side-to-side movement needed for undulatory propulsion with no volume change and no tendency to pull the fish up or down. The basic idea of how this is achieved is shown in Figure 10.98a. It is the bending of a body with a flexible but incompressible backbone by segmented blocks of muscle (Bone and Moore 2008).

In practice, the side-to-side motion requires a musculature with w-shaped myotomes (Figure 10.98b,c) that are anchored in myosepta, partitions consisting of collagen fibers that are in turn attached to the vertebral column and to the skin. The myosepta are deformable but inextensible. When muscle fibers contract, the tension produced is transmitted to the myosepta and in turn to horizontal and median septa (backwards and inwards) which attach to the skin and the spine. Thus, skin is involved in the locomotory process (Bone 1978).

Scombroids have departed from traditional fish undulatory propulsion by using their tail alone as a propulsor and moving the body itself very little (Figure 10.99).

	Swimming type				
	Via trunk	Via tail		Via fins	
	Anguilliform	Thuniform	Carangiform	Labriform	Tetraodontiform
Representative taxa	Eel-like	Tuna-like	Jack-like	Bass-like	Mola mola-like
Propulsive force	Most of body	Caudal region	Posterior half of body	Pectoral fins	Median fins
Propulsive form	Undulation	Undulation	Undulation	Oscillation	Oscillation
Maximum speed (BL s^{-1})	Slow-moderate (2 BL s^{-1})	Fast-very fast (20 BL s^{-1})	Moderate-fast (10 BL s^{-1})	Slow (4 BL s^{-1})	Slow (? BL s^{-1})

Figure 10.97 Form and function in the five basic types of swimming in fishes. The part of the fin that provides propulsion is indicated by cross-hatching. *Source:* Adapted from Helfman et al. (1997).

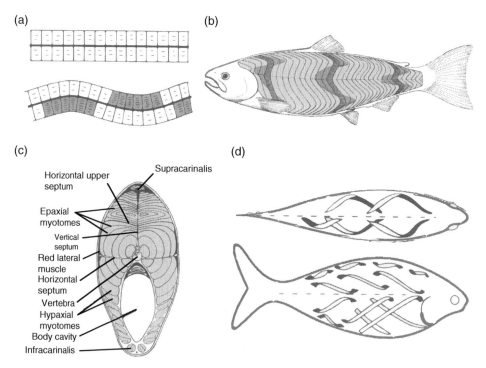

Figure 10.98 Fish musculature form and function. (a) Diagram showing how a body with a flexible but incompressible backbone (thick black line) and segmented muscle blocks (contracted myotomes stippled) can generate undulatory motion; (b) myotome patterns in salmon showing the characteristic W shape; (c) cross-sectional diagram of a salmon's body musculature showing positions of septa; (d) muscle fiber orientation between myotomes in a typical teleost. The helical character of muscle fiber shortening allows smooth undulatory motion with no tendency to pull the fish up or down. *Sources:* (a) Adapted from Bone and Moore (2008), figure 3.1 (p. 62); (b) Adapted from Bond et al. (1996), figure 18-1 (p. 260); (c) Bond et al. (1996), figure 18-2 (p. 261); (d) Adapted from Bone and Moore (2008), figure 3.6 (p. 67).

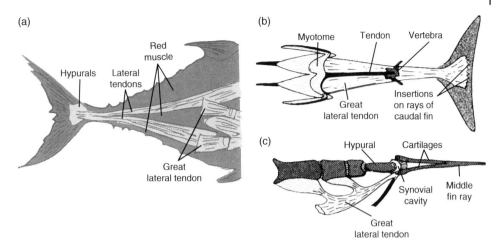

Figure 10.99 The scombrid great lateral tendon. (a) The caudal tendons of a yellowfin tuna showing the great lateral tendon (GLT) linking the posterior myosepta to the hypurals; (b) lateral view showing a single myotome on the left, with smaller tendons (black), and the GLT inserting on the rays of the caudal fin; (c) median horizontal section through a scombrid tail showing two posterior myotomes connecting to the GLT that in turn inserts on the middle fin ray. The anterior end of the fin ray articulates with the hypural via cartilages enclosing a synovial cavity. *Sources:* (a) Adapted from Bone and Moore (2008), figure 3.7 (p. 69); (b, c) Bone et al. (1995), figure 3.5 (p. 50).

They achieve this via tendinous connections between the body musculature and the hypural bone of the tail: the great lateral tendon (GLT) and the smaller lateral tendons. The caudal peduncle is much reduced in size relative to that of a typical teleost, presenting a greatly reduced aspect to the water as the fish oscillates its tail. The effect is that of a body with a propeller rather than the motion of a typical fish where an undulatory wave moves down its body starting either at the head (anguilliform) or in the posterior part of the body (subcarangiform, carangiform) and the myotomal musculature extends to the caudal fin.

Red and White Muscle

Myotomal muscle fibers may be divided into red and white muscle based on the presence of myoglobin. All vertebrates have both "fast" and "slow" types of muscle fibers, but in fishes, the division of labor is most obvious (Table 10.7). Red muscle fibers, which are termed "slow-twitch," are used for sustained swimming (cruising), contain myoglobin, and operate aerobically. White muscle fibers which are termed "fast-twitch," are used for burst swimming, do not contain myoglobin, and operate anaerobically. Cruising fishes using red muscle can swim for extended periods due to the sustainability of the aerobic metabolism of the muscle. In burst swimming, the stores of metabolic fuel used in anaerobic metabolism of the muscle are only sufficient for a few minutes. Thus, the burst is limited to the amount of time it takes to exhaust the biological fuel. Recordings from red and white muscles suggest that white muscle fibers only fire when rapid acceleration is needed, not during normal cruising. Bursting is not all or none; fibers are recruited as needed (Bone 1978).

Distribution of the two fiber types is also quite different, with red muscle distributed axially along the body midline and the

Table 10.7 A comparison of fast and slow muscle fibers in fish. *Source:* Bone (1978), table II (p. 370). Reproduced with the permission of Academic Press.

Slow	Fast
Smaller diameter (20–50% of fast)	Larger diameter (may be more than 300 μm)
Well vascularized	Poorly vascularized
Usually abundant myoglobin, red color	No myoglobin, usually white
Abundant large mitochondria	Few smaller mitochondria with fewer cristae
Oxidative enzyme system	Enzymes of anaerobic glycolysis
Stored lipid and glycogen	Glycogen stored, usually little lipid
Distributed cholinergic innervation	Focal or distributed cholinergic innervation
No propagated muscle action potentials	Propagated muscle action potentials

white muscle forming the complex myotomes described earlier (Figure 10.100). In some fishes, there is a small gradient between the two types, sometimes obvious as a distinct mass of pink fibers in between the two, but in most, the two types appear quite distinct. Red muscle typically makes up from 5 to 15% of the total muscle mass, with the amount of red muscle in the myotome indicating how important cruising is to the lifestyle of the fish. Red muscle in tunas may reach as much as 30% of the muscle mass (Johnston 1981), but in order to supply the red muscles of tuna with needed oxygen, it requires the extensive respiratory and circulatory adaptations described earlier.

Differences in the innervation of the two types of fibers are partially dependent on how advanced the species is in the hierarchy of fishes. In more primitive teleosts such as the herrings, and also among the sharks, the white fibers are focally innervated (Figure 10.100), with one motor endplate at the end of each cell that is served by two motor neurons. The action potential then propagates from the endplate along the length of the fiber. In more advanced fishes such as the scombrids, the white fibers are multiply innervated and, in all fishes, the red fibers are multiply-innervated (Bone and Moore 2008).

Among the vertebrates, the fishes are the only group that can afford to have most of their body mass invested in a musculature that is only used episodically for rapid movement. It is made possible by the buoyancy of the aquatic medium, which is a two-edged sword. It provides buoyancy, but its density makes it difficult to move through. Fishes must cope with the drag it imposes on forward motion: it requires a large musculature to generate the thrust for rapid movement.

Drag and Swimming Costs

Two basic types of drag impede the forward motion of a fish: pressure drag and friction drag. Pressure drag is exactly that. As a fish moves through the dense aqueous medium, it results in pressure at the nose of the fish. Vorticity in the wake directly behind the fish resulting from the motion of its tail and the body shape as it moves through the water results in an area of lower pressure behind the fish. Thus, the fish must cope with pressure at the front and pull at the back as it moves forward. Many fishes (all of the fast movers) have a spindle, or fusiform shape which minimizes the form drag, or the wake created by its shape. Shape of the tail and fins of fast-movers like the tunas and some sharks is designed to minimize vorticity as well.

Skin friction drag is the most important drag component in fish locomotion. It is the friction between the water flowing along the skin of the fish and the stationary boundary layer (Chapter 1) at the fish's skin. For all the fishes of concern here, with the exception of some sharks, scombrids, and billfishes, flow around the body is laminar. Some of the factors that influence friction drag can be modified; these include reducing the wetted surface area by using retractable fins like the pectoral fins of tunas, and having discontinuous fins, like most of the non-eel fish world. Minimizing lateral movements to minimize the amount of fish body presented to the oncoming flow is another. Though impossible to eliminate in undulatory swimmers owing to the side-to-side motion of their tails, a gradient of efficiency can be observed progressing from anguilliform to carangiform swimming and the reduction in how much of the body participates in tail movement. Thunniform swimming is the most efficient because of the design described above. Mucous generation in the skin (barracudas) and denticles in the skin of some fast-moving sharks both are used to stabilize the boundary layer at higher swim speeds and prevent the lateral eddy formation and turbulence that greatly increase drag (Vogel 1994; Bone et al. 1995).

Swimming Costs

The cost of transport is the energy required to transport a unit of mass over a unit of distance (Webb 1997), for example, $kcal\,g^{-1}\,km^{-1}$ (Tucker 1975). Figure 10.101 shows a theoretical curve depicting the power requirements for swimming as speed increases.

The J-shape of the curve comes from the greater contribution of standard metabolism (the ectothermic equivalent of basal or resting metabolism) to the total energy requirements at low speeds as well as the postural costs of hovering. Since neutral buoyancy is approached but rarely completely achieved, the cost of maintaining stability while hovering is part of the energy cost at low speeds (Webb 1997). As speed increases and the fish moves forward, the costs of stability decline until they reach the "bottom of the j" and begin to rise again owing to the increased metabolic demands of propulsion and overcoming drag. The curve depicted in Figure 10.101 represents the gross cost of transport (gcot) because the total energy used by the fish for each swim speed is included. You will encounter two additional terms in summary papers. The first is minimum cost of transport, which is the value at the bottom of the J (Tucker 1975). The other is net cost of transport which is the gcot minus the standard metabolism. Those two are considered to be equivalent.

Beginning with Schmidt-Nielsen (1972), a few papers (e.g. Tucker 1975) have compared cost of transport in the three main locomotory modalities: swimming, flying, and running. All conclude that swimming is the least costly, despite the need for overcoming the resistance of a dense medium.

Maximum Swim Speeds

Maximum swim speeds, particularly of fast-swimming species, are difficult to measure, but a few values have been obtained. Techniques include measuring the line run out from a hooked fish (Walters and Fierstein 1964) or videos of fish swimming at burst speeds (Bone and Moore 2008). Overall conclusions were that maximum speeds for those species tested (all scombroids) varied from 10 to 20 body lengths per second (Wardle et al. 1989).

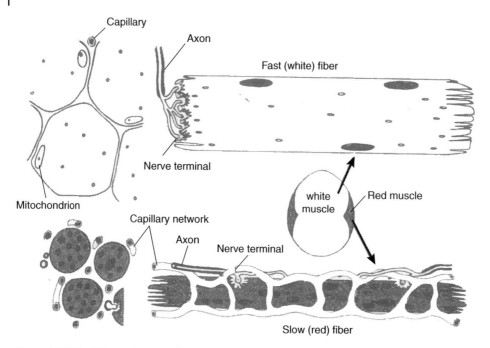

Figure 10.100 Schematic comparison between slow (red) fiber and fast (white) fiber. On the left is a transverse section showing comparative capillary and mitochondrial density. Between the two muscle types is a cross-sectional depiction of their locations at mid-body. *Source:* Bone (1978), figure 5 (p. 371). Reproduced with the permission of Academic Press.

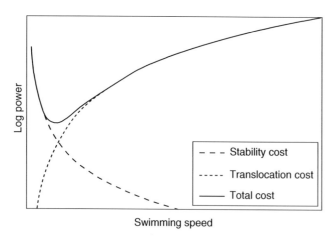

Figure 10.101 Relation between power requirements for swimming and swim speed. Stability costs are postulated to be high at low speed, decreasing with increasing speed. Those costs are due to the energy required for hovering. Costs of translocation (transport) increase with increasing speed. The sum of stability and translocation costs tends toward a J-shape. *Source:* Webb (1997), figure 4 (p. 9). Reproduced with the permission of CRC Press LLC.

Endothermy

During the 1960s, an exciting discovery was made: functional warm-bloodedness in tunas and lamnid sharks (Carey and Teal 1966; Carey et al. 1971). The speed at which muscle fibers contract is a direct function of temperature: the warmer the faster (Hartree and Hill 1921). It follows that sustainable swim speed, a direct function of tail-beat frequency, does as well. During the 1960s, it was discovered that tunas and lamnid sharks are able to maintain their musculature at a higher temperature than ambient by recycling the heat generated from muscle activity using a retial system. This feature substantially enhanced their ability to sustain high tail-beat frequencies and thus speed.

Normally, heat generated by muscular activity is lost to the fish when arterial blood passes through the gills, which are an exceptional heat-exchanger. The minimal diffusion barriers needed for optimizing gas-exchange also rapidly equilibrate the arterial blood to outside temperature. In the tunas and lamnid sharks, heat is trapped by a rete mirabile in the muscle that transfers the heat from venous blood warmed by muscular activity to the incoming arterial blood (Figure 10.102). The efficiency of heat exchange in the retial system and the ability to regulate temperature varies between species, with the bluefin (*Thunnus thynnus*) being the best in this regard. Many tunas, including the albacore, bigeye, skipjack, and yellowfin (*Thunnus alalonga, T. obesus, Katsuwonus pelamis,* and *T. albacares,* respectively), can maintain muscle temperatures at 5–13 °C above ambient, as can several sharks (Carey et al. 1971). The warm-bloodedness exhibited by tunas and lamnid sharks is termed regional endothermy or heterothermy.

Recent studies on the large mesopelagic Lampriform fish, the Opah (*Lampris guttatus*) (Wegner et al. 2015), have revealed a new and novel approach to endothermy in fishes: they have achieved a limited degree (4–6 °C difference between fish and ambient) of whole body endothermy. The biological strategy of heat generation is similar to that of the tunas, in that most of it comes from the muscular activity of their flapping pectoral fins, but the location of the rete that traps the heat is in the gill instead of in the circulation entering and leaving the red muscle. The gill rete retains the heat generated by the muscle so that it can circulate through the entire body of the fish including the heart, giving the fish a physiological boost in cardiac and swimming performance. The fish has fatty connective tissue insulating its swimming muscle and heart as additional mechanisms of conserving metabolic heat. It should be noted that the Opah is a true mesopelagic fish, spending most of its time between 50 and 400 m (Wegner et al. 2015).

Warm-Brained Billfishes

Several of the billfishes, including the swordfish *Xiphias gladius*, the white marlin *Kajikia albida*, the spearfish *Tetrapturus angustirostris*, and the sailfish *Istiophorus albicans*, maintain their eyes and brain at a temperature several degrees (up to 10 °C in swordfish) above ambient, a phenomenon first discovered by Frank Carey of the Woods Hole Oceanographic Institution (Carey 1982). Though most of the billfishes reside primarily in the epipelagic zone in temperate and tropical waters (45°N to 45°S), they often drop into the cooler waters below the thermocline to hunt, particularly the swordfish. Billfishes, unlike the tunas who swim continuously, are considered to be "stalkers and sprinters" owing to the preponderance of white fibers in their musculature (Bone 1966; Carey 1982). They need their eyes and brain to be fully functional to

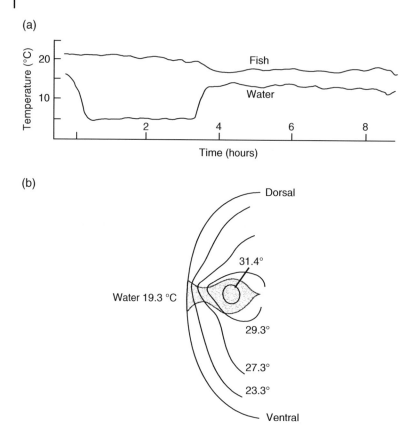

Figure 10.102 Thermoregulation in the bluefin tuna (*Thunnus thynnus*). (a) Records of water temperature and stomach temperature obtained by telemetry from free-swimming bluefin, showing independence of body temperature from changes in water temperature; (b) thermal profile obtained with thermistor probes, red muscle stippled. *Source:* Bone et al. (1995), figure 3.14 (p. 64). Reproduced with the permission of Chapman & Hall.

successfully capture prey, even in the event of a rapid temperature drop such as that encountered at the seasonal thermocline. Unlike the tunas, the billfishes are ectotherms, with body temperatures matching those in the ambient water. However, the billfishes have developed a "brain heater" to offset the slowing effects of cooler temperatures. It is located in apposition to the floor of the cranial cavity, which instead of being bone, is membranous, facilitating the effective transfer of heat to the brain which is cocooned just above it. The heater organ is quite large, many times (about 25 times) larger than the brain itself and has evolved from one of the oculomotor muscles of the eye, the superior rectus. The cells of the heater organ are highly modified. They have lost nearly all their contractile fibrils but are very heavily invested with mitochondria (55–70% of cell volume) and the membrane systems important to the calcium cycling within the cell that normally control muscle contraction. In normal muscle cells, the chemical energy released by ATP hydrolysis during cellular respiration is converted to mechanical work (e.g., muscle contraction). In the modified heater cells, that chemical energy is instead released as heat. Neural stimulation of the heater cells causes the continual release and re-uptake of calcium ions that would normally surround the contractile fibrils, with the attendant hydrolysis of ATP and consequent production of heat. This biochemical process is termed futile cycling.

Oxygen is supplied to the heater cells by a large rete mirabile originating in the carotid artery (Block 1986), and the heater cells are well invested with the myoglobin needed to accept it, allowing the heater cells to function aerobically. The rete traps the heat generated within the heater organ, passing it from the venous to the arterial circulation. For a further description of the complex processes within the heater cells consult Block (1991).

Swimming in Mesopelagic Fishes

Few data are available on swimming in mesopelagic fishes, particularly in the numerically dominant bristlemouths, lanternfishes, and hatchetfishes. Observations of swimming are best obtained in situ and are quite limited. Even submersible observations have their limitations owing to the fact that they must use lights at depth, an unnatural situation, to be able to observe fish behavior.

The observations of Barham (1971) are still some of the most extensive that have been reported, and they showed similarities between locations ranging from the California borderland, to the NW Atlantic, to the warmer Pacific waters off Acapulco and Cabo San Lucas. Most lanternfishes spent part of their day in a quiescent state with their head oriented upward but were much more active at nighttime depths, swimming and maintaining a horizontal attitude. Hatchetfishes maintained a more active horizontal position, rapidly swimming away from the submersible, but were also observed in a motionless state. Overall, most fishes observed were vertical migrators and most showed periods of inactivity during the day with the caveat that they would always flee when they sensed the submersible.

Based on reported and personal observations, the three major midwater taxa can be roughly categorized into the following locomotory types. Sternoptychids are subcarangiform swimmers, using their tail for rapid forward movement but able to use their paddle-like pectoral fins for stabilization and finer scale movement (Janssen et al. 1986). Based on personal observations made during a "night-lighting" operation (using a submerged light to attract and capture animals at night), the myctophid *Tarletonbeania crenularis* was determined to be intermediate between anguilliform and subcarangiform swimmers. *Cyclothone* are very weak swimmers, mainly fluttering their tails for forward movement, but also are capable of weak anguilliform swimming. Personal observations of stomiids, mainly *Stomias atriventer* from the California current and those in respirometers, were mainly tail-flutterers. However, consistent with their white muscle mass, they would be capable of rapid acceleration using whole-body subcarangiform/anguilliform motion. Indeed, this sort of swimming has been observed for both *Stomias* and *Chauliodus* when fleeing from submersibles. Perhaps, tail-fluttering would be a way of maintaining station with little energy expenditure at depth, and the white muscle could be used to "strike" at prey, similar to the action of snakes.

Buoyancy

The universal quest for neutral buoyancy in pelagic fishes may be narrowed down to a few basic strategies. Those are:

1) Use of a gas-filled swimbladder as a source of lift, the function and filling of which was covered in "blood," above. Without question, a gas-filled swimbladder is the most efficient way of achieving

neutral buoyancy, though it also has a metabolic price as noted above, and it must be continually adjusted with changes in depth. The need to continually adjust pressure with depth changes contraindicates its use by vertical migrators.

2) Use of the pectoral fins as hydrofoils to provide upward lift while swimming, analogous to the wings of an aircraft. This is employed by several dense fishes such as the mackerels and tunas (Magnuson 1970), and it was formerly believed to be used in sharks as well. More recent study on the leopard shark, *Triakis semifasciata* (Wilga and Lauder 2000) suggests that lift in sharks is generated by the attitude of the shark body itself. They swim with their nose slightly elevated to the horizontal (11° at slow speeds, less as speed increases). Steady forward swimming is achieved by adjusting the body angle; pectoral fins are mainly used for initiating changes in vertical position and in maneuvering. The leopard shark is a coastal species. In California waters, it is often found swimming in the kelp bed close to shore where maneuverability is particularly important. Whether the same principles hold in the faster swimming mackerel sharks of open waters remains to be seen, but it seems likely.

The use of hydrofoils does have a downside. As with aircraft, there is a minimum forward speed required to maintain the upward lift, the marine equivalent of the stalling speed of aircraft. Thus, they must keep moving or sink (Alexander 1993).

3) Accumulation of positively buoyant substances, usually lipids, either sequestered in one organ such as in the liver of some sharks or distributed in other tissues such as in the Antarctic silverfish *Pleuragramma antarctica*, which carries lipocytes under the skin (Eastman 1998) or even as a buoyant gelatinous layer of glycosaminoglycans such as that found under the skin of stomiids (Yancey et al. 1989). Many vertical migrators have filled their swimbladders with lipid compounds.

Fish densities vary from 1050 to 1090 kg m^{-3}, with a median of about 1075 kg m^{-3} (Alexander 1993). In order to be at neutral buoyancy, the buoyancy organ must give the fish enough lift to match the density of seawater, about 1026 kg m^{-3}. Table 10.8 gives the density of various different lipids used as buoyancy aids by fishes, and the fractional volume required for them to enable a typical (1075 kg m^{-3}) fish to reach neutral buoyancy (Alexander 1993). You will notice that all of the lipid types require a very large fraction of the total volume of the fish to be invested in lipid-rich tissue (23–31%) for them to achieve neutral buoyancy.

Several species of sharks are neutrally buoyant, including the large filter-feeding basking shark *Cetorhinus maximus* (Matthews and Parker 1950) and several species in the order Squaliformes (Corner et al. 1969). The four neutrally buoyant sharks analyzed by Corner et al. (1969) all had very large lipid-rich livers, making up 20–30% of the animal volume and containing over 80% lipid, mostly squalene (Table 10.8); more than enough to confer neutral buoyancy to the fishes. Van Vleet et al. (1984) in a similar study found shark livers heavily invested with wax esters; a different lipid with similar buoyancy properties to squalene (Alexander 2003). The benthopelagic oilfish, *Ruvettus*, also uses wax esters as a source of static lift,

Table 10.8 Fractional volumes and masses of buoyancy aids required to match the density of typical fish to seawater. *Source:* Adapted from Alexander (1993), table 2 (p. 79).

Buoyancy aid	Density of buoyancy aid	Fractional volume	Fractional mass
Gas	Negligible	0.05	—
Squalene	$860\,kg\,m^{-3}$	0.23	0.19
Wax esters	$860\,kg\,m^{-3}$	0.23	0.19
Triglycerides	$930\,kg\,m^{-3}$	0.34	0.31

Note: Density of the fish without buoyancy aids is assumed to be $1075\,kg\,m^{-3}$. Fractional volumes are $V_b/(V+V_b)$, and fractional masses are $m_b(m+m_b)$. V_b and m_b are the volume and mass of the required buoyancy organ, respectively. V and m are the total volume and mass, respectively.

but in the oilfish, the lipid is distributed throughout its tissues (including bone!) instead of localized in one organ as in the sharks (Bone 1978). The ammoniacal squids described in Chapter 8 use ammonium as a source of static lift instead of lipids: the same principle with a different buoyancy aid.

4) Minimization of dense substances within the body, notably muscle protein and bone tissue, resulting in high water contents and lower densities. Reduction of density by lowering protein levels and density of bone has chiefly been exploited by deeper-living species. First described by Denton and Marshall (1958) in the mesopelagic gonostomatid *Gonostoma elongatum* (now *Sigmops elongatum*) and bathypelagic alepocephalid *Xenodermichthys copei*, a systematic study of fish chemical composition by Childress and Nygaard (1973) for 37 species in the California borderland revealed that water levels increased dramatically with depth of occurrence (Figure 10.103) with attendant declines in protein, lipid, and skeletal ash (an indirect measure of bone density) as a function of wet mass. Similar results were obtained for 33 species of midwater fishes in the Gulf of Mexico (Stickney and Torres 1989). Water levels ranged from 60 to 91% of wet mass in the CA fishes and 65 to 88% in the GOM for fishes dwelling in the upper 1200 m of the water column.

Increased water levels and the decreased overall density that results (measured densities in *Gonostoma* and *Xenodermichthys* were 1032 and $1039\,kg\,m^{-3}$ respectively) have obvious limitations, the first of which is that the density of seawater can be approached but never equaled, so neutral buoyancy can never be achieved. The second drawback is that increased water levels are accompanied by a less robust musculature and skeletal structure, compromising locomotory abilities. In the light-limited environment of the mesopelagic zone, particularly at depths below 500 m where water contents are highest, the danger of visual predation and advantages of long-distance pursuit are also reduced, lessening the need for a well-developed locomotory apparatus. Clearly, energetic factors are in play here as well, which will be considered in Chapter 12.

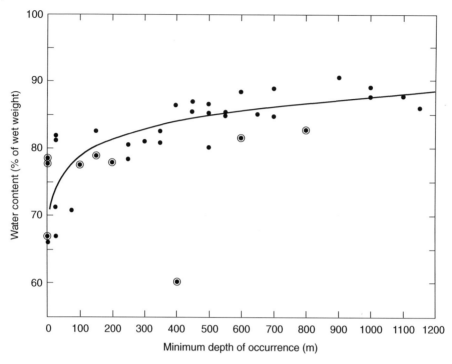

Figure 10.103 Water content as a function of minimum depth of occurrence in a group of midwater fishes. Symbols with circles around them represent species which have well-developed gas-filled swimbladders. The regression line of water content (*W*) as a function of depth (*D*) for fishes without well-developed gas bladders is shown in the figure and is $W = 63.74 D^{0.046 \pm 0.0031}$ (95% confidence; *t*-test). *Source:* Adapted from Childress and Nygaard (1973), figure 1 (p. 1098). Reproduced with the permission of Pergamon Press.

The Nervous System

Anatomy and Basics

The central nervous system (CNS) of fishes is the most advanced and by far the best described of the pelagic taxa addressed thus far. A brief summary of terms (Waterman et al. 1971; Withers 1992) will help in understanding the anatomy and function of the fish nervous system.

Neurons themselves have four structural parts (Figure 10.104a): the dendrites, the cell body, the axon, and the telodendria. The axon and sometimes the dendrites (in sensory neurons) are "insulated" with myelin, a lipid and protein compound that greatly facilitates the speed of neural conduction and is peculiar to the vertebrates. The neural signal effectively jumps between nodes of Ranvier on the axon instead of propagating more slowly along its length. Dendrites convey neural information to the nerve cell body: they are receptive. Axons transmit neural signals from the cell body to their final destination via telodendria, which are short terminal branches of the axon. They are transmissive. The nerve cell body contains the nucleus and cellular organelles.

The telodendria of axons typically end in a chemical synapse which can connect them to a nerve cell body, axon or

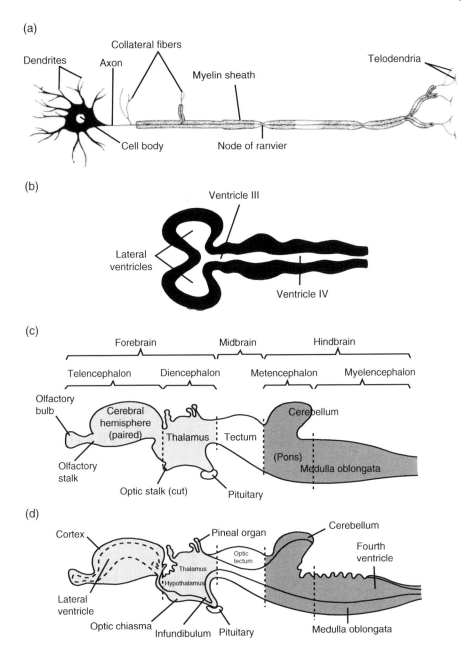

Figure 10.104 Nervous system. (a) Diagram of a motor neuron showing its four major parts; (b) diagram showing the position of the four ventricles within the fish brain. Note that the lateral ventricles are paired; (c) external anatomy of a generalized teleost brain (lateral view); (d) median section of the brain in "(c)" showing internal structures. *Sources:* (a) Adapted from Waterman et al. (1971), p. 423; (b) Adapted from Waterman et al. (1971), p. 439; (c, d) Adapted from Bond et al. (1996), figure 17-2 (p. 245).

dendrite, or a muscle cell. The synapse has a small (c. 200 Å) space between the axon terminus and the target cell termed the synaptic cleft. The neural signal bridges the cleft with neurotransmitters that are released from synaptic vesicles located in the axon terminus. The neurotransmitter (e.g. acetylcholine or serotonin) diffuses across the cleft and depolarizes the target cell.

The central nervous system comprises the brain and spinal cord. The peripheral nervous system comprises the nerves that carry afferent (sensory) input to the CNS and the motor neurons that carry commands from the CNS to the peripheral effectors (e.g. skeletal muscle). Peripheral nerves that connect to the brain are cranial nerves; those that connect with the spinal cord are spinal nerves.

Cranial and spinal nerves may be further subdivided into somatic and visceral types. Somatic (exteroreceptive) sensory neurons relay external sensory information, e.g. taste, touch, and smell. Visceral (interoreceptive) sensory neurons carry information from internal systems such as the gut, glands, and cardiovascular system. Somatic motor neurons lead to the head, limbs, and body wall, including voluntary effectors such as the skeletal musculature. Visceral motor neurons innervate the involuntary internal organs such as the gut. Together, the somatic sensory and motor neurons make up the somatic nervous system. The visceral sensory and motor neurons make up the autonomic nervous system.

Cranial nerves are major arteries of information to and from the brain. Fishes have 10 cranial nerves that serve a variety of functions, correlating with the major subregions of the brain. In early development, the brain arises as a three-lobed structure at the anterior end of the hollow dorsal nerve cord, as is typical of all vertebrates. Later developmental modifications on the same basic plan produce the diversity in brain morphology observed in fishes, supporting their diverse lifestyles.

The Brain

The brain is located in the neurocranium of the skull. It is protected by the skull itself, the membrane (meninges) and cerebrospinal fluid that enclose it, and a matrix of fatty tissue within the cranial cavity (Bond et al. 1996). The three lobes of the embryonic fish brain are the prosencephalon (forebrain), mesencephalon (midbrain), and rhombencephalon (hindbrain). Those three lobes differentiate into five regions (Figure 10.104) each with its own functional responsibilities. Dorsal aspects of shark (*Squalus*) and teleost (*Salmo*) brains are depicted in Figure 10.105a and b; lateral aspects of shark and teleost brains are in Figure 10.105b and c, with labeled cranial nerves. The size of anatomical features within the brain, e.g. the optic lobes or the cerebellum, is dictated by their importance to the lifestyle of the fish (Bond et al. 1996). Thus, strong-swimming visual predators such as the tunas would be expected to have well-developed optic lobes and a large cerebellum to provide the neural integration for vision and motor function.

Four ventricles (hollow spaces) run through the center of the brain (Figure 10.104b). Cerebrospinal fluid (CSF), manufactured in the ventricles, bathes the brain's interior as well as the space surrounding it (Withers 1992). The ventricles continue into the spinal cord as the central canal.

The five distinct regions of the brain begin anteriorly with the telencephalon (forebrain) which contains the olfactory bulbs and is concerned primarily with olfactory senses. Posterior to the telencephalon is the diencephalon ("tween-brain") which contains the pineal organ (pineal gland). The pineal organ is sensitive to light. It appears externally as an unpigmented spot in the skin at the top of the head just posterior to the nares. Its importance varies from species to species and may be useful as a monitor for downwelling light among mesopelagic species such as the hatchetfish *Argyropelecus* (Lagler et al. 1962).

The diencephalon also includes the thalamus, hypothalamus, and pituitary gland. The thalamus and hypothalamus constitute the major processing centers for all sensory neurons except those concerned with olfaction. The pituitary gland is an important regulator within the endocrine system. Finally, the optic nerves enter and cross (optic chiasma) within the diencephalon on their way to the mesencephalon. Together, the telencephalon and diencephalon constitute the cerebrum.

Posterior to the diencephalon is the mesencephalon (mid-brain) which contains the optic tectum of fishes. The optic tectum is visible dorsally as the two optic lobes and is the primary center for visual integration. It is responsible for eye-body coordination (Lagler et al. 1962), connecting visual information with behavioral responses.

Behind the mesencephalon is the metencephalon which constitutes the anterior portion of the hindbrain. Its most prominent feature is the cerebellum, which is responsible for coordinating most "automatic" aspects of swimming including posture, orientation in space, and muscle tone. It does not initiate motor activity but does respond to a variety of sensory inputs (auditory, visual, tactile) coming from other parts of the brain (Withers 1992). In most fishes, it is the largest portion of the brain, particularly in strong swimmers such as the tunas and pelagic sharks. It may also play a role in electroreception as it is well developed in the Mormyrids, well known for their ability to detect prey using electroreception.

The myelencephalon is the posterior part of the hindbrain. It is composed almost entirely of the medulla oblongata (Figure 10.104c and 10.105c). The medulla is an important nerve center, accepting input from all sensory nerves except those of smell and vision. Included are sensory inputs from the acoustico-lateralis system (lateral line). Cranial nerves III–X (see below) arise from the medulla, underscoring its importance as a sensory and motor relay center. Sensory modalities important to the lifestyle of different species show as enlargements in the medulla at the junction of the relevant cranial nerves (Lagler et al. 1962). Besides, its function as a "relay station" to the higher brain areas, some somatic and autonomic functions are governed by the medulla itself including respiratory rhythm and heart rate. The mesencephalon, metencephalon, and myelencephalon are known collectively as the brainstem.

In most of the teleosts, a pair of giant neurons, the Mauthner cells, are found in the medulla at the level of cranial nerve VIII. They receive dendritic input from the cerebellum and optic tectum as well as several of the cranial nerves, and their axons run through the spinal cord to the muscles of the tail. The Mauthner cells are believed to act in a similar way to the giant axons of squids, mediating a rapid escape response (Lagler et al. 1962; Eaton et al. 1977).

Cranial Nerves

Fishes have 11 cranial nerves, 10 of which are designated with roman numerals and one is designated nerve "zero." They serve the same basic functions in fishes as they do in all vertebrates, including man (Waterman et al. 1971).

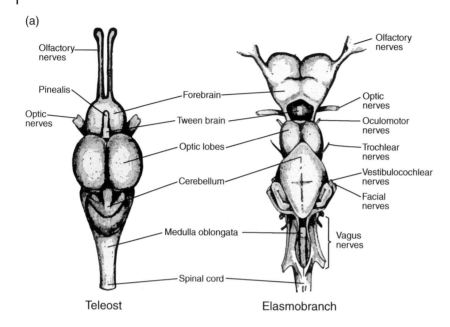
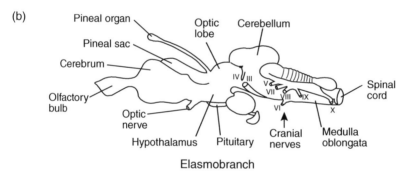
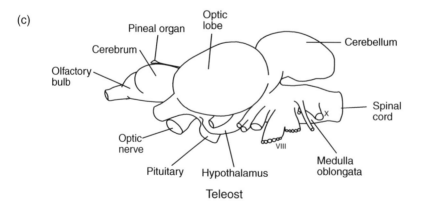

Figure 10.105 Comparison of teleost and elasmobranch brain anatomy. (a) Dorsal views of a teleost (*Salmo*, left) and an elasmobranch (*Squalus*, right) brain; (b) lateral view of the brain of a shark (*Dalatias*); (c) lateral view of the brain of a teleost (*Oncorhynchus*). *Sources:* (a) Adapted from Lagler et al. (1962), figure 3.20 (p. 99); (b, c) Bond et al. (1996), figure 17-1 (p. 243).

i) The olfactory nerve carries sensory input from the olfactory sensory cells to the olfactory bulb in the forebrain.

A small cranial nerve named the terminal nerve and designated "nerve 0" tracks the olfactory nerve but is not connected with the olfactory nerve. It was discovered well after the other cranial nerves were named and hence its non-roman numeral designation. It is believed to be involved with the detection of pheromones and the stimulation of sexual behavior.

ii) The optic nerve is the main sensory nerve of sight. It runs from the retina through the optic chiasma to the optic tectum of the mesencephalon and then to the hypothalamus of the diencephalon.

iii) The oculomotor nerve is a motor nerve that innervates the muscle of the eye.

iv) The trochlear nerve is another motor nerve to the muscles of the eye.

v) The trigeminal nerve is a three-branched nerve that carries sensory information from the head to the metencephalon and motor output to the jaw muscles. It is primarily concerned with the senses of taste and touch.

vi) The abducens nerve is a somatic motor nerve to the lateral rectus muscle of the eye.

vii) The facial nerve is a complex nerve with both sensory and motor functions. It is concerned with sensory inputs from taste and touch receptors from the head and mouth and innervates muscles of the face and hyoid arch.

viii) The acoustic nerve is a sensory nerve from the inner ear.

ix) The glossopharyngeal nerve is a complex nerve with both sensory and motor function. It mainly serves the anterior head region.

x) The Vagus nerve is a highly complex nerve with five branches that provide motor control over most of the branchial muscle and convey sensory input from taste and lateral line receptors. It also has visceral motor and sensory connections to and from the internal organs. It is considered a general visceral efferent nerve and is important in the autonomic nervous system.

The Spinal Cord

The spinal cord is continuous with the medulla oblongata of the hindbrain, leaving the cranial cavity via the foramen magnum. Though considerably less complex than the brain, it is nonetheless a highly sophisticated structure that in most fishes runs the entire length of the cartilaginous or bony vertebral column.

The spinal cord is subdivided into several spinal segments. A bilateral pair of dorsal sensory nerve roots and ventral motor nerve roots emerges from each segment, uniting to form a single spinal nerve on each side (Figure 10.106a, Waterman et al. 1971). In cross section, the spinal cord comprises a small central canal (the remnants of the tubular lumen of the developing embryo), an inner region of grey matter, and an outer region of white matter (Figure 10.106b). The inner grey region consists mainly of unmyelinated nerve cell bodies and their dendrites, giving it its characteristic shade of grey. The outer white matter is made up of myelinated and unmyelinated axons that ascend to or descend from the brain or other segments of the spinal cord. The white shade comes from the myelin insulating the axons.

Sensory Mechanisms

Sensory Modalities

Fishes exhibit a full suite of sensory modalities including:

Photoreception – perception of light, including vision and in some cases the pineal organ.

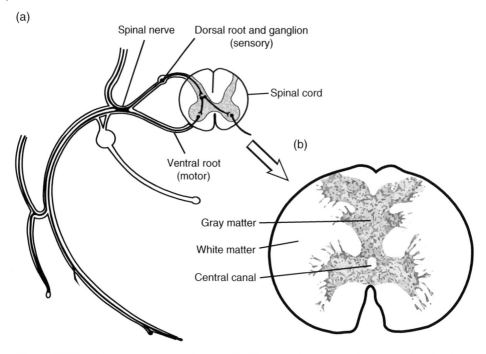

Figure 10.106 Spinal cord and spinal nerves. (a) Diagram of a typical spinal nerve showing dorsal (sensory) and ventral (motor) roots combining to form the spinal nerve; (b) cross section of the spinal cord showing the central canal along with areas of white and gray matter. *Source:* Adapted from Waterman et al. (1971), figures 11-6 and 11-10.

- *Mechano*reception – perception of waterborne vibrations, gravity, and angular acceleration with receptors in the lateral line and inner ear.
- *Electro*reception – perception (and generation) of electric fields
- *Chemo*reception – perception of chemical stimuli (gustation and olfaction)
- *Proprio*ception – awareness of body and limb position, with receptors in the muscle spindles and joints.
- *Magneto*reception – ability to orient to magnetic fields, including the earth's magnetic field for purposes of navigation.

The sensory modality is a property of the receptor itself, but the signal from the receptor to the CNS does not determine the sensory modality: its point of termination in the CNS does. For example, the output from the olfactory epithelium goes to the olfactory lobe where its signal is interpreted as a waterborne chemical, not light or sound. This is known as the "labeled line principle."

Sensory receptors are structured to respond only to the stimuli of their particular modality and may have elaborate accessory structures, such as the lens and cornea of the eye, to facilitate their doing so. When stimulated, the receptor produces an electrical signal termed a "receptor potential." The transformation of a detected stimulus into an electrical signal (receptor potential) is termed "transduction." The receptor potential scales directly with stimulus intensity; that is, it is a graded response. Its electrical signal is transferred to an afferent nerve that delivers the information to the correct area of the brain. The afferent electrical signal is termed a "generator potential."

Photoreception

The rudiments of fish vision, including structure of the retina and lens, were introduced in Chapter 8 in the detailed comparison of cephalopod and fish eyes. What remains to be covered are additional basics of fish vision and adaptations of open-ocean fishes, particularly mesopelagic and bathypelagic fishes and their predators.

The Choroid Body

The retina of teleost fishes is highly demanding of oxygen, yet in most species, it is poorly vascularized, a problem that is solved by a retinal system reminiscent of that in the swimbladder: the choroid rete, also termed the choroid body or choroid gland (Figure 10.107). Located behind the retina, the choroid rete acts as a countercurrent multiplier, generating oxygen partial pressures in the retina and vitreous humor many times that of arterial blood (up to 20 times arterial blood pO_2 depending on the species, Wittenberg and Wittenberg 1974). It is most highly developed in visual predators. The rete utilizes metabolites from the retina to acidify the blood which, due to the Root effect, causes Hb to release oxygen. The released oxygen is then accumulated in the countercurrent exchange of the rete.

Photoreceptors: Rods and Cones

Rods and cones comprise the outermost layer of the retina in the vertebrate eye. Rods and cones contain photoreactive pigments that have a maximum absorbance at a single wavelength of light. The most common photopigment is rhodopsin, which is commonly found in rods that, depending upon the species, have a maximum absorbance of 350–500 nm but are usually about 500 nm (Withers 1992). Rods have a lower threshold for detecting light, providing capable vision in dim light. Cones contain a variety of photopigments, generically termed photopsins, with a variety of absorbance maxima. Shallow-water, freshwater,

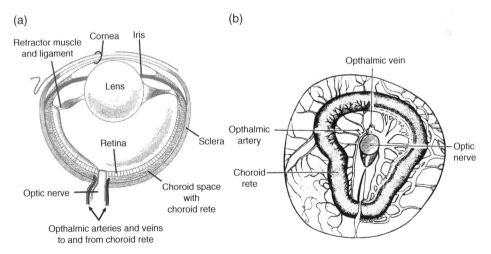

Figure 10.107 The choroid rete. (a) Section through the eye of a teleost showing the choroid rete and its ophthalmic blood supply; (b) the choroid rete of the cod viewed from inside the orbit. The right half of the figure shows the ophthalmic venous sinus and its drainage; on the left the veins have been dissected away to show the arteries. *Sources:* (a) Adapted from Lagler et al. (1962), figure 3.22 (p. 102); (b) Adapted from Bridges et al. (1998), figure 1 (p. 68).

and epipelagic fishes usually have color vision, with duplex retinas containing both rods and cones. In order to see color, at least two populations of cones with two different absorbance maxima are required. Most often, fishes are trichromatic, with three types of cones spanning short, medium, and long wavelengths. Humans also have a trichromatic retina and are able to discriminate 1500 colors (Withers 1992).

Light and Vision

Fishes that encounter daily change in illumination have mechanisms to enhance day and night vision within the retina: retinomotor responses. In a light-adapted retina, cones withdraw from the outer pigmented layer of the retinal epithelium, maximizing their exposure to light (Figure 10.108). At the same time, rods remain within the pigment layer and are shielded from exposure. In dark-adapted eyes, the situation is reversed with rods fully exposed and cones masked by pigment. The masking pigment (melanin) is contained within the cells of the retinal epithelium and migrates toward the photoreceptors in the photopic eye and away in the dark-adapted eye.

The amount of light entering the eye can be adjusted in elasmobranchs and in some teleosts with a contractile iris, but in most teleosts, the pupil is immobile. In deeper-living sharks, which are less likely to experience large changes in light intensity, the ability to adjust pupil diameter is lost as well (Bone and Moore 2008). In those species which are unable to adjust the quantity of incoming light, the retinomotor response is the primary means for protecting the photoreceptors.

In a few deeper-living fishes, e.g. *Argyropelecus*, the spectral quality of light

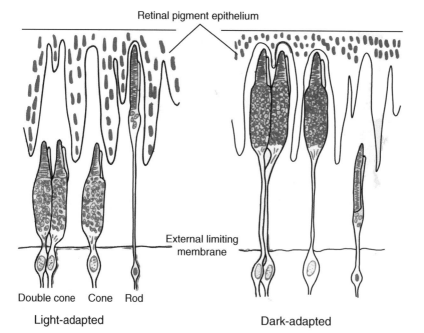

Figure 10.108 Diagram of a light-adapted (left) and a dark-adapted (right) retina of a typical teleost. *Source:* Bone and Moore (2008), figure 10.27 (p. 318). Reproduced with the permission of Taylor and Francis.

entering the eye is adjusted by using a yellow lens (Muntz 1975), which improves the contrast between downwelling ocean light and that emitted by the ventral camouflaging photophores of prey items, likely improving hunting success.

Elasmobranchs and some teleosts have a tapetum lucidum, a reflective layer in the back of the eye (Bond et al. 1996) in either the choroid (elasmobranchs and some mesopelagic teleosts) or in the back of the retina itself. The elasmobranch tapetum is made up of reflective guanine crystals that give them their characteristic green eyeshine when brought on board ship in the dark. The tapetum reflects light passing through the retina back into its photoreceptive layer, improving sensitivity. In elasmobranchs, the tapetum may be occluded with a masking pigment during daylight hours, preventing any eyeshine caused by daylight from making them more conspicuous (Bone and Moore 2008).

Visual and Morphological Adaptations to Deep-Sea Life

The unusual eyes found in many deep-sea fishes are highly functional adaptations to a limited-light environment, as are their shapes, colorations, and photophores.

Tubular Eyes and Retinal Adaptations Given that light intensity is reduced by 90% for each 75 m increase in depth (Denton 1990) and the intensity of downwelling light at any depth in the mesopelagic zone is greatest when looking up, many fishes have upwardly directed eyes, the better to see prey and predators silhouetted against the background light field. Moreover, most of the fishes with upwardly directed eyes have tubular eyes, an adaptational compromise that satisfies (at least partially) the conflicting needs for visual acuity and sensitivity (Locket 1977; Herring 2002).

The lenses of fishes are spherical (Figure 10.109a). Focal length of a spherical lens is about 2½ times its radius. The larger the lens, the more light may enter. However, the bigger the lens, the larger the eye needs to be in order to get a clear image. If the fish retained a spherical eye and increased lens size to accommodate the low light levels at depth, the eyes would be extremely large and take up most of the head. Tubular eyes (Figure 10.109b) allow a larger lens for light gathering and also a clear image by becoming what is essentially the central portion of a normal spherical eye that is laterally reduced. The binocular orientation of tubular eyes further increases sensitivity since both eyes receive light from the same source. They may also provide an advantage in the potential ability to determine distance of objects.

In addition to retinal adaptations to increase the visual field like the accessory retinas of the scopelarchids, several species exhibit retinal adaptations to increase photosensitivity, including changes in the receptors themselves and multibank retinas. First, most deep-sea fishes have all-rod retinas with a rhodopsin absorbing maximally at about 485 nm, close to the wavelength of downwelling light (Douglas et al. 1998) but shifted slightly toward the green to enhance detection of bioluminescence. As noted above, rods are more effective than cones in trapping photons in low-light conditions. The rods of platytroctids and some sternoptychids have exceptionally long outer segments to optimize photon capture, and in some species, such as the deep-living snipe eel, *Serrivomer*, the retina has an extra "tier" of rods between the outer limiting membrane and the pigmented epithelium, forming a multibank retina (Figure 10.109c). Multibank retinas may have up to six tiers of rods (Munk 1966; Locket 1977) and are found in many species

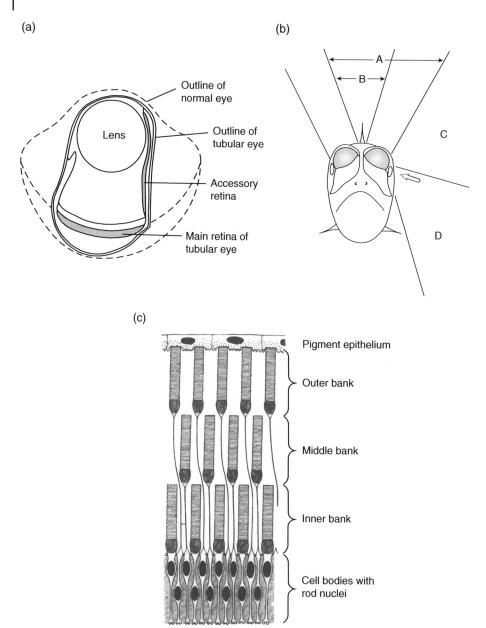

Figure 10.109 Tubular eyes and retinal adaptations. (a) Diagram of the outline of a tubular eye (solid lines) superimposed on the outline of a normal eye (dashed lines). Both contain the same size lens. The main retina of the tubular eye corresponds to the central part of the retina of the normal eye. The reduced visual field of the tubular eye is partly extended by the accessory retina, although it does not receive a focused image; (b) diagram showing the fields of view of the pearleye, *Scopelarchus*, eyes (rostral view). These fields of view are: (a) main retina of one eye (left), (b) overlap of both main retinae giving a dorsal field of binocular vision, (c) accessory retina, and (d) ventral field subserved by dorsal accessory retina. The arrow indicates light passing through the lens pad; (c) diagram of a multibank retina with three tiers of rods. *Sources:* (a) Lockett (1977), figure 12 (p. 101); (b) Lockett (1977), figure 19 (p. 110); (c) Lockett (1977), figure 43 (p. 143). Reproduced with permission of Springer Nature.

of mesopelagic fishes. Any increase in the length of the light path through the eye (in addition to increasing the number of photoreceptors) will increase the probability of photon capture. A tapetum lucidum behind the retina effectively doubles the light path (Herring 2002).

Below 1000 m, visual function tends to decline in fishes. With little or no ambient light available, it is not worth investing the biological currency for increasing specialization (Herring 2002). Many of the bathypelagic fishes, such as the whalefishes and anglerfishes have reduced eyes that are incapable of forming a sharp image but can still function as photometers to detect bioluminescence. In bathypelagic fishes, chemoreceptive and mechanoreceptive senses assume greater importance.

Mechanoreception

The acousticolateralis system of fishes includes the inner ear, the lateral line and head canals, and the superficial neuromasts populating the surface of fishes. The basic mechanoreceptive cell populating all elements of the fish acousticolateralis system is the vertebrate hair cell (Figure 10.110), which is similar in appearance and function to the mechanoreceptive cells of the Crustacea and cephalopods. Hair cells are secondary sense cells, meaning that they are specialized non-neural receptors, in this case designed for detecting water motion. A larger kinocilium and an array (may be a hundred or more, Blaxter 1987) of smaller stereocilia are located at the apex (receptor end) of the cell. As shown in Figure 10.110a, the stereocilia are graded in size and attached to one another at their tips: they are the main players in detecting movement (Bone and Moore 2008). Hair cells are highly polarized; that is, they are preferentially deformed back and forth in one plane. Deflection of the stereocilia in one direction or the other results in depolarization or hyperpolarization and production of an excitatory or inhibitory receptor potential, respectively (Figure 10.110a).

At their base, hair cells are joined by afferent and efferent neurons via chemical synapses, with the afferent neurons carrying information to the brain and the efferent neurons used to modify the sensitivity of the hair cells (Flock 1966). Hair cells are associated in groups to form the free-standing neuromast. Within the neuromast, the hair cell receptors extend into a gelatin-filled wand, the cupula, and are oriented along a common axis (Blaxter 1987) (Figure 10.110b). Neuromasts are found free-standing on the skin of fish, or sub-epidermally in canals on the head and in the lateral line canal along the body.

Sound moving through the water from a point source has a near-field and far-field component, with the former characterized by vibration of the water molecules themselves and the latter by a more orderly propagation of a sound pressure wave as compression and rarefaction. The near-field vibrational region extends to one-sixth the wavelength of the sound (Bone and Moore 2008). Fish mechanoreception is sensitive to both types of aqueous sound propagation, with some division of labor between the free-standing neuromasts, lateral line and head canals (near-field) and the inner ear.

Free-Standing Neuromasts and the Lateral Line
Numbers and locations of free-standing neuromasts vary from species to species. They are believed to be most sensitive to local disturbance and are usually most numerous on the head, with numbers ranging from tens to hundreds (Blaxter 1987; Bone and Moore 2008), with the remainder sprinkled from head to tail over the surface of the fish, including over the caudal fin. In

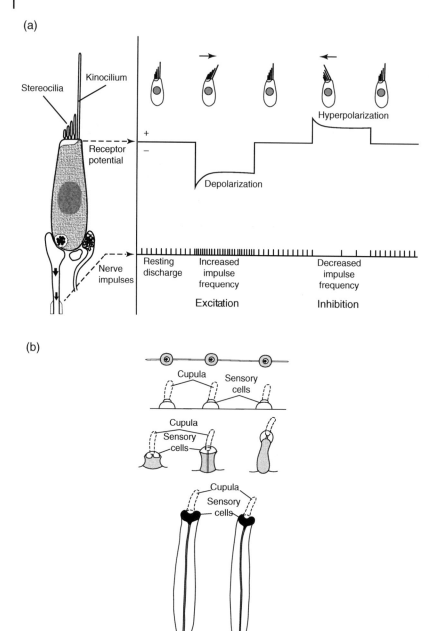

Figure 10.110 Mechanoreception. (a) Neuromast hair cell on the left showing the kinocilium and stereocilia at the apical end and the afferent and efferent nerves at the basal end. The directional properties depend on whether the kinocilium is bent away from or toward the stereocilia by the stimulus (horizontal arrows). If bent away, the hair cell becomes depolarized causing excitation; if bent toward the stereocilia the hair cell becomes hyperpolarized, an inhibitory effect. (b) Free-ending lateral-line organs of three ceratioid anglerfishes: *Dolopichthys* (top) neuromasts on papillae, *Cryptopsaras* (middle) neuromasts on short stalks, and *Neoceratias* neuromasts on long stalks. *Sources:* (a) Adapted from Blaxter (1987), figure 2 (p. 473); (b) Adapted from Marshall (1979), figure 146 (p. 417).

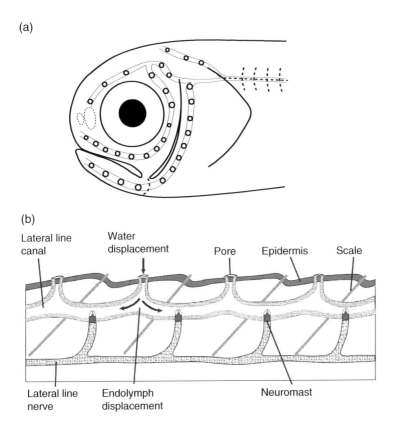

Figure 10.111 Mechanoreception. (a) Lateral-line organs and canals on the head of the lanternfish *Electrona antarctica* (b) diagram of the lateral line canal showing the canal organs and pores. Arrows indicate the direction of water flow into and within the canal.

some species of catfishes, they form patterns in association with the lateral line, presumably to widen the sensory field. Coastal pelagics, notably the clupeids, are well invested with free neuromasts on the head and anterior body. In epipelagic fishes and most coastal species, the cupula of the free neuromasts protrudes slightly or is nearly flush with the skin surface (Marshall 1979).

Deeper-living fishes, particularly bathypelagic groups such as the many families of anglerfishes (Lophiiformes), and the gulper eels and snipe eels (Saccopharyngiformes), have free-standing neuromasts mounted on papillae or stalks, moving them away from the body surface and improving sensitivity.

The gentle water movement typical of the bathypelagic realm minimizes ambient levels of water noise, allowing the stalked neuromasts to function effectively. "Once water is set in motion, the displacement persists for a considerable time. It is surely this aqueous attribute that has prompted the evolution of lateral-line systems that enable users to detect and find the sources of hydrodynamic displacements, whether caused by social partners, prey, or predators" (Marshall 1979). Elegantly stated, and particularly true in the perpetual darkness of the bathypelagic zone.

The subepidermal lateral line canals of fishes include the head canals (Figure 10.111a) and the lateral line itself, running roughly from head to tail down the midline of the fish. The canals are filled with a clear endolymph about equal in consistency with blood plasma or cerebrospinal fluid (Marshall 1979) and communicate with the outside world via a series of pores. The neuromast organs are located within the canals (Figure 10.111b), responding to flow in the surrounding endolymph. In many species, the pores are arranged at intervals of one per scale, either as a tube penetrating the scale itself or capitalizing on the overlap between scales as in the skipjack tuna (Suckling 1967). Lateral lines differ considerably among the fishes, in number, location and morphology, but their mechanoreceptive function is the same.

The endolymph-filled subepidermal canals provide a partial buffer to minimize ambient "noise" generated by currents or by the motion of the fish itself, allowing the fish to identify important mechanoreceptive inputs. Despite any damping effects of the endolymph with the canal, the lateral lines of fish can be exquisitely sensitive to vibrational (near-field) input, as witnessed in the midwater trichiurid *Aphanopus carbo* (700–1300 m – fishbase.org) which can detect its prey at a distance of 32 m (Pumphrey 1950; Bone and Moore 2008). The highly sensitive detection system is reflected in the well-developed acoustico-lateralis lobes of its brain. In most fishes, the maximum distance for near-field detection is about 1 m.

Midwater fishes are usually motionless when first observed in situ, using their operculae or gentle movements of the tail to maintain station, keeping their immediate environment as quiet as possible and maximizing the possibility for hydrodynamic detection of prey. Crustaceans in particular are often negatively buoyant (Childress 1995): they must use their pleopods to keep them moving or they will sink, giving them an obvious acoustic signature. Acoustic and visual cues used together greatly facilitate prey capture and avoidance of predators. Fishes dwelling in the lower mesopelagic and bathypelagic realms, such as the whalefishes and anglerfishes (Figures 10.57, 10.64–10.67, have prominent lateral lines with wide canals to facilitate acoustic reception.

The Inner Ear and Sound Reception
Structure
Unlike terrestrial vertebrates with an outer ear and eardrum for sound reception, fishes have only their inner ears for both hearing and balance. They are located near the brain on either side of the head in regions of the skull known as the otic capsules. Typical of all vertebrate inner ears, they contain a membranous labyrinth with three semicircular canals and otolith organs (Figure 10.112a, b). Both structures contribute to the perception of balance, acceleration, angular momentum, and hearing at different levels depending on species and life style.

Balance, Acceleration, and Angular Momentum
The three semicircular canals are oriented orthogonally, so that motion is detected in the x, y and z planes. Each canal connects to a basal structure: the ampulla. Within the ampulla are hair cells in a gelatinous cupula, similar to the neuromasts within the lateral line. Motion of the fluid in the semicircular canal caused by fish motion bends the cupula, which in turn elicits excitation or inhibition responses from the hair cells depending on the direction of motion. The signals transmitted to the brain are interpreted in terms of balance, acceleration, and angular momentum.

In teleosts, the three otolith organs each contain a single otolith resting atop a

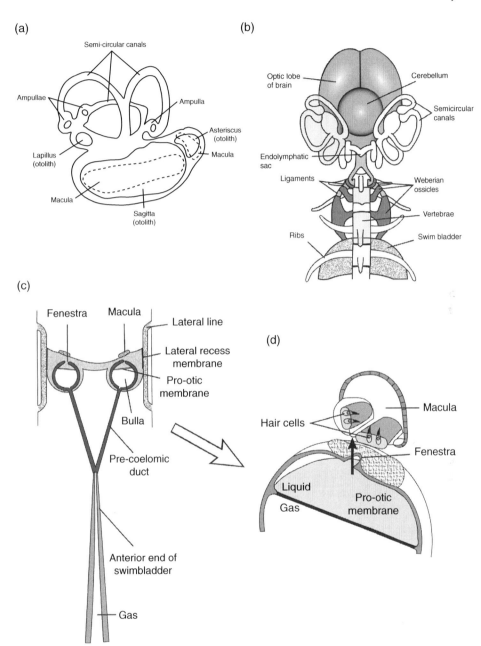

Figure 10.112 Mechanoreception: hearing. (a) Diagram of the inner ear of a typical teleost. The three otoliths and their location within the inner ear are labeled. Two of the maculae upon which the otoliths lie are shown (dashed lines); (b) diagram of the head region of a teleost showing the relationship between the inner ear, the Weberian apparatus, and the swimbladder; (c) diagram of the components of the swimbladder-bulla-lateral line system in clupeid fishes. The parts of the bullae in white contain gas, the shaded parts contain liquid, and the two parts are separated by an elastic pro-otic membrane. An enlargement of the bulla is shown on the right. *Sources:* (a) Adapted from Bone and Moore (2008), figure 10.2 (p. 291); (b) Adapted from Lagler et al. (1962), figure 3.25 (p. 106).

sensory macula composed of hair cells, separated only by the otolith membrane. In elasmobranchs, multiple smaller otoconia perform the same function as the teleost otolith. The otoconia may be formed as calcium concretions by the organism itself (much like the teleost otolith) or, in some bottom-dwellers, may enter as sand grains through the endolymphatic duct, which connects the inner ear to the outside (Lagler et al. 1962). The bed of hair cells within the maculae is polarized in different directions so that any motion of the otolith is conveyed directly to the hair cell and thence to the brain.

Hearing

Two strategies for sound reception are employed by fishes. The first, or direct method, is the most basic, using the difference between the vibration of the otolith and the vibration of the body of the fish to perceive sound. The otolith is about three times denser than the body of the fish and so, when in a sound field, it vibrates with a smaller amplitude and different phase than the body of the fish. Since the otolith is "free-floating" in the endolymph with its only connection to the body of the fish via the otolith membrane, it bends the stereocilia and that motion is perceived as sound (Popper et al. 1988).

The second method employs the swimbladder as a resonator with various levels of structural sophistication, depending on the taxonomic grouping. The key here is that the gas in the swimbladder, even at mesopelagic depths, is far more compressible than water and so is far more sensitive to sound. The sound pressure wave of an acoustic disturbance will cause the walls of the swimbladder to vibrate. These vibrations are then re-radiated to the inner ear where they are interpreted as sound. In two groups of fishes, structural adaptation greatly improves hearing. The fishes in the cohort Otocephala (ear-head), which include the two superorders Clupeomorpha (sardines and herrings) and Ostariophysi (catfishes, carps, characins and knifefishes) both have swimbladder modifications to improve hearing sensitivity. In the Ostariophysi, the swimbladder is connected to the inner ear (Figure 10.112b) by a series of ossicles called the Weberian apparatus, allowing the vibration of the swimbladder to reach the inner ear directly, greatly improving sensitivity and range.

In the clupeids, an even more sophisticated system places anterior extensions of the swimbladder as partially gas-filled bullae (blisters) in direct contact with the inner ear as well as with its lateral recess membrane (Figure 10.112c). Changes in gas pressure within the bullae cause oscillations of the pro-otic membrane, which in turn causes flow of liquid through the fenestra onto the maculae, stimulating the hair cells within it. In addition, the movement of liquid in the bullae is translated into the lateral line canal due to oscillations of the lateral recess membranes. The lateral recess membrane communicates with the entire lateral line system of the fish, giving the clupeids an auditory system that directly stimulates both systems of mechanoreception. The clupeids are the auditory specialists of the fish world with a high sensitivity and broad frequency range. Some representatives are able to detect the echolocation clicks of dolphins, which are in the ultrasound frequency range (Mann et al. 1998, 2001).

Electroreception

Electroreception is another sense absent from the human sensory repertoire, but it is common in elasmobranchs, and it is believed to have originated early in fish evolution (Zakon 1988). The electroreceptors themselves were first described by Lorenzini (1678), but their function was unknown until Kalmijn (1971) showed that they were used by sharks to detect prey. The weak electrical fields produced by nerve and muscle activity are detectable by the ampullae of Lorenzini, which are located in groups on the head and anterior regions of sharks, including pelagic species (Figure 10.113a).

In structure, individual ampullae are modified neuromasts sitting at the bottom of a long, jelly-filled canal Figure 10.113b). In each group of ampullar receptors, the canals lead to pores on the surface of the fish that are sufficiently far apart from one another to allow comparisons of different sites along the body, making them useful in locating prey (Bone and Moore 2008). The canals may be several centimeters in length (Bennet 1971). The jelly within the canals has an ionic concentration equivalent to that of seawater and so is highly conductive, acting like an electrical cable. An external voltage at the electroreceptor pore induces a current flow in the ampulla, which is translated into a generator potential and an afferent sensory signal for integration by the brain. Worthy of note is that the hair cells making up the sensory portions of the ampullae, unlike those of lateral line

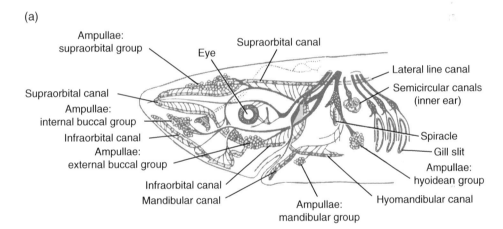

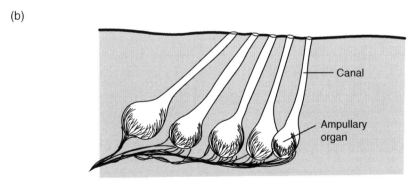

Figure 10.113 Electroreception. (a) Diagram of the distribution of ampullae of Lorenzini and sensory canals on the head of a shark; (b) a group of ampullary organs and sensory canals from the snout of a shark. *Sources:* (a) Lagler et al. (1962), figure 11.23 (p. 390); (b) Adapted from Bone and Moore (2008), figure 10.14 (p. 305).

neuromasts only have afferent innervation. Like the neuromast hair cells, they have a continuous (tonic) afferent output that is altered by detection of their particular sensory stimulus: sound or voltage.

The ampullae are exquisitely sensitive, able to detect voltage gradients of $10^{-8}\,V\,cm^{-1}$ ($0.01\,uV\,cm^{-1}$) (Withers 1992), a major feat, even for a well-equipped neurophysiology lab. Experiments by Kalmijn (1971) showed that dogfish sharks could detect a buried flatfish (plaice: *Pleuronectes platessa*) that was buried 15 cm beneath the surface of a sandy bottom using only their electroreceptors.

Within the teleosts, electroreception is most important to freshwater species living in murky waters such as the mormyrids and gymnotiforms. They have a highly sophisticated electrosensory system (including ampullary and other receptors) that allows them to interrogate the water surrounding them for prey and predators. No electrosensory abilities have been reported for pelagic teleosts, though magnetoreception has been directly demonstrated in rainbow trout (Walker et al. 1997) and doubtless underpins the navigational abilities of tunas and other species (Bone and Moore 2008).

Chemoreception: Olfaction and Gustation
Though easily differentiated in terrestrial species as distance and contact chemoreception, the difference between olfaction (smell) and gustation (taste) in aquatic species is less obvious, since in both it is a direct interaction with waterborne chemicals. Unlike in the Crustacea, where asthetascs on the carapace must serve in both roles, the fishes have a division of labor between the olfactory epithelium of the external nares and gustatory receptors (taste buds) of the mouth and pharynx, and in some cases the skin, gill arches, fins and external barbels, but *not* on the tongue (Bond et al. 1996; Bone and Moore 2008).

Olfaction is important to fishes in prey detection and predator avoidance as well as in reproductive and homing behavior. Gustation is important in food handling, particularly acceptance or rejection of food items. For benthic species with chin barbels such as those in the Ostrariophysi and those species with gustatory receptors on the skin, the gustatory sense can aid in detecting buried and nearby prey.

The olfactory organs in the elasmobranchs and most of the teleosts are paired and are located within external nares (Figure 10.114a). In teleosts, the nares are found on the head, usually anterior and medial to the eyes (Figures 10.75a, and 10.114a, c), whereas in sharks, they are located on the ventral portion of the snout. In both cases, their location allows them to sense directionality of a scent trail as they swim. Water flow through the nares is generated in most pelagic species by forward motion, but in less active species, water motion through the sensory sac can be generated by ciliary activity within it or by activity of facial muscles pumping water through the nares by compressing and relaxing the sensory sac (Bond et al. 1996).

The sensory epithelium of the olfactory organ is composed of lamellae that contain high densities (up to 5×10^5 mm^{-2}, Withers 1992) of receptor cells, some with a high level of specificity. Odorant molecules diffuse into the receptor cells where they bind with receptor proteins. The binding process created an electrical potential which is transmitted through sensory neurons to the olfactory lobe of the brain for interpretation.

Sexual dimorphism exists in the olfactory lamellae of several groups of mesopelagic and bathypelagic fishes. The lamellae can be quite large and take the form of a rosette, as in the male anglerfishes or may be arranged in parallel as in the bathypelagic

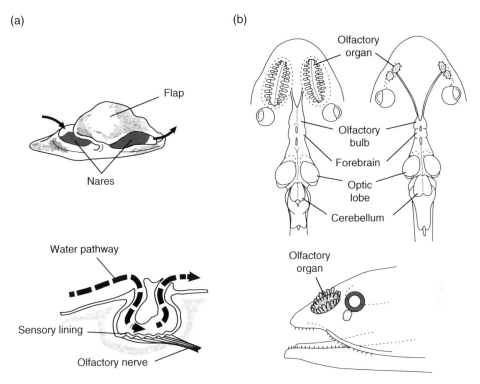

Figure 10.114 Chemoreception. (a) Diagram of the nares of a typical teleost; (b) water pathway and its relation to the sensory lining (olfactory epithelium) supplied by the olfactory nerve; (c) olfactory organs and brains of a male (top left) and female (top right) *Cyclothone microdon*. The bottom figure is the head of a male *C. microdon* showing the macrosmatic olfactory organs. *Sources:* (a, b) Adapted from Lagler et al. (1962), figure 3.21 (p. 101); (c) Marshall (1979), figure 143 (p. 406).

Cyclothone microdon (Figure 10.114c), both of which are considered "macrosmatic." In contrast, the females of most anglers and those of *C. microdon* are microsmatic, with small olfactory organs. A general rule of thumb is "the more lamellae, the more sensitive the olfactory system," with the well-developed sensory epithelia of macrosmatic species complemented by large olfactory lobes in the brain (Figure 10.114c). The sensitive macrosmatic olfactory systems of male bathypelagic species are believed to aid in locating receptive females by following their pheromone trail through the darkness of the bathypelagic zone to its source. Sexual dimorphism in olfactory organs of mesopelagic species is less common, though it is exhibited in mesopelagic *Cyclothone* (e.g. *braueri and signata*) as well as the bathypelagic *C. microdon* (Marshall 1979).

Local concentrations of waterborne chemical stimuli, such as blood from a wounded fish or pheromones released by an ovulating female, persist longer than those in air, particularly in still waters, providing a longer-lasting scent trail as the chemicals dissipate. Sharks in particular, but also most teleosts, are quite capable of following a scent trail to its source using a figure-eight or circular search pattern (Lagler et al. 1962).

Gustatory cells within the taste buds of fishes are similar in function to the olfactory receptors and, in most fishes, are found primarily in the oral cavity, pharynx, and gill arches (Hara 1993). Taste buds themselves are groups of 30–100 receptor cells arranged in a tiny pear-shaped organ (20–50 μm in width, 30–80 μm in height; Sorenson and Caprio 1998) with a pore at its apex. The pore may protrude from, be flush with, or retracted from the surface of the epithelium. As in the olfactory system, stimulus chemicals enter the receptor cells by diffusion and combine with proteins to create afferent signals to the medulla.

Camouflage, Bioluminescence, Photophores

Camouflage

It is hard to hide in a well-lit environment with no structure, like the open-ocean epipelagic zone. However, invertebrates and fishes have developed mechanisms that provide some protection against visual predators. Two of the most prominent methods are transparency and countershading. Except for some larval forms, transparency is not an option for fishes. Instead, they have relied on countershading: a dark dorsum matching the deep-blue color of the ocean below the fish and a silvery underside to match the silvery surface of the ocean above it. By matching their background, countershaded fishes will effectively disappear when viewed by a predator from either above or below. Good examples of countershading include large predators such as dophinfish, tunas, billfishes, and pelagic sharks. Smaller epipelagic fishes such as herrings, sardines, and anchovies which are prey for the large predators also employ countershading. Furthermore, some species of epipelagic fishes such as herrings and anchovies have a knifelike ventral surface to minimize their silhouette when viewed from below. Some mesopelagic species, both migrators (e.g. myctophids) and non-migrators (hatchetfishes), are countershaded. However, the numbers of mesopelagic species with countershading declines with depth of occurrence, with lower mesopelagic and bathypelagic species mostly uniform in color. In addition to countershading, many mesopelagic species utilize photophores to mask their silhouettes.

A recent discovery speaking directly to the importance of camouflage, even in the three-dimensional darkness of the deep-sea, is the ultra-black integument of many deep-sea fishes (Davis and Fielitz 2010). Eighteen species representing eight deep-sea fish families, including vertical migrators as well as mesopelagic and bathypelagic residents, exhibited a near-total (99–99.9%) absorbance of the light impinging on their integument. Species included stomiid predators as well as common prey fishes such as myctophids and Cyclothone. The broad spectrum of species examined suggests that ultra-black integument is the norm and not the exception for deep-sea fishes: one more arrow in the quiver of stealth!

Bioluminescence and Photophores

"The anatomical and physiological expression of bioluminescent abilities reaches its zenith in fishes" (Herring and Morin 1978). A bewildering array of light-producing organs are found in fishes, but they come in two basic types: those that utilize luminescent bacterial symbionts in "culture" and those that utilize a fish's own biochemical system of luciferin and luciferase in light-producing cells, i.e. photophores.

Most midwater fishes have photophores with intrinsic light production that serve a variety of purposes and which vary in their degrees of sophistication. The three numerically dominant families of midwater fishes (Gonostomatidae, Sternoptychidae, and Myctophidae) covered in the taxonomic survey earlier in this chapter provide a good sampler of photophore types and arrays. Two additional families will round out the mix (Stomiidae and Platytroctidae).

The simplest types of photophores are seen in the gonostomatids. They consist of a light-emitting cell (photocyte) in a cup-shaped reflective housing with a simple lens. *Cyclothone* and its close relatives *Gonostoma* and *Sigmops* have these simple photophores arranged serially on the ventrum and along the ventrolateral surface (Figures 10.27 and 10.28). Genera like *Sigmops* and *Margrethia* (Figure 10.28) inhabiting mid-depths in the mesopelagic zone combine half-silvery sides with the ventral and ventrolateral photophores. The mix provides camouflage when viewed from oblique angles or when viewed from directly below. Deeper-living (>500 m) *Cyclothone* such as *Cyclothone microdon* have reduced photophores combined with a completely dark body, a trend also observed in the myctophids.

Myctophids also possess simple cuplike serial photophores. However, myctophid photophores are classified into primary and secondary types with the primary photophores located ventrally and the secondaries, slightly smaller, found throughout the body. The primaries have a lens, and the secondaries do not. The secondary photophores provide counterillumination and camouflage from the side, and their patterns are species-specific (Figures 10.48–10.50; Case et al. 1977).

Luminous tissue is often found on the caudal peduncle of sexually mature lanternfish, often with differences between the sexes in size or location. Those relatively large luminous patches, termed "stern chasers" after the stern-mounted cannons of old fighting ships, are highly variable within the myctophids, with a few genera (e.g., *Tarletonbenia* and *Notoscopelus*) lacking them altogether. Most commonly males have supracaudal organs and females infracaudal, but sometimes the organs are sexually dimorphic in size only, or not sexually dimorphic at all (Herring and Morin 1978). When stimulated, stern chasers flash more brightly than the serial photophores (Barnes and Case 1974), which may point to a predator distracting function as well as the more obvious role of sexual signaling.

The simple cup-like serial photophores of gonostomatids and myctophid are eclipsed by the highly sophisticated photophores of the sternoptychids (Figure 10.115a, b). Like *Sigmops*, sternoptychids also have silvery sides with their ventrum completely covered by photophores (Figure 10.32), but their photophores are tubular and engineered quite differently. The photocytes are contained within highly silvered tubular chambers. In most sternoptychids, the light entering the photophore tube is adjusted for spectral quality by a filter located in the ventral aperture of the tube. The resulting light matches downwelling radiance, not only in spectral quality but also in angular distribution (Denton and Land 1971; Herring 2002). In addition, the photophores also match the intensity of the downwelling through the use of a biological monitoring system. A small (pre-orbital) photophore shines directly into the hatchetfish eye (Figures 10.32 and 10.115b), allowing the fish to directly monitor the output of its photophore array. The fish simply matches the output of the photophore shining into its eye with the intensity of the downwelling daylight. In this way, the fish can assure a continuous match in intensity with changes in depth and time of day (Herring 2002)

Photophores are well vascularized and well invested with nervous tissue, suggesting a plentiful supply of oxygen as well as a high level of neural control (Anctil and Case 1977). Strong neural control is certainly evident in the countershading system of the hatchetfishes described above, and it is present in the serial photophores and stern chasers of myctophids as well (Herring and Morin 1978).

Members of the subfamilies Melanostomiinae and Malacosteinae have a very interesting adaptation to their photophore system: some have suborbital photophores that produce red light. Since red light is well outside the visual spectrum detectable by most mesopelagic species the light is invisible. However, the retinae of the dragonfishes that utilize a red-light photophore (e.g., *Pachystomias*, Figure 10.116a) have complementary visual pigments with absorbance maxima in the red range (Figure 10.116b) (Widder et al. 1984;

Denton et al. 1985; Douglas et al. 1998, 1999). The red light is used to illuminate prey items without them being aware of it. The red light produced by the suborbital photophore is the result of a red fluorescent protein that absorbs the energy of the blue light produced by the photocytes to fluoresce red, and a filter that screens out the shorter wavelengths (Herring 2002). Like most of the other stomiids, *Pachystomias* also has a postorbital photophore that emits in the "normal" blue range (absorption max of 475 nm), which has the greatest penetration in clear seawater and is useful in illuminating prey outside the range of the red photophore.

Luminescent bacteria in culture are best known in the many families of deep-sea anglerfishes, which employ them in the esca of their lures (see Figures 10.64–10.66). Lures of living anglers produce a steady

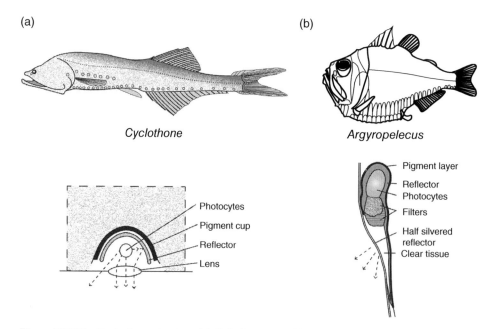

Figure 10.115 Photophore structure. (a) *Cyclothone* sp. and its simple cup-shaped photophore; (b) *Argyropelecus* sp. and its more complex, filtered, and multi-reflector-equipped photophore. *Sources:* (a) photophore adapted from Herring (1985), figure 2 (p. 329); (b) photophore adapted from Denton et al. (1985), figure 2 (p. 67).

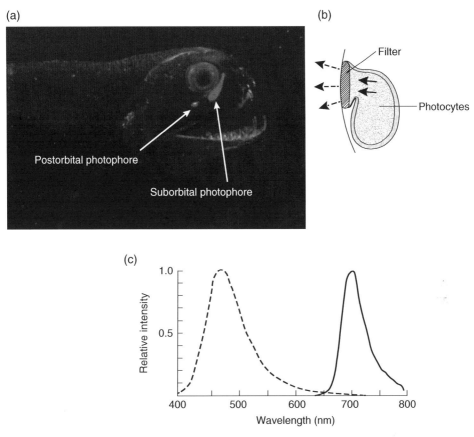

Figure 10.116 Photophores. (a) Close-up of the head of *Pachystomias microdon* showing the blue postorbital and red suborbital photophores; (b) vertical section of the red suborbital photophore of *Malacosteus*. The red light produced by the photocytes is further modified by the filter before leaving the photophore; (c) the resulting spectral emission of the suborbital photophore has a maximum at 708 nm (solid line), very different from the typical blue emission (dotted line) of the postorbital photophore. *Sources:* (a) © Dante Fenolio, reproduced with permission. (b) Denton et al. (1985), figure 8 (p. 74); (c) Adapted from Widder et al. (1984), figure 2 (p. 512). See color plate section for color representation of this figure.

glow, presumably to attract prey to within reach of their jaws, whereupon a remarkably fast lunge is used to engulf them. A dense colony of luminescent bacteria is concentrated in a central cavity within the esca, nourished by the fish's circulation. The glow can be turned on and off to produce pulses or flashes of light, but the mechanism by which the luminescence is turned on and off is unclear (Herring and Morin 1978).

Ventrally directed luminous organs that use bacterial symbionts are found in several species of midwater fishes. Most often the bacteria are housed in a diverticulum of the gut and the generated light is reflected downward with accessory structures, some quite sophisticated. Perhaps, the most famous is that of the opisthoproctid *Opisthoproctus*, which has a rectal luminous organ that shines rostrally into a light guide backed by a reflector that runs the

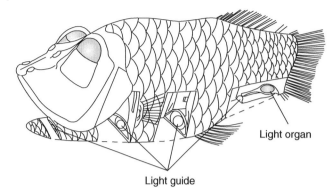

Figure 10.117 Luminous organs. The ventral light-guide system in *Opisthoproctus* shown in section at various positions along the body. *Source*: Herring and Morin (1978), figure 9.6. Reproduced with the permission of Academic Press.

length of the fish (Figure 10.117). The light guide and reflector system direct the luminescence downward into the "sole" (ventral surface) of the fish, resulting in a luminescence "evenly spread over the whole ventral surface of the animal," greatly reducing its silhouette when viewed from beneath (Herring and Morin 1978; Bone and Moore 2008).

The platytroctids (Figure 10.26), or "tubeshoulders," release a cloud of brilliant blue–green luminescence from the tube-like papillae located behind their head when disturbed. The cloud is composed of modified photocytes that form sheets or balls, allowing the cloud to briefly persist (Nicol 1958). The cloud has two likely functions, the first of which is to temporarily blind a predator while making good an escape. The second is the "burglar alarm" theory, whereby the bright cloud advertises the presence of the tubeshoulder and its predator, attracting a second predator to dispatch the first, again allowing the tubeshoulder to escape. Those two strategies apply to the crustacean "spewers," e.g. the oplophorid shrimp, as well.

Bioluminescence is not as widespread in the elasmobranchs as it is in the teleosts, but it is found in several small squaliform sharks in the families Dalatiidae (the kitefin sharks: *Isistius*, *Euprotomicrus* and *Squalilolus*) and Etmopteridae (the lantern sharks: *Etmopterus and Centroscyllium*). In all cases, luminescence is produced by thousands of tiny photophores located on the ventral and ventrolateral aspects of the sharks. Because of their location it is presumed that they have a function in counterillumination and camouflage.

Isisitius brasiliensis, the "cookie-cutter shark" is perhaps the most famous of the small luminescent sharks owing to its curious habit of removing a cookie-shaped plug of flesh from its victims, which include tunas, swordfish and other sharks. Widder (1998) postulated a function for its luminescent display, which completely covers its ventral surface with the exception of a wide, non-luminescent collar, located at the back of the head just behind the jaw. The collar resembles a small fish, acting as bait for upward scanning predators who see only the "collar-fish" silhouetted against the downwelling light. The rest of the little shark is hidden by its countershading luminescence. As the predators swim up to strike the fish, the shark evades and latches onto the body of the predator as it goes by, using its powerful suction and cookie-cutter teeth to remove a plug of tissue from the predator.

References

Aldridge, R.J. and Briggs, D.E.G. (1989). A soft body of evidence. *Natural History*, 6–11 May.

Alexander, R.M. (1993). Buoyancy. In: *The Physiology of Fishes* (ed. D.H. Evans). Boca Raton: CRC Press.

Alexander, R.M. (2003). *Principles of Animal Locomotion*. Princeton: Princeton University Press.

Anctil, M. and Case, J.F. (1977). The caudal luminous organs of lanternfishes: general innervation and ultrastructure. *American Journal of Anatomy* 149: 1–22.

Andriashev, A.P. (1953). *Ancient Deep-Water and Secondary Deep-Water Fishes and Their Importance in Zoogeographical Analysis. Notes of Special Problems in Ichthyology. Moscow, Leningrad: Akademie, Nauk, SSSR, Ikhiol.Kom, Fish and Wildlife Translation Series 6*. Washington, DC: U.S. National Museum.

Applegate, V.C. and Moffett, J.W. (1955). The Sea Lamprey. *Scientific American* 192: 36–41.

Astor, Y., Muller-Karger, F., and Scranton, M.I. (2003). Seasonal and interannual variation in the hydrography of the Cariaco Basin: implications for basin ventilation. *Continental Shelf Research* 23: 125–144.

Badcock, J. (1970). The vertical distribution of mesopelagic fishes collected on the SOND cruise. *Journal of the Marine Biological Association of the United Kingdom* 50: 1001–1044.

Badcock, J. and Merrett, N.R. (1976). Midwater fishes in the eastern North Atlantic - I. Vertical distribution and associated biology with developmental notes on certain myctophids. *Progress in Oceanography* 7: 3–58.

Baird, R.C. and Hopkins, T.L. (1981a). Trophodynamics of the fish *Valencienellus tripunctulatus*. II. Selectivity, grazing rates and resource utilization. *Marine Ecology Progress Series* 5: 11–19.

Baird, R.C. and Hopkins, T.L. (1981b). Trophodynamics of the fish *Valencienellus tripunctulatus*. III. Energetics, resources and feeding strategy. *Marine Ecology Progress Series* 5: 21–28.

Baird, R.C., Milliken, D.M., and Wilson, D.F. (1973). Observations on *Bregmaceros nectabanus* Whitley in the anoxic water of the Cariaco Trench. *Deep-Sea Research* 20: 503–504.

Baird, R.C., Wilson, D.F., Beckett, R.C., and Hopkins, T.L. (1974). *Diaphus taaningi* Norman, the principal component of a shallow sound-scattering layer in the Cariaco Trench, Venezuela. *Journal of Marine Research* 34: 301–312.

Balbontin, F.S., Desilva, S., and Ehrlich, K.F. (1973). A comparative study of anatomical and chemical characteristics of reared and wild herring. *Aquaculture, Amsterdam* 2: 217–240.

Barham, E.G. (1971). Deep-sea fishes: lethargy and vertical orientation. In: *Proceedings of an International Symposium on Biological Sound Scattering in the Ocean* (ed. G.B. Farquhar). Washington, DC: Maury Center for Ocean Science, Department of the Navy.

Barnes, A.T. and Case, J.F. (1974). The luminescence of lanternfish (Myctophidae): spontaneous activity and responses to mechanical, electrical and chemical stimulation. *Journal of Experimental Marine Biology and Ecology* 15: 203–221.

Baum, J.K. and Myers, R.A. (2004). Shifting baselines and the decline of pelagic sharks in the Gulf of Mexico. *Ecology Letters* 7: 135–145.

Baum, J.K., Kehler, D., and Myers, R.A. (2005). Robust estimates of decline for pelagic shark populations in the northwest Atlantic and Gulf of Mexico. *Fisheries* 30: 27–30.

Beamish, R.J., Leask, K.D., Ivanov, O.A. et al. (1999). The ecology, distribution and abundance of midwater fishes of the Subarctic Pacific gyres. *Progress in Oceanography* 43: 399–442.

Belman, B.W. and Anderson, M.E. (1979). Aquarium observations on feeding by *Melanostigma pammelas* (Pisces: Zoarcidae). *Copeia* 1979: 366–369.

Belman, B.W. and Gordon, M.S. (1979). Comparative studies on the metabolism of shallow-water and deep-sea marine fishes. V. Effects of temperature and hydrostatic pressure in the mesopelagic zoarcid *Melanostigma pammelas*. *Marine Biology* 50: 275–281.

Bennet, M.V.L. (1971). Electroreception. In: *Fish Physiology. Volume 5. Sensory Systems and Electric Organs* (eds. W.S. Hoar and D.J. Randall). London: Academic Press.

Bertelsen, E. (1951). The ceratioid fishes. Ontogeny, taxonomy, distribution and biology. *Dana Report* 39: 1–276.

Beverton, R.J. and Holt, S.J. (1959). A review of the lifespans and mortality rates of fishes in nature and their relationship to growth and other physical characteristics. In: *The Lifespan of Animals*, CIBA Foundation Colloquium on Aging no. 5 (eds. G.E.W. Wolstenholme and M. O'connor). London: Churchill.

Bigelow, H.B. and Schroeder, P.C. (1948). Sharks. In: *Fishes of the Western North Atlantic. Number One. Part One* (eds. J. Tee-Van, C.M. Breder, S.F. Hildebrand, et al.). New Haven: Sears Foundation for Marine Research, Yale University.

Bigelow, H.B. and Schroeder, W.C. (1953). Chimaeroids. In: *Fishes of the Western North Atlantic, Part 2* (eds. H.B. Bigelow and W.C. Schroeder). New Haven: Sears Foundation for Marine Research, Yale University.

Bishop, R.E. and Torres, J.J. (1999). Leptocephalus energetics: metabolism and excretion. *Journal of Experimental Biology* 202: 2485–2493.

Bishop, R.E. and Torres, J.J. (2001). Leptocephalus energetics: assembly of the energetics equation. *Marine Biology* 138: 1093–1098.

Bishop, R.E., Torres, J.J., and Crabtree, R.E. (2000). Leptocephalus energetics: chemical composition and growth indices. *Marine Biology* 137: 205–214.

Blaxter, J.H.S. (1987). Structure and development of the lateral line. *Biological Reviews* 62: 471–514.

Block, B.A. (1986). Structure of the brain and eye heater tissue in marlins, sailfish and spearfish. *Journal of Morphology* 190: 169–189.

Block, B.A. (1991). Evolutionary novelties: how fish have built a heater out of muscle. *American Zoologist* 31: 726–742.

Block, B.A., Dewar, H., Blackwell, S.B. et al. (2001). Migratory movements, depth preferences and thermal biology of Atlantic Bluefin tuna. *Science* 293: 1310–1314.

Bond, C.E., Smith, G., Gharrett, A. et al. (1996). *Biology of Fishes*, 2e. United States: Thomson Learning Inc.

Bone, Q. (1966). On the function of the two types of muscle fibres in elasmobranch fish. *Journal of the Marine Biological Association of the United Kingdom* 46: 321–349.

Bone, Q. (1978). Locomotor muscle. In: *Fish Physiology Volume 7* (eds. W.S. Hoar and D.J. Randall). New York: Academic Press.

Bone, Q. and Moore, R.H. (2008). *Biology of Fishes*, 3e. New York: Taylor & Francis.

Bone, Q., Marshall, N.B., and Blaxter, J.H.S. (1995). *Biology of Fishes*, 2e. London: Blackie Academic and Professional.

Bonfil, R.A.O. (2005). Transoceanic migration, spatial dynamics and population linkages of white sharks. *Science* 310: 100–103.

Brainerd, E.L. (2001). Caught in the crossflow. *Nature* 412: 387–388.

Branstetter, S. (1990). Early life-history implications of selected carcharinoid and lamnoid sharks of the Northwest Atlantic. In: *Elasmobranchs as Living Resources: Advances in the Biology, Ecology, Systematics and the Status of the Fisheries* (eds. H.L. Pratt, S.H. Gruber and T. Taniuchi). Washington, DC: U.S. Department of Commerce.

Breder, C.M.J. (1926). The locomotion of fishes. *Zoologica* 4: 159–297.

Bridges, C.R., Berenbrink, M., Muller, R., and Waser, W. (1998). Physiology and biochemistry of the pseudobranch: an unanswered question? *Comparative Biochemistry and Physiology* 119A: 67–77.

Briggs, J.C. (1974). *Marine Zoogeography*. New York: McGraw-Hill.

Brightman, R.I., Torres, J.J., Donnelly, J., and Clarke, M.E. (1996). Energetics of premetamorphic larval red drum. Part II. Growth and biochemical indicators. *Fishery Bulletin US* 95: 431–444.

Brill, R.W. and Bushnell, P.G. (2001). The cardiovascular system of tunas. In: *Tuna: Physiology, Ecology and Evolution* (eds. B.A. Block and E.D. Stevens). San Diego: Academic Press.

Brusca, R.C., Moore, W., and Shuster, S.M. (2016). *Invertebrates*, 3e. Sunderland: Sinauer Associates.

Buddington, R.K. and Diamond, J.M. (1986). Aristotle revisited: the function of pyloric caeca in fish. *Proceedings of the National Academy of Science of the USA* 83: 8012–8014.

Burger, J.W. and Hess, W.N. (1960). Function of the rectal gland in the spiny dogfish. *Science* 131: 670–671.

Burgess, G.H., Beekircher, L.R., Cailliet, G.M. et al. (2005). Is the collapse of shark populations in the Northwest Atlantic Ocean and Gulf of Mexico real? *Fisheries* 30: 19–26.

Burghart, S.E., Hopkins, T.L., and Torres, J.J. (2010). Partitioning of food resources in bathypelagic micronekton in the eastern Gulf of Mexico. *Marine Ecology Progress Series* 399: 131–140.

Cailliet, G.M. and Goldman, K.J. (2004). Age determination and validation in Chondrichthyan fishes. In: *Biology of Sharks and Their Relatives* (eds. J.C. Carrier, J.A. Musick and M.R. Heithaus). Boca Raton: CRC Press.

Cailliet, G.M., Love, M.S., and Ebeling, A.W. (1986). *Fishes. A Field and Laboratory Manual on Their Structure, Identification and Natural History*. Prospect Heights: Waveland Press.

Carey, F.G. (1982). A brain heater in the swordfish. *Science* 216: 1327–1329.

Carey, F.G. and Lawson, K.D. (1973). Temperature regulation in free-swimming Bluefin tuna. *Comparative Biochemistry and Physiology* 44A: 375–392.

Carey, F.G. and Teal, J.M. (1966). Heat conservation in tuna fish muscle. *Proceedings of the National Academy of Sciences of the USA* 56: 1464–1469.

Carey, F.G., Teal, J.M., Kanwisher, J.W. et al. (1971). Warm-bodied fish. *American Zoologist* 11: 137–145.

Case, J.F., Warner, J., Barnes, A.T., and Lowenstine, M. (1977). Bioluminescence of lanternfish (Myctophidae) in response to changes in light intensity. *Nature* 265: 179–181.

Castonguay, M. (1987). Growth of American and European eel larvae as revealed by otolith microstructure. *Canadian Journal of Zoology* 65: 875–878.

Cetta, C.M. and Capuzzo, J.M. (1982). Physiological and biochemical aspects of embryonic and larval development of the winter flounder *Pseudopleuronectes americanus*. *Marine Biology* 71: 327–337.

Chapman, A.D. (2009). *Numbers of Living Species in Australia and the World*, 2e. Canberra: Australian Biological Resources Study.

Checkley, D.M.A.O. (2009). Habitats. In: *Climate Change and Small Pelagic Fish* (eds. D.M. Checkley, J. Alheit, Y. Oozeki and C. Roy). Cambridge: Cambridge University Press.

Childress, J.J. (1995). Are there physiological and biochemical adaptations of metabolism in deep-sea animals? *Trends in Ecology and Evolution* 10: 30–36.

Childress, J.J. and Nygaard, M.H. (1973). The chemical composition of midwater fishes as a function of depth of occurrence off southern California. *Deep-Sea Research* 20: 1093–1109.

Childress, J.J., Price, M.H., Favuzzi, J., and Cowles, D.L. (1990). Chemical composition of midwater fishes as a function of depth of occurrence off the Hawaiian Islands: food availability as a selective factor? *Marine Biology* 105: 235–246.

Clarke, T.A. (1973). Some aspects of the ecology of lanternfishes (Myctophidae) in the Pacific Ocean near Hawaii. *Fishery Bulletin US* 71: 401–434.

Clarke, T.A. (1974). Some aspects of the ecology of stomiatoid fishes in the Pacific Ocean near Hawaii. *Fishery Bulletin US* 72: 337–351.

Clarke, T.A. (1978). Diel feeding patterns of 16 species of mesopelagic fishes from Hawaiian waters. *Fishery Bulletin U.S.* 76: 495–513.

Clarke, T.A. (1982). Feeding habits of stomiatoid fishes from Hawaiian waters. *Fishery Bulletin US* 80: 287–304.

Clarke, T.A. and Wagner, P.J. (1976). Vertical distribution and other aspects of the ecology of certain mesopelagic fishes taken near Hawaii. *Fishery Bulletin U.S.* 74: 635–645.

Collette, B.B. and Nauen, C.E. (1983). *Scombrids of the World. An Annotated and Illustrated Catalogue of Tunas, Mackerels, Bonitos and Related Species*. Rome: Food and Agriculture Organization of the United Nations.

Compagno, L.J.V. (1984). *FAO Species Catalogue. Volume 4. Sharks of the World. An Annotated and Illustrated Catalogue of Shark Species Known to Date*. Rome: Food and agricultural organization of the United Nations.

Compagno, L.J.V., Gottfried, M.D., and Bowman, S.C. (1993). Size and scaling of the giant "megatooth" shark, *Carcharodon megalodon* (Lamnidae). *Journal of Vertebrate Paleontology* 13: 31A.

Corner, E.D.S., Denton, E.J., and Forster, G.R. (1969). On the buoyancy of some deep-sea sharks. *Proceedings of the Royal Society of London B* 171: 415–429.

Courtenay-Latimer, M. (1979). My Story of the First Coelacanth. *Occasional Papers of the California Academy of Sciences, San Francisco* 134: 6–10.

Cowles, D.L. and Childress, J.J. (1995). Aerobic metabolism of the anglerfish *Melanocetus johnsoni*, a deep-pelagic marine sit-and-wait predator. *Deep-Sea Research I* 42: 1631–1638.

Crabtree, R.E. (1995). Chemical composition and energy content of deep-sea demersal fishes from tropical and temperate regions of the Western North Atlantic. *Bulletin of Marine Science* 56: 434–449.

Crabtree, R.E. and Sulak, K.J. (1986). A contribution to the life history and distribution of Atlantic species of the deep-sea fish genus Conocara. *Deep-Sea Research* 33: 1183–1201.

Crabtree, R.E., Cyr, E., Bishop, R.E. et al. (1992). Age and growth of tarpon, *Megalops atlanticus*, larvae in the eastern Gulf of Mexico, with notes on relative abundance and probably spawning areas. *Environmental Biology of Fishes* 35: 361–370.

Dalpadado, P. (1988). Reproductive biology of the lanternfish *Benthosema pterotum* from the Indian Ocean. *Marine Biology* 98: 307–316.

Davis, M.P. and Fielitz, C. (2010). Estimating divergence times of lizardfishes and their allies (Euteleostei: Aulopiformes) and the timing of deep-sea adaptations. *Molecular Phylogenetics and Evolution* 57: 1194–1208.

Dean, B. (1906). *Chimaeroid Fishes and Their Development, Publications of the Carnegie Institution 32*. Washington, DC: Carnegie Institution of Washington.

Degnan, K.J., Karnaky, K.J., and Zadunaisky, J.A. (1977). Active chloride transport in the in vitro opercular skin of a teleost (*Fundulus heteroclitus*). *Journal of Physiology* 271: 155–191.

Denton, E.J. (1990). Light and vision at depths greater than 200 metres. In: *Light and Life in the Sea* (eds. P.J. Herring, A.K. Campbell, M. Whitfielld and L. Maddock). Cambridge: Cambridge University Press.

Denton, E.J. and Land, M.F. (1971). Mechanism of reflexion in silvery layers of fish and cephalopods. *Proceedings of the Royal Society of London A* 178: 43–61.

Denton, E.J. and Marshall, N.B. (1958). The buoyancy of bathypelagic fishes without a gas-filled swimbladder. *Journal of the Marine Biological Association of the United Kingdom* 37: 753–767.

Denton, E.J., Herring, P.J., Widder, E.A. et al. (1985). The role of filters in the photophores of oceanic animals and their relation to vision in the oceanic environment. *Proceedings of the Royal Society of London B* 225: 63–97.

Donnelly, J. and Torres, J.J. (2008). Pelagic fishes in the Marguerite Bay region of the Western Antarctic Peninsula Shelf. *Deep-Sea Research, II* 55: 523–539.

Donoghue, P.C.J., Forey, P.L., and Aldridge, R.J. (2000). Conodont affinity and chordate phylogeny. *Biological Reviews* 75: 191–251.

Douglas, R.H., Partridge, J.C., Dulai, K.S. et al. (1999). Enhanced retinal longwave sensitivity using a chlorophyll-derived photosensitiser in *Malacosteus niger*, a de ep-sea dragon fish with far red bioluminescence. *Vision Research* 39: 2817–2832.

Douglas, R.H., Partridge, J.C., and Marshall, N.J. (1998). The eyes of deep-sea fish I: lens pigmentation, tapeta and visual pigments. *Progress in Retinal and Eye Research* 17: 597–636.

Drazen, J.C. and Sutton, T.T. (2017). Dining in the deep: the feeding ecology of deep-sea fishes. *Annual Review of Marine Science* 9: 337–366.

Eastman, J.T. (1998). Lipid storage systems and the biology of two neutrally buoyant Antarctic notothenioid fishes. *Comparative Biochemistry and Physiology* 90B: 529–537.

Eaton, R.C., Bombarderi, R.A., and Meyer, D.L. (1977). The Mauthner initiated startle response in teleost fish. *Journal of Experimental Biology* 66: 65–81.

Ebeling, A.W. (1962). Melamphaidae, I. Systematics and zoogeography of the species in the bathypelagic fish genus *Melamphaes* Gunther. *Dana Report* 58: 1–164.

Ebeling, A.W. and Weed, W.H.I. (1963). Systematics and distribution of the species in the bathypelagic fish genus Scopelogadus Vaillant. *Dana Report* 60: 1–58.

Eckert, R., Randall, D.J., and Augustine, G. (1988). *Animal Physiology. Mechanisms and Adaptations*, 3e. New York: W.H. Freeman and Company.

Ehrlich, K.F. (1974a). Chemical changes during growth and starvation of larval *Pleuronectes platessa*. *Marine Biology* 24: 39–48.

Ehrlich, K.F. (1974b). Chemical changes during growth and starvation of herring larvae. In: *The Early Life History of Fish* (ed. J.H.S. Blaxter). Heidelberg: Springer-Verlag.

Evans, D.H. (1979). Fish. In: *Comparative Physiology of Osmoregulation in Animals* (ed. G.M.O. Maloiy). Orlando: Academic Press.

Fitch, J.E. and Lavenberg, R.J. (1968). *Deep-Water Fishes of California*. Berkeley: University of California Press.

Flock, A. (1966). Transducing mechanisms in the lateral line canal organ receptors. *Cold Spring Harbor Symposia on Quantitative Biology: Proceedings* 30: 133–145.

Florey, E. (1966). *General and Comparative Animal Physiology*. Philadelphia: W.B. Saunders and Co.

Forey, P.L. (1998). *History of the Coelacanth Fishes*. London: Chapman and Hall.

Frost, B.W. and Mccrone, L.E. (1979). Vertical distribution, diel vertical migration, and abundance of some mesopelagic fishes in the eastern Subarctic Pacific Ocean in summer. *Fishery Bulletin US* 76: 751–770.

Forster, R.P., Goldstein, L., and Rosen, J.K. (1972). Intrarenal control of urea reabsorption by renal tubules of the marine elasmobranch, *Squalus acanthias*. *Comparative Biochemistry and Physiology* 42A: 3–12.

Gartner, J.V. (1991a). Life histories of three species of lanternfishes (Pisces: Myctophidae) from the eastern Gulf of Mexico. I. Morphological and microstructural analysis of sagittal otoliths. *Marine Biology* 111: 11–20.

Gartner, J.V. (1991b). Life histories of three species of lanternfishes (Pisces: Myctophidae) from the eastern Gulf of Mexico. II. Age and growth patterns. *Marine Biology* 111: 21–27.

Gartner, J.V. (1993). Patterns of reproduction in the dominant lanternfish species (Pisces: Myctophidae) of the eastern Gulf of Mexico, with a review of reproduction among tropical-subtropical Myctophidae. *Bulletin of Marine Science* 52: 721–750.

Gartner, J.V., Hopkins, T.L., Baird, R.C., and Milliken, D.M. (1987). Ecology of the lanternfishes (Pisces: Myctophidae) of the eastern Gulf of Mexico. *Fishery Bulletin U.S.* 85: 81–98.

Gaskett, A.C., Bulman, C., He, X., and Goldsworthy, S.D. (2001). Diet composition and guild structure of mesopelagic and bathypelagic fishes near Macquarie Island, Australia. *New Zealand Journal of Marine and Freshwater Research* 35: 469–476.

Gerringer, M.E., Andrews, A.H., Huss, G.R. et al. (2018). Life history of abyssal and hadal fishes from otolith growth zones and oxygen isotopic compositions. *Deep-Sea Research I* 132: 37–50.

Gibbs, R.H. and Krueger, W.H. (1987). Biology of midwater fishes of the Bermuda Ocean Acre. *Smithsonian Contributions to Zoology* 452: 1–187.

Gibbs, R.H. and Roper, C.F.E. (1970). Ocean acre preliminary report on vertical distribution of fishes and cephalopods. In: *Proceedings of an International Symposium on Biological Sound Scattering in the Ocean* (ed. G.B. Farquhar). Washington, DC: Department of the Navy.

Gjøsaeter, J. (1984). Mesopelagic fish: a large potential resource in the Arabian Sea. *Deep-Sea Research* 31: 1019–1035.

Gjøsaeter, J. (1987). Primary growth increments in otoliths of six tropical myctophid species. *Biological Oceanography* 4: 359–382.

Gjøsaeter, J. and Kawaguchi, K. (1980). *A Review of the World Resources of Mesopelagic Fish*. Rome: Food and Agricultural Organization of the United Nations.

Gjøsaeter, J. and Tilseth, S. (1988). Spawning behavior, egg and larval development of the myctophid fish *Benthosema pterotum*. *Marine Biology* 98: 1–6.

Graae, M.J.F. (1967). Lestidium bigelowi, a new species of paralepidid fish with photophores. *Breviora* 277: 1–10.

Graham, J.B. and Dickson, K.A. (2001). Anatomical and physiological specializations for endothermy. In: *Tuna Physiology, Ecology and Evolution* (eds. B.A. Block and E.D. Stevens). San Diego: Academic Press.

Gray, I.E. (1954). A comparative study of the gill area of marine fishes. *Biological Bulletin* 107: 219–225.

Gray, I.E. (1957). A comparative study of the gill area of crabs. *Biological Bulletin* 112: 34–42.

Greely, T.M., Gartner, J.V., and Torres, J.J. (1999). Age and growth of *Electrona antarctica* (Pisces: Myctophidae) the dominant mesopelagic fish of the Southern Ocean. *Marine Biology* 133: 145–158.

Greenwood, P.H., Rosen, D.E., Weitzman, S.H., and Myers, G.S. (1966). Phyletic studies of teleostean fishes, with a provisional classification of living forms. *Bulletin of the American Museum of Natural History* 131: 339–456.

Grigg, G.C. (1969). Temperature induced changes in the oxygen equilibrium curve of the blood of the brown bullhead, Ictalurus nebulosus. *Comparative Biochemistry and Physiology* 28: 1203–1223.

Guinot, G., Adnet, S., Cavin, L., and Cappetta, H. (2013). Cretaceous stem chondrichthyans survived the end-Permian mass extinction. *Nature Communications* 4: 2669.

Gunn, J. and Block, B.A. (2001). Advances in acoustic, archival and satellite tagging of tunas. In: *Fish Physiology* (eds. B.A. Block and E.D. Stevens). San Diego: Academic Press.

Haedrich, R.L. (1964). Food habits and young stages of North Atlantic Alepisaurus (Pisces, Iniomi). *Breviora* 201: 1–15.

Haedrich, R.L. and Merrett, N.R. (1988). Summary atlas of deep-living demersal fishes in the North Atlantic Basin. *Journal of Natural History* 22: 1325–1362.

Hansen, K. and Herring, P.J. (1977). Dual luminescent systems in the anglerfish genus Linophryne (Pisces: Ceratioidea). *Journal of Zoology, London* 182: 103–124.

Hara, T.J. (1993). Chemoreception. In: *The Physiology of Fishes* (ed. D.H. Evans). Boca Raton: CRC Press.

Harbison, G.R. and Janssen, J. (1987). Encounters with a swordfish (*Xiphias gladius*) and sharptail mola (*Masturus lanceolatus*) at depths greater than 600 m. *Copeia* 1987: 511–513.

Harold, A.S., Hartel, K.E., Craddock, J.E., and Moore, J.A. (2002). Hatchetfishes and relatives. Family Sternoptychidae. In: *Bigelow and Schoeder's Fishes of the Gulf of Maine* (eds. B.B. Collette and G. Klein-Macphee). Washington, DC: Smithsonian Institution Press.

Hartree, W. and Hill, A.V. (1921). The nature of the isometric twitch. *Journal of Physiology* 55: 389–411.

Heimberg, A.M., Cowper-Salari, R., Sémon, M. et al. (2010). MicroRNAs reveal the interrelationships of hagfish, lampreys, and gnathostomes and the nature of the ancestral vertebrate. *Proceedings of the National Academy of Science of the USA* 107: 19379–19383.

Helfman, G.S., Collette, B.B., and Facey, D.E. (1997). *The Diversity of Fishes*. Oxford: Blackwell.

Helfman, G.S., Collette, B.B., Facey, D.E., and Bowen, B.B. (2009). *The Diversity of Fishes*. Oxford: Wiley-Blackwell.

Herring, P.J. (1985). Bioluminescence in the Crustacea. *Journal of Crustacean Biology* 5: 557–573.

Herring, P.J. (2000). Species abundance, sexual encounter and bioluminescent signaling in the deep-sea. *Philosophical Transactions of the Royal Society of London* 355B: 1273–1276.

Herring, P.J. (2002). *The Biology of the Deep Ocean*. New York: Oxford University Press.

Herring, P.J. and Morin, J.G. (1978). Bioluminescence in fishes. In: *Bioluminescence in Action* (ed. P.J. Herring). London: Academic Press.

Hochachka, P.W. and Somero, G.N. (1973). *Strategies of Biochemical Adaptation*. Philadelphia: Saunders.

Hochachka, P.W. and Somero, G.N. (1984). *Biochemical Adaptation*. Princeton: Princeton University Press.

Holland, P.W.H. (2006). My sister is a seasquirt. *Heredity* 30: 1–2.

Holland, N.D., Holland, L.Z., and Holland, P.W.H. (2015). Scenarios for the making of vertebrates. *Nature* 520: 450–455.

Honebrink, R.R. (2000). *A Review of the Family Carangidae, with Emphasis on Species Found in Hawaiian Waters*, Technical Report 20-01. Honolulu: Hawaiian Department of Land and Natural Resources, Division of Aquatic Resources.

Hopkins, T.L. (1982). The vertical distribution of zooplankton in the eastern Gulf of Mexico. *Deep-Sea Research* 29: 1069–1083.

Hopkins, T.L. and Baird, R.C. (1973). Diet of the hatchetfish *Sternoptyx diaphana*. *Marine Biology* 21: 34–46.

Hopkins, T.L. and Baird, R.C. (1977). Aspects of the feeding ecology of oceanic midwater fishes. In: *Oceanic Sound Scattering Prediction* (eds. N.R. Anderson and B.J. Zahuranec). New York: Plenum Press.

Hopkins, T.L. and Baird, R.C. (1981). Trophodynamics of the fish *Valencienellus tripunctulatus* I. Vertical distribution, diet and feeding chronology. *Marine Ecology Progress Series* 5: 1–10.

Hopkins, T.L. and Gartner, J.V. (1992). Resource-partitioning and predation impact of a low-latitude myctophid community. *Marine Biology* 114: 185–197.

Hopkins, T.L. and Lancraft, T.M. (1984). The composition and standing stock of mesopelagic micronekton at 27 N 86 W in the eastern Gulf of Mexico. *Contributions in Marine Science* 27: 145–158.

Hopkins, T.L., Sutton, T.T., and Lancraft, T.M. (1996). The trophic structure and predation impact of a low latitude midwater fish assemblage. *Progress in Oceanography* 38: 205–239.

Horn, M.H. (1998). Feeding and digestion. In: *Physiology of Fishes*, 2e (ed. D.H. Evans). Boca Raton: CRC Press.

Howell, W.H. and Krueger, W.H. (1987). Family Sternoptychidae. In: *Smithsonian Contributions to Zoology no. 452: Biology of Fishes of the Bermuda Ocean Acre* (eds. R.H. Gibbs and W.H. Krueger). Washington DC: Smithsonian Institution Press.

Hughes, G.M. and Morgan, M. (1973). The structure of fish gills in relation to their respiratory function. *Biological Reviews* 48: 419–475.

Hully, P.A. (1990). Family Myctophidae. In: *Fishes of the Southern Ocean* (eds. O. Gon and P.C. Heemstra). Grahamstown: J.L.B. Smith Institute of Ichthyology.

Janssen, J. and Harbison, G.R. (1981). Fish in salps: the association of squaretail (*Tetragonurus* spp.) with pelagic tunicates. *Journal of the Marine Biological Association of the United Kingdom* 61: 917–927.

Janssen, J., Harbison, G.R., and Craddock, J.E. (1986). Hatchetfishes hold horizontal attitudes during diagonal descents. *Journal of the Marine Biological Association of the United Kingdom* 66: 825–833.

Janssen, J., Gibbs, R.H., and Pugh, P.R. (1989). Association of *Caristius* sp. (Pisces: Caristiidae) with a siphonophore *Bathyphysa conifera*. *Copeia* 1989: 198–201.

Jensen, D. (1966). The Hagfish. *Scientific American* 214: 82–90.

Johnson, G.D., Paxton, J.R., Sutton, T.T. et al. (2009). Deep-sea mystery solved: astonishing larval transformations and extreme sexual dimorphism unite three fish families. *Biology Letters* 5: 235–239.

Johnston, I.A. (1981). Structure and function of fish muscles. In: *Vertebrate Locomotion* (ed. M.H. Day). London: Academic Press.

Jones, E.C. (1971). Isistius brasiiensis, a squaloid shark, the probable cause of crater wounds on fishes and crustaceans. *Fishery Bulletin U.S.* 69: 791–798.

Jordan, D.S. and Evermann, B.W. (1900). The fishes of north and middle America. *Bulletin of the United States National Museum* 47: 1–3313.

Kalmijn, A.J. (1971). The electric sense of sharks and rays. *Journal of Experimental Biology* 55: 371–383.

Karnella, C. (1987). Family Myctophidae, lanternfishes. In: *Biology of Midwater Fishes of the Bermuda Ocean Acre* (eds. R.H. Gibbs and W.H. Krueger). Washington, DC: Smithsonian Institution Press.

Katz, S.L. (2002). Design of heterothermic muscle in fish. *Journal of Experimental Biology* 205: 2251–2266.

Kawaguchi, K. and Marumo, R. (1967). Biology of *Gonostoma gracile* (Gonostomatidae). I. Morphology, life-history and sex-reversal. *Information Bulletin of Planktology in Japan*, Commemoration Number of Dr.MATSUE, 53–69.

Keene, M.J., Gibbs, R.H., and Krueger, W.H. (1987). Family Melamphaidae, Bigscales. In: *Biology of Midwater Fishes of the Bermuda Ocean Acre* (eds. R.H. Gibbs and W.H. Krueger). Washington, DC: Smithsonian Institution Press.

Keyes, J.L. (1974). Blood-gases and blood-gas transport. *Heart and Lung* 3: 945–954.

Keyes, J.L. (1985). *Fluid, Electrolyte, and Acid-Base Regulation*. Monterey: Wadsworth Health Sciences Division.

Kinzer, J., Bottger-Schnack, R., and Schulz, K. (1993). Aspects of horizontal distribution and diet of myctophid fish in the Arabian Sea with reference to the deep water oxygen deficiency. *Deep-Sea Research II* 40: 783–800.

Klimley, A.P. (1994). The predatory behavior of the white shark. *American Scientist* 82: 122–133.

Kohler, N.E. and Turner, P.A. (2001). Shark tagging: a review of conventional methods and studies. *Environmental Biology of Fishes* 60: 191–223.

Krogh, A. (1929). *The Anatomy and Physiology of Capillaries*, 2e. New Haven: Yale University Press.

Lagler, K.F., Bardach, J.E., and Miller, R.R. (1962). *Ichthyology*. New York: Wiley.

Lancraft, T.M., Hopkins, T.L., and Torres, J.J. (1988). Aspects of the ecology of the mesopelagic fish *Gonostoma elongatum* (Gonostomatidae, Stomiiformes) in the eastern Gulf of Mexico. *Marine Ecology Progress Series* 49: 27–40.

Lancraft, T.M., Torres, J.J., and Hopkins, T.L. (1989). Micronekton and macrozooplankton in the open waters near Antarctic ice edge zones (AMERIEZ 1983 AND 1986). *Polar Biology* 9: 225–233.

Lancraft, T.M., Hopkins, T.L., Torres, J.J., and Donnelly, J. (1991). Oceanic micronektonic/macrozooplanktonic community structure and feeding under ice covered Antarctic waters during the winter (AMERIEZ 1988). *Polar Biology* 11: 157–167.

Liem, K.F. (1980). Adaptive significance of intra- and interspecfic differences in the feeding repertoires of cichlid fishes. *American Zoologist* 20: 295–314.

Linkowski, T.B. (1985). Population biology of the myctophid fish *Gymnoscopelus nicholsi* (Gilbert 1911) from the western South Atlantic. *Journal of Fish Biology* 27: 683–689.

Linkowski, T.B. (1987). Age and growth of four species of Electrona (Teleostei,

Myctophidae). In: *Proceedings of V Congress of European Ichthyologists. Congressus Europaeus Ichthyologorum 1985* (eds. S.O. Kullander and B. Fernholm). Stockholm: Swedish Museum of Natural History.

Linley, T.D., Gerringer, M.E., Yancey, P.H. et al. (2016). Fishes of the hadal zone including new species, in situ observations and depth records of Liparidae. *Deep-Sea Research I* 114: 99–110.

Lisovenko, L.A. and Prut'Ko, V.G. (1987). Reproductive biology of Diaphus suborbitalis in the equatorial part of the Indian Ocean 1. Nature of oogenesis and type of spawning. *Journal of Ichthyology* 26: 47–58.

Locket, N.A. (1977). Adaptations to the deep-sea environment. In: *Handbook of Sensory Physiology. The Visual System of Vertebrates* (ed. F. Crescitelli). Berlin: Springer-Verlag.

Locket, N.A. (1985). The multiple bank rod fovea of *Bajacalifornia drakei*, an alepocephalid deep-sea teleost. *Proceedings of the Royal Society of London B* 224: 7–22.

Lorenzini, S. (1678). *Observazioni intorno alle torpedini*. Firenze: l'Onofri.

Love, R.H., Fisher, R.A., Wilson, M.A., and Nero, R.W. (2003). Unusual swimbladder behavior of fish in the Cariaco Trench. *Deep-Sea Research I* 51: 1–16.

Lowe, T.E., Brill, R.W., and Cousins, K.L. (2000). Blood oxygen-binding characteristics of bigeye tuna (*Thunnus obesus*), a high-energy-demand teleost that is tolerant of low ambient oxygen. *Marine Biology* 136: 1087–1098.

Luck, D.G. and Pietsch, T.W. (2008). In-situ observations of a deep-sea ceratioid anglerfish of the genus Oneirodes (Lophiiformes: Oneirodes). *Copeia* 2008: 466–451.

Magnuson, J.J. (1970). Hydrostatic equilibrium of *Euthynnus affinis*, a pelagic teleost without a swimbladder. *Copeia* 1970: 56–85.

Major, P.F. (1978). Predator-prey interactions in two schooling fishes, *Caranx ignobilis* and *Stolephorus purpureus*. *Animal Behavior* 26: 760–777.

Mann, D.A., Hastings, M.C., and Popper, A.N. (1998). Detection of ultrasonic tones and simulated dolphin echolocation clicks by a teleost fish, the American shad, Alosa sapidissima. *Journal of the Acoustical Society of America* 104: 562–568.

Mann, D.A., Higgs, D., Tavolga, W.N. et al. (2001). Ultrasound detection by clupeiform fishes. *Journal of the Acoustical Society of America* 104: 3048–3054.

Marshall, N.B. (1965). *The Life of Fishes*. London: Weidenfeld and Nicolson.

Marshall, N.B. (1971). *Explorations in the Life of Fishes*. Cambridge: Harvard University Press.

Marshall, N.B. (1979). *Deep-Sea Biology, Developments and Perspectives*. Dorset: Blandford Press Ltd.

Marshall, W.S. and Grosell, M. (2006). Ion transport, osmoregulation, and acid-base balance. In: *The Physiology of Fishes*, 3e (eds. D.H. Evans and J.D. Claiborne). Boca Raton: CRC Press.

Matthews, L.H. and Parker, H.W. (1950). Notes on the anatomy and biology of the basking shark (*Cetorhinus maximus* (Gunner)). *Proceedings of the Zoological Society of London* 120: 535–576.

Mcclain, C.R. and Hardy, S.M. (2010). The dynamics of biogeographic ranges in the deep sea. *Proceedings of the Royal Society of London B* 277: 3533–3546.

Mceachran, J.D. and Fechhelm, J.D. (2005). *Fishes of the Gulf of Mexico*. Austin: University of Texas Press.

Mcginnis, R.F. (1982). *Biogeography of Lanternfishes (Myctophidae) South of 30 S, Antarctic Research Series, v. 35, Biology of the Antarctic Seas XII*. Washington: American Geophysical Union.

Mead, G.W. (1963). Observations of fishes caught over the anoxic waters of the Cariaco Trench, Venezuela. *Deep-Sea Research* 10: 733–734.

Merrett, N.R. and Roe, H.S.J. (1974). Patterns and selectivity in the feeding of certain mesopelagic fishes. *Marine Biology* 28: 115–126.

Miya, M. and Nemoto, T. (1985). Protandrous sex reversal in *Cyclothone atraria* (family Gonostomatidae). *Japanese Journal of Ichthyology* 31: 438–440.

Miya, M. and Nemoto, T. (1991). Comparative life histories of the meso- and bathypelagic fishes of the genus *Cyclothone* (Pisces: Gonostomatidae) in Sagami Bay, central Japan. *Deep-Sea Research* 38: 67–89.

Miya, M. and Nishida, M. (1996). Molecular phylogenetic perspective on the evolution of the deep-sea fish genus *Cyclothone*. *Ichthyological Research* 43: 375–398.

Mochioka, N. and Iwamizu, M. (1996). Diet of anguilloid larvae: leptocephali feed exclusively on larvacean houses and fecal pellets. *Marine Biology* 125: 447–452.

Mollett, H.F., Ebert, D.A., Cailliet, G.M. et al. (1996). A review of length validation methods and protocols to measure large white sharks. In: *Great White Sharks, the Biology of Carcharodon carcharias* (eds. A.P. Klimley and D.G. Ainley). San Diego: Academic Press.

Moog, F. (1981). The lining of the small intestine. *Scientific American* 245: 154–179.

Morrow, J.E. (1964). Family Chauliodontidae. In: *Fishes of the Western North Atlantic, Number 1, Part 4* (eds. H.B. Bigelow, C.M. Breder, G.W. Mead, et al.). New Haven: Sears Foundation for Marine Research, Yale University.

Moser, H.G. and Ahlstrom, E.H. (1974). Role of larval stages in systematic investigations of marine teleosts: a case study. *Fishery Bulletin US* 72: 391–413.

Motta, P.J. and Wilga, C.D. (2001). Advances in the study of feeding behaviors, mechanisms, and mechanics of sharks. *Environmental Biology of Fishes* 60: 131–156.

Muir, B.S. and Kendall, J.I. (1968). Structural modifications in the gills of tunas and some other oceanic fishes. *Copeia* 1968: 388–398.

Munk, O. (1966). Ocular anatomy of some deep-sea teleosts. *Dana Report* 70: 1–62.

Munk, O. (1977). The visual cells and retinal tapetum of the foveate deep-sea fish *Scopelosaurus lepidus* (Teleostei). *Zoomorphologie* 87: 21–49.

Muntz, W.R.A. (1975). Behavioral studies of vision in a fish and possible relationships to the environment. In: *Vision in Fishes* (ed. M.A. Ali). New York: Plenum Press.

Musilova, Z., Cortesi, F., Matschiner, M. et al. (2019). Vision using multiple distinct rod opsins in deep-sea fishes. *Science* 364: 588–592.

Nafpaktitis, B.G. (1977). Family Neoscopelidae. In: *Fishes of the Western North Atlantic, Part 7* (ed. R.H. Gibbs). New Haven: Sears Foundation for Marine Research, Yale University.

Nafpaktitis, B.G. and Nafpaktitis, M. (1969). *Lanternfishes (Family Myctophidae) Collected During Cruises 3 and 6 of the R/V Anton Bruun in the Indian Ocean*. Los Angeles County Museum of Natural History.

Nafpaktitis, B.G., Backus, R.H., Craddock, J.E. et al. (1977). Family Myctophidae. In: *Fishes of the Western North Atlantic, Part 7* (ed. R.H. Gibbs). New Haven: Sears Foundation for Marine Research, Yale University.

Nakamura, I. (1985). *FAO Species Catalogue. Volume 5. Billfishes of the World. An Annotated and Illustrated Catalogue of Marlins, Sailfishes, Spearfishes and Swordfishes Known to Date*. Rome: Food and Agricultural Organization of the United Nations.

Nakamura, I. and Parin, N.V. (1993). *Snake Mackerels and Cutlassfishes of the World (Families Gempylidae and Trichiuridae)*. Rome: Food and Agriculture Organization of the United Nations.

Nelson, J.S. (2006). *Fishes of the World*. Hoboken: Wiley.

Nelson, J.S., Grande, T.C., and Wilson, M.V.H. (2016). *Fishes of the World*, 5e. Hoboken: Wiley.

Nicol, J.A.C. (1958). Observations on luminescence in pelagic animals. *Journal of the Marine Biological Association of the United Kingdom* 37: 705–752.

Nicol, J.A.C. (1960). Studies on luminescence. On the subocular light-organs of stomiatoid fishes. *Journal of the Marine Biological Association of the United Kingdom* 39: 529–548.

Olson, K.R. (1996). Secondary circulation in fish: anatomical organization and physiological significance. *Journal of Experimental Zoology* 275: 172–185.

Otake, T.K. and Maruyama, K. (1993). Dissolved and particulate organic matter as possible food sources for eel leptocephali. *Marine Ecology Progress Series* 92: 27–35.

Ozawa, T., Fuji, K., and Kawaguchi, K. (1977). Feeding chronology of the vertically migrating gonostomatid fish *Vinciguerria nimbaria* (Jordan and Williams), off southern Japan. *Journal of the Oceanographical Society of Japan* 33: 302–327.

Patterson, C. (1965). The phylogeny of the chimaeroids. *Philosophical Transactions of the Royal Society of London. Series B, Biological Sciences* 249: 101–219.

Paxton, J.R. (1989). Synopsis of the whalefishes (Family Cetomimidae) with descriptions of four new genera. *Records of the Austraiian Museum* 41: 135–206.

Paxton, J.R., Ahlstrom, E.H., and Moser, H.G. (1984). Myctophidae: relationships. In: *Ontogeny and Systematics of Fishes. Special Publication I* (eds. H.G. Moser, W.J. Richards, D.M. Cohen, et al.), 239–244. ASIH.

Pearcy, W.G. (1964). Some distributional features of mesopelagic fishes off Oregon. *Deep-Sea Research* 24: 223–245.

Pearcy, W.G., Krygier, E.E., Mesecar, R., and Ramsey, F. (1977). Vertical distribution and migration of oceanic micronekton off Oregon. *Deep-Sea Research* 24: 223–245.

Pfeiler, E. (1986). Towards an explanation of the developmental strategy in leptocephalus larvae of marine teleost fishes. *Environmental Biology of Fishes* 15: 3–13.

Pfeiler, E. (1996). Energetics of metamorphosis in bonefish (*Albula* sp.) leptocephali: role of keratan sulfate glycosaminoglycan. *Fish Physiology and Biochemistry* 15: 359–362.

Pietsch, T.W. (1978). The feeding mechanism of *Stylephorus chordatus* (Teleostei: Lampridiformes): functional and ecological implications. *Copeia* 1978: 255–262.

Pietsch, T.W. (2009). *Oceanic Anglerfishes: Extraordinary Diversity in the Deep Sea*. Berkeley: University of California Press.

Pietsch, T.W. and Van Duzer, J.P. (1980). Systematics and distribution of ceratioid anglerfishes of the family Melanocetidae, with description of a new species from the eastern North Pacific Ocean. *Fishery Bulletin U.S.* 78: 59–87.

Piper, R. (2007). *Extraordinary Animals: An Encyclopedia of Curious and Unusual Animals*. Westport: Greenwood Publishing Group.

Popper, A.N., Rogers, P.H., Saidel, W.M., and Cox, M. (1988). Role of the fish ear in sound processing. In: *Sensory Biology of Aquatic Animals* (eds. J. Atema, R.R. Fay,

A.N. Popper and W.N. Tavolga). New York: Springer-Verlag.

Porter, H.T. and Motta, P.J. (2004). A comparison of strike and prey capture kinematics of three species of piscivorous fishes: Florida gar (*Lepisosteus platyrhyncus*), redline needlefish (*Strongylura notata*) and great barracuda (*Sphyraena barracuda*). *Marine Biology* 145: 989–1000.

Potter, I.F. and Howell, W.H. (2011). Vertical movement and behavior of the ocean sunfish Mola mola in the northwest Atlantic. *Journal of Experimental Marine Biology and Ecology* 396: 138–146.

Potts, W.T.W. (1976). Ion transport and osmoregulation in marine fish. In: *Perspectives in Experimental Biology* (ed. P.S. Davies). Oxford: Pergamon Press.

Potts, W.T.W. and Parry, G. (1964). *Osmotic and Ionic Regulation in Animals*. Oxford: Pergamon Press.

Pough, F.H., Heiser, J.B., and Mcfarland, W.N. (1989). *Vertebrate Life*. New York: Macmillan.

Poulsen, J.Y., Byrkjedal, I., Willassen, E. et al. (2013). Mitogenomic sequences and evidence from unique gene rearrangements corroborate evolutionary relationships of myctophiformes (Neoteleostei). *BMC Evolutionary Biology* 13: 111–132.

Pratt, H.L. and Casey, J.G. (1990). Shark reproductive strategies as a limiting factor in directed fisheries, with a review of Holden's method of estimating growth parameters. In: *Elasmobranchs as Living Resources: Advances in Biology, Ecology, Systematics and Status of the Fisheries* (eds. H.L. Pratt, S.H. Gruber and T. Taniuchi), 97–109. Washington, DC: NOAA. Technical Report NMFS 90.

Priede, I.G. (2017). *Deep-Sea Fishes*. Cambridge: Cambridge University Press.

Priede, I.G. and Froese, R. (2013). Colonization of the deep sea by fishes. *Journal of Fish Biology* 83: 1528–1550.

Prosser, C.L. (1973). *Comparative Animal Physiology*. Philadelphia: W.B. Saunders.

Pugh, W.L. (1972). Collections of midwater organisms in the Cariaco Trench, Venezuela. *Bulletin of Marine Science* 22: 592–600.

Pumphrey, R.J. (1950). Hearing. *Symposia of the Society for Experimental Biology* 4: 3–18.

Riley, J.P. and Chester, R. (1971). *Introduction to Marine Chemistry*. London: Academic Press.

Robison, B.H. and Reisenbichler, K.R. (2008). Macropinna microstoma and the paradox of its tubular eyes. *Copeia* 2008: 780–784.

Romer, A.S. and Parsons, T.S. (1986). *The Vertebrate Body*. New York: CBS College Publishing.

Ross, S.W., Quattrini, A.M., Roa-Varon, A.Y., and Mcclain, J.P. (2010). Species composition and distributions of mesopelagic fishes over the slope of the north-central Gulf of Mexico. *Deep-Sea Research II* 57: 1926–1956.

Sanderson, S.L., Cheer, A.Y., Goodrich, J.S. et al. (2001). Crossflow filtration in suspension feeding fishes. *Nature* 412: 439–441.

Schaefer, B. and Williams, M. (1977). Relationships of fossil and living elasmobranchs. *American Zoologist* 17: 293–302.

Schmidt, J. (1925). The breeding place of the eel. *Reports of the Smithsonian Institution* 1924: 279–316.

Schmidt-Nielsen, K. (1972). Locomotion: energy cost of swimming, flying and running. *Science* 177: 222–228.

Schmidt-Nielsen, K. (1990). *Animal Physiology: Adaptation and Environment*, 4e. Cambridge: Cambridge University Press.

Schmitt, R.J. (1982). Cooperative foraging by yellowtail, *Seriola lalandei* (Carangidae) on two species of fish prey. *Copeia* 1982: 714–717.

Schopf, T.J.M. (1980). *Paleoceanography*. Cambridge: Harvard University Press.

Smith, D.G. (1989). Leptocephali: introduction. In: *Fishes of the Western North Atlantic. Part Nine. Volume Two* (ed. E.B. Bohlke). New Haven: Sears Foundation for Marine Research, Yale University.

Smith, M.M. and Heemstra, P.C. (1991). *Smith's Sea Fishes*. Johannesburg: Southern Book Publishers.

Sorenson, P.W. and Caprio, J. (1998). Chemoreception in fish. In: *The Physiology of Fishes*, 2e (ed. D.H. Evans). Boca Raton: CRC Press.

Springer, F.G. and Gold, J.P. (1989). *Sharks in Question*. Washington, DC: Smithsonian Institution Press.

Stiassny, M.L.J. (2000). Skeletal system. In: *The Laboratory Fish* (ed. G.K. Ostrander). San Diego: Academic Press.

Stickney, D.G. and Torres, J.J. (1989). Proximate composition and energy content of mesopelagic fishes from the eastern Gulf of Mexico. *Marine Biology* 103: 13–24.

Suckling, J.A. (1967). Trunk lateral line nerves: some anatomical aspects. In: *Lateral Line Detectors* (ed. P. Cahn). Bloomington: Indiana University Press.

Sutton, T.T. (2005). Trophic ecology of the deep-sea fish *Malacosteus niger* (Pisces: Stomiidae): an enigmatic feeding ecology to facilitate a unique visual system. *Deep-Sea Research I* 52: 2065–2076.

Sutton, T.T. and Hartel, K.E. (2004). New species of Eustomias (Teleostei: Stomiidae) from the western North Atlantic, with a review of the subgenus Neostomias. *Copeia* 2004: 116–121.

Sutton, T.T. and Hopkins, T.L. (1996a). Trophic ecology of the stomiid (Pisces: Stomiidae) assemblage of the eastern Gulf of Mexico: strategies, selectivity and impact of a mesopelagic top predator group. *Marine Biology* 127: 179–192.

Sutton, T.T. and Hopkins, T.L. (1996b). Species composition, abundance, and vertical distribution of the stomiid (Pisces: Stomiiformes) fish assemblage of the Gulf of Mexico. *Bulletin of Marine Science* 59: 530–542.

Sutton, T.T., Wiebe, P.H., Madin, L., and Bucklin, A. (2010). Diversity and community structure of pelagic fishes to 5000 m depth. *Deep-Sea Research II* 57: 2220–2233.

Sverdrup, H.U., Johnson, M.W., and Fleming, R.H. (1942). *The Oceans, Their Physics, Chemistry, and General Biology*. Prentice-Hall: Englewood Cliffs.

Takashima, R., Nishi, H., Huber, B.T., and Leckie, M. (2006). Greenhouse world and the Mesozoic ocean. *Oceanography* 19: 82–92.

Torres, J.J., Belman, B.W., and Childress, J.J. (1979). Oxygen consumption rates of midwater fishes as a function of depth of occurrence. *Deep-Sea Research* 26: 185–197.

Tricas, T.C. and Mccosker, J.E. (1984). Predatory behavior of the white shark (*Carcharodon carcharias*) with notes on its biology. *Proceedings of the California Academy of Sciences* 43: 221–238.

Tseng, W. (1990). Relationship between growth rate and age at recruitment of *Anguilla japonica* elvers in a Taiwan estuary as inferred from otolith increments. *Marine Biology* 107: 75–81.

Tucker, V.A. (1975). The energetic cost of moving about: walking and running are extremely inefficient forms of locomotion. Much greater efficiency is achieved by birds, fish-and bicyclists. *American Scientist* 63: 413–419.

UNESCO (2009). *Global Open Oceans and Deep Seabed (GOODS) – Biogeographic*

Classification, *IOC Technical Series, 84*. Paris: UNESCO-IOC.

Van Vleet, E.S., Candileri, S., Mcnellie, J. et al. (1984). Neutral lipid components of eleven species of Caribbean sharks. *Comparative Biochemistry and Physiology B* 79: 549–554.

Vinogradov, M.E. (1970). *Vertical Distribution of the Oceanic Zooplankton*. Jerusalem: Israel Program for Scientific Translations.

Vogel, S. (1994). *Life in Moving Fluids: The Physical Biology of Flow*, 2e. Princeton: Princeton University Press.

Von Westernhagen, H. (1974). Observations on the natural spawning of *Alectis indicus* (Ruppell) and *Caranx ignobilis* (Forsk.) (Carangidae). *Journal of Fish Biology* 6: 513–516.

Walker, M.M., Deibel, D.E., Haugh, C.V. et al. (1997). Structure and function of the vertebrate magnetic sense. *Nature* 390: 371–376.

Walsh, P.J. (1998). Nitrogen excretion and metabolism. In: *The Physiology of Fishes*, 2e (ed. D.H. Evans). Boca Raton: CRC Press.

Walters, S.V. and Fierstein, H.L. (1964). Measurements of swimming speeds of yellowfin tuna and wahoo. *Nature* 202: 208–209.

Wardle, C.S., Videler, J.J., Arimoto, T. et al. (1989). The muscle twitch and the maximum swimming speed of giant bluefin tuna, *Thunnus thynnus*. *Journal of Fish Biology* 42: 129–137.

Waterman, A.J., Frye, B.E., Johansen, K. et al. (1971). *Chordate Structure and Function*. New York: The Macmillan Company.

Webb, P.W. (1997). Swimming. In: *The Physiology of Fishes*, 2e (ed. D.H. Evans). Boca Raton: CRC Press.

Wegner, N.C., Snodgrass, O.E., Dewar, H., and Hyde, J.R. (2015). Whole-body endothermy in a mesopelagic fish, the opah, *Lampris guttatus*. *Science* 348: 786–789.

Werner, E.E. and Gilliam, J.T. (1984). The ontogenetic niche and species interaction in size-structured populations. *Annual Review of Ecology and Systematics* 15: 393–425.

Widder, E.A. (1998). The predatory use of counterillumination by the squaloid shark *Isistius brasiliensis*. *Environmental Biology of Fishes* 53: 267–273.

Widder, E.A. (2010). Bioluminescence in the ocean: origins of biological, chemical and ecological diversity. *Science* 328: 704–708.

Widder, E.A., Latz, M.F., Herring, P.J., and Case, J.F. (1984). Far-red bioluminescence from two deep-sea fishes. *Science* 225: 512–514.

Wilga, C.D. and Lauder, G.V. (2000). Three-dimensional kinematics and wake structure of the pectoral fins during locomotion in leopard sharks *Triakis semifasciata*. *Journal of Experimental Biology* 203: 2261–2278.

Wilga, C.D., Motta, P.J., and Sanford, C.P. (2007). Evolution and feeding in elasmobranchs. *Integrative and Comparative Biology* 47: 55–69.

Willis, J.M. and Pearcy, W.G. (1980). Spatial and temporal variations in the population size structure of three lanternfishes (Myctophidae) off Oregon, USA. *Marine Biology* 57: 181–191.

Willmer, P., Stone, G., and Johnston, I. (2005). *Environmental Physiology of Animals*. Suffolk: Blackwell.

Wilson, D.F. (1972). Diel migration of sound-scatterers into, and out of, the Cariaco Trench anoxic water. *Journal of Marine Research* 30: 168–176.

Withers, P.C. (1992). *Comparative Animal Physiology*. Orlando: Saunders.

Wittenberg, J.B. and Wittenberg, B.A. (1974). The choroid rete mirabile of the fish eye. I. Oxygen secretion and structure: comparison with the swimbladder rete mirabile. *Biological Bulletin* 146: 116–136.

WoRMS. World Register of Marine Species. (n.d.) www.marinespecies.org.

Wood, C.M. (1993). Ammonia and urea metabolism and excretion. In: *Physiology of Fishes* (ed. D.H. Evans). Boca Raton: CRC Press.

Yancey, P.H., Lawrence-Berry, R., and Douglas, M.D. (1989). Adaptations of mesopelagic fishes. I. Buoyant glycosaminoglycan layers in species without diel vertical migration. *Marine Biology* 103: 453–459.

Yancey, P.H., Gerringer, M.E., Drazen, J.C. et al. (2014). Marine fish may be biochemically constrained from inhabiting the deepest ocean depths. *Proceedings of the National Academy of Science of the USA* 111: 4461–4465.

Yeh, J. and Drazen, J.C. (2009). Depth zonation and bathymetric trends of deep-sea megafaunal scavengers of the Hawaiian Islands. *Deep-Sea Research I* 56: 251–266.

Young, J.W., Bulman, C.M., Blader, S.J.M., and Wayte, S.E. (1988). Age and growth of the lanternfish *Lampanyctodes hectoris* (Myctophidae) from eastern Tasmania, Australia. *Marine Biology* 99: 569–576.

Zakon, H. (1988). The electroreceptors: diversity in structure and function. In: *Sensory Biology of Aquatic Animals* (eds. J. Atema, R.R. Fay, A.N. Popper and W.N. Tavolga). New York: Springer-Verlag.

Zintzen, V., Roberts, C.D., Anderson, M.J. et al. (2011). Hagfish predatory behaviour and slime defence mechanism. *Scientific Reports* 1: 131.

Zintzen, V., Rogers, K.M., Roberts, C.D. et al. (2013). Hagfish feeding habits along a depth gradient inferred from stable isotopes. *Marine Ecology Progress Series* 485: 223–234.

11

Communities

Introduction

Krebs (1972) defines a community as "any assemblage of populations of living organisms in a prescribed area or habitat," a very general definition but quite useful for oceanographic systems whose boundaries are notably difficult to constrain. A population may be defined as "a group of organisms of the same species occupying a particular place at a particular time."

The major taxa of micronekton and macrozooplankton and their physiological characteristics were covered in preceding chapters, along with a brief introduction to physical oceanography and to biological responses to physical challenges such as temperature and pressure. This chapter will examine three ecoregions with disparate physical and climatic characteristics to see how they differ in species composition and abundance. We have picked the Gulf of Mexico (GOM) as a paradigm tropical-subtropical pelagic community, the California Current as a cold-temperate pelagic community, and the Antarctic/Southern Ocean as a polar system.

Thirty-three global mesopelagic ecoregions (Figure 11.1) were recently identified based on their species composition and physical characteristics (Sutton et al. 2017).

Ecoregions are defined as "areas of ocean that contain geographically distinct assemblages of natural communities and species" (Spalding et al. 2007). Further, they are assumed to be stable enough to have shared a common history of coevolution. Besides species composition, five operators were considered critical in determining the regions and their boundaries: water masses, oxygen minimum zones, temperature profiles, primary production, and, where data were available, observed faunal breaks.

Our three ecoregions are representative of three larger oceanic biomes (Longhurst 1998). The GOM and seven other ecoregions are members of the Trade Wind Biome, characterized by small amplitude responses to trade wind variability; all include weak seasonality and low primary production. Diversity in all trade wind ecoregions is very high. The California Current ecoregion and 12 others are in the Distant Neritic Biome, in which circulation is modified by interaction with continental topography and associated coastal winds, producing a strong upwelling season and high annual production. The Antarctic/Southern Ocean ecoregion and two others constitute the Polar Biome, characterized by low temperatures and extreme seasonality in production due to ice cover and light limitation.

Life in the Open Ocean: The Biology of Pelagic Species, First Edition. Joseph J. Torres and Thomas G. Bailey.
© 2022 John Wiley & Sons Ltd. Published 2022 by John Wiley & Sons Ltd.

Communities

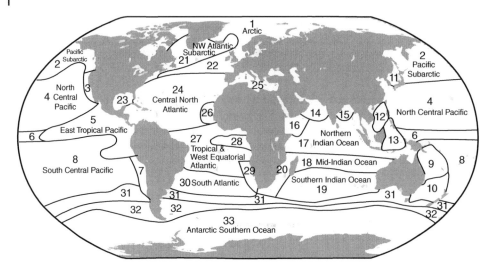

Figure 11.1 The mesopelagic ecoregions of the world's oceans. Key to small ecoregions not labeled: 3 – California Current, 6 – Equatorial Pacific, 7 – Peru Upwelling/Humboldt Current, 9 – Coral Sea, 10 – Tasman Sea, 11 – Sea of Japan, 12 – South China Sea, 13 – Indo-pacific Pocket Basins, 14 – Arabian Sea, 15 – Bay of Bengal, 16 – Somali Current, 20 – Agulhas Current, 22 – North Atlantic Drift, 23 – Gulf of Mexico, 25 – Mediterranean Sea, 26 – Mauritania/Cape Verde, 28 – Guinea Basin and East Equatorial Atlantic, 29 – Benguela Upwelling, 31 – Circumglobal Subtropical Front, 32 – Sub-Antarctic. Depths <200 m are not shown, boundaries are approximate. *Source:* Adapted from Sutton et al. (2017), figure 4 (p. 89).

Ideally, a community approach would include all major taxonomic groupings for the three regions. In practice, one must use the data available, which is why those three regions were chosen. Multi-year studies in each provide data on fishes, crustaceans, and cephalopods. For the GOM and Antarctic systems, basic data on macrozooplankton were also available.

The Gulf of Mexico

The eastern GOM is a subtropical system characterized by warm surface temperatures throughout the year (20–30 °C) and a shallow mixed layer of 25–50 m (Figure 11.2a). The steepest part of the permanent thermocline occurs between 50 and 150 m, dropping to 15–18 °C from sea surface temperatures (Hopkins et al. 1994). The water column is well oxygenated, with the minimum occurring in a broad zone from 450 to 800 m (>2.7 ml l^{-1}) (Torres et al. 2012). The water column can be considered normoxic: O_2 is not a limiting influence on species diversity in this system.

The most dynamic feature of the eastern GOM is the Loop Current (Figure 11.2b), which "connects the Yucatan Current from the Caribbean Sea with the Florida Current and the Gulf Stream by forming an anticyclonic retroflecting current with a looping path through the eastern GOM" (Liu et al. 2016). Penetration of the Loop Current into the GOM is temporally and spatially variable, sometimes reaching as far north as the Mississippi delta (Maul 1977; Sturges and Evans 1983; Sturges and Lugo-Fernandez 2005) but in most years remaining southwest of a primary study location, "Standard Station", at 27 °N 86 °W resides between 150 and 200 m. At intervals, the Loop Current sheds eddies that travel west toward Texas, bringing their resident fauna along. In most years, Standard Station resides in eastern GOM residual water (Hopkins et al. 1994). From a faunal perspective, the chief importance of the Loop Current is as a potential mechanism for transport of Caribbean fauna into the GOM. The faunal composition of the Caribbean, the GOM, and the Sargasso Sea is similar (Hopkins and Lancraft 1984; Gartner et al. 1989; Sutton et al. 2010).

Primary production in the eastern GOM is between <50 and $75\,g\,C\,m^{-2}\,y^{-1}$ (El Sayed 1972; Hopkins 1982), typical of an oligotrophic regime. Zooplankton standing stock was measured at 1.2 g dry mass (DM) m^{-2} in the upper 1000 m of the water column (Hopkins 1982), a higher value than that typical of central gyre systems (ca. $0.498\,g\,DM\,m^{-2}$, Vinogradov 1970), but about equal to that observed in tropical-subtropical boundary currents such as the Kuroshio ($1.3\,g\,DM\,m^{-2}$, Vinogradov 1970) and Canaries Current ($1.2\,g\,DM\,m^{-2}$, Yashnov 1962).

The eastern GOM was sampled annually from 1972 to 2011 by the laboratories of Drs. Tom Hopkins and Jose Torres, producing a very long-term data set and strong familiarity with the suite of species inhabiting the eastern Gulf.

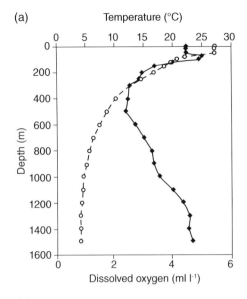
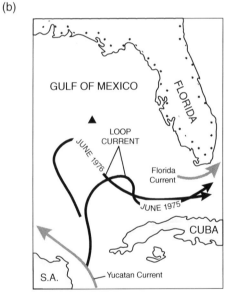

Figure 11.2 The Gulf of Mexico. (a) Profiles of dissolved oxygen (solid line) and temperature (dashed line) in the Gulf of Mexico. (b) Path of the Loop Current in the Gulf of Mexico in June 1975 and 1976. The black triangle marks the location of "Standard Station.". Source: (a) Torres et al. (2012), figure 1 (p. 1909).

(Figure 11.2b). The study site is recognized while at sea by the depth of the 22 °C isotherm, which in Loop Current waters

The Northern California Current

The California Current is the boundary current forming the eastern leg of the North Pacific gyral circulation (Figure 11.3). The waters of the gyral circulation are cooled as they transit the North Pacific latitudes and begin the equatorward journey down the western seaboard of the USA as the Northern California Current. Off the coast

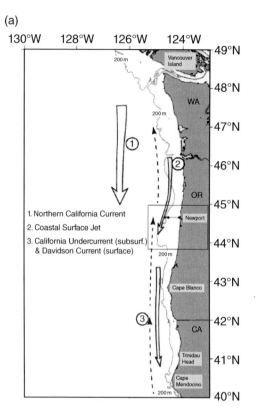
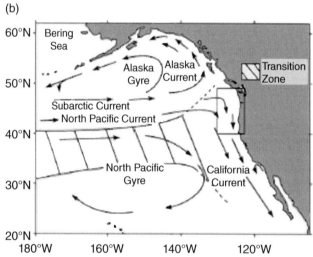

Figure 11.3 Chart of the NE Pacific showing (a) the seasonal circulation in the California Current in relation to (b) the basin-scale circulation. *Source:* Peterson et al. (2017), figure 2 (p. 7269), Journal of Geophysical Research. Copyright © 1969 John Wiley and Sons. Reproduced with the permission of John Wiley & Sons.

of Oregon, sea surface temperatures typically vary from 8 to 15 °C, depending on the location and time of year (Figure 11.4) (Brodeur et al. 2019). The central Oregon coast is an excellent example of a coastal upwelling system, with a fairly straight coastline and well-defined oceanographic seasons. It has been a study site for Oregon State University researchers from the early 1960s until the present day, including a long-term study on its micronekton led by Dr. William G. Pearcy.

The oceanographic seasons off the Oregon coast are governed by wind stress, which determines coastal circulation patterns and the levels of phytoplankton and zooplankton production. During the winter months, strong coastal winds and currents are largely poleward over the shelf and slope, resulting in downwelling. Winter winds and storms deepen the mixed layer, enriching surface waters with nutrients. As winter progresses toward spring, equatorward winds and currents become increasingly important, with episodic upwelling events that provide nutrients for limited phytoplankton blooms as the days grow longer. The summer upwelling season begins in earnest in May, a dramatic transition that affects the whole region. The upwelling is driven by alongshelf equatorward wind stress and is strongest on the inner half of the shelf. Currents on the shelf primarily flow equatorward during the summer months. The poleward flowing California Undercurrent develops as the upwelling season progresses, spreading underneath the equatorward flow of surface waters over the shelf and upper slope. Its main axis is along the upper slope.

Nutrients residing in the deeper water that is episodically upwelled onto the shelf, in concert with the increased daylength of the summer months, drive a robust phytoplankton production. The plentiful phytoplankton in turn produces large standing stocks of zooplankton, particularly the copepod *Pseudocalanus*, that serve as fodder for sardines and anchovies and a variety of other larger species. The high productivity supports a rich coastal fishery.

Shelf circulation patterns are important to the upwelling system off the Oregon coast in a number of ways besides the obvious influence of upwelling on annual productivity. The California undercurrent creates a two-layered flow regime that aids in retaining nutrients and phytoplankton within the system. For example, vertically migrating zooplankton will be naturally retained within the system by feeding on surface productivity during nighttime hours and returning on the undercurrent during the day.

The end of upwelling and return to winter conditions during the fall are as dramatic as the transition to the summer regime, usually accompanying the first major storm of the year. Coastal currents shift once again to primarily poleward, and wind-driven turbulence deepens the mixed layer, carrying phytoplankton away from near-surface waters and solar irradiance, even as the days get shorter (Hickey and Landry 1989).

Estimates of annual primary productivity for the Pacific Northwest vary from 130 to 645 $g\,C\,m^{-2}\,y^{-1}$ (Dortch and Postel 1989; Hickey and Landry 1989) depending on the techniques employed. A generic figure for upwelling systems, presented by Ryther (1969) in his classic paper on phytoplankton and fish production in the sea, is 300 $g\,C\,m^{-2}\,y^{-1}$. Without question, the Pacific Northwest is one of the most productive systems in the global ocean. In the same paper, Dr. Ryther gives a figure for open ocean systems such as the GOM of 50 $g\,C\,m^{-2}\,y^{-1}$, which agrees well with the figures of El Sayed 1972 (50–70 $g\,C\,m^{-2}\,y^{-1}$) cited above.

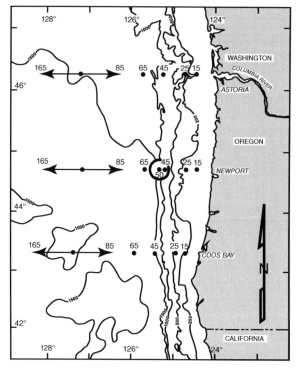

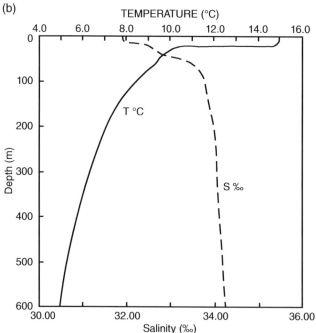

Figure 11.4 Trawling and hydrography studies off the coast of central Oregon. (a) Locations of midwater trawling stations used in the Oregon State University studies. Numbers designate the distance from the coast in nautical miles. The circle includes the area of repeated tows. Depths are in fathoms; (b) temperature and salinity profiles at a station 50 miles off Newport in August 1961. *Sources:* (a) Pearcy (1964), figure 1 (p. 85); (b) Pearcy (1964), figure 5 (p. 92).

As is usually the case with highly productive systems (Gallo and Levin 2016), a well-developed oxygen minimum layer is present in the waters off the Oregon coast (and off California as well), potentially limiting faunal diversity there. A long-term data set on oxygen concentration vs depth has been collected along the same Newport Hydrographic (NH) Line used in Pearcy (1964) (Figure 11.4a). The study (Pierce et al. 2012) covers two disparate 11-year periods off the Oregon coast: 1960–1971 and 1998–2009. It provides an excellent picture of the offshore oxygen environment as well as showing an increase in severity and a shoaling of the minimum in later years. Along the Newport Line, the core (lowest O_2) of the minimum begins at approximately 600 m ($20\,\mu mol\,kg^{-1}$) in the most recent data set, but in the earlier data, it was closer to 700 m (Figure 11.5) (Pierce et al. 2012). Descending from the surface, oxygen begins to decline rapidly from near air saturation of about $272\,\mu mol\,kg^{-1}$, assuming a temperature of 12 °C (Riley and Chester 1971), at a depth of 100 m to $80\,\mu mol\,kg^{-1}$ at 300 m. Below 300 m, it declines more gradually to $20\,\mu mol\,kg^{-1}$ at 600 m.

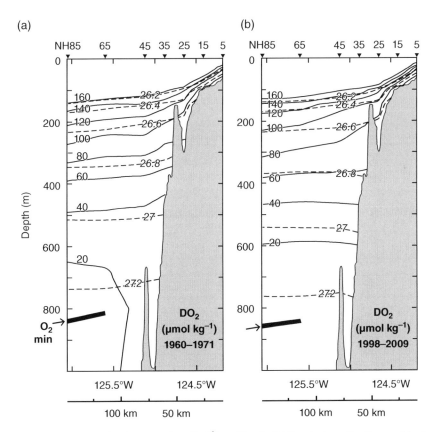

Figure 11.5 Dissolved oxygen ($\mu mol\,kg^{-1}$) profiles (solid lines) along the Newport hydrographic line (NH) off central Oregon during two disparate eleven-year periods. (a) 1960–1971; (b) 1998–2009. Dashed lines represent σ_0 surfaces. *Source:* Pierce et al. (2012), Declining Oxygen in the Northeast Pacific, Journal of Physical Oceanography, vol. 42, 495–501, figure 3 (p. 498). © American Meteorological Society. Used with permission.

Species living in the water column off Oregon must tolerate oxygen concentrations down to 40 µmol kg^{-1} (a PO$_2$ of about 24 mm Hg or 3.2 kPa) if their vertical range exceeds 500 m, and 35 mm Hg between 400 and 500 m (Figure 11.5) (Pierce et al. 2012). Pearcy and Laurs (1966) noted that the vertically migrating fishes off Oregon are most highly concentrated in the 150–500 m depth range during the day, moving into the upper 150 m at night, meaning that the lowest PO$_2$ that they likely encounter is about 35 mm Hg.

When considered in light of their vertical profiles, the directly measured P_c's (see Chapter 2) of important species support an entirely aerobic metabolic strategy. The overwhelmingly biomass-dominant lanternfish off the Oregon coast, *Stenobrachius leucopsaurus*, exhibits a P_c of 12 mm Hg (Torres et al. 1979), well below that needed for an entirely aerobic existence. Similarly, the decapod *Sergestes similis* has a P_c of about 21 mm Hg (Childress 1975), is easily able to live aerobically in the O$_2$ minimum off Oregon, as does the dominant euphausiid *Euphausia pacifica* ($P_c = 18$ mm Hg).

The Antarctic

The Antarctic is an unfamiliar picture to many people. It is remote from populated areas and difficult to reach; few get to experience it firsthand. On most maps of the world, it is portrayed as a continuous band along the bottom of the map, almost an afterthought. As it happens however, the Antarctic is truly a bellwether for the planet (cf. Ainley 2002), an experiment in progress as climate change alters the system. Polar biota cannot find colder water by a change in distribution; they are already at the end of the line (Clarke 1996; Mintenbeck and Torres 2017).

The open ocean of the Antarctic is critically important in the dynamics of the global ecosphere. It is the end-member oceanic system for marine biota, with sea surface temperatures below 50 °S latitude typically ranging from −2 to 2 °C (Gordon and Molinelli, 1982), making it the coldest open-ocean system on the planet. The fauna there must accommodate its low temperatures and extreme seasonality, including the annual advance and retreat of sea ice that covers and uncovers a vast area of sea surface (Figure 11.6). At its maximum extent in September, sea ice covers about 20 million km^2, shading an area twice the size of the continental United States (Gordon and Comiso 1988). The Antarctic continent itself is 13.2 million km^2 in area, considerably larger than the United States (9.83 million km^2). It is the world's fifth largest continent, smaller than South America (17.82 million km^2) but larger than Europe (9.94 million km^2). It covers the Earth's south pole and is the coldest, highest, and most isolated continent. It is surrounded by the windiest and roughest body of water on earth: the Southern Ocean.

The Southern Ocean is officially designated by the International Hydrographic Organization as the circumpolar ocean south of 60S extending to the Antarctic continent (Figure 11.7). It comprises the southernmost portions of the Atlantic, Pacific, and Indian oceans. Moving south toward the continent, it is characterized by an important series of oceanic fronts and water masses. For most scientists, the Antarctic ecosystem begins at the natural boundary known as the Antarctic Polar Front (APF), which is its true northern boundary. Also known as the Antarctic Convergence, it is the location where cold Antarctic surface water meets warmer Sub-Antarctic surface water and, owing to its greater density, slides beneath it to form Antarctic Intermediate Water (Figure 11.8). Heading south, the temperature drop at the

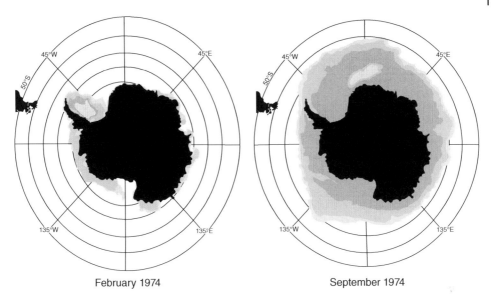

Figure 11.6 Maximum and minimum ice distribution in the Antarctic in 1974. The clear area just offshore in September centered at approximately 0° longitude is a large polynya, an ice-free "oasis." The open-water polynyas serve as refuges for marine mammals and birds. *Source:* Adapted from Zwally et al. (1983), figure B-2b, September 1974 (p. 175); figure B-2a, February 1974 (p. 174).

APF is 3–4 °C and often has a surface manifestation such as sea fog when observed from the deck of a ship. South of the APF at an average latitude of 53 °S and extending south to roughly 65 °S (Nowlin and Klinck 1986; Carmack 1990) is the Antarctic Circumpolar Current (ACC), flowing clockwise around the continent, driven by the prevailing westerly winds. The position of the APF varies with location (Figure 11.7), being nearer to 50 °S in the Atlantic sector, 60 °S in the Pacific and between 50 and 55 °S in the Indian Ocean Sector (Andriashev 1987).

The ACC, formerly known as the West Wind Drift, transports more water than any other current system on the planet, including the Gulf Stream. The reason why is not its speed, which is about $20\,\mathrm{cm\,s^{-1}}$ compared with the Gulf Stream's $200\,\mathrm{cm\,s^{-1}}$, but its enormous size (Carmack 1990). Besides its width, which varies from 200 to 1000 km, the ACC flows from the surface to the bottom (Foster 1984; Carmack 1990). Beneath the cold Antarctic surface, water lies a very large mass of warmer water that forms nearly the entire remainder of the water column: the Circumpolar Deep Water (CDW) (Figure 11.8), also known as Warm Deep Water. Temperatures in the CDW vary from about −0.5 to 2.0 °C (Gordon and Molinelli 1982). The relatively balmy temperatures of the CDW are particularly important to fishes because of their reduced internal ionic concentration relative to the surrounding seawater. Oceanic fishes will freeze at temperatures below −1.0 °C, meaning that they freeze before seawater does (−1.86 °C), making the Antarctic a very dangerous place.

The last water mass relevant to this discussion is Antarctic Bottom Water (ABW). As sea ice forms in the continental margin, brine is excluded from the forming ice into

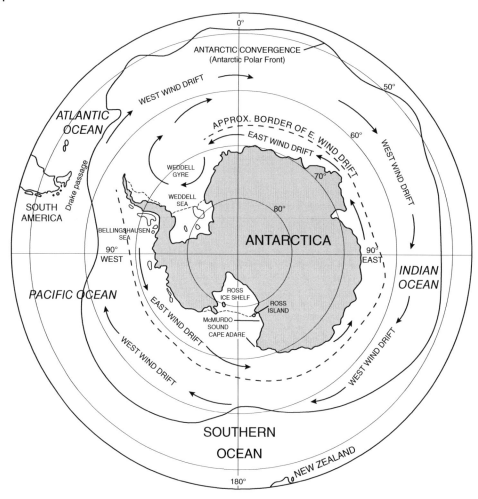

Figure 11.7 Map of Antarctica and the surrounding Southern Ocean. The Antarctic Convergence (Antarctic Polar Front) is depicted with the solid line that meanders between 50 and 60 °S. Major currents are indicated by arrows. The approximate boundary between the West Wind Drift (Antarctic Circumpolar Current) and East Wind Drift (Antarctic Coastal Current) is depicted with a dashed line. The transition between these two currents occurs in the Weddell Sea (Weddell Gyre).

the frigid waters beneath, making them saltier and, with their already cold temperatures, quite dense. The cold dense water sinks, spilling off the shelf down the continental slope and onto the sea floor where it flows beneath the CDW as Antarctic Bottom Water. It is important to recognize that in virtually all Antarctic coastal systems, from the continent out to the shelf break, the entire water column is composed of Ice-Shelf Water, the frigid waters destined to become ABW. A notable exception is the Western Antarctic Peninsula (WAP) shelf where because of the easterly flowing ACC, CDW intrudes onto the shelf via cross-shelf canyons, forming a warmer water column that is more friendly to oceanic fishes. Antarctic Bottom Water spreads northward

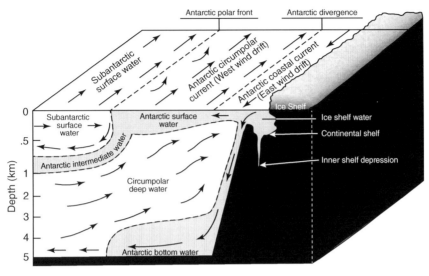

Figure 11.8 Schematic cross section of the water column of the Southern Ocean showing characteristic water masses, currents, and fronts. Note the features of the Antarctic continental shelf including great depth, inner shelf depressions, and relatively shallow shelf break. The coldest water masses are shaded. *Source:* Eastman (1993), figure 1.6 (p. 13). Reproduced with the permission of Academic Press.

well beyond the Southern Ocean, cooling the deep water in all three of the major ocean basins and assuring that much of the world's deep ocean is well oxygenated. Its signature can be seen as far north as 45 °N in the Atlantic and Pacific Oceans (Kennett 1982; Eastman 1993).

The Antarctic system exhibits a considerable amount of diversity. The coastal system on the Western Antarctic Peninsula (WAP) shelf (Parker et al. 2011, 2015) is very different from that on the Ross Sea shelf (Figure 11.7) (Donnelly et al. 2004b), both in terms of the fauna that inhabits it and the temperatures they experience. The study of the Antarctic open-ocean system used here is AMERIEZ (**A**ntarctic **M**arine **E**cosystem **R**esearch in the **I**ce **E**dge **Z**one), a program of expeditions to the Scotia–Weddell Sea regions, vic 60 °S 40 °W, in the spring, fall, and winter seasons spread over three years (1983, 1986, and 1988). The goal of the program was to describe the physics and chemistry of the water column at an Antarctic ice edge at the outset of the spring ice retreat and the biological response to the awakening system. Data were acquired at all trophic levels from phytoplankton on up to seabirds and seals. Fall and winter cruises allowed the AMERIEZ team to describe system changes at the ice edge over the annual cycle. The ice-edge bloom happens each year, and it is responsible for much of the annual production in the Antarctic system. A follow-on expedition during the summer of 1993 to the same general locale allowed a team of scientists to complete the yearly cycle for open-ocean fauna with a summer sampling program.

Early studies on the Southern Ocean considered it to be a highly productive system (El Sayed 1984), due at least in part to seasonal and geographical bias in sampling (Smith and Sakshaug 1990). The large and highly visible stocks of krill, seabirds, and seals were assumed to be supported by a high annual primary production. In fact, the Southern Ocean as a whole is an oligotrophic system with an estimated annual production of $16 g C m^{-2} y^{-1}$ (Smith and Sakshaug 1990), though mesoscale variability is quite high, both in local production and in the biomass of higher trophic levels.

As the Antarctic sea ice retreats in the Austral spring from its northerly limit, the melting ice creates a meltwater lens at its edge (Figure 11.9), stratifying the surface waters that are now exposed to high light levels. The vertical stratification keeps phytoplankton cells in a favorable light environment for photosynthesis, and nutrients released from the ice-melt provide fuel for growth. As the ice continues to retreat, it leaves behind a highly productive region which, in the Weddell Sea, may extend up to 250 km (Smith and Sakshaug 1990) but is highly dependent on location. Life cycles of zooplankton species that include a winter dormant phase (diapause), such as the calanoid copepod *Allantoides acutus*, become active in time to exploit the ice edge bloom (Burghart et al. 1999).

Station locations and thermal profiles for each of the four AMERIEZ ice-edge cruises are given in Figure 11.10a and b. Hydrography of the four study areas was complex, with contributions by Weddell Sea Water, Circumpolar Deep Water (CDW), and Weddell–Scotia confluence water. The spring 1983 (Scotia Sea) and fall 1986 (Weddell Sea) (Lancraft et al. 1989) studies primarily sampled in the open water. They show a surface mixed layer overlying a residual winter water layer, visible in Figure 11.10a as a temperature minimum at about 75 m depth. From 75 m downward, temperatures climb toward a temperature maximum between 200 and 400 m, indicative of CDW. The winter expedition of 1988 (Weddell–Scotia Confluence Line) (Lancraft et al. 1991) shows a temperature profile from within the ice pack with freezing

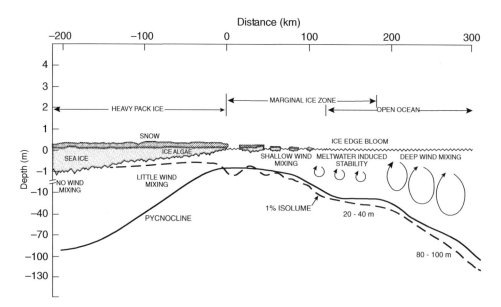

Figure 11.9 Schematic model of conditions necessary for ice edge phytoplankton bloom. *Source:* Sullivan et al. (1988), figure 1 (p. 12488), Journal of Geophysical Research Copyright © 1969 John Wiley and Sons. Reproduced with the permission of John Wiley & Sons.

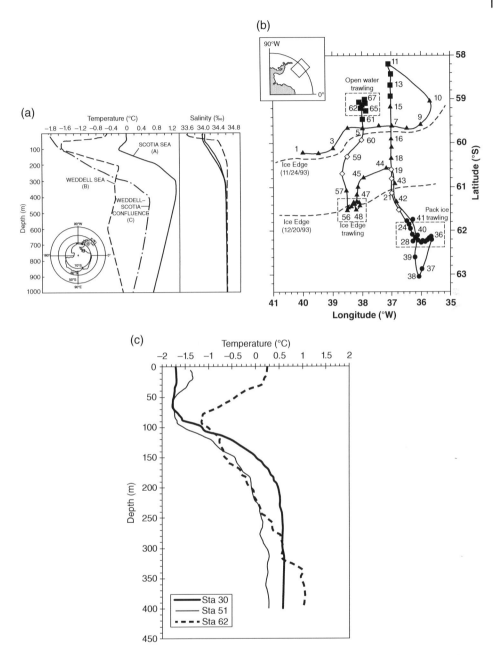

Figure 11.10 Station locations and hydrographic profiles for the AMERIEZ ice-edge cruises. (a) Hydrographic conditions in the Scotia Sea, open waters of the Weddell Sea during Spring, and the Weddell–Scotia confluence during winter in the ice pack. (b) cruise track in the NW Weddell Sea; (filled circle) pack ice stations, (filled triangle) ice edge stations, (filled square) open water stations, (open diamond) transitional stations. Dashed line represents approximate position of ice edge at the beginning and end of the cruise. Dashed boxes show trawling locations during multi-day occupations within each MIZ area. (c) Temperature profiles for representative stations within each MIZ area during the south-north trawling transect; station 30 = pack ice, station 51 = ice edge, station 62 = open water. *Sources*: (a) Lancraft et al. (1991), figure 1 (p. 159); (b) Donnelly et al. (2006), figure 1 (p. 281). (c) Donnelly et al. (2006), figure 2 (p. 283).

temperatures at the surface gradually warming with increasing depth until reaching a temperature maximum of about 0.5 °C at 400 m. The summer 1993 study was focused on micronekton, allowing time for serious trawling at multiple locations within the marginal ice zone. In open waters seaward of the receding ice edge (Station 62), a warm surface mixed layer extended to about 40 m, overlying winter water at about 75 m with a temperature minimum, gradually increasing with depth to a relatively warm 1.0 °C at 400 m. The ice edge station (51) and pack ice station (30) both showed near-freezing surface temperatures increasing with depth. Worthy of note is the pervasive influence of the warm CDW at depths below 200 m.

As the ice edge receded, it left behind a highly productive water column in the open water system with chlorophyll concentrations of $1.7\,\mu g\,l^{-1}$ and primary productivity of $500\text{--}750\,mg\,C\,m^{-2}\,d^{-1}$, extending for at least 150 km behind the core of the bloom. The core of the ice edge bloom exhibited a primary productivity of $1000\,mg\,C\,m^{-2}\,d^{-1}$ and chlorophyll concentration of $3.5\,\mu g\,l^{-1}$. Within the ice pack prior to the development of the bloom, chlorophyll concentrations were quite low, $0.4\,\mu g\,l^{-1}$, but the developing bloom rapidly increased concentrations from 250 to $1000\,mg\,C\,m^{-2}\,d^{-1}$ as the sea surface became exposed to light.

The Weddell–Scotia confluence is a transitional region that separates the cold Weddell Sea waters from the warmer waters of the Scotia Sea. From a physical perspective, it is quite complicated, but interesting. The Antarctic Coastal Current (East Wind Drift) that flows next to the continent from east to west is reflected east by the Antarctic Peninsula. There it meets the westward-flowing Circumpolar current in the vicinity of 60 °S as it rounds the peninsula, creating a diffuse transitional boundary characterized by mesoscale eddies and in some cases density anomalies that transport large bubbles of deeper water, with organisms included, to the surface (cf. Ainley et al. 1986). Very deep-dwelling species like the giant ostracod *Gigantocypris* were found in the guts of flighted seabirds. Originally thought to be a behavioral response by deep-sea species ascending in the water column due to reduced light levels within the pack ice, the bubble transport was later found to be the most likely cause of the unusual dietary data. That discovery is great example of how different disciplines working together can arrive at an interesting new result.

System Comparisons

Tables 11.1–11.7 compare the micronekton taxa of the three systems described above, using species composition in all cases and relative abundance where data were available. The direct comparisons provide a useful example of the similarities and differences between three open ocean systems with widely differing climatic regimes and levels of annual primary production. Perhaps not surprisingly, the great majority of available data on midwater communities come from tropical/subtropical areas, including the waters off Hawaii, Bermuda and in the vicinity of the Canary Islands where weather and sea conditions are more amenable to open ocean sampling programs. Publications resulting from such programs have been cited extensively in the preceding pages. Detailed taxonomic coverage is broadest in the GOM, and the authors of the book contributed to the data acquisition, making it a natural choice. Fewer data are available for the micronekton of temperate cold-water systems. The studies selected came from the long-term program founded by Bill Pearcy of Oregon State

Table 11.1 Species composition of the decapod and mysid assemblage in the upper 1000 m of the eastern Gulf of Mexico, Northern California Current and Antarctic. Data from Hopkins et al. (1989, 1994), Pearcy and Forss (1969), Lancraft et al. (1989, 1991), Donnelly et al. (2006), Burghart et al. (2007), and Brodeur et al. (2019). Species reported as present but lacking abundance data listed with "ND." Species occasionally captured in the mesopelagic zone but primarily found below 1000 m designated with a "B". Abundances reported as number of individuals per square meter in the upper 1000 m.

Gulf of Mexico (27°N 86°W)		California Current (44°N 126°W)		Antarctic (60°S 40°W)					
					Abundance (ind m^{-2})				
Taxon	# m^{-2}	Taxon	# m^{-2}	Taxon	SP	F	SP/F	W	SU
Order Decapoda		Order Decapoda		Order Decapoda					
Family Benthesicymidae		Family Benthesicymidae		Family Benthesicymidae					
Bentheogenemma intermedia	B	Bentheogenemma borealis	0.014	Gennadas kempi	0.0014^a	0.0014^a	0.0014^a	0.0014^a	0.0014^a
Gennadas bouvieri	B	Bentheogenemma burkenroadi	0.05					0.005	0.0053
Gennadas capensis	0.188	Gennadas incertus	0.003						
Gennadas scutatus	0.013	Gennadas propinquus	ND						
Gennadas valens	1.238								
Family Pasiphaeidae		Family Pasiphaeidae		Family Pasiphaeidae					
Parapasiphae sulcatifrons	B	Pasiphaea chacei	0.003	Pasiphaea scotiae			0.03	0.01^a	0.011
Pasiphaea merriami	0.0074	Pasiphaea pacifica	0.005						
		Parapasiphae sulcatifrons cristata	ND						
Family Sergestidae		Family Sergestidae		Family Sergestidae					
Sergestes armatus	0.023	Sergestes similis	7.09	Petalidium foliaceum			0.0014^a	0.01^a	
Sergestes atlanticus	0.097	Sergia tenuiremis	0.01						
Sergestes cornutus	0.032	Petalidium suspiriosum	0.024	Family Nematocarcinidae					

Table 11.1 (Continued)

Gulf of Mexico (27°N 86°W)		California Current (44°N 126°W)		Antarctic (60°S 40°W)					
					Abundance (ind m^{-2})				
Taxon	# m^{-2}	Taxon	# m^{-2}	Taxon	SP	F	SP/F	W	SU
Sergestes curvatus	0.005			*Nematocarcinus lanceopes*			0.085[a]	0.015[a]	0.0283
Sergestes edwardsi	0.036								
Sergestes henseni	0.091			Total abundance			0.116	0.04	0.045
Sergestes paraseminudus	0.044			Total species			4	5	4
Sergestes pectinatus	0.272								
Sergestes sargassi	0.063								
Sergestes vigilax	0.034								
Sergia filicta	0.014								
Sergia grandis	0.004								
Sergia japonicus	0.002								
Sergia robustus	0.069								
Sergia splendens	0.333								
Sergia talismani	0.016								
Sergia tenuiremis	0.005								
Family Olophoridae		Family Oplophoridae							
Acanthephyra acanthitelsonis	0.0067	*Acanthephyra curtirostris*	0.003						
Acanthephyra acutifrons	B	*Hymenodora frontalis*	0.58						
Acanthephyra curtirostris	0.0218	*Hymenodora gracilis*	0.013						
Acanthephyra purpurea	0.0338	*Notostomus japonicus*	ND						

Acanthephyra stylorostrata	0.0068	Systellaspis bravery	0.02
Ephyrina benedicti	B		
Ephyrina ombango	B	Total abundance	7.815
Janicella spinicauda	B	Total species	16
Meningodora mollis	B		
Meningodora vesca	B		
Notostomus elegans	ND		
Notostomus gibbosus	0.0016		
Notostomus westergreni	ND		
Oplophorus gracilirostris	0.00495		
Systellaspis cristata	B		
Systellaspis debilis	0.07		
Family Pandalidae			
Heterocarpus ensifer	ND		
Parapandalus richardi	0.037		
Plesionika polycanthomerus	ND		
Family Penaeidae			
Funchalia villosa	0.008		
Family Bresilidae			
Lucaya bigelowi	B		
Total abundance	2.79505		
Total species	45		

(Continued)

Table 11.1 (Continued)

Gulf of Mexico (27°N 86°W)		California Current (44°N 126°W)		Antarctic (60°S 40°W)				
				Abundance (ind m^{-2})				
Taxon	# m^{-2}	Taxon	# m^{-2}	Taxon	SP	F	SP/F	W	SU
Order Lophogastrida		Order Lophogastrida		Order Lophogastrida					
Family Eucopiidae		Family Eucopiidae		Family Eucopiidae					
Eucopia australis	0.0229	Eucopia unguiculata	ND	Eucopia australis	0.003[a]	ND	ND	0.005[a]	ND
Eucopia unguiculata	0.168								
Eucopia sculpticauda	0.092	Family Gnathophausiidae		Family Gnathophausiidae					
		Neognathophausia gigas	ND	Neognathophausia gigas	ND	ND	ND	0.005[a]	ND
Family Lophogastridae									
Neognathophausia ingens	0.0064								
Total abundance	0.2893								
Total species	4								
		Order Mysida		Order Mysida					
		Family Mysidae		Family Mysidae					
		Boreomysis californica	ND	Boreomysis inermis	0.0014[a]	ND	ND	ND	ND
		Boreomysis rostrata	ND	Boreomysis rostrata	ND	ND	ND	0.01[a]	ND
				Boreomysis sibogae	ND	0.0014	ND	ND	ND
		Total species	4						
				Total abundance	0.0044	0.0014		0.025	
				Total species	5	5	5	5	5

[a] Data were generated by dividing the total numbers captured by the total volume filtered during a given sampling effort. Thus, Spring + Fall numbers combined totals captured during the 1983 and 1986 cruises normalized to the total volume filtered.

Table 11.2 Species composition of the euphausiid assemblage in the upper 1000 m of the eastern Gulf of Mexico, Northern California Current and Antarctic. Data from Kinsey and Hopkins (1994), Gomez-Gutierrez et al. (2005), Lancraft et al. (1989, 1991), and Donnelly et al. (2006). Abundances reported as number of individuals per square meter in the upper 1000 m.

Gulf of Mexico (27°N 86°W)		California Current (44°N 126°W)		Antarctic (60°S 40°W)				
					Abundance (ind m^{-2})			
Taxon	# m^{-2}	Taxon	# m^{-2}	Taxon	Spring	Fall	Winter	Summer
Order Euphausiacea		Order Euphausiacea		Order Euphausiacea				
Family Euphausiidae		Family Euphausiidae		Family Euphausiidae				
Euphausia americana	26.5	Euphausia pacifica	731.5	Euphausia superba	4.34	6.05	11.527	2.54
Euphausia brevis	8.3	Nematoscelis flexipes	1.75	Euphausia triacantha	0.132[a]		0.576	0.019
Euphausia tenera	81.1	Nematoscelis atlantica	13.4	Euphausia frigida			0.021	
Euphausia hemigibba	23	Nematoscelis difficilis	7.8	Thysanoessa macrura	1.88	1.31	2.341	2.143
Euphausia pseudogibba	<0.1	Stylocheiron affine	0.55					
Euphausia mutica	2.9	Stylocheiron longicorne	0.2	Total abundance	6.22	7.36	14.465	4.702
Euphausia gibboides	1.2	Stylocheiron maximum	0.3	Total species	3	2	4	3
Nematoscelis microps	16	Thysanoessa spinifera	290.2					
Nematoscelis atlantica	3.5	Thysanoessa longipes	31.8					
Nematoscelis tenella	1.5	Thysanoessa gregaria	16.8					
Nematobrachion flexipes	0.4	Thysanoessa inermis	0.15					
Nematobrachion sexspinosus	<0.1	Thysanoessa inspinata	0.9					
Nematobrachion boopis	0.3	Thysanoessa inoculatum	8.6					
Stylocheiron abbreviatum	6.5							

(Continued)

Table 11.2 (Continued)

Gulf of Mexico (27°N 86°W)		California Current (44°N 126°W)		Antarctic (60°S 40°W)			
				Abundance (ind m^{-2})			
Taxon	# m^{-2}	Taxon	# m^{-2}	Taxon	Spring	Fall	Winter	Summer
Stylocheiron affine	6.5	Total abundance	1103.95					
Stylocheiron suhmii	3	Total species	13					
Stylocheiron carinatum	28							
Stylocheiron longicorne	10.7							
Stylocheiron elongatum	13.7							
Stylocheiron robustum	0.5							
Stylocheiron maximum	<0.1							
Thysanopoda aequalis	2							
Thysanopoda monacantha	0.4							
Thysanopoda tricuspida	0.1							
Thysanopoda orientalis	0.1							
Thysanopoda obtusifrons	<0.1							
Thysanopoda pectinata	<0.1							
Total abundance	236.2							
Total species	27							

[a] Data were generated by dividing the total numbers captured by the total volume filtered during a given sampling effort.

Table 11.3 Myctophids. Species composition of the myctophid assemblage in the upper 1000 m of the eastern Gulf of Mexico, Northern California Current, and Antarctic. Data from Pearcy (1964), Pearcy and Laurs (1966), Hopkins and Lancraft (1984), Lancraft et al. (1989, 1991), and Donnelly et al. (2006). Species reported as present but lacking abundance data listed with "ND" or "P." Abundances reported as number of individuals per square meter in the upper 1000 m or numbers per 1000 m^{-3}, which are roughly equivalent.

Gulf of Mexico (27°N 86°W)		California Current (44°N 126°W)		Antarctic (60°S 40°W)					
					Abundance (ind m^{-2})				
Taxon	TC/AB	Taxon	AB	Taxon	Spring	Fall	SP/F	Winter	Summer
Order Myctophiformes		Order Myctophiformes		Order Myctophiformes					
Family Myctophidae		Family Myctophidae		Family Myctophidae					
Abundant		**Abundant**		*Electrona Antarctica*	0.18	0.17		0.27	0.11
Benthosema suborbitale	30/0.0311	*Diaphus theta*	1.71	*Electrona carlsbergi*	0.001^a			0.007^a	
Ceratoscopelus warmingii	88/0.0913	*Stenobrachius leucopsaurus*	3.42	*Gymnoscopelus bravery*			0.043^a	0.13^a	0.036
Daphus dumerilii	78/0.0809	*Tarletonbeania crenularis*	0.45	*Gymnoscopelus nicholsi*		P (1)		P (2)	0.011
Lampanyctus alatus	72/0.0747	**Total**	5.58	*Gymnoscopelus opisthopterus*			0.027^a	P (3)	0.014
Lepidophanes guentheri	66/0.0685			*Gymnoscopelus bolini*		P (1)			
Myctophum affine	27/0.0280	**Common**		*Krefftichthys anderssoni*	P (5)			0.007^a	
Notolychnus valdiviae	26/0.0270	*Nannobrachium ritteri*	0.35	*Protomyctophum bolini*	P (2)			0.05^a	0.014
Total	387/0.04015	*Protomyctophum thompsoni*	0.27	*Protomyctophum tenisoni*				0.07^a	
Common		**Total**	0.62	**Total spp./abundance**	4/0.181	3/0.17	2/0.07	8/0.534	5/0.185
Bolinichthys photothorax	2/0.0021								
Centrobranchus nigroocellatus	1/0.0010	**Uncommon**							
Diaphus lucidus	7/0.0073	*Nannobrachium regale*	0.08						

(*Continued*)

Table 11.3 (Continued)

Gulf of Mexico (27°N 86°W)		California Current (44°N 126°W)		Antarctic (60°S 40°W)			
				Taxon	Abundance (ind m^{-2})		
Taxon	TC/AB	Taxon	AB		Spring Fall	SP/F	Winter Summer
D. mollis	25/0.0259	Protomyctophum crockeri	0.05				
D. splendidus	2/0.0021	Stenobrachius nannochir	0.09				
Diogenichthys atlanticus	9/0.0093	Symbolophorus californiensis	0.03				
Hygophum benoiti	42/0.0436	**Total**	**0.25**				
H. taaningi	13/0.0135	**Rare**					
Lampanyctus lineatus	17/0.0176	Ceratoscopelus townsendi	ND				
Total	**118/0.1224**	Lampadena urophaos	ND				
Uncommon							
Bolinichthys supralateralis	2/0.0021	**Total spp./abundance**	**11/6.45**				
Diaphus brachycephalus	4/0.0042						
D. luetkeni	2/0.0021						
D. rafinesqui	12/0.0125						
D. taaningi	1/0.0010						
D. termophilus	1/0.0010						
Gonichthys cocococo	2/0.0021						
Hygophum reinhardtii	2/0.0021						
Lampadena luminosa	3/0.0031						
Lampanyctus ater	6/0.0062						
L. cuprarius	2/0.0021						

L. nobilis	3/0.0031
L. tenuiformis	4/0.0042
Lobianchia gemellarii	2/0.0021
Myctophum nitidulum	1/0.0010
Notoscopelus resplendens	9/0.0093
Taaningichthys bathyphilus	1/0.0010
Total	**57/0.0591**
Rare	
Notoscopelus caudispinosus	1/0.0010
Total	**1/0.001**
Total spp./abundance	**33/0.5841**

SP/F, Spring/Fall; TC, total catch; AB, abundance (# 1000 m^{-3}).

[a] Data were generated by dividing the total numbers captured by the total volume filtered during a given sampling effort. Thus, for Antarctic myctophids, Spring + Fall numbers combined totals captured during the 1983 and 1986 cruises normalized to the total volume filtered. GOM catch data from Gartner et al. (1987) come from the eastern central Gulf (EC) totals divided by the total volume filtered from the Columbus Iselin (CI) cruises.

Table 11.4 Stomiiformes. Species composition and abundance of the pelagic Stomiiformes fish assemblage in the upper 1000 m of the eastern Gulf of Mexico, Northern California Current, and Antarctic. Data from Hopkins and Lancraft (1984), Pearcy (1964), Pearcy and Laurs (1966), Pearcy et al. (1977), Willis and Pearcy (1982), Lancraft et al. (1989, 1991), and Donnelly et al. (2006). Species that were captured during a study, but for which no volume-filtered data were available, are designated with a "P" followed by the total number captured.

Gulf of Mexico (27°N 86°W)		California Current (44°N 126°W)		Antarctic (60°S 40°W)					
						Abundance (ind m^{-2})			
Taxon	TC/AB	Taxon	AB	Taxon	Spring	Fall	SP/F	Winter	Summer
Order Stomiiformes		Order Stomiiformes		Order Stomiiformes					
Family Gonostomatidae		Family Gonostomatidae		Family Gonostomatidae					
Cyclothone acclinidens	296/0.307	Cyclothone acclinidens	0.188	Cyclothone microdon			0.019	0.036	0.023
Cyclothone alba	132/0.137	Cyclothone atraria	0.531	Cyclothone pallida				P (1)	
Cyclothone bravery	127/0.132	Cyclothone microdon		Cyclothone kobayashii			0.002	0.007	0.019
Cyclothone microdon	16/0.017	Cyclothone pallida							
Cyclothone obscura	10/0.010	Cyclothone pseudopallida	0.039						
Cyclothone pallida	1080/1.121	Cyclothone signata	1.086						
Cyclothone pseudopallida	358/0.372								
Cyclothone sp.	37/0.038								
Bonapartia pedialota	1/0.001								
Gonostoma atlanticum	1/0.001								
Sigmops elongatus	39/0.040								
Margrethia obtusirostra	1/0.001								
Maurolicus mulleri	1/0.001								
Polichthys mauli	1/0.001								

Valencienellus tripunctulatus	90/0.093			
Vinciguerria nimbaria	8/0.008			
Vinciguerria poweriae	9/0.009			
Family Sternoptychidae		Family Sternoptychidae		
Argyropelecus aculeatus	21/0.022	Danaphos occulatus		
Argyropelecus affinis	1/0.001	Argyropelecus intermedius		
Argyropelecus gigas	1/0.001	Argyropelecus lychuns		
Argyropelecus hemigyrmnus	50/0.052	Argyropelecus pacificus		
Argyropelecus sladeni	3/0.003	Argyropelecus sladeni	0.049	
Sternoptyx diaphana	51/0.053			
Sternoptyx pseudobscura	43/0.045			
Polyipnus asteroides	1/0.001			
Polyipnus polli	4/0.004			
Polyipnus sp.	4/0.004			
Family Stomiidae		Family Stomiidae		Family Stomiidae
Astronesthes indicus	1/0.001	Chauliodus macouni	0.340	Borostomias antarcticus P (1)
Astronesthes macropogon	1/0.001	Idiacanthus antrostomus		
Astronesthes niger	2/0.002	Aristostomias scintillans		Total abundance 0.021 0.043
Astronesthes similis	1/0.001	Bathophiolus flemingi		
Chauliodus sloani	15/0.016	Tactostoma macropus	1.486	
Aristostomias grimaldi	1/0.001			
Aristostomias tittmanni	1/0.001	Total abundance	3.719	
Aristostomias sp.	5/0.005			
Bathophilus nigerrimus	1/0.001			

(*Continued*)

Table 11.4 (Continued)

Gulf of Mexico (27°N 86°W)		California Current (44°N 126°W)			Antarctic (60°S 40°W)				
						Abundance (ind m^{-2})			
Taxon	TC/AB	Taxon	AB	Taxon	Spring	Fall	SP/F	Winter	Summer
Bathophilus pawneei	3/0.003								
Bathophilus sp.	1/0.001								
Leptostomias bilobatus	1/0.001								
Melanostomias melanops	1/0.001								
Melanostomias valdiviae	1/0.001								
Photostomias guernei	12/0.012								
Total species/abundance	**42/2.525**								

SP/F, Spring/Fall; TC, total catch; AB, abundance (# 1000 m^{-3}).

Table 11.5 Non-myctophid and non-stomiiform fishes. Species composition and abundance of the non-myctophid and non-stomiiformes pelagic fish assemblage in the upper 1000 m of the eastern Gulf of Mexico, Northern California Current, and Antarctic. Data from Hopkins and Lancraft (1984), Pearcy (1964), Pearcy and Laurs (1966), Pearcy et al. (1977), Willis and Pearcy (1982), Lancraft et al. (1989, 1991), and Donnelly et al. (2006). Species that were captured during a study but for which no volume-filtered data were available are designated with a "P" followed by the total number captured.

Gulf of Mexico (27°N 86°W)		California Current (44°N 126°W)		Antarctic (60°S 40°W)				
					Abundance (ind m^{-2})			
Taxon	TC/AB	Taxon	AB	Taxon	Spring	Fall	Spring/Fall	winter	Summer
Order Alepocephaliformes		Order Alepocephaliformes						
Family Alepocephalidae		Family Alepocephalidae						
Photostylus pycnopterus	1/0.001	*Talismania bifurcata*						
Family Platytoctidae	1/0.001	Family Platytroctidae						
Order Tetraodontiformes		*Sagamichthys abei*						
Family Balistidae	2/0.002	Order Argentifiniformes		Order Argentifiniformes				
Order Argentifiniformes		Family Bathylagidae		Family Bathylagidae				
Family Bathylagidae		*Lipolagus ochotensis*	0.241	*Bathylagus antarcticus*			0.130	0.0180	0.068
Melanolagus bericoides	1/0.001	*Pseudobathylagus milleri*	0.114					
Dolicholagus longirostris	8/0.008	*Bathylagus pacificus*	0.090					
Order Ophidiiformes		Family Opisthoproctidae						
Family Brotulidae	3/0.003	*Bathylychnops exilis*						
Order Perciformes		*Macropinna microstoma*	0.027	Family Microstomatidae				
Family Carngidae				*Nansenia Antarctica*	P (4)			
Decapterus punctatus	1/0.001							

(*Continued*)

Table 11.5 (Continued)

Gulf of Mexico (27°N 86°W)		California Current (44°N 126°W)		Antarctic (60°S 40°W)			
					Abundance (ind m^{-2})		
Taxon	TC/AB	Taxon	AB	Taxon	Spring Fall	Spring/Fall	winter Summer
Selar crumenophthalmus	2/0.002						
Order Lophiiformes	6/0.006	Order Lophiiformes	2 spp.				
Order Beryciformes		Order Beryciformes		Order Beryciformes			
Family Melamphaidae		Family Melamphaidae		Family Melamphaidae			
Melamphaes longivelis	1/0.001	*Melamphaes lugubris*	0.042	*Poromitra crassicips*	P (1)		
Melamphaes typhlops	7/0.007	*Poromitra crassicips*	0.065				
Melamphaes sp.	5/0.005						
Poromitra crassiceps	2/0.002						
Scopeloberyx opisthopterus	7/0.007						
Scopeloberyx robustus	3/0.003						
Scopelogadus mizolepis	1/0.001						
Family Cetomimidae	1/0.001	Family Cetomimidae					
Family Rondoletiidae	1/0.001	*Cetostomus regani*					
Rondoletia bicolor	1/0.001						
Order Scombriformes				Order Scombriformes			
Family Chiasmodontidae				Family Gempylidae			
Dysalotus alcocki	1/0.001			*Paradiplospinus gracilis*	P (4)		0.022

Taxon	TC/AB	Taxon	AB	AB	AB
Kali kerberti	1/0.001				
Pseudoscopelus altipinnis	1/0.001				
Order Aulopiformes		Order Aulopiformes			
Family Evermannellidae		Family Scopelarchidae			
Coccorella atlantica	2/0.002	*Benthalbella dentata*	0.057		
Odonostomops normalops	1/0.001				
Family Paralepididae		Family Paralepididae			
Family Omosudidae	2/0.002	*Lestidium ringens*	0.041		
Omosudis lowei	5/0.005				
Family Scopelarchidae					
Scopelarchus analis	2/0.002				
Order Gadiformes		Order Aulopiformes			
Family Melanonidae		Family Scopelarchidae			
Melanonus sp.	3/0.003	*Benthalbella elongata*		P(2)	
Family Bregmacerotidae		Family Paralepididae			
Bregmaceros atlanticus	8/0.008	*Notolepis coatsi*	0.236	0.102	0.014
Order Anguilliformes					0.124
Family Nemichthyidae		Order Gadiformes			
Avocettina infans	4/0.004	Family Macrouridae			
Family Serrivomeridae		*Cyanomacrurus piriei*		P(3)	
Serrivomer brevidentratus	1/0.001	Family Muraenolepididae			
		Muraenolepis microps		0.016	
		Order Trachichthyformes			
		Family Anoplogastridae			
		Anoplogaster cornuta			
Total species/abundance	32/0.087	**System total species/partial abundance**	31/4.396		
		Total species/abundance	4/0.387	8/0.347	4/0.124

TC, total catch; AB, abundance (# 1000 m^{-3}).

Table 11.6 The cephalopods. Species composition of the cephalopod assemblage in the upper 1000 m of the eastern Gulf of Mexico, Northern California Current, and Antarctic. Data for the GOM (Passarella and Hopkins 1991) and NCC (Pearcy 1965) acquired using scientific trawls varying in size from 3.2 to 6.5 m². In the Antarctic, Collins and Rodhouse (2006) summarize data from a variety of expeditions using scientific nets and larger commercial style midwater trawls. Entries marked with an asterisk from Brodeur et al. (2019). A more complete data set for the GOM acquired using a variety of different nets and sources may be found in Judkins et al. (2009). See text.

Gulf of Mexico	California Current	Antarctica
Taxon	Taxon	Taxon
Order Sepioidea	Order Sepioidea	
Family Sepiolidae	Family Sepiolidae	
Heteroteuthis dispar	*Rossia pacifica*	
Order Spirulida		
Family Spirulidae		
Spirula spirula		
Order Myopsida	Order Myopsida	
Family Loliginidae	Family Loliginidae	
	Dorytheuthis opalescens	
Order Oegopsida	Order Oegopsida	Order Oegopsidae
Family Brachioteuthidae		Family Brachioteuthidae
Brachioteuthis riisei		*Brachioteuthis linkovski*
		Slosarczykovia circumantarctica
Family Cranchiidae	Family Cranchiidae	Family Cranchiidae
Bathothauma lyromma	*Galiteuthis armata*	*Galikteuthis glacialis*
Cranchia scabra	*Taonius pavo*	*Mesonychoteuthis hamiltoni*
Galiteuthis armata	*Cranchia scabra*	
Egea inermis		Family Batoteuthidae
Helicocranchia pfefferi		*Batoteuthis skolops*
Leachia sp.		
Liocranchia reinhrdti		Family Psychroteuthidae
Megalocranchia spp.		*Psychroteuthis glacialis*
Family Cycloteuthidae		
Discoteuthis discus		
Family Enoploteuthidae	Family Enoploteuthidae	
Abralia redfieldi	*Abraliopsis felis**	
Abralia veranyl		
Abraliopsis atlantica	Family Gonatidae	
Abraliopsis pfefferi	*Gonatus fabricii*	
Enoploteuthis anapsis	*Gonatus magister*	

Table 11.6 (Continued)

Gulf of Mexico	California Current	Antarctica
Taxon	Taxon	Taxon
Enoploteuthis leptura	*Gonatus onyx**	
Family Pyroteuthidae	*Gonatus anonychus*	
Pterygioteuthis gemmata	*Gonatopsis borealis*	
Pterygioteuthis giardi		
Pyroteuthis margaritifera		
Family Ancistrocheiridae		
Ancistrocheirus lesueuri		
Family Histioteuthidae	Family Histioteuthidae	
Histioteuthis corona	*Meleagroteuthis boleyi*	
Stigmatoteuthis dofleini		
Family Bathyteuthidae		Family Bathyteuthidae
Bathyteuethis abyssicola		*Bathyteuthis abyssicola*
Family Chaenopterygidae		
Chtenopteryx sicula		
Family Lycoteutidae		
Lampadioteuthis megleia		
Family Ommastrephidae		
Ornithoteuthis antillarum		
Sthenoteuthis pteropus		
Family Chrioteuthidae	Family Chiroteuthidae	
Chiroteuthis capensis	*Chiroteuthis calyx**	
Chiroteuthis joubini	*Chiroteuthis veryanyi*	
Family Octopoteuthidae	Family Octopoteuthidae	
Octopoteuthis megaptera	*Octopoteuthis sicula*	
Tanangia danae		
Family Onychoteuthidae	Family Onychoteuthidae	Family Onychoteuthidae
Onykia carriboea	*Onychoteuthis banksi*	*Kondakovia longimana*
Family Thysanoteuthidae	*Moroteuthis robusta*	*Moroteuthis ingens*
Thysanoteuthis rhombus		*Moroteuthis knipovitchi*
Family Mastigoteuthida		Family Mastigoteuthidae
Mastigoteuthis sp.		*Mastigoteuthis psychrophila*
Family Joubinoteuthidae		
Joubinoteuthis portieri		Family Neoteuthidae
		Alluroteuthis antarcticus

(*Continued*)

Table 11.6 (Continued)

Gulf of Mexico	California Current	Antarctica
Taxon	Taxon	Taxon
Order Octopoda	Order Octopoda	
Family Bolitaenidae	Family Bolitaenidae	
Japetella daphana	*Japetella heathi*	
Family Alloposidae		
Haliphron atlanticus		
Family Argonautidae		
Argonauta argo		
Order Vampyromorpha	Order Vampyromorpha	
Family Vampyroteuthidae	Family Vampyroteuthidae	
Vampyroteuthis infernalis	*Vampyroteuthis infernalis*	

University and continuing to this day (cf. Brodeur et al. 2019). The polar studies most comparable to the other two were our own, generated during the AMERIEZ program briefly described above and spanning all four seasons.

An interesting characteristic of open ocean systems is that the taxonomic groupings are similar from place to place, though overall diversity is not. For example, myctophids, gonostomatids, and bathylagids are found in all three regions treated here, but most of the other midwater fish families drop out in the polar system.

The Decapods and Mysids

The species richness of the eastern Gulf of Mexico, the ecological term for total species numbers, dwarfs that of the highly productive Northern California Current and the Antarctic (Table 11.1). Forty-six species of decapods inhabit the GOM resulting in a total decapod abundance of about 2.8 ind m^{-2} in the upper 1000 m. Dendrobranchs dominate the system, contributing the four most abundant species: *Gennadas valens, Sergia splendens, Sergestes pectinatus,* and *Gennadas capensis* as well as half the total species' numbers. *Systellaspis debilis* and *Acanthephyra curtirostris* dominate the Caridea. With respect to decapods, the GOM can be characterized as a high-diversity low biomass system.

The northern California Current (NCC) system is quite different, numbering 20 species in total with two clear dominants and an overall species abundance of about 7.8 ind m^{-2}, nearly three times that of the GOM. One species, *Sergestes similis*, is the overwhelming dominant, with an abundance of 7.09 ind m^{-2}. The other dominant, *Hymenodora frontalis*, is a caridean with the modest abundance of 0.58 ind m^{-2}. The NCC can be classified as a low to modest diversity, high biomass system supported by a high annual primary production.

In the Antarctic pelagial, decapods are nearly absent. Five species were captured during the AMERIEZ program, and all of them were rare. *Pasiphaea scotiae* and *Nematocarcinus lanceopes* were found most often, with *Nematocarcinus* persisting furthest south in the overall study area. The

Table 11.7 Gelatinous zooplankton and amphipods. Species composition of the gelatinous zooplankton assemblage in the upper 1000 m of the eastern Gulf of Mexico and Antarctic system. No comparable data were available for the NCC. Data from Hopkins and Lancraft (1984), Lancraft et al. (1989, 1991), and Donnelly et al. (2006). Species reported as present but lacking the needed data for computing a normalized abundance listed with a "p" followed by the total number captured. Abundances reported as number of individuals 1000 m^{-3} in the upper 1000 m or number of individuals per square meter in the upper 1000 m. Both express the number of individuals 1000 m^{-3} in the upper 1000 m, but they are calculated differently. Number of individuals 1000 m^{-3} were generated by dividing the total numbers captured by the total volume filtered during a given sampling effort. Number of individuals per square meter were calculated by dividing a species' total number in each tow by the total volume filtered and summing all tows vertically. For all Antarctic species but *Salpa thompsoni*, the Spring+Fall numbers combined totals captured during the 1983 and 1986 cruises normalized to the total volume filtered.

Gulf of Mexico		Antarctica – Spring + Fall		Antarctica – Winter		Antarctica – Summer	
Taxon	TC/AB	Taxon	TC/AB	Taxon	TNC/AB	Taxon	AB
Phylum Cnidaria		Phylum Cnidaria		Phylum Cnidaria		Phylum Cnidaria	
Class Hydrozoa		Class Hydrozoa		Class Hydrozoa		Class Hydrozoa	
Order Leptothecata		Order Anthoathecata		Order Anthoathecata		Order Anthoathecata	
Family Aequoreidae	12/0.012	Family Bythotiaridae		Family Bythotiaridae		Family Bythotiaridae	
Class Scyphozoa		*Calycopsis borchgrevinki*	117/0.081	*Calycopsis borchgrevinki*	16/0.116	*Calycopsis borchgrevinki*	0.06
Order Semaeostomeae	11/0.011	Class Scyphozoa		Order Trachymedusae		Class Scyphozoa	
Order Coronatae		Order Coronatae		Family Rhopalonematidae		Order Coronatae	
Family Atollidae		Family Atollidae		*Crossota brunnea*	1/0.007	Family Atollidae	
Atolla sp	5/0.005	*Atolla wyvillei*	34/0.023	Family Halicreatidae		*Atolla wyvillei*	0.016
Family Periphyllidae		Family Periphyllidae		*Halicreas minimum*	7/0.051	Family Periphyllidae	
Periphylla	11/0.011	*Periphylla*	19/0.013	Order Siphonophorae		*Periphylla*	0.01
Order Semaeostomeae		Order Semaeostomeae		Family Diphyidae			
Stygiomedusa gigantea	52/0.054	*Stygiomedusa gigantea*	1/ND	*Diphyes Antarctica*	61/0.441		
Phylum Ctenophora				Class Scyphozoa			
				Order Coronatae			

(Continued)

Table 11.7 (Continued)

Gulf of Mexico		Antarctica – Spring + Fall		Antarctica – Winter		Antarctica – Summer	
Taxon	TC/AB	Taxon	TC/AB	Taxon	TNC/AB	Taxon	AB
				Family Atollidae			
				Atolla wyvillei	8/0.058		
				Family Peripphyllidae			
				Periphylla	1/0.007		
				Order Semaeostomeae			
				Stygiomedusa gigantea	1/0.007		
				Phylum Ctenophora			
				Class Tentaculata			
				Order Cydippida			
				Family Mertensiidae			
				Callianira Antarctica	6/0.043		
				Class Nuda			
				Family Beroidae			
				Beroe sp.	4/0.029		
Phylum Annelida		Phylum Annelida		Phylum Annelida		Phylum Annelida	
Class Polychaeta	7/0.007	Class Polychaeta		Class Polychaeta		Class Polychaeta	
		Order Phyllodocida		Order Phyllodocida		Order Phyllodocida	
		Family Tomopteridae		Family Tomopteridae		Family Tomopteridae	
		Tomopteris carpenteri	24/0.017	*Tomopteris carpenteri*	19/0.137	*Tomopteris carpenteri*	0.046
		Family Typhloscolecidae		Family Typhloscolecidae			
		Travisiopsis coniceps	3/0.002	*Travisiopsis coniceps*	1/0.007		

Taxon	n/value	Taxon	n/value	Taxon	n/value	value
Phylum Mollusca		Family Alciopidae		Family Alciopidae		
Class Gastropoda		*Vanadis Antarctica*	95/0.066	*Vanadis Antarctica*	2/0.014	
Order Littorinimorpha		Phylum Mollusca		Phylum Mollusca		
Heteropods	47/0.049	Class Gastropoda		Class Gastropoda		
		Order Thecosomata		Order Thecosomata		
		Family Cliidae		Family Cliidae		
Pteropods	9/0.009	*Clio pyramidata* (fall only)	953/1.310	*Clio pyramidata*	1/0.007	
Phylum Arthropoda		Phylum Arthropoda		Phylum Arthropoda		
Class Malacostraca		Subphylum Crustacea		Subphylum Crustacea		
Order Amphipoda	25/0.026	Class Malacostraca		Class Malacostraca		
		Order Amphipoda		Order Amphipoda		
		Suborder Hyperiidea		Suborder Hyperiidea		
		Family Cyllopodidae		Family Cyllopodidae		
		Cyllopus lucasii (spring)	43/0.059	*Cyllopus lucasii*	21/0.152	0.063
		Cyllopus lucasii (fall)	158/0.217	Family Hyperiidae		
		Family Phrosinidae		*Themisto gaudichaudii*	7/0.051	0.022
		Primno macropa	16/0.011	Family Megalanceolidae		
		Family Vibiliidae		*Megalanceola stephensi*	1/0.007	0.071
		Vibilia stebbingi	3/0.002	Family Vibiliidae		
		Family Lanceolidae		*Vibilia stebbingi*		0.141
		Lanceola sp. 1986	2/0.003	Suborder Gammaridea		
		Suborder Gammaridea		Family Lysianassidae		
		Family Lysianassidae		*Cyphocaris richardi*	2/0.014	
		Cyphocaris richardi	20/0.014	*Cyphocaris faueri*	1/0.007	
				Parandania boecki	1/0.007	
				Cyphocaris richardi	1/0.007	0.029
				Family Eusiridae		

(*Continued*)

Table 11.7 (Continued)

Gulf of Mexico		Antarctica – Spring + Fall		Antarctica – Winter		Antarctica – Summer	
Taxon	TC/AB	Taxon	TC/AB	Taxon	TNC/AB	Taxon	AB
		Cyphocaris faueri	5/0.003			*Eusirus antarcticus*	0.012
		Parandania boecki	3/0.002				
		Family Eusiridae					
		Eusirus microps 1986	1/0.002				
Phylum Chordata		Phylum Chordata		Phylum Chordata		Phylum Chordata	
Class Thaliacea		Class Thaliacea		Class Thaliacea		Class Thaliacea	
Order Salpida		Order Salpida		Order Salpida		Order Salpida	
Family Salpidae	4/0.004	Family Salpidae		Family Salpidae		Family Salpidae	
Order Pyrosomatida		*Salpa thompsoni* (spring)	82 881/145.33[a]	*Salpa thompsoni*	157/1.135	*Salpa thompsoni*	11.282
Family Pyrosomatidae	1/0.001	*Salpa thompsoni* (fall)	10 523/ 5.62[a]				

TC, total catch; AB, abundance; TNC, total night catch.
[a] Abundances for *Salpa thompsoni* were computed as numbers per square meter in the upper 1000 m.

dendrobranchs *Gennadas* and *Petalidium* were found only in the northernmost portions (vic 60°S) of the study area. The genus *Nematocarcinus* is largely benthic with some demersal representatives. *Nematocarcinus lanceopes* has long spidery pereiopods that would seem best suited to a benthic life, though in the AMERIEZ studies, they were captured at mesopelagic depths in a mouth-closing Tucker trawl. *Nematocarcinus* has been photographed on the bottom in the southeastern Weddell Sea (Gutt et al. 1991) at depths of 800–1200 m and captured in large numbers in benthic trawls (Arntz and Gorny 1991) at depths of up to 4000 m. Its presence in both midwater and benthic trawls is best explained by a demersal lifestyle, where individuals come significantly off the bottom at intervals to forage, or perhaps to release larvae nearer surface production.

The paucity of decapod species in the Antarctic was first described in detail in the 1960s (Yaldwyn 1965) for the benthic and pelagic realms and is now a well-accepted characteristic of the system. The next obvious question is why they are so poorly represented when they are found throughout the globe at lower mesopelagic and bathypelagic depths, which though not as frigid as the Antarctic are cold and food-limited.

Since the most vulnerable point in the development of most species is their early life history, it makes the most sense to look there first. The dendrobranchs are broadcast spawners, producing small planktotrophic larvae that rely on surface productivity for sustenance (see Chapter 7). The carideans are brooders with larger eggs and larvae that hatch at a more advanced state of development. Dendrobranchs have many more larval stages than carideans, who hatch as zooea and can begin feeding soon after. Dendrobranchs are slightly more numerous in the more benign environment of the Western Antarctic Peninsula shelf (Parker et al. 2011, 2015), though the number of species remains the same as in the open waters sampled during the AMERIEZ study.

Both groups must contend with a highly seasonal, short-lived period of primary and secondary production to feed their young larvae. In addition, the frigid temperatures of the Antarctic system result in prolonged larval development times, meaning that as the larvae grow and the productive season winds down their food is likely to become quite scarce. The "boom and bust" character of Antarctic production in concert with slow larval development combines for a difficult early life. Antarctic decapods, notably *Nematocarcinus lanceopes*, are believed to reproduce only once every two years (Thatje et al. 2003, 2005).

To combat the vagaries of a very short productive season, an abbreviated planktotrophic larval development strategy with only two to four instars accompanied by larger energy-rich eggs for increased energy reserves in the Zoea I stage has been observed in Antarctic carideans (Thatje et al. 2003). This confers greater resistance to starvation in the most vulnerable first planktotrophic stage.

The Euphausiids

Euphausiids are an important component of all open ocean systems, feeding on a diversity of smaller prey ranging from phytoplankton cells to smaller zooplankton (see Chapter 7) (Table 11.2). The trends in species' numbers observed with decapods and mysids in Table 11.1 are repeated with euphausiids, though the decline is not quite as pronounced with increasing latitude. Note that abundance numbers for the California Current are not directly comparable to those of the GOM and Antarctic because samples were taken only at the depths of maximum euphausiid abundance (Gomez-Gutierrez et al. 2005) instead of over the entire upper 1000 m of the water column, which inflates total abundance numbers relative to the other two systems. The trends in species dominance hold quite well, however. Two genera hold sway in both the NCC and Antarctic systems, one a suspension feeder (*Euphausia* sp.) specializing in phytoplankton with the other (*Thysanoessa* sp.) more of a plankton picker. The relative abundance numbers show the overwhelming dominance of just two species in the cold-water systems and the more even spread in the GOM.

Like the decapods, the early life history of Antarctic euphausiids must conquer the short "boom and bust" productive season and must do so with an early life history containing more larval instars than the dendrobranchs (Chapter 7). That they do so effectively is manifest in the presence of super-swarms with fabulous numbers of *Euphausia superba* (krill) that episodically appear in the open ocean system and a more persistent, very large (796 ind m^{-2}, Lancraft et al. 2004) resident population off the West Antarctic Peninsula. In the open ocean system, *E. superba* is most abundant at the ice edge itself. Its locomotory abilities and large size (up to 63 mm total length in adults) allow it to seek out productive areas which typically occur in the vicinity of the ice edge (Daly and Macaulay 1988). *E. superba* is a keystone species. It is of critical importance in the dynamics of the whole Antarctic ecosystem through its role as a major prey item for everything from fishes to sea birds, penguins, and marine mammals.

The other dominant Antarctic species, *Thysanoessa macrura*, also has a circumantarctic distribution, making a consistent but smaller contribution to total euphausiid numbers in the region. *Euphausia triacantha* is a deeper-living species (Chapter 7) found associated with Circumpolar Deep Water (Baker 1957), that was captured chiefly in the northernmost stations of the four seasonal studies. *E. frigida* though considered an Antarctic species (Mauchline and Fisher 1969) is chiefly found north of 60°S and was a rare capture.

Significant differences in the life history strategies of the euphausiids in the three systems account for the large disparities in diversity and abundance of populations in the three study areas. Those factors will be discussed to the next chapter.

The Myctophids

Myctophid species in the three regions exhibited the same general trends observed in the decapods and euphausiids (Table 11.3). Species richness was far higher in the GOM than in the cold-water systems, with dominance shared by seven species, in contrast with the three overwhelmingly dominant species off the Oregon coast and the three main species found in the Antarctic system.

Data for the GOM come from a single extended cruise designed to elucidate total abundance and biomass by using 19 repetitive oblique tows in the upper 1000 m (Hopkins and Lancraft 1984). Those results were compared to a more exhaustive study (not shown) combining data from 465 tows over 7 years (Gartner et al. 1987). Total numbers of species were far higher in the more exhaustive study, but dominants remained the same, as did relative numbers of the less abundant species. The study by Hopkins and Lancraft 1984 was chosen for comparison to those in the Northern California Current (NCC) and the Antarctic cruises reported in Lancraft et al. (1989, 1991) and Donnelly et al. (2006) because of its similarity in the type of sampling gear and overall sampling effort.

The three dominant NCC myctophids are found throughout the Northeastern Pacific, though relative numbers vary with location (Frost and McCrone 1979; Beamish et al. 1999). *Stenobrachius leucopsaurus* clearly dominates the myctophid fauna off Washington state and Oregon. Further south off California, its importance begins to wane (Neighbors and Wilson 2006), though it is still abundant. All three of the NCC dominant myctophids are vertical migrators, with daytime distributions between 150 and 500 m and nighttime depths in the upper 150 m. The three species comprise about 52% of the total numbers of mesopelagic fishes in the upper 1000 m of the water column (Pearcy and Laurs 1966). It is worth noting that though the tropical-subtropical GOM has far more species than the NCC, only seven species comprise 72% of the total numbers. Even tropical systems with high diversity have clear dominants.

The Antarctic midwater has far fewer myctophid species altogether, and those listed as present (P) were only captured incidentally, usually at the northernmost stations in the four seasonal studies. The waters in the vicinity of 60 °S form an ecotone, where species with Subantarctic affinities (e.g. *Protomyctophum bolini*, *Krefftichthys andersoni*) are occasionally captured along with the normal complement of Antarctic species. *Gymnoscopelus nicholsi* in particular is found commonly in the waters offshore of the Western Antarctic Peninsula (McGinnis 1982; Parker et al. 2011, 2015). Sea ice dictated the latitudinal extent of sampling which ranged from 57 °S in the spring to 65 °S in the fall. The three clearly dominant species were *Electrona Antarctica*, *Gymnoscopelus bravery*, and *G. opisthopterus*, which was the largest of the three. Smith's Sea Fishes ranks *G. opisthopterus* as the largest myctophid, with lengths reaching greater than 150 mm. In the fall sampling season, *G. opisthopterus* replaced *G. bravery* as the second most numerous lanternfish (behind *Electrona Antarctica*) at the southernmost stations.

Non-myctophid Fishes

Species richness of non-myctophid fishes across the three oceanic systems mirrors the trends observed in the three previous taxonomic groupings (Tables 11.4 and 11.5). However, besides species numbers, the taxa represented showcase one of the more interesting trends observed in meso- and bathypelagic fishes and that is the far-ranging distribution of some species. For example, *Cyclothone microdon*, a deep-dwelling species of *Cyclothone*, is found in all three systems, as is the even deeper-living *C. pallida* (Burghart et al. 2010; Table 11.4). The hatchetfish *Argyropelecus sladeni* is found in the GOM and the NCC (Table 11.4), and the melamphaid *Poromitra crassiceps* is found in all three systems (Table 11.5).

The family Bathylagidae is strongly represented in all three (Table 11.5). In the Antarctic, *Bathylagus antarcticus* is one of the two dominant open ocean species found in the system, not only in the Scotia–Weddell Sea region but also in the open waters of the Eastern Ross Sea (Donnelly et al. 2004) and the WAP shelf (Parker et al. 2011, 2015). Interestingly, during winter, the mesopelagic species reside at greater depth than in the other three seasons (Lancraft et al. 1991), well away from the ice crystals that could spell death, since they lack the antifreezes that allow Antarctic endemics to survive in close proximity to ice (Cullins et al. 2011).

The system comparisons presented in Tables 11.1–11.7 show a clear gradient in species richness, with the highest number of species in the GOM, an intermediate number in the NCC and the fewest in the Southern Ocean. Overall abundance is highest in the NCC, with numbers of individuals varying from two to ten times the totals observed in the GOM. The fewest species and the lowest numbers are both encountered in the Antarctic system, which flies in the face of conventional wisdom based on sampling in regions of high biomass, notably the WAP shelf during the spring and summer seasons. In the great majority of the Antarctic system, resident populations of open ocean species are quite modest, even though in years past it was considered the richest of the three (Smith and Sakshaug 1990).

In any comparison of systems, sampling gear and sampling intensity are important considerations when weighing the data. In Table 11.3, two data sets were mentioned for species composition and abundance of myctophids in the GOM, one with 465 tows and the other with only 19. The major differences between the two were that the total numbers of species were far higher in the Gartner (1987) study, but dominants remained the same in both studies, as did relative numbers of the less abundant species.

For less abundant species, like those listed in Tables 11.4 and 11.5, the disparity is even more pronounced. A comprehensive study of stomiids in the Gulf of Mexico (Sutton and Hopkins 1996) using the catches from 1155 trawl samples resulted in a list totaling 72 species in 18 genera. Compare those results with the 15 stomiid species reported in Table 11.4. The dominants in the two studies are the same, but the numbers of species and total abundance (0.186 vs. 0.049 ind m^{-2}) of the dragonfishes in the GOM are very much greater in the more comprehensive study. That said, the most similar studies are those presented in Table 11.4 giving the most realistic regional comparisons.

Fewer data were available for the non-myctophid fishes of the NCC. The stomiid *Tactostoma macropus* is a clear dominant in the system, as is its confamilial *Chauliodus macouni*. The most abundant group comprises the *Cyclothone* species, also quite important in the GOM and present, but not abundant, in the Antarctic system. Even with the limited available data, the greatest total abundances are exhibited in the NCC system, about double those in the GOM, though with only about half the species of the GOM. It is worth noting the families that are present in all three systems: the myctophids, bathylagids, gonostomatids, melamphaids, paralepidids, scopelarchids, and stomiids (Tables 11.4 and 11.5).

The Cephalopods

The three data sets presented in Table 11.6 were collected using two types of sampling gear. Those of Passarella and Hopkins (1991) (GOM) and Pearcy (1965) (NCC) both used standard scientific trawls (Tucker and IKMT) ranging in mouth area from 3.2 to 9.0 m^2. Scientific trawls are not capable of capturing adults of the highly mobile species such as the loliginids, but they do capture immature stages such as the paralarvae fairly well. The Antarctic data come from Collins and Rodhouse (2006) which were collected using scientific nets as well as larger commercial style midwater trawls. The scientific nets employed in the Antarctic during AMERIEZ were just not capable of catching enough specimens to provide a comparison with the other two systems.

The numbers of cephalopods found in the GOM dwarf those of the NCC and Antarctic systems, continuing the trends seen in the Crustacea and the fishes. The myopsid squids, which include the commercially important loliginids of temperate and tropical systems, completely drop out in the Antarctic system, as do the sepiids and sepiolids (Collins and Rodhouse 2006). Their absence may be attributable to longer development times by their benthic egg masses (see Chapter 8) in very cold water, leaving them more vulnerable to predators and nearshore ice scour (Collins and Rodhouse 2006). The oegopsid squids have planktonic eggs and larvae, rendering them less susceptible to benthic predators, though the "boom or bust" production cycle and temperature effects on development time still pose a problem for reproductive success, as it does for most of the major midwater taxa.

Cephalopods are highly mobile, intermediate to high-level predators in all three systems, playing a role similar to that of fishes. In the Antarctic, the jacks, tunas, and other high-level predatory fish groups are absent and the midwater stomiids, scopelarchids, and gempylids are rarely captured, at least with scientific trawls (e.g. Tables 11.4 and 11.5) and it is likely that the squids fill at least part of that niche. As they grow, squids exhibit a shift in prey from Crustacea (krill and amphipods) to mesopelagic fish, with the largest sizes feeding on cephalopods (Phillips et al. 2003). Data available suggest a longer life in high-latitude squid species such as *Psychroteuthis glacialis* (2–3 years, Piatkowski et al. 1990) and *Galiteuthis glacialis* (2 years, Lu and Williams 1994) than the typical 1 year life span of temperate and tropical species similar to the trend for longer life spans in Antarctic fishes and crustaceans.

Gelatinous Zooplankton and Amphipods

The data presented in Table 11.7 provide a comparison of the major taxa found in the GOM and Antarctic open-ocean systems. There are no similar data for the California current system. Though many taxa in the GOM are not identified to species, the relative contributions of the major groups in each are instructive. A note of caution here is that only the most robust gelatinous taxa are collected by midwater trawls. Many deeper-living species of medusae and ctenophores are quite diaphanous and will pass right through the meshes of a trawl net.

The coronate scyphomedusae *Atolla* and *Periphylla* are found at mesopelagic depths in both systems, despite the profound differences between the two systems in seasonality and in temperatures at depth. However, their relative sizes are quite different. For example, *Periphylla* in the GOM fits easily on the palm of the hand. In the Antarctic, they are typically quite large, reaching sizes that are too large to fit in a gallon jar. Intermediate sizes of the two genera are found in the California current as well (personal observation). Scyphomedusae are truly a ubiquitous component of the mesopelagic fauna. The robust hydromedusa *Calycopsis borchgrevinki* was captured in all four seasons of sampling in the Antarctic and is also commonly captured in the NCC (Brodeur et al. 2019).

Ctenophores were more abundant in the GOM than the Antarctic system, but the absence of taxonomic information for the Gulf comb jellies precludes direct comparisons between the two. The cydippid ctenophore *Callianira Antarctica* captured during the AMERIEZ winter (1988) cruise was also found during a subsequent winter effort (Hofmann et al. 2004) preying on krill larvae and copepods beneath the ice pack (Scolardi et al. 2006), a behavior that would limit its exposure to net capture.

An obvious difference between the GOM and Antarctic systems was in the relative importance of salps. The AMERIEZ spring (1983) study encountered a dense and pervasive bloom of salps within the study area (Table 11.7). Similar blooms have been reported in the vicinity of Elephant Island north of the Antarctic Peninsula (Loeb et al. 1997) and in the peninsular study region of the Palmer Long Term Ecosystem Research program (Ross et al. 2014), both areas were monitored annually during the life of the sampling programs.

Salpa thompsoni overwinters at depth (500–1500 m) as an oozooid ("solitaries") in the warm (ca. 2 °C) temperatures typical of Circumpolar Deep Water (CDW) (Foxton 1966). They migrate to surface waters in the late winter and early spring and begin budding aggregates which, in favorable years ("salp years"), rapidly increase salp populations to the very high densities observed in Lancraft et al. (1989; Table 11.7). Salps proliferate most successfully in warmer oceanic waters with minimum sea ice. In a landmark study using data accumulated over several years, Loeb et al. (1997) found that salp years correlated positively with lower extents of winter sea-ice. In contrast, the krill *Euphausia superba* was most successful in years with high sea-ice and colder water in the spring marginal ice zone. The life histories and mobility of the two species are also factors in determining which of the two dominates at the spring ice edge.

Salpa thompsoni has a seasonal life cycle (see Chapter 9), budding off the aggregates at the surface during the productive season that produce the over-wintering solitaries. Mobility in salps is limited, with their horizontal distributions determined in large measure by prevailing current patterns and their local abundance by reproduction in place. In contrast, adult Antarctic krill are quite large for euphausiids (50–63 mm) and are very capable swimmers (Ross et al. 2014), easily able to move along the spring ice edge to locate areas of peak production.

Multi-year studies of krill abundance along the WAP shelf showed a pattern of episodic recruitment. Two strong year classes in succession were followed by three to four less successful (poor to moderate) recruitment years (Quetin and Ross 2003). The successful cohorts dominated the entire krill population during intervening years leading up to the next successful recruitment event. *E. superba* lives for 5–7 years (Nicol 2000).

Krill are able to feed on ice algae during the winter months and are well situated for exploiting the nascent ice-edge phytoplankton bloom for early spring spawning (cf. Loeb et al. 1997). High ice years, with their more northerly ice edge and colder sea surface temperatures over a wider swath of the open-ocean system, promote larval survival and successful krill recruitment. Conversely, the colder waters accompanying increased sea-ice extent are believed to inhibit the feeding and reproduction of salps, which favor the warmer waters typical of low ice years. The timing of accessibility to production is critical. Once salps reach the surface after their spring ascent, they can reproduce at an extremely rapid rate if conditions are favorable. Observations by Pakhomov et al. (2011) suggest that though they survive in very cold high Antarctic water ($<-1\,°C$), their ability to reproduce requires the warm temperatures typical of the CDW.

Interestingly, a warm water anomaly ("warm blob") that intruded into Oregon waters during 2015 and 2016 caused a shift in macrozooplankton dominance from *Euphausia pacifica*, normally an overwhelming dominant, to *Pyrosoma atlanticum* and *Aequorea victoria* (Brodeur et al. 2019). Pyrosomes littered the beaches of the Oregon coast.

Amphipods captured in the Antarctic system included the free-living gammarids and the free-living (*Themisto, Lanceola*) and salp-associated hyperiids (*Primno, Cyllopus, Vibilia*). The presence of highly dependent genera such as *Vibilia* (see Chapter 7) did coincide with the spring and summer seasons of peak salp abundance, but numbers overall remained low, which is fairly typical of the group. Catches with extremely high salp numbers like those encountered during the austral spring sampling season necessitate sub-sampling the total, likely underestimating the total abundance of amphipods living within the salps themselves. The gammarid *Eusirus antarcticus* not only lives on the underside of the sea ice when sea ice is present (Aarset and

Torres 1989) but is also captured in the midwater. *Cyphocaris* and *Parandania* are panglobal at lower mesopelagic depths. They are captured in the GOM as well as in the Antarctic.

Concluding Observations

The three communities addressed in Chapter 11 differ considerably in their physical and biological characteristics. Without attempting to answer the evolutionary question of why diversity changes so profoundly between the three (cf. McClain and Schlacher 2015), it is possible to identify a few of the important correlates that contribute.

Physical and Biological Factors that Change

Mean Annual Temperature

If there is a single most important factor influencing metabolism, including growth, of ectotherms, temperature is it. Available data on growth and metabolism in midwater species (Chapters 10 and 12) show that growth rate increases with increasing temperature, shortening generation times and thereby population turnover times in the GOM relative to the two other systems. In theory, shorter turnover times may allow the processes of natural selection to proceed more quickly, thereby yielding a higher diversity system, though that is highly speculative. Tropical/subtropical systems throughout the globe do exhibit high diversities (e.g. Briggs 1974, 1995), and the GOM is no exception.

Seasonal Cycling

Factored into the temperature differences between the systems is the seasonal stability. Seasonal change in the GOM is quite muted relative to the NCC and polar systems. In the NCC, the changing seasons bring large but predictable changes in current patterns, ushering in very high productivity during the upwelling season. Temperatures, though considerably cooler than those in the GOM, allow the midwater community to thrive. In the Antarctic, the retreat of the sea ice allows sunlight to reach the surface of the ocean, initiating the system's productive period. For the midwater fishes, sea surface temperatures change from lethal to tolerable and for the invertebrate fauna food is more readily available. Without question, seasonal change is the major driving force defining the character of the polar system.

Annual Production

Annual primary production in the NCC is far higher than that of the GOM and Antarctic systems, and the biomass of its midwater community is far higher as well, with much of that biomass invested in a few very dominant species. It may be characterized as a high biomass, moderate diversity system. With its low level of primary production, the GOM may be characterized as a low biomass, high diversity system, and the oceanic Antarctic as a low biomass, low diversity system. The productivity and biomass in the Antarctic Peninsula region is typically greater (e.g. Parker et al. 2011, 2015), but diversity is still low.

Current Patterns

Current flow in the three regions is quite different, influencing their species composition to varying degrees. The gyral flow across the north Pacific brings its species with it, initially feeding the NCC. The resident midwater community is modified during its flow down the coast as more southerly species make their appearance off California

(Neighbors and Wilson 2006). In the GOM, the Loop Current is a potential vector for more southerly species to enter the GOM, though to what degree is presently unknown. In the Antarctic system, the ACC acts in two ways. From a latitudinal perspective, it isolates the system (Eastman 1993) from more northern influence, though not completely, as there is some exchange from north to south (e.g. McGinnis 1982). From a longitudinal perspective, it keeps the fauna within it fairly homogeneous, which has been demonstrated by sampling programs in different regions within the Antarctic oceanic system as well as by genetic studies on the Antarctic krill (Fevolden and Schneppenheim 1989).

References

Aarset, A.V. and Torres, J.J. (1989). Cold resistance and metabolic responses to salinity variations in the amphipod *Eusirus antarcticus* and the krill *Euphausia superba*. *Polar Biology* 9: 491–497.

Ainley, D.G. (2002). *The Adelie Penguin*. New York: Columbia University Press.

Ainley, D.G., Fraser, W.R., Sullivan, C.G. et al. (1986). Antarctic pack ice structures mesopelagic nektonic communities. *Science*. 232: 847–849.

Andriashev, A.P. (1987). A general review of the Antarctic bottom fish fauna. In: *Fifth Congress of European Ichthyologists, Proceedings, Stockholm, 1985* (eds. S.O. Kullander and B. Fernholm). Swedish Museum of Natural History: Stockholm.

Arntz, W.E. and Gorny, M. (1991). Shrimp (Decapoda, Natantia) occurrence and distribution in the eastern Weddell Sea, Antarctica. *Polar Biology* 11: 169–177.

Baker, A.D.C. (1957). The distribution and life history of *Euphausia triacantha* Holt and Tattersall. *Discovery Reports* 29: 309–340.

Beamish, R.J., Leask, K.D., Ivanov, O.A. et al. (1999). The ecology, distribution, and abundance of midwater fishes of the Subarctic Pacific gyres. *Progress in Oceanography* 43: 399–442.

Briggs, J.C. (1974). *Marine Zoogeography*. New York: McGraw-Hill.

Briggs, J.C. (1995). *Global Biogeography* Amsterdam. Elsevier.

Brodeur, R.D., Auth, T.D., and Phillips, A.J. (2019). Major shifts in pelagic micronekton and macrozooplankton community structure in an upwelling ecosystem related to an unprecedented marine heatwave. *Frontiers in Marine Science* 6: 212.

Burghart, S.E., Hopkins, T.L., Vargo, G.A., and Torres, J.J. (1999). Effects of a rapidly receding ice edge on the abundance, age structure and feeding of three dominant calanoid copepods in the Weddell Sea, Antarctica. *Polar Biology* 22: 279–288.

Burghart, S.E., Hopkins, T.L., and Torres, J.J. (2007). The bathypelagic Decapoda, Lophogastrida, and Mysida of the eastern Gulf of Mexico. *Marine Biology* 152: 315–327.

Burghart, S.E., Hopkins, T.L., and Torres, J.J. (2010). Partitioning of food resources in bathypelagic micronekton in the eastern Gulf of Mexico. *Marine Ecology Progress Series* 399: 131–140.

Carmack, E.C. (1990). Large-scale physical oceanography of polar oceans. In: *Polar Oceanography, Part A: Physical Science* (ed. W.O. Smith). San Diego: Academic Press.

Childress, J.J. (1975). The respiratory rates of midwater crustaceans as a function of depth of occurrence and relation to the oxygen minimum layer off Southern California. *Comparative Biochemistry and Physiology* 50A: 787–799.

Clarke, A. (1996). The influence of climate change on the distribution and evolution of organisms. In: *Animals and Temperature: Phenotypic and Evolutionary Adaptation*, Society for Experimental Biology Seminar Series 59 (eds. I.A. Johnston and A.F. Bennett), 375–407. Cambridge: Cambridge University Press.

Collins, M.A. and Rodhouse, P.G.K. (2006). Southern Ocean Cephalopods. In: *Advances in Marine Biology: Volume 50* (eds. A.J. Southward, C.M. Young and L.A. Fuiman). San Diego: Academic Press.

Cullins, T.L., Devries, A.L., and Torres, J.J. (2011). Antifreeze proteins in pelagic fishes from Marguerite Bay (Western Antarctica). *Deep Sea Research, Part II* 58: 1690–1694.

Daly, K.L. and Macaulay, M.C. (1988). Abundance and distribution of krill in the ice edge zone of the Weddell Sea, austral spring 1983. *Deep-Sea Research* 35: 21–41.

Donnelly, J., Torres, J.J., Sutton, T.T., and Simoniello, C. (2004). Fishes of the eastern Ross Sea, Antarctica. *Polar Biology* 27: 637–650.

Donnelly, J., Sutton, T.T., and Torres, J.J. (2006). Distribution and abundance of micronekton and macrozooplankton in the NW Weddell Sea: relation to a spring ice edge bloom. *Polar Biology* 29: 280–293.

Dortch, Q. and Postel, J.R. (1989). Phytoplankton-Nitrogen Interactions. In: *Coastal Oceanography of Washington and Oregon* (eds. M.R. Landry and B.M. Hickey). Elsevier Science and Technology.

Eastman, J.T. (1993). *Antarctic Fish Biology: Evolution in a Unique Environment*. San Diego: Academic Press.

El Sayed, S.Z. (1972). Primary productivity and standing crop of phytoplankton. In: *Chemistry, Primary Productivity and Benthic Algae of the Gulf of Mexico*. American Geographical Society.

El Sayed, S.Z. (1984). Productivity of Antarctic waters. A reappraisal. In: *Marine Phytoplankton and Productivity* (eds. O. Holm-Hansen, L. Bolis and R. Gilles). Berlin: Springer-Verlag.

Fevolden, S.E. and Schneppenheim, R. (1989). Genetic homogeneity of krill (*Euphausia superba*) in the Southern Ocean. *Polar Biology* 9: 533–539.

Foster, T.D. (1984). The marine environment. In: *Antarctic Ecology*, vol. *2* (ed. R.M. Laws). London: Academic Press.

Foxton, P. (1966). The distribution and life history of *Salpa thompsoni*-Foxton with observations on a related species, *Salpa gerlachei*-Foxton. *Discovery Reports* 34: 1–116.

Frost, B.W. and Mccrone, L.E. (1979). Vertical distribution, diel vertical migration, and abundance of some mesopelagic fishes in the eastern Subarctic Pacific Ocean in summer. *Fishery Bulletin U.S.* 76: 751–770.

Gallo, N.D. and Levin, L.A. (2016). Fish ecology and evolution in the world's oxygen minimum zones and implications of ocean deoxygenation. In: *Advances in Marine Biology* (ed. B.E. Curry). London, San Diego: Academic Press.

Gartner, J.V., Hopkins, T.L., Baird, R.C., and Milliken, D.M. (1987). Ecology of the lanternfishes (Pisces: Myctophidae) of the eastern Gulf of Mexico. *Fishery Bulletin U.S.* 85: 81–98.

Gartner, J.V., Steele, P., and Torres, J.J. (1989). Aspects of the distribution of lanternfishes (Pisces: Myctophidae) from the northern Sargasso Sea. *Bulletin of Marine Science* 45: 555–563.

Gomez-Gutierrez, J., Peterson, W.T., and Miller, C.B. (2005). Cross-shelf life-stage segregation and community structure of the euphausiids off Oregon (1970–1972). *Deep Sea Research, Part II* 52: 289–315.

Gordon, A.L. and Comiso, J.C. (1988). Polynyas in the Southern Ocean. *Scientific American* 258: 90–97.

Gordon, A.L. and Molinelli, E.M. (1982). *Southern Ocean Atlas: Thermohaline Chemical Distributions and the Atlas Data Set*. New York: Columbia University Press.

Gutt, J., Gorny, M., and Arntz, W. (1991). Spatial distribution of Antarctic shrimps

(Crustacea: Decapoda) by underwater photography. *Antarctic Science* 3: 363–369.

Hickey, B.M. and Landry, M.R. (1989). Coastal oceanography of Washington and Oregon: a summary and a prospectus for future studies. In: *Coastal Oceanography of Washington and Oregon* (eds. M.R. Landry and B.M. Hickey). Elsevier Science and Technology.

Hofmann, E.E., Wiebe, P.W., Costa, D.P., and Torres, J.J. (2004). An overview of the Southern Ocean Global Ocean Ecosystems Dynamics program. *Deep Sea Research, Part II* 51: 1921–1924.

Hopkins, T.L. (1982). The vertical distribution of zooplankton in the eastern Gulf of Mexico. *Deep-Sea Research* 29: 1069–1083.

Hopkins, T.L. and Lancraft, T.M. (1984). The composition and standing stock of mesopelagic micronekton at 27 N 86 W in the eastern Gulf of Mexico. *Contributions in Marine Science* 27: 145–158.

Hopkins, T.L., Gartner, J.V., and Flock, M.E. (1989). The caridean shrimp (Decapoda:Natantia) assemblage in the mesopelagic zone of the eastern Gulf of Mexico. *Bulletin of Marine Science* 45: 1–14.

Hopkins, T.L., Flock, M.E., Gartner, J.V., and Torres, J.J. (1994). Structure and trophic ecology of a low latitude midwater decapod and mysid assemblage. *Marine Ecology Progress Series* 109: 143–156.

Judkins, H.L., Vecchione, M., and Roper, C.F.E. (2009). Cephalopoda (Mollusca) of the Gulf of Mexico. In: *Gulf of Mexico Origin, Waters and Biota* (eds. D.L. Felder and D.K. Camp). College Station: Texas A&M University Press.

Kennett, J.P. (1982). *Marine Geology*. Englewood Cliffs, New Jersey: Prentice-Hall.

Kinsey, S.T. and Hopkins, T.L. (1994). Trophic strategies of euphausiids in a low-latitude ecosystem. *Marine Biology* 118: 651–661.

Krebs, C.J. (1972). *Ecology: The Experimental Analysis of Distribution and Abundance*. New York: Harper and Rowe.

Lancraft, T.M., Torres, J.J., and Hopkins, T.L. (1989). Micronekton and macrozooplankton in the open waters near Antarctic ice edge zones (AMERIEZ 1983 and 1986). *Polar Biology* 9: 225–233.

Lancraft, T.M., Hopkins, T.L., Torres, J.J., and Donnelly, J. (1991). Oceanic micronektonic/macrozooplanktonic community structure and feeding under ice covered Antarctic waters during the winter (Ameriez 1988). *Polar Biology* 11: 157–167.

Lancraft, T.M., Reisenbichler, K.R., Hopkins, T.L. et al. (2004). A krill-dominated micronekton and macrozooplankton community in Croker Passage, Antarctica with an estimate of fish predation. *Deep Sea Research, Part II* 51: 2247–2260.

Liu, Y., Weisberg, R.H., Vignudelli, S., and Mitchum, G.T. (2016). Patterns of the loop current system and regions of sea surface height variability in the eastern Gulf of Mexico revealed by self-organizing maps. *Journal of Geophysical Research, Oceans* 121: 2347–2366.

Loeb, V., Siegel, V., Holm-Hansen, O. et al. (1997). Effects of sea-ice extent and krill or salp dominance on the Antarctic food web. *Nature* 387: 897–900.

Longhurst, A.R. (1998). *Ecological Geography of the Sea*. London: Academic Press.

Lu, C.C. and Williams, R. (1994). Contribution to the biology of squid in the Prydz Bay region, Antarctica. *Antarctic Science* 62: 223–229.

Mauchline, J. and Fisher, L.R. (1969). *Advances in Marine Biology*. London: Academic Press.

Maul, G.A. (1977). The annual cycle of the Gulf Loop current. Part I: observations during a one year time series. *Journal of Marine Research* 35: 29–47.

Mcclain, C.R. and Schlacher, T.A. (2015). On some hypotheses of diversity of animal life at great depths on the sea floor. *Marine Ecology* 36: 849–872.

Mcginnis, R.F. (1982). *Biogeography of lanternfishes (Myctophidae) south of 30 S, Antarctic Research Series, v.35, Biology of the Antarctic Seas XII*. Washington: American Geophysical Union.

Mintenbeck, K. and Torres, J.J. (2017). Impact of climate change on the Antarctic silverfish and its consequences for the Antarctic ecosystem. In: *The Antarctic silverfish*, A keystone species in a changing ecosystem. Advances in Polar Ecology, vol. 3 (eds. M. Vacchi, E. Pisano and L. Ghigliotti), 253–286.

Neighbors, M.A. and Wilson, R.R. (2006). Deep sea. In: *The Ecology of Marine Fishes. California and Adjacent Waters* (eds. L.G. Allen, D.J. Pondella II and M.H. Horn). Berkeley: University of California Press.

Nicol, S. (2000). Understanding krill growth and aging: the contribution of experimental studies. *Canadian Journal of Fisheries and Aquatic Sciences* 57: 168–177.

Nowlin, W.D. and Klinck, J.M. (1986). The physics of the Antarctic Circumpolar Current. *Reviews of Geophysics* 24: 469–491.

Pakhomov, E.A., Dubischar, C.D., Hunt, B.P.V. et al. (2011). Biology and life cycles of pelagic tunicates in the Lazarev Sea, Southern Ocean. *Deep Sea Research, Part II* 58: 1677–1689.

Parker, M.L., Donnelly, J., and Torres, J.J. (2011). Invertebrate micronekton and macrozooplankton in the Marguerite Bay region of the Western Antarctic Peninsula. *Deep Sea Research Part II: Topical Studies in Oceanography* 58: 1580–1598.

Parker, M.L., Fraser, W.R., Ashford, J. et al. (2015). Assemblages of micronektonic fishes and invertebrates in a gradient of regional warming along the Western Antarctic Peninsula. *Journal of Marine Systems* 152: 18–41.

Passarella, K.C. and Hopkins, T.L. (1991). Species composition and food habits of the micronektonic cephalopod assemblage in the eastern Gulf of Mexico. *Bulletin of Marine Science* 49: 638–659.

Pearcy, W.G. (1964). Some distributional features of mesopelagic fishes off Oregon. *Journal of Marine Research* 22: 83–102.

Pearcy, W.G. (1965). Species composition and distribution of pelagic cephalopods from the Pacific Ocean off Oregon. *Pacific Science* 19: 261–266.

Pearcy, W.G. and Forss (1969). Depth distribution of oceanic shrimps (Decapoda: Natantia) off Oregon. *Journal of the Fisheries Research Board of Canada* 23: 1135–1143.

Pearcy, W.G. and Laurs, R.M. (1966). Vertical migration and distribution of mesopelagic fishes off Oregon. *Deep-Sea Research* 13: 153–165.

Pearcy, W.G., Krygier, E.E., Mesecar, R., and Ramsey, F. (1977). Vertical distribution and migration of oceanic micronekton off Oregon. *Deep-Sea Research* 24: 223–245.

Peterson, W.T., Fisher, J.L., Strub, P.T. et al. (2017). The pelagic ecosystem of the Northern California Current off Oregon during the 2014–2016 warm anomalies within the context of the past 20 years. *Journal of Geophysical Research, Oceans* 122: 7267–7290.

Phillips, K.L., Nichols, P., and Jackson, G.D. (2003). Size-related dietary changes observed in the squid *Moroteuthis ingens* at the Falkland Islands: stomach contents and fatty acid analyses. *Polar Biology* 26: 474–485.

Piatkowski, U., White, M.G., and Dimmler, W. (1990). Micronekton of the Weddell Sea: distribution and abundance. *Berichte zür Polarforschung* 68: 73–81.

Pierce, S.D., Barth, J.A., Shearman, R.K., and Erofeev, A.Y. (2012). Declining oxygen in the northeast Pacific. *Journal of Physical Oceanography* 42: 495–501.

Quetin, L.B. and Ross, R.M. (2003). Episodic recruitment in Antarctic krill *Euphausia superba* in the Palmer LTER study region.

Marine Ecology Progress Series 259: 185–200.

Riley, J.P. and Chester, R. (1971). *Introduction to Marine Chemistry*. London: Academic Press.

Ross, R.M., Quetin, L.B., Newberger, T. et al. (2014). Trends, cycles, interannual variability for three pelagic species west of the Antarctic Peninsula 1993–2008. *Marine Ecology Progress Series* 515: 11–32.

Ryther, J. (1969). Photosynthesis and fish production in the sea. *Science* 166: 72–76.

Scolardi, K.M., Daly, K.L., Pakhomov, E.A., and Torres, J.J. (2006). Feeding ecology and metabolism of the Antarctic cydippid ctenophore *Callianira antarctica*. *Marine Ecology Progress Series* 317: 111–126.

Smith, W.O. and Sakshaug, E. (1990). Polar phytoplankton. In: *Polar Oceanography Part B: Chemistry, Biology, and Geology* (ed. W.O. Smith). San Diego: Academic Press.

Spalding, M.D., Fox, H.E., Allen, G.R., and Others, A. (2007). Marine ecoregions of the world: a bioregionalization of coastal and shelf areas. *Bioscience* 57: 573–583.

Sturges, W. and Evans, J.C. (1983). On the variability of the loop current in the Gulf of Mexico. *Journal of Marine Research* 41: 639–653.

Sturges, W. and Lugo-Fernandez, A. (2005). *Circulation in the Gulf of Mexico: Observations and Models*. Washington, DC: American Geophysical Union.

Sullivan, C.W., Mcclain, C.R., Comiso, J.C. et al. (1988). Phytoplankton standing crops within an Antarctic ice edge assessed by satellite remote sensing. *Journal of Geophysical Research* 93: 12487–12498.

Sutton, T.T. and Hopkins, T.L. (1996). Species composition, abundance, and vertical distribution of the stomiid (Pisces: Stomiiformes) fish assemblage of the Gulf of Mexico. *Bulletin of Marine Science* 59: 530–542.

Sutton, T.T., Wiebe, P.H., Madin, L., and Bucklin, A. (2010). Diversity and community structure of pelagic fishes to 5000 m depth. *Deep Sea Research, Part II* 57: 2220–2233.

Sutton, T.T., Clark, M.R., Dunn, D.C., and Others, A. (2017). A global biogeographic classification of the mesopelagic zone. *Deep-Sea Research Part I* 126: 85–102.

Thatje, S., Schnack-Schiel, S., and Arntz, W.E. (2003). Developmental trade-offs in Subantarctic meroplankton communities and the enigma of low decapod diversity in high southern latitudes. *Marine Ecology Progress Series* 260: 195–207.

Thatje, S., Bacardit, R., and Arntz, W.E. (2005). Larvae of the deep-sea Nematocarcinidae (Crustacea: Decapoda: Caridea) from the Southern Ocean. *Polar Biology* 28: 290–302.

Torres, J.J., Belman, B.W., and Childress, J.J. (1979). Oxygen consumption rates of midwater fishes as a function of depth of occurrence. *Deep-Sea Research* 26A: 185–197.

Torres, J.J., Grigsby, M.D., and Clarke, M.E. (2012). Aerobic and anaerobic metabolism in oxygen minimum layer fishes: the role of alcohol dehydrogenase. *Journal of Experimental Biology* 215: 1905–1914.

Vinogradov, M.E. (1970). *Vertical Distribution of the Oceanic Zooplankton*. Jerusalem: Israel Program for Scientific Translations.

Willis, J.M. and Pearcy, W.G. (1982). Vertical distribution and migration of fishes of the lower mesopelagic zone off Oregon. *Marine Biology* 70: 87–98.

Yaldwyn, J.C. (1965). Antarctic and subantarctic decapod Crustacea. In: *Biogeography and Ecology in the Antarctic* (eds. J. Van Mieghem and P. Van Oye). The Hague: W. Junk.

Yashnov, V.A. (1962). Vertical distribution of the mass of zooplankton in the tropical region of the Atlantic Ocean. *Oceanology*: 136–141.

Zwally, H.J., Comiso, J.C., Parkinson, C.L. et al. (1983). *Antarctic Sea Ice, 1973–1976: Satellite Passive-Microwave Observations*. Washington, DC: NASA.

12

Energetics

Introduction

Animal energetics is a subdiscipline that describes how animals budget the energy they acquire through feeding. The three basic elements of the budget are growth, metabolism, and excretion, but each of the three can be further subdivided to give greater detail on the fate of ingested energy, forming a complicated balance sheet of energy usage. Animal energetics is the individual counterpoint to ecological energetics, which concerns the amount of energy that is transferred from one trophic level to the next. How efficiently energy moves through the trophic pyramid is a function of how it is used by the species comprising each trophic level. If ingested energy is invested in growth or reproduction, it is available to feed the next trophic level. If it is burned up as metabolism to fuel a high activity level, it is lost to the system.

How species allocate their energy is an important part of their life history. Stearns (1992) lists the following as the principal life history traits.

1) Size at birth
2) Growth pattern
3) Age at maturity
4) Size at maturity
5) Number, size, and sex ratio of offspring
6) Age- and size-specific reproductive investments
7) Age- and size-specific mortality schedules

Clearly, some of the traits in the list (e.g. sex ratio of offspring) are geared toward the more advanced phyla but most apply just as well to cnidarians as they do to fishes. Of particular concern here is "growth pattern." Inherent within a growth pattern is not only how quickly and how large a species grows but the character of its tissues and how much of an energetic investment they represent. An organism like a hydromedusa with tissues that are approximately 95% water is investing a lot less energy in its growth than a crustacean or fish that is between 60 and 80% water, with the remaining 20–40% mainly composed of protein and lipid. Going one step further, the amount of energy that an individual can invest in growth depends on how much it needs for the other demands of life, particularly metabolism.

During the course of daily living, as the chemical energy of food items is combusted to allow for necessities like movement, heartbeat, tissue construction and repair, and maintaining an individual's internal environment, heat is given off by the animal machine. The amount of heat produced is a good

---

Life in the Open Ocean: The Biology of Pelagic Species, First Edition. Joseph J. Torres and Thomas G. Bailey.
© 2022 John Wiley & Sons Ltd. Published 2022 by John Wiley & Sons Ltd.

measure of overall metabolism, particularly for "warm-blooded" or endothermic species like seals. Thus, measurement of heat production has historically been regarded as the gold standard for measurements of metabolism with many different, highly sophisticated devices used to trap and accurately measure animal (including human!) heat production. For aquatic species, measurement of heat production is very difficult for two reasons. First, most experimental subjects are "cold-blooded," or ectotherms, and do not produce much heat. Second, the specific heat of water (see Chapter 1) is very high: it takes a lot of heat energy to get a measurable change in temperature. For that reason, most measurements of metabolism are achieved by measuring the amount of oxygen consumed by the experimental subjects. The amount of oxygen needed to completely combust a substrate such as protein or lipid is directly proportional to the amount of energy liberated, which makes oxygen consumption a highly useful experimental tool for energetics measurements.

A Model Energy Budget

Because of the many different measurements required, energy budgets are difficult to compile. Figure 12.1, from Brett and Groves (1979), is the energy budget of an average carnivorous fish, showing the most important elements. J.R. Brett was a luminary in the field of animal energetics, and in the field of fish physiology in general. It is fitting to use his work as he helped "write the book" on how animals use energy and how they respond to temperature.

At this juncture, the richest source of information available on aquatic species is for fishes. Sufficient data are available to consider two elements of energy usage in a limited number of open-ocean taxa: metabolism and proximate composition.

The first loss term in the energy budget equation is in the digestive process itself. A certain percentage of the total energy value of ingested prey will be refractory to digestion, hard parts like bone and shell as well as some structural proteins and carbohydrates.

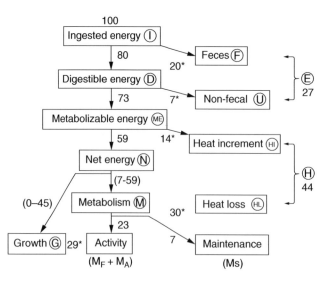

Figure 12.1 Average partitioning of dietary energy for a carnivorous fish. Amounts marked with an asterisk total to 100; figures in brackets indicate the possible range of net energy distributed to metabolism and growth. Source: Brett and Groves (1979), figure 18 (p. 337). Reproduced with the permission of Academic Press.

In the case of carnivorous fish, that loss is about 20% of the initial, or ingested energy (I) value of the prey; the remaining fraction of prey energy available is the digestible energy (D). Clearly, carnivory is going to be the lifestyle with the lowest initial loss of energy due to indigestibility of prey. An herbivore will lose more energy due to the robust and mainly indigestible cell walls of plants. As noted in Figure 12.1, the indigestible fraction of the diet is passed as feces (F). The 80% remaining is known by a variety of terms: apparent assimilated energy, assimilated energy, apparent digestible energy, and digestible energy. To be consistent with Dr. Brett's scheme, we will use the term Digestible Energy (D), and $D = I - F$.

The next loss term is in the nitrogenous waste that makes up the non-fecal excretion, or urinary losses (U). In aquatic species, the most common nitrogenous excretory product is ammonium, sometimes with a small contribution by urea. The source of the nitrogenous waste is mainly in the breakdown of ingested proteins, either because the amount of protein exceeds what is required or the amino acids are poorly balanced. In those cases, the carbon skeletons are deaminated and the nitrogen is excreted as waste. In a typical carnivorous fish, the energy associated with urinary losses is about 7% of the Gross Energy intake, bringing the total losses for excretion (E) to about 27% of the ingested energy. The remaining food energy available is the metabolizable energy (ME): $ME = I - F - U$.

Metabolizable energy is further reduced by the most mysterious of all the loss terms in the energy budget: the heat increment (HI). The heat increment was originally known by a variety of terms, the first of which was "the specific dynamic action" or more properly, since that term was a bad translation from the original German, "the specific dynamic effect (SDE)" (Kleiber 1961). Rubner, one of the early energetics pioneers, first observed that metabolism was elevated during and after feeding in dogs. The effect was thought to be due to protein only (which is what the word specific refers to – specific to protein) but was later found to be associated with feeding in general. "Feeding metabolism" has also been used to describe the same phenomenon. Although heat increment is now the most widely used term, if you peruse the energetics literature you will find all the terms mentioned at the beginning of this paragraph at some point in your reading.

The heat increment is the cost of digestion and metabolic transformation of foodstuffs, which includes muscular activity of the gut, enzymatic breakdown of biomolecules, cost of the active transport of the disassembled nutrients from the gut into circulation, and the beginnings of biosynthesis or reassembly. Since the heat increment involves metabolic processes, it is rightly classified in Dr. Brett's scheme as part of the overall cost of metabolism (M). A loss term, the heat increment energy is lost to the organism prior to its being available for use in maintenance, activity, or growth. It is approximately 14% of the ingested energy. The remaining energy, after subtracting losses due to digestion and molecular transformation, is about 59% of the ingested energy for a carnivorous fish: the net energy (N). The running tally is now $N = I - F - U - HI$.

Net energy has two basic fates. It can be used as a metabolic fuel or it can be used for growth (G), the production of new animal mass. As a metabolic fuel, it allows muscular activity of all kinds, including locomotory muscle as well as that involved in maintenance such as heart and respiratory function. Maintaining ion balance within cells and within the circulatory system, if present, is another energetically demanding process that comes under the general heading of metabolism. The energy combusted for metabolic processes is lost to the system as heat. That which is invested in growth, whether it be the growth of an individual or the production of gametes and offspring to yield the next generation, is energy that is at least partially available to the next trophic level. The overall equation for an energy budget is quite simple: $I = G + M + E$. The basic elements of growth, metabolism, and excretion have been briefly covered, but the devil is in the details for each. The devil is in the lack of much detailed information on energetics in open-ocean taxa. Some data are available for each of the main elements of the energy budget equation, but the complete picture is available for only a very few species. The situation improves with increasing animal complexity and in how easy the species are to manipulate experimentally. Thus, robust taxa like the crustaceans have been studied more than the gelatinous forms comprising the Cnidaria and Ctenophora.

Ingested energy (I). As noted above, the indigestible fraction of ingested food items comprises the first loss term in the energy budget equation: the fecal losses. Organic substrates that are refractory to the digestive process are typically structural compounds such as the chitin in the exoskeleton of crustaceans and the fibrous proteins of connective tissue. Inorganic compounds such as the siliceous testes of marine algae and the calcareous vertebrate skeleton would also resist digestive breakdown.

Digestibility of Biomolecules

Carbohydrates. Carbohydrates vary in size from simple 6-carbon sugars such as glucose to the long-chain polymers typical of the storage polysaccharides such as glycogen and the structural polysaccharides such as chitin. Any loss of carbohydrates occurs in the initial breakdown process and is due to the inability of digestive enzymes to break down the chemical bonds joining the simple sugars that form the long-chain polysaccharide. Once digested, carbohydrates are fully utilized. Worthy of note is that carbohydrates are a vanishingly small percentage of the body composition of pelagic species, less than 1% of the dry weight.

Lipids. As with the carbohydrates, only a few types of lipids are lost to the organism in the digestive process. Once digested, lipids are fully utilized. Examples of lipids that would be refractory to digestion are the waxy cuticles of copepods and other crustaceans. It is worth noting that for virtually every biomolecule in the marine (and terrestrial) environment, if it is present in sufficient quantity, there is a specialized digestive system that has evolved to utilize it. Some fishes, for example, have specialized digestive enzymes that can break down the waxes present in copepod cuticles.

Proteins. Digestion of proteins is more complex than that of carbohydrates and lipids, mainly due to their nitrogen content. The nitrogen present in proteins makes their digestion a multi-step process. On the one hand, the molecular backbone must be broken down into its constitutive parts, the amino acids, and processed. On the other, the nitrogen present in the amino group of amino acids, an integral part of all protein molecular structure as part of the peptide bond that holds the macromolecules together, may form a toxic waste product: ammonium. Thus, proteins are the major source of urinary nitrogen, the second drop in initial ingested energy after fecal losses.

As with the carbohydrates and lipids, some ingested proteins will be refractory to digestion and will be excreted with the remainder of the fecal material. Those proteins that can be digested are broken down into their constitutive amino acids and absorbed through the gut wall. However, if the ingested protein exceeds that which is needed for growth and maintenance or if the amino acids are poorly balanced, the carbon backbones of the amino acids are deaminated. The resulting amino groups are then excreted as nitrogenous waste. In aquatic species, that waste is usually ammonium, but can include some urea as well.

Energy Value of Biomolecules

The energy liberated by a foodstuff when completely combusted with oxygen, as well as the ratio of CO_2 produced to O_2 consumed, is dependent on what foodstuff it is. Thus, for a typical carbohydrate (after Kleiber 1961 and Brett and Groves 1979):

$$C_6H_{12}O_6 + 6O_2 \longleftrightarrow 6CO_2 + 6H_2O$$
$$6\,vols\,CO_2\,/\,6\,vols\,O_2$$

Or a typical lipid such as a triglyceride:

$$2C_{51}H_{98}O_6 + 145O_2 \longleftrightarrow 102CO_2 + 98H_2O$$
$$102\,vols\,CO_2\,/\,145\,vols\,O_2$$

The energy content for each substrate type, the amount of oxygen required to oxidize a gram of substance, the amount of carbon dioxide produced in the oxidation of 1 g of substance, and the RQ values for each type of substrate are presented in the first four data columns of the table below.

	kcal g^{-1}	l O$_2$ g^{-1}	l CO$_2$ g^{-1}	RQ	kcal l O$_2$$^{-1}$	kcal l CO$_2$$^{-1}$
Carbohydrate	4.1	0.75	0.75	1.00	5.0	5.0
Lipid	9.3	2.01	1.40	0.70	4.7	6.6
Protein	4.8	1.07	0.91	0.84	4.5	5.6

kcal g^{-1} = amount of energy in a gram of substance

l O$_2$ g^{-1} = liters of oxygen necessary to completely oxidize a gram of substance

l CO$_2$ g^{-1} = liters of CO$_2$ produced by the oxidation of 1 g of substance

RQ = respiratory quotient = liters of CO$_2$ produced per liter of O$_2$ used

kcal l O$_2$$^{-1}$ = amount of energy released per liter of oxygen used

kcal l CO$_2$$^{-1}$ = amount of energy released per liter of CO$_2$ produced

A few things are worth noticing about these relationships. First, the amount of O$_2$ it takes to completely oxidize each of the substrate molecules. Per unit mass, lipids require about twice the amount of oxygen for complete combustion that carbohydrates and proteins do. This disparity results in an energy yield for complete oxidation for each that is presented in the last two data columns of the above table. Note that the energy yield for the combustion of each of the biomolecules is quite similar in terms of the volume of oxygen needed for combustion. That is why energy metabolism can be measured by measuring oxygen consumption rate.

The ratio of CO$_2$ formed/O$_2$ used is called the respiratory quotient (about 1.0 glucose; 0.70 lipids; 0.80 proteins). If reliable measurements of O$_2$ consumed/CO$_2$ produced can be obtained, the major substrate used for the production of metabolic energy can be deduced. This could be useful for evaluating the diet of an animal recently captured in the field or for looking at the order in which substrates are metabolized in a starved animal. Clearly, an RQ between 0.7 and 1.0 could be a combination of fat and carbohydrate or protein, so there is always some guesswork involved. Other types of data such as the typical diet of a species and O : N ratio can provide a more complete picture.

O : N ratio – another index that is useful for determining the substrate being oxidized is the ratio of the moles of oxygen consumed to the moles of ammonia N excreted – the O : N ratio. If all AAs resulting from protein catabolism are deaminated and excreted as NH_3 and all C – skeletons go to CO_2 and H_2O, then the theoretical O : N minimum is 9.3; generally for carbohydrates 20–30; lipids >30.

If oxygen consumption rate can be used to measure metabolic rate, then theoretically carbon dioxide production could be used in the same way. It is not often used for three reasons. First, CO_2 concentrations are technically difficult to measure. Second, a large pool of CO_2 is present in the body of all animals in the form of HCO_3^-. Acidification of blood through the production of lactate by muscles produces CO_2 that is only indirectly related to metabolism. Third, the different foodstuffs metabolized yield greatly different values of energy for the volume of CO_2 produced, e.g. fat $6.6\,kcal\,1\,CO_2^{-1}$; carbs $5.0\,kcal\,1\,CO_2^{-1}$. That is a large (~25%) difference in energy yield.

To convert a measured oxygen consumption rate to an energy consumption value, an oxycalorific equivalent is often used, which assumes an average diet composed of a mixture of protein, lipids, and carbohydrates. The value most applicable to fishes and invertebrates is $4.63\,kcal\,l^{-1}$ of oxygen consumed ($19.4\,kJ\,l^{-1}$).

Measuring Metabolic Rate

Indirect calorimetry is a measure of energy consumption using a technique that is one step removed from direct heat production. As noted above, the measurement used most often is oxygen consumption rate. There are a variety of different methods for measuring oxygen consumption, including manometric techniques; however, those most commonly employed use sealed vessels or flow-through systems that allow the experimental animal to deplete the oxygen within them. The depletion can be monitored continuously through time with oxygen electrodes or by taking initial and final readings of oxygen concentration using chemical techniques such as the Winkler method (Ikeda et al. 2000).

Oxygen Consumption Rate-Modifying Factors

The oxygen consumption rate, or metabolism, of an animal is a measurement of the energy used for life's basic requirements, such as osmoregulation, locomotory activity, and digestive processes. It commands the lion's share of the net energy for many organisms, the other major portion being devoted to growth or production. It may be considered as energy that is irretrievably lost to the ecosystem, since it is not incorporated into increasing the mass of the animal. Metabolism or O_2 consumption is often expressed as volume of O_2 or moles of O_2/unit of mass (wet, dry, or carbon)/unit of time. Clearly, the rate that most closely approaches the rate in life is that expressed as wet mass. However, rates expressed as dry mass or carbon content are often used for gelatinous animals such as ctenophores because of their very high water content, and the use of carbon content allows for a comparison of rates across a broad range of taxa using a single common denominator.

Metabolic rates are affected by a wide range of variables that may be divided roughly into biotic and abiotic influences. Biotic variables include factors intrinsic to the animal such as nutritional state and activity. Abiotic variables include factors extrinsic to the animal such as temperature, salinity, and ambient oxygen concentration. Abiotic variables are covered in Chapter 2.

The first biotic factor-modifying respiration is nutritional state. Fasting animals have a lower rate of oxygen consumption than fed animals. At least three reasons explain the difference. First, energy costs as a result of digestive processes are eliminated. Second, starvation leads to a reduction in the level of energy-intensive processes, notably locomotion. Third, little energy is being used for anabolism, the processes related to growth or production. Within the energy budget, the transition from metabolizable energy to net energy is a result of the mysterious loss term, the heat increment or specific dynamic effect, that is briefly described above. However, the increase in metabolism associated with feeding is most accurately described as feeding metabolism, and it comprises several elements, the first being the cost of digestive processes.

Digestive processes. The elements of digestion may be broken down as follows:

1) Cost of feeding – mastication and prey capture/excitement.
2) Stomach and intestinal work
3) Digestion: costs associated with digestive enzyme formation and degradation
4) Absorption and transport
5) Amino acid oxidation and formation of nitrogenous waste
6) Biomass formation and growth

Note that that digestive costs may be divided into short- and long-term processes with all being short-term other than (6).

The heat increment is considered by several investigators (e.g. Jobling 1985; Kiorboe et al. 1985) including your authors, to be part of the cost of growth. The effects of feeding have also been described in marine invertebrates, notably the Antarctic krill *Euphausia superba*. On average, an increase of 1.6 times was observed in the krill a result of feeding (Ikeda and Dixon 1984).

Within the budget depicted in Figure 12.1, you'll notice that the drop in energy associated with the heat increment brings the energy available for growth and metabolism to 59% of the ingested energy: the net energy. Further, the loss attributable to heat increment is considered to be part of the total energy used in metabolism or 44%. If we subtract the 14% for HI, the total available for metabolism is 30% and that is divided into maintenance metabolism (7%) and activity (23%)

Activity

Activity is a major influence on metabolism, and it is the next biotic factor to be considered. In fact, it is the single most important one. The reason is that the locomotory musculature is a sizeable fraction of most species' mass – ranging from contractile proteins in the swimming bells of medusae, to the mantles of squid, the swimming legs of shrimp and the swimming muscles of fishes. Muscle contraction is an energy-intensive process, requiring large quantities of adenosine triphosphate (ATP), the biochemical energy currency generated by oxidative metabolism.

A good part of the reason why swimming activity is an expensive enterprise for all taxa is that neither the contraction of muscle (Hill 1950) nor the efficiency of propulsive systems are close to 100% efficient. Energy losses are incurred in the muscle contraction itself as well as during the act of swimming. It requires considerable energy to push through the aqueous environment. Even for the most efficient swimmers, the fishes, propulsive efficiency hovers in the area of 20% (Webb 1975).

Brett (1964) and many investigators to come after him recognized the need to provide a standard for comparing the metabolic rates of ectotherms while accounting for activity. Also, because of the large costs associated with activity, it must be considered when testing the effects of any variable, e.g. temperature, on metabolism. Mammalian physiology, a science with a longer history than that of marine animal physiology, provided some guidelines with its basal metabolic rate: that rate assumed to be due to maintenance processes alone. It is operationally defined as the metabolism of a fasting, resting individual in a thermoneutral environment. For humans, it requires a cooperative subject, for other mammals, a trained one.

Clearly, a true basal metabolic rate measurement is not workable for ectotherms. A viable alternative is standard metabolism, the oxygen consumption rate associated with zero activity, taken by extrapolation of a metabolism vs. activity curve to zero activity. Active metabolism is in turn the oxygen consumption associated with the highest activity able to be sustained for an arbitrary period. An example of how such data may be acquired is detailed below.

The first and most difficult step was to find a means of controlling activity in the experimental subjects, in this case, young sockeye salmon. A swim tunnel respirometer was designed by engineers to allow precise control of swim speed for measurements of metabolism in swimming fishes. Some training or acclimation of the fishes to the apparatus was required. Incentive to maintain position was provided by a gentle shock at the posterior of the animal chamber when the animals did not maintain position by swimming into the current generated within the tunnel. The electrified grid allowed rapid training of the experimental subjects.

Experimental Protocol

Fishes were forced to swim for 75 minutes periods at fixed velocities, starting at $1.0\,\mathrm{ft\,s^{-1}}$ (1 ft = 30.5 cm) and increasing at $0.3\,\mathrm{ft\,s^{-1}}$ for 5–7 hours or until fatigue. Maximum speed was determined as follows:

1) take the highest speed sustainable for 60 minutes, e.g. $2.0\,\mathrm{ft\,s^{-1}}$.
2) take the speed at which fatigue set in, e.g. $2.3\,\mathrm{ft\,s^{-1}}$ and time to fatigue e.g. 20 minutes.
3) then the max speed = $2.0\,\mathrm{ft\,s^{-1}} + (0.3 * 20/60)$
4) the result = $2.1\,\mathrm{ft\,s^{-1}}$
5) this was known as the critical swimming speed (Ucrit)

Results showed that oxygen consumption scaled with swim speed in a logarithmic fashion and that metabolism increased by approximately ninefold as speed increased from 0 to $2.1\,\mathrm{ft\,s^{-1}}$. The difference between standard and active metabolism is the metabolic scope, or scope for metabolism. The active metabolism divided by the standard yields the factorial scope. If we use the numbers for active and standard metabolism from Brett (1964) as an example:

1) Metabolic scope = 800 − 90 = 710 mg $O_2\,\mathrm{kg^{-1}\,h^{-1}}$.
2) Factorial scope = 800/90 = 8.9

Routine Metabolic Rate

Most often, when metabolism is measured on marine species, activity is not controlled. Investigators take precautions to keep activity to as low a level as possible by minimizing visual and mechanical stimulation, resulting in a rate that is above the standard rate, but well below the active rate. The rates thus measured are termed routine rates, and they are by far the most common (Ikeda et al. 2000).

Within aquatic species, factorial scope varies roughly between three for Crustacea (Torres and Childress 1983) and about 10 for fishes (Webb 1975). For gelatinous species, insufficient data exist to do anything but speculate, but it is likely that for most species it is in the same range as that of the Crustacea.

Animal Size as a Modifier of Metabolism

Clearly, oxygen consumption is highly dependent on the size of the animal doing the breathing. A larger individual will consume considerably more oxygen than a smaller one. However, the relationship between animal size and metabolism is not a simple one-to-one scaling with size. It is quite a bit more complicated, and though you might think it would be a straightforward, almost boring variable, it has been the subject of considerable (and currently ongoing) debate for well over 50 years (cf. Hemmingsen 1960).

The relationship between O_2 consumption and body mass is a power function:

- Metabolism $= am^b$ where:
- b = slope
- a = constant
- m = mass
  - Or: $\log m = \log a + b \log W$
  - The log-log plot gives a straight line.

The slope or b-value generally falls between 0.66 and 1.0. A b-value of 1.0 means that respiration with increasing size scales directly with size or mass, that is, a 10-g animal consumes 10 times as much O_2 as a 1 g animal. A b-value of 0.67 indicates that respiration is directly proportional to surface area, because in geometric forms surface area increases at the rate of a linear dimension 2/3 power. It also implies, and the implication is important, that the mass-specific respiration of an animal decreases with increasing size.

The observed metabolic scaling with size was first of interest to those studying mammals and was readily explained: heat ultimately produced by metabolism was dissipated at the body surface. Since the surface to volume ratio is higher in small mammals than large, and as endotherms, they require a constant body temperature, it makes sense that mass-specific metabolism should decline with increasing mass. Unfortunately, the same metabolic scaling holds for ectotherms, where the heat dissipation theory doesn't apply. In point of fact, just about all animals have a b-value between 0.7 and 0.8. The reason for that is a source of ongoing debate.

Scaling effects (b-value) vary slightly with animal grouping. Ectotherm, endotherms, and unicells show different b-values, but they are not significantly different from 0.75. If you combine them all, they go to 0.75. The arguments proposed for why range from the fact that the larger an individual is the greater fraction of its total mass is in the form of slowly respiring connective tissue to a detailed fractal analysis (West et al. 1997, 1999, 2002).

Geometric similarity predicts a b-value of 0.67 for animals with similar morphology.

1) A cube's surface area (SA) scales with its linear dimensions squared (L^2), whereas its mass scales with its volume (L^3). It follows that the surface area scales with mass$^{2/3}$.
2) For animals with similar morphology, if an important or particularly costly metabolic process depends on SA, e.g. heat loss in endotherms or transport processes in unicells – you might expect

metabolism to scale with SA. In fact, it does seem to do that in many unicells and endotherms but not in ectotherms. There is enough variability between species that the overall b-value is 0.75 when the major groups of species are considered over a very large range in mass: a remarkable finding.

A large number of possibilities have been considered for why metabolism scales with mass$^{0.75}$, but all single-cause explanations have been found wanting. For excellent coverage of the arguments and solid rebuttals, we recommend Somero et al. (2017) and Hochachka and Somero (2002) as an excellent place to start.

The bottom line is that when considering data on animal metabolism, mass must be taken into account as a variable both when taking measurements of metabolism and when comparing rates between species. Specifically, mass-specific metabolism declines with increasing mass. The following tables present all metabolism numbers on a mass-specific basis.

Life History Strategies

For pelagic species, data are lacking for many of the life history characteristics listed earlier (Stearns 1992). However, data on metabolism and proximate composition are available for crustacea and fishes from all three of our study systems (Tables 12.1–12.12). The "M" in the energy budget equation is given in Tables 12.1–12.3 for the crustaceans and in Tables 12.7–12.9 for the fishes. Energy budgets over an entire life span have only been estimated for very few species. Examples of such studies include the comparison between two species of euphausiids, *Euphausia pacifica* from the California Current and *E. superba* from the Antarctic (Table 12.13, *see below*) and comparisons between three species of zooplanktivorous fishes from the Gulf of Mexico (GOM) and the Antarctic (Table 12.14, *see below*). Fewer data are available for cephalopods, pteropods, annelids, salps, medusae, and ctenophores, but the data we have are presented in Tables 12.15–12.17.

Table 12.1 Metabolic rates of pelagic euphausiids, decapods, mysids, and amphipods in the eastern Gulf of Mexico. Data from Donnelly and Torres (1988). Data were acquired at the temperatures typical of each species' vertical profile. n = number of replicates, MDO = minimum depth of occurrence: that depth below which 90% of the population lives, MWM = mean wet mass of all experimental subjects (global mean), V_{O_2} = oxygen consumption rate as a function of WM, DM = conversion factor for converting WM to dry mass (DM). Data for *Euphausia tenera* and *Euphausia diomedae* from Ikeda 1974.

	Gulf of Mexico					
Taxon	n	MDO (m)	MWM (g)	T (°C)	V_{O_2} (µlO$_2$ mg WM^{-1} h^{-1})	DM (mg)
Order Euphausiacea						
Family Euphausiidae						
Euphausia tenera	9	0	0.0037	27	0.866	4.85
Euphausia diomedae	2	0	0.0223	27	0.722	4.85
Thysanopoda monacantha	5	100	0.174	7	0.086	4.06
	2			14	0.196	3.06
Order Decapoda						

Table 12.1 (Continued)

					Gulf of Mexico		
Taxon		n	MDO (m)	MWM (g)	T (°C)	V_{O_2} (µlO$_2$ mg WM^{-1} h^{-1})	DM (mg)
Suborder Dendrobranchiata							
Family Benthesicymidae							
Gennadas capensis		2	410	0.265	7	0.098	3.08
Gennadas scutatus		1	135	0.512	14	0.146	3.67
		1	135	0.512	20	0.215	
Gennadas valens		7	150	0.534	7	0.092	4.39
		5	150	0.534	14	0.137	
		8	150	0.534	20	0.244	
Family Penaeidae							
Funchalia villosa		5	70	2.086	7	0.061	3.68
		6	70	2.086	20	0.217	
Family Sergestidae							
Sergestes armatus		2	150	0.029	7	0.085	3.65
		1	150	0.029	14	0.217	
Deosergestes corniculum		2	125	0.284	7	0.09	3.21
		1	125	0.284	20	0.17	
Sergia grandis		1	400	0.713	14	0.159	4.05
		1	400	0.713	20	0.251	
Sergia robustus		1	225	1.635	7	0.094	4.18
		5	225	1.635	14	0.174	
		1	225	1.635	20	0.351	
Sergia splendens		6	195	0.286	7	0.086	3.58
		1	195	0.286	14	0.187	
		5	195	0.286	20	0.3	
Sergia talismani		1	125	0.202	7	0.141	3.99
		3	125	0.202	20	0.263	
Infraorder Caridea							
Family Oplophoridae							
Acanthephyra purpurea		10	325	1.391	7	0.085	3.89
		1	325	1.391	14	0.199	
Oplophorus gracilirostris		4	100	2.01	7	0.095	3.36
		5	100	2.01	20	0.303	
Systellaspis debilis		12	150	1.118	7	0.077	3.37
		3	150	1.118	14	0.156	
		10	150	1.118	20	0.164	
Family Pandalidae							
Parapandalus richardi		3	150	0.284	7	0.094	3.38
		1	150	0.284	14	0.136	

Table 12.2 Metabolic rates of pelagic euphausiids, decapods, mysids, and amphipods in the California Current. Data from Childress (1975). Data were acquired at the temperatures typical of each species' vertical profile. n = number of replicates, MDO = minimum depth of occurrence: that depth below which 90% of the population lives, WM = mean wet mass of all experimental subjects, V_{O_2} = oxygen consumption rate as a function of WM, DM = conversion factor for converting WM to dry mass (DM), ND = no data. Data for *Thysanoessa longipes* from Ikeda (1974), for *Cyphocaris challengeri* from Yamada and Ikeda (2003).

	California Current					
Taxon	n	MDO (m)	WM (g)	T (°C)	V_{O_2} ($\mu l O_2$ mg WM^{-1} h^{-1})	DM (mg)
Order Euphausiacea						
Family Euphausiidae						
Euphausia pacifica	12	0	0.0354	8	0.216	4.85
Euphausia pacifica	9	0	0.0272	12	0.27	4.85
Thysanoessa spinifera	5	0	ND	12	0.224	4.41
Thysanoessa longipes	1	0	0.0417	12	0.38	4.41
Order Decapoda						
Family Benthesicymidae						
Gennadas propinquus	4	400	ND	5.5	0.027	4.29
Family Sergestidae						
Sergestes similis	36	10	ND	10	0.287	4.27
Family Oplophoridae						
	9	500	ND	5.5	0.036	4.15
Hymenodora frontalis	4	400	ND	5.5	0.023	2.76
Family Pasiphaeidae						
Pasiphaea chacei	3	75	ND	7.5	0.112	4.88
Pasiphaea pacifica	1	75	ND	7.5	0.078	5.1
Pasiphaea emarginata	6	600	ND	5.5	0.021	4.29
Order Lophogastrida						
Family Gnathophausiidae						
Neognathophausia gigas	5	1000	ND	4	0.019	4.88
Order Mysida						
Family Mysidae						
Boreomysis californica	6	500	ND	5.5	0.029	5.75
Order Amphipoda						
Family Cyphocarididae						
Cyphocaris challengeri	38	ND	0.029	5	0.059	5.03
Family Scopelocheiridae						
Paracallisoma coecum	1	500	ND	5.5	0.045	3.15
Family Hyperiidae						
Hyperia galba	2	25	ND	10	0.087	8.7
Family Phronimidae						
Phronima sedantaria	4	25	ND	10	0.0262	13.52

Table 12.3 Metabolic rates of pelagic euphausiids, decapods, mysids, and amphipods in Antarctic. Data from Torres et al. (1994a) and Donnelly et al. (2004). Data were acquired at the temperatures typical of each species' vertical profile. n = number of replicates, MDO = minimum depth of occurrence: that depth below which 90% of the population lives, WM = mean wet mass of all experimental subjects, V_{O_2} = oxygen consumption rate as a function of WM, DM = conversion factor for converting WM to dry mass (DM). Data for *Euphausia crystallorophias* from Ikeda and Bruce (1986); for *Themisto gaudichaudii* from Ikeda and Mitchell (1982).

	Antarctica					
Taxon	n	MDO (m)	WM (g)	T (°C)	V_{O_2} (µlO$_2$ mg WM^{-1} h^{-1})	DM (mg)
Order Euphausiacea						
Family Euphausiidae						
Euphausia crystallorophias	11	0	0.211	−1.7	0.118	4.5
Euphausia superba	23	0	1.403	0.5	0.13	4.62
Euphausia triacantha	2	0	0.296	0.5	0.148	4.84
Thysanoessa macrura	7	100	0.072	0.5	0.245	5.05
Order Decapoda						
Family Benthesicymidae						
Gennadas kempi	2	500	1.839	0.5	0.037	3.28
Family Sergestidae						
Petalidium foliaceum	1	500	0.792	0.5	0.036	3.3
Family Nematocarcinidae						
Nematocarcinus lanceopes	3	0	0.308	0.5	0.073	4.5
Family Pasiphaeidae						
Pasiphaea scotiae	3	100	2.928	0.5	0.034	3.06
Order Lophogastrida						
Family Lophogastridae						
Neognathophausia gigas	3	1000	0.456	0.5	0.049	3.47
Family Mysidae						
Boreomysis rostrata	1	550	0.935	0.5	0.055	2.7
Order Amphipoda						
Family Cyphocarididae						
Cyphocaris faueri	6	40	1.275	0.5	0.053	4.24
Cyphocaris richardi	3	340	0.7	0.5	0.038	3.97
Family Eusiridae						
Eusirus antarcticus	26	0	0.047	0.5	0.137	2.56
Family Stegocephalidae						
Parandania boecki	6	500	0.465	0.5	0.041	6.13
Family Cyllopodidae						
Cyllopus lucasii	10	0	0.242	0.5	0.13	4.76
Family Hyperiidae						
Themisto gaudichaudii	8	0	0.0621	−1	0.09	4.73
Family phrosindae						
Primno macropa	6	50	0.129	0.5	0.148	3.4
Family Vibiliidae						
Vibilis stebbingi	2	0	0.117	0.5	0.148	4.8

Table 12.4 Proximate composition of pelagic euphausiids, decapods, and mysids in the eastern Gulf of Mexico. Data from Donnelly et al. (1993). n = number of replicates and MDO = minimum depth of occurrence: that depth below which 90% of the population lives. DVM = migratory pattern where dm = deep migrator, nm = non-migrator, sm = shallow migrator and wm = weak migrator. WM = mean wet mass or wet mass range of all experimental subjects, depending on the study. Compositional attributes are expressed as a function of wet mass (total body composition) and ash-free dry mass (organic matter only). Energy levels are expressed as kcal per 100 grams of WM and AFDM were computed using the compositional attributes and reported values for protein and lipid: 4.8 and 9.3 kcal g^{-1}, respectively (Kleiber 1961; Brett and Groves 1979). To convert to kilojoules (kJ) multiply by 4.19.

Taxon	n	MDO (m)	DVM	WM (g)	Gulf of Mexico							
					Water (%WM)	AFDM (%DM)	Protein (%WM)	Protein (%AFDM)	Lipid (%WM)	Lipid (%AFDM)	kcal·100 g^{-1} WM	kcal·100 g^{-1} AFDM
Order Decapoda												
Family Benthesicymidae												
Gennadas valens	8	150	sm	0.941	77.20	70.6	8.60	54.00	1.00	6.20	62.80	389.80
Family Sergestidae												
Sergestes henseni	11	100	sm	0.476	75.50	82.5	10.90	53.90	1.40	6.80	79.90	395.40
Sergestes paraseminudus	11	100	sm	0.25	67.00	82	14.40	53.00	1.80	6.60	103.90	384.00
Sergia grandis	6	400	dm	1.805	74.50	80.2	11.20	54.60	1.30	6.20	80.70	394.60
Sergia robustus	10	200	dm	1.675	75.50	72.9	10.40	58.80	1.10	6.10	74.80	419.00
Family Oplophoridae												
Acanthephyra curtirostris	4	700	nm	3.618	76.70	79.5	7.90	42.80	3.00	15.10	75.10	405.50
Acanthephyra purpurea	12	300	dm	2.341	73.60	74.8	9.90	50.60	1.50	7.80	78.30	396.60
Systellaspis debilis	13	75	sm	1.312	75.30	78.7	9.30	48.40	1.90	9.20	75.40	387.70
Family Pasiphaeidae												
Pasiphaea merriami	9	75	sm	2.251	75.50	83.2	12.20	62.40	1.20	6.00	89.10	436.90

Species										
Parapasiphaea sulcatifrons	1	800	nm	0.61	72.10	76.3	6.50	30.40	8.90	42.00	117.90	553.90
Order Lophogastrida												
Family Gnathophausiidae												
Neognathophausia ingens	2	700	nm	6.215	85.70	70.5	3.50	35.40	2.50	24.00	47.90	475.50
Family Eucopiidae												
Eucopia sculpticauda	5	600	nm	3.641	93.00	88.2	1.70	27.60	3.20	48.80	36.80	596.60
Eucopia unguiculata	5	600	nm	0.571	82.60	81.4	5.10	36.20	4.90	34.40	74.20	524.10
Mean					77.25		8.58	46.78	2.59	16.86	76.68	443.05

Table 12.5 Proximate composition of pelagic euphausiids, decapods, and mysids in the California Current. Data from Childress and Nygaard (1974). n = number of replicates and MDO = minimum depth of occurrence: that depth below which 90% of the population lives. DVM = migratory pattern where dm = deep migrator, nm = non-migrator, sm = shallow migrator, and wm = weak migrator. WM = mean wet mass or wet mass range of all experimental subjects, depending on the study. Compositional attributes are expressed as a function of wet mass (total body composition) and ash-free dry mass (organic matter only). Energy levels are expressed as kcal per 100 grams of WM and AFDM were computed using the compositional attributes and reported values for protein and lipid: 4.8 and 9.3 kcal·g^{-1}, respectively (Kleiber 1961; Brett and Groves 1979). To convert to kilojoules (kJ) multiply by 4.19.

											California Current	
Taxon	n	MDO (m)	DVM	WM (g)	Water (%WM)	AFDM (%DM)	Protein (%WM)	Protein (%AFDM)	Lipid (%WM)	Lipid (%AFDM)	kcal·100 g^{-1} WM	kcal·100 g^{-1} AFDM
Order Decapoda												
Family Benthesicymidae												
Gennadas propinquus	2	400	dm	0.4–0.8	76.70	90.3	8.73	41.50	6.62	31.50	104.50	496.88
Family Sergestidae												
Sergestes similis	3	10	sm	0.2–2.2	76.60	87.1	10.88	53.40	3.34	16.40	83.82	411.30
Sergestes phorcus	8	350	dm	0.5–5.4	77.50	86.2	12.18	62.80	4.61	23.80	102.08	526.35
Family Oplophoridae												
Acanthephyra curtirostris	3	500	nm	1.7–3.7	75.90	86.4	7.68	36.90	7.52	36.10	107.92	518.27
Systellaspis cristata	2	650	nm	0.5–6.0	72.80	89.3	8.70	35.80	11.90	49.00	154.21	634.89
Family Pasiphaeidae												
Pasiphaea chacei	6	75	sm	0.4–1.5	79.50	84.7	10.50	60.50	2.55	14.70	74.50	429.30
Pasaphaea emarginata	3	600	nm	0.2–0.8	76.70	85.5	7.91	39.70	7.57	38.00	109.50	549.66

	n										
Order Lophogastrida											
Family Gnathophausiidae											
Neognathophausia ingens	17	400	nm	0.2–10.5	72.80	85.7	8.58	36.80	10.54	45.20	140.74	603.78
Neognathophausia gigas	2	1000	nm	1.2–2.1	79.50	81.6	6.27	37.50	5.37	32.10	80.85	483.35
Order Mysida											
Family Mysidae											
Boreomysis californica	4	500	nm	0.07	82.60	81	4.89	34.70	4.64	32.90	67.29	477.46
Order Euphausiacea											
Family Euphausiidae											
Euphausia pacifica	2	0	sm	0.01	79.40	85.5	10.43	59.20	2.29	13.00	71.69	407.00
Mean					77.27		8.80	45.35	6.09	30.25	99.74	503.48

Table 12.6 Proximate composition of pelagic euphausiids, decapods, and mysids in Antarctica. Data from Torres et al. (1994b) and Donnelly et al. (2004). n = number of replicates and MDO = minimum depth of occurrence: that depth below which 90% of the population lives. DVM = migratory pattern where dm = deep migrator, nm = non-migrator, sm = shallow migrator, and wm = weak migrator. WM = mean wet mass or wet mass range of all experimental subjects, depending on the study. Compositional attributes are expressed as a function of wet mass (total body composition) and ash-free dry mass (organic matter only). Energy levels are expressed as kcal per 100 grams of WM and AFDM were computed using the compositional attributes and reported values for protein and lipid: 4.8 and 9.3 kcal g^{-1}, respectively (Kleiber 1961; Brett and Groves 1979). To convert to kilojoules (kJ) multiply by 4.19. A "W" superscript denotes winter data, "F" fall data. Entries without a superscript are spring or summer values.

										Antarctica	
Taxon	n	MDO (m)	DVM	WM (g)	Water (%WM)	AFDM (%DM)	Protein (%WM)	Protein (%AFDM)	Lipid (%WM)	Lipid (%AFDM)	kcal·100 g^{-1} WM kcal·100 g^{-1} AFDM

Taxon	n	MDO (m)	DVM	WM (g)	Water (%WM)	AFDM (%DM)	Protein (%WM)	Protein (%AFDM)	Lipid (%WM)	Lipid (%AFDM)	kcal·100 g^{-1} WM	kcal·100 g^{-1} AFDM
Order Decapoda												
Family Benthesicymidae												
Gennadas kempi	2	500	nm	1.839	69.40	89	9.00	33.10	14.70	54.00	179.90	660.57
Family Sergestidae												
Petalidium foliaceum^w	3	1000	nm	0.689	67.40	93.5	11.50	38.80	14.00	47.50	196.60	523.27
Family Pasiphaeidae												
Pasiphaea scotiae^f	6	100	nm	1.43	63.20	88.9	9.90	30.80	15.20	48.10	200.40	612.56
Pasiphaea scotiae^w	8	1000	nm	1.98	63.30	93.3	7.20	20.60	13.60	40.70	166.40	485.97
Family Nemarocarcinidae												
Nematocarcinus lanceopes	4	150	0	0.28	77.90	85.9	7.10	37.50	6.80	35.00	97.32	512.64
Order Lophogastrida												
Family Gnathophausiidae												

Species									
Neognathophausia gigas	4	1000	nm	0.897	69.40	89.5	4.90	18.30	11.90	44.60	141.90	518.00
Family Eucopiidae												
Eucopia australis[w]	2	1000	nm	0.925	77.80	87.7	6.80	36.30	9.30	49.40	127.00	652.31
Order Mysida												
Family Mysidae												
Boreomysis rostrata[w]	2	550	nm	0.143	75.80	85.9	8.20	40.80	6.30	29.80	105.00	505.10
Order Euphausiacea												
Family Euphausiidae												
Euphausia superba[f]	23	0	nm-sm	0.453	73.30	83.9	10.60	48.40	3.60	15.70	97.10	433.46
Euphausia superba[w]	32	0	nm-sm	0.378	77.30	87.4	8.90	46.80	4.0	19.80	90.70	457.16
Euphausia triacantha[w]	9	50	sm	0.121	76.10	86	8.70	43.10	1.60	2.70	69.60	338.62
Thysanoessa macrura[f]	1	50	sm	0.013	70.40	86.9	8.80	35.00	7.70	30.80	120.30	467.69
Thysanoessa macrura[w]	7	50	sm	0.062	76.90	88	7.00	35.00	5.10	25.40	88.70	436.34
Mean					72.17		8.35	35.73	8.75	34.12	129.30	507.98

Table 12.7 Metabolic rates of mesopelagic fishes in the eastern Gulf of Mexico. Data from Donnelly and Torres (1988). Data were acquired at the temperatures typical of each species' vertical profile. n = number of replicates, MDO = minimum depth of occurrence: that depth below which 90% of the population lives, WM = mean wet mass of all experimental subjects or range of wet mass values depending on the study, V_{O_2} = oxygen consumption rate as a function of WM, DM = conversion factor for converting WM to dry mass (DM), ND = no data.

				Gulf of Mexico		
Taxon	n	MDO (m)	WM (g)	T (°C)	V_{O_2} (µl O_2 mg WM^{-1} h^{-1})	DM (mg)
Order Myctophiformes						
Family Myctophidae						
Lepidophanes guentheri	3	105	1.29	7	0.049	4.05
				14	0.084	
				20	0.229	
Myctophum affine	2	0	1.90	7	0.051	3.19
	2			14	0.198	
Diaphus mollis	2	90	0.17	20	0.292	1.91
Lampanyctus nobilis	2	120	3.67	7	0.043	4.74
Order Stomiiformes						
Family Gonostomatidae						
Sigmops elongatum	4	140	10.192	7	0.039	8.46
	3			14	0.071	
Family Sternoptychidae						
Argyropelecus aculeatus	2	165	2.609	20	0.161	5.01
Order Lophiiformes						
Family Ceratiidae						
Cryptopsaras couesii	2	800	2.676	20	0.122	9.32
Order Trachichthyiformes						
Family Anoplogastridae						
Anoplogaster cornuta	1	600	17.368	7	0.032	8.85
	1			20	0.155	

Table 12.8 Metabolic rates of mesopelagic fishes in the California Current. Data from Torres et al. (1979). Data were acquired at the temperatures typical of each species' vertical profile. n = number of replicates, MDO = minimum depth of occurrence: that depth below which 90% of the population lives, WM = mean wet mass of all experimental subjects or range of wet mass values depending on the study, V_{O_2} = oxygen consumption rate as a function of WM, DM = conversion factor for converting WM to dry mass (DM), ND = no data. Data for *Cyclothone acclinidens* from Smith and Laver (1981).

				California Current		
Taxon	n	MDO (m)	WM (g)	T (°C)	VO_2 (µl o_2 mg WM^{-1} h^{-1})	DM (mg)
Order Myctophiformes						
Family Myctophidae						
Diaphus theta	2	0	0.9–4.4	5	0.107	2.953
	2		1.5–2.7	10	0.208	

Table 12.8 (Continued)

Taxon	n	MDO (m)	WM (g)	T (°C)	VO$_2$ (μl_{O_2} mg WM^{-1} h^{-1})	DM (mg)
			California Current			
Stenobrachius leucopsaurus	6	25	3.4–5.2	5	0.042	3.024
	3		3.8–5.0	10	0.07	
Tarletonbeania crenularis	1	0	3.29	8	0.096	4.37
	2		0.7–2.1	13	0.186	
Nannobrachium ritteri	2	75	1.3–2.3	5	0.041	3.415
	3		1.3–2.9	10	0.059	
Nannobrachium regale	1	500	2.9	5	0.011	7.273
Symbolophorus californiensis	1	0	0.8	5	0.06	4.61
Order Stomiiformes						
Family Gonostomatidae						
Cyclothone acclinidens	3	300	0.543	3	0.024	4.85
Family Stomiidae						
Aristostomias scintillans	1	500	21.1	5	0.01	ND
Stomias atriventer	6	300	17.1–26.6	5	0.017	5.23
	1		9.3	10	0.04	
Order Alepocephaliformes						
Family Alepocephalidae						
Bajacalifornia burragei	3	1000	5.5–44.3	5	0.005	9
Family Platytroctidae						
Sagamichthys abei	1	600	5.7	5	0.016	7.875
Order Argentiniformes						
Family Bathylagidae						
Lipolagus ochotensis	1	0	3.4	10	0.084	ND
Pseudobathylagus milleri	3	550	29.2–53.0	5	0.011	7.727
Order Beryciformes						
Family Melamhaidae						
Poromitra crassiceps	2	400	1.4–32.8	5	0.011	7.636
Order Lophiiformes						
Family Oneirodidae						
Oneirodes acanthias	2	900	3.7–4.7	5	0.008	10.5
Order Trachichthyiformes						
Family Anoplogastridae						
Anoplogaster cornuta	3	550	43.9–57.9	5	0.024	6.667

Table 12.9 Metabolic rates of mesopelagic fishes in the Antarctic. Data from Torres and Somero (1988b), Torres et al. (1994a), and Donnelly et al. (2004). In all studies, data were acquired at the temperatures typical of each species' vertical profile. n = number of replicates, MDO = minimum depth of occurrence: that depth below which 90% of the population lives, WM = mean wet mass of all experimental subjects or range of wet mass values depending on the study, V_{O_2} = oxygen consumption rate as a function of WM, DM = conversion factor for converting WM to dry mass (DM), ND = no data.

Taxon	Antarctic					
	n	MDO (m)	WM (g)	T (°C)	V_{O_2} ($\mu l\, O_2\, mg\, WM^{-1}\, h^{-1}$)	DM (mg)
Order Myctophiformes						
Family Myctophidae						
Electrona antarctica	47	50	4.6	0.5	0.042	3.006
Gymnoscopelus braueri	18	150	12.2	0.5	0.026	2.957
Gymnoscopelus opisthopterus	15	150	19.2	0.5	0.022	3.224
Order Argentiniformes						
Family Bathylagidae						
Bathylagus antarcticus	26	400	10.4	0.5	0.018	7.776
Order Stomiiformes						
Family Gonostomatidae						
Cyclothone microdon	3	500	0.8	0.5	0.016	2.751
Order Aulopiformes						
Family Scopelarchidae						
Benthalbella elongata	1	ND	35.3	0.5	0.037	ND
Order Gadiformes						
Family Macrouridae						
Cyanomacrurus piriei	1	ND	1.8	0.5	0.016	ND
Order Beryciformes						
Family Melamphaidae						
Poromitra crassiceps	1	ND	4.5	0.5	0.028	ND
Order Perciformes						
Family Zoarcidae						
Melanostigma gelatinosum	1	ND	47.2	0.5	0.021	ND

Metabolism and Composition of Pelagic Species

Data were available for many important species residing in the three study regions and when possible, they are presented in the same taxonomic order as the tables in Chapter 11. For gelatinous species, including cnidarians, ctenophores, pteropods, and salps, the focus shifted toward presenting data from subtropical, cold-temperate, and Antarctic waters wherever the studies took place. Metabolism and composition data for California Current species were taken on

Table 12.10 Proximate composition of pelagic fishes in the eastern Gulf of Mexico. Data from Stickney and Torres (1989). n = number of replicates and MDO = minimum depth of occurrence: that depth below which 90% of the population lives. DVM = migratory pattern where dm = deep migrator, nm = non-migrator, sm = shallow migrator, and wm = weak migrator. WM = mean wet mass or wet mass range of all experimental subjects, depending on the study. Compositional attributes are expressed as a function of wet mass (total body composition) and ash-free dry mass (organic matter only) (see text). Energy levels (kcal 100 g^{-1}) are expressed as a function of WM and AFDM and were computed using the compositional attributes and reported values for protein and lipid: 4.8 and 9.3 kcal g^{-1}, respectively (Kleiber 1961; Brett and Groves 1979). To convert to kilojoules (kJ) multiply by 4.19.

											Gulf of Mexico	
Taxon	n	MDO (m)	DVM	WM (g)	Water (%WM)	AFDM (%DM)	Protein (%WM)	Protein (%AFDM)	Lipid (%WM)	Lipid (%AFDM)	kcal·100 g^{-1} WM	kcal·100 g^{-1} AFDM
Order Myctophiformes												
Family Myctophidae												
Benthosema suborbitale	2	25	sm	0.40	66.7	80.4	11.70	46.10	1.60	6.00	81.40	304.04
Diaphus mollis	5	50	sm	0.60	73.3	76.0	12.30	59.80	2.00	9.50	88.30	410.50
Lampanyctus alatus	7	75	sm	0.60	78.9	77.3	13.80	77.30	1.40	7.80	91.70	562.20
Lepidophanes guentheri	11	75	sm	0.80	72.9	68.8	12.30	74.00	1.30	7.30	81.80	438.70
Order Stomiiformes												
Family Gonostomatidae												
Cyclothone pallida	4	500	nm	0.50	81.6	67.6	8.50	75.30	1.00	8.90	57.60	463.08
Sigmops elongatum	10	140	dm	4.10	85.9	67.9	6.90	70.00	1.60	9.90	53.62	560.06
Order Argeentiniformes												
Family Bathylagidae												
Dolicholagus longirostris	6	200	wm	6.80	86.7	76.5	6.40	64.60	0.90	9.10	44.70	359.30

(*Continued*)

Table 12.10 (Continued)

									Gulf of Mexico		
Taxon	n	MDO (m)	DVM	WM (g)	Water (%WM)	AFDM (%DM)	Protein (%WM)	Protein (%AFDM)	Lipid (%WM)	Lipid (%AFDM)	kcal·100 g^{-1} WM	kcal·100 g^{-1} AFDM
Order Trachichthyiformes												
Family Anoplogastridae												
Anoplogaster cornuta	2	600	nm	32.80	88.1	77.1	4.90	52.30	1.50	15.00	41.40	451.23
Order Stephanoberyciformes												
Family Melamphaidae												
Scopelogadus mizolepis	6	500	wm	3.60	88.0	67.7	5.80	71.10	1.40	8.50	45.70	568.40
Mean sm myctophids					73.0		12.5	64.3	1.6	7.7	85.8	428.9
Mean deeper species					86.1		6.5	66.7	1.3	10.3	48.6	480.4
Mean overall					80.2		9.2	65.6	1.4	9.1	65.1	457.5

Table 12.11 Proximate composition of pelagic fishes in the California Current. Data from Childress and Nygaard (1973). n = number of replicates and MDO = minimum depth of occurrence: that depth below which 90% of the population lives. DVM = migratory pattern where dm = deep migrator, nm = non-migrator, sm = shallow migrator, and wm = weak migrator. WM = mean wet mass or wet mass range of all experimental subjects, depending on the study. Compositional attributes are expressed as a function of wet mass (total body composition) and ash-free dry mass (organic matter only) (see text). Energy levels (kcal 100 g^{-1}) are expressed as a function of WM and AFDM and were computed using the compositional attributes and reported values for protein and lipid: 4.8 and 9.3 kcal g^{-1}, respectively (Kleiber 1961; Brett and Groves 1979). To convert to kilojoules (kJ) multiply by 4.19.

										CALIFORNIA CURRENT	
Taxon	n	MDO (m)	DVM	WM (g)	Water (%WM)	AFDM (%DM)	Protein (%WM)	Protein (%AFDM)	Lipid (%WM)	Lipid (%AFDM)	kcal·100 g^{-1} WM	kcal·100 g^{-1} AFDM
Order Myctophiformes												
Family Myctophidae												
Diaphus theta	1	0	sm	1.60	66.1	91.0	12.19	39.50	19.52	63.30	204.80	663.80
Nannobrachium ritteri	3	75	sm	2–5	70.6	90.3	11.18	43.60	11.36	44.30	165.80	646.51
Stenobrachius leucopsaurus	2	25	sm	1–1.5	66.8	89.9	10.59	35.50	13.07	43.80	176.30	590.68
Tarletonbeania crenularis	2	0	sm	0.4–1.2	77.1	80.4	12.02	65.30	3.39	18.40	101.80	552.91
Nannobrachium regale	2	500	nm	2–64	86.3	80.6	5.60	50.70	2.87	26.00	57.70	522.50
Order Stomiiformes												
Family Gonostomatidae												
Cyclothone acclinidens	7	500	nm	0.07–0.6	79.4	82.8	8.72	51.10	6.06	35.50	94.40	553.44
Family Stomiidae												
Stomias atriventer	3	300	wm	5–20	80.9	86.1	5.90	35.90	7.10	43.20	99.60	605.65
Order Argentiniformes												
Family Bathylagidae												

(Continued)

Table 12.11 (Continued)

CALIFORNIA CURRENT

Taxon	n	MDO (m)	DVM	WM (g)	Water (%WM)	AFDM (%DM)	Protein (%WM)	Protein (%AFDM)	Lipid (%WM)	Lipid (%AFDM)	kcal·100 g^{-1} WM	kcal·100 g^{-1} AFDM
Pseudobathylagus milleri Order Trachichthyiformes Family Anoplogastridae	1	550	nm	47.00	87.1	85.2	5.43	49.40	3.13	28.50	59.60	542.27
Anoplogaster cornuta Order Stephanoberyciformes Family Melamphaidae	2	550	nm	27–46	85.0	77.5	6.14	52.60	2.80	24.10	61.70	530.75
Scopelogadus bispinosus	4	450	nm	5–10	85.4	81.3	6.01	50.60	2.72	22.90	68.80	579.62
Mean sm myctophids					70.2		11.5	46.0	11.8	42.5	162.2	613.5
Mean deeper species					84.0		6.3	48.4	4.1	30.0	73.6	555.7
Mean overall					78.5		8.4	47.4	7.2	35.0	109.1	578.8

Table 12.12 Proximate composition of pelagic fishes in the Antarctic. Data from Donnelly et al. (1990). n = number of replicates and MDO = minimum depth of occurrence: that depth below which 90% of the population lives. DVM = migratory pattern where dm = deep migrator, nm = non-migrator, sm = shallow migrator, and wm = weak migrator. WM = mean wet mass or wet mass range of all experimental subjects, depending on the study. Compositional attributes are expressed as a function of wet mass (total body composition) and ash-free dry mass (organic matter only) (see text). Energy levels (kcal 100 g^{-1}) are expressed as a function of WM and AFDM and were computed using the compositional attributes and reported values for protein and lipid: 4.8 and 9.3 kcal g^{-1}, respectively (Kleiber 1961; Brett and Groves 1979). To convert to kilojoules (kJ) multiply by 4.19.

											Antarctic	
Taxon	n	MDO (m)	DVM	WM (g)	Water (%WM)	AFDM (%DM)	Protein (%WM)	Protein (%AFDM)	Lipid (%WM)	Lipid (%AFDM)	kcal·100 g^{-1} WM	kcal·100 g^{-1} AFDM
Order Myctophiformes												
Family Myctophidae												
Electrona antarctica	27	5	sm	3.89	68.7	88.4	10.50	39.30	8.90	31.30	140.00	505.97
Gymnoscopelus braueri	3	160	sm	8.74	66.6	88.6	11.30	39.60	13.70	47.50	189.00	638.70
Protomyctophum boliniw	6	ND	sm	1.51	74.6	85.9	11.50	54.30	4.00	18.80	104.00	476.66
Protomyctophum tenisoniw	3	ND	sm	1.61	72.2	86.5	11.80	50.20	5.00	21.00	113.00	469.90
Gymnoscopelus opisthopterus	6	250	dm	12.29	80.1	86.1	6.00	36.60	8.70	49.10	110.00	642.00
Order Stomiiformes												
Family Gonostomatidae												
Cyclothone microdon	4	500	nm	1.04	67.0	97.1	12.10	39.40	10.40	33.80	167.00	521.00
Order Argentiniformes												
Family Bathylagidae												

(*Continued*)

Table 12.12 (Continued)

								Antarctic				
Taxon	n	MDO (m)	DVM	WM (g)	Water (%WM)	AFDM (%DM)	Protein (%WM)	Protein (%AFDM)	Lipid (%WM)	Lipid (%AFDM)	kcal·100 g^{-1} WM	kcal·100 g^{-1} AFDM
Bathylagus antarcticus	32	170	dm	3.78	85.9	83.9	7.70	66.50	0.90	8.10	53.00	448.00
Order Aulopiformes												
Family Scopelarchidae												
Benthalbella elongataw	2	ND	ND	97.60	77.0	90.8	6.60	34.50	9.80	47.30	127.00	608.12
Order Perciformes												
Family Genpylidae												
Paradiplospinus gracilisw	2	ND	ND	44.95	69.1	90.0	10.00	37.50	14.40	53.40	189.00	679.60
Mean sm myctophids					70.5		11.3	45.9	7.9	29.7	136.5	522.8
Mean deeper species					75.8		8.5	42.9	8.8	38.3	129.2	579.7
Mean overall					73.5		9.7	44.2	8.4	34.5	132.4	554.4

Table 12.13 Euphausiid energetics. Comparison of energy usage for *Euphausia pacifica* and *Euphausia superba*. Data for *E. pacifica* from Ross (1982b) using 8 °C values and a conversion factor of 9.9 kcal g^{-1} Carbon. Length-at-age for *E. superba* from equations in Siegel (1987) and mass from length vs. mass regressions. Yearly energy expended in metabolism estimated from mass values at 75% of age (cf. Ross 1982b). Metabolism for *E. superba* computed assuming eight months of summer metabolism and four months of reduced winter metabolism (Figure 12.3). Energy devoted to somatic growth and to additional production (reproduction + molting) estimated using the ratios in Ross (1982b). Compositional data are from Tables 12.5 and 12.6; metabolism data are from Figure 12.3. Values are approximate and were intended for comparison only.

	California Current	Antarctic
	Euphausia pacifica	*Euphausia superba*
Max. age (years)	1	6
Max. size (g)	0.0359	1.6
Size at 1 year (g)	0.0359	0.054
Energy expended 1 year (kcal)		
Somatic growth	0.0261	0.052
Additional production	0.0486	0.015
Metabolism	0.148	0.161
Total	0.223	0.181
Energy expended lifetime (kcal)		
Somatic growth	0.0261	1.55
Additional production	0.0486	2.75
Metabolism	0.148	19.286
Total	0.223	23.59

Table 12.14 Mesopelagic fish energetics. Brief comparison of energy usage for two migrating zooplanktivores from the GOM and one from the Antarctic. Length-at-age from Lancraft et al. (1988), Gartner (1991a), and Greely et al. (1999). Metabolism for *Lepidophanes* and *Sigmops* generated using 14 °C values (Table 12.7) and an oxycalorific equivalent of 4.63 kcal l^{-1} (Brett and Groves 1979). Average mass used for computation was 44% of the yearly increase in mass (Greely et al. 1999). Metabolism vs. mass curve in Torres and Somero (1988b) was used for *E. antarctica*. Energy invested in growth was calculated from caloric values in Tables 12.10 and 12.12.

	Gulf of Mexico		Antarctic
	Lepidophanes guentheri	*Sigmops elongatum*	*Electrona antarctica*
Max. age (years)	1	1.75	4
Max. size (g)	65	225	107
Size at 1 year (g)	61	137	28.5
Energy expended 1 year (kcal)			
Somatic growth	3.34	9.88	0.414
Metabolism	6.13	23.33	0.227
Total	9.47	33.21	0.641
Energy expended life (kcal)			
Somatic growth	3.52	43.43	24.07
Metabolism	10.22	99.22	25.28
Total	13.74	142.65	49.23

Table 12.15 Metabolic rates of cephalopods in the California Current and Antarctic. Data from Seibel et al. (1997), Donnelly et al. (2004), and Ikeda and Bruce (1986). In all studies, data were acquired at the temperatures typical of each species' vertical profile within their respective ocean system. n = number of replicates, MDO = minimum depth of occurrence: that depth below which 90% of the population lives, WM = mean wet mass of all experimental subjects or range of wet mass values depending on the study, V_{O_2} = oxygen consumption rate as a function of WM, DM = conversion factor for converting WM-specific to dry mass-specific metabolism (DM). Data for *Psychroteuthis glacialis* from Ikeda and Bruce (1986).

Taxon			Gulf of Mexico			
	n	MDO (m)	WM (g)	T (°C)	V_{O_2} ($\mu l O_2$ mg WM^{-1} h^{-1})	
Order Oegopsida						
Family Cranchiidae						
Cranchia scabra	1	10	6.39	5	0.009	
Galiteuthis phyllura	1	300	5.19	5	0.014	
Heliochrancha pfefferi	11	300	0.88	5	0.022	
Family Gonatidae						
Gonatus onyx	1	100	2.3	5	0.197	
Gonatus pyros	5	100	8.58	5	0.098	
Family Histioteuthidae						
Histioteuthis heteropsis	17	150	9.99	5	0.023	
Order Octopoda						
Family Bolitaenidae						
Japetella heathi	11	600	35.19	5	0.004	
Order Vampyromorpha						
Family Vampyroteuthidae						
Vanpyroteuthis infernalis	15	600	223.4	5	0.002	

Taxon			Antarctic			
	n	MDO (m)	WM (g)	T (°C)	V_{O_2} ($\mu l O_2$ mg WM^{-1} h$^{-1)}$)	DM (mg)
Order Oegopsida						
Family Cranchiidae						
Galiteuthis glacialis	6	50	0.772	0.5	0.033	13.94
Family Psychroteuthidae	6	ND	0.276	−0.8	0.085	4.33
Psychroteuthis glacialis						

Table 12.16 Metabolic rates of pteropods, annelids, and salps from tropical–subtropical, cool temperate, and polar ocean systems. Locations and references shown in the table. In all studies, data were acquired at the temperatures typical of each species' vertical profile within their respective ocean system. n = number of replicates, MDO = minimum depth of occurrence: that depth below which 90% of the population lives, WM = mean wet mass of all experimental subjects or range of wet mass values depending on the study, V_{O_2} = oxygen consumption rate as a function of WM, Q_{O_2} = oxygen consumption rate as a function of DM.

Taxon	n	MDO (m)	WM (g)	T (°C)	V_{O_2} (µlO$_2$ mg WM^{-1} h^{-1})	Q_{O_2} (µlO$_2$ mg DM^{-1} h^{-1})	Location	References
Phylum Mollusca								
Order Pteropoda								
Family Limacinidae								
Limacina helicina	22	0	0.00242–0.0149	−2	0.1234	ND	Antarctica	Seibel et al. (2007)
Family Cavoliniidae								
Cavolinia tridentata	3	0	0.013–0.088	18	0.2462	ND	Gulf of California	Seibel et al. (2007)
Cavolinia longirostris	20	26	0.0082	20	0.2752	ND	E. Tropical Pacific	Maas et al. (2012)
Cavolinia longirostris	11	26	0.011	11	0.0712	ND	E. Tropical Pacific	Maas et al. (2012)
Family Creseidae								
Hyalocylis striata	24	20	0.0105	20	0.1644	ND	E. Tropical Pacific	Maas et al. (2012)
Hyalocylis striata	36	20	0.0162	11	0.0495	ND	E. Tropical Pacific	Maas et al. (2012)
Family Cliidae								
Clio pyramidata	13	10	0.0091	20	0.0100	ND	E. Tropical Pacific	Maas et al. (2012)
Family Clionidae								
Clione limacina	31	ND	0.0188–0.262	−2	0.0456	ND	Antarctic	Seibel et al. (2007)
Clione limacina	27	ND	0.2	0.5	0.0641	0.807	Antarctic	Suprenand et al. (2015)
Clione limacina	20	ND	0.055–0.632	10	0.0437	ND	Newfoundland	Seibel et al. (2007)
Family Cliopsidae								

(*Continued*)

Table 12.16 (Continued)

Taxon	n	MDO (m)	WM (g)	T (°C)	V_{O_2} (µlO$_2$ mg WM^{-1} h^{-1})	Q_{O_2} (µlO$_2$ mg DM^{-1} h^{-1})	Location	References
Cliopsis krohni	4	ND	0.550–3.000	5	0.0012	ND	California Current	Seibel et al. (2007)
Family Pneumodermatidae								
Spongiobranchaea australis	25	ND	0.13	0.5	0.0884	2.493	Antarctic	Suprenand et al. (2015)
Phylum Nemertea								
Family Nectonemertidae								
Nectonemertes mirabilis	11	mes	0.152–1.055	5	0.0066	ND	California Current	Thuesen and Childress (1993)
Phylum Annelida								
Class Polychaeta								
Order Phyllodocida								
Family Phyllodocidae								
Pelagobia sp.	4	mes	0.0062–0.0444	5.00	0.0537	ND	California Current	Thuesen and Childress (1993)
Family Tomopteridae								
Tomopteris carpenteri	3	mes	0.355	0.50	0.0970	0.902	Antarctic	Donnelly et al. (2004)
Tomopteris pacifica	2	mes	0.0126–0.0801	5.00	0.0880	ND	California Current	Thuesen and Childress (1993)
Family Typhloscolecidae								
Travisiopsis lobifera	2	mes	0.239–0.268		0.0224	ND	California Current	Thuesen and Childress (1993)
Phylum Chordata								
Class Thaliacea								

Order Salpida							
Family Salpidae							
Salpa thompsoni	12	0	12.52	−1.6	0.0040	0.102	Antarctica	Ikeda and Bruce (1986)
Salpa thompsoni	6	0	32.55	−1.7	0.0033	0.083	Antarctica	Ikeda and Bruce (1986)
Thalia democratica (sol)	3	0	0.074	17	0.0237	0.593	Antarctica	Ikeda (1974)
Salpa fusiformis (sol)	2	0	0.291	17	0.0166	0.415	Antarctica	Ikeda (1974)
Pegea confoederata (sol)	3	0	2.03	26	0.0164	0.41	Antarctica	Ikeda (1974)
Order Pyrosomatida							
Family Pyrosomatidae							
Pyrosomella vencillata	2	0	1.626	26	0.0162	0.405	Antarctica	Ikeda (1974)

Table 12.17 Metabolic rates of medusae and ctenophores from tropical–subtropical, cool temperate, and polar ocean systems. Locations and references shown in the table. In all studies, data were acquired at the temperatures typical of each species' vertical profile within their respective ocean system. n = number of replicates, MDO = minimum depth of occurrence: that depth below which 90% of the population lives, WM = mean wet mass of all experimental subjects or range of wet mass values depending on the study, V_{O_2} = oxygen consumption rate as a function of WM, Q_{O_2} = oxygen consumption rate as a function of DM. Respiration rates for species with an asterisk were measured in situ.

Taxon	n	MDO (m)	WM (g)	T (°C)	V_{O_2} (µlO$_2$ mg WM^{-1} h^{-1})	Q_{O_2} (µlO$_2$ mg DM^{-1} h^{-1})	Location	References
Phylum Cnidaria								
Class Hydrozoa								
Order Leptothecata								
Family Aequoreidae								
Aequorea victoria	45	0	7.25	10.00	0.0044	0.1100	NE Pacific	Larson (1987)
Order Trachymedusae								
Family Rhopalonematidae								
Benthocodon pedunculatus	5	mes	4.2–7.8	8.00	0.0266	0.5594	Caribbean	Bailey et al. (1994b)
Benthocodon pedunculatus	9	mes	1.7–8.4	8.00	0.0085	0.2496	Caribbean	Bailey et al. (1994b)
Crossota rufobrunnea	11	mes	0.062–1.08	5.0	0.0035	0.0863	California Current	Thuesen and Childress (1994)
Family Halicreatidae								
Haliscera bigelowi	5	mes	0.191–1.168	5.00	0.0029	0.0717	California Current	Thuesen and Childress (1994)
Order Narcomedusae								
Family Cuninidae								
Solmissus incisus	4	mes	25.45	6.00	0.0125	0.3132	NW Atlantic	Bailey et al. (1995)
Solmissus incisus	5	mes	25.45	6.00	0.0055	0.1387	NW Atlantic	Bailey et al. (1995)
Family Aeginidae								

Aegina citrea	11	mes	0.366–9.76	5.00	0.0041	0.1036	California Current	Thuesen and Childress (1994)
Order Siphonophorae								
Family Diphyidae								
Diphyes antarctica	8	0	2.085	−1.8	0.0027	0.0672	Antarctic	Ikeda and Bruce (1986)
Class Scyphozoa								
Order Semaeostomeae								
Family Ulmaridae								
Aurelia aurita	18	0	19.6	10.00	0.0056	0.1400	NE Pacific	Larson (1987)
Family Cyaneidae								
Cyanea capillata	4	0	145.91	6.00	0.0032	0.0697	NW Atlantic	Bailey et al. (1995)
Order Coronatae								
Family Atollidae								
Atolla wyvillei	5	mes	0.219–17.2	5.00	0.0030	0.0751	California Current	Thuesen and Childress (1994)
Family Periphyllidae								
Periphylla periphylla	8	mes	2.760–42.39	5.00	0.0021	0.0527	California Current	Thuesen and Childress (1994)
Phylum Ctenophora								
Class Tentaculata								
Order Lobata								
Family Bathocyroidae								
Bathycyroe fosteri	4	mes	24.38	6.00	0.0076	0.2135	NW Atlantic	Bailey et al. (1995)
Bathycyroe fosteri	9	mes	24.38	6.00	0.0022	0.0606	NW Atlantic	Bailey et al. (1995)
Family Ocyropsidae								

(*Continued*)

Table 12.17 (Continued)

Taxon	n	MDO (m)	WM (g)	T (°C)	V_{O_2} (µlO$_2$ mg WM^{-1} h^{-1})	Q_{O_2} (µlO$_2$ mg DM^{-1} h^{-1})	Location	References
Ocyropsis maculata	3	0	ND	25.00	0.0048	0.1200	Caribbean	Kremer et al. (1986)
Family Eurhamphaeidae								
Eurhamphea vexilligera	47	0	12.525	25.00	0.0025	0.0630	Caribbean	Kremer et al. (1986)
Family Bolinopsidae								
Bolinopsis infundibulum	10	mes	50.74	5.0–6.0	0.0021	0.0513	Gulf of Maine	Bailey et al. (1994a)
Bolinopsis vitrea	55	0	7.12	25.00	0.0015	0.0386	Caribbean	Kremer et al. (1986)
Mnemiopsis leidyi	19	0	0.88–1.65	10.00	0.0024	0.0595	NW Atlantic	Kremer (1977)
	21	0	0.90–12.45	25.00	0.0079	0.1964	NW Atlantic	Kremer (1977)
Order Cydippida								
Family Mertensiidae								
Callianira antarctica	10	0	2.5	0.00	0.0074	0.1840	Antarctic	Scolardi et al. (2006)
Mertensia sp.	8	0	2.277	−1.6	0.0053	0.1329	Antarctic	Ikeda and Bruce (1986)
Class Nuda								
Order Beroida								
Family Beroidae								
Beroe ovata	22	0	0.007–0.030	25.00	0.0109	0.2730	Caribbean	Kremer et al. (1986)
Beroe sp.	7	0	15.93	−1.50	0.0010	0.0245	Antarctic	Ikeda and Bruce (1986)

individuals captured off the coast of California, which has a slightly warmer temperature profile in the upper 200 m than does the Oregon Coast. Species composition is quite similar in both regions.

Metabolism of Euphausiids, Decapods, Mysids, and Amphipods

In each of the three study systems as well as one not shown (Hawaii; Cowles et al. 1991; Figure 12.2), metabolism declines with increasing species' depth of occurrence (Torres et al. 1994a), a phenomenon covered in detail in Chapter 2 (Tables 12.1–12.3). The same families (and sometimes the same species) of Crustacea are present in each, though with different levels of representation. In the GOM, sergestids are an important group of vertical migrators occupying the mid-depths of the water column with a large number of species (Hopkins et al. 1994), whereas one species, *Sergestes similis*, is the overwhelming dominant in the California Current. With the exception of *Nematocarcinus*, which is predominantly benthic, the limited number of Antarctic decapods fall into the same families as well.

What is clear from Tables 12.1–12.3 is that, based on oxygen consumption rates, daily life in the GOM is energetically more expensive than that in the California Current or Antarctic. The reason for the regional disparity is partially due to the temperature structure of the water column and partially due to the fact that virtually all of the GOM species vertically migrate into the warm temperatures of the upper 300 m at night (Hopkins et al. 1994; Kinsey and Hopkins 1994). At daytime depths below

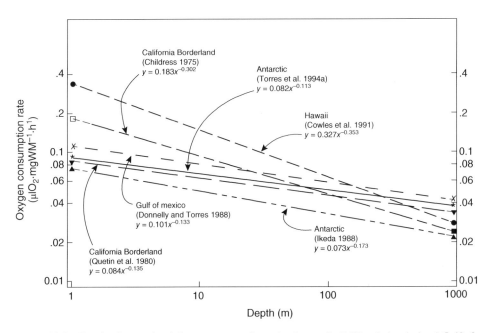

Figure 12.2 Respiration vs. depth in crustaceans from the Antarctic, California borderland, Gulf of Mexico, and central Pacific near Hawaii. All values corrected to a value of 1 g using a *b*-value of 0.75 and to a temperature of 0.5 °C using a Q_{10} of 2.0. *Source:* Torres et al. (1994a), figure 2 (p. 218), Marine Ecology Progress Series Vol 113. ©Inter-Research. Reproduced with the permission of Inter-Research.

500-m rates are quite similar between systems, in most cases differing by a factor of 2 or less at normal habitat temperatures.

A mass- and temperature-corrected composite curve of metabolism vs. depth for six studies treating the subject in different regions of the world ocean (Figure 12.2; Torres et al. 1994a) suggests that the phenomenon of declining metabolism with depth is a universal one for pelagic Crustacea. It underscores the similarity in rates for crustacean fauna dwelling below 500 m. The persistence of the phenomenon in Crustacea from the Antarctic, living in a nearly isothermal water column, supports the idea that the reduced metabolism of deep-living species is an adapted characteristic rather than a response to lowered temperature.

The krill *Euphausia superba* exhibits an interesting metabolic response to the algae-poor winter months in the Antarctic system. Krill show a large (>50%) drop in metabolism during the winter months (Figure 12.3), improving their ability to survive the long dark season of food deprivation (cf. Kawaguchi et al. 1986; Quetin and Ross 1991). None of the other euphausiids exhibited this response, nor did any decapods. The two amphipods, *Cyllopus lucasii* and *Vibilia stebbingi*, both ectocommensals/parasites on salps, also showed a winter drop in metabolism, most likely due to the absence of actively feeding salps at that time of year (Torres et al. 1994a).

Proximate Composition of Pelagic Decapods, Mysids, and Euphausiids

Note that the publications cited here contain more data than are presented in Tables 12.4–12.6. A subset of representative species from each publication was chosen to afford the best comparison between regions. Trends are discussed based on values in the tables, but any reference to statistical significance refers to analyses in the original publications.

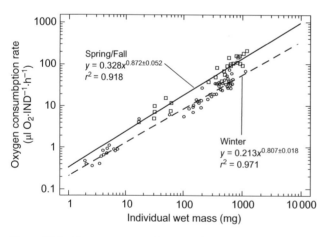

Figure 12.3 Oxygen consumption rate of *Euphausia superba* as a function season. Square symbols represent spring/fall values; circles represent winter values. Slopes are expressed ± standard error. Regressions were fitted using the least squares method; lines are significantly different ($p < 0.01$, ANCOVA). *Source:* Torres et al. (1994a), figure 1 (p. 213), Marine Ecology Progress Series Vol 113. ©Inter-Research. Reproduced with the permission of Inter-Research.

Terminology

Proximate composition is the amount of protein, lipid, and carbohydrate comprising the tissues of an organism (Love 1970). Carbohydrate, though often important in the tissue make-up of benthic species as an energy depot, is a tiny fraction (0.2% or less of WM) of pelagic species' composition and thus was not included in the tables. Ash is the residue left behind when a specimen is combusted at 450 °C, removing all organic matter. It comprises the salts of an individual's blood and any that are incorporated into its carapace. It usually scales directly with its water content. Ash-free dry mass (AFDM) is thus organic matter only.

Trends with Depth of Occurrence

The clearest trends with depth are exhibited by GOM species. Water level increases significantly with increasing species' depth of occurrence. Vertical migrators, the predominant lifestyle in the Gulf, have significantly lower water levels than non-migrators. This trend reflects the more robust structure required by an active migratory lifestyle.

Protein is the principal compositional component in GOM species. Levels of protein decline significantly with depth of occurrence in GOM crustaceans on both a WM and AFDM basis. Protein levels are also significantly higher in migrators than non-migrators. The high protein levels again reflect a more robust locomotory musculature required for diel excursions between 400 and 800 m each day. Lipid levels are generally low in Gulf species and show a significant increase with depth of occurrence. They are significantly lower in migrators than non-migrators.

California current species show a different pattern with respect to water levels, with the mid-depth Crustacea having the lowest values. However, values overall are similar to those in the GOM. Protein levels show a significant decline with depth in the California Current as a function of both WM and AFDM, implying a reduced need for locomotory muscle in deeper living species and likely underpinning the observed decline in metabolism with depth. Lipid levels increase with increasing depth as a function of AFDM, though the trend is not as pronounced as in the GOM.

Representation by decapods and mysids in the Antarctic system is limited, and seasonal changes in species' vertical profiles and behavior complicate the compositional trends with depth. For example, *Pasiphaea scotiae* was not captured in trawls reaching depths less than 1000 m in winter (Lancraft et al. 1991) but was found in the upper 200 m in spring, fall, and summer samples. It showed a significant drop in protein and lipid in winter analyses, suggesting that body substance was being combusted for winter fuel, but unlike *Euphausia superba*, it showed no accompanying drop in metabolism. Worthy of note when considering the Antarctic crustacean fauna is the fact that during all four studies providing the data presented in Tables 12.3 and 12.6, water temperature varied by less than 2 °C between the surface and 1000 m, and the temperature at 1000 m was always greater than that at the surface.

Water levels in Antarctic crustaceans are comparable to those in the other two regions, with decapods having the lowest values. Protein values are generally lower than those in the GOM but are comparable to those in the California Current. Lipid values in Antarctic species are generally comparable, if somewhat higher, than those in the GOM and California Current. Ash level scale significantly with water level in the species examined.

As a function of WM, water, protein, and lipid show no significant correlation with depth. However as a function of AFDM, protein shows a significant decline in both fall and winter. Lipid (%AFDM) has a significant increase with depth in winter but not in fall. Water level scales reciprocally with lipid (%AFDM), suggesting that a low water level implies a high lipid level. The correlation is significant, but not absolute. For example, *Eucopia* shows a high water level and high lipid level, and a drop in winter lipid values. However, in *Pasiphaea scotiae*, the drop in winter lipid values is not accompanied by an increase in water level.

Seasonal Changes

Overwintering mechanisms employed by Antarctic micronektonic Crustacea are important in discerning whether successful species show specialized overwintering strategies or simply adjust those employed by lower latitude species for the more extreme seasonality. At least three mechanisms are possible. The first (Type 1) is an overwintering dormancy cued by photoperiod such as that observed in Antarctic, boreal and cold-temperate copepods (Conover 1988; Miller and Clemons 1988; Hopkins et al. 1993). A Type 1 response includes accumulation of a large lipid reserve and a sinking to midwater depths, typically 300m and below (Hopkins et al. 1993), accompanied by near cessation of all activity (Alldredge et al. 1984). At the other extreme is a "business-as-usual" (Type 3) strategy where metabolic activity is not modulated appreciably and opportunistic feeding coupled with some combustion of tissue carries the species through the winter. In between the two extremes is a Type 2 strategy where metabolic activity is modulated downward either from starvation or by an adapted response to an external cue, accompanied by the depletion of body substance and opportunistic feeding. Evidence points to strategies 2 and 3 as the primary ones employed by Antarctic micronektonic Crustacea. Hyperiid amphipods (not shown) and all species of euphausiids analyzed show winter depletions in compositional attributes (Torres et al. 1994b). In contrast, deeper-living species, including the gammarid amphipods (not shown) and the decapod *Petalidium foliaceum*, show a Type 3 response. The decapod *Pasiphaea scotiae* shows a winter depletion in both protein and lipid but no change in metabolic rate. *Euphausia superba*'s drop in metabolism during the winter months is one of the clearest signals of a Type 2 seasonal response.

Deeper living micronektonic species have access to the diapausing populations of lipid-rich calanoid copepods that sink out of the euphotic zone to overwinter below 300m (Foxton 1956; Hopkins et al. 1993). The majority of zooplankton biomass is found below 300m during the winter season. Since deeper-living micronekton are largely zooplanktivorous, their food supply does not decline appreciably, it simply moves downward (Hopkins et al. 1993).

Trends Across Systems

The data subsets presented in Tables 12.4–12.6 don't include the entire suite of species found in the original publications, but overall trends between systems remain the same. Water levels on average are nearly identical in the GOM and California Current, with both being higher than those in the Antarctic, largely due to the low levels in the Antarctic decapods. Lipid (%WM) is lowest in the GOM, with the two cold water systems showing a greater accumulation of lipid, presumably due to the increased seasonality typical of both systems and greater need for an energy depot, reaching an extreme level in the Antarctic. The general gradient of increasing lipid levels with decreasing environmental temperature is borne out in the caloric values for 100g of wet mass in each system, which show an identical trend. Protein levels (%WM) across all three systems are virtually identical, reflecting similar locomotory demands.

Metabolism of Mesopelagic Fishes

Metabolism of mesopelagic fishes declines with increasing depth of occurrence in all three systems, even in the nearly isothermal water column of the Antarctic (Figure 12.4) (Tables 12.7–12.9). Species richness of mesopelagic fishes is lower in the Antarctic than in the GOM and California Current, but those present belong to the same families inhabiting the midwater in the other two systems and throughout the global ocean. It is important to recognize the fact that Antarctic endemics (Nototheniidae) do not range off the shelf, including the two important pelagics: *Pleuragramma antarctica* (Parker et al. 2015) and *Dissostichus mawsoni*. The open-ocean and shelf faunas do not mix except on the Western Antarctic Peninsula, where the CDW intrudes onto the shelf and provides a more hospitable water column (Hofmann et al. 2004).

Metabolic cold adaptation is evident in Antarctic midwater fishes. Rates at environmental temperature in both systems are quite similar despite the higher environmental temperature in the California Current. When the rates of California Current fishes are adjusted to 0.5 °C, the experimental temperature for Antarctic fishes, their line for respiration vs. depth drops well below that of the Antarctic species (Figure 12.5).

Fewer data are available for GOM species, but myctophids for which data are available show nearly identical values to those of the California Current and Antarctic at temperatures typical of their daytime depths in two of four cases. In the remaining two, the highly active California Current migrators *Diaphus theta* and *Tarletonbeania crenularis* have rates about twice those of the GOM species at similar (5 °C CC vs. 7 °C GOM) temperatures

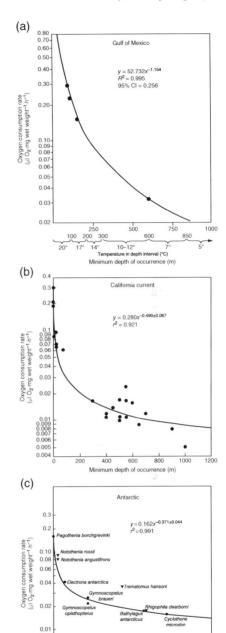

Figure 12.4 Oxygen consumption as a function of minimum depth of occurrence in mesopelagic fishes from the Gulf of Mexico (a), California Current (b), and Antarctic (c). *Sources:* Adapted from (a) Donnelly and Torres (1988), figure 7 (p. 490); (b, c) Torres and Somero (1988a), figure 3 (p. 525).

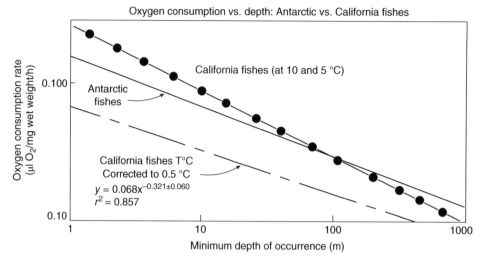

Figure 12.5 The influence of temperature on oxygen consumption vs depth. *Source:* Adapted from Torres and Somero (1988a), figure 4 (p. 525). Reproduced with the permission of Pergamon Press.

(see Tables 12.7 and 12.8). The one species inhabiting both the California Current and GOM systems, *Anoplogaster cornuta*, has very similar rates at habitat temperature (5 °C CC and 7 °C GOM) in both systems.

Proximate Composition of Pelagic Fishes

Available compositional data offer more even coverage of the entire suite of commonly captured species for all three systems, subsets of which are presented in Tables 12.10–12.12. It is mute testament to the fact that it is far easier to capture and freeze delicate specimens for later analysis than it is to capture them alive, bring them to the surface at near-habitat temperature, and then perform experiments on them at sea. Means for shallow migrators and deeper-living species are calculated for values in the table to better show trends with depth. Statements on statistical significance refer to analyses in the relevant publications.

Trends with Depth of Occurrence

Water level increases significantly with increasing depth of occurrence in both the GOM and CC, but not in the Antarctic where low water levels in deeper dwelling species confer enough variability to erase the trend. A significant decline in protein (%WM) and lipid (%WM) with depth occurs in the California Current as a consequence of the increasing water level, but the GOM shows a wet mass decline in protein only. Protein and lipid levels as a function of AFDM show no significant trends with depth in any of the three systems.

Total caloric content (%WM) shows a significant decline with depth in the GOM and California Current, but not in the Antarctic. In all three systems, the fishes are high value food items for marine and aerial predators.

Trends Across Systems

Trends in fish composition mirror those observed in the micronektonic Crustacea. Protein is the largest overall component fraction of WM and AFDM in all three systems, with the highest values in each system contributed by the shallow dwelling myctophids. Lipid is lowest in the GOM, as a function of both WM and AFDM. The cold-temperate and polar systems exhibit higher lipid values and higher overall caloric values than their subtropical brethren, despite the considerable overlap in the families represented in each. Lowest water levels and highest lipid levels are exhibited in the Antarctic species. The larger size of fishes as well as their higher caloric values per unit wet mass make them an important prey item for piscine and mammalian predators as well as swimming and flighted seabirds (cf. Ainley et al. 1986).

Energy and Life History in the Midwater Fauna

Even rough estimates of energy utilization through the life of a species require some knowledge of longevity, and preferably some details on age and growth. Such data are quite rare for midwater species, though it exists for important coastal pelagics such as the Pacific sardine (Lasker 1970) and many other species of commercial importance. Fishes have a time recorder in their ear bones (Chapter 10), but crustaceans have no analogous structure, so any knowledge of their length of life or growth rate must come from repetitive seasonal trawls or laboratory rearing, or both. Fortunately, two important species from two of the systems have been studied in such a way: *Euphausia pacifica* and *Euphausia superba*.

Table 12.13 compares energy utilization in *Euphausia pacifica* in the California Current and *E. superba* from the Antarctic. In these studies, sufficient data were available to estimate energy used in growth and metabolism over the lives of the species. The most complete data set is from a remarkable study which describes the entire energy budget of *E. pacifica* (Ross 1982a,b). Information on *E. superba* comes from a variety of sources cited in the table and figure legends and in the text below.

Though *Euphausia pacifica* and *E. superba* are congeners and the clear dominants in their respective ocean systems, their life histories differ considerably on nearly every level. At, 8°C *E. pacifica* completes its life cycle in just under a year (335 days – Ross 1982b), reproducing as a repetitive batch-spawner in the last quarter of its life. A strong vertical migrator, *E. pacifica*, is primarily herbivorous and a competent swimmer, though not in the league of *E. superba*. Somatic growth commands about 10.5% of its energy usage with the remaining devoted to reproduction, molting, and metabolism. Because of its smaller size and warmer habitat temperature, its routine metabolic rate is about twice that of *E. superba* on a mass-specific basis (Table 12.13).

Euphausia superba lives for six years, reaching a size of 60 mm in the final year of its life (Siegel 1987; Quetin and Ross 1991; Hofmann and Lascara 2000). It is a highly competent swimmer, easily comparing to a small fish in swim speed and avoidance capability (Hamner and Hamner 2000, personal observations). It usually occurs in schools during the productive season that are capable of cruising the ice edge for pockets of high phytoplankton concentrations. It reproduces as a repetitive batch-spawner for the last three years of its life and though not reported in recent years, in the right conditions can form enormous "super-swarms," particularly in the vicinity of the Antarctic Peninsula. A critical element in the success of *E. superba* is its reduced winter metabolism. The energy saved is sufficient to fuel its hypothetical reproductive and molting costs for the year.

The remaining two high Antarctic euphausiids (abundant south of 60°S) are *Thysanoessa macrura* and *Euphausia crystallorophias*. *E. crystallorophias* is a neritic (inshore) species and difficult to sample except during high summer and early fall due to the early formation of inshore pack ice. *T. macrura* is ubiquitous, and though considerably smaller than *E. superba* (35 vs. 60 mm max), it also has a multi-year life cycle, living for about 4.5 years and reproducing over three productive seasons (Siegel 1987). *E. crystallorophias* lives for four to five years and reaches a size of 35–40 mm. Males reproduce over two productive seasons and females for three (Siegel 1987). Overall, Antarctic euphausiids have adopted an iteroparous reproductive strategy: reproducing annually over multiple years of life. The uncertainty of reproductive success in the highly seasonal Antarctic system apparently favors multiple tries. Worthy of note is the fact that all three are broadcast spawners with small larvae that must feed soon after they are developmentally capable of doing so, as opposed to the caridean decapods that produce larger eggs and brood their young. Though counterintuitive, the broadcast spawning strategy is clearly the most successful one for the larger pelagic Crustacea. Life histories in copepods are more abbreviated, about one year for three of the four biomass dominants depending on latitude and including a dormant winter phase.

The comparison of energy utilization and reproductive strategies between *E. pacifica* and *E. superba* in Table 12.13 is particularly interesting. Though the overall growth trajectories are quite different, with *E. superba* reaching a much larger maximum size, the total energy expended in their first year is similar. How it is used is quite different though. In *E. superba*, most of the growth energy is devoted to increasing body mass, in *E. pacifica* most is invested in reproduction. Energy burned in metabolism is similar. *E. pacifica* gets the biological task of reproduction complete in one year. In *E. superba*, it takes three years minimum. All the Antarctic euphausiids have multi-year life histories with reproduction occurring in the last two to three years (Siegel 1987). Reproduction over multiple years and the larger gametic outputs of larger size have proven to be a successful formula for Antarctic euphausiids.

Midwater Fishes

The study of energy utilization and life history in midwater fishes is complicated by the uncertainties involved with the determination of growth. The established standard for growth studies in commercial species, including the Pacific sardine and anchovy, has been the annular rings of sagittal otoliths, which have been experimentally validated in several species. Unfortunately, the rings of myctophid sagittae don't seem to correlate with time in the same way that they do in fishes dwelling in a more conventional day–night cycle. Studies comparing validated microincrements (daily rings) and "annular bands" in migrating lanternfishes show that the life span of lanternfishes is actually much shorter than that what is indicated by the annular rings (Greely et al. 1999).

For midwater fishes, growth of three dominant myctophids in the GOM has been described: *Benthosema suborbitale*, *Diaphus dumerilii*, and *Lepidophanes guentheri* (Gartner 1991a, b), as has growth in the Antarctic-dominant *Electrona antarctica* (Greely et al. 1999) Table 12.14 compares energy utilization in two midwater fish species from the Gulf of Mexico (*L. guentheri* and the gonostomatid *Sigmops elongatum*) and *E. antarctica* from the Antarctic. In all cases, sufficient data were available to estimate energy used in growth and metabolism over the lives of the species.

The three fish species literally compare polar opposites, with two residing in the tropical–subtropical waters of the GOM and one in the Antarctic. All three are zooplanktivorous vertical migrators and dominants in their respective systems. The GOM myctophid *Lepidophanes guentheri* lives for about a year, reaching a length of 61 mm and a mass of 4.3 g, reproducing by repetitive batch-spawning for the last four to six months of its life (Gartner 1991b, 1993). In contrast, the Antarctic myctophid *Electrona antarctica* lives for four years, reaching a length of 101 mm, but also reproduces by repetitive batch-spawning at the end of its life. Of the three dominant GOM lanternfishes examined by Gartner (1991b), two of the three were annuals (*L. guentheri* and *Benthosema suborbitale*), and one, *Diaphus dumerilii*, was believed to live for two years. A life span of a single year or less for small tropical–subtropical myctophids seems to be a common life history (Gjøsæter 1984; Gjøsæter 1987), with larger species requiring more than a year to reach maximum size (Clarke 1973).

Sigmops elongatum is an important member of the GOM micronektonic fish fauna, reaching 225 mm maximum size in the Gulf and increasing the size of its preferred zooplankton prey spectrum with increasing length (Lancraft et al. 1988). *Sigmops* is a protandric hermaphrodite, changing sex from male to female in the middle of its size range (86–117 mm) (Fisher 1983). It resides deeper in the water column than the myctophids reported above and has a higher water content. Reproductively mature fish were found throughout the year, as were juveniles, suggesting year-round breeding for this species. However, in the GOM, each female was believed to spawn only once at the end of its life.

The capsule comparison between GOM, California Current, and Antarctic species suggests that seasonality and temperature itself have an important influence on energy apportionment and general life histories. Though the considerable literature on cold adaptation focuses on metabolism, which is a readily (if not easily) measured characteristic of energy usage in marine species, it is growth that is the major operator in shaping the adapted characteristics of Antarctic biota. Metabolic cold adaptation elevates metabolism to be closer to, but lower than, temperate and tropical species at their respective habitat temperatures (e.g. Kawall et al. 2002). Growth is a more integrative process but behaves in a similar way. For Antarctic species, growth during the productive season must allow sufficient energy accumulation to survive the winter with more limited rations, eventually accruing enough to assure reproductive success.

Comparisons of GOM and Antarctic pelagic fishes show a similar pattern in growth and reproductive timing to the euphausiids presented in Table 12.13, though estimates of energy invested in reproduction is lacking. Energy invested in somatic growth for the two GOM species is about a third of the total first year's expenditures, with the remainder devoted to metabolism. For *Electrona*, most of its first year's expended energy is devoted to growth, and over its life it is nearly a 50–50 split. This is a remarkable growth efficiency, which is partially due to the significantly higher percentage of high-calorie lipid in its body mass (Table 12.12). For *Electrona*, the higher gametic output associated with a larger mass along with semelparous reproduction is a successful life history strategy.

Fewer data on metabolism and chemical composition are available for the gelatinous groups (cnidaria, ctenophora, and salps), or for the pelagic nemerteans, annelids, and molluscs. This is partly because of the difficulty in capturing them in good shape and partly due to the difficulty of manipulating them experimentally. In the case of cephalopods, capturing them at all can be a challenge. They are both intelligent and excellent swimmers, capable of raiding a midwater trawl catch, while the net is fishing and swimming out again before it is returned to the surface, leaving the unlucky scientists with a partially eaten catch. Data presented in Tables 12.15–12.17 are for slower moving mesopelagic species that use stealth and ambush rather than superior locomotory abilities to capture their prey. Also, most are at the small end of the species' size spectrum. Capturing the delicate gelatinous species requires a dedicated effort, including a very slow tow speed and nets designed to capture them gently. Ideally, blue water SCUBA or submersibles are used.

Data presented in Tables 12.15–12.17 give a representative subset of the data available for jellies, worms, pteropods, and cephalopods, both for comparison to the data on crustaceans and fishes and for comparison to each other.

The Cephalopods

The cephalopods shown in Table 12.15 include active migrators in the family Gonatidae as well as deep-living species with a wide-ranging geographic distribution (*Vampyroteuthis infernalis*). Metabolism declines significantly with increasing depth of occurrence in these visual predators (Seibel et al. 1997), supporting the visual interactions hypothesis described in Chapter 2. Superimposed on the overall decline is an additional factor that steepens the relationship with depth, making it more profound than that observed in the fishes and crustaceans: a change in the locomotory modality. Deeper living species may employ medusoid swimming using the webbing between the tentacles to swim like a jellyfish, or in some cases, such as *Vampyroteuthis*, fin-based swimming, both of which are less costly in energy than the jet propulsion utilized by shallow dwellers. The drawback is that neither of the locomotory modes employed by deep-living species is nearly as fast as jet propulsion. But, they don't have to

be. The limited light field makes a stealthy approach a little easier, eliminating much of the need for rapid pursuit.

The range of metabolic rates reported in Table 12.15 is similar to that of fishes (Tables 12.7–12.9), particularly in the deepest-living species, though a bit lower than the crustaceans. Cephalopods have well-developed ammoniacal buoyancy mechanisms (Chapter 8) in the cranchiids and histioteuthids, allowing them to expend less energy for maintaining station.

Rates for juvenile *Galiteuthis* and *Psychroteuthis* from the Antarctic are also shown, with the cranchiid *Galiteuthis* slightly higher than its confamilial *Heliocranchia*, and *Psychroteuthis*, an active squid that gets fairly large, similar to the equally active Gonatids.

Pteropods, Nemerteans, Annelids, Salps, and Pyrosomes

Table 12.16 is a taxonomic sampler of metabolism in the pteropods, nemerteans, annelids, salps, and pyrosomes. Most striking are the high metabolic rates exhibited by the pteropods, equaling or exceeding those of fishes and crustaceans on a per unit mass basis. They have an active lifestyle and are fairly diminutive in size, similar to euphausiids whose metabolic rates are also similar. Pteropods are considered to be gelatinous forms, but the smaller representatives of the group, such as the cavoliniids, are the equals of their crustacean counterparts in activity and metabolism.

Annelids, some of whom are also active swimmers, particularly the tomopterids, exhibit metabolic rates similar to mesopelagic fishes and crustaceans residing at similar depths. Not surprisingly, the more active tomopterids show the highest rates within the group. Though never highly abundant, the genera shown in Table 12.16 are commonly captured in midwater trawls and have a large vertical range.

Nectonemertes mirabilis, a deep-living pelagic nemertean, has a rate an order of magnitude lower than the annelids as a group, but nonetheless higher than the bathypelagic alepocephalid fish *Bajacalifornia burragei* (0.005 ul O_2 mgWM^{-1} h^{-1}), which to be fair is also very much larger (see Table 12.8). *Nectonemertes* shows a mix of quiescence and active swimming in the laboratory which may reflect its lifestyle at depth (Thuesen and Childress 1993).

The chordates have the lowest rates of the several taxonomic groupings in Table 12.16, with the numbers in the group showing a large influence by measurement temperature. For the salps and for *Thalia*, all determinations were done on solitaries. When conditions are favorable, salps can reproduce quite quickly using the reproductive strategy described in Chapter 9. *Salpa thompsoni* in particular can reach high enough numbers (Lancraft et al. 1989) to be competitive with krill (Atkinson et al. 2004). Salps are 96% water (Ikeda and Bruce 1986) and a summer krill (*Euphausia superba*) is 73% water (Table 12.6). Per unit wet mass, it takes a lot more energy to make a krill than a salp. Likewise, a krill is a better meal. Flighted seabirds, for example, only eat the stomachs of salps (Ainley et al. 1986).

The Cnidaria and Ctenophora

The medusae and ctenophores show the lowest rates per unit mass of any of the taxa covered thus far (Table 12.17). Only the most active taxa, the Trachymedusae and Narcomedusae, exhibit rates in the same order of magnitude as the salps and annelids (and crustaceans and fishes). The low rates reflect the general foraging strategy of medusae which is to entrap their prey rather than actively hunt it down. The passive hunting strategy and high (95%) water levels of medusae result in low rates of metabolism per unit wet mass.

The ctenophores also show low rates of metabolism, reflective of their high water content, but have a more diverse array of feeding strategies than the Cnidaria. *Beroe* is an active predator that chases down and engulfs its prey (usually other ctenophores) and the cydippids use tentacles to entrap them. Despite the spectrum of feeding strategies exhibited within the group, rates are uniformly low.

Because of their low swimming speeds and limited evasive ability, medusae and ctenophores can be captured at depth and incubated for respiratory measurements in situ. Though technologically difficult, a number of species have been studied in this way, comparing rates in situ with rates taken at the surface using the identical methods of capture and incubation (marked by asterisks in the table). In all cases, metabolism at depth was higher than that in the lab by a factor of 2-4, an interesting and as yet incompletely understood result. The only variable in the set of experiments was the site of incubation: at habitat depth or in the lab, in effect, the absence of pressure. Yet, pressure studies on several gelatinous species showed no effect (Childress and Thuesen 1993).

Conclusions

One of the main themes discussed here is the idea that metabolism declines with species' depth of occurrence, which has been demonstrated in micronektonic Crustacea, fishes, and cephalopods, all of whom are visual predators, whether it be on other crustaceans, fishes, and cephalopods or on phytoplankton. Gelatinous forms, like the medusae and ctenophores, don't exhibit the same trend, nor do the annelids and a group not treated in the book, the chaetognaths (Thuesen and Childress 1993). There is little doubt that locomotory abilities are most important when visibility is high, for predators and prey. The metabolic costs of forming and using a robust musculature are profound and the relaxation of a selective pressure to maintain them as ambient light declines would benefit the energy budget of deeper-living species: less energy needed for an active metabolism and more for growth and reproduction.

Gelatinous species employ a low-energy lifestyle whether at the surface or at depth. Their watery tissues are less costly to form, and energy expended in metabolism is low. Yet their strategy is a winning one. In the Black Sea, the introduced *Mnemiopsis* nearly wiped out the fishery there by consuming fish larvae, and seasonal increases of *Aequorea* in the coastal waters off British Columbia prey heavily on herring larvae (see Chapters 3 and 4). Though considered to be at the low end of the evolutionary scale, gelatinous species are exceptionally effective predators, most often dining on representatives of the more advanced phyla.

References

Ainley, D.G., Fraser, W.R., Sullivan, C.G. et al. (1986). Antarctic pack ice structures mesopelagic nektonic communities. *Science* 232: 847–849.

Alldredge, A.L., Robison, B.H., Fleminger, A. et al. (1984). Direct sampling and in situ observation of a persistent copepod aggregation in the mesopelagic zone of the Santa Barbara Bain. *Marine Biology* 80: 75–81.

Atkinson, A., Siegel, V., Pakhomov, E.A., and Rothery, P. (2004). Long-term decline in krill stock and increase in salps within the Southern Ocean. *Nature* 432: 100–103.

Bailey, T.G., Youngbluth, M.J., and Owen, G.P. (1994a). Chemical composition and oxygen consumption rates of the ctenophore *Bolinopsis infundibulum* from the Gulf of Maine. *Journal of Plankton Research* 16: 673–689.

Bailey, T.G., Torres, J.J., Youngbluth, M.J., and Owen, G.P. (1994b). Effect of decompression on mesopelagic gelatinous zooplankton: a comparison of in situ and shipboard measurements of metabolism. *Marine Ecology Progress Series* 113: 13–27.

Bailey, T.G., Youngbluth, M.J., and Owen, G.P. (1995). Chemical composition and metabolic rates of gelatinous zooplankton from midwater and benthic boundary layer environments off Cape Hatteras, North Carolina, USA. *Marine Ecology Progress Series* 122: 121–134.

Brett, J.R. (1964). The respiratory metabolism and swimming performance of young sockeye salmon. *Journal of the Fisheries Research Board of Canada* 21: 1183–1226.

Brett, J.R. and Groves, T.D.D. (1979). Physiological energetics. In: *Fish Physiology*, vol. 8 (eds. W.S. Hoar, D.J. Randall and J.R. Brett). New York: Academic Press.

Childress, J.J. (1975). The respiratory rates of midwater crustaceans as a function of depth of occurrence and relation to the oxygen minimum layer off Southern California. *Comparative Biochemistry and Physiology* 50A: 787–799.

Childress, J.J. and Nygaard, M.H. (1973). The chemical composition of midwater fishes as a function of depth of occurrence off southern California. *Deep-Sea Research* 20: 1093–1109.

Childress, J.J. and Nygaard, M.H. (1974). Chemical composition and buoyancy of midwater crustaceans as a function of depth of occurrence off southern California. *Marine Biology* 27: 225–238.

Childress, J.J. and Thuesen, E.V. (1993). Effects of hydrostatic pressure on metabolic rates of six species of deep-sea gelatinous zooplankton. *Limnology and Oceanography* 38: 665–670.

Clarke, T.A. (1973). Some aspects of the ecology of lanternfishes (myctophidae) in the Pacific Ocean near Hawaii. *Fishery Bulletin U.S.* 71: 401–434.

Conover, R.J. (1988). Comparative life histories in the genera *Calanus* and *Neocalanus* in high latitudes of the northern hemisphere. *Hydrobiologia* 167/168: 127–142.

Cowles, D.L., Childress, J.J., and Wells, M.E. (1991). Metabolic rates of midwater crustatceans as a function of depth of occurrence off the Hawaiian islands: food availability as a selective factor? *Marine Biology* 110: 75–83.

Donnelly, J. and Torres, J.J. (1988). Oxygen consumption of midwater fishes and crustaceans from the eastern Gulf of Mexico. *Marine Biology* 97: 483–494.

Donnelly, J., Torres, J.J., Hopkins, T.L., and Lancraft, T.M. (1990). Proximate composition of Antarctic mesopelagic fishes. *Marine Biology* 106: 13–23.

Donnelly, J., Stickney, D.G., and Torres, J.J. (1993). Proximate and elemental composition and energy content of mesopelagic crustaceans from the Eastern Gulf of Mexico. *Marine Biology* 115: 469–480.

Donnelly, J., Kawall, H.G., Geiger, S.P., and Torres, J.J. (2004). Metabolism of Antarctic micronektonic Crustacea across a summer ice edge. *Deep Sea Research, Part II* 51: 2225–2245.

Fisher, R.A. (1983). Protrandric sex reversal in *Gonostoma elongatum* (Pisces: Gonostomatidae) from the Eastern Gulf of Mexico. *Copeia* 1983: 554–557.

Foxton, P. (1956). The distribution of the standing crop of zooplankton in the Southern Ocean. *Discovery Reports* 28: 191–236.

Gartner, J.V. (1991a). Life histories of three species of lanternfishes (Pisces: Myctophidae) from the eastern Gulf of Mexico. I. Morphological and microstructural analysis of sagittal otoliths. *Marine Biology* 111: 11–20.

Gartner, J.V. (1991b). Life histories of three species of lanternfishes (Pisces: Myctophidae) from the eastern Gulf of Mexico. II. Age and growth patterns. *Marine Biology* 111: 21–27.

Gartner, J. (1993). Patterns of reproduction in the dominant lanternfish species (Pisces: Myctophidae) of the eastern Gulf of Mexico, with a review of reproduction among tropical-subtropical Myctophidae. *Bulletin of Marine Science* 52: 721–750.

Gjøsæter, J. (1984). Mesopelagic fish: a large potential resource in the Arabian Sea. *Deep-Sea Research* 31: 1019–1035.

Gjøsæter, H. (1987). Primary growth increments in otoliths of six tropical myctophid species. *Biological Oceanography* 4: 359–382.

Greely, T.M., Gartner, J.V., and Torres, J.J. (1999). Age and growth of *Electrona antarctica* (Pisces: Myctophidae) the dominant mesopelagic fish of the Southern Ocean. *Marine Biology* 133: 145–158.

Hamner, W.M. and Hamner, P.P. (2000). Behavior of Antarctic krill (*Euphausia superba*): schooling, foraging and anti-predatory behavior. *Canadian Journal of Fisheries and Aquatic Sciences* 57 (Suppl 3): 192–202.

Hemmingsen, A.M. (1960). Metabolism in relation to body size. *Report of Steno Memorial Hospital and Nordisk Insulin Laboratory* 9: 1–110.

Hill, A.V. (1950). The dimensions of animals and their muscular dynamics. *Science Progress* 38: 209–230.

Hochachka, P.W. and Somero, G.N. (2002). *Biochemical Adaptation: Mechanism and Process in Physiological Evolution.* New York: Oxford University Press.

Hofmann, E.E. and Lascara, C.M. (2000). Modeling the growth dynamics of

Antarctic krill *Euphausia superba*. *Marine Ecology Progress Series* 194: 219–231.

Hofmann, E.E., Wiebe, P.W., Costa, D.P., and Torres, J.J. (2004). An overview of the Southern Ocean Global Ocean Ecosystems Dynamics program. *Deep Sea Research, Part II* 51: 1921–1924.

Hopkins, T.L., Lancraft, T.M., Torres, J.J., and Donnelly, J. (1993). Community structure and trophic ecology of zooplankton in the Scotia Sea marginal ice zone in winter (1988). *Deep-Sea Research* 40: 81–105.

Hopkins, T.L., Flock, M.E., Gartner, J.V., and Torres, J.J. (1994). Structure and trophic ecology of a low latitude midwater decapod and mysid assemblage. *Marine Ecology Progress Series* 109: 143–156.

Ikeda, T. (1974). Nutritional ecology of marine zooplankton. *Memoirs of the Faculty of Fisheries Hokkaido University* 22: 1–97.

Ikeda, T. (1988). Metabolism and chemical composition of crustaceans from the Antarctic mesopelagic zone. *Deep Sea Research* 35: 1991–2002.

Ikeda, T. and Bruce, B. (1986). Metabolic activity and elemental composition of krill and other zooplankton from Prydz Bay, Antarctica, during early summer. *Marine Biology* 92: 545–555.

Ikeda, T. and Dixon, P. (1984). The influence of feeding on the metabolic activity of Antarctic krill (*Euphausia superba*). *Polar Biology* 3: 1–9.

Ikeda, T. and Mitchell, A.W. (1982). Oxygen uptake, ammonia excretion and phosphate excretion of krill and other Antarctic zooplankton, in relation to their body size and chemical composition. *Marine Biology* 71: 283–298.

Ikeda, T., Torres, J.J., Hernandez-Leon, S., and Geiger, S.P. (2000). Metabolism. In: *Zooplankton Methodology Manual* (eds. R. Harris, P. Wiebe, J. Lenz, et al.). San Diego: Academic Press.

Jobling, M.J. (1985). Growth. In: *Fish Energetics: New Perspectives* (eds. P. Tytler and P. Calow). Beckenham: Croom Helm Ltd.

Kawaguchi, K., Ishikawa, S., and Matsuda, O. (1986). The overwintering strategy of Antarctic krill (*Euphausia superba* Dana) under the coastal fast ice off the Ongul Islands in Lutzow-Holm Bay, Antarctica. *Memoirs of National Institute of Polar Research (Special issue)* 44: 67–85.

Kawall, H.G., Torres, J.J., Sidell, B.D., and Somero, G.N. (2002). Metabolic cold adaptation in Antarctic fishes: evidence from enzymatic activities of brain. *Marine Biology* 140: 279–286.

Kinsey, S.T. and Hopkins, T.L. (1994). Trophic strategies of euphausiids in a low-latitude ecosystem. *Marine Biology* 118: 651–661.

Kiorboe, T., Mohlenberg, F., and Hamburger, K. (1985). Bioenergetics of the planktonic copepod Acartia tonsa: relation between feeding, egg production and respiration, and composition of specific dynamic action. *Marine Ecology Progress Series* 26: 85–97.

Kleiber, M. (1961). *The Fire of Life, West Sussex*. Wiley.

Kremer, P. (1977). Respiration and excretion by the ctenophore *Mnemiopsis leidyi*. *Marine Biology* 44: 43–50.

Kremer, P., Canino, M.F., and Gilmer, R.W. (1986). Metabolism of epipelagic tropical ctenophores. *Marine Biology* 90: 403–412.

Lancraft, T.M., Hopkins, T.L., and Torres, J.J. (1988). Aspects of the ecology of the mesopelagic fish *Gonostoma elongatum* (Gonostomatidae, Stomiiformes) in the eastern Gulf of Mexico. *Marine Ecology Progress Series* 49: 27–40.

Lancraft, T.M., Hopkins, T.L., Torres, J.J., and Donnelly, J. (1991). Oceanic micronektonic/macrozooplanktonic community structure and feeding under ice covered Antarctic waters during the winter (AMERIEZ 1988). *Polar Biology* 11: 157–167.

Lancraft, T.M., Torres, J.J., and Hopkins, T.L. (1989). Micronekton and macrozooplankton in the open waters near Antarctic ice edge zones (AMERIEZ 1983 and 1986). *Polar Biology* 9: 225–233.

Larson, R.J. (1987). Respiration and carbon turnover rates of medusae from the NE Pacific. *Comparative Biochemistry and Physiology* 87A: 93–100.

Lasker, R. (1970). Utilization of zooplankton energy by a Pacific sardine population in the California Current. In: *Marine Food Chains* (ed. J.H. Steele). Edinburgh: Oliver and Boyd.

Love, M. (1970). *The Chemical Biology of Fishes*. London: Academic Press.

Maas, A.E., Wishner, K.F., and Seibel, B.A. (2012). Metabolic suppression in thecosomatous pteropods as an effect of low temperature and hypoxia in the eastern tropical North Pacific. *Marine Biology* 159: 1955–1967.

Miller, C.B. and Clemons, M.J. (1988). Revised life history analysis of large grazing copepods in the Subarctic Pacific Ocean. *Progress in Oceanography* 20: 293–313.

Parker, M.L., Fraser, W.R., Ashford, J. et al. (2015). Assemblages of micronektonic fishes and invertebrates in a gradient of regional warming along the Western Antarctic Peninsula. *Journal of Marine Systems* 152: 18–41.

Quetin, L.B. and Ross, R.M. (1991). Behavioral and physiological characteristics of the Antarctic krill *Euphausia superba*. *American Zoologist* 31: 49–63.

Quetin, L.B., Ross, R.M., and Uchio, K. (1980). Metabolic characteristics of midwater zooplankton: ammonia excretion, O:N ratios and the effects of starvation. *Marine Biology* 59: 201–209.

Ross, R.M. (1982a). Energetics of *Euphausia pacifica* I. Effects of body carbon and nitrogen and temperature on measure and predicted production. *Marine Biology* 68: 1–13.

Ross, R.M. (1982b). Energetics of *Euphausia pacifica* II. Complete carbon and nitrogen budgets at 8 and 12 C throughout the life span. *Marine Biology* 68: 15–23.

Scolardi, K.M., Daly, K.L., Pakhomov, E.A., and Torres, J.J. (2006). Feeding ecology and metabolism of the Antarctic cydippid ctenophore *Callianira antarctica*. *Marine Ecology Progress Series* 317: 111–126.

Seibel, B.A., Thuesen, E.V., Childress, J.J., and Gorodezky, L.A. (1997). Decline in pelagic cephalopod metabolism with habitat depth reflects differences in locomotory efficiency. *Biological Bulletin* 192: 262–278.

Seibel, B.A., Dymowska, A., and Rosenthal, J. (2007). Metabolic temperature compensation and coevolution of locomotory performance in pteropod molluscs. *Integrative and Comparative Biology* 47: 880–891.

Siegel, V. (1987). Age and growth of Antarctic Euphausiacea (Crustacea) under natural conditions. *Marine Biology* 96: 483–495.

Smith, K.L. and Laver, M.B. (1981). Respiration of the bathypelagic fish *Cyclothone acclinidens*. *Marine Biology* 61: 261–266.

Somero, G.N., Lockwood, B.L., and Tomanek, L. (2017). *Biochemical Adaptation. Response to Environmental Challenges from Life's Origins to the Anthropocene*. Sunderland, MA USA: Sinauer Associates, Inc.

Stearns, S.C. (1992). *The Evolution of Life Histories*. New York: Oxford University Press.

Stickney, D.G. and Torres, J.J. (1989). Proximate composition and energy content of mesopelagic fishes from the eastern Gulf of Mexico. *Marine Biology* 103: 13–24.

Suprenand, P.M., Ombres, E.H., and Torres, J.J. (2015). Metabolism of gymnosomatous pteropods in waters of the western Antarctic Peninsula shelf during austral fall. *Marine Ecology Progress Series* 518: 69–83.

Thuesen, E.V. and Childress, J.J. (1993). Metabolic rates, enzyme activities and chemical compositions of some deep-sea pelagic worms, particularly *Nectonemertes mirabiliis* (Nemertea; Hoplonemertinea) and *Poeobius meseres*. *Deep Sea Research, Part I* 40: 937–951.

Thuesen, E.V. and Childress, J.J. (1994). Oxygen consumption rates and metabolic enzyme activities of oceanic California medusae in relation to body size and habitat depth. *Biological Bulletin* 187: 84–98.

Torres, J.J. and Childress, J.J. (1983). Relationship of oxygen consumption to swimming speed in *Euphausia pacifica* 1. Effects of temperature and pressure. *Marine Biology* 74: 79–86.

Torres, J.J. and Somero, G.N. (1988a). Vertical distribution and metabolism in Antarctic mesopelagic fishes. *Comparative Biochemistry and Physiology* 90B: 521–528.

Torres, J.J. and Somero, G.N. (1988b). Metabolism, enzymic activities, and cold adaptation in Antarctic mesopelagic fishes. *Marine Biology* 98: 169–180.

Torres, J.J., Belman, B.W., and Childress, J.J. (1979). Oxygen consumption rates of midwater fishes as a function of depth of occurrence. *Deep-Sea Research* 26A: 185–197.

Torres, J.J., Aarset, A.V., Donnelly, J. et al. (1994a). Metabolism of Antarctic micronektonic Crustacea as a function of depth of occurrence and season. *Marine Ecology Progress Series* 113: 207–219.

Torres, J.J., Donnelly, J., Hopkins, T.L. et al. (1994b). Proximate composition and overwintering strategies of Antarctic micronektonic Crustacea. *Marine Ecology Progress Series* 113: 221–232.

Webb, P.W. (1975). Hydrodynamics and energetics of fish propulsion. *Bulletin of the Fisheries Research Board of Canada* 190: 1–158.

West, G.B., Brown, B.H., and Enquist, B.J. (1997). A general model for the origin of allometric scaling laws in biology. *Science* 276: 122–126.

West, G.B., Brown, B.H., and Enquist, B.J. (1999). The fourth dimension of life: fractal geometry and allometric scaling of organisms. *Science* 284: 1677–1679.

West, G.B., Woodruff, W.H., and Brown, J.H. (2002). Allometric scaling of metabolic rate from molecules and mitochondria to cells and mammals. *Proceedings of the National Academy of Science of the USA* 99: 2473–2478.

Yamada, Y. and Ikeda, T. (2003). Metabolism and chemical composition of four pelagic amphipods in the Oyashio region, western subarctic Pacific. *Marine Ecology Progress Series* 253: 233–241.

Appendix A

Classification of the Chordata

The backbone of the classification system presented here (classes, orders, families) is from the California Academy of Sciences Catalog of Fishes (Eschemeyer's Catalog of Fishes: calacademy.org). Intermediate and higher-level classifications, e.g. subphylum, grade, sub-class, superorder, and cohort, are from Nelson et al. (2016), Helfman et al. (2009), and the World Register of Marine Species (WoRMS). Numbers of species in each taxonomic category, noted in parentheses, are from WoRMS. Classes, orders, and suborders are arranged from most primitive to most advanced, but families are presented alphabetically.

Marine orders and families are in italics; ***those with deep-sea representatives are in bold italics.***

Estimates of deep-sea species are from Priede (2017), fishbase.org, and Helfman et al. (2009) and are in brackets following the family common name. Freshwater orders are in normal typeface. The total number of marine fish species is estimated at 20 000 (WoRMS n.d.).

 PHYLUM CHORDATA
  SUBPHYLUM TUNICATA – The ascidians, thaliaceans, and appendicularians (3070)
  SUBPHYLUM CEPHALOCHORDATA – The lancelets
   CLASS LEPTOCARDII
    Family Branchiostomatidae (30)
  SUBPHYLUM VERTEBRATA
   SUPERCLASS AGNATHA
    CLASS MYXINI
     Order Mixiniformes (hagfishes) (1 family – 85 species)
      Family Myxinidae (85) hagfishes [>21]
    CLASS PETROMYZONTI
     Order Petromyzontiformes (lampreys) (3 families – 13 species)
      Family Geotriidae (1) pouched lampreys
      Family Mordaciidae (2) southern topeyed lampreys
      Family Petromyzontidae (10) sea lamprey [1]
   SUPERCLASS GNATHOSTOMATA
    GRADE CHONDRICHTHYES
    CLASS ELASMOBRANCHII
  SUBCLASS SELACHII (537 spp)
   Order Hexanchiformes (cow and frill sharks) (2 families – 6 species)
    Family Chlamydoselachidae (2) frill sharks [2]
    Family Hexanchidae (4) cow sharks [2]
   Order Heterodontiformes (bullhead and horn sharks) (1 family – 9 species)
    Family Heterodontidae (9)
   Order Orectolobiformes (whale and nurse sharks) (7 families – 45 species)

Family Brachaeluridae (2) blind sharks
Family Ginglymostomatidae (5) nurse sharks
Family Hemiscylliidae (16) bamboo sharks
Family Orectolobidae (12) wobbegongs
Family Parascylliidae (8) collard carpet sharks
Family Rhincodontidae (1) whale sharks
Family Stegostomatidae (1) zebra sharks

Order Lamniformes (mackerel sharks) (7 families – 16 species)
Family Alopiidae (3) thresher sharks [1]
Family Cetorhinidae (1) basking shark [1]
Family Lamnidae (5) mackerel sharks [deep dives only]
Family Megachasmidae (1) megamouth shark [1]
Family Mitsukurinidae (1) goblin shark [1]
Family Odontaspididae (4) sand tiger sharks [2]
Family Pseudocarchariidae (1) crocodile shark [1]

Order Carchariniformes (requiem sharks) (9 families – 290 species)
Family Carcharhinidae (58) (requiem sharks)
Family Hemigaleidae (8) (weasel sharks)
Family Leptochariidae (1) (barbeled houndsharks)
Family Pentanchidae (88) (deepwater cat sharks)[88]
Family Proscylliidae (7) (finback cat sharks) [3]
Family Pseudotriakidae (4) (false catsharks) [2]
Family Scyliorhinidae (68) (cat sharks) [40]

Family Sphyrnidae (10) (hammerhead sharks)
Family Triakidae (46) (houndsharks) [2]

Order Squaliformes (dogfish sharks) (6 families – 137 species)
Family Centrophoridae (16) (centrophorid sharks) [16]
Family Dalatiidae (9) (kitefin sharks) [9]
Family Etmopteridae (51) (lantern sharks) [51]
Family Oxynotidae (6) (prickly dogfish) [6]
Family Somniosidae (18) (sleeper sharks) [5]
Family Squalidae (37) (dogfish sharks) [4]

Order Pristiophoriformes (saw sharks) (1 family – 8 species)
Family Pristiophoridae (8) (saw sharks) [4]

Order Squatiniformes (angel sharks) (1 family – 24 species)
Family Squatinidae (24) (angel sharks) [2]

Order Echinorhiniformes (bramble sharks) (1 family – 2 species)
Family Echinorhinidae (2) (bramble sharks) [2]

Order Pristiophoriformes (saw sharks) (1 family – 8 species)
Family Pristiophoridae (8) (saw sharks) [4]

Order Squatiniformes (angel sharks) (1 family – 24 species)
Family Squatinidae (24) (angel sharks) [2]

SUBCLASS BATOIDEA (668 spp)
Order Torpediniformes (electric rays) (5 families – 68 species)
Family Hypnidae (1) (coffin rays)
Family Platyrhinidae (thornbacks)
Family Narcinidae (30) (numbfishes) [2]

Family Narkidae (12) (sleeper rays) [1]
Family Torpedinidae (25) (torpedo rays) [1]
Order Rhinopristiformes (guitarfishes) (5 families – at least 70 species)
    Family Glaucostegidae (6) (giant guitarfishes)
    Family Pristidae (7) (sawfishes)
    Family Rhinidae (8) (bowmouth guitarfishes/wedgefishes)
    Family Rhinobatidae (48) (guitarfishes)
    Family Trygonorhinidae (banjo rays)
Order Rajiformes (skates) (4 families – at least 303 species)
    Family Anacanthobatidae (17) (smooth skates)
    Family Arhynchobatidae (104) softnosed skates [104]
    Family Gurgesiellidae (pygmy skates)
    Family Rajidae (182) hardnosed skates [at least 14]
Order Myliobatiformes (stingrays) (12 families – at least 230 species)
    Family Aetobatidae (pelagic eagle rays)
    Family Dasyatidae (101)(stingrays)
    Family Gymnuridae (14)(butterfly rays)
    Family Hexatrygonidae (1) sixgill stingray [1]
    Family Mobulidae (10)(devil rays)
    Family Myliobatidae (43) eagle rays (deep-divers)
    Family Plesiobatidae (1) deepwater stingray [1]
    Family Potamotrygonidae (american stingrays)
    Family Rhinopteridae (8)(cownose rays)
    Family Urolophidae (30) (stingarees) [4]
    Family Urotrygonidae (20)(round rays)
    Family Zanobatidae (2) (panrays)
CLASS HOLOCEPHALI
    Order Chimaeriformes (chimaeras) (3 families – 55 species)
    Family Callorynchidae (3) (plownose chimaeras)
    Family Chimaeridae (44) shortnose chimaeras [42]
    Family Rhinochimaeridae (8) longnose chimaeras [8]
GRADE TELEOSTOMI
  CLASS CLADISTII – (FW)
    Order Polypteriformes (bichirs)
    Family Polypteridae
CLASS ACTINOPTERYGII
SUBCLASS CHONDROSTEI
    Order Acipenseriformes (23) (sturgeons and paddlefishes)(FW)
    Family Acipenseridae (23) sturgeons
    Family Polyodontidae –paddlefishes
NEOPTERYGII
SUBCLASS HOLOSTEI
    Order Lepisosteiformes (4) (gars) (FW)
    Family Lepisosteidae (4) gars
    Order Amiiformes (1) (bowfins) (FW)
    Family Amiidae (1) bowfins
SUBCLASS TELEOSTEI
COHORT ELOPOMORPHA
    Order Elopiformes (9) (tarpons and ladyfishes)
    Family Elopidae (ladyfishes)
    Family Megalopidae (Tarpon)
    Order Albuliformes (bonefishes) (1 family – 13 species)
    Family Albulidae (13) bonefishes and pterothrissins [2]
Order Notacanthiformes (halosaurs and spiny eels)(2 families – 27 species)
    Family Halosauridae (16) halosaurs [16]
    Family Notacanthidae (11) spiny eels [11]

Order Anguilliformes (eels) (16 families – 933 species)
Family Anguillidae (16) freshwater eels [spawning only]
Family Chlopsidae (25) false morays [1]
Family Colocongridae (10) shorttail eels [at least 5]
Family Congridae (194) conger eels [at least 18]
Family Derichthyidae (3) longneck eels [3]
Family Heterenchelyidae (mud eels)
Family Muraenesocidae (15) pike congers [at least 5]
Family Muraenidae (200) (moray eels)
Family Moringuidae (spaghetti eels)
Family Myrocongridae (myroconger eels)
Family Nemichthyidae (9) snipe eels [9]
Family Nettastomatidae (42) duckbill eels [at least 10]
Family Ophichthidae (335) snake and worm eels [4]
Family Protanguillidae (1) (primitive cave eels)
Family Serrivomeridae (11) sawtooth eels [11]
Family Synaphobranchidae (41) cutthroat eels [29]
Order Saccopharyngiformes (gulper eels) (5 families – 46 species)
Family Cyematidae (2) bobtail eels [2]
Family Eurypharyngidae (1) umbrellamouth gulpers [1]
Family Monognathidae (15) onejaw gulpers [15]
Family Neocyematidae – bobtail snipe eels
Family Saccopharyngidae (28) swallowers [28]

COHORT OSTEOGLOSSOMORPHA
Order Hiodontiformes (mooneyes) (FW)
Order Osteoglossiformes (bonytongues and elephant fishes) (FW)

COHORT OTOCEPHALA
Superorder Clupeomorpha
Order Clupeiformes (sardines and herrings) (6 families – 334 species)
Family Chirocentridae (2) wolf herrings
Family Clupeidae (153) (herrings, shads, sardines)
Family Denticipitidae (denticle herrings)
Family Dussumieridae (10) (round herrings)
Family Engraulidae (136) (anchovies)
Family Pristgasteridae (33) (longfin herrings)

Superorder Alepocephali
Order Alepocephaliformes
Family Alepocephalidae (98) slickheads [98]
Family Platytroctidae (39) tubeshoulders [39]

Superorder Ostariophysi
Order Gonorhynchiformes (milkfishes) (2 marine families – 6 species)
Family Chanidae (1) milkfishes
Family Gonorhynchidae (5) (beaked sandfishes)
Order Cypriniformes (suckers, loaches and carps) (FW)
Order Characiformes (characins and tetras) (FW)
Order Gymnotiformes (knifefishes) (FW)
Order Siluriformes (catfishes) (39 families – mainly FW)
Family Ariidae (122) (sea catfishes)

COHORT EUTELEOSTEI
Order Lepidogalaxiiformes (FW)
Family Lepiodgalaxiidae (salamanderfishes)

Superorder Protacanthopterygii
Order Esociformes (pikes and mudminnows) (FW)

Order Salmoniformes (salmons, trouts and whitefishes – FW and anadromous)
    Family Salmonidae (48) salmons and trouts

Superorder Osmeromorpha
    Order Argentiniformes (marine smelts) (4 families – 92 species)
        Family Argentinidae (28) herringsmelts [7]
        Family Bathylagidae (23) deep-sea smelts [23]
        Family Microstomatidae (20) pencil smelts [20]
        Family Opisthoproctidae (21) spookfishes [21]
    Order Galaxiiformes (galaxiids) (FW)
    Order Osmeriformes (freshwater smelts) (4 families – 276 species)
        Family Osmeridae (13) smelts
        Family Plecoglossidae (6) (ayu fishes)
        Family Retropinnidae (6) (New Zealand smelts)
        Family Salangidae (13) (icefishes)
    Order Stomiiformes (dragonfishes) (4 families – 415 species)
        Family Gonostomatidae (30) bristlemouths [30]
        Family Phosichthyidae (24) lightfishes [24]
        Family Sternoptychidae (74) hatchetfishes [74]
            Subfamily Maurolycinae (31) pearlsides
            Subfamily Sternoptychinae (43) hatchetfishes
        Family Stomiidae (287) barbeled dragonfishes [287]
            Subfamily Astronesthinae (59) snaggletooths
            Subfamily Chauliodontinae (9) viperfishes
            Subfamily Idiacanthinae (3) black dragonfishes
            Subfamily Malacosteinae (14) loosejaws
            Subfamily Melanostomiinae (193) scaleless black dragonfishes
            Subfamily Stomiinae (9) scaly dragonfishes

Superorder Ateleopodimorpha
    Order Ateleopodiformes (jellynose fishes) (1 family – 13 species)
        Family Ateleopodidae (13) jellynose or tadpolefishes [13]

Superorder Cyclosquamata
    Order Aulopiformes (lizardfishes) (16 families – 269 species)
        Family Alepisauridae (2) lancetfishes [1]
        Family Anotopteridae (3) daggertooths [3]
        Family Aulopidae (13) Aulopus [deep-sea]
        Family Bathysauridae (2) deep-sea lizardfishes [2]
        Family Bathysauroididae (1) largescale deepsea lizardfishes) [1]
        Family Chlorophthalmidae (19) greeneyes [19]
        Family Evermannellidae (9) sabertooth fishes [9]
        Family Giganturidae (2) telescope fishes [2]
        Family Ipnopidae (30) tripod fishes [30]
        Family Notosudidae (17) waryfishes [17]
        Family Omosudidae (1) omosudids [1]
        Family Paralepididae (60) barracudinas [60]
        Family Paraulopidae (14) cucumber fishes [some deep-sea]
        Family Pseudotrichonotidae (2) sand-diving lizardfish
        Family Scopelarchidae (18) pearleyes [18]
        Family Synodontidae (76) Lizardfish

Superorder Scopelomorpha
   Order Myctophiformes (lanternfishes) (2 families – 257 species)
      Family Myctophidae (251) lanternfishes [251]
         Subfamily Diaphinae (headlightfishes)
         Subfamily Gymnoscopelinae (luminous lanternfishes)
         Subfamily Lampanyctinae (toothy lampfishes)
         Subfamily Myctophinae (lanternfishes)
         Subfamily Notolychninae (topside lanternfishes)
      Family Neoscopelidae (6) blackchins [6]

Superorder Lamprimorpha
   Order Lampriformes (oarfishes and ribbonfishes) (6 families – 22 species)
      Family Lampridae (2) opahs [2]
      Family Lophotidae (4) crestfishes [at least 1]
      Family Radiicephalidae (1) tapertails [1]
      Family Regalecidae (3) oarfishes [3]
      Family Trachipteridae (10) ribbonfishes [some mesopelagic]
      Family Veliferidae (2)(velifers)

Superorder Paracanthopterygii
   Order Percopsiformes (cavefishes and trout-perches) (FW)
   Order Zeiformes (dories) (6 families – 34 species)
      Family Cyttidae (3) lookdowns [3]
      Family Grammacolepididae (3) tinselfishes [3]
      Family Oreosomatidae (11) oreos [11]
      Family Parazenidae (4) smooth dories [4]
      Family Zeidae (6) dories [3]
      Family Zeniontidae (7) armoreye dories [7]

Order Stylephoriformes (tube-eyes)(1 family – 1 species)
   Family Stylephoridae (1) tube-eye [1]
Order Gadiformes (cods, grenadiers and hakes) (10 families – 629 species)
   Family Bregmacerotidae (14) codlets [14]
   Family Euclicthyidae (1) eucla cod [1]
   Family Gadidae (25) cods and haddocks [1]
   Family Lotidae (24) hakes and burbots [3]
   Family Macrouridae (411) grenadiers [411]
   Family Melanonidae (3) pelagic cods [3]
   Family Merlucciidae (22) merluccid hakes [at least 8]
   Family Moridae (109) deepsea cods [109]
   Family Muraenolepididae (9) eel cods [at least 4]
   Family Phycidae (11) phycid hakes [at least 2]
Order Polymixiiformes (beardfishes) (1 family – 10 species)
   Family Polymyxiidae (10) beardfishes [10]

Superorder Acanthopterygii
   Order Beryciformes (bigscales) (2 suborders – 8 families – 111 species)
      Suborder Berycoidei (2 families – 71 species)
      Family Berycidae (10) alfonsinos [10 upper slope]
      Family Melamphaidae (61) bigscales [61]
      Suborder Stephanoberycoidei (whalefishes) (6 families – 40 species)
      Family Barbourisiidae (1) velvet whalefishes [1]

Family Cetomimidae (30) whalefishes [30]
Family Gibberichthyidae (2) gibberfishes [2]
Family Hispidoberycidae (1) spinyscale pricklefishes [1]
Family Rondeletiidae (2) redmouth whalefishes [2]
Family Stephanoberycidae (4) pricklefishes [4]

Order Trachichthyiformes (fangtooths and roughies) (5 families – 68 species)
Family Anomalopidae (9) flashlight fishes [1]
Family Anoplogastridae (2) fangtooths [2]
Family Diretmidae (4) spinyfins [4]
Family Monocentridae (4) pinecone fishes
Family Trachichthyidae (49) roughies [at least 15]

Order Holocentriformes (1 family – 87 species)
Family Holocentridae (87) squirrelfishes and soldierfishes [1]

Order Ophidiiformes (cusk eels) (4 families – at least 539 species)
Family Bythitidae – (232?) livebearing brotulas [at least 60]
Family Carapidae (37) pearlfishes [16 – many invertebrate symbionts]
Family Ophidiidae (267) cusk eels [194]
Family Dinematichthyidae – dinematichthyids

Order Batrachoidiformes (toadfishes) (1 family – 80 species)
Family Batrachoididae (80) toadfish [1]

Order Scombriformes (3 suborders – 10 families – 200 species)
Suborder Scombroidei (tunas, cutlassfishes) (3 families – 121 species)
Family Gempylidae (25) snake mackerels [25]
Family Trichiuridae (45) cutlassfishes [28]
Family Scombridae (51) mackerels and tunas
Suborder Scombrolabracoidei (1 family – 1 species)
Family Scombrolabracidae (black mackerels)
Suborder Stromateoidei (6 families – 78 species)
Family Amarsipidae (1) bagless glassfish
Family Ariommatidae (8) ariomids [3]
Family Centrolophidae (31) medusafishes [10]
Family Nomiidae (17) driftfishes [3]
Family Stromateidae (18) butterfishes
Family Tetragonuridae (3) squaretails [3, thaliacean specialists]

Order Syngnathiformes (pipefishes and gurnards) (7 families – 314 species)
Suborder Syngnathoidei (pipefishes) (5 families – 302 species)
Family Aulostomidae (3) trumpetfishes
Family Centriscidae (12) snipefishes [7]
Family Fistulariidae (4) cornetfishes
Family Solenostomidae (6) ghost pipefishes
Family Syngnathidae (277) pipefishes and seahorses
Suborder Dactylopteroidei (2 families – 12 species)
Family Dactylopteridae (7) helmet gurnards [1]
Family Pegasidae (5) seamoths

Order Kurtiformes (nurseryfishes)
Family Kurtidae (2)

Order Gobiiformes (gobies) (9 families – 1735 species)
   Family Eleotridae (77)
   Family Gobiidae (1536) gobies [5] 5 species live at 400 m or greater
   Family Kraemeriidae (8) sand darts
   Family Microdesmidae (85) wormfishes
   Family Odontobutidae (4) FW sleepers
   Family Rhyacicthyidae (3) loach gobies
   Family Schindleriidae (4) infantfishes
   Family Thalasselotrididae (3) ocean sleepers
   Family Xenisthmidae (15) collared wrigglers
Order Synbranchiformes (spiny eels) (FW)
Order Anabantiformes (gouramis and fighting fishes) (FW)
Order Istiophoriformes (swordfishes and marlins) (2 families – 12 species)
   Family Istiophoridae (11) billfishes and marlins
   Family Xiphiidae (1) swordfishes
All species capable of diving to great depth but are not deep-sea residents
Order Pleuronectiformes (flounders and soles) (11 families – 759 species)
   Family Achiridae (23) American soles
   Family Achiropsettidae (4) southern flounders [3]
   Family Bothidae (166) lefteye founders [at least 24]
   Family Citharidae (7) large-scale flounders [4]
   Family Cynoglossidae (141) tonguefishes [28]
   Family Paralichthyidae (109) sand flounders [at least 4]
   Family Paralichthodidae – peppered flounders
   Family Pleuronectidae (103) righteye flounders [at least 19]
   Family Poecilopsettidae – bigeye flounders
   Family Psettodidae (3) spiny turbots
   Family Rhombosoleidae – South Pacific flounders
   Family Samaridae (27) Crested flounders [at least 4]
   Family Scophthalmidae (9) turbots [2]
   Family Soleidae (166) soles [7]
Order Cichliformes (cichlids) (FW)
Order Atheriniformes (silversides) (10 families – 131 species)
   Family Atherinidae (64) (Old world silversides)
   Family Atherinopsidae (33) (New world silversides)
   Family Bedotiidae Malagasy rainbowfishes
   Family Dentatherinidae (1) tusked silversides
   Family Isonidae (5) surf sardines
   Family Melatotaeniidae (2) rainbowfishes
   Family Notocheiridae (1) surf silversides
   Family Phallostethidae (16) priapum fishes
   Family Pseudomugilidae (6) blue eyes
   Family Telmatherinidae (3) sailfin silversides
Order Cyprinodontiformes (killifishes and kin) (11 families – 85 species)
   Suborder Aplocheiloidei
   Family Aplocheilidae (6) old world rivulins
   Family Notobranchiidae – African rivulins
   Family Rivulidae – New world rivulins
   Suborder Cyprinodontidae
   Family Anablepidae (4) four-eyed fishes
   Family Aphaniidae – Oriental killifishes

Family Cyprinodontidae (15) killifishes
Family Fluviphylacidae – American lampeyes
Family Fundulidae (15) topminnows
Family Goodeidae (1) splitfins
Family Pantanodontidae spine killifishes
Family Poeciliidae (44) poecilliids (many FW)
Family Profundulidae middle American killifishes
Family Procatopodidae – African lampeyes
Family Valenciidae European killifishes

Order Beloniformes (flyingfishes and halfbeaks)(6 families – 185 species)
Family Adrianichthyiidae (8) adrianichthyiids
Family Belonidae (26) needlefishes
Family Exocoetidae (67) flyingfishes
Family Hemiranphidae (54) halfbeaks
Family Scomberesocidae (4) sauries
Family Zenarchopteridae (26) internally fertilized halfbeaks

Order Mugiliformes (mullets)(1 family – 82 species)
Family Mugilidae (82)

Order Gobiesociformes (clingfishes) (1 family – 157 species)
Family Gobiesocidae (157) clingfishes [2]

Order Blenniiformes (blennies) (6 families – 942 species)
Family Blenniidae (401) blennies
Family Chaenopsidae (96) pikeblennies
Family Clinidae (92) klipfishes
Family Dactyloscopidae (48) sand stargazers
Family Labrisomidae (126) labrisomids
Family Trypterygiidae (179) threefin blennies

Order Lophiiformes (anglers and frogfishes) (18 families – 368 species)
*** = Ceratioid**
Family Antennariidae (49) frogfishes
Family Brachionichthyidae (14) handfishes
*Family Caulophrynidae (5) fanfins [5]
*Family Centrophrynidae (1) prickly seadevil [1]
*Family Ceratiidae (4) seadevils [4]
Family Chaunacidae (22) sea toads [10]
*Family Diceratiidae (6) double anglers [6]
*Family Gigantactinidae (23) whipnoses [23]
*Family Himantolophidae (22) footballfishes [22]
*Family Linophrynidae (28) leftvents [28]
Family Lophichthyidae (1) lophichthyid frogfishes
Family Lophiidae (29) goosefishes [at least 8]
*Family Melanocetidae (6) black seadevils [6]
*Family Neoceratiidae (1) spiny seadevils [1]
Family Ogcocephalidae (78) batfishes [45]
*Family Oneirodidae (68) dreamers [68]
Family Tetrabrachiidae (2) tetrabrachiid frogfishes
*Family Thaumatichthyidae (9) wolftrap anglers [9]

Order Tetraodontiformes (puffers) (3 suborders – 10 families – 420 species)

Suborder Triacanthodoidei
Family Triacanthidae (7) triplespines
Family Triacanthodidae (23) spikefishes [23]

Suborder Tetraodontoidei
Family Diodontidae (19) porcupine fishes
Family Molidae (5) molas [1]
Family Tetraodontidae (174) puffers [2]
Family Triodontidae (1) three-tooth puffers

Suborder Balistoidei
Family Aracanidae (13) deepwater boxfishes
Family Balistidae (43) triggerfishes
Family Monacanthidae (110) filefishes
Family Ostraciidae (25) boxfishes

Order Scorpaeniformes (6 suborders – 35 families – 1570 species)

Suborder Scorpaenoidei (17 families – 537 species)
Family Apistidae (3) wasp scorpionfishes
Family Aploactinidae (48) velvetfishes
Family Bembridae (9) deepwater flatheads [>2]
Family Eschmeyeridae (1) cofishes
Family Gnathacanthidae (1) red velvetfishes
Family Hoplichthyidae (15) ghost flatheads [1]
Family Neosebastidae (18) gurnard scorpionfishes
Family Parabembridae (2) spratlike flatheads [1]
Family Pataecidae (3) prowfishes
Family Perryenidae (1) whitenose pigfishes
Family Platycephalidae (flatheads)
Family Plectrogenidae (2) stinger flatheads [2]
Family Scorpaenidae (233) scorpionfishes [23]
Family Sebastidae (135) rockfishes [27]
Family Setarchidae (7) deep-sea bristly scorpionfishes [4]
Family Synanceiidae (37) stonefish
Family Tetrarogidae (42) waspfishes [1]
Family Triglidae (125) searobins [20]

Suborder Congiopodoidei (1 family – 8 species)
Family Congiopodidae (8) racehorses [3]

Suborder Anoplopomatoidei (1 family – 2 species)
Family Anoplopomatidae (2) sablefishes [2]

Suborder Hexagrammoidei (1 family – 12 species)
Family Hexagrammidae (12) greenlings [occasional to 500 m]

Suborder Normanichtyoidei (1 family – 1 species)
Family Normanichthyidae (1) barehead scorpionfishes

Suborder Cottoidei (12 families – 733 species)
Family Abyssocotiidae (FW) deepwater Baikal sculpins
Family Agonidae (46) poachers [9]
Family Bathylutichthyidae (2) Antarctic sculpin [1]
Family Comephoridae (FW) Baikal oilfishes
Family Cottidae (197) sculpins [15]
Family Cottocomephoridae (1) (FW) Lake Baikal Sculpins
Family Cyclopteridae (28) lumpfishes [2]
Family Ereunidae (3) deepwater bullheads [3]

Family Hemitripteridae (8) sea ravens [1]
Family Liparidae (406) snailfishes [250]
Family Psychrolutidae (41) fatheads [41]
Family Rhamphocottidae (1) horsehead sculpins
Order Centrarchiformes
   Suborder Percichthyoidei
      Family Percichthyidae (8) temperate perches
   Suborder Tetrapontoidei
      Family Terapontidae (18) tigerfishes
      Family Kuhliidae (12) Aholeholes
      Family Dichistiidae (2) galjoens
      Family Oplegnathidae (7) knifejaws
      Family Kyphosidae (57) sea chubs
      Family Girellidae nibblers
      Family Scorpididae halfmoons
      Family Microcanthidae stripeys
      Family Parascorpididae (1) jutjaws
   Suborder Centrarchoidei
      Family Centrarchidae (sunfishes and FW basses)
      Family Elassomatidae – pygmy sunfishes
      Family Enoplosidae (1) oldwives
      Family Sinipercidae Chinese perches
   Suborder Cirrhitoidei
      Family Cirrhitidae (38) hawkfishes
      Family Chironemidae (6) kelpfishes
      Family Aplodactylidae (5) marblefishes
      Family Latridae (5) trumpeters
Order Acropomatiformes
   Family Leptoscopidae (5) southern sandfishes
   Family Champsodontidae (13) gapers [>3]
   Family Hemerocoetidae Indo-Pacific duckbills
   Family Creediidae (19) sand burrowers
   Family Glaucosomatidae (4) pearl perches
   Family Pempheridae (69) sweepers
   Family Lateolabracidae (2) Asian seaperches
   Family Synagropidae (splitifin ocean basses)
   Family Dinolestidae (1) long-finned pikes
   Family Malakichthyidae (temperate ocean basses)
   Family Polyprionidae (4) wreckfishes [3]
   Family Bathyclupeidae (7) deepsea herrings [7]
   Family Banjosidae (3) banjofishes
   Family Pentacerotidae (14) armorheads [7]
   Family Ostracoberycidae (3) shellskin alfonsinos [3]
   Family Epigonidae (44) deepwater cardinalfishes [44]
   Family Symphysanodontidae (12) slopefishes [12]
   Family Acropomatidae (33) lanternbellies [33]
   Family Howellidae (9) oceanic basslets [9]
   Family Scombropidae (3) gnomefishes [3]
Order Perciformes (11 suborders – 130 families – at least 5757 species)
   Suborder Percoidei (82 families – 3393 species)
      Family Ambassidae (26) Asiatic glassfishes
      Family Apogonidae (401) cardinalfishes
      Family Arripidae (4) Australian salmon

Family Bramidae (20) pomfrets [1]
Family Caesionidae (23) fusiliers
Family Callanthiidae (16) splendid perches [2]
Family Carangidae (150) jacks – deep-divers only
Family Caristiidae (19) veilfins [19]
Family Centrogenyidae (1) false scorpionfishes
Family Cepolidae (26) bandfishes [6]
Family Centropomidae (12) snooks
Family Cepolidae (26) bandfishes
Family Chaetodontidae (131) butterflyfishes [1]
Family Coryphaenidae (3) dolphinfishes
Family Datnioididae (2)
Family Dinopercidae (2) cavebasses
Family Drepaneidae (3) sicklefishes
Family Echeneidae (8) remoras
Family Emmelichthyidae (16) rovers [>6]
Family Gerreidae (54) mojarras
Family Grammatidae (13) basslets [3]
Family Haemulidae (143) grunts
Family Hapalogenyidae (barbeled grunters)
Family Lactariidae (1) false trevailies
Family Latidae (6) giant perches
Family Leiognathidae (50) ponyfishes
Family Leptobramidae (1) beachsalmons
Family Lethrinidae (43) emperors
Family Lobotidae (2) tripletails
Family Lutjanidae (114) snappers [4]
Family Malacanthidae (46) tilefishes [>2]
Family Menidae (1) moonfishes

Family Monodactylidae (6) moonies
Family Moronidae (4) white basses
Family Mullidae (87) goatfishes
Family Nandidae (3)
Family Nemastiidae (1) roosterfishes
Family Nemipteridae (69) threadfin breams [3]
Family Opistognathidae (82) jawfishes [2]
Family Percidae (12) perches and darters
Family Perciliidae – southern basses
Family Plesiopidae (50) roundheads
Family Polycentridae (1) leaffishes
Family Polynemidae (37) threadfins
Family Pomacanthidae (94) angelfishes
Family Pomatomidae (1) bluefishes
Family Priacanthidae (19) bigeyes
Family Pseudochromidae (157) dottybacks
Family Rachycentridae (1) cobias
Family Sciaenidae (270) croakers [3]
Family Serranidae (561) groupers and seabasses [17]
Family Sillaginidae (35) sillagos
Family Sparidae (149) porgies [2]
Family Toxotidae (3) archerfishes
Suborder Labroidei (5 families – 1095 species)
Family Embiotocidae (23) surfperches
Family Labridae (552) wrasses
Family Odacidae (12) cales
Family Pomacentridae (407) damselfishes
Family Scaridae (101) parrotfishes
Suborder Zoarcoidei (14 families – at least 413 species)

Family Anarhichadidae (5) wolffishes [2]
Family Bathymasteridae (7) ronquils [1]
Family Cebidichthyidae
Family Cryptacanthodidae (4) wrymouths
Family Eulophiidae
Family Lumpenidae
Family Neozarcidae
Family Opisthocentridae
Family Pholidae (16) gunnels
Family Ptilichthyidae (1) quillfishes
Family Scytalinidae (1) graveldrivers
Family Stichaeidae (84) pricklebacks [3]
Family Zaproridae (1) prowfishes
Family Zoarcidae (294) eelpouts [176]
Suborder Notothenioidei (8 families – 156 species)
  Family Artedidraconidae (33) barbelled plunderfishes [33]
  Family Bathydraconidae (18) Antarctic dragonfishes [12]
  Family Bovichtidae (11) bovichtids
  Family Channichthyidae (25) crocodile icefishes [25]
  Family Eleginopsidae (1) Patagonian blennies
  Family Harpagiferidae (11) plunderfishes
  Family Nototheniidae (54) notothens [54]
  Family Pseudaphritidae (3) congollis
Suborder Trachinoidei (9 families – 277 species)
  Family Ammodytidae (31) sandlances
  Family Cheimarrichthyidae (1) New Zealand torrentfishes
  Family Chiasmodontidae (32) swallowers [32]
  Family Percophidae (51) duckbills [23]
  Family Pinguipedidae (88) sandperches
  Family Trachinidae (9) weeverfishes
  Family Trichodontidae (2) sandfishes
  Family Trichonotidae (10) sand divers
  Family Uranoscopidae (53) stargazers [>1]
Suborder Icosteoidei (1 family – 1 species)
  Family Icosteidae (1) ragfish [1]
Suborder Callionymoidei (2 families – 218 species)
  Family Callionymidae (203) dragonets [25]
  Family Draconettidae (15) slope dragonets [15]
Suborder Acanthuroidei (6 families – 138 species)
  Family Acanthuridae (88) surgeonfishes
  Family Ephippidae (15) spadefishes
  Family Luvaridae (1) luvars
  Family Scatophagidae (4) scats
  Family Siganidae (29) rabbitfishes
  Family Zanclidae (1) Moorish idols
Suborder Sphyraenoidei (1 family – 28 species)
  Family Sphyraenidae (28) barracudas
Suborder Caproidei (1 family – 18 species)
  Family Caproidae (18) boarfishes
Suborder Gasterosteroidei (3 families – 13 species)
  Family Aulorhynchidae (1) tubesnouts
  Family Gasterosteidae (10) sticklebacks
  Family Hypotichidae (2) sand-eels

CLASS COELACANTHI
Order Coelacanthiformes (1 family – 2 species)
Family Latimeriidae (2) coelacanths [2]

CLASS DIPNEUSTI
Order Ceratodontiformes (3 families – 6 species)
Family Lepidosirenidae (1) South American lungfishes
Family Neoceratodontidae (1) Australian lungfishes
Family Protopteridae (4) African lungfishes

References

HELFMAN, G.S., COLLETTE, B.B., FACEY, D.E., and BOWEN, B.B. (2009). *The Diversity of Fishes*. Oxford: Wiley-Blackwell.

NELSON, J.S., GRANDE, T.C., and WILSON, M.V.H. (2016). *Fishes of the World*, 5ee. Hoboken: John Wiley and Sons.

Priede, I.G. (2017). *Deep-Sea Fishes*. Cambridge: Cambridge University Press.

WoRMS (n.d.). World Register of Marine Species. www.marinespecies.org

Glossary

Aboral End of the gut opposite the side of the oral opening or mouth.

Absolute temperature The absolute or Kelvin temperature scale begins at absolute zero, −273 °C, where all molecular motion ceases. It is expressed in degrees Kelvin. A Kelvin degree is the same as a Celsius degree, but e.g. a temperature of 0 °C equals a temperature of 273 K.

Acclimation An adjustment in temperature tolerance or physiological rate functions elicited in the laboratory by maintenance at constant temperature.

Acclimatization An adjustment in temperature tolerance or physiological rate functions elicited in the wild by seasonal temperature change.

Activation energy The energy required to initiate a chemical reaction. In any population of molecules, only a subset of the total has enough energy to exceed the activation energy threshold.

Acute measurements Physiological rate determinations with no period of acclimation.

Anaerobiosis Living without oxygen.

Archaea A kingdom of protists considered to be intermediate between bacteria (prokaryotes) and eukaryotes. Like the prokaryotes they have no nucleus, but their biochemistry includes resemblances to eukaryotes. First noted as extremophiles capable of living in forbidding environments, later research shows they are truly ubiquitous in distribution.

Avogadro's number and the mole concept Avogadro's number defines the number of molecules in the mole of a substance: 6.022×10^{23}. A mole of a solid substance is equal to its molecular or atomic weight in g. Thus, a mole of NaCl is equal to the atomic mass of Na (23) plus the mass of Cl (35) or 58 g. A mole of a gas at standard temperature and pressure (0 °C and 101 kPa) contains 6.022×10^{23} molecules and occupies a volume of 22.4 l.

Benthic Bottom dwelling.

Benthopelagic Swimming species living in close proximity to the bottom.

Circumoral Surrounding the mouth.

Climatic adaptation Evolutionary adaptation to temperature through natural selection: Metabolic Cold Adaptation or MCA.

Collagen Tightly wound fibrous proteins usually composed of repeated tripeptide units. Important in cartilage and a very wide variety of other structures.

Copepod Any member of the extremely diverse crustacean subclass Copepoda, including free-living and parasitic forms and about 12 000 species.

Cryopelagic Polar species that alternate between the undersurface of fast ice and the water column beneath it, e.g. *Pagothenia borchgrevinki*.

Desaturases Enzymes that introduce double bonds into fatty acid chains, lowering their melting point and increasing fluidity.

Ectotherm Body temperature determined by ambient environmental temperature: "cold-blooded."

Endotherm Regulated constant body temperature using the internal heat generated by metabolic processes: "warm-blooded."

Epifaunal Living *on* the bottom such as a crab or scallop, but not *in* the bottom such as a tube-dwelling worm.

Estuarine Of or referring to coastal regions where freshwater and ocean water meet, resulting in a shoreward gradient from fully marine to lower salinities closer to shore. May be embayments or river mouths and adjacent waters.

Euphausiid Any member of the Crustacean order Euphausiacea. Shrimp-like zooplankton/micronekton ranging from about 1 to 10 cm in size (Chapter 7).

Eurythermal Capable of living over a wide range of environmental temperature.

Fluid dynamics The science of fluid behavior, especially that of fluids in motion.

Gastroderm Epithelium lining the gut.

Gustation Sense of taste in terrestrial biota. In aqueous species, chemoreceptive detection of dissolved molecules.

Heterocercal tail Tail with a pronounced upper or lower lobe, not crescent shaped, an extreme example of which is the tail of the thresher shark.

Homeotherm Constant body temperature. Formerly a widely used term to describe mammals and birds. Has fallen out of favor because an ectotherm can be a homeotherm when it dwells at constant temperature, like those found in the deep sea.

Homeoviscosity The concept that cell membranes must maintain an optimal fluidity to allow transport proteins that must change conformation within the membrane, to function.

Homocercal tail Crescent-shaped tail with equal upper and lower lobes, an example of which is the tail of the great white shark.

Homonomous segmentation All segments the same.

Hydrophobic Literally "fear of water." Not readily soluble in water.

Hyperosmotic Having an ionic concentration greater than that of the external milieu.

Hyporegulation Maintenance of an internal ion concentration below that of the external environment. Typical of marine bony fishes.

Infaunal Burrowing or tube-dwelling species.

Latent heat of fusion The amount of heat energy required per unit mass to change the state of a substance from a solid to a liquid; which is equal to the amount of energy *released* when the same substance changes from a liquid to a solid. In the case of water, $80 \, \text{cal} \, \text{g}^{-1}$ is released when water solidifies into ice.

Life table A summary of the mortality rates operating on a population as a function of the age of its members.

Littoral The sea nearshore, including waters from the high water mark out to several meters depth. Sometimes used as a synonym for the intertidal region of a shoreline.

Maxillary segment The anterior crustacean body segment containing the maxillae, paired limbs located just posterior to the maxillules. The most posterior of the five segments considered part of the head or cephalon.

Maxillulary segment The anterior crustacean body segment containing the maxillules, a crustacean limb located just posterior to the mandibles and usually associated with feeding.

Mechanoreception Perception of sound, vibration and motion, including equilibrium. Receptors function by mechanical displacement.

Mesenchyme or parenchyma A loose aggregate of unspecialized cells within a deformable and jellylike matrix or mesoglea.

Metabolic suppression The drop in metabolic rate accompanying exposure to low oxygen.

Michaelis Constant (K_m) In an enzymic reaction, the K_m is the substrate concentration at which the reaction proceeds at 50% of its maximum velocity (V_{max}).

Mucopolysaccharide A polysaccharide with repeating paired units of two different sugars, such as *n*-acetylglucosamine and glucuronic acid. Important structural component of eel larvae (leptocephali). Now known by its synonym glycosaminoglycan (GAG).

Mucoprotein or proteoglycan A polypeptide joined to a mucopolysaccharide or other heteropolysaccharide. A heteropolysaccharide is a polysaccharide with repeating units of more than one sugar. The polysaccharide is normally the larger portion of a proteoglycan.

Myogenic hearts Hearts where the beat originates in the heart muscle itself (myo = of or pertaining to muscle; genic = generated or initiated) rather than neurally (neurogenic) as it is in the Arthropoda. Secondary regulation (pulse rate or volume expelled per stroke) is effected neurally.

Myotomes V- or W- shaped muscle bundles comprising the swimming musculature of fish.

Neoteny Retention of formerly juvenile characteristics in adult descendants. Paedomorphosis.

Net cost of transport The metabolic cost of swimming at a given speed minus the standard (basal) metabolism.

Normoxia Oxygen concentrations at or near air saturation.

Olfaction Sense of smell in terrestrial biota. In aqueous species, sometimes used to describe chemoreceptive detection of dissolved molecules at very low concentrations.

Osmoconformity Maintenance of an internal ion concentration equal to that of the external environment. Typical of marine invertebrates.

Osmolytes Osmotically active particles. May be charged such as the sodium ion, or uncharged, such as urea.

Osmoregulation Regulating the internal concentrations of water and ions within an organism.

Otoconia the elasmobranch equivalent to the teleost otolith. Multiple small concretions that stimulate the sensory maculae of the inner ear to yield information on the motion of the fish.

Pelagic Of or referring to the region between the surface and bottom in a body of water, usually in the open ocean but can apply to lakes and rivers as well: midwater, water column.

Phagocytic cells Cells capable of engulfing particles by enveloping them with their plasma membrane and conveying them into the cell's interior as a vesicle, where they can be further broken down. Important in the digestive processes of several invertebrate taxa as well as constituting the main line of defense in vertebrate immune systems.

Pleuston Organisms living at the air–sea interface, such as the cnidarian *Velella* and the gastropod *Janthina*. Distinguished from neuston which dwell in the upper few centimeters of the sea, though sometimes the terms are used interchangeably.

Poikilotherm Variable body temperature. The former equivalent to ectotherm. It is not as descriptive and has fallen out of favor.

Porphyrin ring A highly reactive compound formed from four pyrroles covalently joined into a ring. A pyrrole itself is a five-membered ring with the formula C_4H_4NH. In a heme, the ferrous iron atom is positioned at the center of the ring, joined with the nitrogen atoms forming the apex of each of the pentagonal pyrrole rings.

Proostracum In a belemnoid cephalopod, a partially calcified plate extending the dorsal portion of the shell from the phragmocone toward the head.

Protist Single-celled organism.

Protostomate A type of embryonic development where the blastopore of the embryo becomes the mouth of the young organism.

Q_{10} The rate of increase or decrease in reaction velocity with a 10 °C rise or fall in temperature, typically falling between 2 and 3. Most commonly applied to measurements of metabolism, but to other physiological rates as well.

Quaternary structure Refers to a protein having multiple subunits, in the case of hemoglobin: 4.

Rete mirabile Literally, "wonderful net." A capillary bed that facilitates the exchange of heat or oxygen by having thousands of tiny vessels in close proximity. Found in the heat exchangers of tuna and swordfish and in all fishes with functioning swim bladders.

Salinity Salt concentration.

Sessile Unmoving or fixed in place. Used to describe anchored species such as barnacles as well as tube-dwelling worms and bivalves.

Specific heat The amount of heat energy it takes per unit mass to raise the temperature of a substance by 1 °C.

Spiral cleavage A type of embryonic development where the cleavage planes defining the new cells in the early embryo describe a spiral pattern with respect to the axis of the animal and vegetal poles of the embryo. The type of development exhibited by all phyla in the major clade Spiralia.

Standing crop The numbers or biomass of a taxonomic or general biological grouping, e.g. lanternfishes, fish larvae, zooplankton, phytoplankton, at a particular point in time.

Stenothermal Capable of living over a narrow range of environmental temperature.

Surface tension The tendency of liquid surfaces to shrink into the minimum surface area possible: why droplets are formed.

T-S diagram A plot of temperature vs salinity, the characteristics of which define a water mass.

Venturi tube A constriction in the cross-sectional area of a pipe that results in an increase in velocity of the fluid flowing through it.

Water mass A large, readily identified body of oceanic water with a common origin and distinctive characteristics of temperature, salinity, and density.

Index

Note: page numbers in *italics* indicate figures, those in **bold** indicate tables and boxes.

a

abyssal zone   670
abyssopelagic
   definition   670
Acanthopterygii
   pelagic representatives   735
ACC
   size and water transport   853
aciculae   237, *237*
Acipenseriformes   677
acoelomate body plan
   195, *195*
acousticolateralis system
   definition   815
Actinopterygii
   ray-finned fishes   677
activation energy   42
active metabolism
   definition   900
activity metabolism
   experimental
      determination of   900
   and locomotory muscle   899
   muscle efficiency   899
   propulsive efficiency   900
aesthetascs   311, 323, 351, *312*
agnathans
   paleohistory   678
Alepisaurus
   diet and habits   719, *719*
alepocephalids
   distribution, reproduction,
      and feeding   702, *703*
Alicella gigantea   397
Alitta   245
allozymes and isozymes   47
AMERIEZ   855
   sample timing and
      locations   856, *857*
   water column profiles
      857
Amiiformes   676, *676*
ammocoetes   682, *682*
ammoniacal   528
ammonium toxicity   785
ammonotelic   785
ammonotely   303
Amphineura
   origin of name   440
Amphionidacea   379, *286*
   classification and history   379
   development and natural
      history   380
amphipod eyes   305, *307*
amphipods
   Barnard and Vinogradov
      402
   benthopelagic gammarids
      420
   bioluminescent photocytes
      405, *407*
   blood flow   405, *404*
   caprellids, cyamids and
      ingolfiellids   397
   copulation   408
   Cystisoma digestive tract
      408, *409*
   diets
      gut contents   412, **413**
      methods of study   411
   gammaridean digestive
      tract   408
   gammarids
      and gelatinous
         zooplankton   416
      and hyperiid   416
      trends with depth   419, *419*
   gonads   408
   hyperiid families and their
      hosts   **417**
   hyperiid reproduction
      importance of gelatinous
         host   409
      larval deposition and
         development   409
      Phronima sedentaria
         411, *411*
      Themisto   410
   hyperiids
      vs gammarids   403, *405*
      and gelatinous
         zooplankton   397
      vertical range   419

amphipods (cont'd)
  hyperiids and gelatinous
    species
      types of association   416
  importance in pelagic
    communities   416
  nervous system   403, *406*
  pelagic vs benthic   397, *290*
  species numbers   397
ampoule complex   357, 358
anabiosis   278
anamorphic development
  definition and description
    319
anatomy in fish taxonomy   758
Andriashev
  ancient deep water fishes
    672
anecdysis   294
anglerfishes
  family morphologies   747
anhydrobiosis   278
animal energetics
  introduction   893
annelid morphology
  general   231, *233, 234*
Annelida
  class Polychaeta   224
  introduction   219
  polychaete subclass
    Errantia   226, 229
annelid(s)
  antennae   236
  current classification   222
  history of study   221
  nuchal organs   236
  proboscis eversion   236
  recent changes in
    classification   219
  segmentation   219
annular rings   936
Anoplogaster
  size, diet, and distribution
    738, 739
Anoplogaster cornuta   738
Antarctic
  physical characteristics
    852, *853*
  planetary bellwether   852
  sea ice retreat   853, 856, *856*
  Southern Ocean and oceanic
    fronts   852–855, *854*
Antarctic Bottom Water
  (ABW)   19, 20, *23*,
    853, *855*
Antarctic Circumpolar Current
  (ACC)   11, 16, *854, 855*
Antarctic Coastal
  Current   *854, 855*
Antarctic coastal systems
  WAP and Ross Sea   855
Antarctic Convergence
  852–853
Antarctic Intermediate
  Water   19, *19*, 22, *23*,
    852, *855*
Antarctic Polar Front (APF)
  23,852, *854, 855*
antennal glands   297, *302*,
  326, 354
Anthomedusae   110, *113*
Anuropus bathypelagicus
  association with
    Deepstaria   420
  pelagic lifestyle   289, *289*
aphotic zone   32
Aplacophora
  discovery and classification
    440, *441*
Aplysia
  and neural circuitry   439
apodemes   296, *298*
appendicularians
  bioluminescence
    lumisomes   653
  brain and behavior   643
  buoyancy   638
  circulation   626
  differences with other
    tunicates   625
  digestion
    role of stomach lobes
      626
  digestive system   626
  discarded houses
    importance   652
  luminescence   654
  epipelagic distributions
    660
  escape swimming   644
  field distributions   659
  generation times   659
  hermaphroditism   629
  houses
    basic structure and
      function   631
    detailed design and
      function   632
    filters   633
    flow and filtration
      632, *632*
    importance   631
    replacement and
      expansion   *631*
    rudiments   629
    secretion   629, 630
  kowalevskiids   *627*, 629
  Langerhans cells
    644, *644*
  meso-and bathypelagic
    species   660
  as microbial loop short-
    circuit   **649**, 651, **651**
  nano-and picoplankton
    feeding   648, **651**
  nervous system
    miniaturization   643
  oikoblastic epithelium
    629, *630*
  as prey   *652*, 654
  reproductive cycle   629, 630
  statocyst   643
  swimming patterns   638
  tail
    nerves and circulation
      626, *627*
    structure and function
      626, *627*
  trunk
    basic anatomy   626, *628*
    familial differences
      628
  vertical distributions
    trends with depth   660

appendix masculina 351, 358
apposition eyes 305, *306*
  ommatidia 305, *306*
aragonite
  compensation depth 505
Argyropelecus aculeatus 710
  life history 711
  vertical and geographic distribution 711
Argyropelecus and *Sternoptyx*
  latitudinal range 710, *710*
arthrobranch 315, *315*
arthropod-annelid relationship 275
Arthropoda
  groups
    origins and relatedness, 276–277
    size range 278
    universal characteristics 278
arthropod(s)
  classification
    history 273
    major groups 273
    recent classification system 275
ascidians
  as hermaphrodites 610
  blastozooids 610
  feeding 610, *611*
  reproduction 610, *612*
  solitary, social, and compound *605*, 610
ascidiozooids 612
ash
  definition 931
ash-free dry mass
  defintion 931
Astronesthinae
  feeding 712
  geographic and vertical distributions 712, *713*
atrial cavity 610, *611*
atrial siphon 610, *611*
August Krogh 42
Aulopiformes

hermaphroditism 719
Aysheia 277, *278*

b

bacterioplankton
  trophic role 351
barnacles 283, 422, *423*
barreleyes 702
basal metabolism
  definition 900
bathyal zone 670
Bathychordaeinae 626, 645, 653,
bathylagids 702, *704*
  distribution, morphology, feeding 703–705
bathypelagic zone 33
belemnoids 526–528, *527*
Beloniformes
  habitats
    halfbeaks and sauries 745, *745*
    halfbeaks, flying fishes, and sauries 745, 746
Beryciformes
  bigscales, whalefishes, and alfonsinos 736
bichirs 677
  systematic importance 695
billfishes 743, *744*
  brain heater
    circulation to 801
    location, structure, function 743–745, 799
  foraging habits 745
  need for brain heater 799–800
  reproduction 745
  size and distributions 743
biogeographic range 36
biomineralization 506
biting
  definition 763
black jack 753
blood basics 775
  carrying capacity 775
  oxygen solubility 775

volume %
  definition 775
blue-water diving
  importance in studies of gelatinous species 185
branchiostegite 313
breeding dress 358
buccal siphon *605*, 610, *611, 641*
buoyancy mechanisms
  heavy ion replacement 144, 635
  low density tissue 783
burglar alarm 354, 828
By-the-Wind Sailors 150

c

Calanus pacificus 345
calcite
  compensation depth 505
California Undercurrent *848*, 849
calyptopis 319, 328, *329, 333*
cameral fluid 571, 572
cannonball jellies 104
carangids
  body morphs 753, *754*
  classification 753
  cooperative hunting 753–754
  feeding habits 753
  global distribution 755
  spawning habits 754
  species, size, habits 753
Carcharodon megalodon 683
  size. See elasmobranchs
Cariaco Basin 734
  Bregmaceros and anoxic zone 735
  fish fauna 735
carideans
  brood chamber 358–359
  brooding 359
  copulation 358
  egg incubation and temperature 359

966 | Index

carideans (*cont'd*)
  egg size and fecundity 359, **360**
  hatching and post-larval development 355, *356*
  luminescent systems 352
  reproductive anatomy 358
  reproductive sequence 358
  saddle 350
  sex change 359
  spawning 359
  spawning and depth 359
  successive broods 359
caridoid facies 346, *348*
caristiids
  association with siphonophores 755
  feeding 755
  morphology 755, *755*
carpus 297, *299*
catfishes
  marine representatives 700
caudal peduncle *646*, *759*
cephalenteron 623
cephalochordates 669, 670
  basic characteristics 603, *604*
cephalopod circulation
  arterial flow 544, *545*
  hemocyanin and blood oxygen 544
  venous flow 544, *545*
cephalopod digestion
  assimilation 544
  digestive enzymes 542
  digestive gland *540*, 542, movement of food summary 543
  role of caecum 542
  role of esophagus 541–542
  role of the intestine 543
  stomach and caecum 542
  stomach and food breakdown 542
  total time for digestion 543
cephalopod fisheries 583
cephalopod gas exchange
  oxygen extraction **548**
  % utilization
    influence of jet propulsion 546
  ventilatory flow 546
cephalopod habitats 585
  benthic and benthopelagic 584
  coastal 582
  deep benthopelagic 587
  epipelagic ommastrephids 586
  mesopelagic squids 587
  pelagic coastal 583
  selective pressures and depth 586
  shelf/slope to open ocean 583
  vertical migrators 586–587
cephalopod nervous systems
  brain
    lobes and their function 552
    octopods vs decapods 554
    resemblance to gastropods 552
    structure 552, *553*
  cranial brain protection 552
  giant fibers and neural function 555
  introduction 552
  octopus brain and behavior 554, *554*
  role of giant fibers 555, *556*
  squid curiosity 555
  squid rapid escape 555, *556*
cephalopod photoreception
  cephalopod and fish eyes
    pupil and iris 562–563, *563*
  cephalopod and fish lenses
    depth of field 562, *563*
    refractive index 562
  coleioid eyes
    general 562
    importance of spherical lenses 562, *563*
  Nautilus pinhole eyes 561, *561*
  octopus eye
    anatomy 562, *563*
    vs teleost eyes 562
  receptor density
    cephalopods and fishes 565
  retinal structure
    cephalopod vs fish 563, *564*
  size and types of eyes 561
  spectral sensitivity of cephalopods and fishes 565
cephalopod sensory mechanisms 557
  acceleration detectors 557, *558*, *559*
  equilibrium
    statoliths and hair cells 557, *558*, *559*
  hearing
    evidence for 560
  lateral line 560, *560*
  mechanoreception
    definition and included senses 557
  olfaction and the olfactory pit 557
  receptors
    acceleration vs gravity 559
  sensory cells in lips and suckers 557, *540*
  statocysts
    gravity and acceleration 559, *559*
  taste
    discrimination with suckers 557

Index | 967

touch
  skin and suckers 557
cephalopods
  ammoniacal squid
    ammonium sequestration 575
    cranchiids 574
  anatomy
    Argonauta 537, *539*
    general 529, *530*
    myopsids and oegopsids 532, *532*
    nautilids 529, *533*
    sepiids and spirulids 531, *534*
    shift in anteroposterior axis 529, *531, 532*
  appearance in the fossil record 525
  beak 541
  belemnoids
    K-T extinction 526, *527*
  bioluminescence
    octopods and squid 567
  body-color patterns 566
  body pattern function 567
  body pattern library 566
  buoyancy and evolution 525
  buoyancy mechanisms 571
  buoyancy of ammoniacal squid 574
  chromatophore structure and function 566, *566*
  chromatophores 565
  circulation
    introduction 544
  classification
    introduction 565
  coleoid color change 565
  coleoid skin 565
  contrast with other molluscs 529
  counterillumination 569
  courtship and copulation 581
  digestive tract 541, *540*
    buccal mass 541
  egg capsules 581
  egg deposition 578, *580*
  evolution
    immediate ancestor 525
  excretion
    fundamentals 549
  excretory system 549, *550, 551*
    secretion and reabsorption 552
    system function 549
    ultrafiltration 549, *550*
  excretory system function
    evidence for 549
  fecundity 581
  feeding and digestion 540
  female reproductive anatomy 578, *580*
  four major groups anatomy 529
  gas exchange
    introductory summary 544
  general characteristics 525
  jet propulsion mechanics 570, *570*
  jet propulsion vs undulatory swimming 570–571
  larval development 582
  life histories
    coleoids 575–578
    Illex illecibrosus 578
    Loligo opalescens 576
    models 577, *577*
    nautiloids 575
    Octopus bimaculatus 576–577
    stages 582
  locomotory types 569
  luminous organs and secretions 567, *568*
  male reproductive anatomy 579
  metabolism and depth 938
  myopsids vs oegopsids differences 536
  nautilids
    prey capture 541
    shell 529–530
  nautilus buoyancy system 571–572, *572*
  Octopodiformes
    classification 536
  octopods
    Cirrata vs Incirrata 537
    internal anatomy 536, *537*
    prey capture 540
    sucker morphology *535*, 536
  open ocean predators 588
  paleohistory
    belemnoids 526, *527*
    coleoids 526
    nautiloids and ammonoids 525–526
    Plectronoceras cambria 525, *526*
  pelagic taxa 442
  photophore structure 569
  Plectronoceras
    resemblance to nautiloids 525
  prey capture
    squid and cuttlefish 540
  prey handling 541
  salivary glands 541
  seminal receptacle 581
  sepiid cuttlebone
    function 572, *573*
  sepiids and spirulids
    recent classification 531
  shelf–break predators and prey 584
  size range 525
  spermatophoric organ 579, *580*
  Spirula buoyancy mechanisms 572, *573*
  squids vs cuttlefishes
    anatomical differences 532–533
  vertical distributions 584, 585
  vertical range 582

cephalopods and jellies
   metabolism and proximate composition   938
cephalotoxins   541
cerata   516, *519*
chaetae   237
   function   238
Challenger Expedition   60, 420
Chauliodontinae   712, *714*
   diet and feeding chronology   713
   population age structure   714
   vertical distribution asynchronous   712
   viperfishes   712
chelate   350
chiasmodontids
   species, distribution, habits   712
chimaeras   677, *678*, 693
   depth range   691
   feeding   693
   gas exchange   693
   locomotion   693
   paleohistory   677–678
   reproduction   693
chimaeras and elasmobranchs
   similarities   691
Chondrichthyes   677, 693
chondrocranium   758
Chondrophora   150, *150*
   classification   150
   evolution   151, *152*
   feeding   151, **152**
   locomotion   151–153
   medusae   151
Chondrostei   696
Chordata
   introduction and subphyla   603
chordates
   defining characteristics   603
   deuterostomate development   604
   history of classification   604

metabolism   **923**, 939
ciliated organ   542
cincinnuli   351
Circumpolar Deep Water (CDW)   653
cladistics   222, 605
   defining goal   607
Cladistii   *676*, 695
Cladocera   283
cladophore   621, *621*
Cladoselachimorpha   677, *678*
Class Actinopterygii
   species numbers   695
Class Cladistii
   the bichirs   695
Class Coelacanthi-
   the coelacanths   693–695, *695*
Class Dipneusti
   the lungfishes   695
class Elasmobranchii-sharks and rays   682–691
Class Gastropoda   445
   recent classification   449
   traditional classification   450
class Holocephali-
   chimaeras   691
Class Myxini-hagfishes   678–679, *680*
class Petromyzonti-lampreys   679–682, *681*, *682*
Clitellata   220, 223, 224
cnidae   107
   venom   108, *108*
Cnidaria   89
   classification   89
   history   89
   interaction with prey   109
Cnidaria and Ctenophora
   metabolism   **926–928**, 939
   metabolism in-situ   939
Cnidaria foraging strategies
   attracting prey   116
   direct interception   110, *111*
   encounter zone   110, **112**, *113*
   water flow and swimming   115, *116*

cnidarian venoms   108
   toxic effects   109
cnidocil   107
cnidocyte   107
cnidocyte discharge   107
CO_2 yield
   why not used   898
coelacanths   *695*
   buoyancy and ion regulation   694–695
   as fossil group   693
   historical importance   676
   length of life   695
   re-discovery   693–694
   reproduction   695
   skeleton and fins   694
coelom   194
Cohort Elopomorpha
   tarpon, ladyfish and eels   696
colloblasts   161, 163, 173
communities
   definition   845
compensation depth   31–32, 505
compound eyes   305, *306*
   euphausiids and mysids   306
   pelagic decapods   307
   reduced size in deep-living species   308
conchiolin   464
convex eyes   305
cookie-cutter shark   763
cooperativity   777
cor frontale   311, *289*, *314*, *384*, *388*
coriolis force   13
Coronatae   101
coryphaenids
   size and habits   756
   species and distributions   756, *756*
cost of transport   797, *798*
critical swimming speed (Ucrit)
   determination of   900
cross-flow filtration
   definition   762, *762*
crown teleosts   671, 677

Crustacea 273
  central nervous system 303, *304*
  circulatory and respiratory systems
    introduction 311, *314*
  Class Branchiopoda 283
  Class Branchiura 283
  Class Cephalocarida *282*, 283
  Class Copepoda 283
  Class Malacostraca 284
  Class Mystacocarida 283
  Class Ostracoda 284
  Class Remipedia *282*, 283
  Class Tantulocarida 283
  Class Thecostraca 283
  classification systems 273
  digestive system
    basic anatomy 318, *318*
  head or cephalon
    segments 281
  introduction to 273
  major subdivisions 283
  marine
    species numbers 273
  order Decapoda 346
  Pentastomida
    tongue worms 284
  Subclass Eumalacostraca 284
  Superorder Eucarida 284
  Superorder Peracarida 288
  trunk segments 281, *281*
  types of development 319
  unifying characteristic 281
Crustacea central nervous system
  brain 303, *304*
  circumesophogeal connectives 303, *304*
  giant fibers 305
  segmental ganglia 303, *304*
  subesophogeal ganglion 303–304
  supraesophogeal ganglion 303, *304*
crustacean
  joints and appendages 296, *298, 299*
  X-organ 294–295, *295*
  Y-organ 295, *295*
crustacean antennal gland
  four sections 301, *302*
crustacean appendages
  basic limb morphology 297, *299*
Crustacean brain
  protocerebrum, deuterocerebrum, tritocerebrum 304
crustacean chemoreception
  introduction 310
  sensitivity 311, *313*
  sensory hairs 311, *312*
crustacean circulation
  blood flow in gills 316, *316*
  cor frontale 311, *314*
  gills 313, *315*
  heart 311
  infrabranchial sinus 311–312, *314*
  oxygen transport in the blood
    hemocyanin 317
  pericardium 311, *314*
crustacean development
  anamorphic 319
  basic stages 319, *320*
  epimorphic 319
  teloblastic budding 319
crustacean digestive system *314*, 318
  cardiac stomach 318, *318*
  hepatopancreas 318
  midgut gland 314, 318
  pyloric stomach 318, 318
  stomach function 318
crustacean excretory systems 297–303, *300*, **301**, *302*, **303**
  extra-renal mechanisms 301
  nitrogen excretion 303
  system summary 303
crustacean exoskeleton 292–296
  molting 293
  surface setae 293
  basic structure 292
  epicuticle 293
  inner cuticle 293
  molt cycle stages *292*, 294
crustacean joints
  hydraulics and levers 296, *296, 298*
crustacean limbs
  endopod or endopodite 297, *299*
  flexors and extensors 296, *298*
  hinges 296
  protopod *299*
    coxa and basis 297, *299*
  protopod or protopodite 297, *299*
crustacean mechanoreception
  antennae *309*, 310
  Dendrobranchiata 310, *310*
  hair cells 308, *309*
  introduction 308
  statocyst function 309
  statocysts 308
    locations 308, *309*
crustacean molt cycle *292*
  accompanying metabolic changes 294
  hormonal control 294
crustacean photoreception 305
crustacean respiration
  gill anatomy 314, *315*
  water flow in branchial chamber 313
Crustacean sensory modalities
  introduction 305
Crustacean systems
  integument and molting 292

crustacean tail-flip   297
cryptobiosis   278
crystalline cone   305, *306*
Ctenocanthomorpha   677, *678*
ctenoglossate   451
Ctenophora
  bioluminescence   159
  classification history   159
  classification schemes   160–161
  common names   159
  introduction   159
ctenophore foraging
  the beroids   175
  the cestids   175
  the cydippids   173, *174*
  diets, feeding rates, predation impacts   **176**, 178
  feeding specialists   178
  the lobates   *167*, 174, *177*
  the platyctenids   178
  the thalassocalycids   178
ctenophore foraging strategies
  general   173
ctenophore morphology
  auricles   165, *167*
  Beroida   168, 170
  Cestida   168, 169
  Cydippida   160, 162–165, *164*, *165*, *166*
  Ganeshida   171, *172*
  Lobata   165, *167*
  Platyctenida   168, *171*
  Thalassocalycida   172, *173*
ctenophores
  basics   161
  biradial symmetry   163
  digestion   181
  and evolution   186
  gastrovascular canals   163
  global distributions   185, **186, 187**
  as invasive species   179–181, *180*
  locomotion   182–185, *183*, *184*

nerves and sense organs   181
pharyngeal canals   163
pharynx or stomodeum   163
statocyst location   *164*, *165*, 165
stomach or infundibulum   163
tentacular and stomodeal planes   163
Cubomedusae   105, *106*
  feeding   120, *120*
  life history   105
currents
  surface   10
cuttlebone   532
Cuvier   275
cyanobacteria   648
Cyclosquamata
  Aulopiformes
    the lizardfishes   719
Cyclothone
  as zooplankivore   707
  life histories vs depth   707
Cystisoma   305, 408, *409*
  eyes   305

d

dactyl   297, *299*
daily rings   728–730, **729**, 730
dead-end sieve   762
decapods and mysids   **859–862**
  Antarctic   876
    paucity of species   876, 881
  climate and early life history   881
  GOM   876
  NCC   876
deep ocean
  circulation   16
  vertical structure   16
deep-scattering layer   321, 336, *338*
deep-sea anglerfishes
  bioluminescence   751

chin barbel luminescence   *747*, 751
diets   750
esca
  structure and function   *748*, 751
escal luminescence   751
fatal feeding encounters   750
gape and suck feeding   750
in-situ observations   751
jet propulsion   750
locomotion   750
lure   751
non-parasitic males   749
parasitic males   748
parasitic males per female   748
pectoral fin function   750
reproductive strategies
  males   748
sexual dimorphism   747
spawning and development   749
deimatic displays   567
demarsupiation   409
Dendrobranchiata   315, 348, 349
dendrobranchs
  vs carideans, differences in vertical profiles   370
  copulation   357
  cuticular photophores   352
  egg fertilization   357
  egg shedding   357
  larval stages and characteristics   354
  larval vertical distribution   357
  organs of Pesta   352
  recognizing larval stages   355, *355*
  reproductive sequence   357
  seasonality of reproduction   357

depth
  adaptations to   77
  decline in metabolism
    why   80
  decline in metabolism
    with   78
  of field   562
diffusion
  definition   787
diploblastic   194
direct development   92, 95, 215, 389, 409, 582
disphotic zone   32
distal
  definition   297
Distant Neritic Biome   845
Dolichopteryx longipes   702, *703*
doliolids
  basic anatomy   620, *621*
  brain and nervous system   639
  branchial and atrial siphons   620
  budding
    gastrozooids   620–621, *622*
    gonozooids   622–623
    phororozooids   620–622, *622*
  buoyancy   638
  changes with age   620
  development   623, *624*
  dorsal appendix   620, *621*
  duration of reproductive cycle   623, *625*
  feeding filter   620
  field distributions   659
  gill cilia
    rhythmic cessation   641
  jet propulsion
    mechanics   637–638
  larval morphology   623
  life history   620–625
  local blooms   623
  locomotion
    escape response   638
    and feeding   620

normal swimming   637
old nurse vs. zooids   638
muscle bands   620
old nurse   620
reproduction   622
sensory receptors   639–641, *641*
dormant states
  terminology   278
dorsal lamina   610
dorsal process   621
Dosima fascicularis
  float and its manufacture   422
  the pleustonic barnacle   422
Drach
  molt cycle stages   294
drag
  coefficient   8
  friction   7
  pressure   7, *8*
  total   8

e

East Wind Drift   854, 855
ecdysis   *292*, 294
Ecdysozoa   275, **276**, *608*
Echiura   219, *220, 224, 225, 227*
ecoregions
  definition   845, *846*
ecoregions and boundaries
  5 determining factors   845
ectocochleates   525
ectotherm   9, 38
Edward Forbes
  azoic zone   59
eelpouts. *See* zoarcids
Ekman transport   14, *15*
elasmobranchs
  Batoidea-rays   684–691, *690, 692*
  filter feeding structures   691, *692*
  jaw apparatus   760, *760*
  paleohistory   677
  rays
    shared characteristics   684, 690, 691

shark-ray shared characteristics   682–683
sharks
  coastal and oceanic   683–684, **685–687**
  size range   683
sharks vs rays
  species numbers   684
suction feeding   763
urea
  role in osmoregulation   683
vertebral column   760
viviparity
  mechanisms   691
white shark size   683
electivity vs prey spectrum   485
Elopomorpha   677
elopomorphs
  the five orders   698, *698*
  the leptocephalus larva   696–698, *697*
  saccopharynx and eurypharynx   699, *699*
  one-jaw gulpers   699
  snipe eels   699, *699*
endites   297
  role in food manipulation   297
endostyle   605, *606*, 610, *611*
endotherm   9, 799
energetics   *894*
  feeding metabolism   895
  growth (G)   895
  growth pattern   893
  heat increment (HI)   895
  measuring energy usage   893–894
  measuring heat production   894
  metabolism (M)   895
  metabolizable energy (ME)   895
  net energy (N)   895
  O_2 and energy yield   897, **897**
  specific dynamic action   895

energetics (cont'd)
    specific dynamic effect
        (SDE)   895
    urinary losses (U)   895
energy and life histories
    Euphausia pacifica and
        E. superba   935
    GOM, CC, Antarctic   937
energy budget   *894*
    components of   894
    digestible energy (D)   895
    indigestible fraction (F)
        895
    ingested energy (I)   895
enzyme Kcat   50
enzymes
    efficiency and stability
        52–53
ephyrae   101
epimorphic development
    definition   319
epipelagic zone   33
epipodites or exites
    297, *299*
    role in gas exchange   297
Erythropinae   386
esca
    definition   751
eucoelomate body
    plan   *196*, 197
Euphausia pacifica
    energy and life history
        935
    spawning   330
Euphausia superba
    egg-sinking and larval
        development   332
    energy and life history
        935
    spawning   330
    winter feeding   334–335
    winter metabolism drop
        930, *930*
Euphausiacea
    introduction   321
euphausiid photophores
    number and locations
        324, *324*

structure   325–326, 328
wavelengths emitted
    324, 327
euphausiids
    anatomy   322
    bi-lobed eyes   324, 326
    bioluminescence and
        photophores   324
    broadcast spawners vs
        brooders   330
    circulation and
        respiration   326
    classfication   322
    climate and early life
        history   882
    development   328
    developmental sequence
        328, *329*
    digestive system   328
    distribution of genera
        335
    egg buoyancy   332
    excretory system   326
    feeding morphotypes   332
    food and feeding   332
    geographic distribution
        335
    gill structure   326
    gonad development and
        spawning   330
    history of classification
        321
    larval stages   328
    mating   330
    as members of deep scattering
        layer   336
    nervous system   322
    photoreceptors   323–324
        spectral sensitivity
            324, *327*
    pleopods   322, *325*
    seasonality of egg
        production   330
    seasonality of spermatophore
        production   331–332
    sensory mechanisms
        introduction   323
    size and vertical range   321

suspension feeding
    332–334, *334*
vertical distribution and
    vertical migration
        336–340, *337, 338, 339*
euphotic zone   31
eupyrene sperm   477, *478*
eurybathic   419
eurypterids   279
eurythermal   38
Euteleostei   677
    superorders comprising
        700
euthecosomes
    anatomy
        shell morphology
            492, *493*
Evermanellidae
    diet and habits   720
excretion
    definition   785
exopods
    role in swimming   297

f

factorial scope   900
fairy shrimp   283
fangtooths   738
feeding metabolism
    digestive costs   899
    HI and growth   899
fish ammonia excretion
    role of gills   785
fish anatomy
    composite elasmobranch
        758, 760
    composite teleost   758, *759*
fish bioluminescence
    anglerfish luminescent
        bacteria   826
    cookie-cutter shark   828
    countershading with bacteria
        Opisthoproctus
            827–828, *828*
    elasmobranch
        luminescence   828
    hunting with red
        photophores   826, *827*

matching downwelling
light   825
myctophid photophores
primary and
secondary   825
myctophid stern chasers   825
photophore structure
826, *826*
simplest photophore design
gonostomatids   825
sternoptychid photophores
825, *826*
tubeshoulder luminescent
cloud   828
the two types   824
fish blood
ATP and Hb function   779
Bohr effect   778–779, *780*
carbamino compounds
781
carbonic anhydrase
importance   781
CO_2 solubility and
dissociation   780
CO_2 transport summary   782
dissociation curves
types   777
dissolved oxygen levels
775, **777**
hemoglobin structure
777, *778*
high and low affinity Hb's
778
O_2 affinity   777–778
oxygen dissociation curves
777
red cells and Hb function
779
role of hemoglobin   777
Root effect   779, *780*
seasonal affinity changes
779–780
fish camouflage
countershading   824
ultra-black integuments
824
fish chemoreception
gustation   824

importance of olfaction
822
macrosmatic species
822–823, *823*
olfaction
deep-sea sexual
dimorphism   822–823
vs gustation   822
olfactory organ structure
822, *823*
olfactory organs   822, *823*
scent trails   823
fish circulation
and continuity principle
768
secondary circulation
769–770
function   770, *771*
shark pericardium   768
summary   767, *767*
teleost pericardium   768
vascular system
branching and
coalescence   768, *769*
fish CNS
adult brain anatomy   *805*,
806, *808*
brain diencephalon
806–807
brain location and
protection   806
brain mesencephalon   807
brain metencephalon   807
brain myelencephalon
807
brain telencephalon   806
brain ventricles   *805*, 806
cerebellum   807
chemical synapses
804–806, *805*
cranial and spinal nerves
806
cranial nerves I-X   807
definition   806
embryonic fish brain   806
Mauthner cells   807
medulla oblongata   807
spinal cord   809

grey and white matter
809, *810*
segments   809
the 10 cranial nerves   806
fish digestion
4 sections of tract   765, *765*
esophagus   765
intestinal enzymes   766
intestinal length   766
microphagous feeders
765–766
midgut   766
nutrient absorption   766
peristalsis   765
pyloric caecae   766
pyloric valve   765
spiral valve   766
stomach   765
fish electroreception   820
ampullae sensitivity   822
ampullae structure and
function   821, *821*
ampullae of Lorenzini   821
elasmobranchs   820
in freshwater species   822
fish feeding
biters
morphology   763
fish gills
arches and subunits
771, *772*
lamellar structure   773, *772*
multifunctionality   772
fish hearing
clupeids   820
using otoliths only   *819*, 820
using swimbladder
resonance   *819*, 820
Weberian apparatus   *819*, 820
fish hearts
location and structure
768, *769*
fish history
Conodonta   674, *674*
fish lice   283
fish locomotion
cost of transport
definition   797, *798*

fish locomotion (cont'd)
  endothermy
    function 799
    tunas and lamnid
      sharks 799
  gross cost of transport 797
  heterothermy
    species efficiencies 799
  minimizing friction drag 797
  minimum cost of
    transport 797, 798
  muscle and drag 796
  muscle fiber distribution
    795–796
  muscle fiber innervation
    796, 798
  muscle requirements 793
  myotomes
    structure and function 793
  net cost of transport 797
  opah
    whole body endothermy
      799
  pressure drag
    definition 796
  red and white muscle
    function 795
  regional endothermy
    how achieved 799
  scombroids
    great lateral tendon
      793–795, 795
  skin friction drag
    definition and
      importance 796
  the 5 locomotory
    types 793, 794
  the two types of drag 796
fish mechanoreception
  ampulla 818
  aquatic sound movement
    815
  bathypelagic fish
    neuromasts 817
  deep-sea prey detection
    818
  elasmobranch otoconia
    820
  free-standing
    neuromasts 815
  hair cells
    innervation 815, 816
  inner ear structure 818, 819
  lateral line 815
  lateral line canals
    817–818, 817
  lateral line sensitivity 818
  neuromasts 815, 816
  semicircular canals 818
  teleost otolith organs
    818–819, 819
  vertebrate hair cell 815, 816
fish nervous system
  neuron structure 804, 805
  peripheral nervous system
    (PNS) 806
fish osmoregulation
  elasmobranchs
    hydromineral balancing
      system 790
    role of the teleost
      esophagus 791
    role of the teleost gut
      summary function 791
    role of the teleost
      intestine 793
fish osmotic regulation
  elasmobranchs
    basic strategy 787
    gills 790
    rectal gland 790
    systems involved 787
    the basic problem 790
    urine formation 790
    role of body wall 789
    role of drinking 789, 789
    role of skin and scales
      787–789, 788
    role of the gills 790
    role of the kidneys 789
    teleost kidneys 789
  teleosts
    systems involved 787
fish photoreception
  adjusting spectral
    quality 812–813
  advantage of tubular
    eyes 813, 814
  all-rod retinas 813
  bathypelagic fishes 815
  binocular orientation
    813, 814
  choroid and retinal
    oxygen 811, 811
  color vision 811–812
  contractile iris
    teleosts and
      elasmobranchs 812
  day and night visiom
    812, 812
  light and depth 813
  multibank retinas 813, 814
  photoreactive pignents 811
  retinal adaptations
    deep-sea fishes 813
  spherical lenses 813, 814
  tapetum lucidum 721,
    813, 815
fish PNS
  somatic sensory neurons 806
  visceral sensory neurons 806
fish respiration
  5 governing factors 771
  buccal pumping 773, 774
  counter-current
    exchange 774
  gill surface area (GSA)
    774, **776**
  ram ventilation 774
  ram ventilator gills
    774, 775
fish sensory mechanisms
  senses and introduction
    809–810
fish swimbladders 784
  gas addition 783
  gas retention
    role of the rete 783
  rete and gas gland 784
  size and structure 782–783
  types 782
  volume and pressure 783
fishes
  benthic 671

benthopelagic   671
deep-sea colonization   672
demersal   672
epicercal tail
    definition   674
jaws
    origin and importance   676
    living ancient groups   676
paired fins
    importance   676
teleosts
    hypothetical lineage   676–677, 676
    total species   669
flame bulbs   209
flying fishes
    as prey   746
    habitats and gliding   746
fovea   562, 565
fresh water-% of total   2
furcilia   328
fusiform
    definition   797

g

Gadiformes
    Bregmacerotids
        codlets   734, *734*
    the cods
        importance   734
        Macrouridae and Moridae   734
        pelagic   *734*, 735
gastropod
    diet diversity   451
gastropod alimentary canal
    basic anatomy in toto   451, *454*
    crystalline style   452–453
    food conduction   452, *452*
    midgut gland   452, *454*
    radular apparatus   451
    salivary glands   452
gastropod circulation
    general   455, *455*
gastropod excretion
    kidney structure and function   464, *465*
    modification of ultrafiltrate   464
    ultrafiltration   462–464
        role of the auricle   462–464
gastropod nervous systems
    comparison
        generic vs actual anatomy   467, *468*
        the gymnosomes   469
        the heteropod Pterotrachea   469
        role of the major ganglia   467
        shelled pteropods   469
    torted gastropods
        Janthina   466, *468*
        the visceral loop   467
gastropod radula
    basic structure and function   452, *452*
    teeth   452, *453*
gastropod respiration
    gills and gill morphology
        janthinids   457
        heteropod gills   457, *459, 460*
        monopectinate or pectinibranch gills   457, *458*
        naked pteropods   457, *462*
        nudibranchs
            cerata   459–461, *463, 464*
            Phylliroidae   460–461, *464*
        pteropod gills   457, *461*
        water circulation at gill surface   457, 458
gastropod sensory mechanisms   469
    cephalic tentacles   469, *472*
    chemoreception
        osphradium   471
        Janthina   456, 471,
    photoreception
        heteropods   472
        pteropods   473, *474*
        pteropods and nudibranchs   472
    statocysts
        heteropods   472
gastropod shells
    pelagic forms   466
    presence in pelagic taxa   464
    shell formation   466
    shell structure   464–466, *466*
gastropod systems and structures   451
Gaussia princeps
    vertical distribution and bioluminescence   424
Gempylidae
    snake mackerels   740
gempylids
    distribution and diet   740
    two body morphs   740
generator potential   810–811
geostrophic currents   11, 15
germ layers and body cavities   194
Gigantocypris   284, *422*
    swimming   421
    unusual size   421
    vertical and geographic distribution   421
Giganturidae
    developmental transformation   720
gill bars   610
gill rakers
    and suspension feeding   761
Glycera   250
Gnathophausia ingens   72
    accumulation of matter and energy   390
    size frequencies   390, *391*
gonocoel   549
gonopores   351
gonostomatids
    4 body morphs   705–707, *706, 707*
    Cyclothone
        distribution and morphology   705
        sex change   707
        vertical distribution   707

Great Ocean Conveyor
    (AMOC)   10, *12,* 16–24
great white shark. *See*
    elasmobranchs
Gulf of Mexico (GOM)
  description   362
  mesopelagic vs bathypelagic
    decapods
    362, 363
  physical characteristics
    846
  physical differences
    mesopelagic vs
      bathypelagic   362
  primary and secondary
    production   847
gymnosomes
  alimentary canal   507
  buccal apparatus
    506–508, *507, 508, 509*
  chromatophores   506
  hatching and development
    513, *514*
  Hydromyloidea
    brooding   513
    global distribution   513
  locomotion
    basics   510
    fast and slow
      swimming   510
    swimming
      kinematics   510, *511*
  prey capture   507, *509*
  prey handling and
    digestion   507, *509*
  reproduction
    copulation   512
    egg production
      510–512, *512*
    protandry   510
    seasonality   513
    sperm production   512
  selective feeding   508

h

hadal zone   670
hadopelagic   33, 670
  defnition   670
Haeckel   89, 128, 137

hagfishes   *680*
  blood ion concentration   679
  feeding   679
  notochord   679
  reproduction   679
  slime production   679
  vertical distribution   679
hair cells   557, *558*
hammerjaws   722
Hartline   305
hemocyanin   311, 544
  function   311, 317
heteronomous
    segmentation   236
heteropod radula
  structure and function   452
heteropods
  abundance
    general   487, *488,* **489, 490**
    seasonal variation   487
  abundance vs other taxa   487
  Atlantidae   477
  atlantids
    shell and size   477
  buoyancy
    carinariids and
      pterotracheids   483
  carinariids
    genera   478
    shell morphology   478
    size and general
      anatomy   478
  egg production and egg
    strings   490–491
  feeding chronology
    485–487
  fertilization   490
  global distributions   490
  larvae   491
  as a legacy name   477
  locomotion
    atlantid buoyancy   482
    atlantid swimming   480
    carinariids and
      pterotracheids   483
    role of the fin   480
  prey capture
    atlantids   484
    carinariids   484

  prey preferences   484
    carinariids   484
    pterotracheids   485
  pterotracheids
    general anatomy
      480, *481*
    seasonality of
      reproduction   491
    secondary sexual
      chacteristics   490
    vertical distributions
      atlantids   487
      general   487
heterotherm   9
Holocephali   677
  families and species
    691–693, *693*
Holostei   677
Homarus americanus   279
homeostatic control   785
homeothermy   38
homeoviscosity   56
homonomous segmentation
    236
Hybodonta   677, *678*
hydroids on Georges Bank   119
Hydrolagus affinis   678, *678*
hydromedusae
  exumbrellar and subumbrellar
    surfaces   92–93, *93*
  morphology   92, *93, 94, 95*
  stomach   *93,* 94
hydrostatic skeletons
    196, *199*
hyoid arch   758, *760*
hyperbenthos   396
hyperosmotic regulator
  definition   787
hyposmotic regulator
  definition   787
hydromedusae
  radial and ring canals
    93–94

i

Ice-Shelf Water   854
Ichthyococcus   709
Idiacanthinae
  blackdragons   714, *715*

geographic distribution 714
Idiacanthus
　diet and vertical
　　　distribution 714–715
　life history 715
　male-female dichotomy 715
illicium
　definition 751
independent effectors 107
indirect calorimetry 898
ingested energy (I) 894
　cabohydrate digestibility
　　896
　lipid digestibility 896
　protein digestibility 896
interbrachial webbing 540
inverse retina 564, 565
iridophores 565–566
ischium 297, 299
Ischyodus 678, 678
isolume
　definition 324, 337
Istiophoriformes
　the billfishes
　　introduction 743
iteroparity and semelparity
　215
Ivlev electivity index 485

j

jacks. *See* carangids
janthinid radula
　structure and function
　　452
jawless fishes 678
jellynose fishes 719

k

Kelvin degrees 42
Kings of the Sea 711
kinocilium 815, 816

l

labeled line principle 810
Lamarck 90, 221, 275
Lambert's law 31
lampreys 679, 681
　ammocoete larvae 682
　invasion of great lakes 681

paleohistory 679
resemblance to jawed
　　fishes 679
sea lamprey 680
spawning and early
　life 681, 682
Lampriformes
　body morphs 731
Lamprimorpha
　Lampriformes
　　Opahs and Oarfishes
　　　731–732, 732
lancelets 603, 604
Larvacea
　as legacy name for
　　Appendicularia 625
Lepisosteiformes 676, 677
leptocephali 696, 697
　developmental strategy 697
　diet 697
　duration as larvae 696–697
　phase I
　　accumulation of
　　　energy 697
　phase II of development 697
　proteoglycans and
　　locomotion 697
　two phases of
　　development 697
Leptomedusae 96
Leptostraca 284, 285
Leuckart 90, 160, 275
leucophores 565
life history traits 893
light 29, 30
　absorption and
　　scattering 31, 31
　variability with depth 31
Limnomedusae 96
Linneaeus' Systema
　Naturae 221, 273
liparids
　as hadal species
　　752–753, 752
　geographic and vertical
　　distribution 753
lipids and temperature 53
lizardfishes 719
Loop Current

description and
　importance 846
Lophiiformes
　anglerfishes and frogfishes
　　introduction 746–747
　deep-sea anglerfishes
　　morphology
　　　747, 746–749
lophogastrid eyes
　spectral sensitivity 386
lophogastrids
　general characteristics 381
lophogastrids and mysids
　anatomy 382, 383, 384
　abdomen 384–385
　carapace 382
　thoracic limbs 382
　bioluminescence of
　　Gnathophausia
　　　ingens 386
　circulation and respiration
　　386–387, 388
　classification 382
　classification history 381
　digestive system 387
　　alimentary canal 387, 389
　　digestive ceca 387–388,
　　　389
　　nutrient absorption 389
　excretory systems 387
　eye morphology 385
　　changes with depth
　　　385–386
　larval stages 389
　life histories **392**
　　coastal mysids 389
　　Gnathophausia
　　　ingens 390, 391
　lophogastrid diets
　　392–393
　lophogastrid swimming
　　394, 394, 395
　mesopelagic vs bathypelagic
　　importance 396
　mysid diets 392
　mysid swimming 394, 394
　nervous system 385, 385
　reproduction and
　　development 389

lophogastrids and mysids (*cont'd*)
   sensory mechanisms 385
   shallow vs deep life
      histories 390, **392**
luminous roughies 740

m

macrozooplankton
   definition 89
macula
   definition 559
Malacosteinae
   diets
      GOM and Hawaii
         718–719
      feeding mechanics
         717, *718*
      loosejaws 715
manta rays
   feeding 691, *692*
   horns 691, *692*
manubrium 99
Marjorie Courtenay-Latimer
   693
marsupium 281
mass-specific metabolism
   vs mass 901–902
Mathiessen's ratio
   definition 562
Maurolycinae
   pearlsides
      distribution
         709
maxillary glands 297
mechanoreception
   definition and included
      senses 557
Meckel's cartilage 758
medusae
   diets and feeding rates 117
   feeding efficiency 115
   foraging strategies 105
   locomotion 121
   mesoglea 123
   nerve nets and
      swimming 124
   senses and sensory
      mechanisms 128
   swimming and hunting
      behavior 115
   swimming muscles 121
Meganyctiphanes norvegica
   spawning 330
Melamphaidae *736*
   age and reproduction 737
   the bigscales 736
   diets 736–737
   size range and
      abundance 736
   vertical migrations 736
Melanostigma pammelas
   756, *757*
Melanostomiinae
   scaleless black dragonfishes
      715, *716*
   species and geographic
      distribution 715
   vertical distribution and
      diet 715
membranes
   desaturases 56
   fluidity 54
   primer 54, *55*
   short term change 56
meristic characteristics 758
merus 297, *299*
mesial
   definition 297
mesopelagic decapods
   vertical migration
      patterns *364*
mesopelagic ecoregions
   845, *846*
mesopelagic fish communities
   dominant taxa 700–701
mesopelagic fish growth
   annular vs daily rings 936
   GOM vs Antarctic 936
mesopelagic fish life histories
   GOM vs Antarctic 936
   Sigmops elongatum 937
mesopelagic fish swimming
   lethargy and vertical
      orientation 801
   locomotory types 801
mesopelagic fishes
   calories and depth 934
   composition and depth
      water levels 934
   growth measurements 936
   metabolic decline with
      depth 933, *933*
   metabolism 933
      regional similarities 933
   proximate composition
      934
mesopelagic zone 33, *33*
mesoscale
   definition 659
metabolic cold adaptation
   "MCA" 44
metabolic scaling
   b values
      explanations for 901–902
   mammals 901
   mass vs surface area 901
metabolic scope 425
metabolic suppression 74
metabolism
   and activity. *See* activity
      metabolism
   biotic vs abiotic
      influences 899
   feeding metabolism
      elements of. *See* feeding
         metabolism
   influence of size (mass)
      901
   J.R. Brett
      standard and basal 900
   vs mass (equation) 901
   metabolic and factorial
      scope 900
   and nutritional state 899
   and organismal
      requirements 898
   routine metabolism
      900–901
   units 898
metabolism vs mass
   b values 901
metagenesis
   definition 89
metagenetic life history 613
metameric segmentation 197
metamorphic development
   definition 319
metanephridium 241, 243, *243*

metanephromixium   241
Michaelis constant
    "Km"   47
Michaelis-Menten kinetics   51
microbial loop   648
    importance of
        picoplankton   648
micronekton
    definition   319
micronektonic Crustacea
    introduction   319
micronektonic crustaceans
    trophic position   319–320
microphagous feeding
    definition   765–766
Milne-Edwards   321
    role in gastropod
        classification   445
minimum cost of transport   635
Mnemiopsis in the
    Black Sea   179
modern deep-sea fish fauna
    how it came to be   673
molas
    locomotion, diet,
        distribution   752
    ocean sunfishes
        size and habits   752
molecular phylogenetics   222,
    605–607, **609**
Mollusca
    Class Bivalvia-the
        bivalves   444
    Class Caudofoveata
        441, 442
    Class Cephalopoda-
        squids and octopods   445
    classes illustrated   445, *449*
    Class Gastropoda-
        snails and sea
        butterflies   445
    Class Monoplacophora
        442, *443*
    Class Polyplacophora-
        the chitons   442, *443*
    Class Scaphopoda-
        the tooth shells   443, *444*
    Class Solenogastres
        441, 442

Introduction   439
origin of name   440
molluscan circulation
    arteries   455, *455*
    ctenidium   *455, 456*, 457,
    general   455
    pericardial cavity   455
    sinuses
        return flow to the
            heart   455–457, *456*
    systemic heart   455, *455*
molluscan coelom   455
molluscan excretion
    kidney or nephridium
        461, *465*
molluscan nervous systems
    basic pattern   446, *448*
molluscs
    basic body plan   439
    importance to
        neurophysiology
            439
    success as a phylum   439
molt frequency
    influence of external
        stimuli   295
molting hormone   294
molt-inhibiting hormone   294
Monoplacophora   440
morphometric characteristics
    758
Myctophidae
    distribution and importance
        724
myctophids
    age and growth   728
    Antarctic   883
    batch spawning   730
    countershading   727
    diel vertical migration
        726, *728*
    GOM   882
    growth
        Antarctic vs GOM
            729, **729**
        growth vs epipelagics
            729–730, *730*
    morphology   725, *725*,
        726, 727

NCC   882
photophore function   727
photophore patterns
    725, *725*
photophores vs depth
    726, 727
spawning
    regional variablity   731
    vertical distributions   726
zooplanktivory   727
Myllokunmingia fengjiaoa   674
myosepta   793
mysids
    general characteristics   381

n

nanoplankton
    definition   647–648, **648**
Narcomedusae   97–98, *99*
NCC
    downwelling   849
    oceanographic seasons
        849
    oxygen and vertical
        ranges   852
    oxygen minimum   851, *851*
    primary productivity   849
    shelf circulation and vertical
        migrators   849
    upwelling   849
Nectonemertes mirabilis   939
nematocysts   107
Nemertea
    classification   200
    classification history   193
    evolutionary importance   194
    introduction   192
nemertean morphology
    proboscis apparatus   202
nemerteans
    blood   209
    circulatory system   207, *210*
    color   205
    development   215
    digestion   213
    digestive systems
        211, *212*
    excretory systems
        209, *211*

nemerteans (cont'd)
  foraging strategies 215
  nervous system 205, 206
  pelagic body form and locomotion 204, 205
  reproduction 213
  reproductive behavior 214
  reproductive organs 213, *214*
  sense organs 207, 208
  vertical and geographic distributions 215–216, *216*
Neopilina galathea
  discovery and importance 440–441, *443*
Neoscopelidae
  morphology and relationships 723–724, *724*
Neoscopelus 723
neoteny 625
neurocranium
  definition 758
neuropodium *234*, 237
nitrogen excretion
  forms of excreted N 785
  sources of excreted N 785
normal curve 42
North Atlantic Deep Water 20
northern California Current (NCC) 847
  physical characteristics 847–848, *848*
notacanthiformes *698*, 699
notochaetal fans 256
notochord 603
notopodium *225, 234*, 237, *240*
Notosudidae
  vertical distribution 720
  visual adaptations 720, *721*
nudibranchs,
  Cephalopyge
    anatomy and swimming. See nudibranchs: Phylliroidae
    feeding 521–522
    reproduction 522
  cerata
    cnidosacs 516, 518, *519*

external anatomy summary 516
Fiona *520, 522*
  buoyancy and diet 521
  mating and fertilization 521
  rapid growth 521
  reproductive tract *520*, 521
  size and feeding 518–520
Fionidae
  as members of the pleuston 518
glaucids
  habits and diet 526
  stinging ability 516
Glaucus *519*
  buoyancy and coloration 518
  copulation, fertilization, and spawning 518
  diet 516
  translocation of nematocysts 516–518
  hermaphroditism 518
  pelagic families 516
Phylliroe *523*
  diet 522
  mating and spawning 522
  reproductive system 522
  and Zanclea 522, *523*, *524*
Phylliroidae
  luminescence and distribution 524, *525*
  species 521
phylliroids
  morphology and swimming 521–522
  reproductive system 522
  swimming vs creeping 516

O
$\delta^{18}O$ 505
O:N ratio 898
ocean circulation 10
ocean gyres 15

ocean paleohistory 672–673, *673*
ocean surface area 2
ocean volume 2
oceanic biomes 845
oceans and ocean basins 9
odontophore 451, 452, *452*
oikoblastic epithelium 629, *630*
Olfactores 669
olfactory pit 557, *561*
oligopyrene sperm 477, *478*
ommatidia 305, *306*
  lens or cornea 305
  retinular cells 305
  screening pigment 305
Omosudidae
  distribution and diet 722, *722*
oophagy 691
opah
  endothermy 799
open circulatory system
  flow 313
opisthobranchs
  definition 445
orange roughy 740
Order Tetraodontiformes
  puffers and molas 752
orders Lophogastrida and Mysida
  introduction 380
  organs of Pesta 352
Osmeriformes 702
  alepocephalids (slickheads) 702, 703
  Bathylagidae (deep-sea smelts) 703, 704
  mesopelagic families 702
  opisthoproctids (spookfishes) 702, *703*
  platytroctids (tubeshoulders) 705, *705*
Osmeromorpha
  mesopelagic importance 702
osmoconformity 75–76
osmole
  definition 787

osmosis
  definition 787
osmotic pressure
  definition 787
Osteoglossomorpha 677, 696
Osteostracomorphs 674, *675*
ostia 311
ostracoderms 674
Otocephala 677
  groups included 677
  structures behind name 677
otoconia 820
ovoviviparity 691
oxycalorific equivalent 898
oxygen 24
oxygen minima 24
  aerobic adaptations 71–72
  and dead zones 71
  formation 69
  vs intertidal 71
  and Pc 71–72
  residents 69

P

P$_{50}$
  definition 777–778
packhorse lobster 278
Paleonisciformes *676*, 677
Panarthropoda 275, **276**
Panarthropodan phyla
  Onychophora 277
  Tardigrada 277–278
paper nautilus 537–540, *539*
PAR 29
Paracanthopterygii 700
  taxonomic composition 732
Paralepididae
  barracudinas 722
  distribution and diet 722
parapodia 225, *225*
parturial molt 357, 358, 408, 411
Pc
  definition 71

pectoral-fin propulsion 750
pedalium 105, *120*
pelagic 1
pelagic Crustacea
  superorder Peracarida 380
  swimming limb structure and function 380–381, *380*
pelagic crustaceans
  composition and depth
    Antarctic 931
    CC 931
    GOM 931
  metabolic decline with depth 929
  metabolism 929
  overwintering strategies 932
  regional metabolic differences 929–930
pelagic decapods *350*
  anatomy 350
    abdomen 351
    secondary sexual characteristics 351
    thoracic limbs 350–351
  arterial circulation 354
  bathypelagic vs mesopelagic diets 377
  bioluminescence and photophores 352
  branchial chamber 350
  circulatory system 354
  classification 349
  countershading 353
  development 354–357, *355, 356*
  diet overlap and niche separation 376
  feeding guilds 370
  food and feeding 370
  functions of bioluminescence 353
  geographic distribution 377
  history of classification 348
  luminescent types 352

luminous secretions 352–353
marine snow in diet 377
mechano-and chemo-sensory abilities 351
morphological characteristics 346
nervous system 351
photophore patterns 353
principal prey items 370, **371–375**
reproduction 357
sensitivity to UV 352
sensory mechanisms 351–352
spectral sensitivity of eyes 352
statocysts 351
tail fan 351
two major subdivisions 348
vertical distribution patterns (Omori) 361
vertical distributions 361
pelagic fish buoyancy
  lowering density 803, *804*
  swimbladder 801–802
  use of hydrofoils 802
  use of lipids 802, **803**
pelagic gastropods
  heteropods. *See* heteropods
  janthinids
    egg capsules 477, *478*
    external anatomy 475, *476*
    flotation 475
    hermaphroditism 476–477
    larval development 477
    mechanics of reproduction 477
    novel sperm 477, *478*
    oceanic distribution 475
    prey and prey capture 475
  taxa represented 441
pelagic polychaetes
  excretory organs 243
  new species 228

pelagic red crabs 378
    affinity with oxygen
        minima 379
    benthic vs pelagic habits 379
    classification 378
    feeding habits 379
    global distribution 378
    reproduction 379
pelagic sampling gear
    IKMT (Isaacs-Kidd Midwater
        Trawl) 702
    MOCNESS 702
    Tucker trawl 702
pelagic sampling programs 700
pelagic tunicates
    major groups 603–604
Peracarida
    number of marine
        species 380
    pelagic orders 380
    universal characteristics
        380
Perciformes
    introduction 753
    Percoidei
        Caristiidae 755, *755*
        Coryphaenidae 756, *756*
    Trachinoidei
        Chiasmodontidae
            757–758, *757*
    Zoarcoidei
        Zoarcidae 756–757,
            756
Percoidei
    Carangidae-the jacks
        753, *754*
pereon 405
pereopods 350
    definition 297
peripharyngeal bands *613*, 614,
    620, 626, *646*,
peristomial cirri *226, 233,* 236
peristomium 236
peritrophic membrane
    387–388
petasma 330, 351
phantom bottom 336
pharyngeal basket 610, *611,*
    632, 645

pharyngeal jaws 764, *764*
    and diet 764
    description 764
Pholidophoriformes
    676, 677
phosichthyids *709*
    distribution and diets 709
Photostomias guernei 715–719
    vertical distribution 717
phragmocone 525, 526, *527*
Phronima eyes 305–306, *307*
phyllobranchiate
    315–316, *317*
Phyllodocida 228
phylogenetic systematics 605
Phylum Annelida 224
    Class Clitellata 226
    Subclass Hirudinea 228
    Subclass Oligochaeta 227
physostomous swimbladders
    782
Piccard
    Auguste and Jacques 60
picoplankton
    definition 648
    importance in oligotrophic
        systems 648
placental viviparity 691
Platynereis megalops
    mating dance 254
platytroctids
    distribution, bioluminescence
        705, *705*
Pleocyemata 287, 348, 349
pleopods 351
plesiomorphies and
    apomorphies 223
pleura 403
pleurobranch 314–315, *315*
plicate gills
    definition 457, *461*
pneumatic duct 765, 782
podomere
    definition 296, 297, *299*
podopericardial channel
    404, 407
Poeobius *232*, 259
    affinity with Pacific
        sub-arctic 259

poikilothermy 38–39
Polar Biome 845
polychaete
    bioluminescence 267
    body wall 239
    circulatory systems 246
        pelagic species 249
    coelom 239
    cuticle 239
    external anatomy 233, *235*
    development
        Tomopteris 255
    digestive systems 250
    epidermis 239
    excretory systems 241
    foraging strategies 256
        hunters 256
    gas exchange 249
        pelagic species 249–250
    geographic distributions
        259, **260–266**
        cosmpolitan 259
    gut morphology 250
    head region *226,* 234–236
    hearts 247
    hydrostatic skeleton 240
    internal anatomy 239
    locomotion 255
        role of muscle 255
        role of parapodia
            255–256
    mechanoreceptors,
        chemoreceptors
        244, *246*
    metameres 237
    nervous systems 244, *245*
    pelagic 228
    photoreceptors
        245–246, *247*
    proboscides
        role in feeding 250
    protonephridium
        241
    respiratory pigments
        247–249
    sense organs-general 244
    sites of luminous tissue 268
    supraesophogeal ganglion
        244

trunk or metastomial
region 236
ventral nerve cord 244, *245*
vertical distributions
267, *268*
vertical range **264–266**, 267
polychaete diets
Alicopini 258
Loporrhynchidae 258
Tomopterids 258
polychaete foraging strategies
suspension feeders 258
polychaete nervous systems
giant fibers 244
polychaete photoreceptors
spectral sensitivity
245–246
polychaete reproduction
251
behavior 251
epitoky 252, *253*
hermaphroditism 251
lunar influence 252
pelagic species
Aliciopini and
Lopadorrhynchidae
254
Tomopteris 254
synchronicity 252
Polyipnus *710*
vertical distribution 711
Polypteriformes 676, 677
population
definition 845
pressure
and depth 25
effects on nerves 61
effects on pelagic species 63
effects on rates 61
and gas-filled spaces 59
Mariana Trench 59
and survival 60
pressure adaptation
molecular mechanisms 64
pressure effects
on enzyme systems 65
on membranes 68
pressure research
history 60

primary consumers
definition 320
primary growth rings 729
principal life history traits 575
prochlorophytes 648, *652*
proecdysis 294
proostracum 526–527
propodus 297
prosobranchs
definition 445
prostomium 233
proteins
as N source 896
protonephridia 211, *211,*
241, *242*
protonephromixium
241, *242*
proximal
definition 297
proximate composition
definition 931
pseudocoelomate body
plan 195–197, *196*
pseudoconch 492–494
pseudothecosomes
anatomy
differences with
euthecosomes 492
Cymbulidae
anatomy 493
Desmopteridae
anatomy 494–495, *497*
Peraclidae
anatomy 493
pteropods
biogeography 513
classification 492
as climatic indicators 505
copulation
spermatophore tranfer
502–503
diversity with latitude
513–515
egg masses 503
feeding *501–502*
field observations 500
gut contents 500–501
historical descriptions 500
gymnosomes

anatomy 506
hermaphroditism
transition to female
503, *503*
history of classification
491
Limacina retroversa
annual reproductive
cycle 505
limacinids
male anatomy 502
locomotion
butterfly resemblance
495, *498*
cavolinioids 496–497
Limacina wing
movement 496
limacinids 495
pseudothecosomes
497, *499*
swim speeds 497
and ocean pH 505
and paleotemperature
505
reproduction
cavolinioids and
pseudothecosomes
504
seasonality 504
reproduction and
development
thecosomes 502
reproductive anatomy
Limacina 502–503, *503*
suborders and characteristics
491–492
vertical distributions
515–516
vertical migration
euthecosomes 515
viviparity 503
pteropods, worms, and
chordates
metabolism 938
pterygoquadrate 758
pulmonates
definition 445–457
pycnocline 16
pygidium 233, *235*

pyrosomes
  bioluminescence
    characteristics 635, 639
    light organ 639
  blastozooids 611–612
  ciliary arrest potential 639
  development 613
  feeding 611
  as hermaphrodites 612–613, 613
  locomotion 611, 634–635
  locomotory mechanisms 634–635
  nervous system
    function 639
    structure 639, 640
  oozooids and cyathozooids 611–612
  reproduction 611–613
  vertical migration 635, 659

q
Q_{10} approximation 41

r
rabbitfish 691, 693–694
ram feeding
  definition 760–761
  ram capture 762
  suspension feeding 761, 762
ratfish 691, 693–694
ray-finned fishes 695
rays
  diversity in body form 684, 690, 690
  pelagic 690–691, 692
rays and sharks
  filter feeding structures 692
receptaculum seminalis 518, 520
receptor potential 810
Regalecus glesne
  oarfish 731, 732
Regne Animale 275
reproductive patterns
  GOM and Antarctic fishes 937

respiratory cells 790
respiratory quotient 897, **897**
rete mirabile
  definition 783, 784
retinomotor responses
  definition 812, 812
retinula cells
  cephalopod 563–565, 564
Reynolds number 6
rhabdome 305
Rhizostomae
  oral arms 104, 104
rhodopsin 811
rhopalia 125, 126
roughies 740

s
sabertooth fishes 720
salinity 1
  adaptations to 75–76
  elasmobranch adaptations 76–77
  hypo-regulation 76, 77
salp chains 615, 616
salp morphs
  internal anatomy 614, 618
salps
  aggregate or blastozooid morph 613–614
  alternation of generations 613
  body wall 615
  brain and nervous system 642–643, 642
  development 617
  differential migratory behavior 656
  epithelial conduction, OSP's 643
  eye and photoreception 642, 643
  feeding web
    mesh size 614
  gill bar and endostyle 614
  heart, brain, and eye 615
  jet propulsion
    basics 635
    efficiency 637, **637**
    mechanics 636–637, 636

life history 615–618, 619
locomotion
  aggregate swimming 616, 636
  locomotory rhythm
    origination 642–643, 642
  mechanoreception 641, 643
  migratory behavior and field conditions 656
  mucous feeding web 614
  neural and epithelial interaction 641–643
  peripharyngeal band 614
  as prey 654
  protozoan parasites 654
  sexual reproduction 616–617
  solitary or oozooid morph 613–614
  stolon 615, 618, 619
  swim speeds 635
  swimming agility 635
  vertical migration 656
sardines and herrings
  regional abundance 700, 701
scaphognathite 314, 316
sclerotization 293
Scombridae
  mackerels and tunas
    introduction 741
  subfamily Gasterochismatinae 741
  subfamily Scombrinae
    four tribes 741
Scombriformes
  introduction 740
Scombrinae
  diets and vertical distributions 741
  tribe Thunnini
    heterothermy 741–743
scope for metabolism 900
Scopelarchidae
  pearleyes 722
  visual adaptations 722–723, 723

Scopelomorpha
  Myctophiformes
    lanternfishes and blackchins   723
Scorpaeniformes
  Liparidae
    snailfishes   752–753, *752*
scyphistoma   99
Scyphomedusae   99
  vs Hydromedusae   99
  tetramerous symmetry   99
Scyphozoa
  canals   99
  morphology and life histories   101
  strobilation   101, *103*
sea wasp   120
secondary consumers
  definition   320
secondary sense cells
  definition   815
seed shrimps   284
Semaeostomae   102
semelparous   390, 576, 937
semi-permeable membrane
  definition   787
Sergestes similis   341, 357, 361, 876
sessile
  definition   305, 962, *290*
sharks
  blue and tiger
    pups   684
  coastal
    nurseries   684
    and oceanic   683
  declining numbers   684
  early life histories   684
  gestation periods   684
  growth and sexual maturity   684
  livebearers
    pups   684
  oceanic whitetip
    pups   684
  and rays. See elasmobranchs
siphonophore feeding
  digestion   139

fishing behavior   138–139, *138*
siphonophore pneumatophores
  gas composition   133–134
siphonophores
  buoyancy   143–144
  diets and selectivity   139
  diurnal vertical migration   146–147
  ecological importance   141
  epithelial conduction   147–148
  geographic distribution   147
  introduction   127
  life histories   137, *137*
  locomotion   141–143, **144**
  nectophores and bracts   131
  organization   134–136, *135, 136*
  sensory mechanisms   147
  terminology   127
  vertical distribution   144–146, **146**
  zooids   128–134, *131–134*
siphuncle   571, *572*
Sipuncula   108, 110
Skinner
  Dr. Dorothy   293
snailfishes. *See* liparids
Sorberacea   607
sound   26
  propagation   26–27
  speed in water and air   28
  variability with depth   28
Southern Ocean   9
  productivity   855–856
specific heat   2, 894, 962
spermatophores   330
spermatozeugma   477
spherical lenses
  properties and importance   562, *563*
spinyfins   740
Spiralia   275, *276*
squid giant axon   439, 555, 557
standard metabolism
  definition   900

Standard Station   846–847, *847*
statoconia   557
stem and crown groups   670, *671*
stenothermal   38
Stephanoberycoidei
  Barbourissiidae, Cetomimidae, Rondeletiidae   737, *737*
stereocilia   815
stern-chasers   825
Sternoptychidae
  Maurolycinae (pearlsides)   709, *709*
  Sternoptychinae
    hatchetfishes   709–710, *710*
Sternoptychinae
  three genera   710
stigmata   **606**, 610, *611*, 641
Stomatopoda   284
Stomiidae
  Astronesthinae
    snaggletooths   712
stomiids
  geographic distribution   712
  identification   711
  suborbital photophores
    prey sighting   711
Stomiiformes
  Gonostomatidae (bristlemouths), 705–709, *706, 707*
  importance   705
  Phosichthyidae   709
  Sternoptychidae   709
  Stomiidae   711
    the barbeled dragonfishes   711
    Chauliodontinae   712, *714*
    Idiacanthinae   714, 715
    Malacosteinae   715, 717
    Melanostomiinae   715, 716
    Stomiinae   718, 719
Stomiinae
  distribution and diet   719

sturgeons
   maturity and length of
      life   696
   reproduction and
      fecundity   696
   skeleton   696
   vulnerability   696
Stylephoriformes
   Stylephorus chordatus
      732, *733*
Stylephorus
   unique feeding
      mechanism   733
   vision   733
Subclass Chondrostei
   sturgeons and paddlefishes
      696
Subclass Holostei
   gars and bowfins   696
Subclass Teleostei
   the bony fishes   696
Suborder Stromateoidei
   medusafishes and
      squaretails   743
subphylum Chelicerata   277
subphylum Crustacea   276
subphylum Hexapoda   277
subphylum Myriapoda   277
suction feeding
   as ancestral mode   763
superposition eye   306
suspension feeders,
      definition   610
suspensorium   758–759, *761*
Svante Arrhenius   42
swim tunnel respirometer
   900
Swima   231, *232*, 268
swordfish
   brain heaters   743–744
symmetry
   biradial and tetramerous   92
   radial   92, *93*
sympagic   397
system comparisons
   Antarctic
      amphipods   886–887
      Salpa thompsoni   885–886

      salps vs krill   886
   Bathylagidae   883
   between system
      trends   883–884
   cephalopods
      GOM, NCC, Antarctic   884
      role in Antarctic   885
      sampling gear   884
   ctenophores   885
   euphausiids
      intoduction   881
   gelatinous zooplankton and
      amphipods
      introduction   885
   medusae   885
   myctophids
      introduction   882
   NCC
      jellies and the warm
         blob   886
   non-myctophid fishes
      introduction   883
      NCC   884
   salp and krill distributions
      886
   salps   885
   sampling gear   884
System Comparisons
   introduction   858

t

taenioglossate   451, *453*
tapetum lucidum   721, 813, 815
tardigrades
   dormancy   278
Tasmanian giant crab
   278
taste and smell   310
taxonomic ranks   670
tegumental glands   293
teleost
   axial skeleton   760, *761*
   orobranchial cavity   759
   pectoral and pelvic fins
      skeletal support   760
   pectoral girdle   760
teleost gills
   cell types   790

chloride cells
   ion pumps   791, *792*
   location   790
   morphology   790–791,
      792
   smooth tubular system
      (STS)   791
MR cell function   791
MR cells   790
   chloride transport   791
   paracellular pathway
      791
   sodium transport   791
respiratory cells   790
teleosts
   axial skeleton   760, *761*
   the four major radiations
      696
   orobranchial cavity   759
   pectoral and pelvic fins
      skeletal support   760
   pectoral girdle   760
   suction
      how generated   763
   vomerine and glossohyal
      teeth   763–764
telescope fishes   720
teloblastic development   275
telodendria   804, *805*
temperature   9
   acclimation   39
   acclimatization   39
   climatic adaptation   43
   incipient lethal   39–41, *40*
   limits to life   37
   and rate processes   41
   and speciation   50
   and survival   39
   tolerance   39
temperature compensation
   qualitative strategy   47
   quantitative strategy   47
temperature responses
   acute and acclimated   42
tentaculate predation
   the model   114
tentilla   163
terrestrial surface, volume   2

the siphonphore
	conundrum 137
thecosomes
	anatomy
		the mantle 492
	hermaphroditism 502
thelycum   330, *331*, 35*1*, 357
thermal acclimation   43, *44*
thermal tolerance
		polygon 39, *40*
thermohaline circulation   10,
		12, 16–24
thoracomeres, definition   322
Thysanoessa macrura and
		Euphausia crystallorophias
	life histories 935
Thysanoessa raschii
	spawning 330
Torrea candida   245–246
Trachichthyiformes
	Diretmidae
		spinyfins 740
	fangtooths, roughies, and
		pinecone fishes
		738, *739*
	Trachichthyidae
		roughies 740
Trachymedusae   97, *98*
Trade-Wind Biome   845
traditional depth zones
		33, *33*
transduction, definition   810
Trichiuridae
	cutlassfish 741
trichobranchiate
		315–316, *317*
Trieste   60
trilobites   276
trochophore   233–234, 235,
		477, 479
tubular excretory
		organs 241–243
	four basic functions
		298–301, *300*, **301**
	filtration 145, 298
	osmoconcentration or
		dilution 299
	reabsorption 299

secretion   298
tunas
	archival tagging studies
		743
Tunicata
	Appendicularia 625
	classification 607
tunicate systems
	locomotory types 634
tunicates
	ascidians
		anatomy and life history
			609–610, *611, 612*
	bioluminescence 653
		appendicularians 653
		in doliolids 654
		pyrosomes 653
	circulation 646
		appendicularia 647
		doliolids 645
		heart reversal 645
		salps 645
	doliolids
		introduction 618–619
		excretion 647
		uric acid 647
		filtration rates 647, **649–650**
		gas exchange 645
		as hosts 654
		importance in carbon
			flux 647
	nerves and senses
		general 638–639
	particle ingestion
		sizes 647–651, **651**
		as prey 654, **655**
	pyrosomes
		introduction 610–611
	salp swarms 647, *648*
	salps 613
	trophic importance 647
	as vertebrate relatives
		669
	zoogeography 656

u

ultrafiltration
	definition 300

role of blood pressure
		300–301
upwelling   16, *18*
urea excretion
	importance to
		elasmobranchs 785
Urechis caupo   225, *227*
uropods   348, *348,* 351
uterine milk. *See* elasmobranchs

v

Valencienellus tripunctulatus
	vertical distribution and
		diet 710–711, *710*
Vanadis formosa   245–246,
		247
veilfins. *See* caristiids
velvet worms   277, *278*
vertebrates
	closest relatives 669
vertical migration
	acoustic detection 340
	adaptive significance
		and environment 340
		horizontal displacement
			345
		metabolic models 344
	ecological vs adaptive
		reasons for 337
	energy costs 341
	estimates of energy
		costs 343–344
	importance of visual
		predation 345–346
	influence of solar
		eclipses 340
	influence of temperature and
		oxygen 341
	light as proximal cue 339
	predator avoidance
		field-based support
			346
vesicula seminalis   518, *520*
Vinciguerria   709
viscosity   5
	dynamic 5
	kinematic 6
viviparity   691

W

Walsh, Don Lt. 60
waryfishes 720
water
  bears 277
  density 4
  physical properties 2
water masses 12, *See also* T-S diagrams
Weddell-Scotia Confluence 858
  density anomalies and deep-sea species 858
West Wind Drift 853
Western Antarctic Peninsula (WAP) shelf 854, 855
whale lice 397
whalefishes
  bathypelagic importance 737
  larval and juvenile morphology 739, *737*
  lateral line 738
  sexual dimorphism 738
  size range 737–738
  unique reproductive strategy 738

X

X- and Y-organs
  origin of names 295–296
Xenacanthomorpha 677, *678*

Z

zebra display 567
zoarcids
  and hydrothermal vents 757
  pelagic 757
  species and habits 756–757, *756*
zygoneuries 468–469, *468*